Chronomics and Continuous Ambulatory Blood Pressure Monitoring

Kuniaki Otsuka • Germaine Cornelissen
Franz Halberg

Chronomics and Continuous Ambulatory Blood Pressure Monitoring

Vascular Chronomics: From 7-Day/24-Hour to Lifelong Monitoring

Springer

Kuniaki Otsuka
Department of Chronomics & Gerontology
Tokyo Women's Medical University
Medical Center East
Arakawa-ku, Tokyo, Japan

Franz Halberg
University of Minnesota
Halberg Chronobiology Center
Minneapolis, MN, USA

Germaine Cornelissen
University of Minnesota
Halberg Chronobiology Center
Minneapolis, MN, USA

ISBN 978-4-431-56634-2 ISBN 978-4-431-54631-3 (eBook)
DOI 10.1007/978-4-431-54631-3

Springer Tokyo Heidelberg New York Dordrecht London

Softcover reprint of the hardcover 1st edition 2016

Printed on acid-free paper

Springer Japan KK is part of Springer Science+Business Media (www.springer.com)

This book is dedicated to Dr. Franz Halberg

With great sorrow and a heavy heart we inform our readers that on Sunday June 9, 2013, Professor Franz Halberg passed away. Franz had so much energy that we all thought he would live forever. Even during the last few days of his life, he still wanted to go ahead with our project on writing a book and give it his utmost support. We shall miss him deeply.

1987

Franz Halberg's passing shy of his 94th birthday leaves a void that cannot be filled. He will be remembered for founding the fields of chronobiology, chronomics, chronoastrobiology and chronobioethics. His accomplishments are summarized in his over 3,600 scientific publications, in cooperation with colleagues from around the world.

He coined the term circadian (about 24-hour), after documenting "free-running", as proof of the endogenicity of circadian rhythms. As evidence for their importance, Franz showed that biologic rhythms tip the scale between health and disease and even between life and death. Many around the world call him their mentor and turned to him for advice, from study design and data analysis to the interpretation of results in the time dimension.

His work earned him numerous awards. Apart from holding professorships in Laboratory Medicine and Pathology, Physiology, Biology, Bioengineering and Oral Medicine at the University of Minnesota, he received honorary doctorates from the University of Montpellier (France), Ferrara (Italy), Tyumen (Siberia), Brno (Czech Republic), L'Aquila (Italy), and People's Friendship University of Russia (Moscow, Russia).

Born on July 5, 1919 in Romania, Franz Halberg was brought by the US government from post-war Austria to Harvard Medical School, where he held a World Health Organization fellowship. In 1949, he moved to the University of Minnesota, which saw his breakthrough experiments that led to the important discovery of circadian rhythms, and then built-in cycles with many other periods, such as the week, month, year, and even others reflecting signatures of space-terrestrial weather.

Franz discovered that these cycles are partly innate and can be manipulated by environmental synchronizers. He went on to document their importance in clinical medicine, concentrating on preventive cardiology and individually timed treatment in cardiology and oncology.

At 93 years of age and still active 7 days a week in the Halberg Chronobiology Center at the University of Minnesota, he was one of the last two recipients of a lifetime career award from the National Institutes of Health. He was a trail blazer who leaves a remarkable legacy. Some of the concepts he introduced are illustrated in this book, dedicated to his memory that will forever be cherished.

September 24, 2014
Germaine CORNELISSEN, Professor, University of Minnesota
Kuniaki OTSUKA, Professor, Tokyo Women's Medical University

Preface

Chronobiology

Cycles in biology have been reported by keen observers since antiquity. Aschoff [1] suggested that the history of rhythm research began in 650 BC, when the Greek poet Archilochus of Paros wrote "Recognize which rhythms govern man". As noted by Halberg et al. [2], a later fragment by Archilochus refers to the solar eclipse of 6 April 647 BC, which authenticates the fragment on rhythm as dating from the seventh century BC, and identifies Archilochus as a lyric link to both the study of rhythms and to that of the Sun. In Halberg's view, the different fragments from the same poet link rhythms, chronobiology more broadly, and the transdisciplinary mapping of time structures (chronomes) to space physics.

The opening by day and closing by night of the leaves of the tamarind tree was reportedly [3] described as the "nyktitropic movement" (now called nyctinasty) during the fourth century BC by Androsthenes of Thasos, who in 324 BC commanded a ship with 30 oars in the campaigns of Alexander the Great, explored Bahrain and its flora, and the Arabian coast, and circumnavigated India. His book, now apparently lost, was cited by Theophrastus [4]. It first reports on the fact that plants are capable of movement, a characteristic that had previously been attributed only to the animal world. Androsthenes believed that the daily changes from light to darkness rather than changes in environmental temperature were responsible for the leaf movements [3].

Other ancestors singled out by Aschoff [1] include Hippocrates (c. 460–c. 370 BC), Aristotle (384–322 BC), and Galen (130–200). In examining the occurrence of critical days determining a crisis or lysis in the course of illness, Avicenna (980–1037) found, as Hippocrates and Galen did before him, that the week was an important unit of biological time, usually elapsing between the start and resolution of disease [5].

Physiological monitoring started in the seventeenth century. In 1603, William Harvey (1578–1657) noted that "The movement of blood occurs constantly in a circular manner and is the result of the beating of the heart" [6]. Santorio Santorio

(1561–1636), an Italian physician, kept a 30-year record of his bodily functions [7]. His studies of basal metabolism, using instruments of precision, introduced quantitative experimental procedure into medical research. Credit goes to Thomas Sydenham (1624–1689) for advocating the administration of "Peruvian bark", with quinine as its active ingredient, with a specific timing in the evening for the treatment of pain [8]. Julien-Joseph Virey (1775–1846) was first to write a doctoral thesis in medicine devoted to biological rhythms [9]. The physicist Jean-Jacques d'Ortous de Mairan (1678–1771) incidentally reported that he moved a "sensitive" heliotrope plant (*Mimosa pudica*?) to a place where sunlight could not reach it, finding that the plant still opened its leaves during the day and folded them during the night [10]. His finding can be interpreted as demonstrating the persistence of what we now call circadian rhythms [11].

Only in the 1920s, however, did the study of biological rhythms start in earnest. Around this time, a critical mass of integrated contributions from several investigators appeared, contrasting with the earlier publications that were isolated and sporadic [12, 13]. To the mere description of periodicities, analyses of their structure and possible underlying mechanisms started emerging, recognizing the partly endogenous nature of rhythms. Erwin Bünning deserves credit for finding de Mairan's physiological contribution and for interpreting the observation on the plant from a "clock" viewpoint [14–16]. This was also the time when the study of rhythms became institutionalized. For instance, the Internationale Gesellschaft für biologische Rhythmusforschung was initiated in 1937 in Ronneby, Sweden, later becoming known as the Society for Biological Rhythms (SBR).

One of the organizers of the SBR foundation meeting, the German physician Arthur Jores, stated on this occasion that the study of biological rhythms constituted a new working direction within the life sciences insofar as they followed laws of their own [17]. As founding member and future president of the Society, Jores, a physician interested primarily in practical aspects of rhythms from a therapeutic perspective, still viewed rhythms as being exogenous. The endogenous nature of rhythms, essential for the development of a scientific discipline in its own right, was only considered by the two biologists present at the meeting.

Cambrosio and Keating [12, 13] discerned two stages post-WWII, one between the third congress in 1949 held in Hamburg and the ninth meeting in 1967 in Wiesbaden, during which the SBR no longer monopolized events related to biological rhythms, and the other, officially starting at the tenth conference held in Little Rock (Arkansas, USA) in 1971. This was the first meeting in North America, during which the Society assumed the new name of the International Society for Chronobiology (ISC). For the first time, in 1953, American researchers attended the fourth meeting held in Basel, including Franz Halberg, who would play a critical role in the transformation of the SBR into the ISC. Until 1960, the SBR remained primarily European with a strong interest in medicine.

The year 1960 is considered a turning point in the history of chronobiology, not in view of meetings from the SBR but of the Symposium on Biological Clocks held in Cold Spring Harbor [18], organized by Pittendrigh with Aschoff, Bünning, and

Bruce. Together with earlier meetings, notably one in 1957 in Gatlinburg, Tennessee, on photoperiodism and related phenomena in plants and animals [19], it attests to the growing research on rhythms in the USA starting in the 1950s. This effort was essentially centered on circadian rhythms in a biological rather than medical perspective. In the following years, the dichotomy between the medical and biological approaches sharpened. In 1967 at the meeting in Wiesbaden, Jores, the president of the SBR, named Halberg as his successor to the presidency. At the next meeting in 1971, as the SBR became the ISC in Little Rock, the Society explicitly adopted a disciplinary approach to the study of rhythms. Thus, in 1979 at the 14th Congress, the statement was formally made that "chronobiology should become an academic discipline in its own right" and was voted on as part of the Society's constitution.

Cambrosio and Keating [12, 13] provided an objective account of the two contrasting views regarding the study of biological rhythms. On the one hand, Pittendrigh considered that rhythms should be studied within the scope of their original scientific discipline, focusing exclusively on biological aspects of rhythms and their "clock" mechanism modeled by self-sustaining oscillators. On the other hand, Halberg took a multidisciplinary approach, including problems of growth, development, and senescence of interest to pediatricians and gerontologists within the scope of chronobiology, which he viewed as an integrated discipline like genetics. Both men considered 1950 as the time of divergence between their respective approaches, albeit for different reasons. This was the time when Halberg uncovered the circadian adrenal cycle and identified the adrenal as an anatomical entity, with known biochemical factors, capable of certain effects of a physiological nature, namely the depression of blood eosinophil counts [20]. Whereas Pittendrigh's school used a descriptive approach to the study of clocks, Halberg introduced quantitative methods to study rhythms from a statistical viewpoint. His cosinor method [21] made possible the microscopic analysis of rhythmic phenomena with broad applications in all fields of medicine and biology more generally.

According to Pauly and Scheving [22], Franz Halberg was regarded by many scientists as the father of modern chronobiology and as its undisputed leader, particularly as it related to biomedical research and therapy. He served for more than 10 years as president of the SBR, and for another 15 years as president of the ISC. During his tenure and thereafter, numerous scientists trained in his laboratory in Minnesota, thus creating a critical mass disseminated worldwide. His endeavors were further supported by the journal *Chronobiologia*, published between 1974 and 1994 (then the official journal of the ISC) for which he served as editor-in-chief.

In its first issue, Halberg paid tribute to Arthur Jores for his pioneering interest in chronobiological topics several decades before they came to the forefront [23]. He also invited Kalmus to recount the Foundation Meeting of the International Society for the Study of Biological Rhythms [24]. Additional information related to the historical development of chronobiological concepts and its later extension to chronomics can be found in two autobiographical records [25, 26] and in a review of his contributions to the chronobiology of nutrition [27].

Blood Pressure

Blood pressure was first directly measured in an artery in 1733 by the English naturalist the Reverend Stephen Hales [28] (Fig. 1). An Austrian physician, Samuel Siegfried Karl Ritter von Basch, then developed a convenient and simple way of measuring blood pressure by connecting a small balloon to either a mercury column

Fig. 1 First direct measurement of arterial blood pressure in 1733 by the English naturalist the Reverend Stephen Hales (http://www4.ncsu.edu/~msolufse/bpmeasurement.pdf)

or sometimes an aneroid manometer [29] (Fig. 2). The balloon was placed over the artery in the wrist and compressed until the pulsation of the artery in the wrist below the balloon was obliterated. The pressure at which the pulsation disappeared gave a reasonably accurate measurement of systolic blood pressure. Soon, von Basch realized that the higher the systolic pressure, the greater was the risk of stroke and kidney disease [30].

The sphygmomanometer, from the Greek "sphygmos" (pulse) and the scientific term "manometer" (pressure meter), was invented by von Basch in 1881 [29] (Fig. 3). His instrument was extensively tested by Ignaz Zadek [31, 32]. By 1881, data from Zadek were available from four subjects to assess components with periods of about 24, 84, and 168 hours [33] (Fig. 4).

Blood pressure first became easy to measure in 1896 when an Italian physician, Scipione Riva Rocci, developed what we would now recognize as a conventional mercury sphygmomanometer with a cuff around the arm, which was inflated until the pulsation of the artery could no longer be felt [34] (Fig. 5). A few years later, Nicolai Korotkoff, a Russian army surgeon, realized that by listening with a stethoscope below the cuff over the artery at the elbow, characteristic sounds were heard at the systolic pressure (when the heart contracts), but also importantly at the lower (diastolic) pressure (when the heart relaxes). It became very easy to measure both systolic and diastolic pressure accurately with a stethoscope [35].

In 1904, Theodore C. Janeway of Johns Hopkins University (Baltimore), an opinion leader of his time, wrote that "it is essential that a record of the pressure be made at frequent intervals at some time previous (presumably to an examination), to establish the normal level and the extent of the periodic variations. When this is done, it may be possible to demonstrate changes of small extent, which, lacking this standard, for comparison, would be considered within the limits of normal variation" [36].

By 1974, Frederic C. Bartter (of Bartter syndrome fame, while he headed the clinical center at the U.S. National Institutes of Health) described a patient whose

Fig. 2 Development of a convenient and simple way of measuring blood pressure by von Basch (1876), an Austrian physician (http://www4.ncsu.edu/~msolufse/bpmeasurement.pdf)

226 DISEASES OF CIRCULATORY ORGANS.

of a metal ring with an elastic bag inside large enough to admit a finger. The bag is connected by a tube with a mercurial or aneroid manometer on the one hand and with

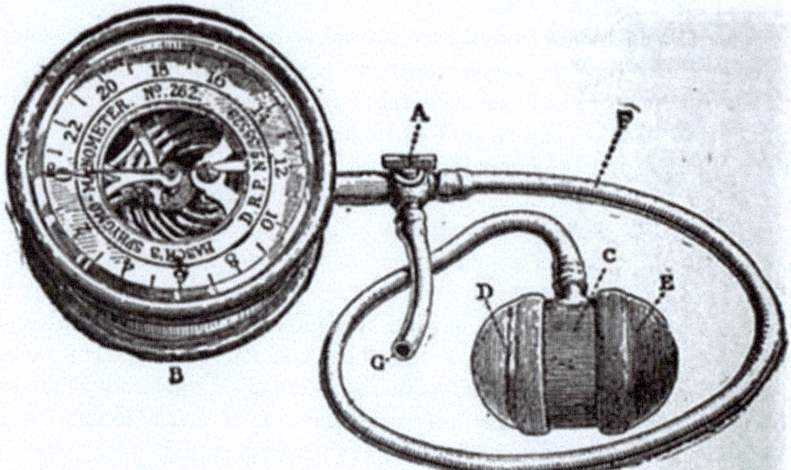

Fig. 29.—Von Basch's sphygmomanometer, from a specimen in the possession of the author. A is a three-way stopcock, by which the aneroid B can be put in communication either with the bulb C or with the outer air. C is a metal ring with caps of india rubber D and E. F is the india rubber tube connecting the bulb and aneroid. This is not the first form of von Basch's sphygmomanometer. The bulb which he first used was included in a metal case somewhat resembling Marey's cardiograph. My instrument needed repair, so I sent it to Vienna, and it was returned by the maker with the bulb figured above.

a pressure bag on the other. After the ring has been placed on the finger the blood is squeezed up from the tip by pressing upward over it a small strong rubber ring or by winding over it a piece of elastic tube, which renders the end of

Fig. 30.—Most recent form of von Basch's sphygmomanometer. (Messrs. Down Bros., 21 St. Thomas's street, S. E.)

Fig. 3 von Basch's sphygmomanometer (1881). Excerpt from: *Diseases of the respiratory organs* (Copyright 1909 by the Year Book Publishers)

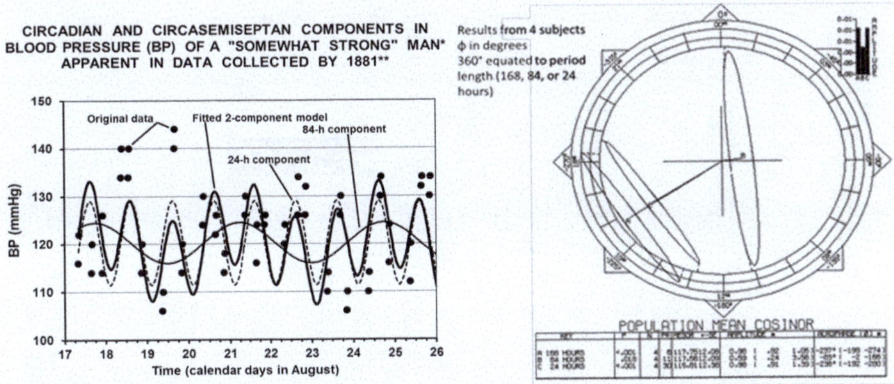

* Case II, p.520 from I.Zadek, Zeitschr f klin Med, 1881; 2: 509-551.

Fig. 4 Historical data collected by Zadek with von Basch's sphygmomanometer, analyzed chronobiologically (© Halberg Chronobiology Center)

Fig. 5 Conventional mercury sphygmomanometer with an arm cuff developed by Riva-Rocci (1896) (http://www4.ncsu.edu/~msolufse/bpmeasurement.pdf)

blood pressure was diagnosed differently by two physicians who saw him at different times of day (Fig. 6): "By conventional standards, this patient is clearly normotensive every morning. But the blood pressure determined each day at 6 in the afternoon provides especially convincing evidence that this patient is a hypertensive. … My plea today is that information contained in [data curves compiled under

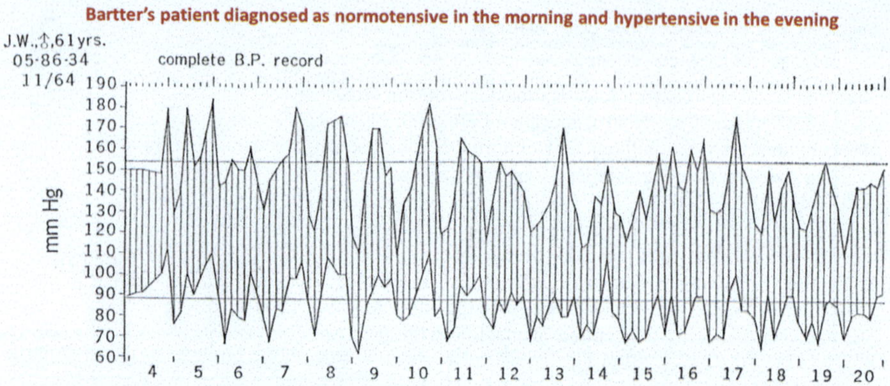

Fig. 6 Historical data collected by Bartter illustrating how circadian rhythms, if ignored, account for contradictory diagnoses reached by measuring blood pressure in the morning or evening (© Halberg Chronobiology Center)

differing circumstances, such as 24 hours a day/7 days a week] become a routine minimal amount of information accepted for the description of a patient's blood pressure. The analysis of this information by cosinor should become a routine. It is essential that enough information be collected to allow objective characterization of a periodic phenomenon, to wit, an estimate of M [the time structure or chronome-adjusted mean, or MESOR] … an estimate of [the amplitude] A itself, and finally an estimate of acrophase, ϕ [a measure of timing]. In this way, a patient can be compared with himself at another time, or under another treatment, and the patient can be compared with a normal or with another patient" [37].

The following chapters review what has been learned since then, using a chronobiologic approach. From the prominent circadian variation in blood pressure and heart rate observed in health and disease, to the vascular variability disorders predictive of adverse outcomes, and extra-circadian cycles shared between physiology and the environment, topics covered herein should be of equal interest to practicing clinicians and to scholars interested in basic science.

Table 1 may best summarize what chronobiology is all about.

Table 1 Minnesota contributions leading to chronobiology (Annu Rev Physiol 1969; 31: 675–725) resolving biological time structures, including circadian (Z Vitamin-, Hormon-u Fermentforsch 1959; 10: 225–96) and other rhythms

1	The partly genetic basis of circadian rhythm characteristics and their uncertainties initially documented by differences in extent of within-day change of circulating eosinophil cell counts, among inbred strains of mice (1950) (Proc Soc Exp Biol 1950; 75: 846–7) and by periods in core temperature, differing from their environmental 24-hour counterparts after blindness, "free-running" (Am J Physiol 1954; 179: 229–55). Human hereditary aspects of about (~) 24-hour and ~7-day cycles documented on twins (Biomed & Pharmacother 2001; 55 [Suppl 1]: 32 s–50s).
2	The importance of the adrenal cortical cycle: documented by a "remove" (and replace) approach (1951: Journal-Lancet [Minneapolis] 1951; 71: 312–19; Journal-Lancet [Minneapolis] 1953; 73: 20–23; Postgrad Med 1958; 24, 349–58).
3	Light and feeding schedules, competing environmental synchronizers of circadian rhythms, mealtime dominating on a restricted diet (1953: Am J Physiol 1953; 174: 109–22).
4	Timing in the cell: RNA and DNA are strongly circadian periodic; RNA synthesis precedes DNA formation in the cell cycle, describing what really happens in time in cellular transcription. A rhythm in cell membrane phospholipid peaks before RNA formation (1958; Proc Soc exp Biol (NY) 1958; 97, 897–900). First hint that RNA may also precede DNA in evolution.
5	Importance of hypothalamus traced from the eyes as transducers of lighting regimen effects on circadian period of adrenal cortex over the pituitary to the CNS (1954: In: Withrow RB., Ed. Photoperiodism and Related Phenomena in Plants and Animals. Ed. Publ. No. 55. Washington DC: AAAS; 1959. p. 803–78; many subsequent articles) and to the cell. The suprachiasmatic nucleus, part of the hypothalamus, was identified later, and its then-claimed importance as sole timer was debunked by 1958 and subsequently (Chronobiologia 1979; 6: 405–24; Anat Rec 1974; 180: 47–52). The long-denied timing mechanisms in the periphery, denied for decades, are also now accepted.
6	Timing for the body as a whole: different, including opposite effects of the same hormones (growth hormone, ACTH, TSH) or (e.g., anti-cancer) drugs, or of the same number of calories, depending only on the circadian timing of administration (1973; Experientia [Basel] 1973; 29: 909–34; Peptide chronomics. In: Kastin A., Ed. Handbook of Biologically Active Peptides. Amsterdam: Elsevier; 2006. p. 1529–64).
7	Circadian rhythms tip the scale between life and death in response to many stimuli: physical stimulus (noise) (1955: Proc Soc exp Biol (NY) 1955; 88: 169–73); whole-body irradiation (Haus E, Halberg F, Loken MK, in Scheving LE, Halberg F, Pauly JE., Eds. Chronobiology, Proc. Int. Soc. for the Study of Biological Rhythms, Little Rock, Ark. Stuttgart: Georg Thieme Publishers/Tokyo: Igaku Shoin Ltd.; 1974. p. 115–22); chemical (alcohol) (J Appl Physiol 1959; 14, 878–80); biochemical (endotoxin) (J Clin Endocrinol 1955; 15: 887); drug (ouabain) (Fed Proc 1959; 18: 63), culminating in the doubling of the 2-year survival rate in patients with advanced perioral cancers (Guest Lecture, Proc. 30th Ann. Cong. Rad., January 1977, Post-Graduate Institute of Medical Education and Research, Chandigarh, India, 8 pp; J Exp Therapeutics Oncol 2003; 3: 223–60).
8	Circadian rhythmicity in a prokaryote (without a nucleus), originally strongly contradicted by a committee on the molecular basis of circadian rhythms (Report of the Dahlem Workshop on the Molecular Basis of *Circadian Rhythms, Berlin, November 3–7, 1975. Berlin: Dahlem Konferenzen, 1976: 462 pp); that partly endogenous rhythms are found in all life (Proc MN Acad Sci 1961; 29: 227–39) is now universally accepted.

(continued)

Table 1 (continued)

9	A broad set of statistical methods and programs for detecting, validating, quantifying, and comparing (always with uncertainties estimated) rhythms, trends and chaos starting with the linear and nonlinear cosinor (in von Mayersbach H., Ed. The Cellular Aspects of Biorhythms, Springer Verlag, NY, 1967, pp. 20–48; cf. Acta med rom 1980; 18: 399–440).
10	Discovery of an (eventually long) list of periods in addition to about-daily (circadian) periods, starting with the biological near-week, followed by discovery of the corresponding geomagnetic near-week followed by numerous other infradian periods (1965: Acta Endocrinol 1965; 50 [Suppl 103]: 5–54; University of Minnesota/Medtronic Chronobiology Seminar Series, #1, December 1991), leading from chronobiology, a figurative microscopy in time, to chronomics, a figurative telescopy in time.

Arakawa-ku, Tokyo, Japan Kuniaki Otsuka
Minneapolis, MN, USA Germaine Cornelissen
Minneapolis, MN, USA Franz Halberg

A Note to the Reader

The duplication of some figures, references and passages of text from one chapter to another is done so that individual chapters can be downloaded and read without needing to cross-reference to other chapters. For the same reason, each chapter has its own bibliography instead of the authors assembling one long bibliography for the whole book.

References

1. Aschoff J. Speech after dinner. 1974 Capri Symposium on timing and toxicity. In: Aschoff J, Ceresa F, Halberg F. (Eds.) Chronobiological Aspects of Endocrinology. Chronobiologia 1974; 1 (Suppl. 1): 483–95.
2. Halberg F, Cornelissen G, Otsuka K, Katinas G, Schwartzkopff O. Essays on chronomics spawned by transdisciplinary chronobiology: Witness in time: Earl Elmer Bakken. Neuroendocrinol Lett 2001; 22: 359–84.
3. Bretzl H. Botanische Forschungen des Alexanderzuges. Leipzig: B.G. Teubner; 1903. p. 120–32.
4. Theophrastus (Hort A. trans.). Enquiry into plants, and minor works on odours and weather signs. Cambridge, Mass.: Harvard University Press / London: W. Heinemann; 1980.
5. Hildebrandt G, Bandt-Reges I. Chronobiologie in der Naturheilkunde: Grundlagen der Circaseptanperiodik. Heidelberg: Haug; 1992. 102 pp.
6. Sutcliffe J, Duin N. A History of Medicine. New York: Barnes & Noble; 1992. 256 pp.
7. Santorio S. De Statica Medicina. Hagae-Comitis, ex typographia A. Vlaco, 1657.
8. SydenhamT (Swan J, trans). The Entire Works of Dr. Thomas Sydenham, Newly made English from the Originals: Wherein the History of acute and chronic Diseases, and the safest and most effectual Methods of treating them, are faithfully, clearly, and accurately delivered. London: Edward Cave, St. John's Gate, 1742. The Life of Dr. Sydenham, pp. v–xi.
9. Virey JJ. Ephémérides de la vie humaine ou recherches sur la révolution journalière et la périodicité de ses phénomènes dans la santé et les maladies. Thesis, University of Paris, 1814, 39 pp.

10. de Mairan JJD. Observation botanique. Histoire de l'Académie Royale des Sciences, 1729, pp. 35–6.
11. Halberg F. Physiologic 24-hour periodicity; general and procedural considerations with reference to the adrenal cycle. Z Vitamin-, Hormon- u Fermentforsch 1959; 10: 225–96.
12. Cambrosio A, Keating P. The disciplinary stake: the case of chronobiology. Social Studies of Science 1983; 13: 323–53.
13. Cambrosio A, Keating P. L'étude des rythmes biologiques: notes socio-historiques. Bulletin du GERB 1988; 21: 9–20
14. Bünning E. Die endogene tagesrhythmik als grundlage der photoperiodischen reaktion. Ber Dtsch Bot Ges 1936; 54: 590.
15. Bünning E. Circadian rhythms and the time measurement in photoperiodism. Cold Spr Harb Symp quant Biol 1960; 25: 249–56.
16. Bünning E, Chandrashekaran MK. Pfeffer's views on rhythms. Chronobiologia 1975; 2: 160–7.
17. Jores A. The origins of chronobiology: an historical outline. Chronobiologia 1975; 2: 155–9.
18. Cold Spring Harbor Symp quant Biol 1960; 25, 524 pp.
19. Withrow RB. (Ed.) Photoperiodism and Related Phenomena in Plants and Animals. Proceedings of the Conference on Photoperiodism, Gatlinburg, Tennessee, October 29-November 2, 1957, sponsored by the Committee on Photobiology, NAS-NRC, and supported by the National Science Foundation. Ed. Publ. No. 55. Washington DC: American Association for the Advancement of Science; 1959.
20. Halberg F, Visscher MB. Regular diurnal physiological variation in eosinophil levels in five stocks of mice. Proc Soc Exp Biol Med 1950; 75: 846–7. doi: 10.3181/00379727-75-18365.
21. Halberg F, Tong YL, Johnson EA. Circadian system phase – an aspect of temporal morphology; procedures and illustrative examples. Proc. International Congress of Anatomists. In: Mayersbach H v. (Ed.) The Cellular Aspects of Biorhythms. Symposium on Rhythmic Research, Sponsored by the VIII International Congress of Anatomy, Wiesbaden, 8–14. August 1965. New York: Springer-Verlag; 1967. pp. 20–48.
22. Pauly JE, Scheving LE. Dedication. Progress in Clinical and Biological Research 1987; 227: xxiii–xxvii.
23. Halberg F, Engel R. Arthur Jores, in appreciation. Chronobiologia 1974; 1: 113–7.
24. Kalmus H. The foundation meeting of the International Society for Biological Rhythms. Chronobiologia 1974; 1: 118–24.
25. Halberg F, Cornelissen G, Katinas G, Syutkina EV, Sothern RB, Zaslavskaya R, Halberg F, Watanabe Y, Schwartzkopff O, Otsuka K, Tarquini R, Perfetto P, Siegelova J. Transdisciplinary unifying implications of circadian findings in the 1950s. J Circadian Rhythms 2003; 1: 2. 61 pp. www.JCircadianRhythms.com/content/pdf/1740-3391-1-2.pdf
26. Halberg F, Cornelissen G, Katinas GS, Hillman D, Otsuka K, Watanabe Y, Wu J, Halberg Francine, Halberg J, Sampson M, Schwartzkopff O, Halberg E. Many rhythms are control information for whatever we do: an autobiography. Folia anthropologica 2012; 12: 5–134. http://ttk.nyme.hu/blgi/Knyvek%20kiadvnyok/FOLIA%20ANTHROPOLOGICA/folia12.pdf
27. Cornelissen G. When you eat matters: 60 years of Franz Halberg's nutrition chronomics. The Open Nutraceuticals Journal 2012; 5 (Suppl 1-M1): 16.
28. Hales S. An account of some hydraulic and hydrostatical experiments made on the blood and blood vessels of animals. London: W. Innys & R. Manby; 1733.
29. Lawrence C. Physiological apparatus in the Wellcome Museum. 3. Early sphygmomanometers. Medical History 1979; 23: 474–8 (+4 figures).
30. von Basch S. On the measurement of blood pressure in humans [in German]. Zeitschr Klin Med 1880; 2: 79–96.
31. Zadek I. Die Messung des Blutdrucks am Menschen mittelst des Basch'chen Apparates. Berlin, med. F., Diss., 25. Nov 1880. Berlin: Schumacher; 1880. 48 p.
32. Zadek I. Die Messung des Blutdrucks am Menschen mittelst des Basch'chen Apparates. Z klin Med 1881; 2: 509–51.

33. Halberg F, Schwartzkopff O, Cornelissen G, Hardeland R, Müller-Bohn T, Katinas G, Revilla MA, Beaty L, Otsuka K, Jozsa R, Zeman M, Csernus V, Hoogerwerf WA, Nagy G, Stebelova K, Olah A, Singh RB, Singh RK, Siegelova J, Dusek J, Fiser B, Czaplicki J, Kumagai Y, Chibisov SM, Frolov VA. Vaskuläres Variabilitäts-Syndrom (VVS) und andere Chronomik 2005–2007. In: Hardeland R. (Ed.) Sonderdruck aus Abhandlungen der Leibniz-Sozietät der Wissenschaften, Band 23: Facetten der Chronobiologie. Berlin: trafo verlag; 2008. p. 89–154.
34. Riva-Rocci S. Un nuovo sfigmomanometro. Gazzetta Medica di Torino 1896; 47: 981–96.
35. Korotkoff NC. On the question of methods of determining the blood pressure. Reports of the Imperial Military Medical Academy, St. Petersburg 1905; 11: 365.
36. Janeway TC. The clinical study of blood pressure. New York: D. Appleton & Co.; 1904. 300 pp.
37. Bartter FC. Periodicity and medicine. In: Scheving LE, Halberg F, Pauly JE. (Eds.) Chronobiology. Tokyo: Igaku Shoin Ltd.; 1974. p. 6–13.

A Brief Overview of Professor Franz Halberg's Life and Work

Franz Halberg was born in Bistritz (Beszterce, Beszterce-Naszód County, Transylvania, now Romania) on July 5, 1919. His enthusiasm about poetry was apparent from his early years. However, his father, a renowned international lawyer who loved medicine, dissuaded him from becoming a poet which he originally wanted to be. (Halberg still practiced poetry on occasion in a contribution to a scientific academy or on birthdays of friends). Thus, in 1936–1937 he started to study mathematics and physics.

Although he never gave up his love for physics, it soon turned spontaneously into interest in a unified science. So he enrolled in medical school and received his degree in 1943 from the Royal Hungarian Ferenc József University in Kolozsvár (Kolozs County, Hungary, which became Romania after World War II).

The contemporary Hungarian government made all efforts in order to revitalize the Kolozsvár University after the Second Vienna Award (1940) and to return it to a high position in international scientific life. Some of the newly appointed professors were really magnificent physicians, famous scientists and brilliant tutors, while others, unfortunately, were not so outstanding.

Franz Halberg was, to all intents and purposes, not so lucky with his laboratory professor, the other internist of the Kolozsvár Medical School at that time, Dr. vitéz váradi Haynal Imre. As a medical student Halberg had a laboratory from Dr. vitéz kibédi Varga Lajos in the department of medicine of the university in Kolozsvár but achieved nothing.

After World War II, he moved to Austria hoping he could find a place with a more inspiring scientific climate and better opportunities than in post-war Romania. He stopped in Innsbruck, the capital of the Tyrol (Austria), where initially he worked as a scientific assistant and later as a university assistant (1946–1948) in the Department of Anatomy of the Innsbruck University.

Franz met his first wife Erna in Minnesota. In Innsbruck, he had met and danced with Erna's sister, Lotte. Contact with a group of leading U. S. scientists teaching in post-war Europe resulted in an invitation to come to the United States where Halberg could find a significantly broader scope of research opportunities.

He left Europe and went to the U. S. in 1948 on a World Health Organization fellowship, where first he became a Research Fellow at Harvard Medical School and an Assistant in Medicine at the Peter Bent Brigham Hospital in Boston, Massachusetts.

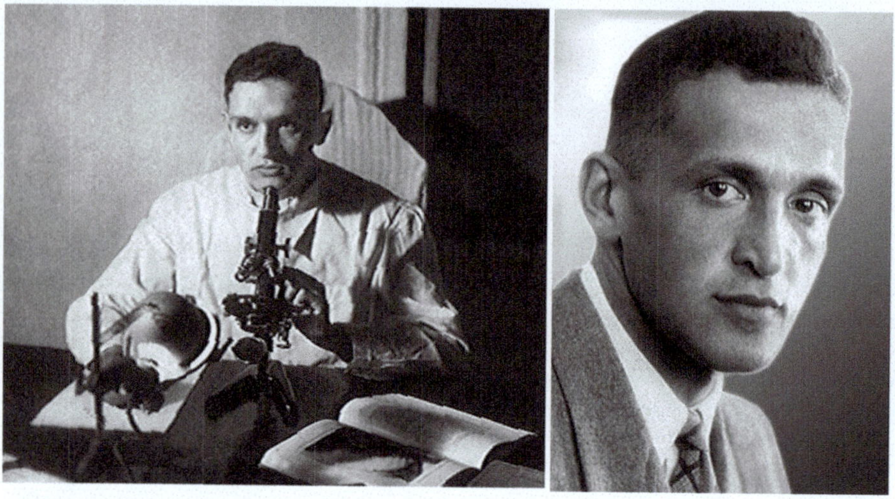

Young Franz Halberg in Tyrol (Austria)

In 1949 he moved to Minneapolis, Minnesota, known as "the land of sky-blue waters". There he joined the staff of the University of Minnesota Medical School. The inspiring scientific atmosphere of Minneapolis and the role model of both parents guided the Halberg daughters, Julia and Francine, towards medical and scientific careers.

Franz Halberg was the founder of the Halberg Chronobiology Center in Minneapolis, where he has been, practically even after his retirement, continuously working, teaching, tutoring and researching for the past sixty years.

I visited Franz for the first time in 1987. It was my first time to talk with him although I had several chances to see him at several meetings in Japan. We discussed for a long time, without sleep, what chronobiology is and why it is needed for medical practice. I remember his strong words clearly: medical doctors all over the world have to learn proper chronobiology heading for chronomedicine. To explore "what is chronomedicine", for the first step, we started long-term continuous ambulatory blood pressure monitoring for over 7 days. For developing it, Franz visited our hospital many times and educated and trained our students, i.e., in 1988, 1991, 1994, 2006, 2007 and I visited him in 1989 and 2000. We have now monitored over 600 subjects including town citizens in Japan as shown in this book.

First visit to Halberg Chronobiology Center, University of Minnesota in 1987

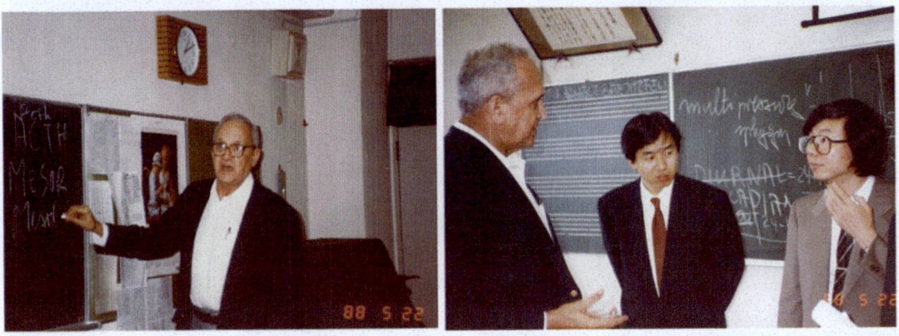

Franz educated us at our University hospital in 1988

Franz trained us at our University hospital in 1991

Franz's third visit, in 2000, when we planned a community-based study. We started the city monitoring in Japan, including comprehensive medical assessment and 7-day/24-hour ABP monitoring

In 2000, we began a community-based study at Uraus town in Hokkaido and at Tosa town in Kochi prefecture, Japan. Franz visited there twice with us and we negotiated with the town mayor and started 7-day/24-hour ambulatory BP monitoring for middle-aged subjects, aged 40 to 80 years, while also evaluating the cardiovascular and neurobehavioral functions of elderly people. Our goal is the prevention of stroke and myocardial infarction and of the decline in cognitive function of the elderly in a community.

We have already monitored 7-day/24-hour ambulatory BP of 686 middle-aged and elderly citizens in both towns from 2000 to 2014. We observed circaseptan components for the average SBP value during the waking span, for the 24-hour average of HR, for the standard deviation (SD) of HR and for the day-night (dipping) ratio of SBP. Specifically, SBP during waking was lowest on Sunday. The BP average during waking shows a circaseptan variation. Thus, we point out that ambulatory BP monitoring limited to 24 hours is not sufficient and propose a routine 7-day screen, which is continued when abnormality is found.

We have started several kinds of interventions, including life-style guidance, education from the viewpoint of clinical nutrition, and consultation regarding drugs for hypertension and dyslipidemia, among other conditions.

Two towns in Japan where Halberg visited and monitored 7-day/24-hour ambulatory BP of over 600 citizens

Halberg's first visit to Uraus city hall, Hokkaido, Japan, 2000. Mr. Kamame Yamamoto (*right*) is the town mayor. He is monitoring his own 7-day/24-hour ambulatory blood pressure

On October 28th, 2006, together with town mayor Mr. Kaname Yamamoto. Franz's second visit to Uraus city hall, Hokkaido, Japan

Franz Halberg attended the 2nd World Congress of Chronobiology, held in Tokyo on November 6th, 2007

Lectures on chronoastrobiology in 2007 at meetings in Vienna and Kyoto

The magnetic activity of the sun has visible photic and invisible non-photic effects on life on earth. Effects of non-photic solar cycles upon plants were recognized by 1838 and championed by economists and transdisciplinary scientists like A.L. Chizhevsky. Halberg discovered new non-daily and non-yearly, i.e. non-photic cycles in both magnetism and corresponding physiological cycles in humans and other organisms. Reports of statistical documentation of a variety of effects upon organisms, including self-monitoring humans over decades, were reviewed at meetings in Vienna and Kyoto in 2007.

There is no place and no need to detail the numerous honorary doctorates, professorships, awards, medals and diplomas Franz received during his long and productive scientific career. At the end of this laudation, a brief scientific biographical sketch lists Professor Halberg's major research experiences, honors, scientific functions, publications and contributions.

However, some pivotal items in Franz's life-work remain to be highlighted. He undisputably placed time as a fourth dimension in life sciences, a new dimension, without which life on planet Earth could not exist. Thus, Professor Halberg is considered worldwide as the founder, father and doyen of chronobiology. More than half a century ago he also coined the term circadian which was referred to by "chronosceptics" as "Halberg's paranoia". He contributed chronobiology, the computer-aided science of the body's time structure, and framed the term chronomics as well. Three entities have been identified to constitute a broad time structure (the chronome): multifrequency rhythms, trends (as a function of age and also as a function of disease risk or response to treatment), and the as-yet unresolved variation (noise). More recently, he coined the term "chronobioethics" as well, in an attempt to quantify endeavors aiming at the hard-to-define "good".

The first moment when even laypersons could realize that chronobiology and circadian rhythms are worth studying was at the time when Halberg and Gupta succeeded in doubling the 2-year survival of patients with oral cancer by timing their radiotherapy according to a circadian marker rhythm, tumor temperature.

Another scientific breakthrough with great practical healthcare impact is based on around-the-clock blood pressure monitoring; Franz identified *c*ircadian *h*yper-*a*mplitude *t*ension, now called "CHAT," an excessive swing (amplitude) of daily changes in blood pressure that develops before a rise in mean blood pressure readings. Surprisingly, CHAT was found to be a greater risk factor for ischemic stroke than family history, obesity, high cholesterol, being male, regular alcohol consumption, smoking and having an elevated mean blood pressure.

Additionally, it was found that patients with diastolic CHAT have a 8.2 greater risk of having a hard event, such as an ischemic stroke, within six years than those whose circadian blood pressure amplitude is within a normal range. As it also turned out, taking blood pressure medication at an undesirable time can cause CHAT. Nonetheless, the importance of treatment timing is usually underestimated by family practitioners and cardiologists.

The results of studies led to the creation of the famous "Womb to Tomb" project more than 20 years ago, which involved monitoring the blood pressure of pregnant women and neonates. The initiative has since been renamed the BIOCOS (an acronym for *bio*sphere and *cos*mos) project and broadened its scope. The purpose of the initiative is to monitor heart rate, blood pressure and other variables *for preventive reasons in maintaining good health* and also to *better understand the effects of the cosmos on our physiology and pathology*.

The mapping of multifrequency rhythms and of other components of the chronome may be the keys to maintain good health and to recognize elevated risk of certain diseases in time so that preventive measures could be carried out before the onset of actual disease. Franz was convinced that measuring and interpreting *chronobiologically*, for example, blood pressure series will sooner or later enable mankind to prevent stroke and also to evade other cataclysms.

Halberg realized that the more constant we try to make our proximal environment, the better we recognize virtually spontaneous variability that reflects the changes around us, near and far. This seemingly chaotic variability has its own laws which can only be discovered and described by analyzing huge series of data and figures. Professor Halberg, perhaps one of the last Renaissance men is, however, the man of the figures as well. "*Quod non est in actis, non est in vita*." What cannot be found in or supported by files (data, figures, etc.), does not exist at all. And, conversely, if we listen carefully enough to the periodic signals in us and around us near and far in the cosmos, we may find new associations and new solutions and broaden our horizons.

Franz's challenging endeavors during his long and prosperous scientific career should be an incentive for us all.

A Note to the Reader

Dedicated, as are all my endeavors, to the memory of my father Dr. Julius Halberg, to Dr. Walter Nelson, with thanks for a lifetime of his cooperation, and to the memory of his mentor Cyrus P. Barnum, professor of biochemistry and, in his time, a leading specialist in nucleic acids. When I asked the strait-laced Barnum for a technician's help with DNA and RNA determinations to map anticipated changes along the 24-hour scale, he replied that the idea of periodicity in nucleic acid – then deemed the most constant of living matter – was so unexpected that he felt he could not assign help paid from a government grant (his technicians were supported by NIH). It made sense to him, however, that if mitoses are periodic, so DNA and RNA should be as well. He replied "I'll be your technician"; a better tech could not have been found.

Franz Halberg

In order to better convey complementary information reported in different chapters but using the same illustrations, this book includes duplicates figures and tables.

Acknowledgments

We extend our thanks to the staff of two cities, Uraus (Hokkaido) and Tosa (Kochi) without whom our project on a community-based health-watch could not have succeeded: Yasuko Taniwaki, Yuhko Ueta-Takahashi, Ayako Kokaji, Mitsue Ito, Kazuki Kitamura, Akihiro Uemura, Shinji Yamashita, (Health and Welfare Center, Tosa District, Kochi), Takushi Nishimura (Tosa town mayor); Yoshie Saito, Yuko Sugiyama, Kensaku Komatsu, (Health and Welfare Center, Uraus town, Kabato District, Hokkaido), Kaname Yamamoto (Uraus town mayor). Our thanks also go to all participants in these two communities who allowed us new insight into healthy aging and the early identification of cardiovascular disease risk, the essential foundation for primary prevention.

For cooperation on long-term continuous blood pressure monitoring, special thanks are due to Shigehiro Ishizuka, Nobuhiko Yasui, Yoshiteru Nozoe (Research Center, and A & D Medical Company, Saitama, Japan). For cooperation on community-based comprehensive medical assessment, we thank Gaku Yamanaka, Shogo Murakami, Kiyotaka Okajima, Junko Sasaki, Tomoko Kikuchi, Sachiko Oinuma, Takashi Yamanaka (Medical Center East, Tokyo Women's Medical University, Tokyo); Kiyohito Okumiya, and Kozo Matsubayashi (Center for Southeast Asian Studies, Kyoto University, Kyoto).

Thanks are also extended to Mary Sampson, Principal Office and Administrative Specialist at the Halberg Chronobiology Center, for her editorial and proof-reading help.

Co-authors in publications underlying the contents of this book include, among others, members of the International Project on the BIOsphere and the COSmos (BIOCOS):

Austria: Thomas Kenner, Physiologisches Institute, Karl-Franzens-Universität, Graz
Azerbaijan: Elchin Khalilov, InterGeo-Tethys, Baku
Belgium: Alain Delcourt, Christian Deruyck, and Guy Toussaint, private practice physicians, Brussels; Christian Muller, Olga Quadens, Aerospace Institute, Brussels
Brazil: Ana Amelia Benedito Silva and Luiz Menna Barreto, University of São Paulo.

Bulgaria: Krasimira Hristova, Department of Noninvasive Functional Dynamic and Imaging, University National Heart Hospital, Sofia

Canada: Karin Olson, University of Alberta Extension Centre, Edmonton, Alberta

China: Zhengrong Wang and Jinyi Wu, West China University of Medical Sciences, Chengdu; Ziyan Zhao, Shandong Academy of Medical Sciences, Jinan

Czech Republic: Jarmila Siegelova, Jiri Dusek, Alena Havelkova, Pavel Homolka, Pavel Prikryl, Masaryk University, Brno; Jaroslav Strestik, Geophysical Institute, Academy of Sciences of the Czech Republic, Prague

France: Christine Amory-Mazaudier, CETP, Observatoire de Saint-Maur, Saint-Maur-des-Fosses; Jerzy Czaplicki, Institut de Pharmacologie et de Biologie Structurale CNRS, Université Paul Sabatier, Toulouse

Germany: Sigrid Berger, University of Heidelberg; Karl Hecht, Charité Hospital, Humboldt University, Berlin

Hong Kong: Brian Tomlinson, Department of Medicine and Therapeutics, Chinese University of Hong Kong, Prince of Wales Hospital, Shatin

Hungary: Rita Jozsa, University Pécs

India: R.B. Singh, Medical Hospital and Research Centre, Centre for Nutrition and Heart Research, Moradabad; R.K. Singh, Department of Biochemistry, King George's Medical College, Lucknow; Atanu Kumar Pati, Pandit Ravishankar Shukla University, Raipur

Italy: Franca Carandente, Department of Chronobiology, and Cristina Maggioni, Department of Obstetrics, University of Milan; Giacomo Laffi, Federico Perfetto, Carlo Rostagno, Roberto Tarquini, Department of Medicine; Roberto Salti, Mayer Pediatric Hospital, University of Florence; Pietro Cugini, University of Rome

Japan: Kuniaki Otsuka, Department of Chronomics and Gerontology, and Gen Mitsutake and Yoshihiko Watanabe, Department of Medicine, Tokyo Women's Medical University, Medical Center East, Tokyo; Keiko Uezono, University of Kyushu, Fukuoka; Akio Fujimura and Yuji Kumagai, Jichi Medical School, Tochigi

Kazakhstan: Lyazzat Gumarova, Department of Biophysics and Biomedicine, Al-Farabi Kazakh National University, Almaty; Alma T. Mansharipova, Kazakhstan Russian Medical University, Almaty

Norway: Andi Weydahl, Finnmark College, Alta

Peru: Julio Chirinos, Santa Maria Catholic University, Arequipa

Poland: Fabien de Meester, Agnieszka Wilczynska, Tsim-Tsoum Institute, Krakow

Russia: Rina M. Zaslavskaya, Moscow Hospital #60, Moscow; Elena V. Syutkina, Inna Malkova, Maria Mitish, Tatiana Turti, Institute of Pediatrics, Academy of Medical Sciences, Moscow; Anatoly Masalov, Lebedev Physics Institute, Moscow; Tamara K. Breus, Space Research Institute, Moscow; Sergey M. Chibisov and Mikhail B. Blagonravov, Russian People's Friendship University, Moscow; George S. Katinas (deceased), I.P. Pavlov First State Medical University, St. Petersburg; Mikhail and Olga Blank and Olga Denisova, N.N. Petrov Oncology Research Institute, Ministry of Health, St. Petersburg; Denis Gubin, Medical Academy, Tyumen

Saudi Arabia: Abdullah A. Al-Abdulgader, Prince Sultan Cardiac Center, Al Ahsa
Slovakia: Anna Gvozdjakova, Miroslav Mikulecky (deceased), and Michal Zeman, Comenius University, Bratislava
Spain: Miguel Revilla, Department of Mathematics and Computer Science, University of Valladolid
Switzerland: Waldemar Ulmer, Varian International, Zug; Vincenzo Valenzi, Libera Universita degli Studi di Scienze Umane e Tecnologiche, Lugano
Taiwan: Chen-Huan Chen, Division of Cardiology, National Yang-Ming University School of Medicine, Taipei
Ukraine: Vladimir Bogdanov and Yuri Gorgo, Kiev National Taras Shevchenko University, Kiev; Anatoly Delyukov, National Academy of Sciences of Ukraine, Kiev
UK: Douglas Wilson, School of Medicine, Pharmacy and Health, Durham University, Durham, England; Hugh Simpson, Department of Surgery, University of Glasgow, Scotland
USA: Germaine Cornélissen, Cathy Lee Gierke, Dewayne Hillman, Othild Schwartzkopff and Robert B. Sothern, Halberg Chronobiology Center, University of Minnesota, Minneapolis; Christopher Bingham, Douglas Hawkins, Department of Applied Statistics, University of Minnesota, Minneapolis; Dana Johnson, Department of Pediatrics, University of Minnesota, Minneapolis; Earl Bakken, Medtronic Inc., Minneapolis, Minnesota; the Phoenix Study Group of the Twin Cities Chapter of the Institute of Electrical and Electronics Engineers, including Christopher Adams, Larry Beaty, Jim Holte, Ellis Nolley and Bob Schlentz (deceased); Katarina Borer, Division of Kinesiology, Department of Movement Science, University of Michigan, Ann Arbor; Mark Engebretson, Department of Space Physics, Augsburg College, Minneapolis, Minnesota; Dan Holley, Department of Biological Sciences, San Jose State University, San Jose, California; Sasikumar Sundaram, University of Arkansas for Medical Sciences, Fort Smith, Arkansas

References of publications relevant to the major topics (1. chronomedicine, 2. chronoecology, 3. chronoastrobiology, and 4. vascular variability syndromes) of this book follow.

Chronomedicine

1981

Cugini P, Scavo D, Cornélissen G, Lee JY, Meucci T, Halberg F. Circadian rhythms of plasma renin, aldosterone and cortisol on habitual and low dietary sodium intake. Hormone Res 1981; 15: 7–27.

Cugini P, Scavo D, Halberg F, Sothern RB, Cornélissen G, Meucci T, Salandi E, Massimiani F. Ageing and circadian rhythm of plasma renin and aldosterone. Maturitas 1981; 3: 173–82.

1982

Otsuka K, Yanaga T, Ichimaru Y, Seto K. Observations of the effect of arrhythmias on the cerebral function by recordings of 24-hour continuous electrocardiograms. Comparison between the sick sinus syndrome and atrio-ventricular block. Jpn Heart J 1982; 23(4): 469–78.
Otsuka K, Yanaga T, Ichimaru Y, Seto K. Sleep and night-type arrhythmias. Jpn Heart J 1982; 23 (4): 479–86.

1983

Saito H, Otsuka K, Sato T, Yoshimatsu K, Kaba H, Seto K, Ichimaru Y, Sato Y, Yanaga T. Arrhythmogenic properties of paradoxical sleep. Am Heart J 1983; 105(5): 875–7.
Otsuka K, Ichimaru Y, Yanaga T, Sato Y. Studies of arrhythmias by 24-hour polygraphic recordings: relationship between atrioventricular block and sleep states. Am Heart J 1983; 105(6): 934–40
Otsuka K, Otsuka Ke, Seto K, Saito H, Kaba H, Sato Y, Ichimaru Y, Yanaga T. Ambulatory ECG-respiration monitoring system. Am Heart J 1983; 106(5 Pt 1): 1173–6.

1984

Halberg F, Cornélissen G, Haus E, Tarquini B, Benvenuti M, Cagnoni M. Timing of medical diagnosis and treatment: clino-circadian quantification of suppression by dexamethasone of the adrenal cortical cycle in healthy men. Chronobiologia 1984; 11: 43–50.
Halberg F, Drayer JIM, Cornélissen G, Weber MA. Cardiovascular reference data base for recognizing circadian mesor- and amplitude-hypertension in apparently healthy men. Chronobiologia 1984; 11: 275–98.
Kawakami T, Otsuka K, Saito H, Seto K, Ozawa T, Yanaga T. Arrhythmogenesis during sleep-induced apnea. Am Heart J 1984; 108(6): 1591–3.

1985

Cugini P, Murano G, Lucia P, Letizia C, Scavo D, Halberg F, Cornélissen G, Sothern RB. Circadian rhythms of plasma renin activity and aldosterone: changes related to age, sex, recumbency and sodium restriction. Chronobiologic specification for reference values. Chronobiology Int 1985; 2: 267–76.
Otsuka K, Sato T, Saito H, Kaba H, Otsuka K, Seto K, Ogura H, Ozawa T. Circadian rhythm of cardiac bradyarrhythmia episodes in rats. Chronobiologia 1985; 12(1): 11–28.

1986

Otsuka K, Ikari M, Ichimaru Y, Saito H, Kawakami T, Otsuka K, Kaba H, Seto K. Experimental study on the relationship between cardiac arrhythmias and sleep states by ambulatory ECG-EEG monitoring. Clin Cardiol 1986; 9(7): 305–13.
Otsuka K, Saito H, Kawakami T, Otsuka K, Kaba H, Seto K, Nojima K, Yanaga T, Kaji M. Effects of limbic-hypothalamic function on arrhythmia occurrence and sleep-induced apnea. Am Heart J 1986; 109(6): 1411–5.

Otsuka K, Sato T, Saito H, Kaba H, Seto K, Yanaga T, Ogura H, Ozawa T. Role of the suprachiasmatic nuclei of the hypothalamus on diurnal rhythm in cardiac arrhythmias. Heart Vessels 1986; 2(1): 15–22.

Cavallini M, Halberg F, Cornélissen G, Enrichens F, Margarit C. Organ transplantation and broader chronotherapy with implantable pump and computer programs for marker rhythm assessment. J Controlled Release 1986; 3: 3–13.

Liu T, Cavallini M, Halberg F, Cornélissen G, Field J, Sutherland DER. More on the need for circadian, circaseptan and circannual optimization of cyclosporine therapy. Experientia 1986; 42: 20–2.

Halberg F, Cornélissen G, Bingham C, Tarquini B, Mainardi G, Cagnoni M, Panero C, Scarpelli P, Romano S, März W, Hellbrügge T, Shinoda M, Kawabata Y. Neonatal monitoring to assess risk for hypertension. Postgrad Med 1986; 79: 44–6.

Cornélissen G, Broda H, Halberg F. Does Gonyaulax polyedra measure a week? Cell Biophysics 1986; 8: 69–85.

1987

Otsuka K, Sadakane N, Ozawa T. Arrhythmogenic properties of disordered breathing during sleep in patients with cardiovascular disorders. Clin Cardiol 1987; 10(12):771–82.

1988

Otsuka K, Watanabe H. Circadian amplitude of diastolic blood pressure gauges family history of high blood pressure in nurses. Chronobiologia 1988; 15:266–7.

Otsuka K, Yanaga T, Watanabe H. Variant angina and REM sleep. Am Heart J 1988; 115(6): 1343–6.

1989

Halberg F, Lakatua D, Lodeiro C, Garcia L, Hermida R, Ayala DE, Tarquini B, Haus E, Cornélissen G. Chronobiology, growth hormone and healthy and malignant growth. J Endocr Invest 1989; 12(Suppl. 3): 41–7.

Otsuka K, Kawakami T, Saito H, Watanabe H. REM sleep and bradyarrhythmia episodes in rats. J Electrocardiol 1989; 22(3): 235–40.

1990

Otsuka K, Watanabe H. Experimental and clinical chronocardiology. Chronobiologia 1990; 17: 135–63.

Halberg FE, Cornélissen G, Halberg F, Halberg E, Halberg J, Kessler T, Mason BH, Glazer ER. About-yearly pattern of cancer deaths, 1942–1987, in patients of a San Francisco hospital. Prog Clin Biol Res 1990; 341A: 567–75.

Cornélissen G. Challenge to the statistician interested in the individualized assessment of health in the inner cities and urban areas: a tribute to Franz Halberg. (Health of Inner Cities and Urban Areas, International Conference, Cardiff, Wales, September 4–7, 1989.) The Statistician 1990; 39: 105–9.

Cornélissen G, Bakken E, Delmore P, Orth-Gomér K, Åkerstedt T, Carandente O, Carandente F, Halberg F. From various kinds of heart rate variability to chronocardiology. Am J Cardiol 1990; 66: 863–8.

1991

Cornélissen G, Halberg F, Prikryl P, Dankova E, Siegelova J, Dusek J, International Womb-to-Tomb Chronome Study Group: Prophylactic aspirin treatment: the merits of timing. JAMA 1991; 266: 3128–9.
Halberg F, Cornélissen G, Carandente F. Chronobiology meets the need for integration in a reductionist climate of biology and medicine. Chronobiologia 1991; 18: 93–103.
Halberg F, Cornélissen G, Carandente F. On with the human chronome initiative: the legacy of Norberto Montalbetti. Chronobiologia 1991; 18: 105–6.

1992

Cornélissen G, Halberg F. Broadly pertinent chronobiology methods quantify phosphate dynamics (chronome) in blood and urine (editorial). Clin Chem 1992; 38: 329–33.
Cornélissen G, Halberg F. Chronobiologic response modifiers and breast cancer development: classical background and chronobiologic tasks remaining. In Vivo 1992; 6: 371–86.
Halberg E, Cornélissen G, Halberg F. Optimization of the chronotherapeutic index in the experimental animal laboratory. In Vivo 1992; 6: 387–402.
Halberg F, Cornélissen G, Bingham C, Fujii S, Halberg E. From experimental units to unique experiments: chronobiologic pilots complement large trials. In Vivo 1992; 6: 403–28.

1993

Wang ZR, Sun XC, Cornélissen G, Wu JY, Meng L, Cai DG, Xue ZN, Halberg F. Doppler flowmeter-assessed circadian rhythms in neonatal cardiac function, family history and intrauterine growth retardation. Am J Perinatol 1993; 10: 119–25.
Cornélissen G, Breus TK, Bingham C, Zaslavskaya R, Varshitsky M, Blank M, Teibloom M, Bakken E, Halberg F. Beyond circadian chronorisk: circadian-circasemiseptan-circaseptan patterns of myocardial infarctions, other vascular events, and emergencies. Chronobiologia 1993; 20: 87–115.
Halberg F, Cornélissen G. More on chronomes: circaseptans and circasemiseptans in Marburg, Germany, and 47 other locations. Chronobiologia 1993; 20: 119–22.

1994

Halberg F, Cornélissen G, Kumagai Y, Bingham C, Saito Y, Tamura K, Otsuka K, Breus T, Rapoport, Komarov F, Zaslavskaya R, del Pozo F, Rodriguez MJ, Arredondo MT, Gomez E, Portela A, Ebihara A, Haus E, Halberg Francine, Halberg J, Fujii S, Delmore P, Bakken E. Telehygiene system for preventive chronopharmacology in space and remote areas on earth. Chronobiologia 1994; 21: 33–43.

Otsuka K, Cornélissen G, Halberg F. Broad scope of a newly developed actometer in chronobiology, particularly chronocardiology. Chronobiologia 1994; 21: 251–64.

Otsuka K, Yamanaka T, Kubo Y. Disruption of fractals of heart rate variability in different types of pathophysiological settings. J Ambul Monit 1994; 7(3): 219–24.

1995

Blank MA, Cornélissen G, Halberg F. Circasemiseptan (about half-weekly) and/or circaseptan (about-weekly) pattern in human mitotic activity? In Vivo 1995; 9: 391–4.

Cornélissen G, Portela A, Halberg F, Bolliet V, Falcón J. Toward a chronome of superfused pike pineals: about-weekly (circaseptan) modulation of circadian melatonin release. In Vivo 1995; 9: 323–9.

Focan C, Cornélissen G, Halberg F. Chronometaanalysis: circasemiseptan (3.5-day) pattern in mitotic activity of murine sarcoma after treatment with cyclophosphamide. In Vivo 1995; 9: 401–5.

Portela A, Cornélissen G, Halberg Franz, Halberg Francine, Halberg J, Hofman MA, Swaab DF, Ikonomov OC, Stoynev AG, Metachronanalysis of circannual and circasemiannual characteristics of human suprachiasmatic vasopressin-containing neurons. In Vivo 1995; 9: 347–58.

Freedman RR, Norton D, Woodward S, Cornélissen G. Core body temperature and circadian rhythm of hot flashes in menopausal women. J Clin Endocrinol Metab 1995; 80: 2354–8.

1996

Halberg F, Cornélissen G, Haus E, Northrup G, Portela A, Wendt H, Otsuka K, Kumagai Y, Watanabe Y, Zaslavskaya R. Clinical relevance of about-yearly changes in blood pressure and the environment. Int J Biometeorol 1996; 39: 161–75.

Nojima K, Otsuka K, Saito H: Ambulatory ECG-EEG monitoring in third generation rats bred under a constant light schedule. J Ambul Monit 1996; 9(4): 365–74.

1997

Tarquini B, Perfetto F, Tarquini R, Cornélissen G, Halberg F. Endothelin-1's chronome indicates diabetic and vascular disease chronorisk. Peptides 1997; 18: 119–32.

Otsuka K, Cornélissen G, Halberg F. Circadian rhythmic fractal scaling of heart rate variability in health and coronary artery disease. Clin Cardiol 1997; 20: 631–8.

Otsuka K, Cornélissen G, Halberg F. Age, gender and fractal scaling in heart rate variability. Clin Sci 1997; 93: 299–308.

Kazuma N, Otsuka K, Matsuoka I, Murata M. Heart rate variability during 24 hours in asthmatic children. Chronobiol Int 1997; 14(6): 597–606.

Halberg F, Cornélissen G, Otsuka K, Watanabe Y, Wood MA, Lambert CR, Zaslavskaya R, Gubin D, Petukhova EY, Delmore P, Bakken E. Rewards in practice from recycling heart rate, ectopy, ischemia, and blood pressure information. J Med Eng Technol 1997; 21: 174–84.

Tarquini B, Cornélissen G, Perfetto F, Tarquini R, Halberg F. About-half-weekly (circasemiseptan) component of the endothelin-1 (ET-1) chronome and vascular disease risk. Peptides 1997; 18: 1237–41.

Otsuka K, Nishimura Y, Kubo Y, Cornélissen G, Halberg F. Chronomes (rhythms, chaos, and age trends) of human heart rate variability in both genders. Comput Cardiol 1997; 24: 49–52.

Anderson Bernadas C, Cornélissen G, Turner CM, Koukkari WL. Rhythmic nature of thigmomorphogenesis and thermal stress of Phaseolus vulgaris L. shoots. J Plant Physiol 1997; 151: 575–80.

Tarquini B, Cornélissen G, Perfetto F, Tarquini R, Halberg F. Chronome assessment of circulating melatonin in humans. In Vivo 1997; 11: 473–84.

1998

Thaela M-J, Jensen MS, Cornélissen G, Halberg F, Nöddegaard F, Jakobsen K, Pierzynowski S.G. Rhythms in pancreatic secretion: circadian and ultradian variation in pancreatic secretion of meal-fed pigs after weaning. J Animal Sci 1998; 76(4): 1131–9.

1999

Herold M, Cornélissen G, Loeckinger A, Koeberle D, Koenig P, Halberg F. About 8-hourly variation of circulating human endothelin-1 (ET-1) in clinical health. Peptides 1999; 19: 821–5.

Cornélissen G, Gubin D, Halberg Francine, Milano G, Halberg Franz. Chronomedical aspects of oncology and geriatrics (review). In Vivo 1999; 13: 77–82.

Cornélissen G, Halberg F. [Book review: Hildebrandt G, Moser M, Lehofer M. Chronobiologie und Chronomedizin: Biologische Rhythmen/Medizinische Konsequenzen.] Chronobiol Int 1999; 16: 223.

Denisov M, Denisova O, Cornélissen G, Halberg F. Toward a heart rate and axillary temperature chronome in schizophrenia: gauge of disease severity. Functional Neurol 1999; 14: 13–20.

Blank M, Denisova O, Cornélissen G, Halberg F. Enhanced circasemiseptan (about 3.5-day) variation in the heart rate of cancer patients ? Anticancer Res 1999; 19: 853–66.

Halberg F, Sothern RB, Cornélissen G, Haus E. The chronobiology of prostate-specific antigen (PSA): case report and chrono-metaanalysis. Anticancer Res 1999; 19: 857–62.

Tarquini B, Tarquini R, Perfetto F, Cornélissen G, Halberg F. Genetic and environmental influences on human cord blood leptin concentration. Pediatrics 1999; 103: 998–1006.

Stoynev AG, Penev PD, Peneva AV, Cornélissen G, Halberg F, Ikonomov OC. Blood pressure and heart rate rhythmicity: differential effects of late pregnancy. Physiol Behav 1999; 66: 269–75.

Halberg F, Cornélissen G, Sampson M, Katinas G, Schwartzkopff O. Season's appreciations 1998. Neuroendocrinol Lett 1999; 20: 31–43.

Tarquini B, Cornélissen G, Tarquini R, Perfetto F, Halberg F. General and unspecific damping by malignancy of the circadian amplitude of circulating human melatonin ? Neuroendocrinol Lett 1999; 20: 25–8.

Maggioni C, Cornélissen G, Antinozzi R, Ferrario M, Grafe A, Halberg F. A half-yearly aspect of circulating melatonin in pregnancies complicated by intrauterine growth retardation. Neuroendocrinol Lett 1999; 20: 55–68.

Halberg F, Cornélissen G, Schwartzkopff O, Cagnoni M, Perfetto F, Tarquini R. Pineal mythology and chronorisk: the swan song of Brunetto Tarquini (*February 8, 1938 - †December 10, 1998). Neuroendocrinol Lett 1999; 20: 91–100.

Halberg F, Cornélissen G, Schwartzkopff O, Cagnoni M, Perfetto F, Tarquini R. Chronome/chrone. Neuroendocrinol Lett 1999; 20: 101.

Halberg F, Cornélissen G, Schwartzkopff O, Cagnoni M, Perfetto F, Tarquini R. Chronorisk/circadian-circannual (Macey, 1994). Neuroendocrinol Lett 1999; 20: 102–4.

Otsuka K, Cornélissen G, Shinagawa M, Nishimura Y, Kubo Y, Hotta N, Fujii C, Ishii T, Omori K, Watanabe Y, Nunoda S, Ohkawa S, Halberg F. Circadian reference values for different endpoints of heart rate variability. Comput Cardiol 1999; 26: 587–90.

Suda Y, Otsuka K, Ban T, Ichikawa S, Higasita R, Takeuchi Y. Heart rate variability after coronary artery bypass grafting. Comput Cardiol 1999; 26: 587–90.

Halberg F, Cornélissen G, Otsuka K, Schwartzkopff O. Chronodiagnosis-based chronotherapy: is blood pressure monitoring for weeks too costly to prevent long-term care for years? Geronto-Geriatrics 1999; 2: 20–30. [In English and Spanish.]

Otsuka K, Cornélissen G, Zhao ZY, Weydahl A, Delyukov A, Gorgo Y, Wang ZR, Perfetto F, Tarquini R, Kubo Y, Shinagawa M, Hotta N, Ishii T, Omori K, Watanabe Y, Nunoda S-i, Ohkawa S-i, Halberg F. Rhythm and trend elements in the time structure, chronome, of heart rate variability. Geronto-Geriatrics 1999; 2: 31–48. [In English and Spanish.]

2000

Halberg F, Cornélissen G, Burioka N, Katinas G, Sampson M, Schwartzkopff O. Circasemicentennial season's appreciations. Neuroendocrinol Lett 2000; 21: 59–68.

Halberg F, Smith HN, Cornélissen G, Delmore P, Schwartzkopff O, International BIOCOS Group. Hurdles to asepsis, universal literacy, and chronobiology—all to be overcome. Neuroendocrinol Lett 2000; 21: 145–60

Salti R, Galluzzi F, Bindi G, Perfetto F, Tarquini R, Halberg F, Cornélissen G. Nocturnal melatonin patterns in children. J Clin Endocr Metab 2000; 85: 2137–44

Cornélissen G, Halberg F, Burioka N, Perfetto F, Tarquini R, Bakken EE. Do plasma melatonin concentrations decline with age? [Letter to the editor re: Zeitzer JM, Daniels JE, Duffy JF, et al. Do plasma melatonin concentrations decline with age? Am J Med 1999; 107: 432–6.] Am J Med 2000; 109: 343–4 (Sept 1).

Simpson HW, Cornélissen G, Katinas GS, Halberg F. Meta-analysis of sequential luteal-cycle-associated changes in human breast tissue. Breast Cancer Res Treat 2000; 63: 171–3.

Herold M, Cornélissen G, Rawson MJ, Katinas GS, Alinder C, Bratteli C, Gubin D, Halberg F. About-daily (circadian) and about-weekly (circaseptan) patterns of human salivary melatonin. J Anti-Aging Med 2000; 3: 263–7.

Kazuma N, Otsuka K, Miyakawa M, Shirase E, Matsuoka I, Murata M. Seasonal variation in heart rate variability in asthmatic children. Chronobiol Int 2000; 17(4): 503–11.

2001

Halberg F, Cornélissen G, Katinas G, Hillman D, Schwartzkopff O. Season's Appreciations 2000: Chronomics complement, among many other fields, genomics and proteomics. Neuroendocrinol Lett 2001; 22: 53–73.

Katinas GS, Halberg F, Cornélissen G, Hawkins D, Bueva MV, Korzhevsky DE, Sapozhnikova LR, Rhodus N, Schaffer E. About 8- and ~84-hour rhythms in endotheliocytes as in endothelin-1 and effect of trauma. Peptides 2001; 22: 647–59.

Halberg F, Cornélissen G, Watanabe Y, Otsuka K, Fiser B, Siegelova J, Mazankova V, Maggioni C, Sothern RB, Katinas GS, Syutkina EV, Burioka N, Schwartzkopff O. Near 10-year and longer periods modulate circadians: intersecting anti-aging and chronoastrobiological research. J Gerontol A Biol Sci Med Sci 2001; 56: M304–M324.

Schaffer E, Cornélissen G, Rhodus N, Halhuber M, Watanabe Y, Halberg F. Blood pressure outcomes of chronobiologically normotensive dental patients: a 7-year follow-up. J Am Dental Assn 2001; 132: 891–9.

Aguilar-Roblero R, Arechiga H, Ashkenazi I, Burioka N, Cipolla Neto J, Cornélissen G, Markus R, Marques N, Menezes AAL, Monk TH, Ralph M, Valdez-Ramirez P, Menna-Barreto L. The brain decade in debate: IV. Chronobiology. Braz J Med Biol Res 2001; 34: 831–41.

Burioka N, Cornélissen G, Halberg F, Kaplan DT. Relationship between correlation dimension and indices of linear analysis in both respiratory movement and electroencephalogram. Clin Neurophysiol 2001; 112: 1147–53.

Delyukov A, Gorgo Yu, Cornélissen G, Otsuka K, Halberg F. Natural environmental associations in a 50-day human electrocardiogram. Int J Biometeorol 2001; 45: 90–9.

Halberg F, Cornélissen G, Otsuka K, Katinas G, Schwartzkopff O. Essays on chronomics spawned by transdisciplinary chronobiology: Witness in time: Earl Elmer Bakken. Neuroendocrinol Lett 2001; 22: 359–84.

Otsuka K, Cornélissen G, Halberg F. Preface: Circadian rhythms and clinical chronobiology. Biomedicine and Pharmacotherapy 2001; 55: 7–18.

Kazuma N, Otsuka K. Seasonal variation in 1/f fluctuations of heart rate in asthmatic children. Biomed Pharmacother 2001; 55(Suppl 1): 102s–5s.

Saitoh T, Watanabe Y, Kubo Y, Shinagawa M, Otsuka K, Ohkawa SI, Watanabe T. Intragastric acidity and circadian rhythm. Biomed Pharmacother 2001; 55(Suppl 1): 138s–41s.

Suda Y, Otsuka K, Niinami H, Ichikawa S, Ban T, Higashita R, Takeuchi Y. Changes in ultra-low and very low frequency heart rate variability after coronary artery bypass grafting. Biomed Pharmacother 2001; 55(Suppl 1): 110s–4s.

Halberg F, Cornélissen G, Otsuka K, Katinas G, Delmore P, Schwartzkopff O. Earl Elmer Bakken: Pacing chronomics. Biomed Pharmacother 2001; 55: 19–20.

Cornélissen G, Engebretson M, Johnson D, Otsuka K, Burioka N, Posch J, Halberg F. The week, inherited in neonatal human twins, found also in geomagnetic pulsations in isolated Antarctica. Biomed Pharmacother 2001; 55: 32–50.

Delyukov A, Gorgo Y, Cornélissen G, Otsuka K, Halberg F. Infradian, notably circaseptan testable feedsidewards among chronomes of the ECG and air temperature and pressure. Biomed Pharmacother 2001; 55: 84–92.

Mitsutake G, Otsuka K, Cornélissen G, Herold M, Günther R, Dawes C, Burch JB, Watson D, Halberg F. Circadian and infradian rhythms in mood. Biomed Pharmacother 2001; 55: 94–100.

Shinagawa M, Kubo Y, Otsuka K, Ohkawa S, Cornélissen G, Halberg F. Impact of circadian amplitude and chronotherapy: relevance to prevention and treatment of stroke. Biomed Pharmacother 2001; 55: 125–32.

Halberg F, Cornélissen G, Otsuka K, Schwartzkopff O, Halberg J, Bakken EE. Chronomics. Biomed Pharmacother 2001; 55: 153–90.

2002

Johnson-Alvares D, Cornélissen G, Portela A, Halberg E, Rudney JD, Yeh C-k, Dodds M, Haus E, Schaffer E, Blank M, Halberg F. Salivary CA130 with and without unilateral autonomic parotid denervation of rats fed different diets. In Vivo 2002; 16: 25–8.

Halberg F, Cornelissen G, Bingham C, Hillman D, Katinas G, Sampson M, Revilla M, Prikryl P Sr, Prikryl P Jr, Sanchez de la Pena S, Gonzalez C, Amory-Mazaudier C, Bouvet J, Barnwell F, Maggioni C, Sothern RB, Wang ZR, Schwartzkopff O, Bakken E. Season's Appreciations 2001. Neuroendocrinol Lett 2002; 23: 170–87.

Appenzeller O, Cornélissen G, Halberg F, Wallace J, Costa MA. Biological rhythms and behavior – then and now. Med Sci Monitor 2002; 8: SR27–30.

Halberg F, Cornélissen G, Katinas G, Appenzeller O, Otsuka K, Sothern RB, Tarquini R, Perfetto F, Maggioni C, Wilson D, Schröder W, Schwartzkopff O, Kessler T, Wang Z, Burioka N, Watanabe Y, Bakken E. System times and time horizons for biospheric near-matches of primarily non-photic environmental cycles. Biomed Pharmacother 2002; 56(Suppl. 2): 266s–72s.

Yamanaka T, Cornélissen G, Halberg F, Katinas G, Hörz Helga, Hörz Herbert, Otsuka K, Bakken EE. Marriage and divorce over a century in Japan: Social biomedicine, not yet societal therapy. Biomed Pharmacother 2002; 56(Suppl 2): 314s–8s.

Watanabe Y, Cornélissen G, Hellbrügge T, Watanabe F, Otsuka K, Schwartzkopff O, Halberg F. Partial spectral element in the chronome of a human neonatal heart rate at term. Biomed Pharmacother 2002; 56(Suppl 2): 374s–8s.

Halberg F, Cornélissen G, Schwartzkopff O. [Book review of Deppert W, Köther K, Kralemann B, Lattmann C, Martens N, Schaefer J, editors. Selbstorganisierte Systemzeiten: Ein interdisziplinärer Diskurs zur Modellierung lebender Systeme auf der Grundlage interner Rhythmen.] Neuroendocrinol Lett 2002; 23: 262–5.

Kazuma N, Otsuka K, Wakamatsu K, Shirase E, Matsuoka I. Heart rate variability in normotensive healthy children with aging. Clin Exp Hypertens 2002; 24(1–2): 83–9.

Mori H, Nakamura N, Tamura N, Sawai M, Tanno T, Narita T, Singh RB, Otsuka K. Circadian variation of basal total vascular tone and chronotherapy in patients with vasospastic angina pectoris. Biomed Pharmacother 2002; 56 Suppl 2:339s–44s.

Saitoh T, Watanabe Y, Kubo Y, Shinagawa M, Otsuka K, Ohkawa SI, Watanabe T. Effect of H2 blockers on the circadian rhythm of intragastric acidity. Biomed Pharmacother 2002; 56(Suppl 2): 349s–52s

Otsuka K, Halberg F. Preface. Quo vadis basic and clinical chronobiology. Clin Exper Hypertens 2002; 24(1&2), vii–ix.

2003

Singh RB, Cornélissen G, Weydahl A, Schwartzkopff O, Katinas G, Otsuka K, Watanabe Y, Yano S, Mori H, Ichimaru Y, Mitsutake G, Pella D, Fanghong L, Zhao Z, Rao RS, Gvozdjakova A, Halberg F. Circadian heart rate and blood pressure variability considered for research and patient care. Int J Cardiol 2003; 87: 9–28.

Burioka N, Cornélissen G, Halberg F, Kaplan DT, Suyama H, Sako T, Shimizu E. Approximate entropy of human respiratory movement during eye-closed waking and different sleep stages. Chest 2003; 123: 80–6.

Qu Y, Wang ZR, Huang X, Wan CW, Yang CL, Liu BL, Cornélissen G, Halberg F. Circadian telomerase activity and DNA synthesis for timing peptide administration. Peptides 2003; 24: 363–9.

Watanabe Y, Cornélissen G, Watanabe M, Watanabe F, Otsuka K, Ohkawa S-i, Kikuchi T, Halberg F. Effects of autogenic training and antihypertensive agents on circadian and circaseptan variation of blood pressure. Clin Exp Hypertens 2003; 25: 405–12.

Cornélissen G, with Halberg F, Schwartzkopff O, Katinas G, Johnson D, Otsuka K, Watanabe Y, Wang ZR, Wan CW, Perfetto F, Tarquini R, Maggioni C, Syutkina EV, Masalov A, Siegelova J, Zhao ZY, Singh RB, Singh RK, Delyukov A, Gorgo Y, Zaslavskaya RM, Gubin GD, Gubin DG, Kumagai Y, Uezono K, Wilson D, Weydahl A, Bakken E. Editor's foreword: What Gesell wished, Hellbrügge accomplished: Chronomics of child development. Neuroendocrinol Lett 2003; 24(Suppl 1): 14–24.

Halberg F, Schwartzkopff O, Cornélissen G. Chronomik ergänzt Genomik: Rechnergestützte Wissenszweige treffen sich in München, November 29–30, 2002. Neuroendocrinol Lett 2003; 24(Suppl 1): 27–59.
Cornélissen G, Reinberg A, Haen E, Schwartzkopff O, Halberg F. Laudatio: Theodor Hellbrügge — Clarus-pons: Utinam multos annos diutius vivas ad rehabilitationis præhabilitationisque scientias perficiendas ! Neuroendocrinol Lett 2003; 24(Suppl. 1): 1–7.
Schwartzkopff O, Cornélissen G, Katinas G, Sampson M, Halberg F. (Helena's Minnesota team): Present history: Helena Raskova Nonageneriana: Ad multos annos. Neuroendocrinol Lett 2003; 24(Suppl 1): 64–5.
Halberg F, Wang ZR, Schwartzkopff O, Cornélissen G. Chronomik für die grosse Öffentlichkeit. Neuroendocrinol Lett 2003; 24(Suppl 1): 74–83.
Kenner T, Cornélissen G, Katinas G, Schwartzkopff O, Kenner B, Halberg F. Population cycle in sudden infant death syndrome (SIDS) ? Neuroendocrinol Lett 2003; 24(Suppl 1): 96–100.
Maggioni C, Cornélissen G, Syutkina EV, Johnson D, Halberg F. Pharmacovigilance: betamimetic drug exposure in pregnancy enhances cardiovascular disease risk of offspring. Neuroendocrinol Lett 2003; 24(Suppl 1): 102–4.
Marazzi A, Ruffieux C, Cornélissen G, Syutkina EV, Johnson D, Halberg F. Circadian and circaseptan patterns of natality and perinatal mortality of infants with different birth weights. Neuroendocrinol Lett 2003; 24(Suppl 1): 105–10.
Siegelova J, Cornélissen G, Schwartzkopff O, Halberg F. Time structures in the development of children. Neuroendocrinol Lett 2003; 24(Suppl 1): 126–31.
Syutkina EV, Cornélissen G, Yatsyk G, Studenikin M, Baranov A, Halberg F. Over a decade of clinical chrononeonatology and chronopediatrics in Moscow. Neuroendocrinol Lett 2003; 24(Suppl 1): 132–8.
Watanabe Y, Nintcheu-Fata S, Katinas G, Cornélissen G, Otsuka K, Hellbrügge T, Schwartzkopff O, Bakken E, Halberg F. Methodology: partial moving spectra of postnatal heart rate chronome. Neuroendocrinol Lett 2003; 24(Suppl 1): 139–44.
Kumagai Y, Kumagai Eu, Kumagai Er, Cornélissen G, Halberg E, Halberg Francine, Schwartzkopff O, Halberg F. Circaseptan axillary temperature rhythm development in a chronomic and familial context. Neuroendocrinol Lett 2003; 24(Suppl 1): 145–51.
Perfetto F, Tarquini R, Salti R, Laffi G, Bubenik G, Kocharyan S, Aslanyan N, Cornélissen G, Katinas GS, Schwartzkopff O, Halberg F. Far from melatonin's time horizon: Halting steps from pediatric ultradians toward a chronome. Neuroendocrinol Lett 2003; 24(Suppl 1): 165–70.
Tarquini R, Perfetto F, Laffi G, Mello G, Cornélissen G, Johnson D, Halberg F. Circadian and circannual aspects of leptin chronome in cord blood. Neuroendocrinol Lett 2003; 24(Suppl 1): 171–4.
Edmunds L, Balzer I, Cornélissen G, Halberg F. Toward a chronome of Euglena gracilis (Klebs). Neuroendocrinol Lett 2003; 24(Suppl 1): 192–9.
Katinas G, Berger S, Cornélissen G, Hillman D, Woolum JC, Engebretson M, Syutkina EV, Masalov A, Wang ZR, Wan CM, Bakken EE, Schwartzkopff O, Halberg F. Acetabularia acetabulum: Unicellular model of neonatal vascular chronome. Neuroendocrinol Lett 2003; 24(Suppl 1): 201–7.
Shinoda H, Ohtsuka-Isoya M, Cornélissen G, Halberg F. Putative circaseptans or other infradians in murine dentin accretion and the suprachiasmatic nuclei. Neuroendocrinol Lett 2003; 24 (Suppl 1): 208–11.
Woolum JC, Cornélissen G, Hayes DK, Halberg F. Spectral aspects of Acetabularia's light transmission chronome: effects of enucleation. Neuroendocrinol Lett 2003; 24(Suppl 1): 212–5.
Zhou B, Wang ZR, Wan CM, Sothern RB, Cornélissen G, Halberg F. Diazepam and melatonin effects upon circadian variation of cultured murine myocardiocytes. Neuroendocrinol Lett 2003; 24(Suppl 1): 216–22.

Burioka N, Cornélissen G, Otsuka K, Shimizu E, Halberg F. Linear and nonlinear indices of variability in respiratory movement, the electroencephalogram and the electrocardiogram. Neuroendocrinol Lett 2003; 24(Suppl 1): 223–30.

Zaslavskaya RM, Makarova LA, Shakarova AN, Komarov F, Wang ZR, Wan CM, Katinas G, Cornélissen G, Halberg F. Individualized time series-based assessment of melatonin effects on blood pressure: model for pediatricians. Neuroendocrinol Lett 2003; 24(Suppl 1): 238–46.

Halberg F, Cornélissen G, Delmore P, Schwartzkopff O, International Womb-to-Tomb Chronome Group: The costs of ignoring a long-overdue chronomedicine: the chronome initiative emerging from the catacombs. Neuroendocrinol Lett 2003; 24(Suppl 1): 247–56.

Halberg F, Cornélissen G, Bingham C, Witte H, Ribary U, Hesse W, Petsche H, Engebretson M, Geissler H-G, Weiss S, Klimesch W, Rappelsberger P, Katinas G, Schwartzkopff O. Chronomics: Imaging in time by phase synchronization reveals wide spectral-biospheric resonances beyond short rhythms. Neuroendocrinol Lett 2003; 24: 355–80.

Cornélissen G, Halberg F, Tarquini R, Perfetto F, Salti R, Laffi G, Otsuka K. Point and interval estimations of circadian melatonin ecphasia in Smith-Magenis syndrome. Biomed Pharmacother 2003; 57(Suppl 1): 31s–4s.

Lee MS, Lee JS, Lee JY, Cornélissen G, Otsuka K, Halberg F. About 7-day (circaseptan) and circadian changes in cold pressor test (CPT). Biomed Pharmacother 2003; 57 (Suppl 1): 39s–44s.

Cornélissen G, Halberg F, Pöllmann L, Pöllman B, Katinas GS, Minne H, Breus T, Sothern RB, Watanabe Y, Tarquini R, Perfetto F, Maggioni C, Wilson D, Gubin D, Otsuka K, Bakken EE. Circasemiannual chronomics: half-yearly biospheric changes in their own right and as a circannual waveform. Biomed Pharmacother 2003; 57(Suppl 1): 45s–54s.

Lee JS, Lee JY, Cornélissen G, Otsuka K, Halberg F. Effects of diaphragmatic breathing on ambulatory blood pressure and heart rate. Biomed Pharmacother 2003; 57(Suppl 1): 87s–91s.

Kubo Y, Murakami S, Matsuoka O, Notta N, Oinuma S, Shinagawa M, Omori K, Nunoda S-i, Otsuka K, Ohkawa S-i, Cornélissen G, Halberg F. Toward chronocardiologic and chronomic insights: dynamics of heart rate associated with head-up tilting. Biomed Pharmacother 2003; 57(Suppl 1): 110s–5s.

Maschke C, Harder J, Cornélissen G, Hecht K, Otsuka K, Halberg F. Chronoecoepidemiology of "strain": infradian chronomics of urinary cortisol and catecholamines during nightly exposure to noise. Biomed Pharmacother 2003; 57(Suppl 1): 126s–35s.

Otsuka K, Cornélissen G, Schwartzkopff O, Bakken EE, Halberg F, Burioka N, Katinas GS, Kane R, Regal PJ, Schaffer E, Sonkowsky R, Patterson R, Engebretson M, Brockway B, Wang ZR, Delmore P, Halpin C, Sarkozy S, Wall D, Halberg J. Clinical chronobiology and chronomegeriatrics: At variance with recommendations of subsequent guidelines, yet focusing indeed on pre-hypertension in the physiological range. Biomed Pharmacother 2003; 57(Suppl 1): 164s–98s.

Halberg F, Cornélissen G, Stoynev A, Ikonomov O, Katinas G, Sampson M, Wang ZR, Wan CM, Singh RB, Otsuka K, Sothern RB, Sothern SB, Sothern MI, Syutkina EV, Masalov A, Perfetto F, Tarquini R, Maggioni C, Kumagai Y, Siegelova J, Fiser B, Homolka P, Dusek J, Uezono K, Watanabe Y, Wu JY, Sonkowsky R, Schwartzkopff O, Hellbrügge T, Spector NH, Baciu I, Hriscu M, Bakken E. Season's Appreciations 2002 and 2003. Imaging in time: The transyear (longer-than-the-calendar year) and the half-year. Neuroendocrinol Lett 2003; 24: 421–40.

Singh Ranjana, Singh Rajesh K, Mahdi AA, Singh RK, Kumar A, Tripathi AK, Rai R, Singh U, Cornélissen G, Schwartzkopff O, Halberg F. Circadian periodicity of plasma lipid peroxides and other anti-oxidants as putative markers in gynecological malignancies. In Vivo 2003; 17: 593–600.

Borer KT, Cornélissen G, Halberg F, Hughes C. Health impact of training intensity in older individuals. Kinesiology 2003; 35: 210–18.

Otsuka K, Cornélissen G, Matsuoka O, Kubo Y, Murakami S, Shinagawa M, Yamanaka T, Nishimura Y, Hotta N, Nunoda S-i, Ohkawa S-i, Halberg F. Mapping of blood pressure and heart rate variability: a model for pediatricians. Neuroendocrinol Lett 2003; 24(Suppl 1): 157–64.

2004

Löckinger A, Köberle D, St. König P, Saria A, Herold M, Cornélissen G, Halberg F. Neuropeptide chronomics in clinically healthy young adults: circaoctohoran and circadian patterns. Peptides 2004; 25: 533–42.

Hrushesky W, Wood P, Levi F, Roemeling R v, Bjarnason G, Focan C, Meier K, Cornélissen G, Halberg F. A recent illustration of some essentials of circadian chronotherapy study design [letter]. J Clin Oncol 2004; 22: 2971–2.

Chen LG, Wang ZR, Wan CM, Xiao J, Guo L, Guo HL, Cornélissen G, Halberg F. Circadian renal rhythms influenced by implanted encapsulated hANP-producing cells in Goldblatt hypertensive rats. Gene Therapy 2004; 11: 1515–22.

Wang ZR, Chen LG, Wan CM, Qu Y, Cornélissen G, Halberg F. In vitro circadian ANP secretion by gene transferring cells encapsulated in polycaprolactone tubes: gene chronotherapy. Peptides 2004; 25: 1259–67.

Perfetto F, Tarquini R, Cornélissen G, Mello G, Tempestini A, Gaudiano P, Mancuso F, Halberg F. Circadian phase difference of leptin in android versus gynoid obesity. Peptides 2004; 25: 1297–306.

Halberg F, Cornélissen G, Schack B. Self-experimentation chronomics for health surveillance and science; also transdisciplinary civic duty? Behav Brain Sci 2004; 27(2): 267–9.

Halberg F, Otsuka K, Katinas G, Sonkowsky R, Regal P, Schwartzkopff O, Jozsa R, Olah A, Zeman M, Bakken EE, Cornélissen G. A chronomic tree of life: ontogenetic and phylogenetic 'memories' of primordial cycles - keys to ethics. Biomed Pharmacother 2004; 58(Suppl 1): S1–11.

Turtoi C, Cornélissen G, Wilson D, Halberg F. About 21-year Hale cyclic signature in agriculture complement those in biomedicine. Biomed Pharmacother 2004; 58(Suppl 1): S12–4.

Chibisov SM, Cornélissen G, Halberg F. Magnetic storm effect on the circulation of rabbits. Biomed Pharmacother 2004; 58(Suppl 1): S15–9.

Cornélissen G, Halberg F, Bakken EE, Singh RB, Otsuka K, Tomlinson B, Delcourt A, Toussaint G, Bathina S, Schwartzkopff O, Wang ZR, Tarquini R, Perfetto F, Pantaleoni GC, Jozsa R, Delmore PA, Nolley E. 100 or 30 years after Janeway or Bartter, Healthwatch helps avoid "flying blind". Biomed Pharmacother 2004; 58(Suppl 1): S69–86.

Gobello C, Cornélissen G, Corrada YA, Goya RG, Otsuka K, Halberg F. Circannual and/or transannual variations in growth hormone and beyond -- call for long data series. Biomed Pharmacother 2004; 58(Suppl 1): S87–90.

Komlos J, Cornélissen G, Woitek U, Otsuka K, Halberg F. Time structures, chronomes, of soldiers' stature mimicking Hale cycle in neonatal body length. Biomed Pharmacother 2004; 58(Suppl 1): S135–9.

Cornélissen G, Axelrod DE, Halberg F. About-weekly variations in nocturia. Biomed Pharmacother 2004; 58(Suppl 1): S140–4.

Halberg F, Cornélissen G, Sothern RB, Katinas GS. Zhennan Xue -- ad multos transannos. Neuroendocrinol Lett 2004; 25(4): 310–6.

Halberg F, Cornélissen G, Bakken EE, Sothern RB, Schwartzkopff O, Hamburger C. Transyears: new endpoints for gerontology and geriatrics or confusing sources of variability? J Gerontol A Biol Sci Med Sci 2004; 59: 1344–7.

2005

Burioka N, Miyata M, Cornélissen G, Halberg F, Takeshima T, Kaplan DT, Suyama H, Endo M, Maegaki Y, Nomura T, Tomita Y, Nakashima K, Shimizu E. Approximate entropy in the electroencephalogram during wake and sleep. Clin EEG Neurosci 2005; 36: 21–3.

Liu Y, Wang Y, Wan C, Zhou W, Peng T, Liu Y, Wang Z, Li G, Cornélissen G, Halberg F. The role of mPer1 in morphine dependence in mice. Neuroscience 2005; 130: 383–8.

Katinas GS, Halberg F, Cornélissen G, Otsuka K, Bakken EE. Time-microscopy for all kinds of data including circadian clock biology. Biomed Pharmacother 2005; 59(Suppl 1): S20–3.

Jozsa R, Halberg F, Cornélissen G, Zeman M, Kazsaki J, Csernus V, Katinas GS, Wendt HW, Schwartzkopff O, Stebelova K, Dulkova K, Chibisov SM, Engebretson M, Pan W, Bubenik GA, Nagy G, Herold M, Hardeland R, Hüther G, Pöggeler B, Tarquini R, Perfetto F, Salti R, Olah A, Csokas N, Delmore P, Otsuka K, Bakken EE, Allen J, Amory-Mazaudier C. Chronomics, neuroendocrine feedsidewards and the recording and consulting of nowcasts -- forecasts of geomagnetics. Biomed Pharmacother 2005; 59(Suppl 1): S24–30.

Cornélissen G, Halberg F, Kovac M, Mikulecky M, Otsuka K, Bakken E. Geographic and extraterrestrial aspects of morbidity and/or mortality patterns from myocardial infarction and stroke. Biomed Pharmacother 2005; 59(Suppl 1): S68–75.

Maggioni C, Cornélissen G, Otsuka K, Halberg F, Consonni D, Nicolini U. Circadian rhythm of maternal blood pressure and fetal growth. Biomed Pharmacother 2005; 59(Suppl 1): S86–91.

Halberg F, Cornélissen G, Panksepp J, Otsuka K, Johnson D. Chronomics of autism and suicide. Biomed Pharmacother 2005; 59(Suppl 1): S100–8.

Jozsa R, Olah A, Cornélissen G, Csernus V, Otsuka K, Zeman M, Nagy G, Kazsaki J, Stebelova K, Csokas N, Pan W, Herold M, Bakken EE, Halberg F. Circadian and extracircadian exploration during daytime hours of circulating corticosterone and other endocrine chronomes. Biomed Pharmacother 2005; 59(Suppl 1): S109–16.

Cornélissen G, Halberg F, Mikulecky M, Florida P, Faraone P, Yamanaka T, Murakami S, Otsuka K, Bakken EE. Yearly and perhaps transyearly human natality patterns near the equator and at higher latitudes. Biomed Pharmacother 2005; 59(Suppl 1): S117–22.

Schwartzkopff O, Cornélissen G, Otsuka K, Halberg F. India revisited: a new hospital and center culminating multidecadal cooperation: From cancer chronotherapy to routine cardiovascular chronotheranostics and further joint Indian-US and international BIOCOS research. Biomed Pharmacother 2005; 59(Suppl 1): S123–31.

Prikryl P, Cornélissen G, Otsuka K, Halberg F. Plasma catecholamines: follow-up on 10-year study in health and cardiovascular disease. Biomed Pharmacother 2005; 59(Suppl 1): S180–7.

Halberg F, Cornélissen G, Faraone P, Poeggeler B, Hardeland R, Katinas G, Schwartzkopff O, Otsuka K, Bakken EE. Prokaryotic and eukaryotic unicellular chronomics. Biomed Pharmacother 2005; 59(Suppl 1): S192–202.

Stebelova K, Zeman M, Cornélissen G, Bubenik G, Jozsa R, Hardeland R, Poeggeler B, Huether G, Olah A, Nagy G, Csernus V, Kazsaki J, Pan W, Otsuka K, Bakken EE, Halberg F. Chronomics reveal and quantify circadian rhythmic melatonin in duodenum of rats. Biomed Pharmacother 2005; 59(Suppl 1): S209–12.

Zeman M, Jozsa R, Cornélissen G, Stebelova K, Bubenik G, Olah A, Poeggeler B, Huether G, Hardeland R, Nagy G, Csernus V, Pan W, Otsuka K, Halberg F. Chronomics: circadian lead of extrapineal vs. pineal melatonin rhythms with an infradian hypothalamic exploration. Biomed Pharmacother 2005; 59(Suppl 1): S213–9.

Poeggeler B, Cornélissen G, Huether G, Hardeland R, Jozsa R, Zeman M, Stebelova K, Olah A, Bubenik G, Pan W, Otsuka K, Schwartzkopff O, Bakken EE, Halberg F. Chronomics affirm extending scope of lead in phase of duodenal vs. pineal circadian melatonin rhythms. Biomed Pharmacother 2005; 59(Suppl 1): S220–4.

Tarquini R, Mazzoccoli G, Dolenti S, Gaudiano P, Comuni C, Laffi G, Perfetto F, Otsuka K, Cornélissen G, Halberg F. Circasemidian rather than circadian variation of circulating osteoprotegerin in clinical health. Biomed Pharmacother 2005; 59(Suppl 1): S225–8.

Singh R, Singh RK, Singh RK, Tripathi AK, Cornélissen G, Schwartzkopff O, Otsuka K, Halberg F. Chronomics of circulating plasma lipid peroxides and anti-oxidant enzymes and other related molecules in cirrhosis of liver: In the memory of late Shri Chetan Singh. Biomed Pharmacother 2005; 59(Suppl 1): S229–35.

Halberg F, Cornélissen G, Otsuka K, Fiser B, Mitsutake G, Wendt HW, Johnson P, Gigolashvili M, Breus T, Sonkowsky R, Chibisov SM, Katinas G, Siegelova J, Dusek J, Singh RB, Berri BL, Schwartzkopff O. Incidence of sudden cardiac death, myocardial infarction and far- and near-transyears. Biomed Pharmacother 2005; 59(Suppl 1): S239–61.

Singh RB, Singh V, Kulshrestha SK, Singh S, Gupta P, Kumar R, Krishna A, Srivastav SS, Gupta SB, Pella D, Cornélissen G. Social class and all-cause mortality in an urban population of North India. Acta cardiologica 2005; 60 (6): 611–7.

2006

Zheng H, Liu YY, Li W, Yang B, Chen DB, Wang XJ, Jiang Z, Wang HX, Wang ZR, Cornélissen G, Halberg F. Beneficial effects of exercise and its molecular mechanisms on depression in rats. Behav Brain Res 2006; 168: 47–55.

Halberg F, Cornélissen G, Katinas G, Tvildiani L, Gigolashvili M, Janashia K, Toba T, Revilla M, Regal P, Sothern RB, Wendt HW, Wang ZR, Zeman M, Jozsa R, Singh RB, Mitsutake G, Chibisov SM, Lee J, Holley D, Holte JE, Sonkowsky RP, Schwartzkopff O, Delmore P, Otsuka K, Bakken EE, Czaplicki J, International BIOCOS Group. Chronobiology's progress: season's appreciations 2004–2005. Time-, frequency-, phase-, variable-, individual-, age- and site-specific chronomics. J Applied Biomed 2006; 4: 1–38.

Halberg F, Cornélissen G, Katinas G, Tvildiani L, Gigolashvili M, Janashia K, Toba T, Revilla M, Regal P, Sothern RB, Wendt HW, Wang ZR, Zeman M, Jozsa R, Singh RB, Mitsutake G, Chibisov SM, Lee J, Holley D, Holte JE, Sonkowsky RP, Schwartzkopff O, Delmore P, Otsuka K, Bakken EE, Czaplicki J, International BIOCOS Group. Chronobiology's progress: Part II, chronomics for an immediately applicable biomedicine. J Applied Biomed 2006; 4: 73–86.

Czaplicki J, Cornélissen G, Halberg F. GOSA, a simulated annealing-based program for global optimization of nonlinear problems, also reveals transyears. J Applied Biomed 2006; 4: 87–93.

Salti R, Tarquini R, Stagi S, Perfetto F, Cornélissen G, Laffi G, Mazzoccoli G, Halberg F. Age-dependent association of exposure to television screen with children's urinary melatonin excretion? Neuroendocrinol Lett 2006; 27(1–2): 73–80.

Cornélissen G, Halberg F. Chronomics of suicides and the solar wind. Br J Psychiatry 2006; 189: 567–8.

Hua H, Wang YQ, Wan CM, Liu YY, Zhu B, Yang CL, Wang XJ, Wang ZR, Cornélissen-Guillaume G, Halberg F. Circadian gene mPer2 overexpression induces cancer cell apoptosis. Cancer Sci 2006; 97(7): 589–96.

2007

Refinetti R, Cornélissen G, Halberg F. Procedures for numerical analysis of circadian rhythms. Biol Rhythm Res 2007; 38(4): 275–325.

Hoogerwerf WA, Hellmich HL, Cornélissen G, Halberg F, Shahinian VB, Bostwick J, Savidge TC, Cassone VM. Clock gene expression in the murine gastrointestinal tract: endogenous rhythmicity and effects of a feeding regimen. Gastroenterology 2007; 133: 1250–60.

2008

Otsuka K, Cornélissen G, Norboo T, Takasugi E, Halberg F. Chronomics and "glocal" (combined global and local) assessment of human life. Prog Theoretical Phys 2008; Suppl. 173: 134–52.

Halberg F, Cornélissen G, Sothern RB, Katinas GS, Schwartzkopff O, Otsuka K. Cycles tipping the scale between death and survival (= "life"). Prog Theoretical Phys 2008; Suppl. 173: 153–81.

Olah A, Jozsa R, Csernus V, Sandor J, Muller A, Zeman M, Hoogerwerf W, Cornélissen G, Halberg F. Stress, geomagnetic disturbance, infradian and circadian sampling for circulating corticosterone and models of human depression? Neuro Res 2008; 13(2): 85–96.

Halberg F, Sothern RB, Cornélissen G, Czaplicki J. Chronomics, human time estimation, and aging. Clin Interventions Aging 2008; 3(4): 749–60.

Halberg F, Cornélissen G, Katinas GS, Siegelova J, Watanabe Y, Sanchez de la Peña S. Introduction: Internet- and self-help-based, scientific, preventive and curative chronobiologic health care: much more for less. Geronto-Geriatrics: Int J Gerontology-ChronomeGeriatrics 2008; 11(14): 110–8.

Halberg F, Cornélissen G, Beaty LA. Pre-habilitation by women to save the cost of rehabilitation by detecting and treating a pre-metabolic syndrome. Geronto-Geriatrics: Int J Gerontology-ChronomeGeriatrics 2008; 11(14): 159–83.

2009

Halberg F, Cornélissen G, Wilson D, Singh RB, De Meester F, Watanabe Y, Otsuka K, Khalilov E. Chronobiology and chronomics: detecting and applying the cycles of nature. Biologist 2009; 56(4): 209–14.

2010

Hoogerwerf WA, Shahinian VB, Cornélissen G, Halberg F, Bostwick J, Timm J, Cassone VM. Rhythmic changes in colonic motility are regulated by period genes. Am J Physiol Gastrointest Liver Physiol 2010; 298: G143-G150. doi:10.1152/ajpgi.00402.2009. PMID: 19926812.

He Y, Wu X, Khan RS, Kastin AJ, Cornélissen-Guillaume GG, Hsuchou H, Robert B, Halberg F, Pan W. IL-15 receptor deletion results in circadian changes of locomotor and metabolic activity. J Molec Neurosci 2010; 41(2): 315–21.

Lista G, Castoldi F, Compagnoni G, Maggioni C, Cornélissen G, Halberg F. Neonatal and maternal concentrations of hydroxil radical and total antioxidant system: protective role of placenta against fetal oxidative stress. Neuroendocrinol Lett 2010; 31(3): 319–24.

Cornélissen G, Galli C, Halberg F, De Meester F, Risé P, Wilczynska-Kwiatek A, Singh RB, Guillaume F. Circadian time structure of fatty acids and vascular monitoring. J Appl Biomed 2010; 8: 93–109.

Bostwick J, Nguyen D, Cornélissen G, Halberg F, Hoogerwerf WA. Effects of acute and chronic STZ-induced diabetes on clock gene expression and feeding in the gastrointestinal tract. Mol Cell Biochem 2010; 338(1–2): 203–13.

Gupta AK, Cornélissen G, Greenway FL, Dhoopati V, Halberg F, Johnson WD. Abnormalities in circadian blood pressure variability and endothelial function: pragmatic markers for adverse cardiometabolic studies in asymptomatic obese adults. Cardiovasc Diabetol 2010; Sept 24; 9: 58(12 pp.) doi:10.1186/1475-2840-9-58.

Cornélissen G, Halberg F, Syutkina EV, Yatsyk GV, Masalov AV, Schwartzkopff O, Johnson D. Assessing variability in neonatal blood pressure, notably in hypotension. J Appl Biomed 2010; 8: 209–11.

Yamamoto N, Kubo Y, Ishizawa K, Kim G, Moriya T, Yamanouchi T, Otsuka K. Detrended fluctuation analysis is considered to be useful as a new indicator for short-term glucose complexity. Diab Technol Therap 2010; 12: 775–83.

2011

Halberg F, Cornélissen G, Grambsch P, McCraty R, Beaty L, Siegelova J, Homolka P, Hillman DC, Finley J, Thomas F, Kino T, Revilla M, Schwartzkopff O. Personalized chronobiologic cybercare; other chronomics' progress by transdisciplinary cycles' congruences: Season's Appreciations 2009. J Appl Biomed 2011; 9: 1–34.

Pan W, Wu X, Kastin AJ, Zhang Y, Hsuchou H, Halberg F, Chatu F, Khan RS, Robert B, Cornélissen-Guillaume GG. Potential protective role of IL15Ra during inflammation. J Mol Neurosci 2011; 43: 412–23.

2012

Costella JF, Halberg F, Hillman D, Mikulecky M, Cornélissen G. Four circadian and two circasemidian periods in sleep-wakefulness of a man on a self-selected routine. Folia anthropologica 2012; 11: 51–3.

Halberg F, Cornélissen G, Katinas GS, Hillman D, Otsuka K, Watanabe Y, Wu J, Halberg Francine, Halberg J, Sampson M, Schwartzkopff O, Halberg E. Many rhythms are control information for whatever we do: an autobiography. Folia anthropologica 2012; 12: 5–134.

Buda BL, Cornélissen G, Toth GA, Buda BKI, Szolnoki V, Veress A, Halberg F. Timing in balneology – outlines of the Sárvár spa study. Folia anthropologica 2012; 12: 141–5.

Cornélissen G, Burioka N, Halberg F. Other-than-annual variations in morning determinations of uric acid. Clin Chem Lab Med 2012; 50(5): 765–70.

Pan W, Hsuchou H, Cornélissen-Guillaume GG, Jayaram B, Wang Y, Tu H, Halberg F, Wu X, Chua SC Jr, Kastin AJ. Endothelial leptin receptor mutation provides partial resistance to diet-induced obesity. J Appl Physiol 2012; 112: 1410–8.

2013

Gumarova L, Cornélissen G, Hillman D, Halberg F. Geographically selective assortment of cycles in pandemics: Meta-analysis of data collected by Chizhevsky. Epidemiol Infect 2013; 141: 2173–84, doi:10.1017/S0950268812002804.

Jayaram B, Pan W, Wang Y, Hsuchou H, Mace A, Cornélissen-Guillaume GG, Mishra PK, Koza RA, Kastin AJ. Astrocytic leptin receptor knockout mice show partial rescue of leptin resistance in diet-induced obesity. J Appl Physiol 2013; 114(6): 734–41.

Rossi Squarcini CF, Nogueira Pires ML, Lopes C, Benedito-Silva AA, Esteves AM, Cornélissen-Guillaume G, Matarazzo C, Garcia D, Peccin da Silva MS, Tufik S, de Mello MT. Free-running circadian rhythms of muscle strength, reaction time, and body temperature in totally blind people. Eur J Appl Physiol 2013; 113: 157–65.

Halberg F, Düll-Pfaff N, Gumarova L, Zenchenko TA, Schwartzkopff O, Freytag EM, Freytag J, Cornélissen G. 27-day cycles in human mortality: Traute and Bernhard Düll. Hist Geo Space Sci 2013; 4: 47–59.

Hsuchou H, Wang Y, Cornélissen-Guillaume GG, Kastin AJ, Jang E, Halberg F, Pan W. Diminished leptin signaling can alter circadian rhythm of metabolic activity and feeding. J Appl Physiol 2013; 115: 995–1003.

Cornélissen G, Halberg F, Halberg J, Schwartzkopff O, Cugini P. Remembering the father of chronobiology and chronomics: Franz Halberg, MD (5 July 1919–9 June 2013). Clin Ter (Rome) 2013; 164(4): i–vi.

Cornélissen G, Halberg Francine, Halberg J, Schwartzkopff O. In memoriam Prof. Franz Halberg (1919–2013). J Appl Biomed 2013; 11: 251.

Cornélissen G, Halberg Francine, Halberg J, Schwartzkopff O. In remembrance: Franz Halberg (5 July 1919–9 June 2013). Neuroendocrinol Lett 2013; 34(6): 107–10.

Cornélissen G, Halberg Francine, Halberg J, Schwartzkopff O. Franz Halberg, MD (5 July 1919–9 June 2013): in appreciation. Chronobiol Int 2013; 30(10): 1205–7.

Cornélissen G, Halberg Francine, Halberg J, Schwartzkopff O. Remembering Franz Halberg, MD (5 July 1919–9 June 2013). J Exp Ther Oncol 2013; 10(3): 239–42. https://www.oldcitypublishing.com/JETO/JETOv10n3 Halberg_obit_2.pdf 4 pp.

2014

Hristova K, Shiue I, Pella D, Singh RB, Chaves H, Basu TK, Ozimek L, Rastogi SS, Takahashi T, Wilson D, DeMeester F, Cheema S, Garg M, Buttar HS, Milovanovic B, Kumar A, Handjiev S, Cornélissen G, Petrov I. Sofia Declaration on transition of prevention strategies for cardiovascular diseases and diabetes mellitus in developing countries: a statement from the International College of Cardiology and International College of Nutrition. Nutrition 2014; 30(9): 1085–9. doi: 10.1016/j.nut.2013.12.013.

Cornélissen G. Cosinor-based rhythmometry. Theor Biol Med Model 2014; 11: 16.

Chronoecology

2002

Otsuka K, Mitsutake G, Yano S. Depression, quality of life, and lifestyle: chronoecological health watch in a community. Biomed Pharmacother 2002; 56(Suppl 2): 231s–42s.

2003

Ho HK, Matsubayashi K, Wada T, Kimura M, Yano S, Otsuka K, Fujisawa M, Kita T, Saijoh K: What determines the life satisfaction of the elderly ? Comparative study of residential care home and community in Japan. Geriatrics and Gerontology International 2003; 3:79–85.

Otsuka K, Murakami K, Kubo Y, Yamanaka T, Mitsutake G, Ohkawa S, Matsubayashi K, Yano S, Cornélissen G, Halberg F. Chronomics for chronoastrobiology with immediate spin-offs for life quality and longevity. Biomed Pharmacother 2003; 57(Suppl 1): 1s–18s.

2004

Wada T, Ishine M, Sakagami T, Okumiya K, Fujisawa M, Murakami S, Otsuka K, Yano S, Kita T, Matsubayashi K. Depression in Japanese community-dwelling elderly – prevalence and association with ADL and QOL. Arch Gerontol Geriatr 2004; 39: 15–23.

Murakami S, Otsuka K, Yamanaka G, Kubo Y, Matsuoka O, Yamanaka T, Shinagawa M, Nunoda S, Ohkawa S, Wada T, Matsubayashi K, Yano S, Cornélissen G, Halberg F. Positive impact of social intervention on disturbed neurobehavioral function in an elderly community-dwelling population: Longitudinal Investigation for Longevity and Aging in Hokkaido County (LILAC). Biomed Pharmacother 2004; 58(Suppl 1): 45s–7s.

Otsuka K, Yamanaka G, Shinagawa M, Murakami S, Yamanaka T, Shibata K, Yano S, Ishizuka S, Singh RB, Cornélissen G, Halberg F. Chronomic community screening reveals about 31 % depression, elevated blood pressure and infradian vascular rhythm alteration. Biomed Pharmacother 2004; 58(Suppl 1): 48s–55s.

2005

Hotta N, Otsuka K, Murakami S, Yamanaka G, Kubo Y, Matsuoka O, Yamanaka T, Shinagawa M, Nunoda S, Nishimura Y, Shibata K, Saitoh H, Nishinaga M, Ishine M, Wada T, Okumiya K, Matsubayashi K, Yano S, Ichihara K, Cornélissen G, Halberg F. Fractal analysis of heart rate variability and mortality in elderly community-dwelling people – Longitudinal Investigation for the Longevity and Aging in Hokkaido County (LILAC) study. Biomed Pharmacother 2005; 59(Suppl 1): 45s–8s.

Matsuoka O, Otsuka K, Murakami S, Hotta N, Yamanaka G, Kubo Y, Yamanaka T, Shinagawa M, Nunoda S, Nishimura Y, Shibata K, Saitoh H, Nishinaga M, Ishine M, Wada T, Okumiya K, Matsubayashi K, Yano S, Ichihara K, Cornélissen G, Halberg F, Ozawa T. Arterial stiffness independently predicts cardiovascular events in an elderly community – Longitudinal Investigation for the Longevity and Aging in Hokkaido County (LILAC) study. Biomed Pharmacother 2005; 59(Suppl 1): 40s–4s.

Murakami S, Otsuka K, Hotta N, Yamanaka G, Kubo Y, Matsuoka O, Yamanaka T, Shinagawa M, Nunoda S, Nishimura Y, Shibata K, Takasugi E, Nishinaga M, Ishine M, Wada T, Okumiya K, Matsubayashi K, Yano S, Ichihara K, Cornélissen G, Halberg F. Common carotid intima-media thickness is predictive of all-cause and cardiovascular mortality in elderly community-dwelling people: Longitudinal Investigation for the Longevity and Aging in Hokkaido County (LILAC) study. Biomed Pharmacother 2005; 59(Suppl 1): 49s–53s.

Otsuka K, Norboo T, Otsuka Y, Higuchi H, Hayajiri M, Narushima C, Sato Y, Tsugoshi T, Murakami S, Wada T, Ishine M, Okumiya K, Matsubayashi K, Yano S, Chogyal T, Angchuk D, Ichihara K, Cornélissen G, Halberg F. Effect of aging on blood pressure in Leh, Ladakh, a high-altitude (3524 m) community, by comparison with a Japanese town. Biomed Pharmacother 2005; 59(Suppl 1): 54s–7s.

Otsuka K, Norboo T, Otsuka Y, Higuchi H, Hayajiri M, Narushima C, Sato Y, Tsugoshi T, Murakami S, Wada T, Ishine M, Okumiya K, Matsubayashi K, Yano S, Chogyal T, Angchuk D, Ichihara K, Cornélissen G, Halberg F. Chronoecological health watch of arterial stiffness and neuro-cardio-pulmonary function in elderly community at high altitude (3524 m), compared with Japanese town. Biomed Pharmacother 2005; 59(Suppl 1): 58s–67s.

Yamanaka G, Otsuka K, Hotta N, Murakami S, Kubo Y, Matsuoka O, Takasugi E, Yamanaka T, Shinagawa M, Nunoda S, Nishimura Y, Shibata K, Saitoh H, Nishinaga M, Ishine M, Wada T, Okumiya K, Matsubayashi K, Yano S, Ishizuka S, Ichihara K, Cornélissen G, Halberg F. Depressive mood is independently related to stroke and cardiovascular events in a community. Biomed Pharmacother 2005; 59(Suppl 1): 31s–9s.

2006

Shirai K, Utono J, Otsuka K, Takata M. A novel blood pressure-independent arterial wall stiffness parameter; cardio-ankle vascular index (CAVI). J Atherosclerosis Thrombosis 2006; 13:101–7.

2007

Fujisawa M, Ishine M, Okumiya K, Otsuka K, Matsubayashi K. Trends in diabetes. Lancet 2007; 369: 1257.

Otsuka K. Impact of stress monitoring and lifestyle in the development of cardiovascular disease. Hypertens Res 2007; 30: 1005–6.

Yamanaka T, Takasugi E, Hotta N, Kubo Y, Otsuka K. Daily living functions of the elderly requiring home visits: A study at a comprehensive assessment clinic for the elderly. Geriat Gerontol Int 2007; 7:388–92.

2008

Wada T, Ishine M, Ishimoto Y, Hirosaki M, Kimura Y, Kasahara Y, Okumiya K, Nishinaga M, Otsuka K, Matsubayashi K. Community-dwelling elderly fallers in Japan are older, more disable, and more depressed than nonfallers. J Am Geriatr Soc 2008; 56: 1570–1.

Ishine M, Okumiya K, Hirosaki M, Sakamoto R, Fujisawa M, Hotta N, Otsuka K, Nishinaga M, Doi Y, Matsubayashi K. Prevalence of hypertension and its awareness, treatment, and satisfactory control through treatment in elderly Japanese. J Am Geriatr Soc 2008; 56: 374–5.

Ishine M, Okumiya K, Kimura Y, Kasahara Y, Wada T, Yamanaka G, Hotta N, Otsuka K, Murakami S, Matsubayashi K. Subjective sleep disturbances were closely associated with comprehensive geriatric functions in dose-responsive manner in the community-dwelling elderly people in Japan. J Am Geriatr Soc 2008; 56: 1571–3.

Okumiya K, Ishine M, Wada T, Fujisawa M, Otsuka K, Matsubayashi K. Lifestyle changes after oral glucose tolerance test improve glucose intolerance in community dwelling elderly people after one year. J Am Geriatr Soc 2008; 56: 767–9.

Otsuka K, Hotta N, Takasugi E, Yamanaka G, Kubo Y, Cornélissen G, Halberg F. Impaired cardiovascular function predicts mortality from respiratory disorders in the elderly [editorial comment]. Hypertens Res 2008; 31: 589–91.

2009

Hirosaki M, Ishimoto Y, Kasahara Y, Kimura Y, Konno A, Sakamoto R, Nakatsuka M, Ishine M, Wada T, Okumiya K, Fujisawa M, Otsuka K, Matsubayashi K. Community-dwelling elderly Japanese people with hobbies are healthier than those lacking hobbies. J Am Geriatri Soc 2009; 57: 1132–3.

Ishimoto Y, Wada T, Hirosaki M, Kasahara Y, Kimura Y, Konno A, Nakatsuka M, Ishine M, Okumiya K, Fujisawa M, Otsuka K, Matsubayashi K. Health-related differences between participants and nonparticipants in community-based geriatric examinations. J Am Geriatri Soc 2009; 57(2): 360–2.

Ishimoto Y, Wada T, Hirosaki M, Kasahara Y, Kimura Y, Konno A, Nakatsuka M, Sakamoto R, Ishine M, Okumiya K, Fujisawa M, Otsuka K, Matsubayashi K. Age and sex significantly influence fall risk in community-dwelling elderly people in Japan. J Am Geriatri Soc 2009; 57: 930–2.

Kimura Y, Wada T, Ishine M, Ishimoto Y, Kasahara Y, Hirosaki M, Konno A, Nakatsuka M, Sakamoto R, Okumiya K, Otsuka K, Matsubayashi K. Community-dwelling elderly with chewing difficulties are more disabled, depressed and have lower quality of life scores. Geriatr Gerontol Int 2009; 9: 102–4.

Kimura Y, Wada T, Ishine M, Ishimoto Y, Kasahara Y, Konno A, Nakatsuka M, Sakamoto R, Okumiya K, Fujisawa M, Otsuka K, Matsubayashi K. Food diversity is closely associated with activities of daily living, depression, and quality of life in community-dwelling elderly people. J Am Geriatri Soc 2009; 57: 922–4.

Yamamoto N, Yamanaka G, Ishikawa M, Takasugi E, Murakami S, Yamanaka T, Ishine M, Matsubayashi K, Hanafusa T, Otsuka K. Cardio-ankle vascular index as a predictor of cognitive impairment in community-dwelling elderly people: Four-year follow-up. Dementia Geriatr Cogn Disord 2009; 28: 153–8.

Yamamoto N, Yamanaka G, Takasugi E, Ishikawa M, Yamanaka T, Murakami S, Hanafusa T, Matsubayashi K, Otsuka K. Lifestyle intervention reversed cognitive function in aged people with diabetes mellitus: Two-year follow up. Diabetes Res Clin Pract 2009; 85(3): 343–6.

Yamamoto N, Hotta N, Takasugi E, Ishikawa M, Murakami S, Yamanaka T, Yamanaka G, Hanafusa T, Matsubayashi K, Otsuka K. The effect of glucose tolerance on central and brachial pressure in elderly people. J Am Geriatri Soc 2009; 57: 1120–2.

2010

Matsubayashi K, Wada T, Ishine M, Sakamoto R, Okumiya K, Ishikawa M, Yamanaka G, Otsuka K, Murakami S, Fujisawa M, Yano S. Community-based geriatric assessment and preventive intervention lowered medical expenses for the elderly. J Am Geriatri Soc 2010; 58: 791–3.

Okumiya K, Sakamoto R, Kimura Y, Ishimoto Y, Wada T, Ishine M, Ishikawa M, Nakajima S, Hozo R, Ge R-L, Norboo T, Otsuka K, Matsubayashi K. Strong association between polycythemia and glucose intolerance in elderly high-altitude dwellers in Asia. J Am Geriatri Soc 2010; 58: 609–11.

Okumiya K, Sakamoto R, Kimura Y, Wada T, Ishine M, Ishikawa M, Nakajima S, Hozo R, Ge R-L, Norboo T, Otsuka K, Matsubayashi K. Diabetes mellitus and hypertension in elderly highlanders in Asia. J Am Geriatri Soc 2010; 58: 1193–5.

Yamamoto N, Yamanaka G, Ishizawa K, Ishikawa M, Murakami S, Yamanaka T, Okumiya K, Ishine M, Matsubayashi K, Otsuka K. Insomnia increases insulin resistance and insulin secretion in elderly people. J Am Geriatri Soc 2010; 58: 801–4.

2011

Matsubayashi K, Ishine M, Wada T, Ishimoto Y, Hirosaki M, Kasahara Y, Kimura Y, Fukutomi E, Ling CW, Sakamoto R, Fujisawa M, Otsuka K, Okumiya K. "Field medicine" for reconsidering "optimal aging". J Am Geriatrics Soc 2011; 59: 1568–70.

Hirosaki M, Ishimoto Y, Kasahara Y, Konno A, Kimura Y, Fukutomi E, Ling CW, Nakatsuka M, Fujisawa M, Sakamoto R, Ishine M, Okumiya K, Otsuka K, Wada T, Matsubayashi K. Self-rated happiness is associated with functional ability, mood, quality of life and income, but not with medical condition in community-dwelling elderly in Japan. Geriatr Gerontol 2011; 11: 531–3.

Sakamoto R, Okumiya K, Ishine M, Lhadon K, Lhamo S, Rinchen S, Dorji N, Dorji T, Fujisawa M, Otsuka K, Wada T, Matsubayashi K. Subjective quality of life in older community-dwelling adults in the kingdom of Bhutan and Japan. J Am Geriatr Soc 2011; 59: 2157–9.

2012

Imai H, Wada T, Sakamoto R, Kasahara Y, Ishimoto Y, Kimura Y, Fukutomi E, Chen WL, Okumiya K, Otsuka K, Matsubayashi K. Activities of daily living rather than depressive symptoms increase the risk of mortality in Japanese community-dwelling elderly people: a 4-year longitudinal survey. J Am Geriatrics Soc 2012; 60: 1191–3.

Chen W, Ishimoto Y, Kimura Y, Fukutomi E, Imai H, Matsubayashi K, Wada T, Fujisawa M, Sakamoto R, Okumiya K, Otsuka K. Association between geriatric function and subjective economy in elderly adults living in a rural community in Japan. J Am Geriatrics Soc 2012; 60: 802–3.

Yamamoto N, Ishizawa K, Ishikawa M, Yamanaka G, Yamanaka T, Murakami S, Hiraiwa T, Okumiya K, Ishine M, Matsubayashi K, Otsuka K. Cognitive function with subclinical hypothyroidism in elderly people without dementia: one year follow up. Geriatr Gerontol Int 2012; 12: 164–5.

Kimura Y, Wada T, Okumiya K, Ishimoto Y, Fukutomi E, Kasahara Y, Chen W, Sakamoto R, Fujisawa M, Otsuka K, Matsubayashi K. Eating alone among community-dwelling Japanese elderly: association with depression and food diversity. J Nutr Health Aging 2012; 16: 728–31.

Okumiya K, Sakamoto R, Kimura Y, Ishimoto Y, Fukutomi E, Kasahara Y, Chen WL, Ishine M, Wada T, Fujisawa M, Imai H, Ishikawa M, Yamamoto N, Otsuka K, Matsubayashi K. J-curve association between economic status and diabetes independent of functional disability in Japanese elderly. Geriatr Gerontol Int 2012; 12: 755–6.

Ishimoto Y, Wada T, Kasahara Y, Kimura Y, Fukutomi E, Chen W, Hirosaki M, Nakatsuka M, Fujisawa M, Sakamoto R, Ishine M, Okumiya K, Otsuka K, Matsubayashi K. Fall Risk Index predicts functional decline regardless of fall experiences among community-dwelling elderly. Geriatr Gerontol Int 2012; 12: 659–66.

Ishikawa M, Yamamoto N, Yamanaka G, Suwa K, Nakajima S, Hozo R, Norboo T, Okumiya K, Matsubayashi K, Otsuka K. Disaster-related psychiatric disorders among survivors of flooding in Ladakh, India. Int J Soc Psychiatr 2012. DOI: 10.1177/0020764012440677.

Kawasaki T, Okumiya K, Norboo T, Otsuka K, Matsubayashi K, De Meester F, Wilson D, Singh RB, Cornelissen G, Halberg F. Chronoecological "glocal" (global and local) health-watch: men better adapted to ambient hypoxia than women. World Heart J 2012; 4: 165–75.

2014

Otsuka K, Cornelissen G, Yamanaka T, Oinuma S, Sasaki J, Yamanaka G, Okumiya K, Matsubayashi K. Comprehensive geriatric assessment reveals sleep disturbances in community-dwelling elderly adults associated with even slight cognitive decline. J Am Geratric Soc 2014; 62: 571–3.

Otsuka K, Cornelissen G, Yamanaka T, Oinuma S, Sasaki J, Yamanaka G, Okumiya K, Matsubayashi K. Time estimation predicts improvement in cognitive function in elderly community-dwelling adults. J Am Geratric Soc 2014; 62: 974–6.

Chronoastrobiology

1994

Breus TK, Mikulecky M, Cornélissen G, Halberg F. Chronobiology and the cosmos: an era of helio-seleno-geophysical chrono-physiology and -pathology. Chronobiologia 1994; 21: 165.

Cornélissen G, Zaslavskaya RM, Kumagai Y, Romanov Y, Halberg F. Chronopharmacologic issues in space. J Clin Pharmacol 1994; 34: 543–51.

Halberg F, Cornélissen G, Bingham C, del Pozo F, Arredondo MT, Otsuka K, Quadens O, Breus T, Rapoport S, Komarov F. Telehygiene system for preventive chronopharmacology in space. J Clin Pharmacol 1994; 34: 552–7.

1995

Breus T, Cornélissen G, Halberg F, Levitin AE. Temporal associations of life with solar and geophysical activity. Ann Geophys 1995; 13: 1211–22.

1996

Cornélissen G, Halberg F, Wendt HW, Bingham C, Sothern RB, Haus E, Kleitman E, Kleitman N, Revilla MA, Revilla M Jr, Breus TK, Pimenov K, Grigoriev A, Mitish MD, Yatsyk GV, Syutkina EV. Resonance of about-weekly human heart rate rhythm with solar activity change. Biologia (Bratislava) 1996; 51: 749–56.

Portela A, Northrup G, Halberg F, Cornélissen G, Wendt H, Melby JC, Haus E. Changes in human blood pressure with season, age and solar cycles: a 26-year record. Int J Biometeorol 1996; 39: 176–81.

1998

Cornélissen G, Halberg F, Sothern RB, Nikityuk BA, Garcia Alonso L, Syutkina EV, Grafe A, Bingham C. Toward a chronoastrobiology: sunspot cycles and geomagnetism as well as sunshine may modulate human morphology. Russ Morphol Newsl [v. 5] N 1998; 3(4): 133–7.

1999

Halberg F, Cornélissen G, Schwartzkopff O, Syutkina EV, Grigoriev AE, Mitish MD, Yatsyk GV, Studenikin MY, Gubin D, Gubin G, Siegelova J, Fiser B, Dusek J, Homolka P, Watanabe Y, Otsuka K, Perfetto F, Tarquini R, Delmore P. Spin-offs from blood pressure and heart rate studies for health care and space research (review). In Vivo 1996; 13: 67–76.

2000

Otsuka K, Yamanaka T, Cornélissen G, Breus T, Chibisov SM, Baevsky R, Halberg F, Siegelova J, Fiser B. Altered chronome of heart rate variability during span of high magnetic activity. Scripta Medica (Brno) 2000; 73: 111–6.

Otsuka K, Ichimaru Y, Cornélissen G, Weydahl A, Holmeslet B, Schwartzkopff O, Halberg F, on behalf of the ICEHRV Working Group. Dynamic analysis of heart rate variability from 7-day Holter recordings associated with geomagnetic activity in subarctic area. Comput Cardiol 2000; 27: 453–6.

Halberg F, Cornélissen G, Otsuka K, Watanabe Y, Delyukov A, Gorgo Y, Zhao ZY, Weydahl A, Sothern RB, Siegelova J, Fiser B, Dusek J, Syutkina EV, Perfetto F, Tarquini R, Singh RB, Rhees B, Lofstrom D, Lofstrom P, Sonkowsky R, Johnson PWC, Schwartzkopff O, International BIOCOS Study Group. Cross-spectrally coherent ~10.5- and 21-year biological and physical cycles, magnetic storms and myocardial infarctions. Invited presentation, NATO Advanced Study Institute on Space Storms and Space Weather Hazards, Crete, Greece, June 19–29, 2000. Neuroendocrinol Lett 2000; 21: 233–58.

2001

Otsuka K, Cornélissen G, Weydahl A, Holmeslet B, Hansen TL, Shinagawa M, Kubo Y, Nishimura Y, Omori K, Yano S, Halberg F. Geomagnetic disturbance associated with decrease in heart rate variability in a subarctic area. Biomed Pharmacother 2001; 55(Suppl 1): 51s–6s.

Otsuka K, Oinuma S, Cornélissen G, Weydahl A, Ichimaru Y, Kobayashi M, Yano S, Holmeslet B, Hansen TL, Mitsutake G, Engebretson MJ, Schwartzkopff O, Halberg F. Alternating-light-darkness-influenced human electrocardiographic magnetoreception in association with geomagnetic pulsations. Biomed Pharmacother 2001; 55(Suppl 1): 63s–75s.

Weydahl A, Sothern RB, Cornélissen G, Wetterberg L. Geomagnetic activity influences the melatonin secretion at latitude 70°N. Biomed Pharmacother 2001; 55: 57–62.

Watanabe Y, Cornélissen G, Halberg F, Otsuka K, Ohkawa S-I. Association by signatures and coherences between the human circulation and helio- and geomagnetic activity. Biomed Pharmacother 2001; 55: 76–83.

Halberg F, Cornélissen G, Otsuka K, Syutkina EV, Masalov A, Breus T, Viduetsky A, Grafe A, Schwartzkopff O. Chronoastrobiology: neonatal numerical counterparts to Schwabe's 10.5 and Hale's 21-year sunspot cycles. In memoriam Boris A. Nikityuk. Int J Prenat Perinat Psychol Med 2001; 13: 257–80.

2002

Oinuma S, Kubo Y, Otsuka K, Yamanaka T, Murakami S, Matsuoka O, Ohkawa S, Cornélissen G, Weydahl A, Holmeslet B, Hall C, Halberg F, on behalf of the "ICEHRV" Working Group. Graded response of heart rate variability, associated with an alteration of geomagnetic activity in a subarctic area. Biomed Pharmacother 2002; 56(Suppl. 2): 284s–8s.

Otsuka K, Singh RB, Weydahl A, Ichimaru Y, Holmeslet B, Hansen TL, Mitsutake G, Murakami S, Kubo Y, Ohkawa S, Yano S, Engebretson MJ, Cornélissen G, Schwartzkopff O, Halberg F, on behalf of the ICEHRV Working Group. Ultra-low frequency geomagnetic pulsations and cardiovascular biological systems in humans. Int J Cardiol 2002; 82(Suppl 1): S26–7.

Cornélissen G, Halberg F, Breus T, Syutkina EV, Baevsky R, Weydahl A, Watanabe Y, Otsuka K, Siegelova J, Fiser B, Bakken EE. Non-photic solar associations of heart rate variability and myocardial infarction. J Atmos Solar-Terrestrial Phys 2002; 64: 707–20.

Cornélissen G, Hillman D, Katinas GS, Rapoport S, Breus TK, Otsuka K, Bakken EE, Halberg F. Geomagnetics and society interact in weekly and broader multiseptans underlying health and environmental integrity. Biomed Pharmacother 2002; 56(Suppl 2): 319s–26s.

Breus TK, Pimenov KYu, Cornélissen G, Halberg F, Syutkina EV, Baevsky RM, Petrov VM, Orth-Gomer K, Åkerstedt T, Otsuka K, Watanabe Y, Chibisov SM. The biological effects of solar activity. Biomed Pharmacother 2002; 56(Suppl 2): 273s–83s.

Starbuck S, Cornélissen G, Halberg F. Is motivation influenced by geomagnetic activity? Biomed Pharmacother 2002; 56(Suppl 2): 289s–97s.

Weydahl A, Sothern RB, Cornélissen G. Non-linear relation of heart rate variability during exercise recovery with local geomagnetic activity. Biomed Pharmacother 2002; 56(Suppl 2): 298s–300s.

2003

Otsuka K, Murakami K, Kubo Y, Yamanaka T, Mitsutake G, Ohkawa S, Matsubayashi K, Yano S, Cornélissen G, Halberg F. Chronomics for chronoastrobiology with immediate spin-offs for life quality and longevity. Biomed Pharmacother 2003; 57(Suppl 1): 1s–18s.

Nintcheu-Fata S, Katinas G, Halberg F, Cornélissen G, Tolstykh V, Michael HN, Otsuka K, Schwartzkopff O, Bakken E. Chronomics of tree rings for chronoastrobiology and beyond. Biomed Pharmacother 2003; 57(Suppl 1): 24s–30s.

Watanabe Y, Cornélissen G, Katinas G, Sothern RB, Halberg F, Watanabe M, Watanabe F, Otsuka K. Non-photic, non-thermic circadecadal solar cycle interaction with cardiovascular circannual and circasemiannual variation in heated air-conditioned habitat. Biomed Pharmacother 2003; 57(Suppl 1): 55s–7s.

Halberg F, Cornélissen G, Schack B, Wendt HW, Minne H, Sothern RB, Watanabe Y, Katinas G, Otsuka K, Bakken EE. Blood pressure self-surveillance for health also reflects 1.3-year Richardson solar wind variation: spin-off from chronomics. Biomed Pharmacother 2003; 57(Suppl 1): 58s–76s.

2004

Mitsutake G, Otsuka K, Oinuma S, Ferguson I, Cornélissen G, Wanliss J, Halberg F. Does exposure to an artificial ULF magnetic field affect blood pressure, heart rate variability and mood? Biomed Pharmacother 2004; 58(Suppl 1): S20–7.

Halberg F, Cornélissen G, Regal P, Otsuka K, Wang ZR, Katinas GS, Siegelova J, Homolka P, Prikryl P, Chibisov SM, Holley DC, Wendt HW, Bingham C, Palm SL, Sonkowsky RP, Sothern RB, Pales E, Mikulecky M, Tarquini R, Perfetto F, Salti R, Maggioni C, Jozsa R, Konradov AA, Kharlitskaya EV, Revilla M, Wan CM, Herold M, Syutkina EV, Masalov AV, Faraone P, Singh RB, Singh RK, Kumar A, Singh R, Sundaram S, Sarabandi T, Pantaleoni GC, Watanabe Y, Kumagai Y, Gubin D, Uezono K, Olah A, Borer K, Kanabrocki EA, Bathina S, Haus E, Hillman D, Schwartzkopff O, Bakken EE, Zeman M. Chronoastrobiology: proposal, nine conferences, heliogeomagnetics, transyears, near-weeks, near-decades, phylogenetic and ontogenetic memories. Biomed Pharmacother 2004; 58(Suppl 1): S150–87.

2005

Cornélissen G, Otsuka K, Halberg F. Near-transyear in solar magnetism. Biomed Pharmacother 2005; 59(Suppl 1): S5–9.

Mitsutake G, Otsuka K, Hayakawa M, Sekiguchi M, Cornélissen G, Halberg F. Does Schumann resonance affect our blood pressure? Biomed Pharmacother 2005; 59(Suppl 1): S10–4.

2006

Chibisov SM, Eremina LZ, Strelkov DG, Romanova EA, Kharlickaya EV, Halberg F, Cornélissen G, Syutkina EV. Physiological data variation in rabbits under extreme environmental conditions (magnetic storm). Eur J Natural History 2006; (2): 54–8.

2009

Otsuka K, Izumi R, Ishioka N, Ohshima H, Mukai C. Chronomics of heart rate variability on earth and in space. Respir Physiol Neurobiol 2009; 169(suppl.1): 69s–72s.

Otsuka K, Cornélissen G, Halberg F. Chronomics of tree rings gauge climate change. Geophys Process Biosphere 2009; 8(1): 63–72. [In Russian with English summary.]

Cornélissen G, Tarquini R, Perfetto F, Otsuka K, Gigolashvili M, Halberg F. Investigation of solar about 5-month cycle in human circulating melatonin: signature of weather in extraterrestrial space? Sun Geosphere 2009; 4(2): 55–9.

2010

Otsuka K, Cornelissen G, Halberg F. Chronomics of climate variations of tree ring width. Izv Atmos Ocean Phys 2010; 46: 61–6.

Halberg F, Cornélissen G, Bernhardt K-H, Sampson M, Schwartzkopff O, Sonntag D. Egeson's (George's) transtridecadal weather cycling and sunspots. Hist Geo Space Sci 2010; 1: 49–61.

Halberg F, Cornélissen G, Sothern RB, Czaplicki J, Schwartzkopff O. Thirty-five-year climatic cycle in heliogeophysics, psychophysiology, military politics, and economics. Izv Atmos Ocean Phys 2010; 46(7): 844–64.

2011

Cornélissen G, Grambsch P, Sothern RB, Katinas G, Otsuka K, Halberg F. Congruent biospheric and solar-terrestrial cycles. J Appl Biomed 2011; 9: 63–102. DOI 10.2478/v10136-009-0023-7.

2012

Cornélissen G, Wendt H, Halberg F. Earthquakes, features of the dynamics of the lithosphere, mirror the heliosphere, as do physiopathologic aspects of the biosphere. Proceedings, Natural Cataclysms and Global Problems of the Modern Civilization, Istanbul, 19–21 September 2011. London: SWB International Publishing House; 2012, pp. 301–6.

Halberg F, Hillman D, Schwartzkopff O, Breus T, Syutkina EV, Masalov A, Cornélissen G. Putative solar signatures can differ among variables and circadian endpoints and in individuals vs. populations. Proceedings, Natural Cataclysms and Global Problems of the Modern Civilization, Istanbul, 19–21 September 2011. London: SWB International Publishing House; 2012, pp. 289–95.

2013

Halberg F, Cornélissen G, Otsuka K, Watanabe Y, Beaty L, Gumarova L, Revilla M, Schwartzkopff O, Siegelova J, Singh RB. Monitoring chronoosphere for knowing oneself and one's environs. Geophys Process Biosphere 2013; 12(4): 5–35.

2014

Zeng W, Liang X, Wan C, Wang Y, Jiang Z, Cheng S, Cornélissen G, Halberg F, Wang Z. Patterns of mortality from cardiac-cerebral vascular disease and influences from the cosmos. Biological Rhythm Research 2014; 11 pp. doi: 10.1080/09291016.2013.876189.

Yamamoto N, Otsuka K, Kubo Y, Hayashi M, Ohshima H, Mukai C. Effects of long-term microgravity exposure in space on circadian rhythms of heart rate variability. Chronobiology International Early Online 2014; 1–14. DOI: 10.3109/07420528.2014.979940.

Vascular Variability Syndromes

1990

Halberg E, Delmore P, Finch M, Cornélissen G, Halberg F. Chronobiologic assessment of deviant human blood pressure: an invitation for improvements. Prog Clin Biol Res 1990; 341A: 305–18.

Hillman D, Otsuka K, Halberg F. Ambulatory blood pressure monitor and "analyzer". Chronobiologia 1990; 17: 305–13.

Otsuka K, Watanabe H, Cornelissen G, Shinoda M, Uezono K, Kawasaki T, Halberg F. Gender, age and circadian blood pressure variation of apparently healthy rural vs. metropolitan Japanese. Chronobiologia 1990; 17: 253–65.

1991

Otsuka K, Cornélissen G, Watanabe H, Hunt SC, Halberg F. Time-varying limits for single blood pressures and heart rates of group-synchronized healthy women. Heart Vessels 1991; 6: 107–11.

Cugini P, Kawasaki T, Di Palma L, Otsuka K, Uezono K. Innovative criteria for diagnosing arterial hypertension via blood pressure monitoring: the chronodiagnosis. J Health Sci 1991; 13: 23–34.

Kuwajima I, Suzuki Y, Shimosawa T, Otsuka K, Kawamura H, Kuramoto K. Effect of nifedipine tablets on ambulatory blood pressure in patients aged less than 60 and greater than 65 years with systemic hypertension. Am J Cardiol 1991; 68 (13): 1351–6.

1992

Nakajima S, Otsuka K, Yamanaka T, Omori K, Kubo Y, Toyoshima T, Watanabe Y, Watanabe H. Ambulatory blood pressure and postprandial hypotension. Am Heart J 1992; 124(6): 1669–71.

1993

Suzuki M, Otsuka K, Guilleminault C. Long-term nasal continuous positive airway pressure administration can normalize hypertension in obstructive sleep apnea patients. Sleep 1993; 16(6): 545–9.

Kuwajima I, Suzuki Y, Kuramoto K, Kawamura H, Otsuka K. Circadian blood pressure change after carteolol chloride in younger and older hypertensive patients. Clin Ther 1993; 15(2): 338–46.

Kuwajima I, Suzuki Y, Otsuka K, Kawamura H, Kuramoto K. Effects of alpha,beta-blocker, arotinolol chloride, on 24-h blood pressure–difference between elderly and younger hypertensive patients. J Clin Pharm Ther 1993; 18(4): 275–80.

1994

Otsuka K, Halberg F. Circadian profiles of blood pressure and heart rate of apparently healthy metropolitan Japanese. Front Med Biol Eng 1994; 6(2): 149–55.

Staessen JA, O'Brien ET, Amery AK, Atkins N, Baumgart P, De Cort P, Degaute J-P, Dolenc P, Gaudemaris RD, Estrom I, Fagard R, Gosse P, Gourlay S, Hayashi H, Imai Y, James G, Kawasaki T, Kuschnir E, Kuwajima I, Lindholm L, Liu L, Macor F, Mancia G, McGrath B, Middeke M, Ming J, Omboni S, Otsuka K, Palatini P, Parati G, Pieper C, Verdecchia P, Zachariah P, Zhang W. Ambulatory blood pressure in normotensive and hypertensive subjects: results from an international database. J Hypertens 1994; 12(Suppl 7): S1–S12.

Cornélissen G, Guillaume F. A chronobiologic approach to blood pressure management. Chronobiologia 1994; 21: 57–60.

1995

Otsuka K, Ishii T, Omori K. Alteration of circadian periodicity in BP, heart rate and physical activity following transmeridian flights. J Ambul Monit 1995; 8(3): 219–30.

Staessen JA, Fagard R, Thijs L, Amery A, Otsuka K, the Participants in The Fourth International Consensus Conference on 24–Hour Ambulatory Blood Pressure Monitoring: A consensus view on the technique of ambulatory blood pressure monitoring. Hypertension 1995; 26: 912–8.

Thijs L, Staessen JA, O'Brien ET, Amery AK, Atkins N, Baumgart P, DeCort P, Degaute J–P, Dolenc P, DeGaudemaris R, Enstrom I, Fagard R, Gosse P, Gourlay S, Hayashi H, Imai Y, James G, Kawasaki T, Kuschnir E, Kuwajima I, Lindholm L, Liu L, Macor F, Mancia G, McGrath B, Middeke M, Ming J, Omboni S, Otsuka K, Palatini P, Parati G, Pieper C, Verdecchia P, Zachariah P, Zhang W. The ambulatory blood pressure in normotensive and hypertensive subjects: results from an international database. Netherlands J Med 1995; 46: 106–14.

1996

Otsuka K, Cornélissen G, Halberg F. Predictive value of blood pressure dipping and swinging with regard to vascular disease risk. Clin Drug Invest 1996; 11: 20–31.

Suzuki M, Guilleminault C, Otsuka K, Shiomi T. Blood pressure "dipping" and "non-dipping" in obstructive sleep apnea syndrome patients. Sleep 1996; 19(5): 382–7.

Kawasaki T, Cugini P, Itoh K, Uezono K, Ogaki T, Yoshimizu Y, Cornélissen G. Circadian rhythm of blood pressure and lifestyle: a study of clinically healthy subjects living in rural and industrialized countries. J Human Hypertension 1996; 10: 281–5.

Cornélissen G, Halberg F. Impeachment of casual blood pressure measurements and the fixed limits for their interpretation and chronobiologic recommendations. Time-dependent Structure and Control of Arterial Blood Pressure, Portaluppi F, Smolensky MH (eds.). Ann NY Acad Sci 1996; 783: 24–46.

Kawasaki T, Itoh K, Cugini P, Uezono K, Cornélissen G. Circadian blood pressure pattern in clinically healthy young Nepalese. Time-dependent Structure and Control of Arterial Blood Pressure, Portaluppi F, Smolensky MH (eds.). Ann NY Acad Sci 1996; 783: 326–7.

Kawasaki T, Uezono K, Cugini P, Itoh K, Cornélissen G. Ambulatory blood pressure monitoring in normotensive smokers and nonsmokers. Time-dependent Structure and Control of Arterial Blood Pressure, Portaluppi F, Smolensky MH (eds.). Ann NY Acad Sci 1996; 783: 328–9.

Portela A, Cornélissen G, Halberg F, Menéndez JC. Can blood pressure variability be reconciled with the interpretation of single measurements versus fixed thresholds? EuroRehab 1996; 2: 33–48.

Halberg F, Cornélissen G, Halpin C, Burchell H, Watanabe Y, Kumagai Y, Otsuka K, Zaslavskaya R. Fleeting "monitor-", "conflict-" or "grief-associated" blood pressure disorders: MESOR-hypertension and circadian hyperamplitudetension (CHAT). EuroRehab 1996; 6: 225–40.

1997

Otsuka K, Cornelissen G., Halberg F. Circadian amplitude-hypertension as a risk factor of ischemic stroke and nephropathy. Vrach issue 1997; 4: 9–11 [In Russian.]

Otsuka K, Cornélissen G, Halberg F, Oehlert G. Excessive circadian amplitude of blood pressure increases risk of ischemic stroke and nephropathy. J Med Eng Technol 1997; 21: 23–30.

Staessen JA, Bieniaszewski L, O'Brien E, Gosse P, Hayashi H, Imai Y, Kawasaki T, Otsuka K, Palatini P, Thijs L, Fagard R, on behalf of the "Ad Hoc" Working Group. Nocturnal blood pressure fall on ambulatory monitoring in a large international database. Hypertension 1997; 29 (1 Pt 1): 30–9.

Cornélissen G, Halberg F, Hawkins D, Otsuka K, Henke W. Individual assessment of antihypertensive response by self-starting cumulative sums. J Med Eng Technol 1997; 21: 111–20.

Cornélissen G, Halberg F, Bingham C, Kumagai Y. Toward engineering for blood pressure surveillance. Biomed Instrum Technol 1997; 31: 489–98.

Cornélissen G, Halberg F, Bingham C, Kumagai Y. More on software chronobioengineering for blood pressure surveillance. Biomed Instrum Technol 1997; 31: 511–3.

Gubin D, Cornélissen G, Halberg F, Gubin G, Uezono K, Kawasaki T. The human blood pressure chronome: a biological gauge of aging. In Vivo 1997; 11: 485–94.

1998

Halberg F, Cornélissen G, Halberg J, Fink H, Chen C-H, Otsuka K, Watanabe Y, Kumagai Y, Syutkina EV, Kawasaki T, Uezono K, Zhao ZY, Schwartzkopff O. Circadian Hyper-Amplitude-Tension, CHAT: a disease risk syndrome of anti-aging medicine. J Anti-Aging Med 1998; 1: 239–59.

1999

Kawasaki T, Uezono K, Cugini P, Ueno M, Abe I, Ashida T, Doi M, Ebihara A, Hayashi H, Katayama S, Matsuoka H, Ogihara T, Otsuka K, Saito N, Shimamoto K. Rationale for time-qualified reference standards for 24-hour blood pressure values and their circadian rhythms in Japanese normotensive adults: a study by the Ambulatory Blood Pressure Monitoring Research Group. Jpn Circ J 1999; 63(10): 744–51.

Otsuka K, Cornélissen G, Shinagawa M, Halberg F. Blood pressure variability assessed by semi-automatic and ambulatorily functional devices for home use. Clin Exp Hypertension 1999; 21: 729–40.

Halberg F, Cornélissen G. Commentary: Chronomes: Time structures within the physiological range identify early disease risk aiming at primary prevention. J Gerontol Med Sci 1999; 54A: M309–11. [Commentary on Cugini P, Gori MC, Petrangeli CM, Tisei P, Giubilei F. Maintained blood pressure and heart rate circadian rhythm in early stage Alzheimer's disease. J Gerontol Med Sci 1999; 54A: M304–8.]

2000

Halberg F, Cornélissen G, Siegelova J, Fiser B, Dusek J, Otsuka K, Delyukov AA, Gorgo Y, Gubin DG, Gubin GD, Schwartzkopff O. The Cosmos and CHAT, prompting blood pressure and heart rate monitoring for Dérer's week. Bratislavske lekarske listy 2000; 101: 260–71.

Halberg F, Cornélissen G, Chen C-H, Katinas GS, Otsuka K, Watanabe Y, Herold M, Loeckinger A, Kreze A, Kreze E, Perfetto F, Tarquini R, Maggioni C, Sothern RB, Schwartzkopff O. Chronobiology: time structures, chronomes, gauge aging, disease risk syndromes and the cosmos. J Anti-Aging Med 2000; 3: 67–90.

Halberg F, Cornélissen G, Katinas GS, Watanabe Y, Otsuka K, Maggioni C, Perfetto F, Tarquini R, Schwartzkopff O, Bakken EE. Feedsidewards: intermodulation (strictly) among time structures, chronomes, in and around us, and cosmo-vasculo-neuroimmunity. About ten-yearly changes: what Galileo missed and Schwabe found. Ann NY Acad Sci 2000; 917: 348–76.

2001

Shinagawa M, Kubo Y, Otsuka K, Ohkawa S, Cornélissen G, Halberg F. Impact of circadian amplitude and chronotherapy: relevance to prevention and treatment of stroke. Biomed Pharmacother 2001; 55(Suppl 1): 125s–32s.

Cornélissen G, Schwartzkopff O, Halberg F, Otsuka K, Watanabe Y. 7-day ambulatory monitoring for adults with hypertension and diabetes [letter]. Am J Kidney Diseases 2001; 37: 878.

2002

Shinagawa M, Otsuka K, Murakami S, Kubo Y, Cornélissen G, Matsubayashi K, Yano S, Mitsutake G, Yasaka Ki, Halberg F. Seven-day (24-h) ambulatory blood pressure monitoring, self-reported depression and quality of life scores. Blood Pressure Monitor 2002; 7(1): 69–76.

Borer KT, Cornélissen G, Halberg F, Brook R, Rajagopalan S, Fay W. Circadian blood pressure overswinging in a physically fit, normotensive African American woman. Am J Hypertens 2002; 15: 827–30.

Stinson SM, Cornélissen G, Scarpelli PT, Halberg F. Self-measurement and ambulatory monitoring of blood pressure: a subject's chronobiological perspective. Biomed Pharmacother 2002; 56(Suppl 2): 333s–8s.

2003

Otsuka K, Cornélissen G, Schwartzkopff O, Bakken EE, Halberg F, Burioka N, Katinas GS, Kane R, Regal PJ, Schaffer E, Sonkowsky R, Patterson R, Engebretson M, Brockway B, Wang ZR, Delmore P, Halpin C, Sarkozy S, Wall D, Halberg J. Clinical chronobiology and chronomegeriatrics: At variance with recommendations of subsequent guidelines, yet focusing indeed on pre-hypertension in the physiological range. Biomed Pharmacother 2003; 57(Suppl 1): 164s–98s.

Cornélissen G, Rigatuso J, Wang ZR, Wan CM, Maggioni C, Syutkina EV, Schwartzkopff O, Johnson DE, Halberg F, International Womb-to-Tomb Chronome Group. Case report of an acceptable average but overswinging blood pressure in Circadian Hyper-Amplitude-Tension, CHAT. Neuroendocrinol Lett 2003; 24(Suppl 1): 84–91.

Watanabe Y, Cornélissen G, Halberg F. Thousands of blood pressure and heart rate measurements at fixed clock hours may mislead. Neuroendocrinol Lett 2003; 24: 339–40.

Katinas G, Halberg F, Cornélissen G, Otsuka K, Tarquini R, Perfetto F, Maggioni C, Schwartzkopff O, Bakken E. Transient circadian hyper-amplitude-tension (CHAT) may be intermittent: case reports illustrating gliding spectral windows. Biomed Pharmacother 2003; 57(Suppl 1): 104s–09s.

Halberg F, Cornélissen G, Spector NH, Sonkowsky RP, Otsuka K, Baciu I, Hriscu M, Schwartzkopff O, Bakken EE. Stress/strain/life revisited. Quantification by blood pressure chronomics: benetensive, transtensive or maletensive chrono-vasculo-neuro-immuno-modulation. Biomed Pharmacother 2003; 57(Suppl 1): 136s–63s.

2004

Otsuka K, Halberg F, Cornélissen G, Katinas G, Singh RB. Preface. Chronomics based on long-term vascular monitoring leads to scientific and clinical findings. Clin Exper Hypertens 2004; 27(2–3): 1–5.

Murakami S, Otsuka K, Kubo Y, Shinagawa M, YamanakaT, Ohkawa S, Kitaura Y. Repeated ambulatory monitoring reveals a Monday morning surge in blood pressure in a community-dwelling population. Am J Hypertens 2004; 17: 1179–83.

Sabath LD, Cornélissen G, Halberg F, Stoynev A, Ikonomov O, Otsuka K. Circadian hypo- and hyper-amplitude-tension (CHAT) associated with putative pheochromocytoma and 12-hourly phenoxybenzamine treatment. Biomed Pharmacother 2004; 58(Suppl 1): S99–106.

Sarabandi T, Cohn J, Cornélissen G, Halberg F, Katinas G, Holte J, Schwartzkopff O, Otsuka K. No baseline for blood pressure in ordinary life: another case of transient chronome alterations. Biomed Pharmacother 2004; 58(Suppl 1): S107–10.

Zaslavskaya RM, Lilitsa GV, Dilmagambetova GS, Halberg F, Cornélissen G, Otsuka K, Singh RB, Stoynev A, Ikonomov O, Tarquini R, Perfetto F, Schwartzkopff O, Bakken EE. Melatonin, refractory hypertension, myocardial ischemia and other challenges in nightly blood pressure lowering. Biomed Pharmacother 2004; 58(Suppl 1): S129–34.

2005

Otsuka K, Halberg F, Cornélissen G, Katinas G, Singh RB. Preface. Clin Exp Hypertens 2005; 2 & 3: 113–8.

Prikryl P, Cornélissen G, Neubauer J, Prikryl P Jr, Karpisek Z, Watanabe Y, Otsuka K, Halberg F. Chronobiologically explored effects of telmisartan. Clin Exp Hypertens 2005; 2 & 3: 119–28.

Sundaram B, Holley DC, Cornélissen G, Naik D, Hanumansetty R, Singh RB, Otsuka K, Halberg F. Circadian and circaseptan (about-weekly) aspects of immigrant Indians' blood pressure and heart rate in California, USA. Biomed Pharmacother 2005; 59(Suppl 1): S76–85.

Katinas GS, Cornélissen G, Otsuka K, Haus E, Bakken EE, Halberg F. Why continued surveillance? Intermittent blood pressure and heart rate abnormality under treatment. Biomed Pharmacother 2005; 59 (Suppl 1): S141–51.

Cornélissen G, Delcourt A, Toussaint G, Otsuka K, Watanabe Y, Siegelova J, Fiser B, Dusek J, Homolka P, Singh RB, Kumar A, Singh RK, Sanchez S, Gonzalez C, Holley D, Sundaram B, Zhao Z, Tomlinson B, Fok B, Zeman M, Dulkova K, Halberg F. Opportunity of detecting pre-hypertension: worldwide data on blood pressure overswinging. Biomed Pharmacother 2005; 59(Suppl 1): S152–7.

Halberg F, Cornélissen G, Otsuka K, Sanchez de la Peña S, Schwartzkopff O, Watanabe Y, Singh RB, Pati AK, Wall DG, Delmore P, Borer K, Beaty LA, Nolley ES, Adams C, Siegelova J, Prikryl P, Homolka P, Dusek J, Fiser B. Why and how to implement 7-day/24-hour blood pressure monitoring? Int J Geronto-Geriatrics 2005; 8(1): 1–31.

2006

Cornélissen G, Chen CH, Halberg F. Predictive value of blood pressure variability: merits of circadian parameters versus dipping patterns. N Engl J Med 2006; 355;8: 850.

2007

Cornélissen G, Halberg F, Otsuka K, Singh RB, Chen CH. Chronobiology predicts actual and proxy outcomes when dipping fails. Hypertension 2007; 49: 237–9. doi:10.1161/01.HYP.0000250392.51418.64.

Halberg F, Cornélissen G, Halberg J, Schwartzkopff O. Pre-hypertensive and other variabilities also await treatment. Am J Med 2007; 120: e19–20. doi:10.1016/j.amjmed.2006.02.045.

2008

Gupta AK, Greenway FL, Cornélissen G, Pan W, Halberg F. Prediabetes is associated with abnormal circadian blood pressure variability. J Human Hypertens 2008; 22: 627–33. doi:10:1038/jnh.2008.32.

Halberg F, Cornélissen G, Otsuka K, Siegelova J, Fiser B, Dusek J, Homolka P, Sanchez de la Peña S, Singh RB, BIOCOS project. Extended consensus on need and means to detect vascular variability disorders (VVDs) and vascular variability syndromes (VVSs). Geronto-Geriatrics: Int J Gerontology-ChronomeGeriatrics 2008; 11(14): 119–46.

Halberg F, Katinas GS, Cornélissen G, Sanchez de la Peña S. Continued self-surveillance for treating a frequent vascular variability syndrome. Geronto-Geriatrics: Int J Gerontology-ChronomeGeriatrics 2008; 11(14): 147–54.

2010

Jiang Z, Li W, Liu Y, Wang Y, Wang Z, Xiao J, Guo H, Cornélissen G, Halberg F. Circadian blood pressure and heart rate characteristics in haemorrhagic vs ischaemic stroke in Chinese people. J Hum Hypertens 2010; 24: 165–74.

2011

Murakami S. Otsuka K. Kono T. Soyama A. Umeda T. Yamamoto N. Morita H. Yamanaka G. Kitaura Y. Impact of outdoor temperature on prewaking morning surge and nocturnal decline in blood pressure in a Japanese population. Hypertens Res 2011; 34: 70–3.

2012

Otsuka K, Cornelissen G. Alcohol consumption and vascular variability anomalies (VVAs) and disorders (VVDs) detected by C-ABPM. Open Nutraceuticals J 2012; 5(Suppl 1-M1): 81–9.

Halberg F, Cornélissen G, Hillman D, Beaty L, Hong S, Schwartzkopff O, Watanabe Y, Otsuka K, Siegelova J. Chronobiologically interpreted ambulatory blood pressure monitoring in health and disease. Global Adv Health Med 2012; 1(2): 64–88.

Cornélissen G, DeMeester F, Wilczynska-Kwiatek A, Halberg F. Relative prominence of weekly-to-daily amplitude ratio of blood pressure changes with age. Proceedings, Natural Cataclysms and Global Problems of the Modern Civilization, Istanbul, 19–21 September 2011. London: SWB International Publishing House; 2012, pp. 297–300.

Cornélissen G, Otsuka K, Halberg F. Week-long ABPM records are a prerequisite for a reliable diagnosis of vascular variability disorders (VVDs). Proceedings, Natural Cataclysms and Global Problems of the Modern Civilization, Istanbul, 19–21 September 2011. London: SWB International Publishing House; 2012, pp. 263–68.

2013

Halberg F, Powell D, Otsuka K, Watanabe Y, Beaty LA, Rosch P, Czaplicki J, Hillman D, Schwartzkopff O, Cornelissen G. Diagnosing vascular variability anomalies, not only MESR-hypertension. Am J Physiol Heart Circ Physiol 2013; 305: H279–94.

Gumarova LZ, Cornélissen G, Halberg F, Mansharipova AT, Otsuka K, Syutkina EV, Masalov AV, Chibisov SM, Frolov VA. Duration of ABPM as an important prerequisite for a reliable diagnosis of vascular variability disorders (VVDs). Vestnik PFUR, seria Meditsina, 2013; No. 1: 27–33.

Watanabe Y, Halberg F, Otsuka K, Cornélissen G. Toward a personalized chronotherapy of high blood pressure and a circadian overswing. Clin Exp Hypertens 2013; 35(4): 257–66. doi: 10.3109/10641963.2013.780073. PMID: 23541237.

2014

Umishio W, Ikaga T, Otsuka K, Ando S. Multivariate analysis of the rise in home blood pressure by personal factors: A field survey on the effect of indoor thermal environment on blood pressure in winter. J Environ Eng (Japan), 2014; 79(701):571–7.

Otsuka K, Okajima K, Yamanaka T, Oinuma S, Sasaki J, Taniwaki Y, Otsuka Kei, Cornelissen G. Aging and the novelty pressor effect in men on the first day of 7-day/24-hour ambulatory blood pressure monitoring. J Am Geratric Soc 2014; 62: 1602–5.

Otsuka K, Okajima K, Oinuma S, Sasaki J, Yamanaka T, Takahashi Y, Kokaji A, Kitamura K, Cornelissen G. Aging and circadian disruption of blood pressure observed using 7-day/24-hour ambulatory blood pressure monitoring. J Am Geriatrics Soc 2014; 62: 2213–5.

Okajima K, Otsuka K, Oinuma S, Sasaki J, Yamanaka T, Cornelissen G. Aging and within- and between-day variability assessed using 7-day/24-hour ambulatory blood pressure monitoring. J Am Geriatrics Soc 2014; 62: 2440–2.

Contents

Chapter 1
From Chronobiology to Chronomedicine: Early Days

Abstract Franz Halberg (1919–2013) developed chronobiology and founded the field of chronomedicine including chronomics, chronoastrobiology, and chronobioethics. He coined the term circadian, after documenting that biologic rhythms tip the scale between health and disease and even between life and death. His work is

---

In approaching any problem, stress and strain in particular, it is tempting to use a "norm" as an initial single value or a daily or yearly average and to refer to it as a "baseline." Over 60 years ago, Halberg wrote [25], citing Zbigniew Z. Godlowski (J. Endocrinol 1952; 8: 102): "… it is not surprising to encounter reference to 'the great variation in circulating eosinophil counts, even in normal conditions, which makes it impossible to establish a base-line' for the study of the physiopathology of these cells. It is indeed impossible to establish a base-line, at least a straight base-line, for eosinophil counts in normal conditions, because there is none in normal conditions." Nor are there baselines for other variables, except in the wishful thinking of too many investigators, including the best, such as Hans Selye, as we will demonstrate. As of March 2012, 63,740 publications referring to "circadian" attest to the ubiquity of these variations; there are many extracircadian changes as well, all constituting control information to anything we do along the scale of time. We appeal to all authors and editors to debunk "baselines" and to use, whenever available, historical and concomitant controls in science (and art, most broadly in all "humanities").

An earlier paper by Halberg was labeled by the editor of the *Journal of Gerontology A*: Biological Sciences and Medical Sciences "Future history." This title applies to this historical sketch; it ends as an implied, here explicit recommendation for the future, to account for a complementary system of cosmic cycles in and around us. This ever-present transdisciplinary spectrum will have to be considered in focus upon anything singled out as a partial system, as the study of time structure in living matter (chronobiology) investigated in biological data aligned with space and other "weather" series (by chronomics). The invariably present complementary system, a sphere of the mind (noös), Vernadsky's noösphere, undergoes aeolian cosmic cycles mirrored in human affairs, as in living matter broadly, in the chronousphere (Gk chronos, time + Attic Gk nous, mind + Gk sphairos, sphere, globe), the nonstationary glocal (global and local) diversity in space and time of the universe which happens to be our home.

We qualify Pierre Charron (1541–1603), who wrote that "The true science and true study of man is man" (Traité de la Sagesse: Preface du premier Livre), and Alexander Pope (1688–1744), who wrote "The proper study of Mankind is Man" (An Essay on Man, Epistle II), adding, of course, the need for the backtracking of human periodicities over an archaeon to the cosmos. This sketch, originally written for a meeting on the history of chronobiology, was later extended in response to a request by Dr. Botond Buda for Folia anthropologica.

K. Otsuka et al., *Chronomics and Continuous Ambulatory Blood Pressure Monitoring*, DOI 10.1007/978-4-431-54631-3_1

summarized in his over 3600 scientific publications, in cooperation with colleagues from around the world. Progress of the chronomedicine exactly depends on his history of research. Thus, herein we introduce him briefly.

Beyond blood pressure and heart rate, the biomedically interested scientist will find an all-too-brief introduction to what students of sociology, biology, and medicine regard as a new science from the molecular to the sociologic and the epidemiologic time structure of life. We show that the large extent of cardiovascular variations can be exploited in the form of dynamic and other endpoints derived for each individual, for use in preventive as well as curative health care.

Keywords Chronobiology • Circadian rhythm • Chronobiologically interpreted ambulatory blood pressure monitoring • Vascular variability disorders • Linear-nonlinear cosinor spectrum

1.1 From Chronobiology to a Unified Science

Living matter is variable in time and space, as is weather. The more constant we try to make our proximal environment, the better we recognize seemingly spontaneous variability that reflects the changes around us, near and far. I describe a journey that started with counts of circulating blood cells and action potentials of our brain and the then-surprising periodicities even in RNA and DNA formation and led to the recognition of the drastic importance of timing, among very many other stimuli, exposure to noise first (and X-irradiation and drugs afterward) in the 1950s. What I had then dubbed "circadian stage" made the difference between ~70 or 80 % death and ~70 or 80 % survival from the same drug, ouabain, or an adrenocortical inhibitor, respectively, or other stimulus. Eventually, I learned that extracircadian rhythms also needed to be mapped; infradians can determine whether we die suddenly because of cardiac malfunction, by our own hand or by that of others. The cycles of the cosmos are found transdisciplinarily and are a challenge as we try to account for them beyond the molecule, at the levels of atom and quantum, with respect to the ever-present associations with space weather.

A perspective in space gains from new complementary imaging tools, chronobiology and chronomics leading (this is humanity's challenge) to chronobioethics. In having fun toward that goal, we can use a set of linear-nonlinear cosinor methods that, whenever possible, should be applied glocally in time and space: nonstationarities are found not only in geo- (and cosmo-)graphy but also in time, in each case requiring the analyses of the longest available time series as a whole (globally) and in sections varied systematically in length (locally). The resulting atlas, when completed, will, we trust, serve a unified science, providing new preventive and therapeutic marker endpoints for variability anomalies among which coexisting multiple circadian and multiple about-7-day cycles are already mapped and await application for strain relief during wellness. We resonate with the solar flares about 5 months and the solar wind speeds over 1-year-long periods, as well as with decadals (about 10 years), didecadals (about 20 years), paratridecadals (about 35 years), and trans-semicentennials (about 600 years) of space weather. These contribute to more than blackouts in and around us, including war and terrorism. The study of space weather is thus relevant to human affairs, including an effect on our mood, long anticipated by Alexander Leonidovich Chizhevsky and indirectly documented by Joseph Vallot, validated herein in "the language of shared frequencies" and by remove-and-replace approaches whose application has just begun.

1.2 Variability: Foe in 1948–1949, Friend Thereafter

Late in 1948, when I arrived at the Peter Bent Brigham Hospital of the Harvard Medical School in Boston, word was that the adrenocortical hormone cortisone, isolated by Edward C. Kendall of the Mayo Clinic (manufactured by the Merck Co.'s Lewis H. Sarett, the likely source of information) given by Philip S. Hench to patients with severe rheumatoid arthritis [1–3], had reproduced the effect of Lourdes: figuratively speaking, patients who came in wheelchairs left walking. This was more impressive than the wheelbarrows in William Withering's (1749–1794) garden, with which patients with congestive heart failure were said to have brought their ascites and presumably left them after receiving Digitalis purpurea [4]. Cortisone was then scarce and a substitute with similar activity was highly desirable. Hence, at Harvard I was assigned to test different substances for cortisone-like activity. Variability was great, so that occasionally the count was spontaneously zero in one or at most two consecutive samples. I was given a few mice for testing. One solution, with tests on one or a few mice, was to make sure to bring the count to zero for 24 hours, something I did not find occurring spontaneously; I used a large dose of 2.8 mg of cortisone, the truck on the right of Fig. 1.1 [5] as a reference standard for eosinopenic activity, then a cautious, but wasteful, approach warranted only in view of great variability in count and the limited numbers of mice available for testing and even though the reference standard was scarce. The dose response curve on the left and in the middle of Fig. 1.1 shows that 1 μg of cortisone was effective in inducing a relative eosinopenia in the ascending stage of the circadian rhythm in eosinophil count in the mice investigated (when a spontaneous decrease in count was most unlikely to occur).

1.3 Epinephrine Test of Adrenocortical Function?

Others in the department of medicine at the Peter Bent Brigham Hospital tested human adrenocortical function: Recant, Hume, Forsham, and Thorn [6–8] had postulated, in keeping with C.N.H. Long [9], that epinephrine (from the adrenal

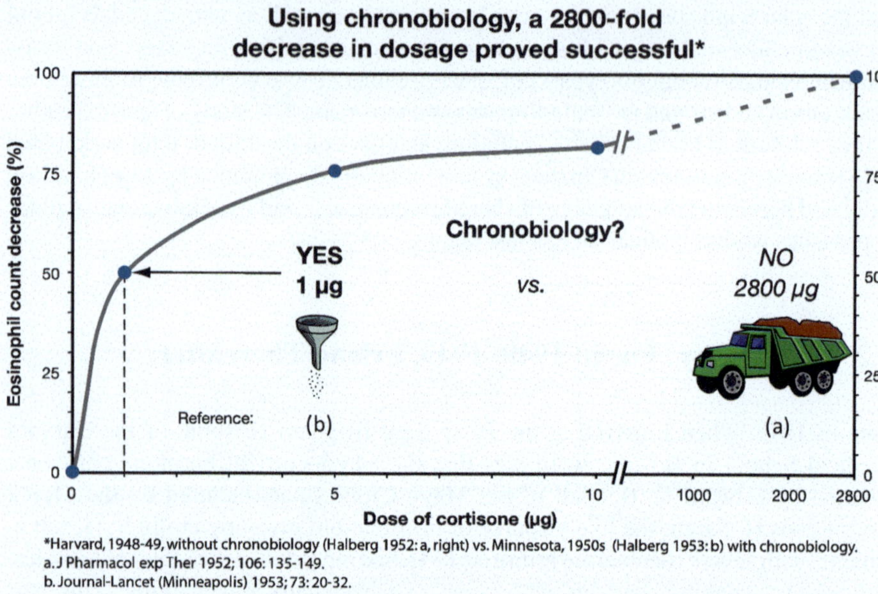

Fig. 1.1 Gain in sensitivity from taking rhythms into account. Circulating eosinophil cell-depressing activity was detected with 1 μg of cortisone when tested in the ascending phase of the circadian rhythm in the count of these blood cells [25]. The increase in sensitivity of the assay is depicted by comparing the minuscule dose on the left, taking into account the rhythm, with the dose depicted as a truckload on the right, when the rhythm was "eliminated" by depressing the cell count to zero for 24 hours by 2.8 mg of cortisone used as the standard in my original bioassay in Boston [5]. Using chronobiology, a 2800-fold decrease in dosage proved successful (© Halberg Chronobiology Center)

medulla) acts upon the hypothalamus, which in turn, via corticotropin-releasing factor, stimulates the pituitary [7, 8] to produce adrenocorticotropin (ACTH), which stimulates the adrenal cortex to secrete corticosteroid which in turn suppresses circulating eosinophil counts. Accordingly a test was recommended for the clinic based on epinephrine-induced eosinopenia, thought to depend upon the presence of the adrenal cortex and which would not occur as a ≥50 % fall in the absence of cortical adrenal tissue. It seemed somewhat surprising that epinephrine from the adrenal medulla would have to act via the hypothalamus and the pituitary to stimulate its next-door neighbor, the adrenal cortex. Hence, I tested the effect of epinephrine in adrenalectomized and gonadectomized mice with ectopic cortical tissue removed, as best I could, along both sides of the spinal column, in the large ligaments of females and the scrotal fat of males. The eosinophil count decreased after my epinephrine injections by ≥50 % when I happened to test. I found eosinopenia in response to epinephrine in mice without adrenals. I reported my inability to confirm the basis of the epinephrine test in mice, presumably deprived of all adrenal cortical tissue (for epinephrine to act upon), to my department head, the late George W. Thorn, Hersey Professor of Medicine at Harvard. Thorn admired my "sticking to my guns," as he put it, but added that all his staff members and distinguished fellows could not be wrong. In the case of the epinephrine test, however, his staff at the time was hardly right [10]: in 1952, William R. Best et al. reported on the "clinical and statistical analyses of 702 4-hour eosinophil response tests to corticotropin, ephedrine, epinephrine and placebos in 284 normal and miscellaneous medical subjects" and concluded that "Greater than 50 % drop of eosinophils has been noted following epinephrine or ephedrine in patients with pituitary tumors and in patients adrenalectomized and receiving small doses of cortisone. … Tests with these substances are therefore of little value in the diagnosis of adrenal, hypothalamic, or pituitary disease, and do not accurately assess the functional capacity of these organs at the time of examination."

1.4 Vascular Variability Disorders (VVDs)

Very many still treat variability as a foe, or ignore it. Typical is a former head of cardiology at the University of Minnesota who was given a week's half-hourly around-the-clock data on one of his patients. He certainly "honestly" added that he would gladly check the results (on a variability disorder, with his *single* measurement!). He is not alone. A family practice department head who monitored himself for a week, in whom we found altered variability, went to his care provider in Rochester, Minnesota, and we never saw him again. These instances are the rule. A respected friend, a former cardiology department head at the Mayo Clinic, who monitored himself repeatedly for Franz Halberg's sake only, as he put it, could not be convinced of the merit of monitoring, even when his own data showed that it was helpful [11].

It took more than a generation for the profession to accept the use of the blood pressure cuff at home (not yet to remove it from the care provider's office), a very great progress, re-advocated in the 1970s [12, 13] and implemented in the third millennium AD [14].[1] It may take another generation to develop a chronobiologically

[1] Self-measurements enabled even the detection of the after-effect upon blood pressure of an earthquake [14]. They did not suffice to detect the same event's antecedents and certainly miss all changes occurring during sleep. Automatic records during sleep can provide essential information.

The blood pressure cuff in the care provider's (physician or nurse) office is no substitute for, and no check of, chronobiologically analyzed serial measurements. The cuff has a place only in the homes of each individual. In the 1960s and 1970s, it was worthwhile to plead [12], as Frederic C. Bartter, of Bartter syndrome, also did, for moving the blood pressure cuff from the providers' offices to everybody's (including his own) workplace by day and taking it home by night. This recommendation holds only until a system of C-ABPM becomes affordably available to everybody, as it is now via BIOCOS. As head of the Hypertension-Endocrine Branch at the US National Institutes of Health (NIH), and later as head of the NIH's Clinical Center, Bartter advocated the interpreted ambulatory blood pressure and heart rate monitoring (C-ABPM) system (Fig. 1.2a–f [15]), which may allow even the mapping of infradian, e.g., paratridecadal cycles [16] (Fig. 1.3) and perhaps some new information thereby concerning the health not only of individuals but also of societies (Fig. 1.4) [17], and may even detect the antecedents of earthquakes (Fig. 1.5) [18]. If we can have sensors in the tires of our cars and computer chips that continuously monitor pressure over the life of the tire, we should be able to measure, as-one-goes, all rhythms that can form the basis of diagnostic and therapeutic measures. When known or assessed, such rhythms resolve effects of aging and are particularly indicated in view of the epidemic of noncommunicable diseases, referred to as a "slow-motion disaster" [19; cf. 20].

In this context, an immediate reward can be anticipated and has been obtained on a small but worldwide scale by recognizing, for instance, vascular variability anomalies (VVAs) by the cosinor method (Fig. 1.2a–g) ([21]; cf. [12, 22, 23]. When these VVAs (Fig. 1.2d) [15] persist in several automatic 7-day around-the-clock records and become vascular variability disorders (VVDs) [15], treatment is indicated and is sometimes as simple as changing the schedule of hypotensive medication (Fig. 1.2e bottom left) [24]. We try not to repeat the mistakes of the past, such as the failure to scrub before surgery. Measuring and interpreting chronobiologically blood pressure series may seem cumbersome, like scrubbing for antisepsis. Nonetheless, in a computer era, self-surveillance could soon be implemented by everyone, monitoring continuously and affordably. VVAs would gauge loads and teach us how to avoid VVDs as a feature of universal preventive health care. The merit of work with vs. without rhythms is apparent quantitatively in Fig. 1.1, showing the dose reduction for corticosteroid-induced eosinopenia by an order of magnitude (from 2.8 mg to one or a few μg [5] vs. [25]).

Qualitatively opposite results obtained in the case of differences in circadian phase started chronobiology in Minnesota (Fig. 1.6a, b) to where the physiologist Maurice B. Visscher (whom I, as an assistant to the dean, had met in Innsbruck before coming to the USA, while he was lecturing as a member of a Unitarian medical mission) offered me a chance to move in 1949. Visscher also gave me the task to study adrenocortical function indirectly by eosinophil counts in two groups of mice, with a high and low breast cancer rate, respectively. I compared their counts without realizing at first that they were under the influence of different synchronizers, light for a fully fed group (feeding in the daily dark span, as rodents do when food is freely available) and the meal time for a group on a calorie-restricted diet ([26], cf. [27]) (Fig. 1.6) that happened to be given in the morning and, since limited in calo-

need for serial measurements before a physical examination since he, like his predecessors, recognized the unreliability of the best single measurement in individuals who need multiple measurements because of variable blood pressure.

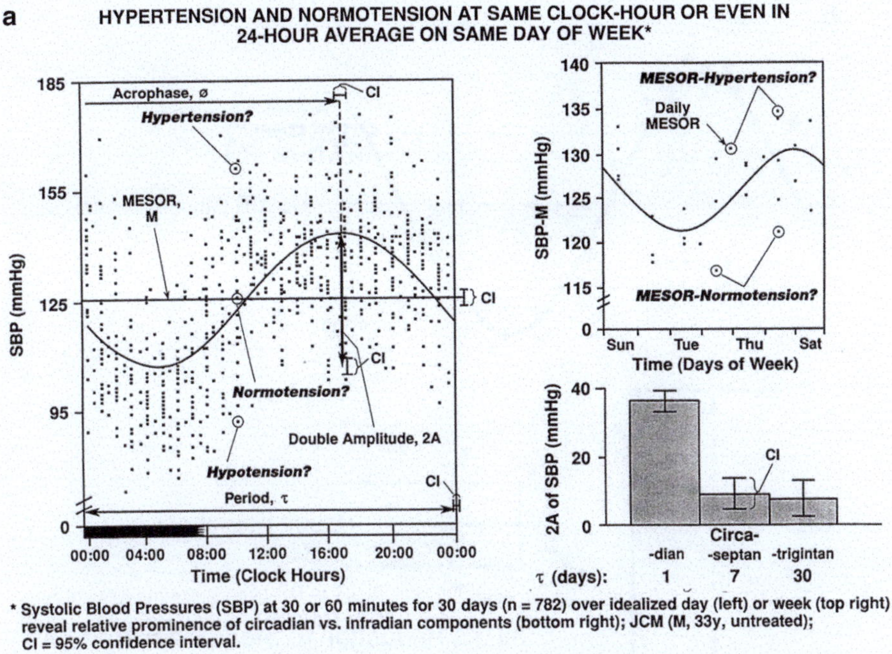

Fig. 1.2 (**a**) Hypertension and normotension at same clock-hour on different days or even "prehypertension" or "normotension" in 24-hour average on same day of the week in different weeks (© Halberg Chronobiology Center). (**b**) *Top*: the MESOR is usually more accurate than the arithmetic mean, as it is less biased from sampling at non-equidistant intervals. *Middle*: the MESOR is usually more precise than the arithmetic mean, as it tends to have a smaller standard error (SE) since the variability accounted for by the usually prominent rhythmic pattern does not enter the error term. *Bottom*: illustration of a cosine function identifying four parameters of the oscillation: MESOR, period, amplitude, and acrophase (© Halberg Chronobiology Center). (**c**) Illustrative parametric (*left*) and nonparametric (*right*) approach bracket a sphygmochron (*middle*) from a MESOR-normotensive man with CHAT, a first tentative diagnosis requiring additional monitoring. After data covering preferably at least 7 days of blood pressure (BP) and heart rate (HR) are downloaded from the e-mail into a computer for analysis, the following results are provided (since the 1990s and currently cost free from corne001@umn.edu) for the patient as well as for the care provider: (1) A list of actual measurements and the times at which they were obtained. (2) A plot of data as a function of time, shown together with the time-specified prediction intervals (PIs) of acceptability for systolic (S) and diastolic (D) BP and HR characteristics. (3) A data summary and a report of any BP and/or HR excess in consecutive 3-hour intervals. This part of the report may be accompanied by a "rhythmometric summary," which is just a more technical form from which the information is derived to prepare the: (4) "Sphygmochron." A sample "sphygmochron" (*center*) illustrates how results are being reported. First, above, the participant's name is kept confidential; a codename is used instead. Gender and age are listed, along with the date and time at which monitoring started, and for how long data were collected. The numerical report consists of two parts labeled "Characteristics" (parametric results) and "Indices of Deviation" (nonparametric results). In each case, results are shown for SBP (when the heart contracts) on the *left*, DBP (when the heart relaxes) in the *middle*, and HR on the *right*. Under "parametric results," a mathematical model of a smooth curve is fitted to the data to assess their circadian variation, which is primarily characterized by four numbers shown in the left-hand section of the graph, one of which, the period, covers with its uncertainty the precise 24 hours, so that the other three numbers are given from the fit of a

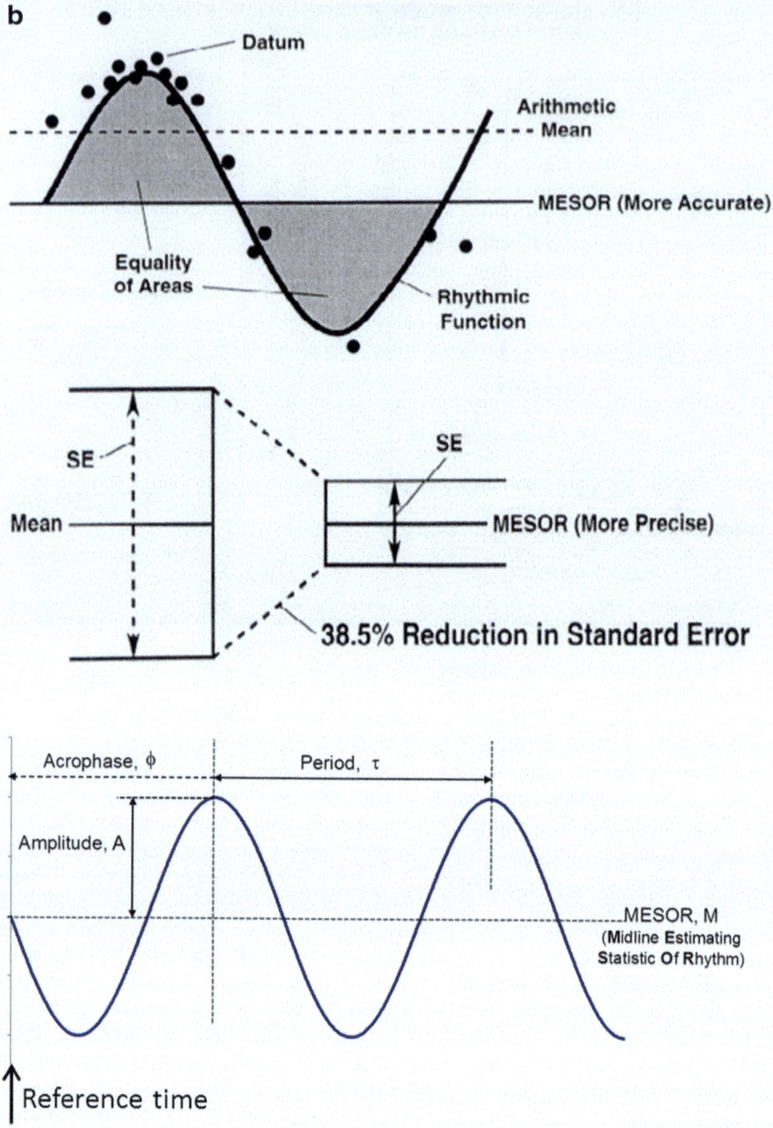

Fig. 1.2 (continued) 24-hour cosine curve. One characteristic, called the "MESOR," is the average value around which values fluctuate. It is very similar to the mean value, but yields more reliable results when the data are not collected at precisely regular intervals, and has a smaller error when the data are equally spaced. Another characteristic, called the "double amplitude," is a measure of the predictable change occurring within a day, from the overall low values found usually during sleep to the high values during the daily active span. The third characteristic, called the "acrophase," is a measure of the time when overall high values are likely to recur each day. For each of the three characteristics ("parameters"), the participant's value is compared to a range of acceptable values, derived from data provided by clinically healthy people of the same gender and age group as the participant. For instance, in the example shown here in the sphygmochron, all parameters are within the range of acceptable values (rectangles), except for the double amplitudes of SBP and DBP. Under "nonparametric results," the participant's data are compared by computer

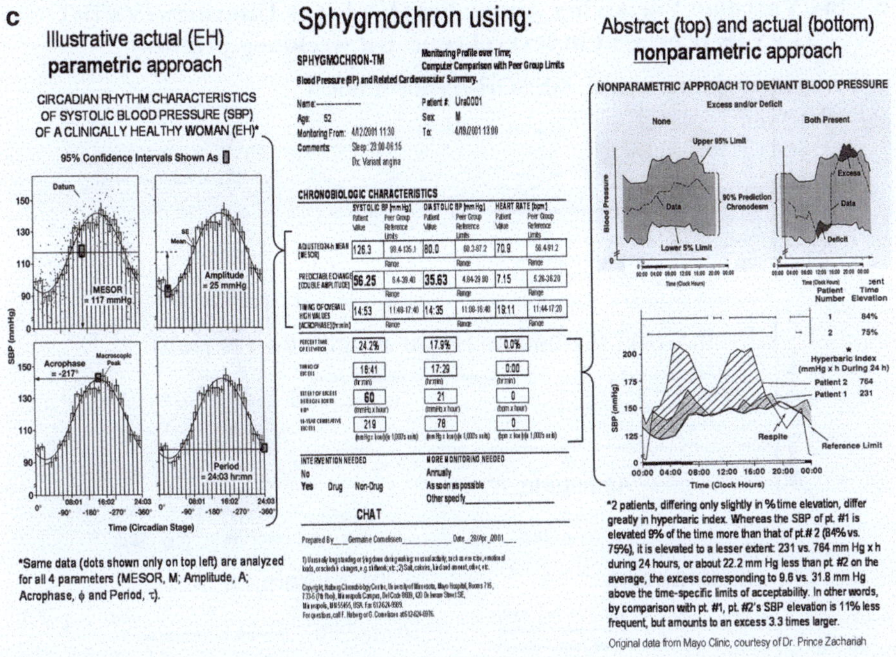

Fig. 1.2 (continued) with time-specified reference values, also derived from chronobiologic archives on clinically healthy subjects matched by gender and age. For this analysis, all data are stacked over an idealized 24-hour day. Whenever a given person's profile exceeds the limits of acceptability of peers, the data are marked as being excessive or deficient. The "percentage time of elevation" reports the relative incidence of excessive values during a 24-hour day. It is common to have occasional high values, but in the example herein, there is reason for concern. The next item, the time of excess, becomes useful when drug treatment should be timed prior to the peak in excess. Excessive values may either be barely above the limit or in turn can be very much higher than the limit. It is therefore important to express the extent of deviation by the "area under the curve," that is, the area between the values when they exceed the limit and this limit itself. Empirically, it has been shown that excess up to about 50 (mmHg × hour during 24 hours) may still be acceptable and accountable for by daily worries and/or physical activities. In the case summarized, the HBI is 60, in *bold*, and if confirmed in the next 7/24 profile, a reason for treatment. On the *top right*, an abstract illustration of excess and deficit is accompanied below by two cases that are similar in terms of percent time elevation. They are very different in terms of hyperbaric index. In patient #2, although the percent time elevation is 9 % smaller than that in patient #1, the hyperbaric index is several times larger. The "timing of excess" can be used as a guide to time the administration of nondrug or, if need be, of drug treatment once there is blood pressure excess above 50 (mmHg × hour during 24 hours) and/or an elevation in MESOR, taking into account the chronopharmacokinetics of the drug prescribed. When, e.g., a tentative diagnosis of MESOR-normotension with CHAT is made, with insight into information provided on the questionnaire given to the participant with the monitor, as a first step, additional analyses may be carried out. Additional monitoring is recommended to check on any abnormality detected during the first monitoring, and if confirmed, the need for intervention is reported to the person monitored so that it can be reported to the health-care provider. In one case summarized elsewhere, the follow-up 7-day monitoring showed that CHAT persisted for both SBP and DBP, while the MESORs were again acceptable. Thus, the diagnosis of CHAT with MESOR-normotension was confirmed. Consultation with a health-care provider was strongly and urgently recommended. In two cases of CHAT without an elevation of the blood pressure MESOR, when such recommendations were ignored, catastrophic disease and high cost occurred, a myocardial infarction in a man or eclampsia

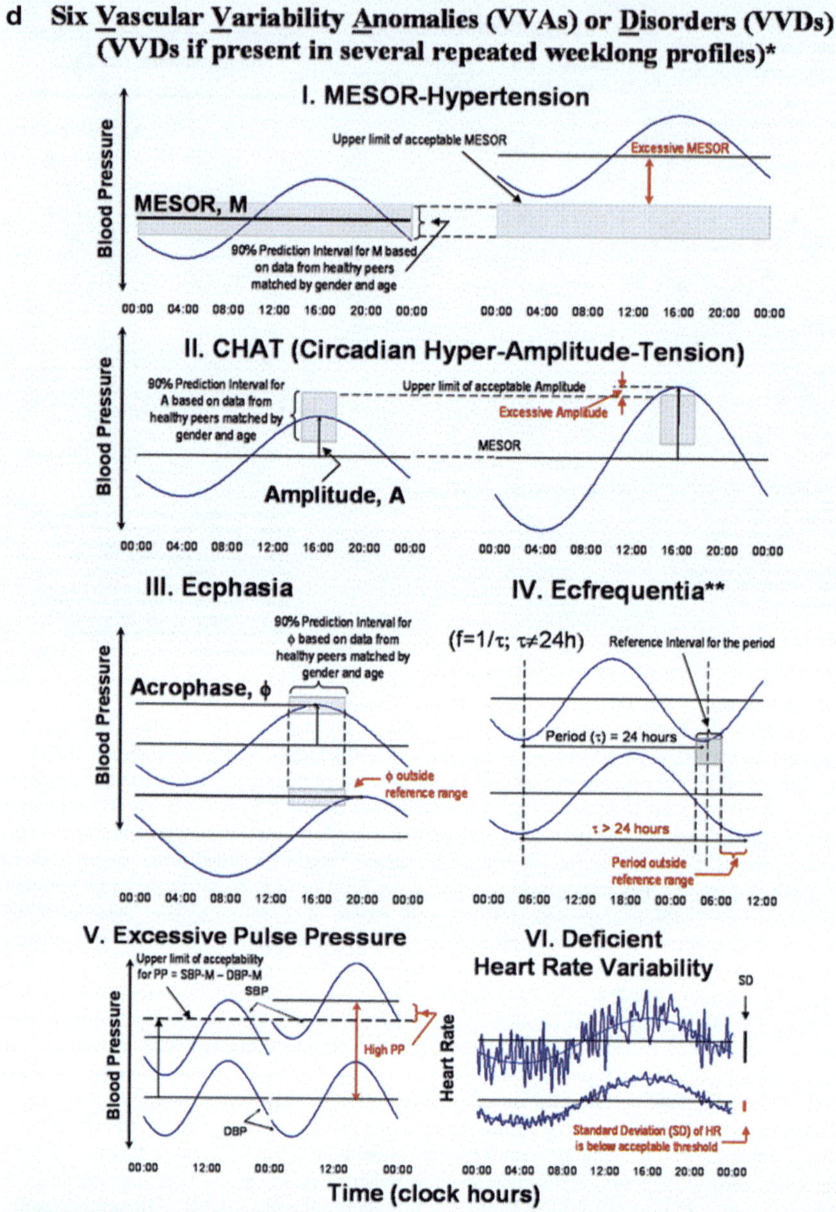

* Validated by chronobiologic analysis of around-the-clock 7-day/24-hour records of measurements at 1-hour or shorter intervals, interpreted in the light of time-specified reference standards qualified by gender and age. **Ecfrequentia: short for frequency (f) alteration (e.g., desynchronization) that can be Dysfrequentia when associated with symptoms and/or persisting in repeated consecutive 7-day records.

Fig. 1.2 (continued) in a pregnant woman with pressures of 115/67 mmHg (SBP/DBP), leading to the delivery of a very premature boy hospitalized for 26 months at a cost of $1 million US.(© Halberg Chronobiology Center). (**d**) Abstract limits for acceptable I. MESOR and II. amplitude (*left*, first 2 rows and *upper curves* in *bottom* two rows) for III. acrophase, IV. frequency (period),

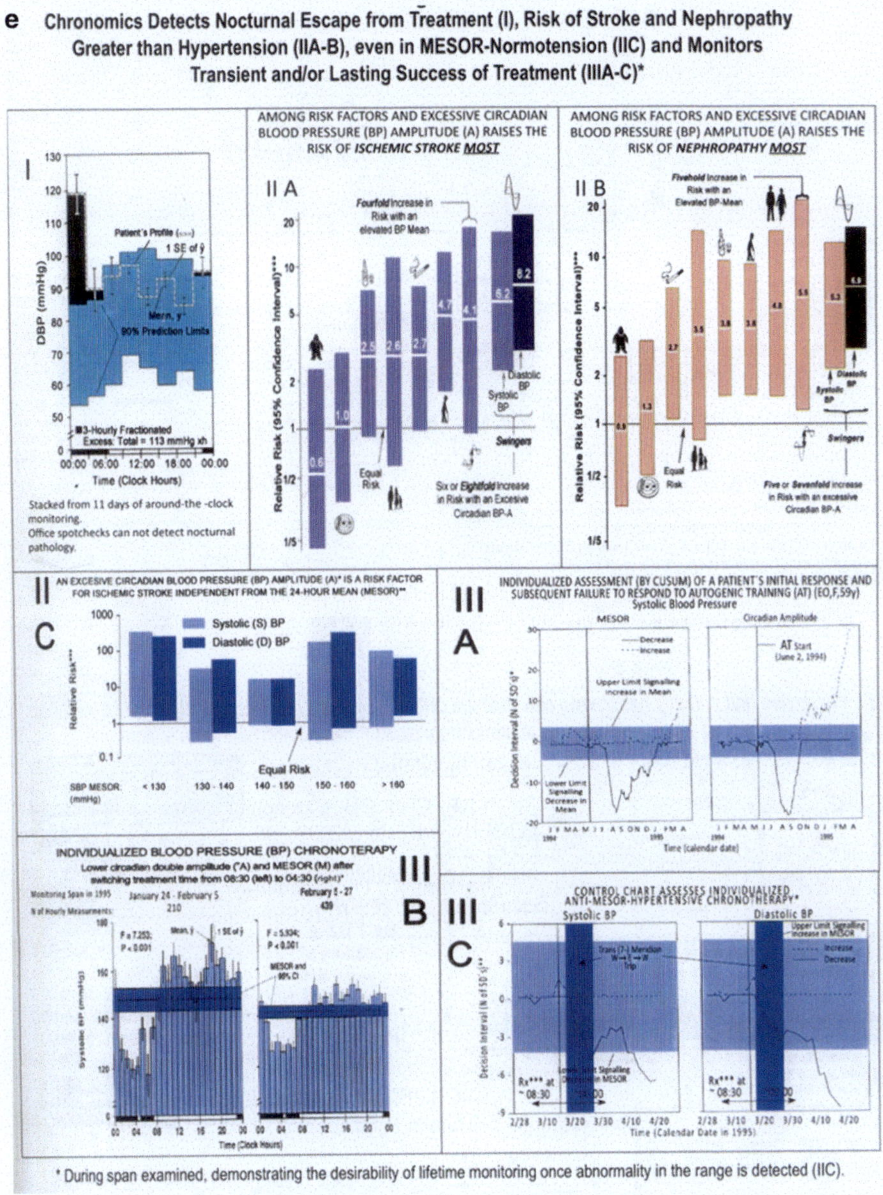

Fig. 1.2 (continued) V. pulse pressure, and VI. heart rate variability, associated with blood pressure and heart rate surveillance and their corresponding vascular variability anomaly (VVA) descriptions. I. MESOR-hypertension (MH) can be systolic (S) (S-MH), diastolic (D) (D-MH), or both (SD-MH), or mean arterial (MA-MH), demonstrated parametrically (by cosine fitting). II. Circadian hyper-amplitude-tension (CHAT), which can also be systolic (S-CHAT), diastolic (D-CHAT), both (SD-CHAT) or mean arterial (MA-CHAT), etc. (by cosine fitting). III. SBP, DBP, or SDBP ecphasia (odd timing of the circadian rhythm of BP but not of that in HR) (by cosine fitting). IV. Ecfrequentia (altered period of the circadian rhythm)

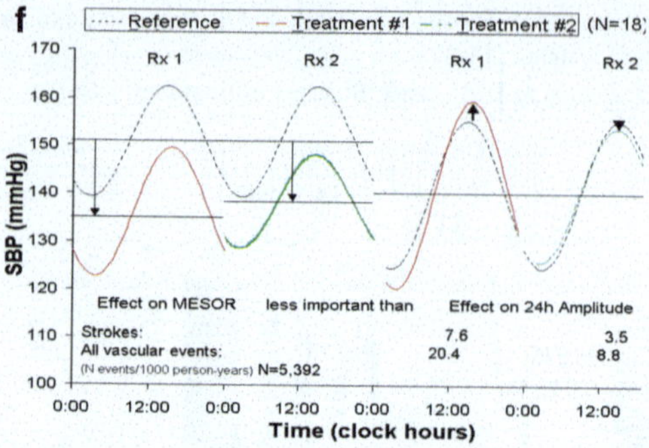

Does treating CHAT reduce morbidity?
(18 patients in double blind placebo controlled study)
(Shinagawa M et al. Biomed Pharmacother 2002; 55: 125-32)

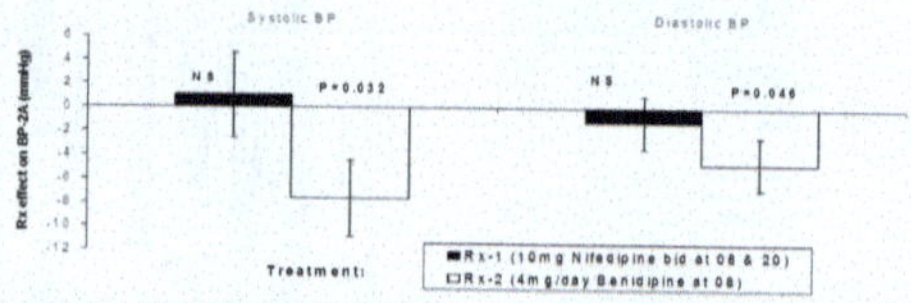

CHAT: Circadian Hyper-Amplitude-Tension (24h-A above upper 95% prediction limit)
Outcomes of 2-year calcium antagonist trials on 5,392 patients. Over 50% reduction in strokes (*left*) and of all severe vascular events (*right*) by treatment (*Rx*) that reduces (*Rx2, white bars*) vs. one that does not reduce (*Rx1, black bars*) the BP-A. Rx1 vs. Rx2 comparison: P<0.001.

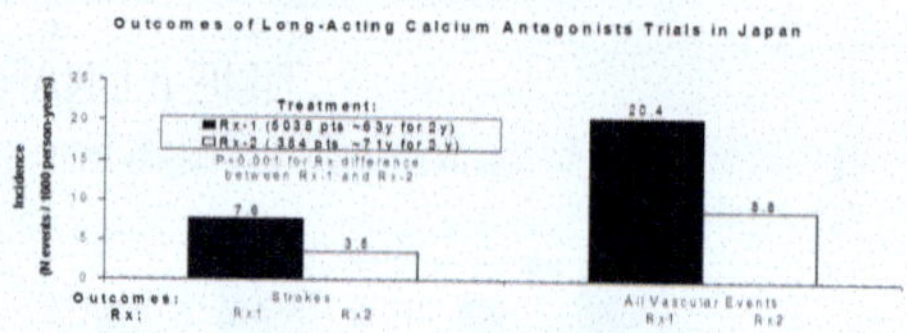

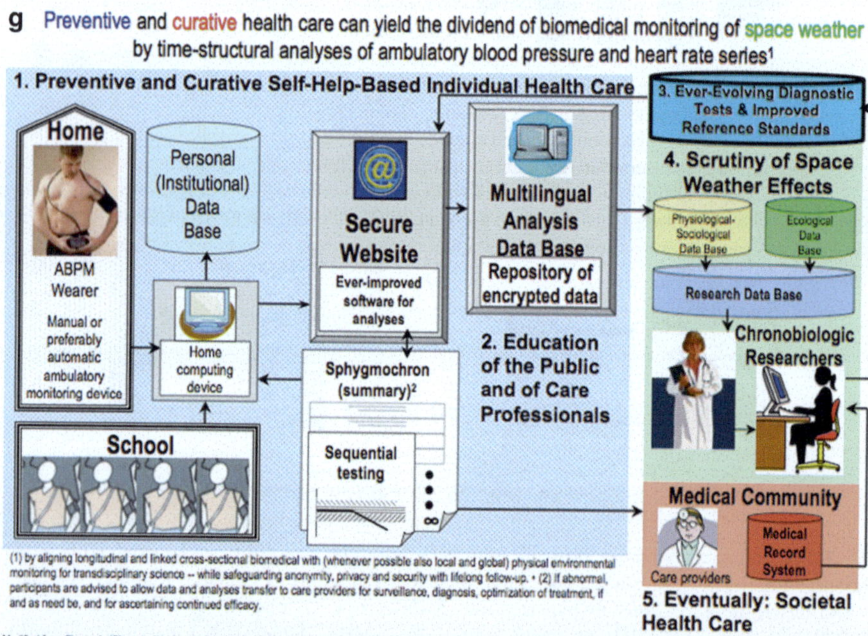

Modified from Figure 1 (Phoenix Architecture) in Adams C Privacy requirements for low-cost chronomedical systems. Int Conf on the Frontiers of Biomedical Science: Chronobiology, Chengdu, China, September 24-26, 2006, p. 64-69 , originally with Larry A. Beaty (www.sphygmochron.org) of the Phoenix Project (www.phoenix.tc-ieee.org/).

Fig. 1.2 (continued)

ries (from carbohydrate and fat only), was promptly consumed. These initially startling opposite results led us, by the early 1950s, to find the synchronization of a circadian rhythm by timing a calorie-restricted diet when the restriction was by 50 %, yet did not involve any changes in the intake of protein and vitamins and thus

Fig. 1.2 (continued) (by cosine fitting). V. Excessive pulse pressure (EPP), when the difference in the MESORs of SBP and DBP for adults exceeds 60 mmHg, a threshold that remains to be replaced by reference values from clinically healthy peers (eventually) with disease-free long-life outcomes specified further by gender, age, ethnicity, and geography. VI. A deficient HR variability (DHRV), defined as a standard deviation of HR less than 7.5 beats/min, a threshold that remains to be replaced by reference values from clinically healthy peers (eventually with disease-free long-life outcomes) specified further by gender, age, ethnicity, and geography (© Halberg Chronobiology Center). (**e**) Illustrative results supporting the need for continued surveillance and for a chronomic analysis of blood pressure series. I: Nocturnal hypertension: data stacked from 11 days of around-the-clock monitoring. Office spot checks cannot detect nocturnal pathology. II A: Among risk factors, an excessive circadian BP amplitude (A) raises the risk of ischemic stroke most. II B: Among risk factors, an excessive circadian BP-A raises the risk of nephropathy most. II C: An excessive circadian BP-A is a risk factor for ischemic stroke independent from the 24-hour mean (MESOR). III A: Individualized assessment (by CUSUM) of a patient's initial response and subsequent failure to respond to autogenic training (AT) (EO, F, 59y). III B: Individualized BP chronotherapy. Lower circadian BP-2A and MESOR after switching treatment time from 08:30 (*left*) to 04:30 (*right*). III C – Control chart assesses individualized anti-MESOR-hypertensive chronotherapy. Chronomics detects nocturnal escape from hypotensive treatment taken in the morning (I above) and conditions such as CHAT, associated with a risk of stroke and nephropathy greater than hypertension (IIA, B), even in MESOR-normotension (IIC), and monitors transient and/or lasting success of treatment (IIIA–C). Merits are as follows: – Detection of abnormality during the night when the dose of medication taken in the morning may no longer be effective in certain patients, facts not seen during office visits in the afternoon but revealed as consistent abnormality by around-the-clock monitoring; – Detection of abnormal circadian pattern of blood pressure (CHAT, "overswinging") associated with a risk of cerebral ischemia and nephropathy larger than other risks (including "hypertension") assessed concomitantly (IIA, B); – Finding that CHAT carries a very high risk even among MESOR-normotensives who do not need antihypertensive medication (IIC); – Availability of statistical procedures such as a self-starting cumulative sum (CUSUM) applicable to the individual patient to determine whether an intervention such as autogenic training is effective and if so for how long it remains effective (IIIA); – N-of-1 designs for the optimization of treatment timing: the same dose of the same medication can further lower the same subject's blood pressure MESOR and circadian amplitude when the timing of daily administration is changed (IIIB, C), as ascertained by as-one-goes (sequential) testing and parameter tests, procedures applicable to the given individual (© Halberg Chronobiology Center). (**f**) Benidipine (taken once a day upon awakening, Rx2) was found in large Asian clinical trials to be associated with better outcomes than nifedipine (taken twice a day, in the morning and in the evening, Rx1). Reducing the incidence of CHAT may be the reason accounting for the difference (almost by a factor 2) in outcomes, whether strokes or all cardiovascular events are considered (© Halberg Chronobiology Center). (**g**) Concept of an international multilingual website serving all comers worldwide, as a project on The Biosphere and the Cosmos, BIOCOS, now does on a small scale (left half of graph) with the data available to care providers (*right bottom*) and (after aligning with epidemiological data on natality, morbidity, and mortality and on crime and terrorism as well as with philanthropy and physical environmental data) for research on medical and broader problems with special reference to effects of space weather. The Phoenix Group of volunteering electrical and electronic engineers from the Twin Cities chapter of the Institute of Electrical and Electronics Engineers (http://www.phoenix.tcieee.org) is planning on developing an inexpensive, cuffless automatic monitor of blood pressure and on implementing the concept of a website (www.sphygmochron.org) for collection and analysis of data collected with these instruments (© Halberg Chronobiology Center)

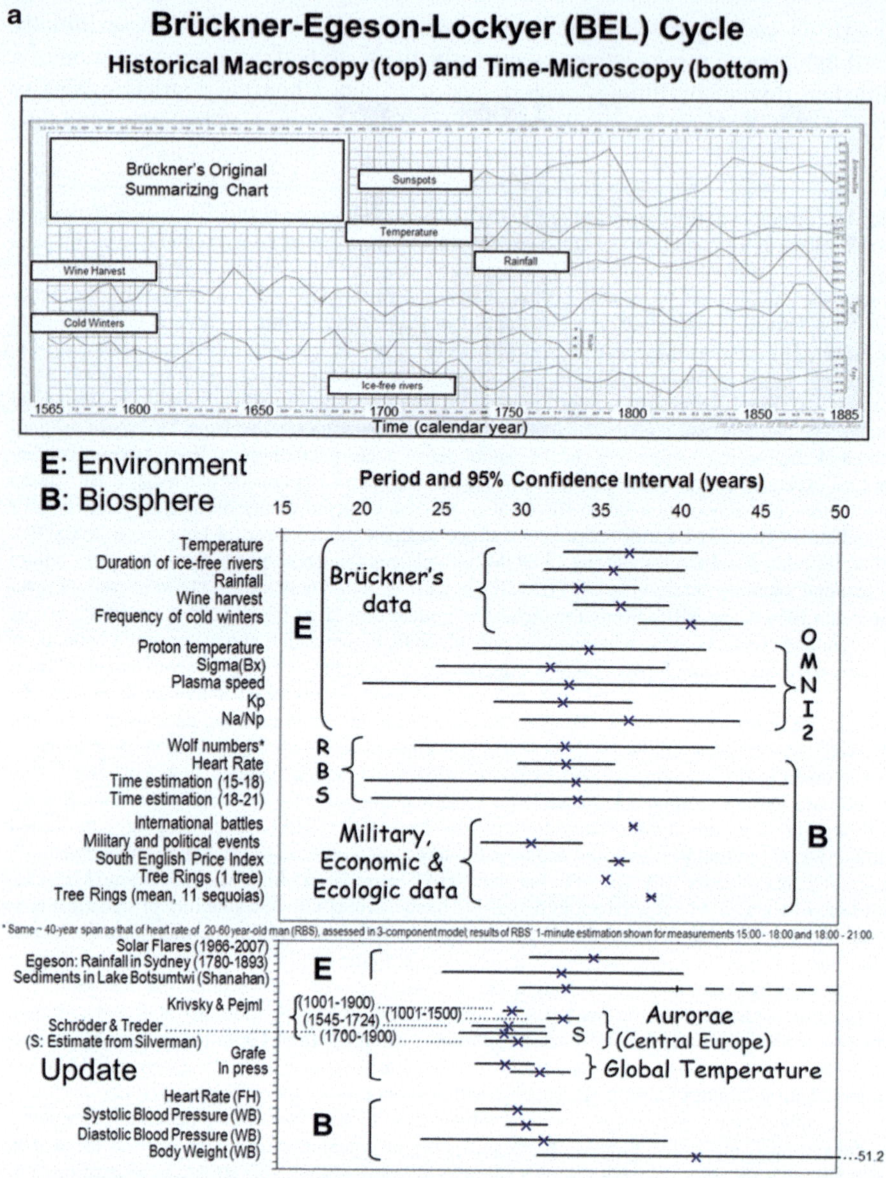

Fig. 1.3 (**a**) Display of original climate data (*top*) and map of environmental-biospheric paratridecadal Brückner-Egeson-Lockyer (BEL) cycles (*bottom*). The BEL cycle originally found in Brückner's data (first five point-and-interval estimates of the period) is also detected by spectral analysis of military, economic, and ecological data and in physiological data from one man (RBS) and from two other men (FH and WB), as well as in many other population phenomena, the mechanisms of which become available to study on individuals for investigators who try to plan beyond their life-spans. An about-33-year cycle was reported as a spectral peak by Shanahan who studied sediments in Lake Bosumtwi in Africa. The spectral coherence peak covers periods of 30–50 years (full length of uncertainty bar shown in graph) but includes a double peak, the second peak corresponding to another spectral peak at 42 years. The tick mark at 40 years corresponds to the small trough on the published coherence graph between the two peaks. Silverman (Rev Geophys 30: 333–51, 1992)

b

BEL Cycle in Human Heart Rate during 40 Years*

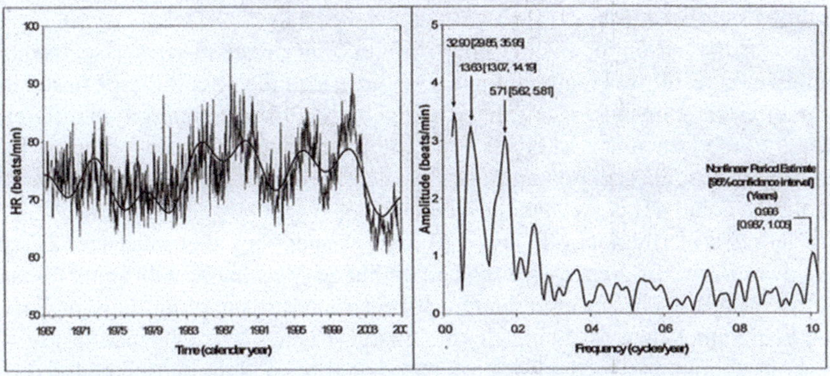

Wolf Numbers During a 40-Year Span of Physiological Monitoring Show an About 33-Year Cycle (top left), Also Seen in Heart Rate Data (Latter Not Shown)*

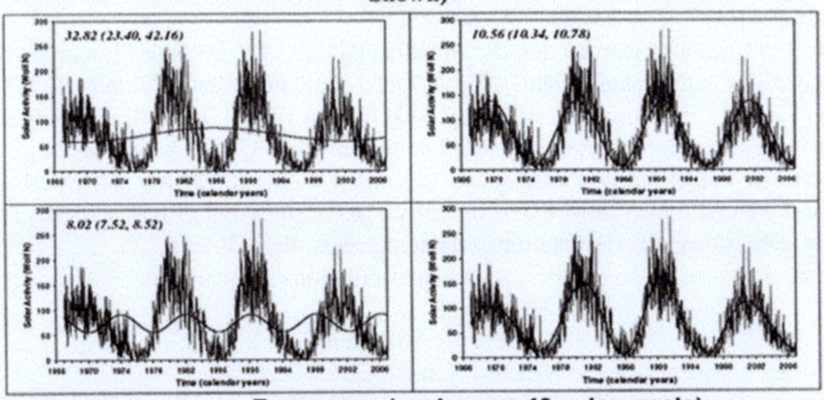

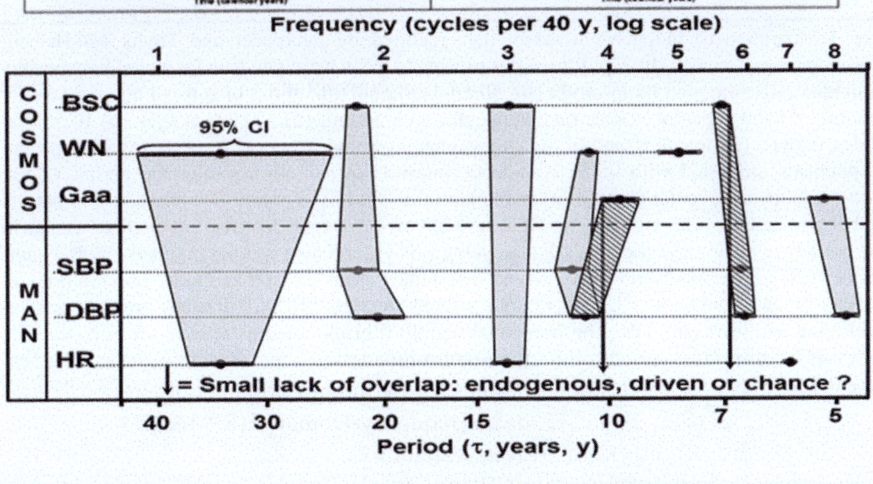

Fig. 1.3 (continued) was first to report an about-33-year spectral peak in aurorae (*S in graph*) but did not provide uncertainty estimates. The latter are derived for aurorae observed in Central Europe as compiled by Krivsky and Pejml during 1001–1900, analyzed globally and during different spans corresponding to varying average numbers of reported aurorae as technology to detect them improved. The span from 1545 to 1724 was of special interest as it allowed a comparison with an

to find its dominance over the synchronization with that by the lighting regimen, found by us earlier [28].[2]

This was the reason for an intergroup difference in circadian acrophase. Its major generalizable result was methodological; it illustrated the possibility of false intergroup differences in spot checks that resulted largely from an intergroup difference in phase of the very many variables that exhibit circadian and many other rhythms. The difference in circadian amplitude revealed the real intergroup difference that spot checks could not detect and thus led to the also generalizable inference that the characteristics of rhythms (cf. Fig. 1.6) are the endpoints that must replace spot checks. Similarly, differences in circadian timing can play havoc with aging research [29], as well as with cancer research [30] restricted to sampling daily at a fixed clock-hour rather than assessing rhythm characteristics. The evidence in Fig. 1.6 was my first hint that a medical science based on time- and rhythm-unqualified spot checks as a whole must eventually be replaced by one based on time series for continuous surveillance, a view validated by the ubiquity of circadian rhythms ([26, 31]; cf. [32]).

By 1953, we also learned that the circadian pattern of convulsions (recorded for prior decades in the same patient!) (Fig. 1.7 left) could be shifted in humans [33, 34] by a change in the timing of sleep and wakefulness (Fig. 1.7 right), a finding for which we subsequently encountered an animal model [35, 36] which revealed an increase in vulnerability during phase-shifting [36] (Fig. 1.8 top left), perhaps because we could also show that different variables shifted at different speeds and hence were transiently desynchronized among each other [36–40].

Fig. 1.3 (continued) independent set of data reported by Schroeder and Treder (© Halberg Chronobiology Center). (**b**) *Top*: time series of weekly mean heart rate data for a clinically healthy individual, RBS (**a**), and its spectrum (**b**). *Middle*: Original time series of sunspot number (Wolf's number, *W*) and superposed spectral components with periods (**a**) 32.82 (*top left*), (**b**) 10.56 (*top right*), (**c**) 8.02 (*bottom left*), and (**d**) its three-component model (*bottom right curve*). The spectral components are listed with 95 % confidence intervals in parentheses after the period values. *Bottom*: influence of solar activity on the human cardiovascular system: the congruence of natural and physiological cycles with periods from several years to several decades. (*1*) Cycle of change in the polarity of solar magnetic field (Hale's cycle), BSC; (*2*) relative sunspot numbers (Wolf's numbers), WN; (*3*) geomagnetic index *aa*, as determined from data of antipodal observatories in Greenwich and Melbourne, Gaa; (*4*) systolic arterial pressure, SBP; (*5*) diastolic arterial pressure, DBP; and (*6*) heart rate, HR. The horizontal length of black columns reflects 95 % confidence intervals for respective periods. Thin near-vertical linking lines and shading indicate congruent periods (© Halberg Chronobiology Center)

[2]The temporal placement of the 24-hour rhythm in eosinophil count of mice on a regimen of light (L) alternating every 12 hours with darkness (D) can be shifted by offering a diet restricted in calories at the beginning of the daily L-span [28].

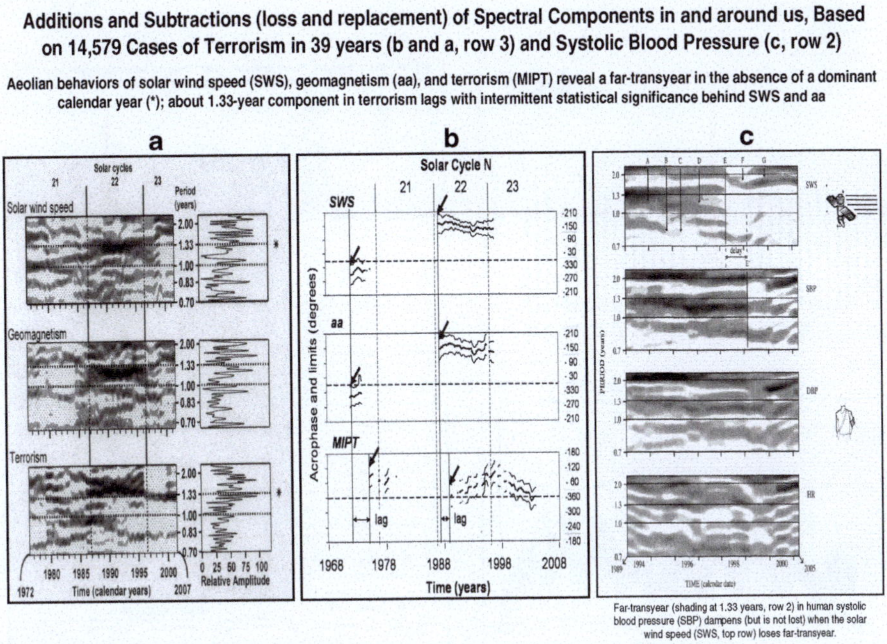

Fig. 1.4 Just as removal and replacement of a gland led to endocrinology, so biological consequences of the loss of environmental spectral components are critical to chronomics, the study of chronomes (time structures) in and around us. The drifting, bi- or trifurcation, disappearance and reappearance of a component with a given frequency and, when present, the waxing and waning of its amplitude, visualized in a and c, is in keeping with external driving of an individual (**c**) and of a population (**a** and **b**). Note components with neighboring frequencies waxing at or near a transyear component in solar wind speed (top of **a** and of **c**), geomagnetism (second row of **a**), and in 39 years of terrorist activity (bottom of **a**), an association supported by an independent method in **b** (by the statistical significance of the fit of a 1.3-year far-transyear to interplanetary [SWS] and terrestrial [aa] magnetism and to terrorism). Analysis in **a** is glocal insofar as it is based on the entire series in the spectral window on the right (global) and in sections of the series in the gliding windows on the left (local). Analyses only on sections of the time series are local in **b** (complementing the global counterpart, as seen on the right of **a**) and local in **c**. (**c**) shows time courses of the frequency structures of the speed of the solar wind (SWS) (*top*) and of an elderly man's (FH) systolic and diastolic blood pressure and heart rate (rows 2–4, respectively), examined by gliding spectral windows. Human systolic blood pressure selectively resonates with solar wind speed (SWS) (top 2 sections). No obvious resonance, only minor coincidence of apparent change, is apparent to some in diastolic blood pressure or heart rate (*bottom* 2 sections). Aeolian rhythms in gliding spectra of SWS and SBP change in frequency (smoothly [A] or abruptly [B, C, D], bifurcating [D, F], and rejoining [G]); they also change in amplitude (B) (up to disappearing [C,E] and reappearing) (© Halberg Chronobiology Center)

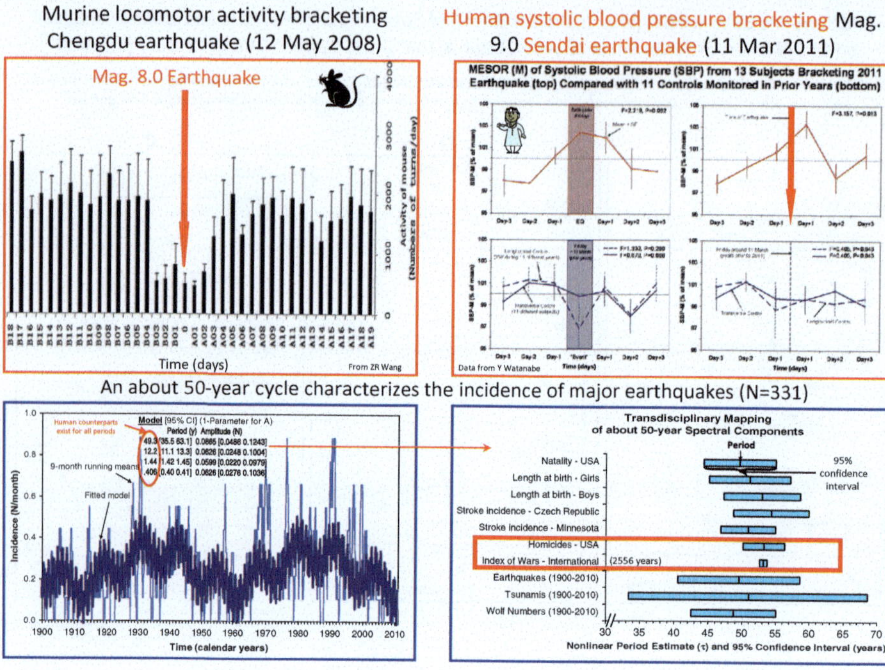

Fig. 1.5 Biospheric contributions to the understanding, if not prediction of earthquakes. *Upper left*: Locomotor activity of some of the mice telemetered around the clock was statistically significantly decreased starting 3 days prior to the magnitude 8.0 earthquake in Chengdu, China, on 12 May 2008 (data of Zhengrong Wang). *Upper right*: Human systolic blood pressure started increasing 2 days prior to the magnitude 9.0 earthquake in Sendai, Japan, on 11 March 2011, documented on the basis of weeklong records of around-the-clock ambulatorily obtained data from 13 Japanese (data of Yoshihiko Watanabe). Similar records from longitudinal and transverse controls differ in their time course, suggesting that the trend observed before the earthquake was related to it rather than being a feature of an anticipated weekly pattern. *Lower left*: the monthly incidence of major earthquakes since 1900 is characterized by the presence of cycles with periods of about 49.3, 12.2, 1.44, and 0.41 year(s), given with their uncertainties in parentheses. *Lower right*: The prominent about-50-year cycle is also documented in physiology, pathology, societal upheavals, and space weather. Nonlinearly estimated periods are displayed with their 95 % confidence intervals shown as the length of corresponding horizontal bars (© Halberg Chronobiology Center)

1.5 "Norms", No Substitute for Concomitant Controls: The Adrenal Cycle, Loads, Baselines, and Spot Checks

By the late 1940s, diurnal variation was well-documented [41], yet in studies other than those focusing on rhythms, it was largely ignored, as it is today. The concept of stress was much discussed from the viewpoint of Hans Selye's discovery that many loads, including toxic influences, involved the adrenal cortex [42]. Although an inverse relation had been considered between the adrenal and the eosinophil count

a Confusing results, that could wrongly be interpreted as "stress" or "allergy", are accounted for by the action of food (offered mornings) and light as competing synchronizers of circulating eosinophils in C3H mice with high breast cancer incidence that can be drastically reduced by calorie restriction

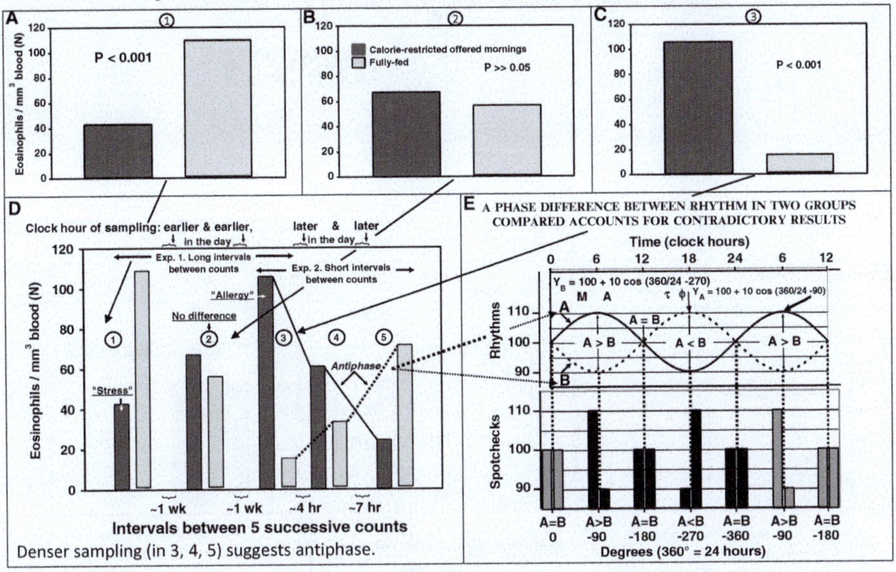

Fig. 1.6 (**a**) Importance of rhythms in assessing intervention effects, illustrated in relation to stress or allergy. *A*. Eosinophil counts seem to be lowered by fasting (and/or stress), when a 50 % reduction in dietary carbohydrates and fats (with proteins, vitamins, and minerals similar to control group) was fed in the morning to C_3H mice (*dark column*). (In this model, the naturally high incidence of breast cancer is lowered by a diet reduced in calories and by ovariectomy, not shown.) The result could have been interpreted as an adrenocortical activation and then assessed by eosinophil depression, with applications for treating breast cancer and for prolonging life. Steroids that depress eosinophil cell counts and perhaps mitoses could be a mechanism through which caloric restriction and ovariectomy act in greatly reducing cancer incidence. This tempting inference was never published. *B*. In view of the importance of this finding for the etiology of cancer, results were replicated on a larger group of animals; 1 week later, a follow-up study with more animals started at an earlier clock-hour yielded confusing results, showing no statistically significant difference between the two groups of mice. *C*. After another week, another study starting at an even earlier clock-hour yielded results opposite to those in the first experiment when considered alone. These findings in C in themselves could have been interpreted as an allergic response, certainly contrary to the "stress" response in *A*. *D*. Sampling at intervals of a few hours in the third study, the stages called 4 and 5, hinted at the reason for the confusion: by sampling at different clock-hours, two groups of mice were found to be characterized by a circadian rhythm with different phases. Opposite effects thus became predictable. *E*. Abstract illustration of two circadian rhythms in antiphase. Differences in opposite direction or no effect are then anticipated from sampling at different clock-hours (© Halberg Chronobiology Center). (**b**) Effect of food restriction on circulating eosinophil counts in mice. Follow-up study on Fig. 1.6a, with a phase difference greatly reduced by offering the restricted diet in the evening. *A*. Even after $\log_{10}$ transformation of the data expressed as percentage of mean, great interindividual variability is apparent in the raw data. *B*. Plots of timepoint mean for mice in each group reveal different circadian patterns. *C–E*. Parameter tests quantify differences, indicating that calorie restriction is associated with a lower MESOR (*C–D*), a larger circadian amplitude (*D–E*), and only a slight difference in acrophase (*D*). The difference in acrophase in this study, where calorie-restricted mice were fed in the evening, is much smaller than the almost-antiphase observed in prior studies (Fig. 1.6a), where calorie-restricted mice were fed in the evening (© Halberg Chronobiology Center)

b Calorie-restricted meal offered evenings only slightly advances phase (no antiphase versus controls, as seen with morning meals), lowers MESOR and amplifies circadian eosinophil rhythm in C_3H mice with high breast cancer incidence that can be reduced by calorie restriction (CR)

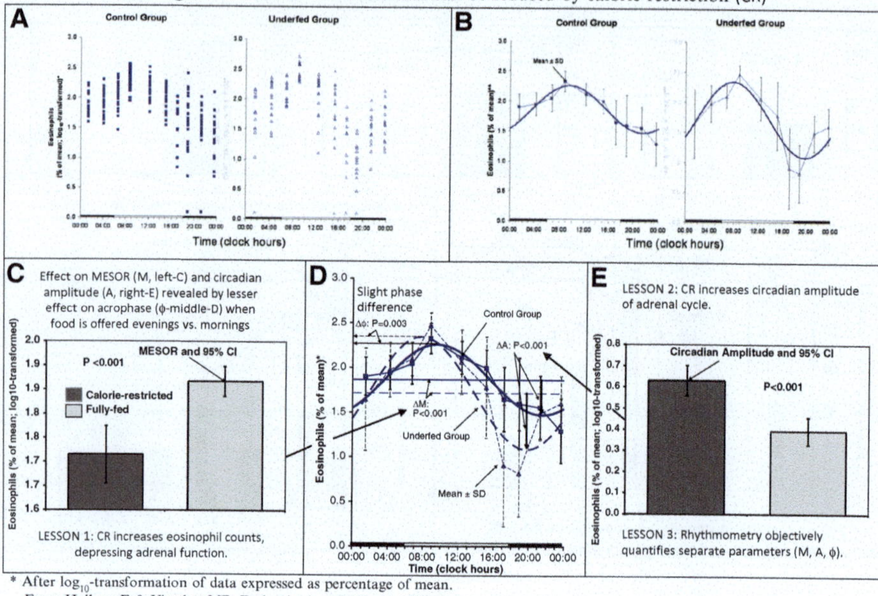

* After $\log_{10}$-transformation of data expressed as percentage of mean.
From Halberg F & Visscher MB. Endocrinology 1952; 51: 329-35.

Fig. 1.6 (continued)

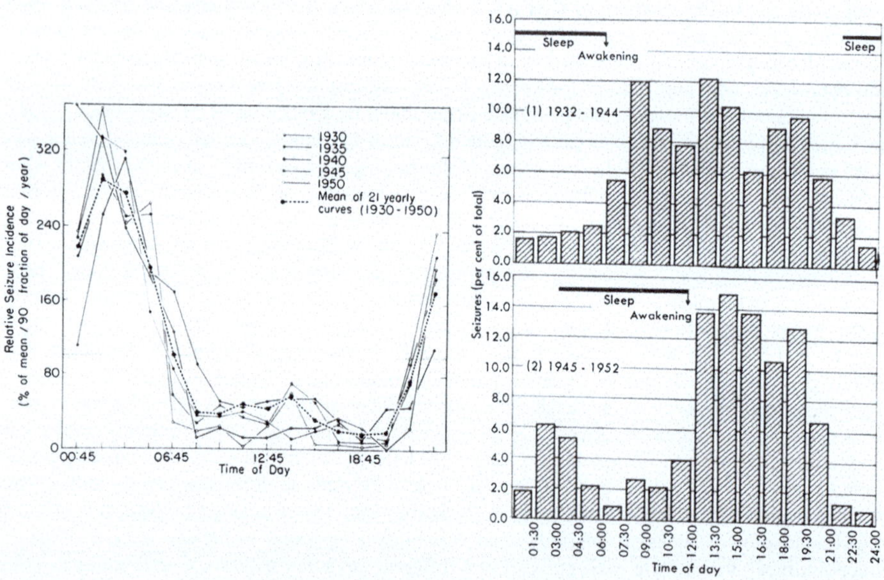

Fig. 1.7 *Left*: unequal distribution of epileptic seizures during the hours of the day in a case observed for 20 years. Night-type. *Right*: unequal distribution of epileptic seizures during the hours of the day on two routines, maintained for many years. Day-type (note change in seizure distribution following change in routine). Daily sleep-wake routine may be regarded as the environmental synchronizer of seizure rhythm; clock-hour as such does not matter in determining times of occurrence of seizures (© Halberg Chronobiology Center)

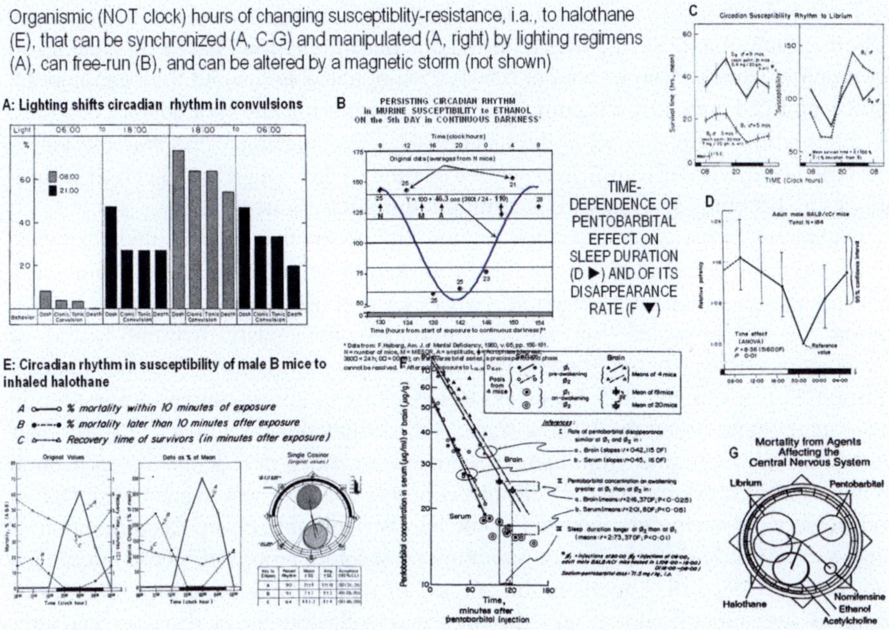

Fig. 1.8 Part of the history of susceptibility rhythms in the response of the mammalian central nervous system (extreme "phase responses"). By 1955, the response to noise (>104 decibels [dB] >0.0002 dyn/cm^2 rms pressure) was shown to be circadian stage-dependent in mice of the susceptible I strain, so that the convulsive risk by day was 11 % and by night 65 % in one group of young mice and 0 % by day and 85 % by night in another group [35]. There were more convulsions and more deaths from exposure to bells in a tub at the beginning of the daily dark span as compared to exposure of animals of the same inbred strain, gender, and age, at the start of the daily light span in mice kept in light from 06:00 to 18:00, alternating with darkness (not shown). This was confirmed in a follow-up study, section A [36]. Again, convulsions and death occurred in a stimulator with exposure to ~117 dB >0.0002 dyn/cm^2 rms pressure at 21:00, but no deaths and fewer convulsions with exposure at 08:00, as seen in the left half of section A. The *right side* of section A shows not only that the timing of this susceptibility rhythm (on the fourth day after reversal of the lighting regimen) can be shifted along the 24-hour scale (can apparently be reversed) and that, at the time investigated (perhaps as a transient), there is an overall increase in susceptibility. Section B, a 6-timepoint phase response curve in continuous darkness shows the persistence of a response rhythm of mice, gauged by the survival time of mice after injection of a fixed dose of ethanol. Sections C and D show further phase response curves to Librium and pentobarbital, i.e., drug-related results under conditions of a 12-hourly alternation of light and darkness. Time-macroscopic (E, *left* and *middle*) and time-microscopic (E, *right*) assessment of parameters of a rhythm in susceptibility to halothane given as an amplitude by the length of a vector and as the acrophase by the vector's direction, with their elliptical 95 % confidence regions as indications of uncertainties. For the acrophase, its 95 % confidence limits (CL) are obtained by tangents to the error ellipse. Section F shows pentobarbital concentration in brain and serum of mice as a function of time after injection at circadian phases of short sleep duration (φ1) and long sleep duration (φ2). Section G shows, with their uncertainties, differences in the timing of phase responses to different agents affecting the mammalian central nervous system (© Halberg Chronobiology Center)

by Emil Schwarz [43],[3] interest in these cells in North America was triggered by Albert J. Dalton and Selye, who reported an "alarm reaction" based on changes in eosinophils from a "norm," e.g., at 0.5, 1, 2, 4, 6, and 8 hours and then again at 24, 30, 48, and 72 hours after a treatment, a schedule ignoring any spontaneous changes, yet compatible with a good night's sleep [44]. The starting value was the single reference count, the "Norm," irrespective of time of day, and was to serve (presumably as a "baseline") in lieu of concomitant untreated controls.

Dalton and Selye summarize their studies: "It appears that the so-called unspecific leukocytosis preceded by a decrease and followed by an increase in eosinophils, is a constant feature of the alarm reaction" (what they proposed as a result of their investigation was not separated from an ever-present, spontaneously rhythmic feature of everyday physiology before it could be dubbed a special "alarm"). They postulated a "response" to the agent they happened to test, without consideration of variation in eosinophil counts along the 24-hour scale, well-documented in humans [45]. By their time (1939), a 24-hour rhythm had not yet been described in rodents; a modification of the solution for staining these cells in a counting chamber was needed [46]. Dalton and Selye were seemingly unaware of the hardly negligible change as a function of time, subsequently documented by sampling on separate groups of inbred mice, each sampled only once [47], to avoid the "stress" of repeated sampling [48]. Such sampling at two clock-hours (Fig. 1.9) [47] was followed more densely thereafter (Fig. 1.10) [49].

The variability of eosinophil counts in mouse strains was particularly great, but hardly the exception [49]. Systematic studies at 4-hour intervals on separate groups of animals at each timepoint were originally carried out to avoid the load (stress) of repeated handling, not only for venisections but also for sampling ear pinna, a point appreciated by Jürgen Aschoff: "Halberg's investigations are so important because they are some of the very few experiments available at this time on the endocrine control of 24-hour periodicity … that consider to a sufficient extent the possible effects of disturbance and have led for the first time to clear results" [50, 51].[4] I also learned that this 6/day equidistant sampling scheme, which I introduced, was

[3] Schwarz broadly views the autonomic nervous system as the "apparatus leading to eosinophilia." Thus, an increase in autonomic tone by direct stimulation (by pilocarpine) or the loss of inhibition of the sympathetic nervous system led to eosinophilia and vice versa, sympathetic stimulation, or the loss of autonomic inhibition (atropine) lower the curve of the count of eosinophil cells. In carrying this thesis to the clinic, the lack of adrenal and thyroidal function leads to eosinophilia and hypertrophy of the chromatin apparatus and of the thyroid to eosinopenia; hence, eosinophilia is found in Addison's disease and in myxedema and eosinopenia in Basedow's (Graves') thyroidal hyperfunction. The 2,758 references he discussed, however, are contradictory, often based upon a percentage of eosinophil cells in a smear rather than upon a direct count in a chamber. The changes of eosinophils along the 24-hour scale were not yet known when Schwarz wrote his 652-page opus nor were the hormones of the adrenal cortex. The isolation of adrenal cortical hormones by Tadeus Reichstein, Edward A. Kendall, and Philip S. Hench (for which they received the Nobel Prize in Physiology or Medicine in 1950) postdated Dalton and Selye's paper [44].

[4] Aschoff's remark [50] reveals that by 1954 [51] others had recognized the need for controls in the study of time-structured variables, apart from the importance of rhythms in themselves, a topic to which Aschoff contributed throughout his lifetime, substantially along naturalistic lines. His comment bears directly upon studies using an initial time-unspecified value as the norm [44], the flaw of very much research then as in 2012.

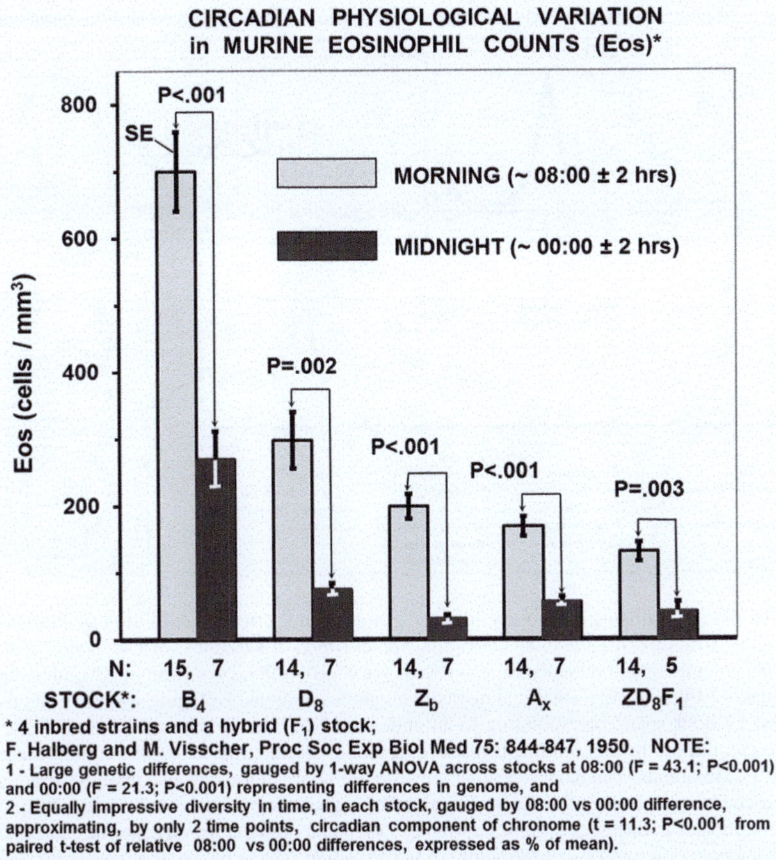

Fig. 1.9 Circadian physiological variation in murine eosinophil counts (Eos). In four inbred strains and a hybrid (F1) stock [47]. Note large genetic differences anchored in the genome, namely, (1) across stocks at 08:00 gauged by one-way ANOVA ($F=43.1$; $P<0.001$) and 00:00 ($F=21.3$; $P<0.001$) and (2) equally impressive diversity in time, in each stock, gauged by 08:00 vs. 00:00 difference, approximating, by only two timepoints, the circadian component of chronome ($t=11.3$; $P<0.001$ from paired t-test of relative 08:00 vs. 00:00 differences, expressed as percent of mean). The ever-present within-day difference can differ among stocks of mice (© Halberg Chronobiology Center)

satisfactory for the fit of a 24-hour cosine curve to assess rhythm parameters analyzed in a series of complementary methodologic developments from the population-mean [21] over the single [12] cosinor and the linear-nonlinear cosinor spectrum to a glocal approach in space and time with gliding spectra and serial sections ([22, 23]; cf. [16, 52]).[5] Four-hourly sampling also became routine when

[5] The development of methodology, a major concern, had a vast field of time series analyses to build on, explored not only by statisticians and mathematicians but also by physicists, among others, including Germaine Cornélissen, whose thesis on the topic prepared her for many subsequent developments. Periodograms and variance (power) spectra were in the forefront in the 1950s and 1960s;

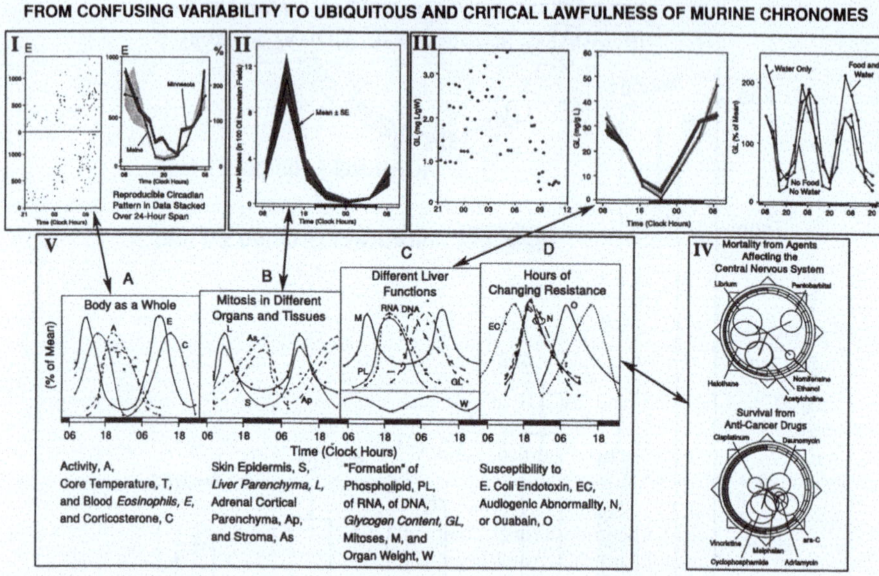

Fig. 1.10 From confusing variability to the ubiquitous and critical lawfulness of murine chronomes (time structures). Time plots of original values (*dots*) of variables such as blood eosinophils (E) (I, *left*) or glycogen content (III, *left*) reveal great variability, confusing at first, until the data are processed by relatively simple statistical techniques such as averaging and stacking over an idealized 24-hour span corresponding to an anticipated 24-hour synchronized periodicity. Once this is done and the results are displayed as a function of time, they show the time-macroscopic ubiquity of circadians. The data averaging for different hours of the day also reveals differences in the time course (in phase) of different functions of a given organ such as the liver (VC), of cell division (mitosis) in different organs and tissues (VB), and of different variables at the level of the body as a whole (VA and D). The structure of the circadian system becomes apparent by the application of the methods of chronobiometry: the circadian variation in blood eosinophils determined years apart in two laboratories as far apart as Minnesota and Maine is closely reproduced (I, *right half*). The lawfulness of the circadian variation yielded by the application of chronobiologic techniques is also revealed for the drastic changes in liver glycogen content (III, *middle*). In this case, a circadian rhythm in the liver's glycogen is seen to persist under conditions of starvation and dehydration, with little if any alteration in the dynamic rhythm characteristics, as compared to usual ad lib conditions, once the data are expressed as a percentage of the overall mean, III, *right*. After a prominent circadian rhythmicity was found at different levels of organization, several series of experiments were carried out under rigorously standardized laboratory conditions in order

---

we used the former on desk computers and the latter on electronic computers. In view of the nature of the short serial data encountered in the clinic, with replications of a cycle of a given individual sought in others, the population-mean cosinor was developed first [21]. The single cosinor followed [12] and, thanks to invaluable help from the late Donald Marquardt, was nonlinearly extended [52]. The gliding spectral window by least squares, developed by the late George S. Katinas, remains invaluable and reveals what may be smeared by a wavelet, but the latter may also reveal what was missed by a linear-nonlinear cosinor. The chronobiologic serial section was also introduced earlier, as soon as longitudinal data covering many cycles became available in larger numbers to complement the analysis of a long series as a whole by those of its sections systematically varied in length. The need for glocality in time was thus recognized early. The glocal approach is illustrated in [16].

the animals had to be killed to remove organs of interest. It often allowed the unaided eye to guesstimate by inspection what curve-fitting tries to approximate with uncertainties whose estimation admittedly remain to be improved, because nonstationarities must be taken into account.

1.6 Telemetry

It was cumbersome to work around the clock for days, or even years in the case of the 4-hourly around-the-clock rectal temperature measurements in mice that lead to the finding of shorter-than-24-hour periods, dubbed by Earl Bakken as "free-running" [53]. We introduced telemetry [54] at least for some variables like body core temperature and eventually kept groups of rats on 6 different schedules of light and darkness, alternating at 12-hour intervals, staggered by 4 hours, so that freely feeding rodents could be sampled at six different circadian stages, at a single clock-hour (Fig. 1.11) [55].

Off and on, I returned to the study of stress and strain and the adrenal cortex on the one hand and to methodology on the other, notably after we found susceptibility rhythms first along the 24-hour scale (Fig. 1.8) [35, 36] and then along the scale of a week (Fig. 1.12a [56] and b [57]). At a meeting at the submarine base in New London, Connecticut [58], Selye indicated to Henri Laborit that the demonstration that circadian stage made the difference between life and death in response to the same stimulus ([35, 36]; cf. [26]) (Fig. 1.8 top left) was an important finding in the

Fig. 1.10 (continued) to investigate the effect of a single physical stimulus such as exposure to noise. Outcomes were as different as no response, convulsion, or even death, as a function of the circadian stage at which the organism was exposed to noise. Whether the stimulus was audiogenic or the exposure to an endotoxin, or to a drug such as ouabain, or to whole-body irradiation, or to other physical or chemical agents, predictable changes were found as a function of the circadian stage at which the stimulus was applied, albeit with differences in the timing of these susceptibility-resistance rhythms to different agents. The hours of changing resistance were thus uncovered, and the times of overall largest response by the organism to a fixed stimulus applied at different rhythm stages mapped (Figure I/VD). Applications followed after cosinor analysis to attempt to assess the uncertainties involved in timing, obtained by drawing tangents to the 95 % confidence ellipses (IV). Prominent susceptibility rhythms were documented in the experimental laboratory, as illustrated here for the case of the mortality from agents affecting the central nervous system and for the case of the survival from (tolerance of) toxic doses of anticancer drugs. In each case, the nonoverlap by the elliptical 95 % confidence region of the center of the circular plot (pole) can be interpreted as the presence of a statistically significant circadian rhythm in the susceptibility of the organism to each of these different agents. The orientation of the directed line (vector) indicates the time of acrophase that is the time of the largest anticipated response. Such charts are helpful in guiding the timing of the administration of the various agents so mapped. The chronotherapy of cancer, guided by the marker tumor temperature, is one critical application resulting from this work (© Halberg Chronobiology Center)

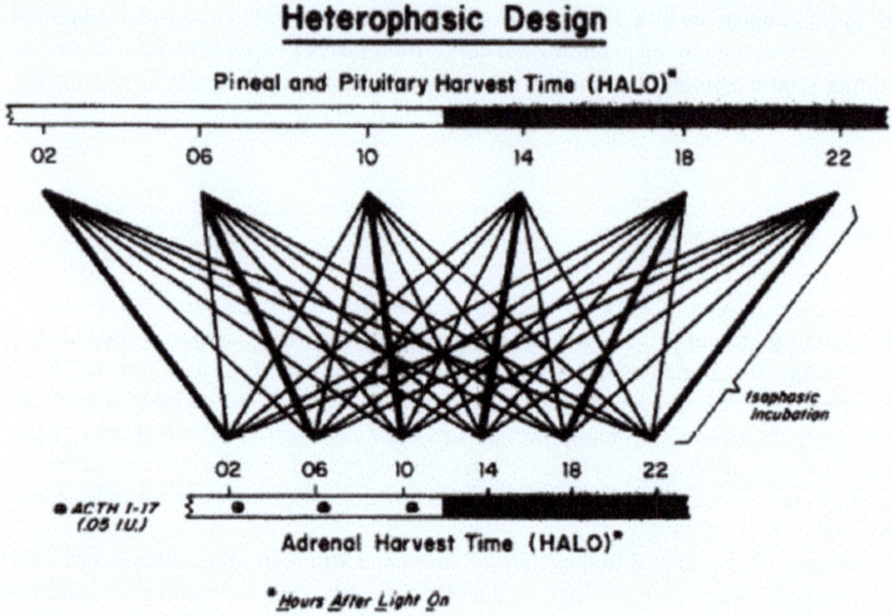

Fig. 1.11 Work concomitantly at each of 6 circadian stages 4 hours apart became possible by separate chambers housing rodents standardized for at least 3 weeks on 6 separate lighting regimens also staggered by 4 hours, so that light, alternating with darkness every 12 hours, went on at midnight in one chamber and at 04:00, 08:00, 12:00, 16:00, or 20:00 in five other chambers, each containing rodents with core temperature telemetry so that circadian stage could be validated. Tissues not only from a given stage but also from the five other stages could thus be obtained, e.g., for incubation studies (© Halberg Chronobiology Center)

history of medicine. I was the more interested to credit his genius of discovering, documenting, and communicating the involvement of the adrenal cortex in the response to stimuli and found particularly drastic changes as a function of injection time after the injection of certain doses of an adrenal cortical inhibitor, SU-4885 (Fig. 1.13a): the overall difference was between >80 % healthy survivors and ~80 % deaths as a function of timing [59], impressive and more than matching the findings we made earlier on susceptibility to ouabain (Fig. 1.13b) involving the difference, as a function only of circadian stage, in one experiment between ~70 % survivors and ~70 % deaths from the same dose (cf. [60–63]). Because of such differences, in the 1960s and for several decades thereafter, the word "stress" was not to be used in my lab when rhythms were not controlled as in the Dalton and Selye report [44]. I trust Selye would be pleased if he could see the infradian charts that now complement the circadians and render at least some vascular and psychophysiological aspects of stress and strain measurable: alterations of novel rhythm parameters gauge the risks of disease in Osler's wear and tear or Selye's conceptual general adaptation syndrome that required the mapping of spontaneous dynamics (cf. [45]).

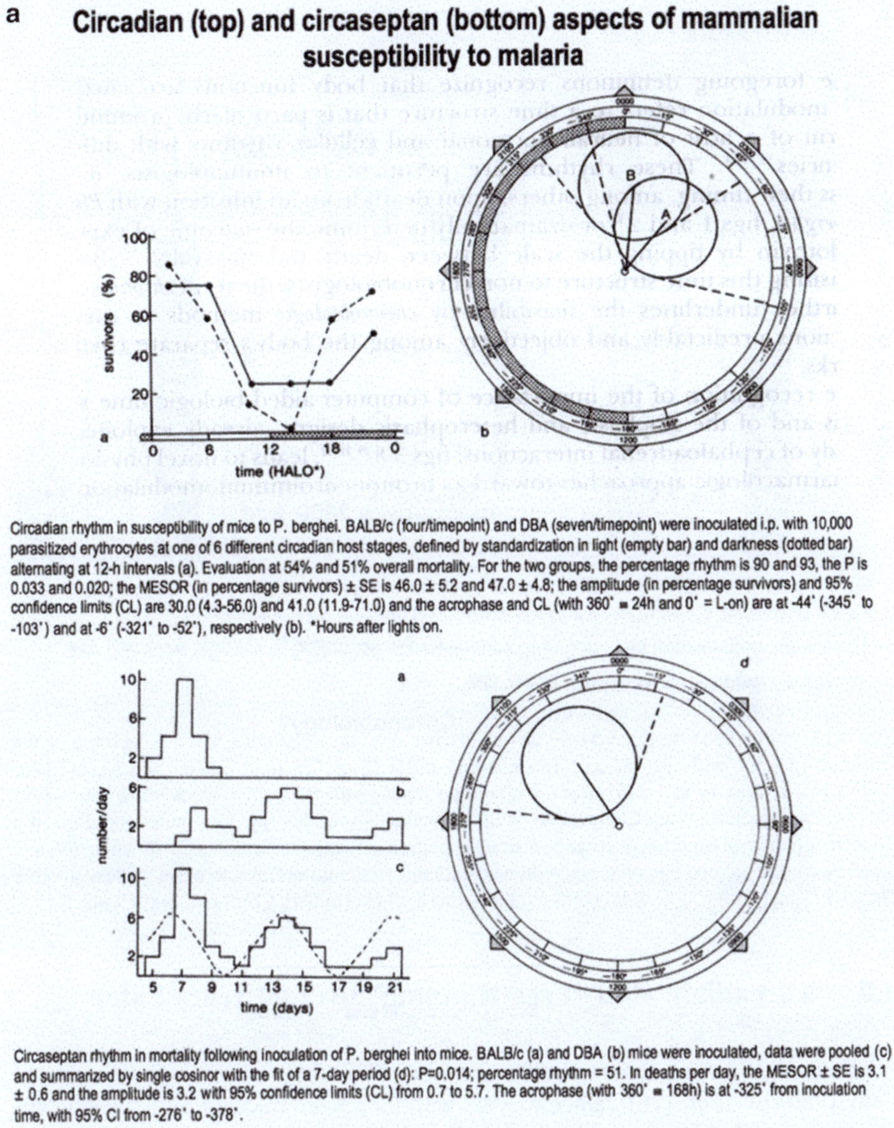

Fig. 1.12 (**a**) *Top*: circadian rhythm in rodent susceptibility to P. berghei. BALB/c mice (at four/timepoint) and DBA mice (at seven/timepoint) were inoculated i.p. with 10,000 parasitized erythrocytes at one of 4 or 6 different circadian host stages, defined by standardization in light (*empty bar*) and darkness (*dotted bar*) alternating at 12-hour intervals (*a*). Evaluation at 54 % and 51 % overall mortality. For the two groups, the percentage rhythm is 90 % and 93 %, the P is 0.033 and 0.020; the MESOR (in percentage survivors) ± SE is 46.0 ± 5.2 and 47.0 ± 4.8; the amplitude (in percentage survivors) and 95 % confidence limits (CL) are 30.0 (4.3–56.0) and 41.0 (11.9–71.0) and the acrophase and CL (with 360° ≡ 24 hours and 0° = L-on) are at −44° (−345° to −103°) and at −6° (−321° to −52°), respectively (*b*). *Hours after lights on. *Bottom*: circaseptan rhythm in mortality following inoculation of P. berghei into mice. BALB/c (*a*) and DBA (*b*) mice were inoculated, data were pooled (*c*) and summarized by single cosinor with the fit of a 7-day cosine curve (*d*): *P*=0.014; percentage

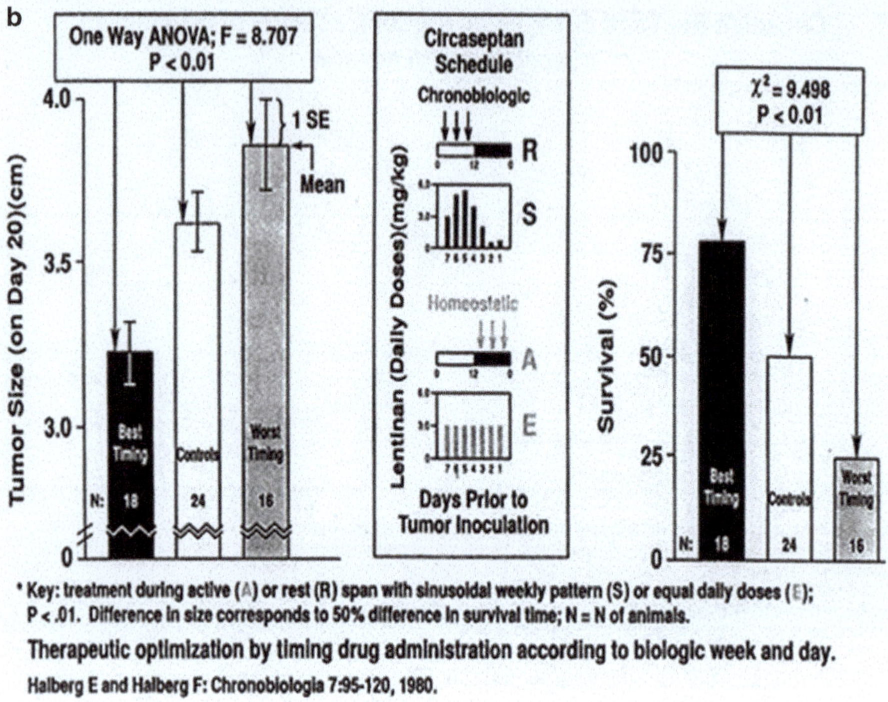

Fig. 1.12 (continued) rhythm = 51 %. In deaths per day, the MESOR ± SE is 3.1 ± 0.6 and the amplitude is 3.2 with 95 % confidence limits (CL) from 0.7 to 5.7. The acrophase (with 360° ≡ 168 hours) is at −325° from inoculation time, with 95 % CI from −276° to −378° (© Halberg Chronobiology Center). (**b**) Multifrequency chronotherapy. The same total dose of a drug (lentinan) as a function of its administration pattern can enhance tumor growth when given in equal doses daily at one circadian stage (homeostatically) and inhibits growth when given at another circadian stage with a sinusoidal pattern (chronotherapy) (© Halberg Chronobiology Center)

1.7 "Circadian" and "Free-Running" over 60 Years Later

"Circadian" and "free-running" were coined in 1950; they were only in internal use until 1959 ([64]; cf. [26]), since to start with they were rejected by nomenclature committees. These terms were intended to indicate both the variability and the partly built-in nature of daily rhythms insofar as the periods of inbred mice differed not only from 24.0 hours but some also differed among individuals (Fig. 1.14-I-C [Eigenfrequenz]) [53]. Endogenicity in organisms, always partial in open systems, was based on (a) differences in the extent of day/night variation of eosinophil blood cell counts among inbred strains of mice, Fig. 1.9, when they had eyes, and (b) when eyes were removed (by surgery if not genetics), Fig. 1.14-IC. Periods (documented by periodograms by 1954 [65], in the desk computer era) in body core temperature were invariably slightly shorter than 24 hours, so that the temporal location of within-day differences scanned the 24-hour scale.

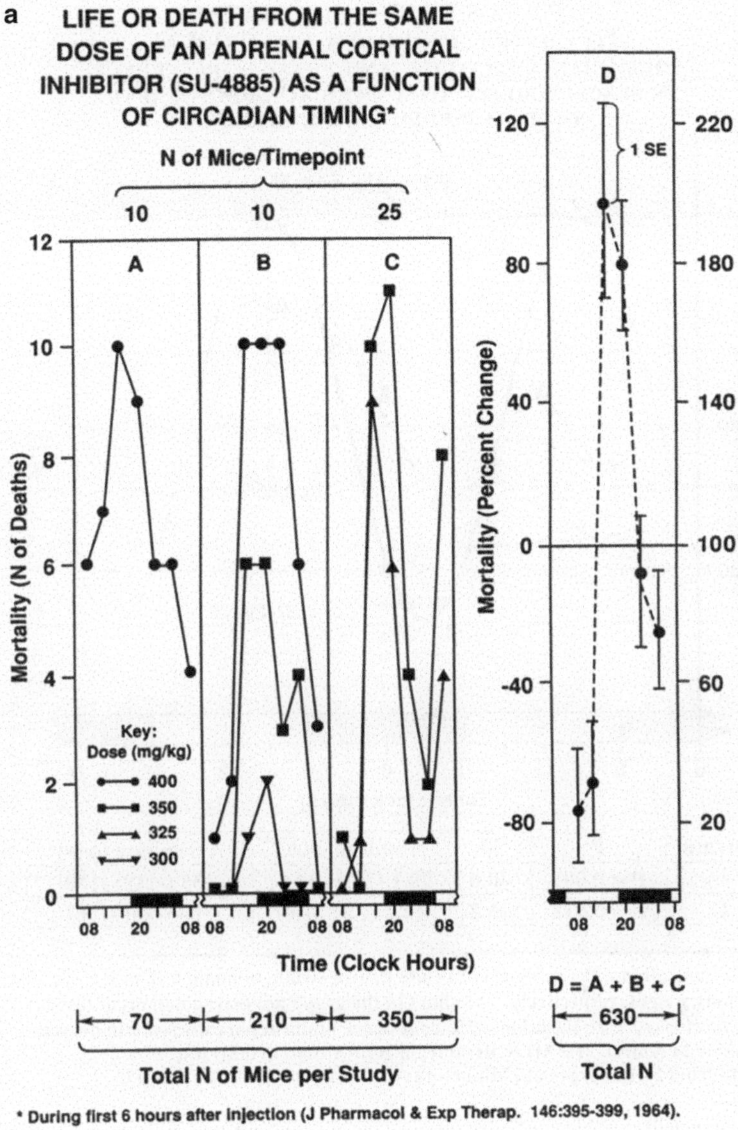

Fig. 1.13 (**a**) Circadian susceptibility-resistance cycle to SU-4885 (© Halberg Chronobiology Center). (**b**) Circadian susceptibility-resistance cycle to ouabain (© Halberg Chronobiology Center)

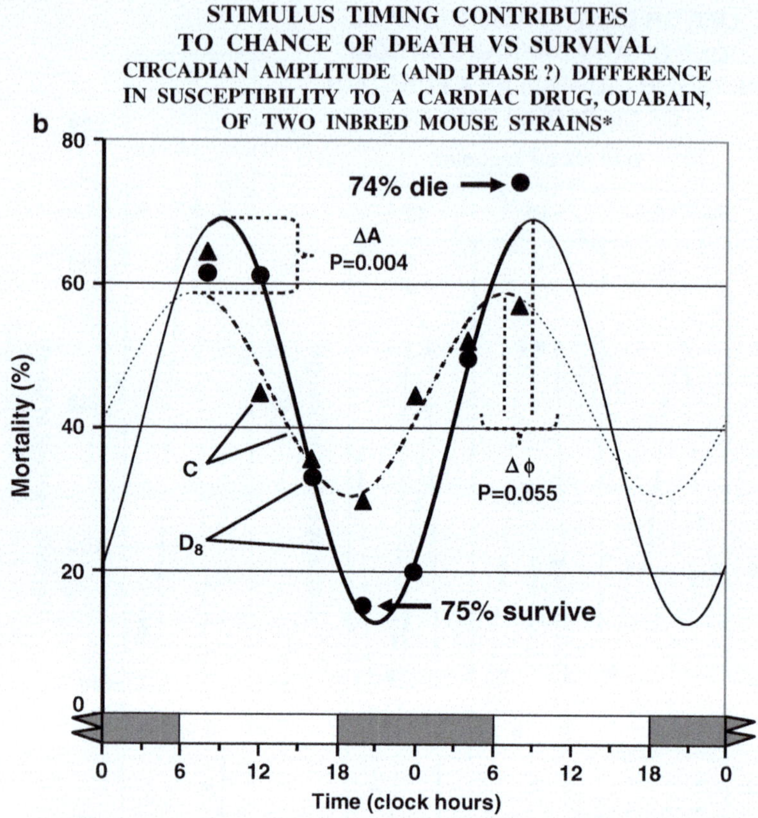

Strain	P	M	Amplitude (A)	Acrophase (ϕ**)
D_8	P < 0.001	40.0	28.1 (21.5; 34.8)	-135 (-119; -150)
C	P = 0.011	45.2	14.2 (7.3; 21.0)	-104 (-73; -136)

*** C: Bagg albino, D_8: subline 8 of Dilute Brown strain; ouabain 0.15 or 0.5 mg/20 g of body weight, respectively. Thin lines in the graph are extrapolations of fitted cosine curves. **In the table: 95% confidence limits in parentheses, ϕ in degrees (360° ≡ 24 hours). M = MESOR; P from zero-A (no-rhythm) test.**
Data from F.Halberg et al., Minnesota Acad Sci Proc 1959; 27: 139-43.

Fig. 1.13 (continued)

In 2012, for the purpose of seeking coexisting neighboring circadian periods, we have learned in indispensable months- or years-long around-the-clock data that a circadian free-running can be but one of coexisting (multiple) circadian periods, including socidian 24.0-hour periods, wrangling with the ~24.8-hour or other components of earth tides in one variable, sleep-wakefulness in health, Table 1.1 [66], or in several (including vascular and endocrine) variables under certain conditions in a disease of 21 years standing, Table 1.2 and Fig. 1.15a–m on JF [67]. In this case, focus on the adrenal cortex is prompted by the adynamia, a symptom of Addison's

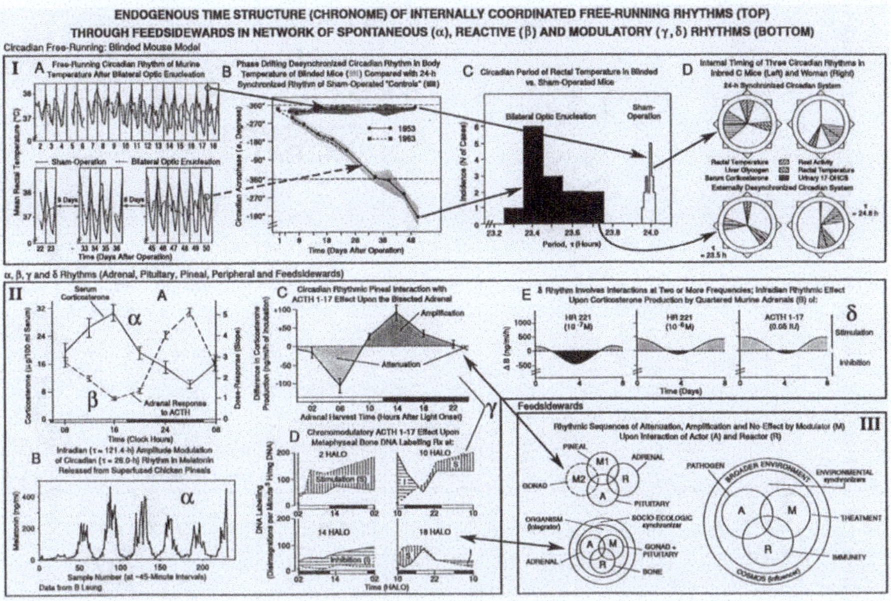

Fig. 1.14 *Top*, section I: desynchronization of circadian rhythm in core temperature of mice after blinding, seen time macroscopically in IA, here leads, in IB, to time microscopy with a chronobiologic serial section showing a different time course of the core temperature acrophases, ϕ, with early separation of the two groups by nonoverlapping 95 % confidence intervals of ϕ in IC, to a summary of individual periodograms that form two separate distributions, and in ID to time relations among three variables in a 24-hour synchronized (*top*) or free-running (*bottom*) system (of mice, *left*, and of a human, *right*). Section II shows a spontaneous (α) rhythm in circulating corticosterone of mice in antiphase with the slope of an in vitro response rhythm to ACTH, a reactive (β) rhythm. The components of the chronome (time structure) are internally coordinated through feedsidewards in a network of rhythms that are more or less spontaneous (α), others primarily reactive (β) or modulatory at a single mapped frequency, such as a circadian (γ), IIC and IID, or at multiply mapped (δ) frequencies, IIE. The effect of one entity (the actor) upon a second (the reactor), such as the pituitary acting upon the adrenal cortical corticosterone production, may be influenced, predictably insofar as rhythmically, by a third entity such as melatonin (the modulator) at the level of the pituitary and the same melatonin may act directly upon the adrenal. Reproducible sequences of attenuation, no effect, and amplification, the time-qualified feedsidewards, replace time-unqualified feedbacks and feedforwards (IIC–E). In sections II and III, feedsidewards include the interaction of a modulator (such as ACTH) upon an actor (such as adrenocortical corticosterone production) acting upon DNA labeling in bone (the reactor). The roles played by endocrines can and do change in various feedsidewards that replace time-unqualified feedbacks and feedforwards (see Fig. 1.15o). Chronomolecular mapping of circadian acrophases has also begun (© Halberg Chronobiology Center)

disease. Focus on the selenosensitivity claimed by JF is in keeping with the double tidal 24.8-hour period in the analyses in a computer output (not in the input), not encountered previously in several variables of the same person. The possibility of a beat with cortisol of vigor must be considered as a mechanism underlying JF's adynamia, as visualized in Fig. 1.15n–p.

Table 1.1 On a self-selected schedule (JFC), a double tidal pull competes with 24-hour environment while initially coexisting free-running circadian fades (global analysis)

	Calendar date		Period	Amplitude	DT-A/S-A (%)	
	All (3 year) data	Double tidal (DT)	24.836 (24.833, 24.838)	0.10606 (0.0854, 0.1268)		*Global*
	1990/02/19 to 1993/03/11	Compromise	24.432 (24.427, 24.436)	0.05643 (0.0357, 0.0772)		
		Free-running	24.260 (24.252, 24.268)	0.03253 (0.0118, 0.0533)		
		Societal (S)	24.001 (24.000, 24.001)	0.25443 (0.2337, 0.2751)	*0.41685*	
		Tidal	12.414 (12.410, 12.418)	0.01728 (–0.0034, 0.038)		
		Semidian	11.999 (11.997, 12.001)	0.03204 (0.0113, 0.0527)		
Local	1: 1990/02/19 to 1991/04/28	Double tidal	24.872 (24.830, 24.914)	0.02513 (–0.0070, 0.0573)		
		Compromise	24.409 (24.384, 24.435)	0.04117 (0.0089, 0.0734)		
		Free-running	24.222 (24.210, 24.235)	0.07921 (0.0468, 0.1116)		
		Societal	24.008 (24.004, 24.011)	0.31605 (0.2838, 0.3483)	*0.07951*	
		Tidal	12.421 (12.411, 12.431)	0.02615 (–0.0059, 0.0582)		
		Semidian	12.007 (12.001, 12.013)	0.04105 (0.0090, 0.0731)		
	2: 1991/04/28 to 1991/09/23	Double tidal	24.898 (24.832, 24.964)	0.12744 (0.0707, 0.1842)		
		Compromise	24.707 (24.671, 24.743)	0.25208 (0.1973, 0.3068)		
		Compromise	24.488 (24.434, 24.541)	0.10274 (0.0552, 0.1502)		
		Societal	24.011 (23.992, 24.030)	0.22758 (0.1815, 0.2736)	*0.55998*	
		Tidal	Did not converge	–		
		Semidian	–	–		
	3: 1991/09/23 to 1992/03/08	Double tidal	24.862 (24.852, 24.872)	0.34858 (0.3089, 0.3882)		
		Compromise	24.523 (24.468, 24.579)	0.06108 (0.0212, 0.1009)		
		Free-running	Did not converge	–		
		Societal	23.997 (23.984, 24.011)	0.22362 (0.1841, 0.2631)	*1.55881*	
		Tidal	12.432 (12.416, 12.448)	0.04969 (0.0106, 0.0888)		
		Semidian	Did not converge	–		

	4: 1992/03/08 to 1992/07/14	Double tidal	24.942 (24.896, 24.989)	0.15133 (0.0960, 0.2067)		
		Compromise	24.451 (24.394, 24.508)	0.11847 (0.0630, 0.1739)		
		Free-running	–	–		
		Societal	23.987 (23.967, 24.006)	0.32099 (0.2660, 0.3760)	*0.47145*	
		Tidal	12.432 (12.398, 12.465)	0.04960 (–0.0051, 0.1042)		
		Semidian	11.982 (11.951, 12.012)	0.05028 (–0.0044, 0.1049)		
	5: 1992/07/14 to 1993/03/11	Double tidal	24.789 (24.782,24.797)	0.31836 (0.2798, 0.3569)		
		Compromise	24.474 (24.438, 24.510)	0.06218 (0.0236, 0.1008)		
		Free-running	–	–		
		Societal	24.009 (23.997, 24.022)	0.17892 (0.1405, 0.2173)	*1.77934*	
		Tidal	12.352 (12.328, 12.376)	0.02358 (–0.0147, 0.0619)		
		Semidian	12.001 (11.988, 12.013)	0.04128 (0.0030, 0.0796)		

The entire available data span was analyzed as a whole (top 6 rows) and then divided into five subspans according to visual inspection of characteristics of 24-hour and 24.84-hour fits in serial sections, suggesting similarities within the spans and differences among them. Initial period guesses for the six nonlinearly searched periods were 24.84, 24.43, 24.26, 24.00, 12.42, and 12.00. Results were omitted if they did not converge to a period fitted to the data, in which case the analysis was redone after removal of these components from the model, to yield results summarized herein. Some periods converged well, but the CI (95 % conservative confidence interval) of the amplitude, A (the CI given in parentheses), covered zero, as apparent from a negative sign inside the last column's second, fifth, and sixth sections, for the case of double tidal and tidal associations in the second section (top) where the societal 24-hour day predominates. The double tidal component converges invariably thereafter. Note that overlaps of zero by the CI of the A are very small. In 2 of 5 sections, the double tidal cycle has the largest A, the ratio of amplitudes of DT over S becomes greater than 1, suggesting that the moon/sun dominate over society in clinical health. These sections are not consecutive, possibly a chance observation, yet in keeping with the possibility that in health there is a wrangling between the double tide and society, as also seen in a patient with twice-yearly depression. From Costella et al. [66]

Table 1.2 Recorded recurrent adynamia. Start-end dates and length (weeks) of spans of adynamia when recorded in winter (W) and summer (S)

	2002	2003	2004	2005	2006	2007	2008	2009	2010	2011	2012
W	–	01.18 f 02.17 f (4)	01.12 02.20 n (5)	01.05 02.15 (5)	12.25 n 02.14 n (7)	01.03 f 03.03 f (8)	01.08 n 02.21 f (6)	12.26 n 02.23 n (8)	01.02 f 02.28 f (8)	01.04 n 03.05 n (8)	01.04[b]
S	06.09 08.07 (8)	06.15 f 08.12 f (8)	06.17 n 08.18 n (8)	06.20 f 08.09 (7)	06.04 08.28 (11)	05.31 f 08.12 n (10)	? 08.18 f (?)	06.09 f 08.08 f (8)	07.08 09.16 (10)	07.07 09.13[a] (10)	

f: date of full moon, *n*: date of new moon

[a]"I have concluded that Sept 1 was a false end to the downtime. I had five functional days, but the last eight [i.e., 09.06–09.13] have been horrible again!" (JF, personal communication, 14 September 2011)

[b]"Although there have been many failed attempts to regain a normal sleep schedule (which has in the past been a clearly noted evidence that the downtime is over), I have not been able to do so" (JF, personal communication, 15 April 2012)

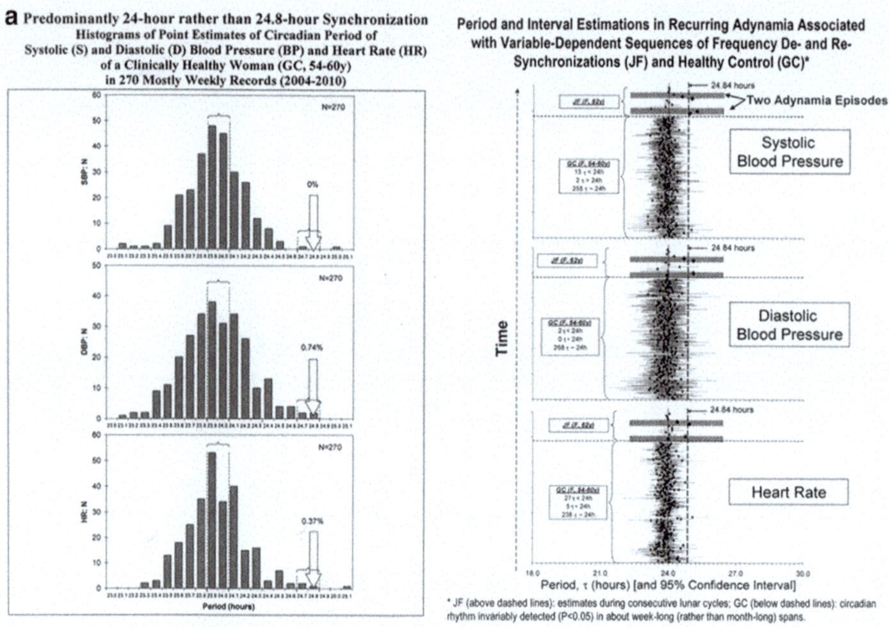

Fig. 1.15 (**a**) In the blood pressure and heart rate of a clinically healthy woman studied at 30-min intervals around the clock over several years, the frequency of one cycle in (a double earth tidal average of) 24.8 hours is extremely low and was not yet detected twice in succession in the same person (GC) in this case or any other studied over the past half-century by ourselves, in health, under conditions of life on a 24-hour average schedule, information that sets the stage for the following slides on JF, a patient with a history of over 20 years of twice-yearly depression (Table 1.2) (© Halberg Chronobiology Center). (**b**) The double tidal average is seen in the first lunar month of two consecutive adynamic episodes investigated in JF – in the computer output of the nonlinearly extended cosinor – in data on self-rated vigor (*top row*). In the second row, the lower curve is the weekly average (MESOR) which summarizes the original data on vigor on top, more clearly

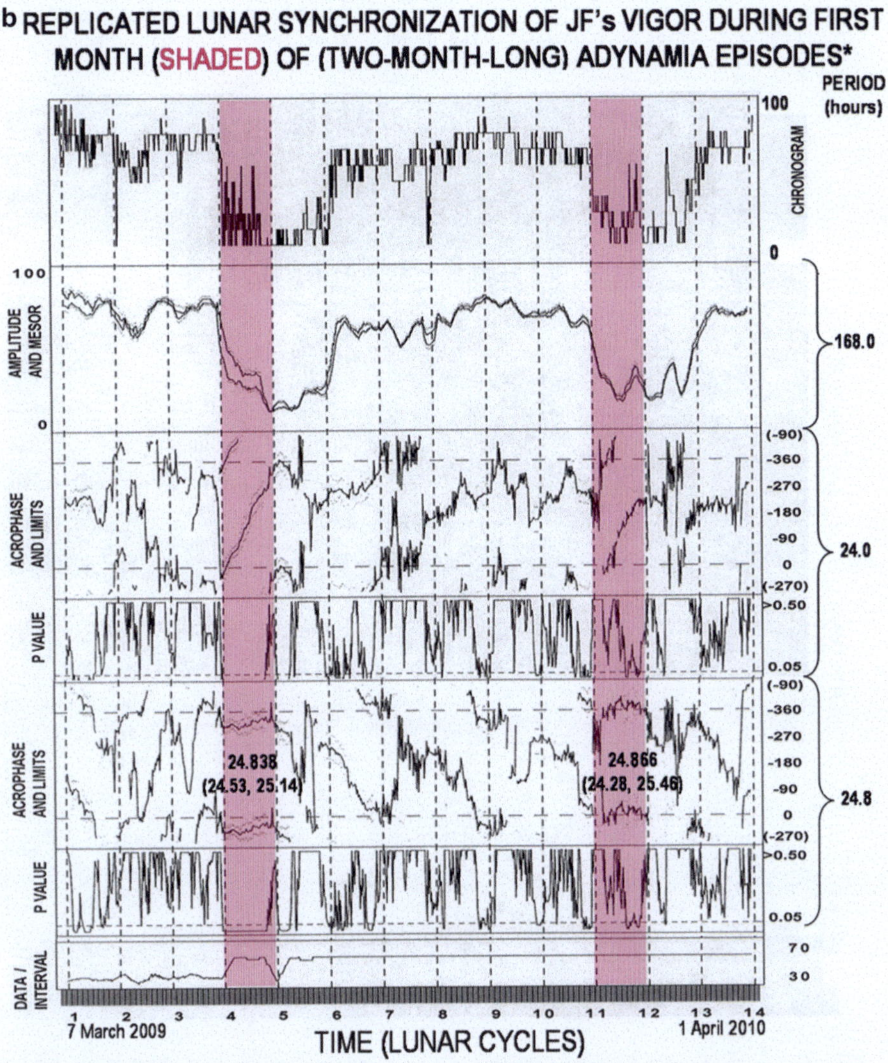

*Dashed vertical lines: full moons (JF reports sensitivity to the moon). N data: 2820; interval: 168 hours; increment: 12 hours.

Fig. 1.15 (continued) revealing the twice-yearly episodes of 2–3 months, of which the first lunar month is shown by vertical pink columns. The difference between the two curves in the second row is the very small circaseptan amplitude, which increases notably during the first episode. The third row shows that, during the first month of each episode, the phase of a 24-hour cosine fit delays and is statistically significant below the dashed horizontal line indicating the 5 % probability level (*in row 4*) during that month in the first episode. With the fit of a 24.8-hour cosine, the phases are more or less horizontal during the first month of each episode indicating a good fit of the double tidal period (*in row 5*). The numbers inserted in row 5 are from a nonlinearly extended cosinor with the CIs (95 % confidence intervals) of the period given in parentheses. None of these CIs covers 24.0 hours; both cover 24.8 hours (© Halberg Chronobiology Center). (**c**) Time course of a 24-hour cosine fit to JF's systolic blood pressure, revealing alternating apparent dominance of 24-hour synchronization and

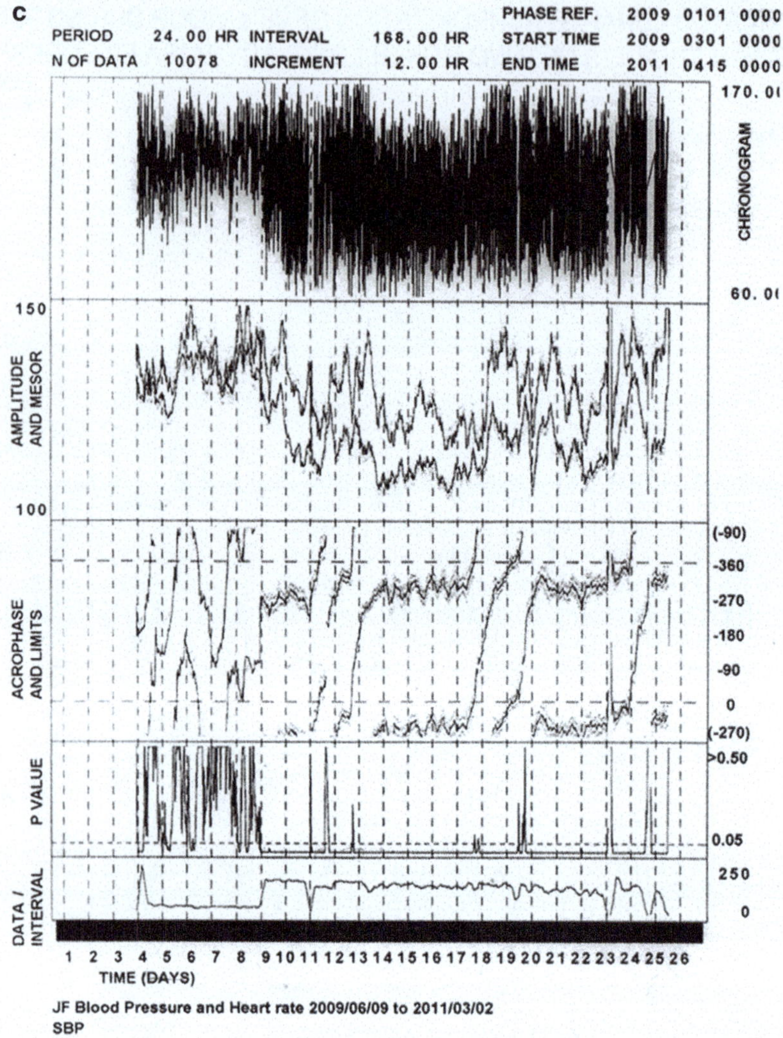

Fig. 1.15 (continued) desynchronization (© Halberg Chronobiology Center). (**d**) Time course of a 24-hour cosine fit to JF's diastolic blood pressure, revealing alternating apparent dominance of 24-hour synchronization and desynchronization (© Halberg Chronobiology Center). (**e**) Time course of a 24-hour cosine fit to JF's heart rate, revealing alternating apparent dominance of 24-hour synchronization and desynchronization (© Halberg Chronobiology Center). (**f**) Coexisting separate circadian periods in JF's heart rate, during an episode of depression with adynamia. A component with a 24.8-hour period dominates with a 95 % confidence interval that does not overlap 24.0 hours. The second largest but numerically smaller amplitude than that of the 24.8-hour peak (of the average double tide) has a CI overlapping 24.0 hours. The more detailed time course of the heart rate's periods is seen in section G (© Halberg Chronobiology Center). (**g**) Time course of two separate coexisting circadian periods in JF's circulation, alternating in (and wrangling for) dominance, as seen by the differences in their amplitudes (© Halberg Chronobiology Center). (**h**) Chronobiologic serial sections of JF's urine volume, with the fit of a 24-hour cosine in the *top row* and of a 24.8-hour cosine curve in the *second row*. The *top row* suggests 24-hour synchrony during wellness in a first lunar cycle and a delaying time course,

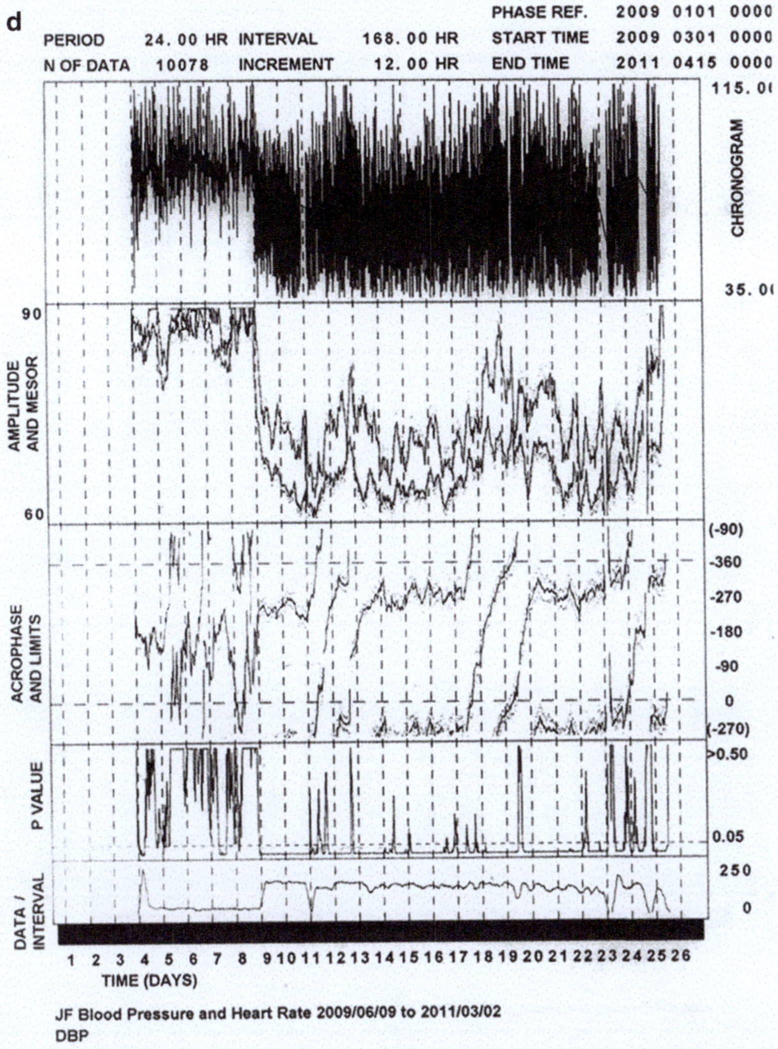

Fig. 1.15 (continued) i.e., a lengthening of period during the remainder of the time of adynamic depression indicating desynchronization from society's 24-hour schedule. In the *second row*, during the first lunar month, the phase advances because the 24-hour period is shorter than the 24.8-hour period fitted in this row. This fit is appropriate during the second lunar month when depression starts, as seen by a horizontal time course, and again near the end of the record, when the 24.8-hour fit again has a horizontal time course of phases. After the second lunar month, wrangling follows with a shortening of the period suggested by an again-diagonal time course of advancing phases, ϕ, until the ϕ again becomes horizontal, i.e., the fitted 24.8-hour period is mostly dominant. During lunar months 2–5, behavioral depression occurs, during which the ϕ is pulled with alternating dominance of the 24.8-hour and 24.0-hour periods, as is also the case in other variables in the first completely investigated episode in section J below, in keeping with the possibility of wrangling by the tides and society, the former dominating usually during the first month of depression and regaining control again during another subspan of a depressive episode (© Halberg Chronobiology Center). (**i**) Coexisting circadian (~24.0-hour and longer) periods in the urine volume (*left*, *top*) and

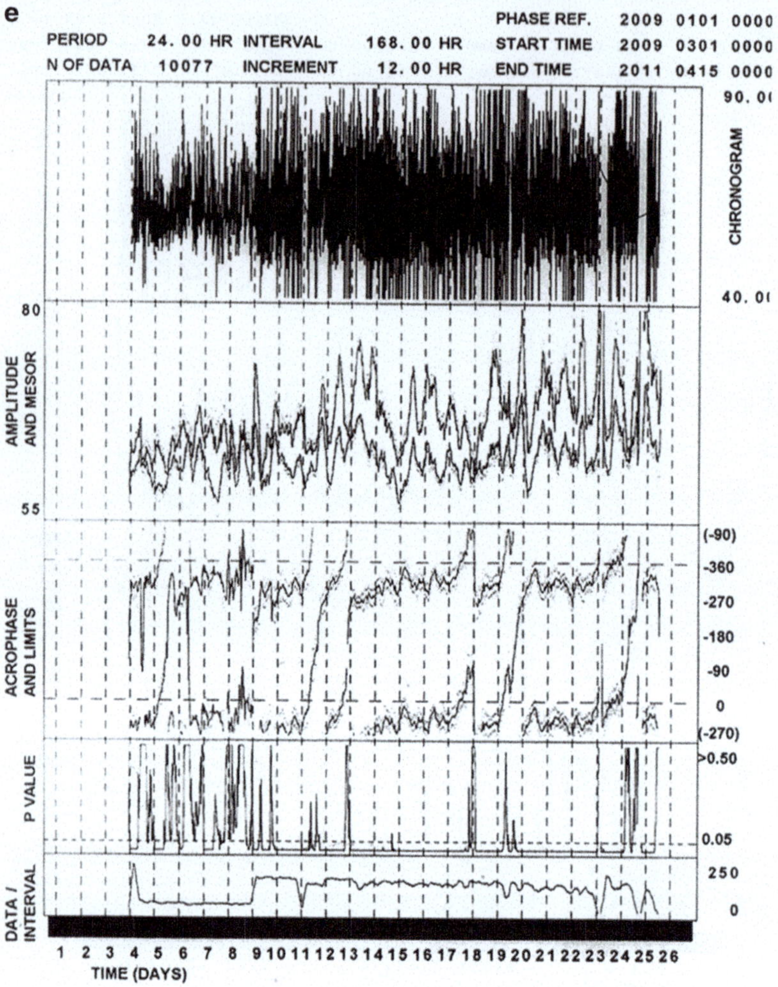

Fig. 1.15 (continued) excretion rate (*right, top*) dominate in alternation, as seen from their amplitudes (*bottom*) (© Halberg Chronobiology Center). **(j)** Wrangling between the tides (the gravitational pulls of moon/sun) on the one hand and society on the other hand in JF's endocrines and in her circulation, with the pull of the tides lengthening the dominant period during spans of depression shaded as vertical blue-shaded columns (note, for more than one variable, two separate dominances of 24.8 hours by the periods plotted during the same first completely investigated episode, the first a clear overshoot) (© Halberg Chronobiology Center). **(k)** Diagonal (desynchronized) time course of JF's phases in salivary cortisol and DHEA during depression (*red*) is prominent, yet slightly present during wellness as well (*blue*) when the period still differs (is free-running) from precisely 24.00 hours ($P<0.001$) (© Halberg Chronobiology Center). **(l)** Diagonal (desynchronized) time course of JF's phases in salivary melatonin and estradiol during depression (*red*) is much less prominent yet slightly present during wellness as well (*blue*), with a period differing (free-running) from precisely 24.00 hours ($P<0.001$) (© Halberg Chronobiology Center). **(m)** Diagonal (desynchronized) time course of JF's phases in testosterone and aldosterone,

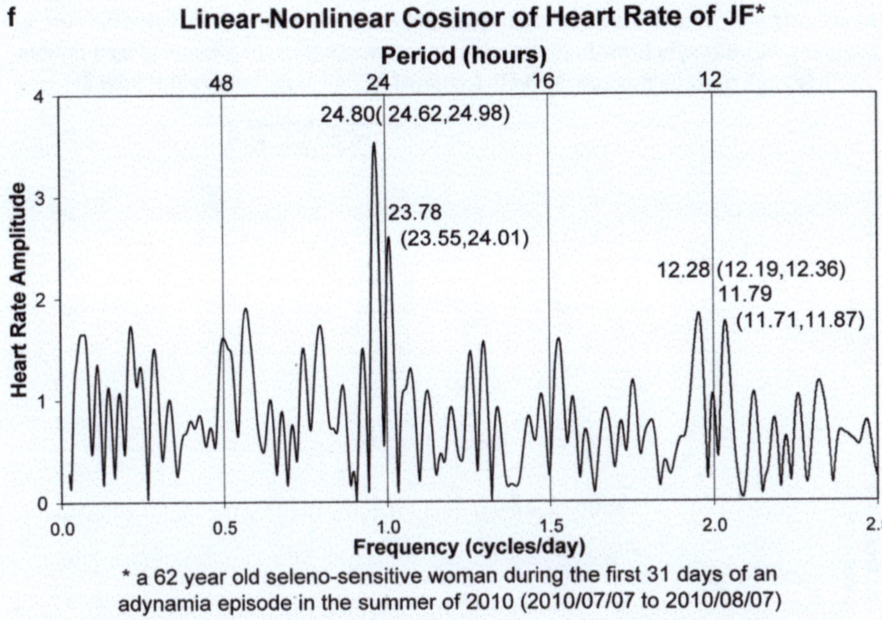

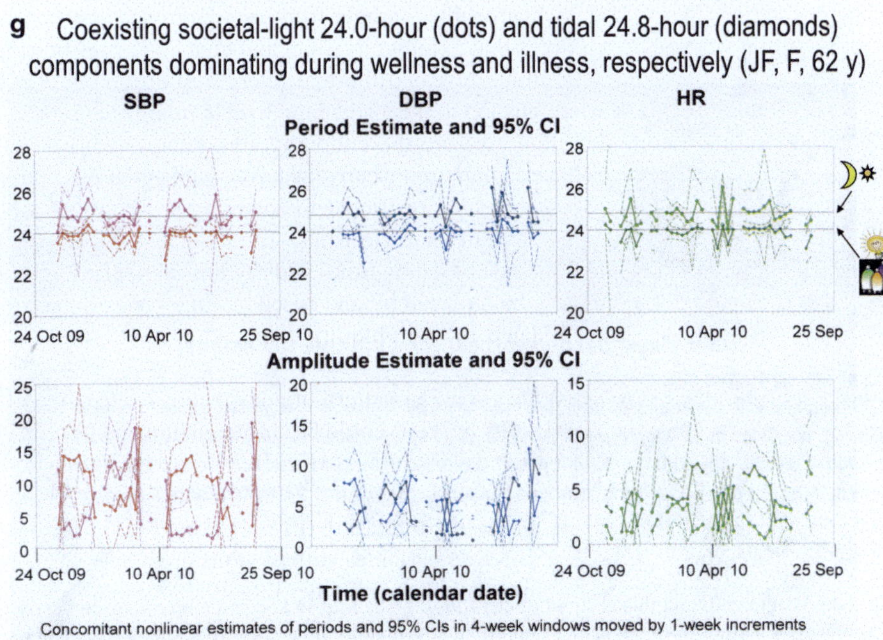

Fig. 1.15 (continued) prominently present during depression (*red*), is not seen in testosterone and aldosterone during wellness (*blue*), presumably because of selective synchronization by twice-daily spironolactone treatment (© Halberg Chronobiology Center). (**n**) Abstract visualization of a hypothetical factor underlying vigor/wellness beating with some other factor (cortisol?). If the twice-yearly episodes of adynamia in JF should come about as a beat of neighboring periods, the nearest beat period to 1 year in the data that happened to be available at the time of this analysis involves cortisol,

h Near-horizontal time course of circadian acrophases, ϕs, when the fit is right (shaded) suggests 24-hour synchronization in a first lunar month (top row) followed by a double tidal pull of water metabolism with transient 24.84-h synchronization (row 2)*

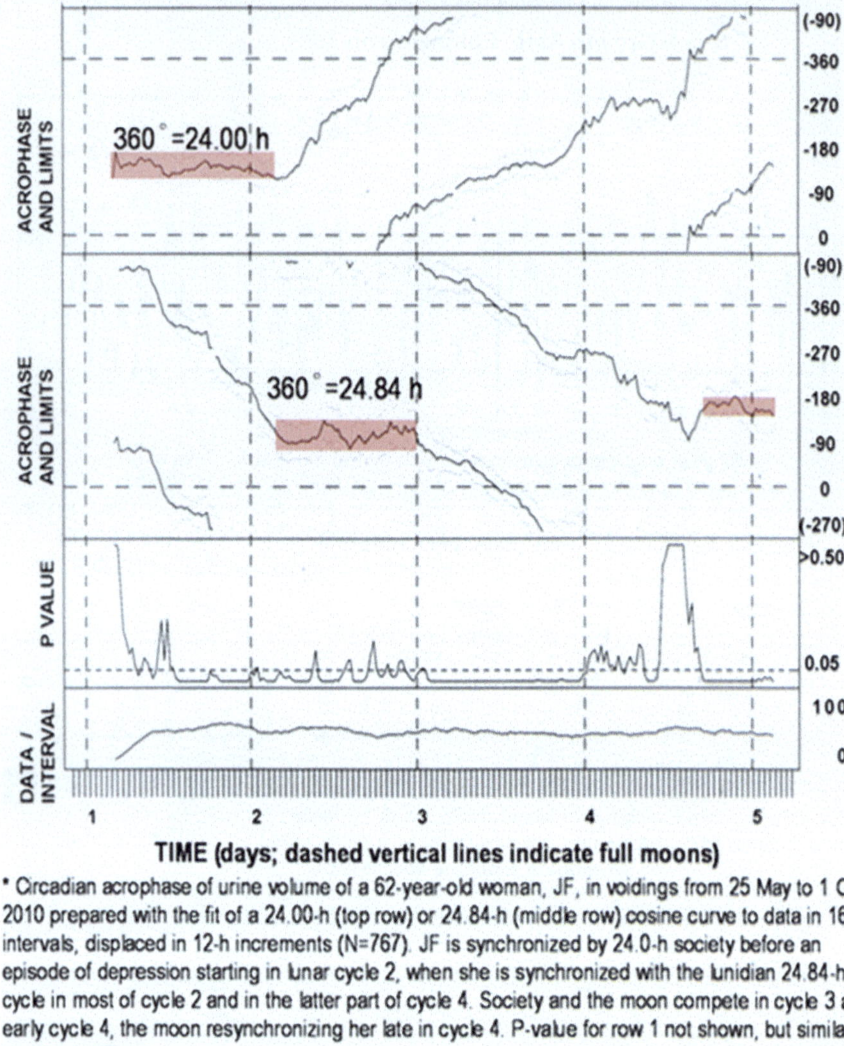

TIME (days; dashed vertical lines indicate full moons)

* Circadian acrophase of urine volume of a 62-year-old woman, JF, in voidings from 25 May to 1 Oct 2010 prepared with the fit of a 24.00-h (top row) or 24.84-h (middle row) cosine curve to data in 168-h intervals, displaced in 12-h increments (N=767). JF is synchronized by 24.0-h society before an episode of depression starting in lunar cycle 2, when she is synchronized with the lunidian 24.84-h cycle in most of cycle 2 and in the latter part of cycle 4. Society and the moon compete in cycle 3 and early cycle 4, the moon resynchronizing her late in cycle 4. P-value for row 1 not shown, but similarly mostly significant.

Fig. 1.15 (continued) and this abstract model (*above*) was thus in keeping with the twice-yearly recurrence (in section **b**) of adynamia (a prominent symptom of adrenocortical insufficiency) (© Halberg Chronobiology Center). (**o**) JF reports that she is sensitive to the moon and to helio- and geomagnetism. The double tidal period and its wrangling with the societal schedules are in keeping with her claim. Underlying mechanisms can be explored by the following survey in the laboratory: (*A*) Observations on the equatorial geomagnetic index Dst (*top*), on the hypothalamus, Ht (*middle*) that may be activated via the storm directly and/or via the pineal, Pin (*bottom*). Concerning an effect of the storm (*top*) upon the neuroendocrine network (*below*), also showing adrenocortical

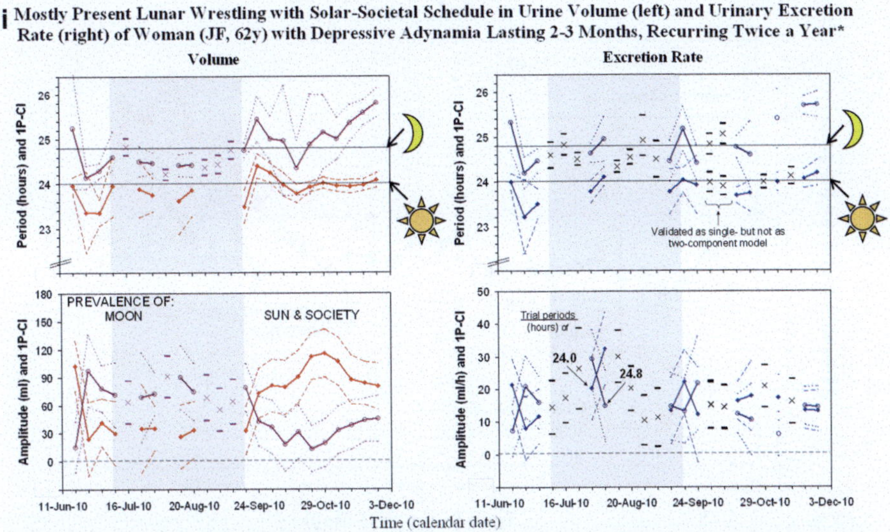

*with thus documented seleno-sensitivity, detected by the concomitant nonlinear fit of a 2-component model consisting of cosine curves with trial periods of 24.0 (filled diamonds) and 24.8 (open circles) hours to data in a 4-week interval displaced by 1 week, yielding estimates for each period (top) and corresponding amplitude (bottom), shown with their 1-parameter confidence intervals (1P-CI) derived by Marquardt's algorithm. Some 95% conservative confidence intervals of the amplitude also do not cover zero (not shown). Parameters could not be estimated in some intervals because model was then reduced to single major component (crosses).

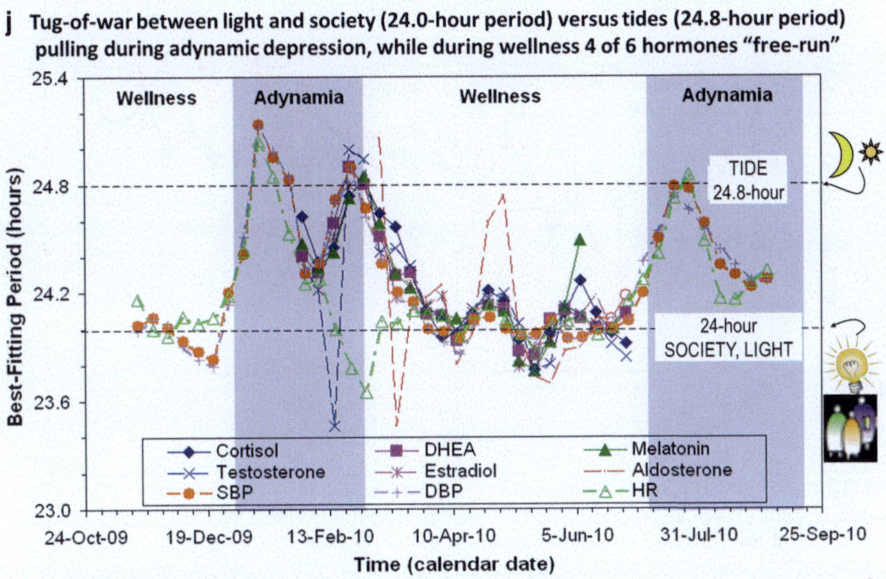

N=11,700 salivary hormonal assays (JF: F, 61-62y; 20y of adynamic episodes lasting 2-3 months and recurring half-yearly).

Fig. 1.15 (continued) involvement. These unplanned observations on rats can be aligned with observations of a decreased human urinary excretion of melatonin during magnetic storms in keeping with a damping of pineal function, including its dynamics, gauged by the lower circadian amplitude during the storm. This "experiment of nature" offers a putative mechanism that cannot be dissected in vivo in the pineal and/or hypothalamus of humans. (*B*) Lack of effect, attenuation, or amplification by aqueous pineal homogenate (APH) of corticosterone production by bisected

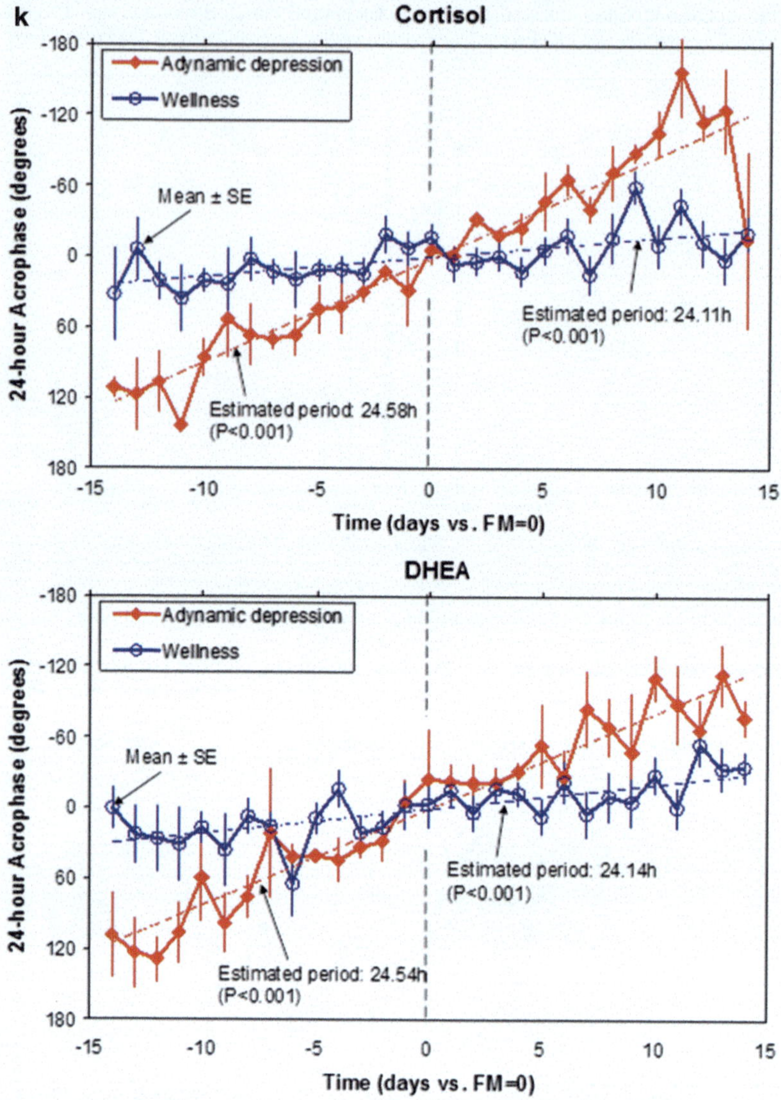

Fig. 1.15 (continued) adrenals in response to a synthetic ACTH 1–17, Synchrodyn (Sy). When magnetic storms reach an organism, their effect will depend upon the network's stage at exposure times, e.g., in the case of a double storm shown in (**a**) (© Halberg Chronobiology Center). (*C*) Results such as those in *A* and *B* are summarized in *C* (*top*). The second row from the top in A illustrates the effect of a double magnetic storm on circadian amplitudes and acrophases in the suprachiasmatic nuclei (SCN) of the hypothalamus and the third row upon the circadian characteristics of the pineal. The consistent findings after SCN removal are changes in circadian amplitude and/or acrophase, and these are consistent with those in *A*. (*D*) Chronomodulation: on the right, a modulator, representing magnetic planetary and interplanetary solar and galactic factors (their effect on geomagnetics is shown in *A*), is conceived as normally modulating or, as storms, altering the (socio-)ecological conditions in the habitat, e.g., the synchronizing effect of the lighting regimen (the actor in the animal room), acting upon the organismic hypothalamic-pineal-

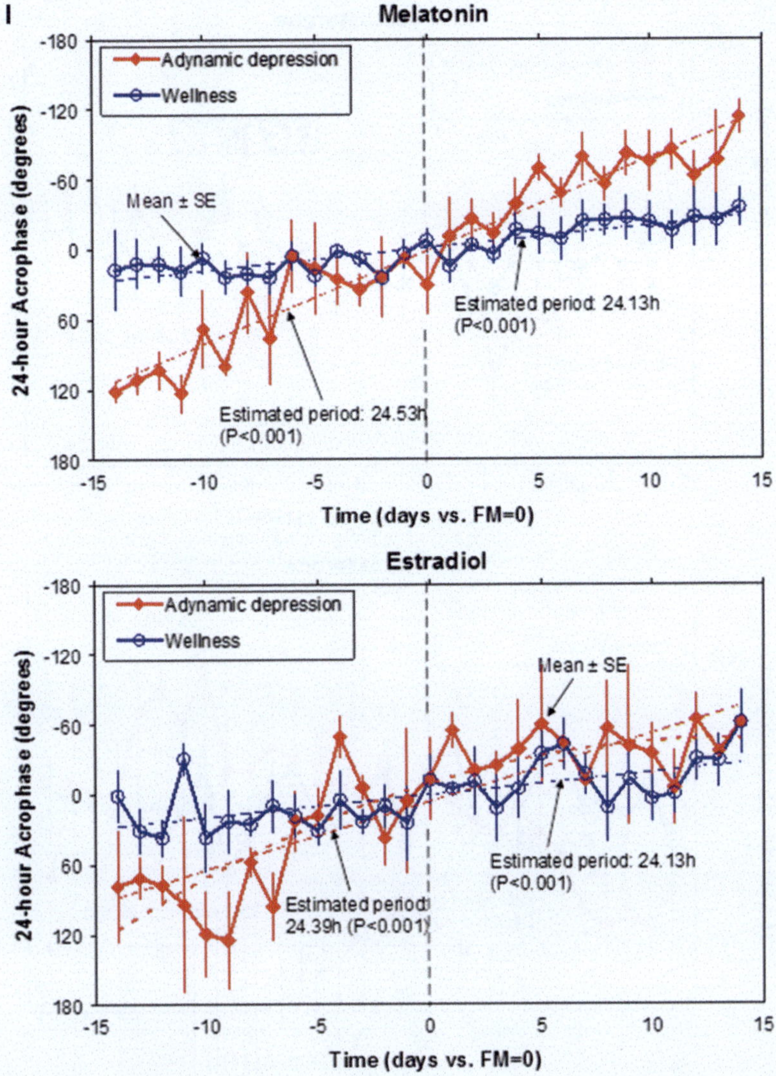

Fig. 1.15 (continued) adrenocortical-vascular network (© Halberg Chronobiology Center). (**p**) That the pull of the tides influences the human circulation, as found by Miroslav Mikulecky for the case of his arrhythmias, is supported by the repeated detection of the average double tidal period in JF and may act on melatonin in both the pineal and the hypothalamus, in keeping with data on the rat for the case of a magnetic storm (© Halberg Chronobiology Center)

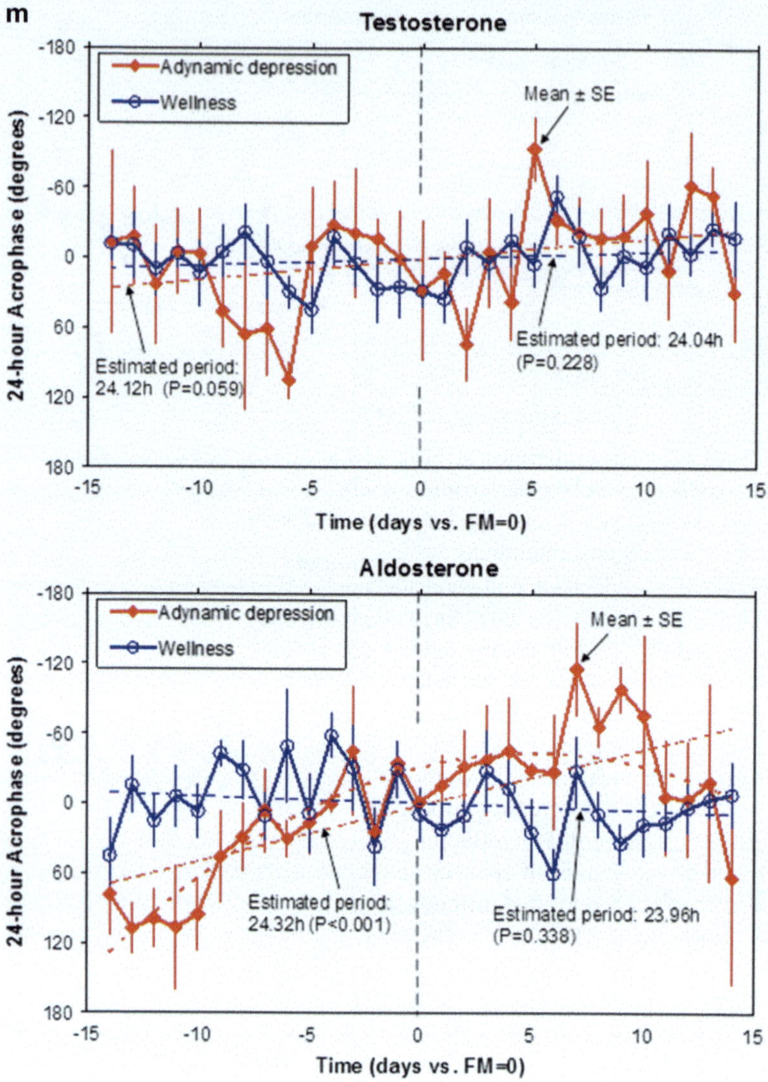

Fig. 1.15 (continued)

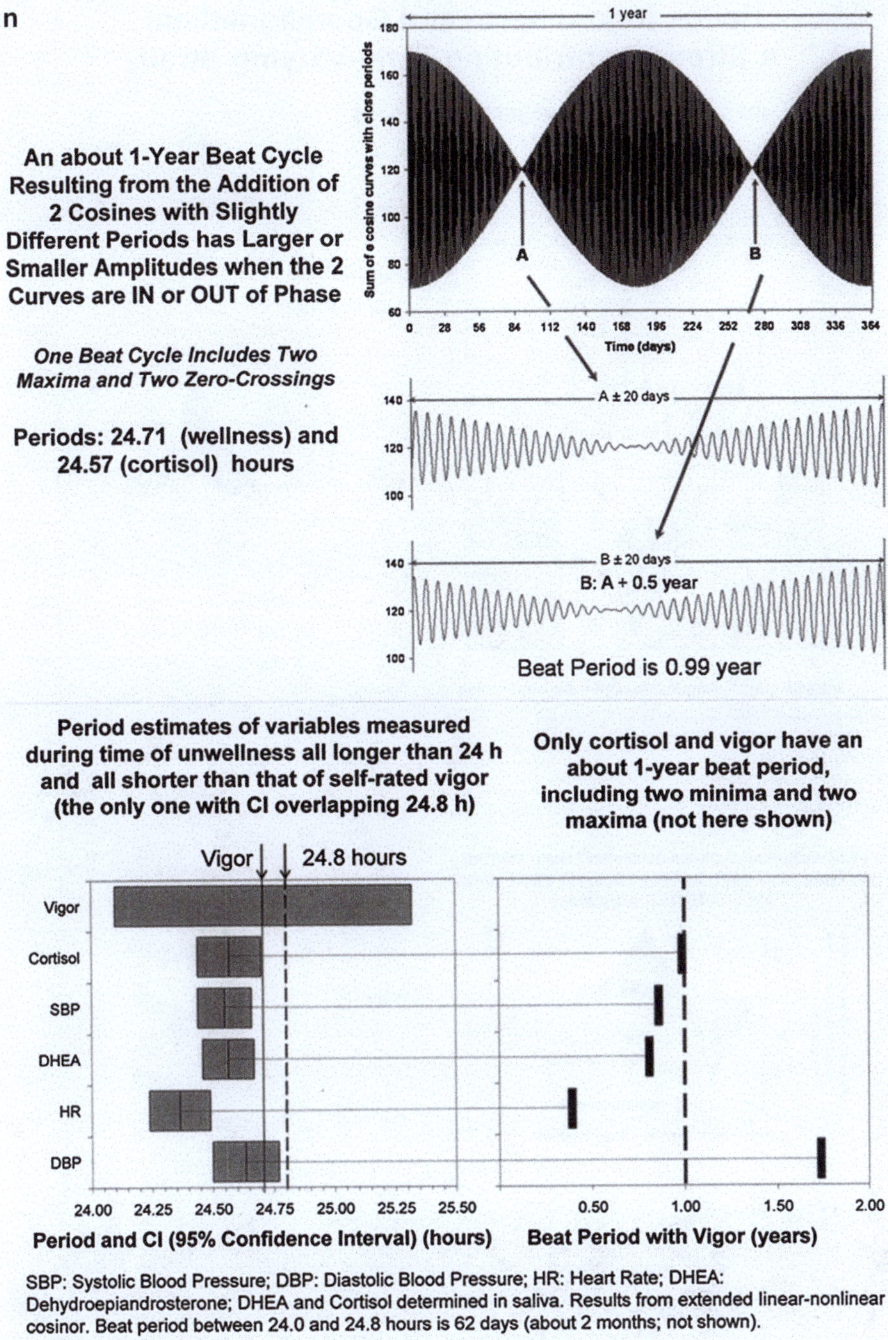

Fig. 1.15 (continued)

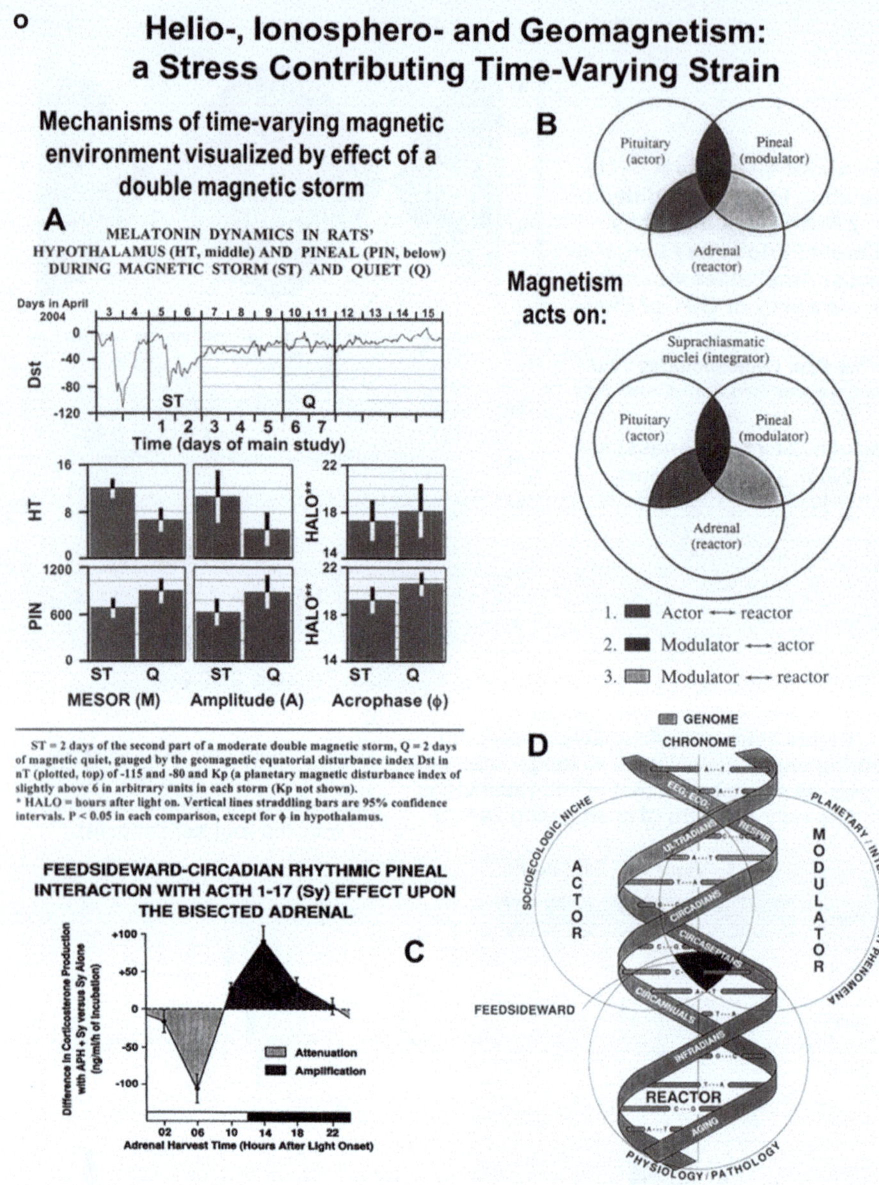

Fig. 1.15 (continued)

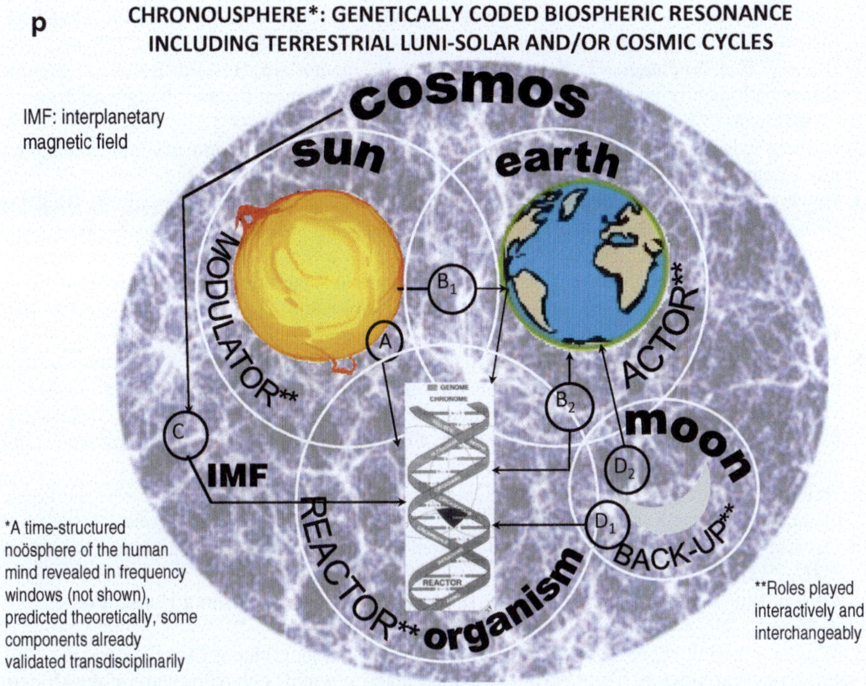

Fig. 1.15 (continued)

References

1. Hench PS. Cortisone and ACTH in clinical medicine. Proc Staff Mtg Mayo Clin 1950;25:474–6.
2. Hench PS. The present status of cortisone and ACTH in general medicine. Proc Roy Soc Med 1950;43:769–73.
3. Hench PS, Kendall EC, Slocumb CH, Polley HF. The effect of a hormone of the adrenal cortex (17-hydroxy-11-dehydrocorticosterone: compound E) and of pituitary adrenocorticotropic hormone on rheumatoid arthritis. Ann Rheum Dis 1949;8:97–104.
4. Withering W. An account of the foxglove and some of its medical uses, with practical remarks on dropsy and other diseases. Birmingham: Swinney; 1785.
5. Halberg F. Some correlations between chemical structure and maximal eosinopenia in adrenalectomized and hypophysectomized mice. J Pharmacol Exp Ther 1952;106:135–40.
6. Recant L, Hume DM, Forsham PH, Thorn GW. Studies on the effect of epinephrine on the pituitary-adrenocortical system. J Clin Endocrinol 1950;10:187–229.
7. Hume DM, Wittenstein GJ. The relationship of the hypothalamus to pituitary-adrenocortical function. In: Soskin S, editor. Progress in clinical endocrinology. New York: Grune & Stratton; 1950. p. 213.
8. Hume DM. The role of the hypothalamus in pituitary-adrenal cortical response to stress. J Clin Invest 1949;28:790.
9. Long CNH, Fry EG. Effect of epinephrine on adrenal cholesterol and ascorbic acid. Proc Soc Exp Biol Med 1945;59:67–8.

10. Best WR, Muehrcke RC, Kark RM. Studies on adrenocortical eosinopenia: a clinical and statistical evaluation of four-hour eosinophil response tests. J Clin Invest 1952;31:733–42.
11. Halberg F, Cornélissen G, Schwartzkopff O. In memoriam Howard Burchell: lifetime chronobiologically-interpreted (C-) ABPM for strain assessment for everybody with diagnostic dividends. World Heart J 2010;2(3):177–96.
12. Halberg F, Johnson EA, Nelson W, Runge W, Sothern R. Autorhythmometry—procedures for physiologic self-measurements and their analysis. Physiol Tchr 1972;1:1–11.
13. Halberg F, Smith HN, Cornélissen G, Delmore P, Schwartzkopff O. International BIOCOS Group. Hurdles to asepsis, universal literacy, and chronobiology – all to be overcome. Neuroendocrinol Lett 2000;21:145–60.
14. Satoh M, Kikuya M, Ohkubo T, Imai Y. Acute and subacute effects of the great East Japan earthquake on home blood pressure values. Hypertension 2011;58:e193–4. doi:10.1161/HYPERTENSIONAHA.111.184077.
15. Halberg F, Cornelissen G, Otsuka K, Siegelova J, Fiser B, Dusek J, Homolka P, Sanchez de la Pena S, Singh RB, BIOCOS project. Extended consensus on need and means to detect vascular variability disorders (VVDs) and vascular variability syndromes (VVSs). Int J Gerontol-Chronome Geriatr 2008;11(14):119–46. Cf. Leibniz-Online Nr. 5, 2009. (http://www2.hu-berlin.de/leibniz-sozietaet/journal/archiv_5_09.html). 35 pp.
16. Halberg F, Cornélissen G, Sothern RB, Czaplicki J, Schwartzkopff O. Thirty-five-year climatic cycle in heliogeophysics, psychophysiology, military politics, and economics. Izv Atmos Oceanic Phys 2010;46(7):844–64 (Backtranslation from Geophysical Processes and Biosphere 2009; 8 [2]: 13–42.).
17. Lord of time. London: Science without Borders/International Publishing House SWB; 2011. p. 45.
18. Watanabe Y, Halberg F, Otsuka K, Cornelissen G. Physiological changes in relation to the 2011 East Japan earthquake. World Forum "Natural Cataclysms and Global Problems of the Modern Civilization", 19–21 Sept 2011, Istanbul, Turkey. London: SWB; 2011. p. 113–4.
19. Rosenbaum L, Lamas D. Global health: facing a "slow-motion disaster" – the UN meeting on noncommunicable diseases. N Engl J Med 2011;365:2345–8.
20. Halberg F, Cornélissen G, Schwartzkopff O, Khalilov E, Khalilov T, Damirov F, Wang Z, Watanabe Y, Otsuka K, Siegelova J, al-Abdulgader AA. Preventive cardiology concerns avoidance of personal and societal health-related and natural cataclysms. In: Halberg F, Kenner T, Fiser B, Siegelova J, editors. Noninvasive methods in cardiology. Brno: Faculty of Medicine, Masaryk University; 2011. p. 33–44.
21. Halberg F, Tong YL, Johnson EA. Circadian system phase—an aspect of temporal morphology; procedures and illustrative examples. Proceedings, International Congress of Anatomists. In: Mayersbach H v, ed. The cellular aspects of biorhythms. Symposium on Rhythmic Research, Sponsored by the VIII International Congress of Anatomy, Wiesbaden, 8–14 August 1965. New York: Springer; 1967. p. 20–48.
22. Cornélissen G, Halberg F. Chronomedicine. In: Armitage P, Colton T, editors. Encyclopedia of biostatistics. 2nd ed. Chichester: Wiley; 2005. p. 796–812.
23. Refinetti R, Cornélissen G, Halberg F. Procedures for numerical analysis of circadian rhythms. Biol Rhythm Res 2007;38(4):275–325. doi:10.1080/09291010600903692.
24. Halberg F, Cornélissen G, International Womb-to-Tomb Chronome Initiative Group. Resolution from a meeting of the International Society for Research on Civilization Diseases and the Environment (New SIRMCE Confederation), Brussels, 17–18 Mar 1995, Fairy tale or reality? Medtronic chronobiology seminar #8, April 1995, 12 pp. text, 18 figures. http://halbergchronobiologycenter.umn.edu (click "Resolution" link on left).
25. Halberg F. Some physiological and clinical aspects of 24-hour periodicity. Journal-Lancet (Minneapolis) 1953;73:20–32.
26. Halberg F, Cornélissen G, Katinas G, Syutkina EV, Sothern RB, Zaslavskaya R, Halberg F, Watanabe Y, Schwartzkopff O, Otsuka K, Tarquini R, Perfetto P, Siegelova J. Transdisciplinary

unifying implications of circadian findings in the 1950s. J Circadian Rhythms 2003;1:2. 61 pp. www.JCircadianRhythms.com/content/pdf/1740-3391-1-2.pdf.

27. Halberg F, Visscher MB. A difference between the effects of dietary calorie restriction on the estrous cycle and on the 24-hour adrenal cortical cycle in rodents. Endocrinology 1952;51: 329–35.
28. Halberg F, Visscher MB, Bittner JJ. Eosinophil rhythm in mice: range of occurrence; effects of illumination, feeding and adrenalectomy. Am J Physiol 1953;174:109–22.
29. Halberg F, Sothern RB, Cornélissen G, Czaplicki J. Chronomics, human time estimation, and aging. Clin Interv Aging 2008;3(4):749–60. DOI http://dx.doi.org/10.2147/CIA.S4194.
30. Halberg F, Cornélissen G, Ulmer W, Blank M, Hrushesky W, Wood P, Singh RK, Wang Z. Cancer Chronomics III. Chronomics for cancer, aging, melatonin and experimental therapeutics researchers. J Exp Ther Oncol 2006;6:73–84.
31. Halberg F. *Quo vadis* basic and clinical chronobiology: promise for health maintenance. Am J Anat 1983;168:543–94.
32. Halberg F, Sothern RB, Katinas G, Cornélissen G, Watanabe Y, Chibisov SM, Shastun SA, Frolov VA. Why decades-long chronobiologically interpreted blood pressure and heart rate monitoring (C-ABPM) coperiodisms toward a chronousphere? Vestnik PFUR, seria Meditsina 2012 (#1):27–36.
33. Halberg F, Halberg E, Gully RJ. Effects of modifications of the daily routine in healthy subjects and in patients with convulsive disorder. Epilepsia (Third Series) 1953;2:150.
34. Halberg F, Howard RB. 24-hour periodicity and experimental medicine. Example and interpretations. Postgrad Med 1958;24:349–58.
35. Halberg F, Bittner JJ, Gully RJ, Albrecht PG, Brackney EL. 24-hour periodicity and audiogenic convulsions in I mice of various ages. Proc Soc Exp Biol Med (NY) 1955;88:169–73.
36. Halberg F, Jacobson E, Wadsworth G, Bittner JJ. Audiogenic abnormality spectra, 24-hour periodicity and lighting. Science 1958;128:657–8.
37. Halberg F, Bittner JJ, Smith D. Belichtungswechsel und 24-Stundenperiodik von Mitosen im Hautepithel der Maus. Z Vitamin-, Hormon- u Fermentforsch 1957;9:69–73.
38. Halberg F, Albrecht PG, Barnum Jr CP. Phase shifting of liver-glycogen rhythm in intact mice. Am J Physiol 1960;199:400–2.
39. Cornélissen G, Halberg J, Halberg F, Sanchez de la Pena S, Nelson W, Schwartzkopff O, Stoynev A, Haus E. Schedule shifts, cancer and longevity: good, bad or indifferent? J Exp Ther Oncol 2008;7(4):263–74.
40. Halberg F, Halberg E, Barnum CP, Bittner JJ. Physiologic 24-hour periodicity in human beings and mice, the lighting regimen and daily routine. In: Withrow RB, editor. Photoperiodism and related phenomena in plants and animals. Ed. Publ. No. 55. Washington, DC: AAAS; 1959. p. 803–78.
41. Kleitman N. Biological rhythms and cycles. Physiol Revs 1949;29:1–30.
42. Selye H. The general adaptation syndrome and the diseases of adaptation. J Clin Endocrinol 1946;6:117–230.
43. Schwarz E. Die Lehre von der allgemeinen und örtlichen "Eosinophilie". Ergebn d allg Path u path Anat 1913;17(Part 1):137–789.
44. Dalton AJ, Selye H. The blood picture during the alarm reaction. Folia Hematologica 1939;62:397–407.
45. Rud F. The eosinophil count in health and in mental disease: a biometrical study. Acta psychiat et neurol 1947; (Suppl 40) 443 pp.
46. Speirs RS, Meyer RK. The effects of stress, adrenal and adrenocorticotropic hormones on the eosinophil levels of mice. Endocrinology 1949;45:403–29.
47. Halberg F, Visscher MB. Regular diurnal physiological variation in eosinophil levels in five stocks of mice. Proc Soc Exp Biol Med (NY) 1950;75:846–7.
48. Halberg F. Changes in eosinophil count of mice with venisections repeated at intervals of several days. Proc Soc Exp Biol Med (NY) 1953;82:160–2.

49. Cornélissen G, Halberg F. Introduction to chronobiology. Medtronic Chronobiology Seminar #7, April 1994, 52 pp. (Library of Congress Catalog Card #94-060580; http://halbergchronobiologycenter.umn.edu)
50. Aschoff J. Speech after dinner. Capri Symposium on Timing and Toxicity. In: Aschoff J, Ceresa F, Halberg F, editors. Chronobiological aspects of endocrinology. Stuttgart: F.K. Schattauer Verlag; 1974/Chronobiologia 1974;1(Suppl. 1):483–95.
51. Halberg F. Beobachtungen über 24 Stunden-Periodik in standardisierter Versuchsanordnung vor und nach Epinephrektomie und bilateraler optischer Enukleation, 20th meeting of the German Physiological Society, Homburg/Saar, September, 1953. Berichte über die gesamte Physiologie und experimentelle Pharmakologie (Berichte über die gesamte Biologie, Abteilung B) 1954;162:354–5. [English translation of Aschoff's remark in Halberg F +. Transdisciplinary unifying implications of circadian findings in the 1950s. J Circadian Rhythms 2003;1:2. 61 pp. www.JCircadianRhythms.com/content/pdf/1740-3391-1-2.pdf, p. 23].
52. Halberg F. Chronobiology: methodological problems. Acta Med Rom 1980;18:399–440.
53. Halberg F, Cornélissen G, Otsuka K, Katinas G, Schwartzkopff O. Essays on chronomics spawned by transdisciplinary chronobiology: witness in time: Earl Elmer Bakken. Neuroendocrinol Lett 2001;22:359–84.
54. Halberg F, Nelson W, Runge WJ, Schmitt OH, Pitts GC, Tremor J, Reynolds OE. Plans for orbital study of rat biorhythms. Results of interest beyond the Biosatellite program. Space Life Sci 1971;2:437–71.
55. Sanchez de la Peña S, Halberg F, Halberg E, Ungar F, Cornélissen G, Sanchez E, Brown G, Scheving LE, Yunis EG, Vecsei P. Pineal modulation of ACTH 1–17 effect upon murine corticosterone production. Brain Res Bull 1983;11:117–25.
56. Sanchez de la Peña S, Halberg F, Schweiger H-G, Eaton J, Sheppard J. Circadian temperature rhythm and circadian-circaseptan (about 7-day) aspects of murine death from malaria. Proc Soc Exp Biol Med 1984;175:196–204.
57. Halberg E, Halberg F. Chronobiologic study design in everyday life, clinic and laboratory. Chronobiologia 1980;7:95–120.
58. Halberg F. Physiologic 24-hour rhythms: a determinant of response to environmental agents. In: Schaefer KE, editor. Man's dependence on the earthly atmosphere. New York: Macmillan; 1962. p. 48–98.
59. Ertel RJ, Halberg F, Ungar F. Circadian system phase-dependent toxicity and other effects of methopyrapone (SU-4885) in mice. J Pharmacol Exp Ther 1964;146:395–9.
60. Halberg F. Organisms as circadian systems; temporal analysis of their physiologic and pathologic responses, including injury and death. In: Symposium on Medical Aspects of Stress in the Military Climate, Walter Reed Army Institute of Research (Col. William D. Tigertt, Medical Corps, Director and Commandant), Walter Reed Army Medical Center (Maj. Gen. A.L. Tynes, Medical Corps, Commanding), Washington, DC, 22–24 Apr 1964. Washington, DC: U.S. Government Printing Office 1965-778-714; 1965. p. 1–36.
61. Halberg F, Cornélissen G, Spector NH, Sonkowsky RP, Otsuka K, Baciu I, Hriscu M, Schwartzkopff O, Bakken EE. Stress/strain/life revisited. Quantification by blood pressure chronomics: benetensive, transtensive or maletensive chrono-vasculo-neuro-immuno-modulation. Biomed Pharmacother 2003;57 Suppl 1:136s–63s.
62. Maschke C, Harder J, Cornélissen G, Hecht K, Otsuka K, Halberg F. Chronoecoepidemiology of "strain": infradian chronomics of urinary cortisol and catecholamines during nightly exposure to noise. Biomed Pharmacother 2003;57 Suppl 1:126s–35s.
63. Halberg F, Cegielski N, Cornélissen G, Ilyia E, Rosch P, Hillman D, Schwartzkopff O, Singh RB, Revilla M, El-Khoury M, De Meester F. Timing nutrition makes the difference concerning body weight and survival: sampling requirements for a 7-day/24-hour circadian endocrine stress–strain test for nutritionists interested in a premetabolic syndrome. In: Proceedings, 15th Congress of Clinical Nutrition, El Sakhna, 19–22 Sept 2010. p. 26–34.
64. Halberg F. Physiologic 24-hour periodicity; general and procedural considerations with reference to the adrenal cycle. Z Vitamin-, Hormon-u Fermentforsch 1959;10:225–96.

65. Halberg F, Visscher MB, Bittner JJ. Relation of visual factors to eosinophil rhythm in mice. Am J Physiol 1954;179:229–35.
66. Costella JF, Halberg F, Hillman D, Mikulecky M, Cornélissen G. Four circadian and two circasemidian periods in sleep-wakefulness of a man on a self-selected routine. Folia Anthropologica 2012;11:51–3.
67. Halberg F, Cornélissen G, Hillman D, Ilyia E, Cegielski N, el-Khoury M, Finley J, Thomas F, Brandes V, Kino T, Pappadoupoulou A, Chrousos GP, Costella JF, Mikulecky M. Multiple circadian periods in a lady with recurring episodes of adynamic depression: case report. In: Halberg F, Kenner T, Fiser B, Siegelova J, editors. Noninvasive methods in cardiology. Brno: Faculty of Medicine, Masaryk University; 2011. p. 45–67.

Chapter 2
From Chronomics to Chronoastrobiology: Many Rhythms Are Control Information for Whatever We Do

Abstract Franz Halberg (1919–2013) developed chronobiology and founded the field of chronomedicine including chronomics, chronoastrobiology, and chronobioethics. His work is summarized in his over 3,600 scientific publications, in cooperation with colleagues from around the world. Progress in chronomedicine depends on how his research developed.

Several international meetings have revealed an accumulating body of reference values for well-established about-daily and about-yearly rhythms of photic origin. Evidence was also provided for about-7-day, about-27-day, about-half-yearly, about-10.5 and about-21-yearly, and even about-50-yearly rhythmicities in us as well as around us, as invisible nonphotic heliogeophysical signatures possibly built into individuals and/or populations. In time series, biological or other, the characteristics of rhythms, chaos, and trends can all be quantified as elements of structures, called chronomes. Chronome mapping with outcomes could eventually serve an individualized optimization of lifestyle, for chronoprevention and chronotherapy as well as for inquiries into the evolution and future of life. The historical path of Franz Halberg's research leading to chronomedicine is presented herein.

Keywords Chronotherapy • Remove-and-replace approach • Chronome • Chronomics • Environmental counterpart

2.1 Back to My First Encounter with the Adrenocortical Cycle

At only one time in history, the earliest 1950s, adrenalectomy could be carried out on patients [1; cf. 2], in the wake of the euphoria from cortisone, making lame people walk again. Thus, a remove-and-replace approach taught us (Fig. 2.1) that the 24-hour rhythm in serum iron was altered but persisted (ACI: adrenocortical insufficiency), while the rhythm in circulating blood eosinophil counts [1] (and in phospholipid labeling and cell division of adrenalectomized mice) seemed to be eliminated or altered by removing the adrenal (in a 2-timepoint approach) [1]. Kaine et al. [2] found that a periodic administration of corticosteroid re-induced the rhythm in humans without adrenals.

K. Otsuka et al., *Chronomics and Continuous Ambulatory Blood Pressure Monitoring*, DOI 10.1007/978-4-431-54631-3_2

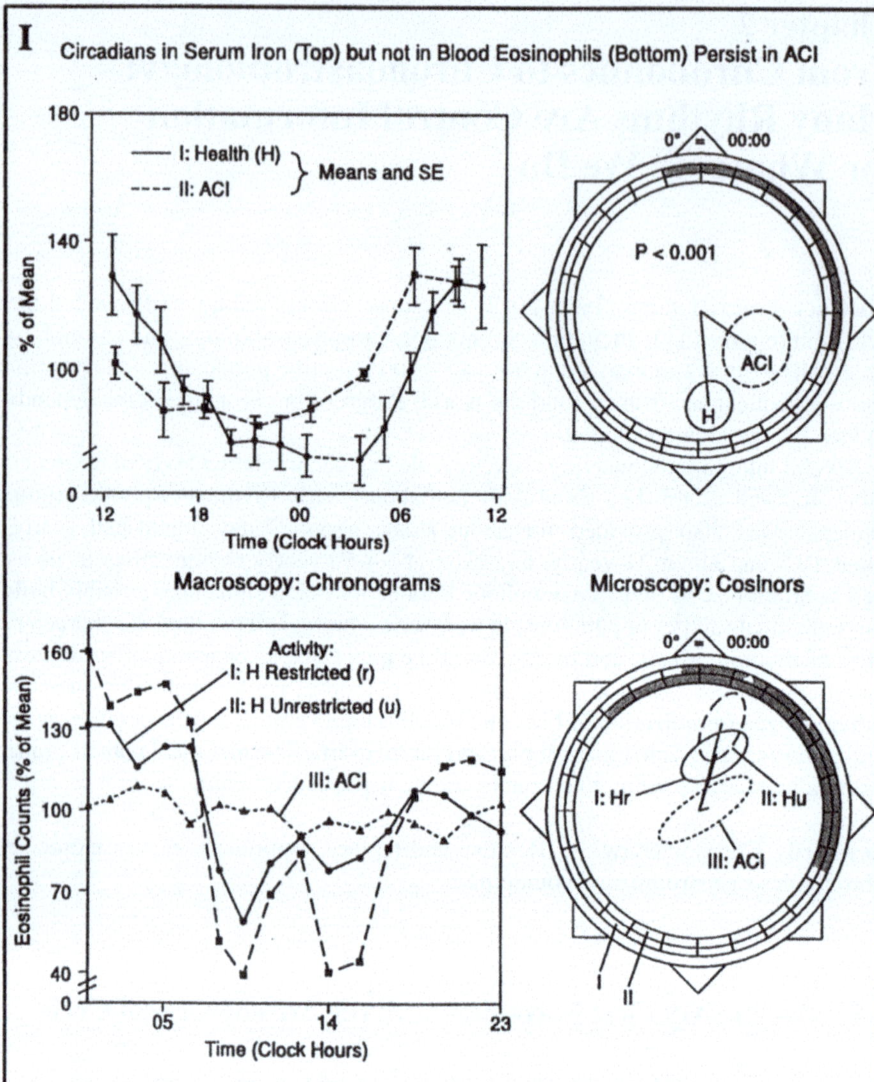

Fig. 2.1 Chronobiologic self-experimentation extended to groups of medical students and staff served to find that the adrenal glands are an important source of circadian rhythmicity in the circulating eosinophil count. For instance, a circadian component of variation could not be demonstrated for eosinophil counts of patients with adrenocortical insufficiency (*bottom*), whereas in healthy people with either restricted or unrestricted activity, this rhythmicity was not only demonstrable but also amplified by enhanced motor activity. Whereas for patients with Addison's disease, a circadian rhythm in eosinophil counts could not be statistically validated, circadians persisted, albeit altered, for the case of serum iron ($P<0.001$) (*top*). As subsequently observed in the case of bilateral, histologically validated ablations of the suprachiasmatic nuclei, some rhythms persist; others may be altered but are not lost in remove-and-replace approaches studied thus far (© Halberg Chronobiology Center)

2.2 Life or Death from the Same Stimulus, Including Drugs: Chronotherapy

Circadian susceptibility rhythms accounting for the difference between survival and death from the same stimulus, e.g., bell ringing, by 1955, and then X-irradiation in mice [3] and a long series of drugs, became the basis of chronotherapy. By 1977, with B.D. Gupta and Akhil Deka, the 2-year disease-free survival rate of patients with perioral cancers (receiving radiation treatment at the time of their peak tumor temperature) could be doubled, Fig. 2.2a, b [4–6]. Susceptibility underwent rhythmicity along both the 24-hour and the 7-day (circaseptan) scale in vivo and in vitro (Fig. 2.3), in the hands of Waldemar Ulmer, who accounted for the circaseptan cycle

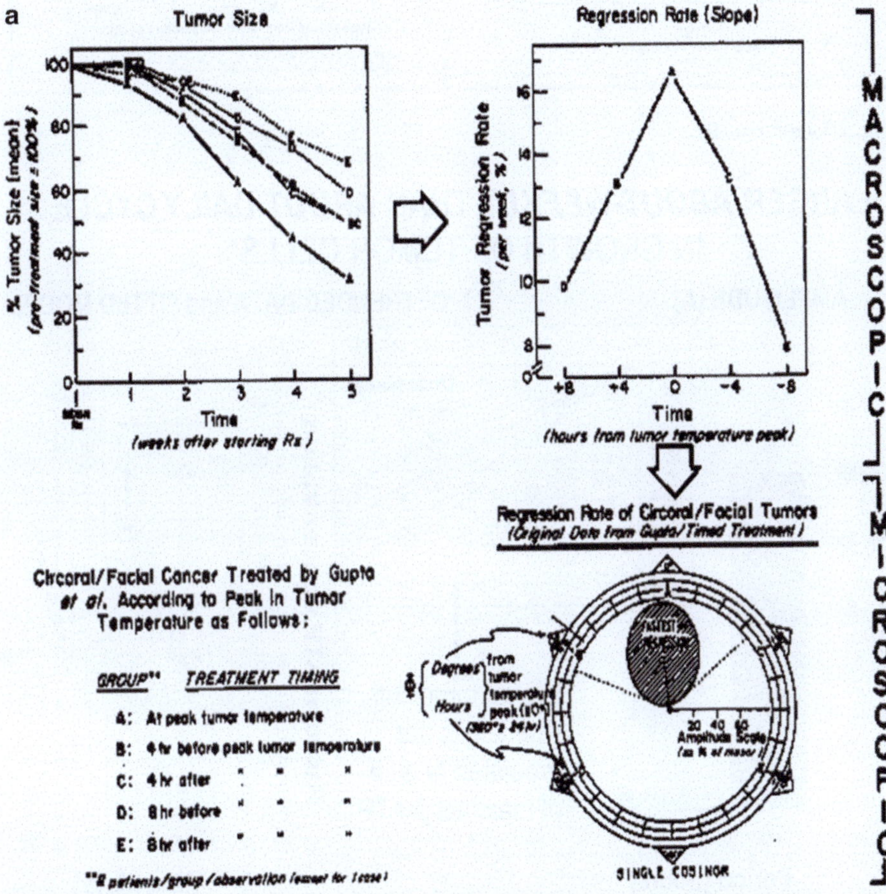

Fig. 2.2 (**a**) Tumor regression rate is circadian stage-dependent in response to X-irradiation (© Halberg Chronobiology Center). (**b**) Treatment timing at peak tumor temperature leads to faster tumor regression (*left*) and to more than doubling of the 24-month disease-free survival rate (0 on abscissa, *right*), as compared to reference groups treated 4 or 8 hours before or after the tumor temperature peak (+8, +4, –4, –8 on abscissa, *right*) or "as usual" (last column, *right*). Studies implemented by B.D. Gupta and Akhil Deka, designed during a PL-480 program sponsored by the USA (© Halberg Chronobiology Center)

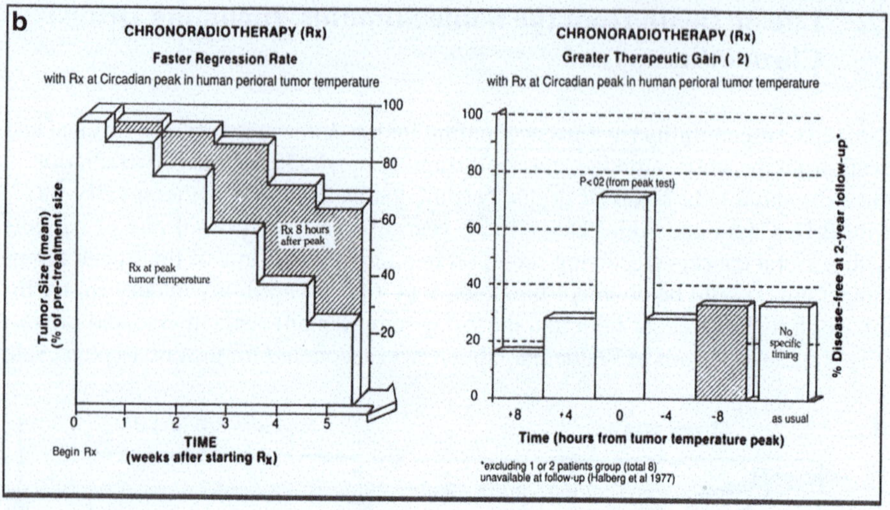

Fig. 2.2 (continued)

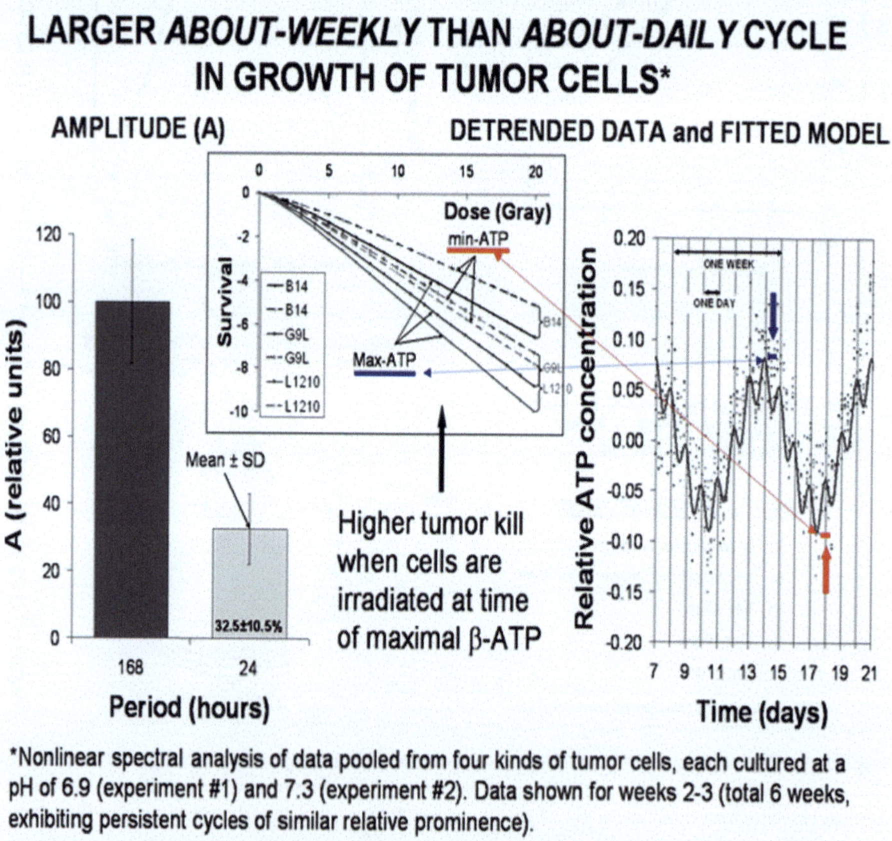

Fig. 2.3 Circaseptan optimization in a 9 L glioma rat tumor model may be even more important than circadian chronotherapy (© Halberg Chronobiology Center)

at the atomic [7] and quantum [8] as well as macroscopic levels, the latter on the right of Fig. 2.3, while therapeutic benefit from treating at maximal β-ATP used as a marker is seen in the middle of this figure (in vitro).

2.3 Cellular Timing

By 1958, a combination of methods, including radioactive tracers, differential centrifugation, and wet chemistry, enabled the demonstration of circadian rhythms in the formation of RNA and DNA [9], with RNA synthesis preceding that of DNA, and growth hormone acting in just one stage of this first mapped circadian cell cycle [10, 11]. In 1960, this cell cycle was reported and depicted as a basic complement to the hypothalamus, Fig. 2.4 [3], as accepted today but denied at the time and for several decades thereafter. Thus, against the prevailing bias of the vast majority of a Naito meeting published in 1979 [12], it was reemphasized that circadian timing resides in peripheral cells as well as in the suprachiasmatic nuclei (SCN) of the hypothalamus [13, 14]. Unless the data were interpreted by visual inspection alone [15, 16], SCN removal (as in the case of adrenal removal) eliminated some rhythms (in motor activity and water drinking) but not those in core temperature, alcohol drinking, corneal mitosis, or thymidine labeling in the gastrointestinal tract, Fig. 2.5a, b. A report on the loss of the circadian rhythm in core temperature was

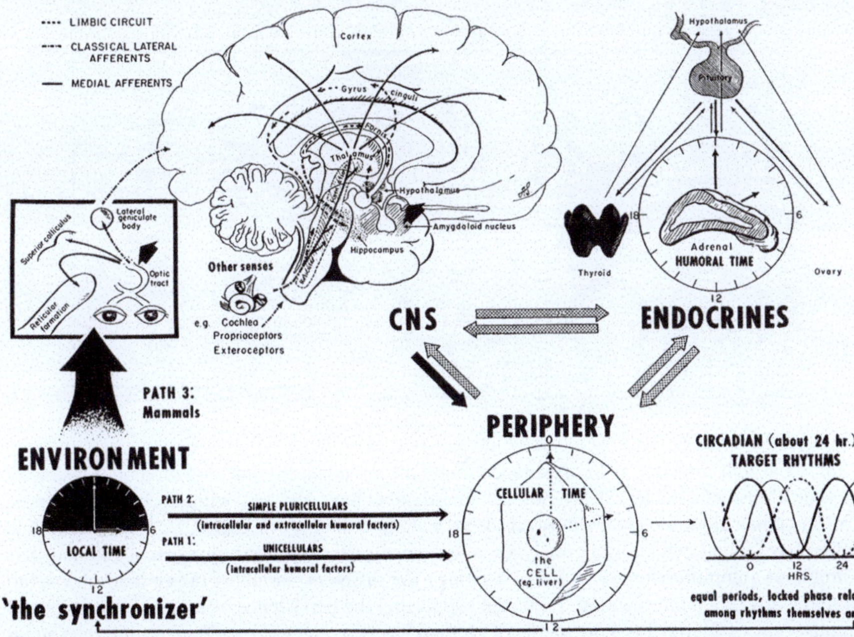

Fig. 2.4 Sketch of factors and pathways known or hypothesized to participate in frequency synchronization *among* circadian rhythms themselves, as well as in synchronization *between* rhythm(s) and environmental synchronizer(s). Emphasis upon the cell as a documented source of rhythmicity by 1960 [3, 9] (© Halberg Chronobiology Center)

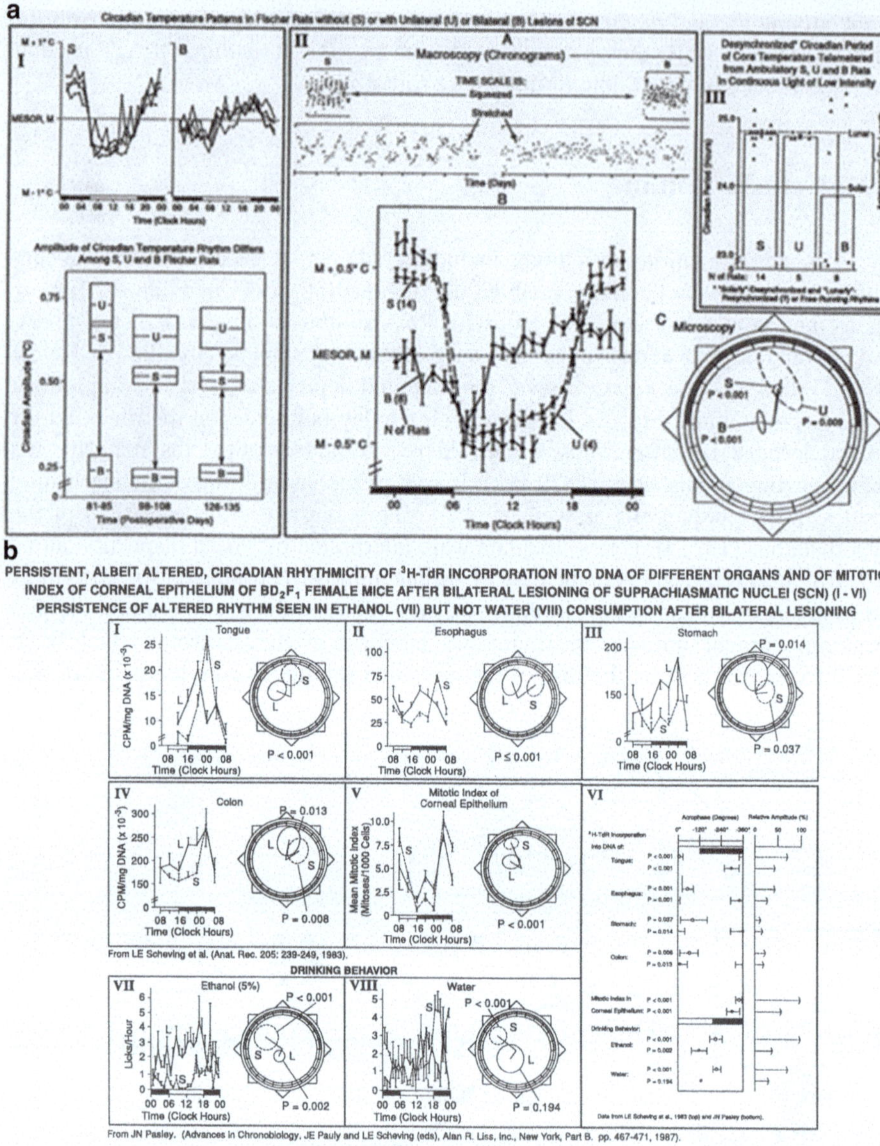

Fig. 2.5 (**a**) Much controversy has revolved around the circadian rhythm alteration after lesioning of the suprachiasmatic nuclei (SCN). Several investigators have claimed, without cosinor analyses, that all circadian rhythms are obliterated after bilateral lesioning of the SCN; others including ourselves have found, with cosinor analyses, a rhythm alteration to a differing extent and in different directions (amplification as well as damping) for different variables rather than an overall obliteration after such intervention. Evidence supporting the persistence of rhythms after bilateral lesioning of the SCN illustrates the merits of chronobiometric analyses to obtain a valid interpretation of the data. This figure shows the circadian temperature patterns in Fischer rats, without or with unilateral or bilateral lesions of the SCN. A visual inspection of the original core temperature

revised when the data were reanalyzed [15 vs. 16]. An amplification of thymidine labeling was found in the stomach after bilateral SCN ablation, Fig. 2.5b, and in core temperature after unilateral ablation, Fig. 2.5a (lower row), the latter finding interpreted as subtractive coupling between the two nuclei [13].

Along a different line of thought, removal of the SCN suggests its mediation of earth tidal effects by the loss of the 24.8-hour (double tidal) period in core temperature of rats kept in continuous light of low intensity, Fig. 2.5a-III. This 24.8-hour component was also found in the self-selected sleep pattern of a healthy man during 3 years, Table 2.1, and in the longest (>260-day) human isolation in a cave study [17, 18], Fig. 2.6, among many others, shorter but important studies of human isolation in shielded and unshielded bunkers by Ruetger Wever [19]. He recorded periods that could have been interpreted, as close to 24.8 hours.

2.4 Prokaryotes

By 1961, a circadian cycle shorter than 24 hours (of ~21 hours) was found in growth or colony advance of a prokaryote [20], Fig. 2.7a–d, and extended in scope [21], but was ignored for subsequent decades by those "naturalists" for whom periodograms and power spectra were (in the words of a personal communication by a sympathetic committee member) "too much mathematics." Not realizing that ruling out a rhythm is difficult with the unaided eye (it requires time series analysis), a committee erroneously decreed, and its decision was accepted by many others for many

Fig. 2.5 (continued) data (IIA) may erroneously convey the impression that rhythmicity clearly seen in the series on the left is abolished in the series on the right. The use of simple chronobiologic techniques such as a plexogram (i.e., the stacking of the data over an idealized anticipated cycle of 24 hours) shows that there is a persisting circadian pattern, albeit an altered one, with a reduced amplitude and an advanced phase, IIB and C. This is also shown in curves obtained from several different animals (IC and D). Whereas the circadian amplitude of core temperature of rats with a histologically validated bilateral lesioning of the SCN is much reduced, it is noteworthy that in rats with a unilateral lesion of the SCN, the amplitude is actually larger than that of sham-operated rats. Evidence supporting the influence of planetary-interplanetary phenomena is provided by the fact that sham-operated rats and rats with a unilateral lesion of the SCN kept in continuous light of low intensity have a circadian period of core temperature that is desynchronized from 24 hours, with an average period close to the "lunar" day (or rather one of the earth's double tidal periods); this is not the case for rats with a bilateral lesioning of the SCN which can have either a shortened or lengthened circadian period (III) (© Halberg Chronobiology Center). (**b**) Persistence of the circadian rhythmicity of the ^{3}H-TdR incorporation into DNA of different organs and of the mitotic index of the corneal epithelium of BD_2F_1 female mice after bilateral lesioning of the SCN has also been extensively demonstrated by Scheving (I-VI) [14] and Pasley (VI and VII). For animals with a bilateral lesion of the SCN, a cosinor on Pasley's data could not validate a circadian rhythm in the drinking of water, but demonstrated it ($P=0.002$) for the drinking of 5 % ethanol (VI and VII). The SCN is not the sole or primary pacemaker of the entire circadian system (© Halberg Chronobiology Center)

Table 2.1 On a self-selected schedule (JFC), a double tidal pull competes with 24-hour environment while initially coexisting free-running circadian fades (global analysis)

	Calendar date		Period	Amplitude	DT-A/S-A (%)	
	All (3-years) data	Double tidal (DT)	24.836 (24.833,24.838)	0.10606 (0.0854,0.1268)		*Global*
	02/19/1990	Compromise	24.432 (24.427,24.436)	0.05643 (0.0357,0.0772)		
	to 03/11/1993	Free-running	24.260 (24.252,24.268)	0.03253 (0.0118,0.0533)		
		Societal (S)	24.001 (24.000,24.001)	0.25443 (0.2337,0.2751)	*0.41685*	
		Tidal	12.414 (12.410,12.418)	0.01728 (–0.0034,0.038)		
		Semidian	11.999 (11.997,12.001)	0.03204 (0.0113,0.0527)		
Local	1: 02/19/1990	Double tidal	24.872 (24.830,24.914)	0.02513 (–0.0070,0.0573)		
	to 04/28/1991	Compromise	24.409 (24.384,24.435)	0.04117 (0.0089,0.0734)		
		Free-running	24.222 (24.210,24.235)	0.07921 (0.0468,0.1116)		
		Societal	24.008 (24.004,24.011)	0.31605 (0.2838,0.3483)	*0.07951*	
		Tidal	12.421 (12.411,12.431)	0.02615 (–0.0059,0.0582)		
		Semidian	12.007 (12.001,12.013)	0.04105 (0.0090,0.0731)		
	2: 04/28/1991	Double tidal	24.898 (24.832,24.964)	0.12744 (0.0707,0.1842)		
	to 09/23/1991	Compromise	24.707 (24.671,24.743)	0.25208 (0.1973,0.3068)		
		Compromise	24.488 (24.434,24.541)	0.10274 (0.0552,0.1502)		
		Societal	24.011 (23.992,24.030)	0.22758 (0.1815,0.2736)	*0.55998*	
		Tidal	Did not converge	–		
		Semidian	–	–		
	3: 09/23/1991	Double tidal	24.862 (24.852,24.872)	0.34858 (0.3089,0.3882)		
	to 03/08/1992	Compromise	24.523 (24.468,24.579)	0.06108 (0.0212,0.1009)		
		Free-running	Did not converge	–		
		Societal	23.997 (23.984,24.011)	0.22362 (0.1841,0.2631)	*1.55881*	
		Tidal	12.432 (12.416,12.448)	0.04969 (0.0106,0.0888)		
		Semidian	Did not converge	–		

	4: 03/08/1992	Double tidal	24.942 (24.896,24.989)	0.15133 (0.0960,0.2067)		
	to 07/14/1992	Compromise	24.451 (24.394,24.508)	0.11847 (0.0630,0.1739)		
		Free-running	–	–		
		Societal	23.987 (23.967,24.006)	0.32099 (0.2660,0.3760)	*0.47145*	
		Tidal	12.432 (12.398,12.465)	0.04960 (–0.0051,0.1042)		
		Semidian	11.982 (11.951,12.012)	0.05028 (–0.0044,0.1049)		
	5: 07/14/1992	Double tidal	24.789 (24.782,24.797)	0.31836 (0.2798,0.3569)		
	to 03/11/1993	Compromise	24.474 (24.438,24.510)	0.06218 (0.0236,0.1008)		
		Free-running	–	–		
		Societal	24.009 (23.997,24.022)	0.17892 (0.1405,0.2173)	*1.77934*	
		Tidal	12.352 (12.328,12.376)	0.02358 (–0.0147,0.0619)		
		Semidian	12.001 (11.988,12.013)	0.04128 (0.0030,0.0796)		

The entire available data span was analyzed as a whole (top six rows) and then divided into five subspans according to visual inspection of characteristics of 24-hour and 24.84-hour fits in serial sections, suggesting similarities within the spans and differences among them. Initial period guesses for the six nonlinearly searched periods were 24.84, 24.43, 24.26, 24.00, 12.42, and 12.00 hours. Results were omitted if they did not converge to a period fitted to the data, in which case the analysis was redone after removal of these components from the model, to yield results summarized herein. Some periods converged well, but the CI (95 % conservative confidence interval) of the amplitude A (the CI given in parentheses) covered zero, as apparent from a negative sign inside the last column's second, fifth, and sixth sections, for the case of double tidal and tidal associations in the second section (top) where the societal 24-hour day predominates. The double tidal component converges invariably thereafter. Note that overlaps of zero by the CI of the A are very small. In 2 of 5 sections, the double tidal cycle has the largest A, the ratio of amplitudes of DT over S becomes greater than 1, suggesting that the moon/sun dominate over society in clinical health. These sections are not consecutive, possibly a chance observation, yet in keeping with the possibility that in health there is a wrangling between the double tide and society, as also seen in a patient with twice-yearly depression. From Costella et al. [66]

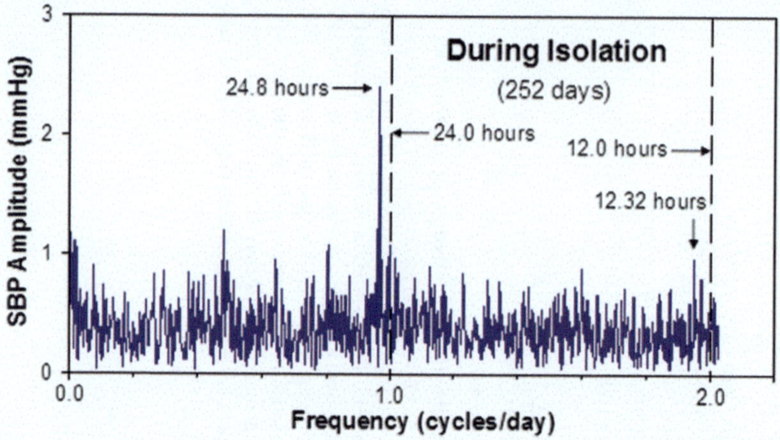

Fig. 2.6 During the longest span of comfortable isolation from a 24-hour schedule of a healthy young woman in a special underground laboratory, but not before or thereafter, a period of 24.8 hours dominantly characterizes her blood circulation (among other peaklets also resolved) upon meta-analysis of original data as a whole (globally) and in serial sections (locally) (© Halberg Chronobiology Center)

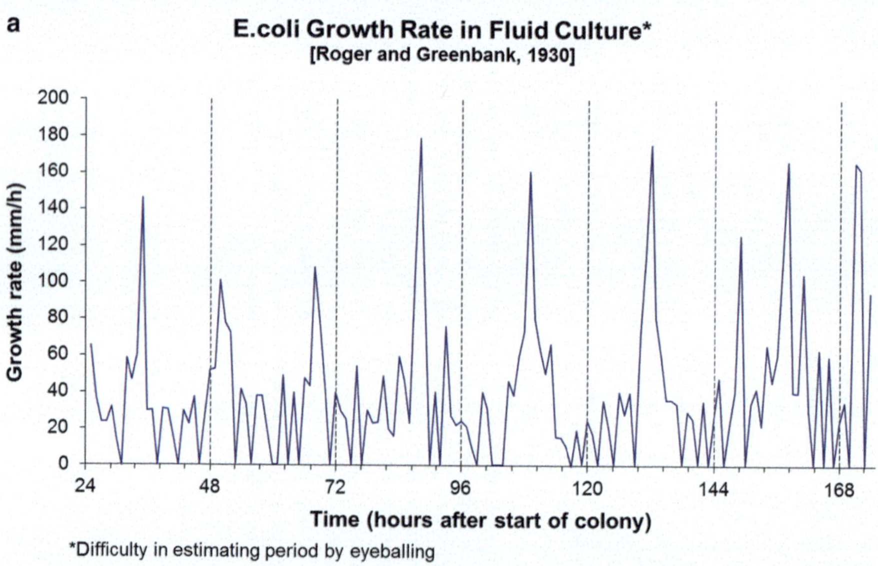

Fig. 2.7 (**a**) Timeplot of the 1930 data of Rogers and Greenbank. Lore Rogers was a noted bacteriologist described by a Cosmos Club (Washington, DC) Vignette of December 1967 as "the bright star in the [US Department of Agriculture] scientific horizon before World War II." In 1961, the data on growth rate (actually colony advance) of *E. coli* in a spiral glass tube, assessed at hourly intervals for slightly over 6 days, were taken off the graph published by Rogers and Greenbank in 1930. Some regularity in the time series is apparent from the spikes in colony growth occurring at intervals shorter than 24 hours, as seen by reference to the vertical dashed lines drawn at daily intervals, yet eyeballing can hardly serve for an estimate of a period and certainly forestalls any evaluation of uncertainties. Rogers and Greenbank wrote: "which is representative of a number of experiments,

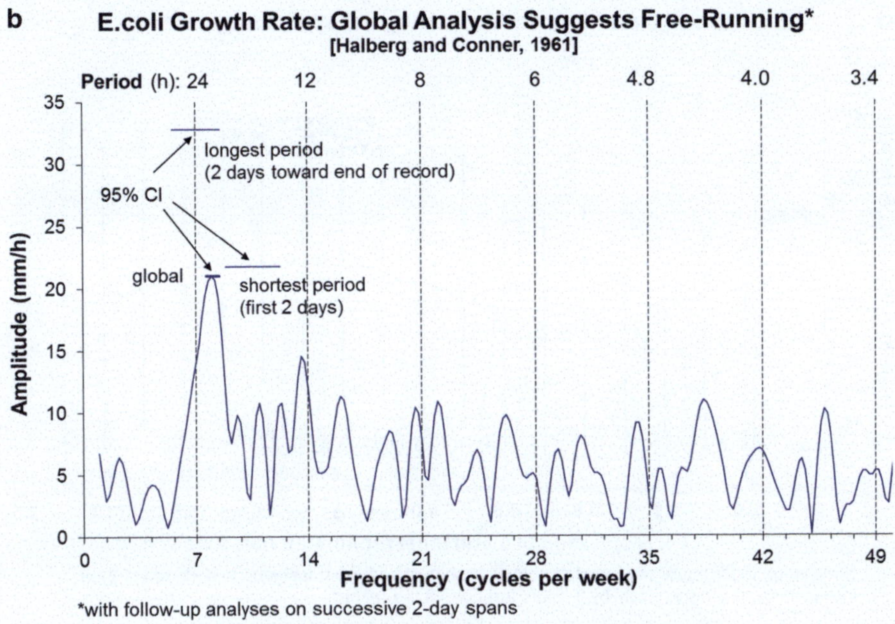

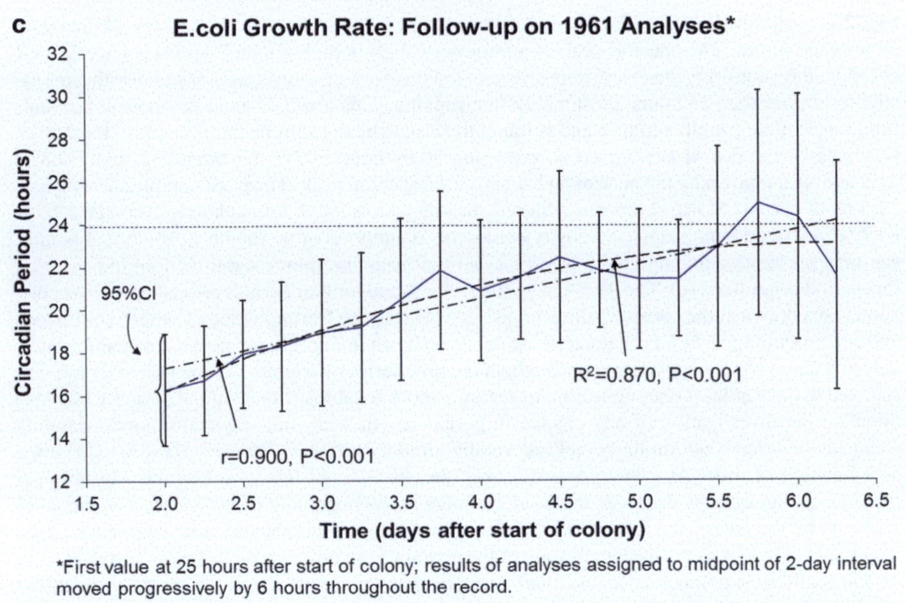

Fig. 2.7 (continued) shows that there is a considerable degree of periodicity to the alternation of rapid and slow growth. In considering this graph, it should be kept in mind that the curve represents not extent of growth, but rate of advance of growth for definite time periods. The number of bacteria in the different periods could not be determined" (© Halberg Chronobiology Center). (**b**) By 1961, Rogers' data on growth rate of *E. coli* (**a**) had been analyzed by a global periodogram and power spectra to find what the naked eye may miss, i.e., what was then interpreted as a free-running circadian rhythm, with a period of about 20.75 hours by Halberg and Conner [20]. An infradian modulation was subsequently reported. Herein, a result of linear-nonlinear rhythmometry

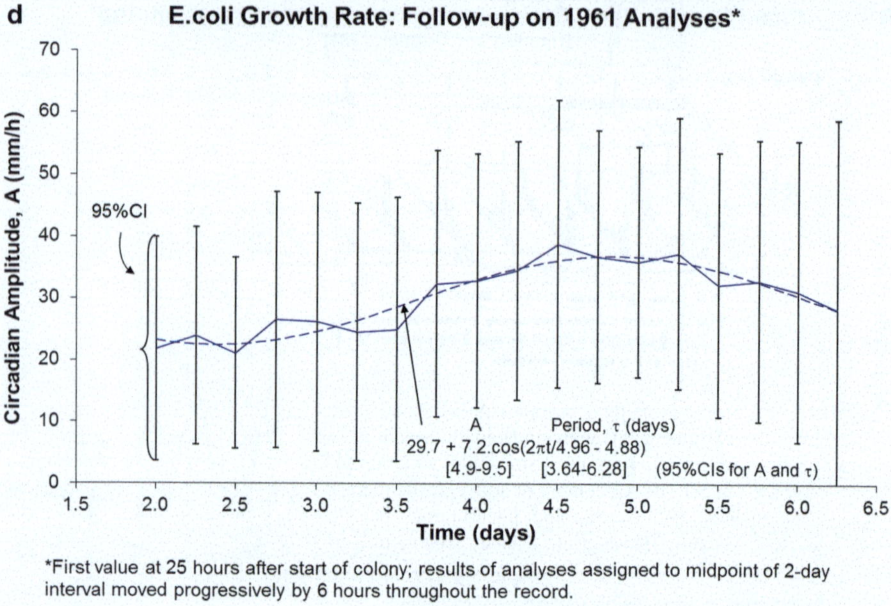

Fig. 2.7 (continued) summarizes period estimates for 2-day sections as well as for all data with their uncertainties. The spectral peak at a frequency slightly higher than 7 cycles per week (or 1 cycle per day) confirms the visual impression of the presence of a circadian variation with a period slightly shorter than 24 hours, presumably free-running. This about-21-hour component accounts for 14.4 % of the overall variance and is found to be statistically significant by cosinor ($P<0.001$). Nonlinearly, the period is resolved as averaging 20.09 hours (95 % CI extending from 19.8 to 21.9 hours), as shown by the horizontal bar above the spectral peak. The peak's amplitude averages 52.6 % of the MESOR (chronome-adjusted mean), with a 95 % CI extending from 19.4 % to 85.7 %, not overlapping zero (thereby rejecting the assumption of no rhythm). Additional nonlinear analyses applied to a moving 2-day interval (nonlinear chronomic serial section) indicate that this period is not fixed (**c**). The 95 % Cls of the shortest and longest periods are also shown as horizontal bars drawn at the corresponding amplitudes (© Halberg Chronobiology Center). (**c**) Periods and corresponding 95 % CIs characterizing the data shown in Fig. 21a are here estimated for intervals of 2 days progressively moved throughout the time series by 6 hours. An interval of 2 days was selected to have at least one replication to secure a more reliable estimate of the period while also allowing the investigation of any change in period length with time. Separate, nonoverlapping 1-day intervals analyzed similarly yield acceptable similar estimates. The period is seen to lengthen as a function of time. As a first approximation, the *dash-dotted line* is in keeping with a linear increase. Being only of the order of about 18 hours at the beginning of the series, it approaches 24 hours by the end of the first week of monitoring. With the qualification that consecutive estimates are not independent, the lengthening of the period has an ordering $P<0.001$ ($r=0.900$). The fit of a second-order polynomial (*solid line*) suggests that the period levels off. In this model, the quadratic term significantly reduces the residual variance (ordering $P<0.02$) beyond the linear term. The results could be interpreted as an originally free-running circadian component that may be pulled by (if not become synchronized with) the 24-hour day, but there is a possible alternative, shown in D (© Halberg Chronobiology Center). (**d**) Amplitude estimates and 95 % CIs corresponding to the changing periods (in **c**) are shown herein. With the same qualification that consecutive amplitude estimates are not independent, further liner-nonlinear analyses of the amplitude sequence indicate the presence of an infradian component with a period of about 5 days (*dashed curve*). The "circadian" amplitude is quite large (of the order of 50 % of the MESOR), being larger in the middle of the

years that circadians were limited to eukaryotes [22] until eventually a flood of papers on rhythms in prokaryotes set the record straight [23]. In 2010, we learned about a rhythm in archaeon, Fig. 2.8 [24].

2.5 Eukaryotes

In eukaryotes, the circadian cycle could be tracked from the adrenal to the cell, the pituitary, and the hypothalamus by 1959 and eventually to the suprachiasmatic nuclei and the pineal, resulting in rhythms of different orders explored with novel designs leading to feedsideward interactions, replacing, whenever feasible, or complementing time-unqualified feedbacks and feedforwards [25].

2.6 Infradians: Chronomics

Free-running repeated itself for the biological week, Fig. 2.9, which in turn led to its physical counterpart in geomagnetics, Fig. 2.10a ([26]; cf. [27, 28]) (as well as to an earlier reported week in rainfall, Fig. 2.10b), starting chronomics with the postulate that for each biological cycle there should be an environmental cycle with a similar period (in the sense of overlapping CIs, 95 % confidence intervals, of their period), and vice versa [29]. A unique set of self-ratings and self-measurements for over 40 years of Robert B. Sothern [30], and shorter records of others, allowed the demonstration of paratridecadal Brückner-Egeson-Lockyer (BEL) cycles with a CI of period near or in the about-30-40-year range. A BEL cycle in environmental temperature could be a first harmonic of the about-60-year climate temperature cycle, described by Enfield and Trimble as a 65–80-year cycle with an 0.4 °C range [31], while Schlesinger and Ramankutty report for 11 geographic regions that "the 65-70-year oscillation is the statistical result of 50-80-year oscillations for the North Atlantic Ocean and the broader Northern Hemispheric continents" ([32; cf. 33]). It is the more interesting to find these periods in neonatal anthropometry, in the length,

Fig. 2.7 (continued) series than at the start. The change in amplitude could be in keeping with a synchronization of the originally free-running component to a 24-hour cycle, but in that case, there should have been a more or less continuous increase until the end of the record. Results from the nonlinear analysis are in keeping with the alternative hypothesis of an infradian modulation of the circadian amplitude with a period of 4.96 days (95 % CI: 3.64–6.28 days). It should be noted that the data were collected under presumably constant conditions, except for the hourly illumination of the colony for taking the measurements. In his 90s, Lore Rogers (the senior author) first wrote in response to an inquiry that he could not recall the conditions of the study. Later he wrote to report that his technician and coauthor G.R. Greenbank (in Florida) had sent him a Christmas card. Rogers subsequently learned from Greenbank about the lack of any known external 24- or 20.75-hour cycle (the latter the average cycle length in a global analysis) (© Halberg Chronobiology Center)

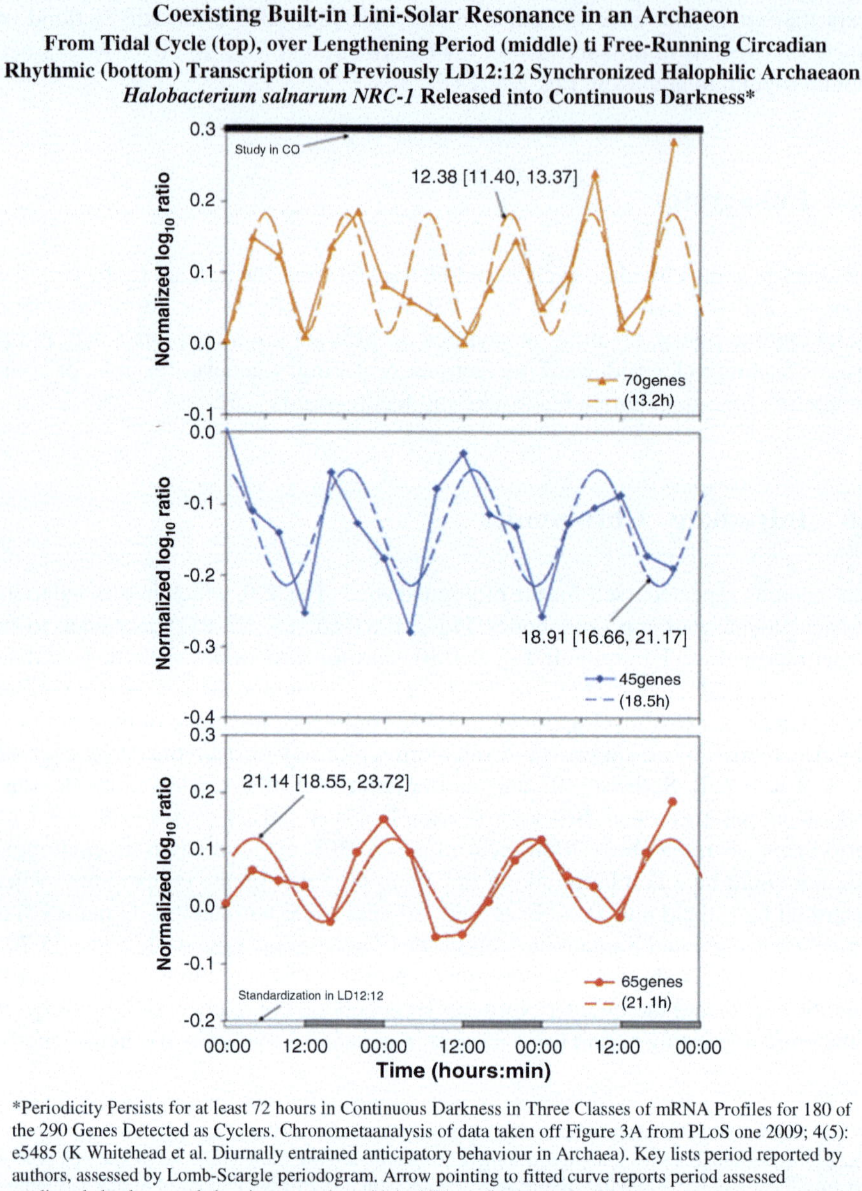

Fig. 2.8 Assessment by the extended cosinor of periods and their 95 % confidence intervals of three classes of mRNA profiles for 180 of the 290 genes detected as cyclers by Whitehead K. et al. (Diurnally entrained anticipatory behavior in Archaea. PLoS ONE 4[5]: e5485. doi:10.1371/journal.pone.0005485) in Archaea kept for at least 72 hours in continuous darkness (© Halberg Chronobiology Center)

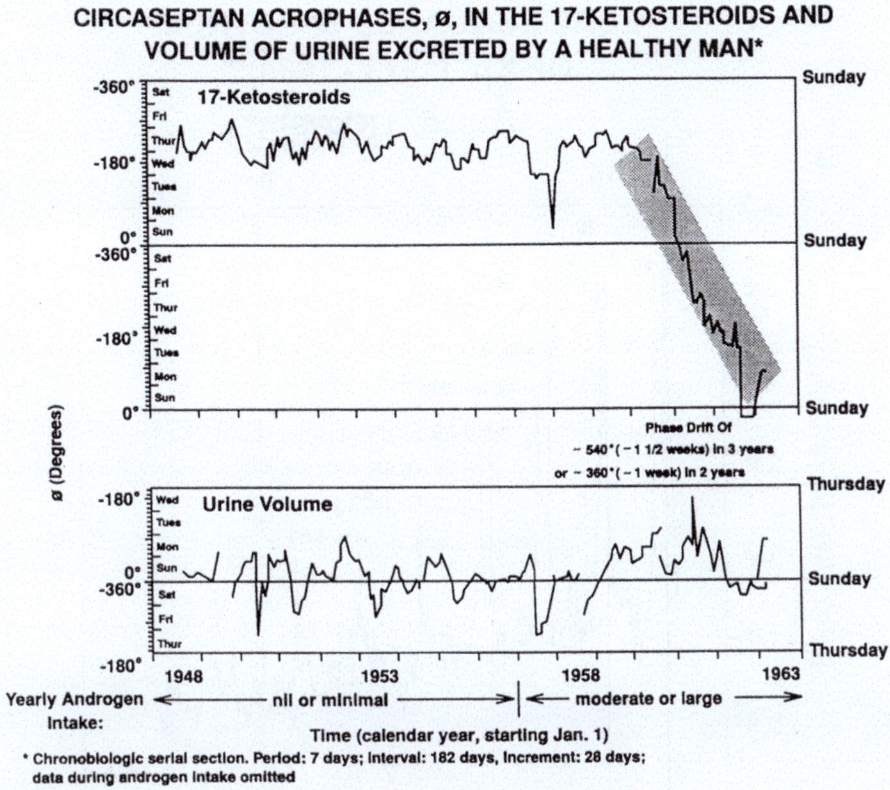

Fig. 2.9 Circaseptan acrophases, ϕ, in the 17-ketosteroids (*top*) and volume of urine excreted by a healthy man, CH (© Halberg Chronobiology Center)

weight, and head circumference at birth of the human neonate [34], Table 2.2, a sensitive magneto- and broader cosmo-meter [35, 36], showing ~50- and ~60-year cycles and one ~80-year cycle, all congruent with the global multidecadal temperature cycle as trans-semicentennians.

These decadal, didecadal, and paratridecadal cycles in longitudinal records of ~6 daily self-measurements (by now for well over 4 decades) [30; cf. 37] were later complemented by automatically collected around-the-clock half-hourly and decades-long blood pressures and heart rates with gaps [38]. Like the manually recorded self-surveillances, the automatic self-surveillances mirrored cycles in space and/or earth weather, Figs. 2.11, 2.12, 2.13, 2.14, and 2.15 (as noted, with their ever-present uncertainties). Moreover, the set of records as a whole suggested a selective assortment of paired periods, coperiodisms, in and around us, that were congruent by virtue of their overlapping CIs (95 % confidence intervals). Whether this assortment is a matter of chance cannot be ruled out, yet it seems important to note the fact, in any event, that different variables and even different characteristics such as the circadian MESOR vs. circadian amplitude of the same variable in the

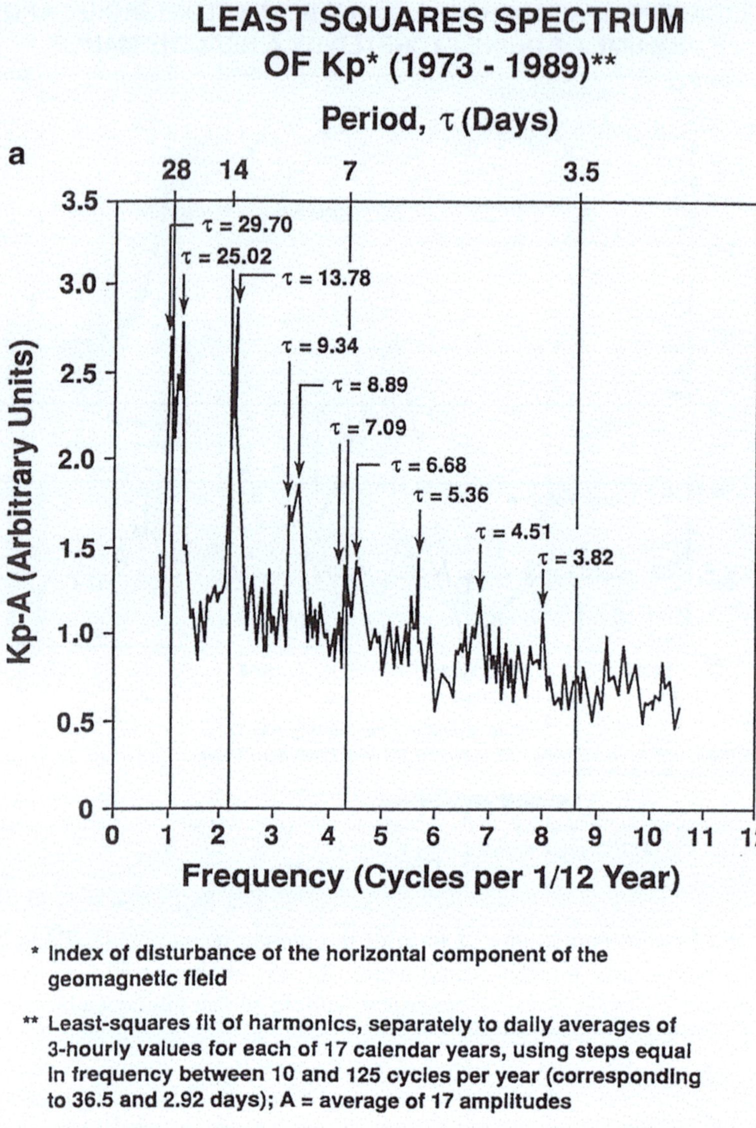

Fig. 2.10 (**a**) Least-squares spectrum of the geomagnetic index (Kp) recorded every 3 hours for 57 years between 1932 and 1990, obtained by a population-mean cosinor spectrum, summarizing results over consecutive years. Large-amplitude peaks at periods of about but not exactly 28 days, 14 days, and 7 days can readily be seen. Although the about-28-day component has long been associated with the relative motion of the moon around the earth, when it comes to geomagnetism, the about-27-day rotation of the sun around its axis has to be considered as a major source of variability in the circatrigintan range. The finding of this natural near week of 6.68 days, and other near weeks (Fig. 2.10b), provided an environmental coperiodism of the societal week (© Halberg Chronobiology Center). (**b**) "We discovered a well-marked period … of 6.6456 days (later improved to 6.6485 days) and also the half of it in the weather of New York and Washington" (Abbot CG. Solar variation and weather, a summary of the evidence, completely illustrated and documented. Washington DC: Smithsonian Miscellaneous Collections 146, No. 3 [Publ. 4545]; 1963; p. 47. Figure 41 on p. 50.)

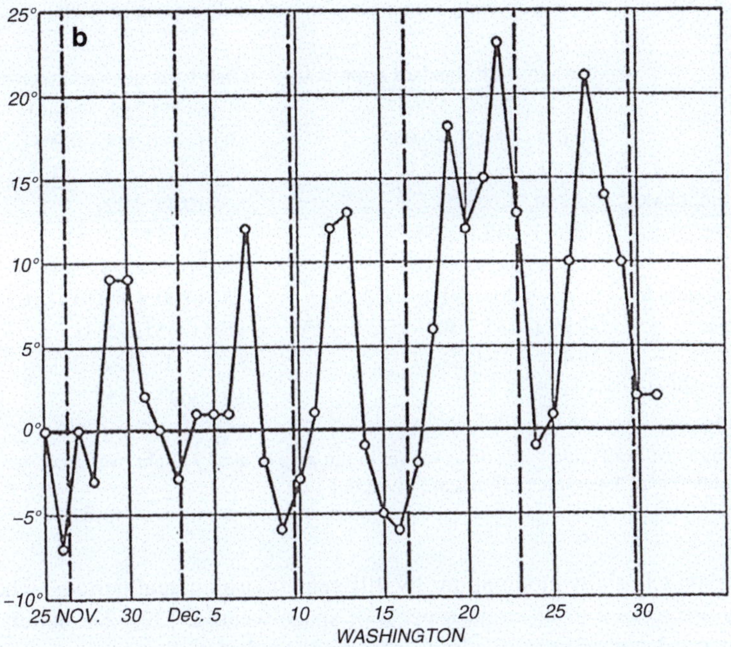

FIG. 41a.—Period of 6.6485 days in Washington, D.C., weather.

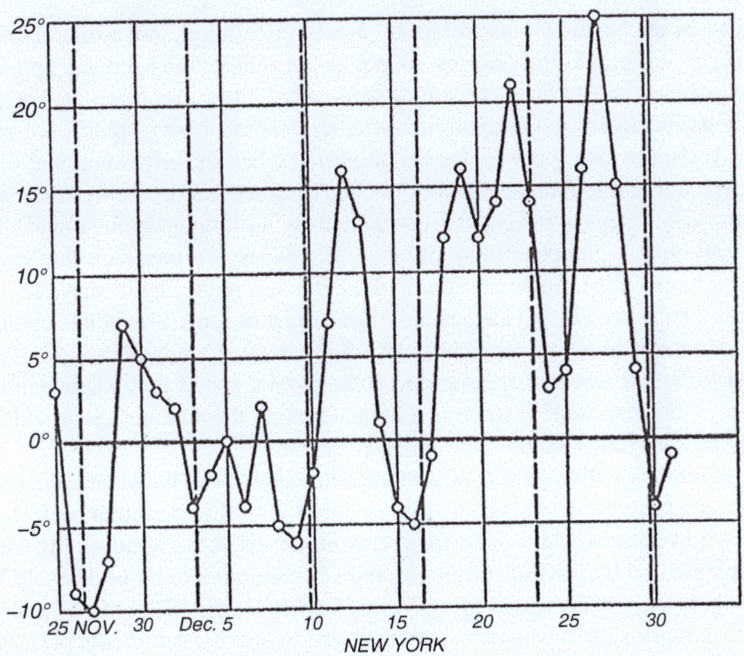

Fig. 2.10 (continued)

Table 2.2 Period(s) in years in body length (cm) at birth in Russia 1874–1985: data of Boris Nikityuk[a]

	Boys (overall standard error: 0.44)	Girls (overall standard error: 0.47)
MESOR	52.224 (51.962, 52.687)	51.564 (51.288, 51.840)
Slope	−0.017 (−0.027, −0.008)	−0.015 (−0.025, −0.005)
Period 1	53.671 (46.12, 61.23)	53.666 (42.00, 65.33)
Double amplitude	1.52 (0.74, 2.28)	1.32 (0.42, 2.22)
Acrophase	−276° (−247, −305)	−274° (−237, −311)
Period 2	32.404 (29.35, 35.13)	33.487 (25.88, 41.09)
Double amplitude	0.76 (−0.00, 1.50) [0.42, 1.08]	0.86 (0.00, 1.74) [0.48, 1.24]
Acrophase	−243° (−187, −300)	−207 (−151, −263)
Period 3	20.441 (18.31, 22.57)	21.217 (19.18, 23.25)
Double amplitude	0.84 (0.56, 1.10)	0.94 (0.62, 1.24)
Acrophase	−184° (−129, −240)	−158° (−109, −208)

[a]95 % confidence limits in ( ) are conservative; in the case of a very slight overlap of zero beyond the second decimal, 1-parameter limits are given in [ ]

same person can show modulations by different environmental periods, whether or not they are related to different aspects of space weather [39]. The latter has too many mapped components to draw inferences to direct causal connections from congruence of periods only.

Congruent cycles in and around us are noteworthy, but do not validate relations. Whenever possible and the cosmos cooperates, a remove-and-replace approach is indicated where the sun does the subtraction and addition (replacement) of a component in its spectrum, and we examine whether the biosphere responds accordingly [37–39]. If the physiological cycles are amplified when the environmental counterpart is detected and is increasing, we conclude that we are driven from outside the organism. When the environmental component is no longer detected and the biospheric component dampens in amplitude yet persists, we conclude that it is built into living matter [38]. Such environmental behavior, including disappearance of a component, be it because it is transiently lost or just obscured by noise, could actually be documented in several cases [37–39] where, to paraphrase Sir William Herschel [40], the sun (the measurable solar wind's speed or sunspot number or area) was "well-disposed" insofar as it cooperated by the subtraction or addition of a weekly or transyearly component in its spectrum.

Myriadennians with a period of 200 million years (MY) between ice ages ([41; cf. 42]) come to mind when cycles of the order of 260 MY are found by Rohde and Muller [43] in the extinction of species repeating itself at the bottom of the sea and are complemented by their uncertainties and Cornélissen's cycle of 37.81 MY with a 95 % confidence region extending from 36.64 to 38.97 MY [44], which corresponds to a major step in water cooling, revealed by marine sediments [41]. Periods in MY in species diversity show the width of the spectrum of biospheric cycles, all sought to be paired eventually with environmental ones, dubbed the Cornélissen-series [45] that awaits much more aeolian mapping. Physical cycles that were

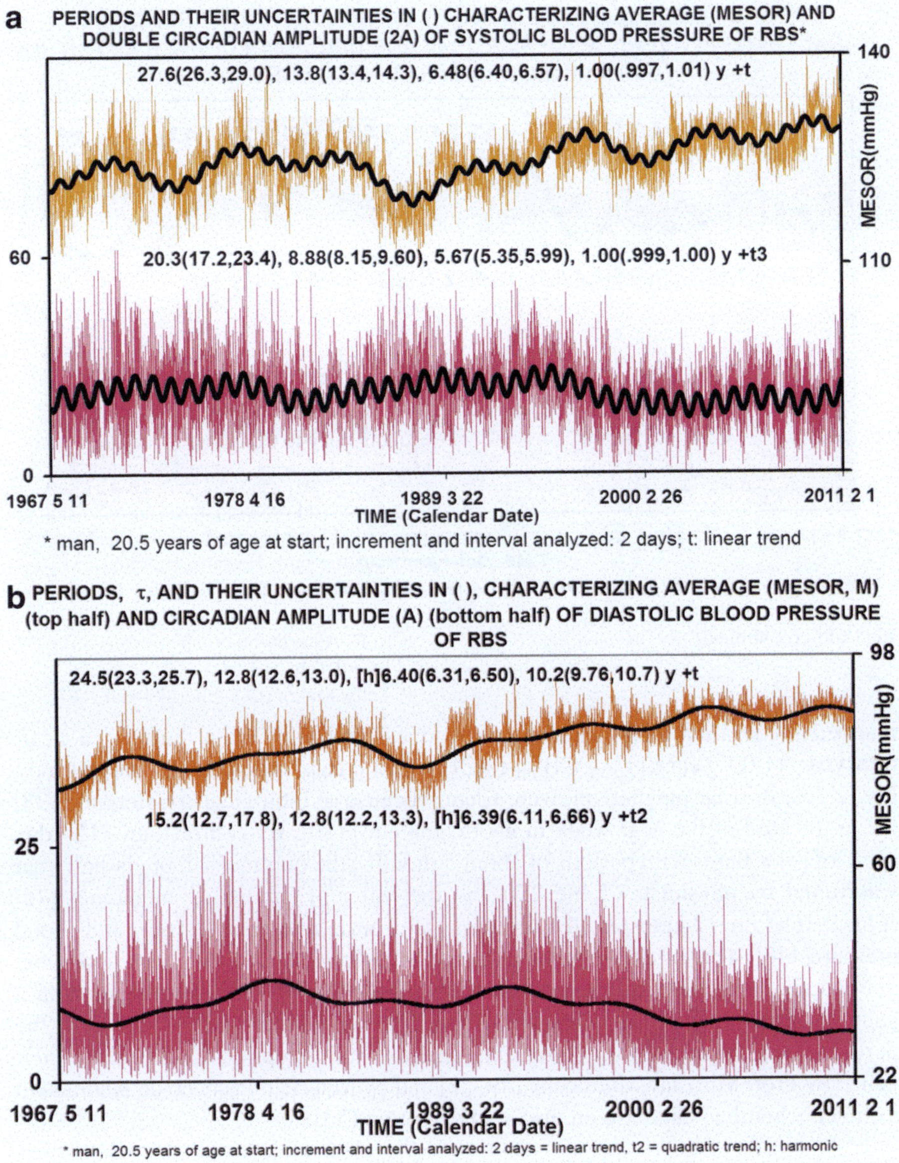

Fig. 2.11 (**a**) When analyzed by the extended cosinor, these two new sets of time series of circadian MESORs, Ms, and double amplitudes, 2As, for systolic blood pressure reveal separate sets of periods during the span investigated in the data of a clinically healthy man, RBS (© Halberg Chronobiology Center). (**b**) When analyzed by the extended cosinor, these two sets of time series of Ms and 2As for diastolic blood pressure of RBS reveal separate sets of periods during the span investigated (© Halberg Chronobiology Center). (**c**) When analyzed by the extended cosinor, these two sets of time series of Ms and 2As for heart rate of RBS reveal separate sets of periods during the span investigated (© Halberg Chronobiology Center)

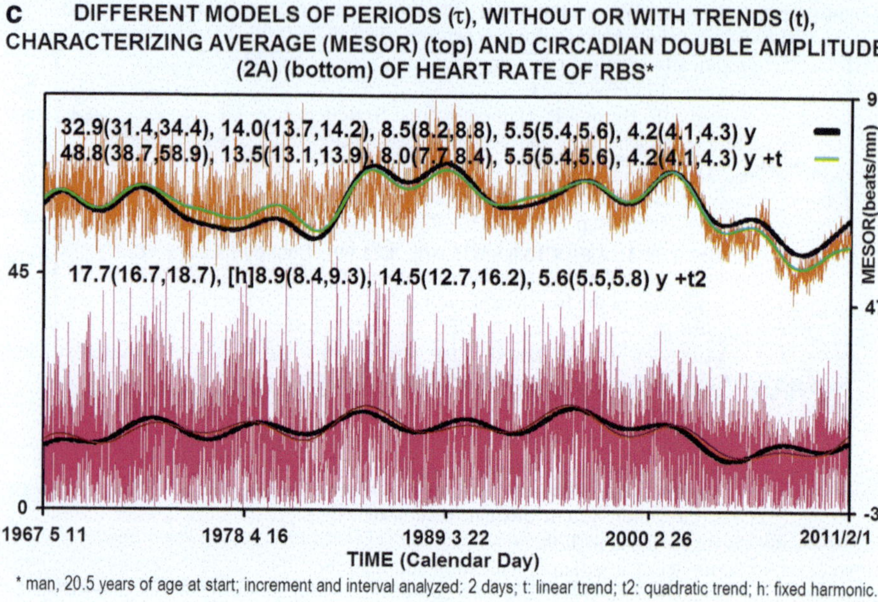

Fig. 2.11 (continued)

mapped in us after their discovery around us, and vice versa, namely, a near-transyear (1.00 year<[τ – CI]<[τ+CI]<1.20 years where τ is the period), Fig. 2.16a, in solar magnetism, were found because of biological counterparts [38] (as in the case of the near week in geomagnetism) [26]. By contrast, an ~154-day cycle of solar flares with periods of about 5 months, the quinmensal, or cis-half year was found by physicists (Table 2.3) and thereafter in circulating melatonin [46] (Fig. 2.16b), in 17-ketosteroid excretion [47], heart rate (Fig. 2.16c), and blood pressure [48], (Fig. 2.16d). Nonstationary infradian periods in space and time, dubbed aeolian, found in the biosphere also include periods much longer than a year, far-transyears (1.20 years≤[τ – CI]<[τ+CI]<1.90 years where τ is the period) ([38]; Fig. 2.17), the latter extended to natality, epilepsy, and stroke incidence by Prof. Miroslav Mikulecky [49–52] and by Mauricio Pacheco de Andrade to a clinical chemical database on uric acid excretion [53].

2.7 When to Eat

Another Minnesota lesson is that the timing of a meal restricted in calories can make the difference between survival vs. death [54] in laboratory mice (without ample fat reserves; but not in rats), in cool temperature with single housing to prevent cuddling. This study was prompted by our experience in India where, among palm trees in Varanasi, in cool (not cold) temperature, I read a newspaper report on

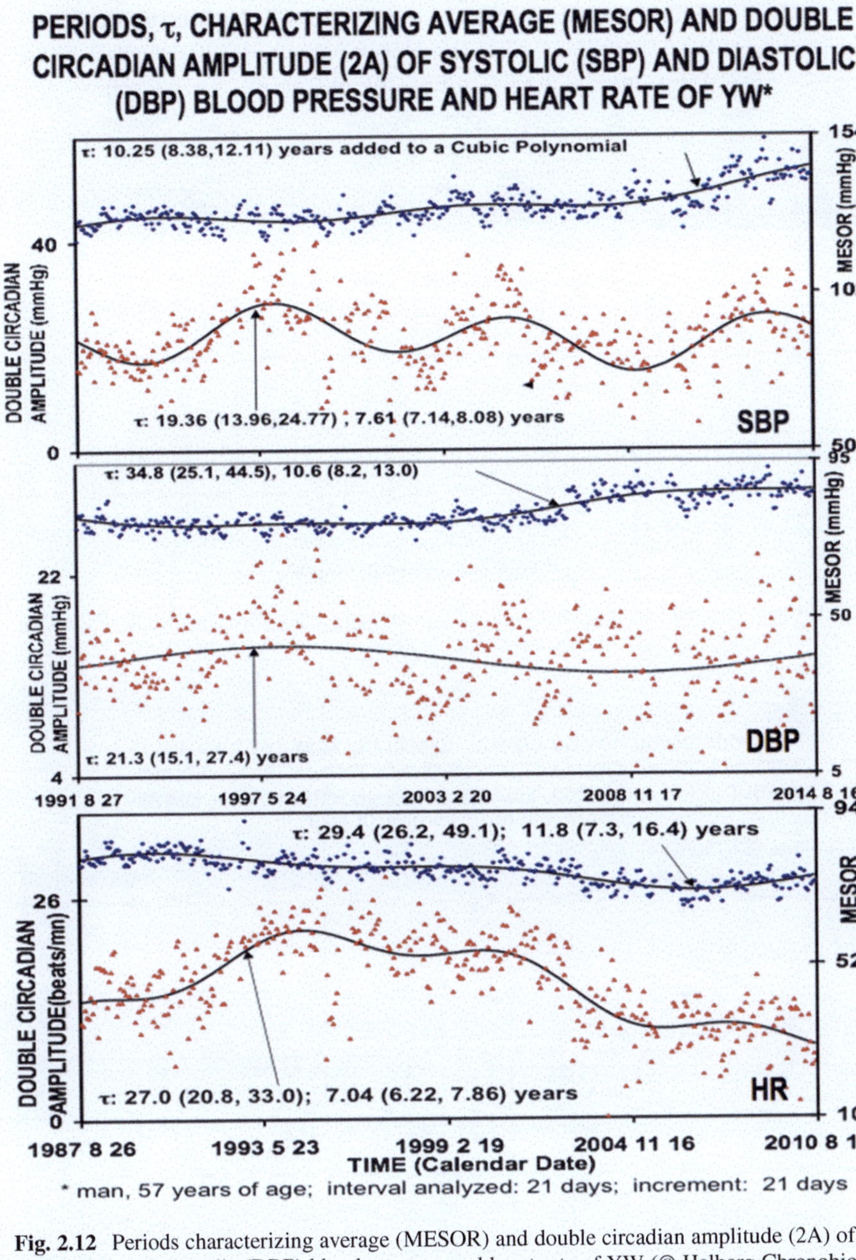

Fig. 2.12 Periods characterizing average (MESOR) and double circadian amplitude (2A) of systolic (SBP) and diastolic (DBP) blood pressure and heart rate of YW (© Halberg Chronobiology Center)

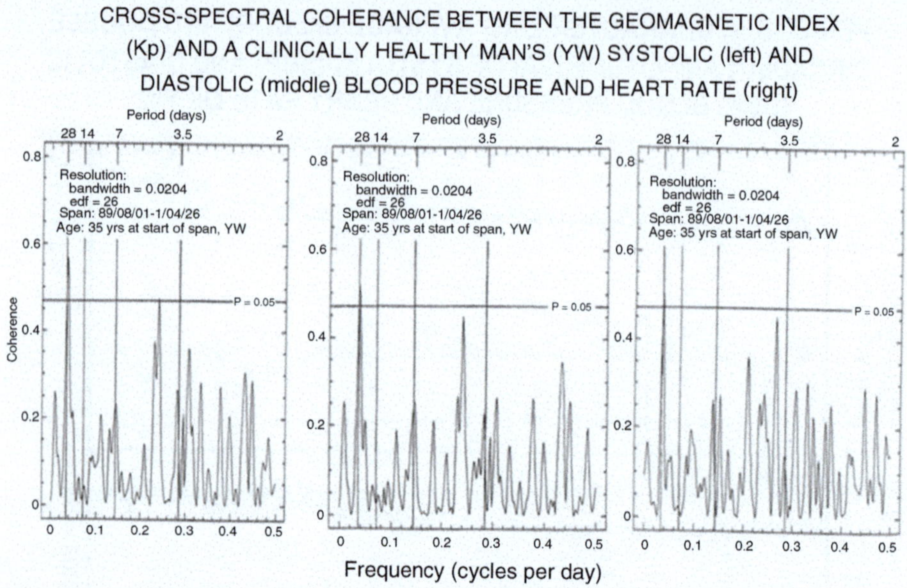

Fig. 2.13 Cross-spectral coherence between the geomagnetic index (Kp) and a clinically healthy man's (YW) systolic (*left*) and diastolic (*middle*) blood pressure and heart rate (*right*) (© Halberg Chronobiology Center)

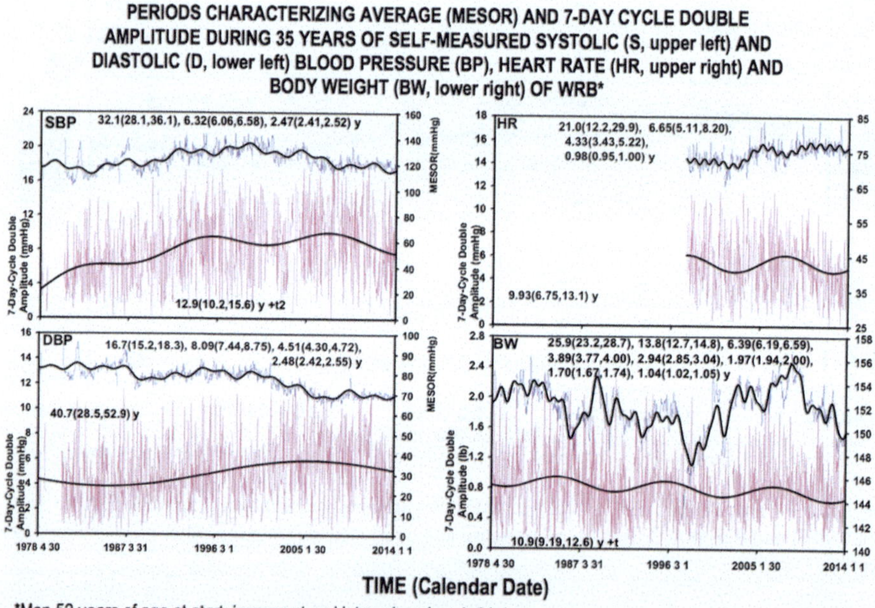

Fig. 2.14 Periods, characterizing 7-day average (MESOR) and 7-day cycle double amplitude during 35 years of self-measured systolic (S, *upper left*) and diastolic (D, *lower left*) blood pressure (BP), heart rate (HR, *upper right*), and body weight (BW, *lower right*) of WRB, a man on hypotensive medication (© Halberg Chronobiology Center)

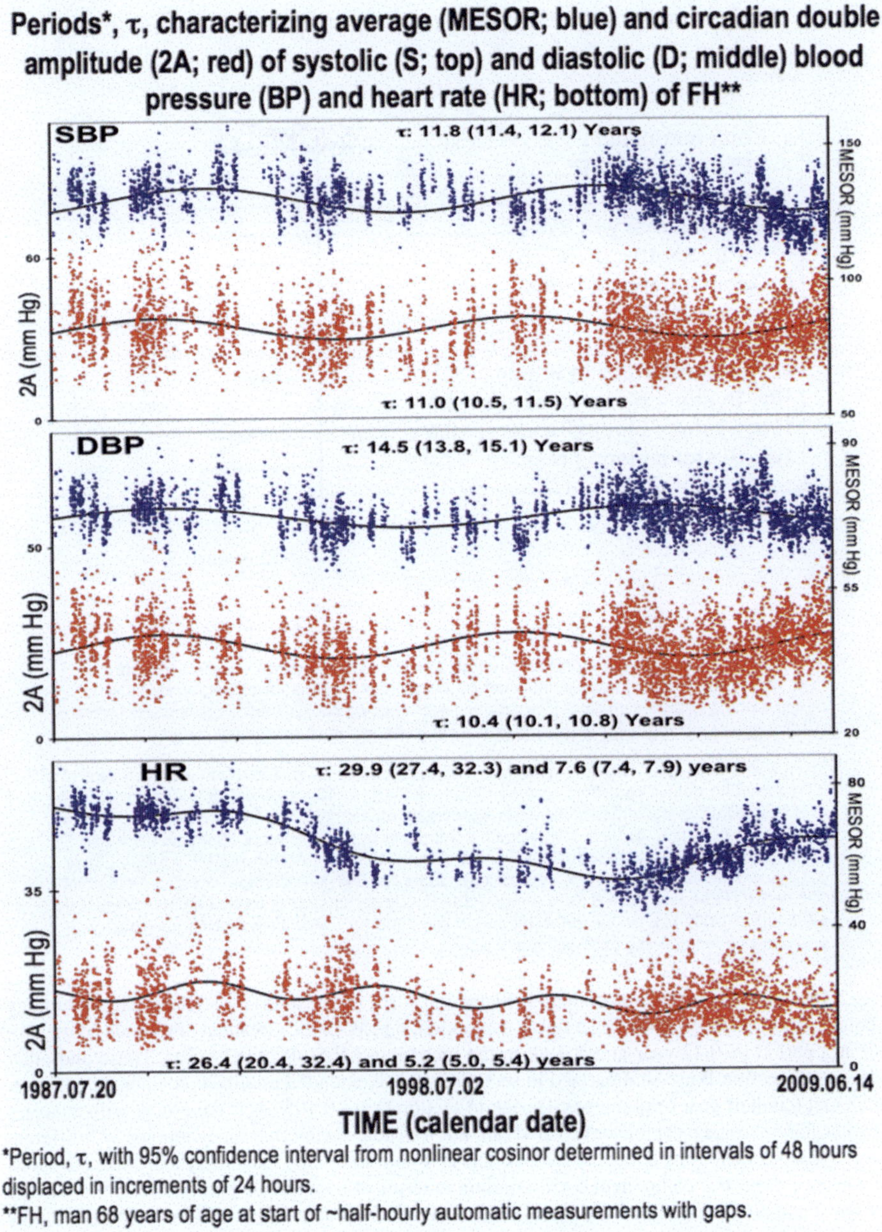

Fig. 2.15 Periods characterizing average (MESOR, *blue*) and circadian double amplitude (2A, *red*) of systolic (S, *top*) and diastolic (D, *middle*) blood pressure (BP) and heart rate (*bottom*) of an elderly man, FH (© Halberg Chronobiology Center)

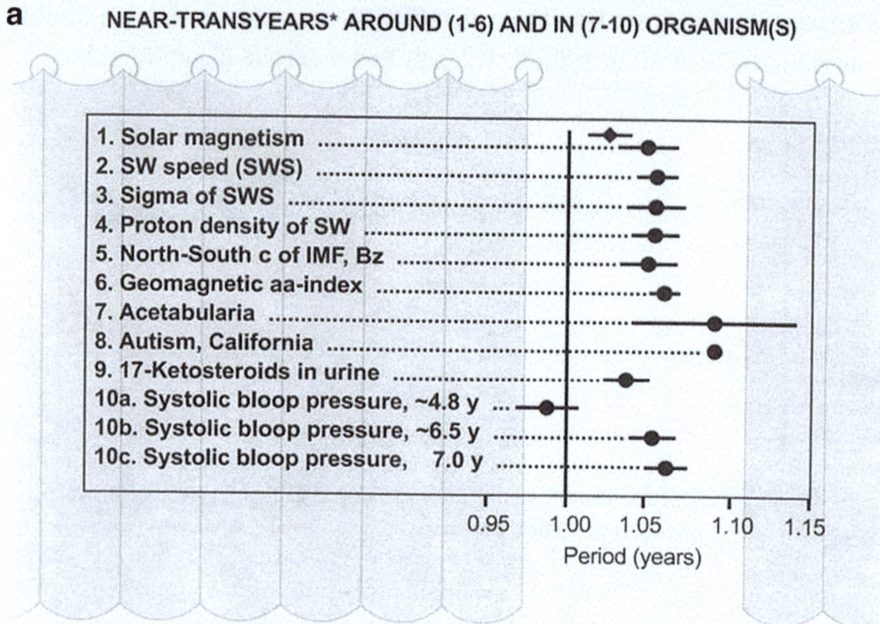

* as separate components and probably not as sidelobes of circannual variation (not stemming from amplitude- and/or phase-modulation by components with a lower frequency). Circles = period lengths (point estimations); horizontal lines crossing circles = 95% confidence intervals (CIs) of spectral components (when the zero-amplitude assumption was rejected nonlinearly).

From www.noaa.gov **1**: Stanford data, May 1975 - Apr 2002 (27 y) [Diamond: longer record of 37 y including earlier data from Crimea (1968 - 2004)]; **6**: 131 y (1868-1998); From OMNI2 (1963-2003) (ftp://nssdcftp.gsfc.nasa.gov/spacecraft_data/omni/) **2-4**: SW = Solar wind; Sigma = Standard deviation; **5**: Bz = North-South component of Interplanetary Magnetic Field (IMF); From BIOCOS **7**: Oxygen production (1980-1994); **8**: Quarterly incidence, 10.5 y (1994-2004); **9**: Daily excretion for 15 y of urinary steroidal metabolites (17-ketosteroids, CH, Oct 1948 - Oct 1963); **10**: Half-hourly record from a man (GSK), 72 y old at start of measurements from Mar 1998 to Feb 2003 (a); to Dec 2004 (b); and to Apr 2005 (c). An initially lacking near-transyear appears with lengthening of series and/or age.

Fig. 2.16 (**a**) Near-transyears around [1–6] and in [7–10] organism(s) (© Halberg Chronobiology Center). (**b**) Section A shows, by a linear estimation, a peak in the circadian amplitude of melatonin at a period of 0.43 year and a peak in the melatonin MESOR at 0.39 year. These and other analyses in section B are summarized in section C by the extended nonlinear cosinor. The parasemiannual (cis-half year or quinmensal) peak in section D at ~0.40 year is congruent with analyses of solar flares. Chance can never be ruled out, but it is noteworthy that many among 172 different people (each contributing data only for 24 hours at 4-hour intervals) must be synchronized to show, as a group, a spectral component corresponding in terms of congruence to a component predicted by Wolff and documented by Rieger et al., and many others (Table 2.2) (© Halberg Chronobiology Center). (**c**) Spectral window of solar flares from 1966 to 2007 (*top left*, 1) shows, with a prominent ~10.38-year Horrebow-Schwabe and a lesser ~34.73-year Brückner-Egeson-Lockyer (BEL) peak (see Fig. 3), an ~0.43-year quinmensal cycle (*top* 2 rows) with coexisting ~0.71- and ~1.72-year components. The statistical significance of these components, as shown for the quinmensals in flares (*row* 2) and in the human heart rate (*bottom row*), and elsewhere for the BEL, waxes and wanes, as shown by *filled* (P<0.05) and *empty* (P>0.05) *symbols* for the rejection of the zero-amplitude assumption (© Halberg Chronobiology Center). (**d**) Global (spectral window of the entire series; top row) and local (gliding spectral windows with intervals of different lengths; rows below) analyses of diastolic blood pressure of an elderly man, GSK, show aeolian quinmensals (© Halberg Chronobiology Center)

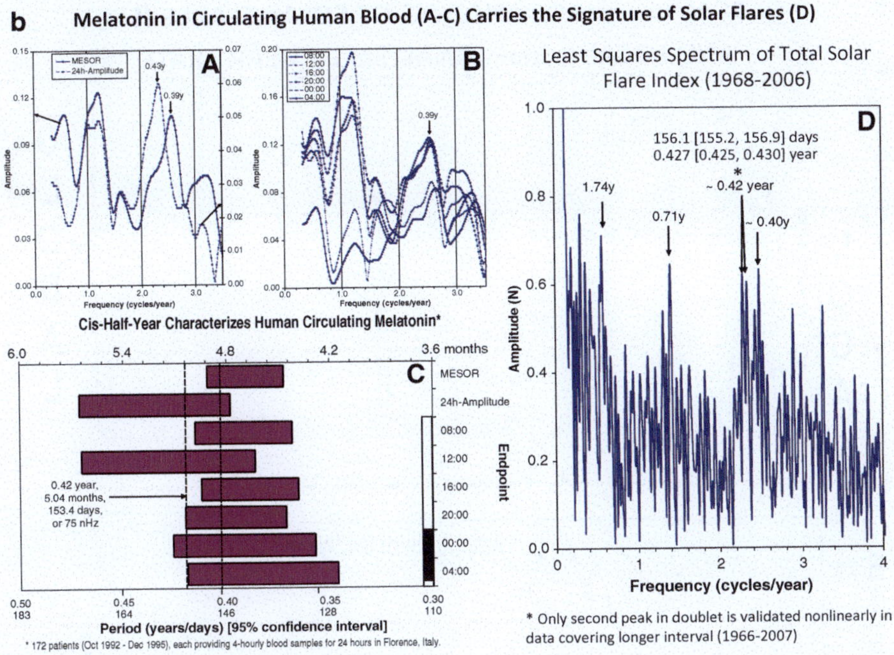

Fig. 2.16 (continued)

a number of deaths among poorly dressed, undernourished people. We were also surprised to learn on humans that clinically healthy students on 2000 cal in a single-meal/day experiment on "breakfast-only" lost body weight, while on "dinner-only" they (and other staff members) had a (relative) gain in weight, Fig. 2.18a–g [55].

2.8 When to Treat

Like timed radiotherapy, chemotherapy guided by salivary marker rhythms also awaits entry into everyday clinical oncotherapy ([11]; [56, 57, 58] and [59]). Much to-us convincing laboratory investigation, Fig. 2.19, in cancer, Fig. 2.19a–i [60], and in hypotensive therapy ([61]; cf. [62]), Fig. 2.19j–l, and clinical evidence based on marker rhythms is not yet utilized in oncology, preventive cardiology, psychiatry, and health care generally. Timing treatment is also a challenge in other fields, notably in cardiocerebrovascular disease. Immediately applicable cost-effective, urgent, and most promising, in view of the slow-motion disaster of noncommunicable disease in very many millions of people in developing areas, is a hypotensive treatment that does not fly blind, as it currently does between visits to a care provider [63; cf. 39]. Chronobiologic self-surveillance optimally clarifies when stress becomes strain by gauging partly novel vascular variability anomalies (VVAs), disorders (VVDs),

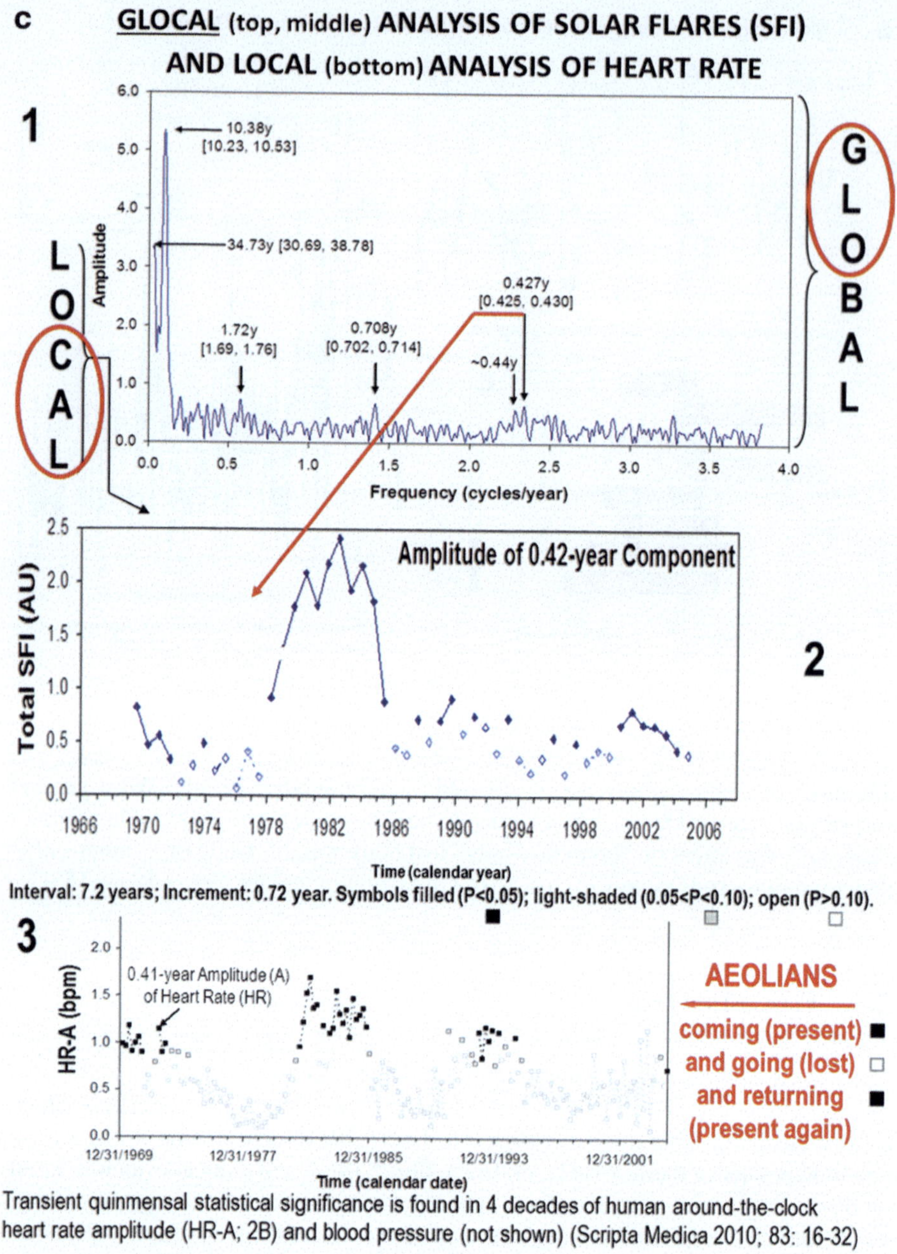

Fig. 2.16 (continued)

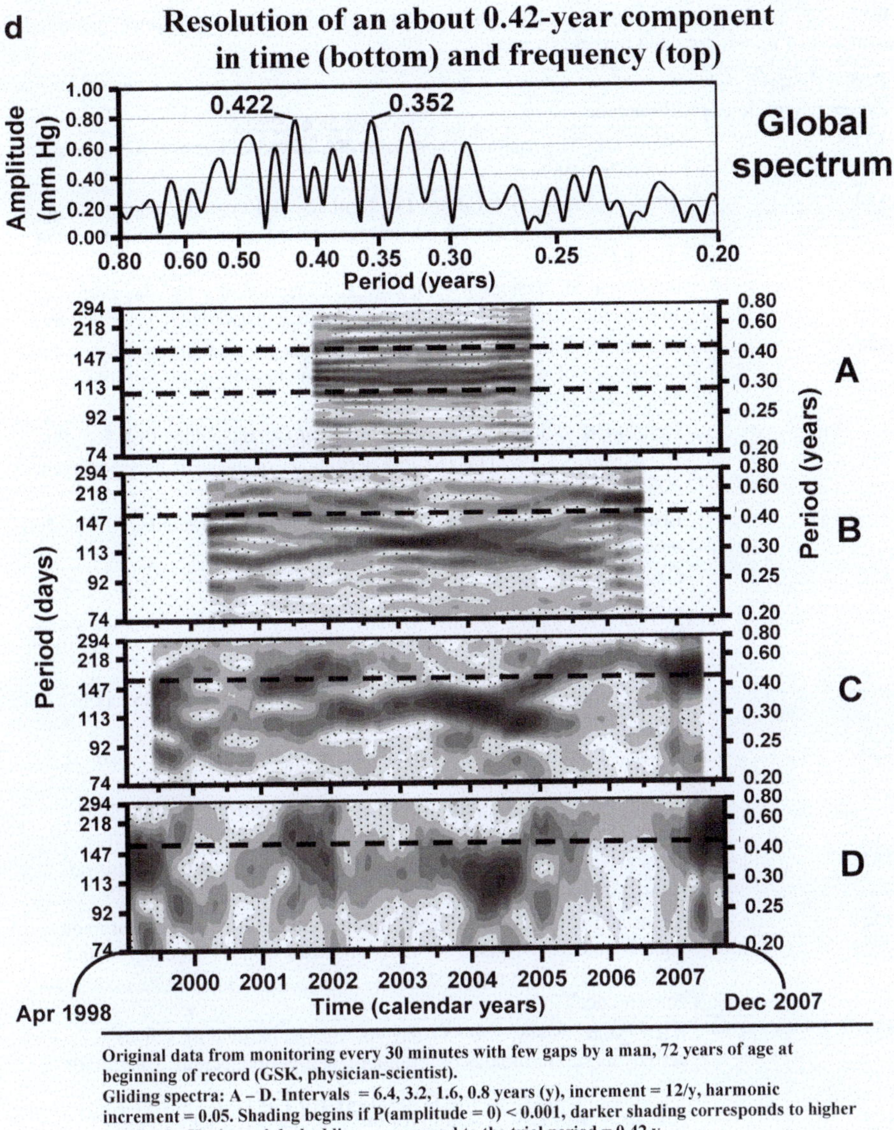

Fig. 2.16 (continued)

Table 2.3 Point estimates of a cis-half year (~154-day) periodicity, sometimes given with estimates of uncertainty and hypothesis testing

Period (in days except where noted)	Reference
154.3	Wolff CL. The rotational spectrum of g-modes in the sun. Astrophys J 1983; 264: 667–76
154	Rieger A, Share GH, Forrest DJ, Kanbach G, Reppin C, Chupp EL. A 154-day periodicity in the occurrence of hard solar flares? Nature 1984; 312: 623–25
158	Kiplinger AL, Dennis BR, Orwig LE. Detection of a 158-day periodicity in the solar hard X-ray flare rate. Bull Am Astronom Soc 1984; 16: 891
152	Bogart RS, Bai T. Confirmation of a 152-day periodicity in the occurrence of solar flares inferred from microwave data. Astrophys J 1985; 299: L51–L55
152–158	Dennis BR. Solar hard X-ray bursts. Solar Physics 1985; 100: 465–90
152	Bai T, Sturrock PA. The 152-day periodicity of the solar flare occurrence rate. Nature 1987; 327: 601–4
Near 155	Lean JL, Brueckner GE. Intermediate-term solar periodicities – 100–500 days. Astrophys J 1989; 337: 568–78
152	Özgüç A, Ataç T. Periodic behavior of solar flare index in solar cycles 20 and 21. Solar Physics 1989; 123: 357–65
154 (±0.6)	Bai T, Cliver EW. A 154-day periodicity in the occurrence rate of proton flares. Astrophys J 1990; 363: 299–309
Near 155	Carbonell M, Ballester JL. A short-term periodicity near 155 day in sunspot areas. Astron Astrophys 1990; 238: 377–81
153	Dröge W, Gibbs K, Grunsfeld JM, Meyer P, Newport BJ, Evenson P, Moses D. A 153 day periodicity in the occurrence of solar flares producing energetic interplanetary electrons. Astrophys J Suppl Ser 1990; 73: 279–83. Applying Rayleigh test for periodicity
155	Lean J. Evolution of the 155-day periodicity in sunspot areas during solar cycles 12 to 21. Astrophys J 1990; 363: 718–27
155	Silverman SM. The 155-day solar period in the sixteenth century and later. Nature 1990; 347: 365–7. "[A]lthough values between 150 and 160 days have been reported, I refer to it here as the 155-day period, for convenience"
154	Bai T, Sturrock PA. The 154-day and related periodicities of solar activity as subharmonics of a fundamental period. Nature 1991; 350: 141–3
154	Kile JN, Cliver EW. A search for the 154 day periodicity in the occurrence rate of solar flares using Ottawa 2.8 GHz burst data, 1955–1990. Astrophys J 1991; 370: 442–8
152–158	Verma VK, Joshi JC, Uddin W, Paliwal DC. Search for a 152–158 days periodicity in the occurrence rate of solar flares inferred from spectral data of radio bursts. Astron Astrophys Suppl Ser 1991; 90: 83–7
154	Bai T. The 77 day periodicity in the flare rate of cycle 22. Astrophys J 1992; 388: L69–L72
Near 155	Carbonell M, Ballester JL. The periodic behaviour of solar activity: the near 155-day periodicity in sunspot areas. Astron Astrophys 1992; 255: 350–62

(continued)

Table 2.3 (continued)

Period (in days except where noted)	Reference
152	Verma VK, Joshi JC, Paliwal DC. Study of periodicities of solar nuclear gamma ray flares and sunspots. Solar Physics 1992; 138: 205–8
154	Bai T, Sturrock PA. Evidence for a fundamental period of the sun and its relation to the 154 day complex of periodicities. Astrophys J 1993; 409: 476–86
~153	Cane HV, Richardson IG, von Rosenvinge TT. Interplanetary magnetic field periodicity of 153 days. Geophys Res Lett 25: 4437–40
158	Oliver R, Ballester JL, Baudin F. Emergence of magnetic flux on the Sun as the cause of a 158-day periodicity in sunspot areas. Nature 1998; 394: 552–3 | doi:10.1038/29012
Near 158	Ballester JL, Oliver R, Baudin F. Discovery of the near 158 day periodicity in group sunspot numbers during the eighteenth century. Astrophys J 1999; 522: L153–L156
151–155	Lou YQ. Rossby-type wave-induced periodicities in flare activities and sunspot areas or groups during solar maxima. Astrophys J 2000; 540: 1102–8
153.9	Hady AA. Analytical studies of solar cycle 23 and its periodicities. Planetary and Space Science 2002; 50: 89–92
156	Krivova NA, Solanki SK. The 1.3-year and 156-day periodicities in sunspot data: wavelet analysis suggests a common origin. Astron Astrophys 2002; 394: 701–6.
Near 160	Ballester JL, Oliver R, Carbonell M. The near 160 day periodicity in the photospheric magnetic flux. Astrophys J 2002; 566: 505–11
Near 5-month	Han Yanben, Han Yonggang. Time variation of the near 5-month period of sunspot numbers. Chinese Sci Bull 2002; 47 [23]: 1967–73. "Many scholars … found similar periods of solar activity from other observations of the Sun. However, these periods are different, such as about 152-, 154-, 156-day, etc. Here we name it a near 5-month period (N5MP) since it is not definite"
154	Prabhakaran Nayar SR, Radhika VN, Revathy K, Ramadas V. Wavelet analysis of solar, solar wind and geomagnetic parameters. Solar Phys 2002; 208: 359–73. "The global wavelet spectra of these parameters … exhibit the presence of a variety of prominent quasi periods around 16 years, 10.6 years, 9.6 years, 5.5 years, 1.3 years, 180 days, 154 days …"
153	Bai T. Periodicities in solar flare occurrence analysis of cycles 19–23. Astrophys J 2003; 591: 406–15
Near 160	Ballester JL, Oliver R, Carbonell M. Return of the near 160 day periodicity in the photospheric magnetic flux during solar cycle 23. Astrophys J 2004; 615: L173–L176
151–158	Knaack R, Stenflo JO, Berdyugina SV. Evolution and rotation of large-scale photospheric magnetic fields of the Sun during cycles 21–23: Periodicities, north–south asymmetries and r-mode signatures. Astron Astrophys 2005; 438: 1067–82
~152	Chowdhury P, Ray PC. Periodicities of solar electron flare occurrence: analysis of cycles 21–23. Mon Not Roy Astronom Soc 2006; 373: 1577–89

(continued)

Table 2.3 (continued)

Period (in days except where noted)	Reference
155	Chowdhury P, Ray PC, Ray S. Periodicity of 155 days in solar electron fluence. Ind J Phys 2008; 82: 95–104
150–160	Dimitropoulou M, Moussas X, Strintzi D. Enhanced Rieger-type periodicities' detection in X-ray solar flares and statistical validation of Rossby waves' existence. Mon Not R Astron Soc 2008; 386: 2278–84
158	Javaraiah J, Ulrich RK, Bertello L, Boyden JE. Search for short-term periodicities in the sun's surface rotation: a revisit. Solar Phys 2009; 257: 61–9
155	Vaquero JM, Trigo RM, Vazquez M, Gallego MC. 155-day periodicity in solar cycles 3 and 4. New Astronomy 2009. doi: 10.1016/j.newast.2009.11.004
155–160	Zaqarashvili TV, Carbonell M, Ballester JL, Oliver R. Magnetic Rossby waves in the solar tachocline and Rieger-type periodicities. arXiv:0911.4591v1 [astro-ph.SR] 24 Nov 2009
154	Fischbach E, Jenkins JH, Buncher JB, Gruenwald JT, Sturrock PA, Javorsek D II. Evidence for solar influences on nuclear decay rates. arXiv:1007.3318v1 [hep.ph] 20 Jul 2010

This incomplete list (added contributions are cited in the references provided) suffices to indicate the variable nature of the period being discussed. Specifications of the decadal solar (Horrebow-Schwabe) cycle in which they are found point indirectly to the intermittency of the components being discussed

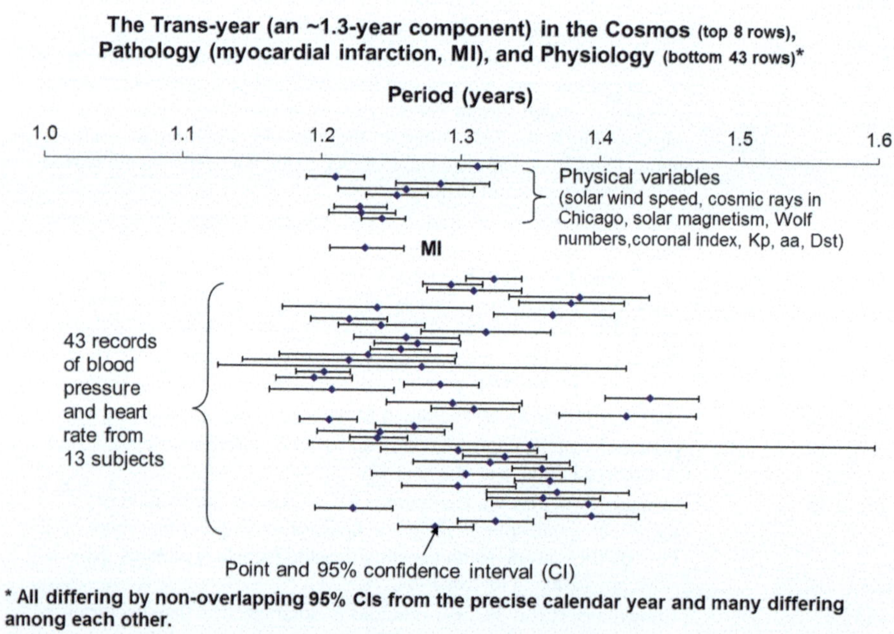

Fig. 2.17 The transyear (an ~1.3-year component) in the cosmos (*top* 8 rows), pathology (myocardial infarction, MI) and physiology (*bottom* 43 rows), all differing by nonoverlapping 95 % CIs from the precise calendar year and some differing among each other (© Halberg Chronobiology Center)

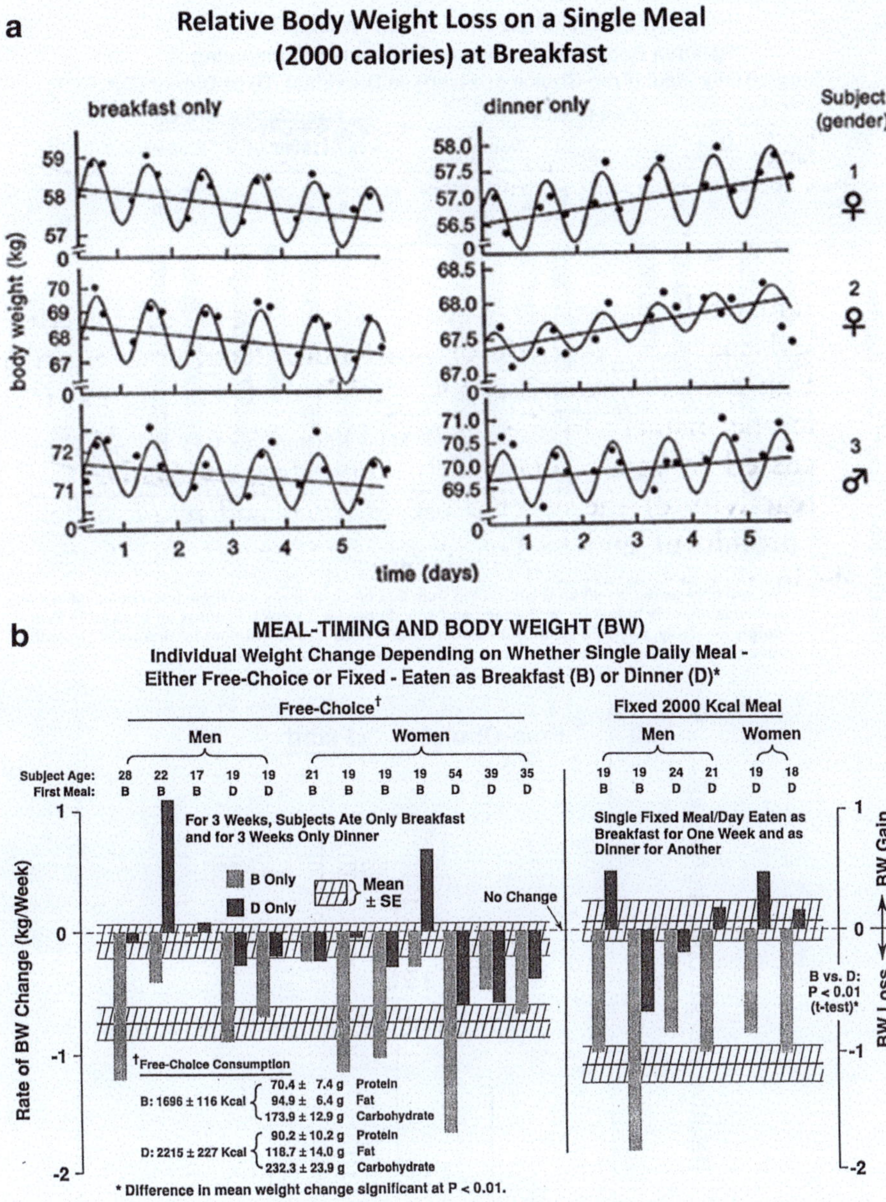

Fig. 2.18 (**a**) Body weight loss on breakfast-only vs. body weight gain on dinner-only in 3 young clinically healthy subjects consuming a single daily meal of 2,000 cal. In this limited study, a 24-hour cosine function and a linear trend fitted concomitantly to the data were statistically significant 5 times out of 6; $P<0.05$ for nonzero-slope 4 times out of 5. Results confirmed by extensive follow-up studies on ad libitum breakfast-only vs. dinner-only for 3 weeks, (© Halberg Chronobiology Center). (**b**) Meal timing and body weight. In two separate studies of the effect of meal timing on body weight, 9 men and 9 women consumed either a fixed 2,000-cal meal or a single free-choice meal as breakfast (B) or dinner (D) (for 1 week on a fixed-calorie meal or

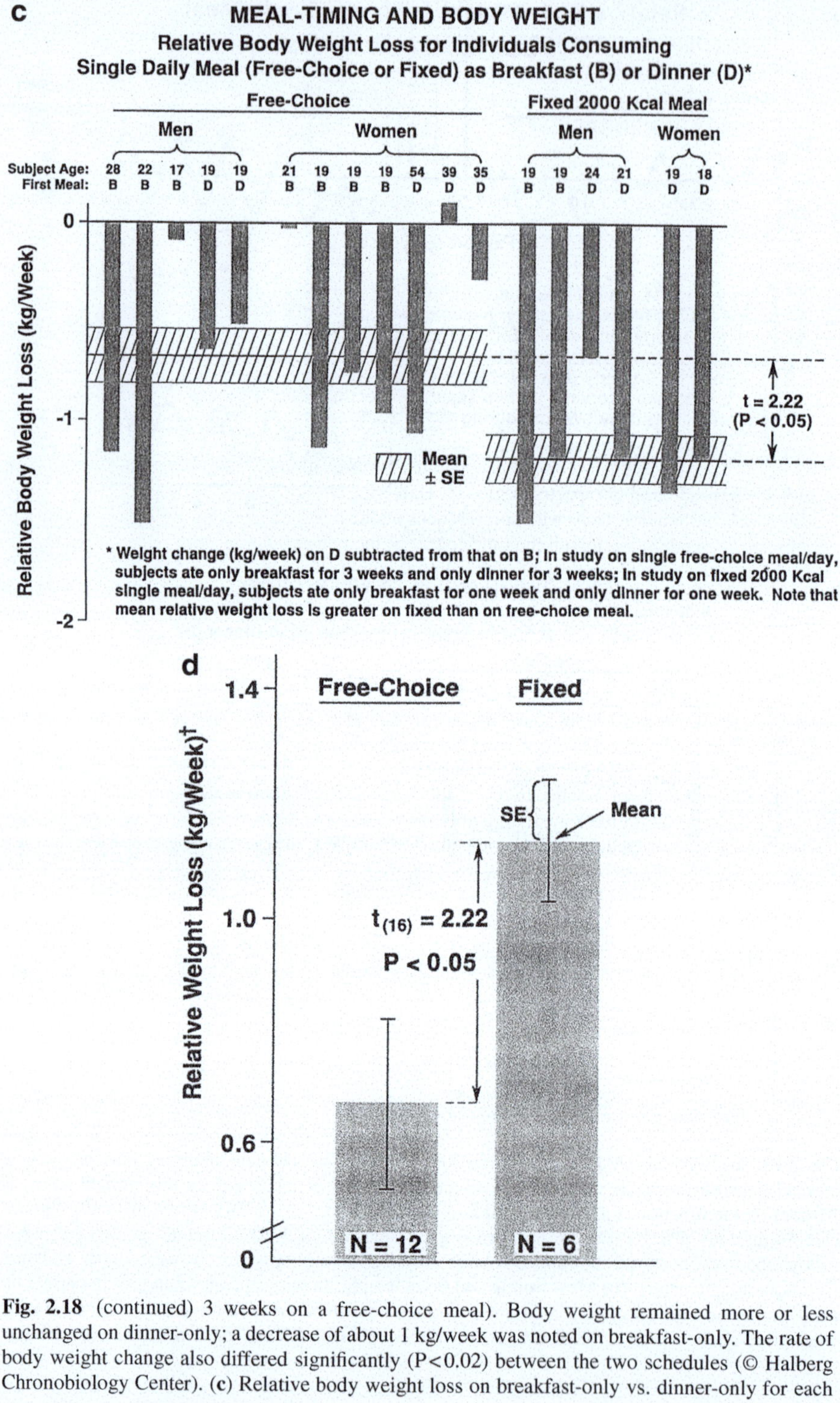

Fig. 2.18 (continued) 3 weeks on a free-choice meal). Body weight remained more or less unchanged on dinner-only; a decrease of about 1 kg/week was noted on breakfast-only. The rate of body weight change also differed significantly ($P<0.02$) between the two schedules (© Halberg Chronobiology Center). (**c**) Relative body weight loss on breakfast-only vs. dinner-only for each

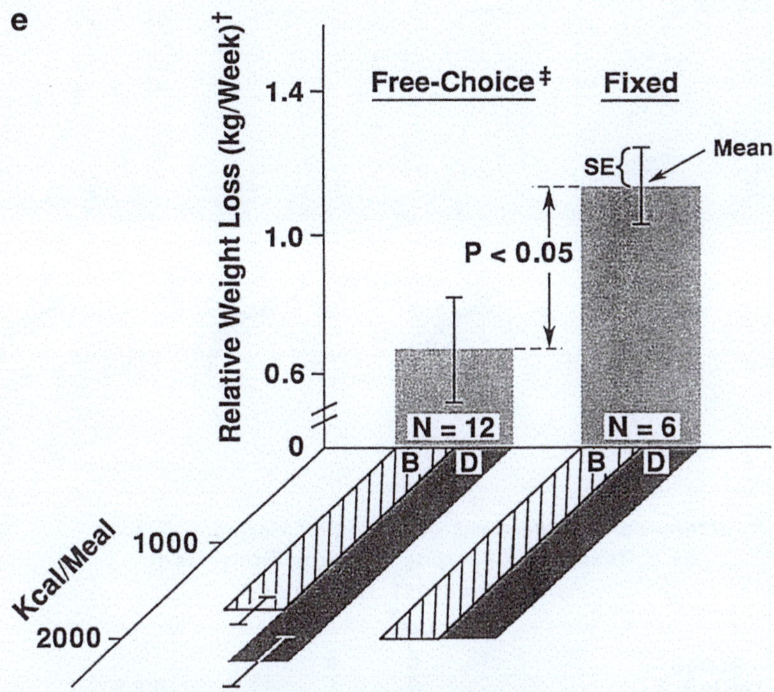

* Here defined as choice of kind and amount of food.

† Weight change (kg/week) on B subtracted from that on D.

‡ Average free-choice intake (± SE) on B = 1696 ± 116 Kcal/meal, on D = 2215 ± 227 Kcal/meal; former value significantly different from 2000 (t = 2.62; P < 0.05), latter not. Food consumed consisted on the average of

g		Protein	Fat	Carbohydrate
Free-Choice	B	70.4 ± 7.4	94.9 ± 6.4	173.9 ± 12.9
	D	90.2 ± 10.2	118.7 ± 14.0	232.3 ± 23.9
Fixed Meal, B & D		77.0	75.0	250.0

Fig. 2.18 (continued) subject participating in the two studies. Only one volunteer gained weight on breakfast vs. dinner. Overall, the difference in relative body weight loss on breakfast (B) vs. dinner (D) is statistically significant (P<0.05), whether a fixed 2,000-cal meal (*right*) or a single free-choice meal (*left*) is consumed. Weight change (kg/week) on D subtracted from that on B. In the study on one free-choice meal per day, subjects ate only breakfast for 3 weeks and only dinner for 3 weeks. In the study on one fixed 2,000-cal meal per day, subjects ate only breakfast for 1 week and only dinner for another 1 week. Note that mean relative weight loss is greater on fixed than on free-choice meal (© Halberg Chronobiology Center). (**d**) Appetite (here defined as choice and amount of food) modifies the effect of meal timing on body weight. Relative body weight loss on breakfast-only (B) as compared to dinner-only (D) is less when meal is free-choice rather than fixed. An overall summary of relative body weight loss on breakfast-only vs. dinner-only in the two studies indicates that the decrease in relative body weight was more pronounced when a fixed 2,000-calorie meal was imposed than when volunteers could choose what they ate (© Halberg Chronobiology Center). (**e**) Appetite (here defined as choice and amount of food) modifies the effect of meal timing on body weight. The lesser body weight loss on breakfast (B) vs. dinner (D) observed on a free-choice vs. a fixed 2,000-cal meal occurred, while calorie consumption on the free-choice meal was less (not more) than 2,000 calories per meal

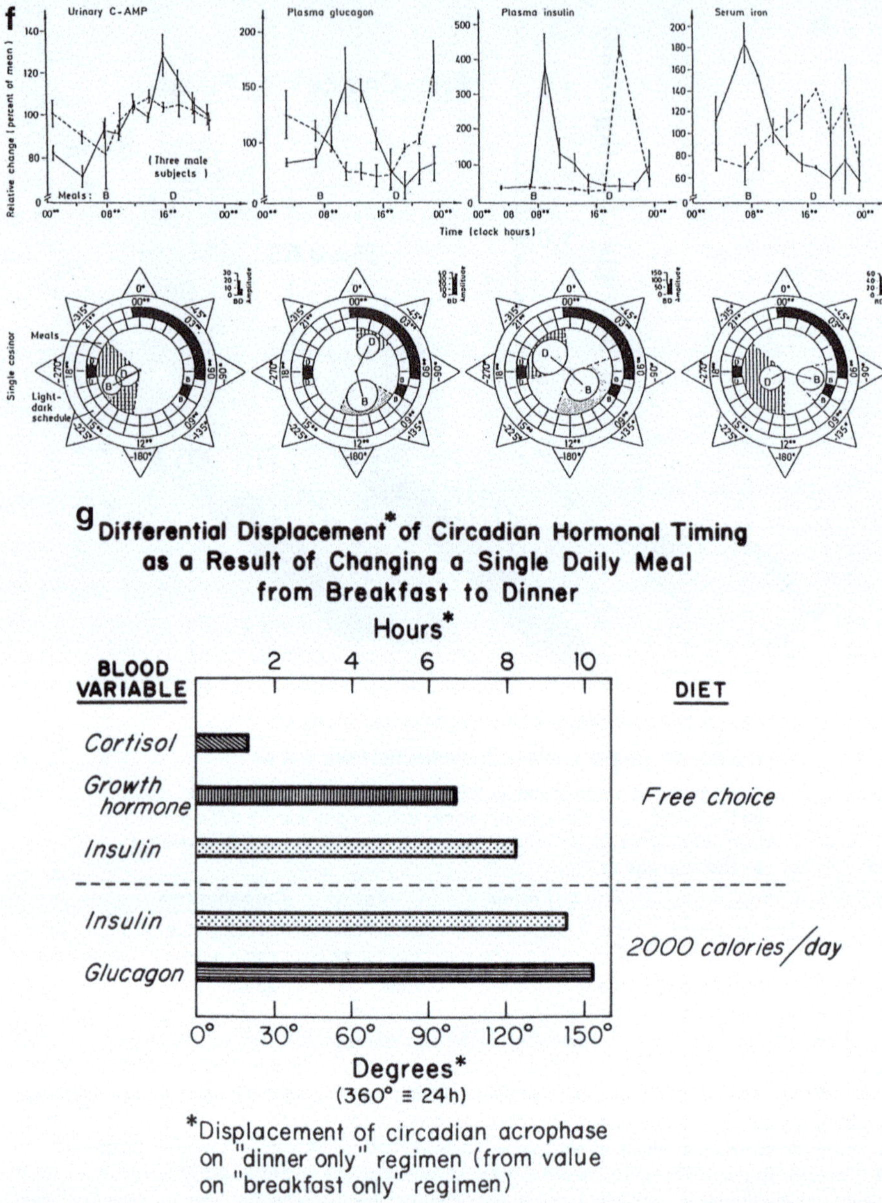

Fig. 2.18 (continued) (© Halberg Chronobiology Center). (**f**) Consuming a single daily meal as breakfast-only vs. dinner-only affects the timing of the circadian chronome component differently for different physiological variables in humans. The single daily meal was 2,000 calories given for 7-day spans. Breakfast (B), *solid line*. Dinner (D), *broken line* (© Halberg Chronobiology Center). (**g**) Differential displacement of circadian hormonal timing as a result of changing a single daily meal from breakfast to dinner. Whereas the circadian rhythm of circulating cortisol is only slightly affected by the timing of a single daily meal, considerable phase-shifts are observed for the case of growth hormone, insulin, and glucagon, resulting in a different internal timing of the latter three hormones vs. cortisol (© Halberg Chronobiology Center)

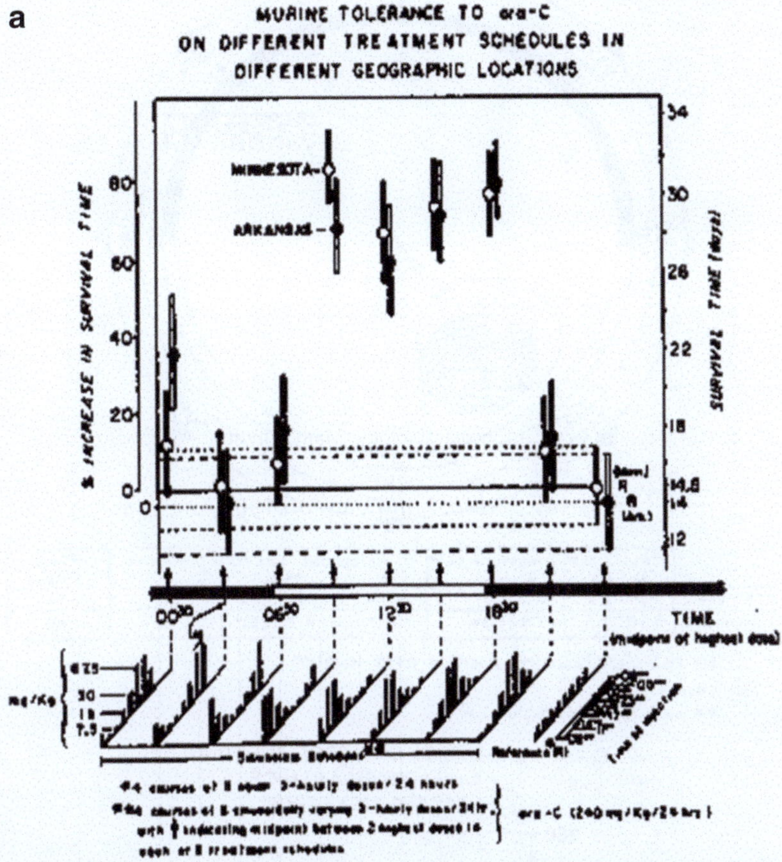

Fig. 2.19 (**a**) Chronotoxicology: comparison of equal dose (homeostatic) ara-C treatment of L1210 leukemia (extreme right) with sinusoidal doses in Arkansas and Minnesota reveals similar gain in survival from certain timings of sinusoidal variation of the same total dose [11] (© Halberg Chronobiology Center). (**b**) Agreement of ara-C chronotherapy timing in Arkansas and Minnesota revealed by cosinor with overlapping uncertainties (© Halberg Chronobiology Center). (**c**) Investigators in Alabama (Rose et al., Cancer Treatment Rep 1978; 62: 1337–49) did not find a difference (in survival times only of those mice that died) between their homeostatic treatment and chronotherapy (cured mice were ignored). Accounting for cures in turn shows a clear benefit from a sinusoidal treatment, actually a much bigger gain in Alabama than in Arkansas and Minnesota (© Halberg Chronobiology Center). (**d**) Circadian variation in toxicity of doxorubicin (© Halberg Chronobiology Center). (**e**) Circadian variation in effectiveness of doxorubicin (adriamycin) (© Halberg Chronobiology Center). (**f**) Gains from cancer chronotherapy: experimental and clinical (© Halberg Chronobiology Center). (**g**) Gains from cancer chronotherapy: experimental and clinical (© Halberg Chronobiology Center). (**h**) Timing cancer chronotherapy depends on the drug used (© Halberg Chronobiology Center). (**i**) Chronotolerance of whole body irradiation (© Halberg Chronobiology Center). (**j**) Circadian stage dependence of antihypertensive response in SHR-SP rats (Original data of Akio Fujimura [61]. © Halberg Chronobiology Center). (**k**) Summary of circadian stage dependence of antihypertensive response in SHR-SP rats (Original data of Akio Fujimura [61]. © Halberg Chronobiology Center). (**l**) Chronotherapeutic gain in laboratory and clinic (© Halberg Chronobiology Center)

b

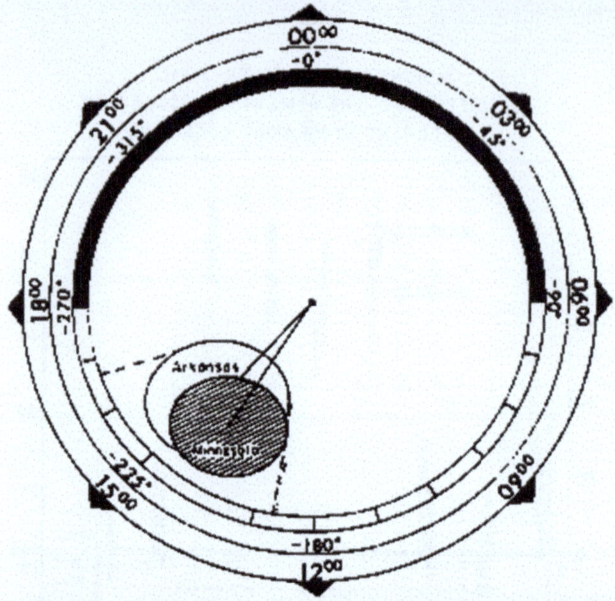

	Mesor (M)* M ± SE	Amplitude (A)* A ± SE	Acrophase (Ø)** Ø	95% CI	% Rhythm	P-value F-test
Minnesota	5.54 ± 2.2	2.06 ± 3.1	-211°	-188°, -233°	22	.01
Arkansas	5.60 ± 2.4	1.67 ± 3.4	-221°	-191°, -251°	13	.01

N = 180 mice (20/time point) in each location

*in hours

**from local midnight

C

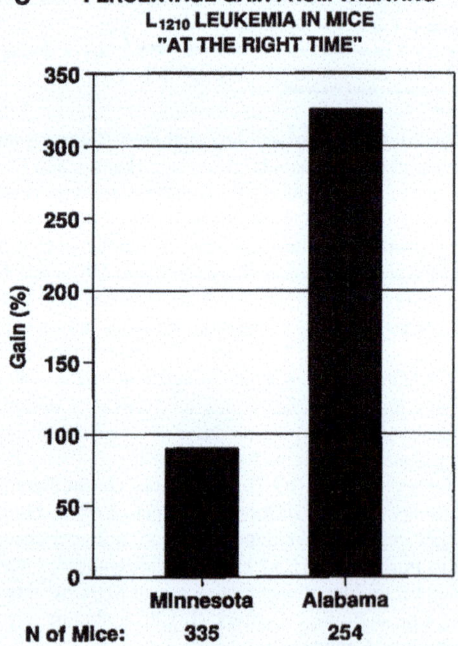

Fig. 2.19 (continued)

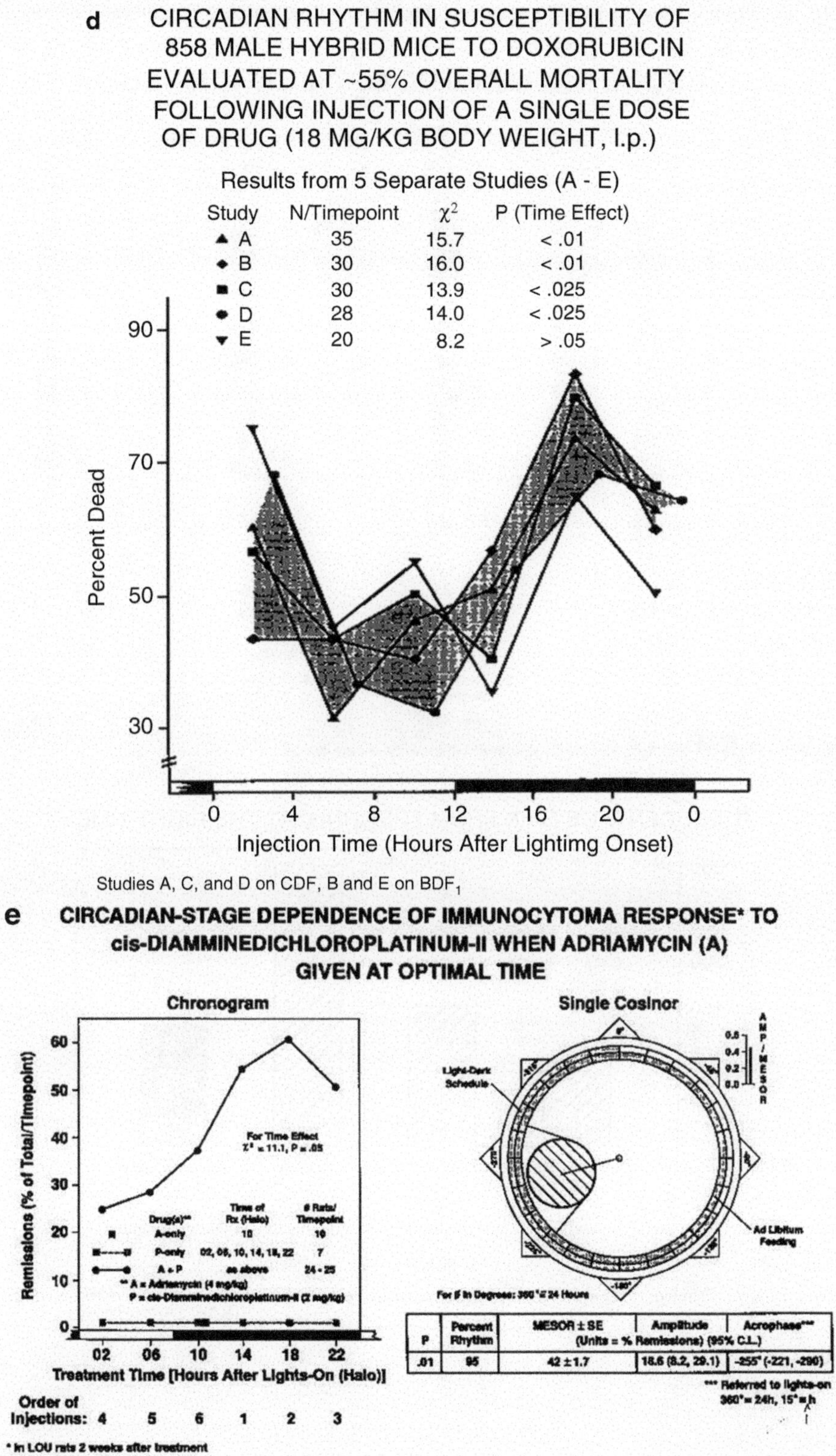

Fig. 2.19 (continued)

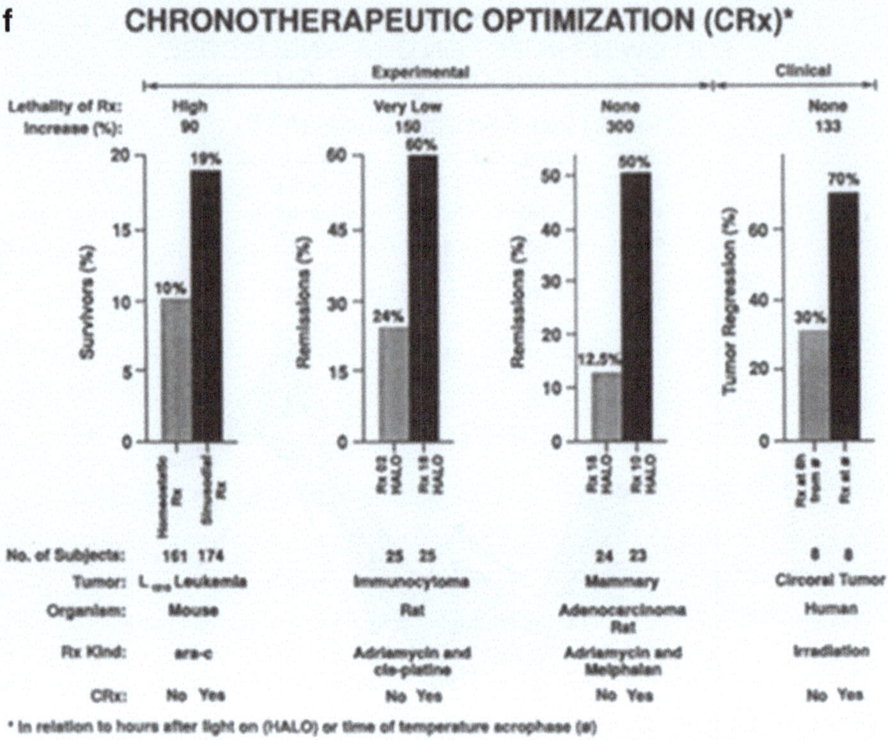

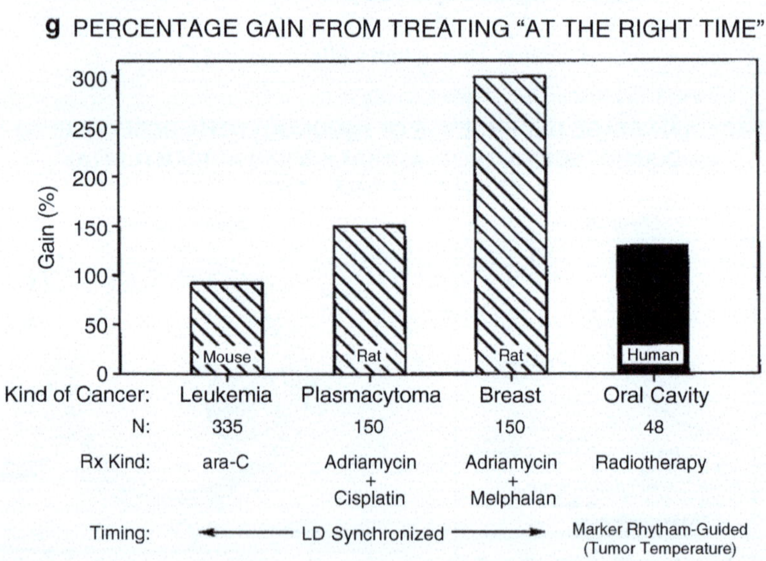

Fig. 2.19 (continued)

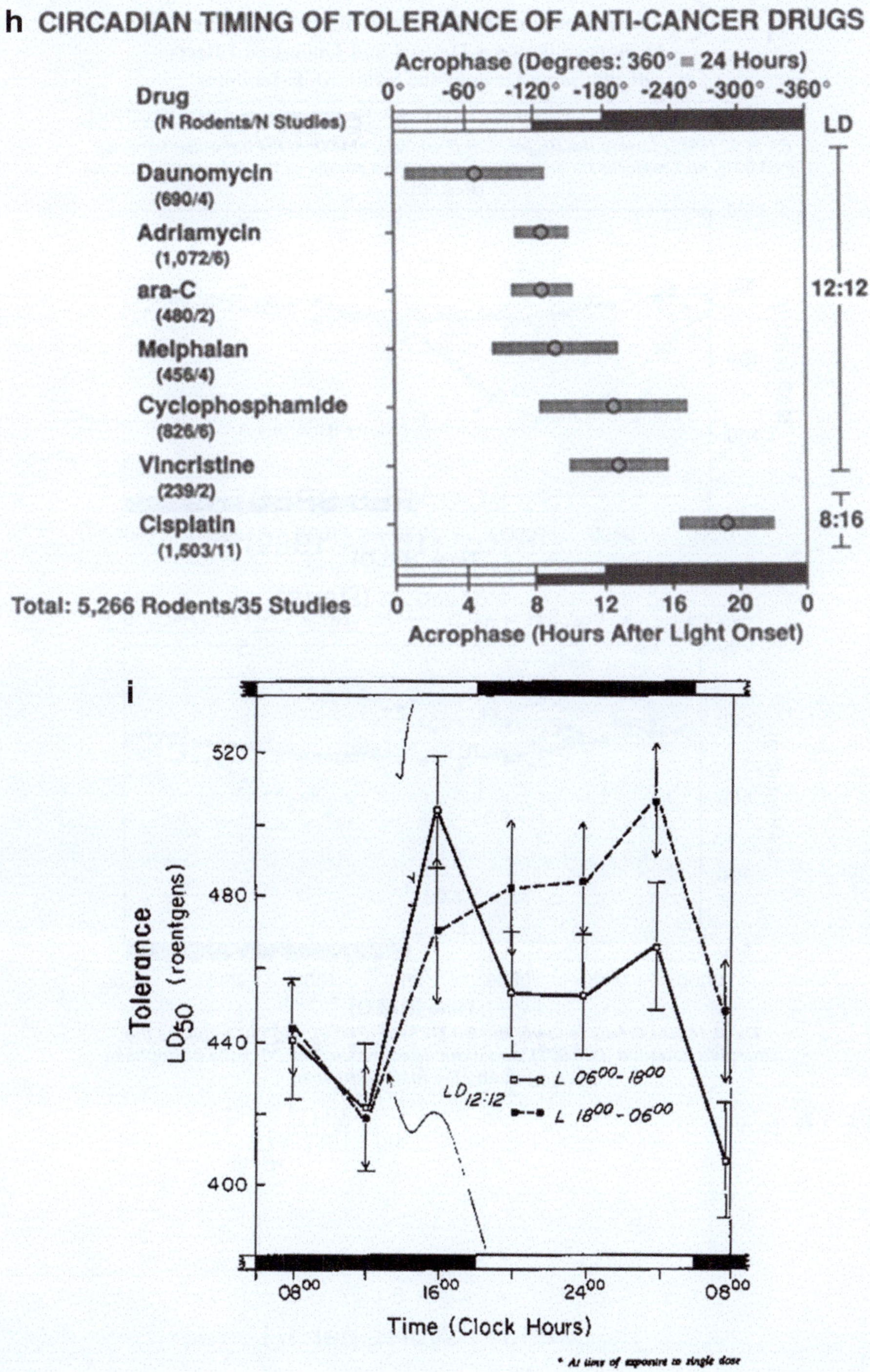

Fig. 2.19 (continued)

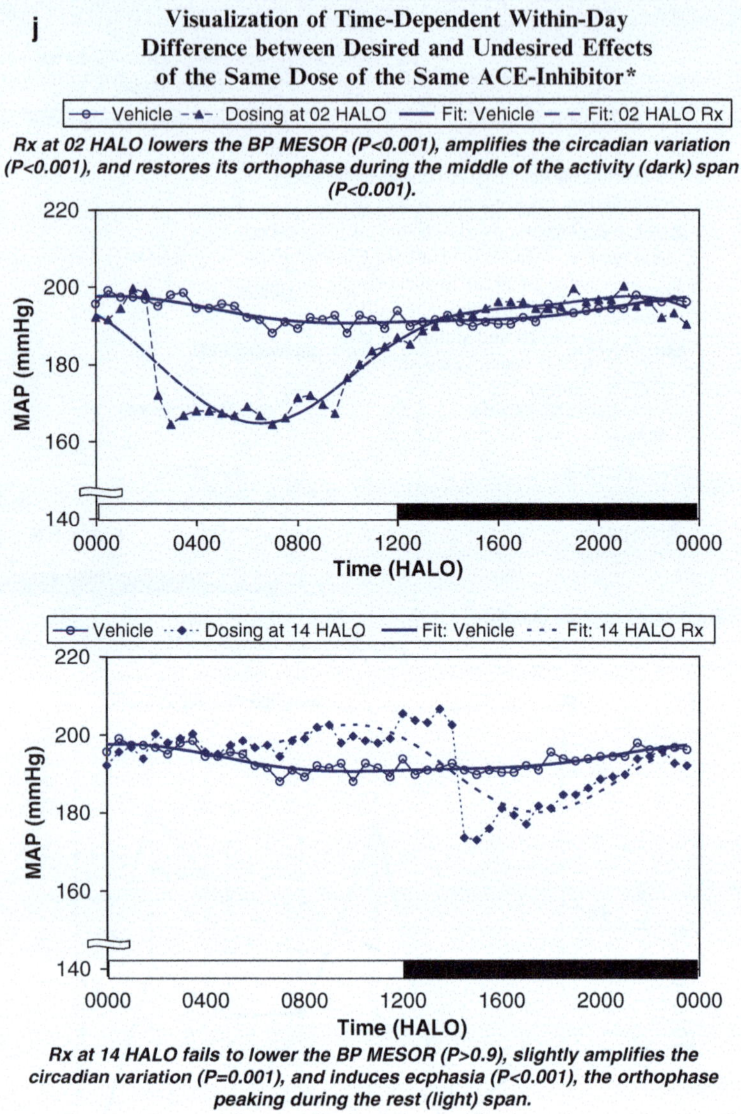

Fig. 2.19 (continued)

k **Survival of Salt-Loaded Okamoto SHR-SP Rats Treated at Different Circadian Stages#**

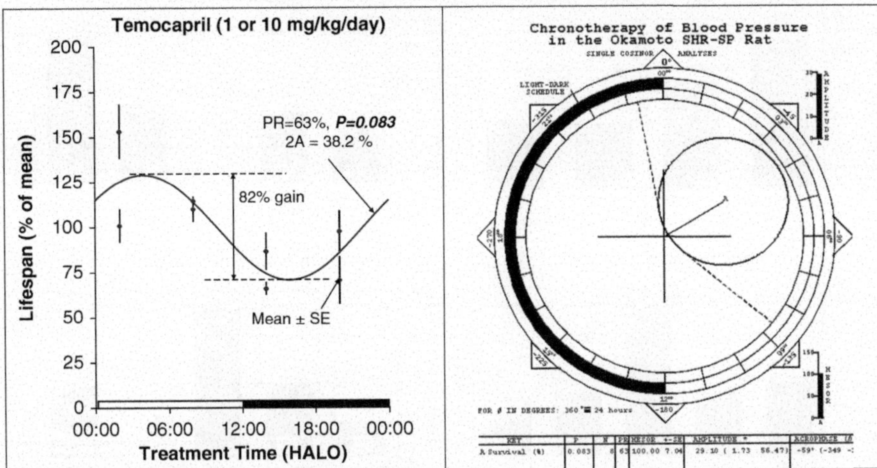

#with an angiotensin converting enzyme inhibitor. Circadian rhythm demonstrable with borderline statistical significance after pooling data at two doses (after expressing data as a percentage of respective overall mean values). Note that best treatment time may occur at time other than actual test times (slightly later than 02 HALO in this case), albeit 95% confidence limits of acrophase include

Fig. 2.19 (continued)

and syndromes (VVSs) [63], providing the means to reach Selye's goal of "stress without distress" [64]. That chronobiologically interpreted ambulatory monitoring detects antecedents of natural disasters such as earthquakes, for possible evasion, as well as those of personal cataclysms such as strokes is all part and parcel of a unified science and art without disciplinary borders; it starts relating to health care when music therapy rather consistently alleviates all symptoms, at least for a while, in JF, a patient with adynamic depression who was not helped by drugs during her 20-year history of adynamia and still depends on palliatively successful music therapy [65]. The same case documents sensitivity to earth tides by the demonstration of a (double tidal) 24.8-hour period in several variables in the computer output of analyses, complementing an earlier cosinor demonstration of a tidal effect on cardiac arrhythmia by Miroslav Mikulecky [66].

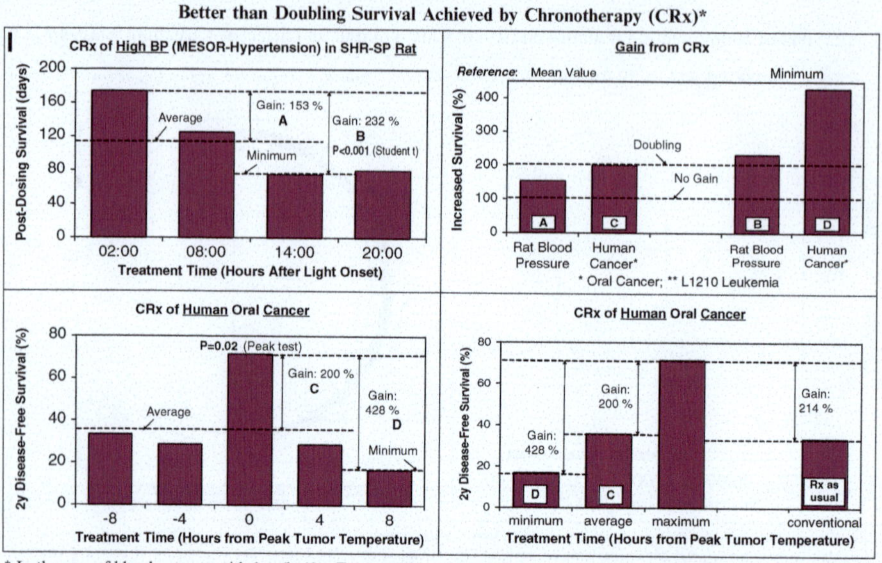

* In the case of blood pressure with 1mg/kg/day Temocapril in salt-loaded stroke-prone Okamoto (SHR-SP) rats (*top left*) and by radiation of human cancers of the oral cavity timed by marker rhythmometry (of tumor temperature) (*bottom left*); Therapeutic gain assessed by comparing response at best time with average response or response at worst time (*top right*), or in the case of chronoradiotherapy, also with response to conventional treatment administered with timing during "regular hours" by convenience

Fig. 2.19 (continued)

References

1. Halberg F, Visscher MB, Flink EB, Berge K, Bock F. Diurnal rhythmic changes in blood eosinophil levels in health and in certain diseases. Journal Lancet (Minneap) 1951;71:312–9.
2. Kaine HD, Seltzer HS, Conn JW. Mechanism of diurnal eosinophil rhythm in man. J Lab Clin Med 1955;45(2):247–52.
3. Halberg F. Temporal coordination of physiologic function. Cold Spr Harb Symp quant Biol 1960;25:289–310. Discussion on LD50, p. 310.
4. Halberg F. Biological as well as physical parameters relate to radiology. Guest lecture. In: Proceedings of the 30th Annual Congress of Radiology, Chandigarh: Post-Graduate Institute of Medical Education and Research, January 1977, 8 pp.
5. Halberg F, Cornélissen G, Schwartzkopff O. Introduction: time, diagnostics, and therapeutics – beyond circadian marker rhythm-guided treatment. In: Youan BC, editor. Chronopharmaceutics: science and technology for biological rhythm-guided therapy and prevention of diseases. Hoboken: Wiley; 2009. p. xvii–xxxii; Halberg F, Cornélissen G, Schwartzkopff O. Chapter 10: Implications and applications of circadian susceptibility rhythms: chronomics and anesthesia. In: Youan BC, editor. Chronopharmaceutics: science and technology for biological rhythm-guided therapy and prevention of diseases. Hoboken: Wiley; 2009. p. 217–55; Cornélissen G, Halberg F. Chapter 11: Treatment with open eyes: markers-guided chronotheranostics. In: Youan BC, editor. Chronopharmaceutics: science and technology for biological rhythm-guided therapy and prevention of diseases. Hoboken: Wiley; 2009. p. 257–323.
6. Halberg F, Cornélissen G, Wang ZR, Wan C, Ulmer W, Katinas G, Singh R, Singh RK, Singh RK, Gupta BD, Singh RB, Kumar A, Kanabrocki E, Sothern RB, Rao G, Bhatt MLBD, Srivastava

M, Rai G, Singh S, Pati AK, Nath P, Halberg F, Halberg J, Schwartzkopff O, Bakken E, Shastri VK. Chronomics: circadian and circaseptan timing of radiotherapy, drugs, calories, perhaps nutriceuticals and beyond. J Exp Ther Oncol 2003;3:223–60.

7. Ulmer W, Cornélissen G, Halberg F. Physical chemistry and the biologic week in the perspective of chrono-oncology. In Vivo 1995;9:363–74.
8. Ulmer W, Cornélissen G, Halberg F. Interaction among (quantum mechanical) resonance-coupled electromagnetic circuits relevant to a natural week. World Heart J 2012; 4 (1): 35–70.
9. Halberg F, Barnum CP, Silber RH, Bittner JJ. 24-hour rhythms at several levels of integration in mice on different lighting regimens. Proc Soc Exp Biol Med (NY) 1958;97:897–900.
10. Litman T, Halberg F, Ellis S, Bittner JJ. Pituitary growth hormone and mitoses in immature mouse liver. Endocrinology 1958;62:361–4.
11. Halberg F, Haus E, Cardoso SS, Scheving LE, Kühl JFW, Shiotsuka R, Rosene G, Pauly JE, Runge W, Spalding JF, Lee JK, Good RA. Toward a chronotherapy of neoplasia: tolerance of treatment depends upon host rhythms. Exp (Basel) 1973;29:909–34.
12. Suda M, Hayaishi O, Nakagawa H, editors. Biological rhythms and their central mechanisms. Naito Foundation. Amsterdam: Elsevier North-Holland Biomedical Press; 1979. See Halberg F, Halberg E, Halberg J. Collateral-interacting hierarchy of rhythm coordination at different organization levels, changing schedules and aging, p. 421–34 (see also discussion p. 435–8).
13. Halberg F, Lubanovic WA, Sothern RB, Brockway B, Powell EW, Pasley JN, Scheving LE. Nomifensine chronopharmacology, schedule shifts and circadian temperature rhythms in di-suprachiasmatically lesioned rats – modeling emotional chronopathology and chronotherapy. Chronobiologia 1979;6:405–24.
14. Scheving LE, Pauly JE, Burns ER, Halberg F, Tsai S, Betterton HO. Lighting regimen dominates interacting meal schedules and synchronizes mitotic rhythm in mouse corneal epithelium. Anat Rec 1974;180:47–52.
15. Osborne AR, Refinetti R. Effects of hypothalamic lesions on the body temperature rhythm of the golden hamster. Neuro Rep 1995;6:2187–92.
16. Refinetti R, Cornélissen G, Halberg F. Unilateral SCN ablation amplifies while bilateral SCN ablation dampens the circadian rhythm in core temperature of hamsters. In: Proceedings of the international conference on the frontiers of biomedical science: chronobiology, Chengdu; Sept 24–26, 2006, p. 196–8.
17. Halberg F, Cornélissen G, Sonkowsky RP, Lanzoni C, Galvagno A, Montalbini M, Schwartzkopff O. Chrononursing (chronutrics), psychiatry and language. New Trends Exp Clin Psychiatr 1998;14:15–26.
18. Cornélissen G, Halberg F, Siegelova J, Galvagno A. The moon's image in prolonged human isolation. In: Halberg F, Kenner T, Fiser B, Siegelova J, editors. Noninvasive methods in cardiology. Czech Republic, Brno: Masaryk University; 2010. p. 69–74.
19. Wever RA. The circadian system of man: results of experiments under temporal isolation. New York: Springer; 1979, p. 279.
20. Halberg F, Conner RL. Circadian organization and microbiology: variance spectra and a periodogram on behavior of *Escherichia coli* growing in fluid culture. Proc Minn Acad Sci 1961;29:227–39.
21. Sturtevant R. Circadian variability in *Klebsiella* demonstrated by cosinor analysis. Int J Chronobiol 1973;1:141–6.
22. Hastings JW, Schweiger HG, editors. Life sciences research report 1: the molecular basis of circadian rhythms. Report of the Dahlem Workshop on the Molecular Basis of Circadian Rhythms, Berlin, November 3–7, 1975. Berlin: Dahlem Konferenzen; 1976. p. 462.
23. Halberg F, Cornélissen G, Faraone P, Poeggeler B, Hardeland R, Katinas G, Schwartzkopff O, Otsuka K, Bakken EE. Prokaryotic and eukaryotic unicellular chronomics. Biomed Pharmacother 2005;59 Suppl 1:S192–202.
24. Cornélissen G, Halberg F, Schwartzkopff O. Competing tidal and circadian lunisolar resonance in an archaeon. World Forum "Natural Cataclysms and Global Problems of the Modern Civilization". Istanbul: SWB; 2011. p. 127.
25. Halberg F, Cornélissen G, Katinas GS, Watanabe Y, Otsuka K, Maggioni C, Perfetto F, Tarquini R, Schwartzkopff O, Bakken EE. Feedsidewards: intermodulation (strictly) among time

structures, chronomes, in and around us, and cosmo-vasculo-neuroimmunity. About ten-yearly changes: what Galileo missed and Schwabe found. In: Conti A, Maestroni GJM, McCann SM, Sternberg EM, Lipton JM, Smith CC, editors. Neuroimmunomodulation (Proceedings 4th International Congress, International Society for Neuroimmunomodulation, Lugano; Sept 29–Oct 2, 1999). Ann NY Acad Sci 2000; 917: 348–76.

26. Halberg F, Breus TK, Cornélissen G, Bingham C, Hillman DC, Rigatuso J, Delmore P, Bakken E, International Womb-to-Tomb Chronome Initiative Group: Chronobiology in space. Keynote, 37th Ann. Mtg. Japan Society for Aerospace and Environmental Medicine, Nagoya, November 8–9, 1991. University of Minnesota/Medtronic Chronobiology Seminar Series, #1, December 1991, 21 pp. of text, 70 figures.
27. Roederer JG. Are magnetic storms hazardous to your health? Eos, transactions. Am Geophys Union 1995;76:441. 444–5.
28. Vladimirskii BM, Narmanskii VYA, Temuriantz NA. Global rhythmics of the solar system in the terrestrial habitat. Biophysics 1995;40:731–6.
29. Halberg F, Cornélissen G, Otsuka K, Watanabe Y, Katinas GS, Burioka N, Delyukov A, Gorgo Y, Zhao ZY, Weydahl A, Sothern RB, Siegelova J, Fiser B, Dusek J, Syutkina EV, Perfetto F, Tarquini R, Singh RB, Rhees B, Lofstrom D, Lofstrom P, Johnson PWC, Schwartzkopff O, International BIOCOS Study Group. Cross-spectrally coherent ~10.5- and 21-year biological and physical cycles, magnetic storms and myocardial infarctions. Neuroendocrinol Lett 2000;21:233–58.
30. Sothern RB, Katinas GS, Fiser B, Siegelova J, Cornélissen G, Halberg F. A transtridecadal cycle in human heart rate: selective infradian, notably multidecadal solar-physiologic BEL congruences. In: Halberg F, Kenner T, Fiser B, Siegelova J, editors. Proceedings, noninvasive methods in cardiology, Brno, Czech Republic;, Oct 4–7, 2008. p. 204–13. http://web.fnusa.cz/files/kfdr2008/sbornik_2008.pdf
31. Enfield DB, Mestas-Nuñez AM, Trimble PJ. The Atlantic multidecadal oscillation and its relation to rainfall and river flows in the continental U.S. Geophys Res Lett 2001;28(10):2077–80.
32. Schlesinger ME, Ramankutty N. An oscillation in the global climate system of period 65–70 years. Nature 1994;367:723–6.
33. Kerr RA. A north Atlantic climate pacemaker for the centuries. Science 2006;288(5473):1984–6.
34. Halberg F, Cornélissen G, Otsuka K, Syutkina EV, Masalov A, Breus T, Viduetsky A, Grafe A, Schwartzkopff O. Chronoastrobiology: neonatal numerical counterparts to Schwabe's 10.5 and Hale's 21-year sunspot cycles. In memoriam Boris A. Nikityuk. Int J Prenat Perinat Psychol Med 2001;13:257–80.
35. Sello S, Halberg F, Cornélissen G. Human babies: a slow-to-read, sensitive population magnetometer, also read by wavelets. In: Halberg F, Kenner T, Fiser B, Siegelova J, editors. Noninvasive methods in cardiology. Brno: Masaryk University; 2011. p. 123–40.
36. Syutkina EV, Cornelissen G, Mitish M, Narogan MV, Surgyk AV, Krylova OS, Masalov A, Schwartzkopff O, Halberg F. Neonates as particularly sensitive magnetoreceptors? World Forum "Natural Cataclysms and Global Problems of the Modern Civilization". Istanbul: SWB; 2011. p. 114–5.
37. Cornélissen G, Halberg F, Wendt HW, Bingham C, Sothern RB, Haus E, Kleitman E, Kleitman N, Revilla MA, Revilla Jr M, Breus TK, Pimenov K, Grigoriev AE, Mitish MD, Yatsyk GV, Syutkina EV. Resonance of about-weekly human heart rate rhythm with solar activity change. Biologia (Bratisl) 1996;51:749–56.
38. Halberg F, Cornélissen G, Katinas G, Tvildiani L, Gigolashvili M, Janashia K, Toba T, Revilla M, Regal P, Sothern RB, Wendt HW, Wang ZR, Zeman M, Jozsa R, Singh RB, Mitsutake G, Chibisov SM, Lee J, Holley D, Holte JE, Sonkowsky RP, Schwartzkopff O, Delmore P, Otsuka K, Bakken EE, Czaplicki J. International BIOCOS group. Chronobiology's progress: season's appreciations 2004–2005. Time-, frequency-, phase-, variable-, individual-, age- and site-specific chronomics. J App Biomed 2006;4:1–38. http://www.zsf.jcu.cz/vyzkum/jab/4_1/halberg.pdf.

39. Halberg F, Cornélissen G, Wilson D, Singh RB, De Meester F, Watanabe Y, Otsuka K, Khalilov E. Chronobiology and chronomics: detecting and applying the cycles of nature. Biologist 2009;56(4):209–14.
40. Herschel W. Observations tending to investigate the nature of the sun, in order to find the causes or symptoms of its variable emission of light and heat; with remarks on the use that may possibly be drawn from solar observations. Phil Trans Roy Soc Lond 1801;91:265–318.
41. Maasch KA. What triggers ice ages? http://www.pbs.org/wgbh/nova/earth/cause-ice-age.htm Posted 01.01.97.
42. Sigel F, Dreier W, Lerche D, Ubers, Göring H. Wissenschaftl. Red. der deutschsprachigen. Schuld ist die Sonne. Thun/Frankfurt am Main: Harri Deutsch; 1979. p. 215.
43. Rohde RA, Muller RA. Cycles in fossil diversity. Nature 2005;434:208–10.
44. Cornélissen G, Bakken EE, Sonkowsky RP, Halberg F. A 38-million-year cycle among myriadennians in the diversity of oceanic genera. Abstract, III international conference, civilization diseases in the spirit of V.I. Moscow: Vernadsky, People's Friendship University of Russia, Oct. 10–12, 2005, p. 47–9.
45. Halberg F. Historical encounters between geophysics and biomedicine leading to the Cornélissen-series and chronoastrobiology. In: Schröder W, editor. Long- and short-term variability in sun's history and global change. Bremen: Science Edition; 2000. p. 271–301.
46. Cornélissen G, Tarquini R, Perfetto F, Otsuka K, Gigolashvili M, Halberg F. Investigation of solar about 5-month cycle in human circulating melatonin: signature of weather in extraterrestrial space? Sun and Geosph 2009;4(2):55–9.
47. Halberg F, Cornélissen G, Schwartzkopff O. Quo vadis chronomics 2008. Measuring variability in us, among us and around us. In: Halberg F, Kenner T, Fiser B, Siegelova J. editors. Proceedings, Noninvasive Methods in Cardiology, Brno, Czech Republic; Oct 4–7, 2008. p. 16–25. http://web.fnusa.cz/files/kfdr2008/sbornik_2008.pdf.
48. Cornélissen G, Halberg F, Sothern RB, Hillman DC, Siegelova J. Blood pressure, heart rate and melatonin cycles synchronization with the season, earth magnetism and solar flares. Scr med 2010;83:16–32.
49. Mikulecky M, Florida PL. Daily birth numbers in Davao, Philippines, 1993–2003: Halberg's transyear stronger than year. Abstract, 26th Seminar, Man in his Terrestrial and Cosmic Environment, Upice, Czech Republic; May 17–19, 2005.
50. Mikulecky M. Reanalysis of variability in south Brazil: Halberg's paraseasonality dominating again. International Conference on the Frontiers of Biomedical Science: Chronobiology, Chengdu; Sept 24–26, 2006, p. 187–8.
51. Kovac M, Mikulecky M. Time sequence of epileptic attacks from the point of view of possible lunisolar connections. International conference on the frontiers of biomedical science: chronobiology, Chengdu; Sep 24–26, 2006, p. 175–9.
52. Kovac M, Mikulecky M. Secular rhythms and Halberg's paraseasonality in the time occurrence of cerebral stroke. Bratisl Lek Listy 2005;106(2):423–7.
53. Pacheco de Andrade M, Cornélissen G, Burioka N, Halberg F. Transyear and cis-half-year accompany calendar-year in spectrum of human uric acid excretion. Proceedings, international conference on the frontiers of biomedical science: chronobiology, Chengdu; Sep 24–26, 2006, p. 155–9.
54. Nelson W, Cadotte L, Halberg F. Circadian timing of single daily "meal" affects survival of mice. Proc Soc exp Biol Med (NY) 1973;144:766–9.
55. Halberg F, Haus E, Cornélissen G. From biologic rhythms to chronomes relevant for nutrition. In: Marriott BM, editor. Not eating enough: overcoming underconsumption of military operational rations. Washington: National Academy Press; 1995. p. 361–72. http://books.nap.edu/books/0309053412/html/361.html#pagetop.
56. Halberg F, Cornélissen G, Bingham C, Fujii S, Halberg E. From experimental units to unique experiments: chronobiologic pilots complement large trials. In Vivo 1992;6:403–28.
57. Kennedy BJ. A lady and chronobiology. Chronobiologia 1993;20:139–44.
58. Halberg F, Nelson W, Cornélissen G, Haus E, Scheving LE, Good RA. On methods for testing and achieving cancer chronotherapy. Cancer Treat Rep 1979;63:1428–30.

59. Halberg F, Prem K, Halberg F, Norman C, Cornelissen G. Cancer Chronomics I. Origins of timed cancer treatment: early marker rhythm-guided individualized chronochemotherapy. J Exp Ther Oncol 2006;6:55–61.
60. Halberg F, Nelson W, Cornélissen G, Haus E, Scheving LE, Good RA. Chronochemotherapy: L1210 leukemia and beyond. Chronobiologia 1979;6:203–11.
61. Kumagai Y, Cornélissen G, Fujimura A, Halberg F, Kharlitskaya EV, Ikonomov O, Blagonravov ML, Chibisov SM, Radysh IV. Chronotherapy of vascular variability disorders: a challenge for the clinic. Proceedings, 1st International Workshop, Physiology of Adaptation and Quality of Life: Problems of Traditional Medicine and Innovation, People's Friendship University of Russia, Moscow; May 14–16, 2008. p. 404–7.
62. Watanabe Y, Cornélissen G, Halberg F, Beaty L, Siegelova J, Otsuka K, Bakken EE. Harm vs. benefit from losartan with hydrochlorothiazide at different circadian times in MESOR-hypertension or CHAT. In: Halberg F, Kenner T, Fiser B, Siegelova J, editors. Proceedings, Noninvasive Methods in Cardiology, Brno, Czech Republic; Oct 4–7, 2008. p. 149–67. http://web.fnusa.cz/files/kfdr2008/sbornik_2008.pdf.
63. Halberg F, Cornelissen F, Halberg F, Kessler T, Otsuka K. Measuring mental strain by duration of blood pressure overswing (CHAT): case report. World Heart J 2010;2(2):141–67.
64. Selye H. Stress without distress. Philadelphia/New York: J.P. Lippincott; 1974. 171 pp.
65. Brandes V, Cornélissen G, Hillman D, Ilyia E, Cegielska N, El-Khoury M, Strestik J, Finley J, Thomas F, Kino T, Chrousos GP, Singh RB, Mikulecky M, Halberg F. Sanoson music therapy of unwellness in a case of recurrent adynamic depression with circadian difrequentia. World Forum "Natural Cataclysms and Global Problems of the Modern Civilization". Istanbul, Turkey: SWB; 2011. p. 134.
66. Mikulecky M. Paroxysmal tachyarrhythmia and moon. Chronobiologia 1990;17:71–3.

Chapter 3
Chronobioethics: What Is Life?

Abstract Franz Halberg (1919–2013) developed chronobiology and founded the field of chronomedicine including chronomics, chronoastrobiology, and chronobioethics. The progress of chronomedicine is one and the same as his history of research. This chapter introduces his foundation of chronobioethics.

Natural geophysical near-weeks, half-years, the solar wind's near- and far-transyears, etc., have evolved in living matter into a transdisciplinary system of coperiodisms in and around us. These ubiquitous cycles, some in the archaeon and the human, constitute congruent signatures not only of the photic day and year but also of the nonphotic, including gravitational and magnetic cyclic influences upon our mental functions in particular. They leave their signatures in terrorism, military-political affairs, crime, suicide, church memberships, and religious proselytism.

Keywords Glocal methods in time and space • Aeolian original data • Chronousphere • Chronobioethics

3.1 Encounters

Prof. Dr. (med.) Dr. (theol.) Mag. (pharmacist) Gustav Sauser, whom I (FH) met while he practiced as a pharmacist in Wels, Upper Austria, after he lost his professorship during the occupation of Austria in World War II, granted me an opportunity for an academic connection. After the war, he was appointed as the head of the anatomy department at the University of Innsbruck and in succession became its dean and rector. Sauser offered me a room in his bombed-out anatomy department building; as his assistant, I participated in international academic weeks [1, 2] where I met the physicist Arthur March (a friend of Erwin Schrödinger), with whom I developed an interest in physics. There, I also read Galileo and the statement attributed to him that I (and others including physicists and historians I consulted had cited) could never identify in his writings: "Measure what is measurable and render measurable what as yet is not," which became the guiding principle of my endeavors and to which I added qualifications like "in time and hence meaningfully": "Measure, in a partial system, everything pertinent that is measurable and render measurable what is pertinent but as yet is not measurable, as simply as possible, but not simpler, in time (tempestive) and hence meaningfully, taking into account

K. Otsuka et al., *Chronomics and Continuous Ambulatory Blood Pressure Monitoring*, DOI 10.1007/978-4-431-54631-3_3

chronomic maps, a described and quantified complementary system" [Omnia propria ex systemate partiali metire quæcumque licet et propria immensa, quam simplicissime sed non simplicius, ad mensuram tempestive et ergo significative redige, reddens rationem tabulae chronomicae – systematis complementaris, descripti et quantificati].

While in Innsbruck from 1946 to 1948, I also headed the section of venereology and dermatology at the French military hospital, where I encountered my first case in which timing could possibly have played a role. In two identically treated soldiers who contracted gonorrhea from the same prostitute on consecutive nights, the same treatment begun the same day cleared gonococci first from the patient who was infected a day later [3]. I also owe Sauser my World Health Organization fellowship that brought me to Harvard in 1948.

3.2 Opportunities Enjoyed and Yet to Be Followed Up On

In Minnesota in 1950, I met Prof. Alan Treloar, who renewed my interest in the inferential statistical validation of all I learned. In 1961, I began receiving a lifetime career award from the US National Institutes of Health (NIH); at over 92 years of age, I am still receiving it. I may be one of only two still-active recipients of these awards. A contract to study the endocrine associations of breast cancer, also from the NIH data, led to multifrequency models of infradian risk assessment. We found modulations by both about-30-day and about-yearly rhythms of the circadian MESOR and amplitude demonstrating them stepwise as partial spectral structures based on parameters obtained by the least-squares fit of cosine curves of 24 hours and 28 and 365 days to data on plasma prolactin, Figs. 3.1, 3.2, 3.3, and 3.4 [4].[1] These are obtained every 20 min, and data on other hormones are obtained at 100-min intervals for 24 hours, in 4 seasons on 3 age groups of Japanese and Minnesotan women, Fig. 3.1a–f, to derive tentative gauges of breast cancer and other disease risks. We then wrote something still pertinent today:

> Once we realize that a spectrum of rhythms exists [Figures 3.1, 3.2, 3.3, and 3.4] and what its components are, we have a chance to specify how, with very few samples if not with a single sample, at the right time for each pertinent variable, one may [try to] achieve multiple aims. One may quantify certain aspects of health. One may recognize risk — before disease sets in, as a basis for prevention. One may also learn when one should look for what and where in the body — in a study of physiological mechanisms, safeguarding our performance — the neuroendocrine and cellular feedsidewards leading to a yet to be tested hemopsy [4–7]

[1] A much larger circannual amplitude in circulating prolactin of Japanese as compared to American women, based on nearly 3,000 determinations [4], had as a result that in different seasons at different clock hours, timing can certainly make the difference between finding or not finding a higher prolactin in plasma of Japanese vs. Americans, and numerically the results can even happen to be opposite at different clock hours in certain seasons [4]. Based on fewer samples, a comparison of melatonin excretion by the two populations of women at different clock hours can also show opposite results.

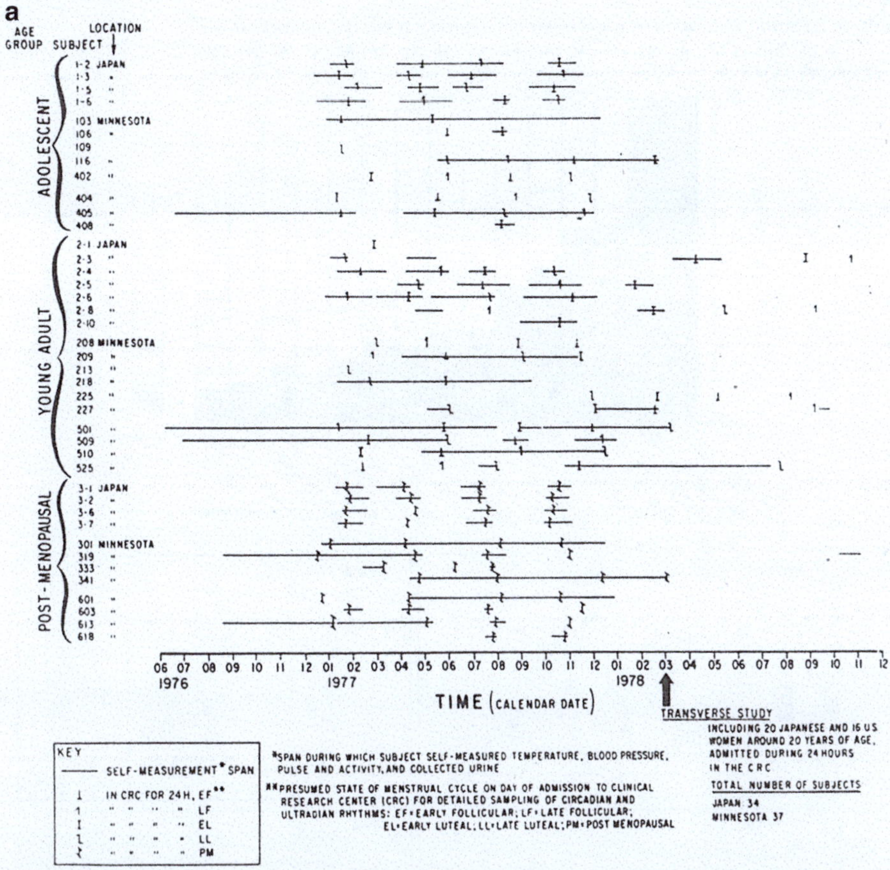

Fig. 3.1 (**a**) Individual sampling schemes (spans) for assessment of multifrequency rhythms in human blood and urine. Experimental design of Minnesota-Japan chronoepidemiologic study on women of 3 age groups: schedule of hospital admission and at-home self-measurements of subjects classified according to age and geographic location, with specification of menstrual cycle stage on day of admission. Limitation: ignorance, at the time, of quinmensals and transyears discovered later in some other hormones that remain to be examined (© Halberg Chronobiology Center). (**b**) Comparison of statistical significance of circadian rhythm, assessed by least-squares fit of a 24-hour cosine curve for 13 hormones in plasma of healthy women sampled for individualized rhythm assessment by cosinor. P from F-test of zero-amplitude hypothesis. Since the hierarchical statistical significance represented by P-values is sampling dependent, prolactin and cortisol, measured every 20 min, were also analyzed at 100-min intervals as for the other hormones. Total number of determinations: 40,765.(© Halberg Chronobiology Center). (**c**) Comparison of plasma prolactin in clinically healthy Japanese and Minnesotan women in four seasons. Samples taken at 20-min intervals over a 24-hour span in Fukuoka City, Japan, and Minneapolis, Minnesota, USA (© Halberg Chronobiology Center). (**d**) Geographic difference in circannual rhythm of human plasma prolactin. Contradictory comparisons of circadian MESOR could be obtained by sampling in different seasons. Moreover, controlling only time of year (or only time of day) may not avoid misleading results due to circannual (or circadian, as the case may be) variations, whereas assessment of circannual (or circadian) rhythm characteristics may reveal important chronoepidemiological or chronoprotopathological results (as would most likely extended focus on rhythms

b

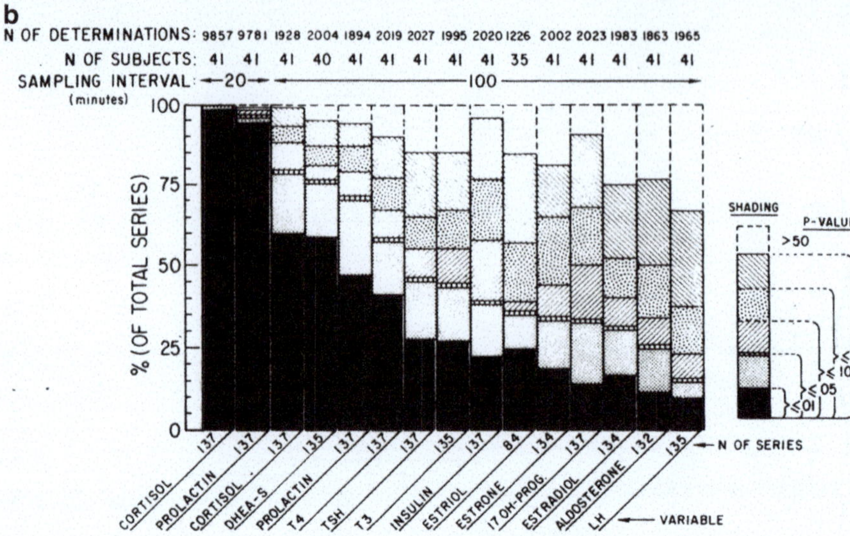

c

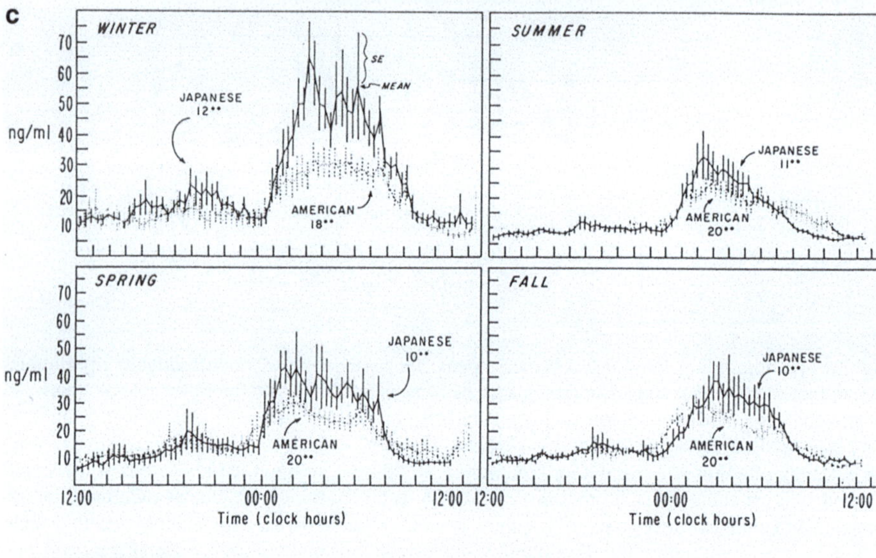

Fig. 3.1 (continued) discovered in the interim). Conclusions from single samples obtained in different seasons (or times of day) can be resolved as differences of circannual (circadian) prolactin rhythm between Japanese and American women. Each subject provided 72 blood samples (8640 determinations, each in duplicate). *W* winter, *Sp* spring, *S* summer, *F* fall (© Halberg Chronobiology Center). (**e**) Circadian rhythm of plasma prolactin in Japanese and North American women in winter. Dense sampling on plasma prolactin suggests both a larger circadian amplitude and a higher rhythm-adjusted mean or MESOR in Japanese as compared to North American women ($P<0.002$ from Hotelling T^2 test) (© Halberg Chronobiology Center). (**f**) Difference of plasma prolactin between clinically healthy Japanese and American women in March 1978 when circannual rhythm is near its maximum, confirmed by complementary study. Fifteen whites of mixed ethnic background (18–24 years old) in Minnesota and 20 Japanese (~20 years old) in Kyushu, Japan (© Halberg Chronobiology Center)

d

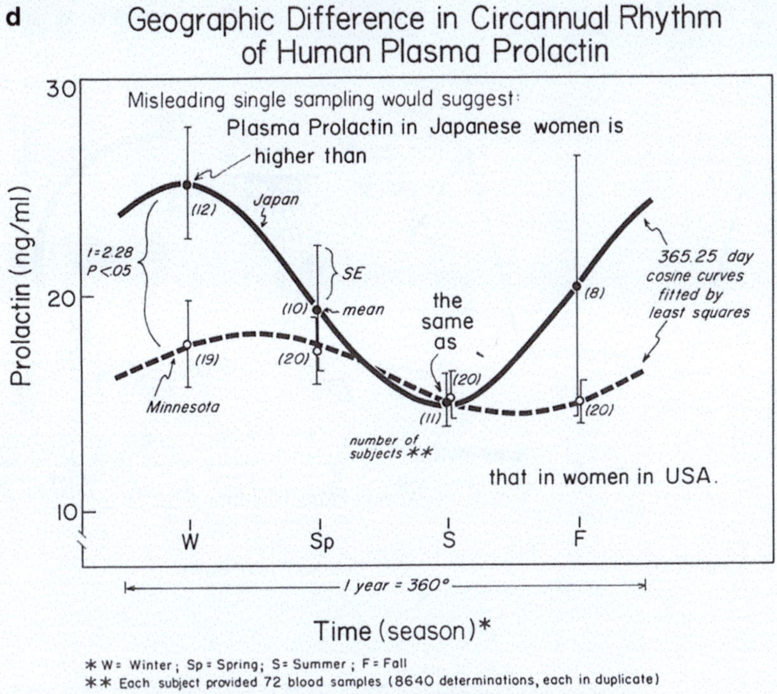

e

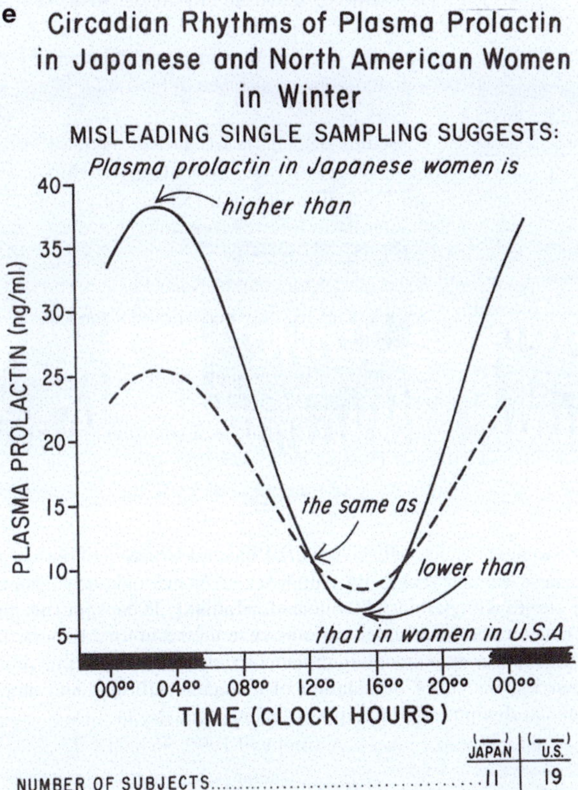

	JAPAN (—)	U.S. (– –)
NUMBER OF SUBJECTS	11	19
NUMBER OF SERIES (SOME SUBJECTS PROVIDED 2 SERIES)	15	26
TOTAL NUMBER OF DETERMINATIONS (IN DUPLICATE)	1070	1824

Fig. 3.1 (continued)

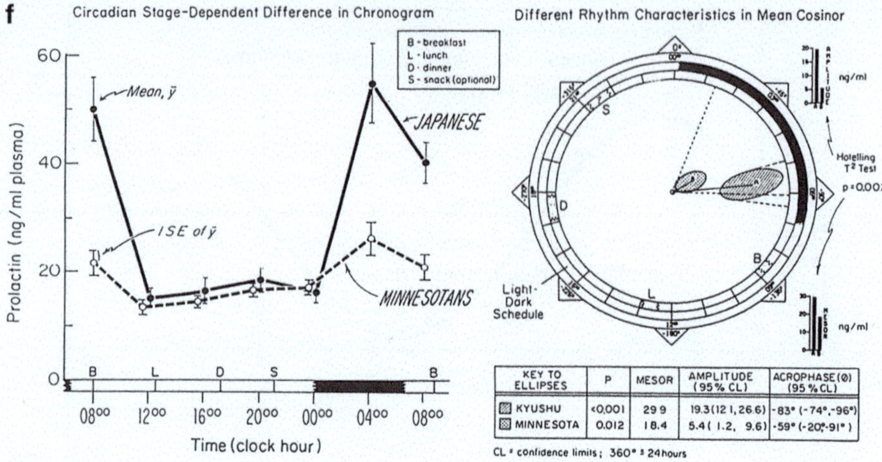

Fig. 3.1 (continued)

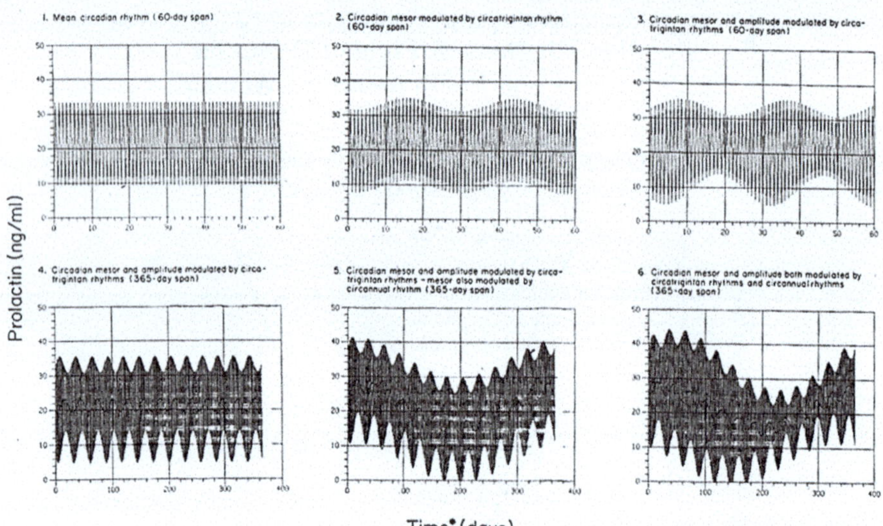

Fig. 3.2 Stepwise theoretical reconstruction of partial spectral structure of human plasma prolactin for group of Japanese women (based on parameter estimates obtained from separate least-squares fits of cosine curves with periods of 24 hours [circadian], 28 days [circatrigintan], and 365 days [circannual] to data on plasma prolactin obtained every 20 min for 24 hours, ~4 times a year on a few women). Didactic example modeling the interaction of human plasma prolactin rhythms with different frequencies, including modulation of circadian MESOR and amplitude by circatrigintan and circannual rhythms (© Halberg Chronobiology Center)

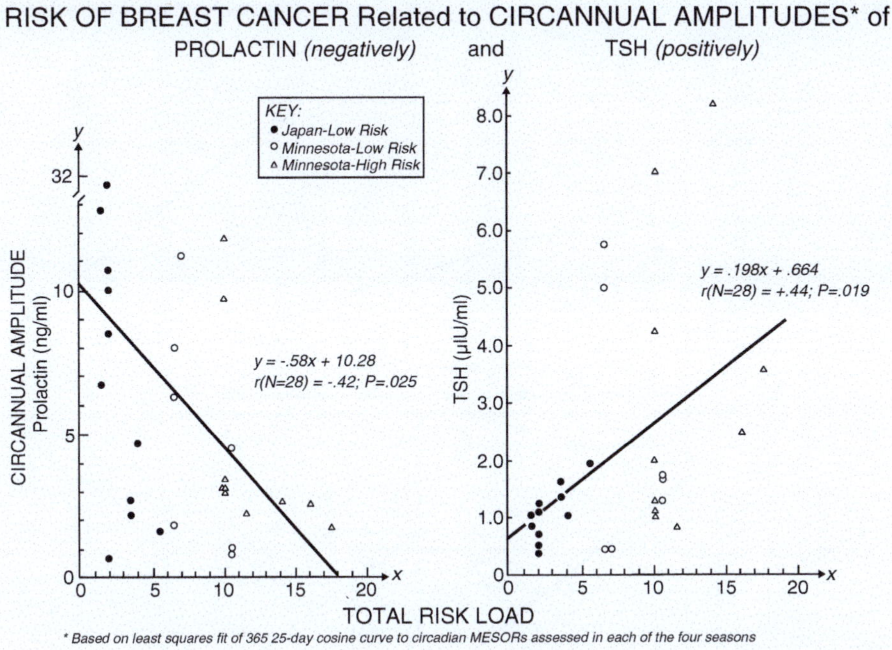

Fig. 3.3 A negative correlation between the total relative breast cancer risk evaluated from epidemiological criteria and the circannual prolactin amplitude (*left*) corroborates the finding that an evaluation of breast cancer risk is associated with a decrease in circannual amplitude (based on least-squares fit of 365.25-day cosine curve to circadian MESORs assessed in each of the four seasons). Note further the positive correlation between epidemiologically assessed breast cancer risk and the circannual amplitude of thyroid-stimulating hormone (TSH) (*right*). Clinical thyroid malfunction has empirically been associated with breast cancer risk. If, then, this topic is still controversial, this may perhaps be accounted for by the circannual stage dependence of the correlation. Note also that in prostatic cancer (a condition characterized by geographic differences in morbidity and mortality similar to those of breast cancer), the extent of circannual variation also changes as a function of risk and/or cancer. In blood sampled with serial independence in the morning at different times of the year, a prominent circannual rhythm in TSH of healthy subjects is lost in prostatic cancer (and perhaps even in men at high risk of prostatic cancer). For prolactin, a circannual rhythm becomes demonstrable in the case of prostatic cancer, while it is not demonstrable with serially independent sampling in healthy men of low or high prostatic cancer risk. Thus, TSH and prolactin show opposite behavior along the 1-year scale in cancers of both breast and prostate (rather than responding in the same way, as is the case along the scale of minutes to hours, following the application of stimuli such as thyrotropin-releasing hormone). The two correlations shown in the figure are just part of a larger correlation matrix. The circumstance is noted that correlations emerged as statistically significant for the very hormones which clinicians have long considered have some relation to breast cancer, yet thus far could not rigorously establish such a relation as biologically significant, perhaps because of too-limited sampling. While these conclusions rest on large samples, they describe only a small number of subjects. Moreover, all conditions required to apply a linear regression between two variables are not satisfied. A test for lack of fit indicates that the model is not adequate for TSH; the error term is not normally distributed. In addition, the assumption of homogeneity of variance is not verified for prolactin as well as for TSH. Finally, in the case of prolactin, there seems to be an age effect on both this circannual amplitude (decreases with age) and the breast cancer risk (increases with age). This may account for the negative correlation illustrated in the figure. Hence, the correlations in the figure are of ordering rather than documenting value. They are intended to emphasize that circannual rhythmicity deserves further study in relation to carcinogenesis. If such correlations can be confirmed and if the circannual rhythms involved should prove to be determinants of carcinogenesis in the human breast, these same correlations will point to the possibility of a chemoprevention of breast cancer (© Halberg Chronobiology Center)

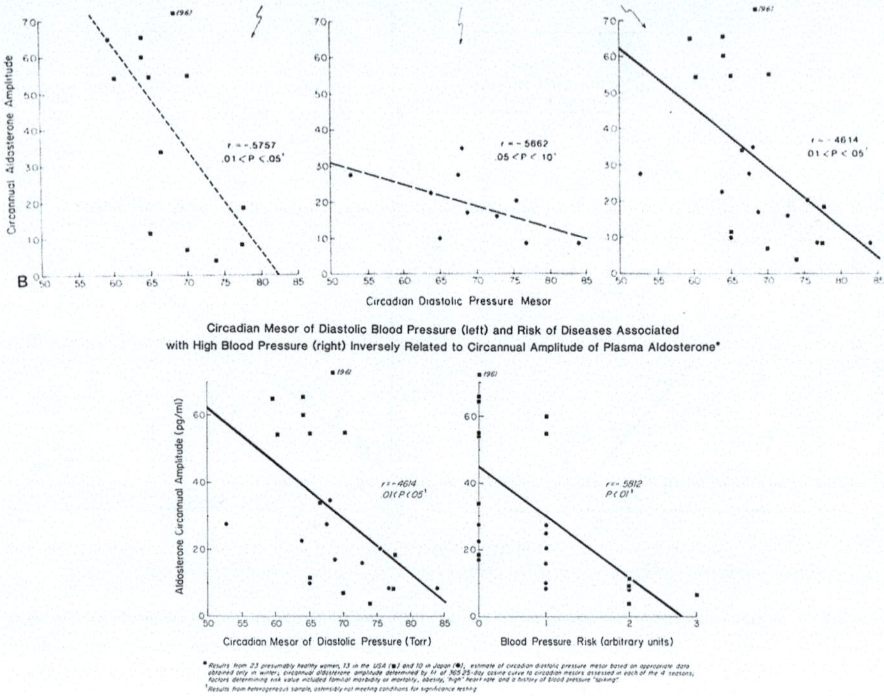

Fig. 3.4 *Top*: negative correlation between the circannual aldosterone amplitude (based on least-squares fit of 365.25-day cosine curve to circadian MESORs assessed in each of the four seasons on each of the women investigated in Minnesota and Japan) and the circadian diastolic blood pressure MESOR (in winter, the season when blood pressure profiles at about 10-min intervals were obtained in both locations) observed in women in Minnesota ($N=13$, *left*), Kyushu, Japan ($N=10$, *middle*), and both ($N=23$, *right*). *Bottom*: negative correlation of the circannual aldosterone amplitude and both the circadian diastolic blood pressure MESOR and the individual risk of diseases associated with high blood pressure. Results from 23 presumably healthy women, 13 in the USA (■) and 10 in Japan (●); estimate of circadian diastolic blood pressure MESOR based on appropriate data obtained only in winter; circannual aldosterone amplitude determined by fit of 365.25-day cosine curve to circadian MESORs assessed in each of the four seasons; factors determining risk value included familial morbidity or mortality, obesity, "high" heart rate, and a history of blood pressure spiking. The same restrictions as outlined for Fig. 3.1g applies to Fig. 3.2 and this figure. For qualification, see legend of Fig. 3.3 (© Halberg Chronobiology Center)

The extension of this model of 1981 [4] to include the periods mapped in the interim of 30 years as coperiodisms in and around us, up to paratridecadals, remains a major challenge for marker rhythmometry.

Whenever possible, for disease risk assessment by infradian modulation, we should prefer the noninvasively collected saliva or urine to plasma. A study of 11,702 samples on six hormones in saliva collected every 4 hours around the clock for many months reveals new facts, including not only multiple distinct frequencies in a very broad spectrum but also, i.e., multiple circadians. Multiple circaseptan frequencies, Fig. 3.5a, b, were also found in cases where, under life in contemporary society, we expected to find only a single circadian or circaseptan frequency, which

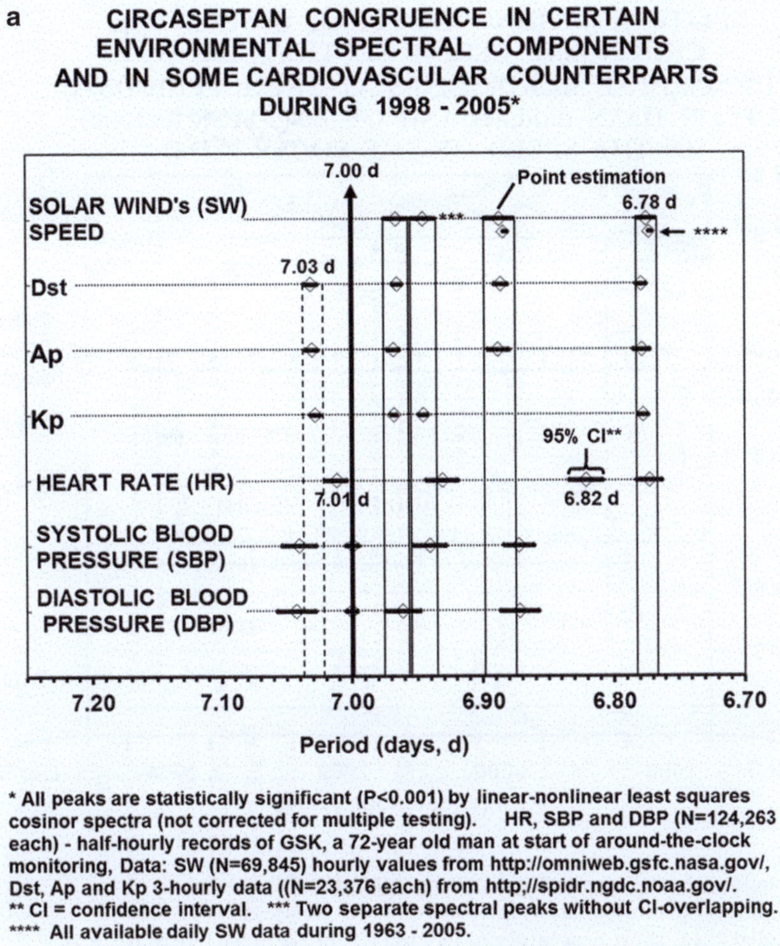

Fig. 3.5 (**a**) Transdisciplinary congruence, defined by overlapping 95 % confidence intervals of the periods in the circaseptan spectral region (© Halberg Chronobiology Center). (**b**) Combination of gliding spectral window (*top*) with special focus on the behavior of two selected periods (serial sections, *middle* and *bottom*), with the time course of the phase validating the 6.77-day period by a more or less horizontal trajectory of phases in only part of the record, but invalidating a precise 7-day periodicity since no dots bracket any horizontal time course of phases, and the small initial section with dots shows a gradual advance (© Halberg Chronobiology Center). (**c**) Fourier power spectral density of rest activity of a clinically healthy man (JFC) on a self-selected sleep-wake schedule (© Halberg Chronobiology Center). (**d**) Sleep-wakefulness on a self-selected schedule analyzed by the fit of cosine curves of 24.00 hours to intervals of 168 hours (top 3 rows) or of 24.84 hours to intervals of 173.88 hours displaced with increments of 12.00 or 12.42 hours, respectively. Nonlinear analyses (not shown here) reveal coexisting circadian periods including a 24.84-hour period (24.828; 24.859) during an initial more or less 24 hour synchronized span seen in row 2. Note, in the first 18 lunar months, in row 2, the dominating, albeit undulating and slightly delaying, not quite horizontal time course of doubly plotted acrophases indicating perhaps the interaction of a 24.00-hour schedule with a competing free-running and/or other mechanism; periods shorter than 24.8 hours are seen in the first 15 lunar months in row 4. This row also shows spans of 24.8-hour (earth tidal, also greatly undulating) synchronization (Original data from John F. Costella © Halberg Chronobiology Center)

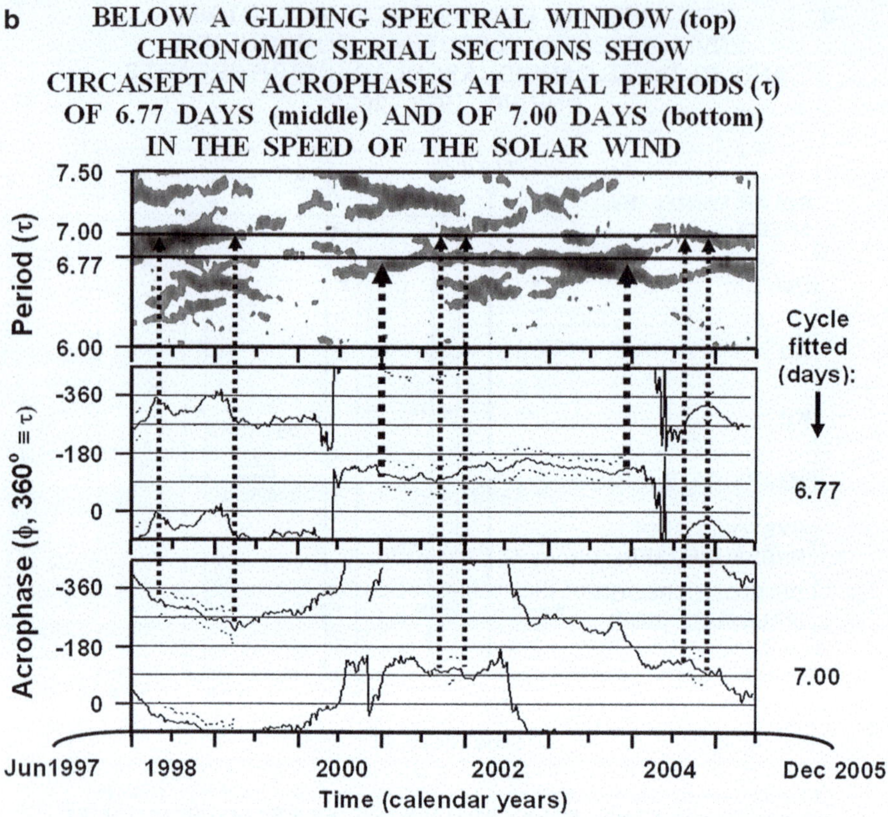

Data from http://omniweb.gsfc.nasa.gov/html/ow_data.html.
In gliding spectral window, interval = 1 year, increment = 1 week;
shaded areas show percentage of rhythm (from 1.2 to 6.8 %). Statistical significance seen as 95% confidence intervals of ϕs shown as dots bracketing curves (middle and bottom). Reference time: 00:00 on Dec 21, 1997. Dotted arrows indicate correspondence between circaseptan components in the phase (middle and bottom) and period (top) domains. 23-rd solar cycle began in May 1996, its maximum was in Apr 2000 and the cycle still continued descending after the end of record.

Fig. 3.5 (continued)

is or is not environmentally synchronized. As noted earlier, there are multiple coexisting circadian frequencies, Table 3.1. Thus, we arrive at new diagnoses; once we rely on time series, we may quantify even the effects of the earth tides on humans [8]. Serial data in turn require glocal methods in time and space, notably if focus is extended beyond the circadian system.

In the early 1960s, representatives from NASA were the first to knock on my door and ask what they could do for Minnesotan chronobiology. NASA, the National Science Foundation (NSF), and NIH enabled the completion of a laboratory for period-

c Fourier power spectral density of a clinically healthy man (JFC) on a self-selected sleep-wake schedule

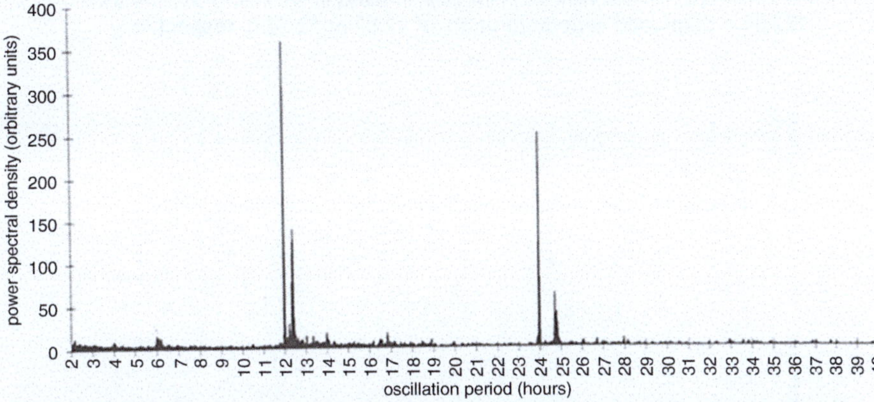

Original analysis by John F. Costella, available from http://assassinationscience.com/johncostella/sleep/sleep-report.pdf (1). Note 4 components with dominant circasemidian component, raising the question of this component being more than a descriptor of the waveform.

Fig. 3.5 (continued)

icity analysis, with standardization, if not full control, of environmental temperature and lighting, and as best we could of sound and smell, since they were the candidate factors accounting for the synchronization of the eosinophil count's circadian rhythm in commonly but not in singly housed mice without eyes (deprived of them by surgery or genetics, ZRD). NASA also supported the introduction of temperature and other telemetry in preparation for a biosatellite study that remains a challenge, as does that of a lunar laboratory [9]. In extraterrestrial space, away from hospitals, it will be especially important to detect undue strain and to alleviate it, a task also critical on earth.

It cannot be overemphasized that cycles, when not considered, can be confounders, or when two cycles being compared differ in phase, or in frequency or for those very many investigators sampling at a single clock hour with an interest in aging. A human mental function, such as the duration of an estimation of one minute, may decrease with age at one time of day, while at another clock hour in the same person in the same longitudinal time series, an increase with age is found, Fig. 3.6. This is a necessary a priori consequence of a decrease in circadian amplitude (or of a change in phase) with age. If other characteristics do not change (they actually do in the case considered), a decrease with age of the circadian amplitude results in an increase at the trough, and a decrease at the peak of a daily rhythm, as readily visualized by a 24-hour cosine fit at the start and at the end (not shown).

Fortunately, the puzzles were solved and nonsensical partial results were not published. Interest in full- and overtime study of the cosmos resulted from our inability to validate a test developed and published by ourselves in Minnesota [10], confirmed abroad for 2 years, yet subject to a modulation by an initially unknown circadecadal rhythm, which could eventually be found in the circadian blood pressure amplitude of newborns. The challenge remains to test other-than-circadian,

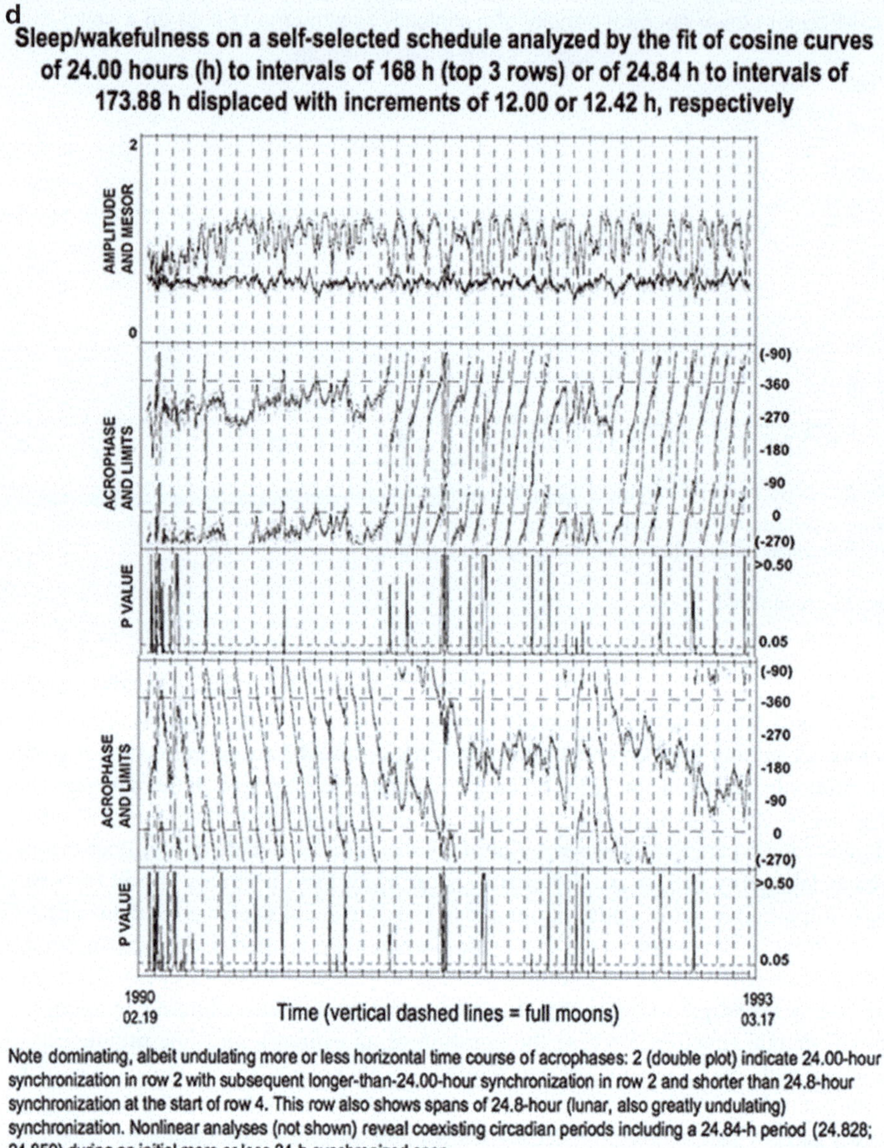

Fig. 3.5 (continued)

e.g., circaseptan amplitudes that may be less confounded by decadal changes for diagnosing the neonatal risk of developing cardiovascular disease later in life, even though an effect of geomagnetism upon them has been demonstrated [11].

Sooner rather than later, we must deal with a broad spectrum of nonphotic, partly gravitational, and/or magnetic novel paired transdisciplinary cycles in and around us in the sphere of the human mind (noös), Vernadsky's noösphere. Mikulecky and

Table 3.1 On a self-selected schedule (JFC), a double tidal pull competes with 24-hour environment while initially coexisting free-running circadian fades (global analysis)

	Calendar date		Period	Amplitude	DT-A/S-A (%)	
	All (3-year) data	Double tidal (DT)	24.836 (24.833, 24.838)	0.10606 (0.0854, 0.1268)	0.41685	Global
	1990/02/19 to 1993/03/11	Compromise	24.432 (24.427, 24.436)	0.05643 (0.0357, 0.0772)		
		Free-running	24.260 (24.252, 24.268)	0.03253 (0.0118, 0.0533)		
		Societal (S)	24.001 (24.000, 24.001)	0.25443 (0.2337, 0.2751)		
		Tidal	12.414 (12.410, 12.418)	0.01728 (–0.0034, 0.038)		
		Semidian	11.999 (11.997, 12.001)	0.03204 (0.0113, 0.0527)		
Local	1: 1990/02/19 to 1991/04/28	Double tidal	24.872 (24.830, 24.914)	0.02513 (–0.0070, 0.0573)	0.07951	
		Compromise	24.409 (24.384, 24.435)	0.04117 (0.0089, 0.0734)		
		Free-running	24.222 (24.210, 24.235)	0.07921 (0.0468, 0.1116)		
		Societal	24.008 (24.004, 24.011)	0.31605 (0.2838, 0.3483)		
		Tidal	12.421 (12.411, 12.431)	0.02615 (–0.0059, 0.0582)		
		Semidian	12.007 (12.001, 12.013)	0.04105 (0.0090, 0.0731)		
	2: 1991/04/28 to 1991/09/23	Double tidal	24.898 (24.832, 24.964)	0.12744 (0.0707, 0.1842)	0.55998	
		Compromise	24.707 (24.671, 24.743)	0.25208 (0.1973, 0.3068)		
		Compromise	24.488 (24.434, 24.541)	0.10274 (0.0552, 0.1502)		
		Societal	24.011 (23.992, 24.030)	0.22758 (0.1815, 0.2736)		
		Tidal	Did not converge	–		
		Semidian	–	–		
	3: 1991/09/23 to 1992/03/08	Double tidal	24.862 (24.852, 24.872)	0.34858 (0.3089, 0.3882)	1.55881	
		Compromise	24.523 (24.468, 24.579)	0.06108 (0.0212, 0.1009)		
		Free-running	Did not converge	–		
		Societal	23.997 (23.984, 24.011)	0.22362 (0.1841, 0.2631)		
		Tidal	12.432 (12.416, 12.448)	0.04969 (0.0106, 0.0888)		
		Semidian	Did not converge	–		

(continued)

Table 3.1 (continued)

	Calendar date		Period	Amplitude	DT-A/S-A (%)	
	4: 1992/03/08 to 1992/07/14	Double tidal	24.942 (24.896, 24.989)	0.15133 (0.0960, 0.2067)	0.47145	
		Compromise	24.451 (24.394, 24.508)	0.11847 (0.0630, 0.1739)		
		Free-running	–	–		
		Societal	23.987 (23.967, 24.006)	0.32099 (0.2660, 0.3760)		
		Tidal	12.432 (12.398, 12.465)	0.04960 (–0.0051, 0.1042)		
		Semidian	11.982 (11.951, 12.012)	0.05028 (–0.0044, 0.1049)		
	5: 1992/07/14 to 1993/03/11	Double tidal	24.789 (24.782, 24.797)	0.31836 (0.2798, 0.3569)	1.77934	
		Compromise	24.474 (24.438, 24.510)	0.06218 (0.0236, 0.1008)		
		Free-running	–	–		
		Societal	24.009 (23.997, 24.022)	0.17892 (0.1405, 0.2173)		
		Tidal	12.352 (12.328, 12.376)	0.02358 (–0.0147, 0.0619)		
		Semidian	12.001 (11.988, 12.013)	0.04128 (0.0030, 0.0796)		

The entire available data span was analyzed as a whole (top six rows) and then divided into five subspans according to visual inspection of characteristics of 24-hour and 24.84-hour fits in serial sections, suggesting similarities within the spans and differences among them. Initial period guesses for the six nonlinearly searched periods were 24.84, 24.43, 24.26, 24.00, 12.42, and 12.00 hours. Results were omitted if they did not converge to a period fitted to the data, in which case the analysis was redone after removal of these components from the model, to yield results summarized herein. Some periods converged well, but the CI (95 % conservative confidence interval) of the amplitude A (the CI given in parentheses) covered zero, as apparent from a negative sign inside the last column's second, fifth, and sixth sections, for the case of double tidal and tidal associations in the second section (top) where the societal 24-hour day predominates. The double tidal component converges invariably thereafter. Note that overlaps of zero by the CI of the A are very small. In 2 of 5 sections, the double tidal cycle has the largest A, the ratio of amplitudes of DT over S becomes greater than 1, suggesting that the moon/sun dominate over society in clinical health. These sections are not consecutive, possibly a chance observation, yet in keeping with the possibility that in health there is a wrangling between the double tide and society, as also seen in a patient with twice-yearly depression. From Costella et al. [66]

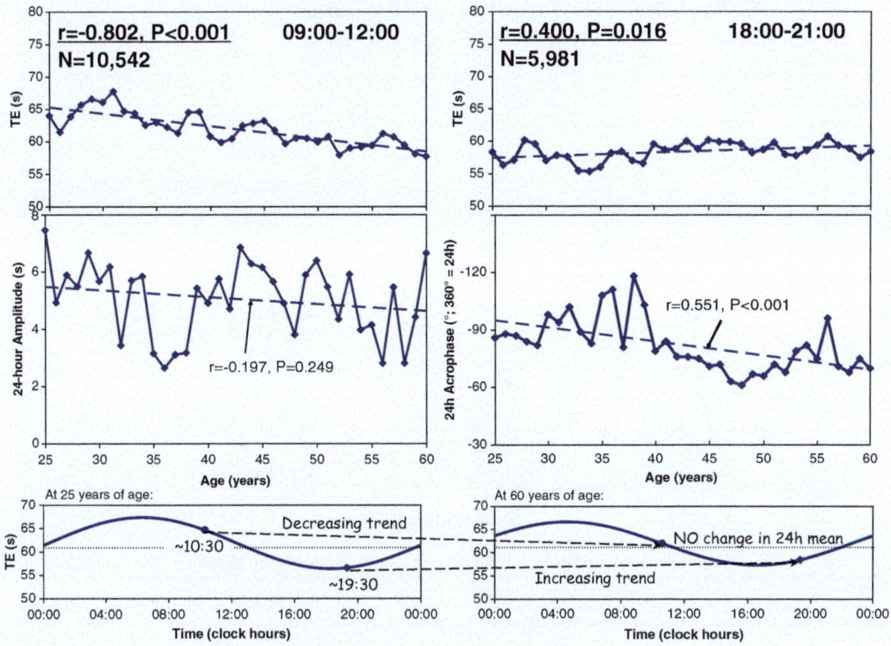

Fig. 3.6 The data of the first and last years of 1-min estimation by RBS were stacked along the 24-hour scale. Whereas parameter tests did not detect any change in the characteristics of the 24-hour synchronized rhythm, a paired *t*-test comparing eight 3-hourly mean values shows a difference ($P<0.05$). Moreover, Student's *t*-test on data in each of eight equidistant bins reveal that at certain circadian stages, 1 min passed faster in the last year than in the first year (year 35 vs. year 1). This difference was statistically significant at all test times between 06:00 and 18:00 (in some of them with $P<0.001$), but was not found between 18:00 and 06:00, suggesting that the change in 1-min time estimation with age is circadian stage dependent, with highly significant differences during part of the daily active phase, but not at other circadian times. In RBS, chronomics demonstrate interaction between the circadian rhythm's stage and age (© Halberg Chronobiology Center)

Pales' ~500-year cycle [12, 13] in the emergence of prominent poets, historians, and physicians, with counterparts in cave temperature and the incidence pattern of international battles [14], Fig. 3.7a–e, show the extent to which the cosmos modulates human affairs, including science and art, leading to a temporally structured chronousphere, coined by portmanteauing the Attic Gk nous (= noos), into Gk chronos (=time) and Gk sphairos (=globe as well as sphere). There is no choice for a unified art and science, whether in the service of health care, anthropology, or a unified science, but to monitor from birth to death with repeated passes over the accumulating data so as to determine loads before they become undue strain and harbingers of disease, to start with in blood pressure and heart rate. Monitoring has to start during wellness for the prevention of illness as soon as affordable unobtrusive instrumentation yields data analyzed via an international multilingual website, the recommendation of a World Forum on "Natural Cataclysms and Global Problems of the Modern Civilization," held September 19–21, 2011 in Istanbul, Turkey.

a

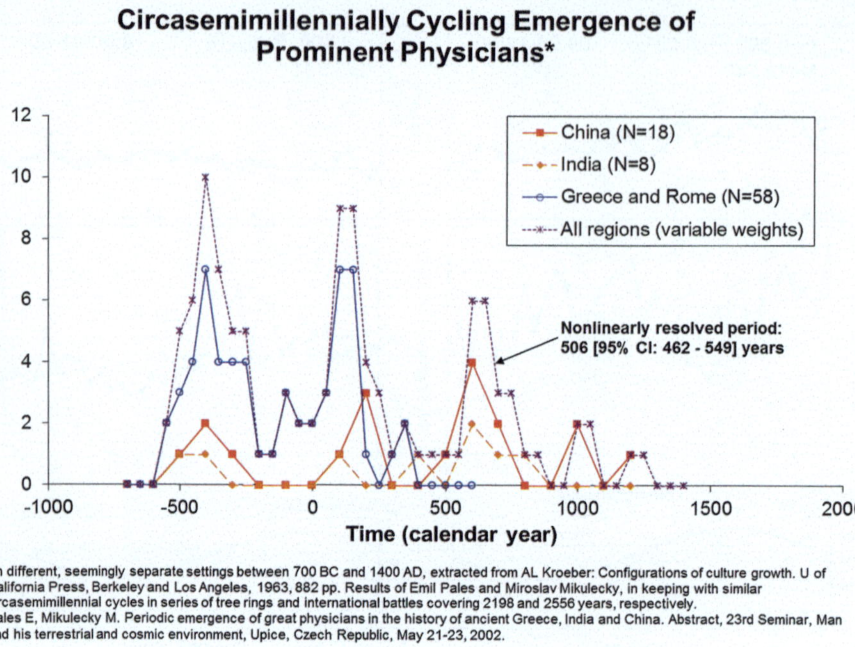

b

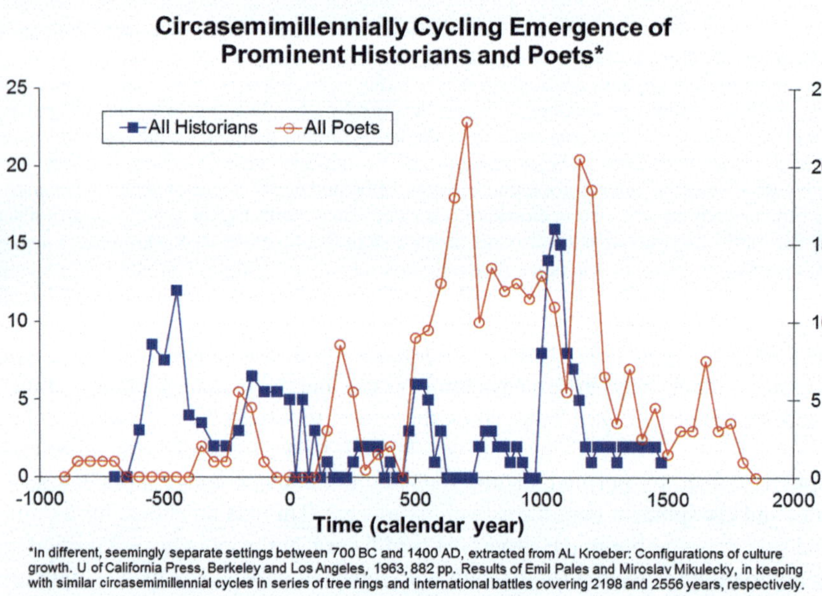

Fig. 3.7 (**a**) About-500-year cycles in the emergence of famous physicians are visible to the naked eye and seem to be synchronized in three completely different geographical regions among which there was originally little if any communication (© Halberg Chronobiology Center). (**b**) About-500-year cycles are also apparent in the emergence of famous historians and poets in three completely different geographical regions among which there was originally little if any communication (© Halberg Chronobiology Center). (**c**) Indirect proxy approximations of solar activity via the effects of climate upon the growth of trees during spans when no other human dynamic indices

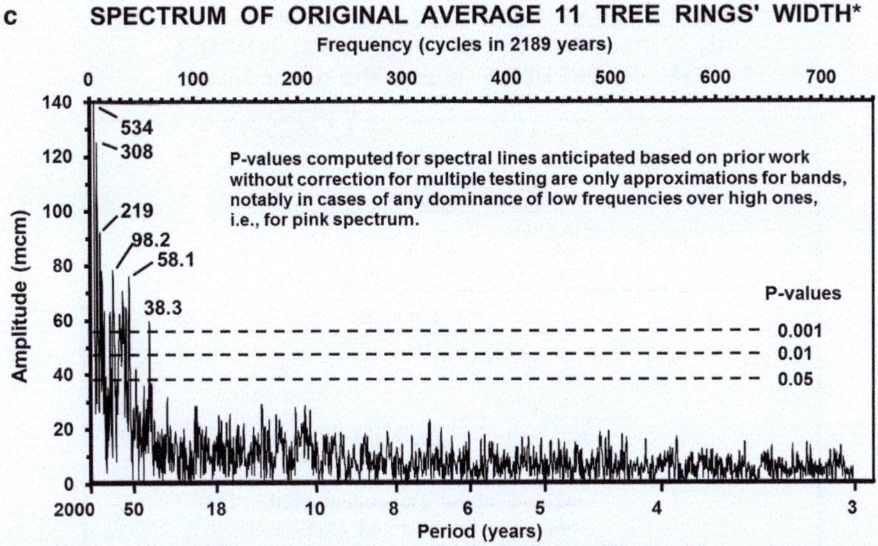

* Averaged values of 11 Californian sequoias. A.E. Douglass. Climatic Cycles and Tree-Growth. Carnegie Institution of Washington, Washington, 1919, pp.119-123.

Fig. 3.7 (continued) exist. A cycle with a period of over 500 years here shown was obtained in the course of studies reported earlier [15]. A similar cycle was also found, among others, in the spectrum of international battles (in log-transformed data) with a period of 499 years and a 95 % CI extending from 459 to 539 years, as also found in human creative cultural growth by Pales and Mikulecky [12, 13] (see Fig. 3.7a, b) (© Halberg Chronobiology Center). (**d**) Chart of about-500-year cycles in the emergence of great historians, physicians, and poets, compared with similar cycles in the Wheeler index of international battles, and in two series likely related to climatic changes, namely, tree ring widths and stalagmite coloring (© Halberg Chronobiology Center). (**e**) Acrophase chart of about-500-year cycles shown in Fig. 3.7d, estimated at average period (© Halberg Chronobiology Center)

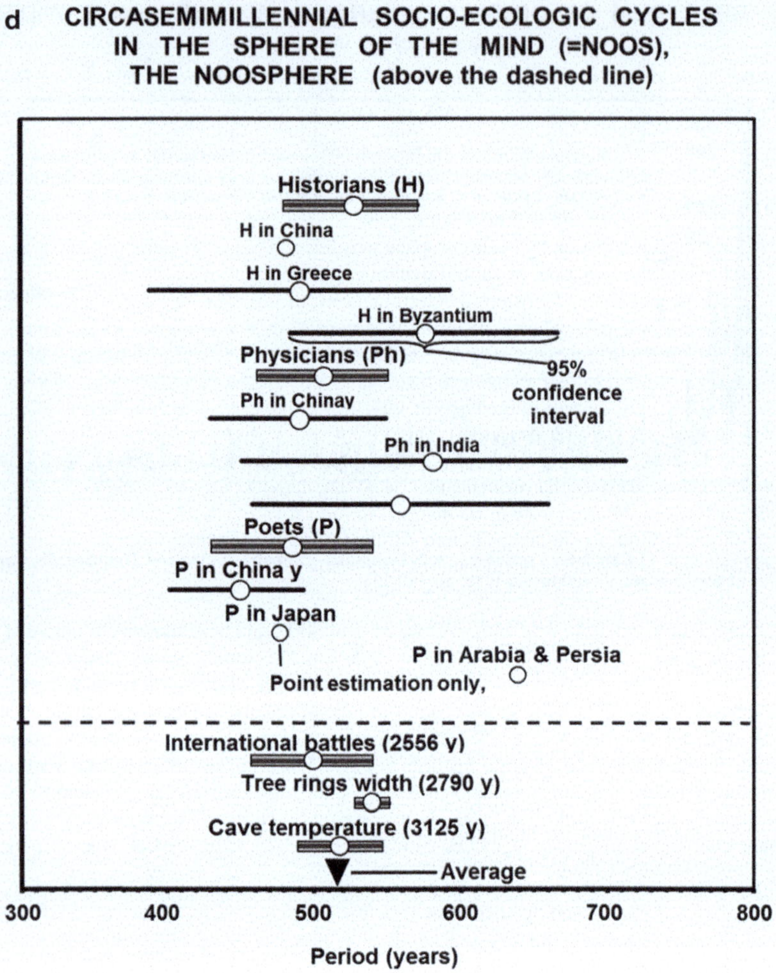

Fig. 3.7 (continued)

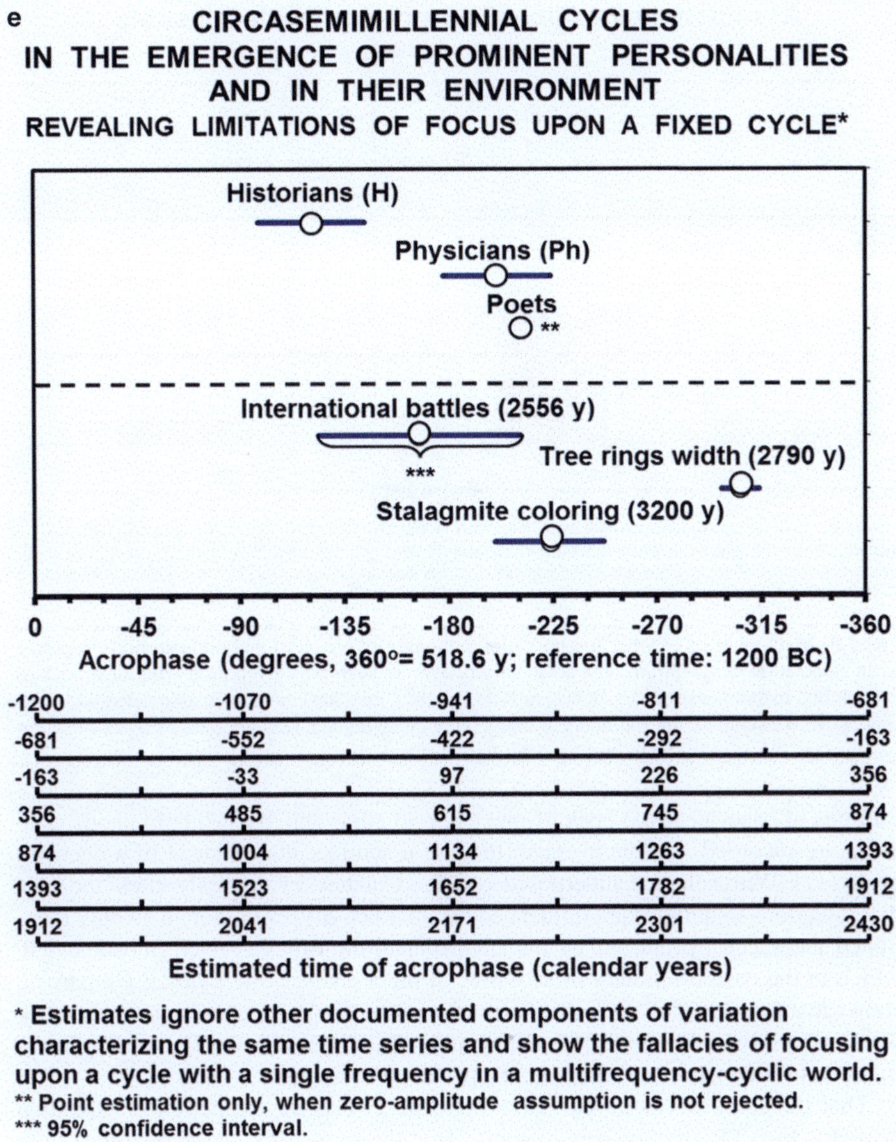

Fig. 3.7 (continued)

3.3 So What?

That aeolian sunspots (actually their cycles) affect many aspects of our physiopathology was documented by Vallot et al., whose data are meta-analyzed in Fig. 3.8. The proposition that infectious diseases such as cholera may be influenced by the cycle in Wolf numbers is one of Alexander Leonidovich Chizhevsky's many lines of evidence, sometimes aptly based on aeolian original data with solar cycles reporting

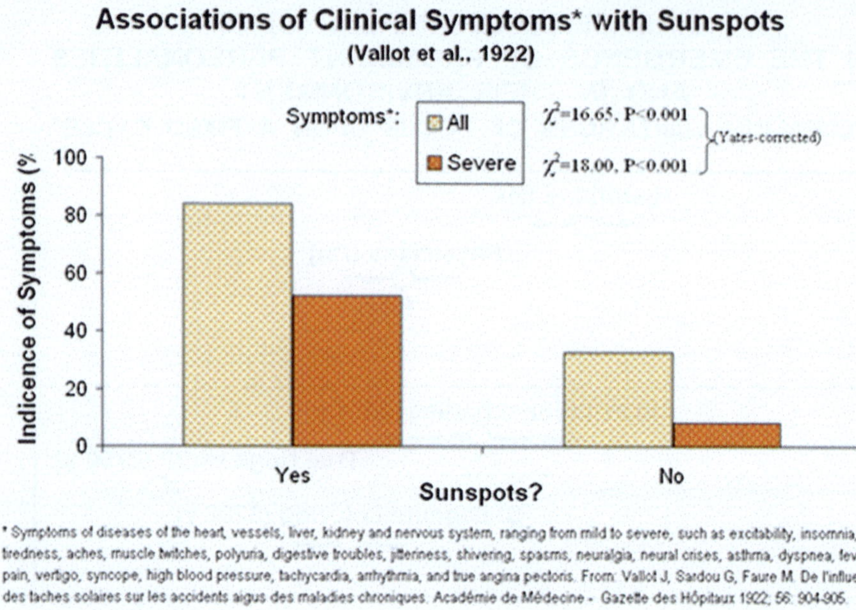

Fig. 3.8 Meta-analysis of what Joseph Vallot, a thoughtful physicist and philanthropist, reported as an association of symptoms with solar activity over 80 years ago. Even mild symptoms, such as excitability, insomnia, tiredness, aches, muscle twitches, etc., have long been associated with sunspots [16]. We stand on the shoulders of many others who noted a lead in phase by a day of symptoms versus sunspots (© Halberg Chronobiology Center)

millions of cases near the peak of one cycle of solar activity, while in another, no cases are recorded. On the average, there is a semidecadal (instead of a decadal) cycle [17]. With selected superposed epochs, Chizhevsky certainly made the case that infections in humans are influenced by solar activity, even though he only mentioned the spectra produced by his friend Vladimir Boleslavovich Shostakovich, which in the case of cholera in Moscow, in meta-analyses by Lyazzat Gumarova, shows that overall (globally), the harmonics of the circaundecennian sunspot cycle were more prominent than the fundamental, which was the one Chizhevsky illustrated as of interest, locally in time: the privilege of genius.

That the same hormone can have opposite effects as a function of circadian cyclicity, Fig. 3.9, may be new to some. That the same 2000 cal at breakfast-only can lead to body weight loss vs. body weight gain on dinner-only may be noteworthy. The doubling of a 2-year survival rate by cancer chronotherapy timed by tumor temperature is just a page of an overdue chrono-oncology based on triangulation guided by marker rhythms [7]. But when we deal with figurative cancers of the human mind (i.e., with man's inhumanity to man), we rest the case for a sphere of the human mind to be optimized, i.e., of a chronousphere, a page of anthropology par excellence with a reference to some of the poems of Chizhevsky (1897–1964) [18], who put his thoughts equally memorably (poetically) in prose: "Peut-être même nos sentiments et nos pensées ne sont-ils qu'un faible écho de ces vibrations

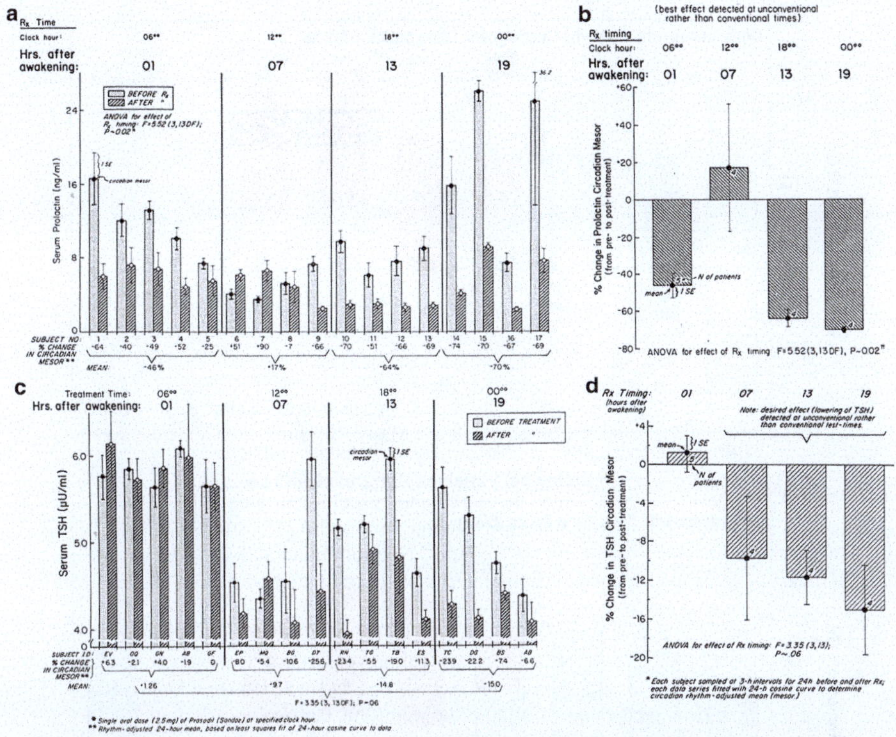

Fig. 3.9 Circadian stage dependence of effect of bromocriptine mesylate on circulating prolactin (**a**–**b**) and TSH (**c**–**d**) of patients with prostatic hypertrophy. Individual responses of patients at one of four different circadian stages are shown on the *left*. A summary by one-way analysis of variance (*right*) indicates that the treatment was effective in lowering prolactin when given at 1, 13, and 19 hours after awakening, but not at 7 hours after awakening. Likewise, the treatment was more effective in lowering TSH at 13 and 19 hours than at 7 hours after awakening, and it was not effective at 1 hours after awakening (© Halberg Chronobiology Center)

du cosmos .... Involontairement, une antique idée nous vient à l'esprit: notre connaissance des phénomènes de la nature ne serait pas autre chose qu'un écho, reçu par nos organes, des processus réels de l'univers" ("Perhaps even our feelings and thoughts are just a weak echo of the vibrations of the cosmos .... Involuntarily an old idea comes to mind: Our knowledge of natural phenomena will not be different from an echo, received by our organs, of the real processes of the universe") [18]. Actually, the analysis of a near 40-year record of self-measurements and self-ratings supports this aspect of his "cosmism" [19] by objective results in Fig. 3.10 [20, 21].

An influence on human psychophysiology, anticipated by Chizhevsky [18], of the antipodal index of geomagnetic disturbance aa (Fig. 3.10, top) and of the non-photic extraterrestrial environment (gauged by solar wind speed, an approximation of interplanetary magnetism; Fig. 3.10, bottom) was assessed by means of the congruence of periods of spectral components (defined by overlap of the 95 % confi-

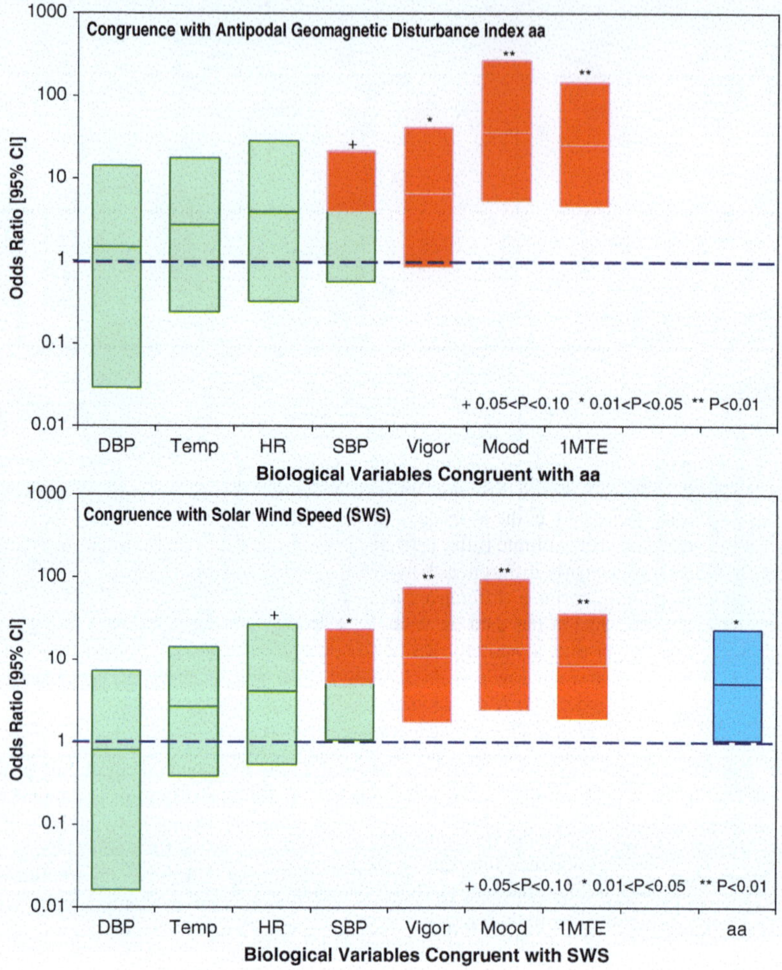

Fig. 3.10 By the criterion of the number of shared frequencies (with overlapping 95 % confidence intervals), odds ratios of human mental functions versus geomagnetics (*top*) or vs. interplanetary magnetism (*bottom*) more than match the accepted association of geo- and interplanetary magnetism (*blue*) (© Halberg Chronobiology Center)

dence intervals of the periods, in the frequency range of one cycle in 2.5 years to 3 cycles per year). The biological data stem from ~40 years of self-measurements (of oral temperature [Temp], systolic blood pressure [SBP], diastolic blood pressure [DBP], and heart rate [HR], and of ratings of mood and vigor and the estimation of 1 min by counting [1MTE], performed about 6 times a day by a clinically healthy man, Dr. Robert B. Sothern). Congruences (assessed by means of odds ratios based on the noncentral hypergeometric distribution) found for 1MTE and for other human mental functions more than equal those of the known association of helio- and geomagnetism (bottom, last column on right of dashed vertical line in blue). Mental functions (full red) show higher congruence than somatic functions (green).

Among the latter, systolic blood pressure (SBP) is responsive, perhaps constituting a seemingly acceptable proxy for the mental functions. P-values are based on the noncentral Fisher hypergeometric distribution, with 95 % confidence intervals computed using Fisher's exact test, used since the null hypothesis was rejected in some, yet not all cases [20, 21]. Much more remains to be learned on how the human mind can optimize its terrestrial, cosmically influenced habitat in the same language of frequencies.

3.4 Maps

The mapping of circadian rhythms with their uncertainties in several species is illustrated in Fig. 3.11a–d. It was a pleasure to comply with a request to publish one of these charts by an opinion-leading naturalist, Colin Pittendrigh, who included it, with its uncertainties, in his swan song [22]. For apparent clarity, it is unfortunately tempting for others to "simplify" such charts by omitting the wide, possibly still liberal 95 % confidence intervals. These, however, are indispensable for all mapping in and around us, as in Figs. 3.11, 3.12, and 3.13a–d, in a complementary monograph on chronoauxology [23], in an introduction to chronobiology, and in the *American Journal of Anatomy*. Infradian mapping must continue on both a population basis and in individuals' circulation. A map of a much broader psychophysiology by Robert B. Sothern [24, 25] already complements for this individual, in Fig. 3.14, by some aspects of infradian temporal coordination, the circadian map shown in Fig. 3.11, in this case only for populations.

Morphology in space and time exhibits multiple coexisting cycles, ranging in length from fractions of seconds to decades in individuals and to still longer periods in populations of living matter, including cycles of 142 [95 % CI: 133–150], 62 [60–63], and 38 [37–39] million years in species diversity. In populations, these cycles are synchronized among individuals; they can be demonstrated by time series to which each individual organism contributes just a single value. I learned this along the circadian scale from eosinophil counts and from RNA and DNA synthesis, in this order, on inbred mice, well over 50 years ago. During the individual human's lifespan, some of the decadal cycles demonstrated in population – across individuals by neonatal anthropometry over 112 years or longer (transverse sampling) – could be reproduced longitudinally [24–26]. Sooner or later, physiological monitoring may be from womb to tomb with technology already available for mice (to test drugs) and for tires (for security). In the case of tires, the pressure sensor technology was apparently developed in the mid-1980s but was not deployed to commercial applications until 2000, after a poorly manufactured tire model from Firestone on Ford trucks [27] was linked to about 140 deaths (Larry A. Beaty, personal communication).[2] The added mortality for the case of untreated vascular

[2] A small-scale ongoing chronobiologically interpreted C-ABPM system, serving for cost-free analyses of self-surveillance data by all comers by the project on The Biosphere and the Cosmos,

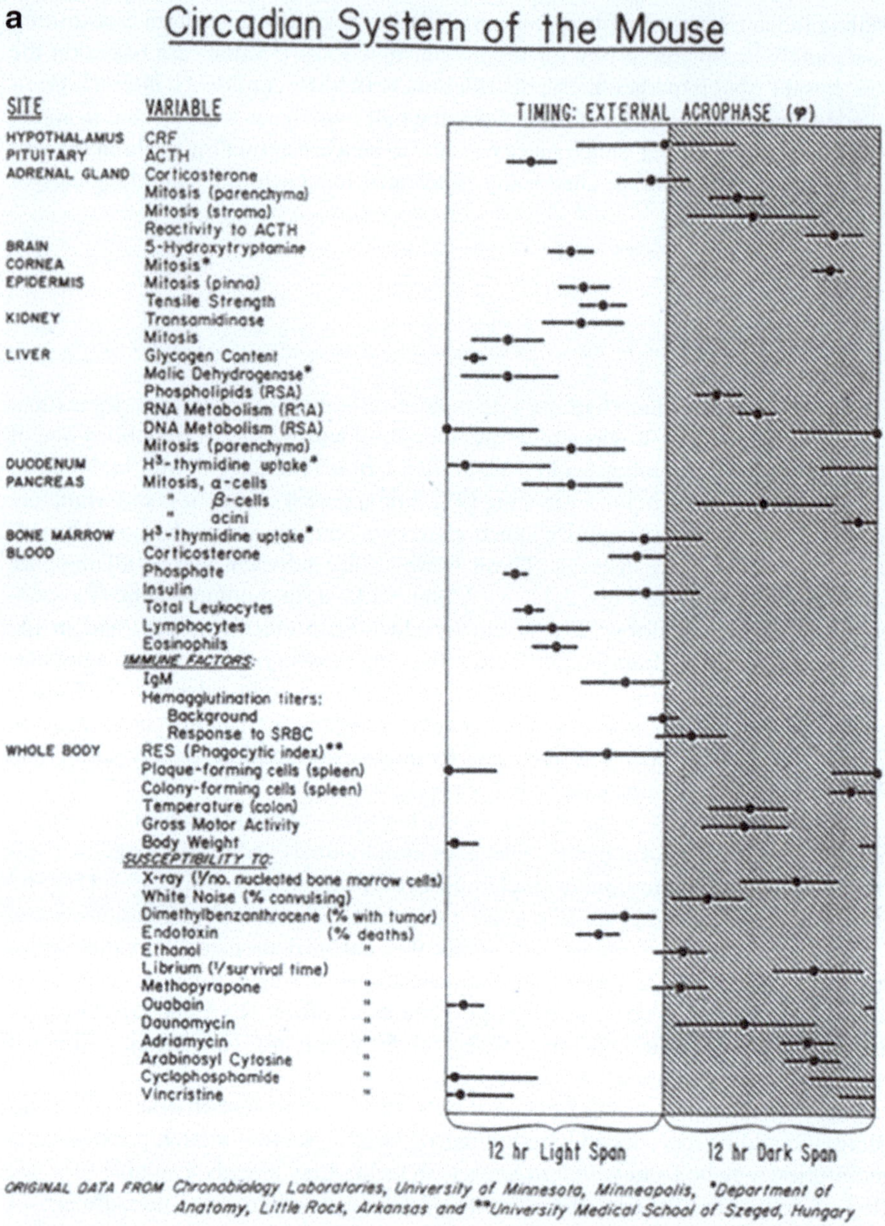

Fig. 3.11 (**a**) Phase chart. Circadian system of the rat (Note differences in phase as a division of labor in time) (© Halberg Chronobiology Center). (**b**) Phase chart. Circadian system of the rat (Note differences in phase as a division of labor in time) (© Halberg Chronobiology Center). (**c**) Phase chart. Human circadian system (Note differences in phase as a division of labor in time) (© Halberg Chronobiology Center). (**d**) Period chart. Human decadal-didecadal system (© Halberg Chronobiology Center). (**e**) Key to period chart. Human decadal-didecadal system (© Halberg Chronobiology Center)

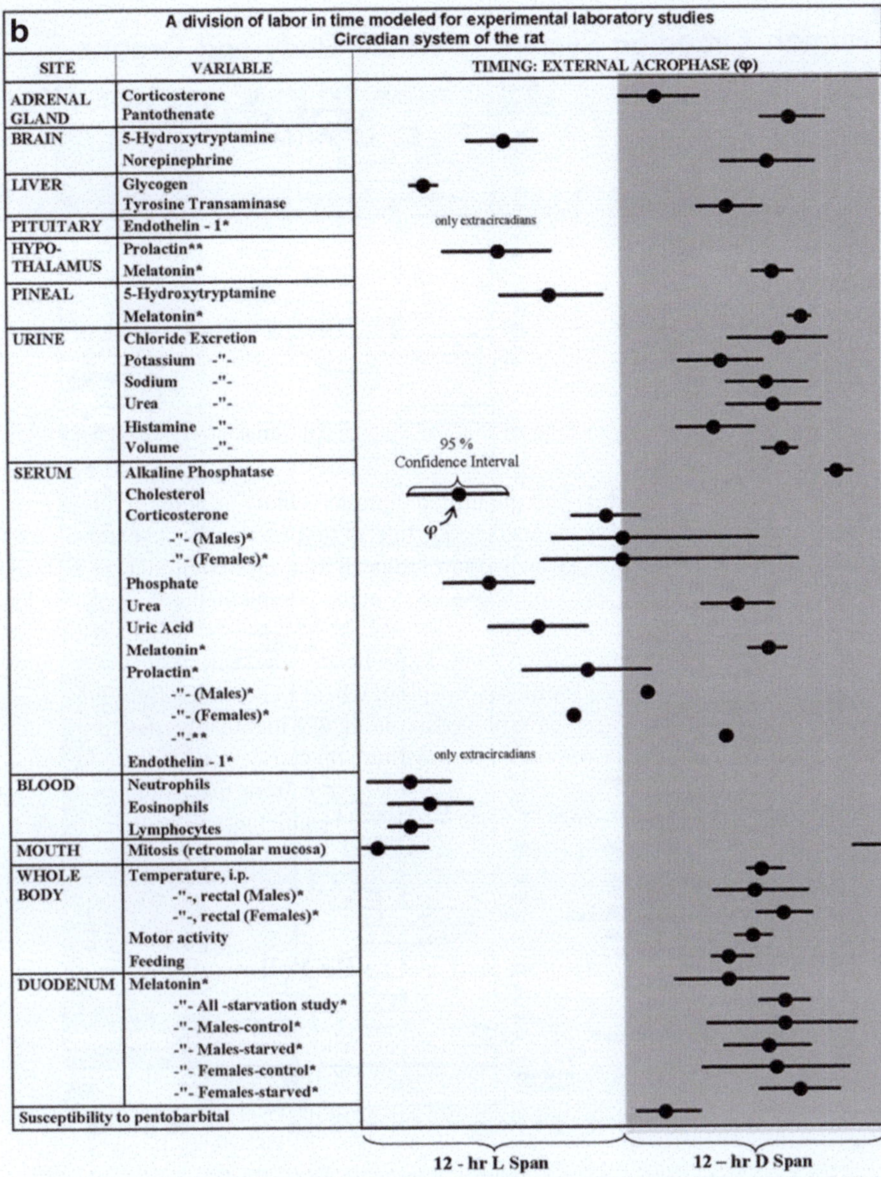

Original data from Chronobiology Laboratories, University of Minnesota, except for those on liver tyrosine transaminase (J. Axelrod), blood leukocytes (J. Pauly and L. Scheving), brain 5-hydroxytryptamine, norepinephrine, and susceptibility to pentobarbital (L. Scheving), on urinary volume and histamine (C. Wilson), and *Pecs study by A. Olah, R. Jozsa, G. Nagy, M. Zeman, V. Csernus, **G. Nagy

Fig. 3.11 (continued)

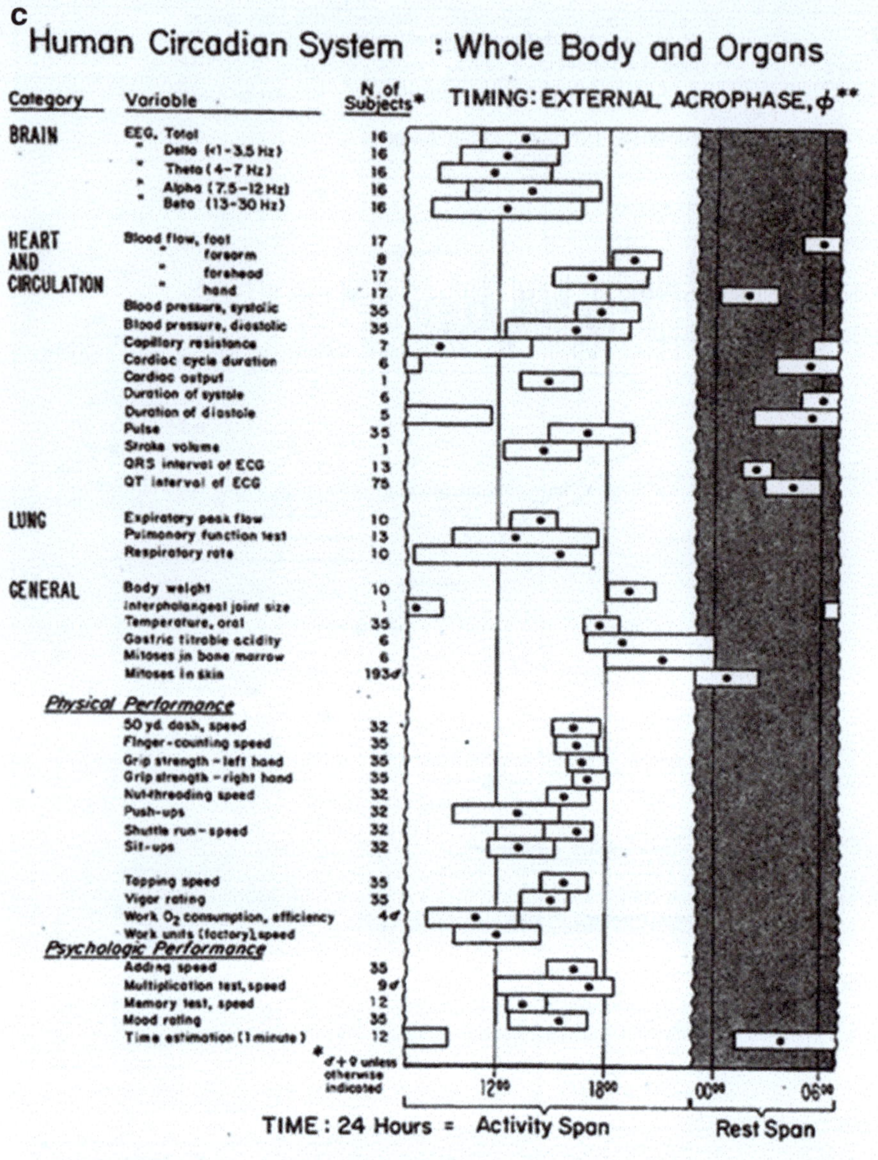

**Cosinor approximation of high values in circadian rhythm. Dots = acrophases; bars = 95 % confidence limits. Non-overlapping bars indicate statistically significant differences in timing.

Fig. 3.11 (continued)

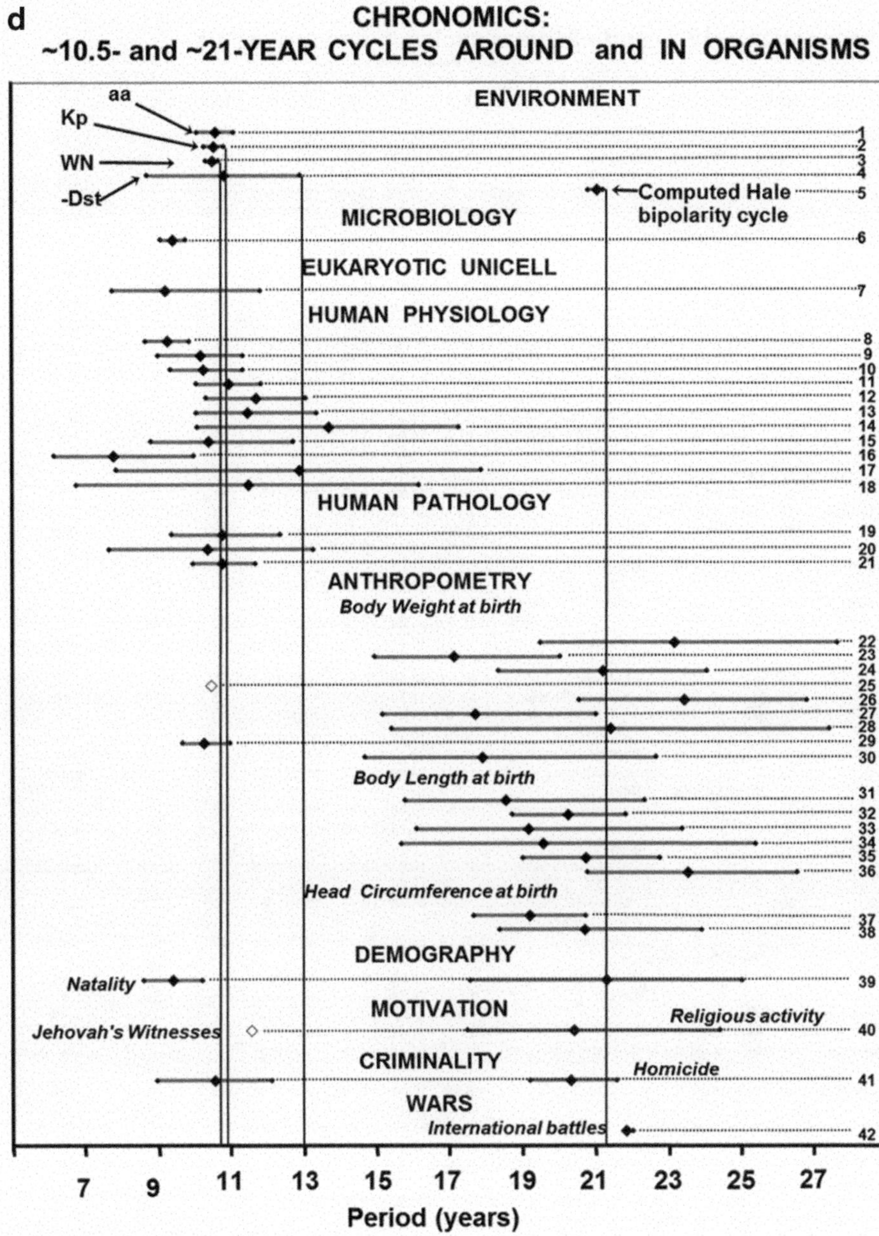

Fig. 3.11 (continued)

e

Chronomics: ~10.5- and ~21-year cycles in and around us

	Line		Period (years) Lower limit*	**Best Fit**	Upper limit*	Series duration Dates	Years	Number of data	Geographic site
Environment	1	aa = Antipodal Geomagnetic Index	10.12	**10.63**	11.13	1890-1999	110	1 / year	
	2	Kp = Planetary Geomagnetic disturbance	10.32	**10.58**	10.85	1932-1999	68	1 / month	
	3	WN = Wolf relative sunspot number	10.37	**10.54**	10.70	1890-1999	110	1 / year	
	4	-Dst = Equatorial geomagnetic disturbance	8.75	**10.85**	12.96	1973-1999	27	"	
			10.48	**10.51**	10.55	1700-1999	300	"	
	5	Bipolarity "Hale Cycle" **	20.86	**21.10**	21.26	1890-1999	110	"	
			21.42	**21.428**	21.43	1700-1999	300	"	
Biology	6	Prokaryotes: Air Bacterial Sectoring	9.12	**9.45**	9.81	1970-1982	13	3,744	Italy
	7	Eukaryotes: Unicellular Algal O_2 Production	7.79	**9.24**	11.87	1980-1991	11	324	Germany
Physiology* **	8	Mood (RBS)	10.11	**11.50**	13.41	1966-1998	33	~5 / day	USA
	9	Time (1-Minute) Estimation (RBS)	9.38	**10.29**	11.37	1966-1998	33	"	"
	10	Urinary 17-ketosteroid excretion (CH)	8.70	**9.30**	9.90	1948-1963	15	1 / day	Denmark
	11	Peak Expiratory Flow (RBS)	10.36	**11.74**	13.11	1966-1998	33	~5 / day	USA
	12	Respiratory Rate (RBS)	10.13	**12.50**	17.32	1966-1998	33	"	"
	13	Systolic Blood Pressure - SBP (RBS)	9.05	**10.21**	11.36	1966-1998	33	"	"
	14	Standard Deviation of SBP (YW)	8.85	**10.43**	12.76	1987-1998	11	~48 / day	Japan
	15	Diastolic Blood Pressure - DBP (RBS)	10.09	**10.98**	11.87	1966-1998	33	~5 / day	USA
	16	Standard Deviation of DBP (YW)	6.18	**7.82**	10.02	1987-1998	11	~48 / day	Japan
	17	Heart Rate - HR (YW)	9.54	**12.93**	17.91	1987-1998	11	"	"
	18	Standard Deviation of HR (YW)	8.27	**11.52**	16.22	1987-1998	11	"	"
Pathology	19	Myocardial Infarction	10.00	**10.80**	11.70	1960-1996	37	129,205	USA
	20	Leptospirosis	9.40	**10.80**	12.40	1949-1995	47	2,907	Slovakia
	21	Diabetes	7.70	**10.40**	13.30	1985-1995	11	1,369	"
Anthropo-metry at birth		***Body Weight*** ***Boys***							
	22	Minnesota	19.53	**23.19**	27.67	1963-1998	36	2,136,745	USA
	23	Alma-Ata Russians	14.99	**17.17**	20.07	1946-1998	53	9,056	Kazakhstan
	24	" Kazakhs	18.39	**21.24**	24.05	1946-1998	53	3,459	"
	25	Moscow		**10.49**		1874-1985	112	5,987	Russia
		Girls							
	26	Minnesota	20.58	**23.46**	26.83	1963-1998	36	1,039,464	USA
	27	Alma-Ata Russians	15.21	**17.75**	21.06	1946-1998	53	9,105	Kazakhstan
	28	" Kazakhs	15.44	**21.45**	27.45	1946-1998	53	3,448	"
	29	Moscow	9.70	**10.29**	11.01	1874-1985	112	5,840	Russia
		Both genders							
	30	Denmark	14.71	**17.94**	22.68	1973-1994	22	1,166,206	Denmark
		Body Length ***Boys***							
	31	Alma-Ata Russians	15.82	**18.58**	22.38	1946-1998	53	9,026	Kazakhstan
	32	Moscow	18.76	**20.28**	21.86	1874-1985	112	5,976	Russia
		Girls							
	33	Alma-Ata Russians	16.13	**19.20**	23.39	1946-1998	53	9,105	Kazakhstan
	34	" Kazakhs	15.72	**19.60**	25.40	1946-1998	53	3,485	"
	35	Moscow	19.05	**20.76**	22.78	1874-1985	112	5,976	Russia
		Both genders							
	36	Denmark	20.81	**23.55**	26.55	1973-1994	22	1,166,206	Denmark
		Head Circumference							
	37	***Boys*** Moscow	17.71	**19.23**	20.75	1874-1985	112	5,976	Russia
	38	***Girls*** "	18.42	**20.73**	23.95	1874-1985	112	5,820	"
Demography	39	Birth rate	8.63	**9.43**	10.23	1940-1996	57	57	USA
			17.61	**21.33**	25.05	1940-1996			
Motivation	41	Religious activity of Jehovah's Witnesses	17.52	**20.44**	24.45	1950-1999	50	328,572[#] 5,653,987[##]	Worldwide
Criminality	40	Homicide	8.99	**10.58**	12.16	1900-1998	99	99	USA
			19.23	**20.35**	21.62				
Wars	42	International battles	21.87	**21.96**	22.06	599BC-1957	2556	2556	Worldwide

* 95% confidence limit; not shown if cycle is not statistically significant.
** Computed by changing the sign of WN at each WN minimum.
***RBS - Dr. Robert B. Sothern, CH - Dr. Christian Hamburger, YW - Dr. Yoshihiko Watanabe.
[#] in 1950, [##] in 1999, pool of 103 plus other unspecified number of sites.

Fig. 3.11 (continued)

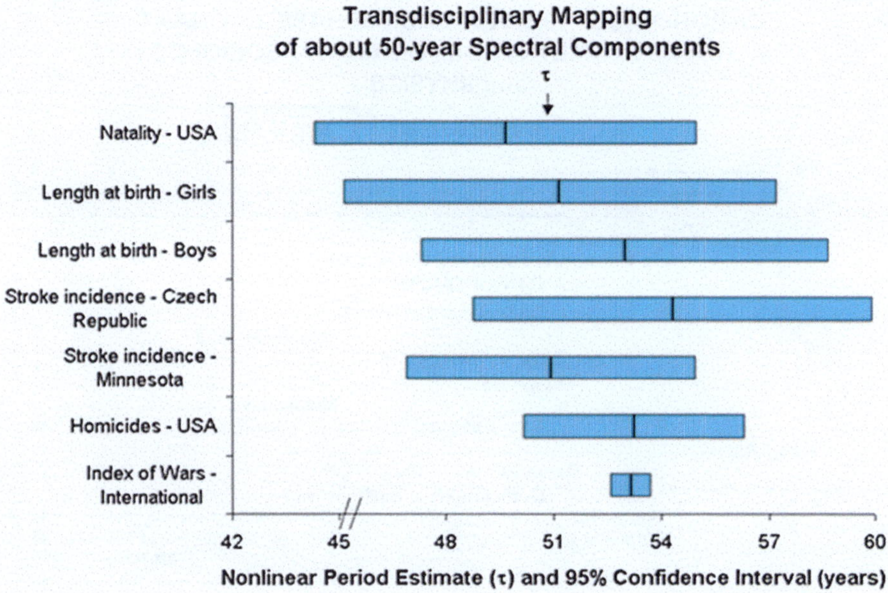

Fig. 3.12 Period chart. Human semicentennial system (© Halberg Chronobiology Center)

variability disorders is also now documented; accordingly, action is overdue by those formulating guidelines for dealing with blood pressure in the clinic.

3.5 Conclusion

Natural geophysical near weeks, aeolian quinmensals (about-5-month cycles Table 3.2) of solar flares, geophysical half years, the solar wind's near- and far-transyears, and decadals, didecadals, paratridecadals, and semicentennians or trans-semicentennians, among others, have evolved in living matter into a transdisciplinary system of coperiodisms in and around us that are congruent by virtue of overlapping

BIOCOS (Halberg et al. [28]), has shown decadal and multidecadal solar periods in the human circulation on the as-yet basic side. From an applied viewpoint, BIOCOS also represents a de facto validation test, on a small scale but worldwide, of C-ABPM. A model for a website has now also been proposed by a World Forum on "Natural Cataclysms and Global Problems of the Modern Civilization," held September 19–21, 2011 in Istanbul, Turkey. The data collected for the prevention of personal cataclysms, such as a massive stroke, may also be used to attempt an understanding (for countermeasures) of the mechanisms whereby space weather contributes to societal disease, such as suicide, crime, homicide, terror and war, and natural disasters such as earthquakes and tsunamis, a task for which the mapping of the time structures of these phenomena has begun, with results all revealing solar-biospheric coperiodisms.

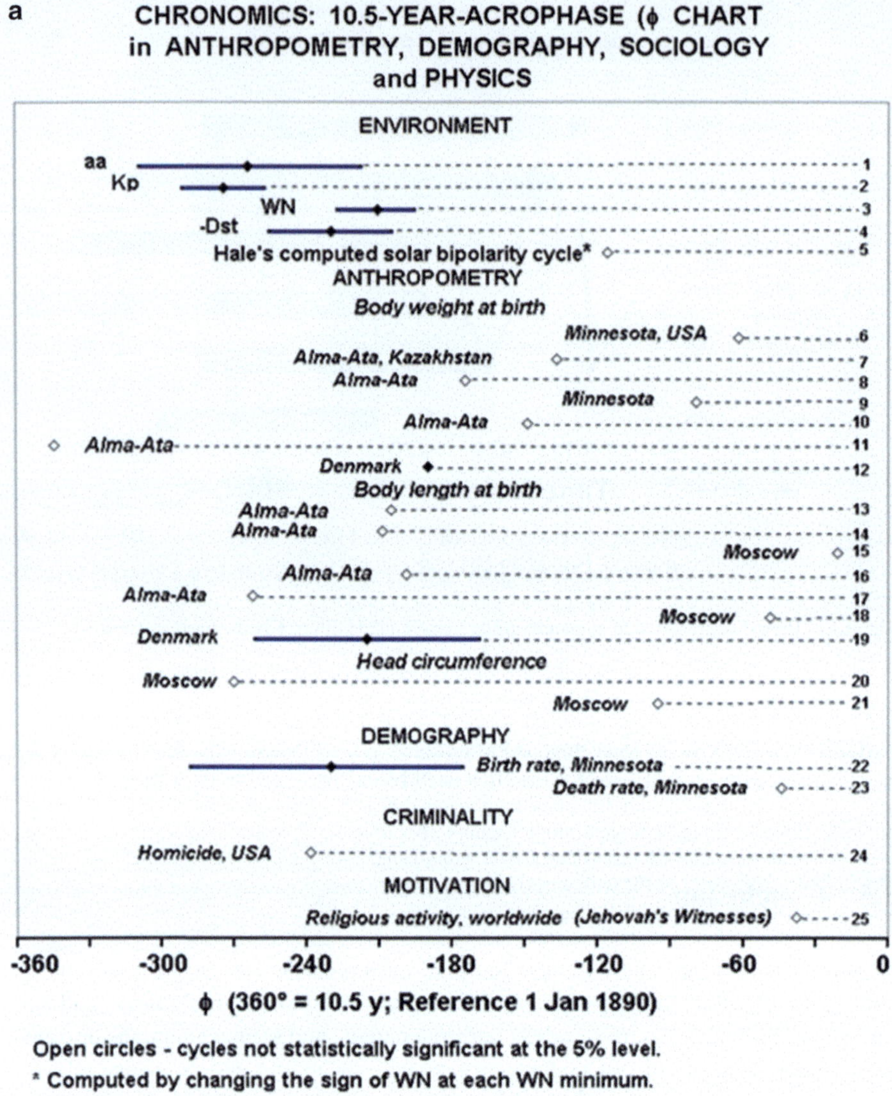

Fig. 3.13 (**a**) Human decadal phase chart (© Halberg Chronobiology Center). (**b**) Key to human decadal phase chart (© Halberg Chronobiology Center). (**c**) Human didecadal phase chart (© Halberg Chronobiology Center). (**d**) Key to human didecadal phase chart (© Halberg Chronobiology Center)

95 % confidence intervals of their periods. Each pair or set of coperiodic cycles was mapped after the rejection of the zero-amplitude (no-cycle) assumption with the uncertainties of each of its characteristics: period, amplitude, phase, and, whenever possible, eventually of waveform (given by the amplitude/phase of harmonics).

b

Chronomics: Acrophases of 10.5-year cycles in and around us

	Line		Acrophase (degrees) Lower limit*	Point estimate	Upper limit*	Series duration Dates	Years	Number of data	Geographic site
Environment	1	aa = Antipodal Geomagnetic Index	-218	**-264**	-309				
	2	Kp = Planetary Geomagnetic disturbance	-257	**-274**	-291	1932-1999	68	1 / month	
	3	WN = Wolf relative sunspot number	-196	**-211**	-227	1890-1999	110	1 / year	
	4	-Dst = Equatorial geomagnetic disturbance	-205	**-230**	-255	1973-1999	27	1 / year	
	5	Bipolarity "Hale Cycle" **		**-116**		1890-1999	110	1 / year	
Anthropometry at birth		***Body Weight*** ***Boys***							
	6	Minnesota		**-62**		1963-1998	36	2,136,745	USA
	7	Alma-Ata Russians		**-137**		1946-1998	53	9,056	Kazakhstan
	8	" Kazakhs		**-175**		1946-1998	53	3,459	"
		Girls							
	9	Minnesota		**-79**		1963-1998	36	1,039,464	USA
	10	Alma-Ata Russians		**-149**		1946-1998	53	9,105	Kazakhstan
	11	" Kazakhs		**-344**		1946-1998	53	3,448	"
		Both genders							
	12	Denmark		**-190**		1973-1994	22	1,166,206	Denmark
		Body Length ***Boys***							
	13	Alma-Ata Russians		**-205**		1946-1998	53	9,026	Kazakhstan
	14	" Kazakhs		**-209**		1946-1998	53	3,325	"
	15	Moscow		**-21**		1874-1985	112	5,976	Russia
		Girls							
	16	Alma-Ata Russians		**-199**		1946-1998	53	9,105	Kazakhstan
	17	" Kazakhs		**-262**		1946-1998	53	3,485	"
	18	Moscow		**-49**		1874-1985	112	5,976	Russia
		Both genders							
	19	Denmark	-169	**-215**	-261	1973-1994	22	1,166,206	Denmark
	20	***Head Circumference***							
		Boys Moscow		**-270**		1874-1985	112	5,976	Russia
	21	***Girls*** "		**-95**		1874-1985	112	5,820	"
Demography	22	Birth rate	-176	**-230**	-288	1940-1996	57	57	USA
	23	Death rate		**-44**		1940-1996	57	57	"
Criminality	24	Homicide		**-238**		1900-1998	99	99	"
Motivation	25	Religious activity of Jehovah's Witnesses		**-38**		1950-1999	50	328,572▪ 5,653,987▪▪	Worldwide

* 95% confidence limit; not shown if cycle is not statistically significant. ** Computed by changing the sign of WN at each WN minimum.
▪ in 1950, ▪▪ in 1999, pool of 103 plus other unspecified number of sites.

Fig. 3.13 (continued)

These multiple coexisting aeolian cycles, seen and unseen, covering a broad range of frequencies, characterize our morphology revealed by anthropometry in space and a partly novel morphology in time, as soon as a curtain of ignorance covering the normal range by such concepts as homeostasis is drawn, Fig. 3.15 [29]. These ubiquitous cycles, some in the archaeon and the human, constitute congruent signatures not only of the photic day and year but also of the nonphotic, including gravitational and magnetic, cyclic influences upon our mental functions in particular. The nonphotics can coexist and can compete with the photics and can even replace the photics, at least for a while, Fig. 3.16. They leave their signatures in terrorism, military-political affairs, crime, suicide, church memberships, and religious proselytism, Fig. 3.17a–f.

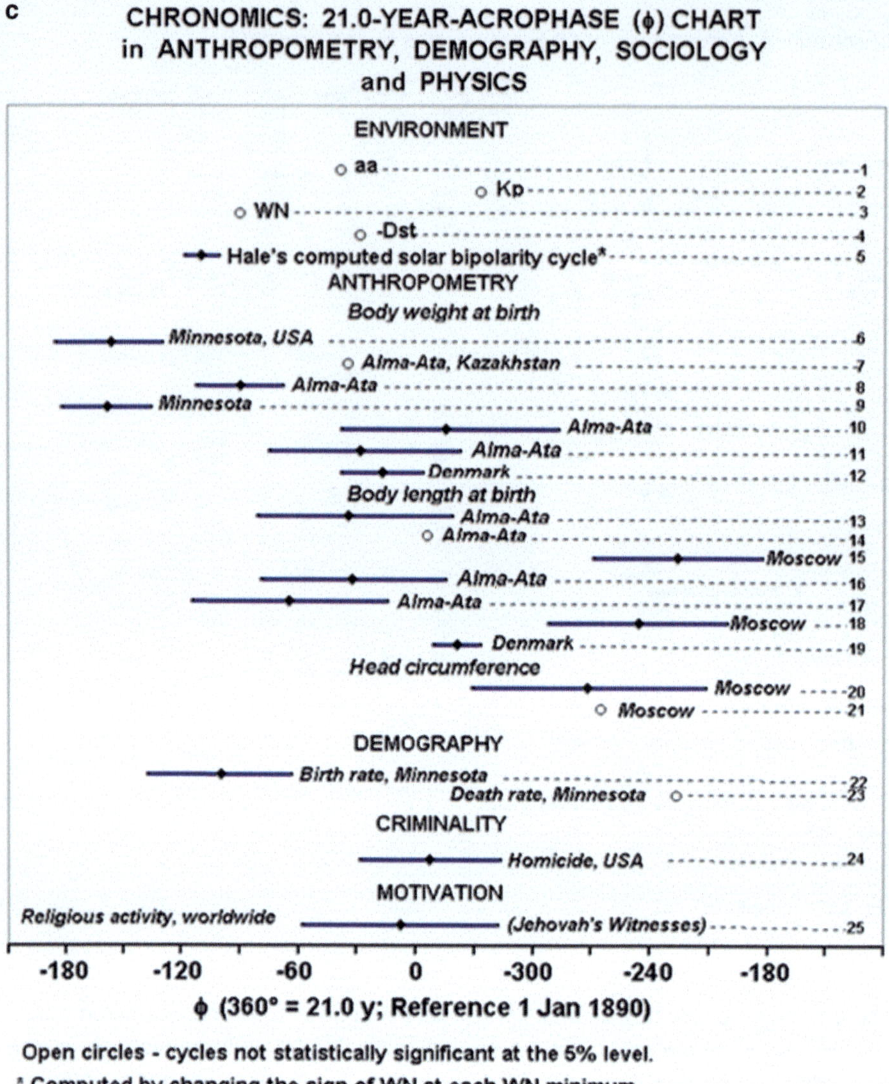

Fig. 3.13 (continued)

As to mechanisms involved, we are still learning about new cycles from old time series, back at the (cortico)steroids, with which my career started, Fig. 3.18a–o. Oscillator theories may be scaffolds, but are not the houses we build with them, or rather the cycles we map in and around us. Mapping the unseen is more challenging than mapping the seen (the photic and thermic day and year). We must learn about the influences of space weather before we can optimize what we regard as desirable –

d

Chronomics: Acrophases of 21.0-year cycles in and around us

	Line		Acrophase (degrees) Lower limit*	Point estimate	Upper limit*	Series duration Dates	Years	Number of data	Geographic site
Environment	1	aa = Antipodal Geomagnetic Index		**-39**					
	2	Kp = Planetary Geomagnetic disturbance		**-327**		1932-1999	68	1 / month	
	3	WN = Wolf relative sunspot number		**-90**		1890-1999	110	1 / year	
	4	-Dst = Equatorial geomagnetic disturbance		**-29**		1973-1999	27	1 / year	
	5	Bipolarity "Hale Cycle" **	-101	**-110**	-118	1890-1999	110	1 / year	
Anthropometry at birth		***Body Weight*** ***Boys***							
	6	Minnesota	-130	**-157**	-185	1963-1998	36	2,136,745	USA
	7	Alma-Ata Russians		**-35**		1946-1998	53	9,056	Kazakhstan
	8	" Kazakhs	-68	**-90**	-112	1946-1998	53	3,459	"
		Girls							
	9	Minnesota	-136	**-159**	-182	1963-1998	36	1,039,464	USA
	10	Alma-Ata Russians	-287	**-345**	-38	1946-1998	53	9,105	Kazakhstan
	11	" Kazakhs	-338	**-29**	-75	1946-1998	53	3,448	"
		Both genders							
	12	Denmark	-357	**-18**	-38	1973-1994	22	1,166,206	Denmark
		Body Length ***Boys***							
	13	Alma-Ata Russians	-342	**-35**	-81	1946-1998	53	9,026	Kazakhstan
	14	" Kazakhs		**-354**		1946-1998	53	3,325	"
	15	Moscow	-183	**-226**	-269	1874-1985	112	5,976	Russia
		Girls							
	16	Alma-Ata Russians	-345	**-33**	-79	1946-1998	53	9,105	Kazakhstan
	17	" Kazakhs	-15	**-65**	-114	1946-1998	53	3,485	"
	18	Moscow	-201	**-246**	-292	1874-1985	112	5,976	Russia
		Both genders							
	19	Denmark	-327	**-339**	-351	1973-1994	22	1,166,206	Denmark
	20	***Head Circumference***							
		Boys Moscow	-212	**-272**	-331	1874-1985	112	5,976	Russia
	21	***Girls*** "		**-265**		1874-1985	112	5,820	"
Demography	22	Birth rate	-64	**-100**	-137	1940-1996	57	57	USA
	23	Death rate		**-226**		1940-1996	57	57	"
Criminality	24	Homicide	-317	**-353**	-28	1900-1998	99	99	"
Motivation	25	Religious activity of Jehovah's Witnesses	-318	**-8**	-58	1950-1999	50	328,572[#] 5,653,987[##]	Worldwide

* 95% confidence limit; not shown if cycle is not statistically significant.
** Computed by changing the sign of WN at each WN minimum.
[#] in 1950, [##] in 1999, pool of 103 plus other unspecified number of sites.

Fig. 3.13 (continued)

love rather than hate, philanthropy rather than crime, and peace rather than war – all carrying mapped signatures of the cosmos. The importance of this task remains a challenge.

3.6 Epilogue

Many other endeavors are apparent from my bibliography, available at http://halbergchronobiologycenter.umn.edu/. Others used their own remove-and-replace approach, independently of us, Fig. 3.19. The help of all coauthors, that of my family and colleagues in particular, whether through their original active involvement or through support that made the studies possible is gratefully acknowledged. To my

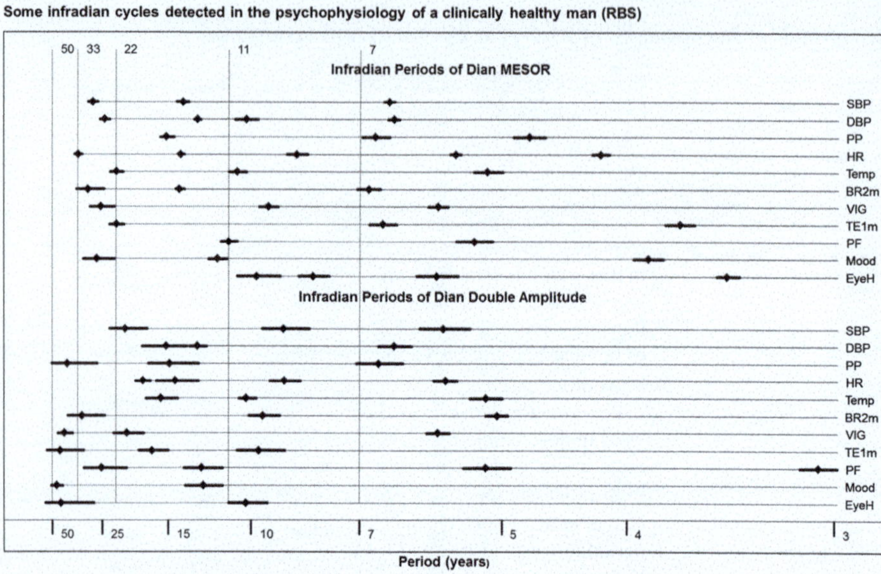

Fig. 3.14 Partial infradian period chart of aspects of the psychophysiology of a healthy man (RBS). *SBP* systolic blood pressure, *DBP* diastolic blood pressure, *PP* pulse pressure, *HR* heart rate, *Temp* oral temperature, *BR2m* breathing rate (2 min), *VIG* self-rated vigor, *TE1m* 1-min time estimation, *PF* peak (expiratory) flow, *Mood* self-rated mood, *EyeH* eye-hand coordination; horizontal bars are 95 % confidence intervals of nonlinearly estimated periods (© Halberg Chronobiology Center)

Table 3.2 Period(s) in years in body length (cm) at birth in Russia 1874–1985: data of Boris Nikityuk[a]

	Boys (overall standard error: 0.44)	Girls (overall standard error: 0.47)
MESOR	52.224 (51.962, 52.687)	51.564 (51.288, 51.840)
Slope	−0.017 (−0.027, −0.008)	−0.015 (−0.025, −0.005)
Period 1	53.671 (46.12, 61.23)	53.666 (42.00, 65.33)
Double amplitude	1.52 (0.74, 2.28)	1.32 (0.42, 2.22)
Acrophase	−276° (−247, −305)	−274° (−237, −311)
Period 2	32.404 (29.35, 35.13)	33.487 (25.88, 41.09)
Double amplitude	0.76 (−0.00, 1.50) [0.42, 1.08]	0.86 (0.00, 1.74) [0.48, 1.24]
Acrophase	−243° (−187, −300)	−207 (−151, −263)
Period 3	20.441 (18.31, 22.57)	21.217 (19.18, 23.25)
Double amplitude	0.84 (0.56, 1.10)	0.94 (0.62, 1.24)
Acrophase	−184° (−129, −240)	−158° (−109, −208)

[a]95% confidence limits in ( ) are conservative; in the case of a very slight overlap of zero beyond the second decimal, 1-parameter limits are given in [ ]

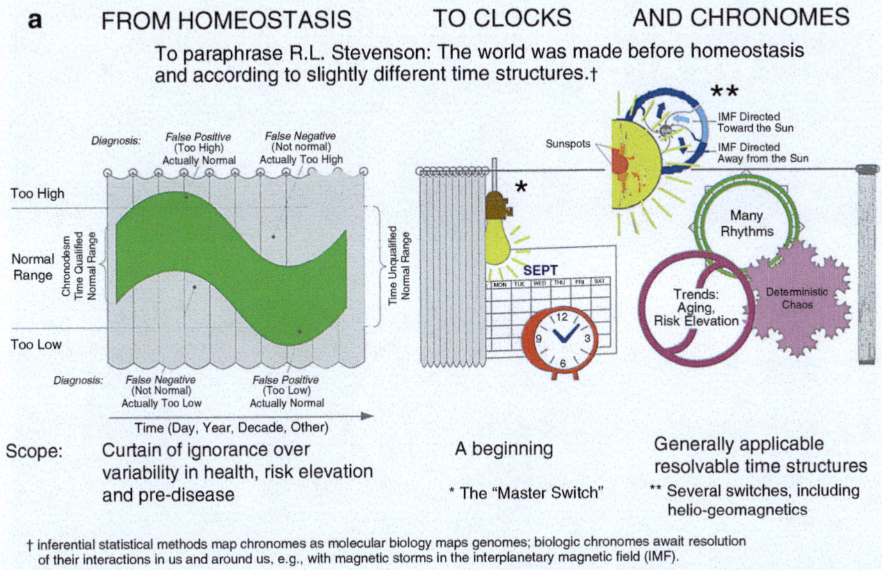

Fig. 3.15 (**a**) Concepts like homeostasis represent an excuse for a curtain of ignorance closed over the range of physiological variation. Clocks and calendars (*middle*) open that curtain only partly, ignoring structures consisting of deterministic and other chaos, trends and the many rhythms other than circadian and circannual components in and around us. Thus they are no substitute for mapped coperiodisms (*right*) in the biosphere and its environment (Fig. 3.15b) (© Halberg Chronobiology Center). (**b**) Multifrequency cycles in us and around us. Biological cycles tend to have environmental counterparts. Some cycles detected in biology prompted their discovery in the environment (© Halberg Chronobiology Center)

colleagues, I owe what became the Cornélissen series of transdisciplinary coperiodisms, a page of anthropology in time. Germaine Cornélissen's contributions since her arrival in Minnesota in the 1970s cannot be separated from mine and vice versa. She and Othild Schwartzkopff described, among others, what I tried to do elsewhere [34–41]. Someday, everyone may have an affordable chance of a life mostly without VVDs, although some of us cannot help but have them transiently and some VVAs more frequently, Fig. 3.20. In pursuit of chronobioethics, Fig. 3.21, we can try to

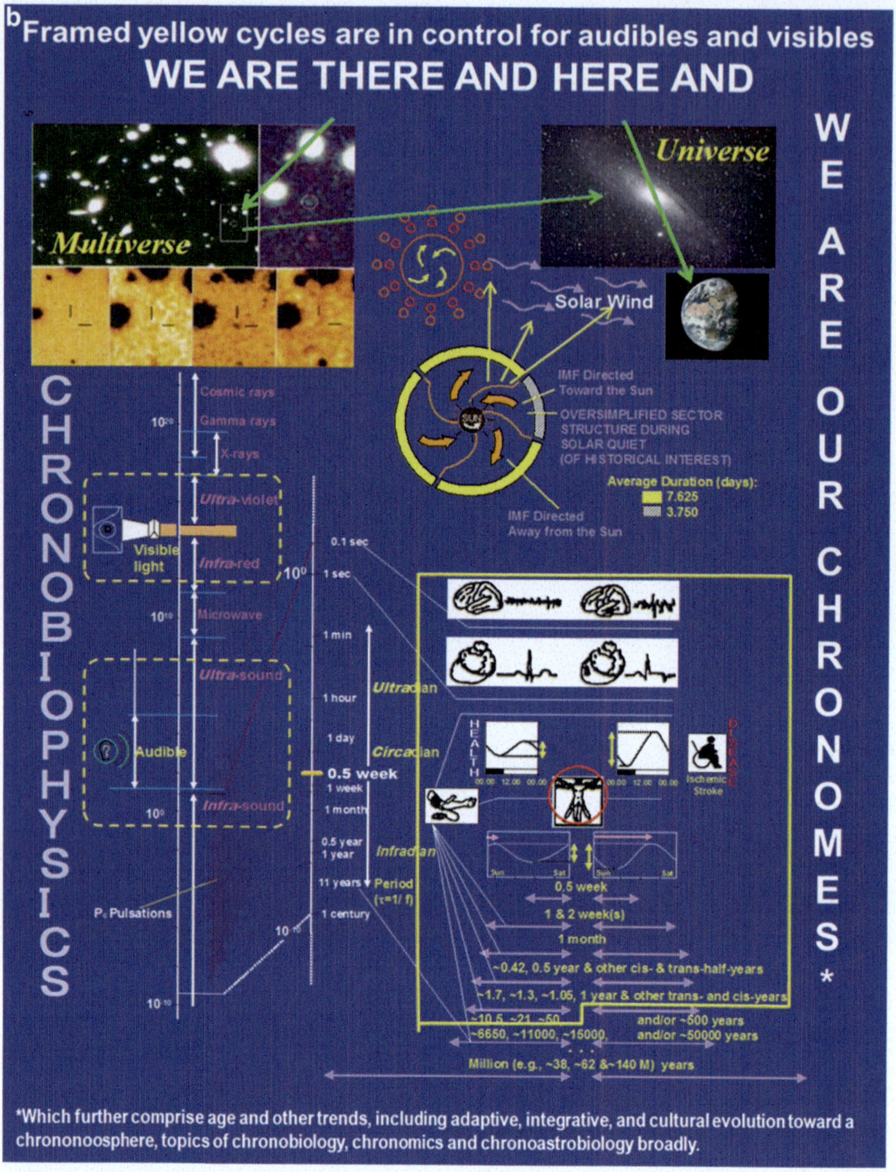

Fig. 3.15 (continued)

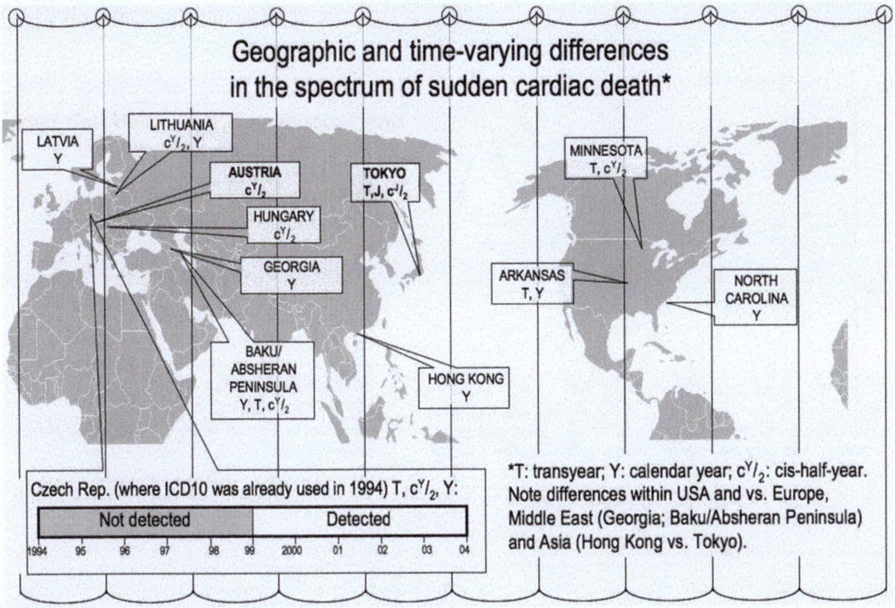

Fig. 3.16 Need for a global and local (glocal) approach in space and time illustrated by incidence pattern in sudden cardiac deaths varying with time. A transyear is found in Minnesota with a cis-half year (cY/2) and both a calendar year and a transyear in Arkansas and the Czech Republic, where a cis-half year, corresponding in length to an also-transient period of hard solar flares, is detected after but not before 1999 (© Halberg Chronobiology Center)

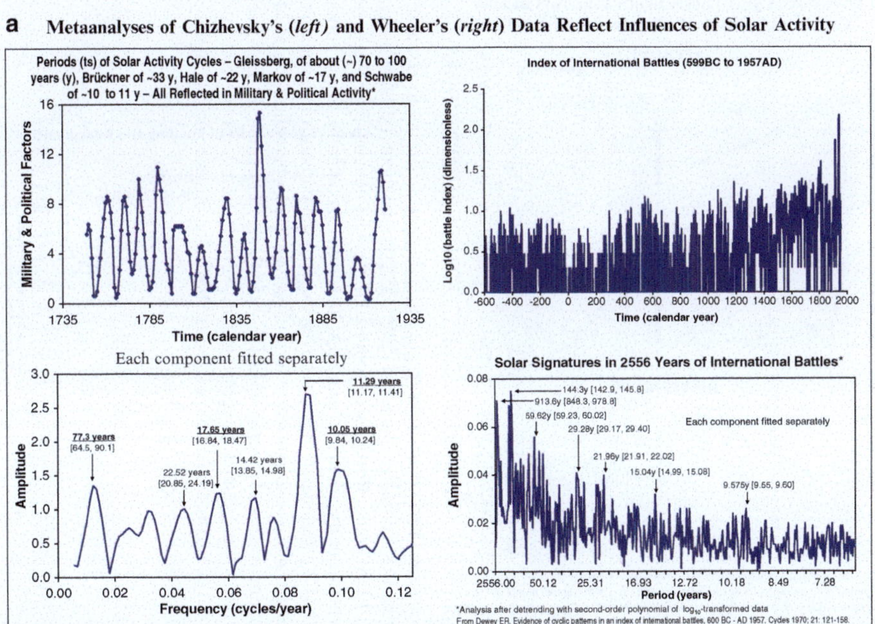

Fig. 3.17 (**a**) Solar periodicities of Chizhevsky's (*left*) and Wheeler's (*right*) data (*top half*) mirrored in our meta-analyses (*bottom half*) (© Halberg Chronobiology Center). (**b**) About

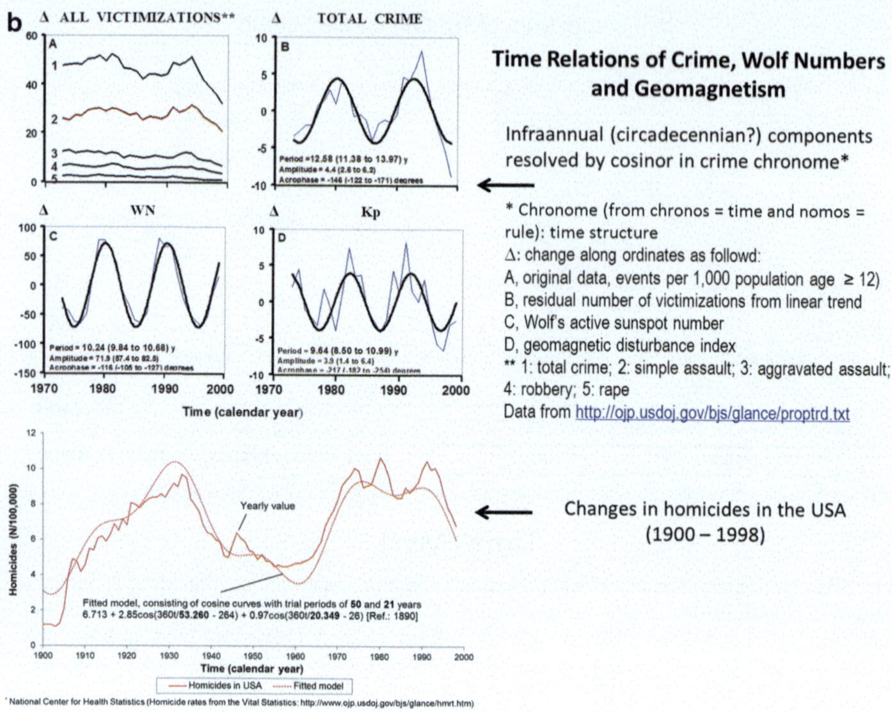

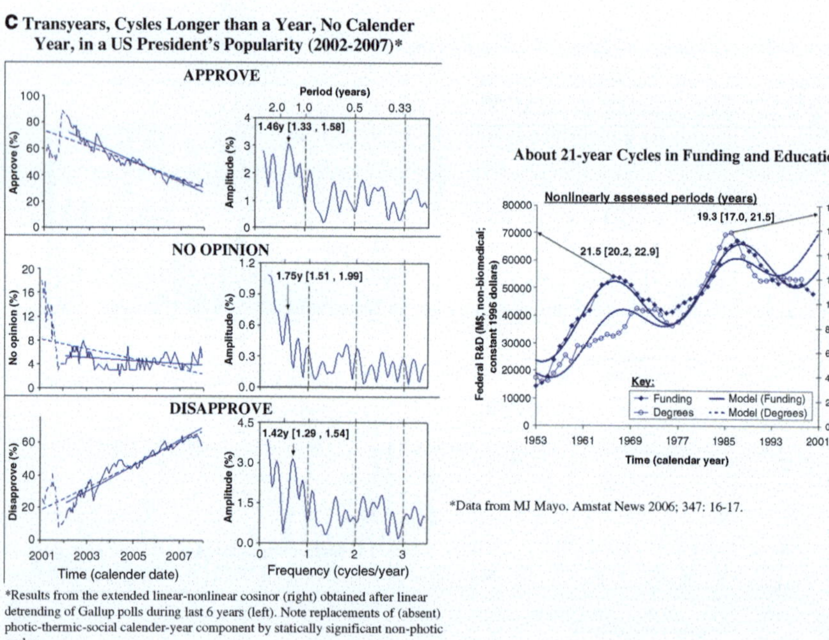

Fig. 3.17 (continued) (~)-10-yearly (*top* and *middle*) and ~50- and 20-yearly periods in the "bad" (morbidity) of society [30] (© Halberg Chronobiology Center). (**c**) The prominence of the magnetic transyears over the calendar year in presidential politics (*left*) and of a near 20-year cycle

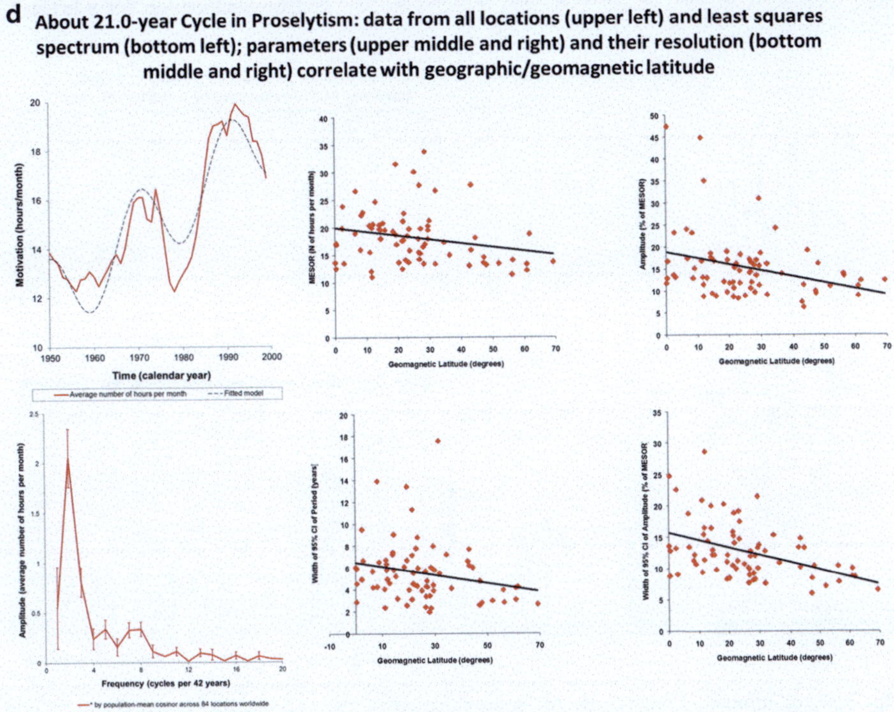

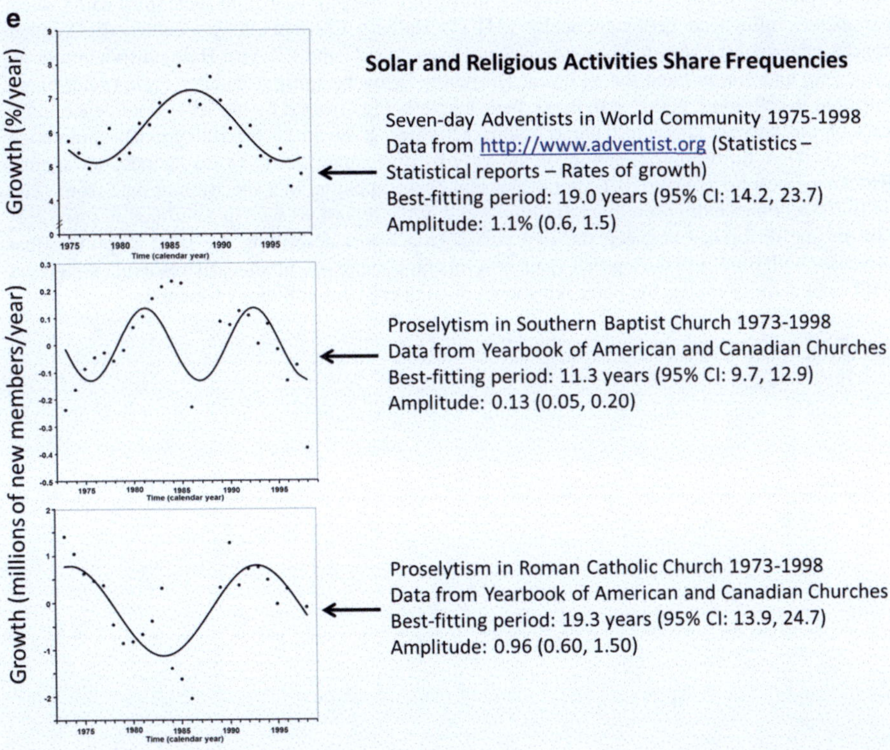

Fig. 3.17 (continued) of sunspot bipolarity in the funding of nonmedical (Defense Department-related) education is in keeping with an association with space weather (© Halberg Chronobiology

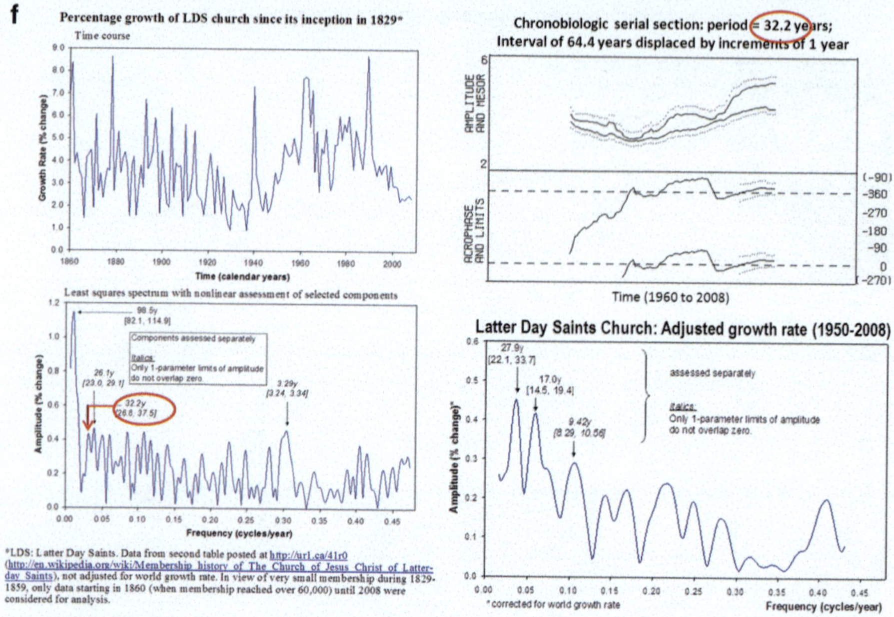

Fig. 3.17 (continued) Center). (**d**) An approximation of the very-hard-to-gauge desire to do "good": proselytism of Jehovah's Witnesses mirrored in the ~21-year Hale cycle (*left*) with cycle's parameters influenced geomagnetically [31] (© Halberg Chronobiology Center). (**e**) Another approximation of the very-hard-to-gauge desire to do "good": the ~21-year Hale sunspot bipolarity cycle (*top* and *bottom*) and the ~11-year Horrebow-Schwabe sunspot number cycle (*middle*) are mirrored in religion ("good" activities), their length being covered by the 95 % confidence intervals of the periods characterizing religious activities given in () (© Halberg Chronobiology Center). (**f**) A third approximation of the very-hard-to-gauge desire to do "good": an aeolian Brückner-Egeson-Lockyer (BEL) paratridecadal cycle, apparent in a spectrum of data from 1860 on (*left*), becomes more prominent during the end of the record, as seen in a chronomic serial section by the difference between the two curves (*top right*) describing the BEL amplitude that increases with time, and during that span, several components are nonlinearly resolved, congruent with solar activity cycles, the latter not shown (© Halberg Chronobiology Center)

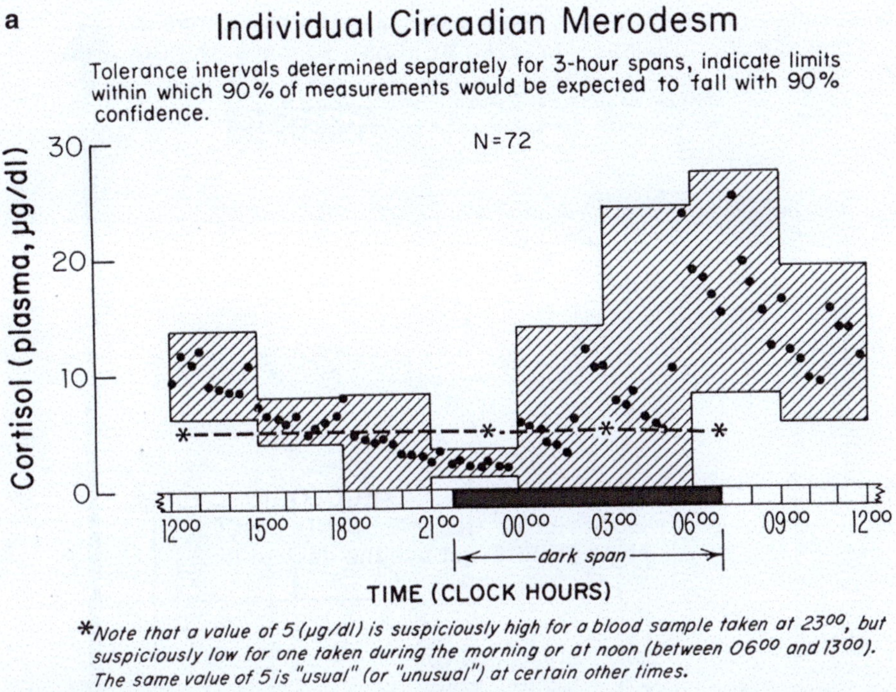

Fig. 3.18 *All.* Cross-validation by cycles in space weather and glocal technique of analysis gradually refines the infradian window of steroid dynamics. A for-us unique time series with few gaps covering 15 years of daily 17-ketosteroid determinations, the analysis of which has been steadily improved over the past half century, resulting in a map of cycles complementing the circadian system and allowing a first view of a bushel basket of novel infradians (with periods longer than 28 hours). On the clinical side, if recognized they can forestall androgen substitution therapy when it is not necessary (in the case examined herein, CH, they did not prevent his androgen use because the numerical analyses were retrospective). There is, of course, other applied and basic interest in learning how the cosmos influences our physiology, as a reference for also-documented pathology, perhaps in part through steroidal mechanisms. (**a**) Didactic demonstration. Variations of circulating corticoid and other hormones can be mapped in the saliva of an individual. It is worthwhile nonetheless to illustrate a personalized chart of a range of acceptable values for time-specified single plasma cortisol samples, a desideratum half a century ago (individual circadian merodesm). Tolerance intervals determined separately for 3-hour spans indicate limits within which 90 % of measurements would be expected to fall with 90 % confidence. On the sleep schedule depicted by a black horizontal bar on the abscissa, a cortisol value such as 5 μg/dl (asterisk) may be too low at one time (e.g., 06:30), too high at another time (e.g., 23:00), or normal at still another time (e.g., 03:00). But at a time when over 11,700 hormone assays could be completed on a motivated subject, reliance on saliva rather than blood and on rhythm characteristics rather than spotchecks can be advocated (© Halberg Chronobiology Center). (**b**) In the 1960s, we obtained a 5-day series of four consecutive 4-hourly and one 8-hourly urine sample/day and a 15-year-long series (with gaps) of daily 17-ketosteroid (KS) excretion by a clinically healthy man, CH. While KS are unspecific, they estimate, among others, the breakdown products of steroids and may convey some aspects of variability in corticosteroids. In the 5-day series, a prominent circadian component is shown by an arrow. In addition, there were much smaller peaks at many frequencies, some of which, admittedly arbitrarily chosen, were studied by cosinor in 4 sections of the 15-year series. Data analyzed were from days when the subject did not self-administer androgen suppositories. Infradian results

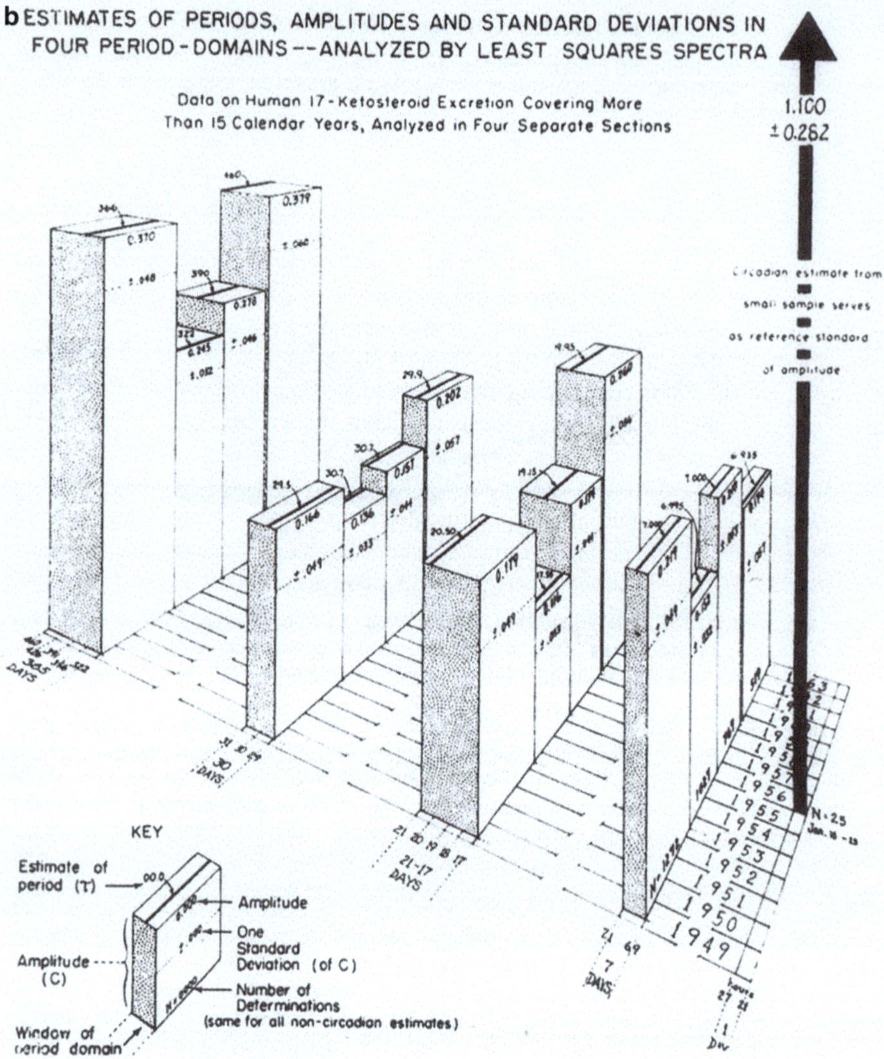

Fig. 3.18 (continued) appeared to be roughly reproducible from one subspan to the next to the unaided eye, an interpretation revised by further analyses (© Halberg Chronobiology Center). (**c**) Three windows were singled out for further analyses: a 24-hour synchronized major component (*right*) stands out prominently, overshadowing the corresponding minor peaklet in urine volume, in which the KSs were determined. Before looking at the about (~)-yearly component (*left*) and an ~ weekly component (*middle*), a change in scale must be kept in mind, as shown by brackets between the windows in the middle and on the right. With this qualification as to their magnitude ("minitude"?), there are peaks in both KS and urine volume (UV) in the circannual region (*left*), with the urine volume peak, but not that in KS peak synchronized to the calendar year: the KS peak is clearly distant from the precise year length. Both KS and UV seem to be overall synchronized by the societal week (*middle*) (© Halberg Chronobiology Center). (**d**) The window of the weekly (circaseptan) region is complemented by a chronobiologic serial section that shows the original data of the long KS series in the top row. Without time series analyses, the unaided eye sees ups

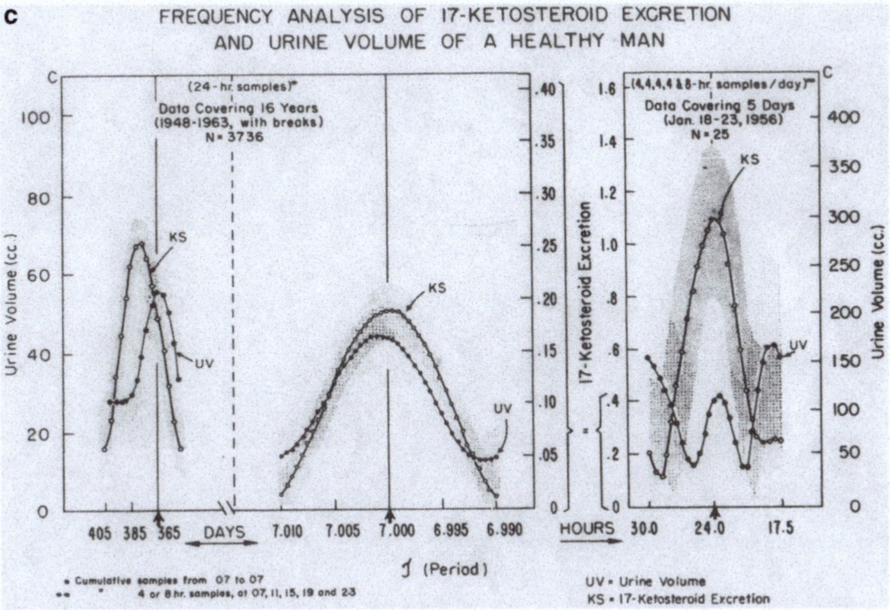

Fig. 3.18 (continued) and downs in the data but can guesstimate neither the precise periodicities in the infradian domain of the time structure of the series nor any CIs (95 % confidence intervals) of the periods. These aeolian components are nonstationary and best assessed with an approximation of their uncertainties. The second row summarizes the time course of changes in the circaseptan extent of change, the amplitude, A, with two vertical lines, labeled SD, being indications of the length of the standard deviation of A. The unaided eye can now detect more clearly than in the top row, high values for the extent of the weekly swing, if not clear peaks about 9 or 10 years apart, possibly a circadecennian modulation of the circaseptan amplitude. It is certainly unable to quantify about-7-day changes and their uncertainties, CIs. The third and fourth rows show changes in the phase, ϕ, of a 7-day cosine curve for analyses made on 0.5- and 1.0-year intervals, respectively, displaced for consecutive computations in increments of 28 days. For the first 12 years, the peaks occur mostly on or near Thursdays, irrespective of the interval length, more smoothly with the longer interval. The corresponding amplitudes remain comparable with earlier ones for the entire span, as shown in row 2. An advance in ϕ (downward trend at A) is also smoothed by using a 1.0- instead of a 0.5-year interval, but the following ϕ advance starting at B and continuing by scanning over more than 1.5 full weekly cycles remains consistent during the last several years. The finding of this ϕ advance indicates an external circaseptan desynchronization of KS in a clinically healthy man, CH. An about (~)-weekly (circaseptan) cycle is synchronized in the first 12 years with the calendar week, while in the last 3 years, perhaps because he self-administered a more massive dose of androgen (as noted, the corresponding data are omitted from the analyses), a circaseptan period of KS was slightly but statistically significantly shorter than a week, desynchronized from the societal calendar week, with which it had been previously synchronized for the preceding 12 years. For the last 3 years, the ϕs follow a diagonal advancing time course. Do KS take a time course of their own at least in part? If so, is there an environmental counterpart as in the case of 24-hour rhythms? The alternation of light and darkness constitutes an environmental cycle to which circadian rhythms constitute an adaptation. In looking for a comparable physical environmental counterpart for the desynchronized week in us, we encountered a near 7-day cycle in geomagnetism, confirmed by others. A cosinor spectrum reveals a dominant natural near week of 6.74 days length in geomagnetism. Some years earlier, Charles Greeley Abbot had "discovered a well-marked period … of 6.6456 days (later improved to 6.6485 days) and also the half of it in the weather of New York and

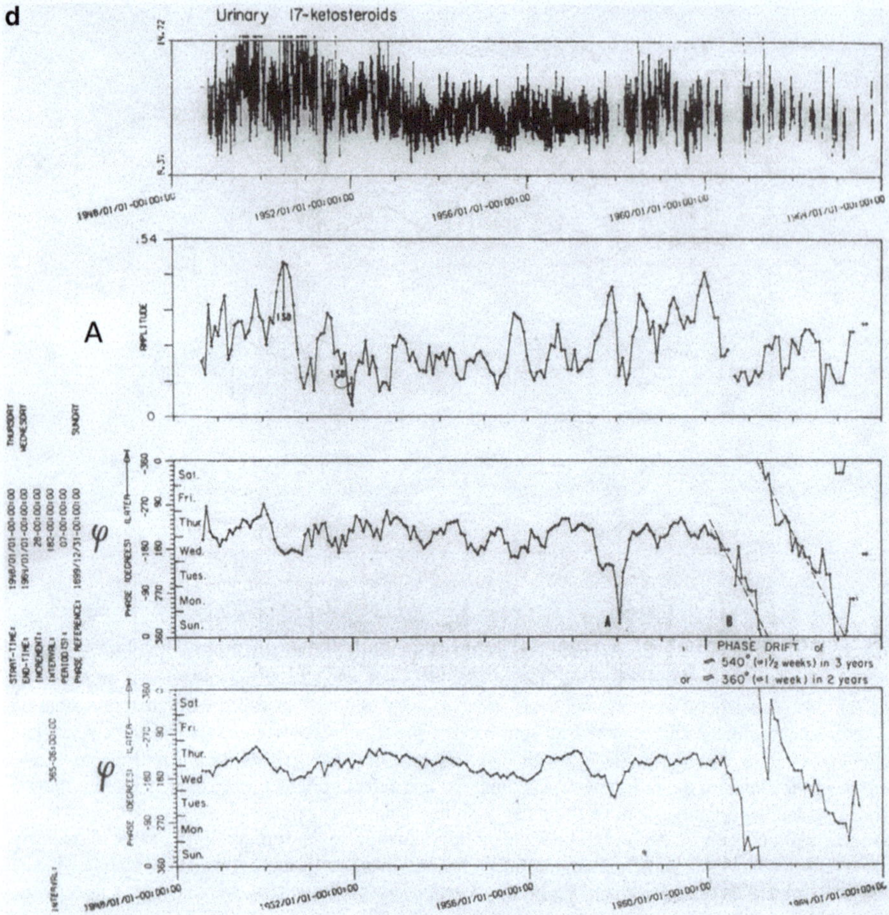

Fig. 3.18 (continued) Washington" (Abbot 1963). Selectively assorted sets of about-weekly geo-, helio-, and biospheric periods were then found in a clinically healthy man [32], Fig. 3.5a (cf. Fig. 3.5b for their aeolian behavior). It was tempting to postulate, on the basis of extensive experience with the week in us and around us, that, for each biological cycle, there should be a corresponding environmental cycle, and vice versa (© Halberg Chronobiology Center). (**e**) A corresponding ϕ advance is not seen for urine volume, which observation renders desynchronization in KS even more remarkable as an indicator of a circaseptan internal desynchronization as well (© Halberg Chronobiology Center). (**f**) It also seemed logical to ask whether the long biochemical 17-KS series actually also had a component of nearly 10 years, possibly associated with the sunspot cycle of Christian Pedersen Horrebow (1718–1776) (Thiele ThN. De Macularum Solis antiquioribus quibusdam observationibus Hafniae institutis. Astronomische Nachrichten 1859; 50: 259–61) and Samuel Heinrich Schwabe (1789–1876). An answer in the affirmative shows an about-10-year component found linearly in the spectrum with a period of 9.28 years which upon nonlinear analyses was longer with a CI covering 10 years [33] (© Halberg Chronobiology Center). (**g**) In seeing just the steady decrease in KS (without knowing that it is part of an about-10-year cycle), CH, a medical professional, indeed considered the need for androgen substitution therapy (*upper left graph*). He may not have realized that a decreasing trend was a transient, followed by an increase (*upper middle graph*), the increase and decrease being but part of a readily seen cycle

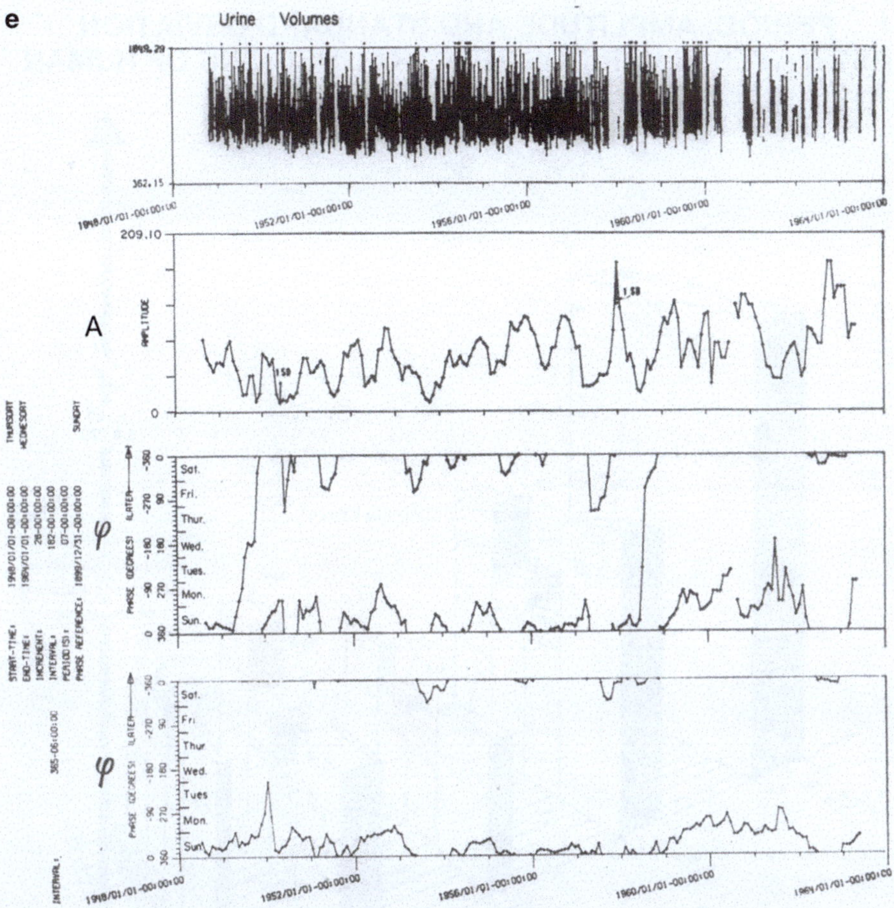

Fig. 3.18 (continued) (*upper right graph*) that when related to age (*lower left*) or to sunspot numbers (*lower middle*) can lead, during different subspans, to contradictory, even opposite correlations (when examined only as a trend, in the absence of prior information). The circadecennian period of 17-KS and the period of urine volume differ (*extreme bottom right*). There are circadecennian, circaseptan, and further transyearly differences between the period of KS and the volume of urine in which it was determined and even differences in φ at the same period, Fig. 3.18o (© Halberg Chronobiology Center). (**h**) On the data as a whole, a spectral window reveals (globally) novel periods (shorter and longer than a year) for KS of CH (© Halberg Chronobiology Center). (**i**) On the data as a whole, a spectral window reveals (globally) novel periods for urine volume of CH (© Halberg Chronobiology Center). (**j**) A glocal analysis, with a global window (*right*) and a gliding window for local study (*middle*), shows a near-transyearly period of KS in CH that can globally (*left*) be separated from a yearly component by a 95 % confidence interval (CI) not overlapping the precise 1-year length, unless the (conservative) CIs are still too liberal. Three far-transyearly periods are also seen globally (*middle and right*) (© Halberg Chronobiology Center). (**k**) Near-transannual (*left*) and far-transannual internal desynchronization of KS from urine volume in CH is suggested by nonoverlapping 95 % confidence intervals of the period estimates (© Halberg Chronobiology Center). (**l**) The chronobiologic serial section shows the urinary KS data (*top row*) recorded over 15 years in CH, a clinically healthy male scientist 49 years of age at start of study. The MESOR, M, is the lower curve, and the quinmensal's (cis-half year's, 0.42 year's) amplitude, A, is the distance

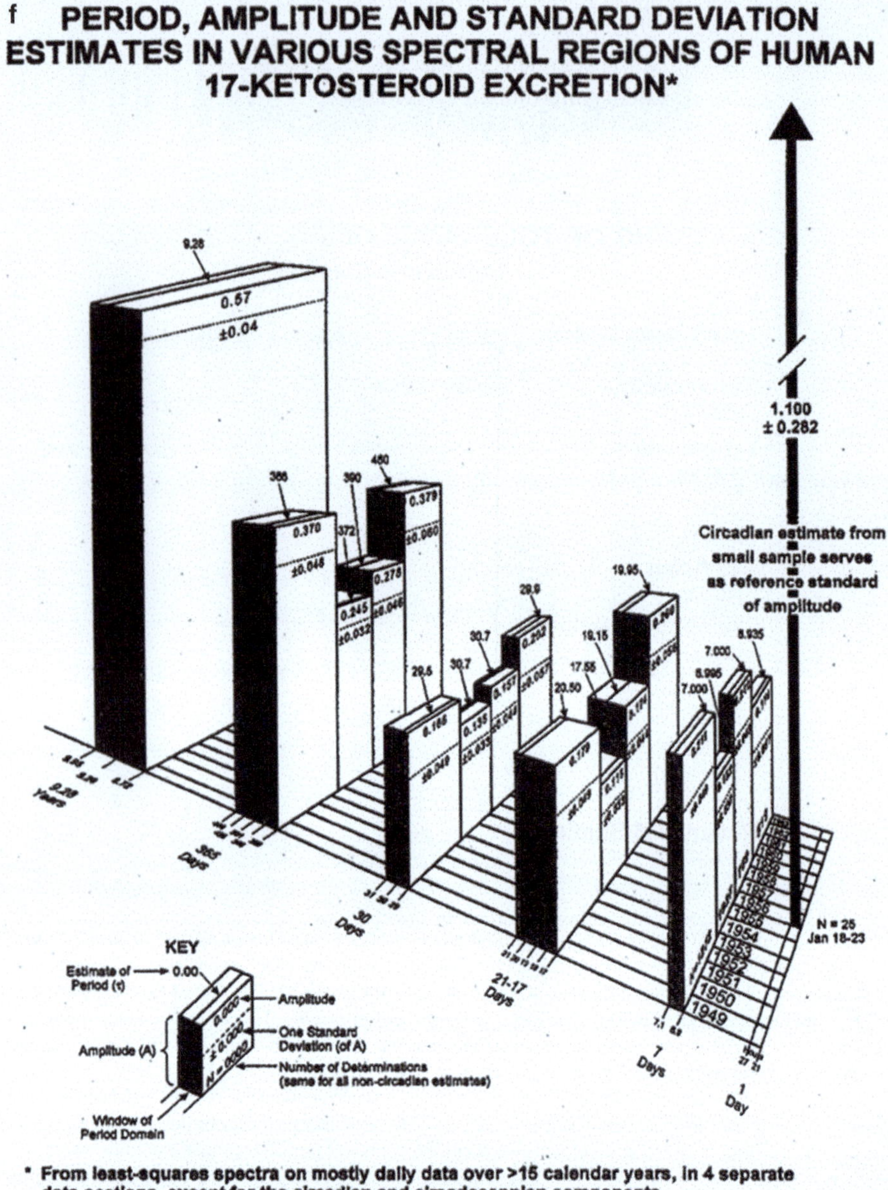

Fig. 3.18 (continued) between the two curves in the second row. *Dots* below the lower and above the *upper curves* indicate the standard errors of the estimates of M and A, respectively. Acrophases (*third row*) are shown with *dots*, corresponding to their CIs (95 % confidence intervals) only when the P-values for the test of the "no 0.42-year" amplitude (*shown in the fourth row*) are below 0.05, the *dashed horizontal line*. The time course of cis-half year acrophases mimics the time course of a 0.42-year component in the planetary geomagnetic index, Kp, Fig. 3.18m. Globally, urine volume is congruent in period with the former two variables, but its phase behavior is congruent with neither; only transiently it seems to mimic the time course of the phase of the relative sunspot

g Contradictory Correlations of Hormonal Metabolites with Age (I, II, IV - 5 of them) or with Solar Activity (V - 5 of them) Stem in part from a Common Years-Long Cycle (III, IV, VI)

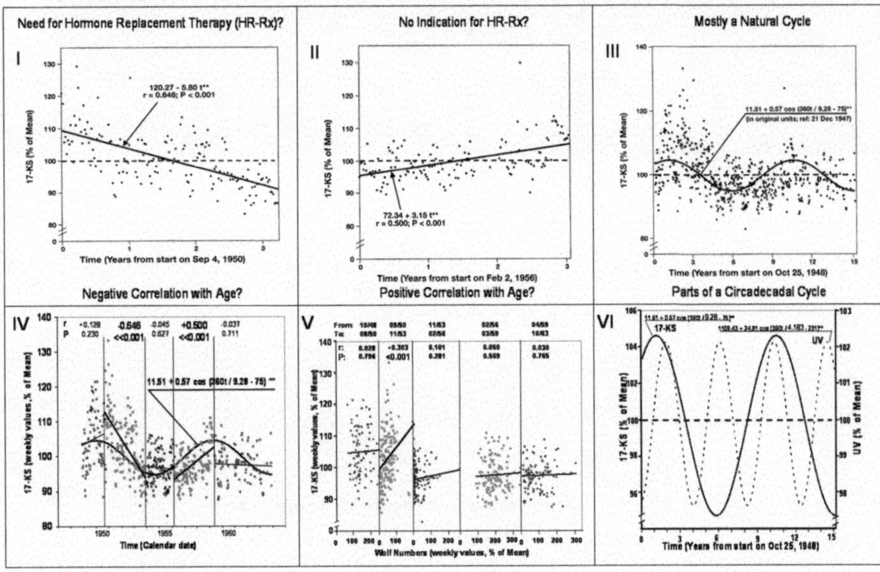

h Partial Least Squares Spectrum of Urinary Excretion of 17-Ketosteroids

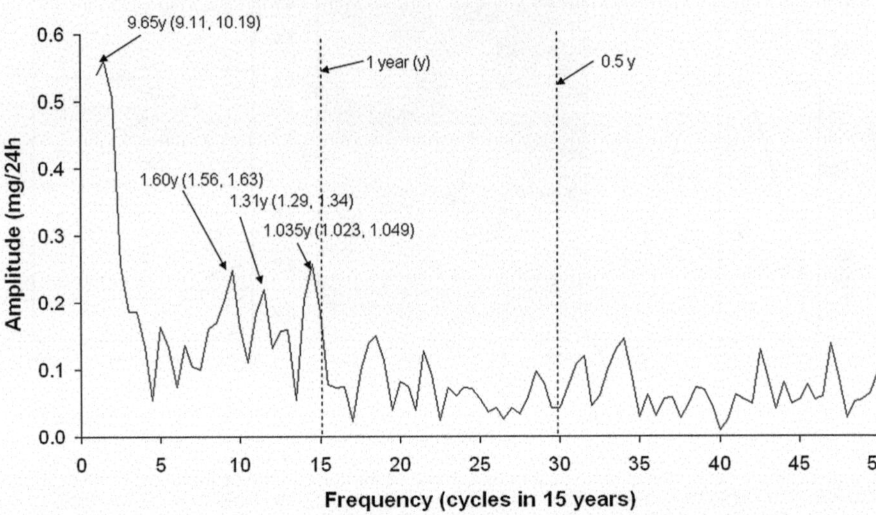

Fig. 3.18 (continued) (Wolf) numbers which differs in its time course from those of KS and Kp (not shown). The two biospheric variables differ in terms of their environmental phase synchronization (© Halberg Chronobiology Center). (**m**) A chronomic serial section of the planetary geomagnetic index Kp from October 23, 1948, to October 29, 1963, matching the KS record of CH, with the fit of a 0.42-year cosine curve shows a similar time course of quinmensal acrophases as that of CH's KS in Fig. 3.18l. Note that statistical significance is lost toward the end of the record

i

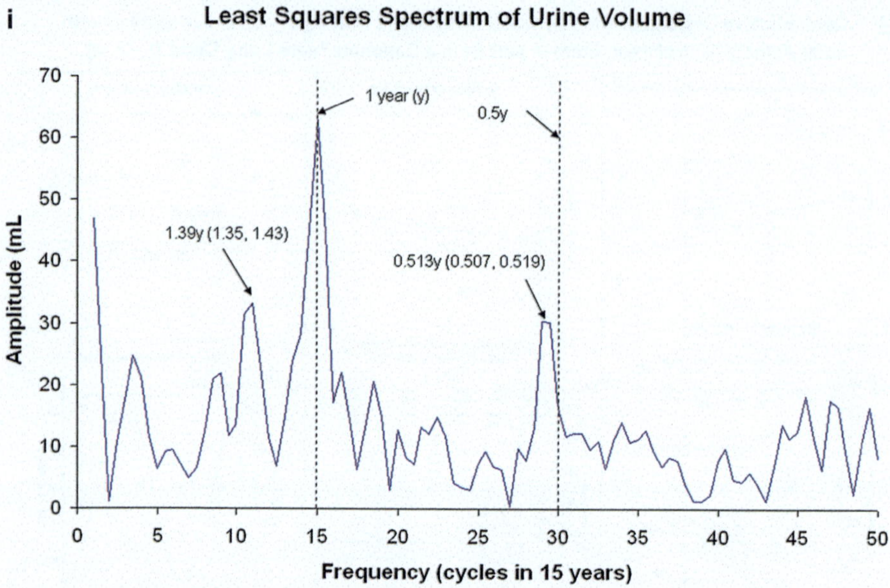

j

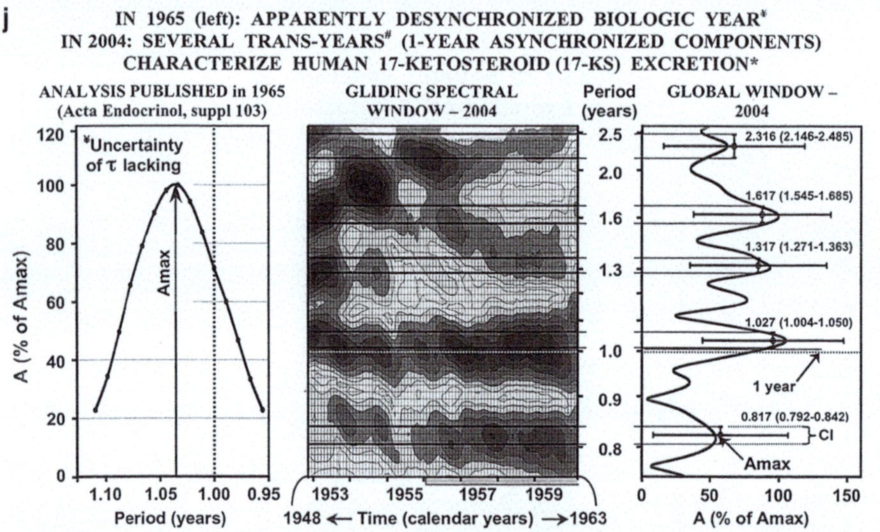

Transyears are oscillations with 95% confidence intervals (CIs) of their period (τ) between 1 and 2 years, overlapping neither of these τs nor one another. Cisyears are oscillations with periods between 1 and 0.5 year with CIs overlapping neither the 1-year nor the 0.5-year τ, as seen for a τ of ~ (but not) = 0.8 y. * ~ 15-year record, daily data (with gaps) from a clinically healthy man (CH, 43 – 58 y), N = 3719. Gliding interval = 8 years (y), increment = 1 month, longest trial τ = 2.265 y, shortest τ = 0.75 y, harmonic increment = 0.05. Darker shading corresponds to larger amplitude (A). Horizontal shaded bar starting in 1956 indicates span of intermittent androgen Rx; data during Rx omitted from analysis.

Fig. 3.18 (continued) for Kp, but not for KS (Fig. 3.18l). Note further at the beginning the lack of statistical significance in Kp, but not in KS. Note the quinmensal amplitudes (As, *second row*) that are quite large for KS at the end of the record, while the quinmensal (cis-half year) is not statistically significant in Kp and its amplitude decreases with time. Time course of only intermittently congruent (possibly driving when present) environmental period resembles that of a biospheric

k

Partial Internal Asynchronization between Urine Volume and 17-Ketosteroids at 2 of 3 Trial Periods Tested*

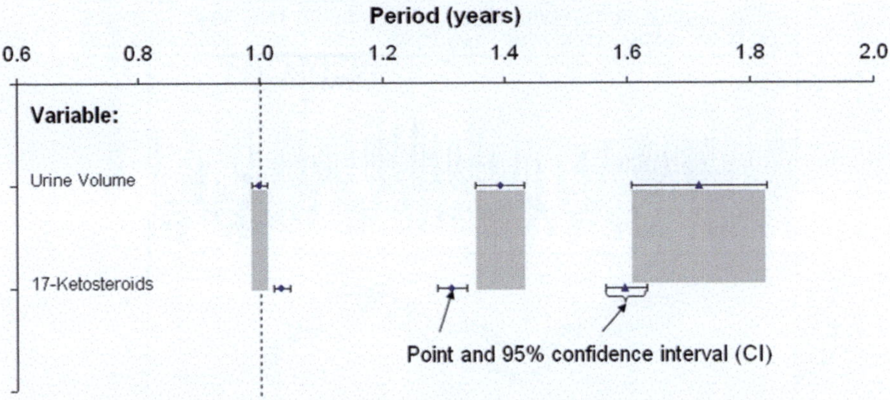

* In clinical health (CH, M, 44y at start) found by linear-nonlinear rhythmometry applied to daily data covering, with gaps, 14 consecutive years.

l

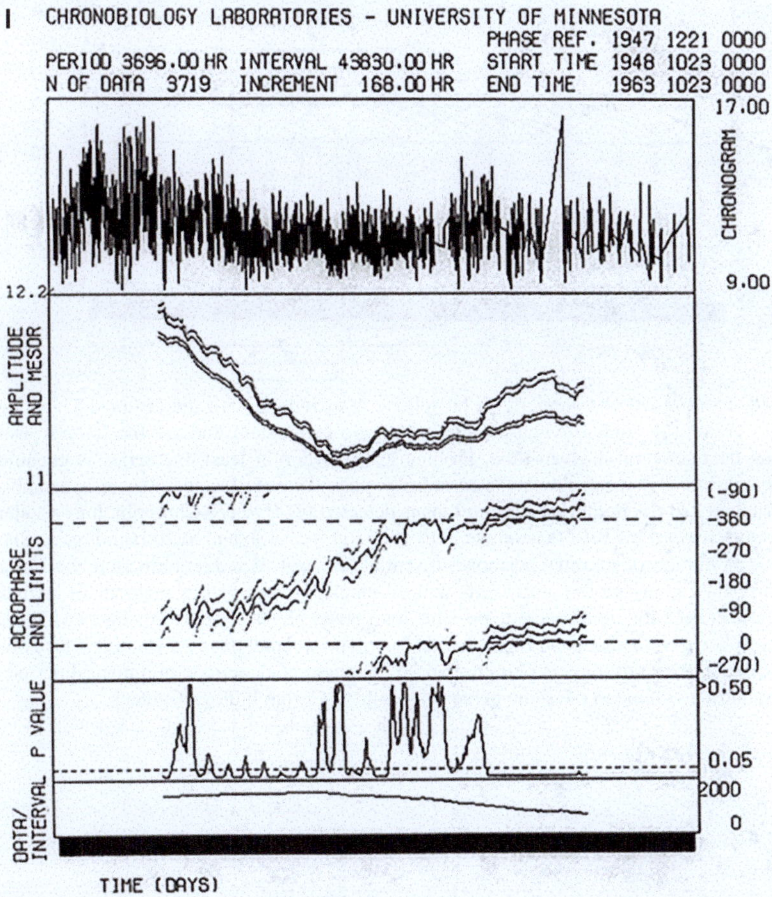

Fig. 3.18 (continued) period. Note, however, the persistence of a quinmensal period in the biosphere (when the previously congruent environmental period is not detected), suggesting a partial

m

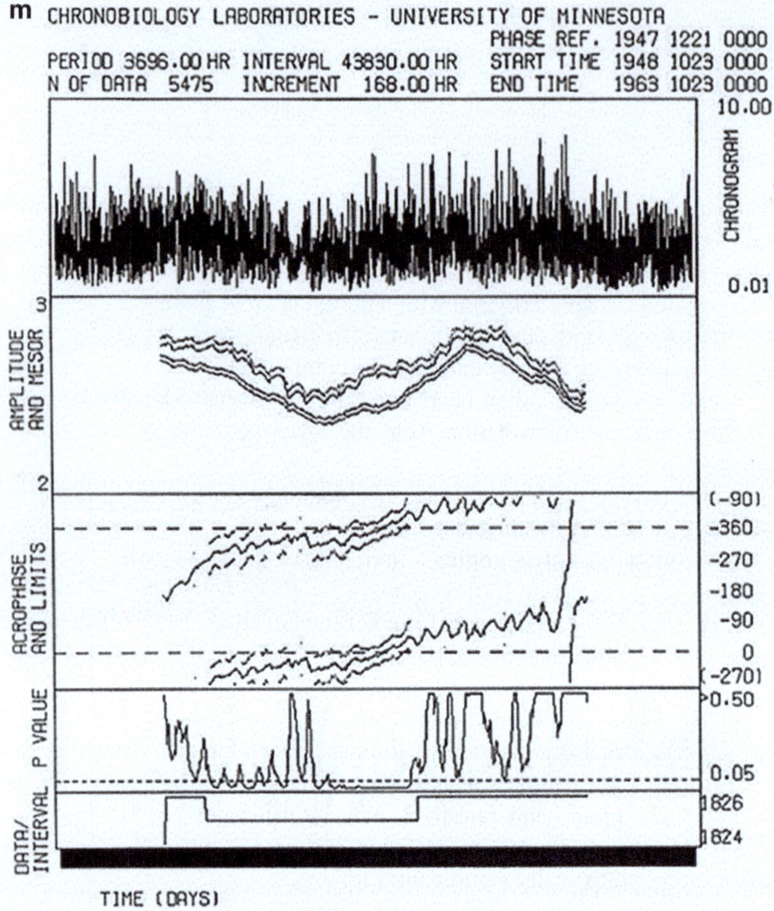

Fig. 3.18 (continued) endogenicity in biospheric associations with the cosmos. These associations are preferably studied when all data series are equidistant and are the longest available, matching data series at a given time. Problems arise when at least one series is unequidistant because of artifacts that can simulate a periodicity when none exists and, vice versa, can obscure a real periodicity to the point that it is no longer detected (© Halberg Chronobiology Center). (**n**) Periodicities revised by 2012 (as compared to 1965 in Fig. 3.18b and to 2001 in Fig. 3.18f) in the breakdown products of steroidal hormones essential for survival and reproduction, the 17-ketosteroids; probably, many more components will eventually be mapped as methods of analyses are refined and cross-validations become possible with novel physical coperiodisms or vice versa, or novel biospheric components prompt a search for environmental ones (© Halberg Chronobiology Center). (**o**) Different drummers seem to lock-in the phase at a common quinmensal period tested of KS and urine volume in the same person (© Halberg Chronobiology Center)

n
Selective congruence of steroid excretion with Earth (Kp) (top half) and of urine volume with Sun (Wolf N) (bottom half) in the same frequency window during 15 years (clinically healthy man, CH)

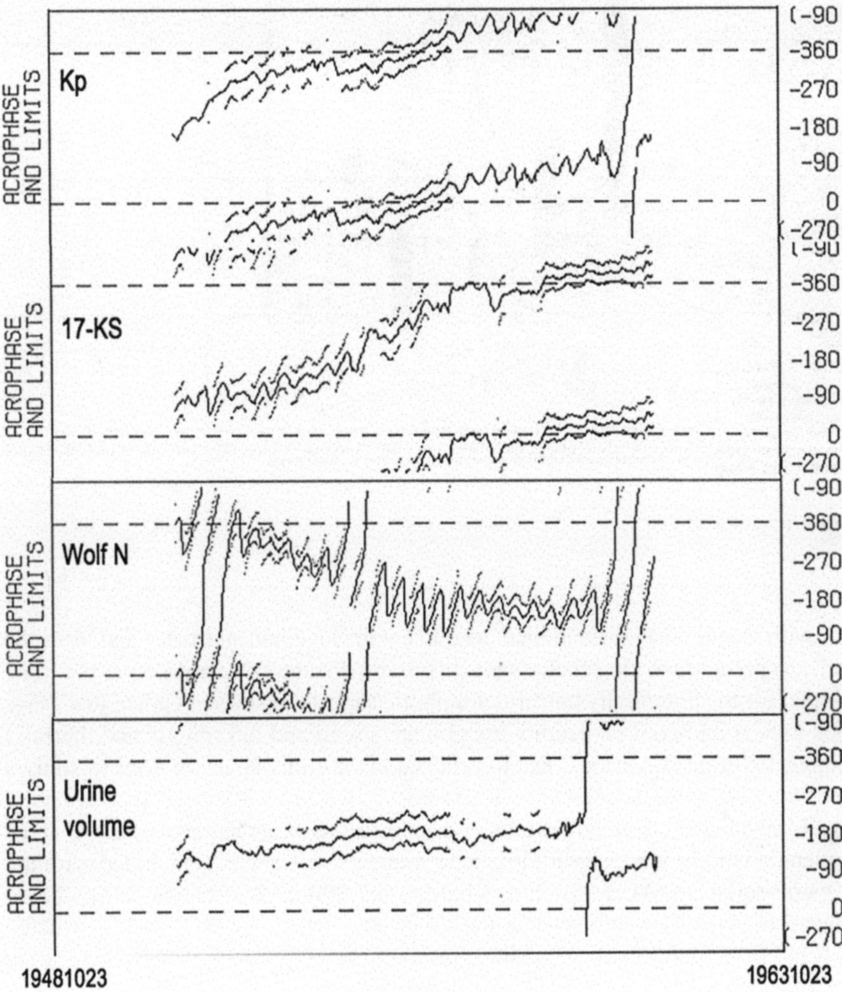

Fig. 3.18 (continued)

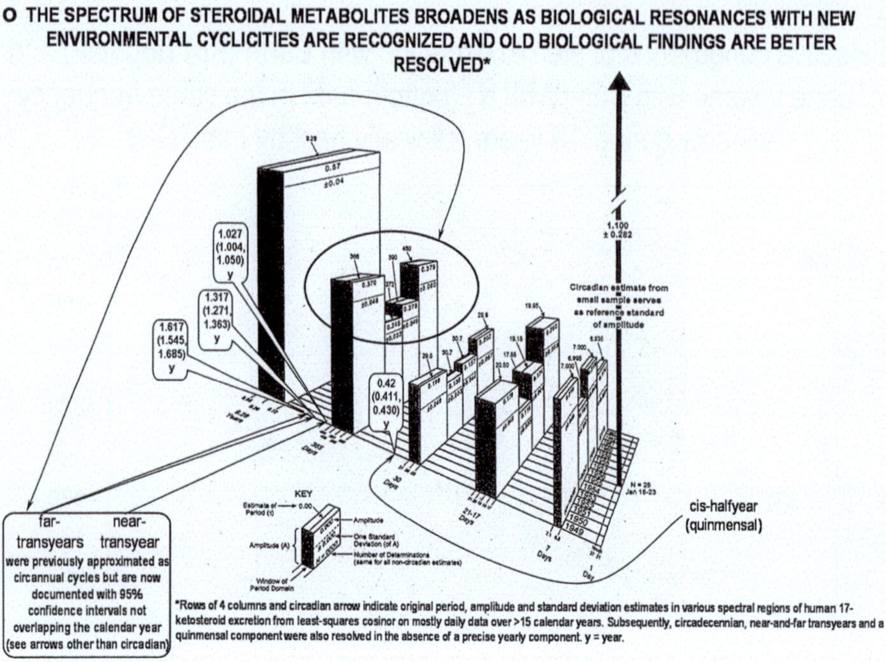

Fig. 3.18 (continued)

arrive, with continuous surveillance, at a life with love and not hate, and with the cosmos, to paraphrase Herschel, "well-disposed." Countermeasures have yet to be developed for the potentially harmful nonphotic aspects of space weather, just as we learned to heat and cool in meeting the climate's seen and felt (photic and thermic) dangerous extreme conditions. But to proceed toward this goal, we need to study a unified science with its aeolian uncertainties. It seems pertinent that apparent physical-biospheric cross-validations continue to occur, as apparent for the three myriadennians in biodiversity finding near matches in Table 3.3, included with the kind permission of Acad. Elchin Khalilov of Baku, Azerbaijan [45]. These biological coperiodisms are seen in the spectrum of Fig. 3.22a, the reconstructed models in Fig. 3.22 (globally), and the chronomic serial sections (locally in time) in Fig. 3.22d, e.

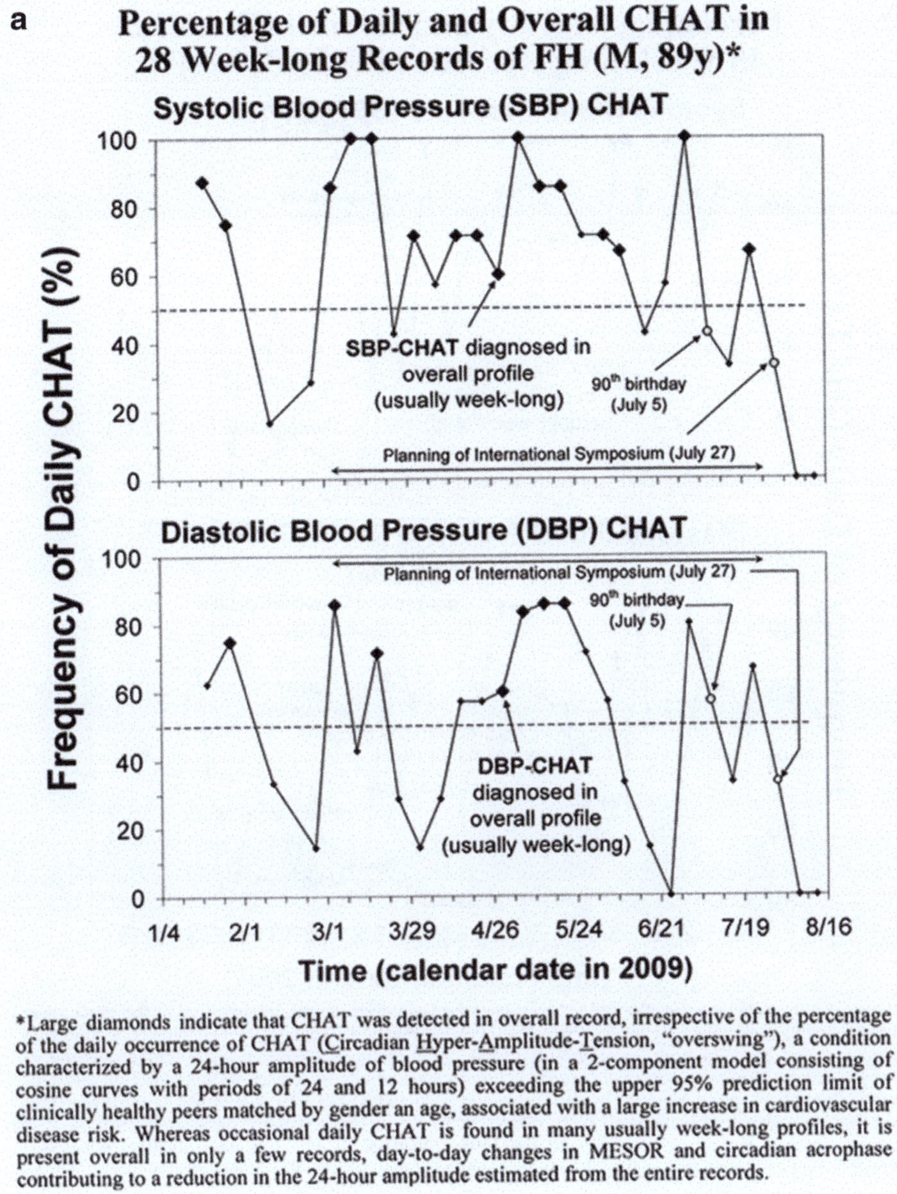

Fig. 3.19 (**a**) CHAT during about 5 months of preparation for a symposium lost in the last 2 post-symposium records, each of 1-week duration. It would seem that after 14 days of normalcy, monitoring could be dispensed with. But consider revising this opinion in view of the following addition in Fig. 3.19b (© Halberg Chronobiology Center). (**b**) After the absence of CHAT in each of two consecutive weekly records, transient CHAT (*small diamonds*) recurs in both the systolic and diastolic blood pressure of FH with weekly CHAT (*bigger diamond*) in one episode in systolic BP and two episodes in diastolic BP. There was no reason to lull oneself into a false sense of safety in Fig. 3.19a based on the last 2 consecutive acceptable 7-day records therein. The full normality in

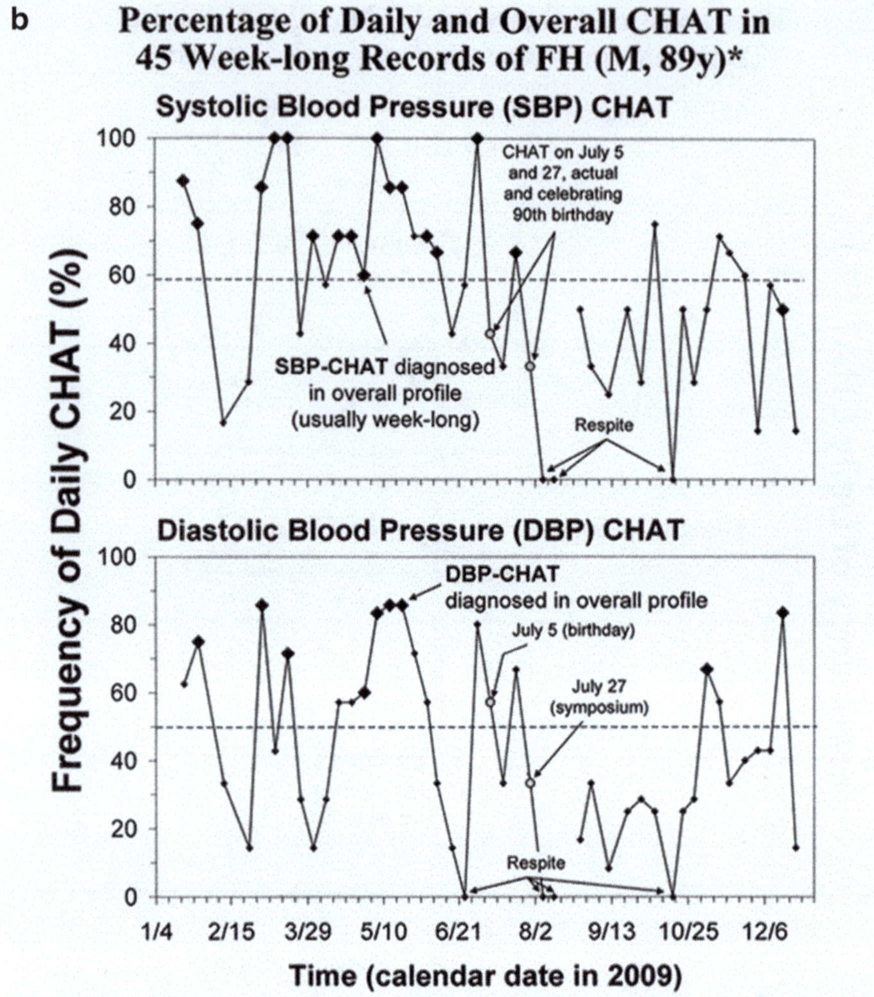

*Large diamonds indicate that CHAT was detected in overall record, irrespective of the percentage of the daily occurrence of CHAT (Circadian Hyper-Amplitude-Tension, "overswing"), a condition characterized by a 24-hour amplitude of blood pressure (in a 2-component model consisting of cosine curves with periods of 24 and 12 hours) exceeding the upper 95% prediction limit of clinically healthy peers matched by gender an age, associated with a large increase in cardiovascular disease risk. Whereas occasional daily CHAT is found in many usually week-long profiles, it is present overall in only a few records, day-to-day changes in MESOR and circadian acrophase contributing to a reduction in the 24-hour amplitude estimated from the entire records.

Fig. 3.19 (continued) the 2 weeks after a successful symposium was followed by transient (daily) CHAT and even occasional weekly CHAT. Take-home lesson: *Once a VVD is diagnosed, continued monitoring is advocated* (© Halberg Chronobiology Center)

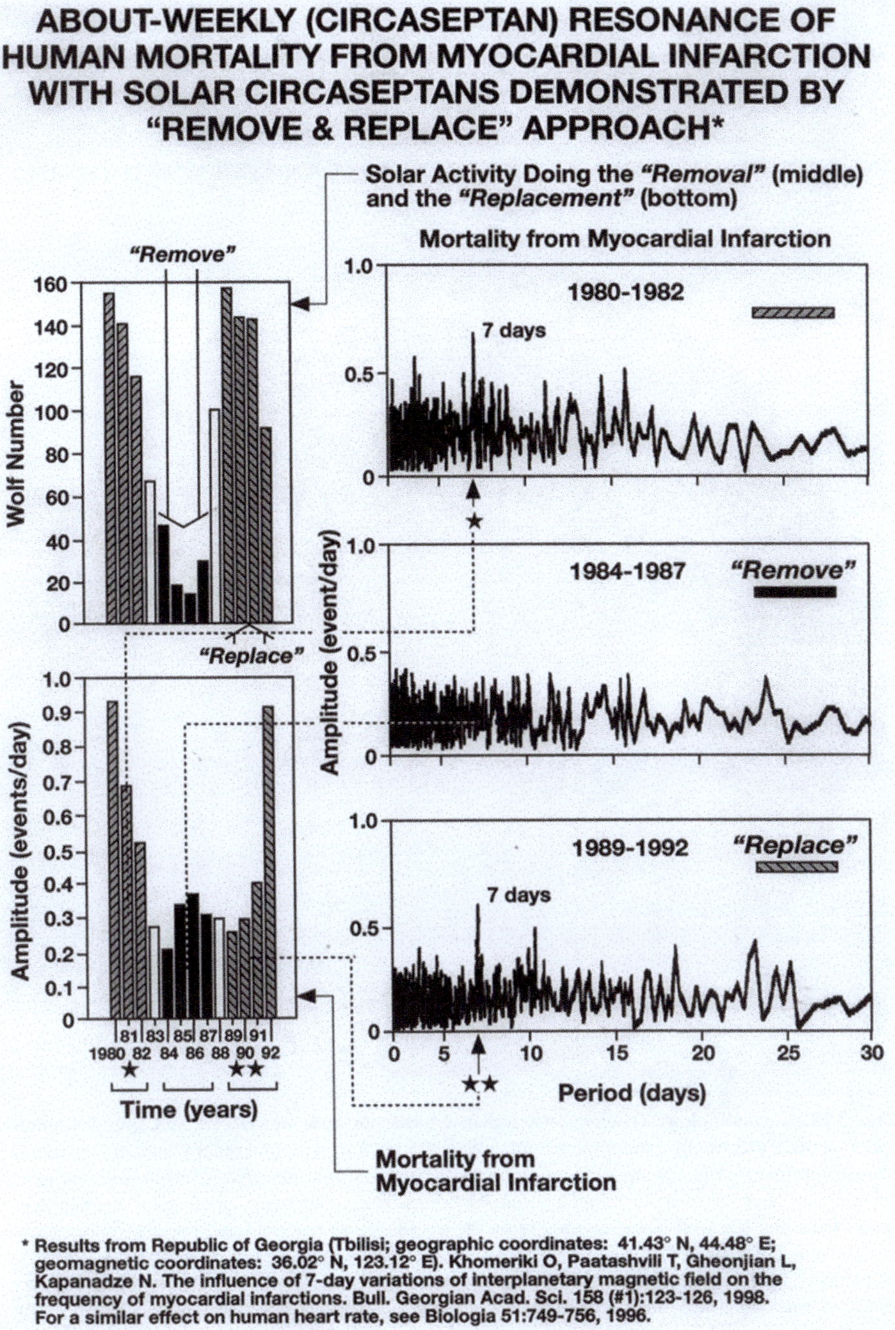

Fig. 3.20 Mortality from cardiopathology in Tbilisi, Republic of Georgia (1980–1992) [42], is characterized by an anticipated weekly variation, a prominent circannual rhythm, and an about-10.5-year cycle similar to that observed for mortality from myocardial infarction in Minnesota (1968–1996) [43]. Corroborating results on human heart rate, the circaseptan component is more prominent during spans of high solar activity than during solar quiet when solar circaseptans also tend to be more prominent [44]. This about-weekly (circaseptan) resonance of human mortality from cardiopathology with sunspot circaseptans was demonstrated by a subtraction-addition procedure by Khomeriki et al. [42] corresponding to our "remove-and-replace" approach (© Halberg Chronobiology Center)

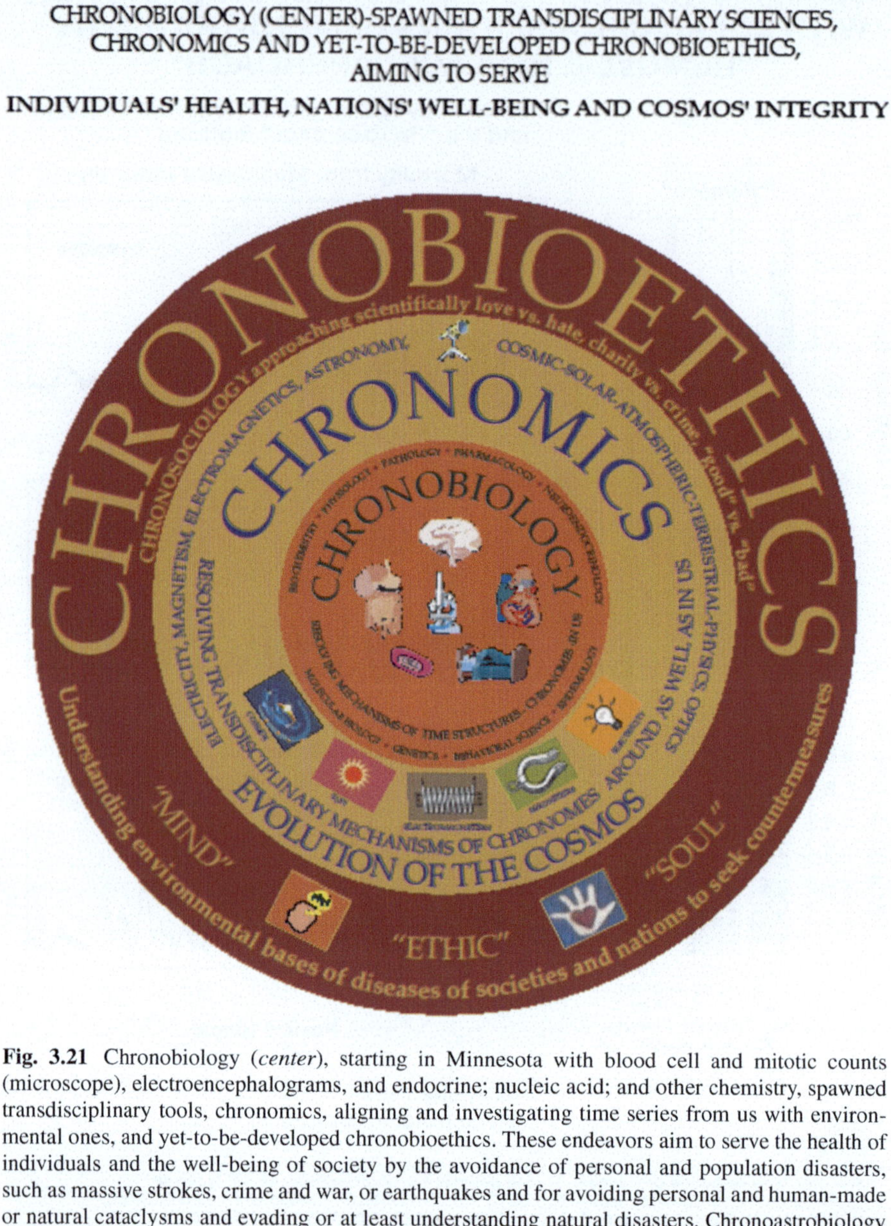

Fig. 3.21 Chronobiology (*center*), starting in Minnesota with blood cell and mitotic counts (microscope), electroencephalograms, and endocrine; nucleic acid; and other chemistry, spawned transdisciplinary tools, chronomics, aligning and investigating time series from us with environmental ones, and yet-to-be-developed chronobioethics. These endeavors aim to serve the health of individuals and the well-being of society by the avoidance of personal and population disasters, such as massive strokes, crime and war, or earthquakes and for avoiding personal and human-made or natural cataclysms and evading or at least understanding natural disasters. Chronoastrobiology is as yet missing in this diagram, introducing remove-and-replace experiments and (Fig. 3.10) documenting an association of human mood with interplanetary and earth magnetism. This may open a chapter in global health care including self-surveillance of populations to prevent or avoid personal and societal or natural environmental cataclysms, respectively (© Halberg Chronobiology Center)

Table 3.3 Major cycles in earth's development

No	Cycle's time span	Cycle name and reference
1	400–500 My	Wilson cycles (V.Y. Khain, 1992), periodic formation and collapse of Pangaeas. These megacycles could be identified with cycles named after the Canadian geophysicist J.T. Wilson who discovered them first on the example of re-closing and opening of the Atlantic.
2	150–200 My	Bertrand cycles (Khain, Seslavinsky, 1991) found at the end of the nineteenth century. In terms of recurrence in the section of multiple-age folded systems of Western Europe and North America of the same sequence of sedimentary formations: shale, flysch and molasse. He singled out four cycles - Huronian, Caledonian, Hercynian and Alpine. Of them, the last three are found in the literature, supplemented, however, with another cycle, Cimmerian, that appeared in the Mesozoic; the Huronian cycle was replaced by N.S.Shatsky with the Baikal cycle.
		Corresponding to one of the cycles of biodiversity mentioned, Shaviv, 2002.
3	140 My	Rohde RA, Muller RA. Cycles in fossil diversity. Nature 2005; 434: 208–210.
	60–63 My	Even periods of large-scale continental volcanism have been shown to have this same 60-million year cycle. Rohde & Muller, 2005
3	30 My	Stille cycles (V.Y. Khain, 1992). These cycles are marked with orogenic phases when fold-thrust deformations were on the rise. Again, these cycles coincided with the periodicity of changes in the intensity of tectonic deformation, island arc volcanism, granite formation, and regional metamorphism. This periodicity is similar to that which is established for great fauna extinctions and regenerations linked in turn, on the one hand, with major transgressions and regressions and on the other, with intensifications of bombardment of the earth with large meteorites and (or) comets.
4	3–5 My	Most pronounced are the stratification of the global stratigraphic scale of the Phanerozoic, and definitely the unconscious empirical basis in developing this scale. In the above-mentioned Vail curve of eustatic sea level fluctuations, this cyclicity corresponds to third-order cycles.
5	400 thousand years	Berger et al. (1989) (changes in the eccentricity of earth's orbit), the periodicity established by M. Milankovitch and successfully used by him to explain the alternation of glacial and interglacial periods
6	100 thousand years	Berger et al. (1989) (changes in the eccentricity of earth's orbit), the periodicity established by M. Milankovitch and successfully used by him to explain the alternation of glacial and interglacial periods
7	41 thousand years	Berger et al. (1989) (changes in the inclination of earth's axis of rotation)
8	19–23 (20) thousand years	Dercourt et al. 1986 (precessions). This cyclicity is also reflected in changes in lithology and thickness of extraglacial deposits, in particular, changes in the calcium carbonate and organic carbon content

Adapted from Khain VE, Khalilov EN [45]

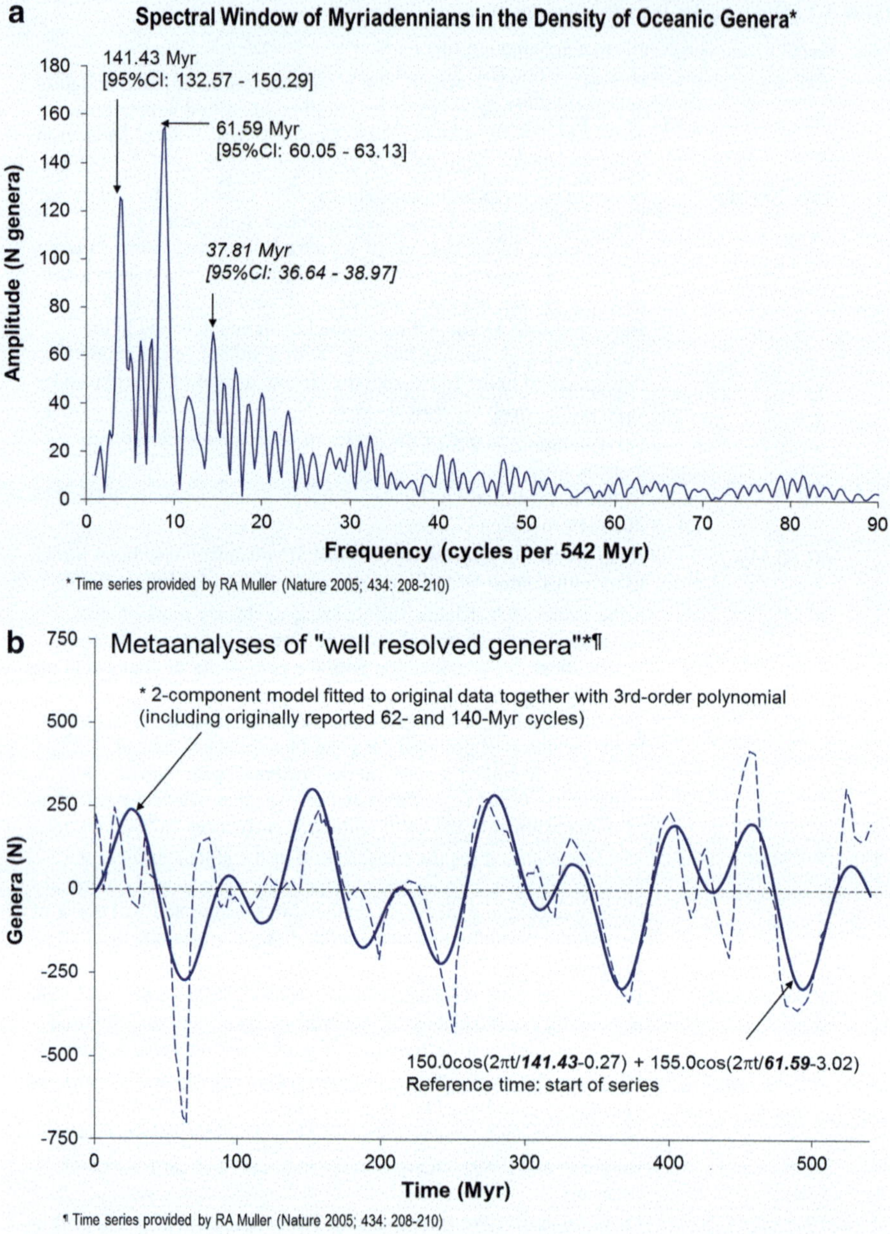

Fig. 3.22 (**a**) A further step toward a tentative cross-validation of biospheric coperiodisms with cycles in (space) weather, a development that started vice versa from the natural biologic near week slightly shorter than 7 days in 17-ketosteroid excretion to a near week of 6.74 days in geomagnetism. Data on fossil diversity during the Phanerozoic, the eon during which hard shells and skeletons left abundant fossils 0 to 542 million years (My) ago, formatted, and previously analyzed by Rohde and Muller who reported about-62- and 140-My cycles based on Raup and Sepkoski's compendium of the first and last stratigraphic appearances of 36,380 marine genera (Science 1982; 215:

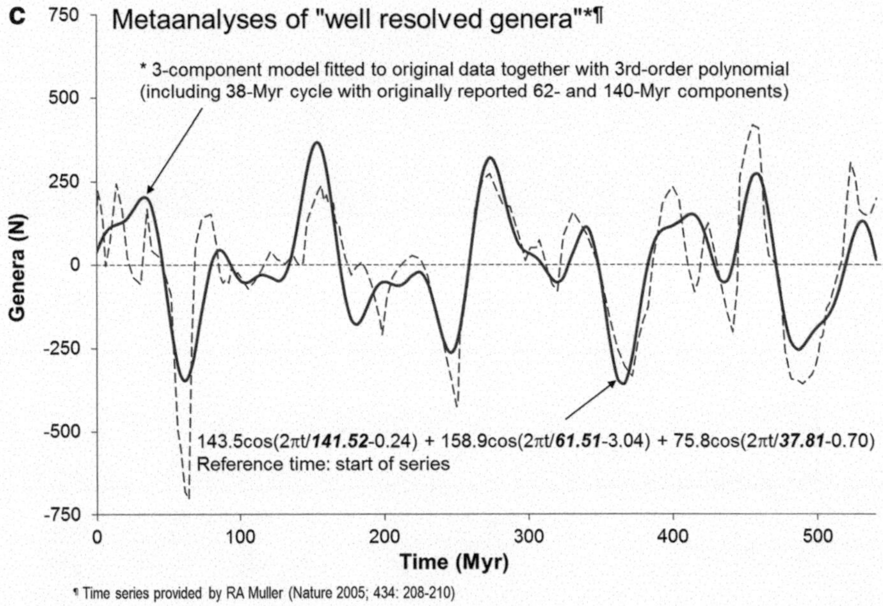

Fig. 3.22 (continued) 1501–1503), were analyzed by linear-nonlinear rhythmometry. In addition to the two major spectral peaks reported by Rohde and Muller, the least-squares spectrum of the residuals from a third-order polynomial trend reveals a third component with a period of about 38 My. Nonlinearly, a model consisting of the two major cycles and a third-order polynomial yields the 95 % confidence intervals (CI) for the periods listed near the spectral peaks. When added to the model, the third component is also resolved with statistical significance, as shown in italics near the spectral peak. The addition of this component to the model results in only minor changes in the estimated periods, amplitudes (in thousands of genera), and their respective 95 % CIs, estimated as 141.52 [132.14, 150.90] My with an amplitude of 143.5 [77.5, 209.5], 61.51 [59.98, 63.05] with an amplitude of 158.9 [133.2, 184.6], and 37.81 [36.64, 38.97] My with an amplitude of 75.8 [13.6, 138.0] (© Halberg Chronobiology Center). (**b**) Plot as a function of time of the well-resolved genera (after removal of a third-order polynomial trend) together with the nonlinearly resolved model, considering only the two major spectral components reported by Rohde and Muller (© Halberg Chronobiology Center). (**c**) Plot as a function of time of the well-resolved genera (after removal of a third-order polynomial trend) together with the nonlinearly resolved model, considering all three spectral components detected by linear-nonlinear rhythmometry (© Halberg Chronobiology Center). (**d**) Chronobiologic serial section of the "well-resolved genera" detrended by means of a third-order polynomial examines the stability of the about-38-My cycle characteristics that may have an environmental coperiodism in the Hans Stille geologic 30-My cycle. The detrended biospheric data on species diversity are shown in the *top row*. Instead of analyzing the time series as a whole, data are fitted with a 37.8 My cosine curve in a 75.6-My interval that is progressively displaced by a 1.26-My increment. The rhythm-adjusted average (MESOR) is the lower curve in the second row, and the 37.8-My amplitude is the distance between the two curves in the second row. *Dots* below the lower curve and above the upper curve are standard errors of the MESOR and myriadennian amplitude, respectively. The 37.8-My acrophases (phases of the maximum) are double plotted in the third row, where 0° corresponds to the beginning of the time series and 360° represents a full 37.8-My cycle. For the first ~150 My, the acrophase remains stable and then undergoes a rapid change around 180 My to resume about the same phase position until about 300 My when a longer lasting disturbance is associated with a shift of the acrophase by almost half a cycle (almost complete reversal). The acrophase returns to its original position after the 400-My mark. Changes in acrophase tend to coincide with times when the amplitude is smaller. Statistical

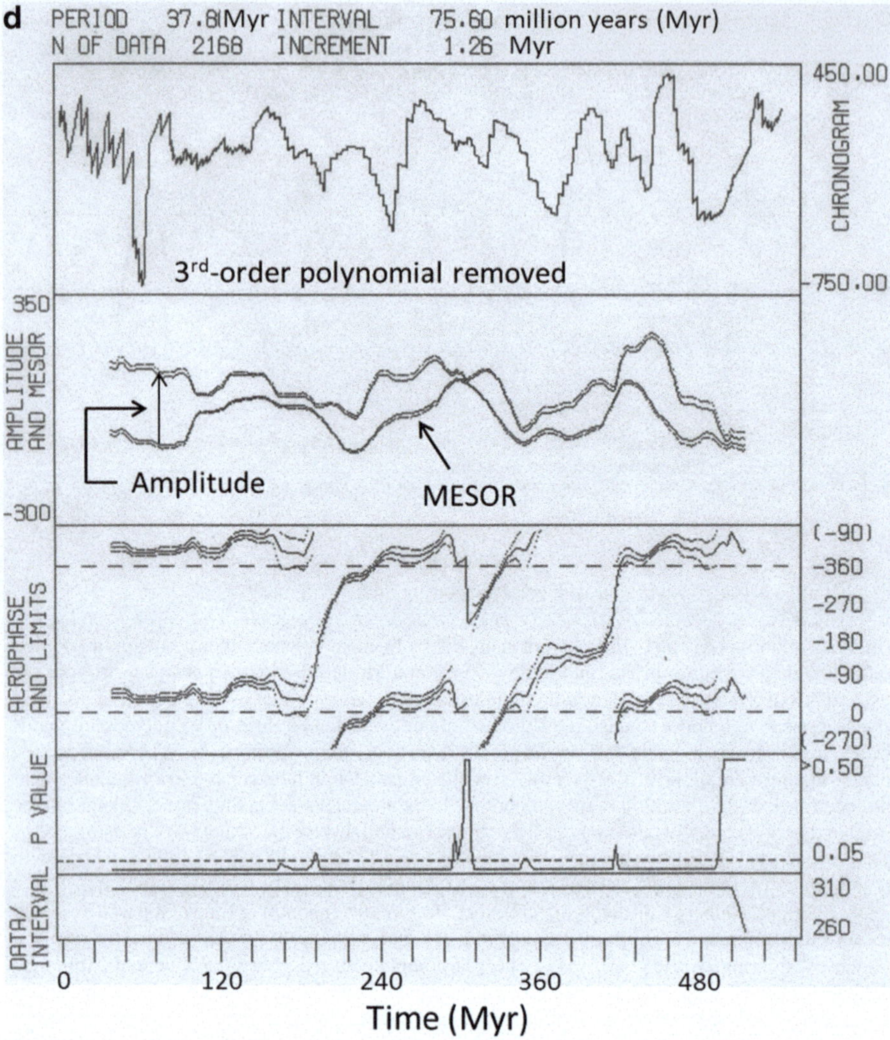

Fig. 3.22 (continued) significance for the 37.8-My fit is shown in the fourth row, indicating that the cycle is detected most of the time, except for short spans when the amplitude is reduced and the acrophase deviates from its original position (© Halberg Chronobiology Center). (**e**) The about-38-My cycle being less prominent than the about 62- and 140-My components, it is likely that the latter influence the estimation of the former in short intervals moved throughout the time series (in Fig. 3.22d). For this reason, another chronobiologic serial section is obtained after removal not only of the third-order polynomial but also of the two largest cycles with periods of about 62 and 140 My, estimated nonlinearly and concomitantly. Residuals are shown in the *top row*. Whereas disturbances in the 37.8-My acrophases are still present, coinciding with spans when the amplitude is reduced, the stability and consistency of the acrophase is improved by comparison with results in Fig. 3.22d (© Halberg Chronobiology Center)

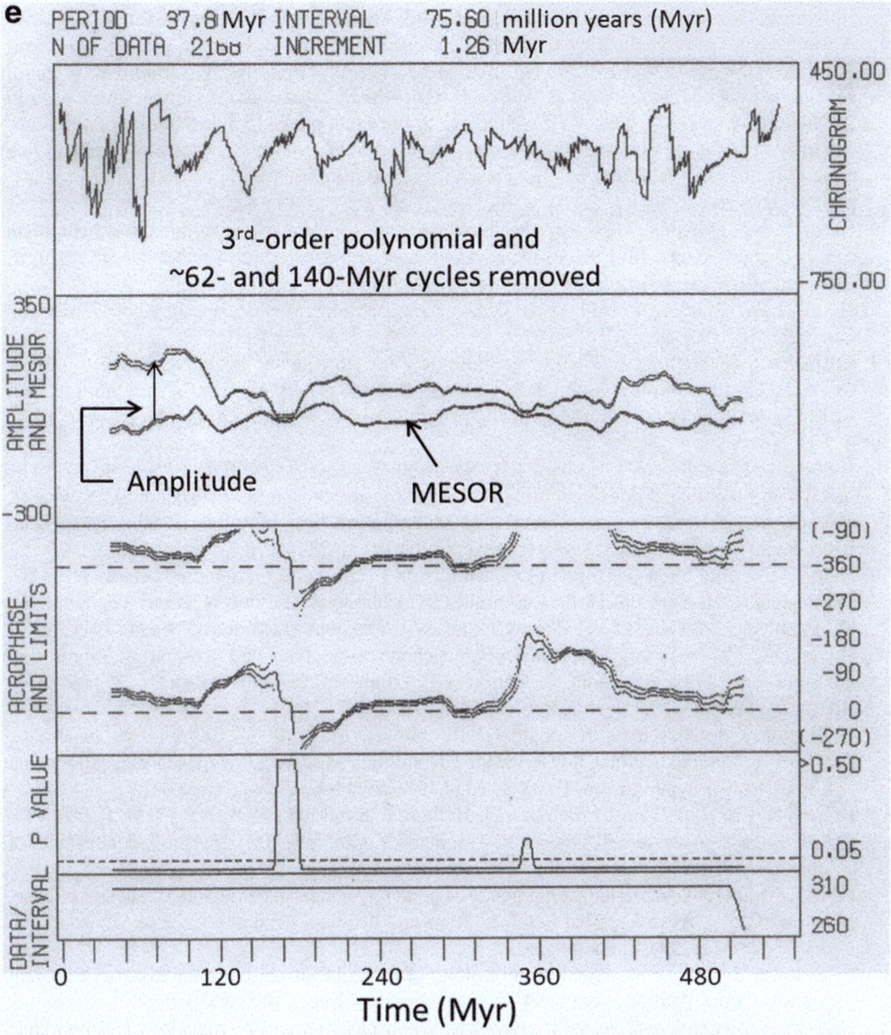

Fig. 3.22 (continued)

References

1. Halberg F. Chapter on "Medizin" in: Jahrbuch der Internationalen Hochschulwochen des Österreichen College. Salzburg: Igonta Verlag; 1946. p. 336–51. [In German.].
2. Halberg F. Neue Fortschritte in der Medizin. Europ med Rdsch 1948;1:21–35 [In German.].
3. Halberg F. A propos du traitement pénicille-sulfamide de la gonococcie. Bull mens Soc Méd mil franc 1948;42:201–3.

4. Halberg F, Cornélissen G, Sothern RB, Wallach LA, Halberg E, Ahlgren A, Kuzel M, Radke A, Barbosa J, Goetz F, Buckley J, Mandel J, Schuman L, Haus E, Lakatua D, Sackett L, Berg H, Wendt HW, Kawasaki T, Ueno M, Uezono K, Matsuoka M, Omae T, Tarquini B, Cagnoni M, Garcia Sainz M, Perez Vega E, Wilson D, Griffiths K, Donati L, Tatti P, Vasta M, Locatelli I, Camagna A, Lauro R, Tritsch G, Wetterberg L. International geographic studies of oncological interest on chronobiological variables. In: Kaiser H, editor. Neoplasms – comparative pathology of growth in animals, plants and man. Baltimore: Williams and Wilkins; 1981. p. 553–96.
5. Hermida RC, Halberg F. From marker bioperiodicities, over marker rhythms, toward human cancer chronotherapy. In: Levine AS, editor, Kaiser H. series editor. Cancer growth and progression, vol. 9, ch. 6. Dordrecht: Kluwer.; 1989. p. 44–56.
6. Hermida RC, Halberg F, Halberg E. Closer to a psychoneuroendocrine hemopsy? Biochim Clin 1986;10:1053–66.
7. Cornélissen G, Halberg F. The chronobiologic pilot study with special reference to cancer research: Is chronobiology or, rather, its neglect wasteful? In: Goldson AL, volume editor, Kaiser H, series editor. Cancer growth and progression, vol. 9, ch. 9. Dordrecht: Kluwer; 1989. p. 103–33.
8. Halberg F, Cornélissen G, Sothern RB, Barnwell F, Cegielski N, Ilyia E, Siegelova J. The moon's and the genes' tides and double tides pulling the biosphere. In: Halberg F, Kenner T, Fiser B, Siegelova J, editors. Noninvasive Methods in Cardiology, 16–17 Sept 2010, Brno Brno: Faculty of Medicine, Masaryk University; 2010. p. 23–45.
9. Halberg F, Cornélissen G, Hillman DC, Bingham C, Halberg E, Guillaume F, Barnwell F, Wu JY, Wang ZR, Halberg FE, Holte J, Schmitt OH, Kellogg PJ, Luyten W, Breus TK, Komarov FI, Mikulecky M, Garcia L, Lodeiro C, Iglesias T, Quadens O, Muller C, Kaada B, Miles L, Hayes DK. Chronobiology in a moon-based chemical analysis and physiologic monitoring laboratory. In: Ponnamperuma C, Gehrke CE, editors. A lunar-based chemical laboratory (LBCAL). Hampton: A. Deepak Publishing; 1992. p. 161–203.
10. Halberg F, Cornélissen G, Bingham C, Tarquini B, Mainardi G, Cagnoni M, Panero C, Scarpelli P, Romano S, März W, Hellbrügge T, Shinoda M, Kawabata Y. Neonatal monitoring to assess risk for hypertension. Postgrad Med 1986;79:44–6.
11. Breus TK, Pimenov KYu, Cornélissen G, Halberg F, Syutkina EV, Baevsky RM, Petrov VM, Orth-Gomer K, Åkerstedt T, Otsuka K, Watanabe Y, Chibisov SM. The biological effects of solar activity. Biomed Pharmacother 2002;56 Suppl 2:273s–83s.
12. Pales E, Mikulecky M. Periodic emergence of great physicians in the history of ancient Greece, India and China. Abstract, 23rd Seminar, Man and his Terrestrial and Cosmic Environment, Upice 21–23 May 2002 (unpaginated).
13. Pales E, Mikulecky Sr M. 500-year periodicity of political instability in the history of ancient Egypt and China. Androgens at work? Neuroendocrinol Lett 2008;29:589–97.
14. Halberg F, Cornélissen G, Regal P, Otsuka K, Wang ZR, Katinas GS, Siegelova J, Homolka P, Prikryl P, Chibisov SM, Holley DC, Wendt RW, Bingham C, Palm SL, Sonkowsky RP, Sothern RB, Pales E, Mikulecky M, Tarquini R, Perfetto F, Salti R, Maggioni C, Jozsa R, Konradov AA, Kharlitskaya EV, Revilla M, Wan CM, Herold M, Syutkina EV, Masalov AV, Faraone P, Singh RB, Singh RK, Kumar A, Singh R, Sundaram S, Sarabandi T, Pantaleoni GC, Watanabe Y, Kumagai Y, Gubin D, Uezono K, Olah A, Borer K, Kanabrocki EA, Bathina S, Haus E, Hillman D, Schwartzkopff O, Bakken EE, Zeman M. Chronoastrobiology: proposal, nine conferences, heliogeomagnetics, transyears, near-weeks, near-decades, phylogenetic and ontogenetic memories. Biomed Pharmacother 2004;58 Suppl 1:S150–87.
15. Nintcheu-Fata S, Katinas G, Halberg F, Cornélissen G, Tolstykh V, Michael HN, Otsuka K, Schwartzkopff O, Bakken E. Chronomics of tree rings for chronoastrobiology and beyond. Biomed Pharmacother 2003;57 Suppl 1:24s–30s.
16. Vallot J, Sardou G, Faure M. De l'influence des taches solaires sur les accidents aigus des maladies chroniques. Gazette des Hôpitaux 1922;56:904–5.
17. Chizhevsky AL. (Gazenko O, foreword; Korzhuyev PA et al. editors) Earth echo of solar storms. Moscow: Thought; 1976. 367 p. with illustrations.

18. Chizhevsky AL. Action de l'ionisation de l'atmosphère et de l'ionisation artificielle de l'air sur les organismes sains et les organismes malades. In: Piéry M, editor. Traité de Climatologie: Biologique et médicale. Tome premier Paris: Masson et Cie; 1934. p. 662–73.
19. Hagemeister M. Russian cosmism in the 1920s and today. In: Rosenthal BG, editor. The occult in Russian and Soviet culture. Ithaca: Cornell University Press; 1997. p. 185–202.
20. Cornélissen G, Grambsch P, Sothern RB, Katinas G, Otsuka K, Halberg F. Congruent biospheric and solar-terrestrial cycles. J Appl Biomed 2011;9:63–102. doi:10.2478/v10136-009-0023-7.
21. Halberg F, Cornélissen G, Grambsch P, McCraty R, Beaty L, Siegelova J, Homolka P, Hillman DC, Finley J, Thomas F, Kino T, Revilla M, Schwartzkopff O. Personalized chronobiologic cybercare; other chronomics' progress by transdisciplinary cycles' congruences: season's appreciations 2009. J Appl Biomed 2011;9:1–34. doi:10.2478/v10136-009-0022-8.
22. Pittendrigh CS. Temporal organization: reflections of a Darwinian clock-watcher. Ann Rev Physiol 1993;55:17–54.
23. Halberg F, Cornélissen G, Salti R, Perfetto R, Tarquini R, Stagi S, Hillman DC, Katinas GS, Hoogerwerf WA, Carandente F, Otsuka K, Czaplicki J, Chibisov SM, Scheving LA, Syutkina EV, Masalov A, Mitsutake G, Wang ZR, Wan CM, Schwartzkopff O, Bakken EE. Chronoauxology. Chronomics: trends and cycles in growth and cosmos rather than secularity. In: Proceedings, 10th Auxology Congress: Human Growth in Sickness and in Health, Florence, 4–7 Jul 2004. Florence: Edizioni Centro Studi Auxologici; 2010. 92 pp.
24. Sothern RB, Halberg F, Hillman D, Cornelissen G. Infradian components gauge aging of the circulation and respiration in self-measurements by a healthy man for over four decades. World Forum "Natural Cataclysms and Global Problems of the Modern Civilization", 19–21 Sept 2011, Istanbul, Turkey. London: SWB; 2011. p. 118–9.
25. Halberg F, Sothern RB, Hillman D, Cornelissen G. Para-tridecadal components characterize psychophysiological variables of a clinically healthy man. World Forum "Natural Cataclysms and Global Problems of the Modern Civilization", 19–21 Sept 2011, Istanbul, Turkey. London: SWB; 2011. p. 117–8.
26. Watanabe Y, Cornélissen G, Hillman D, Otsuka K, Halberg F. Long periods in 23 years of around-the-clock automatic measurements of human blood pressure and pulse. World Forum "Natural Cataclysms and Global Problems of the Modern Civilization", 19–21 Sept 2011, Istanbul, Turkey. London: SWB; 2011. p. 121–2.
27. Hyde J. Tire sensors among new SUV safety initiatives. ABC News, January 9 (year not given but predates 2002). 2 pp. http://abcnews.go.com/US/story?id=94448&page=1
28. Halberg F, Cornélissen G, Otsuka K, Watanabe Y, Katinas GS, Burioka N, Delyukov A, Gorgo Y, Zhao ZY, Weydahl A, Sothern RB, Siegelova J, Fiser B, Dusek J, Syutkina EV, Perfetto F, Tarquini R, Singh RB, Rhees B, Lofstrom D, Lofstrom P, Johnson PWC, Schwartzkopff O. International BIOCOS study group. Cross-spectrally coherent ~10.5- and 21-year biological and physical cycles, magnetic storms and myocardial infarctions. Neuroendocrinol Lett 2000;21:233–58.
29. Halberg F. Chronobiology. Annu Rev Physiol 1969;31:675–725.
30. Halberg F, Otsuka K, Katinas G, Sonkowsky R, Regal P, Schwartzkopff O, Jozsa R, Olah A, Zeman M, Bakken EE, Cornélissen G. A chronomic tree of life: ontogenetic and phylogenetic 'memories' of primordial cycles - keys to ethics. Biomed Pharmacother 2004;58 Suppl 1:S1–S11.
31. Starbuck S, Cornélissen G, Halberg F. Is motivation influenced by geomagnetic activity? Biomed Pharmacother 2002;56 Suppl 2:289s–97s.
32. Halberg F, Cornélissen G, Sothern RB, Katinas GS, Schwartzkopff O, Otsuka K. Cycles tipping the scale between death and survival (= "life"). Prog Theor Phys 2008;120 Suppl 173:153–81.
33. Halberg F, Cornélissen G, Watanabe Y, Otsuka K, Fiser B, Siegelova J, Mazankova V, Maggioni C, Sothern RB, Katinas GS, Syutkina EV, Burioka N, Schwartzkopff O. Near 10-year and longer periods modulate circadians: intersecting anti-aging and chronoastrobiological research. J Gerontol A Biol Sci Med Sci 2001;56:M304–24.

34. Halberg F, Schwartzkopff O, Cornélissen G, Hörz H, Hartung W. Franz Halberg im Treffpunkt Alltagsphysik-Alltagsphysiologie-Alltagsökologie: Autobiographie mit zeitgenössischer Wertung. Leibniz-Online 2010;7. 58 pp. http://www2.hu-berlin.de/leibniz-sozietaet/journal/archiv_07_10.htm
35. Fedor-Freybergh PG. Hommage á Franz Halberg. Neuroendocrinol Lett 1999;20:46–7.
36. Cornélissen G, Halberg E, Halberg F, Halberg J, Sampson M, Hillman D, Nelson W, Sánchez de la Peña S, Wu J, Delmore P, Marques N, Marques MD, Fernandez JR, Hermida RC, Guillaume F, Carandente F. Chronobiology: a frontier in biology and medicine. Chronobiologia 1989;16:383–408.
37. Pauly JE, Scheving LE. Dedication. Prog Clin Biol Res 1997;227A:xxiii–vii.
38. Mikulecky Sr M. Professor Franz Halberg – the grand democrat in the global science. From the molecule to cosmos and back. Neuroendocrinol Lett 2009;30:675–6.
39. Sidorin AYa. Franz Halberg's studies of the heliogeophysical effects on the biosphere. Geophys Processes Biosph 2009;8:5–12 [In Russian with English summary. Dedicated to Franz Halberg on his 90th birthday.].
40. Cornélissen G. Time structures (chronomes) in us and around us: a tribute to Franz Halberg. In: Cornélissen G, Kenner R, Fiser B, Siegelova J, editors. Proceedings, Symposium: Chronobiology in Medicine. Dedicated to the 85th anniversary of Professor Franz Halberg. Brno: Masaryk University; 2004. p. 8–43.
41. Cornélissen G. Celebration of a life's achievements in nutrition by Franz Halberg. The Open Nutraceuticals J. 2012;5(Suppl 1-M1):15, and When you eat matters: 60 years of Franz Halberg's nutrition chronomics. Open Nutraceuticals J 2012;5 (Suppl 1-M1):16.
42. Amiranashvili AG, Cornélissen G, Amiranashvili V, Gheonjian L, Chikhladze VA, Gogua RA, Matiashvili TG, Paatashvili T, Kopytenko YA, Siegelova J, Dusek J, Halberg F. Circannual and circadecennian stages in mortality from cardiovascular causes in Tbilisi, Republic of Georgia (1980–1992). Scripta Medica 2002;75:255–60.
43. Cornélissen G, Halberg F, Breus T, Syutkina EV, Baevsky R, Weydahl A, Watanabe Y, Otsuka K, Siegelova J, Fiser B, Bakken EE. Non-photic solar associations of heart rate variability and myocardial infarction. J Atmos Sol Terr Phys 2002;64:707–20.
44. Cornélissen G, Hillman D, Katinas GS, Rapoport S, Breus TK, Otsuka K, Bakken EE, Halberg F. Geomagnetics and society interact in weekly and broader multiseptans underlying health and environmental integrity. Biomed Pharmacother 2002;56 Suppl 2:319s–26s.
45. Khain VE, Khalilov EN. Cycles in geodynamic processes: their possible nature. Moscow: Scientific World; 2009. 520 p.

Chapter 4
Chronobiology

Abstract This chapter sketches some of the highlights of chronobiology. In a nutshell, considering time as the fourth dimension in biology amounts to changing variability from foe to friend. Akin to the splitting of the "indivisible" atom, chronobiology indeed splits the otherwise neglected seemingly indivisible range into rhythms and other predictable changes, such as trends and chaotic behavior.

Keywords Time structures • Cosinor method • Environmental geophysical cycles • Circadians • Extracircadian oscillations

4.1 Introduction

While counting eosinophils in mouse blood, Franz Halberg was faced with confusing variability [1], Fig. 4.1. Large changes observed in several kinds of blood cell counts persisted when additional precautions were taken not to disturb the animals. He also found large variations in the content of glycogen in liver and in the number of cells in the state of division in the liver and other tissues. After collecting data at different times and pursuing the patterns of change as a function of time, predictable about-24-hour cycles became apparent. Some of these periodicities had large amplitudes and were highly reproducible. Relatively simple statistical methods applied to the data brought about quantification of the underlying patterns, which were to be known as circadian rhythms [2]. The "circa" in "circadian" indicates that the variations recurred "about" every 24 hours and the prefix conveys their statistical rather than purely deterministic causality. By noting that rhythms had periods near ("circa") 24 hours as they recurred in similar ("circa") but not identical sequences and at similar ("circa") intervals, which differed statistically from their precise environmental and societal counterpart, the partly endogenous nature of circadian rhythms is recognized.

Many more periodic changes characterize biological variables, which coexist with trends as a function of development, age, and transition from health to disease. Intermodulations among these different components broadened the scope of time structures (chronomes) beyond the circadian system. "Chronome" is derived from "chronos" (time), "nomos" (rule), and "chromosome" to describe the critical law-

K. Otsuka et al., *Chronomics and Continuous Ambulatory Blood Pressure Monitoring*, DOI 10.1007/978-4-431-54631-3_4

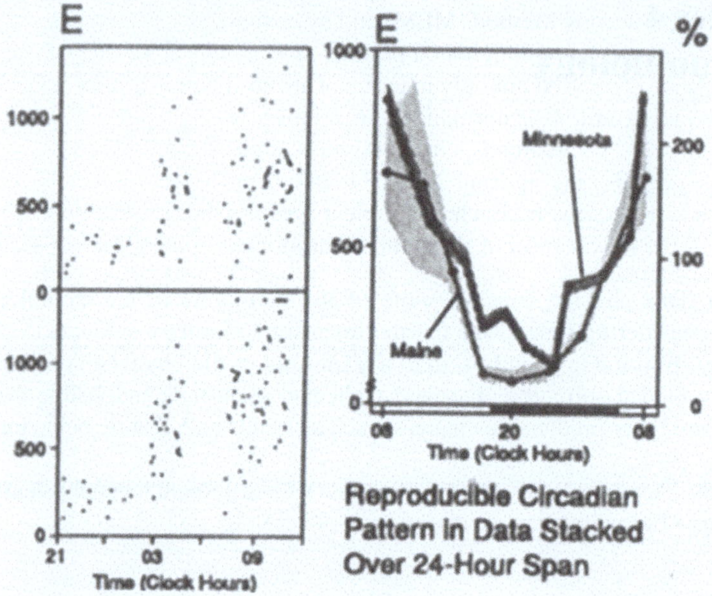

Fig. 4.1 Confusing variability in eosinophil counts partly resolved by large-amplitude circadian rhythm

fulness of biologic structures in time. Mapping the rhythmic structure not only of eosinophil counts but also of the activity of the body as a whole, and its resistance to various stimuli, of mitoses in different organs and tissues, and of different liver functions related to different stages of the cell cycle led to applications to cancer treatment and to the fields of chronopharmacology and chronotherapy, based on the demonstration of the hours of changing resistance to a variety of stimuli [3–6].

As the emerging discipline of chronobiology developed and the ubiquity of circadian rhythms became evident, statistical methods for rhythm detection and quantification of rhythm characteristics with an estimate of uncertainty were devised that could handle non-equidistant data series. The cosinor method Franz Halberg introduced [7] was to play a critical role in providing a solid statistical foundation for the new discipline.

As Franz proceeded with his study of blood eosinophils, mice were kept under different experimental conditions, using different lighting regimens and feeding schedules. Reversing the regimen of 12 hours of light alternating with 12 hours of darkness, Franz found that the timing of rhythms along the 24-hour scale could be moved to any desirable location in time by manipulating the lighting regimen [8], thereby clarifying the role of the environment as a synchronizer of circadian rhythms, Fig. 4.2. Meal timing was also found to affect the timing of circadian rhythms [9].

Environmental geophysical cycles, such as the astronomic day/night cycle, the phases of the moon, the tides, and the seasons, may induce periodic responses in

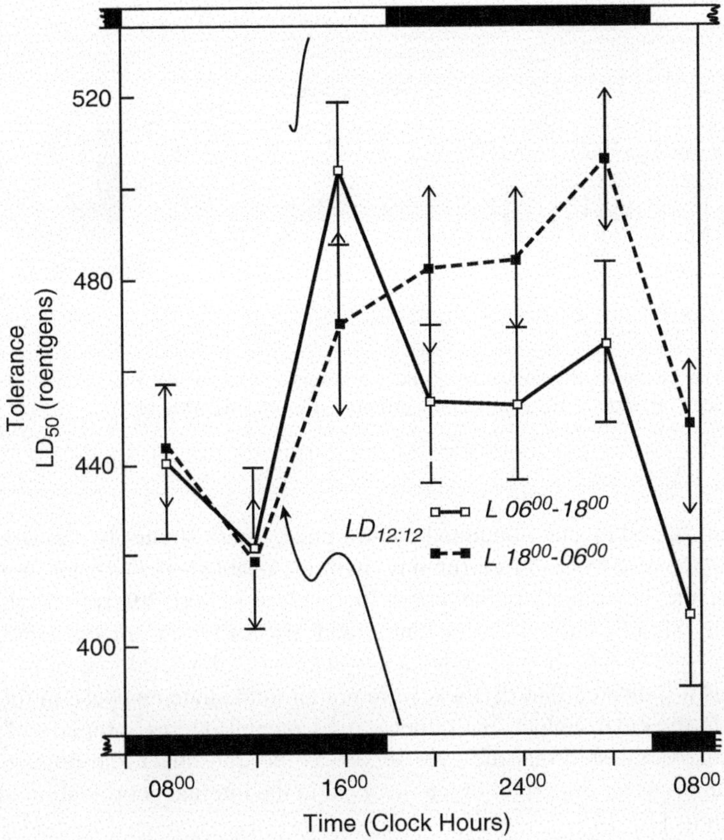

Fig. 4.2 The environment serves as a synchronizer of circadian rhythms. Inverting the LD12:12 lighting regimen is associated with a 12-hour shift in the circadian acrophase

certain species, but in many cases, they rather serve as synchronizers for partly endogenous rhythms [10]. The influence and importance of each of these synchronizers vary among species, depending, for instance, on the degree of exposure to one or the other of these geophysical cycles. An environmentally synchronized endogenous rhythm usually does not undergo an immediate abrupt phase shift of its synchronizer, but rather does so progressively over a number of transient cycles. The time it takes for the phase of a rhythm to adjust to the new environmental schedule depends upon the organism involved, the relative strength of the dominant synchronizer, the extent of the shift in synchronizer phase, the direction of shift (i.e., whether it is a phase advance or a phase delay), and the interaction with other environmental synchronizers (for instance, the lighting and feeding schedules). The rate of phase adjustment also differs among different variables [10].

In order to understand the mechanism through which the alternation of light and darkness affected circadian rhythms, marker variables such as rectal temperature were monitored longitudinally around the clock in mice born anophthalmic, in

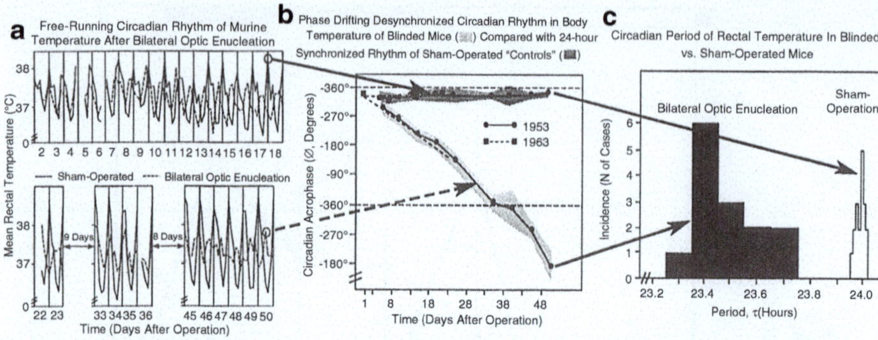

Fig. 4.3 The circadian rhythm persists under constant environmental conditions, assuming a period close to but statistically significantly different from exactly 24 hour

blinded mice, and in sham-operated control mice. These studies by Franz Halberg [11] led to the discovery of free-running rhythms, established on a rigorous statistical basis, the circadian rhythm persisting with a period differing slightly but statistically significantly from 24 hours, and also differing among some of the mice, Fig. 4.3. This finding provided a key element to the development of chronobiology as it indicated that rhythms were not an immediate response to the cyclic changes in the environment, but rather a fundamental internal property of organisms. It also suggested that the eyes served as the transducer and pointed to the hypothalamus as an important organ involved in the internal coordination of circadian rhythms [12–14].

The subsequent discovery of a molecular mechanism underlying circadian rhythms [15, 16] confirmed their partly endogenous nature. Current knowledge considers the suprachiasmatic nuclei in the hypothalamus to play a critical role as "master oscillator" or pacemaker [17]. The view originally held that all circadian rhythms are obliterated after bilateral suprachiasmatic enucleation [18] had to be revised [19], however, and it is now recognized that clock genes are found in almost every cell [20]. Early confirmed work by Franz Halberg [9, 21] had shown the critical role of the adrenal cortex in coordinating circadian rhythms in blood eosinophils, but not in serum iron [22]. Evidence is available, however, showing that some circadian rhythms can persist in one or the other variable after removal of either the suprachiasmatic nuclei [23], and even of the brain as a whole [24], or of both the adrenals [25].

4.2 Rhythms of Different Orders

The lawfulness of the circadian variation, determined by the application of statistical methods, became apparent from the ubiquity of prominent circadian rhythms at different levels of organization. A series of experiments in Franz Halberg's laboratory investigated the effect of a single physical stimulus under rigorously standardized laboratory conditions. Whether the stimulus was audiogenic [3] or due to exposure to an endotoxin [26], or to a drug such as ouabain [27], or to whole-body irradiation [5], predictable changes were found as a function of the circadian stage at which the stimulus was applied, the timing of these susceptibility-resistance rhythms differing from one kind of stimulus to another. This work led to the concept of the hours of changing resistance, maps being prepared to indicate the times of overall largest response by the organism to a fixed stimulus applied at different rhythm stages. Prominent susceptibility rhythms were documented in the experimental laboratory, notably in relation to anticancer drugs [28, 29], Fig. 4.4.

Using staggered lighting regimens, it was possible to examine how neuroendocrine and peripheral systems intermodulate by means of feedsidewards mechanisms

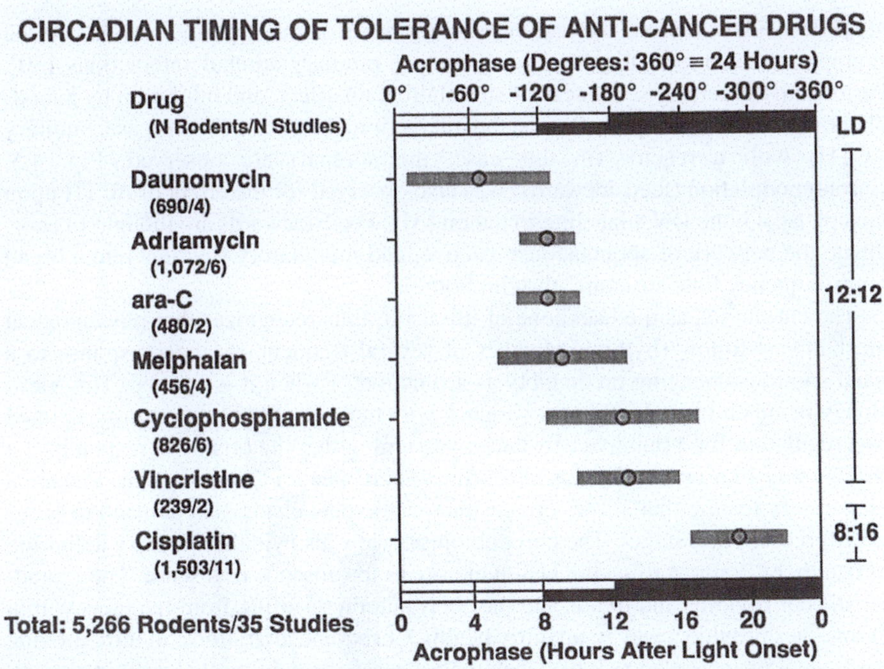

Fig. 4.4 Acrophase chart illustrating the circadian susceptibility-resistance cycle to different anticancer drugs tested in the experimental laboratory

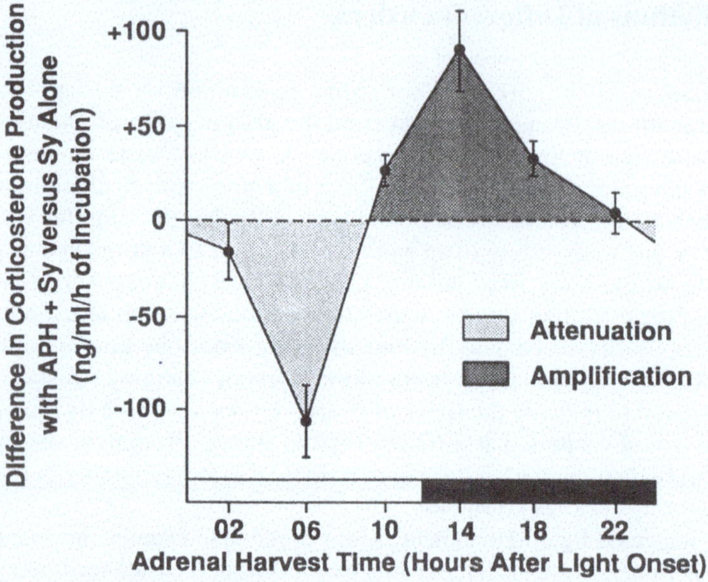

Fig. 4.5 Example of feedsideward-pineal homogenate modifies the effect of ACTH-1-17 on corticosterone production by the adrenals

[14]. Feedsidewards are interactions among three or more periodic entities. One example is the modulation by the pineal of pituitary-adrenal interactions [30]. Periodically recurring sequences of stimulation, no effect, and inhibition by a modulator (in this case, the pineal) upon the interaction of an actor (in this case, pituitary ACTH) with a reactor (in this case, the adrenal) are observed, Fig. 4.5. Chronomodulation (feedsidewards) was also observed for the effect of ACTH upon metaphyseal bone DNA labeling in the rat [31]. Feedsidewards are thought to coordinate the network of spontaneous, reactive, and modulatory rhythms into a broad multifrequency time structure, the chronome.

The chronome, as the backbone of life itself, thus recognizes that physiological mechanisms follow rhythmic changes of several frequencies, their response to a fixed stimulus changing predictably as a function of when it is applied. This view, slowly being embraced [32], is at variance with the concept of homeostasis, defined as a regulation for a putative constancy, presuming that the organism responds in a similar way to a given stimulus at all times. Partly endogenous rhythms include a preparatory feature that allows organisms to anticipate changes in demand to occur at different rhythm stages. The chronobiologic view also yields a positive definition of health, by contrast to seeing health merely as the absence of disease. The characteristics of rhythms and trends and the coordination resulting from relations within chronomes provide a way to quantify health. Chronome alterations, in turn, provide harbingers of increased risk of developing disease, prompting the early institution of prophylactic intervention for a truly primary prevention, while rhythm characteristics also serve for secondary prevention and for the optimization of timing treatment administration when treatment is needed.

4.3 Circadian and Extracircadian Oscillations in Heart Rate

Beyond circadian rhythms are many other oscillations of different frequencies. Corresponding periods range from fractions of seconds in single neurons to seconds in the cardiac and respiratory cycles and a few hours in certain endocrine functions in the ultradian domain. In the infradian domain, about-weekly (circaseptan) and half-weekly (circasemiseptan) changes are also ubiquitous, as are about-monthly (circatrigintan) and circannual variations. Other cycles with periods of about 5 and 16 months and about 5, 10.5, 21, and 50 years and even lower-frequency changes have also been found in biology. Because their environmental counterparts are less stable, additional methodologic tools are needed to study them that can handle nonstationarities. These components will be reviewed in more detail in the chronomics section.

4.3.1 The Cardiac Cycle

The electrocardiogram of a clinically healthy subject reveals a relative complex waveform that repeats itself with some regularity. This basic cycle is composed of several waves, corresponding to the conduction of electrical signals allowing the heart to pump blood into the circulation. An automaticity, outside the body, of the heart as a whole, deprived of its whip and break (its sympathetic and vagal efferents), was found by Aztec priests who sometimes observed beats of the organ after it was ripped out of the body for divination [33]. Their finding has been carried from the single organ to the single cell [34].

4.3.2 Respiratory Sinus Arrhythmia

Respiratory sinus arrhythmia is a naturally occurring cardiopulmonary reflex, the heart rate varying with the phase of the respiratory cycle. Inspiration increases and expiration decreases the rate via vagally mediated reflexes [35]. Expiration favors left ventricular filling by increasing the return to the left heart of blood that was stored in the lungs during inspiration. Respiratory sinus arrhythmia is very noticeable in infants and in children [36]. During voluntary cardiorespiratory synchronization, it becomes less prominent with advancing age. It is estimated to decrease by about 10 % for each decade [37]. Spectral analyses of heart rate variability at rest indicate that sinus arrhythmia declines less with age than lower (vasomotor) frequencies, suggesting that the decline in heart rate variability as a function of age may not be predominantly associated with a decline in frequencies related to respiration [38]. A central cholinergic receptor is thought to be involved in the respiratory modulation of heart rate [39]. Influence of the arterial baroreceptors has been suggested in the generation of respiratory sinus arrhythmia in conscious humans [40]. Heritability of respiratory sinus arrhythmia has been estimated to range from 28 % to 43 % [41].

4.3.3 REM/NREM Cycle

Relying on sleep polygrams, the sympatho-vagal balance, quantified by the spectral power around 10.5 s and 3.6 s (LF/HF), was found to decrease the more so the more synchronized sleep was, whereas this ratio was increased during REM sleep, suggesting a sympathetic predominance during REM sleep [42]. Increases in HF and decreases in LF spectral power of heart rate variability across NREM sleep stages and opposite changes in REM sleep and wake were corroborated in other studies [43, 44]. About-90-min components in heart rate variability observed in our own studies [45, 46] are likely related to the sleep-wake (REM/NREM) cycle.

4.3.4 Other Ultradian Components

Longitudinal recordings of neonatal heart rate during the first weeks of life have indicated the presence of ultradian changes likely related to the feeding schedule. Babies monitored in Florence (Italy), who were fed every 3 hours, have a prominent spectral peak at a frequency of one cycle in 3 hours. A prominent spectral peak at a frequency of one cycle in 4 hours is found in heart rate records from babies in Germany fed every 4 hours [47]. In adults, some neuropeptides can have more prominent ultradian than circadian changes. This is, for instance, the case of endothelin-1, which undergoes a pronounced about-8-hour variation [48–50], Fig. 4.6. An 8-hour rhythmic variation was also found to characterize the urinary excretion of norepinephrine in women with Bartter's syndrome and in clinically healthy volunteers [51].

4.3.5 Circadians

Day-night differences not fully accounted for by the sleep-wake cycle are already demonstrable in early extrauterine life [52]. During the first week of life, a circadian rhythm is detected on a population basis, peaking around 01:00, in almost antiphase with the mother's circadian rhythm [53]. Changes in heart rate within a single day usually exceed the range of change in this variable as a function of age in adulthood (between 20 and 60 years of age) [54]. Under habitual conditions, heart rate usually assumes higher values during the daily active span and lower values during the daily rest span. Following a change in schedule, this pattern adjusts gradually rather than

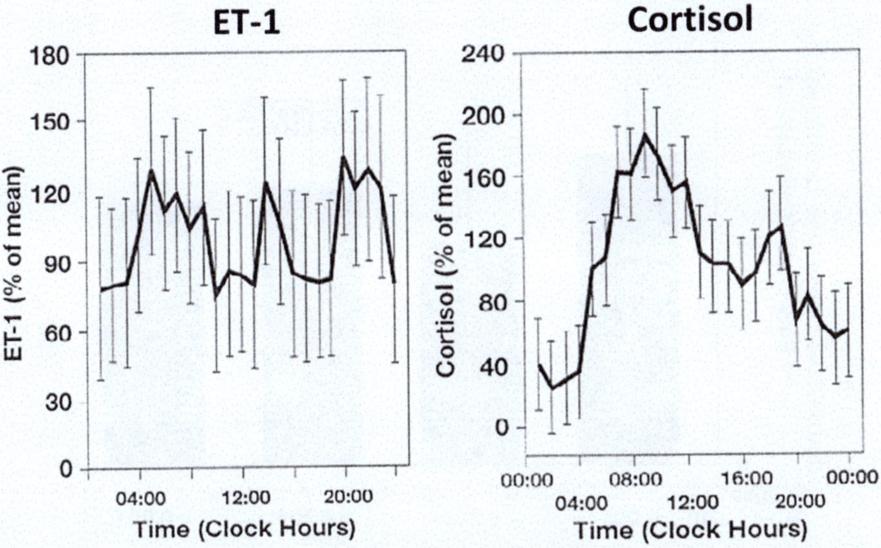

Fig. 4.6 An 8-hour variation characterizes circulatory endothelin-1 (ET-1) in healthy adults, whereas cortisol, determined concomitantly, undergoes a large-amplitude circadian rhythm

immediately to the new routine [55]. The circadian rhythm in heart rate persists during complete bedrest for 36 hours and a 4-hourly hypocaloric diet, indicating that it is not fully dependent upon the rest/activity cycle [56]. In the absence of time clues, the persistence of a circadian rhythm with a period near 24 hours but differing from exactly 24 hours with statistical significance (free-running) is another indirect line of evidence documenting a partly endogenous component. A free-running, presumably partly endogenous circadian heart rate component has been demonstrated to persist under conditions of isolation from the usual physical as well as societal environment [57, 58].

Non-chronobiologic studies on twins accounted by heredity for 54 % of the interindividual variation in ventricular rate and for 34 % in the case of atrial depolarization [59]. A statistically significant heritability of 60 % or more has also been reported for resting heart rate [60] as well as for the activity of heart rate during mental activities [61]. A genetic background of the circadian rhythm in human heart rate was proposed by Barcal et al. [62]. Heritability not only of the MESOR but also of the circadian amplitude was confirmed in studies of twins reared apart [63]. Differences in 24-hour amplitude of heart rate are much smaller within twin pairs than among twin pairs. Intra-class correlation analyses on results from 35 monozygotic and 22 dizygotic twin pairs reared apart document a genetic basis for the circadian rhythm of human heart rate [64]. After adjusting for gender and age, the difference in intra-class correlation coefficient between monozygotic and dizygotic twin pairs remains statistically significant, Fig. 4.7.

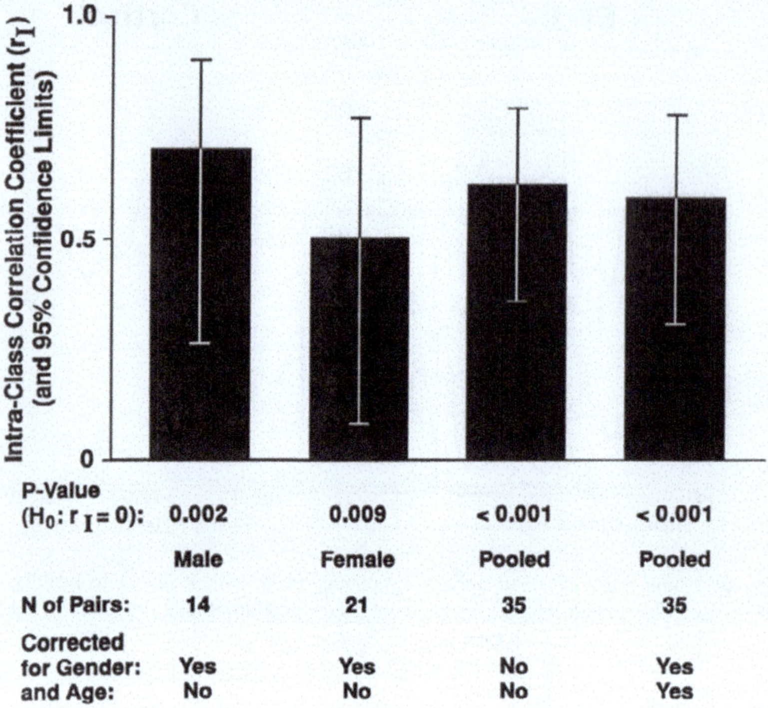

Fig. 4.7 Heritability of the circadian amplitude of heart rate examined in monozygotic twins reared apart

4.3.6 Circaseptans and Circasemiseptans

The biologic week is also ubiquitous and partly endogenous, being amenable to synchronization by social schedules, as evidenced from several lines of research. Free-running was originally shown by Franz Halberg in a 15-year-long record of daily determinations of urinary excretion of 17-ketosteroids. Whereas urine volume remained synchronized for 7 days, 17-KS started to free-run after 12 years, in association with self-administration of testosterone [65], Fig. 4.8. Free-running of the about-7-day and about-3.5-day components has also been observed in isolation studies [66, 67]. Amplification, if not induction of circaseptans after a single stimulus, is another source of indirect evidence for a partly endogenous biologic week. The stimulus can be as mild as balneotherapy [68] or as severe as organ transplantation [69]. Birth is such a single stimulus to which babies respond with a prominent circaseptan component in blood pressure and heart rate [70–73]. When babies are monitored around the clock for several weeks, a circaseptan component is detected with statistical significance on an individualized basis, the circaseptan period, estimated by nonlinear least squares, differing from precisely 7 days as well as among individual babies [72]. In this study, when results are pooled from a large group of

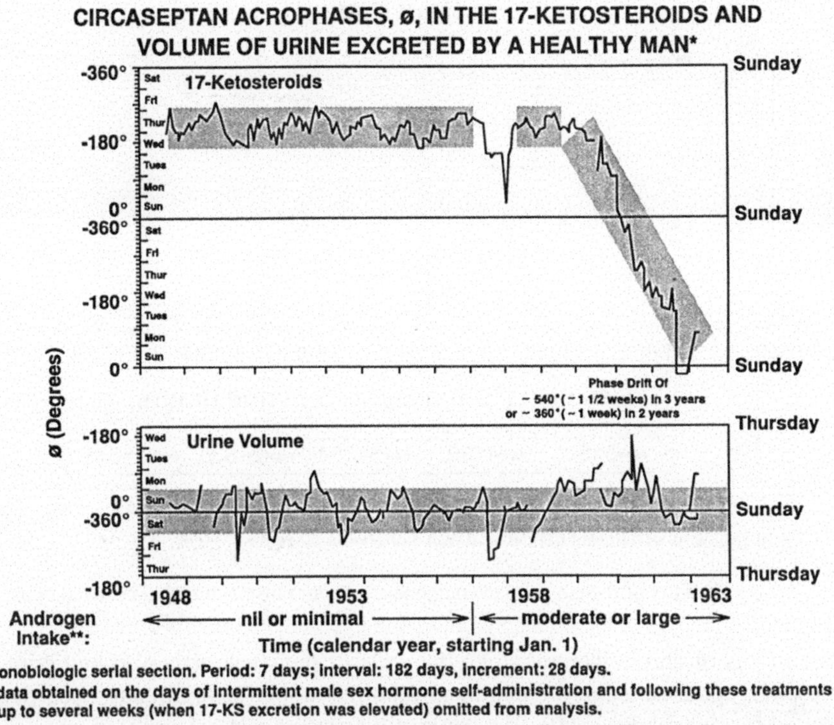

Fig. 4.8 Free-running of the weekly component in the urinary excretion of 17-ketosteroids, but not of urine volume of a clinically healthy man

babies, analyses of variance indicate that the circaseptan component is synchronized by the time of birth (developmental age) rather than by the societal schedule. In a small group of neonatal twins, the circaseptan period was more similar between twins in a pair than among different pairs of twins, as assessed by an intra-class correlation coefficient [74], Fig. 4.9. A circaseptan response to a single stimulus (release into continuous darkness from an LD12:12 regimen) was also observed in a unicell [75]. Additional indirect evidence stems from several studies in *Acetabularia* [76], face flies [77], and springtails [78], where the response variable (growth or lifespan) followed an about-7-day pattern in relation to the interval between successive shifts they were submitted to [for review, see 79]. Circaseptans and circasemiseptans have been particularly prominent in relation to growth, development, and immunology [80].

While circadians are much more prominent in adulthood, circaseptans and circasemiseptans usually remain demonstrable, with a peak usually around mid-week [81–83]. The relative prominence of circaseptans and circasemiseptans as compared to circadians varies as a function of age, regaining importance in the elderly [84, 85]. Circaseptans are also more pronounced in the presence of a depressed mood [86, 87], as it is also the case for negative affect, as compared to positive affect [88].

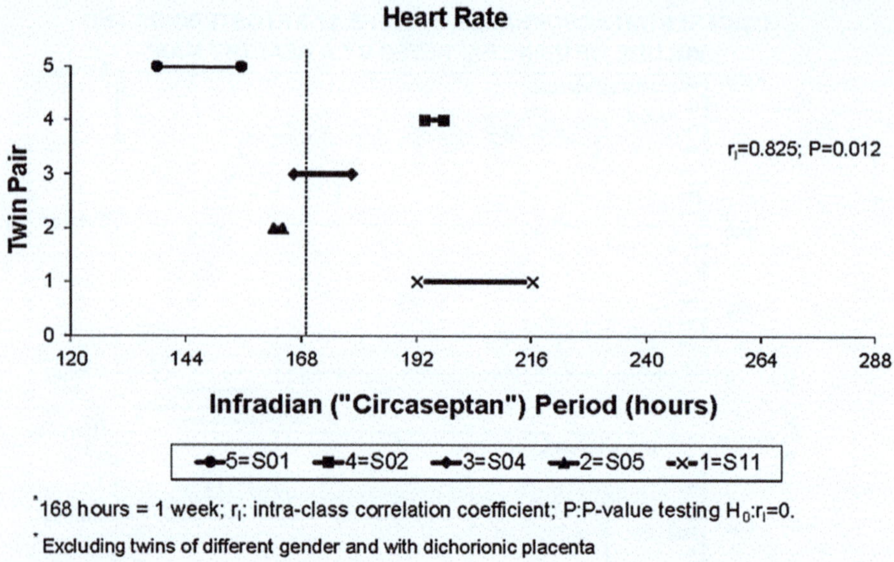

Fig. 4.9 Heritability of the circaseptan component of heart rate examined in neonatal twins

The merits of chronotherapy along the scale of the week, achieved by optimizing the circaseptan schedule of treatment administration, have already been demonstrated for several drugs [89–91], while still awaiting clinical applications.

4.3.7 Circatrigintans

In one isolation study, the circaseptan component of heart rate, which was free-running, did not immediately resynchronize upon return to a societal routine, while the circadian system quickly adjusted to the 24-hour schedule. The circaseptan system failed to do so for weeks, until menstruation was re-induced by hormones [58]. Neurohormonal or other effects of menstruation may be a condition for synchronization of a built-in circaseptan component with societal schedules [58]. About-monthly cycles are observed not only in relation to menstruation. By 1657, Santorio's aphorisms comment on changes in body weight of one or two pounds recurring about once a month in a mature healthy man [92]. The unusual case of a man who bled from his thumb about once a month is recorded [93], as is a purpura of the calf recurring for 6 years at intervals of about 4 weeks in a 60-year-old man with Morbus maculosus Werlhofii [94]. Circatrigintans characterize the frequency of neutrophil leukocytes with androgen-induced nuclear appendages [95]. In a variety of life forms, variables related not only to the reproductive tract but also to other organ systems of individuals of both genders show overt cycles in the range of about 30 days [96]. The about-monthly component of breast surface temperature discriminates between women with breast cancer who will die or survive [97]. The incidence of tachyarrhythmia has also been reported to vary as a function of lunar stage [98]. Differential effects of space and/or terrestrial weather with respect to suicide and other deaths associated with the nervous and sensory systems vs. death from cardiac or respiratory disease as well as overall death by differences in the phase of a common 27-day cycle characterizing these mortality patterns [99].

4.3.8 *Circannuals*

About-yearly features characterize biology broadly in plants and animals. Human growth rate, body weight, metabolic rate, and pineal gland function are all circannual periodic [100, 101]. The importance of human circannuals is seen in morbidity and mortality statistics. In particular, the incidence of cardiovascular and respiratory diseases follows a prominent circannual variation [102, 103], Fig. 4.10. Some about-yearly changes persist and phase-drift under constant environmental conditions. In natural lighting, in continuous light, in continuous darkness, or on a fixed LD12:12 regimen, a circannual variation of large amplitude with little or no damping from year to year was found to characterize the ovarian weight of the Jamuna

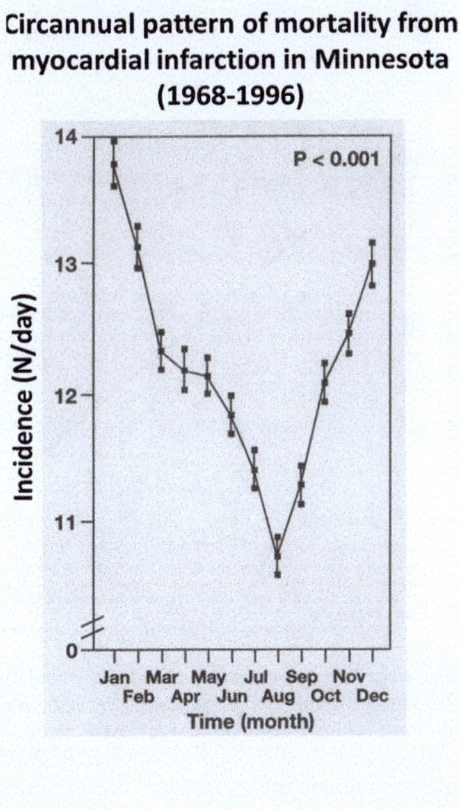

Fig. 4.10 The incidence of cardiovascular diseases is circannual periodic

River catfish [104]. Under natural lighting conditions, the best-fitting circannual period was close to 365 days, whereas under constant environmental conditions, it was longer, extending to 393 to 400 days, suggesting partial endogenicity. The circannual amplitude of diastolic blood pressure in clinically healthy women was found to correlate negatively with the familial and personal risk of developing high blood pressure and/or related cardiovascular disease later in life, as did the circannual amplitude of circulating aldosterone [105], Fig. 4.11. Discrimination and classification techniques applied to these data singled out aldosterone as a classifier of cardiovascular disease risk and corroborated the circannual stage dependence of the discriminating power of aldosterone as a classifier for this condition [14].

A circannual variation in heart rate cannot invariably be demonstrated [106]. By contrast, the circadian amplitude of heart rate was found to be larger in the summer and fall than in the winter and spring in a longitudinal study of 10 women [107], the circadian acrophase peaking later in the winter than in the other seasons. These circannual changes in the circadian amplitude and acrophase as a function of circannual stage are in keeping with opposite circannual acrophases estimated from data collected around the clock by ambulatory monitoring at different clock hours by a clinically healthy man during 15 years [108], Fig. 4.12. While measurements obtained during the daytime assume overall high values in the winter, measurements at night and in the early morning assume overall high values in the summer.

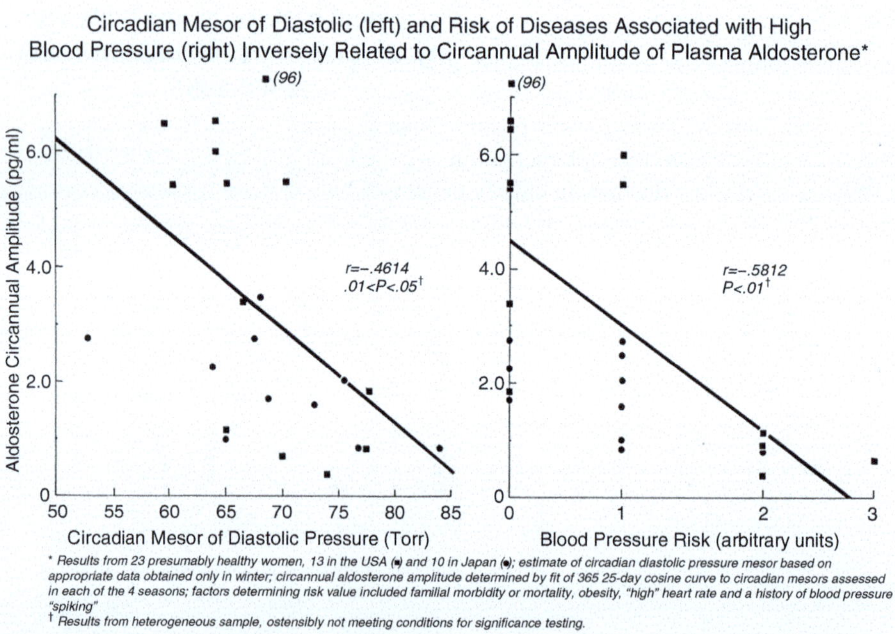

Fig. 4.11 The circannual amplitude of circulating aldosterone correlates negatively with the MESOR of diastolic blood pressure (*left*) and with the risk of developing high blood pressure and/or related cardiovascular disease (assessed by questionnaire)

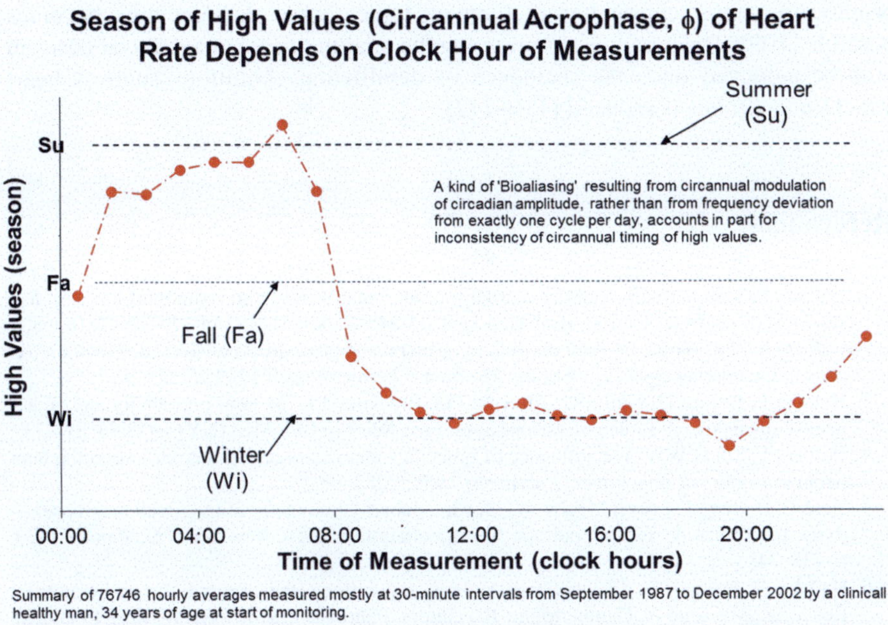

Fig. 4.12 Circannual changes in the circadian amplitude and/or acrophase, but not necessarily in the MESOR, are in keeping with different circannual acrophases of heart rate depending on when during the day it is measured

4.4 Conclusion

As are many other variables, heart rate is characterized by multiple frequency rhythms, from the 1-Hz cardiac cycle to secular variations likely associated with space-terrestrial weather, as discussed elsewhere in this book. These oscillations are to a large extent predictable and cannot solely be accounted for by changes in locomotor activity. They are part of a broad time structure, the chronome. Apart from the element consisting of rhythmic changes, chronomes include two other elements, namely, chaotic behavior and trends in both rhythmic and chaotic endpoints as a function of age and as a function of health status.

These findings have profound implications for the future of health care. First, they point to the need to derive time-specified reference values specified by gender and age. By partitioning a large portion of variability, predictable in terms of rhythms, chaos, and trends, the residual level of uncertainty (noise) can be greatly reduced, thereby allowing for more refined diagnoses. Most importantly, rhythm characteristics and other endpoints representing lawful variations (norms) within the physiological range provide a quantification of health. Deviations from these norms, in turn, can be used as early warning signs for impending risks. Early prophylactic intervention before there is target organ damage thus becomes possible [109]. The paradox of improving the quality of care at a reduced cost can thus be achieved once the incidence of debilitating conditions such as heart attacks and strokes is reduced [110–112].

References

1. Halberg F, Visscher MB. Regular diurnal physiological variation in eosinophil levels in five stocks of mice. Proc Soc Exp Biol Med 1950;75:846–7. doi:10.3181/00379727-75-18365.
2. Halberg F. Physiologic 24-hour periodicity; general and procedural considerations with reference to the adrenal cycle. Z Vitamin-, Hormon-u Fermentforsch 1959;10:225–96.
3. Halberg F, Bittner JJ, Gully RJ, Albrecht PG, Brackney EL. 24-hour periodicity and audiogenic convulsions in I mice of various ages. Proc Soc exp Biol Med (NY) 1955;88:169–73.
4. Halberg F, Spink WW. The influence of brucella somatic antigen (endotoxin) upon the temperature rhythm of intact mice. Lab Invest 1956;5:283–94.
5. Haus E, Halberg F, Loken MK, Kim US. Circadian rhythmometry of mammalian radiosensitivity. In: Tobias A, Todd P, editors. Space radiation biology. New York: Academic; 1973. p. 435–74.
6. Halberg F, Halberg E. Chronopharmacology and further steps toward chronotherapy. In: Benet LZ, Massoud N, Gambertoglio JG, editors. Pharmacokinetic basis for drug treatment. New York: Raven; 1984. p. 221–48.
7. Halberg F, Tong YL, Johnson EA. Circadian system phase – an aspect of temporal morphology; procedures and illustrative examples. Proceedings, International Congress of Anatomists. In: von Mayersbach H, editor. The cellular aspects of biorhythms: Symposium on Biorhythms. New York: Springer-Verlag; 1967. p. 20–48.
8. Halberg F, Barnum CP, Silber RH, Bittner JJ. 24-hour rhythms at several levels of integration in mice on different lighting regimens. Proc Soc Exp Biol (NY) 1958;97:897–900.
9. Halberg F, Visscher MB, Bittner JJ. Eosinophil rhythm in mice: range of occurrence; effects of illumination, feeding and adrenalectomy. Amer J Physiol 1953;174:109–22.
10. Haus E, Cornelissen G, Halberg F. Introduction to chronobiology. In: Scheving LE, Halberg F, editors. Chronobiology: principles and applications to shifts in schedules. Alphen aan den Rijn, The Netherlands: Sijthoff and Noordhoff; 1980. p. 1–32.
11. Halberg F, Visscher MB, Bittner JJ. Relation of visual factors to eosinophil rhythm in mice. Amer J Physiol 1954;179:229–35.
12. Ungar F, Halberg F. Circadian rhythm in the in vitro response of mouse adrenal to adrenocorticotropic hormone. Science 1962;137:1058–60.
13. Ungar F, Halberg F. In vitro exploration of a circadian rhythm in adrenocorticotropic activity of C mouse hypophysis. Experientia (Basel) 1963;19:158–9.
14. Halberg F. Quo vadis basic and clinical chronobiology: promise for health maintenance. Am J Anat 1983;168:543–94.
15. Konopka RJ, Benzer S. Clock mutants of Drosophila melanogaster. Proc Natl Acad Sci USA 1971;68(9):2112–6.
16. Ralph MR, Menaker M. A mutation of the circadian system in golden hamsters. Science 1988;241:1225–7.
17. Bernard S, Gonze D, Cajavec B, Herzel H, Kramer A. Synchronization-induced rhythmicity of circadian oscillators in the suprachiasmatic nucleus. PLoS Comput Biol 2007;3(4), e68.
18. Takahashi JS, Menaker M. Role of the suprachiasmatic nuclei in the circadian system of the house sparrow. Passer domesticus. J Neurosci 1982;2(6):815–28.
19. Halberg F, Lubanovic WA, Sothern RB, Brockway B, Powell EW, Pasley JN, Scheving LE. Nomifensine chronopharmacology, schedule shifts and circadian temperature rhythms in

di-suprachiasmatically lesioned rats – modeling emotional chronopathology and chronotherapy. Chronobiologia 1979;6:405–24.
20. Plautz JD, Kaneko M, Hall JC, Kay SA. Independent photoreceptive circadian clocks throughout Drosophila. Science 1997;278:1632–5.
21. Best WR, Muehrcke RC, Kark RM. Studies on adrenocortical eosinopenia: a clinical and statistical evaluation of four-hour eosinophil response tests. J Clin Invest 1952;31:733–42.
22. Halberg F, Howard RB. 24-hour periodicity and experimental medicine. Example and interpretations. Postgrad Med 1958;24:349–58.
23. Powell EW, Pasley JN, Scheving LE, Halberg F. Amplitude-reduction and acrophase-advance of circadian mitotic rhythm in corneal epithelium of mice with bilaterally lesioned suprachiasmatic nuclei. Anat Rec 1980;197:277–81.
24. Galicich JH, Halberg F, French LA, Ungar F. Effect of cerebral ablation on a circadian pituitary adrenocorticotropic rhythm in C mice. Endocrinology 1965;76:895–901.
25. Lakatua DJ, Haus E, Halberg F. Persistence of circadian periodicity in BALB/c mice after adrenalectomy and during inanition and dehydration. J Steroid Biochem 1974;5:386.
26. Halberg F, Johnson EA, Brown BW, Bittner JJ. Susceptibility rhythm to E. coli endotoxin and bioassay. Proc Soc Exp Biol Med 1960;103:142–4. doi:10.3181/00379727-103-25439.
27. Halberg F, Stephens AN. Susceptibility to ouabain and physiologic circadian periodicity. Proc Minn Acad Sci 1959;27:139–43.
28. Halberg F, Haus E, Cardoso SS, Scheving LE, Kühl JFW, Shiotsuka R, Rosene G, Pauly JE, Runge W, Spalding JF, Lee JK, Good RA. Toward a chronotherapy of neoplasia: tolerance of treatment depends upon host rhythms. Experientia (Basel) 1973;29:909–34.
29. Halberg F, Gupta BD, Haus E, Halberg E, Deka AC, Nelson W, Sothern RB, Cornélissen G, Lee JK, Lakatua DJ, Scheving LE, Burns ER. Steps toward a cancer chronopolytherapy. In: Proceedings of the XIV International Congress of Therapeutics. Montpellier, France: L'Expansion Scientifique Française; 1977. p. 151–196.
30. Halberg F, Guillaume F, Sanchez de la Peña S, Cavallini M. Cornélissen G. Cephalo-adrenal interactions in the broader context of pragmatic and theoretical rhythm models. Chronobiologia 1986;13:137–54.
31. Walker WV, Russell JE, Simmons DJ, Scheving LE, Cornélissen G, Halberg F. Effect of an adrenocorticotropin analogue, ACTH 1–17, on DNA synthesis in murine metaphyseal bone. Biochem Pharmacol 1985;34:1191–6.
32. Schulkin J. Rethinking homeostasis: allostatic regulation in physiology and pathophysiology. Cambridge, MA: The MIT Press; 2003.
33. Leake CD. The historical development of cardiovascular physiology. In: Hamilton WF, Dow P, editors. Handbook of physiology: Section 2: Circulation, Volume 1. Washington: American Physiological Society; 1962. p. 11–22.
34. Han HW, Shao DL, Wu JY, Cornélissen G, Halberg F. Chronobiologic approach to beat-to-beat variations of cultured murine myocardial cells. Cell Biophys 1991;18:217–29.
35. Sokolow M, McIlroy MB. Clinical cardiology. 4th ed. Los Altos: Lange Medical Publications; 1986. 661 pp.
36. Davies HEF. Respiratory change in heart rate, sinus arrhythmia in the elderly. Geront Clin 1975;17:96–100.
37. Hrushesky WJ, Fader D, Schmitt O, Gilbertsen V. The respiratory sinus arrhythmia: a measure of cardiac age. Science 1984;224(4652):1001–4.
38. Jennings JR, Mack ME. Does aging differentially reduce heart rate variability related to respiration? Exp Aging Res 1984;10(1):19–23.
39. Lacroix D, Logier R, Kacet S, Hazard JR, Dagano J, Lekieffre J. Effects of consecutive administration of central and peripheral anticholinergic agents on respiratory sinus arrhythmia in normal subjects. J Auton Nerv Syst 1992;39(3):211–7.
40. Piepoli M, Sleight P, Leuzzi S, Valle F, Spadacini G, Passino C, Johnston J, Bernardi L. Origin of respiratory sinus arrhythmia in conscious humans. An important role for arterial carotid baroreceptors. Circulation 1997;95(7):1813–21.

41. Snieder H, Boomsma DI, Van Doornen LJ, De Geus EJ. Heritability of respiratory sinus arrhythmia: dependency on task and respiration rate. Psychophysiology 1997;34(3):317–28.
42. Scholz UJ, Bianchi AM, Cerutti S, Kubicki S. Vegetative background of sleep: spectral analysis of the heart rate variability. Physiol Behav 1997;62(5):1037–43.
43. Toscani L, Gangemi PF, Parigi A, Silipo R, Ragghianti P, Sirabella E, Morelli M, Bagnoli L, Vergassola R, Zaccara G. Human heart rate variability and sleep stages. Ital J Neurol Sci 1996;17(6):437–9.
44. Bonnet MH, Arand DL. Heart rate variability: sleep stage, time of night, and arousal influences. Electroencephalogr Clin Neurophysiol 1997;102(5):390–6.
45. Mehta P, Cornelissen G, Portela A, Holte J, Halberg F. Heart rate chronome in health assessed by two different instruments. Policlin Sez Med 1995;102:195–208.
46. Isaacson P, Lee R, Cornelissen G, Schwartzkopff O, Siegelova J, Halberg F. Chronomes of oxygen saturation and heart rate in health and after coronary artery bypass grafting. Scr Medica (Brno) 2000;73:77–83.
47. Otsuka K, Cornelissen G. Alcohol consumption and vascular variability anomalies (VVAs) and disorders (VVDs) detected by C-ABPM. Open Nutr J 2012;5(Suppl 1-M1):81–9.
48. Tarquini B, Perfetto F, Tarquini R, Cornelissen G, Halberg F. Endothelin-1's chronome indicates diabetic and vascular disease chronorisk. Peptides 1997;18:119–32.
49. Herold M, Cornelissen G, Loeckinger A, Koeberle D, Koenig P, Halberg F. About 8-hourly variation of circulating human endothelin-1 (ET-1) in clinical health. Peptides 1998;19:821–5.
50. Katinas GS, Halberg F, Cornelissen G, Hawkins D, Bueva MV, Korzhevsky DE, Sapozhnikova LR, Rhodus N, Schaffer E. About 8- and ~84-hour rhythms in endotheliocytes as in endothelin 1 and effect of trauma. Peptides 2001;22:647–59.
51. Simpson HW, Bartter FC, Halberg F, Gill JR, Delea C, Gullner H-G, Haus E. A circaoctohoran rhythm of urine epinephrine excretion in man. Prog Clin Biol Res 1990;341A:157–60.
52. Biriukovich AA. Changes in the frequency of cardiac contractions with the hour of the day in children from birth till 18 years of age. Pediatriia 1972;51(6):27–32.
53. Halberg F, Cornelissen G, Bakken E. Caregiving merged with chronobiologic outcome assessment, research and education in health maintenance organizations (HMOs). Prog Clin Biol Res 1990;341B:491–549.
54. Halberg F, Drayer JIM, Cornelissen G, Weber MA. Cardiovascular reference data base for recognizing circadian mesor- and amplitude-hypertension in apparently healthy men. Chronobiologia 1984;11:275–98.
55. Halberg F, Cornelissen G, Halberg E, Halberg J, Delmore P, Shinoda M, Bakken E. Chronobiology of human blood pressure. Medtronic Continuing Medical Education Seminars. 4th ed. Minneapolis: Medtronic Inc.; 1988. p. 242.
56. Reinberg A, Ghata J, Halberg F, Gervais P, Abulker C, Dupont J, Gaudeau C. Rythmes circadiens du pouls, de la pression artérielle, des excrétions urinaires en 17-hydroxycorticostéroïdes, catécholamines et potassium chez l'homme adulte sain, actif et au repos. Ann Endocrinol (Paris) 1970;31:277–87.
57. Halberg F, Siffre M, Engeli M, Hillman D, Reinberg A. Étude en libre-cours des rythmes circadiens du pouls de l'alternance veille-sommeil et de l'estimation du temps pendant les deux mois de séjour souterrain d'un homme adulte jeune. CR Acad Sci (Paris) 1965;260:1259–62.
58. Halberg F, Tamura K, Cornelissen G. Chronobioengineering toward a cost-effective quality health care. Front Med Biol Eng 1994;6:83–102.
59. Havlik RJ, Garrison RJ, Fabsitz R, Feinleib M. Variability of heart rate, P-R, QRS and Q-T durations in twins. J Electrocardiol 1980;13(1):45–8.
60. Smith TW, Turner CW, Ford MH, Hunt SC, Barlow GK, Stults BM, Williams RR. Blood pressure reactivity in adult male twins. Health Psychol 1987;6(3):209–20.
61. Turner JR. Individual differences in heart rate response during behavioral challenge. Psychophysiology 1989;26(5):497–505.

62. Barcal R, Sova J, Krizanovska M, Levy J, Matousek J. Genetic background of circadian rhythms. Nature 1968;220(5172):1128–31.
63. Hanson BR, Halberg F, Tuna N, Bouchard Jr TJ, Lykken DT, Cornelissen G, Heston LL. Rhythmometry reveals heritability of circadian characteristics of heart rate of human twins reared apart. Cardiologia 1984;29:267–82.
64. Kumagai Y, Cornelissen G, Lykken DT, Bouchard TJ Jr, Tuna N, Delmore P, Zaslavskaya R, Teibloom M, Eckert E, Halberg F. In vitro and in vivo evidence for built-in features of the human heart rate chronome. In: Cornelissen G, Halberg E, Bakken E, Delmore P, Halberg F, editors. Toward phase zero preclinical and clinical trials: chronobiologic designs and illustrative applications. University of Minnesota Medtronic Chronobiology Seminar Series, #6, second extended edition, Feb 1993, Part I, p. 47–83.
65. Halberg F, Engeli M, Hamburger C, Hillman D. Spectral resolution of low-frequency, small-amplitude rhythms in excreted 17-ketosteroid; probable androgen induced circaseptan desychronization. Acta endocrinol (Kbh) 1965;50 Suppl 103:5–54.
66. Sanchez de la Peña S, Halberg F, Galvagno A, Montalbini M, Follini S, Wu J, Degioanni J, Kutyna F, Hillman DC, Kawabata Y, Cornelissen G. Circadian and circaseptan (about-7-day) free-running physiologic rhythms of a woman in social isolation. In: Proceedings 2nd annual IEEE Symposium on Computer-Based Medical Systems, Minneapolis, June 26–27, 1989. Washington DC: Computer Society Press; 1989. p 273–8.
67. Halberg F, Cornelissen G, Sonkowsky RP, Lanzoni C, Galvagno A, Montalbini M, Schwartzkopff O. Chrononursing (chronutrics), psychiatry and language. New Trends Exp Clin Psychiatr 1998;14:15–26.
68. Hildebrandt G, Bandt-Reges I. Chronobiologie in der Naturheilkunde: Grundlagen der circaseptanperiodik. Heidelberg: Haug; 1992. p. 102.
69. DeVecchi A, Halberg F, Sothern RB, Cantaluppi A, Ponticelli C. Circaseptan rhythmic aspects of rejection in treated patients with kidney transplant. In: Walker CA, Winget CM, Soliman KFA, editors. Chronopharmacology and chronotherapeutics. Tallahassee, Florida: Florida A & M University Foundation; 1981. p. 339–53.
70. Cornelissen G, Halberg F, Tarquini B, Mainardi G, Panero C, Cariddi A, Sorice V, Cagnoni M. Blood pressure rhythmometry during the first week of human life. In: Tarquini B, editors. Social diseases and chronobiology: Proceedings III International Symposium Social Diseases and Chronobiology, Florence, Nov. 29, 1986. Bologna: Società Editrice Esculapio; 1987. p. 113–22.
71. Turti T, Syutkina EV, Cornelissen G, Grigoriev AE, Mitish MD, Abramian AS, Siegelova J, Fiser B, Dusek J, Al-Kubati M, Muchova L, Uhlir M, Halberg F. Multiseptan-over-circadian prominence of neonatal blood pressure and heart rate in Moscow, Russia. Scr Medica (Brno) 1996;67 Suppl 2:85–92.
72. Siegelova J, Cornelissen G, Schwartzkopff O, Halberg F. Time structures in the development of children. Neuroendocrinol Lett 2003;24 Suppl 1:126–31.
73. Cornelissen G, Siegelova J, Halberg F. Blood pressure and heart rate dynamics during pregnancy and early extra-uterine life: Methodology for a chrononeonatology. In: Halberg F, Kenner T, Fiser B, editors. Proceedings, Symposium: the Importance of Chronobiology in Diagnosing and Therapy of Internal Diseases. Faculty of Medicine, Masaryk University, Brno, Czech Republic, January 10–13, 2002. Brno: Masaryk University; 2002. p.;58–96.
74. Cornelissen G, Engebretson M, Johnson D, Otsuka K, Burioka N, Posch J, Halberg F. The week, inherited in neonatal human twins, found also in geomagnetic pulsations in isolated Antarctica. Biomed Pharmacother 2001;55 Suppl 1:32s–50.
75. Halberg F, Cornelissen G, Faraone P, Poeggeler B, Hardeland R, Katinas G, Schwartzkopff O, Otsuka K, Bakken EE. Prokaryotic and eukaryotic unicellular chronomics. Biomed Pharmacother 2005;59 Suppl 1:S192–202. PMCID: PMC3082478.
76. Schweiger H-G, Berger S, Kretschmer H, Mörler H, Halberg E, Sothern RB, Halberg F. Evidence for a circaseptan and a circasemiseptan growth response to light/dark cycle shifts in nucleated and enucleated acetabularia cells, respectively. Proc Natl Acad Sci USA 1986;83:8619–23.

77. Hayes DK, Halberg F, Cornelissen G, Shankaraiah K. Frequency response of the face fly, Musca autumnalis (Diptera: Muscidae), to lighting schedule shifts at varied intervals. Ann Entomol Soc Am 1986;79:317–23.
78. Marques MD, Cutkomp LK, Cornelissen G, Marques N, Halberg F. Life span of springtail on 12-h shifts at varying intervals of 12-hourly alternation in ambient temperature. Prog Clin Biol Res 1987;227A:257–64.
79. Cornelissen G, Halberg J, Halberg F, Sanchez de la Pena S, Nelson W, Schwartzkopff O, Stoynev A. Haus E. Schedule shifts, cancer and longevity: good, bad or indifferent? J Exp Ther Oncol 2008;7(4):263–74. PMCID: PMC2646176.
80. Halberg F. Historical encounters between geophysics and biomedicine leading to the Cornelissen-series and chronoastrobiology. In: Schröder W, editor. Long- and short-term variability in sun's history and global change. Bremen: Science Edition; 2000. p. 271–301.
81. Singh RB, Cornelissen G, Siegelova J, Homolka P, Halberg F. About half-weekly (circasemiseptan) pattern in blood pressure and heart rate in men and women of India. MEFA 8th International Fair of Medical Technology and Pharmacy, Brno, Czech Republic, 7–10 Nov 2000, abstract 24.
82. Watanabe Y, Asahi Y, Wu JY, Hillman D, Otsuka K, Cornelissen G, Halberg F. Circadian-infradian aspects of the human adult blood pressure and heart rate chronomes assessed longitudinally. In: Halberg F, Watanabe H, editors. Proceedings Workshop on Computer Methods on Chronobiology and Chronomedicine, Tokyo, Sept. 13, 1990. Tokyo: Medical Review; 1992. p.;233–44.
83. Siegelova J, Cornelissen G, Fiser B, Dusek J, Homolka P, Masek M, Jancik J, Svacinova H, Halberg F. Blood pressure and heart rate monitoring in humans: the circaseptan and circadian rhythms. Clin Auton Res 2003;13:60–1.
84. Gubin D, Cornelissen G, Halberg F, Gubin G, Uezono K, Kawasaki T. The human blood pressure chronome: a biological gauge of aging. In Vivo 1997;11:485–94.
85. Gubin D, Cornelissen G, Halberg F, Gubin GD, Turti T, Syutkina EV, Grigoriev AE, Mitish MD, Yatsyk GV, Ikonomov O, Stoynev A, Madjirova N, Siegelova J, Fiser B, Dusek J. Half-weekly and weekly blood pressure patterns in late human ontogeny. Scr Medica (Brno) 1997;70:207–16.
86. Rawson MJ, Cornelissen G, Holte J, Katinas G, Eckert E, Siegelova J, Fiser B, Halberg F. Circadian and circaseptan components of blood pressure and heart rate during depression. Scr Medica (Brno) 2000;73:117–24.
87. Shinagawa M, Otsuka K, Murakami S, Kubo Y, Cornelissen G, Matsubayashi K, Yano S, Mitsutake G, Yasaka K, Halberg F. Seven-day (24-h) ambulatory blood pressure monitoring, self-reported depression and quality of life scores. Blood Press Monit 2002;7:69–76.
88. Cornelissen G, Watson D, Mitsutake G, Fiser B, Siegelova J, Dusek J, Vohlidalova L, Svacinova H, Halberg F. Mapping of circaseptan and circadian changes in mood. Scr Med 2005;78:89–98. PMCID: PMC2577283.
89. Halberg E, Halberg F. Chronobiologic study design in everyday life, clinic and laboratory. Chronobiologia 1980;7:95–120.
90. Liu T, Cavallini M, Halberg F, Cornelissen G, Field J, Sutherland DER. More on the need for circadian, circaseptan and circannual optimization of cyclosporine therapy. Experientia 1986;42:20–2.
91. Ulmer W, Cornelissen G, Revilla M, Siegelova J, Dusek J, Halberg F. Circadian and circaseptan dependence of the beta-ATP peak of four different cancer cell cultures: implications for chronoradiotherapy. Scr Medica (Brno) 2001;74:87–92.
92. Santorio S. De Statica Medicina. Hagae-Comitis, ex typographia A. Vlaco;,1657.
93. Ash. An extract of the journal of the society at Dublin giving an account of a periodical evacuation of blood at the end of one of the fingers. Philosop Trans 1685;15:989–90.
94. Demmer T. Morbus maculosus werlhofii in regelmassigen vierwochentlichen schuben bei einem 60 jahrigen mann, nebst untersuchungen uber die blutplattchen. Folia Haemat 1920;26:74–86.

95. Månson JC. Cyclic variations of the frequency of neutrophil leucocytes with "androgen induced" nucleus appendages in an adult man. Life Sci 1965;4:329–34.
96. Cornelissen G, Halberg F. Introduction to chronobiology. Medtronic Chronobiology Seminar #7, Apr 1994. pp. 52. (Library of Congress Catalog Card #94-060580) https://sites.google.com/a/umn.edu/halbergchronobiologycenter/.
97. Simpson HW, George D, Sothern RB, Griffith K. The chronobra identifies prevailing mammary vascularity as a candidate variable in breast cancer post-operative outcome prediction. Springer Plus 2013;2:241.
98. Mikulecký M. Paroxysmal tachyarrhythmia and moon. Chronobiologia 1990;17:71–3.
99. Halberg F, Düll-Pfaff N, Gumarova L, Zenchenko TA, Schwartzkopff O, Freytag EM, Freytag J, Cornelissen G. 27-day cycles in human mortality: Traute and Bernhard Düll. Hist Geo Sp Sci 2013;4:47–59. doi:10.5194/hgss-4-47-2013.
100. Garcia Alonso L, Hillman D, Cornelissen G, Garcia Penalta X, Wang ZR, Halberg F. Nature, not solely nurture: chronome as well as season governs growth patterns of infants. In: Otsuka K, Cornelissen G, Halberg F, editors. Chronocardiology and chronomedicine: humans in time and cosmos. Tokyo: Life Science Publishing; 1993. p. 71–5.
101. Portela A, Cornelissen G, Franz H, Francine H, Halberg J, Hofman MA, Swaab DF, Ikonomov OC, Stoynev AG. Metachronanalysis of circannual and circasemiannual characteristics of human suprachiasmatic vasopressin-containing neurons. In Vivo 1995;9:347–58.
102. Smolensky M, Halberg F, Sargent II F. Chronobiology of the life sequence. In: Itoh S, Ogata K, Yoshimura H, editors. Advances in climatic physiology. Tokyo: Igaku Shoin Ltd; 1972. p. 281–318.
103. Cornelissen G, Breus TK, Bingham C, Zaslavskaya R, Varshitsky M, Mirsky B, Teibloom M, Tarquini B, Bakken E, Halberg F, International Womb-to-Tomb Chronome Initiative Group. Beyond circadian chronorisk: worldwide circaseptan-circasemiseptan patterns of myocardial infarctions, other vascular events, and emergencies. Chronobiologia 1993;20:87–115.
104. Sundararaj BI, Vasal S, Halberg F. Circannual rhythmic ovarian recrudescence in the catfish Heteropneustes fossilis (Bloch). In: Takahashi R, Halberg F, Walker C, editors. Toward chronopharmacology, Proceedings 8th IUPHAR Congress and Sat. Symposia, Nagasaki, July 27–28, 1981. Oxford/New York: Pergamon Press; 1982;p. 319–37.
105. Halberg F, Cornelissen G, Sothern RB, Wallach LA, Halberg E, Ahlgren A, Kuzel M, Radke A, Barbosa J, Goetz F, Buckley J, Mandel J, Schuman L, Haus E, Lakatua D, Sackett L, Berg H, Wendt HW, Kawasaki T, Ueno M, Uezono K, Matsuoka M, Omae T, Tarquini B, Cagnoni M, Garcia Sainz M, Perez Vega E, Wilson D, Griffiths K, Donati L, Tatti P, Vasta M, Locatelli I, Camagna A, Lauro R, Tritsch G, Wetterberg L. International geographic studies of oncological interest on chronobiological variables. In: Kaiser H, editor. Neoplasms -- comparative pathology of growth in animals, plants and man. Baltimore: Williams and Wilkins; 1981. p. 553–96.
106. Halberg F, Cornelissen G, Regal P, Otsuka K, Wang ZR, Katinas GS, Siegelova J, Homolka P, Prikryl P, Chibisov SM, Holley DC, Wendt HW, Bingham C, Palm SL, Sonkowsky RP, Sothern RB, Pales E, Mikulecky M, Tarquini R, Perfetto F, Salti R, Maggioni C, Jozsa R, Konradov AA, Kharlitskaya EV, Revilla M, Wan CM, Herold M, Syutkina EV, Masalov AV, Faraone P, Singh RB, Singh RK, Kumar A, Singh R, Sundaram S, Sarabandi T, Pantaleoni GC, Watanabe Y, Kumagai Y, Gubin D, Uezono K, Olah A, Borer K, Kanabrocki EA, Bathina S, Haus E, Hillman D, Schwartzkopff O, Bakken EE, Zeman M. Chronoastrobiology: proposal, nine conferences, heliogeomagnetics, transyears, near-weeks, near-decades, phylogenetic and ontogenetic memories. Biomed Pharmacother 2004;58 Suppl 1:S150–87.
107. Uezono K, Kawasaki T, Otsuka K, Yanaga A. Circannual differences in circadian variations of blood pressure and heart rate in Japanese women. In: Halberg F, Watanabe H, editors. Proceedings Workshop on Computer Methods on Chronobiology and Chronomedicine, Tokyo, Sept. 13, 1990. Tokyo: Medical Review; 1992. p. 233–44.
108. Watanabe Y, Cornelissen G, Halberg F. Thousands of blood pressure and heart rate measurements at fixed clock hours may mislead. Neuroendocrinol Lett 2003;24:339–40.

109. Halberg F, Cornelissen G, Carandente F. Chronobiology leads toward preventive health care for all: cost reduction with quality improvement. A challenge to education and technology via chronobiology. Chronobiologia 1991;18:187–93.
110. Cornelissen G, Otsuka K, Halberg F. Blood pressure and heart rate chronome mapping: a complement to the human genome initiative. In: Otsuka K, Cornélissen G, Halberg F, editors. Chronocardiology and chronomedicine: humans in time and cosmos. Tokyo: Life Science Publishing; 1993. p. 16–48.
111. Cornelissen G, Halberg F, Sonkowsky RP. Chronome. In: Macey SL, editor. Encyclopedia of time. New York: Garland Publishing; 1994. p. 107.
112. Halberg F, Cornelissen G, Otsuka K, Schwartzkopff O, Halberg J, Bakken EE. Chronomics. Biomed & Pharmacother 2001;55 Suppl 1:153s–90s.

Chapter 5
Chronomics of Heart Rate Variability

Abstract Chronomics is a term comparable to genomics, or proteomics, and it can disclose covered signals in the original time series data, which are not visible without being resolved by chronomics. Chronomes are time structures consisting of (1) multifrequency rhythms covering frequencies over 10 orders of magnitude, (2) elements of chaos, (3) trends in chaotic and rhythmic endpoints, and (4) other as yet unresolved variability, which are found in larger and larger systems along longer and longer time scales in us and still longer ones around us.

The circadian aspects of heart rate variability (HRV) have been intensively studied, but there are only few reports on the broader chronome of HRV.

Keywords Heart rate variability • Circadian elements of chronomics • Age trends in heart rate variability • Multifactorial interactions

5.1 Introduction

Heart rate variability (HRV) has become an important tool in cardiology not only as a noninvasive gauge of autonomic nervous system activity but also as an index predicting subsequent morbidity and/or mortality of patients with coronary artery disease. Ambulatory monitoring of the ECG has provided an opportunity to study long-term fluctuations in the autonomic cardiovascular function. HR and HRV in humans are characterized by a spectrum of rhythms, including ultradian, circadian, and infradian changes, as part of an even broader time structure, also including chaos and trends in rhythmic and chaotic endpoints. The overall time structure has been called the CHRONOME [1–7]. The rhythmic features of HR and HRV include not only circadian variations (with a period of about 24 hours) but also circaseptan changes (with a period of about 7 days), circannual components (with a period of about 1 year), and even circadecadal cycles (with a period of about 10 years), among others. In addition to the chronome-adjusted mean value (MESOR), each of these components is characterized by an amplitude and acrophase, measures of the extent and timing of predictable change within a cycle, while their waveform can be determined by amplitude-acrophase pairs of harmonic components that modify the shape of a rhythm with a given fundamental period.

K. Otsuka et al., *Chronomics and Continuous Ambulatory Blood Pressure Monitoring*, DOI 10.1007/978-4-431-54631-3_5

HR and HRV are also characterized by other changes which may be nonlinear in nature, some of them pertaining to what has been called deterministic chaos [8–12]. Endpoints to assess deterministic or other chaotic behavior of HRV include the fractal dimension and complexity, used as imputations, that is, as intermediate computations, whether or not any tests for nonlinearity or more specifically for deterministically chaotic behavior are met. Several endpoints of HRV in the frequency domain are also useful and include the spectral power in different spectral regions. Of interest is the spectral power around 10.5 s (0.04–0.15 Hz; also known as "LF") and the spectral power around 3.6 s (0.15–0.40 Hz; also known as "HF"), which in turn are found to be circadian stage-dependent. These different endpoints also exhibit trends as a function of age and in the presence of disease. Trends as a function of age may differ between men and women. Patients with different cardiovascular disorders have been shown to have an altered time structure (chronome) of HRV. In turn, patients with an altered HRV chronome have been shown to be at an increased vascular disease risk, a concept that has led to the identification of new disease risk syndromes. The circadian aspects of HRV have been intensively studied, but there are only few reports on the broader chronome of HRV.

We have studied the chronomics of HRV, which include the ultradian components (i.e., the circacentuminutan, circaoctohoran, and circasemidian aspects of HRV pertaining to cycles with periods of about 100 minutes, about 8 and 12 hours, respectively) as well as the circadians. Trends as a function of development are also being mapped. We also noticed that there exist associations of helio- and geomagnetics on physiological chronomes, such as that of HRV, which have counterparts in our environment, and that our genetic makeup in time may have evolved from our adaptation to and integration with nonphotic as well as photic aspects of our cosmos.

5.2 Mapping of Chronome Components

Ambulatory around-the-clock 24-hour ECG records were obtained by using a two-channel Holter recorder (SM-28, Fukuda-Denshi, Tokyo) and an analyzing system (SCM-280-3, Fukuda-Denshi, Tokyo), including an interval counter of RR intervals and a built-in A/D converter. The interval resolution was 8 msec.

For the analysis of a trend with development, 19 infants (aged from 25 days to 3 months of age), 22 children (aged 3–9 years), 18 boys and girls (aged 10–14 years), 14 pubertal boys (aged 15–20 years), and 19 young men (aged from 21 to 29 years) were examined.

The mapping of circadian changes in HRV endpoints as a function of age and development was performed in Tokyo on 285 males (aged from 25 days to 92 years) and 117 women (aged 3–91 years), who were randomly selected from a larger group of clinically healthy volunteers. The clinical health status of all subjects investigated was assessed by means of a physical medical examination and laboratory data, including blood and urine examinations, chest X-ray, and ECGs at rest. Adult subjects were found to be normotensive by WHO criteria (SBP and DBP below 140 and 90 mmHg, with no history of major diseases).

Frequency-domain measures were obtained with the MemCalc/CHIRAM (Suwa Trust Co., Ltd., Tokyo) software. Time series of NN intervals covering 5 min were processed consecutively, and the spectral power in different frequency regions was computed, namely, in the "very-low-frequency" ("VLF-1," 0.003–0.04 Hz, "VLF-2," 0.005–0.04 Hz, SAS (90-sec)), "low-frequency" (spectral power centered around 10.5 sec: "LF-1," 0.04–0.15 Hz, "LF-2," 0.05–0.07 Hz, MSNA (15-sec)), and "high-frequency" (spectral power centered around 3.6 sec: "HF," 0.15–0.40 Hz) regions of the MEM spectrum (Fig. 5.1). Next, 180-min NN interval time series were processed consecutively, progressively displaced by 5 min, to estimate the "total spectral frequency" ("TF," 0.0001–0.50 Hz), the "ultralow-frequency" ("ULF-1," 0.0001–0.003 Hz, "ULF-2," 0.0001–0.0003 Hz, circacentuminutan or Kleitman (90 min), "ULF-3," 0.0003–0.005 Hz "extremely low frequency" (ELF) (30 min)) components, and the slope of 1/f fluctuations of NN intervals ("β (beta)"), in the frequency range of 0.0001–0.01 Hz.

The dynamics of HRV measures, assessed every 5 min, were analyzed by both the Maximum Entropy Method (MEM) and by cosinor, involving the least-squares fit of cosine curves with anticipated periods. Spectral components of each MEM spectrum were expressed not only in terms of spectral power or amplitude but also as power ratio (unit, %), normalized in relation to the "total spectral power" (dividing by "TF"). Circadians, circasemidians, and circaoctohorans were defined in terms of periods ranging from 20 to 28 hours, from 10 to 14 hours, and from 6.5 to 9.5 hours, respectively. Each component was assumed to be present (detectable) when the power ratio of these three chronome components was equal to or greater than 10 %.

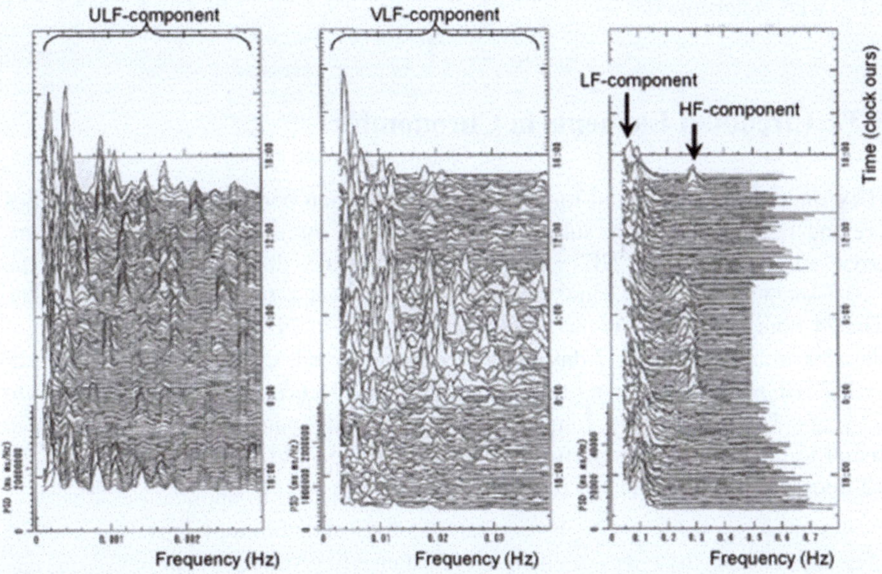

Fig. 5.1 Power spectral array of ULF (*left*), VLF (*middle*), LF, and HF components (*right*)

For assessing any trend with age, time-domain measures, SDNN, pNN50, SDANNs (1, 5, 30, and 60 min), SDNNIDXs (1, 5, 30, and 60 min), r-MSSD, NN50(+), NN50(−), total NN50, and pNN50 were computed. The mean cycle length of the normal-normal RR (NN) intervals over 24 hours and the standard deviation (SD) of NN intervals over 24 hours were calculated as 24-hour NN and SDNN, respectively. SDANNs (1, 5, 30, and 60 min) were calculated as the SD of the 1-min, 5-min, 30-min, and 60-min mean NN intervals over the entire 24-hour ECG recordings, respectively. Similarly, SDNNIDXs (1, 5, 30, and 60 min) were calculated as the mean of the SD of the 1-min, 5-min, 30-min, and 60-min NN intervals over 24 hours, respectively. The difference between adjacent NN intervals was also computed as the root mean square of successive differences (r-MSSD), calculated as the square root of the mean of the sum of the squares of differences between adjacent NN intervals over the entire 24-hour record. NN50 was assessed as follows: successive NN intervals were subtracted from each other and compared with 50 msec. If the difference was larger or smaller than 50 msec, the positive or negative count was incremented by one, with the totals over 24 hours being represented by NN50(+) and NN50(−), respectively. NN50 and pNN50 represent the total count and the percentage of differences between adjacent NN intervals that are greater than 50 msec, computed over the entire 24-hour ECG recording. Frequency-domain measures were determined as follows: The time series of NN intervals were reconstructed using a third-degree spline interpolation method to yield a new smoothed instantaneous time series sampled at 4 Hz. A 10-min span of the reconstructed time series was processed consecutively by the fast Fourier transform (FFT), with successive 10-min spans overlapping by 150 sec. Spectra obtained within a 60-min span were averaged, and the "LF," "HF," and the ~10.5 sec/~3.6 sec spectral power ratio ("LF/HF") were determined.

5.3 Circadian Elements in Chronomics

HRV endpoints were found to be eminently circadian periodic in the 359 profiles previously analyzed (from subjects 10–92 years of age). The spectral power centered around 3.6 sec ("HF"=high frequency) peaks during the night, and the ~10.5 sec/~3.6 sec ("LF=low frequency/HF") power ratio peaks during the day. The 24-hour component is the most prominent one in 292 profiles (81.3 %) out of the 359 around-the-clock 7-day ECG records analyzed. Circadian variation of HF and LF/HF ratio is shown in Fig. 5.2, measured in 218 healthy males aged from 3 to 92 years. The second most frequently observed component is the 8-hour variation, found to be statistically significant in 14 profiles (3.9 %), and the third one is the 12-hour component, detected in 13 profiles (3.6 %).

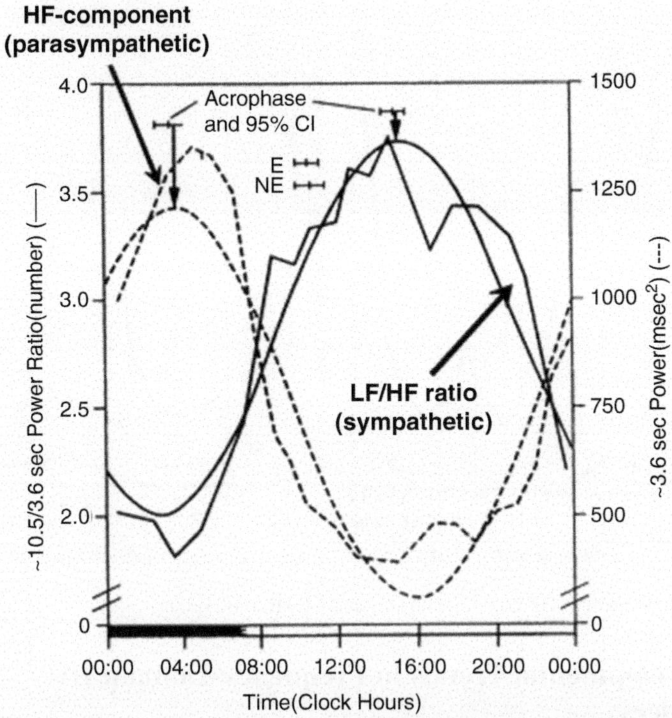

	"Tone"	
	"Sympathetic"	"Parasympathetic"
P:	<0.001	<0.001
MESOR±SE:	2.88±0.04	769±24
Double Amplitude (95% CI):	1.70(1.46, 1.96)	904(763, 1045)
Acrophase (95% CI):	14:51(14:26, 15:24)	03:36(03:00; 04:12)

Fig. 5.2 Circadian variation of HF and LF/HF ratio

5.4 Circadian Reference Values for Different Endpoints of HRV

HRV serves not only as a noninvasive measure of autonomic nervous system activity but also as an index predicting long-term survival in patients with coronary artery disease (CAD). Here we introduce "chronodesms" (Fig. 5.3), that is, time-specified reference values that account for the circadian variation in HRV endpoints. To illustrate their qualified usefulness and to show the need for refinement by outcomes, circadian profiles of CAD patients are interpreted in the light of chronodesms derived from healthy peer groups, revealing nearly deviant features in the presence of disease.

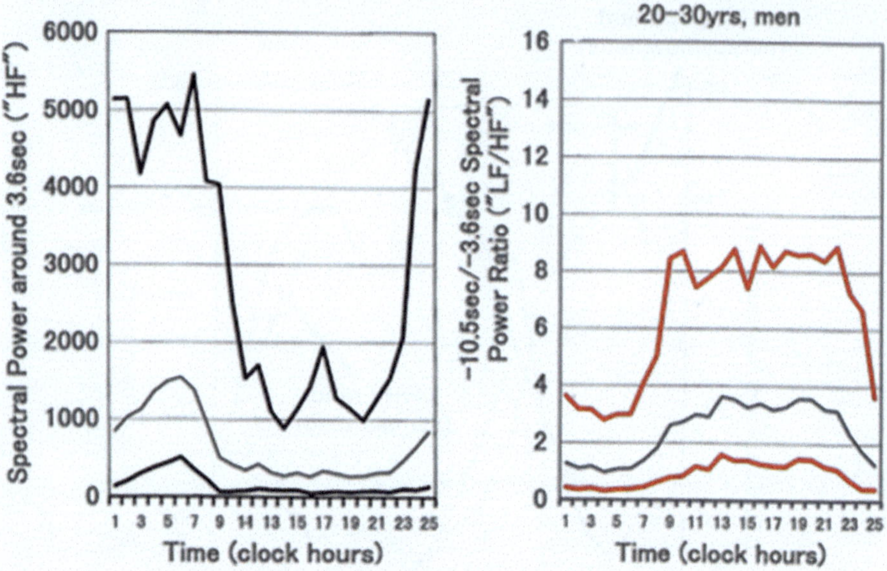

Fig. 5.3 Chronodesm of HF component (*left*) and LF/HF ratio (*right*)

5.5 Developmental Trends in Frequency-Domain HRV Indices

The finding of an 8-hour component in circulating endothelin-1 [13, 14] and also in the population density of endotheliocytes [15] is one of the noteworthy observations in recent years, notably since the hypothesis was made that an 8-hour periodicity may be a key between the linear and nonlinear systems of life, as proposed by Li and Yorke in 1975 [16]. Such mechanisms may reside in endotheliocytes, since circulating cortisol did not show an about-8-hour (circaoctohoran) component [13]. As noted above, another recent finding relates to the slope (β (beta)) over the two-decade band from 0.0001 to 0.01 Hz, known as 1/f behavior. Several authors have proposed that this slope is a better predictor of all-cause mortality than the traditional power spectral bands [8–10].

The development of the chronome of different endpoints of HRV, notably of the circadian, circasemidian, and circaoctohoran components, was investigated. Results are summarized in Table 5.1 in terms of the incidence of detection of each of these three components for six different frequency-domain measures of HRV (1/f slope, "TF," "ULF," "VLF," "LF," and "HF"). The incidence of detection was compared among the following five age groups: 25 days to 3 months ($n=19$), 3–9 years ($n=22$), 10–14 years ($n=18$), 15–20 years ($n=14$), and 21–29 years ($n=19$). The incidence of detection of the circadian component for 1/f changes with age (21.05, 27.27, 61.11, 57.14, and 31.58 %, respectively, $p<0.05$) was mostly present around puberty. Averaged circadian profiles of the 1/f slope are shown in Fig. 5.4. Not

Table 5.1 Developmental trends in the incidence of detection of circadian, circasemidian, and circaoctohoran components of HRV indices

Circardian (20–28 hour) component									
Age		1/f	TF	ULF	VLF	LF	HF	χ^2	p-value
25 days to 3 months	(n=19)	21.05	36.84	31.58	10.53	5.26	15.79	8.77	N.S
3–9 years	(n=22)	27.27	31.82	45.45	63.64	63.64	77.27	17.35	<0.005
10–14 years	(n=18)	61.11	33.33	33.33	50.00	33.33	77.78	12.31	<0.05
15–20 years	(n=14)	57.14	28.57	35.71	35.71	35.71	50.00	3.36	N.S
21–29 years	(n=19)	31.58	42.11	31.58	57.89	36.84	78.95	13.29	<0.05
χ^2		9.98	0.82	1.23	14.18	15.17	24.16		
p-value		<0.05	N.S	N.S	<0.01	<0.005	<0.0001		
Circasemidian (10–14 hour) component									
Age		1/f	TF	ULF	VLF	LF	HF	χ^2	p-value
25 days to 3 months	(n=19)	47.37	31.58	21.05	5.26	0.00	0.00	24.50	<0.0005
3–9 years	(n=22)	31.82	22.73	36.36	4.55	4.55	18.18	12.45	<0.05
10–14 years	(n=18)	27.78	22.22	44.44	5.56	5.56	16.67	12.10	<0.05
15–20 years	(n=14)	35.71	21.43	42.86	21.43	14.29	28.57	3.89	N.S
21–29 years	(n=19)	26.32	26.32	36.84	10.53	15.79	47.37	8.73	N.S
χ^2		2.40	0.69	2.70	3.92	4.66	13.43		
p-value		N.S	N.S	N.S	N.S	N.S	<0.01		
Circaoctohoran (6.5–9.5 hour) component									
Age		1/f	TF	ULF	VLF	LF	HF	χ^2	p-value
25 days to 3 months	(n=19)	52.63	42.11	31.58	10.53	5.26	5.26	21.40	<0.001
3–9 years	(n=22)	54.55	18.18	50.00	4.55	4.55	9.09	32.08	0.0001
10–14 years	(n=18)	38.89	11.11	44.44	5.56	16.67	0.00	18.98	<0.005
15–20 years	(n=14)	50.00	57.14	57.14	7.14	14.29	14.29	17.79	<0.005
21–29 years	(n=19)	57.89	47.37	63.16	15.79	15.79	0.00	29.21	<0.0001
χ^2		1.55	12.12	4.36	2.09	2.79	4.83		
p-value		N.S	<0.05	N.S	N.S	N.S	N.S		

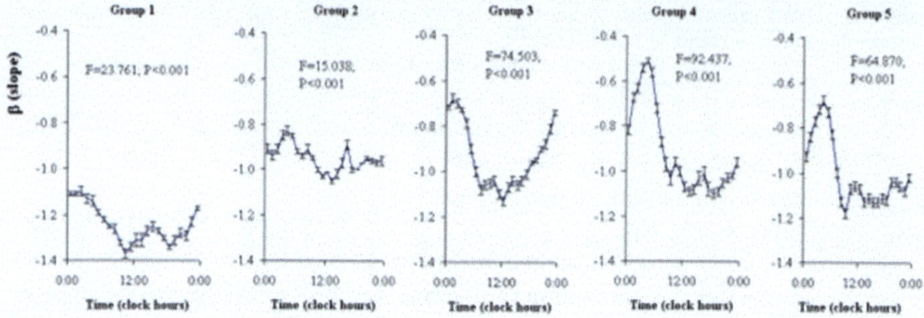

Fig. 5.4 *Average circadian profiles of 1/f behavior of RR intervals*. Group 1, 25 days to 3 months (n = 19); Group 2, 3–9 years (n = 22); Group 3, 10–14 years (n = 18); Group 4, 15–20 years (n = 14); and Group 5, 21–29 years (n = 19)

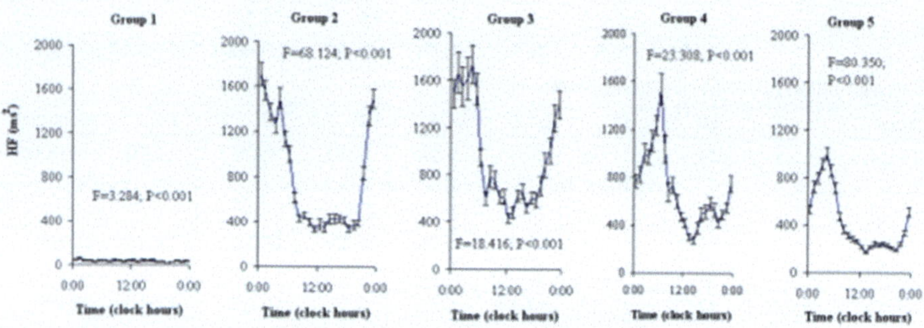

Fig. 5.5 *Average circadian profiles of the "HF" component*. Group 1, 25 days to 3 months (n = 19); Group 2, 3–9 years (n = 22); Group 3, 10–14 years (n = 18); Group 4, 15–20 years (n = 14); and Group 5, 21–29 years (n = 19)

shown in Table 5.1 is an increase with age in the power ratio of the circadian spectrum: 40.8, 39.8, 41.7, 56.1, and 51.1 %, respectively ($p < 0.05$). No statistically significant changes in the circadian period, assessed by MEM spectral analysis, were found as a function of age: 24.5, 21.2, 22.6, 20.9, and 26.0 hours, respectively.

Table 5.1 suggests that the about 35 % circadian rhythm detection for "TF" and "ULF" does not change appreciably with age. By contrast, the detectability of a 24-hour component for "VLF," "LF," and "HF" is much lower early in life than after 3 years of age. Averaged circadian profiles of "HF" are shown in Fig. 5.5. Similar changes with age are not seen for the circaoctohoran component, however. For 1/f, the incidence of detection is 52.63, 54.55, 38.89, 50.00, and 57.89 %, respectively (N.S.), and for the power ratio of the MEM spectrum, it is 27.7, 27.1, 24.9, 26.3, and 22.9 %, respectively (N.S.). Results for the circasemidian component show an increase with age only for "HF" and may reflect the development of the circadian system. Also apparent from Table 5.1 is the larger detection of a circaoctohoran component for 1/f, "TF," and "ULF" as compared to "VLF," "LF," and "HF," reflecting the autonomic nervous activity.

5.6 Age Trends in Heart Rate Variability

Changes with age of circadian components in the spectral element of HRV chronomics are highly statistically significant (Fig. 5.6) in both genders. The relation is not invariably linear, as indicated by the presence of higher-order polynomial terms selected by stepwise regression, using a sixth-order model at the outset. For instance, NN50, SDNNIDX (1-min), and "HF" reach a minimum around 66.5 years (male: 65, female: 68), 72 years (male: 70, female: 74), and 59.5 years (male: 61, female: 58), respectively, and increase thereafter, as shown in Fig. 5.7, raising the question whether the oldest age group represents a category of long-lived subjects, who by this status may have been healthier than the average subject in the younger age groups. Thus, circadian elements in HRV chronomics, obtained with the MemCalc/CHIRAM (Suwa Trust Co., Ltd., Tokyo) software, are shown to vary with age in Appendices 1, 2, 3, and 4, which are shown as an average of 24 hours, sleeping and waking time spans.

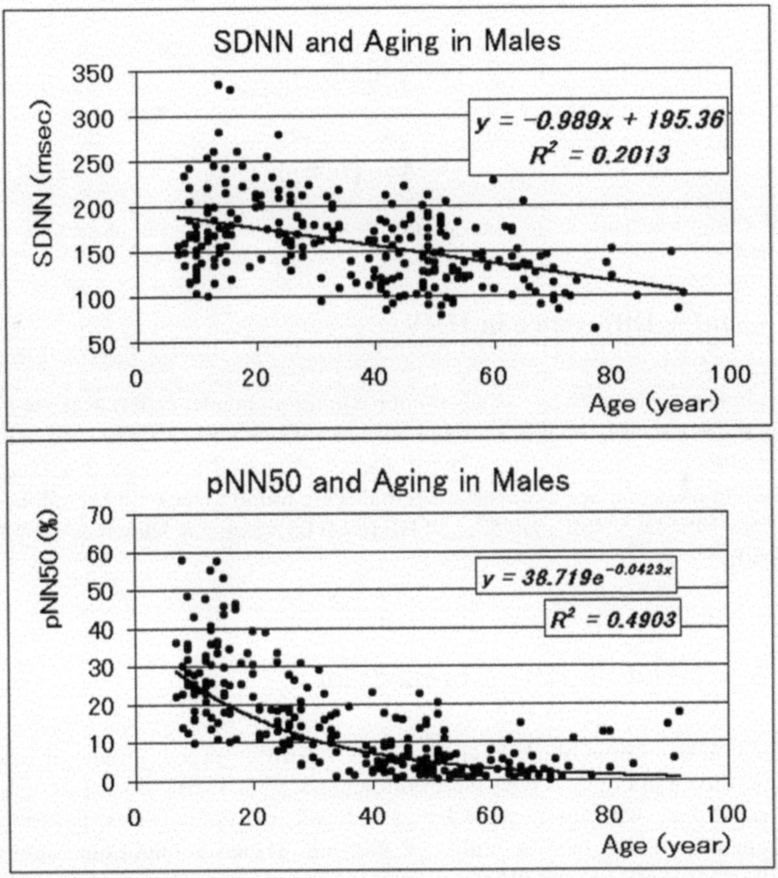

Fig. 5.6 *Top*: Linear decrease of SDNN with age in male subjects. *Bottom*: Exponential decrease of pNN50 with age in male subjects

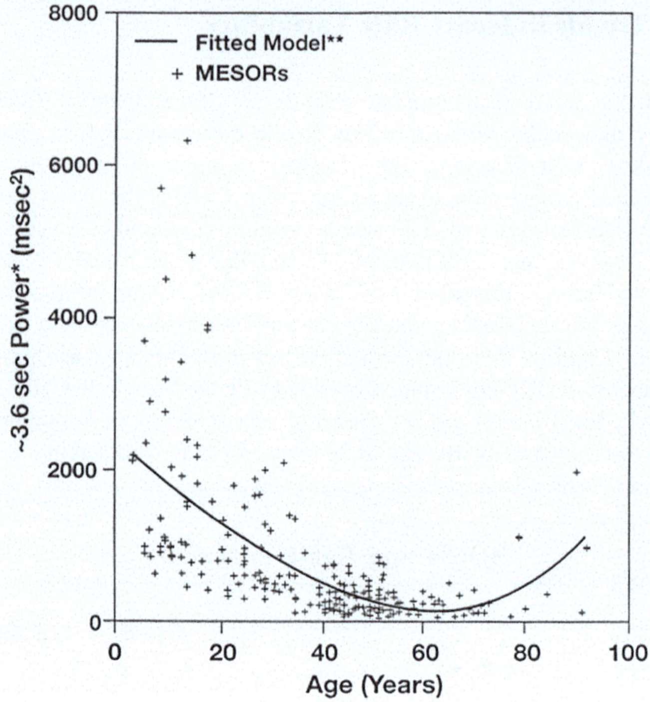

Fig. 5.7 Changes with age in HF component of HRV, suggesting parasympathetic tone, in 218 healthy males

5.7 Gender Difference in HRV

Using a linear model with age as a first approximation, gender differences are found for the slope but not for the intercept of r-MSSD, NN50, pNN50, and HF, the females showing a slower decline with age as compared to males ($p<0.05$). By 1-way ANOVA, using age as covariate, females are found to have higher values than males for r-MSSD, NN50, pNN50, and HF ($p<0.05$) (Fig. 5.8, shown as not NN50, but RR50).

5.8 Age Trends in HRV Chronodesm

Circadian chronodesms of "HF" and LF/HF" for different age groups of healthy males are shown in Fig. 5.9. They were computed as 90 % prediction limits on $\log_{10}$-transformed data, with the mean values and lower and upper limits of variability back-transformed into original units for plotting. These chronodesms show that both HF and LF/HF vary greatly as a function of age.

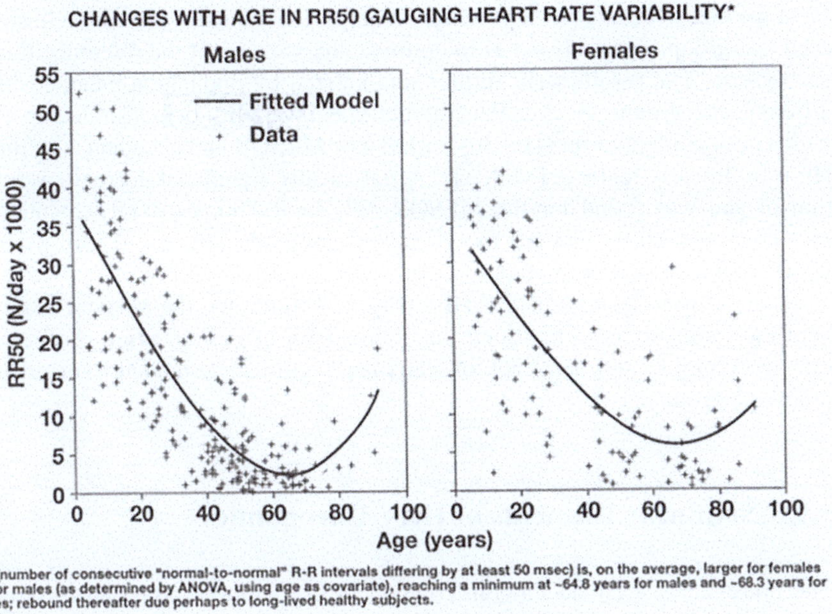

Fig. 5.8 Gender differences in RR50 found for the slope of the regression curve, suggesting cardiac parasympathetic tone in the aged kept better in females compared to males

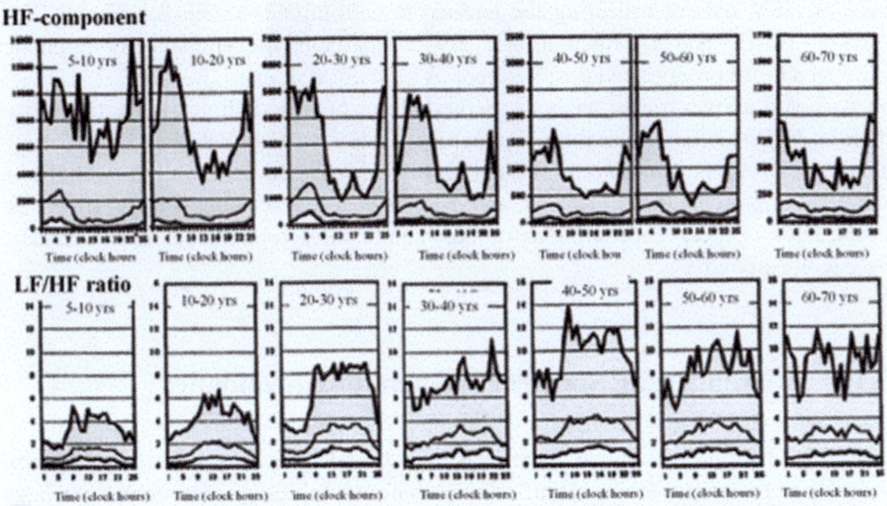

Fig. 5.9 Circadian chronodesms of the HF component (*top*) and the LF/HF ratio (*bottom*) for different age groups of healthy males

In the chronodesms for "HF" (Fig. 5.9, top), the vertical scales are adjusted for each age group separately, so as to optimize the viewing of the reference limits in each group. The results were further summarized by population-mean cosinor in different age groups. A circadian rhythm was invariably demonstrated ($p<0.05$) with a peak at night (around 03:30) for HF. The MESOR and circadian amplitude of HF were linearly regressed with age. A statistically significant decrease as a function of age was found for the MESOR (all: $r=-0.590$, $p<0.001$; 20–60 years: $r=-0.495$, $p<0.001$) and for the circadian amplitude (all: $r=-0.530$, $p<0.001$; 20–60 years: $r=-0.474$, $p<0.001$).

In the chronodesms for "LF/HF" (Fig. 5.9, bottom), the vertical scales are kept the same, so as to facilitate the comparison of the circadian profiles from one age group to another. Lack of statistical significance tended to occur in the older age groups.

5.9 Nonlinear Elements of HRV Chronomics

The presence of a circaoctohoran component of variation, proposed to be a key between the linear and nonlinear systems of life by Li and York in 1975 [16], was found not to differ among the five age groups considered above for the case of HRV indices reflecting the autonomic nervous system activity ("LF" and "HF" components). It was found to be already present in a sizeable proportion of infants for the case of HRV indices reflecting the endocrine or immune system function ("TF," "ULF," "VLF," and 1/f fluctuations). By comparison, the circadian component showed a more pronounced development as a function of age.

As noted above, fractal analysis methods have provided clinically useful information. In particular, the scaling exponent α, calculated with detrended fluctuation analysis (DFA), is available to clinical cardiology. Circadian variation of the two scaling components (α_1; <11 beats, and α_2; >11 beats) describing fractal properties of the HRV is shown as a comparison between healthy young subjects and elderly people in Appendix 5.

5.10 A Broadening Spectral Element of Chronomics

In clinical health, HRV is characterized by a broad time structure, including not only a prominent circadian and infradian variation, including infra-annuals such as circadecadals, but also statistically significant circaoctohoran and circasemidian components, in addition to trends as a function of age. Previously, we have reported that a circadian rhythm is demonstrated in both genders and in different age groups (<10 years, 10–20 years, 20–30 years, 30–40 years, 40–50 years, 50–60 years, 60–70 years, and >70 years) except for the oldest age group. The "HF" component may reflect the parasympathetic activity with a peak during the night. A strong circadian component is also shown for the "LF/HF" spectral power ratio, which is considered to be an indicator of sympatho-vagal interaction, and is characterized by a peak during the day.

A circaseptan component is also demonstrated for most of the HRV indices, including SDNN, observed both in middle aged and in elderly subjects. SDNN is known to be a good predictor of long-term survival, both in patients with myocardial infarction and in a cohort of elderly subjects. A meta-analysis of 47 studies shows the incidence of myocardial infarction to be increased on Mondays with a secondary peak on Thursdays and Fridays [2]. The circaseptan pattern of SDNN for the middle-aged subjects is characterized by a decrease on Mondays, with a secondary trough on Fridays, and may be at least one of the reasons underlying the circaseptan pattern of incidence of cardiovascular events. An important aspect of the circaseptan variability may be the relative prominence of the circasemiseptan (3.5-day) component, which could account for the secondary peak later in the week.

A circannual variation in "LF" and "HF" components has been observed in elderly subjects living in Shandong province, China, where the environmental temperature is hot in summer (up to 40 ° C) and cold in winter (as low as −15 ° C). A physiological interpretation for "LF" still needs to be clarified. "LF" fluctuations reflect combined sympathetic and parasympathetic activity. The "LF" peak is strikingly reduced by atropine. An increase in its power has been observed as a consequence of sympathetic activation. Thus, an increase in "LF" is widely viewed as a marker of sympathetic activation and vice versa. De Boer et al. [17] suggested that the "LF" peak was caused by the rather slow baroreflex response at 2–3 seconds in the beat-to-beat changes in arterial pressure. Thus, the decrease in "LF" in the winter may suggest an increase in arterial pressure related to an alteration in environmental temperature. It is noteworthy that β (the slope of the 1/f fluctuations of RR intervals), noted as one of the most important predictors of adverse cardiovascular outcomes, is characterized by a circannual component. "β (beta)" was most negative in the summer, when an 83-year-old man experienced a minor stroke. Nicolau et al. [18] investigated circannual variations in cardiac mortality and reported a peak in July with a broader peak during the cold season. Previously, Smolensky et al. [19] and Reinberg et al. [20] had also reported a circannual pattern in the incidence of cardiovascular events.

We also have shown previously that a relation exists between HRV endpoints and latitude, suggesting that there may be an effect of sunshine duration on the chronome mechanism that affects HRV measurements [21]. These results may also be interpreted as an association between environmental factors and human beings who have adapted to the surrounding cycles in the cosmos.

5.11 Community-Based "LILAC" Study for Longevity

In 2000, we began a community-based study named "Longitudinal Investigation for Longevity and Aging in Hokkaido County (LILAC study)" and monitored 7-day/24-hour ambulatory BP of middle-aged subjects, aged 40–74 years, while also evaluating the cardiovascular and neurobehavioral functions of elderly people. Our goal was the prevention of stroke and myocardial infarction and of the decline in cognitive function of the elderly in a community.

CHAT (short for circadian hyper-amplitude-tension), a condition defined by an excessive circadian amplitude of blood pressure (BP), above a threshold approximated by the upper 95 % prediction limit of clinically healthy peers matched by gender, age, and ethnicity, is associated with a large increase in vascular disease risk, cerebral ischemic events, and nephropathy in particular [1, 11, 22–32]. We have already monitored 7-day/24-hour ambulatory BP of 218 middle-aged citizens in Hokkaido County from April 2001 to April 2003. We observed circaseptan components for the average systolic (S) BP value during the waking span, for the 24-hour average of HR, for the standard deviation (SD) of HR, and for the day-night (dipping) ratio of SBP. Specifically, SBP during waking was lowest on Sunday. The BP average during waking shows a circaseptan variation. Thus, we point out that ambulatory BP monitoring limited to 24 hours is not sufficient and propose a routine 7-day screen, which is continued when abnormality is found. Among the 218 records of 7-day/24-hours ambulatory BP, we also observed 32 (14.7 %) and 11 (5.0 %) cases of SBP and diastolic (D) BP CHAT, respectively. We have started several kinds of interventions, including lifestyle guidance, education from the viewpoint of clinical nutrition, and consultation of drugs for hypertension and hyperlipidemia, among other conditions.

We also determined in the longitudinal community-based study (LILAC) whether BP, HR, and HRV are predictors of cognitive function in the elderly. We initially examined 115 people older than 75 years (average, 79.6 years). BP was measured at the beginning of the study in a sitting position, and pulse wave velocity (PWV) was measured between the right arm and ankle in a supine position, using an ABI/Form instrument (Nippon Colin Co., Ltd., Komaki, Japan). The first 1-hour of ambulatory ECG recording was obtained during routine medical examination conducted each year in July. The data were processed for HRV using a Fukuda-Denshi Holter analysis system (SCM-280-3). Time-domain measures (SDNN, pNN50, SDANN, and Lorenz plot indices: length (L), width (W), and L/W ratio) and frequency-domain measures (spectral power in the "very low frequency" (VLF), 0.003–0.04 Hz; "low frequency" (LF), 0.04–0.15 Hz; and "high frequency" (HF), 0.15–0.40 Hz regions and the "LF/HF" ratio) were determined. Except for SDNN and HR, calculated over the whole 1-hour record, all indices were computed as averages over consecutive 5-min intervals. Spectral indices were obtained by the Maximum Entropy Method (MEM) with the MemCalc/CHIRAM program (Suwa Trust Co., Ltd., Tokyo, Japan). Using as reference the data obtained in July 2000, the cardiovascular coordination function of each participant was scored as 3, 2, or 1 point for each of the following three indices (SBP, HR, and "VLF" component of the HRV): SBP $>$ 160, 140–159, or $<$ 140 mmHg; HR $>$ 80, 70–79, or $<$ 70 beats/min; and "VLF" $<$ 800, 800–1000, or $>$ 1000 msec^2. Participants were classified into either the normal, mildly disordered, or disordered group when the sum of these indices was $\leq$ 4, 5 or 6, or $\geq$ 7, respectively.

The Japanese version of the Mini-Mental State Examination (MMSE) and the Hasegawa Dementia Scale Revised (HDSR) were used to measure the overall cognitive function, including verbal orientation, memory, and constructional ability. The Up & Go test measured (in seconds) the time it took the subject to stand up from a chair, walk a distance of 3 meters, turn, walk back to the chair, and sit down again. This test is a simple measure of physical mobility and demonstrates the subject's balance, gait speed, and functional ability (Up & Go). A lower time score indicates better physical mobility. Functional reach, used to evaluate balance, represents the maximal distance a subject can reach forward beyond arm's length while maintaining a fixed base of support in the standing position. A higher score indicates better balance. Manual dexterity was assessed using a panel with combinations of ten hooks, ten big buttons, and five small buttons. There were three discrete measurements of time recorded for each participant (ten "hook-ons," ten big "button-on-and-offs," and five small "button-on-and-offs"). Total manual dexterity time in seconds, defined as the button score (Button-S), was calculated by adding the average times for one hook-on and one big or small button-on-and-off. A lower button score indicates better manual dexterity.

We evaluated the effects of several kinds of health consultation, rehabilitation of disordered function, healthy lifestyle modification by promoting complete cessation of smoking, weight reduction, reduction of salt intake, moderation in the consumption of fruits and vegetables and alcohol intake, as well as advising medical prescription for the local general practitioner. The paired *t*-test was used to compare each neurobehavioral endpoint between 2000 and 2002. Results were considered to be statistically significant at $p<0.05$.

In 2000 [33], the cardiovascular coordination score did not correlate with any index of neurobehavioral function, although it showed a negative correlation with SDNN, SDANN, pNN50, "LF," or "HF" components ($p<0.0001$) and a positive correlation with PWV ($p<0.01$). We were able to follow up 72 of the 115 subjects. We found that between 2000 and 2002 the cognitive function, estimated by MMSE and HDSR, was maintained or improved as follows: In the cardiovascular coordination disordered group, MMSE and HDSR improved from 24.6 to 26.0 ($p=0.06$) and from 23.8 to 25.9 ($p=0.04$), respectively. In the mildly disordered group, these indices improved from 23.4 to 25.7 ($p=0.005$) and from 23.4 to 25.1 (N.S.), respectively. In the normal cardiovascular coordination group, MMSE and HDSR were maintained from 25.6 to 26.0 (N.S.) and from 24.9 to 26.4 (N.S.), respectively. There were no statistically significant alterations in activities of daily living (ADL), assessed by Up & Go, functional reach, and button score, in any of the groups.

Although a cross-sectional study did not show any apparent correlation between cardiovascular and neurobehavioral functions in subjects 75 years of age or older, an intervention aimed at preventing stroke and a decline in cognitive function in an elderly community population induced an improvement of the cognitive function, especially in people suffering from hypertension, tachycardia, or decreased HRV. This study demonstrates a positive impact of a simple social intervention, including advising on the implementation of a medical prescription, in improving a disordered cognitive function in elderly people. People with a disordered coordination of cardiovascular function are more sensitive to such an intervention, suggesting that the cardiovascular function is a major factor affecting cognitive function.

5.12 Conclusion

In clinical health, both HR and BP variabilities (HRV and BPV) are characterized by a broad time structure that includes the prominent circadians and also ultradian (notably about 8 hours and about 12 hours) and infradian (notably about-weekly, about-yearly, and about-10-yearly) changes, in addition to undergoing trends with growth, maturation, and aging. Alterations in these HRV and BPV chronomes are observed in the presence of disease and at times of increasing magnetic activity. We postulate that the physiological chronomes, such as those of HRV endpoints, have counterparts in our environment and that our genetic makeup in time may have evolved to adapt to and integrate with our cosmos. Future work should focus on how phenomena in the cosmos, including helio- and geomagnetics, can affect physiological chronomes, those of the HRV in particular.

Human beings have shielded themselves from the unkind features of obvious extremes of weather and climate by building houses and have developed means, e.g., sulfonamides and antibiotics, to deal with at least some of the microbial contagions confronting us. These infections via the mutation of microorganisms causing them and by (steroidal?) changes in the host may be accelerated by solar activity. This is one more reason to explore the possibility of feedsidewards underlying phase response curves along genetically anchored built-in cycles from "then" (a billion years of evolution ago) as well as "now."

In conclusion, there is a need for a transdisciplinary combined archival, physiological, and demographic, always inferential statistical methodology in order to resolve multifactorial interactions underlying "contagions" of the body and of the mind, which later may also be under the nonphotic influence of the sun or the galaxies, at least insofar as their cyclicity is concerned. They may lead to a heightened susceptibility of the masses to demagogues as well as to the demagogues themselves. We face the task of a scientific chronobioethic that should complement chronomics and chronobiology in everybody's service. It may not be coincidence that circadecadal changes in religious motivation [34] and homicide [4] are nearly in antiphase (200° apart). When religiosity reigns from crime one abstains (religio reget, homicidium subsidet).

Appendix 1

HRV measures in healthy men reported for a 24-hour recording span

	25 days to 5 months		3–9 years		10–14 years		15–19 years		20–29 years		40–49 years		50–59 years		60–69 years	
	(n = 19)		(n = 22)		(n = 18)		(n = 14)		(n = 25)		(n = 17)		(n = 21)		(n = 12)	
	Mean	SD	Mean	SD	Mean	SD	Mean	SD	Mean	SD	Mean	SD	Mean	SD	Mean	SD
HR	144.7	11.7	92.3	12.3	80.0	11.4	71.9	10.0	76.2	7.9	79.1	8.4	76.0	7.4	67.3	7.0
NN	417.5	35.5	660.7	86.0	765.3	112.1	848.5	109.1	796.2	89.2	767.3	90.3	796.2	72.4	900.3	93.9
CVRR	19.5	3.3	24.4	2.5	25.6	6.6	25.0	6.3	26.4	4.0	19.6	4.0	18.9	3.8	18.4	5.7
SDNN	52.9	12.4	132.9	33.3	174.9	62.2	190.8	38.3	186.0	33.4	131.8	32.4	134.8	30.8	156.2	42.7
r-MSSD	13.3	4.7	50.3	22.2	57.2	30.0	52.9	23.5	40.3	12.9	25.0	9.0	21.8	5.4	23.7	7.3
NN	203735.9	17866.5	127129.3	21105.7	107271.1	19574.6	98292.6	16525.9	108987.2	13702.1	112754.1	11770.3	107065.1	9676.0	93024.0	13677.8
NN50	872.4	1403.5	21695.1	11354.9	24093.5	14022.6	21903.3	12117.9	15865.6	7614.0	4776.2	5136.7	3361.2	2852.5	3647.6	3966.5
NN50+	754.0	1433.0	10507.7	5836.7	11417.3	6592.3	10720.1	5916.4	7715.2	3604.6	2538.9	2583.5	1646.5	1394.0	717.4	398.7
NN50-	653.7	1274.3	11187.4	5543.4	12676.2	7454.2	11183.1	6212.6	8150.4	4044.3	2820.0	2958.3	1714.7	1484.4	699.0	420.5
pNN50	0.76	1.43	21.01	12.52	25.05	16.59	23.56	14.38	15.26	8.80	5.05	5.49	3.24	2.88	4.35	4.83
pNN50+	0.38	0.76	10.14	6.14	11.84	7.77	11.54	7.03	7.43	4.20	2.38	2.55	1.60	1.41	0.74	0.43
pNN50-	0.34	0.69	10.88	6.38	13.21	8.84	12.01	7.35	7.84	4.64	2.65	2.95	1.65	1.47	0.74	0.54
SDANN5	45.3	10.7	118.2	26.2	154.4	59.2	169.8	39.8	178.5	35.6	120.1	31.9	127.0	31.6	138.3	54.0
SDANN30	38.7	9.2	113.7	25.8	147.8	60.8	161.5	40.8	173.4	36.0	114.1	32.5	121.1	31.3	132.7	55.5
SDmean5	24.7	7.4	65.6	21.1	82.0	29.3	82.8	20.7	71.8	12.9	52.3	12.3	47.4	11.1	42.5	10.7
SDmean30	33.9	9.4	75.7	22.9	98.6	33.6	102.6	26.0	88.7	14.6	67.3	16.2	63.9	16.1	59.8	15.2
N	299.1	32.3	419.4	40.3	409.4	92.9	470.9	142.4	466.0	72.1	505.5	76.8	527.7	63.3	517.4	44.6
X	418.2	40.1	572.9	62.7	681.1	135.4	802.0	175.8	693.0	91.6	729.8	115.9	750.0	88.3	828.9	165.1
M	538.8	68.7	942.0	179.4	1164.3	237.4	1238.1	195.5	1139.1	143.0	1037.8	146.7	1090.9	147.3	1106.6	114.1
TINN	239.8	55.8	522.6	172.8	754.8	282.3	767.0	250.8	673.2	128.0	532.2	134.8	563.1	154.9	588.9	95.5
HRVI	15.0	3.5	32.7	10.8	47.2	17.6	47.9	15.7	42.1	8.0	33.3	8.4	35.2	9.7	36.8	6.0

(continued)

(continued)

	25 days to 5 months		3–9 years		10–14 years		15–19 years		20–29 years		40–49 years		50–59 years		60–69 years	
	(n=19)		(n=22)		(n = 18)		(n=14)		(n =25)		(n=17)		(n=21)		(n=12)	
	Mean	SD	Mean	SD	Mean	SD	Mean	SD	Mean	SD	Mean	SD	Mean	SD	Mean	SD
TI	14.7	3.5	31.4	11.2	45.9	18.1	45.6	17.3	39.4	8.7	31.8	9.0	33.9	10.3	39.2	9.1
90%Len	229.3	51.5	590.9	136.3	779.7	270.1	856.6	160.8	840.8	124.5	604.9	146.6	630.3	142.2	700.0	168.0
90%Wid	30.4	11.4	123.2	62.1	154.9	109.3	128.5	64.3	97.1	34.4	56.9	21.5	48.8	14.3	53.7	19.8
Len/Wid	8.01	1.58	5.50	1.72	6.37	2.56	8.33	4.20	9.47	2.77	11.30	2.94	13.51	3.29	13.91	4.16
Trend	−1.235	0.144	−0.944	0.119	−0.922	0.123	−0.899	0.156	−0.968	0.129	−1.075	0.095	−1.141	0.119	−1.185	0.128
TF	1737.3	928.5	6686.7	3686.6	10933.0	6484.1	11879.8	5637.3	9172.3	2934.7	5874.5	2634.9	5571.9	2517.6	6050.6	2398.9
ULF	1165.4	610.1	2678.5	1331.3	4837.7	2743.2	5724.6	3262.0	4516.7	1515.5	3416.6	1553.2	3536.4	1772.8	3943.0	1637.1
VLF	381.5	194.4	2155.0	1269.8	3860.2	2517.2	4279.5	1797.8	3143.3	1254.7	1701.6	834.9	1486.9	662.7	1660.2	722.8
LF	142.8	115.3	975.7	611.0	1198.4	659.0	1136.4	452.5	1019.7	310.9	606.5	313.9	433.6	194.0	327.8	148.1
HF	33.5	37.5	808.1	641.1	970.3	1079.0	695.3	552.4	469.2	321.3	135.8	94.3	105.4	55.7	110.1	90.0
LF/HF	5.881	1.966	1.493	0.621	2.219	1.790	2.424	1.608	2.774	1.180	5.373	2.408	4.442	1.339	3.759	1.828
ULF/TF	0.679	0.069	0.417	0.080	0.448	0.065	0.473	0.069	0.493	0.071	0.580	0.054	0.626	0.083	0.658	0.065
VLF/TF	0.224	0.052	0.319	0.058	0.351	0.066	0.370	0.077	0.340	0.054	0.289	0.042	0.269	0.050	0.270	0.063
LF/TF	0.076	0.025	0.141	0.020	0.117	0.025	0.100	0.017	0.114	0.029	0.105	0.028	0.083	0.032	0.053	0.009
HF/TF	0.016	0.011	0.112	0.051	0.078	0.044	0.054	0.028	0.050	0.028	0.023	0.010	0.021	0.013	0.018	0.009
Kleitman	641.9	365.3	1163.0	581.1	1776.0	999.9	2182.1	1104.7	1770.2	703.0	1562.1	786.7	1609.3	837.9	1294.8	772.3
ELF	598.8	296.0	1895.2	1121.4	3794.8	2344.7	4460.4	2751.5	3323.9	1258.7	2205.6	1071.1	2238.3	1170.1	2177.4	1077.2
SAS	306.3	164.1	1775.3	990.5	3127.2	2019.1	3361.6	1305.9	2565.9	1050.0	1350.5	687.6	1175.7	528.4	931.5	441.1
MSNA	45.5	36.7	292.9	159.7	366.0	206.4	325.8	110.4	286.6	72.8	183.3	86.3	142.4	54.8	88.9	54.0
Kleitman/TF	0.372	0.074	0.184	0.055	0.168	0.053	0.187	0.049	0.195	0.055	0.267	0.059	0.289	0.069	0.282	0.074
ELF/TF	0.353	0.061	0.288	0.047	0.345	0.033	0.361	0.050	0.361	0.055	0.373	0.062	0.391	0.064	0.458	0.054
SAS/TF	0.178	0.044	0.265	0.049	0.286	0.060	0.295	0.069	0.277	0.046	0.229	0.038	0.214	0.047	0.194	0.045
MSNA/TF	0.024	0.008	0.044	0.005	0.036	0.009	0.029	0.006	0.032	0.006	0.032	0.007	0.027	0.010	0.018	0.004

Glossary: *ELF* means 30-min period, *Kleitman* means 90-min period, *MSNA* means a period showing “muscle sympathetic nerve activity,” *SAS* means a period showing “sleep apnea syndrome”

Appendix 2

HRV measures in healthy men reported during the sleep span at night

	25 days to 5 months		3–9 years		10–14 years		15–19 years		20–29 years		40–49 years		50–59 years		60–69 years	
	(n = 19)		(n = 22)		(n = 18)		(n = 14)		(n = 25)		(n = 17)		(n = 21)		(n = 12)	
	Mean	SD	Mean	SD	Mean	SD	Mean	SD	Mean	SD	Mean	SD	Mean	SD	Mean	SD
HR	133.6	13.2	74.4	9.6	64.3	13.6	56.2	6.1	57.5	5.8	67.0	9.3	63.0	7.0	60.7	5.8
NN	453.5	46.7	818.6	105.5	969.2	185.3	1077.9	101.4	1054.9	114.2	913.9	138.2	963.3	96.9	996.3	86.7
CVRR	13.6	2.5	14.8	3.1	12.4	2.6	11.1	2.0	11.6	2.5	10.4	3.3	9.6	2.4	9.2	2.3
SDNN	41.9	10.9	109.6	27.3	122.4	44.9	124.3	26.7	124.7	23.3	89.5	26.6	91.5	27.9	92.7	31.1
r-MSSD	15.3	6.0	72.6	29.2	80.1	46.0	74.3	34.5	62.2	22.3	29.4	13.3	28.8	8.9	20.8	5.4
NN	52259.4	13196.2	34902.1	5481.3	27106.2	7612.8	21565.4	4843.8	24727.1	5212.3	25318.4	4312.7	23530.8	3349.5	27385.0	2504.0
NN50	543.6	1043.3	12399.4	4545.7	9849.1	4635.6	8534.1	4469.1	8807.9	4322.7	1945.6	2284.4	1792.6	1529.2	657.7	660.7
NN50+	305.3	586.6	6314.0	2456.7	4791.6	2208.1	4109.7	2139.3	4235.0	2052.6	959.1	1046.3	899.4	707.1	339.7	332.8
NN50-	238.3	460.3	6406.2	2203.8	5198.6	2463.8	4424.4	2337.1	4572.9	2316.4	986.5	1242.8	893.2	828.2	318.0	342.7
pNN50	1.07	1.74	39.50	17.12	42.96	23.34	41.39	20.91	36.06	16.73	8.04	9.63	7.67	6.46	2.46	2.49
pNN50+	0.59	0.98	18.94	8.42	19.89	10.28	19.96	10.14	17.28	7.83	3.96	4.48	3.85	3.00	1.27	1.19
pNN50-	0.48	0.78	20.55	8.79	23.07	13.14	21.42	10.83	18.78	9.04	4.06	5.17	3.81	3.50	1.20	1.30
SDANN5	30.2	7.0	64.6	19.5	67.4	24.4	64.9	18.4	73.4	20.2	60.2	24.0	64.8	27.4	74.6	30.5
SDANN30	24.4	5.8	53.6	17.9	49.6	20.0	50.9	18.1	59.4	23.0	57.9	26.4	59.2	28.7	69.0	36.7
SDmean5	25.9	9.2	83.0	23.0	95.1	38.4	96.2	23.8	90.3	19.9	58.5	18.3	56.8	16.1	42.9	12.4
SDmean30	33.4	10.8	93.8	24.3	110.1	43.2	111.4	24.2	105.5	20.7	69.4	19.5	69.3	19.6	59.2	19.9
N	370.4	32.8	579.9	101.8	726.8	151.2	864.7	115.1	826.4	124.5	749.0	144.3	797.6	93.8	827.3	85.3
X	463.0	46.1	835.0	126.6	1003.8	208.3	1088.3	93.9	1084.7	133.6	921.8	155.2	970.6	110.0	1016.3	97.6
M	544.7	71.9	1062.8	155.0	1219.2	269.1	1322.5	161.0	1293.9	125.1	1101.4	178.5	1144.9	130.0	1181.9	143.3
TINN	174.0	63.7	482.8	155.3	492.3	204.0	457.9	167.5	467.4	103.2	352.2	153.2	347.3	99.6	354.4	141.2

(continued)

(continued)

	25 days to 5 months		3–9 years		10–14 years		15–19 years		20–29 years		40–49 years		50–59 years		60–69 years	
	(n = 19)		(n = 22)		(n = 18)		(n = 14)		(n = 25)		(n = 17)		(n = 21)		(n = 12)	
	Mean	SD	Mean	SD	Mean	SD	Mean	SD	Mean	SD	Mean	SD	Mean	SD	Mean	SD
HRVI	10.9	4.0	30.2	9.7	30.8	12.7	28.6	10.5	29.2	6.5	22.0	9.6	21.7	6.2	22.2	8.8
TI	10.3	4.2	29.1	10.1	29.2	12.6	26.7	10.5	27.4	6.5	20.8	9.7	20.4	6.1	20.8	8.6
90%Len	197.6	50.9	499.7	117.8	550.9	206.1	575.0	138.8	585.1	122.8	405.6	125.8	421.8	131.0	412.9	150.2
90%Wid	35.7	15.7	177.0	73.9	213.1	179.2	178.2	85.0	148.4	57.8	66.6	33.2	68.2	24.0	46.9	12.5
Len/Wid	6.07	1.56	3.11	0.88	3.38	1.31	3.99	2.42	4.28	1.11	6.67	2.23	6.53	1.84	9.19	3.48
Trend	−1.123	0.173	−0.880	0.191	−0.765	0.149	−0.624	0.173	−0.715	0.214	−0.782	0.204	−0.819	0.156	−1.006	0.250
TF	1636.7	889.0	9511.0	4309.3	13159.5	8181.6	13080.1	4961.6	11632.7	4308.6	5838.7	3017.7	6021.3	3144.3	4595.4	3471.6
ULF	970.7	466.6	3551.4	1668.1	4965.3	2963.9	4832.6	2022.3	4440.6	1901.4	2465.2	1080.6	2830.7	1725.1	2769.4	2683.9
VLF	430.6	264.9	3095.8	1483.3	5220.5	3438.3	5788.9	2305.9	5082.1	2177.0	2466.8	1698.3	2381.0	1275.9	1528.0	819.9
LF	168.1	142.5	1276.2	686.9	1283.4	804.9	1214.2	615.7	1147.7	435.1	691.4	393.1	613.6	359.8	222.1	131.5
HF	48.1	64.3	1518.8	1013.8	1613.8	1991.0	1190.2	993.7	930.1	737.7	201.4	176.7	187.0	113.3	69.5	30.8
LF/HF	5.833	2.422	1.028	0.493	1.954	2.535	1.871	2.182	1.623	0.811	4.619	2.858	3.804	1.994	3.394	1.629
ULF/TF	0.620	0.091	0.380	0.083	0.383	0.070	0.371	0.076	0.382	0.097	0.439	0.091	0.471	0.104	0.569	0.129
VLF/TF	0.257	0.065	0.327	0.071	0.394	0.083	0.449	0.094	0.437	0.090	0.408	0.084	0.393	0.076	0.358	0.129
LF/TF	0.092	0.031	0.132	0.027	0.108	0.038	0.093	0.027	0.102	0.031	0.117	0.018	0.103	0.040	0.051	0.010
HF/TF	0.023	0.022	0.154	0.068	0.110	0.079	0.083	0.048	0.077	0.043	0.033	0.016	0.032	0.016	0.019	0.012
Kleitman	530.8	242.1	1782.3	956.2	1940.6	1199.9	1863.6	1016.1	1563.5	915.8	1183.9	520.6	1315.6	1041.4	1302.4	1572.6
ELF	516.7	277.6	2276.5	1066.9	3920.8	2659.0	3974.4	1714.9	3660.8	1512.5	1616.1	741.4	1857.3	1002.7	1716.1	1317.4
SAS	353.7	224.6	2588.4	1212.0	4324.5	2765.1	4783.5	1837.6	4298.4	1843.0	2132.1	1532.0	2038.9	1096.2	1278.9	650.9
MSNA	53.4	47.8	407.4	190.1	434.7	293.0	395.0	179.9	376.0	112.5	242.9	137.9	219.3	112.8	83.8	52.8
Kleitman/TF	0.348	0.088	0.188	0.053	0.153	0.061	0.145	0.061	0.135	0.059	0.214	0.055	0.218	0.091	0.250	0.099
ELF/TF	0.320	0.062	0.244	0.052	0.294	0.053	0.302	0.046	0.314	0.059	0.283	0.040	0.308	0.060	0.373	0.057
SAS/TF	0.209	0.056	0.275	0.062	0.330	0.075	0.372	0.078	0.370	0.079	0.350	0.077	0.338	0.073	0.305	0.117
MSNA/TF	0.029	0.010	0.043	0.007	0.037	0.017	0.030	0.009	0.034	0.010	0.041	0.009	0.038	0.014	0.019	0.004

Appendix 3

HRV measures in healthy men reported during the waking span during daytime

	15–19 years		20–29 years		40–49 years		50–59 years		60–69 years	
	(n=14)		(n=25)		(n=17)		(n=21)		(n=12)	
	Mean	SD	Mean	SD	Mean	SD	Mean	SD	Mean	SD
HR	78.1	12.2	84.4	8.7	83.6	9.0	80.9	7.9	77.4	5.6
NN	784.8	114.2	718.9	80.1	726.1	84.0	748.5	70.5	778.8	55.7
CVRR	22.0	5.2	19.7	2.7	17.0	3.9	15.4	2.6	16.7	7.1
SDNN	150.5	40.0	119.8	20.0	105.3	26.9	100.2	22.0	113.1	38.9
r-MSSD	45.2	20.4	30.5	8.5	23.2	8.2	19.2	4.6	18.4	3.1
NN	76727.3	15058.4	84260.1	14284.0	87435.7	8646.5	83534.3	9440.9	71945.3	12458.1
NN50	13369.2	8552.8	7057.7	4181.9	3413.3	3439.3	1568.6	1635.3	758.7	608.4
NN50+	6610.4	4217.3	3480.2	1992.9	1579.8	1632.6	747.1	863.6	377.7	320.1
NN50-	6758.8	4344.6	3577.5	2203.4	1833.5	1834.3	821.5	793.6	381.0	297.8
pNN50	18.7	12.8	9.0	6.4	4.2	4.4	2.0	2.1	1.1	0.8
pNN50+	9.3	6.3	4.5	3.0	1.9	2.0	0.9	1.1	0.5	0.4
pNN50-	9.5	6.5	4.6	3.3	2.3	2.4	1.0	1.0	0.6	0.4
SDANN5	125.5	36.9	103.3	20.1	90.2	24.3	88.0	19.5	103.5	44.1
SDANN30	109.1	37.9	87.6	20.4	78.3	24.5	74.8	18.8	93.6	48.0
SDmean5	77.5	21.6	63.4	11.8	49.9	12.0	43.6	10.7	42.2	11.4
SDmean30	99.1	29.3	81.2	15.2	66.3	17.0	61.6	16.2	59.8	16.2
N	460.6	145.0	465.4	72.5	507.6	63.9	524.1	64.2	521.7	48.9
X	791.7	161.6	691.4	91.3	713.3	119.0	748.9	86.9	768.3	125.8
M	1106.5	182.6	992.6	109.8	959.1	123.6	979.5	120.4	1013.7	102.3
TINN	645.9	240.2	526.9	93.5	451.6	103.7	455.5	124.9	491.7	92.9

(continued)

(continued)

	15–19 years		20–29 years		40–49 years		50–59 years		60–69 years	
	(n=14)		(n=25)		(n=17)		(n=21)		(n=12)	
	Mean	SD	Mean	SD	Mean	SD	Mean	SD	Mean	SD
HRVI	40.4	15.0	32.9	5.8	28.2	6.5	28.5	7.8	30.7	5.8
TI	39.1	15.8	32.0	5.8	27.4	6.5	27.8	8.0	29.4	6.9
90%Len	690.9	174.7	552.8	92.6	471.9	118.6	462.0	106.1	501.0	151.6
90%Wid	109.5	55.3	70.8	21.8	52.6	18.5	43.4	11.1	39.6	5.7
Len/Wid	7.69	3.68	8.33	2.25	9.44	2.31	10.85	1.89	12.78	3.97
Trend	−1.028	0.154	−1.101	0.121	−1.196	0.085	−1.286	0.121	−1.376	0.132
TF	11527.2	6730.9	8064.4	2967.4	5886.0	2775.4	5452.4	2618.0	4765.5	2183.8
ULF	6218.6	4276.9	4570.1	1933.2	3770.7	1805.1	3868.3	2039.9	3364.1	1534.8
VLF	3653.1	1902.8	2267.9	924.7	1418.4	705.6	1136.9	514.3	1082.1	563.6
LF	1110.3	459.7	960.7	318.8	573.8	330.1	363.3	159.0	258.9	148.4
HF	504.1	404.6	246.4	134.2	109.3	74.1	73.9	36.4	51.6	26.1
LF/HF	3.160	1.714	4.490	1.705	6.175	2.429	5.265	1.611	4.939	2.169
ULF/TF	0.529	0.076	0.560	0.071	0.636	0.050	0.696	0.072	0.718	0.059
VLF/TF	0.324	0.071	0.281	0.044	0.242	0.036	0.213	0.041	0.216	0.047
LF/TF	0.103	0.019	0.125	0.038	0.102	0.038	0.073	0.030	0.052	0.018
HF/TF	0.040	0.020	0.031	0.014	0.018	0.008	0.016	0.012	0.012	0.004
Kleitman	2351.4	1516.1	1875.4	1036.0	1694.4	988.7	1760.8	1043.7	1261.0	435.8
ELF	4750.9	3489.5	3179.8	1385.2	2432.5	1240.4	2407.0	1316.7	2432.9	1342.7
SAS	2769.4	1269.7	1782.8	748.6	1062.2	527.4	837.5	374.6	752.3	373.9
MSNA	299.0	106.9	247.1	75.9	161.1	82.8	112.6	45.1	90.5	56.7
Kleitman/TF	0.207	0.055	0.230	0.072	0.284	0.076	0.320	0.081	0.287	0.068
ELF/TF	0.395	0.060	0.390	0.063	0.412	0.081	0.431	0.071	0.495	0.060
SAS/TF	0.252	0.066	0.221	0.038	0.182	0.028	0.158	0.034	0.153	0.034
MSNA/TF	0.028	0.006	0.032	0.007	0.029	0.009	0.023	0.009	0.018	0.007

Appendix 4

HRV measures in elderly men in a community

	75–79 years		80–84 years		85–89 years	
	(n=27)		(n=27)		(n=11)	
	Mean	SD	Mean	SD	Mean	SD
HR	70.7	8.7	71.7	9.4	76.5	13.3
NN	860.9	101.0	849.5	97.2	805.4	135.1
CVRR	6.7	3.2	6.3	1.8	5.9	1.5
SDNN	52.8	22.8	49.4	16.3	44.1	19.9
r-MSSD	22.6	8.6	29.7	24.6	23.2	7.5
NN	3599.9	949.8	3531.4	580.1	4842.6	1842.7
NN50	125.2	194.5	235.6	406.5	138.3	105.5
NN50+	67.0	107.2	111.4	202.2	75.9	66.3
NN50-	58.2	91.3	124.2	205.6	62.4	50.6
pNN50	3.4	4.5	6.4	10.6	2.8	1.9
pNN50+	1.9	2.5	3.0	5.2	1.5	1.0
pNN50-	1.6	2.1	3.4	5.4	1.3	1.0
SDANN5	32.9	20.2	27.4	12.8	26.1	17.5
SDANN30	21.3	14.8	16.6	16.4	14.6	7.2
SDmean5	39.7	15.2	39.1	14.2	33.0	11.7
SDmean30	48.1	19.9	45.9	17.3	39.9	19.2
N	751.8	105.9	743.0	89.7	710.9	112.1
X	857.8	107.5	858.2	100.4	818.4	142.8
M	981.8	122.7	956.1	114.4	900.4	167.5
TINN	230.0	97.8	213.1	70.4	189.3	79.5

(continued)

(continued)

	75–79 years		80–84 years		85–89 years	
	(n = 27)		(n = 27)		(n = 11)	
	Mean	SD	Mean	SD	Mean	SD
HRVI	14.4	6.1	13.3	4.4	11.8	5.0
TI	13.9	6.1	12.8	4.4	11.5	4.9
90%Len	246.4	112.2	222.0	75.4	214.4	93.8
90%Wid	51.1	22.7	80.6	84.8	77.7	113.2
Len/Wid	5.00	1.66	4.32	2.08	4.53	1.84
VLF	1009.6	725.2	806.0	475.7	693.9	695.4
LF	303.0	377.6	234.8	254.9	97.5	73.2
HF	90.1	87.7	161.9	298.4	67.8	39.3
LF/HF	3.781	2.411	2.595	1.903	1.347	0.661

HRV measures in elderly women in a community

	70–74 years		75–79 years		80–84 years		85–89 years		90–97 years	
	(n=9)		(n=45)		(n=39)		(n=17)		(n=10)	
	Mean	SD	Mean	SD	Mean	SD	Mean	SD	Mean	SD
HR	74.6	13.5	76.6	10.3	78.2	12.6	76.0	11.6	79.2	9.9
NN	827.0	143.7	797.4	108.6	784.4	114.1	805.9	119.6	769.1	99.0
CVRR	6.9	2.3	7.0	2.3	6.1	2.1	6.6	2.4	5.8	3.0
SDNN	50.5	15.7	50.7	22.0	41.7	13.4	45.9	11.6	39.5	23.6
r-MSSD	19.8	7.0	22.5	13.0	21.8	10.4	20.7	7.4	26.6	18.1
NN	4454.6	1651.5	3952.6	918.4	4209.2	1328.2	4458.9	1283.2	4642.2	1445.3
NN50	95.0	146.0	170.8	349.4	158.6	264.7	124.7	196.9	329.3	447.0
NN50+	44.7	67.4	88.6	177.5	82.6	138.6	62.5	102.4	163.5	214.6
NN50-	50.3	78.6	82.2	173.8	76.0	128.4	62.2	94.8	165.8	236.8
pNN50	2.4	3.8	4.2	7.8	4.2	7.0	2.6	3.5	6.2	8.4
pNN50+	1.1	1.8	2.2	4.0	2.2	3.7	1.3	1.8	3.1	4.1
pNN50-	1.3	2.1	2.0	3.9	2.0	3.4	1.3	1.7	3.1	4.5
SDANN5	26.1	10.1	28.5	16.0	23.9	11.2	32.1	10.0	24.6	21.2
SDANN30	15.0	17.7	18.0	18.5	11.2	8.2	14.0	10.1	27.5	36.3
SDmean5	41.5	12.6	39.4	16.0	33.2	10.7	30.0	9.5	29.2	12.7
SDmean30	45.8	15.8	47.7	20.8	39.2	12.5	40.5	12.4	34.2	16.0
N	722.7	131.2	694.3	89.0	692.7	106.0	727.7	126.9	689.2	95.8
X	834.4	145.7	802.8	119.3	787.3	120.0	819.2	117.6	783.6	111.2
M	939.9	165.6	911.5	137.6	880.2	126.5	894.0	123.4	848.7	117.1
TINN	217.2	72.6	217.2	81.0	187.5	59.4	166.4	46.1	159.7	78.1
HRVI	13.6	4.5	13.6	5.1	11.7	3.7	10.4	2.9	10.0	4.9
TI	13.1	4.5	13.1	4.9	11.4	3.7	9.7	2.9	9.5	4.7

(continued)

(continued)

	70–74 years		75–79 years		80–84 years		85–89 years		90–97 years	
	(n=9)		(n=45)		(n=39)		(n=17)		(n=10)	
	Mean	SD	Mean	SD	Mean	SD	Mean	SD	Mean	SD
90%Len	237.0	75.2	234.7	108.4	188.4	63.4	213.6	60.7	171.4	99.3
90%Wid	42.7	17.5	52.5	36.5	50.5	30.8	42.6	18.0	64.5	49.5
Len/Wid	5.92	2.03	5.22	2.21	4.44	1.71	5.60	2.25	3.11	1.21
VLF	1303.4	706.8	985.9	779.2	668.8	447.9	504.2	370.9	435.1	425.7
LF	295.4	302.3	205.7	225.8	163.3	165.1	86.2	90.2	89.0	110.6
HF	64.7	58.0	93.7	170.6	85.6	98.3	60.5	55.1	92.9	97.6
LF/HF	4.100	1.249	3.067	1.798	2.573	1.659	1.492	0.671	1.486	1.077

Appendix 5

Circadian variation of two scaling components calculated by DFA: comparison between healthy young and elderly

	α1					α2				
	Young	(n = 10)	Elderly	(n = 11)		Young	(n = 10)	Elderly	(n = 11)	
Clock hour	Mean	SD	Mean	SD	*t*-test	Mean	SD	Mean	SD	*t*-test
0	0.978	0.160	1.333	0.252	N.S.	0.860	0.153	0.974	0.131	N.S.
1	1.044	0.344	1.243	0.364	N.S.	0.913	0.173	0.931	0.186	N.S.
2	1.140	0.181	1.231	0.351	N.S.	1.012	0.152	0.932	0.200	N.S.
3	1.101	0.269	1.176	0.313	N.S.	0.941	0.141	0.969	0.114	N.S.
4	1.120	0.290	1.345	0.282	<0.05	0.908	0.094	0.911	0.095	N.S.
5	0.998	0.319	1.452	0.210	N.S.	0.867	0.199	1.043	0.190	N.S.
6	1.202	0.232	1.352	0.262	<0.01	0.927	0.114	1.174	0.143	<0.001
7	1.185	0.247	1.393	0.153	<0.01	0.915	0.082	1.059	0.165	<0.05
8	1.343	0.147	1.212	0.274	<0.001	1.000	0.087	1.134	0.184	N.S.
9	1.322	0.240	1.187	0.383	<0.005	1.023	0.126	1.058	0.205	N.S.
10	1.291	0.198	1.216	0.362	<0.005	1.026	0.081	1.118	0.207	N.S.
11	1.275	0.159	1.255	0.310	<0.005	0.981	0.181	1.089	0.125	N.S.
12	1.367	0.118	1.270	0.332	<0.001	0.920	0.102	1.128	0.119	<0.001
13	1.337	0.123	1.213	0.291	<0.001	1.047	0.142	1.101	0.172	N.S.
14	1.313	0.201	1.183	0.300	<0.001	0.981	0.110	1.143	0.191	<0.05
15	1.352	0.110	1.268	0.201	<0.001	1.019	0.143	1.074	0.192	N.S.
16	1.383	0.140	1.153	0.282	<0.001	1.034	0.152	1.123	0.142	N.S.
17	1.369	0.087	1.276	0.248	<0.001	0.999	0.121	1.139	0.167	<0.05
18	1.407	0.138	1.298	0.182	<0.001	1.042	0.085	1.096	0.183	N.S.

(continued)

(continued)

	α1					α2				
	Young	(n = 10)	Elderly	(n = 11)		Young	(n = 10)	Elderly	(n = 11)	
Clock hour	Mean	SD	Mean	SD	*t*-test	Mean	SD	Mean	SD	*t*-test
19	1.402	0.091	1.269	0.240	<0.001	0.953	0.149	1.095	0.127	<0.05
20	1.356	0.096	1.302	0.221	<0.001	1.039	0.128	1.102	0.138	N.S.
21	1.345	0.110	1.331	0.178	<0.001	0.975	0.142	1.104	0.208	N.S.
22	1.343	0.112	1.157	0.327	<0.001	1.005	0.162	0.970	0.170	N.S.
23	1.192	0.254	1.099	0.329	<0.05	0.991	0.122	0.927	0.181	N.S.
24 hours	1.257	0.112	1.259	0.179	N.S.	0.974	0.039	1.058	0.062	<0.01
Daytime	1.350	0.104	1.240	0.209	N.S.	1.000	0.054	1.111	0.073	<0.001
Night	1.063	0.227	1.296	0.235	<0.05	0.917	0.065	0.960	0.076	N.S.
t-test	<0.005		N.S.			<0.05		<0.005		

References

1. Cornélissen G, Halberg F, Schwartzkopff O, Delmore P, Katinas G, Hunter D, Tarquini B, Tarquini R, Perfetto F, Watanabe Y, Otsuka K. Chronomes, time structures, for chronobioengineering for "a full life". Biomed Instrum Technol 1999;33:152–87.
2. Cornélissen G, Breus TK, Bingham C, Zaslavskaya R, Varshitsky M, Mirsky B, Teibloom M, Tarquini B, Bakken E, Halberg F, International Womb-to-Tomb Chronome Initiative Group. Beyond circadian chronorisk: worldwide circaseptan-circasemiseptan patterns of myocardial infarctions, other vascular events, and emergencies. Chronobiologia 1993;20:87–115.
3. Cornélissen G, Halberg F. Introduction to chronobiology. Medtronic chronobiology seminar #7, Apr 1994, p. 52 (Library of Congress Catalog Card #94-060580). https://sites.google.com/a/umn.edu/halbergchronobiologycenter/
4. Halberg F, Cornélissen G, Katinas G, Hillman D, Sampson M, Schwartzkopff O. Season's appreciations 2000. Chronomics complement, among many other fields, genomics and proteomics. Neuroendocrinol Lett 2001;22:53–73.
5. Halberg F, Cornélissen G, Otsuka K, Schwartzkopff O, Halberg J, Bakken EE. Chronomics. Biomed Pharmacother 2001;55 Suppl 1:153–90.
6. Halberg F, Haus E, Cornélissen G. From biologic rhythms to chronomes relevant for nutrition. In: Marriott BM, editor. Not eating enough: overcoming underconsumption of military operational rations. Washington, DC: National Academy Press; 1995. p. 361–72.
7. Otsuka K, Cornélissen G, Zhao ZY, Weydahl A, Delyukov A, Gorgo Y, Wang ZR, Perfetto F, Tarquini R, Kubo Y, Shinagawa M, Hotta N, Ishii T, Omori K, Watanabe Y, Nunoda S-i, Ohkawa S-i, Halberg F. Rhythm and trend elements in the time structure, chronome, of heart rate variability. Gerontol Geriatr 1999;2:31–48.
8. Bigger JT Jr, Steinman RC, Rolnitzky LM, Fleiss JL, Albrecht P, Cohen RJ. Power law behavior of RR-interval variability in healthy middle-aged persons, patients with recent acute myocardial infarction, and patients with heart transplants. Circulation 1996;93:2142–51.
9. Huikuri HV, Makikallio TH, Peng CK, Goldberger AL, Hintze U, Moller M for the DIAMOND Study Group. Fractal correlation properties of R-R interval dynamics and mortality in patients with depressed left ventricular function after an acute myocardial infarction. Circulation 2000;101:47–53.
10. Makikallio TH, Huikuri HV, Makikallio A, Sourander LB, Mitrani RD, Castellanos A, Myerburg RJ. Prediction of sudden cardiac death by fractal analysis of heart rate variability in elderly subjects. J Am Coll Cardiol 2001;37:1395–402.
11. Oinuma S, Kubo Y, Otsuka K, Yamanaka T, Murakami S, Matsuoka O, Ohkawa S, Cornélissen G, Weydahl A, Holmeslet B, Hall C, Halberg F. Graded response of heart rate variability, associated with an alteration of geomagnetic activity in a subarctic area. Biomed Pharmacother 2002;56:284s–8.
12. Otsuka K, Cornélissen G, Halberg F. Circadian rhythmic fractal scaling of heart rate variability in health and coronary artery disease. Clin Cardiol 1997;20:631–8.
13. Herold M, Cornélissen G, Loeckinger A, Koeberle D, Koenig P, Halberg F. About 8-hourly variation of circulating human endothelin-1 (ET-1) in clinical health. Peptides 1998;19:821–5.
14. Simpson HW, Bartter FC, Halberg F, Gill JR, Delea C, Gullner H-G, Haus E. A circaoctohoran rhythm of urine epinephrine excretion in man. Prog Clin Biol Res 1990;341A:157–60.
15. Katinas GS, Halberg F, Cornélissen G, Hawkins D, Bueva MV, Korzhevsky DE, Sapozhnikova LR, Rhodus N, Schaffer E. About 8- and ~84-hour rhythms in endotheliocytes as in endothelin-1 and effect of trauma. Peptides 2001;22:647–59.
16. Li T-Y, Yorke JA. Period three implies chaos. Am Math Mon 1975;82:985–92.
17. de Boer RW, Karemaker JM, Strackee J. Hemodynamic fluctuations and baroreflex sensitivity in humans: a beat-to-beat model. Am J Physiol 1987;253(3 Pt 2):H680–9.

18. Nicolau GY, Haus E, Popescu M, Sackett-Lundeen L, Petrescu E. Circadian, weekly and seasonal variations in cardiac mortality, blood pressure and catecholamine excretion. Chronobiol Int 1991;8:149–59.
19. Smolensky M, Halberg F, Sargent FII. Chronobiology of the life sequence. In: Itoh S, Ogata K, Yoshimura H, editors. Advances in climatic physiology. Tokyo: Igaku Shoin Ltd; 1972. p. 281–318.
20. Reinberg A, Gervais P, Halberg F, Gaultier M, Roynette N, Abulker C, Dupont J. Mortalité des adultes: Rythmes circadiens et circannuels dans un hôpital parisien et en France. Nouv Presse Med 1973;2:289–94.
21. Otsuka K, Oinuma S, Cornélissen G, Weydahl A, Ichimaru Y, Kobayashi M, Yano S, Holmeslet B, Hansen TL, Mitsutake G, Engebretson MJ, Schwartzkopff O, Halberg F. Alternating-light-darkness-influenced human electrocardiographic magnetoreception in association with geomagnetic pulsations. Biomed Pharmacother 2001;55 Suppl 1:63–75.
22. Cornélissen G, Schwartzkopff O, Halberg F, Otsuka K, Watanabe Y. 7-day ambulatory monitoring for adults with hypertension and diabetes [letter]. Am J Kidney Dis 2001;37:878.
23. Halberg F, Cornélissen G, Halberg J, Fink H, Chen C-H, Otsuka K, Watanabe Y, Kumagai Y, Syutkina EV, Kawasaki T, Uezono K, Zhao ZY, Schwartzkopff O. Circadian hyper-amplitude-tension, CHAT: a disease risk syndrome of anti-aging medicine. J Anti Aging Med 1998;1:239–59.
24. Halberg F, Cornélissen G, Halpin C, Burchell H, Watanabe Y, Kumagai Y, Otsuka K, Zaslavskaya R. Fleeting "monitor-", "conflict-", or "grief-associated" blood pressure disorders: MESOR-hypertension and circadian hyper-amplitude-tension (CHAT). Euro Rehab 1996;6:225–40.
25. Halberg F, Cornélissen G, Otsuka K, Schwartzkopff O. Chronodiagnosis-based chronotherapy: is blood pressure monitoring for weeks too costly to prevent long-term care for years? Gerontol Geriatr 1999;2:20–30.
26. Halberg F, Cornélissen G, Otsuka K, Watanabe Y, Woods MA, Lambert CR, Zaslavskaya R, Gubin D, Petukhova EY, Delmore P, Bakken E. Rewards in practice from chrono-meta-analyses "recycling" heart rate, ectopy, ischemia and blood pressure information. J Med Eng Technol 1997;21:174–84.
27. Halberg F, Cornélissen G, Wall D, Otsuka K, Halberg J, Watanabe Y, Siegelova J, Homolka P, Fiser B, Dusek J, Sánchez de la Peña S, Maggioni C, Delyukov A, Gorgo Y, Gubin D, Carandente F, Schaffer E, Borer K, Sonkowsky RP, Schwartzkopff O. Engineering and governmental challenge: 7-day/24-hour chronobiologic blood pressure and heart rate screening: part I. Biomed Instrum Technol 2002;36:89–122.
28. Otsuka K, editor. Chronome & Janus-medicine: heart rate variability (HRV) and BP variability (BPV) from a viewpoint of chronobiology and ecology. Tokyo: Medical Review; 1998. 213 pp.
29. Otsuka K, Cornélissen G, Halberg F, Oehlerts G. Excessive circadian amplitude of blood pressure increases risk of ischaemic stroke and nephropathy. J Med Eng Technol 1997;21:23–30.
30. Otsuka K, Cornélissen G, Halberg F. Predictive value of blood pressure dipping and swinging with regard to vascular disease risk. Clin Drug Invest 1996;11:20–31.
31. Otsuka K, Cornélissen G, Halberg F. Circadian rhythms and clinical chronobiology. Biomed Pharmacother 2001;55:7–18.
32. Otsuka K, Cornélissen G, Shinagawa M, Halberg F. Blood pressure variability assessed by semiautomatic and ambulatorily functional devices for home use. Clin Exp Hypertens 1999;21:729–40.
33. Otsuka K, Murakami K, Kubo Y, Yamanaka T, Mitsutake G, Ohkawa S, Matsubayashi K, Yano S, Cornelissen G, Halberg F. Chronomics for chronoastrobiology with immediate spin-offs for life quality and longevity. Biomed Pharmacother 2003;57 Suppl 1:1s–18s.
34. Starbuck S, Cornélissen G, Halberg F. Is motivation influenced by geomagnetic activity? Biomed Pharmacother 2002;56 Suppl 2:289–97.

Chapter 6
Chronomics of Life for Improving the Quality of Health

Abstract Chronomics maps cycles in us and around us. Environmental cycles include the lighting and feeding schedules as well as other non-photic phenomena that are part of space weather. Chronomics can solve the complexity of organisms' interactions with the environment near and far broadly.

For any component in biological spectra, one looks for, and often finds, natural (as well as anthropogenic) physical environmental counterparts. If coherences are found, e.g., by cross-spectra reflecting external and internal interactions, they are best taken into account, as far as possible, on the basis of chronome maps. This was done for the case of a built-in free-running about-weekly (circaseptan), but not exactly 7-day component in spectra of human urinary 17-ketosteroid excretion (17-KS). Another, again not exactly 7-day, natural, slightly shorter than 7-day wobbly average 6.75-day counterpart, stands out clearly in spectra of the planetary geomagnetic disturbance index, Kp.

With this qualification, chronome mapping with outcomes could eventually serve an individualized optimization of lifestyle, for chronoprevention and chronotherapy as well as for inquiries into the evolution and future of life.

Keywords Chronomics • Meal timing • Chronoprevention • Chronotherapy • Chronobioengineering • Disease-risk syndromes

6.1 Introduction

Several international meetings have revealed an accumulating body of reference values for well-established about-daily and about-yearly rhythms of photic origin and evidence also for about-7-day, about-27-day, about-half-yearly, about-10.5- and 21-yearly, and even about-50-yearly rhythmicities in us as well as around us, as invisible nonphotic heliogeophysical signatures possibly built into individuals and/or populations, complementing the biological year and day. In time series, biological or others, which are dense and long enough, the characteristics of rhythms, chaos (deterministic and other), and trends can all three be quantified as elements of structures, called chronomes (Table 6.1). Chronobiologic methodology assesses uncertainties in comparisons of endpoints in all elements of chronomes, before and

K. Otsuka et al., *Chronomics and Continuous Ambulatory Blood Pressure Monitoring*, DOI 10.1007/978-4-431-54631-3_6

Table 6.1 Minnesotan contributions to *chronomics*, resolving transdisciplinary geo-bio-noospheric coperiodisms[a]

1	A set of methods and programs resolving, with *uncertainties, coexisting multiple, sometimes spectrally neighboring periods, with their nonoverlapping CIs (95 % confidence intervals), revealing and mapping nonstationarity in time and space* (including the transient disappearance of an aeolian nonphotic component in certain spectral or geographic locations)
2	*Congruences* defined by overlapping CIs of period and/or phase, e.g., within the organism (division of labor in time), within the environment (e.g., Kp vs. Wolf numbers, geo- vs. heliomagnetism) or external-internal congruences (e.g., blood pressure and geomagnetism) at a given period
3	*New internal-external congruent CIs of periods define coperiodisms*, some found first in living matter and next in the abiotic environment, others vice versa, constituting a transdisciplinary spectrum with component periods of *a near week, an about (~)-2-week, ~1-month, ~5-month (quinmensal), semiannual, circannual, near- and far-transannual, diennian, septennian, decadal, ~15–17-year, paradecadal, paradidecadal, paratridecadal (BEL, Bruckner-Egeson-Lockyer, near 30–40 years), ~50-year (semicentennian), transsemicentennian, semimillennian, and myriadennian components*
4	Partial endogenicity, *genetic coding* supported by periods with nonoverlapping CIs in and around us (free-run) and by "after single stimulus" manifestation ("induction")
5	Brain (pineal and suprachiasmatic nucleus, SCN) (mediating) or heart, circulation and cell (reflecting) nonphotic environmental effects (shown by SCN ablation or magnetic storm effects)
6	Nonphotic spectral components in populations of *eukaryotes and prokaryotes* (decadals in bacterial mutations)
7	*Selective assortment (SA)* of periods (or phases) among and within individuals, in organ systems, variables, cycle characteristics (MESOR vs. amplitude), and in their lock-ins (of phase, e.g., of 17-ketosteroids with Kp and urine volume not with Kp but possibly with Wolf numbers)
8	*SA* of periods in different variables of the same population or of phases at a fixed period in different populations
9	Nonphotics can tip the scale between little or none and many *infections*, *sudden cardiac death*, *suicide*, and *terrorism*
10	Odds ratios for the number of shared frequencies between human mental functions and either helio- or geomagnetism more than match the association of helio- and geomagnetism to each other

Biomed Pharmacother 2004; 58(Suppl 1):S150–S187

[a]Rules learned from decades-long longitudinal studies of populations and of individuals analyzed globally and locally, *glocally in space and time,* as a method as well as a map, updated *progressively* at intervals determined by the cycles detected *and pergressively* surveilled in serial sections and their sequential analysis in repeated passes over the accumulating data for resolution of changes in characteristics of each cycle in individuals, eventually from womb to tomb and in populations beyond the human lifespan

Choices in dealing with respect to personal, societal (human-made), and natural disasters

1. We currently try, but may not be able, to respond to each cataclysm after it occurs, whether it is a massive stroke, a crime, a terrorist attack, or an earthquake

2. We can assess currently conventional risk factors and try to reduce their impact according to visits with a care provider, but fly blind with respect to vascular and other variability disorders that cannot be detected by a single or 24-hour spotcheck, or *alternatively*

3. The individual and society can survey the time structure of personal health by, first, vascular monitoring and society can implement a system using the physiological data aligned with those from epidemiological and social monitoring, with the tools of a temporal microscope (chronobiology) and telescope (chronomics), to assess one's position by comparing a personalized

Table 6.1 (continued)

with a corresponding gender-age-ethnicity-matched chronosphere providing reference standards as glocal maps via an international website
Of particular interest are *far-transyears*, mostly between 1.2 and 1.9 years, which drift in frequency or phase and wax and wane in amplitude to the point of disappearance and reappearance. So do *near-transyears*, longer than 1 year but shorter than 1.2 years, found *after their detection in biology* in solar and terrestrial *magnetism* and other counterparts, just as we found the *geomagnetic* near week in Kp, validated by Roederer and extended to aa by Vladimirsky as a counterpart to a set of biological near weeks, most of which differ from precisely 1 week and may hence be partly built-in

after: (1) changes in lifestyle, such as meal quality, quantity, timing, and salting, (2) preventive non-drug interventions for vascular disease risk lowering, or (3) drug treatments of high risk as well as of actual vascular disease, all on a practicable, individualized, as well as population basis. A collateral hierarchy characterizes molecular to psychosocial aspects of living things that are open to their socio-ecologic environs and thus are synchronizable and/or otherwise manipulable by society, meals, lighting, heating, and nonphotic, nonthermic environmental variations that become predictable to the extent that they appear to be cycles, yet adhere only to a statistical, rather than a deterministic causality. With this qualification, chronome mapping with outcomes could eventually serve an individualized optimization of lifestyle, for chronoprevention and chronotherapy as well as for inquiries into the evolution and future of life.

6.2 Chaos, Trends, and/or Rhythms Constituting Structures in Time

The popular press, and often the journals *Science* [1–3] and *Nature* [4], look upon chronobiology as dealing with biological clocks and calendars, eventually to depict when to eat [2, 5; cf. [6–8], when to treat [2, 6, 8, 9], and when best to perform in school, at work, on hobbies, or in sports. Needlessly, the question of "when" is often raised (restricted) as to clock hours or calendar dates. This is an important intermediate step shown in Fig. 6.1a from the concept of a putative balance or homeostasis, on the left, to the middle of this figure, to Johnson's "exceptionally substantial and durable self-winding and self-regulating physiological clock" [10; cf. 11]. Since there are clocks galore [1], we best take the often-indispensable next step from the middle to the right of Fig. 6.1a. Attractive simple models, middle, cannot account for more complex circumstances in an organism's time structures that contain more built-in rhythms than the circadian and contain further chaos and trends in us, leading to possibly genetically anchored chronomes. Moreover, the rhythms of internal chronomes intermodulate with a host of cycles in the environment, while environments also contain interacting chaos and trends [12]. In this context, the complex right portion of Fig. 6.1a calls to mind Einstein's adage that everything must be made as simple as possible, but not simpler. Concepts such as clocks, calendars, or oscillators have served as useful scaffolds, but inferential statistical methods applied

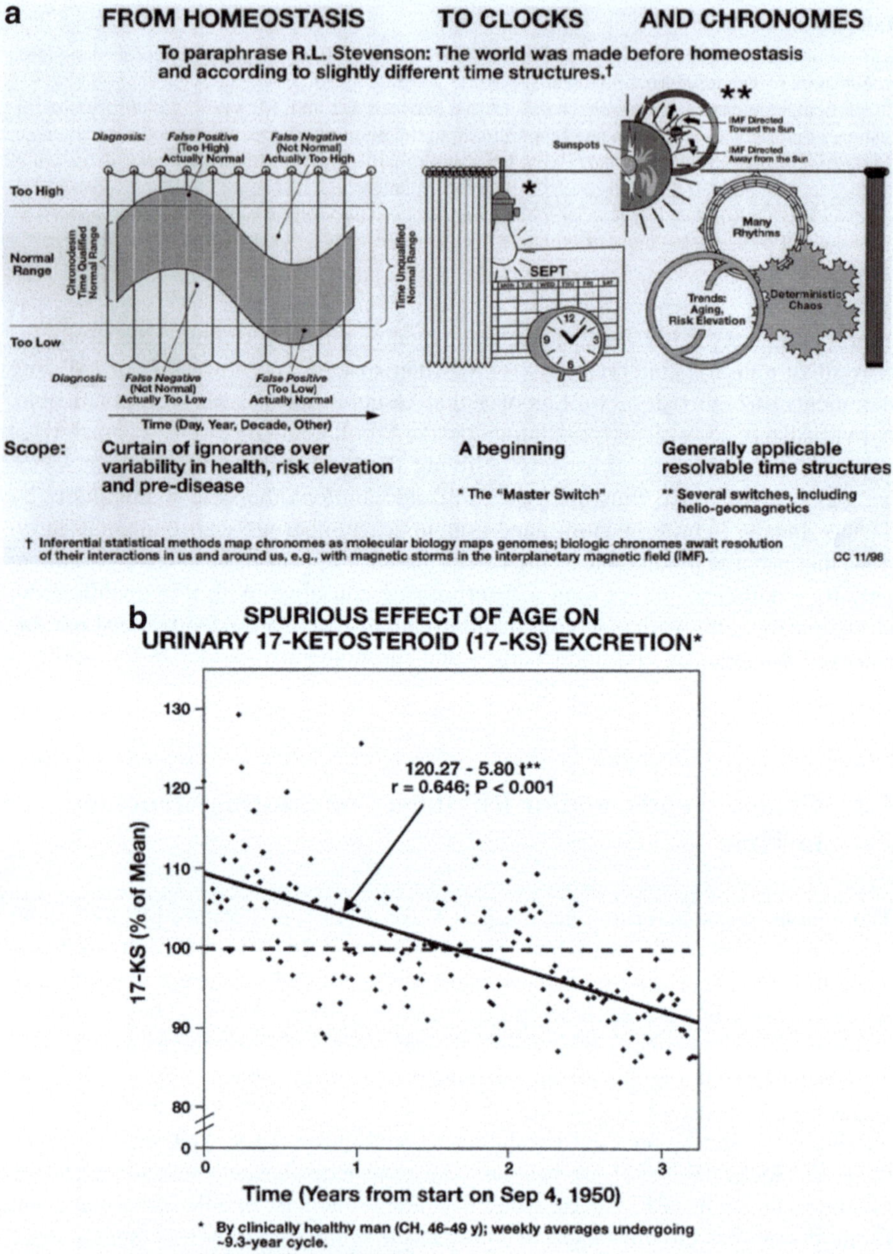

Fig. 6.1 (**a**) *We turn from a master clock serving regulation for constancy to an integrative internal-external collateral hierarchy for physiological coordination.* Homeostasis postulates that physiological processes remain largely within certain ranges in health and seeks departure from such "normal values" to diagnose overt disease. Thereby, variability within the normal range, however, is often dealt with as if it were narrow, random, or trivial, the body striving for at least a relative "constancy." The alternative, rather than complement, to this status quo, on the *left* of this figure,

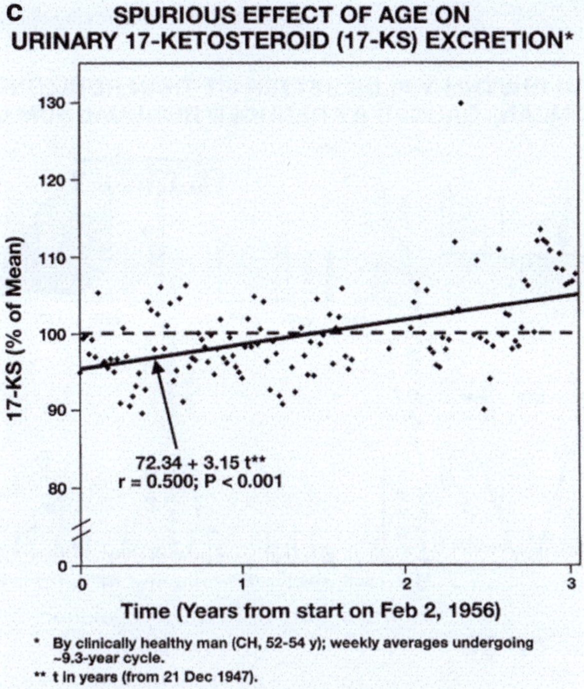

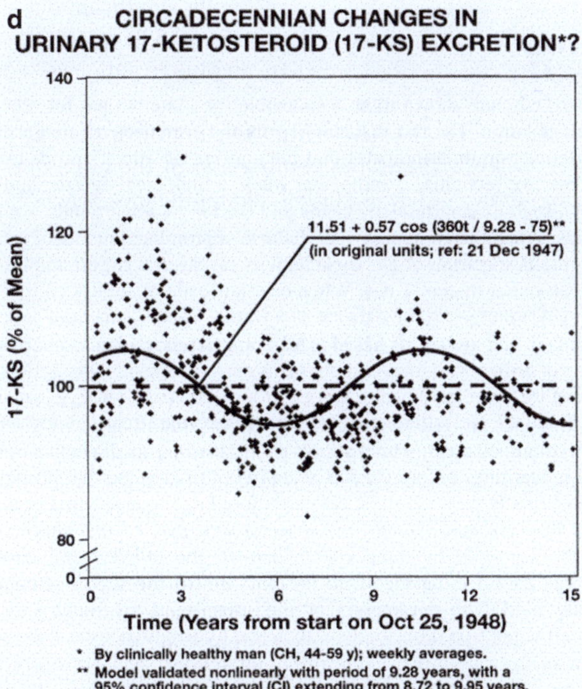

Fig. 6.1 (continued) is learning about the rules of rhythmic and chaotic variations that take place within the "usual value" ranges that led to the postulation of a "biological clock" that would enable the body to keep track of time. By removal and replacement experiments, it was located first in the

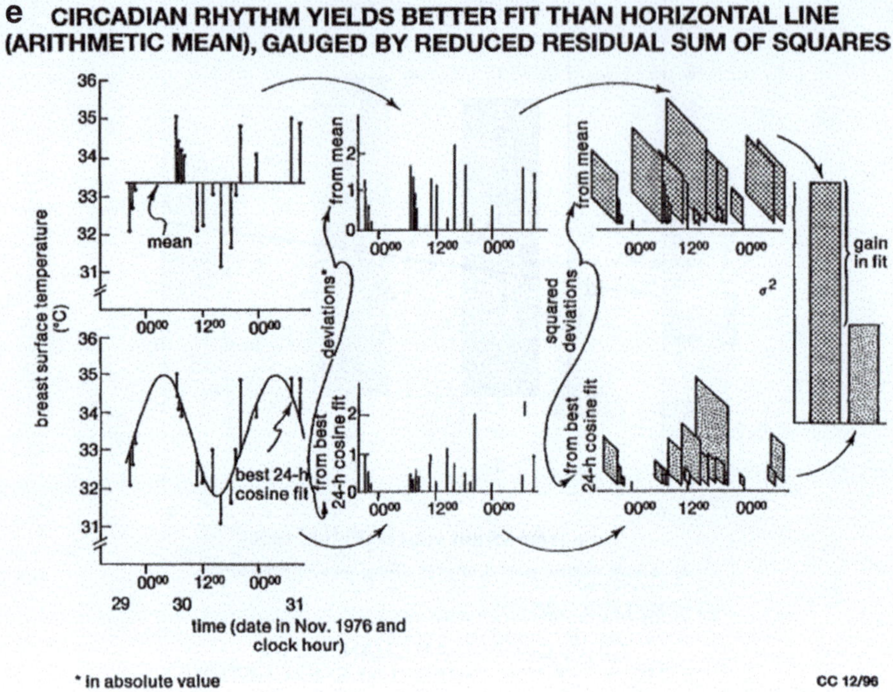

Fig. 6.1 (continued) adrenal, shown to be responsible for some but not for other circadians that persisted after brain ablation. The fact that single cells and even bacteria are genetically coded for a spectrum of rhythmic variation indicates that the concept of "clock" needs extension. Beyond clocks and calendars, we recognize a biological week, a biological decade, and also other rules found in variability, such as deterministic chaos and the long-known trends, some of which may turn out to be cycles. When the giant alga *Acetabularia*, a prominent model of a "clock," is placed into continuous light, its spectrum of electrical activity reveals the largest amplitude for a component of about 1 week rather than of 1 day. When over a decade of studies on this alga are pooled, an about-10-year cycle emerges in the data set as a whole. The alignment of spectral components and chaos and trends in and around us has also begun. Long-term longitudinal, but not yet entire, lifetime monitoring of critical variables complements current linked cross-sectional (hybrid) reference values required for preventive health and environmental care. Changes occurring within the usual value range as longer, and still longer cycles are resolvable as chronomes, with a (predictable multifrequency) rhythmic element, which allows us to measure the dynamics of everyday life, in order to obtain, e.g., warnings before the *fait accompli* of disease, so that prophylactic measures can be instituted in a timely way and to detect heretofore largely undetected or unquantified environmental effects. The abstract sketch of the sector structure of the interplanetary magnetic field (shown on the *top right* by three visible *arrows*, the fourth being covered by another circle showing solar flares) is an idealized presentation of the sector structure in the interplanetary magnetic field. The parameters of the solar wind are much more variable than originally visualized when first described, as sketched by irregular solar flares. Associations of helio- and geomagnetic variability, myocardial infarctions, and strokes are accumulating and are just the tip of the iceberg, with a profound effect of magnetism (recognized by Gilbert in 1600), apparent in the human ECG, notably in the auroral region. In external-internal interactions, a broad spectrum of rhythms (in both the environment and in living matter) organizes deterministic and other chaos and trends. Trends pursued long enough may become low-frequency cycles, e.g., for the detection of any risk elevation and for timely action

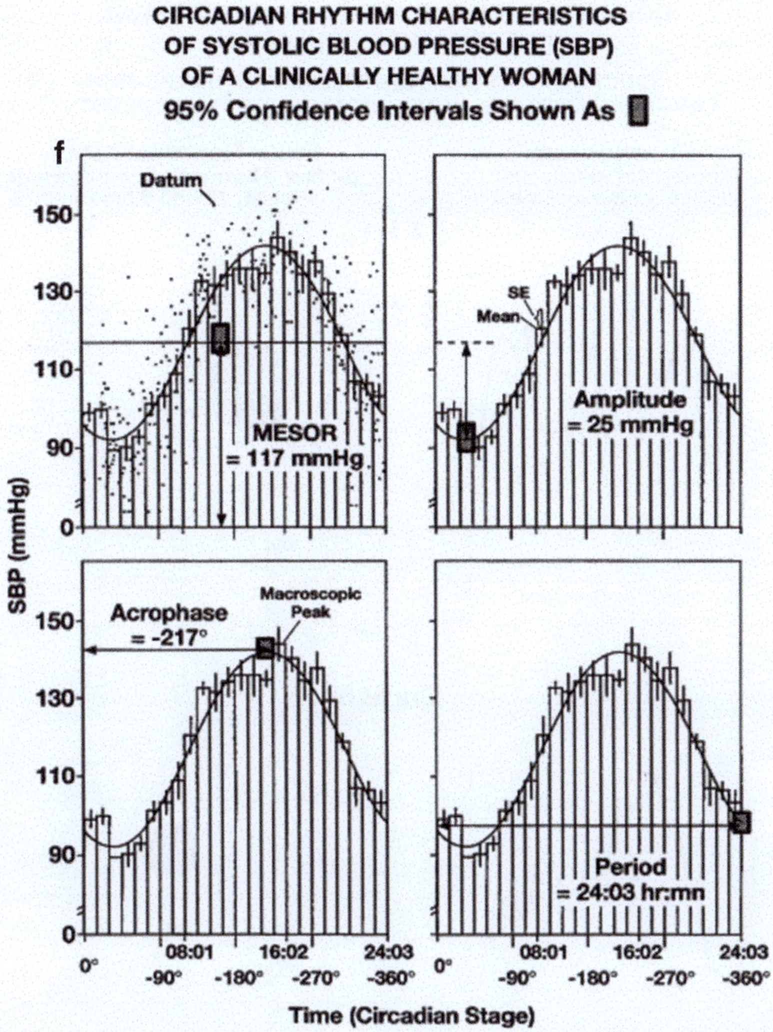

Fig. 6.1 (continued) (© Halberg Chronobiology Center). (**b**) In the absence of systematic chronome mapping, with seemingly ample documentation, a decrease over several years in metabolites related to the adrenal and the male gonad (in the excretion of 17-ketosteroids) may be described as a sign of aging, notably of the declining sex gland activity, and may be published as such in any professional journal (but see Fig. 6.1**c**) [20, 21] (© Halberg Chronobiology Center). (**c**) A surprising increase in steroidal metabolites (gonadal activity?) of a man in his 50s constitutes a statistically but not biologically significant effect which may encourage speculation, perhaps by *post hoc ergo propter hoc* reasoning about how the increase in 17-ketosteroid excretion (rejuvenation?) came about. The authors were tempted to publish Figs. 6.1b and c separately in different professional journals to show that referees are unlikely to question trends and to validate the need for longitudinally and concomitantly mapping all three elements (chaos, rhythms, and trends) of chronomes instead of spotchecks of aging irrespective of rhythms [20, 21] (© Halberg Chronobiology Center). (**d**) By unmasking an about-9-year rhythm, this figure resolves any controversy that may have arisen from the separate viewing of Figs. 6.1b and c. To the extent that in the future such

g

COEXISTING 8-HOUR AND 24-HOUR PATTERNS IN THE SAME CIRCULATION

CIRCAOCTOHORAN ENDOTHELIN-1 (ET-1) VERSUS CIRCADIAN CORTISOL IN 7 CLINICALLY HEALTHY STUDENTS*

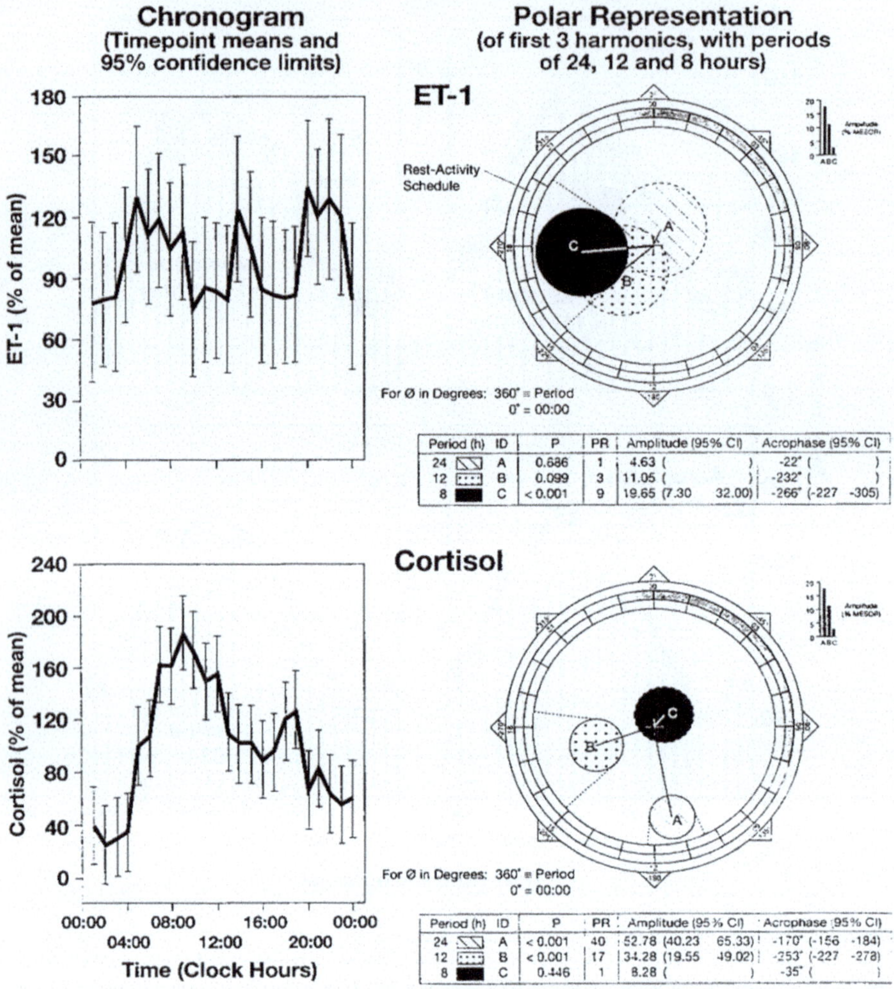

ET-1

Period (h)	ID	P	PR	Amplitude (95% CI)	Acrophase (95% CI)
24	A	0.686	1	4.63 ( )	-22° ( )
12	B	0.099	3	11.05 ( )	-232° ( )
8	C	< 0.001	9	19.65 (7.30 32.00)	-266° (-227 -305)

Cortisol

Period (h)	ID	P	PR	Amplitude (95% CI)	Acrophase (95% CI)
24	A	< 0.001	40	52.78 (40.23 65.33)	-170° (-156 -184)
12	B	< 0.001	17	34.28 (19.55 49.02)	-253° (-227 -278)
8	C	0.446	1	8.28 ( )	-35° ( )

*** 2 women and 7 men, 22-27 years of age; note that 8-hour component (C) is most significant for ET-1 (P < 0.001) but it is not detected for Cortisol (P > 0.4).**

Fig. 6.1 (continued) rhythmicities can be reproduced repeatedly on each of several different persons, the task still ahead of us, and to the extent that such cycles reveal a nonoverlap of the 95 % confidence intervals with those of cycles in solar activity or in geomagnetic disturbance indices, one may postulate that such changes, like the circadians or the about-7-day (circaseptan) rhythms, are built into us by an evolutionary integration of life into its environmental invisible nonphotic chronome in the velocity changes of the solar wind. Corpuscular effects, e.g., from the sun, complement the effect of visible light and readily recognized temperature and can be ascertained beyond similar spectral signatures and cross-spectral coherence by remove-and-replace approaches,

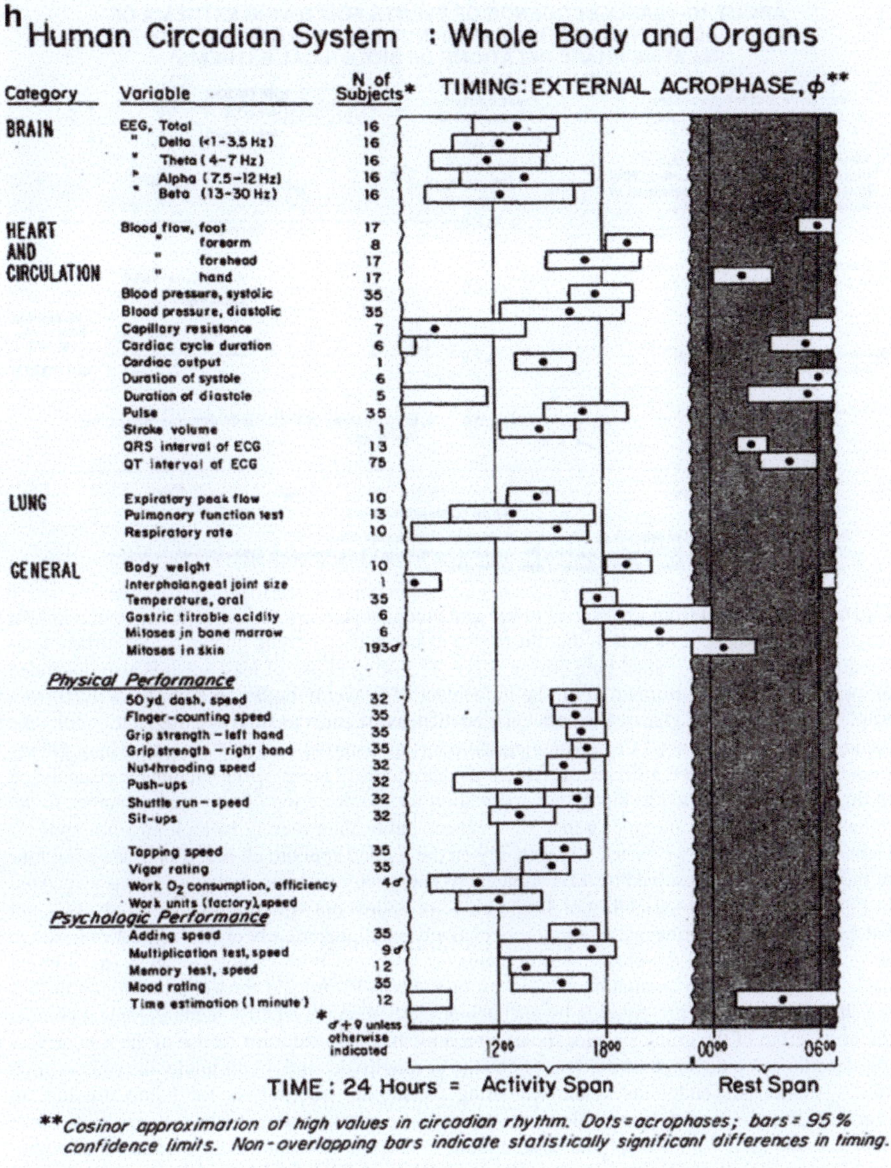

Fig. 6.1 (continued) including superposed epochs (© Halberg Chronobiology Center). (**e**) The same data set is shown, as the case may be, as a deviation (*upward* or *downward*, on the *left* of this figure) either from its mean, shown as a horizontal line (*top*) or as a deviation from the best-fitting 24-hour cosine (*bottom*). In the next column from the *left* to *right*, the deviations are shown by the length of lines, irrespective of sign; on the average they are smaller when related to the cosine curve, as compared to their mean, as best seen in the third pair toward the *right*, where the deviations are squared: the *squares* at the *bottom*, as deviations from the cosine curve, are obviously smaller than those at the *top*, as further summarized by the length of two columns on the extreme *right*. The thus-unqualified method of least squares underlies both the computation of a mean, as a step for computing deviations from the mean or from other standards. A wealth of information

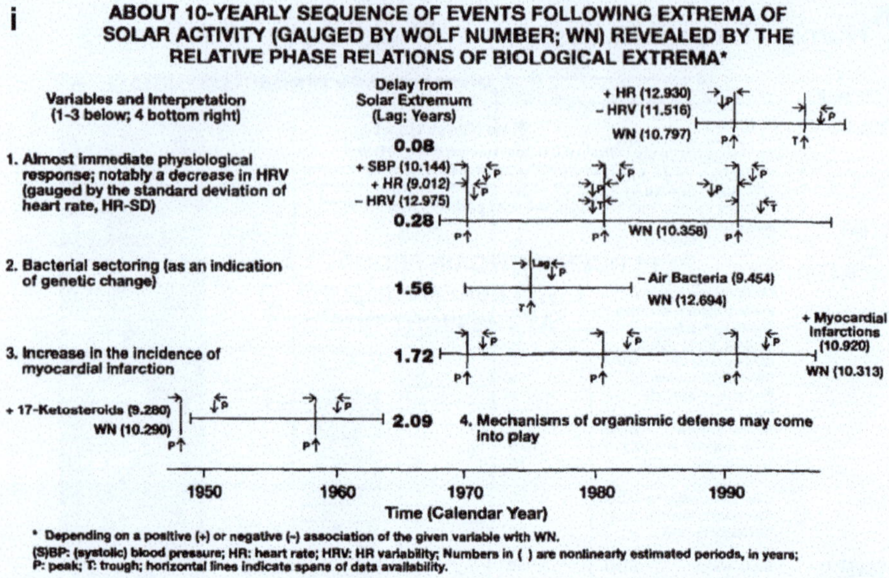

Fig. 6.1 (continued) from time series in life and other science reveals the importance of a known pertinent mathematical model. In the illustrative example used herein, human breast surface temperature can be approximated by a cosine curve. The original use of least squares also depended on prior independent information on the movement of heavenly bodies in ellipses, derived from Kepler and available to Gauss. Gauss developed the least-squares method and used it to predict the location of a lost "planet," Ceres (now classified as an asteroid or dwarf planet), where Olbers found it on December 31, 1801. In the case of the chronome of breast surface temperature analyzed in this figure, the method awaits broad application for the detection of very early changes in the female breast prior to the appearance of a cancer rather than merely to look at a hot spot (© Halberg Chronobiology Center). (**f**) *Dots* only on the *top left* apply to all four graphs showing four of the characteristics resolved by least-squares rhythmometry with the method of Gauss visualized in Fig. 6.1e. The great variability of single blood pressures notwithstanding, the relatively small hatched quadrangles or hexagons show the relatively small uncertainty of characteristics based in several days of around-the-clock measurements (© Halberg Chronobiology Center). (**g**) Original data: Time plots of two circulating substances, endothelin-1 (*top*) and cortisol (*bottom*), on the *left* are shown in a polar representation by the cosinor method on the *right*. A methodological point is the nonoverlap of the graph's center (pole) on *top* by the darkened error ellipse of the 8-hour vector, describing by its length the extent of change or amplitude, and, by its angle, the time of high values, acrophase, endpoints visualized along rectangular coordinates for blood pressure in Fig. 6.1f. The nonoverlap of the pole by the error ellipse corresponding to the 8-hour cosine curve fit rejects the zero 8-hour amplitude assumption, in keeping with an 8-hour rhythm for endothelin-1. This figure also shows that in the same circulation of seven healthy medical students, on the average, cortisol (*bottom*) can reveal statistically highly significant 24-hour and 12-hour components, but no 8-hour component, since the corresponding darkened error ellipse (c) overlaps the pole, the opposite of what is seen on top, where the ellipses of the 24- and 12-hour components include the pole, results also seen from tables accompanying the graph. Endothelin sometimes in some people obeys a "24-hour clock," but at most times in most people, it exhibits a primary 8 hour rhythm. The designation of a "clock" has no heuristic value in this case and many others and need not obscure the search for mechanisms (© Halberg Chronobiology Center). (**h**) Circadian acrophase mapping is much more extensive than shown in this chart that suffices to reveal the principle of synchronization with differences in phase. Major advances beyond this chart include (1). the preparation of such a chart for many more variables of the individual subject and (2). the extension from phases to include amplitude and waveform at all pertinent periods, thereby to

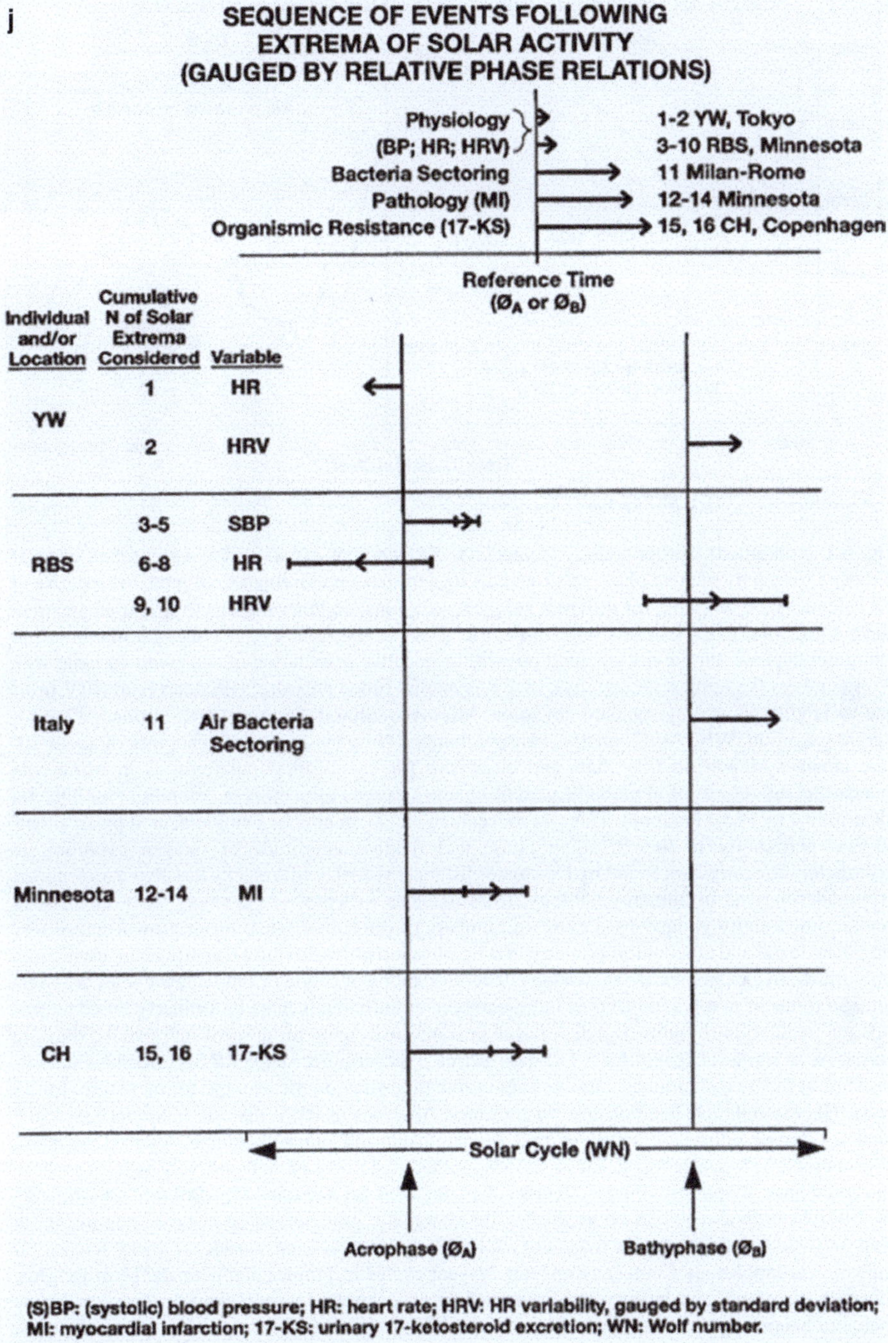

Fig. 6.1 (continued) detect modulations with ever lower frequencies, including those shown in Figs. 6.1i and j, indeed for some individual subjects (© Halberg Chronobiology Center). (**i**) The cycles in geomagnetic disturbance (first two *rows*) and the Schwabe (and Hale) cycles of solar activity have counterparts in the biosphere. For phase relations, see Fig. 6.1j (© Halberg

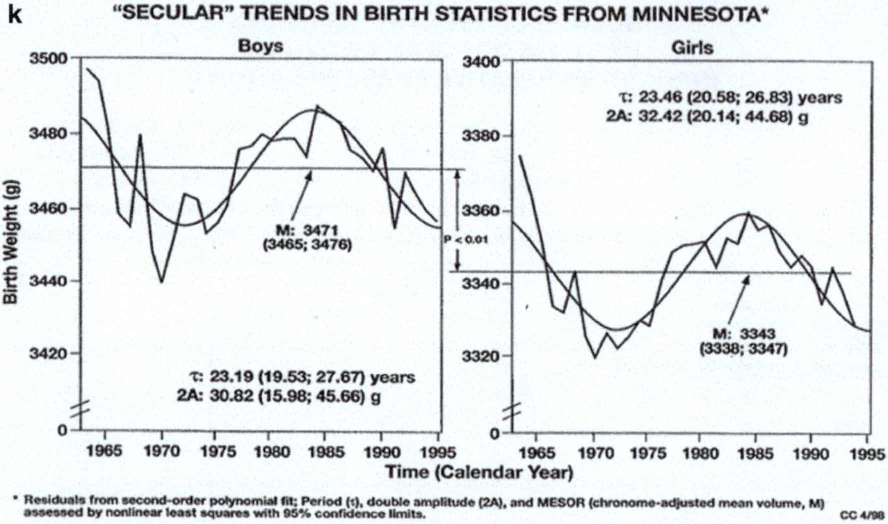

Fig. 6.1 (continued) Chronobiology Center). (**j**) Sequence of events following extrema of solar activity (gauged by relative phase relations). Delays from solar extrema are aligned, irrespective of the calendar date of study, for different subjects, locations, and/or variables to reveal an apparent bunching of biological extrema with respect to those of the solar cycle. Focus on maximum or minimum depends on the documented positive or negative association of the given variable with Wolf numbers (gauging solar activity). (S)BP, (systolic) blood pressure; HR, heart rate; HRV, heart rate variability gauged by standard deviation; MI, myocardial infarction; 17-KS, urinary 17-keto-steroid excretion; WN, Wolf's relative sunspot number. Note the degree of generality of about-10-year changes shown on this chart and elsewhere for events in populations as in Minnesota myocardial infarctions or in probable genetic changes gauged by sectors in colonies of air bacteria (determined by Piero Faraone). This circadecennian chart can now be complemented by an about-10-year periodicity in the circadian phase and relative amplitude of oxygen evolution in *Acetabularia acetabulum*, found by Dewayne Hillman. Miroslav Mikulecky has also mapped a set of circadecennians in human pathology, from diabetes to leptospirosis, along with changes in human productivity gauged by a cycle in numbers of published titles. From growth gauged by human neonatal and other anthropometry to religious motivation, about-21-yearly changes abound. Only illustrative examples are shown here. The limitations of single or only a few cycles are overcome in terms of numbers insofar as large numbers of individuals must be similarly timed to provide the population rhythms depicted, based in some cases upon millions of individuals; they are overcome in terms of length by a 112-year series. Furthermore, subject RBS continues to self-measure over 10 physiological variables about five times/day on the average for most days for 35 years. His example needs emulation for lifetimes. Analyses of RBS' data serve to emphasize the need for denser automatic series so that deterministic and other chaos can also be assessed. Information on circadecennians has already suggested that many more controversies such as those that could have arisen from Figs. 6.1b and c, had the latter been obtained by different investigators on different individuals, can be avoided. The systematic government-sponsored monitoring, at least on some test pilots, seems mandatory for a solid biological and transdisciplinary science (© Halberg Chronobiology Center). (**k**) About-21-year cycles in Minnesotan neonatal birth weights, shown in this graph, also characterize detrended data (not shown) from Denmark, during the same span, but happen to be in antiphase. Moreover, in 112 years of data from the late Boris Nikityuk, results of other neonatal anthropometry show about-20-year cycles, but birth weight is frequency-multiplied to an about-10-year cycle in Moscow but not in Alma-Ata. The near-antiphase between the time course of circavigintunennians in data from Minnesota and Denmark during the same span, documented at each site, with large numbers of individuals as well as a frequency multiplication at one of two other sites, remains a transdisciplinary challenge in a multifactorial situation (© Halberg Chronobiology Center)

to time series are needed for resolving the complex sources of variability, and their underlying chronomes [13, 14; cf. 15].[1]

If one measures densely enough, deterministic and other chaotic endpoints can be computed in sets of environmental and biological variables such as a correlation dimension [14, 16–18] or an approximate entropy or complexity, as one element of chaos in chronomes. A second element of these chronomes consists of trends, which can be approximated by polynomials, to quantify them during development and aging, and also with, e.g., disease-risk elevation, starting at birth [19]. If one samples long enough, trends can turn out to be roughly periodic components (Fig. 6.1b–d) [19, 21, 22]. Indeed, the third and major insofar as predictable element of chronomes is a spectrum of periodicities or near periodicities, in us or around us. It comprises changes with periods ranging from fractions of a second to decades, assessed before or after detrending by linear-nonlinear rhythmometry (Fig. 6.1e–g) including with circadians (Fig. 6.1h) about-10.5-year (Fig. 6.1i and j), about-21-year (Fig. 6.1k), and about-50-year cycles, the latter mainly mapped thus far in populations by neonatal anthropometry in 112 years of data and for strokes in Minnesota and the Czech Republic only for the past half-century [21].

Data inspection (eyeballing) of time plots and analyses of variance, albeit useful, often are not enough for dealing separately or in combination with all characteristics of the rhythmic component of time structure (Fig. 6.2a, b). Section IIA of Fig. 6.2a (in the two swarms of measurements on the right) shows to the naked eye

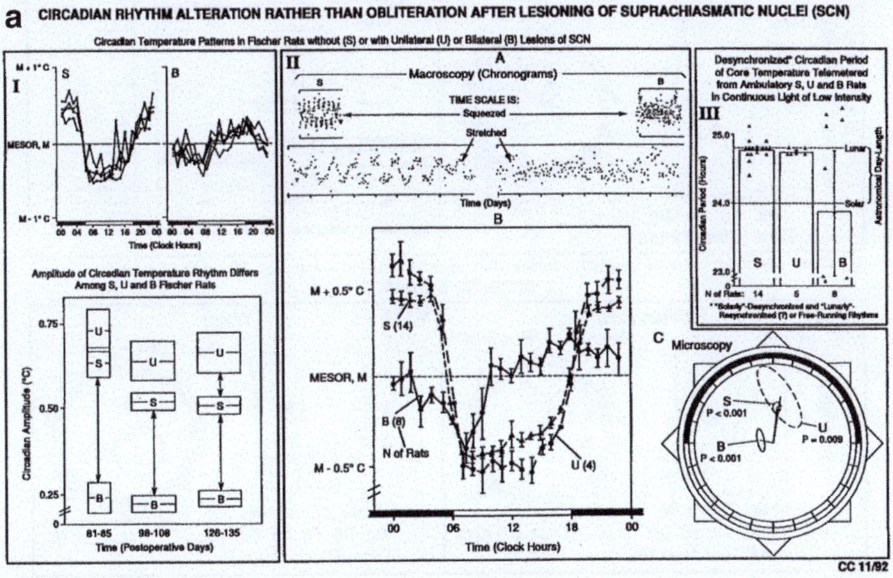

Fig. 6.2 (**a**) By eyeballing alone of Section IIA on the *right*, the circadian rhythm in core temperature seems to be lost in the rat with a SCN lesion. Simple stacking reveals the persistence of a circadian rhythm in telemetered core temperature for individual animals (Section IB *top*), but with a smaller within-day change (IB) as compared to controls (IA). This finding is also seen after

[1] From *chronos*, time, and *nomos*, rule, rather than from alternate derivations [15].

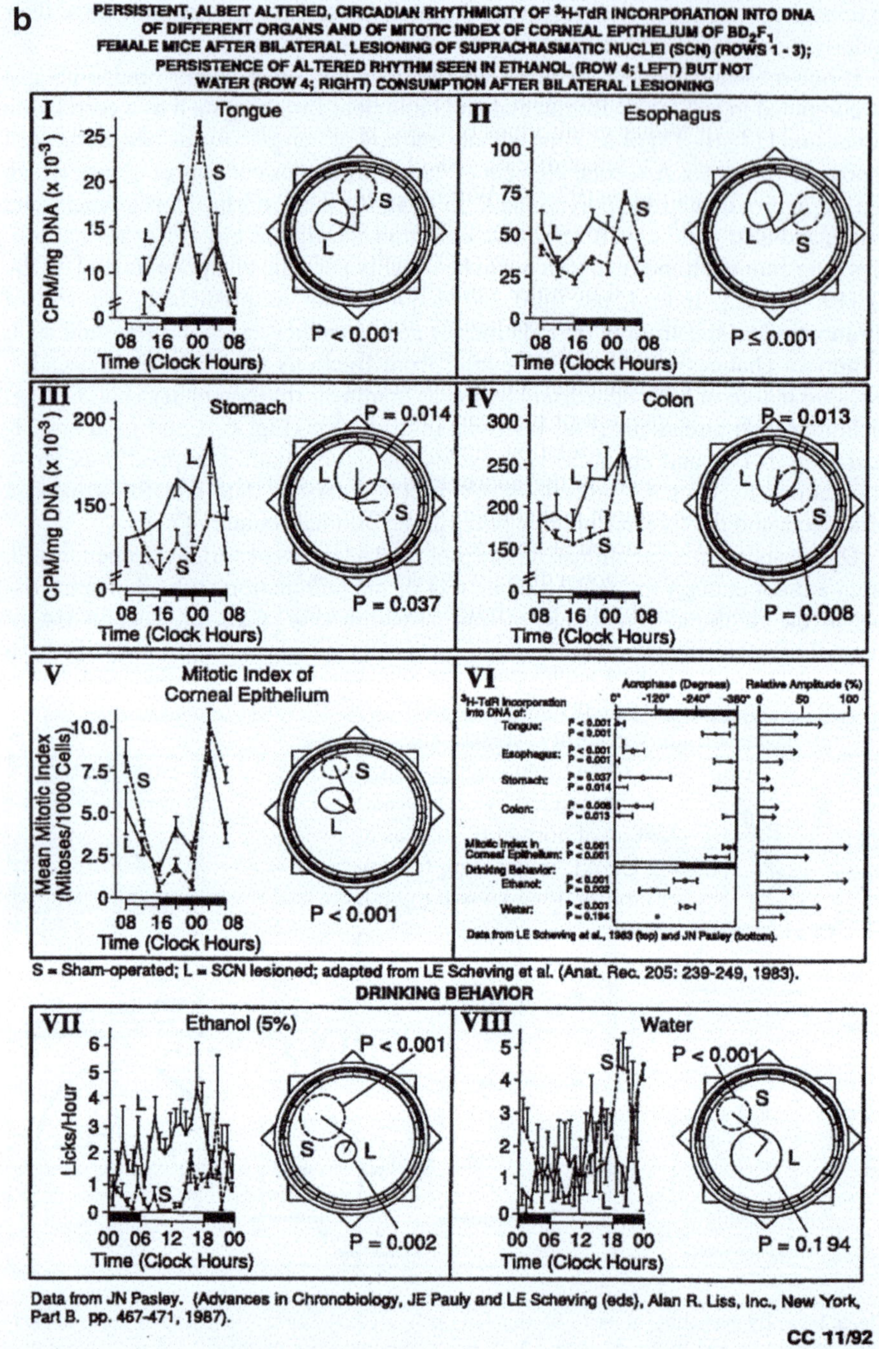

Fig. 6.2 (continued) averaging in Section IIB *bottom*. Microscopy (C), apart from quantifying the rhythm by cosinor, reveals a great amplitude lowering by bilateral SCN ablation and a phase advancement seen as an earlier and shorter vector. Section IIC thus validates by the nonoverlap of the center or pole of the graph, by the error ellipse representing a 95 % confidence region, that the removal of the SCN is compatible with the persistence of a statistically highly significant circadian

why the eyeballing of original data can lead to claims that the circadian temperature rhythm is lost after ablation of the suprachiasmatic nuclei (SCN). Whether the time scale for plotting is squeezed (top of section IIA, Fig. 6.2a) or stretched (bottom of section IIA, Fig. 6.2a), the rhythm clearly seen in the two swarms of dots on the left seems lost on the right. After stacking, eyeballing suffices to "see" the rhythm, e.g., in section I (top) for the SCN-ablated animals separately or averaged in section II B. For quantification, time-microscopic inferential statistical methodology [23, 24], also shown in Fig. 6.2a, b, is essential.

Macroscopy (left) and microscopy (right) in each section are aligned in Sections I–V and VII–VIII of Fig. 6.2b. On the right, in Sections I–V, VII, and VIII, a time-microscopic display in polar coordinates shows the extent of change, amplitude, by the length of a vector, while the vector's angle represents the acrophase, the time of overall high values. The ellipse around the tip of the vector represents the 95 % confidence region of the amplitude-acrophase pair; when it does not overlap the center of the plot (the pole), a rhythm is demonstrated, as it is in all cases except for water drinking (Section VIII) for which the ellipse covers the pole and hence the zero-amplitude assumption tested by the ellipse is not rejected and a rhythm is consequently not demonstrated.

Pole overlap, such as that for the case of water consumption (Section VIII), is not seen for ethanol drinking. For the latter variable and all others, a rhythm is demonstrated, and, except for DNA labeling in the stomach, the circadian amplitude is reduced by ablation of the SCN. In all cases, the phase is advanced, as succinctly summarized in the chronome map in Section VI (Fig. 6.2b). The usual concern about point estimates of period and phase is best complemented by uncertainty estimates, including those of amplitude and, when the density of the data permits, of waveform. Each of these parameters is to be given, whenever possible, with interval estimates. An indication of the uncertainties involved is particularly important when long-term treatment is the issue [6, 18, 23].

Fig. 6.2 (continued) rhythm in core temperature quantified with its parameters and their uncertainties. The polar cosinor displays in Fig. 6.2b also quantify a phase advance of rhythm after histologically validated bilateral SCN ablation in several tissues, with the exception of the stomach, which may respond to food directly rather than via the SCN, again as seen by the shorter vector for the B (SCN-ablated) group, as compared to longer and later vectors for the sham-ablated S group. When the ablation, unintentionally, as discovered at post-mortem, was unilateral (U), the circadian amplitude is enhanced (Section I *bottom*), a finding suggesting a subtractive coupling between the two SCN. Section III is in keeping with the speculation of an effect by lunar factors upon the "free-running" period of about 24.8 hours found in controls or unilaterally ablated animals. If this should be in part a lunar effect, it is lost in animals subjected to bilateral suprachiasmatic lesions (© Halberg Chronobiology Center). (**b**) The SCN coordinates a collateral hierarchy that can be quantified in terms of amplitude and phase: the major effect of bilateral SCN ablation is thus far invariably, comparably to the behavior of core temperature in Fig. 6.2a, an advance in phase for the L animals in eight cases out of eight, with a reduction in amplitude, except for DNA labeling in the stomach (Section III). Section VI shows a microscopic phase and amplitude chart summarizing the findings in the other sections for a number of variables other than core temperature, studied as marker rhythm in Fig. 6.2a, confirming and extending the scope of the lesson learned in Fig. 6.2a: the SCN, rather than being a master clock leading to the abolition of all rhythms when ablated bilaterally, is compatible with their persistence, except for water drinking among the functions investigated. A subjective time-macroscopic interpretation-based impression that led to the master clock illusion (e.g., Fig. 6.2a, Section II *right*) is thus resolved by quantification (© Halberg Chronobiology Center)

The estimation of characteristics from cosine fitting provides endpoints in several fields, from problems of eating and salting to treating disease, and is indispensable for otherwise silent disease-risk recognition and for risk lowering by precautionary intervention. A dividend of curve-fitting is an improved assessment of a time structure-adjusted mean or MESOR, of general interest in science since, as compared to the arithmetic mean, in series characterized by rhythms, the MESOR is usually more accurate in the case of unequidistant data and more precise in the case of equidistant ones (Fig. 6.3). Recent meetings on chronomics [25–28; cf. 29], considering data covering up to somewhat more than a century, have documented time structures characterized by rhythms that replace confounding secularity (Fig. 6.1b–d) [20, 21]. A wealth of detailed information obtained in chronobiology on a collateral hierarchy [30] in the circadian system, ready not only as a basis for further research but some of it also for immediate utilization in everyday life, is here illustrated in a historic context.

6.3 Started Millennia Ago and Persisting Today

The scholarship of the late Jürgen Aschoff [31] traced the root of thoughts on rhythms from a fragment of verse by Archilochus of Paros (c. 680–640 BC): γιγνωσχε δ′οιος ρυθμος ανθρωπον εχει ("Recognize which rhythm governs man"). In finding this fragment from Archilochus, Aschoff demonstrated in him a link between solar physics and human rhythmicity as far back as seven centuries before the Common Era (as validated by the date of an eclipse).[2] Hippocrates, Aristotle, and Galen followed in Aschoff's gallery of ancestors with others, up to his own day, crowned by his implied recognition by 1974 of, as he put it, the need for a "cosinor beast" (read objective quantification) [31]. Rhythm "analysis" carried out time-macroscopically, only with data inspection by the unaided eye, can mislead to decades of erroneous inferences of an absent rhythm, e.g., in Fig. 6.2a, Section IIA (right), whereas the stacking of the same data (Sections I (top) and II B) does reveal the rhythm, and an analysis of variance establishes its statistical significance, yet none of these steps replaces the complementary and necessary computer-aided quantification of universally applicable endpoints, such as amplitudes and phases [7, 13, 23, 24]. The cosinor approach provides objective, generally applicable endpoints whereby much controversy can be avoided (Figs. 6.1b–d and 6.2a, b).

6.4 Centuries Ago and Today

In a thesis published in Paris on April 23, 1814, Julien-Joseph Virey [32; cf. 6] wrote that an individual who during 24 hours eats only once in the evening could find himself with a *nuance of temperament* different from that of somebody who

[2]Archilochus was reputedly inspired by the solar eclipse of April 6, 647 BC, which, he wrote, "made night from mid-day, hiding the light of the shining Sun," which circumstance authenticates his contribution from the seventh century BC.

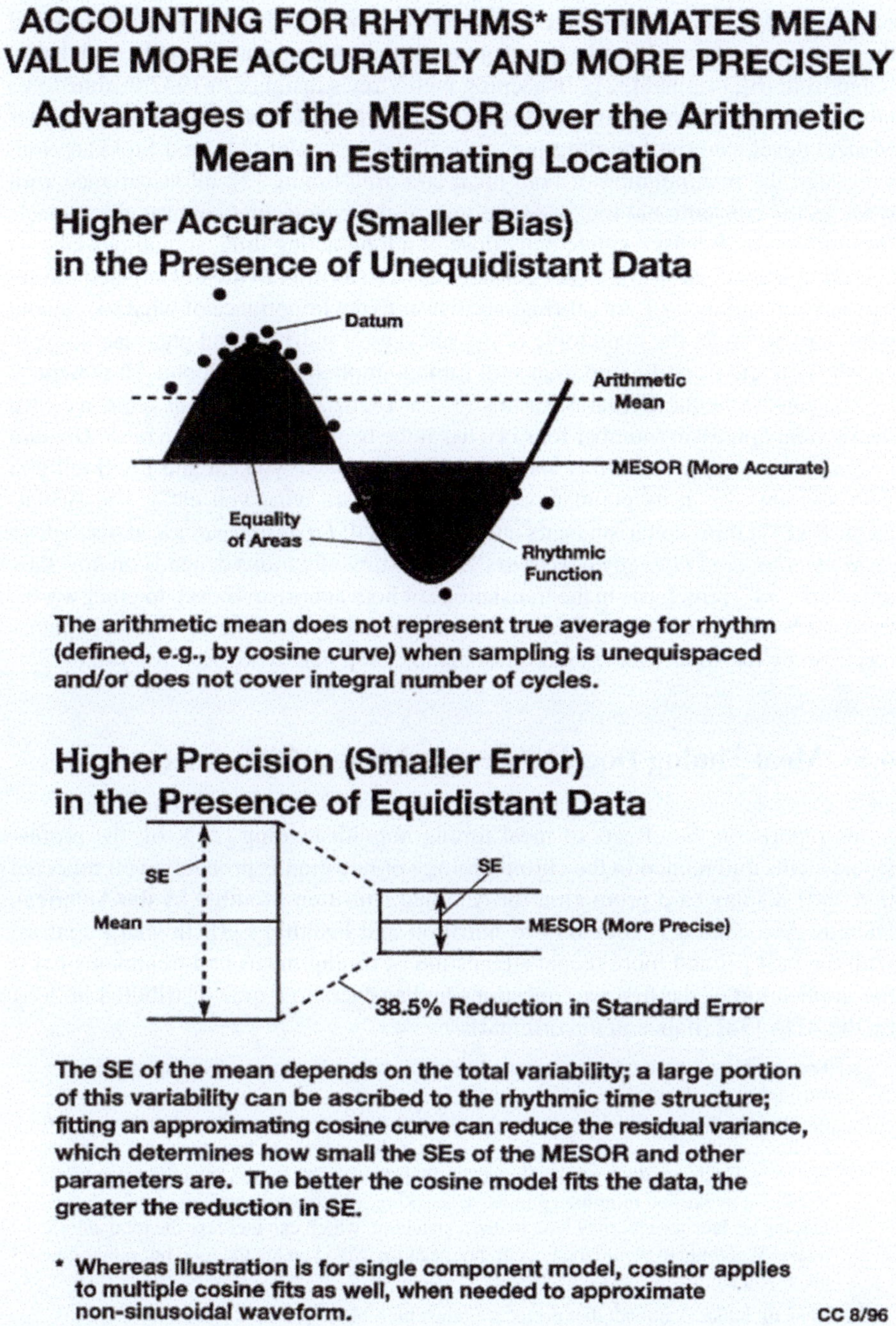

Fig. 6.3 For any data series (whether from physics, chemistry, biology, psychology, or sociology), characterized by a spectrum of rhythms, a MESOR (**m**idline-*e*stimating *s*tatistic *o*f *r*hythm or rather of a chronome-adjusted mean) can be computed and has merits: as compared to the arithmetic mean of the same series, the MESOR is usually more accurate and more precise (© Halberg Chronobiology Center)

eats only in the morning. For him "... the morning meal will be the most salutary and the most rejuvenating." Similarly, any given drug is not indicated equally at all hours (cited from [6]; original [32]). In keeping with Virey's insights, in 1937, Arthur Jores labeled the neglect of timing as "the idiocy of 'three times a day'" [33], Werner Menzel developed the first drug pump for timed treatment [34], and Stokkan et al. suggested the manipulation of both meal and drug timing [2], all at variance with what seems conventional today. At the turn of the twenty-first century, three meals for nutrition or at least a single daily dose of a long-acting drug, sometimes labeled "chronotherapeutics" [35], do not consider any merits of individually targeted timing by marker rhythms [36]. An affirmative answer to the importance of what we eat and with what we treat, the main topic of the sciences of nutrition and pharmacology, in no way detracts from the importance of timing, implicit in Archilochus' fragment.

To Jean-Anthelme Brillat-Savarin's (1755–1826) *Physiologie du goût*, we owe the familiar aphorism (number four in a list at the beginning of volume one), *Dis-moi ce que tu manges, je te dirai ce que tu es* ("Tell me *what* you eat and I will tell you who you are," or in its popular paraphrase "You are what you eat"). But Brillat-Savarin's [37] third aphorism bears attention as well: *La destinée des nations dépend de la manière dont elles se nourissent* ("The destiny of nations depends on *how* they are nourished") [emphasis in the translations of both aphorisms ours]. In other words, "you are *how* you eat." Could this include the possibility, at variance with an explanatory note by the American Dietetic Association [38], that "you are *when* you eat?"

6.5 Meal Timing Doesn't Seem to Make a Difference

Some reports on the effects of meal timing notwithstanding [2, 5, 6], the professional status quo related to the chronobiology of nutrition is probably well reflected in a 2001 display card promotion for National Nutrition Month® by the American Dietetic Association ("Your link to nutrition and health") [38]. In sharp contrast with the past [7] and more recent suggestion of timing meals and treatments [2] is the implication of the first two questions and answers to a quiz distributed in 2001 by the ADA [38] (italics in the original):

1. **True or False**: A late dinner is more likely to cause weight gain than eating the same meal earlier in the day.
   FALSE. The clock really doesn't make a difference. It's not *when* you eat, but *what* you eat that counts. No matter when they're eaten, calories seem to have the same effect on the body. Evidence does suggest however, that regular mealtimes, including breakfast, reduce fat intake and minimize impulsive snacking, which results in fewer total calories. Staying up late at night may lead to *extra* snacking, which can increase the total caloric intake. Weight gain would then occur, not because *when* you ate the food but rather *how much* you ate.
2. **True or False**: A no-salt diet protects against high blood pressure.
   FALSE. If you're sodium-sensitive, then removing the salt shaker from the table does help protect against high blood pressure, or hypertension. However, sodium is the real issue, not just salt, which is 40 percent sodium by weight. Both salt and sodium are used widely in processed foods. Check labels so you consume just a moderate amount, about 2,400 milligrams daily. To help control high blood pressure, maintain a healthy weight, be physically active, and avoid smoking.

Explicit in the answer to question 1 and implicit in the answer to question 2 is the claim, contradicted by the following evidence [7], that timing "doesn't make a difference."

6.6 Body Weight in Lean Individuals Depends on the Timing of the Meal

In a clinical research center, two women and four men consumed a single 2000-kcal (kcal) meal in the morning each day for 1 week, after which the individual consumed the same meal in the evening for another week, on the same limited activity routine. In these studies, the effect of the sequence of meal timing was controlled by a crossover design. "Breakfast-only" meant that the subject ate all of the kcal for the given day within 1 hour of awakening after a 23-hour fast. "Dinner-only" meant that the subject fasted for at least 12 hours after awakening on each day of this study subspan. Although activity was not measured, subjects were sedentary in a clinical research center throughout the study. Figure 6.4a (right) shows that when eating breakfast-only, all six subjects lost weight, whereas four out of six gained weight while eating dinner-only. The remaining two subjects lost weight on both schedules, but lost more weight on breakfast-only than on dinner-only. Using the weight change (kg/week) on dinner-only as a reference, one can subtract it from the change on breakfast-only in order to find a *relative* body weight loss on breakfast of nearly 1.2 kg/week (Fig. 6.4b (right)) [7].

In order to study interactions of meal quality, amount and timing, 12 other lean subjects (five men and seven women) were allowed a single, free-choice meal daily. For the first 3 weeks, seven subjects ate the meal as breakfast, while the other five ate it as dinner. Each subject changed the meal timing between breakfast and dinner for a second 3-week span. Subjects consuming a free-choice meal ate fewer kcal for breakfast, on the average, than for dinner ($P<0.01$), and as could be expected from the controlled study of a fixed 2000-kcal meal, their relative weight *loss* on breakfast-only (1646 ± 116 kcal) was greater than on dinner-only (2215 ± 227 kcal) (Fig. 6.4a left). Their weight loss, however, was smaller when compared to breakfast-only and dinner-only of the subjects eating a fixed 2000-kcal meal ($P<0.05$) (Fig. 6.4a, b left). In addition to the effect of total caloric intake, an influence of meal quality was thus apparent: the extent of relative body weight loss on breakfast was higher when the kind of meal was fixed (and usually disliked) rather than free choice (and less disliked) (Fig. 6.4c) [6, 7, 39].

Body weight gain or loss as a function of meal timing is also found in catfish [40, 41], again at variance with the cautious suggestion of the anonymous writer for the American Dietetic Association ("no matter when they're eaten, calories seem to have the same effect on the body").

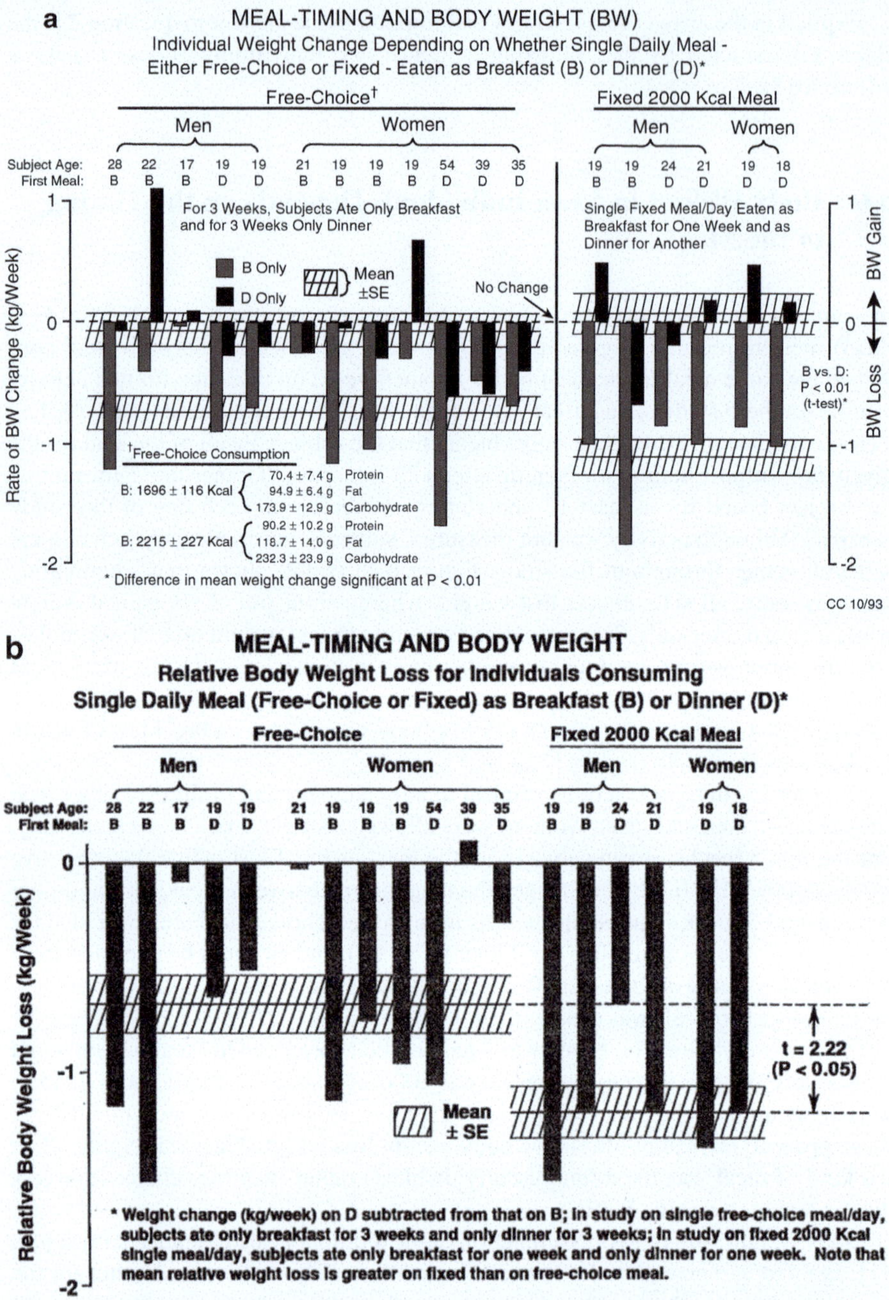

Fig. 6.4 (**a**) Meal timing and body weight. In two separate studies on the effect of meal timing on body weight, a total of nine men and nine women consumed, some a fixed 2000 kcal meal for 1 week, others a single free-choice meal for 3 weeks as breakfast-only (B) or as dinner-only (D). Body weight remained more or less unchanged on dinner-only; a decrease of about 1 kg/week was noted on breakfast-only. When eating just a single meal within 1 hour of awakening (breakfast-only, *gray bars*) or not before 12 hours after awakening (dinner-only, *black bars*), with only one exception

C "APPETITE"* MODIFIES EFFECT OF MEAL-TIMING ON BODY WEIGHT

Relative Body Weight Loss on Breakfast-Only (B) as Compared to Dinner-Only (D) is Less When Meal is Free-Choice (C) Rather than Fixed (F), Even Though Relative Calorie Intake is Smaller on C vs. F

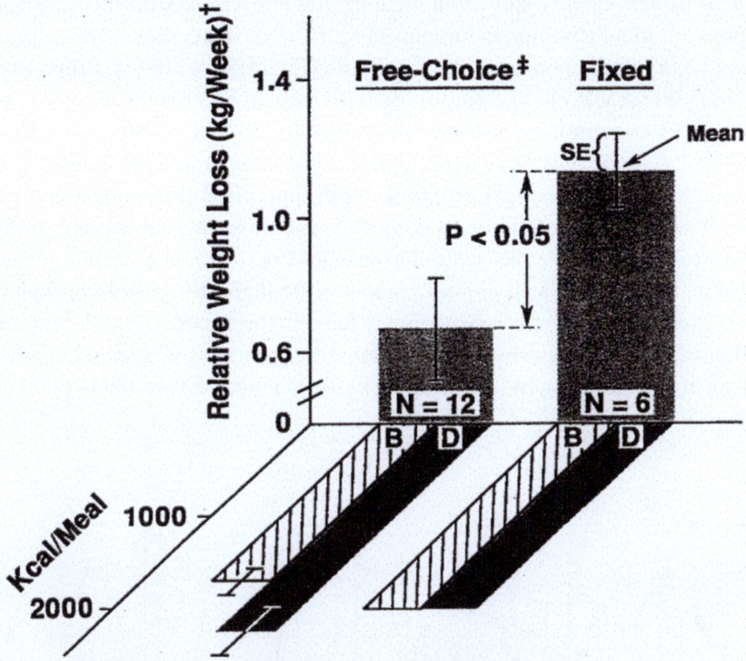

* Here defined as choice of kind and amount of food.

† Weight change (kg/week) on B subtracted from that on D.

‡ Average free-choice intake (± SE) on B = 1696 ± 116 Kcal/meal, on D = 2215 ± 227 Kcal/meal; former value significantly different from 2000 (t = 2.62; P < 0.05), latter not. Food consumed consisted on the average of

		g Protein	Fat	Carbohydrate
Free-Choice	B	70.4 ± 7.4	94.9 ± 6.4	173.9 ± 12.9
	D	90.2 ± 10.2	118.7 ± 14.0	232.3 ± 23.9
Fixed Meal, B & D		77.0	75.0	250.0

CC 10/93

Fig. 6.4 (continued) there is a *relative body weight loss on breakfast-only* as compared to dinner-only (cf. Fig. 6.4b and c) (© Halberg Chronobiology Center). (**b**) The rate of body weight change, i.e., the extent of overall relative body weight loss on breakfast-only vs. dinner-only, also differed significantly (P<0.01) between the two schedules (© Halberg Chronobiology Center). (**c**) Relative body weight loss on breakfast-only vs. dinner-only for each subject participating in the two studies described in Fig. 6.4b. Only one volunteer gained weight on breakfast vs. dinner. Overall, the difference in relative body weight loss on breakfast (B) vs. dinner (D) is statistically significant (P<0.05), whether a fixed 2000-cal meal or a single free-choice meal is consumed. Weight change (kg/week) on D subtracted from that on B. In the study on one free-choice meal per day, subjects ate only breakfast for 3 weeks and only dinner for 3 weeks. In the study on one fixed 2000-cal meal per day, subjects ate only breakfast for 1 week and only dinner for 1 week. Note that mean relative weight loss is greater on fixed than on free-choice meal (© Halberg Chronobiology Center)

6.7 Meal Timing Alters Human and Murine Chronomes

The investigations, which compare a single daily meal consumed as breakfast vs. one eaten as dinner, clearly show that meal timing is a way to shift the time relations among physiological rhythms in humans (Fig. 6.5a–c) in keeping with earlier [6, 7] and recent [2, 5] studies on laboratory animals (Fig. 6.5d, e) [6, 7]. Some rhythms, like that in cyclic AMP, change amplitude with a change in meal timing, but not the acrophase. The concomitant assessment of amplitude and acrophase in Fig. 6.5a reveals, in the cosinor plot, the quantification of an effect upon the circadian amplitude in the absence of an effect on phase, both quantified time-microscopically in Fig. 6.5a. It would be worthwhile to determine the most favorable configuration of rhythms in order to assure the peak alertness and performance during any critical times of day, but more sensitive performance tests than those used previously will have to be used (Fig. 6.5c) [39]. Note that at least in the experimental laboratory, the scrambling of the circadian rhythmic element of chronomes is associated with anti-cancer and life-prolonging (not shortening) effects (Fig. 6.5f) [42–44].

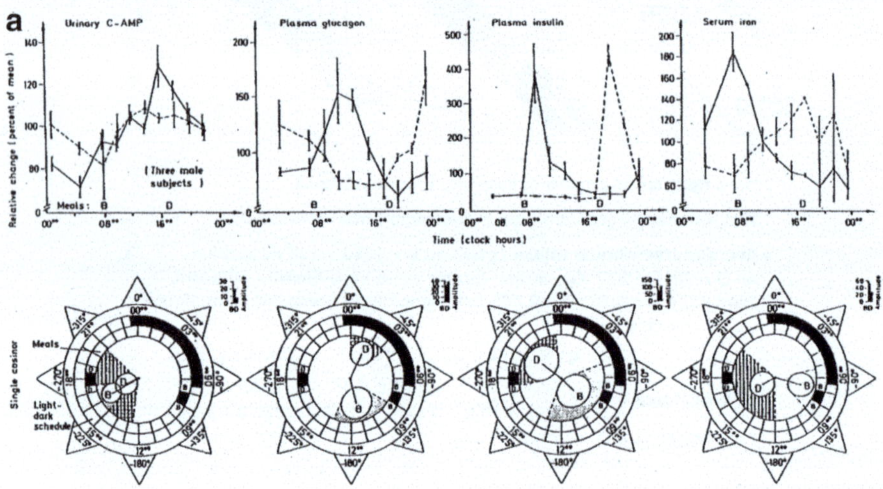

Fig. 6.5 (**a**) A single daily meal as breakfast-only (within 1 hour of getting up) vs. as dinner-only (a meal at least 12 hours after getting up) has very different effects on different human circadian rhythms. A clear difference in amplitude of the circadian rhythm in urinary cyclic AMP excretion is seen with no change in the timing of this rhythm, while the timings of the circadian rhythms in circulating insulin, glucagon, and iron all differ with high statistical significance, as apparent from nonoverlapping 95 % confidence ellipses (© Halberg Chronobiology Center). (**b**) A large difference in the timing of the insulin rhythm is seen on breakfast-only vs. dinner-only, whether the diet involves items of some limited choice or a fixed diet of 2000 cal/day; in the same human circulation, there is just a much lesser difference in the timing of the circadian rhythm of cortisol (© Halberg Chronobiology Center). (**c**) Failure of single meal timing to shift circadian changes in the performance of simple habitual tasks (© Halberg Chronobiology Center). (**d**) Amplification, under

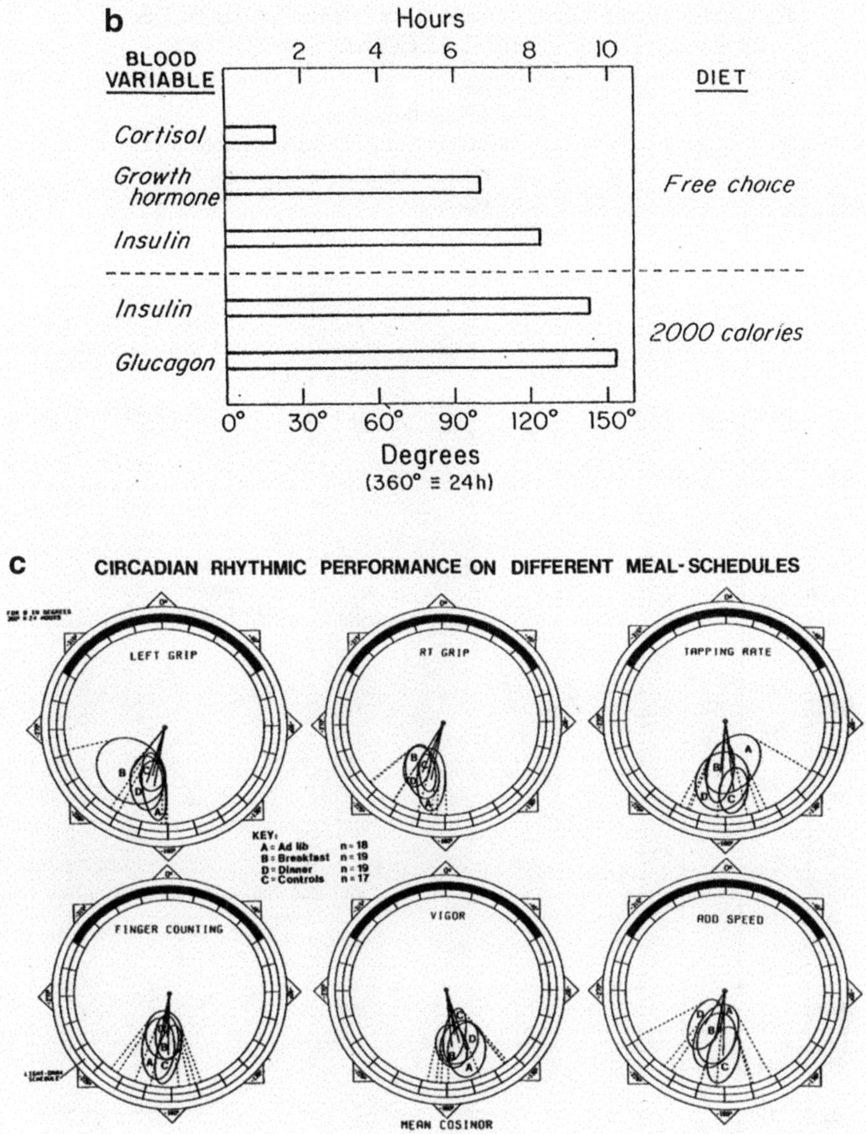

Fig. 6.5 (continued) laboratory conditions, of circadian change by meal timing (© Halberg Chronobiology Center). (**e**) Restricting food accessibility to the first 4 hours of the daily 12 hour light span shifts the circadian rhythm in susceptibility of mice to the carcinostatic drug, adriamycin [6], as hypothesized subsequently [2]. Units of MESOR and amplitude = mean survival time (hours) at 90% overall mortality; units of acrophase = degrees after 00^{00}, with 15° = 1 hour (© Halberg Chronobiology Center). (**f**) When rhythms are manipulated (if not "exercised") by alternating the temporal placement along the 24-hour scale of a regimen of lighting into phase and again out of phase with the availability of food during a restricted span of the day, breast cancer incidence is statistically significantly lowered rather than enhanced by a "stress" effect (© Halberg Chronobiology Center)

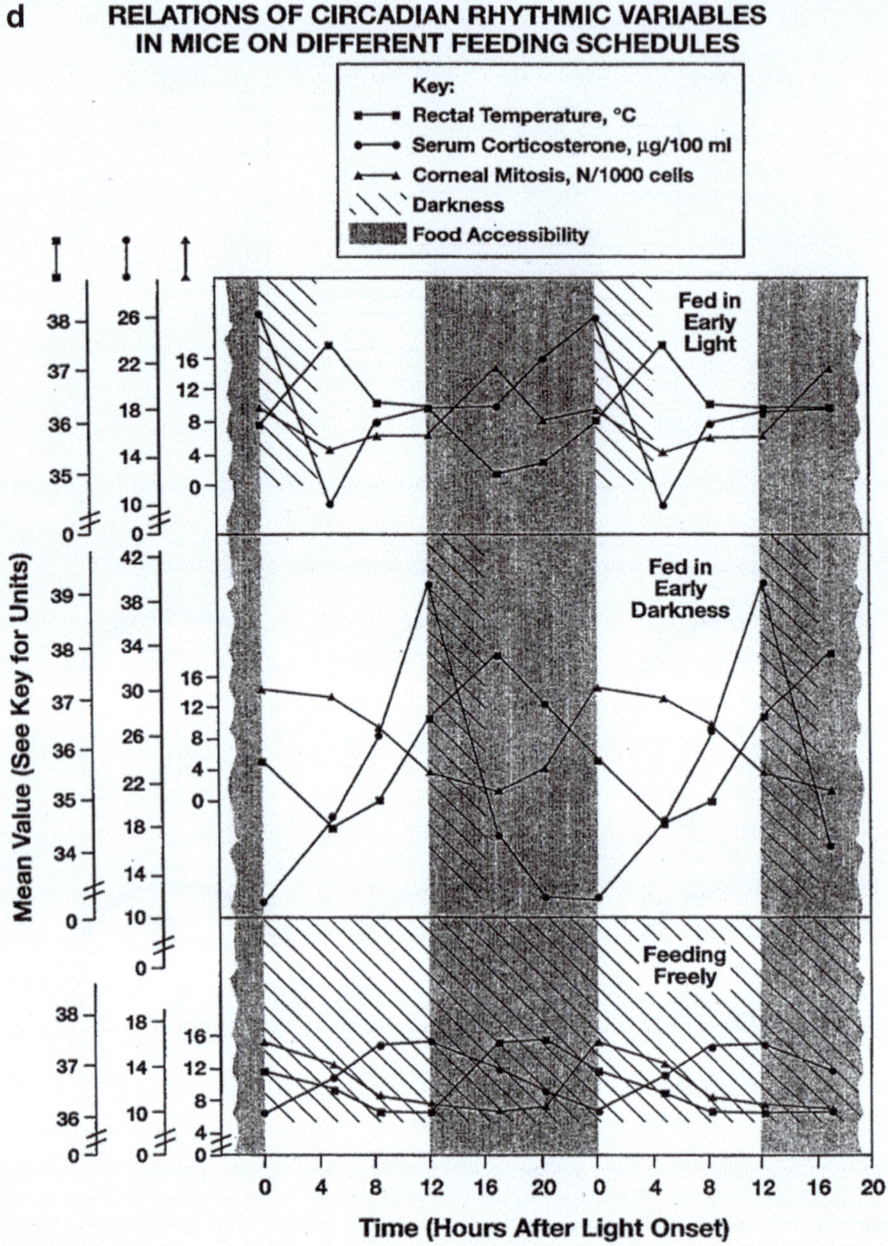

Fig. 6.5 (continued)

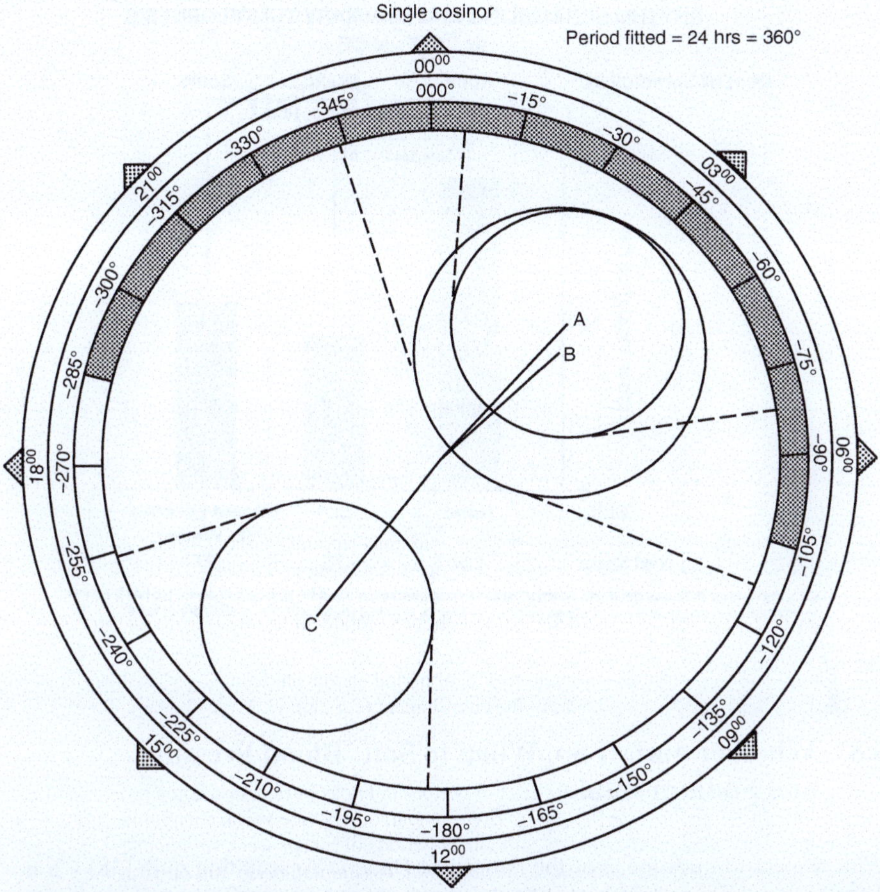

series ident	access to food (dose)	percent rhythm (%)	messor S.E.	amplitude (conserv. 95% CI)	acrophase (conserv. 95% CI)	P F-test
A	L (I)	10	229 6	35 (15 to 56)	−43° (− 7 to −78)	.001
B	L (F)	4	246 6	22 (2 to 42)	−49° (−113 to −345)	.024
C	A (I)	10	225 6	39 (17 to 62)	−218° (−183 to −254)	.001

L = limited to first 4 hrs in 12 hrs light span
A = ad lib.
I = dose by weight at injection time
F = dose fixed
CI = confidence interval

Fig. 6.5 (continued)

f

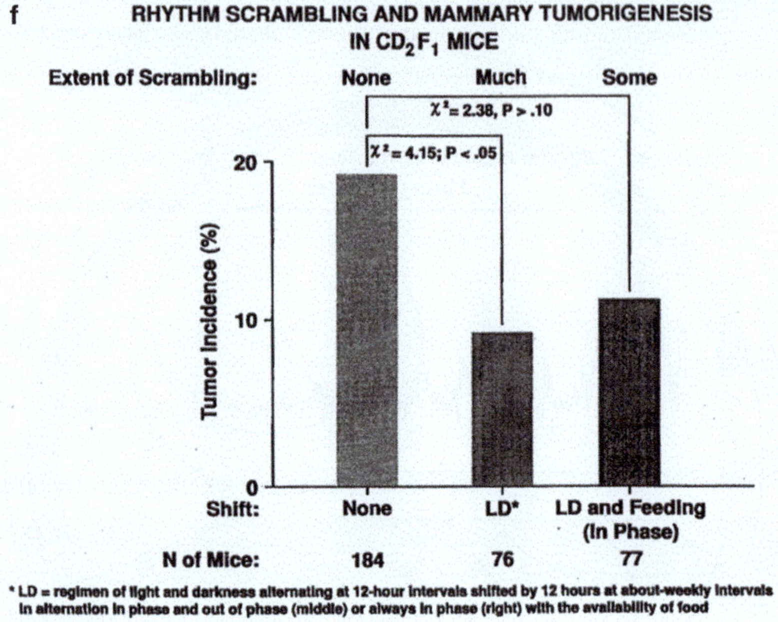

Fig. 6.5 (continued)

6.8 Whether and, If So, When to Salt: Blood Pressure and Sodium Intake

With respect to question 2 of the American Dietetic Association quiz [38] ("a no-salt diet protects against high blood pressure"), the answer ("sodium is the real issue, not just salt") must also be qualified. The assumption prevails that a high dietary sodium intake leads to higher blood pressure than a lower intake, increasing the risk of cardiovascular disease. A basis of this proposition is a cross-sectional study of sodium and blood pressure in 52 populations [45; cf. 46, 47], criticized by Freedman and Petitti [48], who note that of 52 populations, 4 had an association of low salt and low blood pressure, but across the remaining 48 populations, "blood pressures go down as salt levels go up – contradicting the [sodium: our addition] hypothesis. Experimental evidence suggests that the effect of a large reduction of salt intake on blood pressure is modest and health consequences remain to be determined. Funding agencies and medical journals have taken a stronger position favoring the salt hypothesis than is warranted, raising questions upon the interaction between policy, process and science" [48].

The chronobiologic evidence against a sweeping sodium hypothesis is much stronger [49–53]. Instead of the spotchecks of a long series of epidemiological studies [45–47, 54–58], longitudinal study under controlled conditions on small groups reveals that on the average, an extreme reduction of dietary sodium intake indeed can lower the rhythm-adjusted average blood pressure. This is shown in Fig. 6.6a in a dietitian-controlled study under the conditions of a metabolic ward at the NIH. As to individuals, Fig. 6.6b (top) shows that on a reduced sodium intake, the MESOR of blood pressure is statistically significantly lower in 11 of 15 individuals, but in three it remains unchanged (NS) and in at least one other subject, it

a BLOOD PRESSURE (BP) VARIATION WITH CHANGES IN SODIUM INTAKE[1]

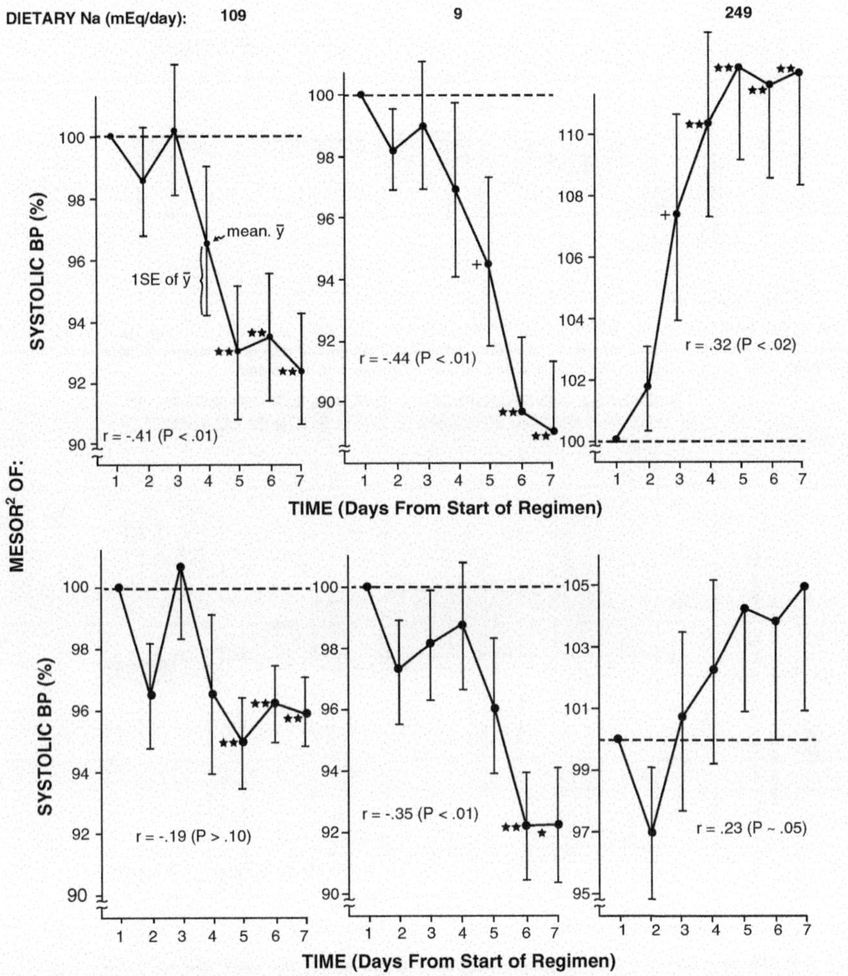

1) Data from 11 subjects in study B:r and P from linear regression: ★★ P < .01: ★ P < .05: +P < .10 from Student t-test (in testing equality of maen mesor to 100%. with mesor at start of study regimen equated to 100%)

2) 24-hour rhythm-adjusted mean (MESOR) expressed relative to mesor of day 1 in each stage of study.

Fig. 6.6 (**a**) In 11 subjects placed on a diet of 109 and then of 9 mEq of sodium/day, blood pressure decreased on the average. Return to 249 mEq of sodium/day was associated with a rise in blood pressure, with the effect ascertained with statistical significance by the third day of this last stage of the study for systolic but not for diastolic blood pressure. Original NIH data of Frederic

b

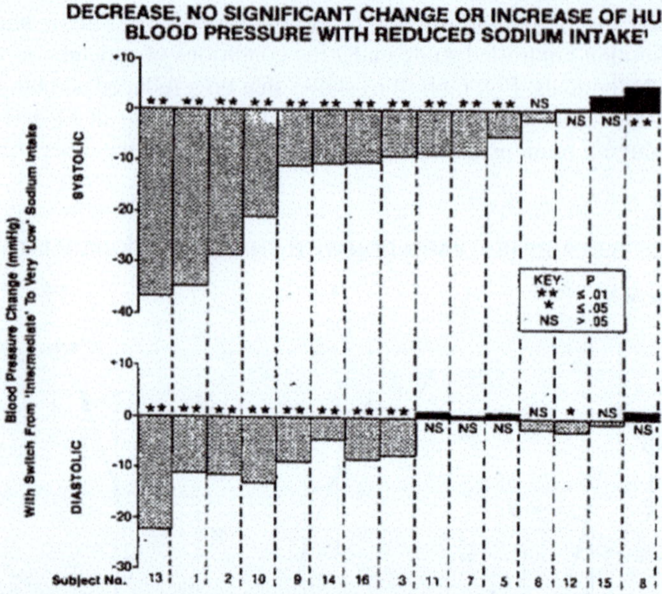

¹From 109 to 9 mEq Na⁺/day, with 70 mEq K⁺/day, each stage lasting one week. Change gauged individually by test of equality (in the two stages) of circadian MESOR (24-h rhythm-adjusted mean). Subject 4 did not contribute data during stage I (intermediate salt intake) and hence is not listed.

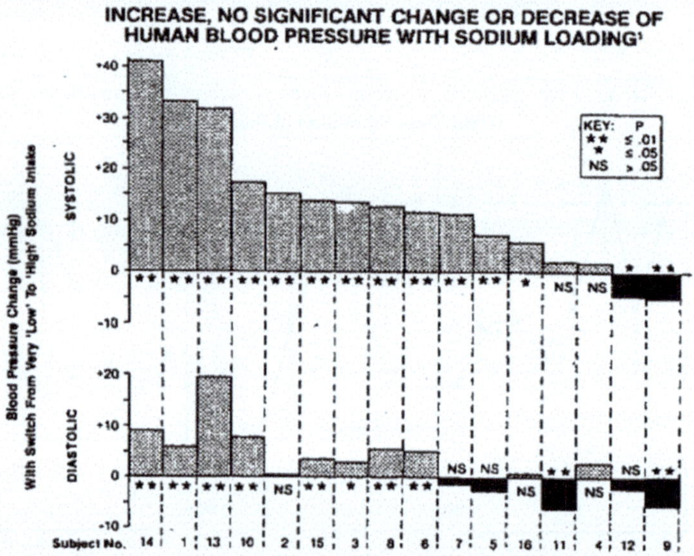

¹From 9 to 249 mEq Na⁺/day, with 70 mEq K⁺/day, each stage lasting one week. Change gauged individually by test of equality (in the two stages) of circadian MESOR (24-h rhythm-adjusted mean).

Fig. 6.6 (continued) C. Bartter, Catherine Delea, and Terukazu Kawasaki sent for analyses to Minnesota in the 1970s (© Halberg Chronobiology Center). (**b**) On an individualized basis, a majority of responses ascertained individually are in keeping with the "sodium hypothesis," yet

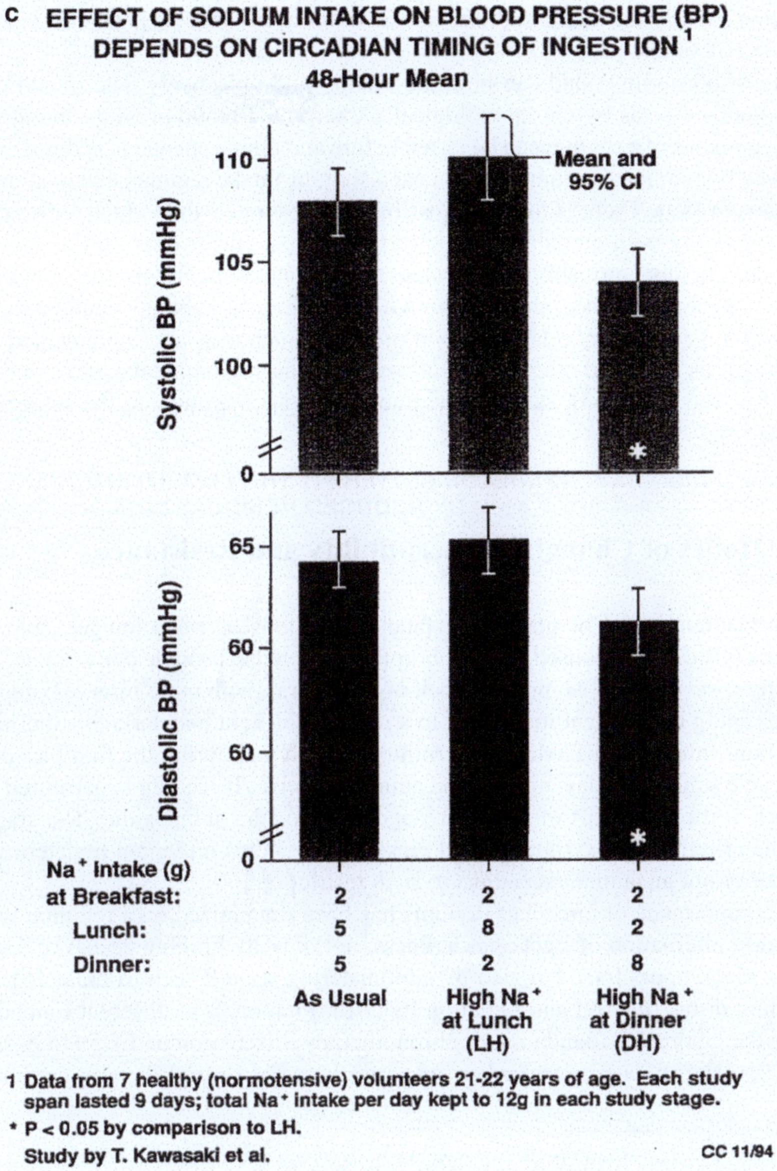

Fig. 6.6 (continued) there are also nonresponders and inverse responders (in *black*) (original data of Frederic C. Bartter, Catherine Delea, and Terukazu Kawasaki) (© Halberg Chronobiology Center). (**c**) *When* to salt also plays a role [49, 50, 52, 53] (© Halberg Chronobiology Center)

is higher (when the daily sodium intake is reduced) [50, 51], as subsequently extended by others [49].[3]

The response must and can be more readily and objectively determined on an individualized basis by chronobiologically interpreted blood pressure monitoring. The hypothesis of a rhythm can be tested before and after a change in sodium intake, followed by parameter estimations on each sodium-intake regimen and a parameter comparison [59]. There is thus the possibility to test one's individualized response cost-effectively with ambulatory monitoring and parameter tests on the data thus obtained. Thereby, probabilities (P-values) used thus far mainly for research can be used for decisions concerning one's own response to a habitual manipulation of sodium intake as it may relate to blood pressure. Such tests are implemented cost-effectively, as indicated elsewhere. But again, timing sodium intake makes a difference (Fig. 6.6c) [52, 53], as seems pertinent to a vast literature on the subject [45, 46, 54–58].

6.9 Hours of Changing Susceptibility and Resistance

A mammalian model of potentially fatal interactions between hunger, cold, and rhythms is the singly housed mouse abruptly restricted to a single daily "meal." The short-term survival of the mouse (albeit not of the rat, with more reserves) depends on the timing of the meal in relation to a regimen of light and darkness alternating at 12-hour intervals. If feeding is permitted for 4 hours during the first part of the dark cycle stage each day, most of the animals survive. If feeding is permitted only 4 hours in the early part of the light stage, most of the animals die. The stage of circadian rhythms at the time of food presentation is thus important in determining whether or not an animal can adjust to such restriction [60].

The importance of circadian rhythms has been dramatized on a regimen with a 12-hourly alternation of light and darkness, not only by limiting access to food to certain clock hours [60], but also by administering stimuli such as noise [61, 62], including drugs [63–69] and radiation [64, 70], to animals at different timepoints, laying the laboratory foundation of chronotherapy (discussion on LD50 to X-ray in [64, 70]) (Fig. 6.7a–e): even when mice are kept for several days in continuous

[3] Figure 6.6 A-B stems from the 1970s, when one of us served as consultant to Frederic C. Bartter, then head of the Hypertension-Endocrine Branch of the NIH. In his data, we found the inverse response, when it was not yet replicated. Later, after Bartter moved to Texas, he called to inform us that he had replicated the finding of the inverse response. Bittle and Molina published the results, posthumously crediting Bartter as a coauthor [49]. Independently, Ludwig Pöllmann, as an occupational physician in Frankfurt, Germany, also found inverse responses, albeit extremely rarely, and had controversies with his colleagues, who dismissed the results as a regression to the mean. Pöllmann relates (albeit anecdotally) his experience with a factory worker who had self-measured his blood pressure, found that it increased when he was salt-deprived (on the low-sodium diet provided by his employer), and thus always kept his own salt handy when he ate.

a

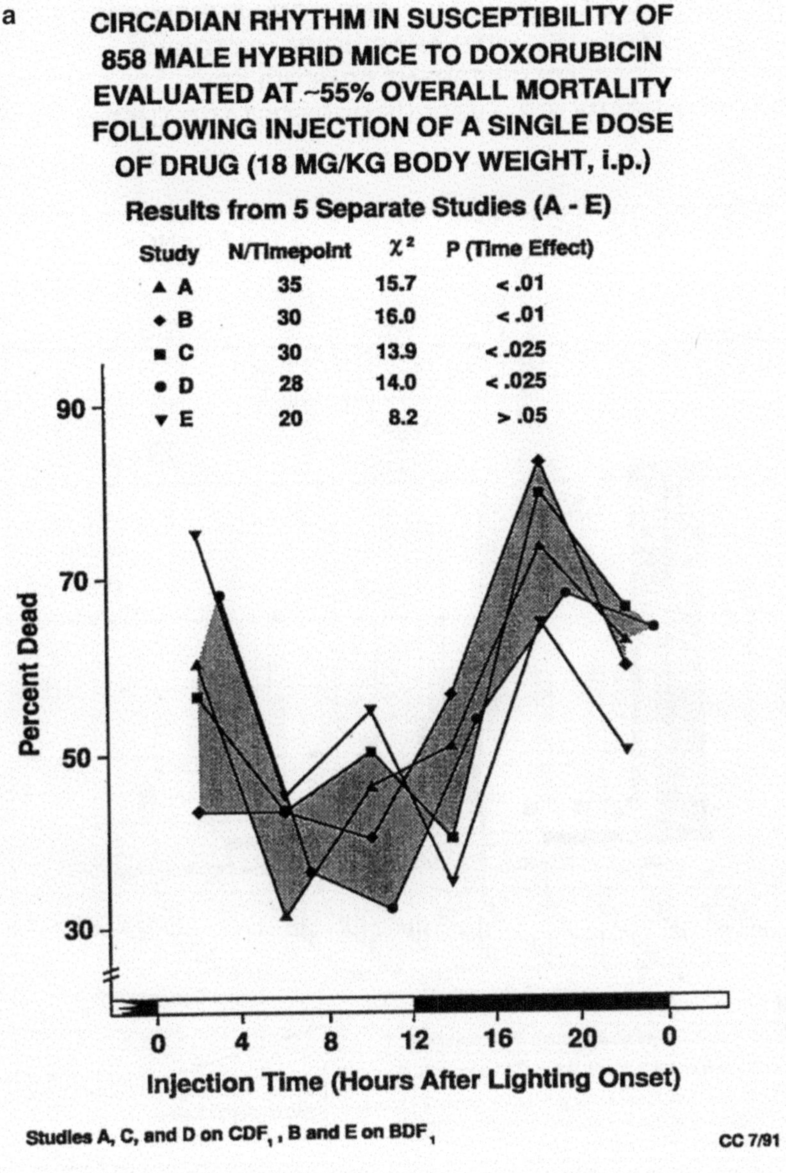

Fig. 6.7 (**a**) Studies in chronotoxicology with ethanol and ouabain in the 1950s, [24, 66; cf. 64, 65, 67] not here shown, revealed that the same dose of the same drug killed most of a group of inbred mice at one time, but was associated with a survival of the majority of animals at another rhythmically alternating and, to that extent, predictable time. Data on the tolerance of doxorubicin followed, showing the times of differential lethality. The extent of variability qualifies circadian susceptibility rhythms and leads to the suggestion of marker rhythmometry (© Halberg Chronobiology Center). (**b**) In the model chosen by contractors for NIH, the effect of ara-C in equidistant equal doses in the treatment of L1210 leukemia was improved by redistributing the same total dose as a sinusoidal treatment pattern with gradually changing doses. Sinusoidal treatment proved to be superior from the viewpoints of tolerance (toxicity) and of effectiveness

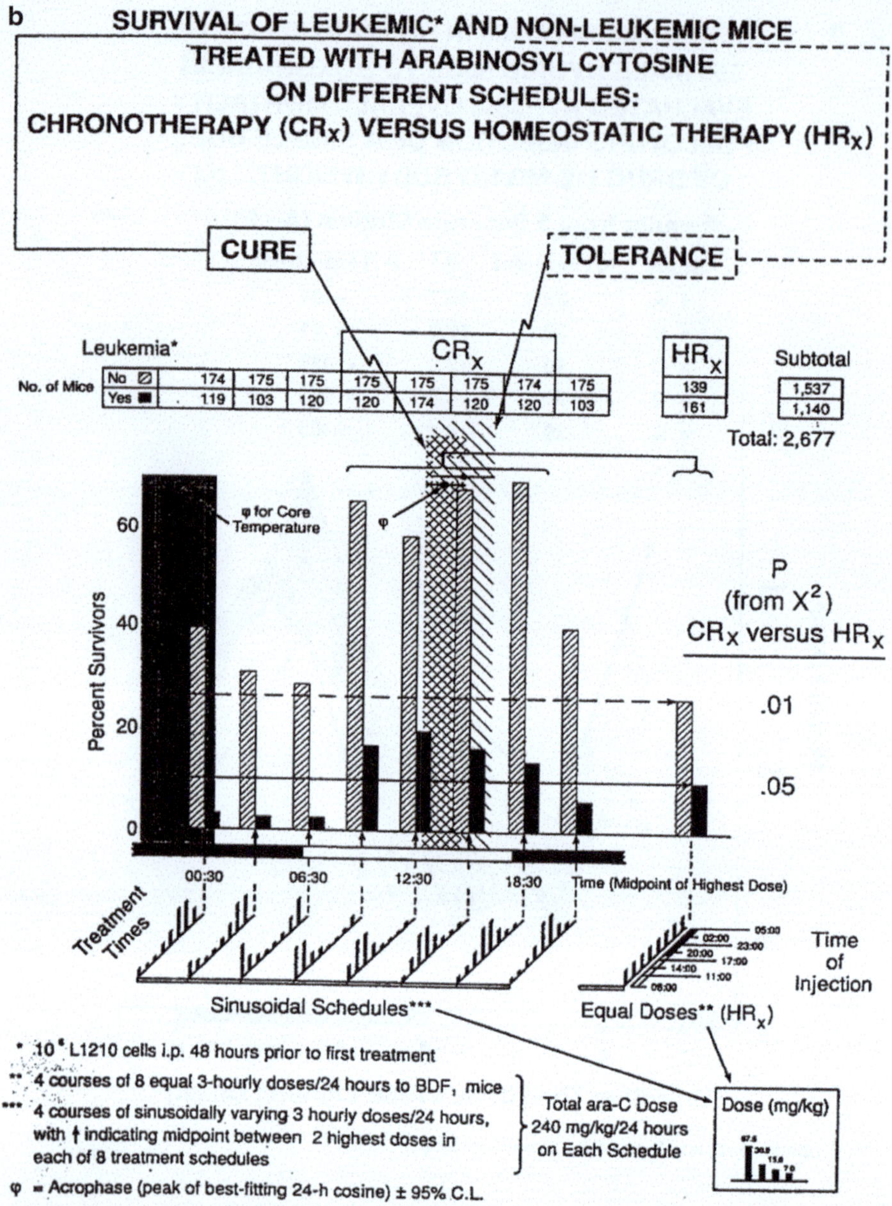

Fig. 6.7 (continued) (survival time) as a step toward cure (© Halberg Chronobiology Center). (**c**) Circadian rhythm in therapeutic index demonstrated in 1973 [85], still awaiting implementation by marker rhythms for chronochemotherapy in the clinic (© Halberg Chronobiology Center). (**d**) Confirmation in data published from Alabama of the merits of sinusoidal treatment, when cures are taken into account. The work in Alabama was mandated by Gordon Zubrod, the late former head of chemotherapy at the NIH, and was first published by comparing survival times, without accounting for cures. It was presented in an invited lecture at the FDA under the title "ignorance of chronotherapy, or rather indolence, and eventually criminal negligence." Subsequently, when a

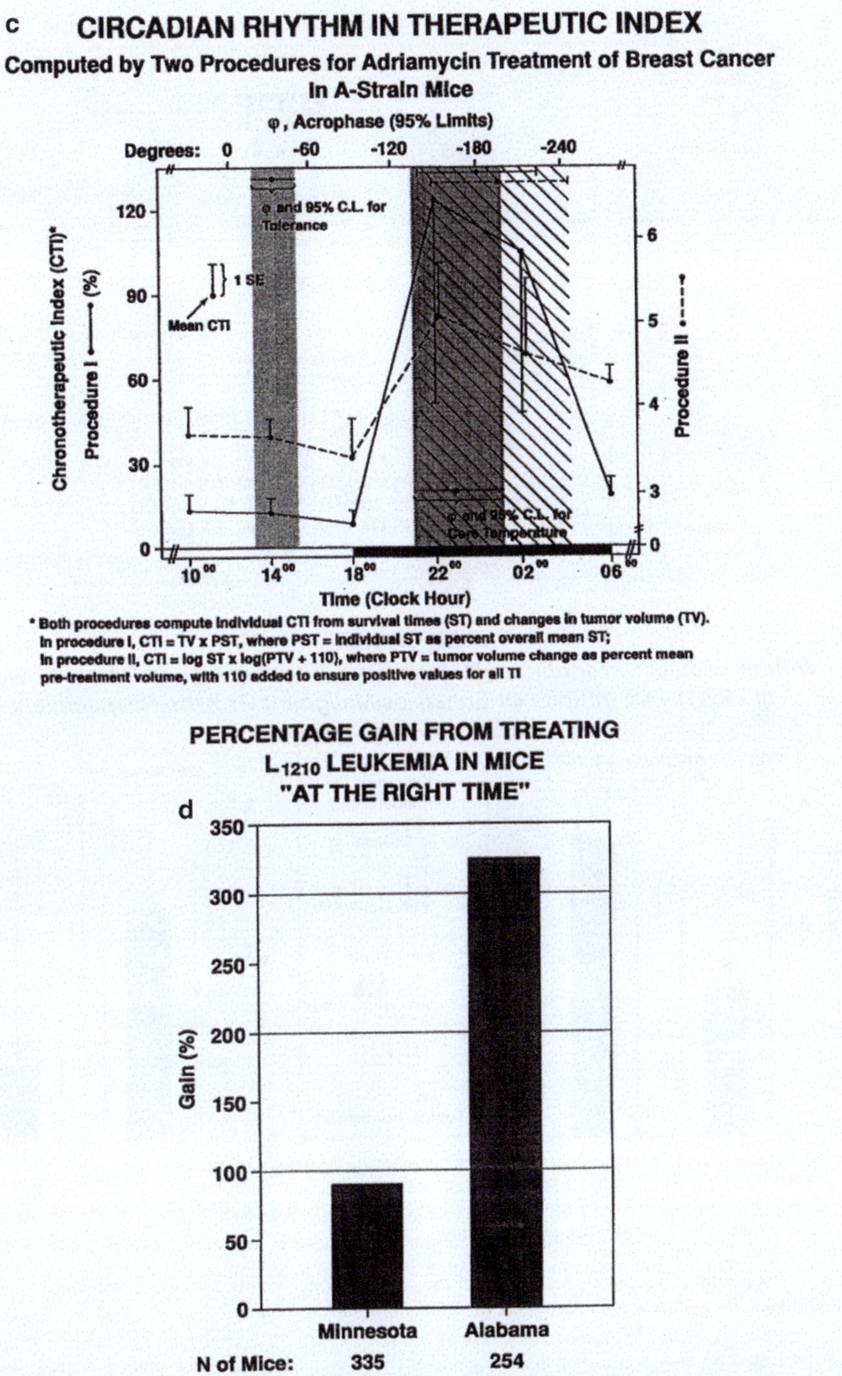

Fig. 6.7 (continued) delegation of chronobiologists including the late Frederic C. Bartter, the late Lawrence E. Scheving, and the senior author presented the results to the commissioner of the FDA for further testing and transfer to the clinic, the commissioner assured the delegation of full

e

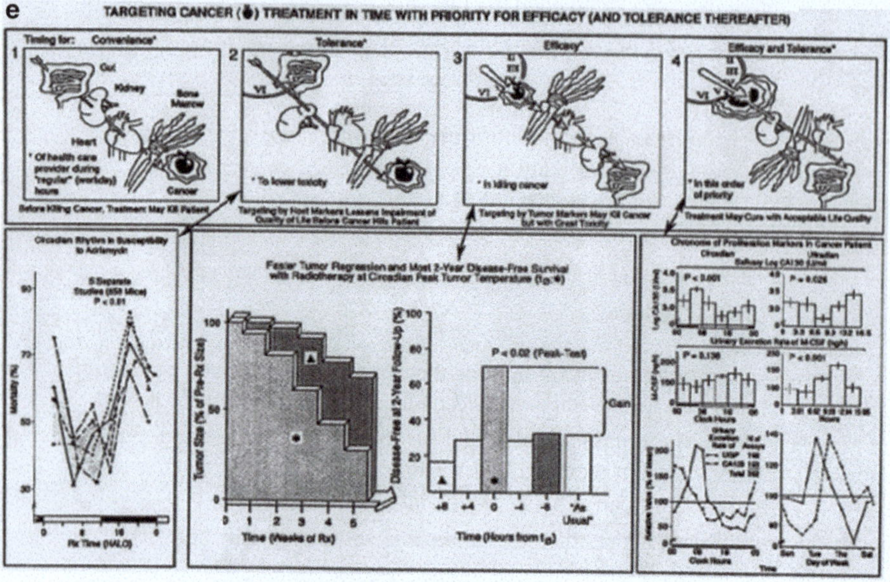

f

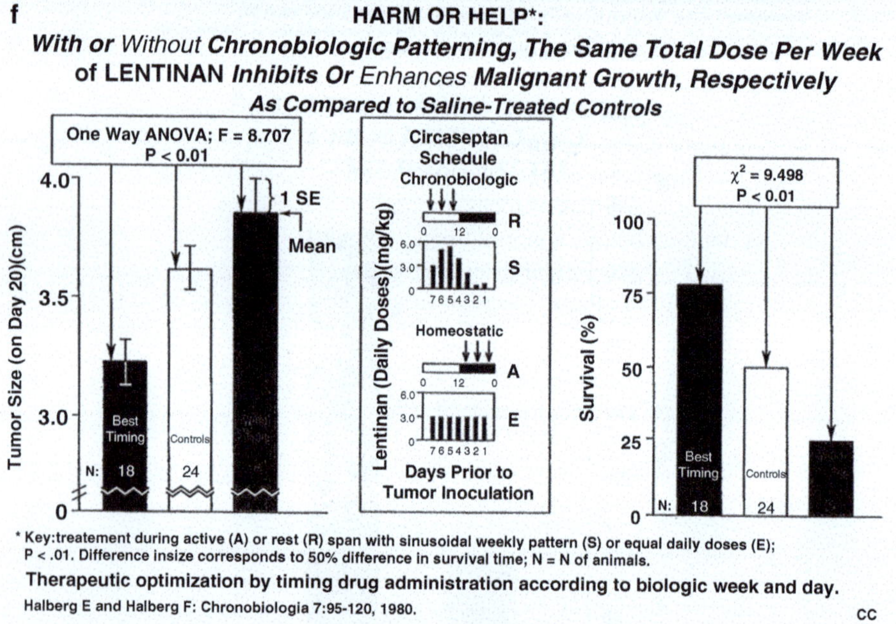

Fig. 6.7 (continued) support. A member of the commissioner's staff who was a colleague in chronobiology reported, however, that after the delegation left, the commissioner called for the pursuit of "business as usual." When the authors advocated further clinical trials, an NIH statistician present protested against action because it would involve including "another sheet of paper" in the protocol (© Halberg Chronobiology Center). (**e**) B.D. Gupta and A.K. Deka carried out the first marker rhythm-guided chronotherapy in Chandigarh, India [36]. Their data show that when perioral temperature is measured and the treatment is administered at the time of peak tumor temperature, the disease-free survival rate 2 years later can be doubled. The tumors are only of historical

darkness in controlled environmental temperature, most of those treated at one circadian time may die, whereas when the same dose of a drug is given at another circadian time, most will live [71]. The optimization by a sinusoidally varying drug administration pattern according to about-7-day [72, 73; cf. 13] as well as other rhythms can account for the difference between enhancing and inhibiting a malignant growth and thus shortening or lengthening survival time [13, 72] (Fig. 6.7f). Just as an organism's response to toxins and drugs varies according to circadian and other, e.g., about-weekly or about-yearly rhythms, so too does its response to nutrients, as indicated by circadian cycles in glucose tolerance, the rate of amino acid metabolism and the respiratory quotient, among others [6, 8, 73–78].

In the broad context of earlier work, the liver's "ability to adapt temporally to changes in the feeding pattern" [2] and that of other organs, notably of the lung [2], heart, kidney, and pancreas [5], is worthy of further study of the manipulability of the circadian system, apparent also for rodents from Fig. 6.5d. Figure 6.5a–c shows for humans that the timing of food intake can shift the circadian rhythm in circulating insulin and glucagon under conditions of meal timing that do not at all or barely change the temporal placement of the circadian rhythms in urinary cyclic AMP excretion or in circulating cortisol, respectively [7, 76]. Another recent paper reports on the resetting of circadian timing in peripheral tissues by glucocorticoid [79]. Such findings [2, 5] can all be viewed in the light of an earlier demonstration of the ability of another glucocorticoid, prednisone [80], just like meal timing, to shift several variables in human patients with asthma, and the added finding not explored recently [2, 5, 79] of a cosinor-assessed phase response curve as a function of the glucocorticoid's timing [80]. These results may prompt further systematic studies around the clock with regard for an ever-broadening spectrum of rhythms, i.e., with a view of other chronome-documented rhythms, rather than at a single time point [2].

As to circadians, we concur with Stokkan et al. [2], who conclude that manipulating the "entrainment of the liver (and perhaps other peripheral organs) to a specific phase might increase both the practicality and the therapeutic efficacy of timed administration of drugs or radiation," since in Fig. 6.5e the possibility of shifting the susceptibility rhythm to doxorubicin has been demonstrated over a quarter century ago [6]. The use of marker rhythms has already proved itself in cancer chronoradio-

Fig. 6.7 (continued) interest now, since with present-day treatment procedures, they never become large enough to interfere with eating or breathing and to be thus sufficiently large to be readily accessible for the measurement of tumor temperature as a marker, as in the case of some of the patients summarized herein. While the kind of tumor studied is only of historical interest, the results for prospective studies are a *historic* demonstration of the use of marker rhythms (© Halberg Chronobiology Center). **(f)** It is useful to focus on a circadian mechanism, exclusive focus may harm the patient if critical about-weekly (circaseptan) modulations and the broader chronome are ignored. In the case summarized in this graph, the about-weekly schedule contributes to the difference between the stimulation and inhibition of a subsequently implanted immunocytoma. This finding is of critical importance for any treatment aimed at timely cancer prevention; it is possibly also pertinent for the treatment of an actual condition, cancer, or others [72]. The gain from a sinusoidal vs. equal-dose treatment pattern is also demonstrated for cyclosporine in the case of pancreas transplantation (not shown) [73] (© Halberg Chronobiology Center) therapy (Fig. 6.7e) [36, 81] and is particularly attractive also for timing drugs and further for assessing their effect upon the malignant growth [82, 13, 83–90].

6.10 Implications of Circadian Rhythms in Therapeutic Index

In the laboratory, a circadian rhythm of the therapeutic index was documented by 1973 [85]. It was experimentally validated by assessing doxorubicin-associated mouse breast cancer shrinkage and survival time augmentation on six groups of inbred A-strain mice with spontaneous mammary cancers, feeding ad libitum in light from 06.00 to 18.00 alternating with darkness. The animals were given a fixed dose of doxorubicin i.p. at different circadian times. The numbers of survivors and caliper measurements of tumor volume were recorded to find the best timing, if any, at which doxorubicin reduced tumor size and increased survival time. On the basis of these data, the chronotherapeutic index (CTI) was defined as CTI = PST x TVC, where PST = an individual's percent survival time relative to overall mean (≡100 %) of all treated individuals, irrespective of treatment time, and TVC = the individual's extrapolated 47-hour posttreatment tumor volume change (a) relative to individual's extrapolated 1-hour pretreatment tumor volume (b) expressed as ratio (a/b). Results thus analyzed showed the highest CTI (=122 %) at 22:00, and the lowest CTI (=8 %) at 18:00. A 24-hour cosine model fitted by least squares showed statistical significance for the original mean survival times, for survival times expressed as percentage of overall mean, and for CTI (Fig. 6.7c) [72]. Also in the laboratory, a sinusoidal administration pattern has already led to an increase in cures (Fig. 6.7d) [13]. In these same studies and in human patients, we started triangulating for maximal efficacy with minimal toxicity, in this order of priority [36, 82] (cf. Fig. 6.7e). Tumor marker rhythms have been tested both for guiding treatment timing and as substitutes for the survival time for gauging the short-term treatment effect.

6.11 Scrambling Rhythms

We cannot overemphasize the need for further research on chronomics, with marker rhythms for both, securing the desired effect and for avoiding undesired toxicities (Fig. 6.7a–e). Figure 6.5f shows that the scrambling of rhythms can serve for cancer prevention. How it relates to treatment optimization remains to be tested. A long series of studies on ara-C (Fig. 6.7b, d), then on doxorubicin (Fig. 6.7c) [85], and cisplatinum [82, 90] underlie a clinical study [82, 88, 89]: in a first stage, we aimed at finding the treatment time associated with least toxicity by alternating doxorubicin in the morning and cisplatinum in the evening on the one hand with cisplatinum in the morning and doxorubicin in the evening on the other hand. The former schedule was associated with lesser toxicity in this first stage of using toxicity markers [88, 89]. Thereafter, the two schedules were compared and the former schedule was found to be superior to the latter, as reported in Science [88]. Most interesting, and a challenge awaiting further research, was that the alternating schedule was better yet [89], perhaps because it covered more, if not all circadian stages, thereby helping the last cancer cell kill. Whether there was also an effect from scrambling rhythms in ways other than alternating the treatment remains an open question for which future study may utilize some mapped aspects of the chronomes of tumor markers [84]. Finally, the observation that some tumors may be free-running may itself render it essential to use marker rhythmometry rather than "flying blind" [91; cf. 92].

6.12 Disease-Risk Syndromes

The major applied purpose for mapping chronomes is the detection of disease-risk syndromes such as blood pressure "overswinging" or heart rate "underswinging" [92]. Weeklong or longer monitorings of blood pressure and heart rate, coupled to time-structure analyses, detect disease-risk elevations, as a warning of the need for a preventive *prehabilitation* [92–94]. Within the normal range of physiological variation, computer methods have already quantified chronomes that can serve as reference values. As compared with a (gender- and age-matched) peer group-derived threshold, too much blood pressure variability (*c*ircadian *h*yper-*a*mplitude-*t*ension (CHAT)) and too little heart rate variability are indicators of great vascular disease risk. *C*hronome *a*lterations of *h*eart *r*ate *v*ariabilities (CAHRVs) consist of a loss of "jitter," e.g., a reduced circadian standard deviation of heart rate or of other alterations in the spectral element of the heart rate variability chronome, such as in the circadian rhythm of the correlation dimension [14, 18], or in other endpoints of deterministic chaos, such as the approximate entropy. The recognition of these alterations can again prompt prehabilitation in health rather than waiting for the fait accompli of disease, as noted elsewhere [14, 92–94].

While it has not been shown directly as yet that treating CHAT removes the risks of stroke and nephropathy, the possibility of eliminating CHAT by autogenic treatment or drugs is documented [92, 93, 95, 96]. Diastolic CHAT statistically significantly increases the risk of nephropathy by 590 %, even when the 24-hour average of systolic/diastolic blood pressure is below 130/80 mmHg (by comparison, a decreased heart rate variability statistically significantly increases the risk of nephropathy by 440 %). CHAT may be transient; hence, a 7-day span of ambulatory monitoring has been recommended by an international consensus meeting [94]. For the reliable diagnosis of an abnormality such as an overall blood pressure elevation as well as for detecting CHAT, monitoring for a week is prompted by day-to-day as well as within-day variability, long known to be large and to characterize circadian and ultradian characteristics.

The 7-day/24-hour monitoring required for the diagnosis of CHAT is useful in any event to detect reliably an overall elevation in blood pressure, the proven etiopathogenic factor. Indeed, a high blood pressure can be detected much more reliably with the 7-day/24-hour approach, which reveals that many measurements below 140/90 mmHg are found during office hours in conventionally diagnosed hypertensives in the absence of antihypertensive treatment. A few patients have over 50 % of their values below the 140/90 mmHg limit during office hours. The merit of weeks-long automatic monitoring of those who may be on medication they do not need, as documented by placebo spans of a week or longer, is also reported [97]. During office hours, normotensives can have occasional hypertensive values and vice versa. Even a 24-hour ambulatory blood pressure average can be high for several days and acceptable for months or years of ambulatory monitoring thereafter. Current practice based on official guidelines needs improvement, and the 7-day automatic ambulatory monitoring is practical and requires no more work on the part of care providers than monitoring for 6 [98] or 24 hours. Instruments for research use by the practitioner are available with a large price reduction through the Halberg Chronobiology Center at the University of Minnesota, along with analyses and recommendations in the light of reference standards from clinically healthy peers, specified by clock hour as well as gender and age for Japanese as well as Caucasians. The disciplined patient with diabetes, used to self-help via self-determinations of blood glucose, can be used as a role model for many other nondiabetic patients with blood pressure problems who can also learn self-help.

6.13 Challenges for Industry

The current disease-cure-oriented health-care system does too little too late, at too great an expense for those who can afford it. Engineering for concomitantly resolving the internal schedules and how they are influenced by external conditions could change the status quo. Chronoengineering in the broad sense could develop a system that recognizes risk elevation early enough to eventually prompt the immediate institution of preventive measures administered according to an optimal schedule [93, 94]. Chronobioengineering for contemporaneous external as well as internal monitoring and decision-helping could improve the unobtrusive resolution of in-part predictable variations that occur within the physiologic range and could secure benefit for the given person by contrast to merits assessed only as an average for the population. The resolution of chronomes usually yields not only a more precise and more accurate estimate of the mean value, but also new dynamic endpoints, useful for screening, diagnosis, and prognosis and, when needed, for the institution of timely and timed treatment (chronotherapy). The implementation of engineering for concomitantly resolving the external and internal schedules depends on telecommunications and telehygiene as well as telemedicine.

Several ingredients are essential for the development of such a system. Even though much progress has already been made in each field separately, the different modules, which have largely evolved as separate units, will need to be integrated [93]. First, information needs to be acquired in the form of data collected at intervals over time. This is a task for the monitoring industry. Second, clinically pertinent information needs to be extracted from the data series, an analytic task for which the principles of windowing, compacting, and recycling by repeated passes over the accumulating data have been advocated. Last but not least is the need for reference values serving for the interpretation of the data and the endpoints derived therefrom.

Archives constructed around the chronomes of blood pressure and heart rate monitored from womb to tomb are already available and are still accumulating, yet they are lacking for several populations in the USA and abroad. Reference values for the diagnosis of CHAT or CAHRVs need to be adjusted as a function not only of gender and age, but also for ethnicity, in order to know the threshold values beyond which risk is increased [93, 94]. The integration of all three ingredients (monitoring, analysis of the given series, and the interpretation of results in the light of reference values) will ultimately allow closing the loop between instrumentation for the monitoring of vital signs and instrumentation for the delivery of timely treatment patterned to maximize its efficacy and minimize its side effects. The need for analysis was visualized a quarter century ago [99] and that for assessing variability before an office visit nearly a full century ago [100].

The systematic monitoring of external rhythms such as those in geo- and/or anthropo-magnetics, along with physiological ones for the concurrent analysis of rhythms with longer and longer periods, detects alterations anywhere around and between the 40 cycles/sec and the 1 cycle/10.5- or 21- or 50-year regions of the spectra of action potentials in humans [101]. Chronobiomimetic engineering for discovering both instantaneous and long-term chronorisk alterations can provide warnings of increased risk associated, for example, with interplanetary magnetic storms, which may bring about blackouts in the grids not only of power stations, but can also trigger myocardial infarctions, strokes, and traffic accidents by acting on the grids of the heart and brain. If risk-lowering therapy is then instituted automatically, health care will be extended beyond the pacemaker-cardioverter-defibrillator, which focuses on the frequency of 1 cycle/sec, yet it is also affected by chronomes as a whole. Instrumentation that automatically detects increased risks in the light of a data base of reference values that can be used for chronodiagnosis is now accumulating. It should be refined to serve both chronoastrobiology and a practical chronotherapy that aims treatment and assesses its effect on the basis of chronobiologically interpreted physiological and physical monitoring in time, the common link of all science.

References

1. Pennisi E. Multiple clocks keep time in fruit fly tissues. Science 1997;278:1560–1.
2. Stokkan K-A, Yamazaki S, Tei H, Sakaki Y, Menaker M. Entrainment of the circadian clock in the liver by feeding. Science 2001;291:490–2.
3. Szuromi P (editor). This week in Science: setting some clocks ahead. Science 2001;291:395.

4. Roenneberg T, Morse D. Two circadian oscillators in one cell [letter]. Nature 1993;362: 362–4.
5. Damiola F, Minh NL, Preitner N, Kornmann B, Fleury-Olela F, Schibler U. Restricted feeding uncouples circadian oscillators in peripheral tissues from the central pacemaker in the suprachiasmatic nucleus. Genes Dev 2000;14:2950–61.
6. Halberg F. From aniatrotoxicosis and aniatrosepsis toward chronotherapy: introductory remarks to the 1974 Capri symposium on timing and toxicity: the necessity for relating treatment to bodily rhythms. In: Aschoff J, Ceresa F, Halberg F, editors. Chronobiological aspects of endocrinology. Stuttgart: F.K. Schattauer Verlag; 1974. p. 1–34.
7. Halberg F, Haus E, Cornélissen G. From biologic rhythms to chronomes relevant for nutrition. In: Marriott BM, editor. Not eating enough: overcoming underconsumption of military operational rations. Washington, DC: National Academy Press; 1995. p. 361–72.
8. Reinberg A. Chronopharmacology in man. In: Aschoff J, Ceresa F, Halberg F, editors. Chronobiological aspects of endocrinology. Stuttgart: F.K. Schattauer Verlag; 1974. p. 157–85.
9. Halberg F. When to treat. Haematologica (Pavia) 1975;60:1–30.
10. Johnson MS. Effect of continuous light on periodic spontaneous activity of white-footed mice *(Peromyscus)*. J Exp Zool 1939;82:315–28.
11. Halberg F. Some physiological and clinical aspects of 24-hour periodicity. J Lancet (Minneapolis) 1953;73:20–32.
12. Cornélissen G, Engebretson M, Johnson D, Otsuka K, Burioka N, Posch J, Halberg F. The week, inherited in neonatal human twins, found also in geomagnetic pulsations in isolated Antarctica. Biomed Pharmacother 2001;55(Suppl 1):32s–50s.
13. Cornélissen G, Halberg F. Introduction to chronobiology. Medtronic chronobiology seminar #7, April 1994, 52 pp. (Library of Congress Catalog Card #94-060580. https://sites.google.com/a/umn.edu/halbergchronobiologycenter/.
14. Otsuka K, editor. Chronome & Janus-medicine: Heart Rate Variability (HRV) and BP Variability (BPV) from a viewpoint of chronobiology and ecology. Tokyo: Medical Review; 1998. 213 p.
15. Lederberg J, McCray AT. 'Ome sweet 'omics—a genealogical treasury of words. The Scientist, April 2, 2001, p. 8.
16. Burioka N, Cornélissen G, Halberg F, Kaplan DT. Relationship between correlation dimension and indices of linear analysis in both respiratory movement and electroencephalogram. Clin Neurophysiol 2001;112:1147–53.
17. Goldberger AL. Fractal variability vs. pathologic periodicity: complexity loss and stereotypy in disease. Perspect Biol Med 1997;40:543–61.
18. Otsuka K, Cornélissen G, Halberg F. Circadian rhythmic fractal scaling of heart rate variability in health and coronary artery disease. Clin Cardiol 1997;20:631–8.
19. Halberg F, Cornélissen G, Bingham C, Tarquini B, Mainardi G, Cagnoni M, Panero C, Scarpelli P, Romano S, März W, Hellbrügge T, Shinoda M, Kawabata Y. Neonatal monitoring to assess risk for hypertension. Postgrad Med 1986;79:44–6.
20. Halberg F, Cornélissen G, Chen C-H, Katinas GS, Otsuka K, Watanabe Y, Herold M, Loeckinger A, Kreze A, Kreze E, Perfetto F, Tarquini R, Maggioni C, Sothern RB, Schwartzkopff O. Chronobiology: time structures, chronomes, gauge aging, disease risk syndromes and the cosmos. J Anti-Aging Med 2000;3:67–90.
21. Halberg F, Cornélissen G, Watanabe Y, Otsuka K, Fiser B, Siegelova J, Mazankova V, Maggioni C, Sothern RB, Katinas GS, Syutkina EV, Burioka N, Schwartzkopff O. Near 10-year and longer periods modulate circadians: intersecting anti-aging and chronoastrobiological research. J Gerontol A Biol Sci Med Sci 2001;56:M304–24.
22. Halberg F, Cornélissen G, Otsuka K, Watanabe Y, Katinas GS, Burioka N, Delyukov A, Gorgo Y, Zhao ZY, Weydahl A, Sothern RB, Siegelova J, Fiser B, Dusek J, Syutkina EV, Perfetto F, Tarquini R, Singh RB, Rhees B, Lofstrom D, Lofstrom P, Johnson PWC, Schwartzkopff O, International BIOCOS Study Group. Cross-spectrally coherent ~10.5- and 21-year biological and physical cycles, magnetic storms and myocardial infarctions. Neuroendocrinol Lett 2000;21:233–58.

23. Cornélissen G, Halberg F. Chronomedicine. In: Armitage P, Colton T, editors-in-chief. Encyclopedia of biostatistics, v. 1. Chichester: John Wiley & Sons Ltd; 1998. p. 642–9.
24. Halberg F. Chronobiology. Ann Rev Physiol 1969;31:675–725.
25. Mikulecky M, editor. The moon and living matter. Kosice, Slovakia, September 23–25, 1993. Bratislava: Slovak Medical Society; 1993. 97 p.
26. Mikulecky M, editor. Sun, moon and living matter. Bratislava, Slovakia, June 28–July 1, 1994. Bratislava: Slovak Medical Society; 1994. 159 p.
27. Mikulecky M, editor. Chronobiology & its roots in the cosmos. High Tatras, Slovakia, September 2–6, 1997. Bratislava: Slovak Medical Society; 1997. 287 p.
28. Otsuka K, organizer. Proceedings, 1st International Symposium, Workshop on Chronoastrobiology & Chronotherapy (Satellite Symposium, 7th Annual Meeting, Japanese Society for Chronobiology), Kudan, Chiyodaku, Tokyo, 11 Nov 2000, pp 170.
29. Halberg F, Cornélissen G, Katinas G, Hillman D, Schwartzkopff O. Season's appreciations 2000: chronomics complement, among many other fields, genomics and proteomics. Neuroendocrinol Lett 2001;22:53–73.
30. Halberg F, Halberg E, Halberg J. Collateral-interacting hierarchy of rhythm coordination at different organization levels, changing schedules and aging. In: Suda M, Hayaishi O, Nakagaga H, editors. Biological rhythms and their central mechanism. Naito Foundation. Amsterdam: Elsevier North-Holland Biomedical Press; 1979. p. 421–34. see also discussion p. 435–38.
31. Aschoff J. Speech after dinner. 1974 Capri symposium on timing and toxicity. Chronobiologia 1974; 1 (Suppl. 1) 483–95.
32. Virey JJ. Ephémérides de la vie humaine ou recherches sur la révolution journalière et la periodicité de ses phénomènes dans la santé et les maladies. Thesis, University of Paris; 1814, 39 pp.
33. Jores A. 3mal täglich. Hippokrates (Stuttgart) 1939;10:1185–8.
34. Menzel W. Menschliche Tag-Nacht-Rhythmik und Schichtarbeit. Basel/Stuttgart: Benno Schwabe & Co. Verlag; 1962.
35. Anis Anwar Y, White WB. Chronotherapeutics for cardiovascular disease. Drugs 1998;55:631–43.
36. Halberg F, Gupta BD, Haus E, Halberg E, Deka AC, Nelson W, Sothern RB, Cornélissen G, Lee JK, Lakatua DJ, Scheving LE, Burns ER. Steps toward a cancer chronopolytherapy. In: Proc. XIV International Congress of Therapeutics; Montpellier: L'Expansion Scientifique Française; 1977. p. 151–96.
37. Brillat-Savarin JA. Aphorismes 3 et 4. In Physiologie du gout, ou, méditations de gastronomie transcendente. Ouvrage théorique, historique et à l'ordre du jour, dedié aux gastronomes parisiens. Paris: A. Soutelet & Co.; 1826: vol. I.
38. American Dietetic Association. National Nutrition Month®: Food & fitness: build a healthy lifestyle. [Display card.] American Dietetic Association; Beltsville, MD, USA, 2001.
39. Graeber RC, Gatty R, Halberg F, Levine H. Human eating behavior: preferences, consumption patterns and biorhythms. NATICK/TR-78/022 Technical reports, U.S. Army; 1978, 287 p. Available while supplies last on request from the University of Minnesota Chronobiology Center, corne001@umn.edu.
40. Halberg J. Catfish anyone? Chronobiologia 1974;1:127–9.
41. Sundararaj BI, Nath P, Halberg F. Circadian meal timing in relation to lighting schedule optimizes catfish body weight gain. J Nutr 1982;12:1085–97.
42. Carlebach R, Ashkenazi IE. Temporal disorder: therapeutic considerations. Chronobiologia 1987;14:162.
43. Halberg F, Sánchez S, Nelson W. Rhythm scrambling and tumorigenesis in $CD2F_1$ mice. In: Tarquini B, Vergassola R, editors. III Int. Symposium, Social Diseases and Chronobiology, Florence, Nov. 29; 1986. p. 59–61.
44. Kort WJ, Zondervan PE, Hulsman LOM, Weijma JM, Westbroek DL. Light-dark-shift stress, with special reference to spontaneous tumor incidence in female BN rats. JNCI 1986;76:439–46.

45. Intersalt Cooperative Research Group. INTERSALT: an international study of electrolyte excretion and blood pressure. Results for 24-hour urinary sodium and potassium excretion. BMJ 1988;297:319–28.
46. Law M. Commentary: evidence on salt is consistent. BMJ 1996;312:1284–5.
47. MacGregor GA, Sever PS. Salt—overwhelming evidence but still no action: can a consensus be reached with the food industry? BMJ 1996;312:1287–9.
48. Freedman DA, Petitti DB. Salt and blood pressure: conventional wisdom reconsidered. Evaluation Review 2001;25:267–87.
49. Bittle Jr CC, Molina DJ, Bartter FC. Salt sensitivity in essential hypertension as determined by the cosinor method. Hypertension 1985;7:989–94.
50. Cornélissen G, Kawasaki T, Uezono K, Delea C, Halberg F. II: Blood pressure rhythms and salt. Ann Ist Super Sanita 1993;29:667–77.
51. Halberg F, Cornélissen G, Halberg E, Halberg J, Delmore P, Shinoda M, Bakken E. Chronobiology of human blood pressure. 4th ed. Medtronic Continuing Medical Education Seminars. Minneapolis: Medtronic Inc.; 1988. 242 p.
52. Itoh K, Kawasaki T, Cugini P. Effect of timing of salt intake to 24-hour blood pressure and its circadian rhythm. In: Portaluppi F, Smolensky MH, editors. Time-dependent structure and control of arterial blood pressure. Ann NY Acad Sci 1996;783:324–5.
53. Kawasaki T, Itoh K, Cugini P. Influence of reapportionment of daily salt intake on circadian blood pressure pattern in normotensive subjects. J Nutr Sci Vitaminol 1994;40:459–66.
54. Midgley JP, Matthew AG, Greenwood CM, Logan AG. Effect of reduced dietary sodium on blood pressure. JAMA 1996;275:1590–7.
55. Stamler J. The INTERSALT study: background, methods, findings, and implications. Am J Clin Nutr 1997;65 Suppl 2:626S–42S.
56. Stamler J, Elliott P, Dyer AR, Stamler R, Kesteloot H, Marmot M, for the INTERSALT Steering and Editorial Committee. Commentary: sodium and blood pressure in the INTERSALT study and other studies: in reply to the Salt Institute. BMJ 1996;312:1285–7.
57. Svetkey LP, Sacks FM, Obarzanek E, Vollmer WM, Appel LJ, Lin PH, Karanja NM, Harsha DW, Bray GA, Aickin M, Proschan MA, Windhauser MM, Swain JF, McCarron PB, Rhodes DG, Laws RL. The DASH diet, sodium intake and blood pressure trial (DASH-sodium): rationale and design. DASH-Sodium Collaborative Research Group. J Am Diet Assoc 1999;99(8 Suppl):S96–S104.
58. Taubes G. The (political) science of salt. Science 1998;281:898–907.
59. Bingham C, Arbogast B, Cornélissen Guillaume G, Lee JK, Halberg F. Inferential statistical methods for estimating and comparing cosinor parameters. Chronobiologia 1982;9: 397–439.
60. Nelson W, Cadotte L, Halberg F. Circadian timing of single daily "meal" affects survival of mice. Proc Soc Exp Biol Med (NY) 1973;144:766–9.
61. Halberg F, Bittner JJ, Gully RJ, Albrecht PG, Brackney EL. 24-hour periodicity and audiogenic convulsions in I mice of various ages. Proc Soc Exp Biol Med (NY) 1955;88: 169–73.
62. Halberg F, Jacobson E, Wadsworth G, Bittner JJ. Audiogenic abnormality spectra, 24-hour periodicity and lighting. Science 1958;128:657–8.
63. Ertel RJ, Halberg F, Ungar F. Circadian system phase-dependent toxicity and other effects of methopyrapone (SU-4885) in mice. J Pharmacol Exp Ther 1964;146:395–9.
64. Halberg F. Temporal coordination of physiologic function. Cold Spr Harb Symp Quant Biol 1960;25:289–310. Cf. also discussion of LD50.
65. Halberg F, Stephens AN. Susceptibility to ouabain and physiologic circadian periodicity. Proc Minn Acad Sci 1959;27:139–43.
66. Haus E. Periodicity in response and susceptibility to environmental stimuli. Ann NY Acad Sci 1964;117:281–91.
67. Haus E, Halberg F. 24-hour rhythm in susceptibility of C mice to a toxic dose of ethanol. J Appl Physiol 1959;14:878–80.

68. Pauly JE. Chronobiology: anatomy in time. Am J Anat 1983;168:365–88.
69. Scheving LE. The dimension of time in biology and medicine: chronobiology. Endeavour 1976;35:66–72.
70. Haus E, Halberg F, Loken MK, Kim US. Circadian rhythmometry of mammalian radiosensitivity. In: Tobias A, Todd P, editors. Space radiation biology. New York: Academic; 1973. p. 435–74.
71. Halberg F. Symposium on "Some current research methods and results with special reference to the central nervous system." Physiological approach. Amer J Ment Defic 1960;65: 156–71.
72. Halberg E, Halberg F. Chronobiologic study design in everyday life, clinic and laboratory. Chronobiologia 1980;7:95–120.
73. Liu T, Cavallini M, Halberg F, Cornélissen G, Field J, Sutherland DER. More on the need for circadian, circaseptan and circannual optimization of cyclosporine therapy. Experientia 1986;42:20–2.
74. Capani F, Carfagnini A, Consoli F, della Loggia A, de Ponte M, di Felice S, Donatelli T, Guagnano I, Guarnieri M, Iezzi M, Sensi S. Chronobiological aspects of energy expenditure in obesity. In: Enzi G, Crepaldi G, Pozza G, Renold AE, editors. Serono Symposium no. 28, Obesity: Pathogenesis and Treatment. London/New York: Academic; 1981. p. 87–103.
75. Caviezel F, Cattaneo AG, Matonti S, Vigano W, Pappalettera AE, Pozza G. Single daily meal treatment in obesity. In: Enzi G, Crepaldi G, Pozza G, Renold AE, editors. Serono Symposium no. 28, Obesity: Pathogenesis and Treatment. London/New York: Academic; 1981. p. 105–15.
76. Goetz FC, Bishop J, Halberg F, Sothern RB, Brunning R, Senske B, Greenberg B, Minors D, Stoney P, Smith ID, Rosen GD, Cressey D, Haus E, Apfelbaum M. Timing of single daily meal influences relations among human circadian rhythms in urinary cyclic AMP and hemic glucagon, insulin and iron. Experientia (Basel) 1976;32:1081–4.
77. Halberg F, Halberg E, Carandente F. Chronobiology and metabolism in the broader context of timely intervention and timed treatment. In: Lindenlaub E, editor. Diabetes research today, meeting of the Minkowski prize winners, Symposia Medica Hoechst 12 (Capri). Stuttgart/New York: F.K. Schattauer Verlag; 1976. p. 45–95.
78. Reinberg A. Biological rhythms and energy balance in man. In: Apfelbaum M, editor. Energy balance in man. Paris: Masson; 1971. p. 285–95.
79. Balsalobre A, Brown SA, Marcacci L, Tronche F, Kellendonk C, Reichardt HM, Schütz G, Schibler U. Resetting of circadian time in peripheral tissues by glucocorticoid signaling. Science 2000;289:2344–7.
80. Reindl K, Falliers C, Halberg F, Chai H, Hillman D, Nelson W. Circadian acrophases in peak expiratory flow rate and urinary electrolyte excretion of asthmatic children: phase-shifting of rhythms by prednisone given in different circadian system phases. Rass Neurol Veg 1969;23:5–26.
81. Halberg F, Halberg E, Barnum CP, Bittner JJ. Physiologic 24-hour periodicity in human beings and mice, the lighting regimen and daily routine. In: Withrow RB, editor. Photoperiodism and related phenomena in plants and animals, Ed. Publ. No. 55. Washington, DC: Am. Assn. Adv. Sci; 1959. p. 803–78.
82. Cornélissen G, Halberg F. The chronobiologic pilot study with special reference to cancer research: is chronobiology or, rather, its neglect wasteful? In: Goldson AL, editor, H. Kaiser (series editor). Cancer growth and progression, vol. 9, ch. 9. Dordrecht: Kluwer Academic Publishers; 1989. p. 103–33.
83. Halberg F, Halberg J, Halberg E, Halberg F. Chronobiology, radiobiology and steps toward the timing of cancer radiotherapy. In: Goldson AL, editor, H. Kaiserseries editor. Cancer growth and progression. vol. 9, ch. 19. Dordrecht: Kluwer Academic Publishers; 1989. p. 227–53.
84. Halberg F. The week in phylogeny and ontogeny: opportunities for oncology. In Vivo 1995;9:269–78.

85. Halberg F, Haus E, Cardoso SS, Scheving LE, Kühl JFW, Shiotsuka R, Rosene G, Pauly JE, Runge W, Spalding JF, Lee JK, Good RA. Toward a chronotherapy of neoplasia: tolerance of treatment depends upon host rhythms. Experientia (Basel) 1973;29:909–34.
86. Halberg F, Nelson W, Cornélissen G, Haus E, Scheving LE, Good RA. On methods for testing and achieving cancer chronotherapy. Cancer Treat Rep 1979;63:1428–30.
87. Hermida Dominguez RC, Halberg F, Langevin TR. Serial white blood cell counts and chronochemotherapy according to highest values (macrophases) or by model characteristics (acrophases). In: Halberg F, Reale L, Tarquini B, editors. Proc. 2nd Int. Conf. Medico-Social Aspects of Chronobiology, Florence, Oct. 2, 1984. Rome: Istituto Italiano di Medicina Sociale; 1986. p. 327–43.
88. Hrushesky WJM. Circadian timing of cancer chemotherapy. Science 1985;228:73–5.
89. Hrushesky WJM. Automatic chronotherapy: an integral part of the future of medicine. In: Hekkens WTJM, Kerkhof GA, Rietveld WJ, editors. Trends in chronobiology. Oxford: Pergamon Press; 1988. p. 281–93.
90. Kennedy BJ, Halberg F, Sothern RB, Halberg E, Zinneman HH, Scheving LE. Circadian and methodologic aspects of toxicity from platinum and adriamycin interaction in rats with immunocytoma. Chronobiologia 1977;4:123–4.
91. Halberg E, Halberg F, Cornélissen G, Garcia-Sainz M, Simpson HW, Taggett-Anderson MA, Haus E. Toward a chronopsy: Part II. A thermopsy revealing asymmetrical circadian variation in surface temperature of human female breasts and related studies. Chronobiologia 1979;6:231–57.
92. Halberg F, Cornélissen G, Halberg J, Fink H, Chen C-H, Otsuka K, Watanabe Y, Kumagai Y, Syutkina EV, Kawasaki T, Uezono K, Zhao ZY, Schwartzkopff O. Circadian hyper-amplitude-tension, CHAT: a disease risk syndrome of anti-aging medicine. J Anti-Aging Med 1998;1:239–59. Editor's Note by Fossel M., p. 239.
93. Cornélissen G, Halberg F, Schwartzkopff O, Delmore P, Katinas G, Hunter D, Tarquini B, Tarquini R, Perfetto F, Watanabe Y, Otsuka K. Chronomes, time structures, for chronobioengineering for "a full life". Biomed Instrum Technol 1999;33:152–87.
94. Halberg F, Cornélissen G, International Womb-to-Tomb Chronome Initiative Group. Resolution from a meeting of the International Society for Research on Civilization Diseases and the Environment (New SIRMCE Confederation), Brussels, Belgium, March 17–18, 1995: Fairy tale or reality? Medtronic Chronobiology Seminar #8, April 1995, 12 pp. text, 18 figures. https://sites.google.com/a/umn.edu/halbergchronobiologycenter
95. Watanabe Y, Cornélissen G, Halberg F, Saito Y, Fukuda K, Otsuka K, Kikuchi T. Chronobiometric assessment of autogenic training effects upon blood pressure and heart rate. Percept Mot Skills 1996;83:1395–410.
96. Watanabe Y, Cornélissen G, Halberg F, Saito Y, Fukuda K, Revilla M, Rodriguez C, Hawkins D, Otsuka K, Kikuchi T. Method and need for continued assessment of autogenic training effect upon blood pressure: case report. New Trends Exp Clin Psychiatr 1996;12:45–50.
97. Little J, Sánchez de la Peña S, Cornélissen G, Abramowitz P, Tuna N, Halberg F. Longitudinal chronobiologic blood pressure monitoring for assessing the need and timing of antihypertensive treatment. Prog Clin Biol Res 1990;341B:601–11.
98. Sheps SG, Canzanello VJ. Current role of automated ambulatory blood pressure and self-measured blood pressure determinations in clinical practice. Mayo Clin Proc 1994;69:1000–5.
99. Bartter FC. Periodicity and medicine. In: Scheving LE, Halberg F, Pauly JE, editors. Chronobiology. Tokyo: Igaku Shoin Ltd; 1974. p. 6–13. "By conventional standards, this patient is clearly normotensive every morning. But the blood pressure determined each day at 6 in the afternoon provides especially convincing evidence that this patient is a hypertensive. …My plea today is that information contained in [data curves compiled under differing circumstances, such as 24 hours a day/7 days a week] become a routine minimal amount of information accepted for the description of a patient's blood pressure. The analysis of this information by cosinor should become a routine. It is essential that enough information be collected to allow objective characterization of a periodic phenomenon, to wit, an estimate of

M as given for the three statuses in this patient, an estimate of A itself, and finally an estimate of acrophase, ϕ. In this way, a patient can be compared with himself at another time, or under another treatment, and the patient can be compared with a normal or with another patient.".
100. Janeway TC. The clinical study of blood pressure. New York: D. Appleton & Co; 1904. 300 p. "… it is essential that a record of the pressure be made at frequent intervals at some time previous [presumably to an examination], to establish the normal level and the extent of the periodic variations. When this is done, it may be possible to demonstrate changes of small extent, which, lacking this standard for comparison, would be considered within the limits of normal variation.".
101. Otsuka K, Ichimaru Y, Cornélissen G, Weydahl A, Holmeslet B, Schwartzkopff O, Halberg F, on behalf of the ICEHRV Working Group. Dynamic analysis of heart rate variability from 7-day Holter recordings associated with geomagnetic activity in subarctic area. Comput Cardiol 2000;27:453–6.

Chapter 7
Chronomics of Tree Rings Gauge Climate Change

Abstract Variations in the average annual tree rings of 11 sequoia for 2189 years are studied. The power spectrum of tree ring variations, calculated by the Maximum Entropy Method, is power-law in character with the coefficient β (beta) close to −1.00, suggesting the fractal character of the considered time series. The calculation of the coefficient β in the 200-year sliding window showed that this coefficient rapidly drops to zero or very minor positive values, including a break in the fractal structure. We identified seven such episodes, two (the latest) of which correspond to Spörer and Maunder solar minima. The other five episodes, which occurred around 100 BC and 500, 820, and 880 AD, i.e., before regular sunspot observations, may correspond to climate change. By combining methods aimed at identifying the specific spectral components, such as the Schwabe cycle and behavior of the 1/f dependence as a chaos characteristic, the chronobiologic (chronomics) approach can be used to study the global climate processes – such as cycles of about 500 years – bearing on global warming.

Keywords Chronomics of tree rings • Chronobiologic serial section • Fractal scaling • Solar activity

7.1 From Chronobiology to Transdisciplinary Chronomics

We consider how chronobiology and chronomics can serve in dealing with civilization and culture. Chronobiology started with circadian rhythms and led to the mapping of chronomes with a spectral element consisting of photic and nonphotic, visible and largely invisible (except for the aurorae) cycles. Photic cycles involve electromagnetic radiation in the visible domain; nonphotic cycles involve corpuscular emissions from the sun and beyond, including helio-, seleno-, and geomagnetics, ultraviolet flux and gravitation, and other stimuli such as anthropogenic ones. Cycles include, at one extreme, those in the action potentials of the major generators of electric and magnetic fields, the heart and the brain, recorded in ECGs and EEGs and in encephalo- and cardiomagnetograms. When such records become long, they reveal decadal cycles found not only in demographic and other data bases but also in much needed longitudinal heart rate data [1].

K. Otsuka et al., *Chronomics and Continuous Ambulatory Blood Pressure Monitoring*, DOI 10.1007/978-4-431-54631-3_7

Statistical studies have also revealed new non-daily, non-yearly, i.e., nonphotic cycles in both magnetism and corresponding physiological cycles in humans and other organisms [2–5]. Thus it has become increasingly clear that invisible nonphotic (involving largely magnetic) influences from the sun affect the cycles of life on earth today and apparently have even left genetic footprints in partly endogenous physiological chronomes (time structures) that are open to the environment [6].

7.2 Why Tree Rings

Whenever a solar origin is postulated in physics, economics, agriculture, or psychophysiology, the postulate of quasi-reciprocal cycles in and around us can be helpful, and indeed some solar rhythms exhibit cycle lengths found in biology and vice versa [1]. But the physicists' records of aurorae and sunspots are not long enough to look for cycles half a millennium in length. It seemed interesting therefore to look again for a proxy marker of solar activity in tree ring widths that cover the past 2,000 years or more, as others had done before us.

7.3 Background on Tree Maps

The journal "Tree-Ring Research" has a multi-decadal history, and for about a century, tree rings have been recognized as indicators of climatic and other environmental, notably solar changes. Douglass's three volumes of "Climatic Cycles and Tree-Growth" [7, 8] were reproduced in 1971 in the series "Historiae naturalis classica" [9]. We had analyzed 11 sequoias that grew on the lower western slopes of the Sierra Nevada and had found cycles with periods of about 6.51, 13.2, 38.3, 58.1, 98.6, 186, and 534 years [10]. In data taken from a published graph, an about 500-year cycle was also found to characterize the coloring of stalagmites, possibly reflecting changes in earth's subsurface temperature [11].

7.4 Fractal Scaling Calculated by the Maximum Entropy Method

Fractal scaling is one of the characteristics of chaos (or complexity). The concept of a fractal is most often associated with irregular geometric objects that display self-similarity. Fractal forms are composed of subunits (and sub-subunits, and so on) that resemble the structure of the overall object [12]. In an ideal model, this property holds on all scales. The real world, however, necessarily imposes upper and lower bounds over which such scale-invariant behavior applies. Many non-Euclidean structures in nature, such as branching trees, wrinkly coastlines, and the rough surfaces of mountains, are fractal. A number of complex anatomic structures also display fractal-like geometry. Application of fractal analysis in biology and medicine may provide a new approach to risk assessment and to forecast sudden death in animals (including humans). Elucidating the fractal nature involved in physiology and in complex signaling networks is emerging as a major challenge in the post-genomic era.

For the analysis presented herein, the average measurements of studies on 11 sequoia trees were taken from reference [9], pp. 119–23, covering 2,189 years. A time plot (chronogram) is presented in Fig. 7.1 (top). The fractal nature of the time series of tree ring width was assessed by computing the power spectrum of the 2,175-year time series by the Maximum Entropy Method (MEM). A robust line fitting algorithm of log (power) on log (frequency) was then applied to the power spectrum between 10^{-4} and 0.50 Hz (where the relationship between these variables was linear), and the slope (β) of this line was calculated.

As a result, the slope (β) of the log(power) vs. log(frequency) spectrum of the 2,189 years of tree ring widths was −1.0017 (Fig. 7.1, bottom). A slope (β) around −1.00 means fractal nature, indicating that the system is under control. The time

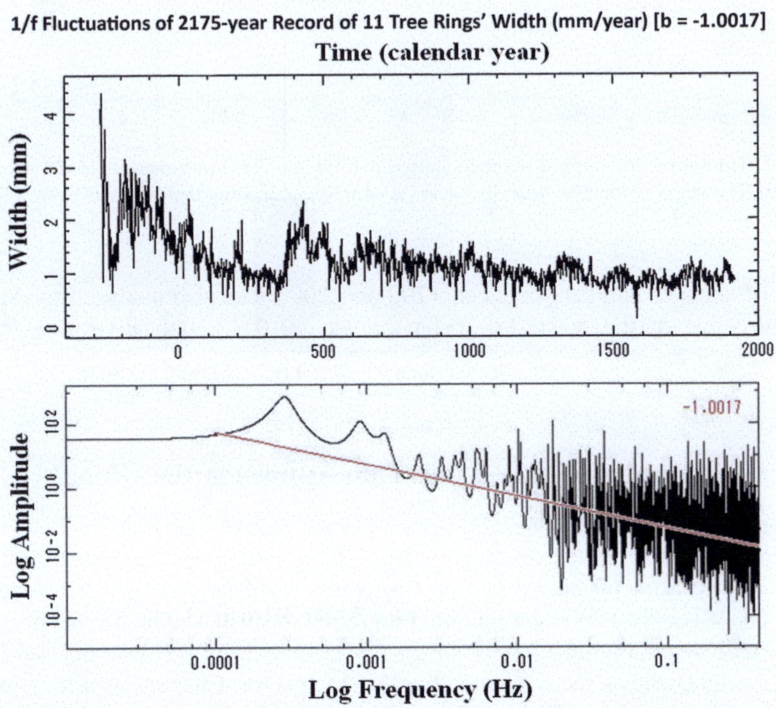

Fig. 7.1 Time series of tree ring widths (*top*) and its power spectrum (*bottom*). *Top*: Average measurements of 11 sequoia trees were taken from Douglass (1971), pp. 119–23, covering 2,189 years. *Bottom*: Power spectrum of the 2,175-year time series by the Maximum Entropy Method (MEM). The slope (β) is −1.0017, indicative of the fractal nature of tree rings

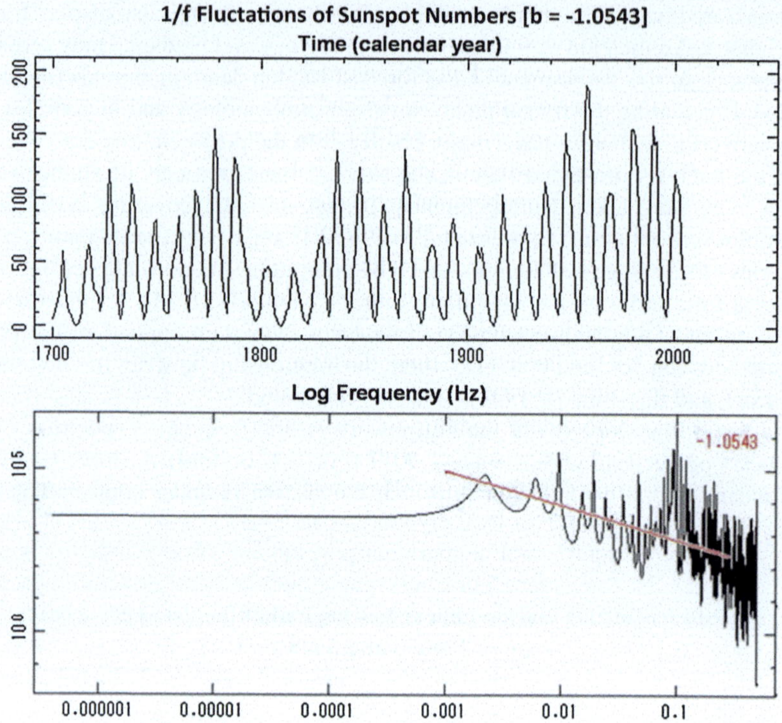

Fig. 7.2 Time series of sunspot numbers between 1700 and 2000 (*top*) and its MEM spectrum (*bottom*). The slope of the power-law behavior is –1.0543, indicating that sunspots show a fractal nature

series of sunspot numbers between 1700 and 2000 was also analyzed by MEM, and the slope of the power-law behavior was –1.0543, indicative of a fractal nature, Fig. 7.2.

7.5 Chronomics of Tree Rings Can Approach the Global Climate Processes

Changes in the fractal nature of tree ring width were assessed by repeating the MEM analysis over a 200-year interval progressively displaced by 5 years throughout the time series. Figure 7.3 (left) shows three examples of 200-year intervals, and Fig. 7.3 (right) shows the corresponding MEM spectra. Changes as a function of time in the slope (β) of the power-law behavior are shown in Fig. 7.4 (bottom), revealing episodic disruptions of the fractal nature. Seven episodes (arrows) are identified in Fig. 7.4. The last two episodes coincide with the Spörer and Maunder minima. Global climate changes may also have happened at times corresponding to the other 5 episodes of disturbed fractal nature.

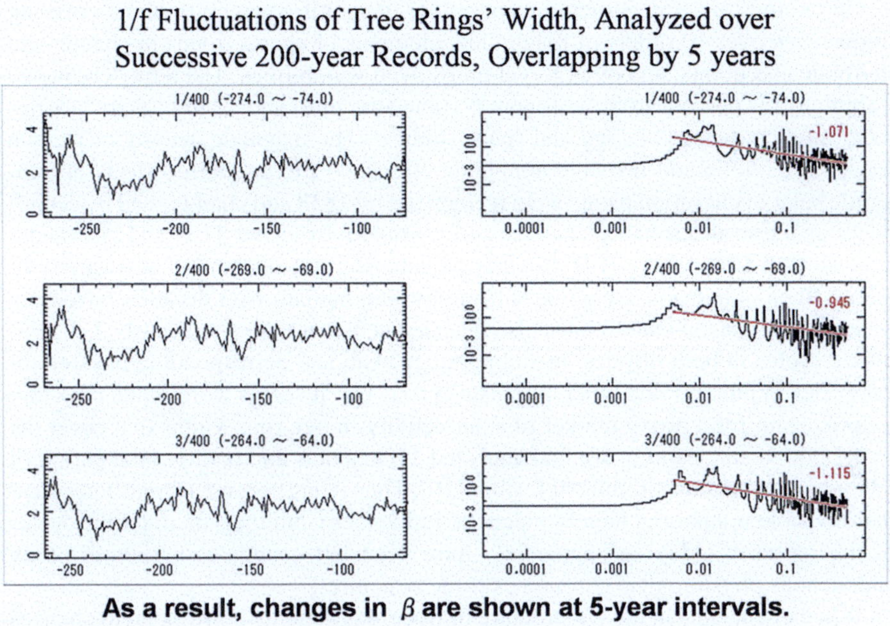

Fig. 7.3 Changes in the fractal nature of tree ring width. Three examples of 200-year intervals (*left*) and their respective MEM spectra, shown as log (power) on log (frequency) (*right*)

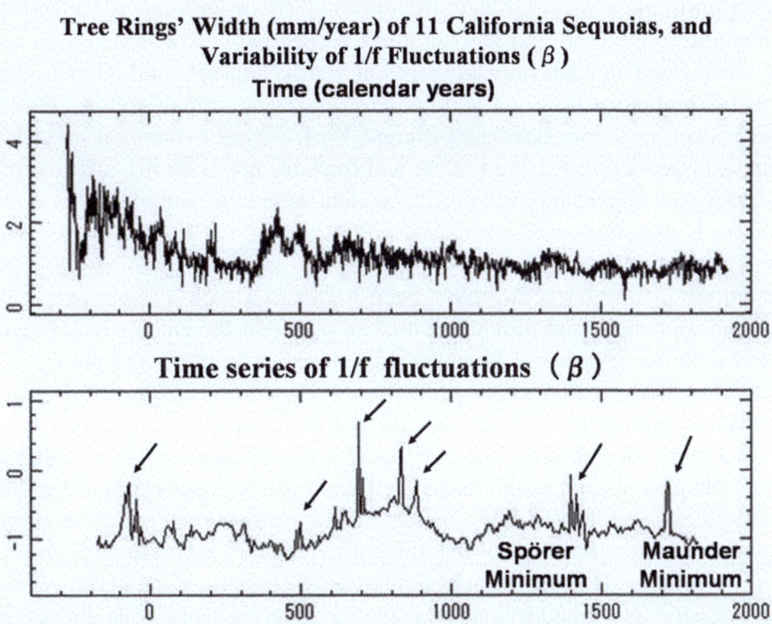

Fig. 7.4 Changes in the slope (β) of the power-law behavior (*bottom*) of tree ring widths, shown again (*top*) for alignment with episodes of altered fractal nature. Note that the last two episodes of altered fractal nature coincide with the Spörer and Maunder minima

Chronomes are time structures consisting of multifrequency rhythms covering frequencies over 10 orders of magnitude, elements of chaos, trends in chaotic and rhythmic end points, and other as yet unresolved variabilities. The term "chronoastrobiology" is derived from "chronome" (time structure) and "astrobiology" (reciprocal implications of biology and space studies). The magnetic activity of the sun has visible photic and invisible nonphotic effects on life on earth. Effects of nonphotic solar cycles upon plants were recognized by 1838 and championed by economists and transdisciplinary scientists like Eduard Brückner [13] and Alexander Leonidovich Chizhevsky [14]. Reports of statistical documentation of a variety of effects upon organisms, including self-monitoring humans over decades, have been reviewed. Statistical studies have also revealed new non-daily, non-yearly, i.e., nonphotic cycles in both physics, including magnetism and corresponding physiological cycles in humans and other organisms [1–5, 11]. It seems worthwhile therefore to look again for a proxy marker of solar activity in tree ring widths that cover the past 2,000 years or more. We had analyzed 11 sequoias and had found a period of 534 years as the most prominent feature [10]. Thus it has become increasingly clear that invisible nonphotic, largely magnetic influences from the sun affect the cycles of life on earth today and apparently have even left genetic footprints in partly endogenous physiological chronomes (time structures).

Whether changes in the 1/f structure of tree rings actually correspond to changes in solar activity or climate also should be scrutinized in auroral sightings or in other long records of reconstructed solar activity such as those based on the ^{10}Be concentration in polar ice [15]. No change in slope was observed during the Dalton/Hallström minimum of solar activity (1790–830) or around 1050 and 1300, corresponding to the Oort and Wolf minima, respectively. A 200-year interval was chosen since tree rings are only available at yearly intervals and the 1/f behavior needs to be documented over a wide enough frequency range and this limitation may have obscured some changes in climate. With respect to the total series length, however, 200 years are less than 10 % and thus are not critically affected by age trends and may compensate for the fact that data were not detrended prior to analysis. Previously obtained results from least squares spectra applied to filtered data are shown in Fig. 7.5 [10].

Alterations in time structure have been associated with a departure from the norm. From this viewpoint, it is interesting to note that the change in 1/f behavior coincided with the Spörer and Maunder minima, even though some physical interpretation of episodes when a change in slope took place is still lacking, a task for future work.

A low-dimensional strange attractor was reported to characterize worldwide climate over the past million years, based on the oxygen isotope record of the equatorial Pacific deep-sea core V28-238, suggesting that long-term climatic evolution may be viewed as the manifestation of deterministic dynamics involving a limited number of key variables [16]. This conclusion has been challenged, however. No sign of such an attractor was found by one investigator studying worldwide climate of the past 1–2 Myr (averages over about 5,000-year spans) and local climate of the past 7,000 years [17]. No evidence of chaotic behavior in 13 tree ring series from the Salt

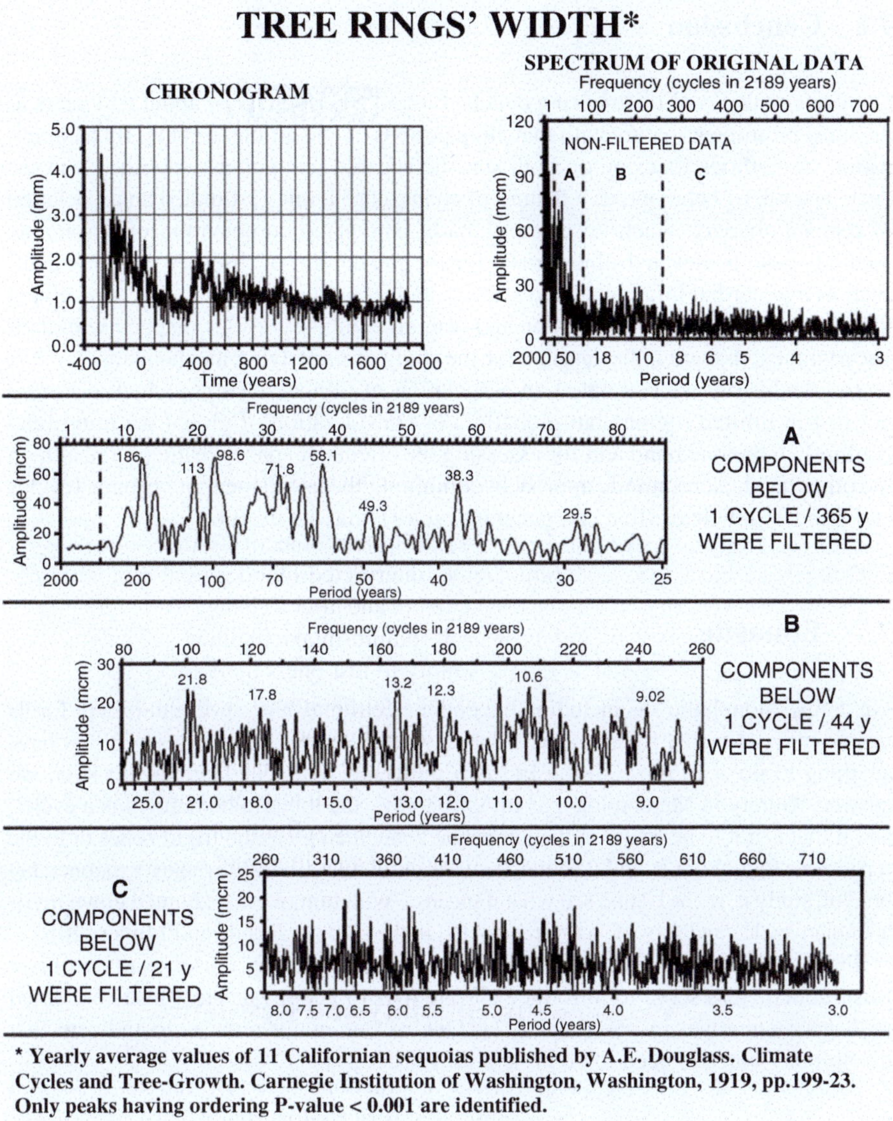

Fig. 7.5 Least squares spectra of tree rings' width over different frequency ranges after different data filtering (© Halberg Chronobiology Center)

and Verde River basins in Arizona was found by other investigators who could model the data by a second-order stationary stochastic process [18]. Whether the tree ring data analyzed herein are characterized by chaotic properties or not, the estimation of the slope (β), overall and in a moving 200-year window, quantifies properties of time structures that were found to be altered at times of changing climatic conditions. As such, fractal analysis can be viewed as being no more than a convenient framework for the description of complex morphologies and other dynamical parameters.

7.6 Conclusion

Our prior analyses of the tree ring data had found a periodicity of about 534 years as the most prominent feature with matching periods in and around us [11]. By a combination of methods focusing on both specific spectral components like the Schwabe cycle and on 1/f behavior, as a feature of chaos, chronomics estimated an association of climate change, which occurred globally. The same combination of chronomic methods detects elevated illness-risks for the prevention of diseases of individuals, such as myocardial infarctions and strokes, and, equally important, aims at resolving illnesses of societies, such as crime and war, all exhibiting cycles mapped with their uncertainties, that are indispensable for the study of underlying mechanisms.

In conclusion, we estimated an association of climate change, which occurred not just in limited regions but globally. For consideration of global warming (and cooling) on the one hand and for assessing the effects of space weather on health on the other hand, a chronomic approach, combining the assessment of chronos (cycles and trends) and chaos (that can generate cycles), may be useful.

7.7 Epilogue

We define "nonphotic" as including the entire electromagnetic spectrum except for its visible range. Finally, rightly, a man asked whether cycles in millions of years have anything to do with those in the last two millennia. With Sir Norman Lockyer, we answer: "Surely in Meteorology, as in Astronomy, 'the thing to hunt down is a cycle', and if that is not to be found in the temperate zone, then go to the frigid zones or to the torrid zone to look for it, and if found, then above all things, and in whatever manner, lay hold of, study it, record it and see what it means." With him, we have hunted down myriadennian cycles and provide a reference [19] in the steps of Rohde and Muller [20].

Fully realizing the limitations of our current knowledge, these results will, we trust, nonetheless serve to introduce chronomics to geophysicists, those concerned with geomagnetism in particular [21]. One of our major aims is to indicate that disciplinary barriers need to be broken down [22, 23].

References

1. Halberg F, Cornélissen G, Otsuka K, Watanabe Y, Katinas GS, Burioka N, Delyukov A, Gorgo Y, Zhao ZY, Weydahl A, Sothern RB, Siegelova J, Fiser B, Dusek J, Syutkina EV, Perfetto F, Tarquini R, Singh RB, Rhees B, Lofstrom D, Lofstrom P, Johnson PWC, Schwartzkopff O. International BIOCOS Study Group. Cross-spectrally coherent ~10.5- and 21-year biological and physical cycles, magnetic storms and myocardial infarctions. Neuroendocrinol Lett 2000;21:233–58.
2. Cornélissen G, Halberg F, Breus T, Syutkina EV, Baevsky R, Weydahl A, Watanabe Y, Otsuka K, Siegelova J, Fiser B, Bakken EE. Non-photic solar associations of heart rate variability and myocardial infarction. JASTP 2002;64:707–20.

3. Otsuka K, editor. Proceedings of the 1st Int Symp Workshop on Circadian Rhythms and Clinical Chronotherapy, Tokyo, 11 Nov 2000. Biomed Pharmacother 2001;55(Suppl 1):7s–190s.
4. Otsuka K, editor. Proceedings of the 2nd Int Symp Workshop on Circadian Rhythms and Clinical Chronotherapy, Tokyo, 17 Nov 2001. Biomed Pharmacother 2002;56(Suppl 2): 231s–328s.
5. Otsuka K, editor. Proceedings of the 3rd Int Symp Workshop on Circadian Rhythms and Clinical Chronotherapy, Tokyo, 9 Nov 2002. Biomed Pharmacother 2003;57(Suppl 1): 1s–198s.
6. Halberg F, Cornélissen G, Otsuka K, Schwartzkopff O, Halberg J, Bakken EE. Chronomics. Biomed Pharmacother 2001;55 Suppl 1:153s–90s.
7. Douglass AE. Climatic cycles and tree-growth, vol. 1. Washington, DC: Carnegie Institution; 1919. p. 128.
8. Douglass AE. Climatic cycles and tree-growth, vol. 2. Washington, DC: Carnegie Institution; 1936. p. 172.
9. Douglass AE. Climatic cycles and tree-growth. New York: Stechert-Hafner Service Agency; 1971 (Historiae naturalis classica, t. 90). https://archive.org/details/climaticcyclestr01douguoft
10. Nintcheu-Fata S, Katinas G, Halberg F, Cornélissen G, Tolstykh V, Michael HN, Otsuka K, Schwartzkopff O, Bakken E. Chronomics of tree rings for chronoastrobiology and beyond. Biomed Pharmacother 2003;57(suppl. 1):S24–S30.
11. Halberg F, Cornélissen G, Regal P, Otsuka K, Wang ZR, Katinas GS, Siegelova J, Homolka P, Prikryl P, Chibisov SM, Holley DC, Wendt HW, Bingham C, Palm SL, Sonkowsky RP, Sothern RB, Pales E, Mikulecky M, Tarquini R, Perfetto F, Salti R, Maggioni C, Jozsa R, Konradov AA, Kharlitskaya EV, Revilla M, Wan CM, Herold M, Syutkina EV, Masalov AV, Faraone P, Singh RB, Singh RK, Kumar A, Singh R, Sundaram S, Sarabandi T, Pantaleoni GC, Watanabe Y, Kumagai Y, Gubin D, Uezono K, Olah A, Borer K, Kanabrocki EA, Bathina S, Haus E, Hillman D, Schwartzkopff O, Bakken EE, Zeman M. Chronoastrobiology: proposal, nine conferences, heliogeomagnetics, transyears, near-weeks, near-decades, phylogenetic and ontogenetic memories. Biomed Pharmacother 2004;58 Suppl 1:S150–87.
12. Mandelbrot BB. The fractal geometry of nature. New York: WH Freeman; 1982. ISBN 0-7167-1186-9.
13. Brückner E. Klimaschwankungen seit 1700 nebst Beobachtungen über die Klimaschwankungen der Diluvialzeit. (Penck A, Hrsg. Geographische Abhandlungen, Band IV.) E. Hölzel, Wien und Olmütz 1890.
14. Chizhevsky [Tchijevsky] AL (de Smitt VP, trans and condensed). Physical factors of the historical process. Cycles 1971;22:11–27. http://cyclesresearchinstitute.org/cycles-history/chizhevsky1.pdf
15. Usoskin IG, Mursula K, Solanki S, Schüssler M, Alanko K. Reconstruction of solar activity for the last millennium using ^{10}Be data. Astron Astrophys 2004;413:745–51.
16. Nicolis C, Nicolis G. Is there a climatic attractor? Nature 1984;311:529–32.
17. Grassberger P. Do climatic attractors exist? Nature 1986;323:609–12.
18. Jeong GD, Rao AR. Chaos characteristics of tree ring series. J Hydrol 1996;182:239–57.
19. Cornélissen G, Bakken EE, Sonkowsky RP, Halberg F. A 38-million-year cycle among myriadennians in the diversity of oceanic genera. In: Abstract, III International Conference, Civilization Diseases in the Spirit of V.I. Vernadsky, People's Friendship University of Russia, Moscow, 10–12 Oct 2005, p. 47–9.
20. Rohde RA, Muller RA. Cycles in fossil diversity. Nature 2005;434:208–9.
21. Persinger MA. Geophysical variables and behavior: LXVI. Geomagnetic storm sudden commencements and commercial air crashes. Percept Mot Skills 1991;72:476–8.
22. Roederer JG. Tearing down disciplinary barriers. Eos Trans AGU 1985;66:681, 684–5.
23. Gregori GP. Galaxy-sun-earth relations: the origin of the magnetic field and of the endogenous energy of the Earth, with implications for volcanism, geodynamics and climate control, and related items of concern for stars, planets, satellites, and other planetary objects. A discussion in a prologue and two parts. Beitraege zur Geschichte der Geophysik und Kosmischen Physik, Band 3, Heft 2, 2002, 471 pp. ISNN: 1615–2824. NE: Gregori, Giovanni P. (c) Science Edition, Arbeitskreiz Geschichte Geophysik/W. Schroeder, Germany; 2002.

Chapter 8
Introduction to Chronobioengineering

Abstract This chapter introduces methods of analyzing the behavior of blood pressure (BP) variability. When many measurements are taken around the clock, one likely finds a few values above or below fixed critical limits. In addition to providing an estimate of such deviant single values, the chronobioengineering approach also addresses new and more pertinent questions such as the following: "To what extent does the BP exceed the threshold at a given time?" "How long does BP remain above a given (time-varying) threshold?" "When does most of the BP excess occur?"

Keywords Sphygmochron • Population-mean cosinor • Gliding spectra • Remove-and-replace approach • Glocalization

8.1 Introduction

Whereas current guidelines [1, 2] include the use of home measurements and ambulatory blood pressure monitoring (ABPM), blood pressure (BP) and heart rate (HR) readings in the clinic remain widely used for diagnosis and as a guide to the treatment of hypertension. An elevated BP is a medical problem receiving much attention since its treatment can prevent major incapacitating diseases: lowering blood pressure by 5 mmHg reportedly reduces the risk of stroke by an estimated 34 % and ischemic heart disease by 21 % from any pretreatment level [3]. Early detection is complicated by several factors, including the facts that the etiology of hypertension remains obscure; the condition is often asymptomatic; measurements may be few and far apart, notably by clinically healthy people who may not visit a physician regularly; BP is so highly variable that isolated readings may be misleading; and once diagnosed, many patients do not adequately follow the steps necessary for a successful treatment, including lifestyle changes such as diet and exercise.

To this list, two important issues need to be added that relate directly to the topic of this book. One issue consists of focus not only on the average BP but also on BP variability. As detailed below, abnormal patterns of BP variability are also indicative of an increased cardiovascular disease risk, beyond the risk associated with an elevated BP itself [4]. As these vascular variability disorders (VVDs) can occur

K. Otsuka et al., *Chronomics and Continuous Ambulatory Blood Pressure Monitoring*, DOI 10.1007/978-4-431-54631-3_8

within the physiological range in apparently normotensive individuals, they are often missed in current practice. The other issue relates to the treatment or rather to treatment timing. Only few clinicians are aware of the fact that even 24-hour formulations can have drastically different effects depending on when (at which circadian stage) they are administered [5–8]. Timing antihypertensive medication thus offers the possibility to optimize treatment, making sure that a larger extent of BP lowering is not obtained at the cost of aggravating other VVDs [9].

Home BP and ABPM are important tools to follow the time course of BP and BP variability, and their use can reduce medical costs [10]. Whereas home BP can be very useful to detect desired or undesired trends as a function of time and thereby prompt adjustments in the treatment plan in a timely fashion, it is not sufficient to rigorously assess the circadian variation based on reliable data obtained during undisturbed sleep as well as during the active daily span [11]. Undue BP elevation by night, as in the case illustrated in Fig. 8.1, can only be detected by ABPM.

Similar issues apply to the monitoring of the electrocardiogram. In both cases, it is important to automatically collect data around the clock, preferably for spans longer than 24 hours [12, 13]. Moreover, information beyond that provided by 24-hour, daytime, and nighttime means is best extracted by a chronobiological analysis of the data, interpreted in the light of time-specified reference values qualified by gender and age [4].

With automatic instrumentation for indirect, noninvasive BP and HR monitoring, beyond the screening for VVDs, several additional pertinent questions can be

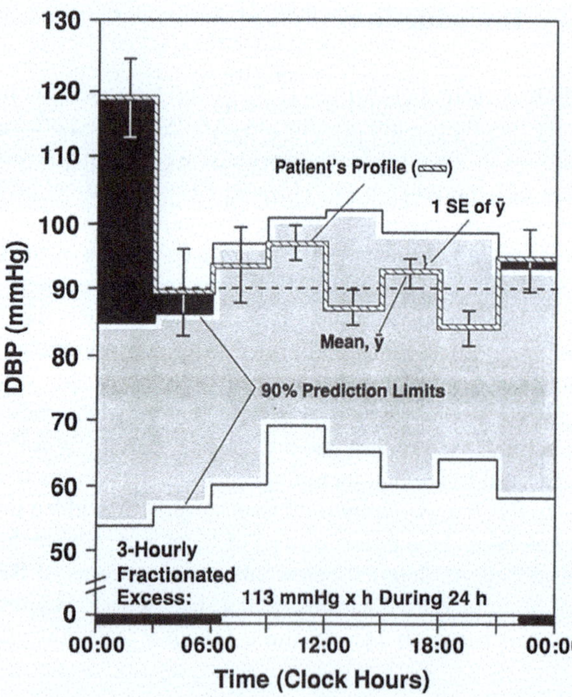

Fig. 8.1 Undue elevation of diastolic blood pressure, recurring over 10 consecutive nights in a patient (DJ, M, 78y) treated in the morning with an ACE-inhibitor (Vasotec, 10 mg), can only be detected by ABPM (© Halberg Chronobiology Center)

addressed, such as (a) how long is BP elevated above acceptable limits within a day, (b) by how much does BP exceed acceptable limits, and (c) at what time is BP excess the greatest [14, 15]. In the next sections, we review how BP and the ECG can be measured, and what are the precision and accuracy of different techniques for their measurement. Sampling requirements are addressed, analytical methods for their analysis are presented, and chronotherapeutic considerations are discussed.

8.2 Instrumentation

Before reviewing methods of measuring BP, it is important to define what kind of BP measurements are considered herein. The origin of BP is the pumping action of the heart. Its value depends on the relationship between cardiac output and peripheral resistance [16]. Venous BP, determined by vascular tone, blood volume, cardiac output, and the force of contraction of the chambers of the right side of the heart, requires invasive methods for its determination [16]. What is commonly measured, as done herein, is arterial BP. Arterial BP is the pressure exerted on the arterial walls when blood flows through the arteries [16]. Arterial BP assumes its highest value, called systolic (S) BP, when the heart contracts and ejects blood to the arteries, and its lowest value, called diastolic (D) BP, when the heart relaxes, which occurs between the ejections of blood from the heart. Pulse pressure (PP) is the difference between SBP and DBP. The period from the end of one heart contraction to the end of the next is called the cardiac cycle. Mean arterial pressure (MAP) is the average pressure during a cardiac cycle and corresponds to the driving force of peripheral perfusion [16].

Working on a variety of principles (e.g., resistance, inductance, and capacitance), BP sensors often detect mechanical signals (such as BP waves) and convert them into electric signals for further processing or transmission. For accurate and reliable measurements, a sensor should have good sensitivity, linearity, and stability [17]. Indirect or noninvasive methods of measuring BP are commonly based on the fact that if a superficial artery is palpated, a rough idea of the pressure within it can be gained by feeling the force necessary to compress it. Only since hydraulic and pneumatic artery-occluding devices were introduced, methods became available to calibrate the occluding force in terms of pressure and to relate it to the pressure within the artery [17]. The upper arm, containing the brachial artery, is the most common site for indirect measurement in view of its closeness to the heart. Usually, an occlusive cuff of appropriate size placed over the upper arm is inflated to a pressure greater than the SBP, usually by about 30 mmHg. The cuff is then gradually deflated, while a detector system (as simple as manual palpation of the radial pulse) determines the point at which blood flow is restored to the limb [16]. Indirect methods most commonly used in ambulatory monitors are auscultation and oscillometry.

The auscultatory method relies on the observation that a completely compressed artery in a normal condition does not produce any sound. Korotkoff [18] noticed that as counterpressure drops below SBP, audible sounds can be heard that can be attributed either to phenomena pertaining to the arterial wall or to those associated with the flow of blood in the artery [17]. Before complete disappearance of the Korotkoff sounds, murmurs are heard, corresponding to a cuff pressure slightly above DBP, while it is slightly below DBP at cessation of the sounds. The auscultatory method may fail in subjects with a low BP or in the presence of environmental noise [17]. Auscultatory measurements may also be inaccurate in conditions such as obesity, pregnancy, arteriosclerosis, and shock [16]. Simple mechanical errors related to mercury sphygmomanometers have been overcome by the use of more elaborate detectors, newer instrumentation relying on sensors using plethysmographic principles, pulse wave velocity sensors, and audible as well as ultrasonic microphones [19].

The oscillometric method relies on the detection of the pulse, the intra-arterial pulsation being transmitted to the occluding cuff. The pressure in the cuff is measured by a sensor. Arterial pressure oscillations are superimposed on the cuff pressure when the blood vessel is no longer fully occluded. Separation of the superimposed oscillations from the cuff pressure is accomplished by filters that extract the corresponding signals. Signal sampling is carried out at a rate determined by the pulse [20]. As counterpressure is reduced from above SBP to below DBP, oscillations appear, increase in magnitude to a maximum, and then decrease in amplitude. Unlike the Korotkoff sounds, the pressure oscillations are detectable throughout the whole measurement, even at cuff pressures higher than SBP or lower than DBP. Oscillometric devices thus use empirical algorithms to determine SBP and DBP from the envelope of the oscillometric pressure curve, based on the observation that MAP corresponds to the lowest cuff pressure of maximum oscillations [21]. Algorithms are usually adjusted by calibrating the SBP and DBP readings against auscultatory measurements at the factory. Electronic pressure and pulse monitors based on oscillometry have recently become popular for their simplicity of use and reliability [16].

Many home BP monitors use the oscillometric method of measuring BP. Some home devices include a memory to store measurements obtained by the press of a button. Some also have a USB connection to upload the data to a computer or have the capability to transfer data by telephone. These devices are much more affordable than ambulatory monitors, which work on the same basic principle but are equipped with a timer to automatically trigger cuff inflation at programmed intervals. Results from studies reported herein were obtained with the ABPM-630 (Colin Medical, Komaki, Japan) or the TM-2421 and TM-2430 (A&D, Tokyo, Japan) monitors. These devices can store data collected at 30-min intervals for 1 week and have been validated [22–24].

Other BP monitors are also becoming more widely available that are based on techniques other that auscultation or oscillometry. Several inexpensive sensor technologies using pulse wave velocity and pulse wave analysis [25] techniques are appealing for their dispensing with the bothersome arm cuff, but face other challenges such as the need for calibration. Local pressure waveforms in large arteries have been measured indirectly by ultrasound, based on pulse wave velocity measurements [26].

While brachial BP is predictive of cardiovascular outcome, interest in assessing central pressure is growing since it may better represent the load imposed on the coronary and cerebral arteries and thereby bear a stronger relationship to vascular damage and prognosis [27]. Central pressures can be estimated by using radial applanation tonometry yielding radial arterial pressure waveforms. The SphygmoCor (AtCor Medical, Sydney, Australia) is a device using this technique, which, however, presents the drawback of needing a calibration based on brachial mean and diastolic pressures obtained immediately prior to its use. The BPro wrist-bound device (HealthSTATS International, Singapore) is a noninvasive pulse wave acquisition system using a modified applanation tonometry approach capable of obtaining arterial radial pulse waves in real time over 24 hours [28]. MEMS technology is also becoming an option for measuring BP in humans. The use of an implantable pressure sensor in proximity with an artery may soon become not only feasible but acceptable as well [29].

Automatic ambulatory monitors for the indirect noninvasive measurement of BP make it possible to follow the time course of BP variation around the clock in large groups of normotensive as well as hypertensive individuals. Monitoring of clinically healthy subjects was useful to derive time-specified reference values, separately for men and women in different age groups [15]. These limits and reference values for circadian rhythm characteristics are useful to identify abnormal patterns early, before there is target organ damage. Primary prevention can thus be achieved by screening every citizen rather than restrict the monitoring to "special cases" for whom an elevated BP is already suspected.

Both in clinical practice and research, ABPM remains largely restricted to 24 hours, or at most 48 hours. Sampling typically varies from intervals of 15–60 min. In view of the large day-to-day variability seen in a large fraction of both normotensive and hypertensive individuals, extending the monitoring span to more than 24 hours is essential. Whereas a 35 % reduction in the uncertainty of circadian parameter estimates can be gained from extending the monitoring session from 1 to 2 days [14], a 7-day session further offers an estimate of the about-weekly variation for an assessment of day-to-day changes in BP and/or HR as they may correlate with differences in daily activities and everyday life events. Habituation to the monitor, work days vs. holidays or weekends, can thus be taken into consideration in making an informed decision regarding the need of treatment and its modality. Whenever monitoring is performed over several days, measurements at 30-min intervals constitute a good compromise with practicability. Changing the sampling interval from 30 to 60 min during rest/sleep may facilitate the monitoring of some people for whom the monitor is less well tolerated at night, but from a time series analysis viewpoint, keeping the sampling rate the same is preferred.

ECG records herein were obtained using an ambulatory two-channel Holter recorder (SM-28, Fukuda-Denshi, Tokyo, Japan), including an interval counter of RR intervals and a built-in A/D converter. The interval resolution was 8 ms [30].

Time-domain measures included SDNN, SDANNs (1, 5, 30, and 60 min), SD means (1, 5, 30, and 60 min), r-MSSD, RR50(+), RR50(−), total RR50, and pNN50. Mean cycle length of the normal-normal RR (NN) intervals for 24 hours and standard deviation (SD) of NN intervals for 24 hours were calculated as 24-hour NN and SDNN, respectively. SDANNs (1, 5, 30, and 60 min) were calculated as the SD of the 1-min, 5-min, 30-min, and 60-min mean NN intervals over the entire 24-hour ECG recordings, respectively. Similarly, SDmeans (1, 5, 30, and 60 min) were calculated as the mean of the SD of the 1-min, 5-min, 30-min, and 60-min NN intervals over 24 hours, respectively. The triangular index (TI) and TINN were also evaluated in some studies. The difference between adjacent NN intervals was computed to derive r-MSSD as the root mean square of successive differences (square root of the mean of the sum of the squared differences between adjacent NN intervals). Successive NN intervals longer or shorter than 50 ms were tallied as RR50(+) and RR50(−), respectively. RR50 represents the total count, and pNN50 the percentage of differences greater than 50 ms, computed over the entire 24-hour ECG record [30].

Frequency-domain measures were obtained by applying a third-degree spline interpolation method to the time series of NN intervals, resampled at 4 Hz. Intervals of 5–10 min of the reconstructed and resampled time series, progressively moved by 150 s, were analyzed by the fast Fourier transform. From the spectra, the power centered around 10.5 s (0.04–0.15 Hz, LH) and around 3.6 s (0.15–0.40 Hz, HF) was determined, and the LF/HF ratio computed. The 1/f behavior of the spectrum was also determined [30].

8.3 Methods of Time Series Analysis for Stationary and Nonstationary (Aeolian) Data

Conceived as a regression problem, cosinor-based methods, first introduced by Franz Halberg [31], are applicable to non-equidistant data (Fig. 8.2). This is a major advantage. Another dividend is the feasibility of being able to readily derive confidence intervals for parameters of rhythmic components of known periods. For prominent components such as the ubiquitous circadian rhythms, prior (external)

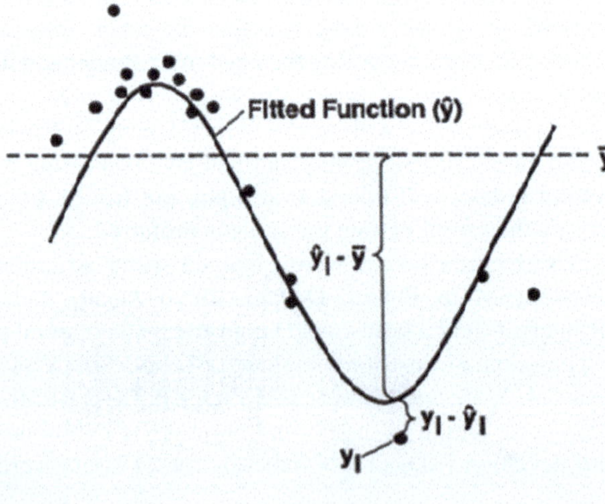

Fig. 8.2 Fit of a cosine curve by least squares to non-equidistant data (© Halberg Chronobiology Center)

information is advantageously exploited to simplify computations. Originally developed for the analysis of short and sparse data series, the extended cosinor has been further developed for the analysis of long time series, focusing both on rhythm detection and parameter estimation, as reviewed below.

8.3.1 *The Single-Component Single Cosinor*

Periodograms and classical spectra originally used in chronobiology [32, 33] required the data to be equidistant and to cover more than a single cycle. Whereas some spectral analysis techniques are available to analyze non-equidistant data [34–36], algorithms available in most software packages remain limited to the case of equidistant data. Least-squares procedures, while having their own limitations [37, 38], are applicable to non-equidistant data. When applied to equidistant data, results at Fourier frequencies are identical to those of the discrete Fourier transform [39].

Whenever a rhythm of known (or approximatively known) period can be assumed, as in the case of the usually prominent circadian rhythms, a cosine curve can be fitted to the data (Y_i, i = 1, 2, …, N) obtained at discrete sampling times (t_i, i = 1, 2, …, N):

$$Y(t) = M + A\cos(2\pi t / \tau + \phi) + e(t) \tag{8.1}$$

where M is the MESOR (midline estimating statistic of rhythm, a rhythm-adjusted mean), A is the amplitude (a measure of half the extent of predictable variation within a cycle), ϕ is the acrophase (a measure of the time of overall high values recurring in each cycle), τ is the period (duration of one cycle), and e(t) is the error term, assumed to be normally distributed with mean zero and constant variance (Fig. 8.3).

When period can be assumed known, using well-known trigonometric angle sum identity, the model can be rewritten as

$$Y(t) = M + \beta x + \gamma z + e(t) \tag{8.2}$$

where $\beta = A\cos\phi$; $\gamma = -A\sin\phi$; $x = \cos(2\pi t/\tau)$; $z = \sin(2\pi t/\tau)$.

M, β, and γ are estimated by solving the so-called normal equations, obtained by expressing that the residual sum of squares (RSS) is minimal when its first-order derivatives with respect to each parameter are zero:

$$RSS = \sum_i [Y_i - (\hat{M} + \hat{\beta} x_i + \hat{\gamma} z_i)]^2 \tag{8.3}$$

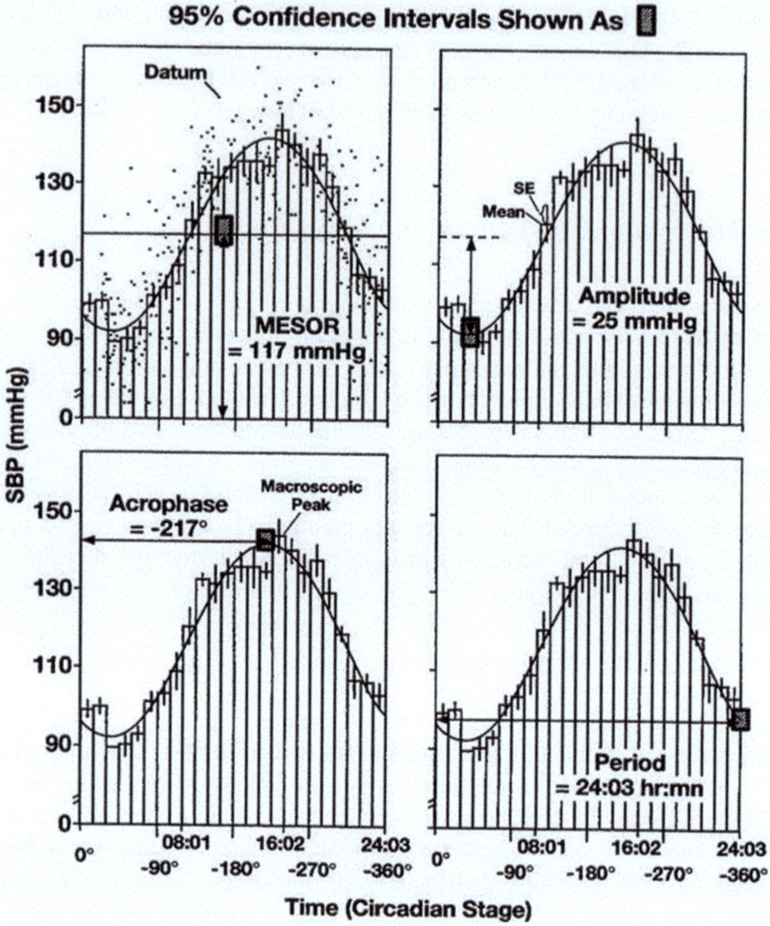

Fig. 8.3 Definition of parameters. The MESOR (midline estimating statistic of rhythm) is a rhythm-adjusted mean value, which is usually more precise and more accurate than the arithmetic mean. The double amplitude is a measure of the extent of predictable change within one cycle. The acrophase is a measure of the timing of overall high values recurring in each cycle, in relation to a reference time. The period is the duration of one cycle. The error term, assumed to be normally distributed with zero mean and constant variance, is used for rhythm detection and for the derivation of confidence intervals (© Halberg Chronobiology Center)

$$\begin{Bmatrix} \sum Y_i \\ \sum Y_i x_i \\ \sum Y_i z_i \end{Bmatrix} = \begin{Bmatrix} N & \sum x_i & \sum z_i \\ \sum x_i & \sum x_i^2 & \sum x_i z_i \\ \sum z_i & \sum x_i z_i & \sum z_i^2 \end{Bmatrix} \begin{Bmatrix} M \\ \beta \\ \gamma \end{Bmatrix} \tag{8.4}$$

Estimates of the amplitude and acrophase can be derived from the estimates of β and γ by the following relations:

$A=[\beta^2+\gamma^2]^{1/2}$ and $\phi=\arctan(-\gamma/\beta)+K\pi$, where K is an integer, the correct value of ϕ depending on the signs of β and γ.

For rhythm detection, the total sum of squares (TSS, the sum of squared differences between the data and the arithmetic mean) is partitioned into the model sum of squares (MSS, the sum of squared differences between the estimated values based on the fitted model and the arithmetic mean) and the residual sum of squares (RSS, the sum of squared differences between the data and the estimated values from the fitted model). The rhythm is statistically significant when the model sum of squares is large relative to the residual sum of squares, as determined by the F test:

$$F=\frac{(MSS/2)}{(RSS/(N-3))} \tag{8.5}$$

where 2 and $N-3$ are the numbers of degrees of freedom for the model ($k=3$ parameters -1) and the error term ($N-k$). The null hypothesis (H_0) that there is no rhythm (the amplitude is zero) is rejected when $F>F_{1-\alpha}(2, N-3)$, where α relates to the probability level chosen for testing H_0.

The same F statistic evaluated at the estimated value of A instead of at $A=0$ yields the equation of the $(1-\alpha)$ confidence ellipse, from which conservative confidence limits can be derived for A and ϕ by drawing concentric circles centered on the pole tangent to the error ellipse and tangents from the pole to the error ellipse, respectively [38, 40].

8.3.2 *The Multiple-Component Single Cosinor*

The single-component cosinor is easily extended to a multiple-component:

$$Y(t)=M+\sum_j A_j \cos(2\pi t/\tau_j+\phi_j)+e(t), j=1,2,\ldots,p \tag{8.6}$$

Instead of solving a system of 3 equations in 3 unknowns, there are $2p+1$ normal equations to estimate M and p pairs of (β_j, γ_j) or (A_j, ϕ_j) when τ_j are assumed known. A confidence ellipsoid can be determined [40] from which approximate confidence intervals can be derived for each component's amplitude and acrophase. Computations are greatly simplified in the case of equidistant data covering an integer number of cycles [41]. A multiple-component model is useful to obtain a better approximation of the signal's waveform when it deviates from sinusoidality. As described below, a 2-component model consisting of cosine curves with periods of 24 and 12 hours usually provides a good fit to approximate the circadian waveform of BP and HR. Reference values based on this model have been derived so that deviation from these norms indicates abnormality. Outcome studies have shown that such abnormal patterns are associated with an increased cardiovascular disease risk [4, 42].

8.3.3 *The Population-Mean Cosinor*

When data are collected as a function of time on three or more individuals, the population-mean cosinor procedure is useful to make inferences concerning a population rhythm, provided the k individuals represent a random sample from that population. Each individual series is analyzed by the single cosinor, using the same single- or multiple-component model. Provided the individual parameter estimates follow a normal distribution and the within-individual variances can be assumed to be equal, the population rhythm can be estimated by computing the arithmetic mean of the individual MESORs and the vectorial averages of the (β, γ) pairs related to each component in the model [38, 40]. The population amplitudes and acrophases can be estimated using the relations

$$\hat{\beta}* = \hat{A}*\cos\hat{\phi}*; \hat{\gamma}* = -\hat{A}*\sin\hat{\phi}* \tag{8.7}$$

where "*" indicates that the expectations are population averages and not averages over the k individuals sampled. In the above conditions and assuming normality of errors and individual parameters, sample variances can be computed as

$$\begin{aligned}&\hat{\sigma}_M^2 = \Sigma_j\left(\hat{M}_j - \hat{M}*\right)^2/(k-1); \hat{\sigma}_{\beta n}^2 = \Sigma_j\left(\hat{\beta}_{nj} - \hat{\beta}*\right)^2/(k-1); \hat{\sigma}_{\gamma n}^2 = \Sigma_j\left(\hat{\gamma}_{nj} - \hat{\gamma}*\right)^2/(k-1);\\ &\hat{\sigma}_{M\beta 1}^2 = \Sigma_j\left(\hat{M}_j - \hat{M}*\right)\left(\hat{\beta}_{1j} - \hat{\beta}_1*\right)/(k-1); \text{ and similarly for the other cross-products}\end{aligned} \tag{8.8}$$

for j = 1, 2, …, k and n = 1, …, 2p + 1 in the case of a p-component model.

Applying the population-mean cosinor for each trial period (p = 1), a confidence interval for M* is given by

$$\hat{M}^* \pm t_{1-\alpha/2}(k-1)\hat{\sigma}_M / k^{1/2} \tag{8.9}$$

A joint 1 − α confidence ellipse for $\left(\hat{\beta}*, \hat{\gamma}*\right)$ consists of all points (β_z*, γ_z*) satisfying

$$\begin{aligned}&\left(\beta_z* - \hat{\beta}*\right)^2/\hat{\sigma}_\beta^2 - 2r\left(\beta_z* - \hat{\beta}*\right)\left(\gamma_z* - \hat{\gamma}*\right)/\hat{\sigma}_\beta\hat{\sigma}_\gamma +\\ &\left(\gamma_z* - \hat{\gamma}*\right)^2/\hat{\sigma}_\gamma^2 \le 2\left(1-r^2\right)(k-1)F_{1-\alpha}(2, k-2)/\left(k(k-2)\right)\end{aligned} \tag{8.10}$$

where $r = \hat{\sigma}_{\beta\gamma}/\left(\hat{\sigma}_\beta\hat{\sigma}_\gamma\right)$.

The null hypothesis of A* = 0 is rejected if

$$\left[\left(k(k-2)\right)/\left(2(k-1)\right)\right]\left[1/\left(1-r^2\right)\right]\left[\hat{\beta}^{*2}/\hat{\sigma}_\beta^2 - 2r\hat{\beta}^*\hat{\gamma}^*/\hat{\sigma}_\beta\hat{\sigma}_\gamma + \hat{\gamma}^{*2}/\hat{\sigma}_\gamma^2\right] > F_{1-\alpha}(2, k-2) \tag{8.11}$$

Approximate confidence intervals for the amplitude and acrophase can be obtained by drawing concentric circles and tangents from the pole to the error ellipse, respectively.

8.3.4 Parameter Tests

To characterize differences in circadian (or other) patterns between two (or more) time series, conventional tests (such as Student's *t*-test) are often applied at each timepoint when replications are available or the time series covers more than a single cycle and data can be stacked over a single idealized cycle. Results from this approach may not always be readily interpreted as differences in opposite direction can be found when the major difference affects the amplitude and/or acrophase but not the MESOR [43, 44]. Adjustment for multiple testing is also often overlooked, which may reduce the statistical power when considered.

Alternatively, test statistics have been developed to test the equality of MESORs, amplitudes, and acrophases considered jointly or separately for both the single and population-mean cosinor [40]. A reduction in circadian amplitude and an advance in circadian acrophase have been documented, for instance, for core temperature in association with the bilateral lesioning of the suprachiasmatic nuclei [45, 46] (Fig. 8.4). An amplitude reduction and acrophase advance are features also found in association with aging for variables such as BP and HR [47, 48].

Parameter tests comparing single data series present the advantage of allowing inferences to be made on an individual basis. This is important, for instance, in order to ascertain that a given intervention has the desired effect, and it has been successfully used to optimize the timing of administration of antihypertensive medication [8, 49] (Fig. 8.5).

The self-starting cumulative sum (CUSUM) control chart [50] is a complementary approach that can be used to determine for the individual patient whether an intervention was effective [51]. In addition, the CUSUM also provides an estimate of when the intervention started having an effect. When this time coincides approximately with the time of intervention, a result as close to causality possible in statistics is reached. An illustrative example is provided in Fig. 8.6 [52].

8.3.5 Least-Squares Spectra

Just as a chronogram (plot of the data as a function of time) usually provides valuable information regarding the presence of trends, outliers, or other features having a bearing on the need for preprocessing the data prior to analysis, viewing the entire time structure of a time series in the frequency domain is also invaluable. Whereas it has to be considered a macroscopic approach, a least-squares spectrum or plot of the amplitudes as a function of frequency over the entire frequency range (from 1/T, where T is the length of the observation span, to $1/2\Delta t$, where Δt is the average

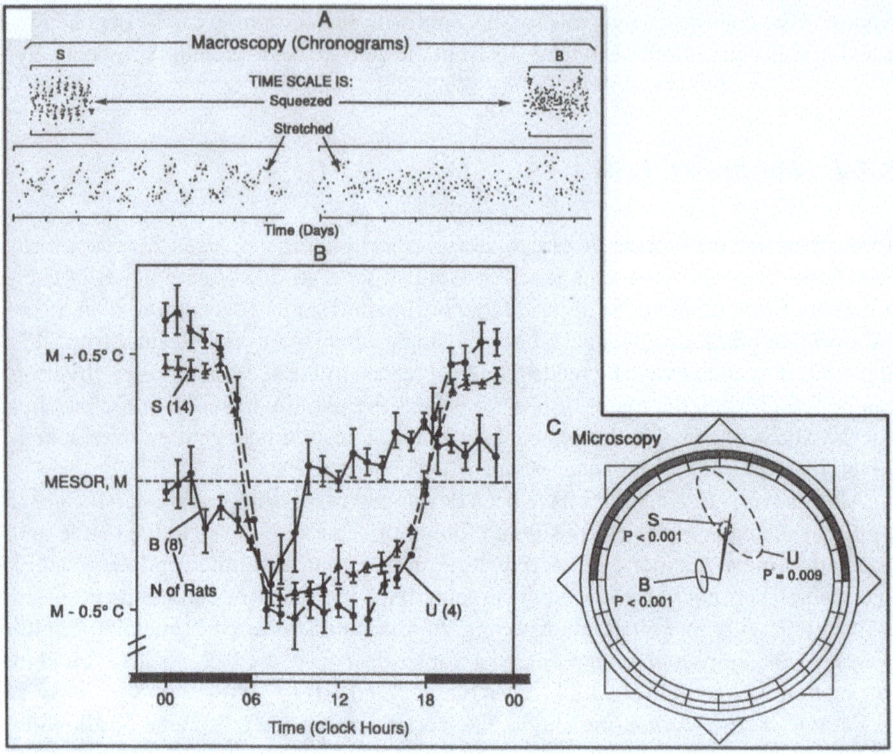

Fig. 8.4 Parameter tests indicate that the circadian amplitude of murine core temperature is reduced and the circadian acrophase is advanced after bilateral lesioning of the suprachiasmatic nuclei (SCN). An increase in circadian amplitude is noted after unilateral lesioning of the SCN (Modified from [46]. © Halberg Chronobiology Center)

sampling interval) is useful to reveal the presence of periodic signals and to provide information regarding the structure of the noise.

A least-squares spectrum is easily computed by applying the single cosinor procedure at trial periods corresponding to the Fourier frequencies. When the data are equidistant, this approach yields results identical to the discrete Fourier transform [39]. When the data are non-equidistant, the procedure remains applicable without the need for interpolation. Another advantage of the least-squares spectrum is the fact that the fundamental frequency can be selected in a judicious way, since it is not constrained to be the exact time series length. Likewise, a harmonic increment less than 1 yields estimates of (A, ϕ) pairs at frequencies intermediate to consecutive Fourier frequencies, thereby providing a better macroscopic view of periods corresponding to peaks in amplitude. It should be noted, however, that unless signals thus detected were anticipated a priori, these cycles cannot be taken at their face value and need additional work to be rigorously assessed. In the want of independent replications, the stability of such cycles can be further examined by chronobiologic serial sections and/or gliding spectra, provided the data series covers a sufficient number of cycles of the periodicity of interest. These methods are described below.

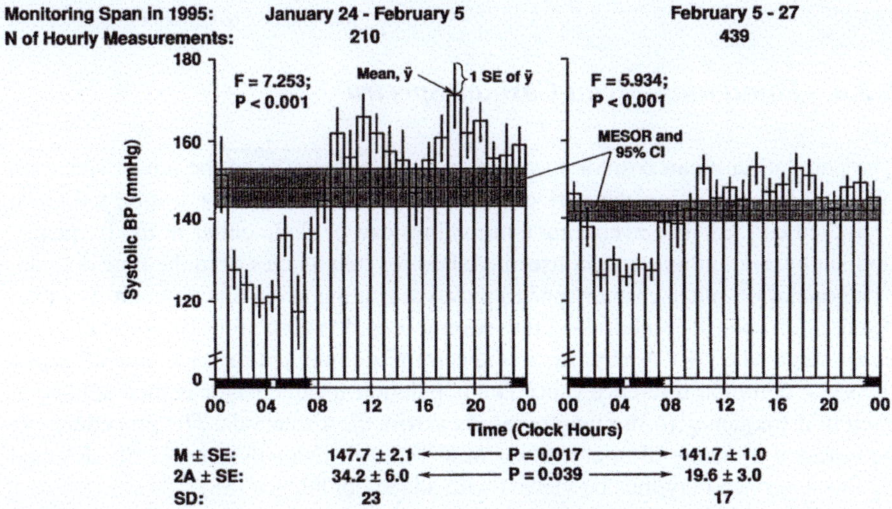

Fig. 8.5 Changing the timing of the same dose of the same medication (diltiazem hydrochloride, 240 mg) to the same patient (FH, M, 75y) from 08:30 to 04:30 is associated with a further lowering of the MESOR and with a reduction in circadian amplitude of SBP. Both effects are validated with statistical significance by parameter tests applied to individual time series (Modified from [8]. © Halberg Chronobiology Center)

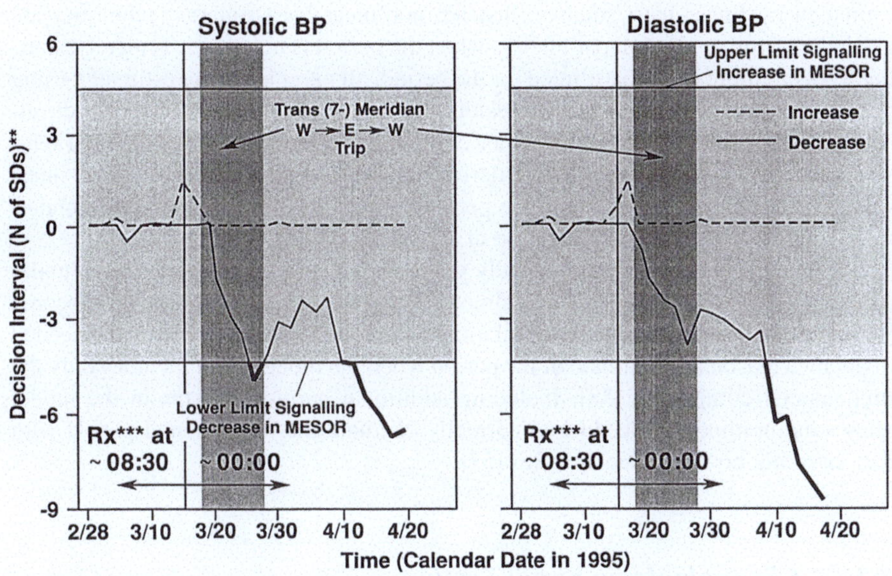

Fig. 8.6 Self-starting cumulative sum (CUSUM) control chart assessing the efficacy of a change in timing (from 08:30 to 00:00) of treatment (diltiazem hydrochloride, 240 mg) in a 75-year-old man (FH). As long as the two lines remain within the decision interval (*shaded bands*), there is no change in the MESOR of SBP or DBP. A decrease below the lower boundary of the decision interval indicates a statistically significant reduction in BP MESOR, observed a few days after the change in treatment timing. The time when the BP MESOR started to decrease is estimated by tracking the line segment, leading to the breakout back to the last occasion on which it lay on the *horizontal zero line*, a time that closely corresponds to when the timing of treatment was changed (Modified from [8]. © Halberg Chronobiology Center)

8.3.6 *Population-Mean Cosinor Spectra*

The population-mean cosinor spectrum is a method similar to the power spectrum obtained by smoothing the periodogram. It is more reliable for testing unknown periodicities [53]. The averaging (smoothing) can be done either in the frequency domain by averaging across consecutive Fourier frequencies or in the time domain. The population-mean cosinor spectrum uses averaging in the time domain. The total observation span T is subdivided into several (e.g., k) subsections (or intervals, I) of equal length (I = T/k). A least-squares spectrum is computed for each interval, using the same common reference time. The population-mean cosinor is then applied at each trial frequency to summarize results across the k intervals. The procedure can be repeated by using different values of k. Unknown signals consistently detected by this approach may thus be viewed with added confidence [38].

8.3.7 *Extended Linear-Nonlinear Cosinor*

As the period is in the argument of the cosine function, this parameter cannot be estimated by linear least squares. Instead, nonlinear least-squares techniques are needed to estimate rhythm parameters when the period is unknown. Typically, starting from an initial (guess) estimate for the period, all parameters are estimated using iterations aimed at minimizing the residual sum of squares. Marquardt [54] developed an algorithm which performs an optimum interpolation between the Taylor series and gradient methods. He also derived a way to approximate confidence intervals for all parameters, including the period [55]. Bingham offers an easily understood approach for the particular case of single-component models [56].

The linear-nonlinear cosinor usually yields an acceptable estimate of the fundamental period of signals with a symmetrical waveform, even from short records [57]. For asymmetrical signals, simulations indicate that about 5 cycles are needed to obtain a reasonable estimate of the period when the model fitted includes only the fundamental component [58]. Including additional harmonic terms in the model allows the nonlinear procedure to correctly estimate the fundamental period with data covering no more than 2 cycles [57].

8.3.8 *Chronobiologic Serial Section*

It has been defined [59] as a method providing analytical results obtained by fitting a fixed-period cosine curve to consecutive overlapping or nonoverlapping data sections of fixed length, called intervals, displaced in increments throughout a time series – displayed with the original data as moving P-values (for rhythm detection) and as moving point and interval estimates for the MESOR, amplitude, and acrophase (for rhythm estimation). Chronobiologic serial sections with nonoverlapping intervals are called fractionated. This approach is recommended whenever estimates thus obtained are used as imputations for further analyses, such as for the assessment of trends in rhythm parameters as a function of time, so as not to violate the underlying assumption of independence. Chronobiologic serial sections with overlapping intervals are called pergressive. From a macroscopic point of view, this approach better conveys smooth changes in rhythm characteristics as a function of time (Fig. 8.7).

The abovementioned methods constitute the core of cosinor rhythmometry. They are applicable in a wide variety of fields in medicine, biology more generally, and beyond. The sections below are devoted to related cosinor techniques with specific applications in chronocardiology and chronomics. One section deals with a special case of the multiple-component cosinor, the sphygmochron. With applications to the analysis of ABPM records, a model consisting of cosine curves with periods of 24 and 12 hours usually approximates the circadian waveform of BP and HR well. Using a fixed model in preference to obtaining the best fitting model for each particular record presents the advantage of being amenable to comparison with reference values derived from records of clinically healthy peers matched by gender and age. A nonparametric analysis of the data complements this approach, as outlined below. Two other sections deal with the analysis of nonstationary time series, gliding spectra, and a remove-and-replace approach that may help understand to some extent the causality of some of the relations between biota and their natural environment. Finally, the case for a glocal viewpoint is made to account for geographic differences observed in the time structure (chronome) of physiopathology. A cartography of chronomes may shed light on possible causes underlying the observed differences.

8.4 The Sphygmochron

The sphygmochron is a computer summary of changes as a function of time in BP and HR measured around the clock, preferably with an automatic ambulatory monitor [15, 60]. The sphygmochron reports on results from a double-barreled (parametric and nonparametric) approach, as illustrated in Fig. 8.8. Together with the sphygmochron form itself, results reported to the subject and/or care provider include the list of actual measurements and the times at which they were obtained, as well as a plot of SBP, DBP, and HR as a function of time. Mean arterial pressure (MAP), pulse pressure (PP = SBP–DBP), and pulse pressure product (PPP = SBP·HR/100) are computed from each set of measurements and included in the listing. Data summaries are prepared in two ways: first, the number of measurements, arithmetic mean, and standard deviation (SD) are listed for each variable, overall as well as for each day (from midnight to the next midnight); second, daytime, nighttime, and 24-hour means and SDs are computed according to the subject's rest-activity schedule (the default rest span is defined as the span from 23:00 to 07:00), from which day-night ratios are computed according to Verdecchia [61].

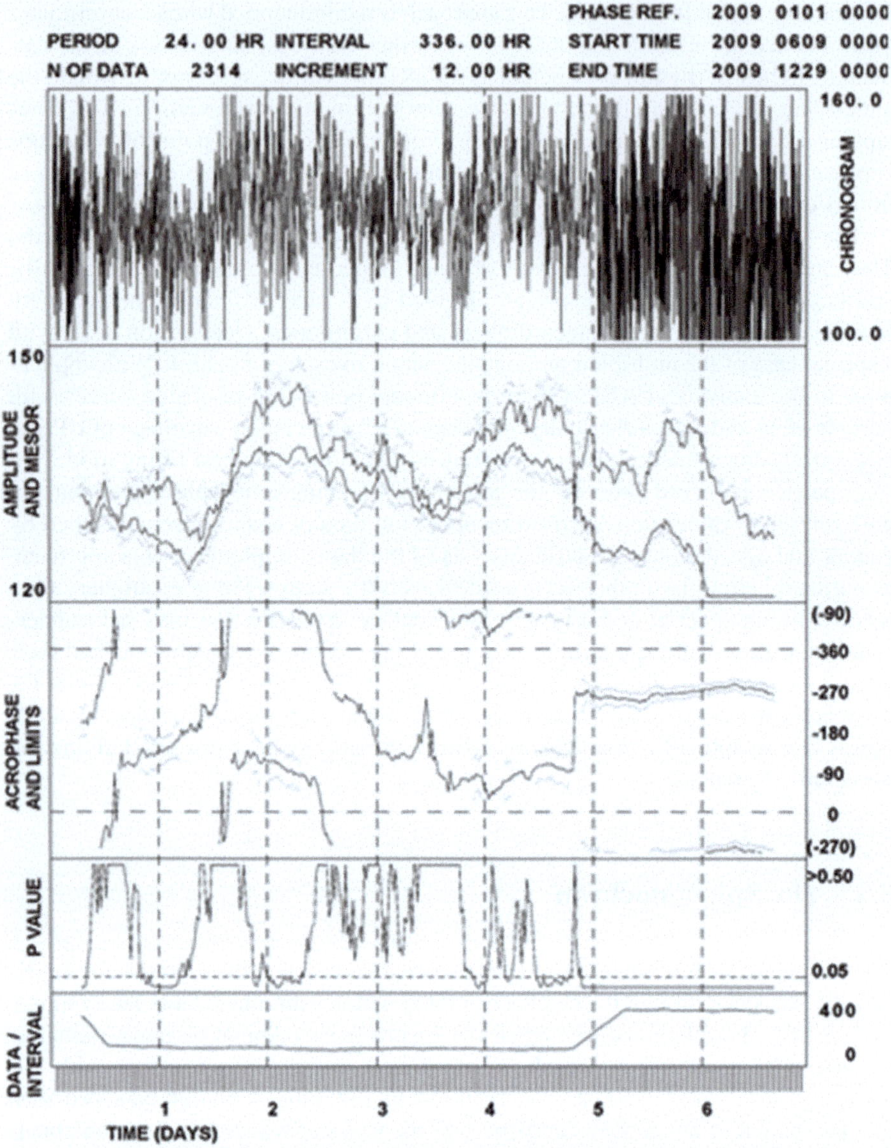

Fig. 8.7 Time course of systolic blood pressure of a woman (JF, 61y at start of monitoring) during the span from June 9 to December 29, 2009, during which 2314 measurements were taken, first manually by self-measurements and then (after *event line 5*) by ambulatory monitoring, including data during rest/sleep. A 24-hour cosine curve is fitted by least squares in 2-week intervals progressively displaced by 12 hours. *Vertical dashed lines* indicate dates of full moons. The data are displayed as a function of time in row 1. The MESOR (*lower curve*) and the 24-hour amplitude (distance

SBP, DBP, and HR data are stacked over an idealized 24-hour day and plotted together with time-specified 90 % prediction limits [62] obtained from clinically healthy subjects of the same gender and age group, as shown in Fig. 8.9.

The SBP, DBP, and HR data are analyzed in two ways. Parametrically, a model consisting of cosine curves with periods of 24 and 12 hours is fitted by least squares to each record, yielding estimates of the MESOR and of the amplitude and acrophase of each component. Nonparametrically, the stacked data are compared to the time-specified upper 95 % prediction limit to determine the percentage time elevation (proportion of the 24-hour day when the subject's data are above the acceptable limit), the amount of BP and/or HR excess (called hyperbaric index, HBI, and tachycardic index, TCI, respectively, defined as the area delineated by the subject's data when they are excessive and the limit itself), and the time of excess (computed as the center of gravity of the area of excess) (Fig. 8.10). Because excess may not always occur as a single episode in a day, excess over consecutive 3-hour intervals (fractionated HBI and TCI) may also be assessed as a guide to timing any needed treatment.

The sphygmochron form summarizes the results from the parametric and nonparametric analyses. As shown in Fig. 8.8, in order to keep the subject's name confidential, a codename is used instead. Gender and age are listed, along with the date and time at which monitoring started and for how long data were collected. The numerical report consists of two parts labeled "Characteristics" (parametric results) and "Indices of Deviation" (nonparametric results). In each case, results are shown for SBP (when the heart contracts) on the left, DBP (when the heart relaxes) in the middle, and HR on the right. The characteristics listed are the MESOR and the 24-hour double amplitude and acrophase, listed together with the 90 % prediction limits from clinically healthy subjects of the same gender and age group. The nonparametric results list the percentage time elevation, amount of excess, and timing of excess, together with a 10-year projection of the amount of excess. Because data are collected under ambulatory conditions, it is common to find occasional excessive values, associated, for instance, with increased activity or emotions [63]. Empirically, excess up to about 50 (mmHg × hour during 24 hours) is considered acceptable and accountable for by daily worries and/or physical activities [64].

Based on the parametric and nonparametric results, a recommendation is made regarding the need for any intervention or further monitoring. MESOR-normotension corresponds to having all rhythm characteristics within the 90 % prediction limits and no undue BP excess. When one or several features of the circadian variation in BP and/or HR deviate from the norms established in clinically healthy peers

Fig. 8.7 (continued) between the two *curves*) are shown in row 2 with their SEs shown as *dots*. The 24-hour acrophases are plotted in row 3 with their 95 % confidence limits. P-values and the number of data per interval are shown in rows 4 and 5, respectively. It can be seen that ambulatory monitoring is associated with rhythm detection, a lowering of the MESOR (nightly data are lower than during waking), an increase in amplitude, and a stable acrophase (© Halberg Chronobiology Center)

SPHYGMOCHRON-TM **Monitoring Profile over Time; Computer Comparison with Peer Group Limits**

Blood Pressure (BP) and Related Cardiovascular Summary.

Name:---------------------		Patient #:	Ura0001
Age: 52		Sex:	M
Monitoring From:	4/12/2001 11:30	To:	4/19/2001 13:00
Comments:	Sleep: 23:00-06:15		
	Dx: Variant angina		

CHRONOBIOLOGIC CHARACTERISTICS

	SYSTOLIC BP (mmHg) Patient Value	SYSTOLIC BP (mmHg) Peer Group Reference Limits	DIASTOLIC BP (mmHg) Patient Value	DIASTOLIC BP (mmHg) Peer Group Reference Limits	HEART RATE (bpm) Patient Value	HEART RATE (bpm) Peer Group Reference Limits
ADJUSTED 24-h MEAN (MESOR)	126.3	98.4-135.1 Range	80.0	60.3-87.2 Range	70.9	56.4-91.2 Range
PREDICTABLE CHANGE (DOUBLE AMPLITUDE)	**56.25**	6.4-39.40 Range	**35.63**	4.84-29.80 Range	7.15	5.26-36.20 Range
TIMING OF OVERALL HIGH VALUES (ACROPHASE) (hr:min)	14:53	11:48-17:40 Range	14:35	11:08-16:48 Range	19:11	11:44-17:20 Range

	STD (MIN; MAX)*	STD (MIN; MAX)*	STD (MIN; MAX)*
PERCENT TIME OF ELEVATION	24.2%	17.9%	0.0%
TIMING OF EXCESS	16:41 (hr:min)	17:29 (hr:min)	0:00 (hr:min)
EXTENT OF EXCESS DURING 24 HOURS HBI*	**60** (mmHg x hour)	21 (mmHg x hour)	0 (mmHg x hour)
10-YEAR CUMULATIVE EXCESS	219 (mmHg x hour)(in 1,000's units)	78 (mmHg x hour)(in 1,000's units)	0 (mmHg x hour)(in 1,000's units)

Individualized bounded indices: (STD = Standard)(Min = Minimum)(Max = Maximum)(HBI = Hyperbanic Index)

INTERVENTION NEEDED			MORE MONITORING NEEDED
No			Annually
Yes	Drug	Non-Drug	As soon as possible
			Other specify____________________
		CHAT	

Prepared By_____Germaine Cornelissen________________ Date__28/Apr_/2001____

1) Unusually long standing or lying down during waking: unusual activity, such as exercise, emotional loads, or schedule changes, e.g. shiftwork; etc.; 2) Salt, calories, kind and amount, other, etc.

Fig. 8.8 Illustrative sphygmochron. This otherwise MESOR-normotensive man has an excessive 24-hour amplitude of both SBP and DBP (CHAT, brief for circadian hyper-amplitude-tension), a condition associated with an increased cardiovascular disease risk in several outcome studies. The 24-hour double amplitude of both SBP and DBP exceeds the upper 95 % prediction limit of clinically healthy peers matched by gender and age, shown next to the subject's results (© Halberg Chronobiology Center)

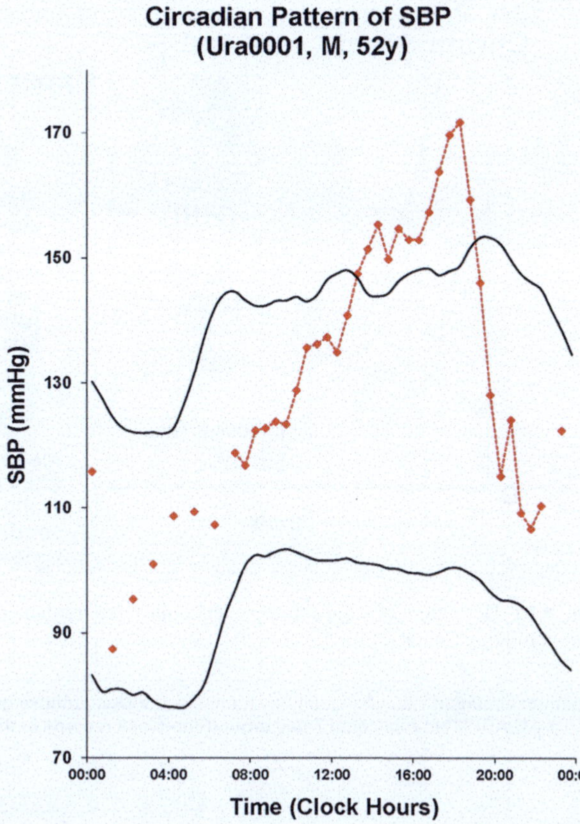

Fig. 8.9 Plot of SBP data collected over 7 days stacked over an idealized 24-hour day, displayed together with time-specified 90 % prediction limits derived from clinically healthy peers matched by gender and age. The circadian overswing in SBP is readily seen, as reported in the sphygmochron of Fig. 8.8 (© Halberg Chronobiology Center)

matched by gender and age, the following vascular variability anomalies (VVAs) or disorders (VVDs when VVAs persist) are identified (Fig. 8.11) [4, 42]:

- MESOR-hypertension (MH) when the MESOR of SBP and/or DBP is excessive
- CHAT (circadian hyper-amplitude-tension) when the 24-hour amplitude of SBP and/or DBP is excessive
- Ecphasia when the 24-hour acrophase of SBP and/or DBP is outside acceptable limits, but the 24-hour acrophase of HR is within acceptable limits
- Excessive pulse pressure (EPP) when the difference in MESOR between SBP and DBP exceeds 60 mmHg
- Deficient HR variability (DHRV) when the overall SD of HR is less than 7.5 beats/min
- Excessive pulse pressure product (EPPP) when the product of the MESORs of SBP and HR is above 10,000 mmHg x beats/min

EPP, DHRV, and EPPP need further specification by gender and age. Another VVA is the deviation of the circadian period from 24 hours, determined by the nonoverlap of 24 hours by the 95 % confidence interval of the circadian period, determined by nonlinear least squares. Thus far, this condition has not been associated with an increased cardiovascular disease risk in outcome studies.

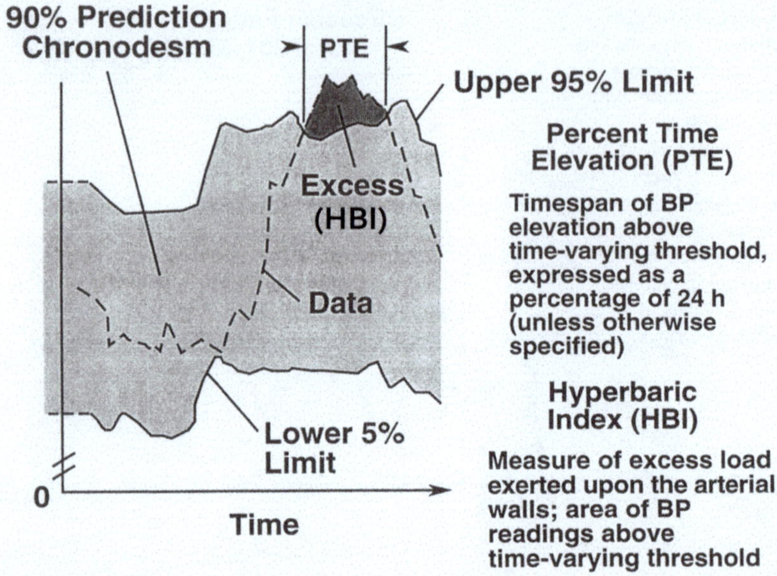

Fig. 8.10 Definition of nonparametric endpoints reported in the sphygmochron (© Halberg Chronobiology Center)

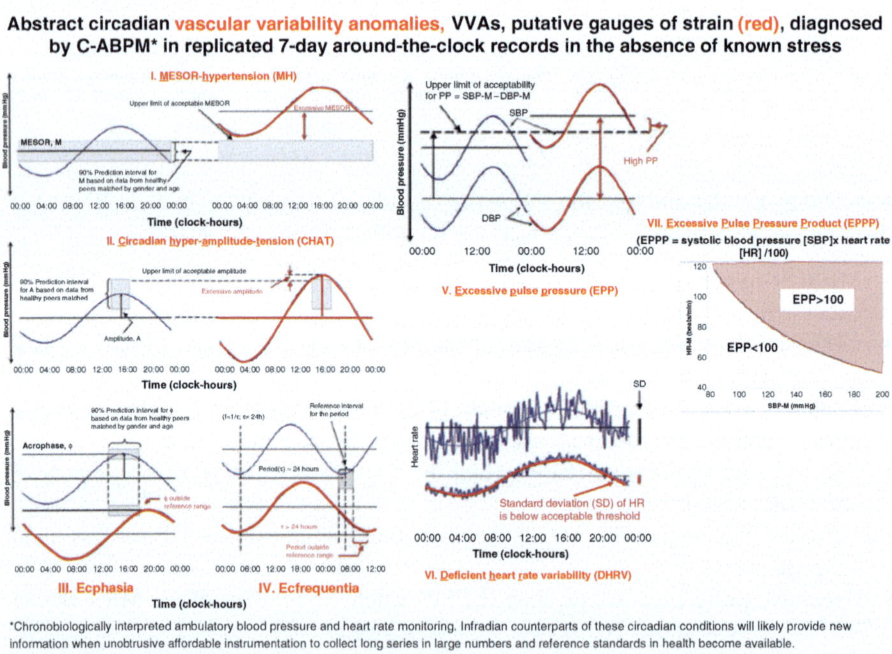

Fig. 8.11 Definition of vascular variability anomalies (VVAs) or, if they persist, vascular variability disorders (VVDs) (© Halberg Chronobiology Center)

8.5 Gliding Spectra

Not all time series are stationary. It is often the rule rather than the exception that characteristics of a signal change as a function of time. Despite reports about the reproducibility of results from repeated 24-hour ABPM sessions [65], the day-to-day variability in circadian rhythm characteristics can be quite large [14, 66–68]. The discrepancy can be accounted for by the observation that many studies rely on regression or correlation analysis to compare results from repeated monitoring, the BP of subjects included in the study varying in a wide range. Whereas regression and correlation methods tend to emphasize similarities, other statistical tests such as assessing the distribution of individual differences shed a better light on the extent of variation present between consecutive BP profiles [14].

Whereas variability is often found in relation to the average (MESOR) and prominence (amplitude), some rhythmic components are relatively stable in period and phase. This is notably the case of circadian rhythms. Once focus is placed on other periodicities, such as the week, or some nonphotic cycles such as the cis-half year or the transyear (components with periods of about 5 and 16 months, respectively, found in space weather), drifts in period and/or phase are also common. Even in the case of circadian rhythms, changes in period and/or phase can be expected, for instance, in relation to travel across time zones [69].

In such cases, a global analysis of the time series as a whole cannot provide information regarding any changes that may have taken place as a function of time. Several approaches are available for this purpose. When data are equidistant or rendered equidistant by averaging and filling data gaps by interpolation, wavelets can be performed [70]. This approach has been useful to uncover components not detected earlier [71]. Short-term Fourier transforms can be used to visualize changes in the spectral structure of the data as a function of time [72]. Alternatively, gliding spectral windows [73] can be computed, as illustrated in Fig. 8.12.

As in the case of the chronobiologic serial section [59], gliding spectra are obtained by performing analyses over a span shorter than the entire time series (called "interval"), the analyses being repeated after the interval is progressively moved by a given time increment throughout the data series. Whereas a single-cosine curve with a given period is fitted in each interval in the chronobiologic serial section, in the case of gliding spectra, a least-squares spectrum is computed over each interval over a specified frequency range. In order to better visualize the time course of changes in frequency and/or amplitude occurring as a function of time, spectral components are estimated not only at the Fourier frequencies but also at intermediate frequencies by choosing a fractional harmonic increment. Likewise, a time increment shorter than the interval used for analysis is selected. The results (usually the amplitude, but the percentage rhythm can also be considered) are then displayed in a 3D graph and/or a surface chart. In such a display, time and frequency are the two horizontal axes, and amplitude (or percentage rhythm) is shown on the vertical axis in a 3D plot or as different shadings in a surface chart [38, 59].

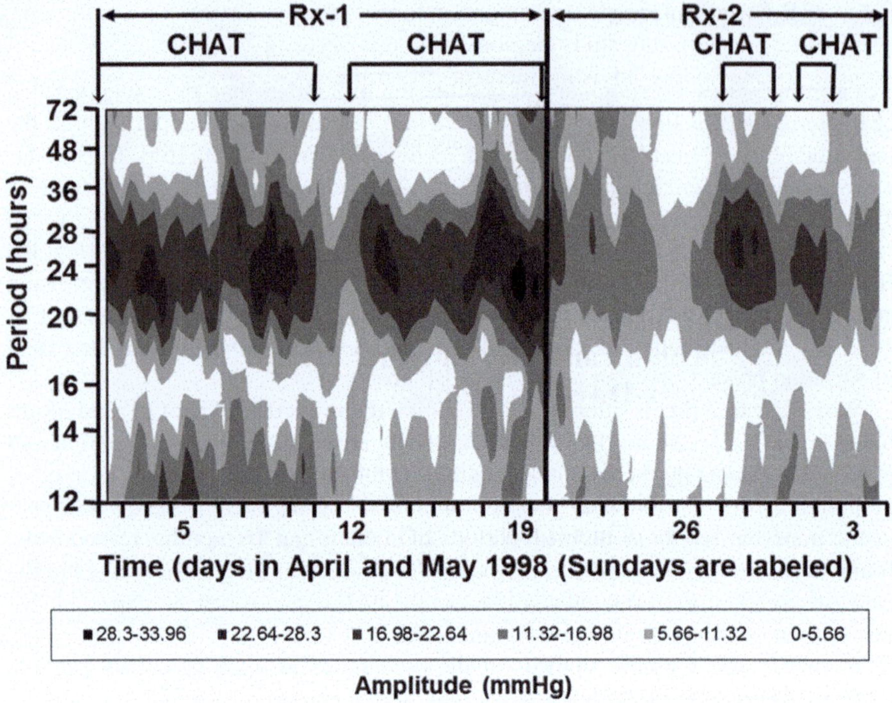

Fig. 8.12 Example of gliding spectral window (analysis by G. Katinas). SBP data are analyzed in 2-day intervals progressively moved by 8 hours. The circadian rhythm is seen prominently by the *darker shading* corresponding to a period of about 24 hours. A change in antihypertensive treatment is associated with transient elimination of CHAT (circadian hyper-amplitude-tension, too large a circadian amplitude of blood pressure), a risk marker of cardiovascular disease
Rx1: nifedipine (10 mg bid at 08:00 and 20:00); Rx2: diltiazem (90 mg tid at 10:00, 15:00, and 20:00) (© Halberg Chronobiology Center)

Whereas the procedure can be performed on non-equidistant data, the interpretation of results is greatly helped when data are equidistant, as changes in sampling rate are also associated with changes in spectral structure appearing on the graph. A judicious choice of the interval length and of the frequency range examined is important in order to minimize sidelobes. The use of a Hanning taper [74] is also helpful in this kind of exploratory analysis [38].

Whenever a signal is present and is not too wobbly in frequency, a complementary analysis by chronobiologic serial section can be very informative, as documented earlier [75]. In the presence of signals varying greatly in period length, a nonlinear serial section can be performed [38]. This approach consists of fitting a model including the component of interest by nonlinear least squares [54] and of plotting the estimated period with its 95 % confidence limits for each interval as it is moved throughout the data series. Changes in the length of the solar activity cycle have thus been assessed by fitting an about-11-year cosine curve to yearly sunspots data in a 35-year interval moved by 5 years [76].

8.6 Remove-and-Replace Approach

The remove-and-replace approach serves for the comparison of the time structure of two or more variables in an attempt to determine whether changes in one variable are related to similar changes in another variable [77]. Usually, one variable will relate to the natural environment and the other to biology, the purpose being to understand how one or the other environmental variable may influence physiology, pathology, or any other biological system. This problem can also be addressed by cross-spectral analysis and spectral coherence [78]. Superposed epoch analysis is another option [79]. Too often, however, investigators limit their analysis to the computation of correlation coefficients [80]. One major pitfall of such an approach is the disregard for periodicities characterizing the variables under consideration. Correlations are notoriously known to lead to spurious results in the presence of rhythms, as the correlation coefficient is largely dependent on the phase difference of a given periodic component between the two variables being compared [81]. In the presence of cycles, and they are ubiquitous, methods other than the Pierson product-moment correlation are advocated.

The remove-and-replace approach was so named after the precedent in endocrinology where a gland or organ (such as the pancreas) is first removed to determine what the resulting health problems are, and its hormone (such as insulin) or other secretion is then administered to find out whether health can thus be restored. Herein, the environment itself is doing the removal and replacement. It can be the presence vs. the absence of a given condition such as a magnetic storm or of a given spectral component (such as the week). Biological states during spans corresponding to the presence or absence of given environmental features can then be compared by an armamentarium of standard statistical tools [77, 82].

The following examples may convey the method better than any theoretical considerations. In one study, an about-weekly variation in the rate of change in sunspot area was detected intermittently in five different spans analyzed by Walsh spectra [83]. The question was asked whether the relative prominence of an about-weekly variation in the sun may have a bearing on human physiology. Longitudinal records of self-measurements of HR available during two or more of these five spans were analyzed by cosinor. Circaseptans in HR were invariably amplified when present in the sun and dampened when absent in the sun (Fig. 8.13) [84]. Resonance of circaseptans in human HR with solar circaseptans was also demonstrated with statistical significance on an individual basis for the longest self-measurement record ($P<0.001$) (Fig. 8.13, *top*) [84]. The HR MESOR, circaseptan amplitude, and circaseptan-to-circadian amplitude ratio further underwent an about-11-year cycle similar to the solar activity cycle [84]. Similarly, in the Republic of Georgia, a 7-day component in mortality from myocardial infarction was detected with statistical significance during 1980–1982 when solar activity was high and circaseptans characterized the interplanetary magnetic field. But during 1984–1987, when solar activity was down, as were circaseptans in the interplanetary field, circaseptans were no longer detected in the spectrum of myocardial infarctions. In 1989–1992,

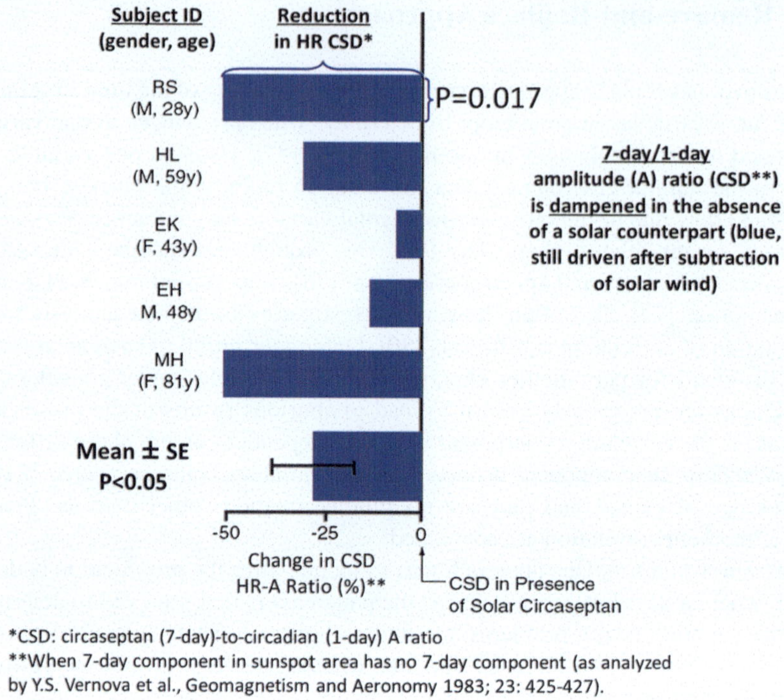

Fig. 8.13 The circaseptan-to-circadian amplitude ratio of HR is compared between spans when an about-weekly variation was or was not detected in the rate of change in sunspot area analyzed by Walsh spectra. In their absence, the relative about-weekly variation in HR is invariably dampened (© Halberg Chronobiology Center)

during the next solar maximum, circaseptans were again detected both in myocardial infarctions and in the environment [85].

In another study, an about-half-monthly component was detected in the daily incidence of sudden cardiac death in Moscow, Russia. Its period was estimated nonlinearly to average 15.2 days [95 % CI: 15.15–15.30 days]. According to the criterion of a nonoverlap of 95 % confidence intervals, this period was statistically significantly different from the half-monthly period characterizing local K analyzed during the same 3-year span (1979–1981). For K, the corresponding period was estimated to be 14.0 [95 % CI: 13.94–14.17] days [86]. Whereas the about-half-monthly period of both sudden cardiac death and K was estimated with a relatively tight 95 % confidence interval, this component showed great wobbliness and was not consistently detected when the data were analyzed by moving spectra over a shorter interval of 3 months, progressively displaced by 1.5 weeks throughout the time series. A possible resonance with occasional frequency trapping between K and sudden cardiac death was suggested by the fact that the half-monthly component was seemingly more prominently expressed in sudden cardiac death when it was detected in the spectrum of K [86, 87].

A remove-and-replace approach was also instrumental in showing that an about-5-month (cis-half year) component was detected in HR self-measured by a clinically healthy man for over 4 decades during spans of maxima in solar flares [88]. Similarly, it showed that an about-1.3-year (transyear) component was detected in the SBP of an elderly man when it was prominently present in solar wind speed and that it dampened but persisted after it was no longer detected with statistical significance in solar wind speed [89]. A similar finding was made for the transyear in the incidence of international terrorist acts, in which case there was not only a communality of cycles between terrorism and space weather, gauged by solar wind speed on the one hand and the antipodal geomagnetic disturbance index aa on the other hand, but also a coincidence in time, the transyear being most prominent in all three variables during solar cycle 22 [75, 77]. In a set of variables self-rated or self-measured for up to 40 years by a clinically healthy man, a congruence test also detected statistically significant associations with solar wind speed and the antipodal geomagnetic index aa, in mental functions, such as 1-min estimation, in mood and vigor, and in SBP [90, 91].

8.7 Glocalization

The lighting regimen and temperature are two important environmental synchronizers of circadian and circannual rhythms. These environmental variables undergo drastic changes as a function of latitude. It is thus not surprising that some biological functions also undergo changes in their characteristics as a function of latitude. A dependence of the circannual acrophase of human natality on latitude was reported as early as 1973 [92]. Dependence on the seasons was demonstrated by the inversion of the circannual rhythm of mortality from respiratory and cardiovascular conditions between the northern and southern hemispheres [93]. Cartographic results from both studies were actually shown on actual world and USA maps, respectively.

Nonphotic environmental influences also are latitude-dependent magnetic storms being felt more strongly near the poles than at midlatitudes [94], ring currents also circling along the equator [95]. These geographic differences may account for the fact that natality is characterized by a 1-year synchronized rhythm in Italy and Japan but follows an about-1.3-year (transyear) variation in the Philippines instead [96]. In a study of religious motivation including data from 103 different geographic locations, an about-21-year component was detected with statistical significance in most sites, its characteristics also showing a strong dependence on latitude [97]. An about-1.3-year (transyear) component, a signature of solar wind speed, was also found to characterize sudden cardiac death in some, but not all geographic locations for which data were available for analysis, and in one country (Czech Republic), this component was detected only during the last half of the record [87, 98]. Important differences in cardiac function are also found in relation to altitude [99–103].

Whether it is to gain a better understanding of environmental influences on human health, to follow the development of the circadian rhythm in BP and HR during the first few weeks after birth, or to examine how vital signs are affected by everyday life events, the temporal mapping of rhythms should be performed both at the global level by analyzing entire time series to explore as wide a frequency range as possible and locally to assess any changes as a function of time within a more specific frequency range. This dual, combined global and local, approach is known as "glocality in time" [101], as a parallel to "glocality in space."

8.8 Discussion and Conclusion

Not discussed above are some other techniques of time series analysis [104–106]. Most of them rely on equidistant data. Among them, the Maximum Entropy Method (MEM) deserves a special place as this approach, based on the principle of maximum entropy, is a method of spectral density estimation yielding a better estimate of the center of a spectral peak, thus improving spectral accuracy. A more accurate estimation of the period can thus be obtained [107, 108].

Another class of methods of data analysis relates to chaos. Different endpoints such as the correlation dimension and the approximate entropy index have been computed from ECG and EEG records [109–120]. In one study, the cosinor method was instrumental in showing a circadian variation in the correlation dimension, with notable difference between health and disease [112]. Other endpoints such as the fractal behavior have been routinely assessed in ECG records in studies focused on heart rate variability [110–113, 115, 121].

Focus was placed specifically on the use of the cosinor and its different extensions herein. The method is fairly simple, and its results lead to meaningful interpretations. Despite some shortcomings related to the difficulty of satisfying all assumptions underlying the use of regression techniques, its wide-ranging applications have played an important role in the development of chronobiology as a quantitative scientific discipline. Applied to ABPM records, it detects VVAs to predict cardiovascular disease risk with a greater accuracy than the day-night ratio [122–124].

References

1. The Seventh Report of the Joint National Committee on Prevention, Detection, Evaluation, and Treatment of High Blood Pressure. US Department of Health and Human Services, National Institutes of Health, National Heart, Lung, and Blood Institute; 2004 Aug. NIH Publication No. 04-5230.
2. James PA, Oparil S, Carter BL, Cushman WC, Dennison-Himmelfarb C, Handler J, Lackland DT, LeFevre ML, MacKenzie TD, Ogedegbe O, Smith Jr SC, Svetkey LP, Taler SJ, Townsend RR, Wright Jr JT, Narva AS, Ortiz E. Evidence-based guideline for the management of high

blood pressure in adults. Report from the Panel Members appointed to the Eighth Joint National Committee (JNC8). JAMA 2014;31(5):507–20. doi:10.1001/jama2013.284427.

3. Law M, Wald N, Morris J. Lowering blood pressure to prevent myocardial infarction and stroke: a new preventive strategy. Health Technol Assess 2003;7(31):1–94.
4. Halberg F, Cornelissen G, Otsuka K, Siegelova J, Fiser B, Dusek J, Homolka P, Sanchez de la Pena S, Singh RB, BIOCOS project. Extended consensus on means and need to detect vascular variability disorders (VVDs) and vascular variability syndromes (VVSs). World Heart J 2010;2(4):279–305.
5. Güllner HG, Bartter FC, Halberg F, Delea C. Circadian temperature and blood pressure rhythms guide timed optimization and gauge antimesorhypertensive prazosin effects. Chronobiologia 1979;6:105.
6. Zaslavskaya RM. Chronodiagnosis and chronotherapy of cardiovascular diseases. 2nd ed. Moscow: Medicina; 1993. p. 397. Translation into English from Russian.
7. Cornelissen G, Zaslavskaya RM, Kumagai Y, Romanov Y, Halberg F. Chronopharmacologic issues in space. J Clin Pharmacol 1994;34:543–51.
8. Halberg F, Cornelissen G, International Womb-to-Tomb Chronome Initiative Group: Resolution from a meeting of the International Society for Research on Civilization Diseases and the Environment (New SIRMCE Confederation), Brussels, Belgium, 1995 Mar 17–18; Fairy tale or reality? Medtronic Chronobiology Seminar #8, 1995 Apr; p. 12. text, 18 figures.
9. Watanabe Y, Halberg F, Otsuka K, Cornelissen G. Toward a personalized chronotherapy of high blood pressure and a circadian overswing. Clin Exp Hypertens 2013;35(4):257–66. doi :10.3109/10641963.2013.780073.
10. Soghikian K, Casper S, Fireman BH, Hunkeler E, Hurley LB, Tekawa I, Vogt TM. Home blood pressure monitoring: effect on use of medical services and medical care costs. Med Care 1992;30(9):855–65.
11. Stinson SM, Cornelissen G, Scarpelli PT, Halberg F. Self-measurement and ambulatory monitoring of blood pressure: a subject's chronobiological perspective. Biomed Pharmacother 2002;56 Suppl 2:333s–8s.
12. Carandente O, Halberg F, Cornelissen G. Two circadian and an infradian rhythm in human cardiac arrhythmia? Prog Clin Biol Res 1987;227B:159–65.
13. Jabaudon D, Sztajzel J, Sievert K, Landis T, Sztajzel R. Usefulness of ambulatory 7-day ECG monitoring for the detection of atrial fibrillation and flutter after acute stroke and transient ischemic attack. Stroke 2004;35:1647–51. doi:10.1161/01.STR.0000131269.69502.d9.
14. Cornelissen G. Instrumentation and data analysis methods needed for blood pressure monitoring in chronobiology. In: Scheving LE, Halberg F, Ehret CF, editors. Chronobiotechnology and chronobiological engineering. Dordrecht, The Netherlands: Martinus Nijhoff; 1987. p. 241–61.
15. Cornelissen G, Otsuka K, Halberg F. Blood pressure and heart rate chronome mapping: a complement to the human genome initiative. In: Otsuka K, Cornelissen G, Halberg F, editors. Chronocardiology and chronomedicine: humans in time and cosmos. Tokyo: Life Science Publishing; 1993. p. 16–48.
16. Webster JG. The measurement, instrumentation, and sensors handbook. Boca Raton, Florida: CRC Press; 1998. p. 2630.
17. Geddes LA. The direct and indirect measurement of blood pressure. Chicago: Year Book Medical Publishers; 1970. p. 196.
18. Korotkoff NS. On the subject of methods of measuring blood pressure. Bull Imp Military Med Acad St Petersburg 1905;11:365–7.
19. Meldrum SJ. Indirect blood pressure measurement. Br J Clin Equip 1976;1:257–65.
20. Yamakoshi K. Non-invasive techniques for ambulatory blood pressure monitoring and simultaneous cardiovascular measurement. J Ambulat Monit 1991;4:123–43.
21. Ramsey M. Non-invasive automatic determination of mean arterial pressure. Med Biol Eng Comput 1979;17:11–8.

22. White WB, Lund-Johansen P, McCabe EJ. Clinical evaluation of the Colin ABPM-630 at rest and during exercise: an ambulatory blood pressure monitor with gas-powered cuff inflation. J Hypertens 1989;7(6):477–83.
23. Keavney B, Bird R, Caiazza A, Casadei B, Conway J. Measurement of blood pressure using the auscultatory and oscillometric methods in the same cuff deflation: validation and field trial of the A&D TM2421 monitor. J Hum Hypertens 2000;14(9):573–9.
24. Palatini P, Frigo G, Bertolo O, Roman E, Da Corta R, Winnicki M. Validation of the A&D TM-2430 device for ambulatory blood pressure monitoring and evaluation of performance according to subjects' characteristics. Blood Press Monit 1998;3:255–60.
25. Baruch MC, Warburton DE, Bredin SS, Cote A, Gerdt DW, Adkins CM. Pulse decomposition analysis of the digital arterial pulse during hemorrhage simulation. Nonlinear Biomed Phys 2011;5:1. doi:10.1186/1753-4631-5-1.
26. Beulen BW, Bijnens N, Koutsouridis GG, Brands PJ, Rutten MC, van de Vosse FN. Toward noninvasive blood pressure assessment in arteries by using ultrasound. Ultrasound Med Biol. 2011;37(5):788–97. doi:10.1016/j.ultrasmedbio2011.01.020.
27. Roman MJ, Devereux RB, Kizer JR, Lee ET, Galloway JM, Ali T, Umans JG, Howard BV. Central pressure more strongly relates to vascular disease and outcome than does brachial pressure: the STRONG Heart Study. Hypertension 2007;50(1):197–203. doi:10.1161/HYPERTENSIONAHA.107.089078.
28. Nair D, Tan SY, Gan HW, Lim SF, Tan J, Zhu M, Gao H, Chua NH, Peh WL, Mak KH. The use of ambulatory tonometric radial arterial wave capture to measure ambulatory blood pressure: the validation of a novel wrist-bound device in adults. J Hum Hypertens 2008;22:220–2. doi:10.1038/sj.jhh.1002306.
29. Franco-Ayala M, Martinez-Pinon F, Reyes-Barranca A, Sanchez de la Pena S, Alvarez-Chavez JA. Implantable blood pressure sensor for analyzing elasticity in arteries. In: Varadan VK. (Ed.). Proc. SPIE 7291, Nanosensors, Biosensors, and Info-Tech Sensors and Systems 2009. San Diego, CA. Conference volume 7291 (31 March 2009). doi:10.1117/12.814666.
30. Otsuka K, Cornelissen G, Zhao ZY, Weydahl A, Delyukov A, Gorgo Y, Wang ZR, Perfetto F, Tarquini R, Kubo Y, Shinagawa M, Hotta N, Ishii T, Omori K, Watanabe Y, Nunoda S, Ohkawa S, Halberg F. Rhythm and trend elements in the time structure, chronome, of heart rate variability. Gerontol Geriatr 1999;2:31–48.
31. Halberg F, Tong YL, Johnson EA. Circadian system phase -- an aspect of temporal morphology; procedures and illustrative examples. Proceedings of the International Congress of Anatomists. In: von Mayersbach H, editor. The cellular aspects of biorhythms, Symposium on Biorhythms. New York: Springer-Verlag; 1967. p. 20–48.
32. Halberg F, Panofsky H. I: Thermo-variance spectra; method and clinical illustrations. Exp Med Surg 1961;19:284–309.
33. Panofsky H, Halberg F. II. Thermo-variance spectra; simplified computational example and other methodology. Exp Med Surg 1961;19:323–38.
34. Marquardt DW, Acuff SK. Direct quadratic spectrum estimation from unequally spaced data. In: Anderson OD, Perryman MR, editors. Applied time series analysis. Amsterdam: North-Holland Publ. Co; 1982. p. 199–227.
35. Brillinger D, Fienberg S, Gani J, Hartigan J, Krickeberg K, editors. Time series analysis of irregularly observed data. New York/Berlin/Heidelberg/Tokyo: Springer-Verlag; 1984.
36. Brillinger D, Fienberg S, Gani J, Hartigan J, Krickeberg K, editors. Robust and nonlinear time series analysis. New York/Berlin/Heidelberg/Tokyo: Springer-Verlag; 1984.
37. De Prins J, Cornelissen G, Malbecq W. Statistical procedures in chronobiology and chronopharmacology. In: Reinberg A, Smolensky M, Labrecque G, editors. Annual review of chronopharmacology, vol. 2. Oxford: Pergamon Press; 1986. p. 27–141.
38. Cornelissen G. Cosinor-based rhythmometry. Theor Biol Med Model 2014;11:16–24. doi:10.1186/1742-4682-11-16.
39. Bloomfield P. Fourier analysis of time series: an introduction. New York: John Wiley & Sons, Inc.; 1976.

40. Bingham C, Arbogast B, Cornelissen Guillaume G, Lee JK, Halberg F. Inferential statistical methods for estimating and comparing cosinor parameters. Chronobiologia 1982;9: 397–439.
41. Bliss CI. Statistics in biology, vol. 1. New York: McGraw Hill; 1967.
42. Halberg F, Powell D, Otsuka K, Watanabe Y, Beaty LA, Rosch P, Czaplicki J, Hillman D, Schwartzkopff O, Cornelissen G. Diagnosing vascular variability anomalies, not only MESOR-hypertension. Am J Physiol Heart Circ Physiol 2013;305:H279–94.
43. Halberg F, Bingham C, Cornelissen G. Clinical trials: the larger the better? Chronobiologia 1993;20:193–212.
44. Cornelissen G, Halberg F. Treatment with open eyes: markers-guided chronotheranostics. In: Youan BC, editor. Chronopharmaceutics: science and technology for biological rhythm-guided therapy and prevention of diseases. Hoboken, NJ: John Wiley & Sons, Inc.; 2009. p. 257–323.
45. Halberg F, Lubanovic WA, Sothern RB, Brockway B, Powell EW, Pasley JN, Scheving LE. Nomifensine chronopharmacology, schedule shifts and circadian temperature rhythms in di-suprachiasmatically lesioned rats—modeling emotional chronopathology and chronotherapy. Chronobiologia 1979;6:405–24.
46. Cornelissen G, Halberg F. Introduction to chronobiology. Medtronic Chronobiology Seminar #7, 1994 Apr; p. 52. Library of Congress Catalog Card #94-060580. http://halbergchronobiologycenter.umn.edu
47. Halberg F. Biologic rhythms, hormones, and aging. Chapter 13. In: Vernadakis A, Timiras P, editors. Hormones in development and aging. New York: Spectrum Publications; 1982. p. 451–76.
48. Froy O. Circadian rhythms, aging, and life span in mammals. Physiology (Bethesda) 2011;26:225–35.
49. Halberg F, Bakken E, Cornelissen G, Halberg J, Halberg E, Wu J, Sanchez de la Peña S, Delmore P, Tarquini B. Chronobiologic blood pressure assessment with a cardiovascular summary, the sphygmochron. In: Meyer-Sabellek W, Anlauf M, Gotzen R, Steinfeld L, editors. Blood pressure measurements. Darmstadt, FRG: Steinkopff Verlag; 1990. p. 297–326.
50. Hawkins DM. Self-starting CUSUM charts for location and scale. The Statistician 1987;36:299–315.
51. Cornelissen G, Halberg F, Hawkins D, Otsuka K, Henke W. Individual assessment of antihypertensive response by self-starting cumulative sums. J Med Eng Technol 1997;21:111–20.
52. Halberg F, Cornelissen G, Katinas G, Tvildiani L, Gigolashvili M, Janashia K, Toba T, Revilla M, Regal P, Sothern RB, Wendt HW, Wang ZR, Zeman M, Jozsa R, Singh RB, Mitsutake G, Chibisov SM, Lee J, Holley D, Holte JE, Sonkowsky RP, Schwartzkopff O, Delmore P, Otsuka K, Bakken EE, Czaplicki J, International BIOCOS Group. Chronobiology's progress: Part II. Chronomics for an immediately applicable biomedicine. J Appl Biomed 2006;4:73–86. http://www.zsf.jcu.cz/vyzkum/jab/4_2/halberg2.pdf
53. Bendat JS, Piersol AG. Random data-analysis and measurement procedures. New York: Wiley Interscience; 1971.
54. Marquardt DW. An algorithm for least squares estimation of nonlinear parameters. J Soc Ind Appl Math 1963;11:431–41.
55. Marquardt DW. Least squares estimation of nonlinear parameters. IBM Share Library Distribution No. 309401, 1966.
56. Bingham C, Cornelissen G, Halberg E, Halberg F. Testing period for single cosinor: extent of human 24-h cardiovascular "synchronization" on ordinary routine. Chronobiologia 1984;11:263–74.
57. Cornelissen G, Grambsch P, Halberg F. Editorial. World Heart J 2011;3:123–34.
58. Halberg F, Cornelissen G, Sothern RB, Wallach LA, Halberg E, Ahlgren A, Kuzel M, Radke A, Barbosa J, Goetz F, Buckley J, Mandel J, Schuman L, Haus E, Lakatua D, Sackett L, Berg H, Wendt HW, Kawasaki T, Ueno M, Uezono K, Matsuoka M, Omae T, Tarquini B, Cagnoni M, Garcia Sainz M, Perez Vega E, Wilson D, Griffiths K, Donati L, Tatti P, Vasta M, Locatelli I, Camagna A, Lauro R, Tritsch G, Wetterberg L. International geographic studies of onco-

logical interest on chronobiological variables. In: Kaiser H, editor. Neoplasms-comparative pathology of growth in animals, plants and Man. Baltimore: Williams and Wilkins; 1981. p. 553–96.
59. Halberg F, Carandente F, Cornelissen G, Katinas GS. Glossary of chronobiology. Chronobiologia 1977;4 Suppl 1:189.
60. Cornelissen G, Halberg F, Bakken EE, Singh RB, Otsuka K, Tomlinson B, Delcourt A, Toussaint G, Bathina S, Schwartzkopff O, Wang ZR, Tarquini R, Perfetto F, Pantaleoni GC, Jozsa R, Delmore PA, Nolley E. 100 or 30 years after Janeway or Bartter, Healthwatch helps avoid "flying blind". Biomed Pharmacother 2004;58 Suppl 1:S69–86.
61. Verdecchia P, Schillaci G, Guerrieri M, Gatteschi C, Benemio G, Boldrini F, Porcellati C. Circadian blood pressure changes and left ventricular hypertrophy in essential hypertension. Circulation 1990;81:528–36.
62. Nelson W, Cornelissen G, Hinkley D, Bingham C, Halberg F. Construction of rhythm-specified reference intervals and regions, with emphasis on "hybrid" data, illustrated for plasma cortisol. Chronobiologia 1983;10:179–93.
63. Halberg F, Cornelissen G, Otsuka K, Katinas GS, Schwartzkopff O, Halpin C, Mikulecky M, Revilla M, Siegelova J, Homolka P, Dusek J, Fiser B, Singh RB. Chronomics detects altered vascular variabilities constituting risks greater than hypertension: with an illustrative case report. In: Mitro P, Pella D, Rybar R, Valocik G, editors. Proceedings of the 2nd Congress on Cardiovascular Diseases, Kosice, Slovakia, 2002 Apr 25–27. Bologna: Monduzzi Editore; 2002. p. 223–58.
64. Cornelissen G, Kawasaki T, Uezono K, Delea C, Halberg F. II: Blood pressure rhythms and salt. Ann Ist Super Sanita 1993;29:667–77.
65. Engfeldt P, Danielsson B, Nyman K, Aberg K, Aberg H. 24-hour ambulatory blood pressure monitoring in elderly normotensive individuals and its reproducibility after one year. J Hum Hypertens 1994;8(8):545–50.
66. Halberg F, Cornelissen F, Halberg F, Kessler T, Otsuka K. Measuring mental strain by duration of blood pressure overswing (CHAT): case report. World Heart J 2010;2(2):141–67.
67. Cornelissen G, Otsuka K, Halberg F. Week-long ABPM records are a prerequisite for a reliable diagnosis of vascular variability disorders (VVDs). In: Proceedings of the Natural Cataclysms and Global Problems of the Modern Civilization, Istanbul, 2011 Sep 19–21. London: SWB International Publishing House; 2012. p. 250–5.
68. Gumarova LZ, Cornelissen G, Halberg F, Mansharipova AT, Otsuka K, Syutkina EV, Masalov AV, Chibisov SM, Frolov VA. Duration of ABPM as an important prerequisite for a reliable diagnosis of vascular variability disorders (VVDs). Vestnik PFUR, seria Meditsina, 2013; No. 1: 27-33.
69. Halberg F, Cornelissen G, Halberg E, Halberg J, Delmore P, Shinoda M, Bakken E. Chronobiology of human blood pressure. Medtronic Continuing Medical Education Seminars. 4th ed. Minneapolis: Medtronic Inc; 1988. p. 242.
70. Daubechies I. The wavelet transform, time-frequency localization and signal analysis. IEEE Trans Inf Theory 1990;36:961–1005.
71. Sello S, Halberg F, Cornelissen G. Human babies: a slow-to-read, sensitive population magnetometer, also read by wavelets. In: Halberg F, Kenner T, Fiser B, Siegelova J, editors. Noninvasive Methods in Cardiology. Brno, Czech Republic: Faculty of Medicine, Masaryk University; 2011. p. 123–40.
72. Czaplicki J, Cornelissen G, Halberg F. Spectrograms recognize multiple circadian periods by blood pressure and heart rate tensiometry. World Heart J 2012;4:9–22.
73. Nintcheu-Fata S, Cornelissen G, Katinas G, Halberg F, Fiser B, Siegelova J, Masek M, Dusek J. Software for contour maps of moving least-squares spectra. Scr Med (Brno) 2003;76:279–83.
74. Nuttall AH. Some windows with very good sidelobe behavior. IEEE Trans Acoust Speech Signal Process 1981;29:84–91.
75. Halberg F, Cornelissen G, Sothern RB, Katinas GS, Schwartzkopff O, Otsuka K. Cycles tipping the scale between death and survival (= "life"). Prog Theor Phys 2008; Suppl 173:153–81.

76. Cornelissen G, Halberg F, Gheonjian L, Paatashvili T, Faraone P, Watanabe Y, Otsuka K, Sothern RB, Breus T, Baevsky R, Engebretson M, Schröder W. Schwabe's ~10.5- and Hale's ~21-year cycles in human pathology and physiology. In: Schröder W, editor. Long- and short-term variability in Sun's history and global change. Bremen: Science Edition; 2000. p. 79–88.
77. Cornelissen G, Otsuka K, Halberg F. Remove and replace for a scrutiny of space weather and human affairs. In: Grigoriev AI, Zeleny LM, editors. Proceedings of the International Conference on Space Weather Effects in Humans; 2013; Space Research Institute, Moscow, Russia; p. 508–38.
78. White LB, Boashash B. Cross-spectral analysis of non-stationary processes. IEEE Trans Inf Theory 1990;36(4):830–5.
79. Chree C, Stagg JM. Recurrence phenomena in terrestrial magnetism. Philos Trans R Soc Lond A 1928;227:21–62.
80. Stoupel E, Birk E, Kogan A, Klinger G, Abramson E, Israelevich P, Sulkes J, Linder N. Congenital heart disease: correlation with fluctuations in cosmophysical activity, 1995–2005. Int J Cardiol 2009;135(2):207–10.
81. Halberg F, Düll-Pfaff N, Gumarova L, Zenchenko TA, Schwartzkopff O, Freytag EM, Freytag J, Cornelissen G. 27-day cycles in human mortality: Traute and Bernhard Düll. Hist Geo Space Sci 2013;4:47–59. doi:10.5194/hgss-4-47-2013. NIHMSID: 485452.
82. Cornelissen G, Watanabe Y, Otsuka K, Halberg F. Influences of space and terrestrial weather on human physiology and pathology. In: Rosch PJ, editor. Bioelectromagnetic and subtle energy medicine. 2nd ed. Boca Raton: CRC Press; 2015, p. 389–400.
83. Vernova YS, Pochtarev VI, Ptitsyna NG, Tyasto MI. Short-period variations in the rate of change of solar activity as a geosensitive parameter. Geomagn Aeron 1983;23:425–7.
84. Cornelissen G, Halberg F, Wendt HW, Bingham C, Sothern RB, Haus E, Kleitman E, Kleitman N, Revilla MA, Revilla Jr M, Breus TK, Pimenov K, Grigoriev AE, Mitish MD, Yatsyk GV, Syutkina EV. Resonance of about-weekly human heart rate rhythm with solar activity change. Biologia (Bratislava) 1996;51:749–56.
85. Amiranashvili AG, Cornelissen G, Amiranashvili V, Gheonjian L, Chikhladze VA, Gogua RA, Matiashvili TG, Paatashvili T, Kopytenko YA, Siegelova J, Dusek J, Halberg F. Circannual and circadecennian stages in mortality from cardiovascular causes in Tbilisi, Republic of Georgia (1980–1992). Scr Med (Brno) 2002;75:255–60.
86. Cornelissen G, Hillman D, Katinas GS, Rapoport S, Breus TK, Otsuka K, Bakken EE, Halberg F. Geomagnetics and society interact in weekly and broader multiseptans underlying health and environmental integrity. Biomed Pharmacother 2002;56 Suppl 2:319s–26s.
87. Halberg F, Cornelissen G, Otsuka K, Fiser B, Mitsutake G, Wendt HW, Johnson P, Gigolashvili M, Breus T, Sonkowsky R, Chibisov SM, Katinas G, Siegelova J, Dusek J, Singh RB, Berri BL, Schwartzkopff O. Incidence of sudden cardiac death, myocardial infarction and far- and near-transyears. Biomed Pharmacother 2005;59 Suppl 1:S239–61.
88. Cornelissen G, Halberg F, Sothern RB, Hillman DC, Siegelova J. Blood pressure, heart rate and melatonin cycles synchronization with the season, earth magnetism and solar flares. Scr Med 2010;83:16–32.
89. Halberg F, Cornelissen G, Katinas G, Tvildiani L, Gigolashvili M, Janashia K, Toba T, Revilla M, Regal P, Sothern RB, Wendt HW, Wang ZR, Zeman M, Jozsa R, Singh RB, Mitsutake G, Chibisov SM, Lee J, Holley D, Holte JE, Sonkowsky RP, Schwartzkopff O, Delmore P, Otsuka K, Bakken EE, Czaplicki J, International BIOCOS Group. Chronobiology's progress: season's appreciations 2004–2005. Time, frequency, phase, variable, individual, age- and site-specific chronomics. J Appl Biomed 2006;4:1–38. http://www.zsf.jcu.cz/vyzkum/jab/4_1/halberg.pdf
90. Halberg F, Cornelissen G, Grambsch P, McCraty R, Beaty L, Siegelova J, Homolka P, Hillman DC, Finley J, Thomas F, Kino T, Revilla M, Schwartzkopff O. Personalized chronobiologic cybercare; other chronomics' progress by transdisciplinary cycles' congruences: season's appreciations 2009. J Appl Biomed 2011;9:1–34. doi:10.2478/v10136-009-0022-8.
91. Cornelissen G, Grambsch P, Sothern RB, Katinas G, Otsuka K, Halberg F. Congruent biospheric and solar-terrestrial cycles. J Appl Biomed 2011;9:63–102. doi:10.2478/v10136-009-0023-7.

92. Batschelet E, Hillman D, Smolensky M, Halberg F. Angular-linear correlation coefficient for rhythmometry and circannual changes of human births at different latitudes. Int J Chronobiol 1973;1:183–202.
93. Smolensky M, Halberg F, Sargent F. II. Chronobiology of the life sequence. In: Itoh S, Ogata K, Yoshimura H, editors. Advances in climatic physiology. Tokyo: Igaku Shoin Ltd; 1972. p. 281–318.
94. Daglis IA, editor. Space storms and space weather hazards. New York: Springer-Verlag; 2001. ISBN 1-4020-0031-6.
95. Kozyra JU, Liemohn MW. Ring current energy input and decay. Space Sci Rev 2003;109(1–4):105–31.
96. Cornelissen G, Halberg F, Mikulecky M, Florida P, Faraone P, Yamanaka T, Murakami S, Otsuka K, Bakken EE. Yearly and perhaps transyearly human natality patterns near the equator and at higher latitudes. Biomed Pharmacother 2005;59 Suppl 1:S117–22. PMCID: PMC2576447.
97. Starbuck S, Cornelissen G, Halberg F. Is motivation influenced by geomagnetic activity? Biomed Pharmacother 2002;56 Suppl 2:289s–97s.
98. Cornelissen G, Halberg F, Rostagno C, Otsuka K. A chronomic approach to cardiac arrhythmia and sudden cardiac death. J Auton Nerv Syst 2007;44:251–4.
99. Otsuka K, Norboo T, Otsuka Y, Higuchi H, Hayajiri M, Narushima C, Sato Y, Tsugoshi T, Murakami S, Wada T, Ishine M, Okumiya K, Matsubayashi K, Yano S, Chogyal T, Angchuk D, Ichihara K, Cornelissen G, Halberg F. Effect of aging on blood pressure in Leh, Ladakh, a high-altitude (3524 m) community, by comparison with a Japanese town. Biomed Pharmacother 2005;59 Suppl 1:S54–7. PMCID: PMC2736910.
100. Otsuka K, Norboo T, Otsuka Y, Higuchi H, Hayajiri M, Narushima C, Sato Y, Tsugoshi T, Murakami S, Wada T, Ishine M, Okumiya K, Matsubayashi K, Yano S, Chogyal T, Angchuk D, Ichihara K, Cornelissen G, Halberg F. Chronoecological health watch of arterial stiffness and neuro-cardio-pulmonary function in elderly community at high altitude (3524 m), compared with Japanese town. Biomed Pharmacother 2005;59 Suppl 1:S58–67. PMCID: PMC2819461.
101. Otsuka K, Cornelissen G, Norboo T, Takasugi E, Halberg F. Chronomics and "glocal" (combined global and local) assessment of human life. Prog Theor Phys 2008; Suppl 173:134–52.
102. Kawasaki T, Okumiya K, Norboo T, Otsuka K, Matsubayashi K, De Meester F, Wilson DW, Singh RB, Cornelissen G, Halberg F. Chronoecological "glocal" (global and local) health-watch: men better adapted to ambient hypoxia than women. World Heart J 2012;4(2/3): 165–75.
103. Kawasaki T, Yatagai A, Nakaoka T, Otsuka Ke, Otsuka Y, Watanabe Y, Otsuka Ku, Okumiya K, Matsubayashi K, Norboo T, Cornélissen G, Halberg F. Astro-glocal spatially and temporally (global and local) comprehensive health watch, especially at high altitude. In: Grigoriev AI, Zeleny LM, editors. Proceedings of the International Conference on Space Weather Effects in Humans; 2013; Space Research Institute, Moscow, Russia; 2013. p. 539–50.
104. Box GEP, Jenkins GM. Time series analysis: forecasting and control. San Francisco: Holden-Day; 1970.
105. Percival DB, Walden AT. Spectral analysis for physical applications. New York: Cambridge University Press; 1993.
106. Kalman RE. A new approach to linear filtering and prediction problems. Trans ASME J Basic Eng 1960;82(1):35–45. doi:10.1115/1.3662552.
107. Burg JP. Maximum entropy spectral analysis. Paper presented at the 37th Annual International Meeting of the Society for Exploration Geophysicists, Oklahoma City, 1967, p. 34–41.
108. Anderson N. On the calculation of filter coefficients for maximum entropy spectral analysis. Geophysics 1974;39:69–72.
109. Bassingthwaighte JB, Liebovitch LS, West BJ. Fractal physiology. New York: Oxford University Press; 1994.

110. Otsuka K, Nishikawa Y, Kubo Y, Cornelissen G, Halberg F. Chronomes (rhythms, chaos, and age trends) of human heart rate variability in both genders. Comput Cardiol 1997;24:49–52.
111. Otsuka K, Cornelissen G, Halberg F. Age, gender and fractal scaling in heart rate variability. Clin Sci 1997;93:299–308.
112. Otsuka K, Cornelissen G, Halberg F. Circadian rhythmic fractal scaling of heart rate variability in health and coronary artery disease. Clin Cardiol 1997;20:631–8.
113. Otsuka K, Nishimura Y, Kubo Y, Shinagawa M, Watanabe H, Cornelissen G, Halberg F. Chronos and chaos: heart rate variability in healthy Japanese. The Autonomic Nervous System (Tokyo) 1998;35:271–9.
114. Cornelissen G, Engebretson M, Johnson D, Otsuka K, Burioka N, Posch J, Halberg F. The week, inherited in neonatal human twins, found also in geomagnetic pulsations in isolated Antarctica. Biomed Pharmacother 2001;55 Suppl 1:32s–50s.
115. Hotta N, Otsuka K, Murakami S, Yamanaka G, Kubo Y, Matsuoka O, Yamanaka T, Shinagawa M, Nunoda S, Nishimura Y, Shibata K, Saitoh H, Nishinaga M, Ishine M, Wada T, Okumiya K, Matsubayashi K, Yano S, Ichihara K, Cornelissen G, Halberg F. Fractal analysis of heart rate variability and mortality in elderly community-dwelling people Longitudinal Investigation for the Longevity and Aging in Hokkaido County (LILAC) study. Biomed Pharmacother 2005;59 Suppl 1:S45–8. PMCID: PMC2820556.
116. Burioka N, Cornelissen G, Halberg F, Kaplan DT. Relationship between correlation dimension and indices of linear analysis in both respiratory movement and electroencephalogram. Clin Neurophysiol 2001;112:1147–53.
117. Burioka N, Cornelissen G, Halberg F, Kaplan DT, Suyama H, Sako T, Shimizu E. Approximate entropy of human respiratory movement during eye-closed waking and different sleep stages. Chest 2003;123:80–6.
118. Burioka N, Cornelissen G, Otsuka K, Shimizu E, Halberg F. Linear and nonlinear indices of variability in respiratory movement, the electroencephalogram and the electrocardiogram. Neuroendocrinol Lett 2003;24 Suppl 1:223–30.
119. Burioka N, Cornelissen G, Maegaki Y, Halberg F, Kaplan DT, Miyata M, Fukuoka Y, Endo M, Suyama H, Tomita Y, Shimizu E. Approximate entropy of the electroencephalogram in healthy awake subjects and absence epilepsy patients. Clin EEG Neurosci 2005;36:188–93.
120. Burioka N, Miyata M, Cornelissen G, Halberg F, Takeshima T, Kaplan DT, Suyama H, Endo M, Maegaki Y, Nomura T, Tomita Y, Nakashima K, Shimizu E. Approximate entropy in the electroencephalogram during wake and sleep. Clin EEG Neurosci 2005;36:21–3. PMCID: PMC2563806.
121. Otsuka K. Chronome & Janus-medicine: heart rate variability (HRV) and BP variability (BPV) from a viewpoint of chronobiology and ecology. Tokyo: Medical Review; 1998. 213 pp. ISBN 4-89600-219-9.
122. Otsuka K, Cornelissen G, Halberg F. Predictive value of blood pressure dipping and swinging with regard to vascular disease risk. Clin Drug Investig 1996;11:20–31.
123. Otsuka K, Cornelissen G, Halberg F, Oehlert G. Excessive circadian amplitude of blood pressure increases risk of ischemic stroke and nephropathy. J Med Eng Technol 1997;21:23–30.
124. Cornelissen G, Halberg F, Otsuka K, Singh RB, Chen CH. Chronobiology predicts actual and proxy outcomes when dipping fails. Hypertension 2007;49:237–9. doi:10.1161/01.HYP.0000250392.51418.64. PMCID: PMC2613012.

Chapter 9
What Community-Based Glocal Comprehensive Assessment Tells Us: Chronoecology Is Far Superior to Evidence-Based Medicine (EBM)

Abstract In this chapter, we focus on another field of chronoecology. In the quest for a better medicine, since 2000 we have examined community-dwelling subjects from the viewpoint of a "glocal" (combined global and local) comprehensive assessment in several villages. We applied "chronomics" to detect disease at an early stage by longitudinal follow-ups of clinical and lifestyle data to assess changes in health status. Namely, a variety of cognitive, neurobehavioral, and neuropsychological as well as cardiovascular functions will need to be investigated to more precisely map their chronomes in space and time, i.e., in communities worldwide. Herein, we introduce some of the major findings.

Keyword Chronoecology • Cognitive function • Community-based comprehensive medical assessment

9.1 Chronoecology

Recently disclosed facts suggest that organisms evolved on Earth acquired many of the visible and invisible cycles of their habitat and/or of their cosmos. While circadian systems are well documented both time-macroscopically [1–7] and time-microscopically [1–3, 8–12], the temporal organization of physiological function is much more extensive. Humans have developed "clock" genes of the circadian, and probably many other components in the spectral element of chronomes, beyond about-yearly (circannual) and about-weekly (circaseptan) features, as a product of adaptation to, or rather integration with cycles in the cosmos. These resolvable time structures, chronomes, in us have counterparts around us, also consisting of rhythms, trends, and chaos, as is increasingly being recognized.

In this section, we focus on another field of chronoecology in medicine [13]. A variety of cognitive, neurobehavioral, and neuropsychological as well as cardiovascular functions will need to be investigated to more precisely map their chronomes in space and time, i.e., in communities worldwide. Fortunately, we have succeeded in investigating several rural Japanese towns, including Uraus and Tosa (Figs. 9.1–9.7), but also old towns in the Himalayan valley (Figs. 9.8–9.10). We have also performed a project on stroke prevention, based on community-based

K. Otsuka et al., *Chronomics and Continuous Ambulatory Blood Pressure Monitoring*, DOI 10.1007/978-4-431-54631-3_9

*c*omprehensive *m*edical *a*ssessment (CMA), including 7-day/24-hour monitoring of blood pressure as a public service, in Uraus and Tosa, towns in Hokkaido and in Shikoku county, Japan, respectively, according to plans originally made in the city of Roseville, a suburb of St. Paul, Minnesota [14]. Mayor Kaname Yamamoto in Uraus and mayor Takushi Nishimura in Tosa have extended to all residents of their cities an offer of community-based CMA, including 7-day/24-hour blood pressure monitoring, which is the first historic chronome mapping for stroke prevention [15–22].

Fig. 9.1 Pictures of officials who made the CMA studies of two rural Japanese towns possible

Fig. 9.2 A rural scene of Uraus. *Top*: winter; *lower left*: spring; *lower right*: summer (Picture copyright, Mr. Akiho Toshiyuki)

Fig. 9.3 Professor Halberg and Professor Cornelissen visited Uraus town and discussed with town mayor, Mr. Kaname Yamamoto, how we should proceed the chronobiological health-watch in November 3, 2006

Fig. 9.4 A rural scene of Tosa (Picture copyright, Ms. Kayo Kawamura)

Fig. 9.5 Professor Halberg and Professor Cornelissen visited Tosa town and discussed with town mayor, Mr. Takushi Nishimura, how we should proceed with the chronobiological health-watch in November 3, 2006

Fig. 9.6 Professor Halberg gave a lecture for citizens in Tosa town

Fig. 9.7 The signing ceremony of a joint project for the chronobiological health-watch between Professor Franz Halberg and Tosa town mayor, Mr. Takushi Nishimura

Fig. 9.8 *Top*: The Ladakh region in the Himalayas of India; *bottom*: Members of the research team that carried out studies of residents of Leh, Ladakh

Fig. 9.9 Health-watch in Ladakh at the Himalayan valley

Fig. 9.10 Festival at Hemis Temple in Ladakh in July 12, 2008

References

1. Halberg F, Visscher MB. Regular diurnal physiological variation in eosinophil levels in five stocks of mice. Proc Soc Exp Biol (NY) 1950;75:846–7.
2. Halberg F, Stephens AN. Susceptibility to ouabain and physiologic circadian periodicity. Proc Minn Acad Sci 1959;27:139–43.
3. Halberg F. Chronobiology. Annu Rev Physiol 1969;31:675–725.
4. Tei H, Okamura H, Shigeyoshi Y, Fukuhara C, Ozawa R, Hirose M, Sakaki Y. Circadian oscillation of a mammalian homologue of the Drosophila period gene. Nature 1997;389:512–6.
5. Casmore AR, Jarillo JA, Wu YJ, Liu D. Cryptochromes: blue light receptors for plants and animals. Science 1999;284:760–5.
6. Aschoff J, editor. Circadian clocks. Proceedings of the Feldafing Summer School, Sept 1964. Amsterdam: North-Holland Publishing Company; 1965.
7. Pittendrigh CS. Temporal organization: reflections of a Darwinian clock-watcher. Annu Rev Physiol 1993;55:17–54.
8. Otsuka K. Chronome & Janus-medicine: heart rate variability (HRV) and BP variability (BPV) from a viewpoint of chronobiology and ecology. Tokyo: Medical Review; 1998. 213 pp.
9. Cornelissen G, Halberg F. Introduction to chronobiology. Medtronic chronobiology seminar #7, Apr 1994, p. 52 (Library of Congress Catalog Card #94-060580). https://sites.google.com/a/umn.edu/halbergchronobiologycenter/
10. Reinberg A, Smolensky MH. Biological rhythms and medicine. Cellular, metabolic, physiopathologic, and pharmacologic aspects. New York: Springer; 1983. 305 pp.
11. Touitou Y, Haus E, editors. Biological rhythms in clinical and laboratory medicine. Berlin: Springer; 1992. 730 pp.
12. Halberg F, Cornelissen G, Otsuka K, Watanabe Y, Katinas GS, Burioka N, Delyukov A, Gorgo Y, Zhao ZY, Weydahl A, Sothern RB, Siegelova J, Fiser B, Dusek J, Syutkina EV, Perfetto F, Tarquini R, Singh RB, Rhees B, Lofstrom D, Lofstrom P, Johnson PWC, Schwartzkopff O, International BIOCOS Study Group. Cross-spectrally coherent ~10.5- and 21-year biological and physical cycles, magnetic storms and myocardial infarctions. Neuroendocrinol Lett 2000;21:233–58.

13. Halberg F, Cornelissen G, Katinas G, Hillman D, Sampson M, Schwartzkopff O. Season's appreciations 2000. Chronomics complement, among many other fields, genomics and proteomics. Neuroendocrinol Lett 2001;22:53–73.
14. Halberg F, Smith HN, Cornélissen G, Delmore P, Schwartzkopff O, International BIOCOS Group. Hurdles to asepsis, universal literacy, and chronobiology – all to be overcome. Neuroendocrinol Lett 2000;21:145–60.
15. Otsuka K, Cornelissen G, Halberg F, editors. Chronocardiology and chronomedicine: humans in time and cosmos. Tokyo: Life Science Pub; 1993. p. 138.
16. Otsuka K, Murakami S, Matsubayashi K, et al. Chronomes, aging and disease. In: Furukawa H, Nishibuchi M, Kono Y, Kaida Y, editors. Ecological destruction, health, and development. Kyoto; Kyoto University Press 2004. p. 367–94.
17. Otsuka K. Chronomics for prehabilitation. Tokyo: Mishimasha Co.; 2007. p. 261 (in Japanese).
18. Otsuka K. Chronobiology & chronomedicine for successful aging. Tokyo: NTT Pub.; 2010. p. 168 (in Japanese).
19. Otsuka K. Checking out clock, always leads us to health. Tokyo: Makino Pub.; 2012. p. 229 (in Japanese).
20. Otsuka K. Notice the mystery of circadian clock, which leads us to health, 101-Shinsho-155. Tokyo: Shogakukan Co.; 2013. p. 204 (in Japanese).
21. Otsuka K. Clinical chronomedicine; glocalization in time and space. Tokyo: Nakayama Shoten Co. Ltd.; 2013. p. 325 (in Japanese).
22. Otsuka K. Chronogerontology & fountain of youth. Tokyo: Mishimasha Co.; 2014. p. 261 (in Japanese).

Chapter 10
Community-Based Comprehensive Geriatric Assessment

Abstract Intervention aimed at preventing stroke induced an improvement of the cognitive function in an elderly community-dwelling population, especially in people suffering from hypertension, tachycardia, or decreased heart rate variability. People with a disordered coordination of cardiovascular function were more sensitive to such an intervention, suggesting that the cardiovascular function was a major factor affecting cognitive function. The comprehensive geriatric assessment involved measurement of arterial stiffness; blood pressure variability as shown in orthostatic hypotension, quality of sleep, and heart rate variability, useful for "prehabilitation" in elderly community-dwelling people.

Keywords Comprehensive geriatric assessment • Cognitive function • Heart rate variability • Blood pressure variability • Prehabilitation

10.1 Positive Impact of Social Intervention on Disordered Neurobehavioral Function of Elderly Community-Dwelling Population: *L*ongitudinal *I*nvestigation for the *L*ongevity and *A*ging in Hokkaido *C*ounty (LILAC)

Several cardiovascular indices, such as blood pressure (BP), heart rate (HR), and heart rate variability (HRV), are predictors of vascular disease risk in the elderly. It has not yet been shown, however, how these indices are associated with maintained cognitive function. In 2000, we began a community-based study to longitudinally investigate the longevity and aging in Hokkaido County (LILAC) [1–5] and to evaluate the population's neurocardiological function. Our goal was the prevention of stroke and the decline in cognitive function of the elderly in a community dwelling.

We initially examined 115 people older than 75 years (average, 79.6 years). BP was measured at the beginning of study in a sitting position, and pulse wave velocity (PWV) was measured between the right arm and ankle in a supine position, using an ABI/form instrument (Nippon Colin Co., Ltd., Komaki, Japan). The first 1 hour

K. Otsuka et al., *Chronomics and Continuous Ambulatory Blood Pressure Monitoring*, DOI 10.1007/978-4-431-54631-3_10

of ambulatory ECG recording was obtained during routine medical examination conducted each year in July. The data were processed for HRV using a Fukuda-Denshi Holter analysis system (SCM-280-3). Time-domain measures (SDNN, pNN50, SDANN, and Lorenz plot indices: length (L), width (W), and L/W ratio) and frequency-domain measures were determined, namely the spectral power in the very low frequency (VLF), 0.003–0.04 Hz; low frequency (LF), 0.04–0.15 Hz; and high frequency (HF), 0.15–0.40 Hz regions and the LF/HF ratio. Except for SDNN and HR, calculated over the whole 1-hour record, all indices were computed as averages over consecutive 5-min intervals. Spectral indices were obtained by the maximum entropy method (MEM) with the MemCalc/CHIRAM program (Suwa Trust Co., Ltd., Tokyo, Japan).

Using as reference the data obtained in July 2000, the cardiovascular coordination function of each participant was scored as 3, 2, or 1 point for each of the following three indices (systolic BP, HR, and VLF component of the HRV): systolic BP > 160, 140–159, or < 140 mmHg; HR > 80, 70–79, or < 70 beats/min; and VLF < 800, 800–1000, or > 1000 msec^2.

Participants were classified into either the normal, mildly disordered, or disordered group when the sum of these indices was ≤ 4, 5 or 6, or ≥ 7, respectively.

The Japanese version of the Mini-Mental State Examination (MMSE) and the Hasegawa Dementia Scale Revised (HDSR) was used to measure the overall cognitive function, including verbal orientation, memory, and constructional ability. The Up & Go test measured, in seconds, the time it took the subject to stand up from a chair, walk a distance of 3 m, turn, walk back to the chair, and sit down again. This test is a simple measure of physical mobility and demonstrates the subject's balance, gait speed, and functional ability (Up & Go). A lower time score indicates better physical mobility. Functional reach, used to evaluate balance, represents the maximal distance a subject can reach forward beyond arm's length while maintaining a fixed base of support in the standing position. A higher score indicates better balance. Manual dexterity was assessed using a panel with combinations of ten hooks, ten big buttons, and five small buttons. Three discrete measurements of time were recorded for each participant (ten "hook-ons", ten big "button-on-and-offs", and five small "button-on-and-offs"). Total manual dexterity time in seconds, defined as the button score (Button-S), was calculated by adding the average times for one hook-on and one big or small button-on-and-off. A lower button score indicates better manual dexterity.

We evaluated the effects of several kinds of health consultation, rehabilitation of disordered function, healthy lifestyle modification by promoting complete cessation of smoking, weight reduction, reduction of salt intake, moderation in the consumption of fruits and vegetables and alcohol intake, as well as advising medical prescription for the local general practitioner. The paired *t*-test was used to compare each neurobehavioral endpoint between 2000 and 2002. Results were considered to be statistically significant at $p<0.05$.

In 2000 (reference), the cardiovascular coordination score did not correlate with any index of neurobehavioral function, although it showed a negative correlation with SDNN, SDANN, pNN50, LF, or HF components ($p<0.0001$) and a positive correlation with baPWV ($p<0.01$). We were able to follow-up 72 of the 115 subjects.

We found that between 2000 and 2002, the cognitive function, estimated by MMSE and HDSR, was maintained or improved as follows: In the cardiovascular coordination disordered group, MMSE and HDSR improved from 24.6 to 26.0 ($p=0.06$) and from 23.8 to 25.9 ($p=0.04$), respectively. In the mildly disordered group, these indices improved from 23.4 to 25.7 ($p=0.005$) and from 23.4 to 25.1 (N.S.), respectively. In the normal cardiovascular coordination group, MMSE and HDSR were maintained from 25.6 to 26.0 (N.S.) and from 24.9 to 26.4 (N.S.), respectively. There were no statistically significant alterations in activities of daily living (ADL), assessed by Up & Go, functional reach, and button score, in any of the groups.

Although a cross-sectional study did not show any apparent correlation between cardiovascular and neurobehavioral functions in subjects 75 years of age or older, an intervention aimed at preventing stroke and a decline in cognitive function in an elderly community-dwelling population induced an improvement of the cognitive function, especially in people suffering from hypertension, tachycardia, or decreased HRV. In conclusion, this study demonstrates a positive impact of a simple social intervention, including advising medical prescription, in improving a disordered cognitive function in elderly people. It is noteworthy that people with a disordered coordination of cardiovascular function are more sensitive to such an intervention, suggesting that the cardiovascular function is a major factor affecting cognitive function.

10.2 Comprehensive Geriatric Assessment in the Field and Morbidity and Mortality in Elderly Community-Dwelling People

We elucidated the predictive value of comprehensive geriatric assessment (CGA) including heart rate variability and arterial stiffness on the risk of cerebral cardiovascular death, decline of motor function, and cognitive impairment in the LILAC study.

The LILAC study was conducted from July 2000 to November 2005 in 272 persons older than 75 years (female ratio was 59.6 %, average age was 79.1±4.7 years, hypertension was 53.0 %, diabetes mellitus was 14.7 %, and dyslipidemia was 47.0 %). Medical examination and CGA involved cognitive function assessments including MMSE, HDSR, and Kohs block test and ADL assessments including the Up & Go test, functional reach test (FR), and button test, and geriatric depression scale consultations were performed every year (Table 10.1). Heart rate variability (HRV) and arterial stiffness measurement including brachial-ankle PWV (baPWV) and ankle-brachial pressure index were evaluated.

During the follow-up span of 1213±403 days, 38 subjects died, and eight of the deaths were found to be attributable to cerebral cardiovascular disease. The motor function decreased in 44 subjects and cognitive impairment occurred in 56 subjects (20.6%) during follow-up (16.2%). A higher baPWV value, presence of orthostatic

Table 10.1 Reference values of comprehensive medical assessment at Uraus town, in Hokkaido county

Variables	Mean ± S.D.	Variables	Mean ± S.D.
N	272	SDNN (mm/sec^2)	38.5 ± 17.7
Female ratio (%)	59.6	pNN50% (%)	5.1 ± 10.4
Age (years)	79.1 ± 4.7	Insomnia (%)	6.4
BMI (kg/m^2)	23.5 ± 3.8	Current smoking (%)	7.2
SBP (mmHg)	141 ± 19	Hypertension (%)	53.0
DBP (mmHg)	75 ± 10	Diabetes mellitus (%)	15.1
HR (beats/min)	71 ± 12	Dyslipidemia (%)	47.0
baPWV (cm/sec)	2092 ± 504	Plasma glucose (mg/dL)	109.6 ± 33.6
ABI (%)	1.01 ± 0.14	AST (IU/L)	25.6 ± 9.1
OH (%)	2.6	ALT (IU/L)	27.3 ± 23.5
Up & Go test (sec)	16.6 ± 6.7	LDH (IU/L)	418.3 ± 91.1
Functional reach (cm)	25.7 ± 7.9	Total protein (mg/dL)	7.2 ± 0.2
Button test (sec)	15.8 ± 7.4	Albumin (mg/dL)	4.2 ± 0.2
MMSE (points)	24.1 ± 4.4	BUN (mg/dL)	18.6 ± 5.7
HDS-R (points)	23.7 ± 5.3	Creatinin (mg/dL)	1.00 ± 0.26
Kohs (points)	19.1 ± 11.2	Total cholesterol (mg/dL)	194.8 ± 30.3
GDS (points)	4.8 ± 6.7	Triglyceride (mg/dL)	127.9 ± 78.0
LF (mm/sec)	219.2 ± 339.6	HDL (mg/dL)	53.0 ± 13.7
HF (mm/sec)	129.8 ± 316.0		
LF/HF	3.09 ± 5.33		

BMI body mass index, *SBP* systolic blood pressure, *DBP* diastolic blood pressure, *HR* heart rate, *baPWV* brachial-ankle pulse wave velocity, *ABI* ankle-brachial index, *OH* orthostatic hypotension, *MMSE* mini-mental state examination, *HDSR* revised Hasegawa Dementia Scale, *GDS* Geriatric Depression Score, *LF* low frequency component, *HF* high frequency component, SDNN standard deviation of normal to normal intervals, *pNN50 %* the proportion derived by diving NN50 by the total number of NN intervals, *BUN* blood urea nitrogen, *HDL* HDL-cholesterol

hypotension (OH), and lower HDSR were independently related to cerebral cardiovascular death (+500 cm/s in baPWV; RR = 2.87, 95 %CI [1.40, 5.90] P<0.01, presence of OH; RR = 1.61, 95 %CI [1.40, 5.90] P<0.01, −5 points in HDSR; RR = 3.00, 95 %CI [1.71, 7.42] P = 0.02) (Table 10.2, Fig. 10.1). Aging, lower HDSR, and presence of insomnia were independently related to decline in motor function (+5 years in age; RR = 1.77, 95 %CI [1.61, 3.17] P<0.00, −5 points in HDSR; RR = 1.36, 95 %CI [1.08, 1.71] P<0.01, presence of insomnia; RR = 5.37, 95 %CI [0.86, 4.00] P = 0.04) (Table 10.3). Aging, lower FR, and longer button test were independently related to decline in cognitive impairment (+5 years in age; RR = 2.26, 95 %CI [1.61, 3.17] P<0.00, −10 cm in FR; RR = 1.64, 95 %CI [1.09, 2.44] P = 0.02, +5 s in longer button test; RR = 1.28, 95 %CI [1.04, 1.58] P = 0.04) (Table 10.4).

Table 10.2 Stepwise regression analysis of reference values on cardiovascular death

Variable	Relative risk	95 % CI	P value
Age (years) +5	1.13	(0.36, 3.51)	0.83
Gender (male vs. female)	1.15	(0.14, 9.78)	0.89
Systolic blood pressure (mmHg)	0.90	(0.71, 1.14)	0.36
Diastolic blood pressure (mmHg)	1.39	(0.82, 2.34)	0.21
baPWV (cm/sec) +500*	2.87	(1.40, 5.90)	<0.01
Triglyceride (mg/dL) +50	1.80	(0.02, 6.80)	0.66
Orthostatic hypotension (yes vs. no)*	1.61	(1.01, 2.62)	<0.05
Snack taking (times/week)	1.02	(0.04, 1.40)	0.91
HDS-R (points) −5*	3.00	(1.21, 7.42)	0.02

baPWV brachial-ankle pulse wave velocity, *HDSR* revised Hasegawa Dementia Scale
**P*<0.05

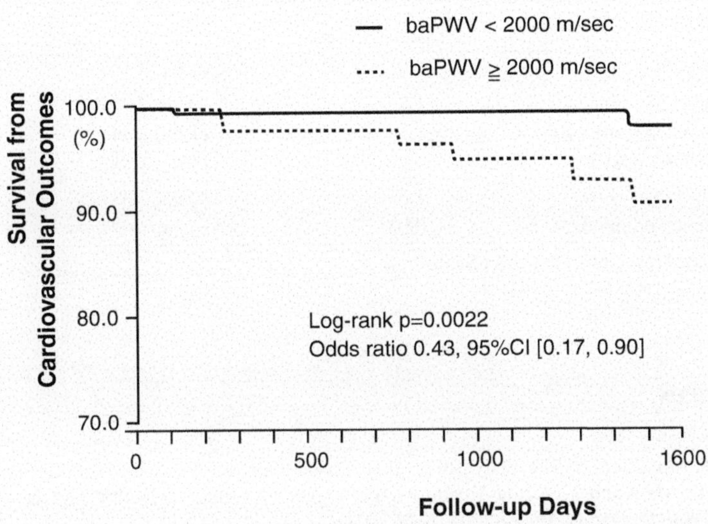

Fig. 10.1 Decreased survival rate from cardiovascular death in citizens with larger baPWV

In conclusion, baPWV was a useful predictor of cerebro- and cardiovascular death. Motor dysfunction and cognitive impairment were correlated. CGA involving arterial stiffness, BP variability as shown in orthostatic hypotension, quality of sleep, and HRV (e.g., SDNN, LF component of HRV, although it was not statistically significant in this analysis) was useful for prehabilitation [2, 6, 7] in elderly community-dwelling people.

Table 10.3 Stepwise regression analysis of reference values on decline in activity of daily living (ADL)

Variable	Relative risk	95 % CI	P value
Age (years) +5*	1.77	(1.87, 3.61)	<0.001
Gender (male vs. female)	0.94	(0.52, 1.66)	0.82
HDS-R (points) −5*	1.36	(1.08, 1.71)	<0.01
MMSE (points) −5	1.14	(0.67, 1.95)	0.63
SDNN (mm/sec^2) +100	0.52	(0.25, 1.08)	0.07
Insomnia (yes vs. no) *	5.37	(1.05, 27.4)	0.04
Diabetes mellitus (yes vs. no)	1.83	(0.86, 4.00)	0.11

HDSR revised Hasegawa Dementia Scale, *MMSE* mini-mental state examination, *SDNN* standard deviation of normal to normal intervals
*$P<0.05$

Table 10.4 Stepwise regression analysis of reference values on decline in cognitive function

Variable	Relative risk	95 % CI	P value
Age (years) +5*	2.26	(1.61, 3.17)	<0.01
Gender (male vs. female)	1.17	(0.58, 2.35)	0.96
Functional reach (cm) −10*	1.64	(1.09, 2.44)	0.02
Button test (seconds) +5*	1.28	(1.04, 1.58)	0.04
Up & Go test (sec) +10	1.51	(0.98, 2.33)	0.06
Low frequency component	0.93	(0.84, 1.02)	0.11

HDSR revised Hasegawa Dementia Scale
*$P<0.05$

References

1. Otsuka K, Murakami S, Matsubayashi K, et al. Chronomes, aging and disease. In: Furukawa H, Nishibuchi M, Kono Y, Kaida Y, editors. Ecological destruction, health, and development. Kyoto: Kyoto University Press; 2004. p. 367–94.
2. Otsuka K. Chronomics for prehabilitation. Tokyo: Mishimasha Co.; 2007. p. 261 (in Japanese).
3. Otsuka K. Chronobiology & chronomedicine for successful aging. Tokyo: NTT Pub.; 2010. p. 168 (in Japanese).
4. Otsuka K. Checking out clock, always leads us to health. Tokyo: Makino Pub; 2012. p. 229 (in Japanese).
5. Otsuka K. Notice the mystery of circadian clock, which leads us to health, 101-Shinsho-155. Tokyo: Shogakukan Co.; 2013. p. 204 (in Japanese).
6. Otsuka K. Clinical chronomedicine; glocalization in time and space. Tokyo: Nakayama Shoten Co. Ltd.; 2013. p. 325 (in Japanese).
7. Otsuka K. Chronogerontology & fountain of youth. Tokyo: Mishimasha Co; 2014. p. 261 (in Japanese).

Chapter 11
Chronomic Community Screening: Depressive Mood, Heart Rate Variability, and Cardiovascular Outcomes in a Community

Abstract Humans have developed "clock" genes of the circadian, and probably many other components in the spectral element of chronomes, beyond about-yearly (circannual) and about-weekly (circaseptan) features, as a product of adaptation to, or rather integration with, cycles in the cosmos. While life originally integrated itself into the cycles of an anthropogenically unpolluted environment, the environmental cycles are now being changed in keeping with the schedules of societal life as in the case of global temperature. Hence, a variety of cognitive, neurobehavioral and neuropsychological as well as cardiovascular functions will need to be investigated to more precisely map their chronomes in space and time, in order to understand chronoastrobiology, based on both the system times and time horizons yielded by chronomes.

We have started a project (chronoecological health-watch) on stroke prevention, based on community-based comprehensive medical assessment, including 7-day/24-hour monitoring of blood pressure (BP) as a public service, in Uraus, a town in Hokkaido, Japan (latitude: 43.45° N, longitude: 141.85° E), according to plans originally made in the city of Roseville, a suburb of St. Paul, Minnesota. In Uraus, Mayor Kaname Yamamoto has extended to all residents of his city an offer of 7-day/24-hour blood pressure monitoring and has started a historic first chronome mapping for stroke prevention.

Chronoecological health-watch was started in 2000 and consultations were repeated every 3 months. Several results of the follow-up on adverse cardiovascular outcomes, or cancer death, are presented here, especially focusing on the relations of depressive mood, 7-day/24-hour BP monitoring, and heart rate variability.

Keywords Depressive mood • Chronomic community screening • Nonphotic transyear components • Community-dwelling elderly

11.1 Depressive Mood Is a Prospective Factor of Stroke and Cardiovascular Events in a Community in Japan

Several lines of evidence suggest that clinical depression is a risk factor for cardiac morbidity and mortality in patients with coronary heart disease, especially after an acute myocardial infarction (MI). Major depression is associated with a 4-fold

K. Otsuka et al., *Chronomics and Continuous Ambulatory Blood Pressure Monitoring*, DOI 10.1007/978-4-431-54631-3_11

increase in the risk of mortality during the first 6 months after an acute MI, and its prognostic significance is comparable to that of left ventricular dysfunction and a history of MI. Depression is not unusual among individuals with coronary heart disease, with studies indicating that between 15 % and 22 % of patients suffer from depression after a cardiac event.

Not only circadian, but also circannual, circaseptan (about 7-day), and other variations characterize the incidence of adverse vascular events, such as acute myocardial infarctions, strokes, and sudden cardiac deaths, carrying important pathophysiological implications. Most studies showed an increased incidence in the morning, peaking between 06:00 and 12:00. Ambulatory blood pressure monitoring (ABPM) has become an important tool in the diagnosis and management of hypertension. Several studies have indicated that target organ damage and cardiovascular morbidity are more strongly associated with circadian BP endpoints than with office BP values. We reported that circadian and other patterns are synchronized by socio-ecologic factors, such as human lifestyles, as well as economic and environmental conditions. We demonstrated the presence of a weekly BP variation in community-dwelling subjects, including a surge in BP on Mondays, and we showed that it was most prominent in depressed people.

In 2000, we began a community-based study to longitudinally investigate the longevity and aging of a population in Hokkaido County (LILAC) [1, 2], including the evaluation of the younger population aged from 40 to 74 years (health-watch) [3]. Our goal is the prevention of cardiovascular events, notably strokes and myocardial ischemic events in a community. Herein, we examine whether a depressive mood is associated with an increased risk of stroke and cardiovascular mortality, and we estimate the predictive value of a depressive mood on the incidence of adverse vascular events, using a multivariate Cox model.

The 224 subjects monitored between April 2001 and March 2004 were between 24 and 79 years of age (88 men and 136 women; mean age: 56.8 ± 11.2 years), living in Uraus in Hokkaido (latitude: 43.45° N, longitude: 141.85° E). All subjects visited and utilized the free health screening, counseling, and educational services of the Uraus office. Subjects with definite neurological diseases, such as Parkinsonism and stroke, and those who were too severely ill to stand without help were excluded. Lifestyle, including nutritional and sleep quality, was investigated by several types of questionnaires. BP was measured at the beginning of the study in a sitting position, and brachial-ankle pulse wave velocity (baPWV) was measured between the right arm and ankle in a supine position, using an ABI/Form instrument (Nippon Colin Co., Ltd., Komaki, Japan). In all studies, baPWV was measured twice after at least a 5-min rest. For the assessment of the predictive value, only baPWV measures from participants showing normal ankle/brachial pressure index (ABI) values (>0.90) were used.

The 7-day/24-hour BP recordings were obtained with a commercially available ambulatory BP recorder (TM-2431, A&D Company, Japan) [4]. Subjects were fitted with the recorder in the morning, usually between 10:00 and 11:30 a.m. of the first day of monitoring. The recorder was set to take measurements at intervals of 30 min between 7:00 a.m. and 10:00 p.m. and of 60 min between 10:00 p.m. and 7:00 a.m.

Subjects were instructed to follow their usual daily routine after they left the office. Subjects were instructed to remain motionless each time a reading was to be taken and then to record their activity on a diary sheet. Stored data were retrieved and analyzed on a personal computer with commercially available software for the device (TM-2430-15, A&D Company, Japan). The data were further analyzed by sphygmochron for the entire 7-day span and separately for each day of monitoring. In addition to the circadian rhythm characteristics, estimates were also obtained for mean values during 24-hour, waking and sleeping spans, and for the day-night ratio (as a gauge of "dipping") for SBP, DBP, and HR and also for mean blood pressure (MBP = 1/3SBP + 2/3DBP), pulse pressure (PP = SBP–DBP), and the double product (DP = SBP × HR/100) ("dip" is defined as the awake-sleep difference divided by awake (×100), where awake and sleep are average values during the awake and sleep span, respectively).

Numerous depression rating scales are currently available, which represent a mixture of observer-rated and self-rating scales. In this investigation, subjects were asked about 15 items on a depression scale, by means of a self-administered questionnaire [5]. When the depression score was ≥5, subjects were considered to have a depressed mood. A depression score was obtained again 1–2 years later. Subjects with a score during the second screening higher by at least 2 points compared to that of the first screening were assigned to the enhanced depressive mood group (irrespective of the score at the start of study). All other subjects were assigned to the control group.

The health-watch study was started on April 12, 2001. One or two doctors of our team visited and provided repeated consultations every 3 months. They offered advice in relation to the rehabilitation of disordered functions and healthy lifestyle modifications (promoting complete cessation of smoking, weight reduction, reduction of salt intake, moderation in the consumption of fruits and vegetables, and alcohol intake). They also advised in terms of medical prescriptions for the local general practitioner.

The follow-up herein ended on November 30, 2004. During this time, 9 subjects died of myocardial infarction or stroke. The follow-up time was defined as the time elapsed between the date of the first (reference) examination and the date of first cardiovascular event or death.

We used Cox's regression analysis to calculate the unadjusted and adjusted relative risks (RRs) and 95 % confidence intervals (CIs) for cardiovascular death. To identify independent predictors of stroke and cardiovascular disease in relation to a depressive mood, we used a multivariate Cox regression analysis with stepwise selection, including as variables age, gender, BMI, lifestyle, sleep quality, QOL, baPWV, and ABP endpoints. The independent correlation of the enhanced depressive mood was determined by means of a logistic regression analysis. Kaplan-Meier event probability curves were computed with two groups, and the cumulative probability of events of two groups was compared by means of the log-rank test. Significance was considered at a value of $p < 0.05$.

The 224 subjects comprised 88 men and 136 women. The mean age of participants at entry was 56.8 years. The mean follow-up time was 1064 days, during which 4 subjects suffered from adverse vascular outcomes (myocardial infarction, 1 man and 1 woman; stroke, 2 men).

As a comprehensive assessment in this chronoecological health-watch, we checked out the following items: indices of environmental conditions, lifestyle, quality of sleep, QOL, and 7-day ABP endpoints, such as the 7-day (from Thursday to Wednesday) and daily averages of SBP, DBP, and HR; the incidence of SBP and DBP CHAT (circadian hyper-amplitude-tension) and of SBP, DBP, and HR "dipping" (SBP-dip, DBP-dip, and HR-dip); estimates of the circadian amplitude and acrophase of SBP, DBP, and HR; as well as estimates of the percentage time elevation (PTE) of SBP, DBP, and HR. Averages over 7 days were also computed (not shown) as SBP 1-7, DBP 1-7, pulse pressure (PP) 1-7, double product (DP) 1-7, 24-hour SD of HR (HRSD1-7), SBP-dip 1-7, DBP-dip 1-7, HR-dip 1-7, MAP-dip 1-7, PP-dip 1-7, and DP-dip 1-7.

Depression scores were obtained for 218 of the 224 citizens at the start of study and from 179 subjects after 1–2 years (177 subjects provided a score during both sessions). Scores obtained during the first and second sessions averaged 4.11 ± 3.11 and 4.42 ± 3.14, respectively. Of the 177 citizens who answered both questionnaires, 39 were assigned to the enhanced depressive mood group. A comparison of characteristics between the two groups is shown in Table 11.1. It can be seen that depressive subjects had a higher body mass index (BMI), they consumed smaller meals (smaller portions of rice), and their total serum cholesterol was higher as compared to the control subjects. The time to fall asleep was also slightly longer in the depressive group ($p = 0.051$). Kaplan-Meier curves for event-free survival revealed a significant difference between the two groups ($p = 0.0124$, log-rank test) (Fig. 11.1). Subjects in the control group had a better event-free survival than depressed subjects.

Among the variables used in Cox proportional hazard models, depressive mood, assessed by GDS, as well as DBP-MESOR and SBP-Amplitude showed a statistically significant association with the occurrence of vascular outcomes. It should be noted that GDS during the second but not during the first session was statistically significantly associated with adverse vascular outcomes. In univariate analyses, an increase by 2 or 3 points of the GDS score during the second session was associated with a RR of 2.121 (95 % CI: 1.237–3.637) or 3.088 (95 % CI: 1.375–6.935), respectively ($P = 0.0063$). A 5 mmHg increase in DBP-MESOR and a 3 mmHg increase in SBP-Amplitude were associated with RRs of 2.143 (95 % CI: 1.232–3.727, $P = 0.0070$) and 0.700 (95 % CI: 0.495–0.989, $p = 0.0430$), respectively. In multivariate analyses, when both the second GDS score and DBP-MESOR were used as continuous variables in the same model, GDS remained statistically significantly associated with the occurrence of cardiovascular death. After adjustment for DBP-MESOR, a 3-point increase in GDS score was associated with a RR of 2.172 (95 % CI: 1.123–4.200).

Monday and/or Tuesday endpoints of the 7-day ABP showed a statistically significant association with adverse vascular outcomes. A 5-mmHg increase in DBP-5 or DBP-6, namely, in the arithmetic mean of DBP from data collected on Mondays and Tuesdays, was associated with RRs of 1.576 (95 % CI: 1.011–2.457, $P = 0.0446$) and 1.666 (95 % CI: 1.052–2.639, $p = 0.0297$), respectively. A 20-unit increase in

Table 11.1 Comparison between the control and enhanced depressive mood groups

		Control			Enhanced depressive mood				
	Units	n	Mean	SD	n	Mean	SD	t-value	p-value
GDS 1	Points	138	4.34	3.13	39	3.26	2.70	1.980	0.049
GDS 2	Points	138	3.75	2.83	39	6.80	3.07	−5.827	<0.001
Difference of GDS	Points	138	−0.61	1.57	39	3.54	1.85	−13.976	<0.001
Age	Years	138	56.7	11.7	39	57.7	10.3	−0.462	N.S.
Gender		138			39			0.896	N.S.
BMI	kg/m^2	138	24.1	2.6	39	25.2	2.9	−2.202	0.029
Smoking	Points	135	9.5	14.9	39	8.3	14.2	0.455	N.S.
Family Hx	Points	138	2.27	2.22	39	2.18	2.02	0.243	N.S.
Alcohol	Points	137	0.77	1.24	39	0.82	0.97	−0.238	N.S.
Vegetables	Times daily	138	2.12	0.80	39	2.27	0.77	−1.019	N.S.
Subjective healthiness	%	138	65.6	17.6	39	65.0	17.5	0.182	N.S.
Subjective mood	%	138	69.8	17.9	39	65.9	19.0	1.219	N.S.
Sense of happiness	%	138	74.3	19.3	39	72.6	17.2	0.514	N.S.
Getting up time	Clock hours	138	6:13	0.03	39	6:02	0.03	1.294	N.S.
Duration of sleep	hours	115	7:50	0.04	33	7:51	0.04	−0.051	N.S.
Time for falling asleep	Min	133	18.5	12.7	39	24.1	23.1	−1.964	0.051
Sleep well ?	Points	135	1.2	0.4	39	1.3	0.5	1.255	N.S.
Sleep quality score	Points	135	19.8	2.4	39	19.5	2.8	0.834	N.S.
Salt intake daily	g	123	13.6	2.2	39	13.3	1.7	0.966	N.S.
Rice	Cups	123	2.60	0.95	39	2.20	0.64	2.442	0.016
rt baPWV	cm/sec	127	1458.1	309.1	35	1520.6	335.4	−1.039	N.S.
lt baPWV	cm/sec	127	1500.6	330.1	35	1555.7	340.2	−0.870	N.S.
rt ABI		127	1.12	0.08	35	1.09	0.08	1.761	0.080
lt ABI		127	1.11	0.08	35	1.09	0.09	1.332	N.S.
ECG (SV1±RV5)	mm	98	28.0	9.3	24	29.7	10.0	−0.794	N.S.
CTR	%	81	47.9	5.5	28	49.4	4.9	−1.310	N.S.
Total cholesterol	mg/dl	131	204.4	30.3	37	215.5	28.0	−2.009	0.046
HDL cholesterol	mg/dl	130	58.8	14.2	37	59.6	13.0	−0.304	N.S.
Triglyceride	mg/dl	131	123.3	85.3	37	128.9	83.9	−0.353	N.S.

(continued)

Table 11.1 (continued)

		Control			Enhanced depressive mood				
	Units	n	Mean	SD	n	Mean	SD	t-value	p-value
Uric acid	mg/dl	128	5.0	1.3	35	5.0	1.6	−0.243	N.S.
Serum creatinine	mg/dl	125	0.88	0.33	34	0.88	0.16	0.046	N.S.
FBS	mg/dl	128	99.5	16.5	36	96.7	12.9	0.939	N.S.
Follow-up days	days	138	1085.8	227.9	39	1128.1	260.5	−0.991	N.S.

GDS scores and difference of GDS scores used for classification. Gender: M = 1, F = 0. rt: right; lt: left.

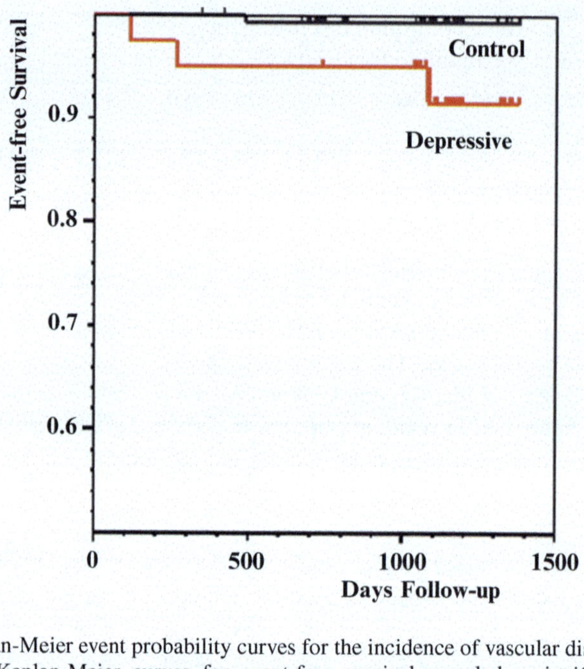

Fig. 11.1 Kaplan-Meier event probability curves for the incidence of vascular disease. Calculation of Kaplan-Meier curves for event-free survival revealed a significant difference between the control and depressive groups (p = 0.0124, log-rank test). A depression score was obtained again after 1–2 years, and when the score during the second screening had increased by more than 2 points compared to the first screening, such subjects were assigned to the enhanced depressive mood group, shown as *depressive*, while the other subjects were assigned to the control group, shown as *control*. It can be seen that subjects in the control group had fewer adverse vascular events (better event-free survival)

DP-5 (double product on Mondays) and a 3 % increase in MAP-dip-5 were associated with RRs of 3.067 (95 % CI: 1.075–8.753, p=0.0362) and 0.824 (95 % CI: 0.683–0.995, P=0.0439), respectively. Saturday endpoints also related to adverse vascular outcomes. A 3 % increase in PP-dip-3 (dipping ratio of pulse pressure on Saturday nights) and a 5-unit increase in DP-dip-3 (dipping ratio of Saturday's double product) were inversely associated with adverse vascular outcome, with RRs of 0.865 (95 % CI: 0.756–0.989, P=0.0337) and 0.752 (95 % CI: 0.596–0.948, P=0.0158), respectively.

Depression is a risk factor for cardiac morbidity and mortality in patients with coronary heart disease [6–8]. Anda et al. [9] followed a cohort of 2832 persons for an average of 12.4 years. During the reference stage, 11.1 % of the study cohort had a depressed affect. The adjusted relative risk of fatal coronary heart disease (CHD) was 1.4 for patients with depressed affect, compared with patients who were not depressed. For nonfatal CHD, the adjusted relative risk was 1.6. For both fatal and nonfatal CHD, the increased risk associated with depression was independent of established risk factors for CHD (e.g., smoking, cholesterol concentration, family history). Barefoot and Schroll [8] reported that high scores on a measure of symptoms of depressed mood were associated with an increased risk for acute myocardial infarction (AMI) and early mortality during a 27-year follow-up and that the impact of a depressed affect on health did not differ between men and women. We also observed in middle-aged community-dwelling people that a depressive mood, assessed by GDS, predicted the occurrence of vascular diseases beyond the prediction provided by age, gender, ABP, lifestyle, and environmental conditions, as assessed by means of a multivariate Cox model. A depressive mood, especially when enhanced for 1–2 years, was associated with adverse vascular outcomes. Results herein suggest the clinical importance of repeated assessment of a depressive mood and of taking sufficient care of depressed subjects.

Another result of the present study is that in community-dwelling people, circadian and circaseptan characteristics of BP variability derived from 7-day/24-hour ABP predicted the occurrence of vascular diseases beyond the prediction provided by age, gender, depressive mood, and lifestyle, as assessed by means of a multivariate Cox model. Ambulatory 24-hour blood pressure monitoring has become quite popular in recent years, offering a number of advantages from the viewpoint of both diagnosis and treatment [10]. One advantage rests on the fact that clinic blood pressure is not representative of blood pressure values outside the clinic. Results obtained herein show that blood pressure is characterized not only by a circadian but also by a weekly variation. The 24-hour averages of SBP and DBP are seen to differ from one day to another in the same citizen in the same 7-day record. In middle-aged community-dwelling people, DBP-MESOR and SBP-Amplitude predicted the occurrence of vascular diseases, assessed by means of a multivariate Cox model. We found that several endpoints from data collected on Mondays predicted adverse outcomes, suggesting that circaseptan as well as circadian characteristics are associated with the occurrence of vascular disease. We reported on weekly BP characteristics earlier [11]. Most citizens in a community showed a novelty effect (the first day effect), a holiday dip, and a Monday morning surge of SBP. A Monday peak has been reported for the incidence of acute myocardial infarction, sudden cardiac death, and stroke [12–16]. Our previous studies showed that the morning BP surge on Mondays was statistically significantly higher compared with other weekdays [17]. Although a direct association between the Monday surge in BP and cardiovascular events could not be demonstrated directly, it is possible that a morning surge of BP on Mondays may trigger cardiovascular events. This possibility is supported by the fact that we found that depressive citizens had a more prominent circaseptan component for SBP, DBP, and DP, compared with nondepressive subjects [18].

Results herein indicate the clinical importance of taking care of depressed subjects. In view of the strong relation between depression and adverse cardiac events, studies should be done to ascertain whether treatment of depression, especially from the standpoint of chronodiagnosis and chronotherapy, can reduce elevated blood pressure values and improve the altered variability in BP and HR, so as to reduce the incidence of adverse cardiac events. Fewer than 7 days of monitoring means a greater chance of a false diagnosis, on which a treatment decision may depend for the long term, including perhaps a potentially unnecessary treatment for decades. Less than 7 days of monitoring can be compared with taking the pulse for less than a cardiac cycle, if the circaseptan aspect of time structures (chronomes) is viewed as one of many new spectral components, with even longer periods, including transyears characterizing series of blood pressure and heart rate covering at least 5 and up to 38 years [19]. When the ECG and EEG, as well as the blood pressure of small rodents, can be monitored continuously for most of their lifetime, this study is but a small step toward universal 7-day and eventually continuous monitoring of BP and HR, or the ECG.

11.2 Chronomic Community Screening Reveals an Association Between Depression and Infradian Vascular Rhythm Alteration

Depression is a risk factor for cardiac morbidity and mortality in patients with coronary heart disease, especially after an acute myocardial infarction [8, 9]. The link between depression and mortality in older persons is now hotly debated. Fried et al. [20] observed an association between depression and mortality, and Schulz et al. [21] followed to show in a large population-based study of older persons, the Cardiovascular Health Study, that high scores of symptoms of depressed mood are an independent risk factor for mortality. Motivational depletion may be a key underlying mechanism for the depression-mortality effect.

Much research has shown that depression predicts total mortality and the prognosis of patients after MI, but evidence concerning the role of depression in the pathogenesis of coronary disease has been less clear. Ambulatory blood pressure monitoring (ABPM) has become an important tool in the diagnosis and management of hypertension. Several previous studies have indicated that target organ damage and cardiovascular morbidity are more strongly associated with circadian variations and the average BP over a 24-hour span than with office BPs.

In recent years, not only circadian, but also circannual, circaseptan (about 7-day) [12–14, 16, 15], and other variations in the incidence of adverse, notably vascular events, such as acute myocardial infarctions, strokes, and sudden cardiac deaths, have been noted in many studies and may carry important pathophysiological implications. Most studies showed an increased onset in the morning with a peak incidence between 06:00 and 12:00, which has been aligned with the concomitant rapid increase in BP, HR, cardiac output and platelet aggregability, and the corresponding decrease in cardiac vagal activity and in fibrinolytic function. These patterns are synchronized by socio-ecologic factors such as human lifestyles and economic and environmental conditions. We have already reported the presence of a weekly variation in BP in community-dwelling subjects. In this investigation, we focus on how depressed mood is associated with abnormal about-weekly and half-weekly variations in BP and HR, including the BP surge on Mondays.

The 217 subjects monitored between April 2001 and April 2003 were between 24 and 79 years of age (85 men and 132 women; mean age, 56.8 ± 11.3 years), living in Uraus in Hokkaido (latitude 43.45° N and longitude 141.85° E). All subjects visited and utilized the free health screening, counseling, and educational services of the Uraus office. Subjects with definite neurological diseases, such as Parkinsonism and stroke, and those who were too severely ill to stand without help were excluded. The 7-day/24-hour BP recordings were obtained from all participants as outlined above.

Numerous depression rating scales are currently available, which represent a mixture of observer-rated and self-rating scales. In this investigation, subjects were asked about 15 items on a depression scale, through the use of a self-administered questionnaire [5]. When the depression score was ≥5, subjects were considered to have a depressed mood.

Results are reported as mean ± SD. Student's t-tests were used to compare subjects with depression scores <5 or ≥5. Analyses of variance (ANOVA) with repeated measures were used to assess any day-of-week and group (depression score) effects. Cosinor methods with parameter tests [22] compared the circadian, circasemiseptan, and circaseptan characteristics of SBP, DBP, and HR of subjects in both groups. A P-value below 0.05 was considered to indicate statistical significance (and below 0.10, borderline statistical significance).

Table 11.2 lists the 24-hour averages and standard deviations (SD) of the 7 consecutive days of monitoring for SBP, DBP, PP, DP, HR-SD, SBP-dip, DBP-dip, PP-dip, DP-dip, and HR-dip. The day-to-day variation is statistically significant for

Table 11.2 Day-to-day variation of the 24-hour average and standard deviation obtained from the 7-day/24-hour ambulatory monitoring

		SBP		DBP		Pulse pressure		Double product		HR-SD	
	n	Mean	SD	Mean	SD	Mean	SD	Mean	SD	Mean	SD
The 1st day	212	129.7	16.5	79.5	9.5	50.5	10.5	90.8	15.2	11.2	4.6
The 2nd day	214	128.4	15.8	78.8	9.1	49.9	9.9	92.4	15.7	11.8	4.2
The 3rd day	213	127.0	14.7	77.9	8.8	49.5	9.7	91.1	14.0	12.2	4.5
The 4th day	214	126.7	15.1	77.6	8.8	49.5	9.7	90.7	14.9	11.9	4.5
The 5th day	209	127.5	15.4	78.4	9.0	49.5	9.8	92.2	15.7	11.4	4.1
The 6th day	203	127.8	15.5	78.4	9.3	49.8	10.0	92.0	15.4	11.7	4.4
The 7th day	172	127.9	15.6	78.5	9.1	49.8	10.2	92.3	14.8	11.8	4.3
	F-value	4.11		5.03		0.91		2.61		3.33	
	P-value	<0.0005		<0.0001		N.S.		<0.05		<0.005	
		SBP dip		**DBP dip**		**Pulse pressure dip**		**Double product dip**		**HR-dip**	
	n	**Mean**	**SD**	**Mean**	**SD**	**Mean**	**SD**	**Mean**	**SD**	**Mean**	**SD**
The 1st day	215	17.5	9.8	18.3	10.7	14.6	17.7	29.7	13.0	14.7	12.3
The 2nd day	215	15.3	10.4	16.8	10.5	12.2	16.7	29.2	12.5	16.0	11.4
The 3rd day	215	14.6	10.7	16.4	10.7	11.4	15.8	29.9	12.8	17.6	10.3
The 4th day	214	15.6	10.6	16.5	11.2	13.1	16.2	29.6	13.7	16.3	12.7
The 5th day	210	15.8	10.4	17.8	10.2	11.4	17.5	30.5	12.2	17.3	10.2
The 6th day	204	16.7	10.1	18.4	10.9	12.5	16.2	31.5	11.8	17.6	10.7
The 7th day	172	16.4	9.2	17.5	10.5	13.8	14.9	31.0	11.9	17.3	10.9
	F-value	3.40		1.74		2.04		0.38		2.44	
	P-value	<0.005		N.S.		N.S.		N.S.		<0.05	

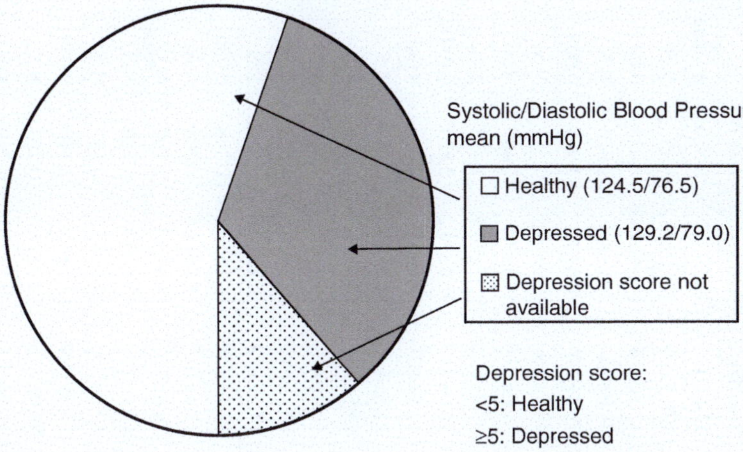

Fig. 11.2 Pie chart showing large proportion of population of depressed citizens (depression score ≥5)

Table 11.3 Difference in 7-day average of ambulatory monitoring measures between healthy and depressed mood groups

	Healthy (n=120)		Depressed (n=72)		Student's t test	
	Mean	SD	Mean	SD	t-value	p-value
7-day SBP	124.50	15.17	129.17	13.99	−2.1310	0.0344
7-day DBP	76.53	8.66	79.03	8.15	−1.9872	0.0483
7-day HR	69.83	7.59	69.55	6.25	0.2645	N.S.
SBP-dip	16.40	7.09	15.21	7.63	1.1046	N.S.
DBP-dip	18.22	7.01	16.30	7.46	1.7965	0.0740
HR-dip	17.37	8.03	16.51	7.42	0.7429	N.S.

the 24-hour average of SBP, DBP, DP, HR-SD, and SBP-dip, which shows a novelty effect. Namely, the 24-hour averages of SBP and DBP and the dipping ratio of SBP were highest and HR-SD and HR-dip were lowest on the first day of the 7-day monitoring. On the other hand, a novelty effect was not detected for DP.

Depression scores were obtained for 192 of the 217 citizens. In 72 subjects, the depression score was ≥5 (31 %), and they were considered to have a depressed mood (Fig. 11.2). Table 11.3 shows the difference in 7-day averages between the healthy and depressed mood groups. SBP and DBP averages were statistically significantly higher in depressed subjects (SBP, 129.2 vs. 124.5 mmHg; $p=0.034$; DBP, 79.0 vs. 76.5 mmHg; $p=0.041$). The 7-day HR average did not differ between the two groups. DBP dipping was smaller in the depressed subjects (16.30 vs. 18.22 %; $p=0.04$), but the dipping ratio of SBP and HR was not statistically significantly different.

Table 11.4 shows differences in the day-to-day variation of ambulatory measures among weekdays for both the healthy and depressed mood groups, analyzed by one-

Table 11.4 Difference of ambulatory BP and HR measures for each weekday between healthy and depressed mood groups

		Thursday	Friday	Saturday	Sunday	Monday	Tuesday	Wednesday	ANOVA	
Number of healthy/depressed group		118/72	120/72	119/72	120/72	118/71	114/70	97/59	F-value	p-value
24h-SBP	Healthy mood	127.3±17.0	126.7±16.9	125.6±15.1	125.4±15.8	126.4±16.5	126.1±16.3	125.0±15.9	1.367	N.S.
(mm Hg)	Depressed mood	132.0±15.1	131.0±14.4	128.3±14.0	128.2±14.3	128.9±14.2	130.0±14.5	131.6±14.8	3.035	0.007
24h-DBP	Healthy mood	78.5±9.7	77.9±9.6	77.4±8.9	77.1±8.9	78.0±9.4	78.0±9.7	77.2±8.8	1.923	N.S.
(mmHg)	Depressed mood	80.8±9.5	80.6±8.9	78.9±8.7	78.4±8.9	79.2±8.7	79.4±9.0	80.5±9.8	2.965	0.008
Pulse Pressure	Healthy mood	48.9±10.5	49.0±10.4	48.4±9.8	48.5±10.1	48.7±10.4	48.5±9.9	48.0±10.5	0.351	N.S.
(mmHg)	Depressed mood	51.9±9.4	50.9±9.4	49.9±8.9	50.4±9.1	50.4±8.7	51.0±10.0	51.5±9.0	1.489	N.S.
Double Product	Healthy mood	89.2±16.1	91.2±16.9	90.7±14.2	90.4±15.8	91.5±16.5	90.8±15.9	90.2±14.7	1.428	N.S.
(mmHgxbpm)	Depressed mood	92.8±13.4	95.1±14.2	92.2±12.9	91.1±13.6	93.4±14.5	94.0±14.7	95.4±14.2	2.825	0.011
HR-SD	Healthy mood	11.4±4.8	12.0±4.1	12.1±4.5	12.1±4.8	11.2±3.9	11.5±4.2	12.0±3.9	2.924	0.008
(bpm)	Depressed mood	11.5±4.4	12.1±4.4	12.6±4.6	11.8±4.1	11.8±4.2	11.8±4.1	11.7±4.7	0.749	N.S.
SBP–dip	Healthy mood	17.8±9.8	16.3±10.2	15.2±11.2	16.5±10.4	16.6±10.4	17.5±9.6	17.5±9.2	2.040	N.S.
(mmHg)	Depressed mood	17.0±9.9	14.9±10.4	14.3±10.0	15.1±11.3	15.4±10.4	16.4±10.5	15.1±9.5	1.612	N.S.
DBP–dip	Healthy mood	19.5±10.7	17.9±10.1	16.9±11.4	17.1±11.4	18.3±10.1	19.8±10.9	18.7±10.5	1.853	N.S.
(mmHg)	Depressed mood	16.2±10.4	15.8±11.2	15.7±10.7	16.3±11.6	17.5±10.3	17.5±10.4	16.2±10.6	0.353	N.S.
HR–dip	Healthy mood	14.7±13.4	16.6±10.7	18.6±10.0	17.7±13.8	17.9±9.7	18.7±9.2	18.1±10.2	2.268	0.036
(bpm)	Depressed mood	14.7±11.3	15.7±12.9	17.4±11.3	13.9±11.2	17.2±10.9	17.4±10.7	17.1±11.4	1.208	N.S.
MAP–dip	Healthy mood	18.9±9.6	17.1±9.6	16.0±10.9	16.9±10.5	17.6±9.8	18.8±9.7	18.3±9.5	2.100	N.S.
(mmHg)	Depressed mood	16.7±9.4	15.5±10.3	15.1±10.0	15.8±11.0	16.6±9.8	17.1±9.9	15.8±9.4	0.720	N.S.
Pulse pressure dip	Healthy mood	13.3±18.1	13.2±16.2	12.2±16.2	14.2±15.9	12.8±16.5	12.2±5.9	14.6±13.5	0.587	N.S.
(mmHg)	Depressed mood	17.0±17.9	12.4±17.3	11.5±13.9	12.5±17.0	10.7±19.0	13.3±16.9	12.3±16.3	2.140	0.049
Double product dip	Healthy mood	29.8±14.6	30.6±11.7	31.1±13.2	31.6±14.1	31.6±12.2	33.0±10.5	32.7±11.5	0.729	N.S.
(mmHgxbpm)	Depressed mood	29.8±11.0	28.6±13.9	29.7±11.5	27.2±13.4	30.1±12.5	31.1±12.4	29.5±12.1	0.760	N.S.

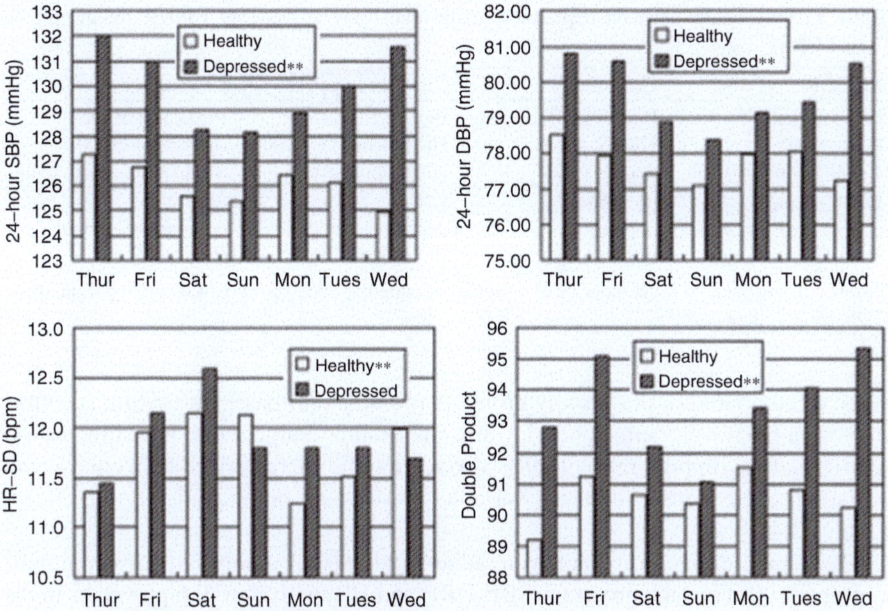

Fig. 11.3 Different weekly variation of ambulatory blood pressure measures and heart rate variability between the healthy and depressed groups.
SBP, DBP, and DP are higher every day of the week in the depressed group as compared to the healthy group. A prominent circaseptan rhythm is detected in the depressed group, overshadowing any novelty effect, holiday dip, and Monday surge seen in the healthy group (Table 11.3). Pulse pressure, the SD of HR, and the dipping ratio of DBP also seem to be characterized by a circaseptan component, even though these differences could not be documented with statistical significance.
* $p<0.05$ by ANOVA, ** $p<0.01$ by ANOVA

way ANOVA. In the healthy group, HR variability, estimated by the SD of HR and the HR dip, shows a weekly variation, with higher values on Saturdays and Sundays (during holidays) and lower values on Mondays. The 24-hour BP measures showed no statistically significant day-to-day variation. In the depressed group, the HR variability measures showed no statistically significant day-to-day variation, but there was a statistically significant weekly variation in BP measures, namely, SBP, DBP, and DP were lower on holidays (holiday dip), and the dipping ratio of PP showed a novelty effect.

Figure 11.3 displays differences in weekly variation of SBP, DBP, DP, and HR-SD between the healthy and depressed groups. A novelty effect and a Monday surge of the 24-hour SBP and DBP as well as a novelty effect and a Monday dip of the HR-SD are demonstrated in the healthy group, which was not statistically significant by ANOVA, but was statistically significant by Student's *t*-test (Table 11.4). In the depressed group, SBP, DBP, and DP assume higher values every day of the

Table 11.5 Relation between depression score and 7-day ambulatory BP/HR monitoring in Uraus, Japan

MESOR	Circadian	Circasemiseptan	Circaseptan
SBP (normal/ depressive): 125.0/129.6 P=0.036	P<0.001 SBP, DBP, HR (normal/ depressive)	P<0.05 SBP, DBP, HR (normal) P<0.001 HR (depressive)	P<0.05 SBP, DBP (normal) P<0.001 SBP, DBP (depressive)
DBP (normal/ depressive): 76.8/79.3 P=0.052	no normal/ depressive difference	HR-A: depressive>normal (P=0.078)	SBP-A: depressive>normal (P=0.027)

Number of profiles: 217, Number of complete assessments: 193, Scores of GDS-15 ≤ 4: healthy; GDS-15 ≥ 5: depressive

week as compared to the healthy group. A prominent circaseptan rhythm accounts in part for the novelty effect, holiday dip, and Monday surge. Pulse pressure, the SD of HR, and the dipping ratio of DBP also seem to be characterized by a circaseptan variation, even though for these variables, results are not statistically significant (Table 11.4).

By population-mean cosinor, a circadian rhythm is invariably statistically significant ($p<0.001$) in each group for SBP, DBP, and HR, with overall high values in the afternoon. A circasemiseptan component is statistically significant for the healthy subjects but not for the BP of the depressed group. A circaseptan component is documented with statistical significance for the BP but not for the HR of both groups. Apart from the difference in mean value reported above, no difference in the circadian pattern is found between the healthy and depressed subjects. The two groups differ, however, in terms of the circaseptan amplitude of SBP ($p=0.027$), whereas the difference in circasemiseptan amplitude of HR is also of borderline statistical significance ($p=0.078$), as shown in Table 11.5.

Ambulatory 24-hour blood pressure monitoring has become quite popular in recent years, offering a number of advantages from the viewpoint of both diagnosis and treatment [10]. One basis of superiority rests on the fact that clinic blood pressure is not representative of blood pressure values outside the clinic. Results obtained herein show that blood pressure is characterized not only by a circadian but also by a weekly variation. The 24-hour averages of SBP and DBP are seen to differ from one day to another in the same citizen in the same 7-day record.

It is also known that increased BP variability on ambulatory BP offers prognostic information regarding adverse cardiovascular outcomes in patients with or without hypertension. Previous reports suggested that a difference in weekly variation according to working status of the citizens was of particular interest and external triggering factors may play a role in the acute causation of the disease. For a long time and again recently, contributions of psychological factors on the pathogenesis of cardiovascular disease and implications for therapy have been noted for primary and secondary prevention [23]. Thus, in this investigation, we focused on the contribution of depressed mood, finding that depression can affect the BP average and HR variability. The 24-hour average of SBP and DBP were statistically significantly higher in the depressed citizens than in the healthy group, but in terms of the 24-hour average of HR, there was no statistically significant difference between the two groups.

We have reported in previous studies on the characteristics of weekly BP variations [11]. Most citizens in a community show a novelty effect (the first day effect), a holiday dip, and a Monday surge of SBP during the waking span. These characteristics were confirmed herein also in the healthy group for the 24-hour average of SBP and DBP, while a novelty effect and a Monday dip of the SD of HR was newly detected. The major finding, however, was that these characteristics disappeared in the depressed citizens, who instead showed a prominent circaseptan component for SBP, DBP, and DP. About-weekly changes in pulse pressure, the SD of HR, and the dipping ratio of DBP were also observed but were not statistically significant. Circaseptan (about 7-day) periodicity was independent of holidays and working days.

In this investigation, we also observed a new fact related to HR variability, namely a likely difference between healthy and depressed subjects in terms of their weekly variation in HR-SD. In citizens without depressed mood, HR variability, estimated by the SD of HR, was lowest on Mondays or on the first day of monitoring, whereas in citizens with depressed mood, it was lower than in the healthy subjects on most days of a week, showing a circaseptan variation independent of holidays. HR variability, such as SDNN (standard deviation of normal cycle R-R normal), has been known to be a good predictor of long-term survival, both in patients with myocardial infarction [24] and in a cohort of elderly subjects [25].

In patients after a myocardial infarction, depression is associated with an increased mortality rate, which is similar to that associated with left ventricular dysfunction or a history of previous myocardial infarction [6, 7]. Depression is an independent risk factor for progression of cardiovascular disease [8]. Little is known about how depression contributes to this increased risk. As a possible explanation for this association, the following factors are discussed: (1) increased platelet aggregation [26]; (2) altered cardiac autonomic tone, including elevated sympathetic nervous activity [27], decreased HR variability [28], increased BP variability [29], and reduced baroreflex cardiac control [30]; (3) increased QT dispersion [31]; (4) increased susceptibility to ventricular arrhythmias [32]; (5) dysregulation of the hypothalamic pituitary-adrenal axis [26]; (6) poor adherence to cardiac treatment regimens; and (7) social isolation and reductions in the cognitive functioning [33]. The present study suggests another new factor consisting of inappropriate biological rhythms, such as a prominent circaseptan rhythm. A prominent circaseptan component was also reported to characterize the negative affect of the PANAS questionnaire in clinically healthy students [34]. This finding provides a new important step in the mechanistic analysis of the nature of cardiovascular disease that accompanies a key component of psychological depression. In view of the strong relation between depression and adverse cardiac events, studies should be done to ascertain whether treatment of depression, especially from a standpoint of chronodiagnosis and chronotherapy, can reduce elevated blood pressure values and improve the altered BP and HR variability, so as to reduce the incidence of adverse cardiac events.

Results in this investigation indicate the clinical importance of taking care of depressed subjects. Fewer than 7 days of monitoring means a greater chance of a false diagnosis on which a treatment decision may depend for the long term, including perhaps a potentially unnecessary treatment for decades. Less than 7 days of monitoring can be compared with taking the pulse for less than a cardiac cycle, if the circaseptan aspect of time structures, chronomes, is viewed as a spectral component, with many new components with even longer periods, including transyears characterizing series of blood pressure and heart rate covering at least 5 and up to 38 years [19, 35, 36]. When the ECG and EEG, as well as the blood pressure of small rodents, can be monitored continuously for most of their lifetime, this study is but a small step toward universal 7-day and eventually continuous monitoring of BP and HR [37, 38], or the ECG [39].

11.3 No Calendar Year but Nonphotic Transyear and Cis-Half-Year Components Characterize Sudden Cardiac Death (SCD) in Minnesota

SCD (ICD10 code I46.1) does not include death from myocardial infarction (MI) since 1999 in Minnesota. The incidence patterns of the two conditions can differ in the same geographic location, namely, Minnesota, USA. There, MIs (1968–1996, 29 years) are characterized by a prominent calendar year, not detected for SCD. Instead, SCDs (1999–2003, 5 years) are characterized by an about 1.3-year transyear similar to variations in the solar wind speed, a component found invariably in all 46 longitudinal records of blood pressure and heart rate analyzed thus far. A transyear was also found for SCDs in Tokyo, in Arkansas, and in the Czech Republic but not in South Carolina, Hungary, Latvia, Lithuania, the Republic of Georgia, or Hong Kong.

Daily incidences of SCD in Minnesota were analyzed for the span from 1999 to 2005 (7 years) by cosinor in the frequency range of one cycle in 7 years to one cycle in about 2 months, using a 0.2 harmonic increment. Point and CI (95 % confidence interval) estimates of the periods corresponding to spectral peaks were determined by nonlinear least squares. The 469 SCDs in 1999–2005 were characterized by a transyear with a period of 1.275y (95 % CI: 1.126, 1.425y) and a cis-half year with an about 0.42-year period of 0.409y (95 % CI: 0.396, 0.422y).

By contrast, the 9,648 cardiac arrests in 1979–1998 were characterized by a 1.0-year synchronized component, the about 1.3-year transyear being much less prominent, reaching only borderline statistical significance. The results for SCD in Minnesota resemble those in Tokyo. They are also in keeping with an about 1.3-year component detected in a 10-year long record of atrial fibrillations, potentially fatal arrhythmia that may underlie SCD. These results imply the possible influence of nonphotic as well as of photic solar effects, at least in some geographic locations.

This possibility is in keeping with the finding of an about-10.5-year cycle characterizing the incidence of MI in Minnesota (1968–1996), accounting for a 5 % difference in mortality between years of maximal vs. minimal solar activity. The fact that nonphotic (e.g., magnetic) effects are latitude-dependent may account in part for geographic differences observed in the patterns of incidence of SCD. Nonphotic influences may also stem from several sources, such as cosmic rays, solar flares, the solar wind, the interplanetary magnetic field, and/or geomagnetic disturbances as measured on earth, to name but a few.

Even for a cycle as prominent as the solar activity cycle averaging about 11 years, it is not invariably found for all variables related to the sun-earth physical environment. For instance, in the OMNI 2 data during 1963–2003 (41 years), Wolf numbers have an about-10.90-year cycle (95 % CI: 10.72, 11.07 years), but solar wind speed is characterized by an about-8.39-year cycle (95 % CI: 8.07, 8.72 years), statistically significantly shorter than the cycle characterizing Wolf numbers, as shown by nonoverlapping 95 % CIs of their periods.

Similar differences are observed among all 40 variables recorded by OMNI 2 for the about-1.3-year component. Congruences in time-varying spectra and cross-spectra between mortality statistics from different causes and natural physical environmental variables may help determine more precisely what factors may trigger an undesired effect or be protective.

For SCD, when data are pooled from all 9 available geographic locations, totaling 5,503 cases during 1998–2003, high geomagnetic activity is associated with a decreased incidence of SCD (partial results shown in Table 11.6). For SCD, underlying factors may be explored in pertinent markers such as the incidence of potentially lethal arrhythmia recorded in some pacemaker-cardioverter-defibrillators, monitored longitudinally for the individual patient, before an untoward event occurs.

Table 11.6 Geomagnetic/geographic differences among cycles with periods in the range of 0.8–2.0 years

Characterizing the incidence of sudden cardiac death (SCD)[a, b]									
Site	Span	T, Δt, N	SCD (N)	Period (y)	(95%CI)	Amplitude	(95%CI)	A(% MESOR)	P-value[b]
				Candidate Transyear detected					
Minnesota	1999–2003	5y, 1d, 1826	343	1.392	(1.173, 1.611)	0.042	(0.00, 0.09)	22.0	0.014
Arkansas	1999–2003	5y, 1d, 1826	273	1.095	(0.939, 1.251)	0.032	(0.00, 0.07)	21.1	0.040
				1.686	(1.293, 2.071)	0.031	(0.00, 0.07)	20.7	0.044
Czech Rep.	1999–2003	5y, 1d, 1826	1006	0.974	(0.856, 1.091)	0.078	(0.00, 0.16)	14.2	0.007
				1.759	(1.408, 2.110)	0.077	(0.00, 0.15)	13.9	0.010
	1994–2003	10y, 1d, 3652	1792	1.726	(1.605, 1.848)	0.074	(0.02, 0.13)	15.1	<0.001
				1.000	(0.944, 1.056)	0.052	(0.00, 0.10)	10.6	0.010
				Candidate Transyear not detected					
North Carolina	1999–2003	5y, 1d, 1826	752	0.929	(0.834, 1.023)	0.069	(0.00, 0.14)	16.9	0.007
Tbilisi, Georgia	Nov 1999–2003	4.1y, 1d, 1505	130	0.988	(0.862, 1.114)	0.035	(0.00, 0.07)	40.7	0.007
Hong Kong	2001–2003	3y, 1m, 36	52	0.843	(0.651, 1.036)	0.022	(NS)	44.9	0.077
Myocardial infarction (MI)									
Site	Span	T, Δt, N	MI (N)	Period (y)	(95%CI)	Amplitude	(95%CI)	A(% MESOR)	P-value[2]
				Coexisting year (circannual) and Transyear					
Czech Rep.	1999–2003	5y, 1d, 1826	52598	1.014	(0.989, 1.038)	2.85	(2.22, 3.48)	9.88	<0.001
				1.354	(1.252, 1.456)	1.35	(0.69, 2.02)	4.68	<0.001
	1994–2003	10y, 1d, 3652	115520	0.998	(0.988, 1.009)	3.03	(2.47, 3.60)	9.58	<0.001
				1.453	(1.417, 1.489)	1.91	(1.34, 2.49)	6.04	<0.001
				1.150	(1.116, 1.184)	1.23	(0.64, 1.82)	3.88	<0.001

[a]With focus on transyears, with periods longer than 1.0 year (underlined, double underline for neartransyear)
[b]International classification of disease (ICD10) Code 146.1 excluding MI and sudden death of unknown or unspecified cause (except before 1999). *T* Length of data series (y = years); Δt: sampling interval (d = day, m = month); *N* number of data (including 0s). Period and 95 % confidence interval (CI) estimated by nonlinear least squares
In longer (10-year) series, a neartransyear (cycle with a period between 1.0 and 1.2 year) is detected for MIs in addition to a fartransyear. Brevity of series and lack of ordering statistical significance qualify results from Hong Kong
Note that transyears are found in 3 of 6 locations ($P<0.05$ by linear least squares) with a relative amplitude exceeding 12 (% of MESOR)
[c]From linear least squares analysis, not corrected for multiple testing. Amplitude expressed in N/day

11.4 Fantasy or Fact? The Relations Between Heart Rate Variability and Cancer Death in Elderly Community-Dwelling People: Longitudinal Investigation for the Longevity and Aging in Hokkaido County (LILAC) Study

Inflammatory signals in the microenvironment of mutated cells such as tumor necrosis factor alpha (TNF-α) enhance the development of transformed cells into tumor cells. Inflammatory cytokines such as interleukin-1 also promote escape from apoptosis, angiogenesis, and metastasis. Gidron et al. recently developed a model according to which the vagus nerve may inform the brain about preclinical tumors and modulates them [40, 41]. This was based on converging evidence showing that the vagus nerve transmits to the brain information about low-dose peripheral inflammation and that vagotomized tumor-bearing mice did not manifest the expected anorexia and a few follow-up studies showing that vagotomized ulcer patients died more than non-vagotomized ulcer patients from various visceral cancers. Furthermore, stimulating the vagus nerve inhibits peripheral cytokine levels including TNF-α, suggesting that the vagus may also modulate tumor progression. However, the relations between noninvasive measures of vagal activity and cancer incidence and cancer death have not been tested.

Heart rate variability (HRV) has been used to evaluate the activity of the parasympathetic tone [42, 43]. HRV represents a fine-tuning of the beat-to-beat control mechanisms. Vagal afferent stimulation leads to reflex excitation of vagal efferent activity and inhibition of sympathetic efferent activity. Efferent vagal activity directed to the sinus node is characterized by discharge largely synchronous with each cardiac cycle that can be modulated by central (respiratory center) and peripheral (oscillation in arterial pressure and respiratory movements) oscillators. These modulatory effects of neural mechanisms on the sinus node are represented by the time- and frequency-domain analyses of HRV. These efferent vagal activities can be measured by the parameters of CVRR, r-MSSD, NN50, pNN50, and the HF component. Since depression has been shown in some studies to predict cancer mortality and to be inversely related with HRV also in the elderly, we also considered the effects of this variable.

The purpose of this study was to test the relations between vagus nerve activity, as indexed by HRV parameters, and cancer death among initially healthy elderly people in a Japanese community. Following the model on the role of the vagus nerve in tumor-to-brain communication and modulation, we expected that vagal activity at the outset would be inversely related to future cancer death.

The LILAC study started on July 25, 2000. We examined 298 people (120 men and 178 women) older than 75 years (average age: 79.6 years). The participants resided in the Hokkaido County, north of Japan. This study was approved at the town office in Uraus by Tokyo Women's Medical University and Kyoto University Review Board (Clinical Research #2912).

One hour of ambulatory ECG recording was obtained on July 25–27, 2000. The data were processed for HRV using a Fukuda-Denshi Holter analysis system (SCM-280-3). Time-domain measures (CVRR, SDANN, r-MSSD, NN50, NN50+, NN50−, pNN50, pNN50+, and pNN50−) and frequency-domain measures (spectral power in the "very low frequency" (VLF), 0.003–0.04 Hz; "low frequency" (LF), 0.04–0.15 Hz; and "high frequency" (HF), 0.15–0.40 Hz regions and LF/HF ratio) were determined. All indices were computed as averages over consecutive 5-min intervals. Spectral indices were obtained by the maximum entropy method (MEM) with the MemCalc/CHIRAM program (Suwa Trust Co., Ltd., Tokyo, Japan).

BP was measured in a sitting position, and the brachial-ankle PWV (baPWV) was measured between the right arm and ankle in a supine position, using an ABI/Form instrument (Nippon Colin Co., Ltd., Komaki, Japan). The baPWV was measured using a volume-plethysmographic method. baPWV was measured in duplicate after at least a 5-min rest. Only baPWV measurements from participants with normal ankle/brachial pressure index (ABI) values (>0.90) were considered. The maximal value among the 4 readings was used for analysis. An echocardiogram and a conventional ECG record were also obtained as usual.

The Japanese version of the Mini-Mental State Examination (MMSE) and the Hasegawa Dementia Scale Revised (HDSR) were used to assess the overall cognitive function, including verbal orientation, memory, and constructional ability (Kohs block test). The Up & Go test measured, in seconds, the time it took the subject to get up from a chair, walk a distance of 3 m, turn, walk back to the chair, and sit down again. This test is a simple measure of physical mobility and demonstrates the subject's balance, gait speed, and functional ability (Up and Go). A lower time score indicates better physical mobility. Functional reach (FR), used to evaluate balance, represents the maximal distance a subject can reach forward beyond arm's length while maintaining a fixed base of support in the standing position. A higher score indicates better balance. Manual dexterity was assessed using a panel with combinations of 10 hooks, 10 big buttons, and 5 small buttons. There were three discrete measurements of time recorded for each participant (10 "hook-ons," 10 big "button-on-and-offs," and 5 small "button-on-and-offs"). The total manual dexterity time in seconds, defined as the button score (button-S), was calculated by adding the average times for one hook-on and one big or small button-on-and-off. A lower button score indicates better manual dexterity.

Numerous depression rating scales are currently available, which represent a mixture of observer-rated and self-rating scales. In this investigation, subjects were evaluated with a 15-item self-administered depression questionnaire [5]. When the depression score was ≥5, subjects were considered to have a depressed mood.

Consultations were repeated every year (end of July or beginning of August). In addition, one or two doctors of our team visited every 3 months and offered several kinds of health consultation regarding the rehabilitation of disordered functions, healthy lifestyle modifications such as the promotion of complete smoking cessation, weight reduction, reduction of salt intake, moderation in the consumption of fruits and vegetables and alcohol intake, and advice in terms of medical prescriptions for the local general practitioner.

In the present investigation, the follow-up ended on October 1, 2005. The follow-up time was defined as the time elapsed between the date of first (reference) examination and the date of cancer death.

We used Cox regression analysis to calculate the unadjusted and adjusted relative risk (RR) and 95 % confidence interval (CI) for all-cause mortality. To identify independent predictors of cancer death, we used multivariate Cox regression analyses with stepwise selection. Variables forced into the multivariate models were age, gender, body mass index (BMI), and heart rate (HR) in addition to significant HRV predictors of cancer death. Significance was considered at a value of $p<0.05$.

The mean age of the participants at entry was 79.6 years. The mean follow-up time was 1430.4 days, during which 38 subjects died (25 men and 13 women). Out of them, 11 participants died of cancer (9 men and 2 women, 4 cases of lung cancer, 2 cases of pancreas cancers, 2 cases of stomach cancers, a gall bladder cancer, a rectum cancer, and a prostate cancer), and their mean follow-up time was 774.8 days. HRV was analyzed in 151 participants out of the 298 participants, excluding those with cardiac arrhythmias such as atrial fibrillation and frequent atrial and ventricular ectopies.

Among the variables used in Cox proportional hazard models, gender, ABI, GDS, and HRV indices (CVRR and pNN50–) showed a statistically significant association with cancer death (Table 11.7). In univariate analyses, gender was significantly associated with cancer death, with a relative risk of 7.268 ($p=0.01119$) for male gender. ABI significantly predicted all-cause mortality, and a 0.1 decrease was associated with a relative risk of 1.476 ($p=0.00870$). GDS was also significantly predicted, and a 2-point increase was associated with a relative risk of 1.722 ($p=0.02478$). The relative risk of developing cancer death predicted by a 1.0 % increase in CVRR and pNN50– was 1.441 and 1.251 ($p=0.00922$ and $p=0.04030$), respectively. The other indices of HRV (SDANN, r-MSSD, NN50–, and pNN50) also showed a trend of significant relative risk predicting cancer death (Table 11.7).

In multivariate analyses (Table 11.8), when age, gender, BMI, systolic blood pressure, and HR were used as continuous variables in the same model, GDS and HRV indices (CVRR, SDANN, pNN50, and pNN50–) remained significantly associated with the occurrence of cancer death. Specifically, a 2-point increase in GDS was associated with a relative risk of 1.801 ($p=0.02510$). Five milliseconds increase in SDANN was associated with a relative risk of 1.420 ($p=0.00788$). One percent increase in CVRR, pNN50, or pNN50– was associated with a relative risk of 1.597, 1.153, or 1.301 ($p=0.01245$, $p=0.04635$, or $p=0.03766$), respectively (Table 11.8).

We found statistically significant relations between HRV indices reflecting vagus nerve activity and cancer mortality in elderly community-dwelling people in North Japan. These relations were independent of participant's age, gender, BMI, SBP, and HR. Previous experimental and clinical investigations [5, 44, 45] suggest that an

Table 11.7 Relative risk of cancer death at Uraus town in Hokkaido

	All subjects (n=298)					
Variables	n	β	z value	p-value	Relative risk	95% CI
Gender	296	1.9834	2.5368	0.01119	7.268	1.570~33.646
Age	298	0.0652	1.1580	N.S.		
BMI	284	−0.0219	0.2494	N.S.		
Systolic BP	284	0.0057	0.3791	N.S.		
Diastolic BP	283	0.0228	0.8385	N.S.		
Heart rate (HR)	284	0.0373	1.6164	N.S.		
PWV	245	0.0005	0.9370	N.S.		
ABI (−0.10)	263	−3.8923	2.6237	0.00870	1.476	1.103~1.974
Postural BP change	282	−0.0071	0.5080	N.S.		
Af	283	0.7609	1.2751	N.S.		
Frequent APCs	283	−0.0600	0.0572	N.S.		
Frequent VPCs	283	0.6841	0.8747	N.S.		
Lown's grade	217	−0.0823	0.4905	N.S.		
GDS (2)	158	0.2717	2.2448	0.02478	1.722	1.071~2.767
Up & Go	293	0.0397	1.2678	N.S.		
Functional reach (FR)	292	0.0188	0.4835	N.S.		
Button test	290	0.0168	0.5084	N.S.		
MMSE	291	−0.0223	0.3192	N.S.		
HDSR	291	−0.0355	0.6459	N.S.		
Kohs	272	0.0013	0.0416	N.S.		
CVRR(1.00)	115	0.3650	2.6037	0.00922	1.441	1.094~1.896
SDANN	115	0.0484	1.8575	0.06324		
r-MSSD	115	0.0520	1.9056	0.05670		
NN50	115	0.0292	1.4775	N.S.		
NN50+	115	0.0593	1.1416	N.S.		
NN50−	115	0.5926	1.7388	0.08207		
pNN50	115	0.1159	1.8186	0.06898		
pNN50+	115	0.2105	1.4821	N.S.		
pNN50− (1.00)	115	0.2238	2.0506	0.04030	1.251	1.010~1.549
SDNN	151	−0.0090	0.2741	N.S.		
VLF	151	0.0001	0.1008	N.S.		
LF	151	−0.0010	0.4028	N.S.		
HF	151	−0.0039	0.4344	N.S.		
LF/HF	151	0.0305	0.1148	N.S.		

Af=atrial fibrillation; APC=atrial premature contraction, VPC=ventricular premature contraction

increase in the CVRR and pNN50 parameters (namely, pNN50, pNN50+, and pNN50−) reflects an increased vagal activity. It is not clear why the frequency-domain index of HF-HRV was not associated with the occurrence of cancer death. We suppose that in elderly participants, the degree of change in HRV is small and spectral analysis in these elderly citizens was largely affected by a contamination of noise including atrial premature contractions.

Table 11.8 Relative risk of cancer death at Uraus town in Hokkaido, adjusted for age, sex, BMI, systolic BP, and heart rate

	All subjects (n=298)					
Variables	n	β	z value	p-value	Relative risk	95% CI
PWV	237	0.0005	0.8438	N.S.		
ABI (−0.10)	255	−2.8620	1.7821	0.07474		
Postural BP change	267	−0.0001	0.0068	N.S.		
Af	267	0.9094	1.3719	N.S.		
Frequent APCs	267	−0.6166	0.5823	N.S.		
Frequent VPCs	267	0.3622	0.4525	N.S.		
Lown's grade	203	−0.2013	1.1947	N.S.		
GDS (2)	147	0.2941	2.2398	0.02510	1.801	1.076~3.013
Up & Go	278	0.0326	0.8465	N.S.		
Functional reach (FR)	277	0.0458	1.0076	N.S.		
Button test	276	0.0057	0.1503	N.S.		
MMSE	277	−0.0260	0.3481	N.S.		
HDSR	277	−0.0220	0.3340	N.S.		
Kohs	258	0.0050	0.1494	N.S.		
CVRR (1.00)	115	0.4684	2.4991	0.01245	1.597	1.106~2.307
SDANN (5.0)	115	0.0702	2.6572	0.00788	1.420	1.096~1.840
r-MSSD	115	0.0616	1.9356	0.05292		
NN50	115	0.0396	1.6827	0.09243		
NN50+	115	0.0724	1.4445	N.S.		
NN50−	115	0.0782	1.8273	0.06766		
pNN50 (1.00)	115	0.1426	1.9922	0.04635	1.153	1.002~1.327
pNN50+ (1.00)	115	0.2822	1.7967	0.07239		
pNN50− (1.00)	115	0.2630	2.0785	0.03766	1.301	1.015~1.667
SDNN	142	−0.0096	0.3043	N.S.		
VLF	142	0.0001	0.2120	N.S.		
LF	142	−0.0009	0.4090	N.S.		
HF	142	−0.1278	1.1613	N.S.		
LF/HF	142	0.0974	0.3247	N.S.		

Af=atrial fibrillation; APC=atrial premature contraction; VPC=ventricular premature contraction

However, the observed direction of relations between HRV and cancer mortality is opposite to the one we hypothesized. We have hypothesized that the afferent route of the vagal nerve transmits information to the brain about precancerous cells via its transmission of tumor-associated proinflammatory cytokines and that both the ascending vagus-nucleus tractus solitarius hypothalamus-pituitary-adrenal axis and the descending cholinergic vagal-dependent anti-inflammatory routes slow or even prevent tumorigenesis by inhibition of tumor-associated proinflammatory cytokines. However, we found that higher vagal activity, as indexed by various measures of HRV, was positively rather than inversely related to cancer mortality.

It is difficult to explain this pattern of results. On one hand, activation of the vagus inhibits proinflammatory cytokines such as TNF-α, and HRV is inversely related to concentrations of proinflammatory cytokines. Since TNF-α may promote tumor growth in its initial stages, vagal activity (and HRV) was thought to be inversely related to future cancer mortality. However, on the other hand, TNF-α has a "two edge" role in cancer and also has antitumor effects in several tumors. Thus, it is possible that higher HRV, which may inhibit TNF-α, thus would reflect less protection from cancer by TNF, possibly leading to greater cancer mortality. The latter chain of events could possibly account for the pattern of results observed in the present study. It is also possible that these findings partly reflect the older age of the participants in the present study and that HRV may be inversely related to cancer mortality if assessed in younger people.

Although no index of arterial stiffness (PWV, ABI, and postural BP changes), ADL activity (Up & Go, Functional Reach and Button tests), and no cognitive function (MMSE, HDS-R, and Kohs cubic tests) showed significant associations with cancer mortality, an increase in depressive mood (GDS score) was significantly predictive of cancer death. This is in line with various studies showing that depressive symptoms predict early mortality [8] and cancer mortality. We have also observed in middle-aged community-dwelling people that a depressive mood, assessed by GDS, predicted the occurrence of vascular diseases beyond the effects of age, gender, 7-day/24-hour ambulatory BP, lifestyle, and environmental conditions [46].

In conclusion, the fact that both depressive symptoms and HRV predicted cancer mortality in the same direction is surprising given that depression is associated with reduced HRV. It is possible that their effects on cancer mortality are indeed independent, as revealed by our multivariate analysis.

However, this study has a few limitations. First, it was not planned to test cancer mortality from its onset; thus we did not present information concerning the type of cancer. This is important since the hypothesized role of the vagus in tumor-to-brain communication and modulation may be mainly relevant to visceral cancers. It is also possible that the observed results which contradict our hypothesis may partly stem from including non-visceral cancers in the outcome variable. Second, the number of cancer deaths and total sample were too small to examine additional more focused questions such as type of cancer. Future studies need to replicate this study addressing these limitations, and testing our hypothesis in other cultures as well, which may have dietary habits known to affect cancer development.

11.5 Appendix: Time-Domain Measures of HRV

CVRR is "the coefficient of variation of the NN intervals over 5 min."

SDANN is "the standard deviation of the average NN intervals calculated over 5 min, which is an estimate of the changes in heart rate due to cycles longer than 5 min."

r-MSSD is "the square root of the mean squared differences of successive NN intervals."

NN50 is the number of interval differences of successive NN intervals greater than +50 ms and shorter than −50 ms.

NN50+ is the number of interval differences of successive NN intervals greater than +50 ms.

NN50− is the number of interval differences of successive NN intervals shorter than −50 ms.

pNN50 is the proportion derived by dividing NN50 by the total number of NN intervals.

pNN50+ is the proportion derived by dividing NN50+ by the total number of the corresponding NN intervals.

pNN50− is the proportion derived by dividing NN50− by the total number of the corresponding NN intervals.

SDNN is "the standard deviation of the NN intervals over 60 min."

References

1. Otsuka K, Murakami S, Kubo Y, et al. Chronomics for chronoastrobiology with immediate spin-offs for life quality and longevity. Biomed Pharmacother 2003;57:1s–18s.
2. Murakami S, Otsuka K, Yamanaka G, et al. Positive impact of social intervention on disturbed neurobehavioral function in an elderly community-dwelling population: longitudinal investigation for longevity and aging in Hokkaido county (LILAC). Biomed Pharmacother 2004;58:45s–7s.
3. Otsuka K, Mitsutake G, Yano S, et al. Depression, quality of life, and lifestyle: chronoecological health watch in a community. Biomed Pharmacother 2002;56:231s–42s.
4. Palatini P, Frigo G, Bertolo O, et al. Validation of the A&D TM-2430 device for ambulatory blood pressure monitoring and evaluation of performance according to subjects' characteristics. Blood Press Monit 1998;3:255–9.
5. Yesavage JA, Brink TL, Rose TL, et al. Development of validation of a geriatric depression screening scale: a preliminary report. J Psychiatr Res 1983;17:37–49.
6. Frasure-Smith N, Lesperance F, Talajic M. Depression following myocardial infarction: determine predictive variables. JAMA 1993;270:1819–61.
7. Frasure-Smith N, Lesperance F, Talajic M. Depression and 18-month prognosis after myocardial infarction. Circulation 1995;91:999–1005.
8. Barefoot JC, Schroll M. Symptoms of depression, acute myocardial infarction, and total mortality in a community sample. Circulation 1996;93:1976–80.
9. Anda R, Williamson D, Jones D, et al. Depressed affect, hopelessness, and the risk of ischemic heart disease in a cohort of U.S. adults. Epidemiology 1993;4:285–94.
10. Cornélissen G, Schwartzkopff O, Halberg F, et al. 7-day ambulatory monitoring for adults with hypertension and diabetes [letter]. Am J Kidney Dis 2001;37:878.
11. Shinagawa M, Otsuka K, Murakami S, et al. Seven-day (24-h) ambulatory blood pressure monitoring, self-reported depression and quality of life scores. Blood Press Monit 2002;7:69–76.
12. Halberg F. Chronobiology. Annu Rev Physiol 1969;31:675–725.
13. Otsuka K, editor. Chronome & Janus-medicine: Heart Rate Variability (HRV) and BP Variability (BPV) from a viewpoint of chronobiology and ecology. Med Rev. Tokyo; 1998. p. 213.

14. Halberg F, Cornélissen G, Katinas G, et al. Season's Appreciations 2000. Chronomics complement, among many other fields, genomics and proteomics. Neuroendocrinol Lett 2001;22:53–73.
15. Cornélissen G, Breus TK, Bingham C, et al. Beyond circadian chronorisk: worldwide circaseptan-circasemiseptan patterns of myocardial infarctions, other vascular events, and emergencies. Chronobiologia 1993;20:87–115.
16. Cornélissen G, Tamura K, Tarquini B, et al. Differences in some circadian patterns of cardiac arrhythmia, myocardial infarction and other adverse vascular events. Chronobiologia 1994;21:79–88.
17. Murakami S, Otsuka K, Kubo Y, et al. Repeated ambulatory monitoring reveals a Monday morning surge in blood pressure in a community-dwelling population. Am J Hypertens 2004;17:1179–83.
18. Otsuka K, Yamanaka G, Shinagawa M, et al. Chronomic community screening reveals about 31 % depression, elevated blood pressure and infradian vascular rhythm alteration. Biomed Pharmacother 2004;58:48s–55s.
19. Cornélissen G, Masalov A, Halberg F, et al. Multiple resonances among time structures, chronomes, around and in us. Is an about 1.3-year periodicity in solar wind built into the human cardiovascular chronome? Hum Physiol 2004;30(2):86–92.
20. Fried LP, Kronmal RA, Newman AB, et al. Risk factors for 5-year mortality in older adults: the Cardiovascular Heart Study. JAMA 1998;279:585–92.
21. Schulz R, Beach SR, Ives DG, et al. Association between depression and mortality in older adults. Arch Intern Med 2000;160:1761–8.
22. Bingham C, Arbogast B, Cornélissen G, et al. Inferential statistical methods for estimating and comparing cosinor parameters. Chronobiologia 1982;9:397–439.
23. Pickering TG. Ambulatory monitoring and blood pressure variability. London: Science Press; 1991.
24. Tsuji H, Venditti FJ, Manders ES, et al. Reduced heart rate variability and mortality risk in an elderly cohort: the Framingham Heart Study. Circulation 1994;90:878–83.
25. Rozanski A, Blumenthal A, Kaplan J. Impact of psychological factors on the pathogenesis of cardiovascular disease and implications for therapy. Circulation 1999;99:2192–217.
26. Grignani G, Soffiantino F, Zucchella M, et al. Platelet activation by emotional stress in patients with coronary artery disease. Circulation 1991;83 Suppl 4:II-128–36.
27. Veith RC, Lewis N, Linares OA, et al. Sympathetic nervous system activity in major depression. Basal and desipramine-induced alterations in plasma noradrenaline kinetics. Arch Gen Psychiat 1994;51:411–22.
28. Carney RM, Blumenthal JA, Stein PK, et al. Depression, heart rate variability, and acute myocardial infarction. Circulation 2001;104:2024–8.
29. Sloan RP, Shapiro PA, Bagiella E, et al. Cardiac autonomic control buffers blood pressure variability responses to challenge: a psychophysiologic model of coronary artery disease. Psychosom Med 1999;61:58–68.
30. Watkins LL, Blumenthal JA, Carney RM. Association of anxiety with reduced baroreflex cardiac control in patients after acute myocardial infarction. Am Heart J 2002;143:460–6.
31. Piccirillo G, Viola E, Nocco M, et al. Autonomic modulation and QT interval dispersion in hypertensive subjects with anxiety. Hypertension 1999;34:242–6.
32. Grippo AJ, Santos CM, Johnson RF, et al. Increased susceptibility to ventricular arrhythmias in a rodent model of experimental depression. Am J Physiol Heart Circ Physiol 2004;286:H619–26. doi:10.1152/ajpheart.00450.2003.
33. DiMatteo MR, Lepper HS, Croghan TW. Depression is a risk factor for noncompliance with medical treatment. Meta-analysis of the effects of anxiety and depression on patient adherence. Arch Intern Med 2000;160:2101–7.
34. Cornélissen G, Watson D, Mitsutake G, Fiser B, Siegelova J, Dusek J, Vohlidalova L, Svacinova H, Halberg F. Mapping of circaseptan and circadian changes in mood. Scr Med 2005; 78:89–98.

35. Halberg F, Cornélissen G, Schack B. Self-experimentation on chronomes, time structures; chronomics for health surveillance and science: also transdisciplinary civic duty? Behav Brain Sci 2004;27(2):267–9.
36. Cornélissen G, Halberg F, Rostagno C, Otsuka K. A chronomic approach to cardiac arrhythmia and sudden cardiac death. Auton Nerv Syst 2007;44:251–4.
37. Cornélissen G, Halberg F, Cornélissen G, Schack B, et al. Blood pressure self-surveillance for health also reflects 1.3-year Richardson solar wind variation: spin-off from chronomics. Biomed Pharmacother 2003;57 Suppl 1:58s–76s.
38. Cornélissen G, Delmore P, Halberg F. Health Watch 3. Minneapolis: Halberg Chronobiology Center, University of Minnesota/Department of Communications, Medtronic Inc.; 2004. 31 pp.
39. Jabaudon D, Sztajzel J, Sievert K, et al. Usefulness of ambulatory 7-day ECG monitoring for the detection of atrial fibrillation and flutter after acute stroke and transient ischemic attack. Stroke 2004;35:1647–51.
40. Gidron Y, Perry H, Glennie M. Does the vagus nerve inform the brain about preclinical tumors and modulate them? Lancet Oncol 2005;6:245–8.
41. Gidron Y, De Couck M, De Greve J. If you have an active vagus nerve, cancer stage may no longer be important. J Biol Regul Homeost Agents 2014;28:195–201.
42. Task Force of the European Society of Cardiology the North American Society of Pacing Electrophysiology. Heart rate variability: standards of measurement, physiological interpretation, and clinical use. Circulation 1996;93:1043–65.
43. Akselrod S, Gordon D, Ubel FA, et al. Power spectrum analysis of heart rate fluctuation: a quantitative probe of beat to beat cardiovascular control. Science 1981;213:220–2.
44. Pomeranz M, Macaulay RJB, Caudill MA, et al. Assessment of autonomic function in humans by heart rate spectral analysis. Am J Physiol 1985;248:H151–3.
45. Malliani A, Pagani M, Lombardi F, et al. Cardiovascular neural regulation explored in the frequency domain. Circulation 1991;84:1482–92.
46. Yamanaka G, Otsuka K, Hotta N, et al. Depressive mood is independently related to stroke and cardiovascular events in a community. Biomed Pharmacother 2005;59:31s–9s.

Chapter 12
Chronoecological "Glocal" (Global and Local) Comprehensive Health-Watch at High Altitude

Abstract Humans adjusted to ambient hypoxia via changes in the expression of hypoxia-related genes. Gender differences, however, have not yet been studied from the viewpoint of a glocal (combined global and local) comprehensive assessment. Using a glocal comprehensive assessment, we studied the effects of high altitude on pulse oximetry (SpO_2), blood pressure (BP), heart rate (HR), aortic stiffness of cardio-ankle vascular index (CAVI), and autonomic cardiovascular and cognitive function. Subjects consisted of 25,211 Japanese (11,845 women) and 1,858 Ladakhis (1,081 women). As compared to Japanese living at low altitude, highland people showed higher CAVI values in both men and women. Highland people also had lower LF and HF components and a higher LF/HF ratio, lower SpO_2, higher diastolic BP, and higher HR. Elderly residents in the high-altitude community had a poorer cognitive function, estimated by the Kohs block design test, and slept less. These findings suggest that the hypobaric hypoxic condition at high altitude induced a decrease in parasympathetic activity and an increase in sympathetic drive, which may result in an imbalance of the cardiopulmonary-vascular functions in Ladakhis. As to gender differences, systolic BP and CAVI measures were higher in Ladakhi women. Ladakhi women showed a steeper negative slope of the regression line between SpO_2 and age and steeper slopes of the regression lines between systolic and diastolic BP vs. age than Ladakhi men.

Keywords High-altitude community • Glocal comprehensive assessment • Cardio-ankle vascular index • Cognitive function • Gender difference

12.1 Introduction

Leh, Ladakh, is a strongly Buddhist district of east Kashmir, adjacent to Tibet, lying at 3,524 m above sea level between the ranges of Karakoram and the Himalayas. It was virtually unknown to the West until the 1970s, and it still has limited contacts with the outside world today. The economy is based on subsistence farmers who grow mainly barley but also legumes. Whereas physiological adjustment to high altitude has been studied in resident populations of the Andes, Tibet, Nepal, North America, and Europe, much less is known about high-altitude natives in India.

K. Otsuka et al., *Chronomics and Continuous Ambulatory Blood Pressure Monitoring*, DOI 10.1007/978-4-431-54631-3_12

The study of human biology in the mountains began about 2,000 years ago. The Jesuit Jose de Acosta, a sixteenth century traveler in the Andes of Peru, was the first to attribute his discomfort when reaching an altitude of 4,802 m to the "thin air" [1]. Numerous physiological studies attempted to understand the effects of ambient hypoxia on humans. We have learned how humans adapted to it via changes in the expression of hypoxia-related genes [2–11]. We also noticed that the adaptation to hypobaric hypoxia at high altitude was more pronounced in men than in women. Gender differences, however, have not yet been studied from the viewpoint of a glocal (combined global and local) comprehensive assessment [12–14].

12.2 Glocal Comprehensive Assessment in High Altitude

This investigation focused on chronoecology in Ladakh, using a sleep-habit questionnaire, endpoints of the circulation, autonomic nervous function, and quality of health as a physiological test system at high altitude, for comparison with chronoecological studies in several Japanese towns (Tosa, Kochi and Uraus, Hokkaido). Individuals living permanently at high altitude must contend with chronic hypobaric hypoxia. Several compensatory mechanisms have evolved to enable them to live and work in this inhospitable environment. We examined whether Ladakhi people have acquired any kind of notable cardiovascular adaptation.

To determine any effects of high altitude on the cardiopulmonary and neurocognitive functions, physical examinations were performed on 1,858 resting Ladakhi subjects living in 119 villages at altitudes between 2,200 and 4,590 m (1,081 women and 777 men, aged from 13 to 92 years, average 51.4 years). They included the measurement of pulse oximetry (SpO_2), blood pressure (BP), heart rate (HR), respiration rate (RR), and body mass index (BMI). BP and HR were measured twice in each of the sitting, supine, and standing positions using a semiautomated BP device (UA-767PC, A&D Co, Ltd, Tokyo, Japan). Orthostatic hypotension (OH) was defined when systolic (S) or diastolic (D) BP decreased by more than 20 or 10 mmHg from the supine to the standing position, respectively. Orthostatic hypertension (OHT) was defined when SBP or DBP increased by more than 20 or 10 mmHg from the supine to the standing position, respectively.

Aortic stiffness of cardio-ankle vascular index (CAVI) of right and left ankles [15, 16] was measured twice using a VaSera instrument (Fukuda Denshi, Tokyo) in the 1,858 Ladakhis as well as in 25,211 Japanese (13,366 men and 11,845 women, aged from 16 to 98 years, average 48.0 years). The average and maximal values from the four measurements were defined as CAVI-mean and CAVI-max, respectively.

Using a glocal comprehensive assessment, including a sleep-habit questionnaire, we studied the effects of high altitude on the autonomic cardiovascular function in Japanese T-town and Uraus and in several villages, including Leh, Ladakh. With continuous monitoring of the pulse-pulse interval using a newly developed pulse oximetry monitor (Konica-Minolta, Japan), we analyzed the autonomic cardiovascular activity.

12.3 Neurocognitive Functions in High Altitude

Of the 1,858 Ladakhis living at an altitude of 2,200–4,590 m, we selected middle-aged and elderly subjects over 40 years of age (1,331 citizens aged from 40 to 92 years, average 58.5 years). Any gender differences in compensatory mechanisms were investigated by comparing the cardiovascular functions, including BP, HR, ECG, and CAVI, between men and women in three groups living at different altitudes, i.e., from 2,500 to 3,000 m (66 men, average 62.7 years; 61 women, average 54.2 years), from 3,200 to 3,720 m (390 men, average 62.6 years; 635 women, average 56.2 years), and from 3,800 to 4,590 m (73 men, average 60.3 years; 110 women, average 55.5 years). We also assessed any gender-dependent effect of altitude by comparing the slopes of regression lines of BP with age between women and men.

The Kohs block design test and the time estimation test were used to assess the overall cognitive function. Subjects were 54 Ladakhi residents older than 70 years(46 men, 73.2 years of age) living at an altitude of 3,300 to 4,600 m and 67 residents of a Japanese town (Uraus) older than 70 years (52 men, 73.2 years of age). Time estimation was performed in a supine position, 10-s and 60-s being estimated with and without counting. The Up & Go test measured, in seconds, the time it took the subject to stand up from a chair, walk a distance of 3 m, turn, walk back to the chair, and sit down again. Functional reach, used to evaluate balance, represents the maximal distance a subject can reach forward beyond arm's length while maintaining a fixed base of support in the standing position. Manual dexterity was assessed using a panel with combinations of ten hooks (hook-on), ten big buttons (button-on-and-off), and five small buttons (button-on-and-off). The total manual dexterity time in sec., defined as the button score (Button), was calculated by adding the average times for one hook-on and one big or small button-on-and-off.

Vascular stiffness and compliance of brachial artery were measured by using a chronos instrument (A&D company, Tokyo) in 80 male Japanese (28–59 years of age, average 44.9 years) and in 45 male Ladakhis (25–67 years of age, average 47.9 years) living at an altitude of 3,524 m. This instrument can measure vascular stiffness and compliance of brachial artery as Pba (pressure of initial point assessed by pressure-area correlation curve, reflecting a difference of volume of right brachial artery vs. 10 mmHg alteration of pressure difference between arm and artery), A40 (cross-sectional area of brachial artery on 40 mmHg cuff pressure measured by color Doppler echogram), eA (estimated cross-sectional area of brachial artery), dAmax (maximum change of cross-sectional area of brachial artery), Kc/V40 (volume compliance of brachial artery normalized by vessel volume), β40 (stiffness on 40 mmHg cuff pressure), hb PWV (heart brachial PWV), and ha PWV (heart ankle PWV).

12.4 Aortic Stiffness of Cardio-Ankle Vascular Index (CAVI)

Ladakhi people showed increased aortic stiffness estimated by CAVI. Japanese and Ladakhi subjects were classified into 13 age groups, in 5-year classes from under 25 years to over 80 years, as shown in Table 12.1. CAVI values increased with age in both Ladakhi and Japanese people. Highlanders showed higher CAVI-max and also higher SBP values than lowlanders.

Table. 12.1 Comparison of age-matched CAVI-max* values between Japanese and Ladakhi people

CAVI-max (women)

Age Group (years)	Japanese			Ladakhi			Comparison	
	n	Mean	SD	n	Mean	SD	t-value	p-value
<24	325	6.37	0.62	27	6.80	0.89	-3.324	<0.001
25-29	564	6.76	0.62	37	7.03	0.88	-2.457	<0.05
30-34	1037	6.96	0.67	46	7.18	0.60	-2.300	<0.05
35-39	1409	7.13	0.64	72	7.27	0.73	-1.776	N.S.
40-44	1424	7.36	0.66	108	7.50	0.89	-2.055	<0.05
45-49	1496	7.62	0.70	102	7.87	1.24	-3.287	<0.005
50-54	1992	7.95	0.71	110	8.22	1.28	-3.581	<0.0005
55-59	1935	8.22	0.77	71	8.31	1.09	-0.961	N.S.
60-64	810	8.56	0.80	82	8.86	1.37	-3.044	<0.005
65-69	353	8.98	0.94	60	9.64	1.46	-4.603	<0.00005
70-74	158	9.30	0.98	52	10.07	1.64	-4.117	<0.0001
75-79	179	9.59	1.27	35	10.70	2.34	-4.031	<0.0001
>80	163	10.19	1.49	13	11.56	2.51	-3.001	<0.005

CAVI-max (men)

Age Group (years)	Japanese			Ladakhi			Comparison	
	n	Mean	SD	n	Mean	SD	t-value	p-value
<24	279	6.30	0.67	11	7.18	0.74	-4.272	<0.00005
25-29	859	6.79	0.67	27	7.40	0.82	-4.586	<0.00005
30-34	1644	7.08	0.69	29	7.41	0.83	-2.559	<0.05
35-39	1656	7.34	0.69	45	7.50	1.11	-1.434	N.S.
40-44	1433	7.60	0.72	73	7.95	0.68	-4.1	<0.00005
45-49	1457	7.87	0.76	59	8.09	0.92	-2.269	<0.05
50-54	1544	8.21	0.82	60	8.17	1.24	0.44	N.S.
55-59	2057	8.56	0.88	59	8.66	1.21	-0.877	N.S.
60-64	1263	8.95	0.85	61	9.06	1.92	-0.857	N.S.
65-69	594	9.31	1.07	50	9.65	1.67	-2.003	<0.05
70-74	273	9.57	1.07	40	10.38	2.30	-3.707	<0.0005
75-79	196	10.24	1.14	23	11.08	3.04	-2.629	<0.01
>80	111	10.59	1.72	24	10.84	2.33	-0.608	N.S.

* maximal value of Cardio-Ankle Vascular Index

12.5 Cardiopulmonary Autonomic Activity in High Altitude

Cardiovascular function was impaired in elderly Ladakhis compared to Japanese subjects. Autonomic cardiovascular activity was compared between 54 Ladakhi residents (46 men, 73.2 years of age) living at an altitude of 3,300–4,600 m and 67 residents of a Japanese Uraus (52 men, 73.2 years of age). High-altitude residents of Leh slept less (6.9 vs. 7.8 hours, $p<0.001$), had a shorter time for falling asleep (15.5 vs. 25.6 min, $p<0.005$), and had a lower SpO_2 (84.4 vs. 96.0 %, $p<0.00001$). Compared to Japanese subjects, they also had lower CVRR ($p<0.05$), VLF ($p<0.00001$), LF ($p<0.05$), and HF ($p<0.001$) components and a higher LF/HF ratio (2.35 vs. 1.69, $p<0.05$), DBP ($p<0.05$), and heart rate ($p<0.00001$).

Next, we checked SpO_2, respiration rate, BP, and ECG findings in the elderly in Ladakh and were surprised. Residents in Ladakh suffered less coronary artery disease, atrial fibrillation, and metabolic syndrome (3 %, 0 %, and 3 %, respectively). In Leh at 3,524 m, elderly residents had a lower SpO_2 and a higher RR than elderly Japanese residents in Tosa (SpO_2, 88.0 ± 4.3 % vs. 96.6 ± 1.2 %, $p<0.0001$; RR, 21.1 ± 4.2 vs. 17.8 ± 4.2, $p<0.0001$). All six measurements of DBP and five of the 6 HR measurements were statistically significantly higher in Leh than in Tosa. Changes in HR from the supine to the standing position were larger in Leh than in Tosa (13.8 ± 8.1 bpm vs. 9.9 ± 6.7 bpm, $p<0.001$). A higher voltage of the QRS complex (SV1 + RV5) was observed in the ECG of Leh's residents, but the so-called lifestyle-related diseases were very rare there.

12.6 Men Better Adapted to Ambient Hypoxia than Women

The Ladakhi people must have obtained some kind of characteristic adaptation to their high-altitude environment. Notably, there was a different cardiovascular adaptation between men and women; men seem to be better adapted to ambient hypoxia than women in Ladakh. Compensatory mechanisms were examined in 1,334 middle-aged and elderly Ladakhis (528 men and 806 women) by comparing cardiovascular functions among three groups living at different altitudes, namely from 2,500 to 3,000 m (66 men and 61 women), from 3,200 to 3,720 m (390 men and 635 women), and from 3,800 to 4,590 m (72 men and 110 women). Results are shown in Tables 12.2 and 12.3.

Ladakhi men showed mostly full adaptations, which include SBP and DBP, SpO_2, and ECG findings, including heart rate, first degree AV block, ST depression, abnormal Q waves and R wave amplitudes, and CAVI (both average and maximum of four measurements), except for an increased OH of DBP in higher-altitude dwellers (Table 12.2). By contrast, adaptations in Ladakhi women were not as complete. SBP in sitting position and average and maximum CAVI measures were higher, and incidences of first degree AV block, ST depression, and abnormal Q waves were more frequent in higher altitude dwellers (Table 12.3).

Table 12.2 Cardiovascular and other characteristics of men at different high altitudes in Ladakh, Kashmir, India (n=528)

Altitude	2,486 m			3,475 m			4,080 m			1-way ANOVA	
Variables	n	Mean	SD	n	Mean	SD	n	Mean	SD	F-value	p-value
Age	66	62.7	11.6	390	62.6	10.8	72	60.3	10.1	1.46	N.S.
BW	66	60.7	11.8	389	61.4	10.5	72	61.6	11.2	0.14	N.S.
Height	66	160.3	6.2	389	162.2	7.4	72	162.3	7.1	2.11	N.S.
BMI	66	23.5	3.7	389	23.4	3.9	72	23.3	3.5	0.07	N.S.
Systolic BP sitting	66	133.7	21.2	388	138.7	25.3	70	134.1	21.8	1.94	N.S.
Diastolic BP sitting	66	84.8	10.9	389	86.4	12.7	69	84.8	11.6	0.88	N.S.
Systolic BP supine	66	133.1	19.0	387	133.4	21.9	70	129.8	20.2	0.78	N.S.
Diastolic BP supine	66	80.2	9.3	387	80.5	12.6	70	81.5	11.4	0.25	N.S.
Systolic BP standing	66	130.9	19.4	386	132.1	24.0	70	126.5	20.6	1.72	N.S.
Diastolic BP standing	66	85.2	13.1	386	85.5	12.9	70	83.6	13.0	0.61	N.S.
⊿Systolic BP	66	−2.2	12.1	386	−1.4	13.8	70	−3.2	11.2	0.62	N.S.
OH Systolic BP (%)	66	4.5		386	7.4		70	10.0		1.55*	N.S.
OHT Systolic BP (%)	66	3.0		386	5.2		70	2.9		1.15*	N.S.
⊿Diastolic BP	66	5.0	9.5	386	4.9	9.2	70	2.1	7.8	3.05	<0.05
OH Diastolic BP (%)	66	3.2		386	3.6		70	14.3		177.7*	<0.001
OHT Diastolic BP (%)	66	25.8		386	24.4		70	17.1		1.91*	N.S.
SpO_2	66	88.2	4.8	377	88.2	4.8	69	87.4	5.6	0.83	N.S.
Pulse	66	78.3	12.9	377	74.5	13.9	68	73.2	11.6	2.89	N.S.
ECG HR	66	70.6	12.3	389	67.2	12.8	69	66.2	12.1	2.46	N.S.
1st degree AV block (%)	66	0		384	2.3		65	0		3.12*	N.S.
ST depression (%)	66	0		375	4.8		65	1.5		4.59*	N.S.
Abnormal Q (%)	66	0		374	1.9		64	0		2.35*	N.S.
SV1+RV5 (mm)	66	26.7	9.3	374	24.4	8.5	64	26.0	10.9	2.41	N.S.
CAVI-mean (m/sec)	66	8.73	1.43	389	8.75	1.72	71	9.08	1.74	1.17	N.S.
CAVI-max (m/sec)	66	9.26	1.66	389	9.26	2.06	71	9.63	2.18	1.02	N.S.

abbreviations: see text. *; χ^2-value

Table 12.3 Cardiovascular and other characteristics of women at different high altitudes in Ladakh, Kashmir, India (n=806)

Altitude	2,500 m			3,472 m			4,084 m			1-way ANOVA	
Variables	n	Mean	SD	n	Mean	SD	n	Mean	SD	F-value	p-value
Age	61	62.7	11.6	635	62.6	10.8	110	60.3	10.1	1.01	N.S.
BW	61	49.0	10.1	635	51.7	9.2	110	52.0	8.1	2.69	N.S.
Height	61	151.4	8.0	635	151.7	6.9	110	152.0	6.0	0.13	N.S.
BMI	61	21.3	3.7	635	22.5	3.8	110	22.5	3.2	2.90	N.S.
Systolic BP sitting	61	127.0	21.7	631	134.8	22.3	109	135.0	24.7	3.38	<0.05
Diastolic BP sitting	61	81.3	12.1	630	85.3	12.5	109	83.7	14.6	3.28	<0.05
Systolic BP supine	61	126.0	20.1	630	129.6	21.6	108	127.8	21.5	0.98	N.S.
Diastolic BP supine	61	76.8	12.2	630	79.5	12.9	108	78.4	12.9	1.40	N.S.
Systolic BP standing	61	124.8	21.9	628	128.1	21.7	108	126.5	22.4	0.83	N.S.
Diastolic BP standing	61	82.4	12.6	627	84.6	12.2	108	81.0	13.7	4.35	<0.05
⊿Systolic BP	61	−1.3	13.9	627	−1.7	13.6	108	−1.3	9.7	0.06	N.S.
OH Systolic BP (%)	61	6.7		627	7.7		108	7.5		0.10*	N.S.
OHT Systolic BP (%)	61	3.3		628	4.6		108	1.9		1.90*	N.S.
⊿Diastolic BP	61	5.7	7.3	626	5.1	8.9	108	2.6	9.6	4.09	<0.05
OH Diastolic BP (%)	61	3.3		626	3.2		109	4.6		0.55*	N.S.
OHT Diastolic BP (%)	61	25.0		626	28		108	19.4		3.55*	N.S.
SpO_2	61	87.8	5.8	620	87.8	5.5	109	88.4	5.4	0.61	N.S.
Pulse	61	83.3	13.7	619	76.9	14.1	109	73.2	11.3	10.51	<0.001
ECG HR	61	74.1	14.3	633	69.4	12.0	108	66.7	12.4	7.18	<0.005
1st degree AV block (%)	60	0		618	0.6		99	3.0		6.02*	<0.05
ST depression (%)	60	0		615	7.0		98	10.2		6.15*	<0.05
Abnormal Q (%)	60	1.7		613	0.5		97	4.1		11.01*	<0.005
SV1+RV5 (mm)	61	24.7	6.7	619	23.9	7.5	102	25.3	9.4	1.61	N.S.
CAVI-mean (m/sec)	61	8.09	2.20	633	8.14	1.35	109	8.58	1.59	4.29	<0.05
CAVI-max (m/sec)	61	8.68	2.44	633	8.57	1.64	109	9.05	1.77	3.62	<0.05

Abbreviations: see text. *; χ^2-value

Relations of BP measured in sitting, supine, and standing positions vs. age, and relations of SpO_2 and CAVI vs. age are shown in Table 12.4. Gender differences in the slope of regression lines are noteworthy. Namely, Ladakhi women showed a steeper negative slope of the regression line between SpO_2 and age than Ladakhi men (−0.1215 vs. −0.0670 %/year, $p<0.0005$) (Fig. 12.1). As compared to men,

Slope from linear regression line with age	Women (n=1,076)		Men (n=773)		Test of equality of slope	
	Correlation coefficient	slope	Correlation coefficient	slope	t-value	p-value
Systolic BP sitting	0.5039	0.7819	0.3508	0.5094	4.31	<0.0001
Diasolic BP sitting	0.3607	0.3253	0.1911	0.1490	4.65	<0.0001
Systolic BP supine	0.4835	0.7009	0.3852	0.5151	3.16	<0.005
Diastolic BP supine	0.4194	0.3730	0.2761	0.2169	4.24	<0.0001
Systolic BP standing	0.4541	0.6643	0.3099	0.4220	3.96	<0.0005
Diasolic BP standing	0.3792	0.3361	0.1830	0.1478	4.95	<0.0001
SpO_2	−0.3296	−0.1215	−0.2232	−0.0670	3.56	<0.0005
CAVI-mean	0.6260	0.0646	0.6173	0.0647	−0.02	N.S.
CAVI-max	0.6124	0.0731	0.5885	0.0736	−0.10	N.S.

Table 12.4 Gender differences in changes with age of cardiovascular endpoints

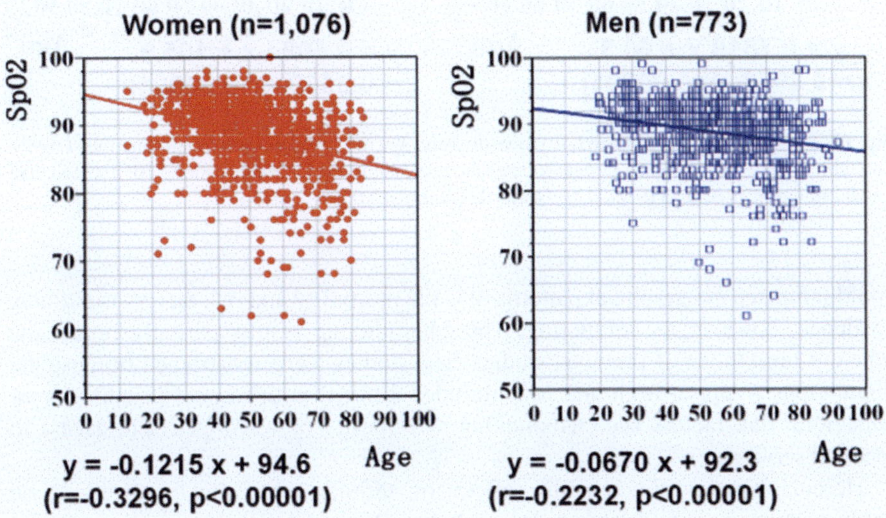

Fig. 12.1 Relations between SpO2 measures and age in Ladakhi women (*Left*) and men (*right*). Ladakhi women (*left*) showed steeper negative slope of the regression line between SpO2 and age than Ladakhi men (*right*); –0.1215 vs. –0.0670 %/year, $p < 0.0005$

Ladakhi women also showed steeper positive slopes of the regression lines vs. age of SBP and DBP measured in the sitting, supine, and standing positions (Fig. 12.2).

12.7 Glocal (Global and Local) Comprehensive Health-Watch

Morbidity and mortality from cardiovascular disease (CVD) constitute a major global health burden. Geographical disparities in CVD morbidity and mortality are substantial, even within countries. Little is known about the role of independent environmental factors in the development of CVD. One environmental factor that

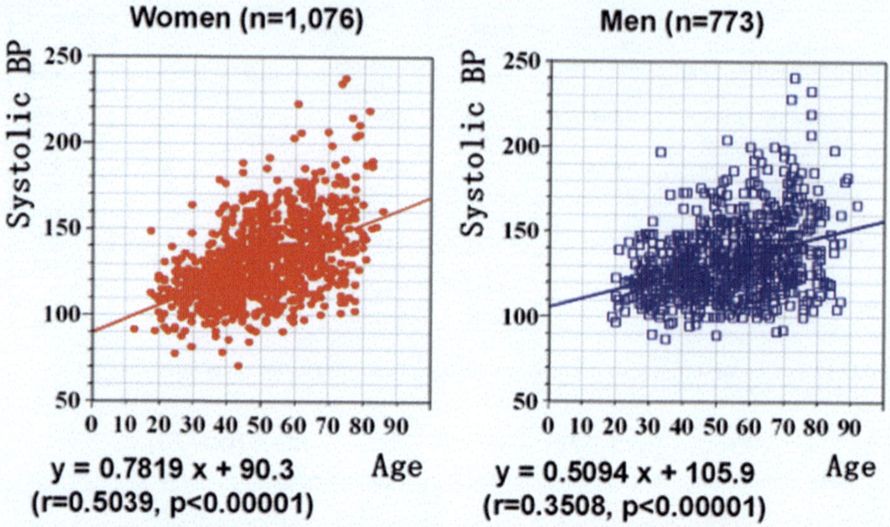

Fig. 12.2 Systolic blood pressure (BP) measures and age in Ladakhi women (*left*) and men (*right*). Ladakhi women (*left*) showed larger slope of the regression line between systolic BP and age than Ladakhi men (*right*); 0.7819 vs. 0.5094 mmHg/year, $p<0.0001$

could influence geographical patterns of CVD morbidity and risk factors is altitude. Evidence, however, is conflicting. The contradicting results partially reflect the inherent limitations of previous studies. The studies were constrained because the populations living at high and low altitudes differed with respect to ethnicity or behavioral risk factors such as smoking and obesity or with respect to access to medical services.

Herein, we investigated the effect of altitude in Ladakhi dwellers on cardiopulmonary-vascular functions, including BP at rest, BP alterations associated with postural change, ECG findings, SpO_2, CAVI, and the autonomic function. Altitude was found to affect cardiovascular functions in Ladakhis. Using a 13 age-group classification from <25 to >80 years in 5-year bins, we observed that CAVI increased with age both in Japanese and Ladakhis. High-altitude Ladakhis showed higher CAVI values than the lowland Japanese people in most of the 13 age groups in both men and women ($p<0.00001$).

As compared to Japanese subjects, Ladakhis also had a lower CVRR ($p<0.05$) and VLF ($p<0.00001$), LF ($p<0.05$), and HF ($p<0.001$) components and a higher LF/HF ratio (2.35 vs. 1.69, $p<0.05$), as well as a higher DBP ($p<0.05$) and heart rate ($p<0.00001$). Residents in the high-altitude community of Leh had a poorer cognitive function, estimated by the Kohs block design test, slept less ($p<0.001$), and had a lower SpO_2 ($p<0.00001$). These findings suggest that a hypobaric hypoxic condition at high altitude induced a decrease in parasympathetic activity, including baroreflex sensitivity reflected by the LF component, and an increase in sympathetic drive, which may result in an imbalance of the cardiopulmonary-vascular functions in Ladakhi people.

The present study also revealed markedly different adaptations to high altitude between men and women. As found by others, the protective effect of altitude was more pronounced in men than in women for CVD. Our principal findings are that although SpO_2 showed a definite adaptation to the ambient hypoxia in both genders, (1) SBP in the sitting position was higher at higher altitude; (2) pulse and ECG HR were lower at higher altitude in women, but not in men; (3) incidences of first degree AV block, ST depression, and abnormal Q waves were more frequent with increasing altitude in women; (4) both average and maximal CAVI measures were statistically significantly larger at higher altitude only in women, as shown in Tables 12.3 and 12.4; and (5) as compared to men, Ladakhi women showed a steeper negative slope of SpO_2 with age ($p < 0.0005$) and a larger slope of the regression line with age of SBP and DBP measured in the sitting, supine, and standing positions (Figs. 12.1 and 12.2).

Gender differences in neuro-cardiopulmonary functions observed herein may be accounted for by the following factors [14]: First, improved lung mechanics may be an important adaptation to indoor biomass and outdoor dust exposure in high-altitude populations [17]. Norboo et al. [18] reported that the prevalence of chronic cough was greater among women than among men. The percentage of villagers over 50 years of age with a forced expiratory volume in one sec/forced vital capacity ratio ($FEV_{1.0}/FVC$) of less than 65 % was 24 % in men and 32 % in women. Older people exposed to environmental dust may develop advanced cardiopulmonary dysfunction. In Ladakh, women are more heavily exposed to dust with the use of chimneys in the kitchen, and to dust storms in the course of their work, and appear to be more commonly affected than men [19]. Second, it has been reported that the urinary sodium/potassium ratio is larger in women than in men in Ladakh [20]. Urinary potassium (nmol/24-hour) is lower in women (37.4 ± 13.2) than in men (56.7 ± 23.8). The urinary sodium/potassium ratio is higher in women (5.52 ± 2.03) than in men (4.24 ± 1.99), which itself is higher than in Japanese (Osaka) women (3.90 ± 1.14). Nutritional imbalances may be associated with unfavorable neuro-cardiopulmonary functions in women, a question being pursued in our ongoing study of how nutrition may affect physiological adaptation to a high-altitude environment. Third, medical services are only now developing in Ladakh, and citizens, especially poor women, have not received sufficiently high-quality care. Lastly, Leh, the present capital of Ladakh, was once the central meeting point for trade caravans from Central Asia and the plains of India. Their religion is Tibetan Buddhism, and many monks practice it in its original form. From a cultural anthropologic standpoint, such original lifestyles, including the educational system, may contribute to the gender differences seen in Ladakh.

Recently, Faeh et al. [21] reported a lower mortality from CAD and stroke in high-altitude dwellers in Switzerland. They postulate a protective effect of altitude that was more pronounced in men than in women and offered the following explanation: First, men are physically more active than women, and exercise under moderate hypoxic conditions may provide more health benefit than at sea level.

Second, it is likely that at higher altitudes, men have professional (tourism industry, farming) and leisure time (sport) outdoor activities and therefore profit more from ultraviolet radiation: vitamin D, omega-3 fatty acids, and other nutrients can have the same protective effect as altitude. Although the protective effect of altitude is more pronounced in men than in women, as presented by Faeh et al. [21] and by our own observations (Tables 12.2 and 12.3 and Figs. 12.1 and 12.2), our study differs insofar as, for instance, Ladakhis have stiffer arteries, as compared to Japanese. The Swiss study may have the advantage of similar ethnicity.

12.8 Conclusion

We confirmed the effect of altitude on cardiopulmonary functions in elderly dwellers. Most were statistically significantly affected by the environmental condition of hypobaric hypoxia. DBP, HR, and changes in HR from the supine to the standing position were higher in Leh than in Tosa. A limitation qualifying our glocal approach is that the ethnicity of the populations in Japan and Kashmir differed. The longitudinal assessment of circadian characteristics is the more important since even circadian stage-specified investigations covering 3.5 decades in the same subject yield opposite results at different circadian stages [22].

The failure to sample systematically around the clock notwithstanding [23], our data indicate the need for a glocal comprehensive chronobiologic assessment to reach a better diagnosis and a more fruitful treatment [24]. It will be important to quantify associations among different environmental variables (weather, inside and outside temperature, hours of sunshine, geomagnetic latitude and geographical altitude, as well as other differences [25, 26], including ethnicity) and lifestyle characteristics (body weight, salt intake, fruit and vegetable intake, alcohol consumption, smoking, physical activity, and duration of sleep), including depression and QOL scores (wellness, mood, family relationships, financial satisfaction, life satisfaction, and sense of happiness), and neuro-behavioral functions and BP changes. A glocal comprehensive assessment should be repeated longitudinally in order to aim at stroke prevention in a community dwelling, especially at high altitude.

References

1. Gilbert DL. The first documented description of mountain sickness: the Andes or Periacaca story. Respir Physiol 1983;52:327–49.
2. Rajput C, Najib S, Norboo T, et al. Endothelin-1 gene variants and levels associate with adaptation to hypobaric hypoxia in high-altitude natives. Biochem Biophys Res Commun 2006;341:1218–24.
3. Charu R, Stobdan T, Ram RB, et al. Susceptibility to high altitude pulmonary oedema: role of ACE and ET-1 polymorphisms. Thorax 2006;61:1011–2.
4. Ahsan A, Charu R, Pash MAQ, et al. eNOS allelic variants at the same locus associate with HAPE and adaptation. Thorax 2004;59:1000–2.

5. Ahsan A, Norboo T, Baig MA, et al. Simultaneous selection of the wild-type genotypes of the G894T and 4B/4A polymorphisms of NOS3 associate with high-altitude adaptation. Ann Hum Genet 2005;69:260–7.
6. Ahsan A, Mohd G, Norboo T, et al. Heterozygotes of NOS3 polymorphisms contribute to reduced nitrogen oxides in high-altitude pulmonary edema. Chest 2006;130:1511–9.
7. Qadar Pasha MA, Khan AP, Kumar R, et al. Angiotensin converting enzyme insertion allele in relation to high altitude adaptation. Ann Hum Genet 2001;65:531–6.
8. Kumar R, Pasha MAQ, Khan AP, et al. Association of high-altitude systemic hypertension with the deletion allele-of the angiotensin-converting enzyme (ACE) gene. Int J Biometeorol 2003;48:10–4.
9. Rajput C, Makhijani K, Norboo T, et al. CYP11B2 gene polymorphisms and hypertension in highlanders accustomed to high salt intake. J Hypertens 2005;23:79–86.
10. Rajput C, Arif E, Vibhuti A, et al. Predominance of interaction among wild-type alleles of CYP11B2 in Himalayan natives with high-altitude adaptation. Biochem Biophys Res Commun 2006;348:735–40.
11. Xing G, Qualls C, Huicho L, et al. Adaptation and mal-adaptation to ambient hypoxia; Andean, Ethiopian and Himalayan patterns. PLoS One 2008;3, e2342.
12. Otsuka K, Cornelissen G, Norboo T. Chronomics and "glocal" (combined global and local) assessment of human life. Prog Theor Phys 2008;Suppl 173:134–52.
13. Otsuka K, Norboo T, Otsuka Y. Effect of aging on blood pressure in Leh, Ladakh, a high-altitude (3524 m) community, by comparison with a Japanese town. Biomed Pharmacother 2005;59:54s–7.
14. Otsuka K, Norboo T, Otsuka Y. Chronoecological health watch of arterial stiffness and neuro-cardio-pulmonary function in elderly community at high-altitude (3524 m), compared with Japanese town. Biomed Pharmacother 2005;59:58s–67.
15. Shirai K, Utino J, Otsuka K, et al. A novel blood pressure-independent arterial wall stiffness parameter: cardio-ankle-vascular index (CAVI). J Atheroscler Thromb 2006;13:101–7.
16. Yamamoto N, Yamanaka G, Ishikawa M, et al. Cardio-ankle vascular index as a predictor of cognitive impairment in community-dwelling elderly people: Four-year follow-up. Dement Geriatr Cogn Disord 2009;28:153–8.
17. Wood S, Norboo T, Lilly M, et al. Cardiopulmonary function in high-altitude residents of Ladakh. High Altitude Med Biol 2003;4:445–54.
18. Norboo T, Yahya M, Bruce NG, et al. Domestic pollution and respiratory illness in a Himalayan village. Int J Epidemiol 1991;20:749–57.
19. Norboo T, Angchuk PT, Yahya M, et al. Silicosis in a Himalayan village population: role of environmental dust. Thorax 1991;46:341–3.
20. The INTERSALT Cooperation Research Group Appendix Tables. Centre-specific results by age and sex. J Hum Hypertens 1989;5:331–407.
21. Faeh D, Gutzwiller F, Bopp M, et al. Lower mortality from coronary heart disease and stroke at higher altitudes in Switzerland. Circulation 2009;120:495–501.
22. Halberg F, Sothern RB, Cornélissen G et al. Chronomics, human time estimation, and aging. Clin Interv in Aging 2008; 3: 749–60. http://www.dovepress.com/articles.php?article_id=2608
23. Otsuka K, Cornélissen G, Schwartzkopff O. Clinical chronobiology and chronome-geriatrics: At variance with recommendations of subsequent guidelines, yet focusing indeed on pre-hypertension in the physiological range. Biomed & Pharmacother 2003;57:164s–98.
24. Halberg F, Cornélissen G, Otsuka K et al. Extended consensus on means and need to detect vascular variability disorders (VVDs) and vascular variability syndromes (VVSs). Leibniz-Online Nr. 5, 2009. pp. 35. (http://www.leibniz-sozietaet.de/journal).
25. Appleton EV. Foreword. In: Stetson HT, editor. Sunspots in Action. New York: Ronald Press Co; 1947. p. iii–vi.
26. Halberg F, Cornélissen G, Katinas G et al. Chronobiology's progress: season's appreciations 2004–2005. Time-, frequency-, phase-, variable-, individual-, age- and site-specific chronomics. J Appl Biomedicine 2006;4:1–38. http://www.zsf.jcu.cz/vyzkum/jab/4_1/halberg.pdf

Chapter 13
Chronoastrobiology

Abstract Life did not emerge spontaneously. Life was created on this planet so that it "matches" with the natural environment and was further evolved through the physical/chemical processes which reflect the state of the planet, including the atmosphere and water, as well as the magnetic field. The present state of the natural environment of the earth has been accomplished so as to balance with the conditions of the solar system and the rest of the universe.

We must realize that what we can see by our human eyes is limited in scope, wavelength, etc. That we cannot see some things does not mean at all that they do not exist in reality. There is a lot we must learn from nature. Life on the earth's surface is, directly or indirectly, influenced or, more explicitly, controlled by how this planet was created and how the present environment has evolved. This is one of the messages the aurora in the polar sky is sending us.

From Professor Yosuke Kamide's words, "Our life is protected by the Earth's atmosphere and magnetic field: what aurora research tells us" (Biomed Pharmacother 55:21s–4s, 2001).

Humans live on the planet in space according to the law of nature

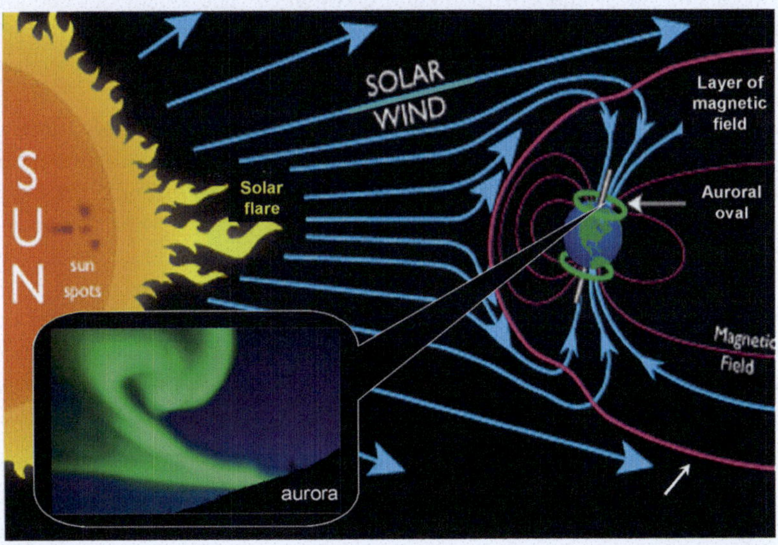

K. Otsuka et al., *Chronomics and Continuous Ambulatory Blood Pressure Monitoring*, DOI 10.1007/978-4-431-54631-3_13

Keywords Chronoastrobiology • Geomagnetic activity • Magnetoreception • Heart rate variability • Cardiovascular outcomes

13.1 Introduction

The term "chronoastrobiology" is derived from "chronome" (time structure) and "astrobiology" (reciprocal implications of biology and space studies). It was the title of a proposal for the development of a chronoastrobiology research initiative on February 8, 1999, by Franz Halberg, Germaine Cornelissen, and Othild Schwartzkopff of the University of Minnesota in Minneapolis; by Daniel C Holley of San Jose State University, California; and, within the US National Aeronautics and Space Administration (NASA), by Charles M Winget, then of the Ames Research Center, Moffett Field, California [1].

This proposal included four specific goals that remain as important today as they were in 1999:

- First, the longitudinal monitoring of psychophysiological variables to compare their time structure with that of space and terrestrial environmental variables. Because some periodic components in biology (such as the circadian rhythm) may be genetically anchored, they may more readily be synchronized and/or resonate with their counterparts in the environment. The sequences of integration of rhythms into the genome, possibly revealed by their relative prominence in ontogeny and phylogeny, may shed light on the sites and time course of life on earth.
- Second, data analyses by complementary approaches for examining associations between biological and environmental chronomes. The extent of concordance between results as a function of time and of their reproducibility across several geographic locations in a glocal approach should lead to a better understanding of environmental influences on biota.
- Third, the mapping of cycles characterizing important physiological functions in humans and other selected species at different stages of development. Such a comparative physiological approach can be followed to examine the extent to which the evolution of chronomes may lead to testable hypotheses concerning the origins of life.
- Fourth, as a dividend from the establishment of reference databases resulting from the above basic work, to identify disease risk syndromes. A better understanding of underlying triggers and the development of countermeasures to lower risk by timely and timed intervention (chronotherapy) aim as much at primary as it does at secondary prevention.

From early in Franz Halberg's career, the mapping of rhythms proved to be very useful as a way to defining health and learning about the origins of life. His work on the circadian cell cycle provides the best illustration of how the study of rhythms may shed light on internal as well as external evolution. Halberg deserves great credit for disproving the then prevailing dogma that the nucleic acid was the most constant feature of the body with his demonstration in the 1950s of a circadian periodic cell cycle [2–4]. Not only did he show that DNA and RNA were characterized by a robust circadian rhythm, Halberg also showed that RNA synthesis precedes that of DNA (Fig. 13.1), contrary to the dominating view of a linear biochemical sequence starting from DNA to RNA and protein [2–4]. Halberg's finding of a lead in phase of the circadian rhythm in RNA formation vs. DNA formation was later reproduced for the case of RNA content, shown to precede DNA content in the unicell Euglena [5].

With the qualification that RNA formation also concerns nondividing cells, the circadian mapping of nucleic acids raised the question of a temporal precedence of RNA vs. DNA. Halberg's finding in growing mouse liver provided a hint that RNA may be more atavistic than DNA, decades prior to the chemical evidence suggesting that an RNA molecule could have existed that "naturally contained the sequence of information for its reproduction through reciprocal-base pairing and could also catalyze a synthesis of more like RNA strands" [6]. For this stage in the development of life, it was hypothesized that "RNA molecules and cofactors sufficed to carry out

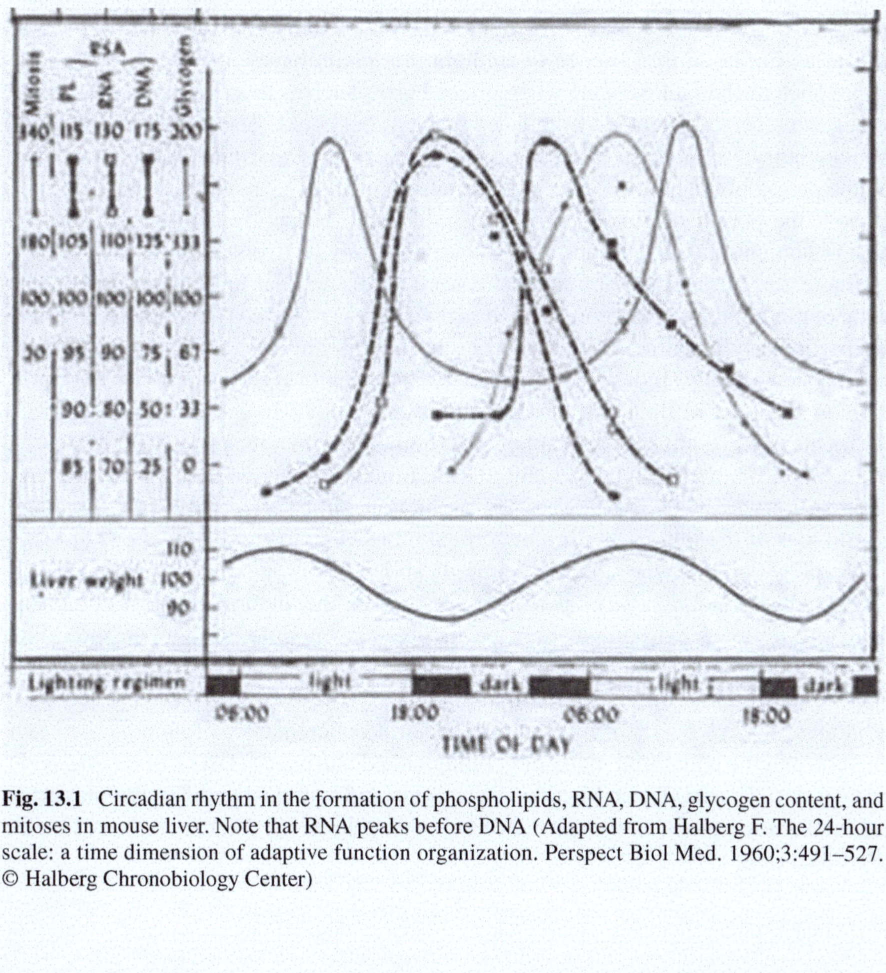

Fig. 13.1 Circadian rhythm in the formation of phospholipids, RNA, DNA, glycogen content, and mitoses in mouse liver. Note that RNA peaks before DNA (Adapted from Halberg F. The 24-hour scale: a time dimension of adaptive function organization. Perspect Biol Med. 1960;3:491–527. © Halberg Chronobiology Center)

all the chemical reactions necessary for the first cellular structures" [6]. Along this line of thought, an RNA world may well have occurred before our DNA world. If the relative phases of their circadian rhythms reflect evolutionary changes, the precedence of RNA synthesis before DNA formation in the given cell cycle [2] can be viewed as an early hint of the chain of events in the course of evolution [7].

Along a similar line of thought, it is tempting to speculate about the role of multiseptans throughout evolution. Multiseptans are components with periods of about 7 days and multiples or submultiples thereof, among which the half week and month are also ubiquitous. A biological week is indeed built into bacteria and unicellular organisms, perhaps another chronobiologic contribution to phylogeny [8–18]. As to ontogeny, the about-weekly change can be more prominent than the about-daily change in human neonatal blood pressure and heart rate [13, 19–28], as it is in the early postnatal life of other species [29–31]. An about-7-day pattern was even observed in DNA labeling under daily STH stimulation in mouse tissues [32], and multiseptans were found to modulate circadian cell division and cell settling patterns in Euglena [33]. As in the case of circadians, circaseptan counterparts are present in the environment. Abbot reported an about-6.65-day cycle in the weather [34]. A near match of the week is also found in the index of geomagnetic disturbance Kp [35–38], likely the fourth harmonic of the about-27-day solar rotation period. Notably, if it developed at the bottom of an ocean, in the absence of sunlight, life could have latched onto nonphotic cycles such as the multiseptans when it faced tasks such as those in immunology and repair, which could not be accomplished in a day, but could be accomplished within a week, a hint of an internal evolution complementary to Darwinian evolution [5, 36]. As the period of the biological week differs from precisely 7 days in human isolation, as does the period of Kp [39–44], it is unlikely to be purely anthropogenic, even though human-made contributions to the weather may be overwhelming [45].

Interestingly, an effect of a magnetic storm at the cellular level and possibly also a site of mammalian magnetoreception in the eye was reported [46]. Mitotic rhythms are particularly prominent as mapped in the urodele, mouse, and rat cornea. They were systematically studied by Lawrence E. Scheving in eight separate 24-hour profiles on rats kept in light and darkness alternating at 12-hour intervals (Fig. 13.2) [47]. One of the eight surveys happened to coincide with a magnetic storm. As seen from Fig. 13.3, the circadian amplitude of corneal mitosis on the day of the magnetic storm was below the lower 5 % prediction limit derived from the circadian amplitudes of the mitotic rhythm in corneal mitoses on the seven quiet days [46].

Work by Faraone who studied seemingly spontaneous and mutagen-induced sectoring in bacteria may also be pertinent. He showed that natural physical influences from the remote environment may be involved in modulating genetic microbial change [48–50]. Any physical environmental associations with the incidence of colonies with one or more differentiated sectors (CSD) were examined for air bacteria and *Staphylococcus aureus*. Cross-spectral coherence was found between air bacteria and the equatorial index of geomagnetic disturbance (Dst) around one cycle in 18.6 days ($P=0.005$) and between *S. aureus* and both Dst and Kp, the planetary index of geomagnetic disturbance, around one cycle in 161.5 days or about 1 week ($P<0.05$) [50].

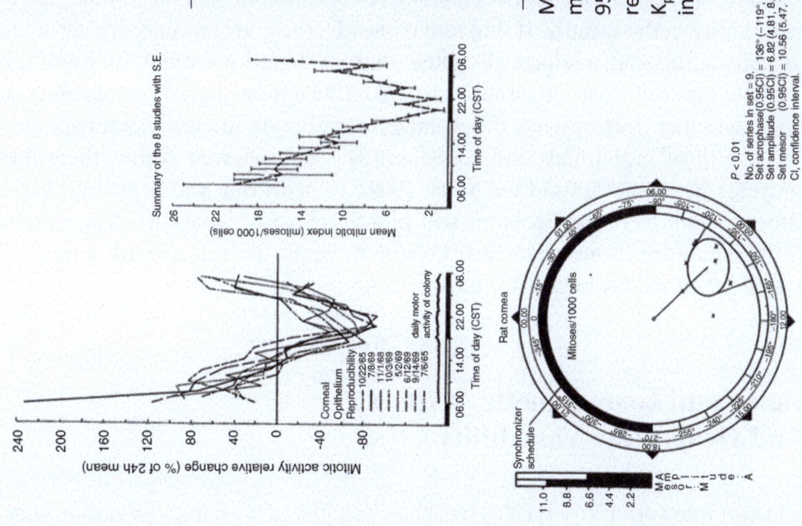

Environmental conditions on day of corneal mitotic profile providing characteristics tabulated in rats from L6:18D18:6

Date	N	P	M±SE	A±SE	φ (95% CI)	K_p	aa	Dst	WN	Cosmic rays	Solar wind speed (km/sec)
1965.07.06	10	0.01	14.90± 0.63	9.81 ± 0.092	-138°(-128,-149)	3.10	34.0	-7.9	17	405,242	—
1965.09.14	14	0.01	12.48± 0.55	7.72 ± 0.101	-119°(-108,-131)	0.66	6.5	3.4	8	413,400	334
1965.10.22	8	0.01	11.61± 0.63	5.64 ± 0.158	-179°(-161,-197)	2.53	27.7	16.3	2.6	401,895	367
1968.11.01	7	0.01	7.56± 0.68	3.78 ± 0.252	-140°(-111,-168)	6.08	134.3	-133.0	89	347,029	907
1969.05.02	8	0.01	7.46± 0.51	5.58 ± 0.129	-125°(-110,-140)	3.51	40.3	-17.4	77	374,416	—
1969.06.12	8	0.01	9.13± 0.75	7.61 ± 0.139	-135°(-119,-151)	3.11	27.0	-23.2	187	360,000	—
1969.07.08	8	0.02	8.59± 1.55	7.96 ± 0.276	-120°(-89,-151)	1.25	11.3	13.7	122	349,837	—
1969.10.03	8	0.01	10.41± 0.52	7.57 ± 0.096	-155°(-146,-166)	3.04	31.1	-43.0	99	409,762	—

M: Midline-Estimating Statistic Of Rhythm, a rhythm-adjusted mean; A: amplitude, a measure of extent of predictable change within one cycle; SE: standard error; 95% CI: 95% confidence interval; φ: acrophase, measure of timing of overall high values recurring in each cycle, expressed in (negative) degrees with 360° ≡ a 24 hours, 0° = 00:00; K_p: planetary disturbance index; aa: geomagnetic index; Dst: equatorial disturbance index; WN: Wolf's equatorial sunspot numbers.

Fig. 13.2 Circadian rhythm in corneal mitoses in rats assessed in eight different studies. One of these studies was performed on a day of a magnetic storm. Quiet conditions prevailed on days of the other seven studies (© Halberg Chronobiology Center)

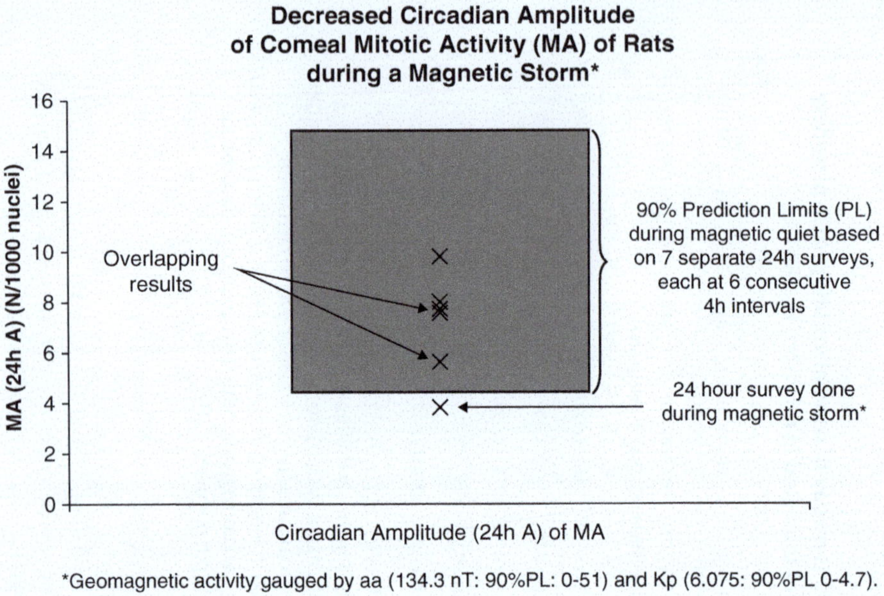

Fig. 13.3 On the day of a magnetic storm, the circadian amplitude of rats' corneal mitotic activity was reduced, by comparison with results obtained during quiet days (© Halberg Chronobiology Center)

Herein, investigations of environmental influences on biota rely on methods other than correlation analysis, an approach too often applied inappropriately. Indeed, the correlation coefficient measures the strength of a relation between two variables, but not the agreement between them. As correlation depends on the range of the true quantity in the sample, if that range is wide, the correlation is greater than when it is narrow. Instead, reciprocal cycles, that is, shared periodicities found in biology and its environment, constitute critical information underlying evolution and genetics, whether dealing with the weather and climate in earth's atmosphere, or with the health of individuals and societies [51]. As reviewed below, there are matching cycles in us and around us. Some relate to heart rate and heart rate variability. Others relate to blood pressure and blood pressure variability. Yet, others deal with the incidence of adverse cardiovascular events, as well as with a host of other variables of broad biomedical interest.

13.2 Solar and Geomagnetic Activity and Heart Rate Variability

Decreased heart rate variability (HRV) has been associated with increased mortality after acute myocardial infarction [52] and after coronary artery bypass grafting surgery [53]. A one-standard deviation (SD) drop in the SD of total normal RR intervals was associated with a hazard ratio of 1.47 (95 % CI: 1.16, 1.86) in the Framingham Heart Study [54]. Together with increased sympathetic tone, it has been related to poor prognosis in several patient populations [55] and an increased risk of sudden cardiac death in patients with myocardial infarction, congestive heart failure, or hypertension [56]. A decreased nighttime HRV was reportedly a strong marker for the development of stroke in apparently healthy subjects [57]. In healthy young adults, reduced HRV indices were independently associated with increased CRP concentrations [58]. Understanding how the natural environment influences HRV is thus important, both on earth and in extraterrestrial space, where crews are spending ever extending journeys on the International Space Station.

In Stockholm, Sweden, the 24-hour SDs of hourly estimates of heart rate were obtained by Holter monitoring from 50 men who had had an acute myocardial infarction or had angina pectoris. Compared with the 24-hour SD of heart rate from 50 clinically healthy men of similar age, it was 20.5 % lower in the patients ($P=0.002$) [59]. Not only is HRV decreased in the presence of ischemic heart disease, a low HRV is also a predictor of cardiovascular disease risk. Indeed, in Tokyo, Japan, the heart rate of 121 normotensive subjects and 176 treated hypertensive patients was measured around the clock by ambulatory monitoring for 2 days [60]. At the time of monitoring, all 297 participants were free of heart disease. The incidence of ischemic heart disease was recorded prospectively for 6 years. As compared to participants free of adverse outcome, patients who had developed coronary artery disease within the 6 years of follow-up have a 20.0 % lower HRV ($P=0.04$). Results from these two studies were aligned with those of a retrospective analysis of archived data on all crews of the Soyuz spacecraft for 1990–1994. Based on the ECG records from 47 male and 2 female cosmonauts, the SD of heart rate was 30 % lower in cosmonauts examined during a magnetic storm, as compared to cosmonauts monitored on quiet days ($P=0.041$) [61], suggesting that exposure to geomagnetic disturbances may increase cardiovascular disease risk (Fig. 13.4).

A reduced HRV on days of high magnetic activity vs. quiet days has also been documented in the longitudinal electrocardiographic (ECG) record of a clinically healthy man, both in terms of the coefficients of variation of R-R intervals and of the total spectral power [62]. In the frequency domain, the decrease in HRV was statistically significant in the very-low-frequency region (periods in the range of 6.5–333 s) but not in the low-frequency (periods of about 10.5 s) or high-frequency (periods of about 3.6 s) region, suggesting an underlying physiological mechanism other than the parasympathetic system as responsible for changes in HRV in response to magnetic activity [62, 63].

To follow up on these results, effects of geomagnetic disturbance on HRV were assessed in adults living at high latitude, in the auroral oval, where magnetic storms are more frequent and more intense than at lower latitudes. Nearly continuous, 7-day beat-to-beat ECG records were obtained between December 1998 and November 2000, on clinically healthy men ($N=15$) and women ($N=4$) from Finnmark College in Alta, Norway (69°56′ N, 23°22′ E). Their age ranged from 21 to 59 years. A geomagnetic record was obtained from the Auroral Observatory of the University of Tromsø, in Tromsø, Norway (69°39′ N, 18°56′ E), including 1-min data of total intensity (F, in nT), declination (D, angle between geographic and magnetic north,

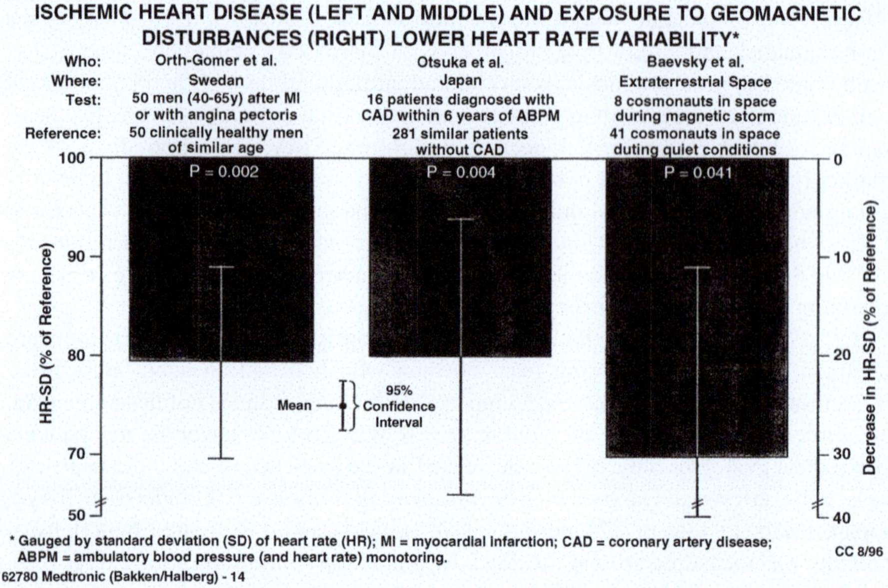

Fig. 13.4 HRV is decreased in patients with ischemic heart disease (*left*). A lower HRV is a predictor of coronary artery disease (*center*). By comparison to an about 20 % decrease in HRV in these two patient populations, HRV is about 30 % lower in healthy cosmonauts who are monitored during a magnetic storm as compared to those studied during magnetic quiet (© Halberg Chronobiology Center)

in degrees), inclination (I, angle between horizontal plane and magnetic direction, in degrees), horizontal intensity (H in nT), and vertical intensity (Z in nT). A decrease in HRV (P=0.002) and an increase in the 24-hour average of heart rate (P=0.020) were documented on days of high geomagnetic disturbance [64] (Fig. 13.5).

As expected, the decrease in spectral power was found primarily at frequencies lower than 0.04 Hz and was not statistically significant around one cycle in 3.6 s. The decrease in HRV was more pronounced in the VLF (21.9 % decrease, P<0.001) than in the ULF (15.5 % decrease, P=0.009) region of the spectrum [64–67]. This effect was more pronounced in spring and autumn, in the presence of a daily light-dark cycle, than in summer (continuous light) or in winter (continuous darkness) [64, 68]. This finding is clinically important because a reduction in the VLF range is associated with an increased risk of cardiac and dysrhythmic death [64]. Several mechanisms have been proposed for the VLF component, namely, thermoregulation [69], the renin-angiotensin-aldosterone system, and other humoral factors [70].

Investigation of 7-day ECG records from five clinically healthy young men living above the arctic circle compared HRV measures assessed over separate 24-hour spans among days of low, middle, and high geomagnetic activity, defined as days with values of the geomagnetic index ap<7, 7–20, and 20–45, respectively. A graded response was observed, the extent of HRV decrease depending on the degree of geomagnetic activity. Specifically, as compared to days when ap<7, TF was decreased

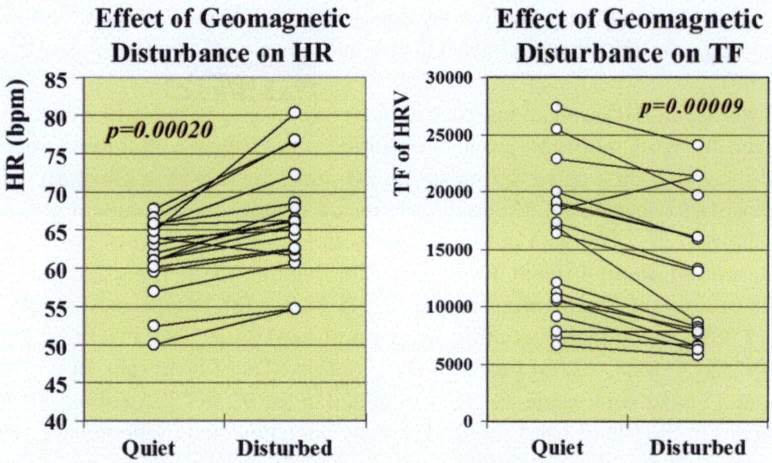

Fig. 13.5 HRV is decreased and heart rate is increased on magnetically disturbed days as compared to quiet days in clinically healthy subjects studied above the arctic circle

by 18.1 % and 31.6 %, ULF by 18.1 % and 27.5 %, and VLF by 12.9 % and 28.6 % on days when 7<ap<20 and 20<ap<45, respectively [64, 71–73]. A graded decrease in HRV to geomagnetic activity suggests the existence of human magnetoreceptors. Phillips et al. [74] proposed a light-dependent magnetoreception mechanism and established a link between magnetic field sensitivity and the visual system in eastern red-spotted newts. The results suggest the operation of a light-darkness-alternation-influenced, if not light-dependent, magnetoreception mechanism in humans.

The existence of such a mechanism is supported by results from cross-spectral analyses. Group-averaged coherences at frequencies of 0.001–0.0001 Hz (0.25–5 hour) differed with statistical significance (P<0.001) among the three lighting conditions, being on the average smaller during long days or long nights than during spring or autumn. Intermittent fields and components in the ultralow-frequency band of the electromagnetic field have been considered to be more bioeffective than the predominantly sinusoidal power line fields in the extremely low-frequency band. The ultralow-frequencies of the electromagnetic field are composed of six kinds of continuous pulsations, including Pc 1 (period range: 0.2–5 s), Pc 2 (5–10 s), Pc 3 (10–45 s), Pc 4 (45–150 s), Pc 5 (150–600 s), and Pc 6 (10 min–5 hours). The cross-spectral analysis results thus suggest that the magnetoreception mechanism may involve the Pc 6 geomagnetic pulsations [64].

Associations between geomagnetic storms and HRV were observed not only on the day of a storm but also for 2 days immediately following a geomagnetically disturbed day. Whereas changes in HR, NN interval, r–MSSD, and pNN50 were found to be statistically significant only during the day of the storm, changes in SDNNIDX (30 min) and in the power-law scaling of the power spectra remained statistically significant for 1 or 2 days following the day with a geomagnetic storm [64]. The 1/f slope was statistically significantly steeper not only on the day of the storm but also on the 2 days following a geomagnetic storm ($P<0.001$), suggesting that geomagnetic disturbances have a lasting influence on the fractal scaling of HRV (Fig. 13.6).

A decrease in HRV and an increase in heart rate in association with solar activity, as gauged by Wolf numbers, are also observed in a 26-year longitudinal record of around-the-clock data collected mostly at 30-min intervals by a clinically healthy man (Figs. 13.7 and 13.8). Magnetic storms are usually more frequent at times of high solar activity. The data are modeled by the least-squares fit of cosine curves with periods of about 1.0 and 10.5 years. Nonlinearly, the periods and their 95 % confidence limits (CI) are estimated as 1.02 [95 %CI: 1.00, 1.04] and 10.60 [95 %CI: 9.21, 11.98] years for the monthly means and as 0.98 [95 %CI: 0.96, 1.00] and 10.44 [95 %CI: 8.85, 12.03] years for the monthly SDs of heart rate. In both cases, the about-11-year solar cycle length is included in the 95 % CI. An about-10.5-year component in the SD of heart rate had already been detected in the first 11 years of this record [63].

The mapping of chronomes, and particularly of cycles shared between physiology and the environment, may thus prove to be a valuable approach to address the goals specified at the beginning of this chapter. For this approach to be successful, reference values from womb to tomb are essential, a project started by the International Chronome Ecology of HRV (ICEHRV) Working Group [64]. For this purpose, around-the-clock ambulatory ECG records have already been obtained from infants (25 days–3months of age), children (3–9 years of age), boys and girls (10–14 years of age), pubertal boys and girls (15–20 years of age), adults of both genders (2–92 years of age), and pregnant women in Japan. In addition, 7-day or

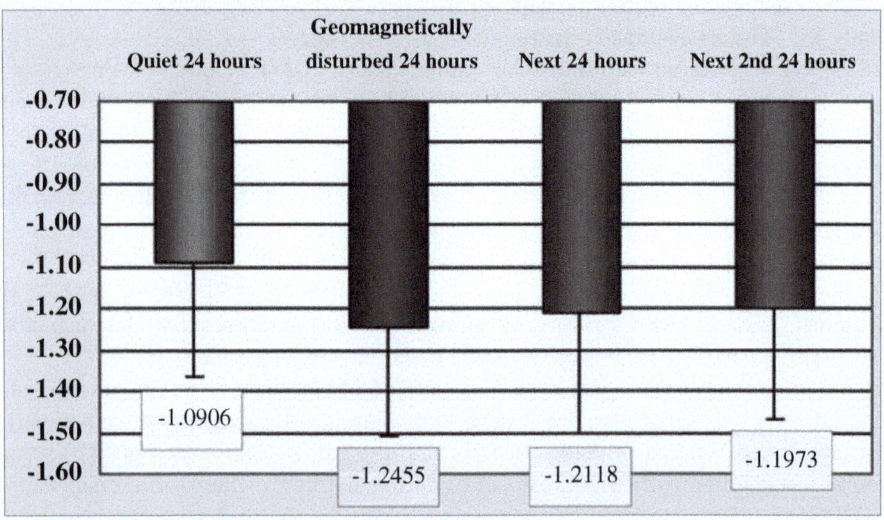

Fig. 13.6 The power-law scaling of the power spectra of HRV remains statistically significant for 1 or 2 days following the day with a geomagnetic storm

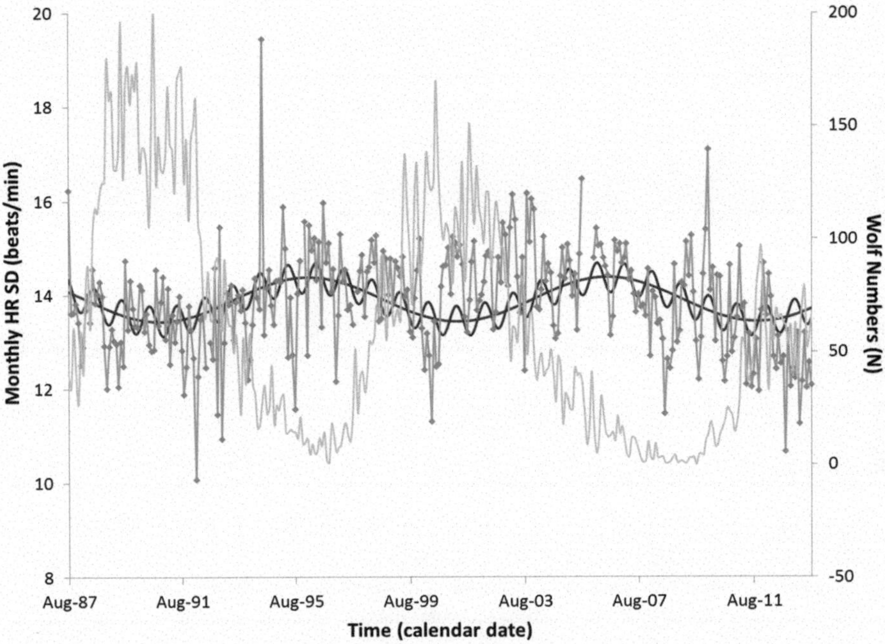

Fig. 13.7 Monthly standard deviations (SD) of heart rate (HR) measured around the clock mostly at 30-min intervals by a clinically healthy man for 26 years are characterized by a prominent about-10.5-year cycle, here shown with the circannual variation. At times of high solar activity, gauged by Wolf numbers (*gray curve*), the HR-SD is decreased (Data from Y Watanabe. © Halberg Chronobiology Center)

longer ECG records were obtained from clinically healthy adults in different geographic locations. Similar records from patients with coronary artery disease and with dilated cardiomyopathy have already found marked differences in their time structure [65, 75, 76].

13.3 Solar and Geomagnetic Activity and Blood Pressure Variability

The influence of space weather on blood pressure (BP) was noticed in a record of weekly measurements taken mostly on Sunday mornings by a 71-year-old psychiatrist [77]. His data were characterized by a large-amplitude circannual variation peaking in the summer. Although some studies reported higher BP values in summer than in winter in elderly populations [78], BP is usually higher in winter than in summer [79–81]. It seems likely, however, that circannual changes in BP are partly endogenous, even if they are statistically significantly correlated with weather and temperature [82, 83] or wind [79].

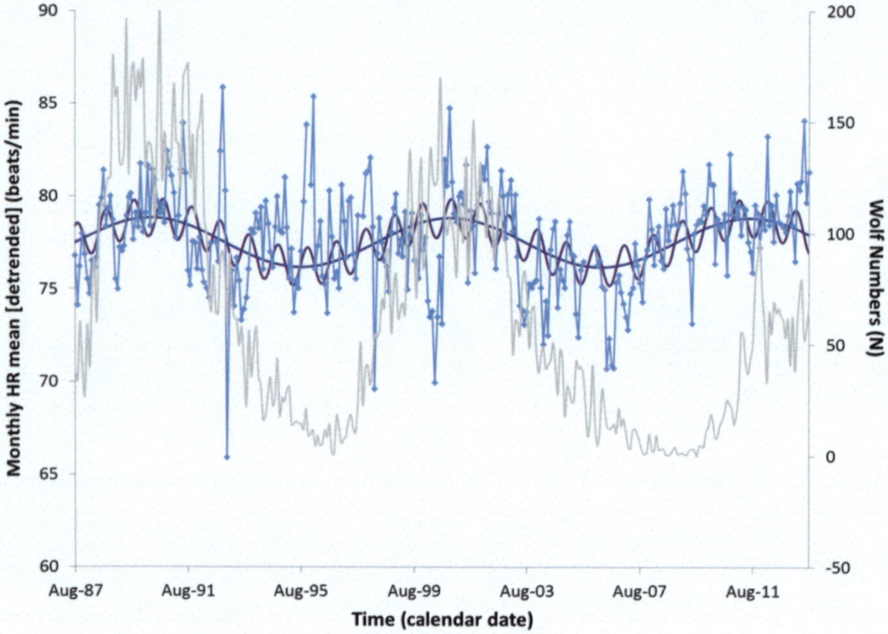

Fig. 13.8 Monthly averages of heart rate (HR) measured around the clock mostly at 30-min intervals by a clinically healthy man for 26 years are characterized by a prominent about-10.5-year cycle, here shown with the circannual variation. At times of high solar activity, gauged by Wolf numbers (*gray curve*), the HR is increased (Data linearly detrended from Y Watanabe. © Halberg Chronobiology Center)

Circannual changes characterize not only BP but also a variety of hormones and their receptors [84–87], electrophysiological and other cardiovascular variables, as well as morbidity and mortality statistics [88–93]. For instance, changes in ventricular cardiac function, both during systole and diastole, exhibit a circannual rhythm in clinically healthy normotensive subjects [84]. The renin-angiotensin system may underlie circannual changes in BP in view of the fact that a very prominent circannual variation characterizes the response of renin release to calcium deprivation in isolated rat glomeruli [85, 86]. The surge in late spring coincides with a time of changes in the human circannual rhythm of circulating aldosterone [85]. A circannual rhythm in circulating aldosterone is a major classifier of risk of developing high BP and/or related cardiovascular disease [85]. Noteworthy is the observation that the circannual MESOR of circulating aldosterone is higher in women at low familial and personal cardiovascular disease risk, as compared to women at high risk [85]. This result is contrary to the view that as an aldosterone-secreting tumor of the adrenal cortex is associated with a high BP, a high cardiovascular disease risk should also be associated with elevated circulating aldosterone [77, 94]. The chronobiologic result in health is in keeping with the report of higher plasma aldosterone concentrations in children of parents with vs. without hypertension [95].

Melatonin may also be involved in the coordination of BP as an endogenous central hypotensive factor [96]. Melatonin administration in the evening has been associated with a decrease in BP [97]. Urinary melatonin excretion was also positively correlated with cardiovascular disease risk assessed by questionnaire [98]. Melatonin has been implicated in a variety of cardiovascular pathophysiological processes, including anti-inflammatory, antioxidant, antihypertensive, and antilipidemic functions [99]. In the circulation, the concentration of melatonin follows a circadian rhythm, with high values at night, thus providing time clues to target tissues endowed with melatonin receptors [99]. It has indeed been suggested that cardiovascular effects of melatonin are at least partly mediated via MT1 and MT2 receptors [100]. It was found that MT1 was expressed in healthy and diseased human coronary arteries and that MT2 was present in human arteries and left ventricles, its expression being altered in coronary heart disease [101]. Reportedly, melatonin is also able to synchronize neuronal firing in the suprachiasmatic nuclei (SCN) via membrane receptors, and clock gene expression is apparently sensitive to melatonin in the pars tuberalis [102].

The pineal being a sensitive receptor that can respond to variations of the order of a few nanoteslas (nT) induced by space weather [103–109] may also be involved indirectly in relation to BP changes and their related cardiovascular complications. In Alta, Norway (70° N), changes in geomagnetic activity above 80 nT/3 hours were associated with a reduced circadian amplitude of salivary melatonin [109]. This result is in keeping with a reduction in a melatonin metabolite excretion in the presence of increased geomagnetic activity, which was greater when the ambient light exposure was also reduced [108]. In this study as well, decreased melatonin secretion was observed when geomagnetic activity exceeded a threshold value of 30 nT. Further evidence in the experimental laboratory supports an effect of magnetic storms on heart function. Indeed, the ultrastructure of cardiomyocytes of rabbits was found to be drastically altered at a time coinciding with strong magnetic storms by comparison with that usually observed during quiet geomagnetic conditions [110].

In addition to the circannual rhythm in BP of the 71-year-old psychiatrist, an about-11-year variation similar to the solar activity cycle was resolved nonlinearly in his 26-year record spanning from 1969 to 1994 [111]. The fit of a model containing both a solar cycle and a 1-year cosine curve was statistically significant for solar cycles 20 (11.6 years), 21 (10.3 years), and 22 (7 years) covered by his record. The solar cycle component was statistically significant during each solar cycle for diastolic BP and during the last two solar cycles for systolic BP, acrophases occurring consistently approximately 2 years after maximal solar activity [111]. These results were compared to about-11-year and about-1-year changes in BP already reported in two other cases by the mid-1990s, one a clinically healthy man, the other an elderly man treated for MESOR-hypertension [111].

Since then, the record of self-measurements from the clinically healthy man was extended to span over four decades, and longitudinal time series from six other subjects covering 10 years or longer became available for analysis. These seven biomedical scientists included five physicians, four of them treated for high

BP. While two of the subjects provided self-measurements, around-the-clock measurements were obtained by ambulatory monitoring from the other five researchers. One more subject contributed data on urinary excretion of 17-ketosteroids for 15 years [112, 113]. Nonlinear analyses of the original data and of the circadian (or in one case, circaseptan) MESORs and amplitudes revealed the presence of cycles with periods clustering around 5 and 10 years [114] (Fig. 13.9). These results strongly suggest that human physiology may be influenced by nonphotic environmental cycles related to solar activity.

Additional evidence stems from the detection in longitudinal BP and HR records of other nonphotic cycles featured in terrestrial-space weather. One such cycle is the transyear detected in solar wind speed [115], which has a period of about 1.3 years. As nonphotic cycles are wobbly, their periods are not stable [116]. While transyears can have periods in a relatively wise range, they are statistically significantly longer than one year. An about-1.3-year transyear was invariably detected in all 43 time series of BP and HR from 13 subjects [117, 118] (Fig. 13.10). When a gliding spectrum was obtained for the SBP record of an elderly man and compared to that of solar wind speed, similarities in the time course of the transyear could be followed between these two variables [119]. Results showed that the transyear in solar wind

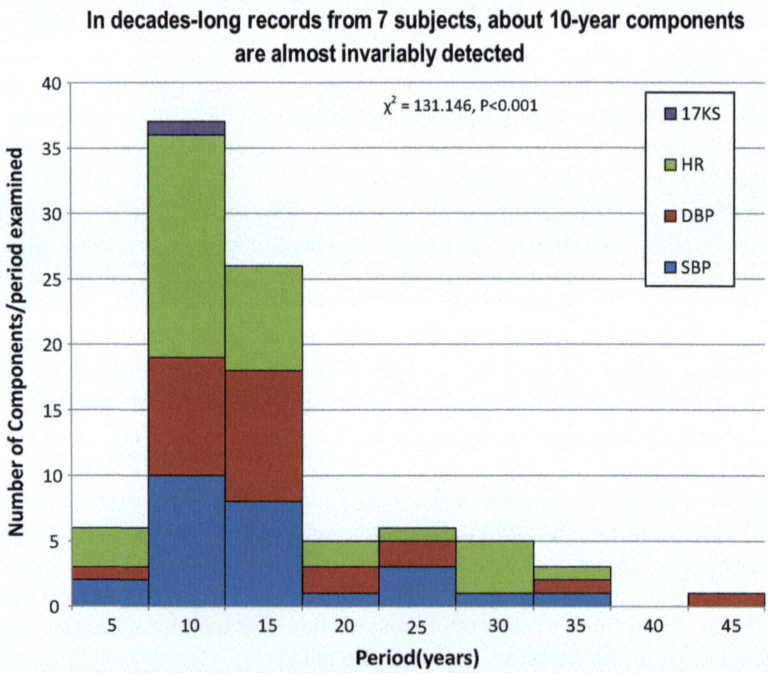

Fig. 13.9 Histogram of periods resolved nonlinearly in records of systolic (S) and diastolic (D) blood pressure (BP) and heart rate (HR) covering decades. One additional subject provided a 15-year series of urinary excretion of 17-ketosteroids. Note cluster around 10 years (© Halberg Chronobiology Center)

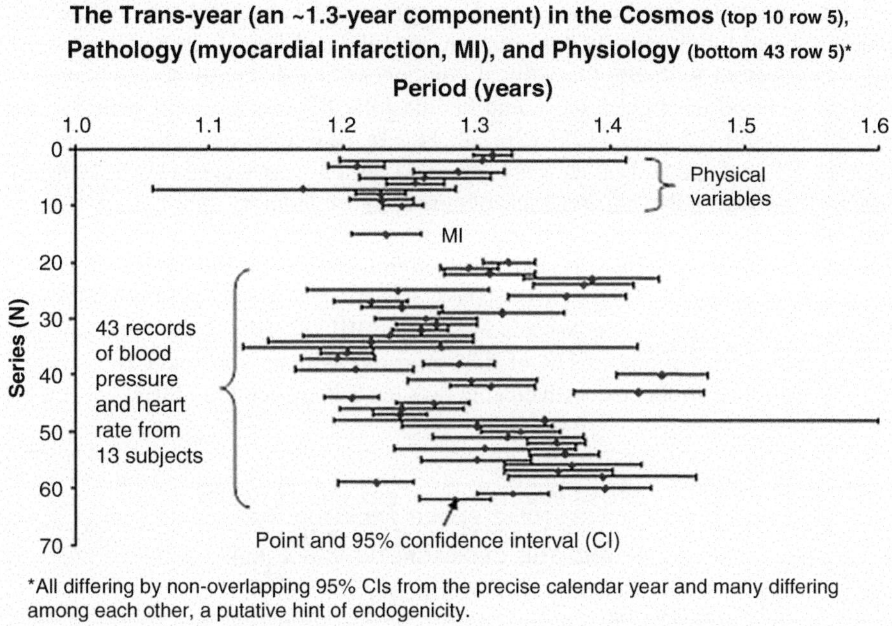

Fig. 13.10 A transyear with a period of about 1.3 years is invariably detected in all 43 time series spanning 10 years or longer, provided by 13 subjects who monitored themselves around the clock by self-measurement or ambulatory monitoring. Periods estimated nonlinearly with their 95 % confidence intervals (© Halberg Chronobiology Center)

speed was expressed prominently between 1989 and 1996, after which it splits into two separate spectral bands until 1997 when the signal fades away and is no longer detectable. In SBP, the transyear remains prominently expressed until 1998, after which its amplitude is decreased but remains detectable until the end of the record in 2005, suggesting that this component, while still resonating with its environment, may also be partly genetically anchored [119].

Transyears have also been documented in transverse data series from neonates, where they are more prominent than the circannual component [120]. Whereas the about-yearly rhythm predominates over transyears in adulthood on the average, transyears regain prominence in the elderly, as gauged by amplitude ratios of these two cycles [120]. A similar time course as a function of age was also found for the about-weekly component in BP and HR [121–123]. The circaseptan period of neonatal BP and HR measured around the clock for the first few weeks after birth, estimated by nonlinear least squares, correlated positively with the corresponding period of the local index of geomagnetic disturbance (K) analyzed during spans matching each neonatal record [124]. These hybrid neonatal data, obtained around the clock for several days from 32 babies monitored during the first month of life, also showed cross-spectral coherence (>0.70) with K at periods of about 3.3 and 8.8 days [24].

Cross-spectral coherence was also documented in a baby boy born prematurely and monitored longitudinally between Jan 26, 1989 and Jan 22, 1990. His systolic BP was coherent with Bz, the vertical component of the interplanetary magnetic field, at a period of about 5 days, and his diastolic BP was coherent with Bz at a period of about 3.5 days [125] (Fig. 13.11). In adulthood as well, associations between BP and space weather were observed. In a 3-year record (1988–1990) of around-the-clock BP measurements obtained by ambulatory monitoring from a clinically healthy man, cross-spectral coherence with the planetary index of geomagnetic disturbance (Kp) was statistically significant for systolic and diastolic BP at periods of about 27.7 days (SBP: 0.567, P=0.010; DBP: 0.516, P=0.025) and 4.2 days [126] (Fig. 13.12). Associations of the BP and HR of another clinically healthy man with Kp and of HR of a 28-year-old woman who spent 267 days in social isolation with Kp and with cosmic rays have also been found [63].

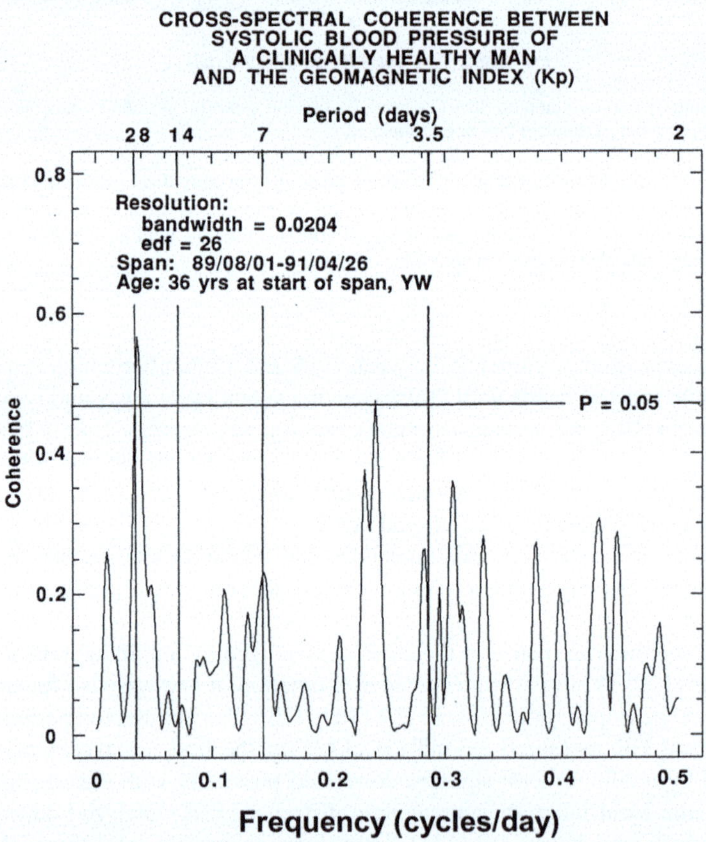

Fig. 13.11 Cross-spectral coherence between diastolic blood pressure of a baby boy born prematurely and Bz, the vertical component of the interplanetary magnetic field, at periods of about 3.5 days (3.56 days; P=0.008), 2.9, and 5.0 days (© Halberg Chronobiology Center)

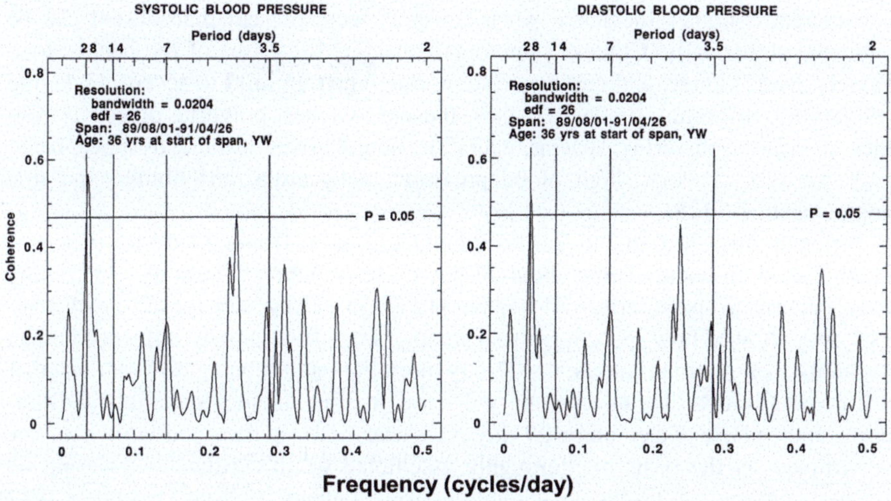

Fig. 13.12 Cross-spectral coherence between systolic (*left*) and diastolic (*right*) blood pressure of a clinically healthy man and Kp, the planetary index of geomagnetic disturbance, at periods of about 27.7 and 4.2 days (© Halberg Chronobiology Center)

Effects of space weather were also observed in the experimental laboratory. The circadian rhythm of systolic and mean intracardiac pressure measured around the clock in the left and right ventricles of male Chinchilla rabbits was found to be markedly altered in September 1984 when the study coincided with the occurrence of two strong geomagnetic storms, as compared to three other studies carried out between June 1984 and March 1985, when quiet geomagnetic conditions prevailed [110]. Considering the involvement of melatonin in relation to blood pressure, it may also be pertinent that in rats sampled around the clock for 7 days, alterations in the circadian rhythm of pineal and hypothalamic melatonin were found following the second extremum of a moderate double magnetic storm [127]. As compared to quiet conditions, during the storm, the MESOR and circadian amplitude were reduced in the case of pineal melatonin and increased in the case of hypothalamic melatonin [127].

13.4 Solar and Geomagnetic Activity and Cardiovascular Outcomes

Already by 1922, there was evidence for an adverse effect of space weather on human health. Two physicians, Sardou and Faure, and Vallot, an astronomer, meteorologist, and generous supporter of research, then published their finding that severe and fatal illnesses as well as common head cold and other aches were more frequent when there were many sunspots [128]. Whether all or only severe symptoms were considered, a higher incidence when sunspots were present than absent can be validated statistically. Clinical symptoms considered diseases of the heart, blood vessels, liver, kidney, and nervous system, ranging from mild to severe, including excitability, insomnia, tiredness, aches, muscle twitches, polyuria, digestive troubles, jitteriness, shivering, spasms, neuralgia, neural crises, asthma, dyspnea, fever, pain, vertigo, syncope, high blood pressure, tachycardia, arrhythmia, and true angina pectoris [128].

Much credit is also due to Bernhard and Gertraud ("Traute") Düll for their pioneering work in documenting about-27-day cycles in human mortality and relating these patterns to space-terrestrial weather [129–135]. Using superposed epoch analysis, they showed how space/terrestrial weather may differentially influence human mortality from various causes, the 27-day mortality pattern being different whether death was from cardiac or respiratory disease or from suicide. Specifically, differences in the phase of a common 27-day cycle characterized these mortality patterns, determined on the basis of thoroughly assembled death certificates covering 68 stacked consecutive 27-day solar rotations from January 1928 to December 1932 (5 years) [136].

An association between the incidence of various diseases, myocardial infarctions and strokes in particular, and geomagnetic storms and other electromagnetic variations has long been claimed by scientists in the former Soviet Union [137–139]. As reviewed by Halberg et al. [140], results of direct experiments in which biological specimens are exposed to artificially induced electromagnetic fields support studies of correlation between geomagnetic activity and human morbidity and mortality [141]. Electromagnetic fields have been reported to affect blood pressure [142], baroreflex function [143], and HRV [144]. Effects of geomagnetic disturbances have also been reported on the capillary blood flow of patients with ischemic heart disease [145].

Such effects upon the cardiovascular system may contribute to observed associations between space/terrestrial weather and the incidence of adverse events, such as myocardial infarction. By superposed epochs, a consistent association of excessive geomagnetic activity with cardiovascular disease and myocardial infarctions particularly was documented in data from the 1960s to 1970s in two geographic locations [146]. Despite other reports of increased incidence of heart attacks in the presence of increased geomagnetic disturbances [147–155], not all studies found such associations [156, 157]. In particular, two large studies in the USA, carried out with impeccable inferential statistics, reported no solar effect in their investigation of associations of solar activity and magnetic disturbances with mortality [158, 159]. While there is no good explanation thus far for the controversy, mostly because underlying mechanisms are not well understood, several possible contributing factors have been proposed. They include the usually higher level of artificial magnetic disturbances from industry and domestic appliances in the USA and the fact that mortality from myocardial infarction does not necessarily coincide with the onset of symptoms, should they have been triggered by the magnetic storm [147]. Not following a sequence of several Horrebow-Schwabe (about 11-year) cycles and studying their variability may be another reason [1]. Lipa et al. [159] cautiously considered the possibility that any causal relationship between geomagnetic disturbances and coronary heart disease may be sensitive to geographic location, to the phase of the solar cycle, or to some other parameter which might distinguish their data from those of Gnevyshev and Novikova [141]. Distinguishing between mortality and morbidity data has also been recommended, as an about 10.5 % increase in mortality from myocardial infarction in association with big geomagnetic storms contrasted with a lack of effect on the incidence of morbid events in one study [152].

In a database of over 6.3 million cases requiring the call for an ambulance during the span from Jan 1, 1979 to Dec 31, 1981 in Moscow, Russia, including 85,819 diagnoses of myocardial infarction, cross-spectral coherence was found in association with the local index K of geomagnetic disturbance and with Bz, the north–south component of the interplanetary magnetic field at the same frequency of one cycle in about 3.15 days [125, 160] (Fig. 13.13). A north-to-south Bz turn triggers aurorae in the upper atmosphere and storms in the magnetosphere, the total current in the magnetosphere increasing up to ten times within a few minutes [161]. Defining a storm as a north-to-south daily mean changing from $\geq$1nT to$<-$1.5nT, there were 32 events during the 3-year span. Superposed epochs analysis indicated that the daily incidence of myocardial infarctions increased by 7.6 % after a Bz turn (P=0.027), followed, however, by a 5.9 % decrease on the following day (P=0.029) (Fig. 13.14). This result was consistently observed during each year considered separately [162]. Using a different method, the result was independently corroborated

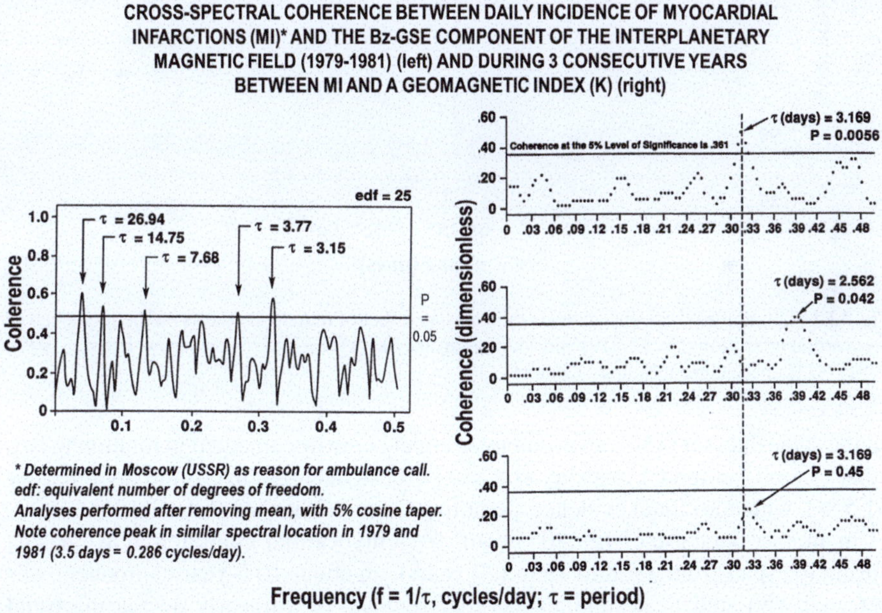

Fig. 13.13 Cross-spectral coherence between the daily incidence of myocardial infarctions in Moscow, Russia (Jan 1, 1979–Dec 31, 1981), and Bz (*left*) and local K (*right*) at similar frequency of about one cycle in 3.15 days (© Halberg Chronobiology Center)

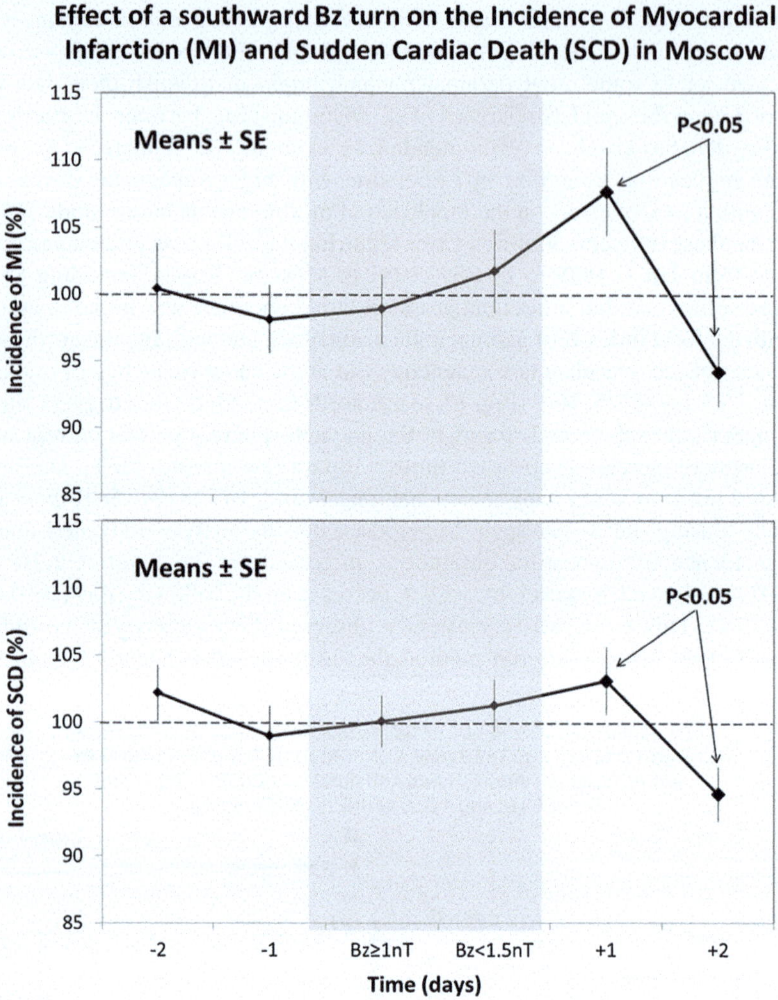

Fig. 13.14 Superposed epochs analysis indicates an effect of magnetic storms on the incidence of myocardial infarction (N = 85,819) and sudden cardiac death (N = 71,525) (Moscow, Russia; 1979–1981) (© Halberg Chronobiology Center)

on the same data set [150], an even larger effect being documented in relation to two other criteria (antipodal index aa > 60 and Forbush decrease in cosmic ray intensity >1.5 %), with additional evidence stemming from another set of data [151].

In another database of 129,205 deaths from myocardial infarction in Minnesota during the span from 1968 to 1996 (29 years), an about-10.5-year component was detected with statistical significance after removal of a linearly decreasing trend (Fig. 13.15). During years of maximal solar activity, there was an excess of 220 deaths per year (about 5 %) as compared to years of minimal solar activity (P = 0.023) [163]. When the data are analyzed separately for each of the three solar cycles,

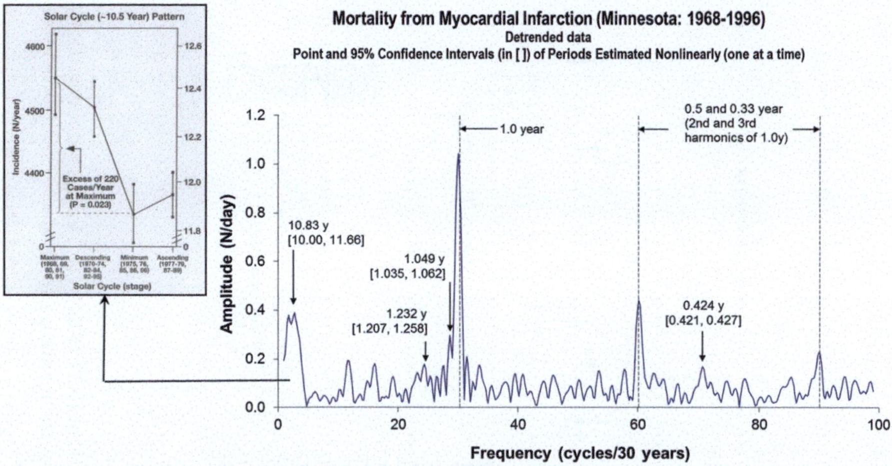

Fig. 13.15 An about-11-year cycle characterizes the daily incidence of mortality from myocardial infarction in Minnesota (1968–1996). In addition to a prominent annual variation, note the presence of a transyear and cis-half year with periods of about 1.3 years and 0.42 year, signatures of solar wind speed and solar flares, respectively (© Halberg Chronobiology Center)

statistically significant differences are found, however, perhaps providing some answer for the different results reported above [140]. Fitting a 10.5-year cosine curve to the detrended data, the model is statistically significant for the spans from 1968 to 1977 ($P=0.006$) and from 1978 to 1986 ($P<0.001$), but not for the span from 1987 to 1996 ($P>0.50$). Whereas acrophases are very similar, differing by only 10° (2.8 % or 3.5 months), amplitudes vary from 131 and 278 deaths/year during the first two cycles to 25 deaths/year during the third cycle. The about-10.5-year cycle can be resolved nonlinearly overall and for the first two cycles, but not for the third, the period and its 95 % confidence interval being estimated as 10.83 [10.00, 11.66] years overall and as 11.48 [3.23, 19.72] and 10.64 [8.60, 13.28] years during the first two cycles, respectively [163]. As seen from Fig. 13.15, mortality from myocardial infarction in Minnesota is also characterized not only by a prominent annual variation, but also by a transyear and a cis-half year with periods of about 1.3 years and 0.42 year. These nonphotic cycles are resolved nonlinearly. They reflect signatures found in solar wind speed [115] and in the solar flare index [164], respectively.

Other diagnoses in the 1979–1981 Moscow database included strokes, hypertensive crises, arrhythmia, and sudden cardiac death, among others [125]. In contrast to the other cardiovascular morbidity statistics, all showing a very prominent weekly pattern, the circaseptan component characterizing sudden cardiac death, although statistically significant, was much less pronounced. Instead, an about-half-monthly component was detected with a period of about 2 weeks in least squares and population-mean cosinor spectra (Fig. 13.16). Nonlinearly, its period is estimated as 15.2 [95 % CI: 15.15–15.30] days. According to the criterion of a nonoverlap of 95 % confidence intervals, this period is statistically significantly different from the

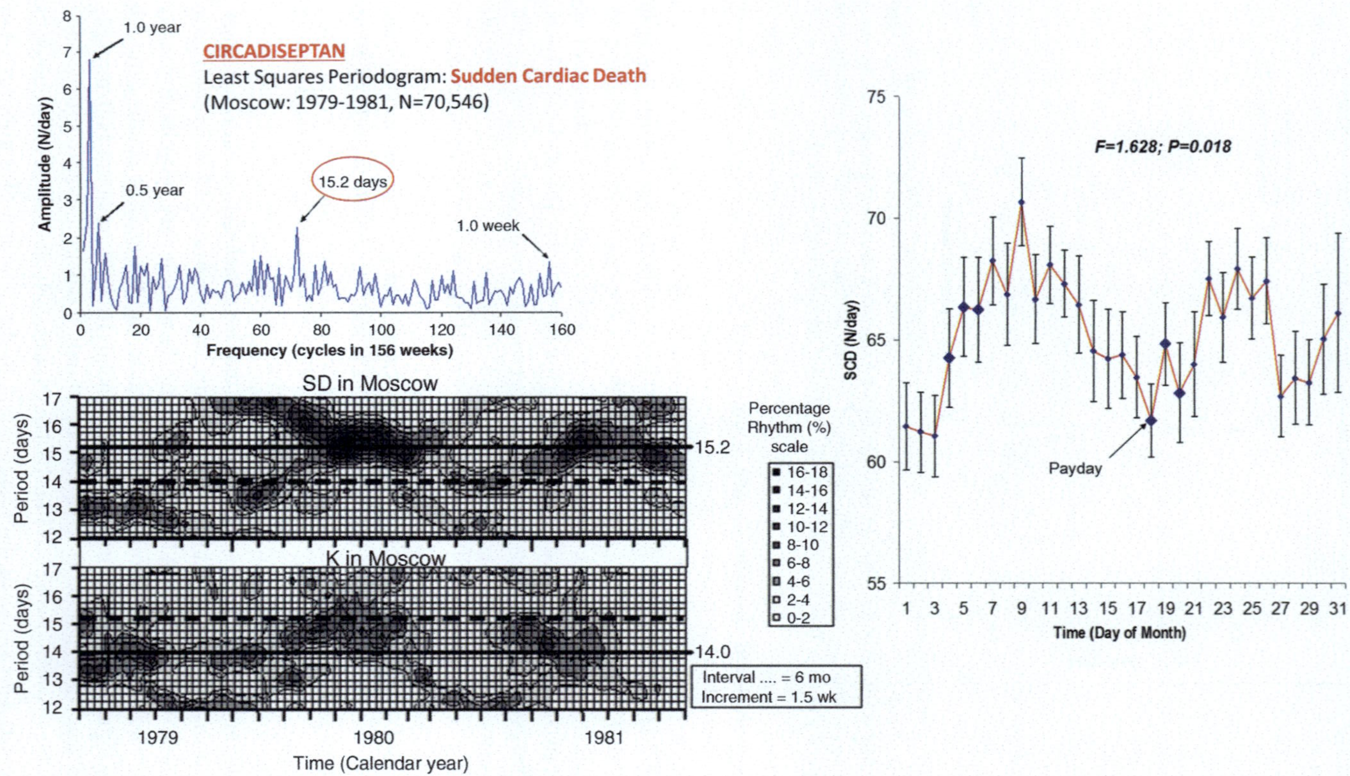

Fig. 13.16 The least-squares spectrum of the daily incidence of sudden cardiac death shows a peak at a period of about 15 days (*top*, *left*), visualized by stacking the data over an idealized double cycle (*right*). Gliding spectral windows (*bottom*, *left*) indicate that this component is expressed more strongly at times when it is also prominent in the local index K of geomagnetic disturbance (© Halberg Chronobiology Center)

half-monthly period characterizing local K analyzed during the same 3-year span. For K, the corresponding period is estimated to be 14.0 [95 % CI: 13.94–14.17] days. Whereas the about-half-monthly period of both sudden cardiac death and local K is estimated with a relatively tight 95 % confidence interval when analyzed over the entire 3-year span, this component's period shows great wobbliness when the data are analyzed by moving spectra over a shorter interval of 3 months, progressively displaced by 1.5 weeks throughout the time series. This component is not consistently detected either for sudden cardiac death or for local K (Fig. 13.16). A possible resonance with occasional frequency trapping between the two variables at this frequency is suggested by the fact that the half-monthly component is seemingly more prominently expressed in sudden cardiac death when it is detected in the spectrum of K [37, 165, 166].

Since the tenth revision of the International Classification of Diseases (ICD10), sudden cardiac death (code I46.1) has been separated from cardiac arrest associated with a myocardial infarction and is defined as unexpected death from cardiac causes within 1–24 hours of the onset of acute symptoms. Daily incidences of sudden cardiac death (ICD10, code I46.1, excluding death from myocardial infarction) in Minnesota during 1999–2003 (5 years) were compared to those of daily incidences of mortality from myocardial infarction during the same span in Minnesota and with those of daily (or monthly) incidences of sudden cardiac death in some other geographic locations.

Whereas the incidence of myocardial infarctions in Minnesota during 1999–2003 is characterized by a prominent about-yearly variation ($P<0.001$) with a 1.009-year period and a 95 % CI [0.965, 1.052] overlapping the precise calendar year, this is not the case for sudden cardiac death, for which a transyear with a period of about 1.3 years ($P=0.014$) yields a better fit (Fig. 13.17). Nonlinearly, its period is estimated to average 1.39 years [95 % CI: 1.17, 1.61 years]. These results resemble those observed in Tokyo, Japan, over a slightly longer span of 8 years (1998–2005; 225 cases) [167] (Fig. 13.17). Whereas a transyear is detected in the absence of a calendar year in Minnesota and Tokyo, Japan, both components are detected in Arkansas and the Czech Republic. A calendar year, in the absence of a transyear, is also detected in North Carolina, Switzerland, and Tbilisi, the Republic of Georgia [119, 166]. Transyears are likely a nonphotic signature of space weather since they have been detected in the geomagnetic disturbance index Ap [168], in the incidence of aurorae [169], in the interplanetary magnetic field [170], in solar wind speed [115, 116], and in helioseismology [171], also leaving their signature in the terrestrial environment [172].

As seen in Fig. 13.17, an about-0.42-year component (a cis-half year) is also resolved nonlinearly in the incidence of sudden cardiac death in Minnesota and Tokyo. In Minnesota, its period is 0.42 [95 %CI: 0.40, 0.44] year, and in Tokyo, it is 0.42 [95 %CI: 0.41, 0.43] year. Albeit of small amplitude, a cis-half year is also detected in the 29-year series of mortality from myocardial infarction in Minnesota (1968–1996; 129,205 deaths), with an estimated period of 0.424 [95 % CI: 0.421, 0.427] year. Cis-half years have been reported to characterize solar flares, physicists estimating the period to be approximately 154 days [164], ranging between about

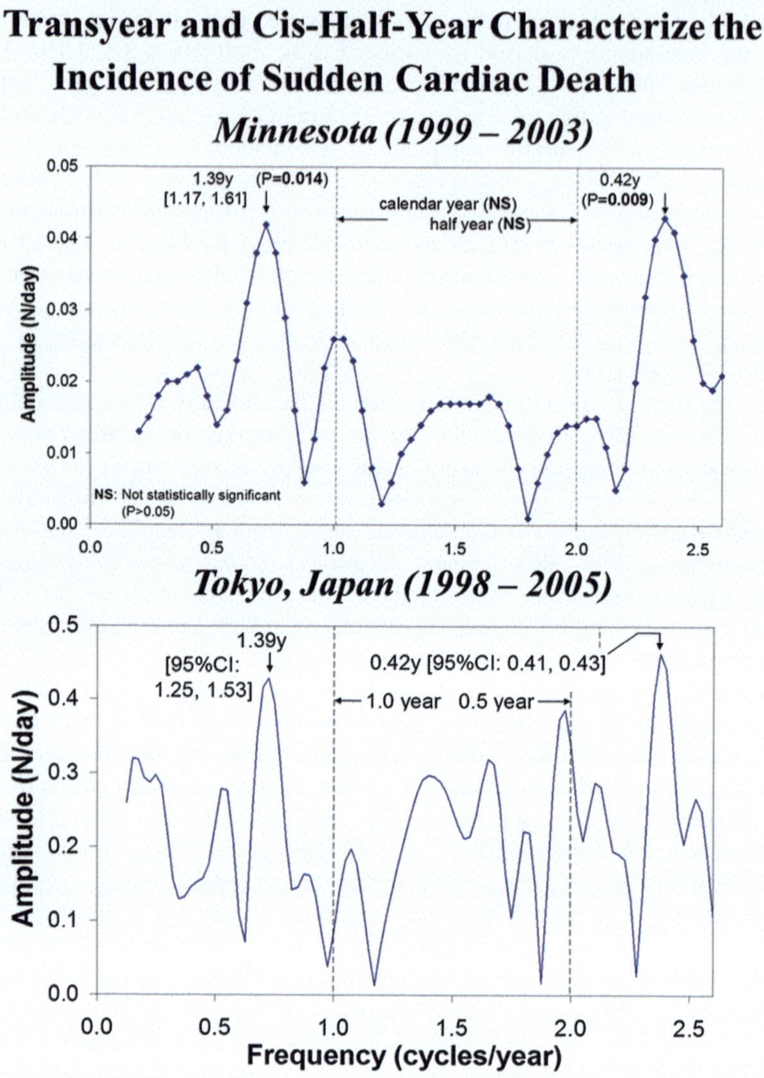

Fig. 13.17 About-1.3-year (transyear) and about-0.42-year (cis-half year) components are detected with statistical significance in the least-squares spectrum of sudden cardiac death in Minnesota (*top*) and Tokyo (*bottom*), but not in all of the other geographic locations investigated (not shown) (© Halberg Chronobiology Center)

150 to 160 days from different studies by different investigators [173]. Our own analysis of the solar flare index for the span from 1968 to 2006 found a double peak around 0.42 year and an additional spectral line around 0.40 year (Fig. 13.18).

Interestingly, a cis-half year is also detected in the longitudinal record of a clinically healthy man spanning four decades, as he aged from about 20–60 years [174]. A multiple-component model consisting of cosine curves with periods of 0.41, 0.50,

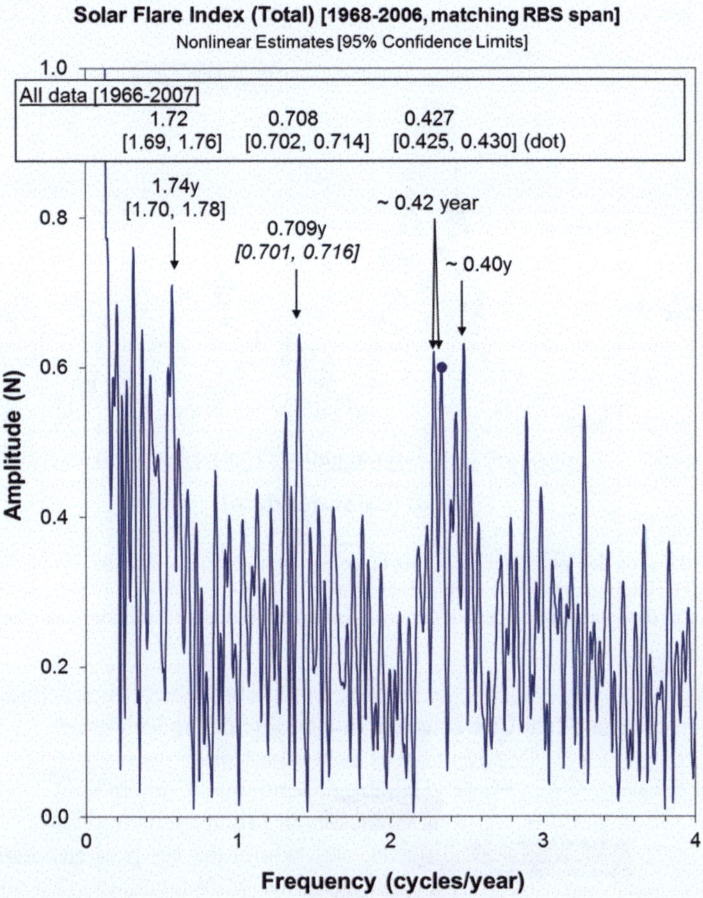

Fig. 13.18 The presence of multiple peaks with periods of about 0.42 year in solar flares may reflect the wobbly nature of nonphotic components (© Halberg Chronobiology Center)

and 1.00 year was fitted to weekly means of heart rate collected about 5–6 times a day over 39 years. As seen from Fig. 13.19, the relative prominence of the cis-half year and its statistical significance change as a function of time in a nonrandom fashion, following the intensity of solar flares [174]. A cis-half year is also detected in serially independent sampling of circulating melatonin measured around the clock for 24 hours in 172 subjects examined between October 1992 and December 1995 in Florence, Italy [175, 176]. An about-0.4-year component was detected and validated by nonlinear least squares for the time series of MESORs and circadian amplitudes, as well as for those of samples collected at each of the six timepoints (at 00:00, 04:00, 08:00, 12:00, 16:00, and 20:00) (Fig. 13.20). The results suggest that melatonin may mediate some of the sun's effects upon the biosphere in certain frequency windows such as the cis-half year of about 5 months.

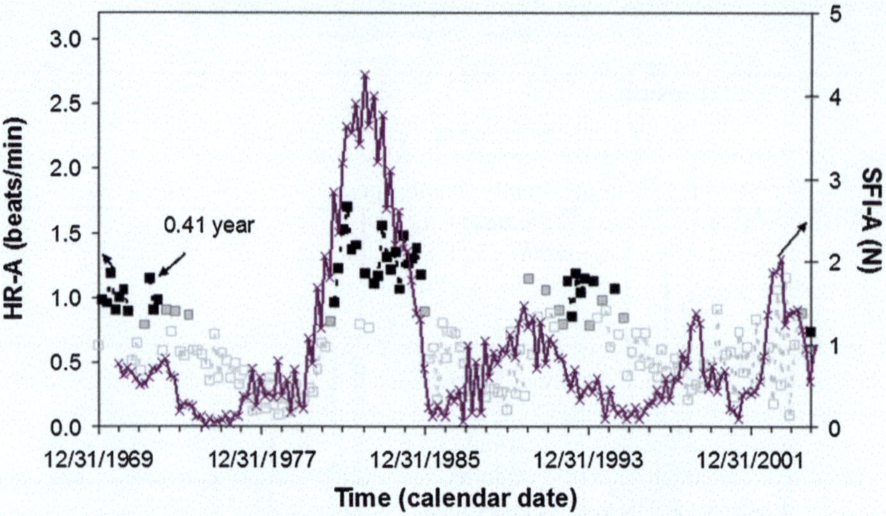

Fig. 13.19 The relative prominence of a 0.41-year component in human heart rate (*dots*, *filled circles* when P<0.05) changes in a nonrandom fashion, following a similar time course as that of a cis-half year in the total solar flare index (*purple curve*) (© Halberg Chronobiology Center)

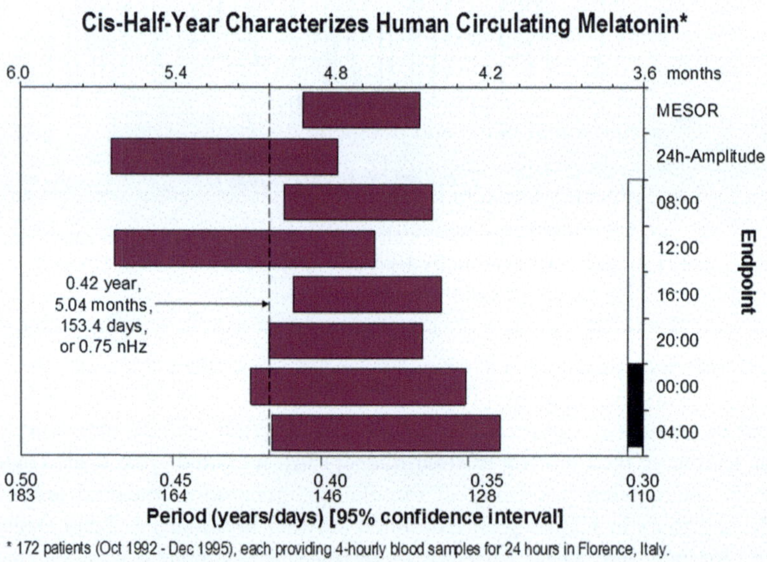

Fig. 13.20 A cis-half year is detected nonlinearly in serially independent MESORs and circadian amplitudes of circulating melatonin and in samples collected at each of six timepoints (00:00, 04:00, 08:00, 12:00, 16:00, and 20:00) from 172 subjects examined between October 1992 and December 1995 (© Halberg Chronobiology Center)

13.5 Conclusion

It has long been known how the photic environment affects circadian and circannual rhythms [2, 3, 177–180]. Circadian and circannual systems may have evolved in biota as mainly photic solar signatures with a clear survival value, documented by the hours of changing resistance. Circadian timing makes the difference between life and death, with the availability of the same amount of food [181] or in response to the same stimulus, whether it is X-irradiation [182], noise [183], ethanol [184], a drug such as ouabain [18] or SU-4885 [185], or bacteriological [182, 186–189]. Early in evolution, circadians are critical and constitute a major component of the time structures of *E. coli* [190] or cyanobacteria [13], a topic of chronoastrobiology inquiring into the origins of life in a phylogenetic perspective [191, 192].

Surprisingly, in some geographic locations, nonphotic features of the sun and space weather can more generally override seasonal effects, even in places such as Minnesota where the contrast between winter and summer is strong. This is, for instance, the case of the incidence of sudden cardiac death during the years 1999–2003 [119, 193]. This is also the case of other conditions, such as the incidence of suicides in Minnesota [194] and in Australia [195].

Results on adverse cardiovascular outcomes presented herein point to two possible underlying mechanisms. On the one hand, an effect on the autonomic nervous system may account for the increased risk of coronary artery disease and a higher incidence of myocardial infarctions in association with a decreased HR variability, notably in vulnerable people who may be more susceptible to the influence of geomagnetic storms, as also suggested by others [196]. On the other hand, changes seen in melatonin point to the possible involvement of the pineal gland. Geomagnetic disturbances may lead to disruptions of hormonal systems, perhaps mediated by the cryptochrome genes, known primarily as key components of the circadian pacemaker [197]. In one study, the administration of melatonin was used for prophylaxis, to prevent the influence of geomagnetic storms in patients with cardiovascular disease [198]. In the experimental laboratory, exposure of rats for 2 days before birth to 7 days after birth to 0.5-Hz pulsed-square wave magnetic fields in the nT range (designed to imitate geomagnetic continuous pulsations), when applied in a specific direction, resulted in a reduction in the number of neurons within the parasolitary nucleus [199]. The authors concluded that cell migration in structures within the brain stem modulating vestibular-related arousal and respiratory or cardiovascular stability may thus be affected. Studies on rabbits also indicate that geomagnetic field disturbances affect brainstem and higher cardiovascular regulatory centers modulating BP and HR variabilities associated with the arterial baroreflex [200]. It is possible that natural geomagnetic fields affect baroreceptor sensory transduction, modulating calcium channel permeability of the baroreceptor membranes, as infusion of verapamil blocked the effect of geomagnetic fields on baroreflex sensitivity [201].

Life on earth has been shaped by the sun. Data from observatories on earth and in space as well as from satellites have considerably broaden our knowledge of solar dynamics occurring on a wide range of timescales, from the 5-min oscillations to the 22-year magnetic cycle and beyond [202, 203]. The earth's magnetic field has also been critical to the evolution of life, as it provides a protective shield against cosmic rays, preventing the atmosphere from being eroded by the solar wind [204]. The proposition that variations on the sun and in the earth's magnetic field can affect human health has important implications, notably for those living at high geomagnetic latitudes where the geomagnetic fluctuations have larger amplitudes [202]. Such latitude dependence has been reported for melatonin [205]. It was also observed for the about-22-year cycle in data on religious motivation available over 50 years from 103 different geographic locations [206] (Fig. 13.21).

Having evolved in the presence of varying natural magnetic and electric fields, it seems inevitable that humans resonate with many of the cycles characterizing terrestrial-space weather. For each validated periodicity with a given frequency, a corresponding component may be sought in the time structure of one or several natural physical environmental variables and, likewise, that for each periodic component in the environment, a near-matching organismic component may be sought [191]. This view then prompts the need to map cycles in broad time structures (chronomes) to resolve standards of life [191, 207]. Selective congruence, namely, a pairing of various biospheric cycles of certain frequencies with different environmental cycles of similar frequencies, characterizes a broad transdisciplinary spectrum. Congruence has been defined as the total or partial overlap of 95 % confidence intervals of periods of anticipated cycles between two (or more) variables [208, 209]. An overall assessment of congruence over a specific spectral region can be defined by the proportion (out of the total) of statistically significant spectral peaks with overlapping and/or overlying 95 % confidence intervals of their periods. As illustrated in Fig. 13.22, in a set of variables self-rated or self-measured for up to 40 years by a clinically healthy man, a test of congruence detects statistically significant associations with solar wind speed and with the antipodal geomagnetic index aa, in mental functions, such as a 1-min time estimation, mood, and vigor and in systolic blood pressure [208, 209]. In the spectral range from one cycle in 2.5 years to three cycles per year, the number of congruences found for the estimation of 1 min and for mood more than equals that of the known association of helio- and geomagnetism.

Other possible mechanisms relate to the Schumann resonance signals. Schumann hypothesized that electromagnetic signals could resonate in the cavity between the earth's surface and the ionosphere [210]. The so-called Schumann resonance signals are electromagnetic resonances of the global earth-ionosphere (quasi) spherical-shell cavity [211]. The Schumann resonance signals may constitute a plausible mechanism for the human health effects of solar and geomagnetic activity because they share the same frequency range as brain waves [212]. Another proposed mechanism implicates wind eddies carrying ionized air that can produce oscillatory signals in the range 3–6 Hz [213]. These could be associated with thunderstorm activity or with winds such as the Fohn wind in Austria and could have localized biological effects that may account for geographic differences. Wind speed as well as fluctuations of atmospheric pressure in the range of 0.01–0.10 Hz showed statistically sig-

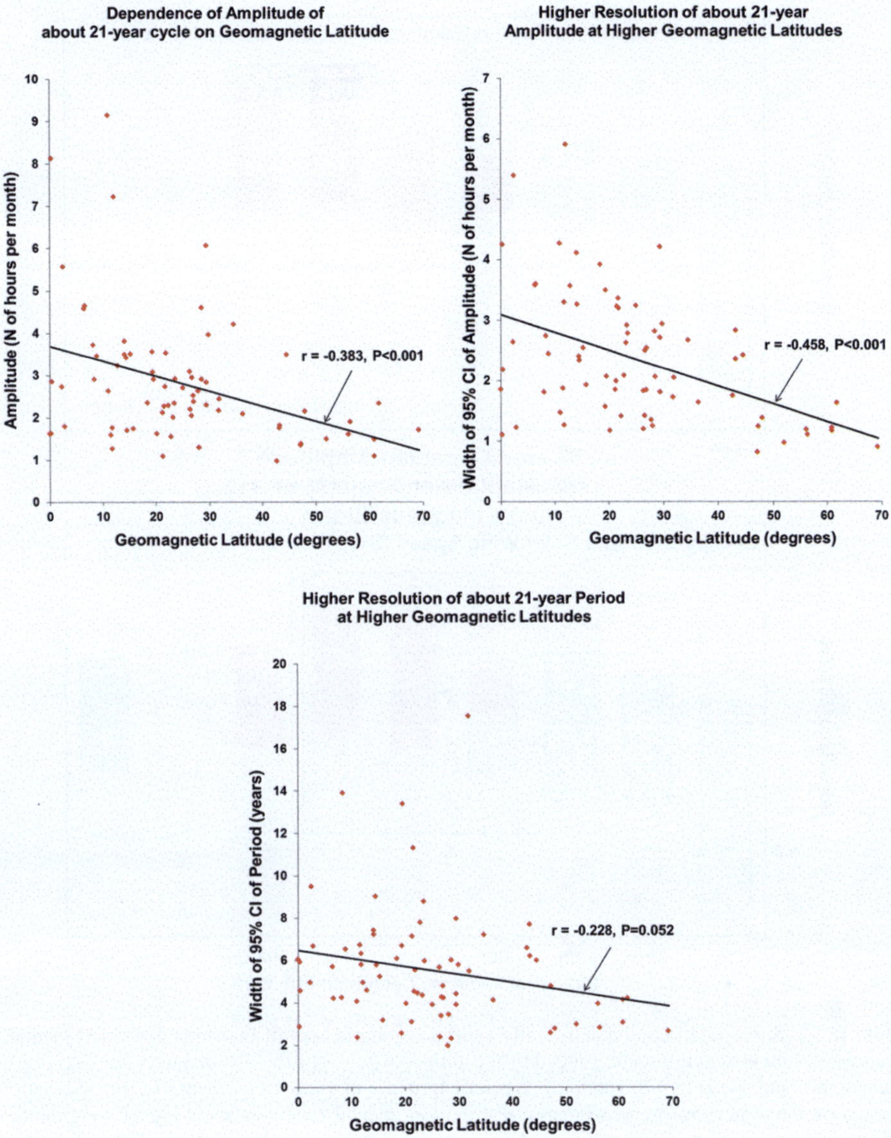

Fig. 13.21 Latitude dependence of amplitude of about-21-year cycle characterizing religious motivation recorded in 103 different geographic locations (record length sufficient for assessment of Hale cycle in 73 sites reported herein) (© Halberg Chronobiology Center)

nificant cross-spectral coherence (0.70–0.98) with corresponding oscillations of heart rate variability in a 50-day around-the-clock electrocardiographic record from a clinically healthy man [214, 215].

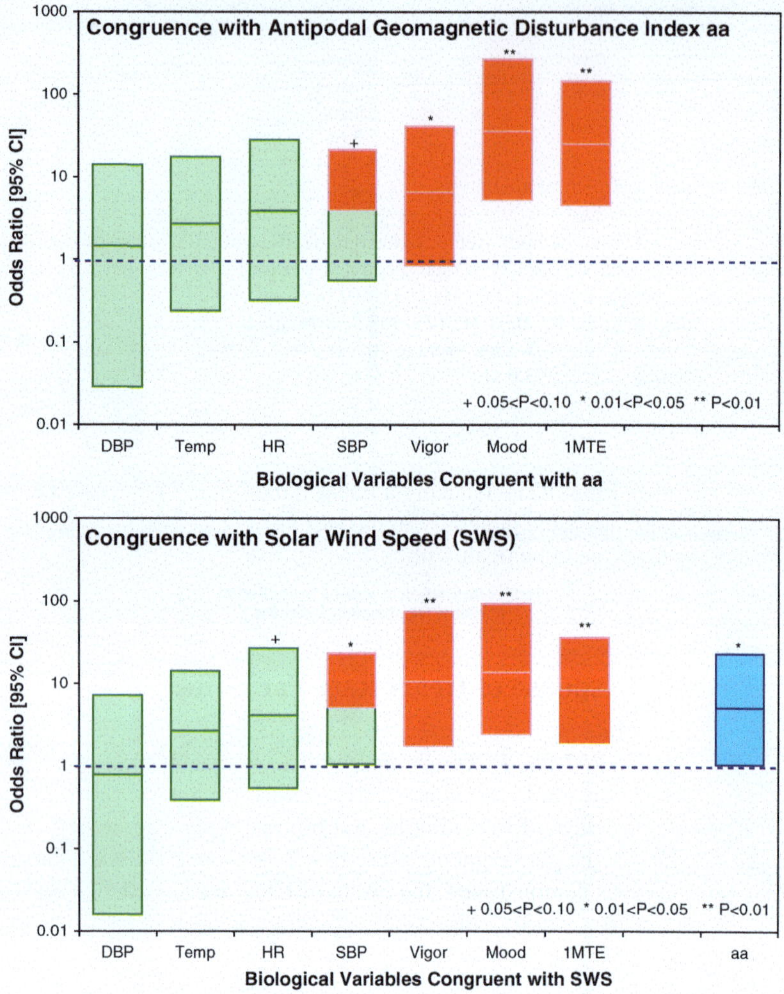

Fig. 13.22 Some inferentially statistically validated congruences of human somatic and mental functions with the environment, gauged by the antipodal geomagnetic disturbance value (aa) (*top*) and solar wind speed (SWS) (*bottom*). When congruence is assessed by means of the odds ratio based on the noncentral hypergeometric distribution, mental functions show higher congruence than somatic functions. *DBP*: diastolic blood pressure, *Temp*: oral temperature, *HR*: heart rate, *SBP*: systolic blood pressure, *1MTE*: 1-min time estimation, variables self-rated or self-measured several times a day by a clinically healthy man as he aged from 20 to 60 years of age. SBP approximates higher congruence with environment of mental (*red*) (vs. somatic, *green*) functions that compare favorably with congruence of aa vs. SWS (*blue*, *bottom right*) (© Halberg Chronobiology Center)

Magnetoreception based on magnetite has been proposed [216]. Semm and Beason [217] obtained clear recordings of responses to weak magnetic stimuli. They found that single units in the superficial ophthalmic branch and ganglion cells of the trigeminal nerve system responded to changes in the intensity of the earth's magnetic field as small as 200 nT. They also showed that the firing rates of units increased with the logarithmic function of field intensity. Magnetite biomineralization and magnetoreception found in different organisms, including humans [218], suggest that biosystems may have adapted to variations in the geomagnetic field induced by variations in the solar wind and the IMF [219]. Magnetocytes, loaded with crystals of biogenic magnetite, have been found in the human brain and may characterize the adrenal glands [218, 220]. These biological functions, while unknown, may detect the geomagnetic field, as do magnetostatic bacteria, protozoans, migratory fish, and birds [221]. Magnetic crystals could be biochemically important, for example, in spin states of reaction intermediates, controlling the decay path of a triplet state, as suggested by McLauchlan [222, 223]. According to his theoretical estimation, fields of a few mTs generated by solar activity can affect the rate of chemical reactions, when some bonds break and reform.

There has also been some investigation into the role of calcium ions as a potential link between geomagnetic variations and human health issues [224]. Cyclotron resonance and parametric resonance are different ways in which ELF electric and magnetic fields could be absorbed by biological systems [225]. Interaction of ions in the body with external magnetic fields is another possible mechanism [226]. On a theoretical basis, the resonant frequency of ions diffusing in a weak magnetic field was estimated to be near one cycle per week [227], a ubiquitous spectral component found to characterize melatonin [228–230]. An important and interesting concept in studying the potential health effects of electromagnetic fields is that of a "biological window," introduced by Ross Adey [231], who showed that calcium output of brain cells in rabbits could only be triggered using very low magnetic field intensities and a specific low frequency. More generally, a biological window is an electromagnetic signal that only has a biological effect if it is of the right intensity in the pertinent frequency range.

Irrespective of the underlying mechanisms, the systematic mapping of cycles in physiopathology and in the near and far environment, as planned in an atlas of chronomes [207], should provide a better understanding of which physiological variables are congruent with which environmental variables at which frequencies during what time spans in which geographic locations. But focus on period length is not sufficient, and an added critical step will be to examine phase relations with respect to a given environmental cycle taken as reference, such as the about-11-year solar activity cycle. Considering that the solar activity cycle length can vary greatly from one cycle to another, phase relations may best be assessed for each cycle separately, as already done in a few instances [214, 215, 232, 233].

References

1. Halberg F, Cornelissen G, Regal P, Otsuka K, Wang ZR, Katinas GS, Siegelova J, Homolka P, Prikryl P, Chibisov SM, Holley DC, Wendt HW, Bingham C, Palm SL, Sonkowsky RP, Sothern RB, Pales E, Mikulecky M, Tarquini R, Perfetto F, Salti R, Maggioni C, Jozsa R, Konradov AA, Kharlitskaya EV, Revilla M, Wan CM, Herold M, Syutkina EV, Masalov AV, Faraone P, Singh RB, Singh RK, Kumar A, Singh R, Sundaram S, Sarabandi T, Pantaleoni GC, Watanabe Y, Kumagai Y, Gubin D, Uezono K, Olah A, Borer K, Kanabrocki EA, Bathina S, Haus E, Hillman D, Schwartzkopff O, Bakken EE, Zeman M. Chronoastrobiology: proposal, nine conferences, heliogeomagnetics, transyears, near-weeks, near-decades, phylogenetic and ontogenetic memories. Biomed Pharmacother 2004;58 Suppl 1:S150–87.
2. Halberg F, Barnum CP, Silber RH, Bittner JJ. 24-hour rhythms at several levels of integration in mice on different lighting regimens. Proc Soc Exp Biol Med (Maywood) 1958;97:897–900. doi:10.3181/00379727-97-23915.
3. Halberg F, Halberg E, Barnum CP, Bittner JJ. Physiologic 24-hour periodicity in human beings and mice, the lighting regimen and daily routine. In: Withrow RB, editor. Photoperiodism and related phenomena in plants and animals. Proceedings of the Conference on Photoperiodism, Gatlinburg, Tennessee, October 29–November 2, 1957. Ed. Publ. No. 55. Washington, DC: American Association for the Advancement of Science; 1959. p. 803–78.
4. Halberg F. The 24-hour scale: a time dimension of adaptive function organization. Perspect Biol Med 1960;3:491–527.
5. Edmunds LN, Halberg F. Circadian time structure of Euglena: a model system amenable to quantification. In: Kaiser H, editor. Neoplasms—comparative pathology of growth in animals, plants and man. Baltimore: Williams and Wilkins; 1981. p. 105–34.
6. de Duve C. The beginnings of life on earth. Am Sci 1995;83(5):428–37.
7. Halberg F, Cornelissen G. Chronoastrobiology. In: Proc. 3rd International Symposium of Chronobiology and Chronomedicine, Kunming, China, October 7–12, 1998. Chengdu: Chinese Society for Chronobiology and Chronomedicine; 1998. p. 128–42.
8. Cornelissen G, Broda H, Halberg F. Does Gonyaulax polyedra measure a week? Cell Biophys 1986;8:69–85.
9. Schweiger H-G, Berger S, Kretschmer H, Mörler H, Halberg E, Sothern RB, Halberg F. Evidence for a circaseptan and a circasemiseptan growth response to light/dark cycle shifts in nucleated and enucleated Acetabularia cells, respectively. Proc Natl Acad Sci USA 1986;83:8619–23. PMCID: PMC386982.
10. Cornelissen G, Halberg F. Do "protists" talk "circaseptan" as well as "circadian'?. Abstract, conference on biochemistry, cell biology and protists, in honor of George G. Holz Jr. and Robert L. Conner: Satellite meeting, 43rd Ann. Mtg. Society of Protozoologists, Luray, June 22–24; 1990. p. 30.
11. Halberg F, Cornelissen G. The spectrum of rhythms in microorganisms revisited. Abstract, conference on biochemistry, cell biology and protists, in honor of George G. Holz Jr. and Robert L. Conner: Satellite meeting, 43rd Ann. Mtg. Society of Protozoologists, Luray, June 22–24; 1990. p. 31.
12. Halberg F, Cornelissen G. The spectrum of rhythms in microorganisms revisited. Biochim Clin 1991;15:163–4.
13. Cornelissen G, Halberg F. Introduction to chronobiology. Medtronic Chronobiology Seminar #7, April 1994; p. 52 (Library of Congress Catalog Card #94-060580. https://sites.google.com/a/umn.edu/halbergchronobiologycenter/
14. Hardeland R, Cornelissen G, Halberg F. Coexisting circaseptan and circadian patterns in the motility of Euglena gracilis Klebs in continuous darkness. Abstract, 6° Convegno Nazionale de Cronobiologia, Chianciano, November 27–28; 1998. p. 66.
15. Katinas G, Hillman D, Siegelova J, Dusek J, Svacinova H, Cornelissen G, Halberg F. About-weekly changes in electrical potential, chloroplast migration and oxygen production in

Acetabularia grown under continuous exposure to light. Scripta Medica (Brno) 2002;75:309–14.
16. Edmunds L, Balzer I, Cornelissen G, Halberg F. Toward a chronome of Euglena gracilis (Klebs). Neuroendocrinol Lett 2003;24 Suppl 1:192–9.
17. Katinas G, Berger S, Cornelissen G, Hillman D, Woolum JC, Engebretson M, Syutkina EV, Masalov A, Wang ZR, Wan CM, Bakken EE, Schwartzkopff O, Halberg F. Acetabularia acetabulum: unicellular model of neonatal vascular chronome. Neuroendocrinol Lett 2003;24 Suppl 1:201–7.
18. Woolum JC, Cornelissen G, Hayes DK, Halberg F. Spectral aspects of Acetabularia's light transmission chronome: effects of enucleation. Neuroendocrinol Lett 2003;24 Suppl 1:212–5.
19. Cornelissen G, Halberg F, Tarquini B, Mainardi G, Panero C, Cariddi A, Sorice V, Cagnoni M. Blood pressure rhythmometry during the first week of human life. In: Tarquini B, editor. Social Diseases and Chronobiology: Proc. III Int. Symp. Social Diseases and Chronobiology, Florence, Nov. 29, 1986. Bologna: Società Editrice Esculapio; 1987. p. 113–22.
20. Wrbsky P, Mills M, Cornelissen G, Johnson D, Halberg F. Circadian and circaseptan variations of systolic blood pressure (SBP) and heart rate (HR) in preterm babies. Chronobiologia 1993;20:135–6.
21. Halberg F, Cornelissen G, Wrbsky P, Johnson D, Rigatuso J, Tarquini B, Mainardi G, Breus T, Syutkina EV, Grigoriev AE, Abramian A, Mitish M, Wakasugi K, Tamura K. About 3.5-day (circasemiseptan) and about 7-day (circaseptan) blood pressure features in human prematurity. Chronobiologia 1994;21:146–51.
22. Siegelova J, Dusek J, Fiser B, Nekvasil R, Muchova M, Cornelissen G, Halberg F. Circaseptan rhythm in blood pressure and heart rate in newborns. Scripta Medica (Brno) 1996;67 Suppl 2:63–70.
23. Turti T, Syutkina EV, Cornelissen G, Grigoriev AE, Mitish MD, Abramian AS, Siegelova J, Fiser B, Dusek J, Al-Kubati M, Muchova L, Uhlir M, Halberg F. Multiseptan-over-circadian prominence of neonatal blood pressure and heart rate in Moscow, Russia. Scripta Medica (Brno) 1996;67 Suppl 2:85–92.
24. Syutkina EV, Cornelissen G, Grigoriev AE, Mitish MD, Turti T, Yatsyk GV, Pimenov K, Breus TK, Studenikin MY, Siegelova J, Fiser B, Dusek J, Johnson D, Halberg F. Neonatal intensive care may consider associations of cardiovascular rhythms with local magnetic disturbance. Scripta Medica (Brno) 1997;70:217–26.
25. Cornelissen G, Engebretson M, Johnson D, Otsuka K, Burioka N, Posch J, Halberg F. The week, inherited in neonatal human twins, found also in geomagnetic pulsations in isolated Antarctica. Biomed Pharmacother 2001;55 Suppl 1:32s–50s.
26. Cornelissen G, Siegelova J, Halberg F. Blood pressure and heart rate dynamics during pregnancy and early extra-uterine life: Methodology for a chrononeonatology. In: Halberg F, Kenner T, Fiser B, editors. Proceedings, Symposium: The importance of chronobiology in diagnosing and therapy of internal diseases. Faculty of Medicine, Masaryk University, Brno, Czech Republic, January 10–13, 2002. Brno: Masaryk University; 2002. p. 58–96.
27. Watanabe Y, Cornelissen G, Hellbrügge T, Watanabe F, Otsuka K, Schwartzkopff O, Halberg F. Partial spectral element in the chronome of a human neonatal heart rate at term. Biomed Pharmacother 2002;56 Suppl 2:374s–8s.
28. Siegelova J, Cornelissen G, Schwartzkopff O, Halberg F. Time structures in the development of children. Neuroendocrinol Lett 2003;24 Suppl 1:126–31.
29. Díez-Noguera A, Cambras T, Cornelissen G, Halberg F. A biological week in the activity chronome of the weanling rat: a chrono-meta-analysis. Abstract, 4° Convegno Nazionale, Società Italiana di Cronobiologia, Gubbio (Perugia), June 1–2, 1996. p. 81–2.
30. Thaela M-J, Cornelissen G, Halberg F, Rantzer D, Svendsen J, Pierzynowski SG. Extra-circadian (about-weekly, half-weekly and 8-hourly) variation in pancreatic secretion of piglets. Proc. XXXIII Int. Cong. International Union of Physiological Sciences, St. Petersburg, June 30–July 5; 1997. abstract P041.36.

31. Fanjul Moles ML, Cornelissen G, Miranda Anaya M, Prieto Sagredo J, Halberg F. Larger infradian vs. circadian prominence of locomotor activity in young vs. older crayfish. Abstract, 6° Convegno Nazionale de Cronobiologia, Chianciano, November 27–28; 1998. p. 65.
32. Halberg F, Cornelissen G. About-7-day (circaseptan) pattern of DNA labelling under daily STH stimulation in mouse tissues: a metachronanalysis. Abstract, Perspectives in Immunology and Medicine 1944–1994: A Symposium in Honor of Robert A. Good, St. Petersburg, May 20–21; 1994.
33. Cornelissen G, Halberg F, Edmunds L. Multiseptan modulations of circadian cell division and cell settling patterns in Euglena. Abstract, 2nd Int. Symp. of Chronobiology and Chronomedicine, Shenyang, Sept. 28–Oct. 2; 1996. p. 63–4.
34. Abbot CG. Solar variation and weather, a summary of the evidence, completely illustrated and documented. Smithsonian Miscellaneous Collections 146, No. 3 (Publ. 4545), Washington, DC; 1963. p. 67. + 4 plates.
35. Roederer JG. Are magnetic storms hazardous to your health? Eos, Trans Amer Geophys Union 1995;76(441):444–5.
36. Vladimirskii BM, Narmanskii VY, Temuriantz NA. Global rhythmics of the solar system in the terrestrial habitat. Biophysics 1995;40:731–6.
37. Cornelissen G, Hillman D, Katinas GS, Rapoport S, Breus TK, Otsuka K, Bakken EE, Halberg F. Geomagnetics and society interact in weekly and broader multiseptans underlying health and environmental integrity. Biomed Pharmacother 2002;56 Suppl 2:319s–26s.
38. Halberg F, Marques N, Cornelissen G, Bingham C, Sanchez de la Peña S, Halberg J, Marques M, Wu J, Halberg E. Circaseptan biologic time structure reviewed in the light of contributions by Laurence K. Cutkomp and Ladislav Dérer. Acta entomol bohemoslov 1990;87:1–29.
39. Lucas EA, Halberg F, Straub KD, Carter J, Redmond DP, Scrima L, Scheving LE. About 7-day (circaseptan) free-running rhythms in urinary norepinephrine and systolic and diastolic blood pressure during human social isolation studies. Sleep Res 1985;14:305.
40. Montalbini M, Galvagno A, Follini S, Sanchez de la Peña S, Hillman DC, Wu J, Degioanni J, Kawabata Y, Kutyna F, Bungo M, Halberg F. Cardiovascular about-24-hour (circadian) and about-7-day (circaseptan) rhythms in human social isolation. Abstract, Health of Inner Cities and Urban Areas, International Conference, Cardiff, September 4–7, 1989, The Institute of Statisticians/SIRMCE; 1989. p. 101.
41. Sanchez de la Peña S, Halberg F, Galvagno A, Montalbini M, Follini S, Wu J, Degioanni J, Kutyna F, Hillman DC, Kawabata Y, Cornelissen G. Circadian and circaseptan (about-7-day) free-running physiologic rhythms of a woman in social isolation. Proc. 2nd Ann. IEEE Symp. on Computer-Based Medical Systems, Minneapolis, June 26–27, 1989. Washington DC: Computer Society Press; 1989. p. 273–8.
42. Galvagno A, Montalbini M, Sanchez de la Peña S, Scheving LE, Degioanni J, Kutyna F, Halberg F. About 7-day (circaseptan) cardiovascular in the social isolation of a cave. Abstract, 1st International Design for Extreme Environments Assembly, Houston, Nov. 12–15; 1991. HF 2.3.2.
43. Hillman DC, Siffre M, Milano G, Halberg F. Urinary about-84-hour (circasemiseptan) variations of a woman isolated in a cave and cosmic ray effects. New Trends Exp Clin Psychiatr 1994;10:173–8.
44. Halberg F, Cornelissen G, Sonkowsky RP, Lanzoni C, Galvagno A, Montalbini M, Schwartzkopff O. Chrononursing (chronutrics), psychiatry and language. New Trends Exp Clin Psychiatr 1998;14:15–26.
45. Cerveny RS, Balling RC Jr. Weekly cycles of air pollutants, precipitations and tropical cyclones in the coastal NW Atlantic region. Nature 1998;394:561–3.
46. Halberg F, Cornelissen G, Schwartzkopff O, Katinas GS, Chibisov SM, Khalitskaya EV, Mitsutake G, Otsuka K, Scheving LA, Bakken EE. Chronometaanalysis: magnetic storm associated with a reduction in circadian amplitude of rhythm in corneal cell division. Proceedings, International Conference on the Frontiers of Biomedical Science: Chronobiology, Chengdu, September 24–26; 2006. p. 40–2.

47. Halberg F, Haus E, Cardoso SS, Scheving LE, Kühl JFW, Shiotsuka R, Rosene G, Pauly JE, Runge W, Spalding JF, Lee JK, Good RA. Toward a chronotherapy of neoplasia: tolerance of treatment depends upon host rhythms. Experientia (Basel) 1973;29:909–34.
48. Faraone P. La frequenza delle colonie a settore differenziato (C.S.D.), fra i batteri in sospensione nell'aria esterna, in tre anni di osservazione (sua correlazione con vari parametri: numeri di Wolf, flusso solar, attivita' geomagnetica, fenomeni fluttuanti). Ann Sclavo 1973;15:207–24.
49. Faraone P. The CSD frequency variation with the solar activity and with the altitude, after twenty years researches. Proceedings, International Medical Congress of Mountain Climatology, Roccaraso (L'Aquila), June 7–9; 1991. p. 1–18.
50. Faraone P, Cornelissen G, Katinas GS, Halberg F, Siegelova J. Astrophysical influences on sectoring in colonies of microorganisms. Scripta Medica (Brno) 2001;74:107–14.
51. Halberg F, with Bakken EE, Katinas GS, Cornelissen G, Zaslavskaya RM, Blank MA, Syutkina EV, Breus TK, Watanabe Y, Masalov A, Chibisov SM. Chronoastrobiology: Vernadsky's future science? Benefits from spectra of circadians and promise of a new transdisciplinary spectrum of near-matching cycles in and around us. Opening keynote, proceedings, III International Conference, Civilization Diseases in the Spirit of V.I. Vernadsky, People's Friendship University of Russia, Moscow, Oct 10–12; 2005. p. 4–22.
52. Kleiger RE, Miller JP, Bigger Jr JT, Moss AJ, Multicenter Post-Infarction Research Group. Decreased heart rate variability and its association with increased mortality after acute myocardial infarction. Am J Cardiol 1987;59:256–62.
53. Lakusic N, Mahovic D, Sonicki Z, Slivnjak V, Baborski F. Outcome of patients with normal and decreased heart rate variability after coronary artery bypass grafting surgery. Int J Cardiol 2013;166(2):516–8.
54. Tsuji H, Larson MG, Venditti Jr FJ, Manders ES, Evans JC, Feldman CL, Levy D. Impact of reduced heart rate variability on risk for cardiac events. The Framingham Heart Study. Circulation 1996;94(11):2850–5.
55. Jensen-Urstad K, Reichard P, Jensen-Urstad M. Decreased heart rate variability in patients with type I diabetes mellitus is related to arterial wall stiffness. J Int Med 1999;245(1):57–61.
56. Wolk R. Central origin of decreased heart rate variability in patients with cardiovascular diseases. Med Hypotheses 1996;46(5):479–81.
57. Binici Z, Mouridsen MR, Kober L, Sajadieh A. Decreased nighttime heart rate variability is associated with increased stroke risk. Stroke 2011;42(11):3196–201.
58. Haarala A, Kahonen M, Eklund C, Jylhava J, Koskinen T, Taittonen L, Huupponen R, Lehtimaki T, Viikari J, Raitakari OT, Hurme M. Heart rate variability is independently associated with C-reactive protein but not with serum amyloid A. The cardiovascular risk in Young Finns Study. Eur J Clin Invest 2011;41(9):91–957.
59. Orth-Gomér K, Halberg F, Sothern R, Åkerstedt T, Theorell T, Cornelissen G. The circadian rhythm of ventricular arrhythmias. In: Takahashi R, Halberg F, Walker C, editors. Toward chronopharmacology, Proc. 8th IUPHAR Cong. and Sat. Symposia, Nagasaki, July 27–28, 1981. Oxford/New York: Pergamon Press; 1982. p. 191–202.
60. Otsuka K, Cornelissen G, Halberg F, Oehlert G. Excessive circadian amplitude of blood pressure increases risk of ischemic stroke and nephropathy. J Med Eng Technol 1997;21:23–30.
61. Baevsky RM, Petrov VM, Cornelissen G, Halberg F, Orth-Gomér K, Åkerstedt T, Otsuka K, Breus T, Siegelova J, Dusek J, Fiser B. Meta-analyzed heart rate variability, exposure to geomagnetic storms, and the risk of ischemic heart disease. Scripta Medica (Brno) 1997;70:199–204.
62. Otsuka K, Yamanaka T, Cornelissen G, Breus T, Chibisov SM, Baevsky R, Halberg F, Siegelova J, Fiser B. Altered chronome of heart rate variability during span of high magnetic activity. Scripta Medica (Brno) 2000;73:111–6.

63. Cornelissen G, Halberg F, Schwartzkopff O, Delmore P, Katinas G, Hunter D, Tarquini B, Tarquini R, Perfetto F, Watanabe Y, Otsuka K. Chronomes, time structures, for chronobioengineering for "a full life". Biomed Instrum Technol 1999;33:152–87.
64. Otsuka K, Murakami K, Kubo Y, Yamanaka T, Mitsutake G, Ohkawa S, Matsubayashi K, Yano S, Cornelissen G, Halberg F. Chronomics for chronoastrobiology with immediate spin-offs for life quality and longevity. Biomed Pharmacother 2003;57 Suppl 1:1s–18s.
65. Otsuka K, Cornelissen G, Zhao ZY, Weydahl A, Delyukov A, Gorgo Y, Wang ZR, Perfetto F, Tarquini R, Kubo Y, Shinagawa M, Hotta N, Ishii T, Omori K, Watanabe Y, Nunoda S-i, Ohkawa S-i, Halberg F. Rhythm and trend elements in the time structure, chronome, of heart rate variability. Geronto-Geriatrics 1999;2:31–48.
66. Otsuka K, Ichimaru Y, Cornelissen G, Weydahl A, Holmeslet B, Schwartzkopff O, Halberg F, on behalf of the ICEHRV Working Group. Dynamic analysis of heart rate variability from 7-day Holter recordings associated with geomagnetic activity in subarctic area. Comput Cardiol 2000;27:453–6.
67. Otsuka K, Cornelissen G, Weydahl A, Holmeslet B, Hansen TL, Shinagawa M, Kubo Y, Nishimura Y, Omori K, Yano S, Halberg F. Geomagnetic disturbance associated with decrease in heart rate variability in a subarctic area. Biomed Pharmacother 2001;55 Suppl 1:51s–6s.
68. Otsuka K, Oinuma S, Cornelissen G, Weydahl A, Ichimaru Y, Kobayashi M, Yano S, Holmeslet B, Hansen TL, Mitsutake G, Engebretson MJ, Schwartzkopff O, Halberg F. Alternating-light-darkness-influenced human electrocardiographic magnetoreception in association with geomagnetic pulsations. Biomed Pharmacother 2001;55 Suppl 1:63s–75s.
69. Fleisher LA, Frank SM, Sessler DI, Cheng C, Matsukawa T, Vannier CA. Thermoregulation and heart rate variability. Clin Sci 1996;90(2):97–103.
70. Tripathi KK. Very low frequency oscillations in the power spectra of heart rate variability during dry supine immersion and exposure to non-hypoxic hypobaria. Physiol Meas 2011;32(6):717–29.
71. Oinuma S, Kubo Y, Otsuka K, Yamanaka T, Murakami S, Matsuoka O, Ohkawa S, Cornelissen G, Weydahl A, Holmeslet B, Hall C, Halberg F, on behalf of the "ICEHRV" Working Group. Graded response of heart rate variability, associated with an alteration of geomagnetic activity in a subarctic area. Biomed Pharmacother 2002;56 Suppl 2:284s–8s.
72. Otsuka K, Murakami S, Oinuma S, Kubo Y, Shinagawa M, Weydahl A, Cornelissen G, Halberg F. Altered heart rate variability on geomagnetically disturbed day and the following two days in subarctic. In: Mitro P, Pella D, Rybar R, Valocik G, editors. Proceedings, 2nd Congress on Cardiovascular Diseases, Kosice, 25–27 April 2002. Bologna: Monduzzi Editore; 2002. p. 271–83.
73. Otsuka K, Singh RB, Weydahl A, Ichimaru Y, Holmeslet B, Hansen TL, Mitsutake G, Murakami S, Kubo Y, Ohkawa S, Yano S, Engebretson MJ, Cornelissen G, Schwartzkopff O, Halberg F, on behalf of the ICEHRV Working Group. Ultra-low frequency geomagnetic pulsations and cardiovascular biological systems in humans. Int J Cardiol 2002;82 Suppl 1:S26–7.
74. Phillips JB, Deutschlander ME, Freake MJ, Borland SC. The role of extraocular photoreceptors in newt magnetic compass orientation: parallels between light-dependent magnetoreception and polarized light detection in vertebrates. J Exp Biol 2001;204(Pt 14):2543–52.
75. Otsuka K, Cornelissen G, Halberg F. Circadian rhythmic fractal scaling of heart rate variability in health and coronary artery disease. Clin Cardiol 1997;20:631–8.
76. Otsuka K, editor. Chronome & Janus-medicine: heart rate variability (HRV) and BP variability (BPV) from a viewpoint of chronobiology and ecology. Tokyo: Medical Review; 1998. p. 213.
77. Halberg F, Cornelissen G, Haus E, Northrup G, Portela A, Wendt H, Otsuka K, Kumagai Y, Watanabe Y, Zaslavskaya R. Clinical relevance of about-yearly changes in blood pressure and the environment. Int J Biometeorol 1996;39:161–75.

78. Nicolau G, Haus E, Bogdan C, Plinga L, Robu E, Ungureanu E, Sackett-Lundeen L, Petrescu E. Circannual rhythms of systolic and diastolic blood pressure, in relation to plasma aldosterone and urinary norepinephrine in elderly subjects and in children. Endocrinologie 1986;24:97–107.
79. Seguro C, Siddi PP, Sau F, Giardina G, Scano G, Mura O, Arru A, Zedda N, Cherchi A. The effect of antihypertensive therapy on the seasonal variations in arterial pressure in hypertensive patients. Cardiologia 1992;37:51–8.
80. Verdon F, Boudry JF, Chuat M, Studer JP, Truong CB, Jacot E. Seasonal variations in arterial pressure in hypertensive patients. Schweiz Med Wochenschr 1993;123:2363–9.
81. Woodhouse PR, Khaw KT, Plummer M. Seasonal variation of blood pressure and its relation to ambient temperature in an elderly population. J Hypertension 1993;11:1267–74.
82. Brennan PJ, Greenberg G, Miall WE, Thompson SG. Seasonal variation in arterial blood pressure. Br Med J. (Clin Res Ed) 1982;285:919–23.
83. Allan TM, Douglas AS. Seasonal variation in health and disease: with sections on effects of weather and temperature. A bibliography. London: Mansell; 1994.
84. Spelsberg TC, Halberg F. Circannual rhythms in steroid receptor concentration and nuclear binding in the chick oviduct. Endocrinology 1980;80:1234–44.
85. Halberg F, Cornelissen G, Sothern RB, Wallach LA, Halberg E, Ahlgren A, Kuzel M, Radke A, Barbosa J, Goetz F, Buckley J, Mandel J, Schuman L, Haus E, Lakatua D, Sackett L, Berg H, Wendt HW, Kawasaki T, Ueno M, Uezono K, Matsuoka M, Omae T, Tarquini B, Cagnoni M, Garcia Sainz M, Perez Vega E, Wilson D, Griffiths K, Donati L, Tatti P, Vasta M, Locatelli I, Camagna A, Lauro R, Tritsch G, Wetterberg L. International geographic studies of oncological interest on chronobiological variables. In: Kaiser H, editor. Neoplasms – comparative pathology of growth in animals, plants and man. Baltimore: Williams and Wilkins; 1981. p. 553–96.
86. Halberg F, Lagoguey JM, Reinberg A. Human circannual rhythms in a broad spectral structure. Int J Chronobiol 1983;8:225–68.
87. Zucker I. Neuroendocrine substrates of circannual variation. In: Kupfer DJ, Monk TH, Barchas JD, editors. Biological rhythms and mental disorders. New York: Guilford Press; 1988. p. 219–51.
88. Smolensky M, Halberg F, Sargent II F. Chronobiology of the life sequence. In: Itoh S, Ogata K, Yoshimura H, editors. Advances in Climatic Physiology. Tokyo: Igaku Shoin Ltd; 1972. p. 281–318.
89. Reinberg A, Gervais P, Halberg F, Gaultier M, Roynette N, Abulker C, Dupont J. Mortalité des adultes: Rythmes circadiens et circannuels dans un hôpital parisien et en France. Nouv Presse Med 1973;2:289–94.
90. Miyatake K, Okamoto M, Kinoshita N, Owa M, Nakasone I, Sakahibara H, Nimura Y. Augmentation of atrial contribution to left ventricular inflow with aging as assessed by intracardiac Doppler flowmetry. Am J Cardiol 1984;53:586–9.
91. Nicolau GY, Haus E, Popescu M, Sackett-Lundeen L, Petrescu E. Circadian, weekly and seasonal variations in cardiac mortality, blood pressure and catecholamine excretion. Chronobiol Int 1991;8:149–59.
92. Saito Y, Tamura K, Hillman DC, Cornelissen G, Halberg F. Circannual changes in ventricular function assessed by echocardiography. Chronobiologia 1992;19:59–65.
93. Cornelissen G, Breus TK, Bingham C, Zaslavskaya R, Varshitsky M, Mirsky B, Teibloom M, Tarquini B, Bakken E, Halberg F, International Womb-to-Tomb Chronome Initiative Group. Beyond circadian chronorisk: worldwide circaseptan-circasemiseptan patterns of myocardial infarctions, other vascular events, and emergencies. Chronobiologia 1993;20:87–115.
94. Halberg F, Haus E, Halberg E, Cornelissen G, Scarpelli P, Tarquini B, Cagnoni M, Wilson D, Griffiths K, Simpson H, Balestra E, Reale L. Chronobiologic challenges in social medicine: illustrative tasks in cardiology and oncology. In: Halberg F, Reale L, Tarquini B, editors. Proceedings of the 2nd International Conference on Medico-Social Aspects of Chronobiology, Florence, Oct 2, 1984; Rome: Istituto Italiano di Medicina Sociale; 1986. p. 13–42.

95. Studer A, Lüscher T, Greminger P, Epstein FH, Grimm J, Leumann EP, Tenschert W, Siegenthaler W, Vetter W. Blutdruck, Renin-Angiotensin-Aldosteron-System und andere kardiovaskuläre Risikofaktoren bei Kindern essentieller Hypertoniker. Klin Wochenschr 1982;60:275–84.
96. Kawashima K, Nagakura A, Wurzburger RJ, Spector S. Melatonin in serum and the pineal of spontaneously hypertensive rats. Clin Exp Hypertens A 1984;6:1517–28.
97. Zaslavskaya RM, Makarova LA, Shakarova AN, Komarov F, Wang ZR, Wan CM, Katinas G, Cornelissen G, Halberg F. Individualized time series-based assessment of melatonin effects on blood pressure: model for pediatricians. Neuroendocrinol Lett 2003;24 Suppl 1:238–46.
98. Wetterberg L, Halberg F, Halberg E, Haus E, Kawasaki T, Ueno M, Uezono K, Cornelissen G, Matsuoka M, Omae T. Circadian characteristics of urinary melatonin from clinically healthy women at different civilization disease risk. Acta Med Scand 1986;220:71–81.
99. Dominguez-Rodriguez A. Melatonin in cardiovascular disease. Expert Opin Investig Drugs 2012;21(11):1593–6.
100. Paulis L, Simko F, Laudon M. Cardiovascular effects of melatonin receptor agonists. Expert Opin Investig Drugs 2012;21(11):1661–78.
101. Ekmekcioglu C, Thalhammer T, Humpeler S, Mehrabi MR, Glogar HD, Holzenbein T, Markovic O, Leibetseder VJ, Strauss-Blasche G, Marktl W. The melatonin receptor MT2 is present in the human cardiovascular system. J Pineal Res 2003;35(1):40–4.
102. Zeman M, Herichova I. Melatonin and clock genes expression in the cardiovascular system. Front Biosci 2013;5:743–53.
103. Wilson BW, Anderson LE, Hilton DI, Phillips RD. Chronic exposure to 60-Hz electric fields: effects on pineal function in the rat. Bioelectromagnetics 1981;2:371–80.
104. Semm P. Neurobiological investigations on the magnetic sensitivity of the pineal gland in rodents and pigeons. Comp Biochem Physiol 1983;76:683–9.
105. Wilson BW, Chess EK, Anderson LE. 60-Hz electric field effects on pineal melatonin rhythms: time course for onset and recovery. Bioelectromagnetics 1986;7:239–42.
106. Reiter RJ, Anderson LE, Buschbom RL, Wilson BW. Reduction of the nocturnal rise in pineal melatonin levels in rats exposed ro 60 Hz electric fields in utero and for 23 days after birth. Life Sci 1988;42:2203–6.
107. Burch JB, Reif JS, Yost MG, Keefe TJ, Pitrat CA. Reduced excretion of a melatonin metabolite in workers exposed to 60 Hz magnetic fields. Am J Epidemiol 1999;150(1):27–36.
108. Burch JB, Reif JS, Yost MG. Geomagnetic disturbances are associated with reduced nocturnal excretion of a melatonin metabolite in humans. Neurosci Lett 1999;266(3):209–12.
109. Weydahl A, Sothern RB, Cornelissen G, Wetterberg L. Geomagnetic activity influences the melatonin secretion at latitude 70°N. Biomed Pharmacother 2001;55:57–62.
110. Chibisov SM, Cornelissen G, Halberg F. Magnetic storm effect on the circulation of rabbits. Biomed Pharmacother 2004;58 Suppl 1:S15–9.
111. Portela A, Northrup G, Halberg F, Cornelissen G, Wendt H, Melby JC, Haus E. Changes in human blood pressure with season, age and solar cycles: a 26-year record. Int J Biometeorol 1996;39:176–81.
112. Halberg F, Engeli M, Hamburger C, Hillman D. Spectral resolution of low-frequency, small-amplitude rhythms in excreted 17-ketosteroid; probable androgen induced circaseptan desynchronization. Acta Endocrinol (Kbh) 1965;50 Suppl 103:5–54.
113. Halberg F, Cornelissen G, Watanabe Y, Otsuka K, Fiser B, Siegelova J, Mazankova V, Maggioni C, Sothern RB, Katinas GS, Syutkina EV, Burioka N, Schwartzkopff O. Near 10-year and longer periods modulate circadians: intersecting anti-aging and chronoastrobiological research. J Gerontol A Biol Sci Med Sci 2001;56:M304–24.
114. Halberg F, Cornelissen G, Schwartzkopff O, Khasigawala P, Hillman D, Sothern RB, Katinas G, Hong S, Siegelova J. Decadal and multidecadal cycles in the cardiovascular system relating to diagnosis and treatment? In: Kenner T, Cornelissen G, Siegelova J, Dobsak P, editors. Noninvasive methods in cardiology 2013. Brno: Masaryk University; 2013. p. 69–78.

115. Richardson JD, Paularena KI, Belcher JW, Lazarus AJ. Solar wind oscillations with a 1.3-year period. Geophys Res Lett 1994;21:1559–60.
116. Mursula K, Zieger B. The 1.3-year variation in solar wind speed and geomagnetic activity. Adv Space Res 2000;25:1939–42.
117. Halberg F, Cornelissen G, Schack B, Wendt HW, Minne H, Sothern RB, Watanabe Y, Katinas G, Otsuka K, Bakken EE. Blood pressure self-surveillance for health also reflects 1.3-year Richardson solar wind variation: spin-off from chronomics. Biomed Pharmacother 2003;57 Suppl 1:58s–76s.
118. Cornelissen G, Masalov A, Halberg F, Richardson JD, Katinas GS, Sothern RB, Watanabe Y, Syutkina EV, Wendt HW, Bakken EE, Romanov Y. Multiple resonances among time structures, chronomes, around and in us. Is an about 1.3-year periodicity in solar wind built into the human cardiovascular chronome? Hum Physiol 2004;30(2):86–92.
119. Halberg F, Cornelissen G, Katinas G, Tvildiani L, Gigolashvili M, Janashia K, Toba T, Revilla M, Regal P, Sothern RB, Wendt HW, Wang ZR, Zeman M, Jozsa R, Singh RB, Mitsutake G, Chibisov SM, Lee J, Holley D, Holte JE, Sonkowsky RP, Schwartzkopff O, Delmore P, Otsuka K, Bakken EE, Czaplicki J, International BIOCOS Group. Chronobiology's progress: season's appreciations 2004–2005. Time-, frequency-, phase-, variable-, individual-, age- and site-specific chronomics. J Appl Biomed 2006;4:1–38.
120. Cornelissen G, Syutkina EV, Johnson D, Otsuka K, Halberg F. Near-transyears and transyears in geomagnetics and biology. In: Matsubayasi K, editor. Proceedings of the 5th International Workshop on Chronoastrobiology and Chronotherapy; Nov 6, 2004; Division of Human-Nature Dynamics, Center for Southeast Asian Studies. p. 40–3.
121. Gubin D, Cornelissen G, Halberg F, Gubin G, Uezono K, Kawasaki T. The human blood pressure chronome: a biological gauge of aging. In Vivo 1997;11:485–94.
122. Gubin D, Cornelissen G, Halberg F, Gubin GD, Turti T, Syutkina EV, Grigoriev AE, Mitish MD, Yatsyk GV, Ikonomov O, Stoynev A, Madjirova N, Siegelova J, Fiser B, Dusek J. Half-weekly and weekly blood pressure patterns in late human ontogeny. Scripta Medica (Brno) 1997;70:207–16.
123. Cornelissen G, Sothern RB, Halberg F. Age and circaseptan-to-circadian prominence of blood pressure in a normotensive clinically healthy man. In: Eriguchi M, editor. Abstract 11 Proceedings of the 3rd International Symposium: Workshop on Chronoastrobiology and Chronotherapy; Nov 9, 2002; Research Center for Advanced Science and Technology, University of Tokyo.
124. Syutkina EV, Cornelissen G, Yatsyk G, Studenikin M, Baranov A, Halberg F. Over a decade of clinical chrononeonatology and chronopediatrics in Moscow. Neuroendocrinol Lett 2003;24 Suppl 1:132–8.
125. Halberg F, Breus TK, Cornelissen G, Bingham C, Hillman DC, Rigatuso J, Delmore P, Bakken E, International Womb-to-Tomb Chronome Initiative Group: Chronobiology in space. University of Minnesota/Medtronic Chronobiology Seminar Series; December 1, 1991; p. 21 of text, 70 figures.
126. Watanabe Y, Hillman DC, Otsuka K, Bingham C, Breus TK, Cornelissen G, Halberg F. Cross-spectral coherence between geomagnetic disturbance and human cardiovascular variables at non-societal frequencies. Chronobiologia 1994;21:265–72.
127. Jozsa R, Halberg F, Cornelissen G, Zeman M, Kazsaki J, Csernus V, Katinas GS, Wendt HW, Schwartzkopff O, Stebelova K, Dulkova K, Chibisov SM, Engebretson M, Pan W, Bubenik GA, Nagy G, Herold M, Hardeland R, Hüther G, Pöggeler B, Tarquini R, Perfetto F, Salti R, Olah A, Csokas N, Delmore P, Otsuka K, Bakken EE, Allen J, Amory-Mazaudier C. Chronomics, neuroendocrine feedsidewards and the recording and consulting of nowcasts forecasts of geomagnetics. Biomed & Pharmacother 2005;59 Suppl 1:S24–30.
128. Vallot J, Sardou G, Faure M. De l'influence des taches solaires: sur les accidents aigus des maladies chroniques. J Med Lyon 1922;9(62):904–5.

129. Düll T, Düll B. Über die Abhängigkeit des Gesundheitszustandes von plötzlichen Eruptionen auf der Sonne und die Existenz einer 27tägigen Periode in den Sterbefällen. Virchows Archiv für pathologische Anatomie und Physiologie 1934;293(2):272–319.
130. Düll T, Düll B. Gleichgewichtsstörungen in unserem Leben durch Eruptionen auf der Sonne. Deutsche Rundschau 1934;61(3):161–8.
131. Düll T, Düll B. Zusammenhänge zwischen Störungen des Erdmagnetismus und Häufungen von Todesfällen. Dtsch Med Wochenschr 1935;3:95.
132. Düll T, Düll B. Neue Untersuchungen über die Beziehungen zwischen der Zahl der täglichen Todesfälle und dem erdmagnetischen Störungscharakter. Bioklimatische Beiblätter der Meteorologischen Zeitschrift 1935;2(1):24–31.
133. Düll T, Düll B. Statistique sur la dépendance de la mortalité des influences géophysique et cosmiques. La Cote d'Azur Médicale 1937;18(12):308–9.
134. Düll B, Düll T. Erd- und sonnenphysikalische Vorgänge in ihrer Bedeutung für Krankheits- und Todesauslösung. Nosokomeion (Vierteljahresschrift für das Krankenhauswesen) 1938;9(2):103–21.
135. Düll T, Düll B. Les influences physiques du soleil sur les phénomènes biologiques. Biodynamisme (Paris) 1938;3(4):83–8.
136. Halberg F, Düll-Pfaff N, Gumarova L, Zenchenko TA, Schwartzkopff O, Freytag EM, Freytag J, Cornélissen G. 27-day cycles in human mortality: Traute and Bernhard Düll. Hist Geo Space Sci 2013;4:47–59. 10.5194/hgss-4-47-2013. NIHMSID: 485452.
137. Chizhevsky AL. Cosmobiologie et Rythme du Milieu extérieur. Verhandlungen, Zweite Konferenz der Internationalen Gesellschaft für Biologische Rhythmusforschung, am 25. und 26. August 1939, Utrecht, Holland, Holmgren Hj, editors. Acta Med Scand 1940; 103(Suppl S108):211–26.
138. Chizhevsky AL. The earth in the universe. Fedynsky VV, editor. NASA TT F-345 TT 66–51025; 1968. p. 280.
139. Novikova KF, Gnevyshev MN, Tokareva NV, Ohl AI, Panov TN. The effect of solar activity on the development of myocardial infarction morbidity and mortality. Cardiology, Moscow 1968;4:109.
140. Halberg F, Cornelissen G, Otsuka K, Watanabe Y, Katinas GS, Burioka N, Delyukov A, Gorgo Y, Zhao ZY, Weydahl A, Sothern RB, Siegelova J, Fiser B, Dusek J, Syutkina EV, Perfetto F, Tarquini R, Singh RB, Rhees B, Lofstrom D, Lofstrom P, Johnson PWC, Schwartzkopff O, International BIOCOS Study Group. Cross-spectrally coherent ~10.5- and 21-year biological and physical cycles, magnetic storms and myocardial infarctions. Neuroendocrinol Lett 2000;21:233–58.
141. Gnevyshev MN, Novikova KF. The influence of solar activity on the earth's biosphere. Part I. J Interdiscipl Cycle Res 1972;3:99–104.
142. Gmittov J, Gmitrova A. Geomagnetic field and artificial 0.2T static magnetic field combined effect on blood pressure. Electro- and Magnetobiology 1994;13:117–22.
143. Gmitrov J, Ohkubo C, Yamada S, Gmitrova A, Xu S. Static magnetic field effects on sinocarotid baroreceptors in rabbits exposed under conscious conditions. Electro- and Magnetobiology 1995;14:217–28.
144. Bortkiewicz A, Gadzicka E, Zymslony M. Heart rate variability in workers exposed to medium-frequency electromagnetic fields. J Auton Nerv Syst 1996;59:91–7.
145. Gurfinkel I, Liubimov VV, Oraevsky VN, Parfenova LM, Iur'ev AS. The effect of geomagnetic disturbances in capillary blood flow in ischemic heart disease patients. Biofizika 1995;40:793–9.
146. Stoupel E. Forecasting in cardiology. New York: John Wiley & Sons; 1976. 141 pp.
147. Malin SRC, Srivastava BJ. Correlation between heart attacks and magnetic activity. Nature 1979;277:646–8.
148. Otto W, Hempel WE, Wagner CU, Best A. Einige periodische und aperiodische Variationen der Herzinfarktsterblichkeit in der DDR. Zeitschrift für die Gesamte Innere Medizin und Ihre Grenzgebiete 1982;37:756–63.

149. Sitar J. The causality of lunar changes on cardiovascular mortality. Casopis Lekaru Cesych 1990;129:1425–30.
150. Villoresi G, Breus TK, Iucci N, Dorman LI, Rapoport SI. The influence of geophysical and social effects on the incidences of clinically important pathologies (Moscow 1979–1981). Phys Med 1994;10:79–91.
151. Villoresi G, Kopytenko YA, Ptitsyna NG, Tyasto MI, Kopytenko EA, Iucci N, Voronov PM. The influence of geomagnetic storms and ma0made magnetic field disturbances on the incidence of myocardial infarction in St. Petersburg (Russia). Phys Med 1994;10:107–17.
152. Villoresi G, Ptitsyna NG, Tyatso MI, Iucci N. Myocardial infarct and geomagnetic disturbances: analyses of data on morbidity and mortality. Biofizika 1998;43:623–31.
153. Gurfinkel I, Kuleshova VP, Oraevskii VN. Assessment of a geomagnetic storm on the frequency of appearance of acute cardiovascular pathology. Biofizika 1998;43(4):654–8.
154. Rapoport SI, Breus TK, Kozyreva OV, Malinovskaia NK. Geomagnetic pulsations and myocardial infarctions. Ter Arkh 2006;78(4):56–60.
155. Vencloviene J, Babarskiene R, Slapikas R, Sakalyte G. The association between phenomena on the sun, geomagnetic activity, meteorological variables, and cardiovascular characteristic of patients with myocardial infarction. Int J Biometeorol 2013;57(5):797–804.
156. Knox EG, Armstrong E, Lancashire R, Wall M, Haynes R. Heart attacks and geomagnetic activity. Nature 1979;281(5732):564–5.
157. Messner T, Haggstrom I, Sandahl I, Lundberg V. No covariation between the geomagnetic activity and the incidence of acute myocardial infarction in the polar area of northern Sweden. Int J Biometeorol 2002;46(2):90–4.
158. Feinleib M, Rogot E, Sturrock PA. Solar activity and mortality in the United States. Int J Epidemiol 1975;4:227–9.
159. Lipa BJ, Sturrock PA, Rogot E. Search for correlation between geomagnetic disturbances and mortality. Nature 1976;259:302–4.
160. Breus T, Cornelissen G, Halberg F, Levitin AE. Temporal associations of life with solar and geophysical activity. Annales Geophysicae 1995;13:1211–22.
161. Arnoldi RL. Signature in the interplanetary medium for substorms. J Geophys Res 1971;76:5189–201.
162. Cornelissen G, Wendt HW, Guillaume F, Bingham C, Halberg F, Breus TK, Rapoport S, Komarov F. Disturbances of the interplanetary magnetic field and human pathology. Chronobiologia 1994;21:151–4.
163. Cornelissen G, Halberg F, Breus T, Syutkina EV, Baevsky R, Weydahl A, Watanabe Y, Otsuka K, Siegelova J, Fiser B, Bakken EE. Non-photic solar associations of heart rate variability and myocardial infarction. J Atmos Solar-Terr Phys 2002;64:707–20.
164. Rieger A, Share GH, Forrest DJ, Kanbach G, Reppin C, Chupp EL. A 154-day periodicity in the occurrence of hard solar flares? Nature 1984;312:623–5.
165. Halberg F, Cornelissen G, Otsuka K, Fiser B, Mitsutake G, Wendt HW, Johnson P, Gigolashvili M, Breus T, Sonkowsky R, Chibisov SM, Katinas G, Siegelova J, Dusek J, Singh RB, Berri BL, Schwartzkopff O. Incidence of sudden cardiac death, myocardial infarction and far- and near-transyears. Biomed Pharmacother 2005;59 Suppl 1:S239–61.
166. Cornelissen G, Otsuka K, Halberg F. Remove and replace for a scrutiny of space weather and human affairs. In: Grigoriev AI, Zeleny LM. editors. Proceedings of the International Conference, Space Weather Effects in Humans: In Space and on Earth, Space Research Institute, Moscow, Russia; 2012 June 4–8; Moscow: IKI RAN;2013. p. 508–38.
167. Hamamatsu A, Cornelissen G, Otsuka Ku, Halberg F, Chibisov S (presenter). Linear-nonlinear rhythmometry documents a transyear and a cis half year in sudden cardiac death (ICD 10, code I46.1) in Tokyo. In: Proceedings of the International Symposium, Problems of ecological and physiological adaptation, People's Friendship University of Russia, Moscow; 2007 Jan 30–31; Moscow: People's Friendship University of Russia; 2007. p. 542–5.
168. Fraser-Smith AC. Spectrum of the geomagnetic activity index Ap. J Geophys Res 1972;77(23):4209–20.

169. Silverman SM, Shapiro R. Power spectral analysis of auroral occurrence frequency. J Geophys Res 1983;88(A8):6310–6.
170. Gonzalez ALC, Gonzalez WD. Periodicities in the interplanetary magnetic field polarity. J Geophys Res 1987;92:4357–75.
171. Howe R, Christensen-Dalsgaard J, Hill F, Komm RW, Larsen RM, Schou J, Thompson MJ, Toomre J. Dynamic variations at the base of the solar convection zone. Science 2000;287:2456—60.
172. Prabhakaran Nayar SR. Periodicities in solar activity and their signature in the terrestrial environment. ILWS workshop Goa, Feb 19–24, 2006, 9 pp.
173. Cornelissen G, Burioka N, Halberg F. Other-than-annual variations in morning determinations of uric acid [editorial]. Clin Chem Lab Med 2012;50(5):765–70.
174. Cornelissen G, Halberg F, Sothern RB, Hillman DC, Siegelova J. Blood pressure, heart rate and melatonin cycles synchronization with the season, earth magnetism and solar flares. Scripta Medica (Brno) 2010;83:16–32.
175. Tarquini B, Cornelissen G, Perfetto F, Tarquini R, Halberg F. Chronome assessment of circulating melatonin in humans. In Vivo 1997;11:473–84.
176. Cornelissen G, Tarquini R, Perfetto F, Otsuka K, Gigolashvili M, Halberg F. Investigation of solar about 5-month cycle in human circulating melatonin: signature of weather in extraterrestrial space? Sun and Geosphere 2009;4(2):55–9.
177. Halberg F, Jacobson E, Wadsworth G, Bittner JJ. Audiogenic abnormality spectra, 24-hour periodicity and lighting. Science 1958;128:657–8.
178. Halberg F, Barnum CP. Continuous light or darkness and circadian periodic mitosis and metabolism in C and D¬8 mice. Amer J Physiol 1961;201:227–30.
179. Batschelet E, Hillman D, Smolensky M, Halberg F. Angular-linear correlation coefficient for rhythmometry and circannual changes of human births at different latitudes. Int J Chronobiol 1973;1:183–202.
180. Halberg F, Haus E, Tarquini B, Cagnoni M, Cornelissen G, Lakatua D, Kawasaki T, Wallach LA, Halberg F, Omae T. Circannual and circadian variations in some blood hormones, notably prolactin. In: Condorelli L, Teodori U, Sangiorgi M, Neri Serneri R, editors. Internal medicine, Proc. XIV Int. Cong. Internal Medicine ISM, Rome, Italy, October 15–19, 1978. Amsterdam/Oxford: Excerpta Medica; 1979. p. 5–11.
181. Nelson W, Cadotte L, Halberg F. Circadian timing of single daily "meal" affects survival of mice. Proc Soc Exp Biol Med (Maywood) 1973;144:766–9. doi:10.3181/00379727-144-37678.
182. Halberg F. Temporal coordination of physiologic function. Cold Spr Harb Symp Quant Biol 1960;25:289–310.
183. Halberg F, Bittner JJ, Gully RJ, Albrecht PG, Brackney EL. 24-hour periodicity and audiogenic convulsions in I mice of various ages. Proc Soc Exp Biol Med (Maywood) 1955;88:169–73. doi:10.3181/00379727-88-21526.
184. Nelson W, Halberg F. Effects of a synchronizer phase-shift on circadian rhythms in response of mice to ethanol or ouabain. Space Life Sci 1973;4:249–57.
185. Ertel RJ, Halberg F, Ungar F. Circadian system phase-dependent toxicity and other effects of methopyrapone (SU-4885) in mice. J Pharmacol Exp Ther 1964;146:395–9.
186. Bick H, Halberg F, Spink WW. Investigation on polyvinylpyrrolidone in mice given brucella endotoxin. Proc Soc Exp Biol Med (Maywood) 1956;93:7–10. doi:10.3181/00379727-93-22646.
187. Halberg F, Spink WW. The influence of brucella somatic antigen (endotoxin) upon the temperature rhythm of intact mice. Lab Invest 1956;5:283–94.
188. Halberg F, Johnson EA, Brown BW, Bittner JJ. Susceptibility rhythm to E. coli endotoxin and bioassay. Proc Soc Exp Biol Med (Maywood) 1960;103:142–4. doi:10.3181/00379727-103-25439.
189. Kovats T, Nelson W, Berner D, Halberg F. Endotoxin chronotolerance. In: Reinberg A, Halberg F, editors. Chronopharmacology, Proc. Satellite Symp. 7th Int. Cong. Pharmacol., Paris 1978. Oxford/New York: Pergamon Press; 1979. p. 223–30.

190. Halberg F, Conner RL. Circadian organization and microbiology: variance spectra and a periodogram on behavior of Escherichia coli growing in fluid culture. Proc Minn Acad Sci 1961;29:227–39.
191. Halberg F, Cornelissen G, Otsuka K, Katinas G, Schwartzkopff O. Physiological monitoring from bacteria and eukaryotic unicells to humans for chronoastrobiology and chronomedicine. Proceedings, 1st International Symposium, Workshop on Chronoastrobiology & Chronotherapy (Satellite Symposium, 7th Annual Meeting, Japanese Society for Chronobiology); Kudan, Chiyodaku, Tokyo; 2000. p. 67–106.
192. Halberg F, Cornelissen G, Sothern RB, Katinas GS, Schwartzkopff O, Otsuka K. Cycles tipping the scale between death and survival (= "life"). Prog Theor Phys 2008;Suppl. 173:153–81.
193. Cornelissen G, Halberg F, Rostagno C, Otsuka K. A chronomic approach to cardiac arrhythmia and sudden cardiac death. Auton Nerv Syst 2007;44:251–4.
194. Halberg F, Cornelissen G, Panksepp J, Otsuka K, Johnson D. Chronomics of autism and suicide. Biomed Pharmacother 2005;59 Suppl 1:S100–8.
195. Halberg F, Cornelissen G, Berk M, Dodd S, Henry M, Wetterberg L, Nolley E, Beaty L, Siegelova J, Fiser B, Wolff C, BIOCOS project. Solar signatures in Australian suicide incidence: gender differences in prominence of photic vs. nonphotic spectral components. In: Halberg F, Kenner T, Fiser B, Siegelova J. editors. Proceedings, Noninvasive Methods in Cardiology, Brno, Czech Republic, October 4–7, 2008. p. 44–62.
196. Stoupel E, Petrauskiene J, Kalediene R, Domarkiene S, Abramson E, Sulkes J. Distribution of deaths from ischemic heart disease and stroke. Environmental and aging influences in men and women. J Basic Clin Physiol Pharmacol 1996;7:303–19.
197. Close J. Are stress responses to geomagnetic storms mediated by the cryptochrome compass system? Proc R Soc B Biol Sci 2012;279(1736):2081–90.
198. Rapoport SI, Smirnova AV, Naumcheva NN, Gaidash SP. The first experience in application of melatonin (melaxen) for prophylaxis of the effects of magnetic storms on patients with cardiovascular pathology. Klin Med 2007;85(8):33–6.
199. Dupont MJ, McKay BE, Parker G, Persinger MA. Geophysical variables and behavior. XCIX. Reductions in numbers of neurons within the parasolitary nucleus in rats exposed perinatally to a magnetic pattern designed to imitate geomagnetic continuous pulsations: implications for sudden infant death. Percept Motor Skills 2004;98(3 Pt 1):958–66.
200. Gmitrov J, Gmitrova A. Geomagnetic field effect on cardiovascular regulation. Bioelectromagnetics 2004;25(2):92–101.
201. Gmitrov J, Ohkubo C. Verapamil protective effect on natural and artificial magnetic field cardiovascular impact. Bioelectromagnetics 2002;23(7):531–41.
202. Palmer SJ, Rycroft MJ, Cermack M. Solar and geomagnetic activity, extremely low frequency magnetic and electric fields and human health at the earth's surface. Surv Geophys 2006;27:557–95. doi:10.1007/s10712-006-9010-7.
203. Lang KR. The sun. In: Beatty JK, Collins Petersen C, Chaikin A, editors. The new solar system. 4th ed. Cambridge: Cambridge University Press; 1999. p. 23–38.
204. Van Allen JA, Bagenal F. Planetary magnetospheres and the interplanetary medium. In: Beatty JK, Collins Petersen C, Chaikin A, editors. The new solar system. 4th ed. Cambridge: Cambridge University Press; 1999. p. 39–58.
205. Wetterberg L, Bratlid T, Knorring LV, Eberhard G, Yuwiler A. A multinational study of the relationships between nighttime urinary melatonin production, age, gender, body size, and latitude. Eur Arch Psychiatr Clin Neurosci 1999;249:256–62.
206. Starbuck S, Cornelissen G, Halberg F. Is motivation influenced by geomagnetic activity? Biomed Pharmacother 2002;56 Suppl 2:289s–97s.
207. Cornelissen G, Beaty LA, Siegelova J, Lee Gierke C, Otsuka K, Watanabe Y, Syutkina EV, Masalov A, Gumarova L, Halberg F. An atlas of chronomes to map broad time structures. In: Kenner T, Cornelissen G, Siegelova J, Dobsak P. editors. Noninvasive methods in cardiology; 2013 May 3–4 and October 21; Brno, Czech Republic. Brno: Faculty of Medicine, Masaryk University; 2013. p. 102–16.

208. Halberg F, Cornelissen G, Grambsch P, McCraty R, Beaty L, Siegelova J, Homolka P, Hillman DC, Finley J, Thomas F, Kino T, Revilla M, Schwartzkopff O. Personalized chronobiologic cybercare; other chronomics' progress by transdisciplinary cycles' congruences: season's appreciations 2009. J Appl Biomed 2011;9:1–34. doi:10.2478/v10136-009-0022-8.
209. Cornelissen G, Grambsch P, Sothern RB, Katinas G, Otsuka K, Halberg F. Congruent biospheric and solar-terrestrial cycles. J Appl Biomed 2011;9:63–102. doi:10.2478/v10136-009-0023-7.
210. Schumann WO. Uber die strahlungslosen Eigenschwingungen einer leitenden Kugel, die von einer Luftschicht und einer Ionospharenhulle umgeben ist. Z Naturforsch 1952;7a:149–54.
211. Barr R, Jones DL, Rodger CJ. ELF and VLF radio waves. J Atmos Sol-Terr Phys 2000;62(17–18):1689–718.
212. Cherry NJ. Schumann resonances, a plausible biophysical mechanism for the human health effects of solar/geomagnetic activity. Nat Hazards 2002;26(3):279–331.
213. Hainsworth LB. The effect of geophysical phenomena on human health. Speculat Sci Technol 1983;6(5):439–44.
214. Delyukov A, Gorgo Y, Cornelissen G, Otsuka K, Halberg F. Natural environmental associations in a 50-day human electrocardiogram. Int J Biometeorol 2001;45:90–9.
215. Delyukov A, Gorgo Y, Cornelissen G, Otsuka K, Halberg F. Infradian, notably circaseptan testable feedsidewards among chronomes of the ECG and air temperature and pressure. Biomed Pharmacother 2001;55 Suppl 1:84–92.
216. Kirschvink JL, Walker MM, Diebel CE. Magnetite-based magnetoreception. Curr Opin Neurobiol 2001;11:462–7.
217. Semm P, Beason RC. Responses to small magnetic variations by the trigeminal system of the bobolink. Brain Res Bull 1990;25:735–40.
218. Kirschvink JL, Jones DS, McFadden B, editors. Magnetite biomineralization and magnetoreception in organisms: a new biomagnetism. New York: Plenum Press; 1985. p. 682 + xxi.
219. Breus TK, Pimenov KY, Cornelissen G, Halberg F, Syutkina EV, Baevsky RM, Petrov VM, Orth-Gomer K, Åkerstedt T, Otsuka K, Watanabe Y, Chibisov SM. The biological effects of solar activity. Biomed Pharmacother 2002;56 Suppl 2:273s–83s.
220. Kirschvink JL, Kobayashi-Kirschvink A, Diaz Ricci J, Kirschvink SJ. Magnetite in human tissues: a mechanism for the biological effects of weak ELF magnetic fields. Bioelectromagnetics 1992;13 Suppl 1:101–14.
221. Campbell WH. Geomagnetism applications. US Geological Survey Circular 1109. Washington, DC: US Government Printing Office; 1995. 32 pp.
222. McLauchlan K. Are environmental magnetic fields dangerous? Phys World 1992;5(1):41–5.
223. Coghlan A. Are power lines bad for you? New Sci 1992;134:22–3.
224. Blackman CF, Benane SG, House DE, Elliott DJ. Importance of alignment between local DC magnetic field and an oscillating magnetic field in responses of brain tissue in vitro and in vivo. Bioelectromagnetics 1990;11(2):159–67.
225. Cleary SF. Biophysical mechanisms of interaction. In: Stone WR, editor. Review of radio science 1990–1992. Oxford: Oxford University Press; 1993. p. 717–35.
226. Ulmer W. On the role of the interactions of ions with external magnetic fields in physiologic processes and their importance in chronobiology. In Vivo 2002;16:32–6.
227. Ulmer W, Cornelissen G, Halberg F. Physical chemistry and the biologic week in the perspective of chrono-oncology. In Vivo 1995;9:363–74.
228. Leung B, Cornelissen G, Hillman D, Wang ZR, Binkley S, Bingham C, Halberg F. Halting steps toward a circadian-infradian pineal melatonin chronome. In: Halberg F, Watanabe H, editor. Proc. Workshop on Computer Methods on Chronobiology and Chronomedicine, Tokyo, Sept. 13, 1990. Tokyo: Medical Review; 1992. p. 263–85.

229. Herold M, Cornelissen G, Rawson MJ, Katinas GS, Alinder C, Bratteli C, Gubin D, Halberg F. About-daily (circadian) and about-weekly (circaseptan) patterns of human salivary melatonin. J Anti-Aging Med 2000;3:263–7.
230. Cornelissen G, Halberg F, Zeman M, Jozsa R, Tarquini R, Perfetto F, Salti R, Bakken EE. Toward a chronome (time structure) of melatonin. In: Pandi-Perumal SR, Cardinali DP, editors. Melatonin: from molecules to therapy. New York: Nova Biomedical Books; 2007. p. 135–76.
231. Bawin SM, Adey WR. Sensitivity of calcium binding in cerebral tissue to weak electric fields oscillating at low frequency. Proc Natl Acad Sci USA 1976;73:1999–2003.
232. Halberg F. Historical encounters between geophysics and biomedicine leading to the Cornelissen-series and chronoastrobiology. In: Schröder W, editor. Long- and short-term variability in sun's history and global change. Bremen: Science Edition; 2000. p. 271–301.
233. Cornelissen G, Halberg F, Gheonjian L, Paatashvili T, Faraone P, Watanabe Y, Otsuka K, Sothern RB, Breus T, Baevsky R, Engebretson M, Schröder W. Schwabe's ~10.5- and Hale's ~21-year cycles in human pathology and physiology. In: Schröder W, editor. Long- and short-term variability in sun's history and global change. Bremen: Science Edition; 2000. p. 79–88.

Chapter 14
Heart Rate Variability Reflecting the Dynamics of Solar Activity

Abstract Geomagnetic variations of partly interplanetary origin, with cyclic signatures in human affairs and pathology, include the incidence of various diseases, for which this study of healthy subjects seeks a mechanism eventually by worldwide archival and physiological monitoring, notably of heart rate variability (HRV).

In the past half-century, the possible health and other hazards of natural, solar variability-driven temporal variations in the magnetic field have become a controversial subject in view of inconsistent results. Some well-documented claims of associations between geomagnetic storms and myocardial infarction or stroke have been rejected by a study based on ample data, analyzed by rigorous methods. It seems possible that inter-solar cycle and geographic variability, if not geographic differences, account for discrepancies. Herein, we examine the start of a planetary study on any influence of geomagnetic disturbances that are most pronounced in the auroral oval, on human HRV.

Evidence is provided herein that an alteration of HRV is most apparent in the circaminutan ("VLF") region, which is clinically important, because a reduction in its power is a predictor of morbidity and mortality from cardiovascular disease. The circadecasecundan ("LF") component of HRV also decreased in association with geomagnetic disturbance, which may reflect an episodic alteration of arterial pressure related to changes in geomagnetic activity. Lastly, our study suggests the existence of a light-dark-influenced magnetoreception mechanism in humans involving mainly the Pc 6 band of the magnetic field.

Keywords Geomagnetic storms in subarctic • Heart rate variability • Graded response to geomagnetic activity • Transfer function from geomagnetics • Light-dark alteration induced magnetoreception

14.1 Prologue

When religious motivation prevails, homicide abstains (religio reget, homicidium subsidet): homicide and religious motivation are in antiphase along the scale of an ~10.5-year Schwabe cycle. Chronomics aims to increase life expectancy and life quality, as it strives for the stars by a timed and timely chronoastrobiology [1–37];

K. Otsuka et al., *Chronomics and Continuous Ambulatory Blood Pressure Monitoring*, DOI 10.1007/978-4-431-54631-3_14

by information thus obtained, it also aims at improving everyday life on earth by prehabilitation, to minimize the need for rehabilitation. Chronomic detection of elevated illness-risks aims at the prevention of diseases of individuals, such as myocardial infarctions and strokes, and, equally important, chronomics resolves illnesses of societies, such as crime and war, all exhibiting mapped cycles, which are indispensable for the study of underlying mechanisms.

14.2 Introduction

Effects of geomagnetic disturbance on heart rate variability (HRV), the 1/f fractal scaling in particular, are being assessed in adults living at high latitude, where magnetic storms are more frequent and more intense than at lower latitudes. The latter may constitute a signal or a proxy, and possibly a mechanism underlying both undesirable and desirable effects, depending upon circumstances yet to be elucidated. Any circadecadal stage dependence of morbidity and/or mortality from certain conditions such as myocardial infarctions remains to be studied in both adult and pediatric populations. Further work could thus examine whether any associations of geomagnetic disturbances may account, at least in part, through effects upon the circulation, for long-term infra-annual changes, possibly anchored in the population's gene pool, observed in a number of anthropologic measurements at birth as well as in other population statistics.

14.3 Clinical Chronoastrobiology

In the past century, reports of possible health and other hazards of natural, solar variability-driven temporal variations in the earth's magnetic field have remained largely controversial, in part because of the lack of rigorous studies based on prospectively collected data, which, in turn, were interpreted mostly by visual inspection rather than by inferential statistics. More recent studies [1–17, 23, 35–57] have provided new information, as reviewed herein, but do not fully account for some discrepant results, which rejected associations between geomagnetic storms and myocardial infarction or stroke based on ample data, analyzed by rigorous methods [33]. A reconciliation of inconsistencies may perhaps be reached by a further scrutiny of possibly contributing factors, such as the solar cycle stage and geographic location of the different studies. Toward this goal, a planetary study on any influence of geomagnetic disturbances on human HRV was started, with focus on the auroral oval, where magnetic storms are most pronounced. The magnetic field variations exhibit complex spectra and include the frequency band between 0.001 and 10 Hz, which is regarded as "ultralow frequency" by physicists. Since the "ultralow-frequency" range, like other endpoints used in cardiology, refers to much higher frequencies than about-yearly and other changes playing a role in environmental-organismic interactions also in relation to HRV, the current designations used in cardiology are all placed in quotation marks to indicate the need for possible revision of nomenclature. Whether or not this suggestion has an immediate response, we point to a need for the development of instrumentation and software that renders the assessment of circadian, infradian, and even infra-annual (truly low frequency) modulations routinely feasible.

14.4 Geomagnetic Influences on Heart Rate Variability in the Subarctic

Associations between geomagnetic storms and myocardial infarction or stroke are of particular interest and constitute active areas of current research. In Minnesota, mortality from myocardial infarction is increased by 5 % at times of maximal solar activity, when geomagnetic disturbances occur more frequently [6]. Susceptible individuals may then be at an increased cardiovascular disease risk. Although there is mounting evidence for such associations, they are far from being fully understood and have remained partly controversial [10]. HRV is known as a powerful predictor of vascular disease risk in apparently healthy people, as evidenced in an elderly cohort, as well as in patients suffering from coronary artery disease, valvular heart disease, and congestive heart failure [3, 4, 8, 14, 24, 28, 31, 34, 37, 40, 43, 52, 53]. During the past several years, we focused on any effect of geomagnetic variations on HRV.

Nearly continuous, 7-day beat-to-beat ECG records were obtained between December 10, 1998, and November 2, 2000, on clinically healthy men (N = 15) and women (N = 4) from Finnmark College in Alta, Norway. Their age ranged from 21 to 59 years. Alta is located at a latitude of 69° 56 min north, a longitude of 23° 22 min east, and an altitude of 3 m above sea level. Mean atmospheric pressure in Alta was 1007.7 hPa. The yearly mean outside temperature was 2.3°C, and the lowest recorded temperature was −30.8°C, with temperatures below 0°C for almost half of the year. Because of the high latitude, the sun was not visible above the horizon for 52 days in winter (from November 26 to January 17) and did not set for 72 days in summer (from May 17 to July 27). During the remaining 241 days (8 months), there was an alternation between light and darkness (day-night) (Fig. 14.1). A geomagnetic record of the following variables was obtained at 1-min intervals from the Auroral Observatory of the University of Tromsø, in Tromsø, Norway (latitude 69° 39 min north, longitude 18° 56 min east) (Fig. 14.2): total intensity (F in nT), declination (D, angle between geographic and magnetic north, in degrees), inclination (I, angle between horizontal plane and magnetic direction, in degrees), horizontal intensity (H in nT), and vertical intensity (Z in nT).

One of the 19 subjects was excluded from this analysis, because all 7 days of the monitoring session were geomagnetically active. HRV was compared in the other 18 subjects between the two geomagnetic conditions (presence vs. absence of geomagnetic storm). Abbreviations and designations in frequency terms are shown in

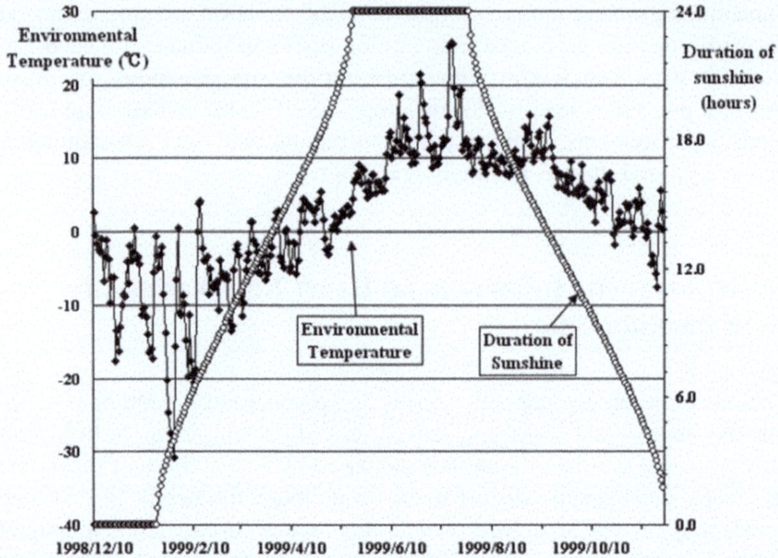

Fig. 14.1 Duration of sunshine (*open circles*) and environmental temperature (*black quadrangles*) in Alta, Norway (70°N), from December10, 1998, to November 23, 1999

Fig. 14.2 Geomagnetic recording at Auroral Observatory, University of Tromso, in 2002

Table 14.1. HRV was suppressed during a geomagnetically disturbed day (Fig. 14.3). Alterations of HRV associated with geomagnetic activity are summarized in Fig. 14.4. An increase in the 24-hour average of heart rate (HR) (P=0.020) and a decrease in HRV (P=0.002) were documented on days of high geomagnetic disturbance, findings fully corroborated as additional profiles were collected. The

Table 14.1 Abbreviations and designations in frequency terms awaiting transdisciplinary solutions

Abbrev.	Denomination ("F" = frequency)	Frequency range	Period (s) Range		Midpoint
Electromagnetic field					
EHF	Extremely high F	30–300 GHz	$3.3\ 10^{-12}$–$3.3\ 10^{-11}$		$6\ 10^{-12}$
SHF	Super high F	3–30 GHz	$3.3\ 10^{-11}$–$3.3\ 10^{-10}$		$6\ 10^{-11}$
UHF	Ultra high F	300–3000 MHz	$3.3\ 10^{-10}$–$3.3\ 10^{-9}$		$6\ 10^{-10}$
VHF	Very high F	30–300 MHz	$3.3\ 10^{-9}$–$3.3\ 10^{-8}$		$6\ 10^{-9}$
HF	High F	3–30 MHz	$3.3\ 10^{-8}$–$3.3\ 10^{-7}$		$6\ 10^{-8}$
MF	Medium F	300–3000 KHz	$3.3\ 10^{-7}$–$3.3\ 10^{-6}$		$6\ 10^{-7}$
LF	Low F	30–300 KHz	$3.3\ 10^{-6}$–$3.3\ 10^{-5}$		$6\ 10^{-6}$
VLF	Very low F	3–30 KHz	$3.3\ 10^{-5}$–$3.3\ 10^{-4}$		$6\ 10^{-5}$
ELF	Extremely low F	3–3000 Hz	$3.3\ 10^{-4}$–$3.3\ 10^{-1}$		$6.7\ 10^{-4}$
ULF	Ultra low F	0.001–5 Hz	$3.3\ 10^{-1}$–10^{3}		$6.7\ 10^{-1}$
Geomagnetic pulsations (IAGA classification)					
cont.:	Pc 1	0.2–5.0 Hz	0.2	–5	0.38
	Pc 2	0.1–0.2 Hz	5	–10	6.7
	Pc 3	22–100 mHz	10	–45	16.4
	Pc 4	6.7–22 mHz	45	–150	69.7 (–1.15 min)
	Pc 5	1.7–6.7 mHz	150	–600	241 (–4 min)
	Pc 5	0.055–1.7 mHz	600	–18,000 (10 min–5 hour)	1140 (–19 min)
irreg.:	Pi 1	0.025–1 Hz	1	–40	2.0
	Pi 2	6.7–25 mHz	40	–150	63.1 (–1.05 min)
Cardiology ("heart rate variability") [1]					
ULF	Ultra-low F	0.0001–0.003 Hz	3.33	–10,000 (–5 min to –3 hour)	645 (–10.75 min)
VLF	Very low F	0.003–0.04 Hz	25	–333	–46.5
LF	Low F	0.04–0.15 Hz	6.7	–25	–10.5
HF	High F	0.15–0.4 Hz	2.5	–6.7	–3.6

"cont." = continuous, "irreg." = irregular. Please note the difference in frequency ranges of ULF, VLF, LF, and HF between terminology used in relation to the electromagnetic field vs. heart rate variability

decrease in spectral power was found primarily at frequencies lower than 0.04 Hz and was not statistically significant around one cycle in the HF component (3.6 sec) (Fig. 14.4). The physiological mechanism involved may be other than the parasympathetic, usually identified with spectral power centered around one cycle in 3.6 s, a spectral region wherein no statistically significant differences were found either in this or in an earlier study. The decrease in HRV was more pronounced in the circaminutan ("VLF") (21.9 % decrease, $p<0.00001$) than in the "ULF" (15.5 % decrease, $p=0.00865$) region of the spectrum.

In this table, we propose, whenever possible, the use of midpoints of frequency ranges, whether we deal with the purely physical or the biological spectra. From a biological viewpoint, it is important that one should not call "low/very low/ultralow"

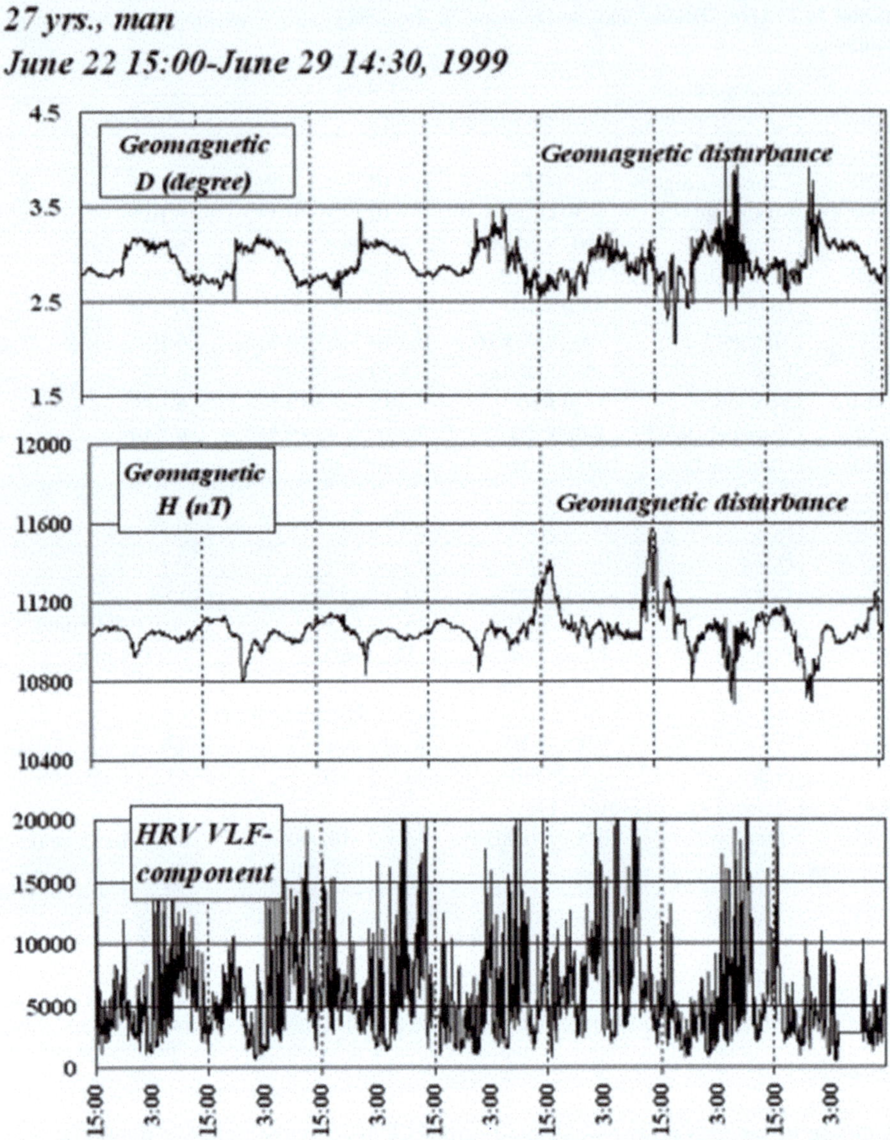

Fig. 14.3 A week's record of geomagnetic D- and H-components in Tromso (near Alta) and VLF-component of HRV in Alta. The VLF-component is apparently suppressed during a geomagnetically disturbed day. Alterations of HRV associated with a geoimagnetic storm are summarized in Table 14.3. A 21.9 % decrease was observed in the 24-hour average of the VLF chronome component ($P<0.000001$)

frequencies that complete one cycle in seconds or minutes when there are also components with frequencies that complete one cycle in about 10 years [2, 3].

Because of earlier work [1], but only for the purpose of this table, the midpoint was computed as the period corresponding to the middle of the frequency range

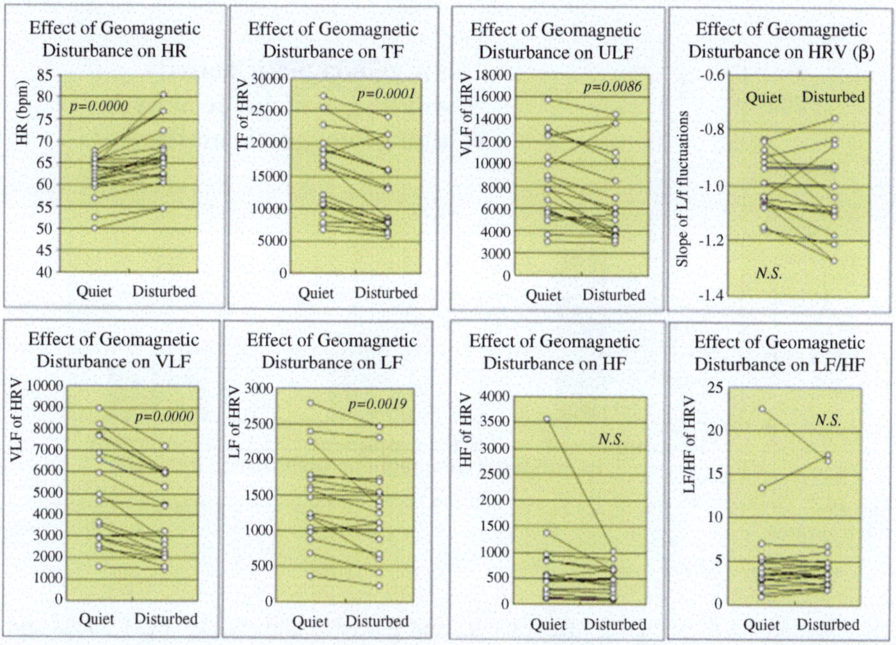

Fig. 14.4 Comparison of the 24-hour average of heart rate (HR) and HRV measures between quiet and geomagnetically disturbed days TF is total spectral power from 5-min epochs; ULF is spectral power in frequency range 0.0001–0.003 Hz (333–10,000 s, or 5 min–3 hour); VLF is spectral power in frequency range 0.003–0.04 Hz (25–333 s, average 46.5 s); LF is spectral power in frequency range 0.04–0.15 Hz (6.7–25 s, average 10.5 s); HF is spectral power in frequency range 0.15–0.40 Hz (2.5–6.7 s, average 3.6 s)

(midpoint = ((Fmin + Fmax)/2) − 1), using a linear rather than a log frequency scale, which latter is more desirable generally in physics as in biology [4] when wide frequency ranges may have to be dealt with.

14.5 Graded Response of Heart Rate Variability

Another investigation of 7-day ECG records from five clinically healthy young men living above the arctic circle compared HRV measures assessed over separate 24-hour spans among days of low, middle, and high geomagnetic activity, defined as days with values of the geomagnetic index ap <7, 7–20, and 20–45, respectively. A graded response was demonstrated, the extent of HRV decrease depending on the degree of geomagnetic activity (Fig. 14.5). Specifically, "TF" was decreased by

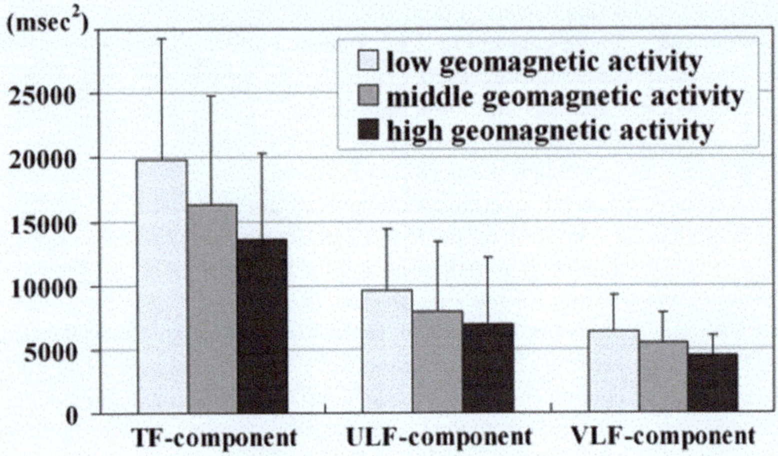

Fig. 14.5 Graded response of HRV, associated with an alteration of geomagnetic activity in a subarctic area

18.1 % and 31.6 %, "ULF" by 18.1 % and 27.5 %, and "VLF" by 12.9 % and 28.6 %, on days when 7<ap<20 and 20<ap<45 (vs. days when ap<7), respectively (Table 14.2). A graded response in HRV to geomagnetic activity suggests the existence of human magnetoreceptors. Phillips and Borland [46] proposed a light-dependent magnetoreception mechanism and established a link between magnetic field sensitivity and the visual system in eastern red-spotted newts. The results observed in this study suggest the operation of a light-darkness-alternation-influenced, if not light-dependent magnetoreception mechanism in humans.

14.6 Transfer Function from Geomagnetics to HRV

Transfer function from geomagnetics (input) to heart rate variability (output) was analyzed in six healthy young subjects in Alta, Norway. Average and standard deviation of each frequency is shown in Fig. 14.6, which shows that the frequency response is larger for the VLF-component of HRV compared with LF- or HF-components of HRV. It is one of the reasons why geomagnetic disturbance affects HRV not in HF- or LF-component, but in VLF- or ULF-components as we have observed in this investigation.

Table 14.2 Heart rate variability parameters on days of low, middle, and high geomagnetic activity in healthy subjects (n=5)

	Low		Middle		High		2-way ANOVA
	Mean	SD	Mean	SD	Mean	SD	
ap	5.35	1.37	12.20	4.95	28.65	8.16	p-value
Average HR	63.18	6.89	63.89	7.07	67.73	5.43	N.S.
NN	1005.13	108.72	995.92	110.12	934.05	76.29	N.S.
CVNN	10.97	1.10	10.20	1.31	10.17	1.55	N.S.
SDNNIDX	107.16	22.64	97.89	20.44	89.01	13.63	0.0144
r-MSSD	72.55	29.02	63.31	18.81	55.38	11.23	N.S.
NN50	87.89	26.70	81.04	17.83	72.71	15.65	N.S.
%NN50	36.86	12.67	33.01	9.36	27.52	5.75	N.S.
Length 90 %	498.56	111.08	457.23	98.36	419.68	75.99	0.0102
Width 90 %	197.90	97.61	164.71	63.13	147.45	48.18	N.S.
Length/width	3.1075	0.9197	3.3072	0.6979	3.4054	0.3501	N.S.
β (1/f)	−1.0133	0.1053	−0.9641	0.1099	−0.9236	0.1670	N.S.
TF	19870.99	9433.06	16300.58	8548.13	13594.99	6750.35	0.0013
ULF	9619.19	4764.94	7874.47	5507.11	6969.53	5184.13	0.0102
VLF	6302.54	2869.74	5489.85	2288.11	4497.39	1597.39	0.0209
LF	2080.00	998.54	1874.63	693.81	1576.80	430.25	N.S.
HF	1307.31	1350.66	950.17	790.46	692.18	304.77	N.S.
LF/HF	3.5943	1.6091	4.0682	1.6339	3.9495	1.2558	N.S.
ULF/TF	0.4950	0.0612	0.4706	0.0597	0.4827	0.0941	N.S.
VLF/TF	0.3213	0.0229	0.3391	0.0411	0.3265	0.0507	N.S.
LF/TF	0.1100	0.0110	0.1261	0.0245	0.1259	0.0322	N.S.
HF/TF	0.0710	0.0633	0.0614	0.0408	0.0618	0.0387	N.S.

Average HR=24-hour average of heart rate (bpm); β(1/f)=slope of power-law regression line of HR variability between 10^{-2} and 10^{-4}, CVNN=coefficient of variation of NN intervals (%); HF; high-frequency spectral component (msec^2); Length90 %=90 % length of Lorenz plot (msec); L/W=length 90 % to width 90 % ratio; LF=low-frequency spectral component (msec^2); NN50=number of difference between adjacent normal cycle intervals >50 msec computed over the entire 24-hour ECG recordings; pNN50=percent of difference between adjacent normal cycle intervals >50 msec computed over the entire 24-hour ECG recordings (%); r-MSSD=root mean square successive difference (msec); SDNNIDX=mean of the SDs of all NN intervals for all 5-min segments of a 24-hour ECG recording (msec); TF=total frequency spectral component (msec^2); ULF=ultralow-frequency spectral component (msec^2); VLF=very-low-frequency spectral component (msec^2); width 90 %=90 % width of Lorenz plot (msec)

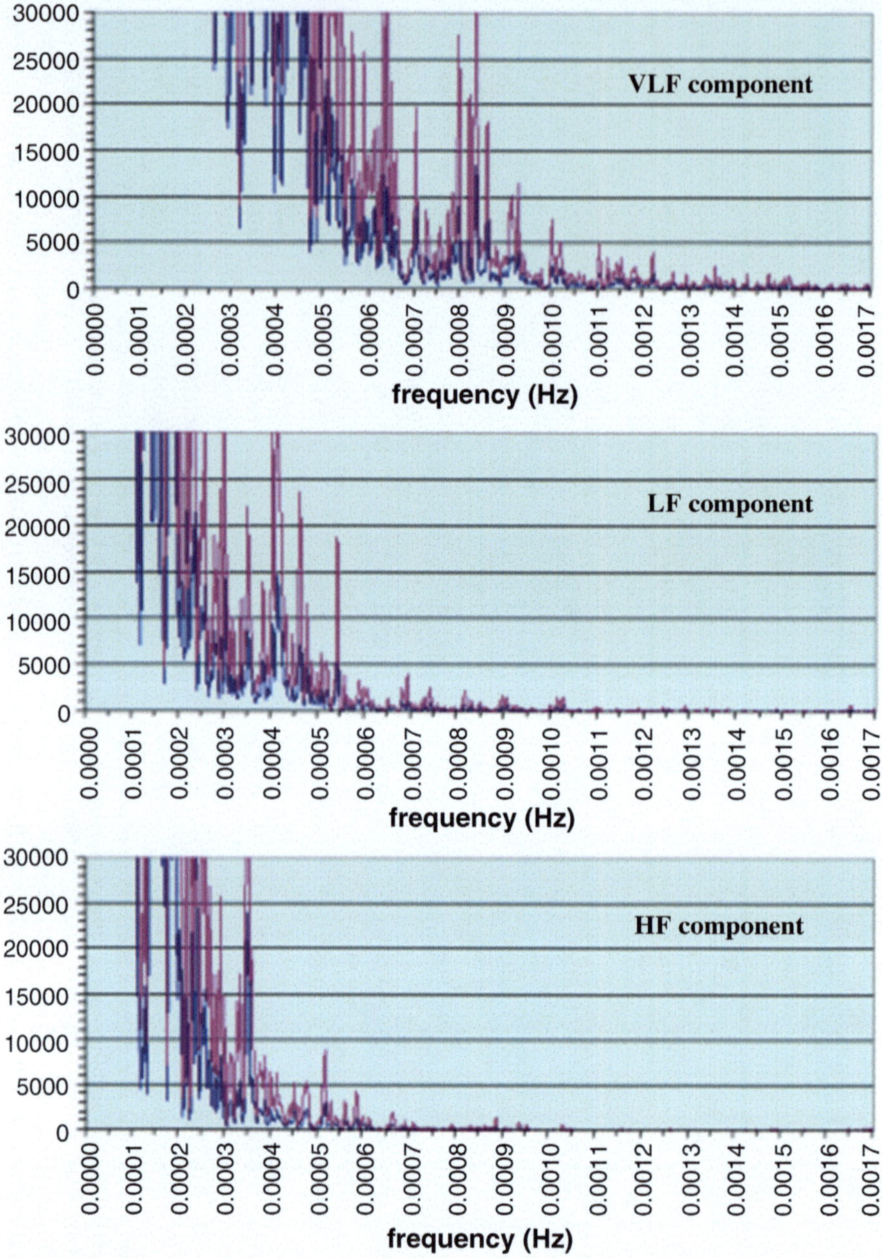

Fig. 14.6 Heart rate variability response to fluctuations in geomagnetic H-component: group-average transfer functions in healthy subjects (n = 6)

14.7 Light-Darkness-Alternation-Influenced Magnetoreception

Circadian characteristics of physiological functions undergo circannual changes in humans, but much less prominently in heart rate as compared to blood pressure, among several other variables. An effect of sunshine duration (from sunrise to sunset) on the magnetoreception of frequency-domain measures of HRV was investigated. Six of the 19 subjects were monitored under the dark-light sunshine condition between February 11 and October 28, with an average duration of sunshine of 625.2 ± 176.6 min (428–920 min), classified as the D/L group. Two subjects were monitored under the continuous, albeit slightly changing in intensity, light (sunshine) condition from May 14 to June 30, and the four other subjects were studied under the continuous dark condition (except for artificial lighting) from December 10 to January 13, classified as the L/L group and D/D group, respectively.

Alterations of HRV associated with geomagnetic disturbances differed in the D/L vs. the L/L and D/D groups, which was more pronounced in the presence of a daily light-dark alternation (D/L) than in continuous light (L/L) or in continuous darkness (D/D), as shown in Table 14.3. The decrease in "TF," "ULF," and "LF" components of HRV was statistically significant only in the D/L group, while the decrease in the "VLF" component was observed in both groups.

This finding is clinically important because a reduction in the circaminutan ("VLF") range is associated with an increased risk of cardiac and dysrhythmic death. Several mechanisms have been proposed for the circaminutan ("VLF") component, namely, thermoregulation, the renin-angiotensin-aldosterone system, and other humoral factors.

14.8 Coherence Coefficient by Cross-Spectral Analysis

Light-darkness-alternation-influenced magnetoreception was confirmed by cross-spectral analysis. An example of cross-spectral analysis from H-component of the geomagnetics to VLF-component of the HRV is shown in Fig. 14.7.

Group-averaged coherences between the geomagnetic H and the HRV "VLF" component are shown in Fig. 14.8, including the D/L group ($n = 3$), the L/L group ($n = 3$; the two subjects under the continuous light (sunshine) condition and another subject under the 1195 min of daily sunshine average condition monitored from May 3 to May 11), and the D/D group ($n = 3$).

The average coherences at frequencies coincident with Pc 5′ (with periods ranging from 10 minutes to 5 hours) were 0.4572 ± 0.1371 (D/L), 0.3720 ± 0.1334 (L/L), and 0.3002 ± 0.1182 (D/D). The difference among groups was statistically significant ($F = 363.0$, $p < 0.000001$), the average during D/D being smaller than those during D/L ($p = 0.000002$) and L/L ($p = 0.000072$) (Fig. 14.9, left). No statistically

Table 14.3 Geomagnetic disturbance-associated decreases in HR variability during the alternation of darkness (D) and light (L), i.e., DL vs. continuous darkness (DD) or light (LL)

Endpoint	Lighting regimen: Continuous light	Magnetic quiet		Geomagnetic disturbance		Student t-value	p-value
		Mean	SD	Mean	SD		
Heart rate	D/L	61.42	5.14	66.05	7.32	−3.5721	0.01601
	D/D or L/L	61.06	6.11	66.38	7.67	−5.6800	0.00236
NN interval	D/L	983.0	88.6	917.8	105.83	4.6307	0.00568
	D/D or L/L	1010.9	101.0	935.3	99.57	6.0950	0.00172
β (slope of 1/f)	D/L	−1.03	0.12	−1.08	0.15	0.9356	N.S.
	D/D or L/L	−0.94	0.10	−0.98	0.17	0.8011	N.S.
"TF"	D/L	16851.0	7550.3	13610.1	7928.5	7.8010	0.00056
	D/D or L/L	16426.0	3080.7	13391.5	4962.8	2.0461	N.S.
"ULF"	D/L	9233.2	4037.0	7708.1	5017.7	2.6359	0.04621
	D/D or L/L	8413.7	2882.8	7362.6	3775.7	1.0318	N.S.
"VLF"	D/L	5664.1	2923.0	4245.0	2453.3	6.8679	0.00100
	D/D or L/L	5409.9	1294.0	4185.5	1458.1	4.7584	0.00507
"LF"	D/L	1410.0	624.9	1272.1	692.2	4.1032	0.00933
	D/D or L/L	1492.9	504.9	1275.0	324.2	1.4299	N.S.

"HF"	D/L	511.0	497.1	355.8	290.0	1.4011	N.S.
	D/D or L/L	1045.1	1248.3	574.1	263.1	1.1216	N.S.
"LF/HF"	D/L	5.390	4.489	6.087	5.780	−1.0361	N.S.
	D/D or L/L	2.973	1.386	3.490	1.567	−2.4620	N.S.
"ULF/TF"	D/L	0.5578	0.0606	0.5585	0.0876	−0.0278	N.S.
	D/D or L/L	0.5070	0.1000	0.5291	0.1061	−0.8503	N.S.
"VLF/TF"	D/L	0.3266	0.0529	0.3132	0.0636	0.8387	N.S.
	D/D or L/L	0.3297	0.0537	0.3061	0.0468	1.0825	N.S.
"LF/TF"	D/L	0.0881	0.0306	0.1040	0.0520	−1.7132	N.S.
	D/D or L/L	0.0959	0.0334	0.1110	0.0452	−1.8897	N.S.
"HF/TF"	D/L	0.0259	0.0181	0.0225	0.0086	0.6418	N.S.
	D/D or L/L	0.0649	0.0598	0.0510	0.0299	1.0665	N.S.

D/L (n=6), D/D or L/L (n=6)
"TF" is total spectral power from 5-min epochs. "ULF" is spectral power in frequency range 0.0001–0.003 Hz (333–10,000 s or ~5 min to ~ 3 hour; ~10.75 min or 645 s). "VLF" is spectral power in frequency range 0.003–0.04 Hz (25–333 s; ~46.5 s). "LF" is spectral power in frequency range 0.04–0.15 Hz (6.7–25 s; ~10.5 s). "HF" is spectral power in frequency range 0.15–0.40 Hz (2.5–6.7 s; ~3.6 s)

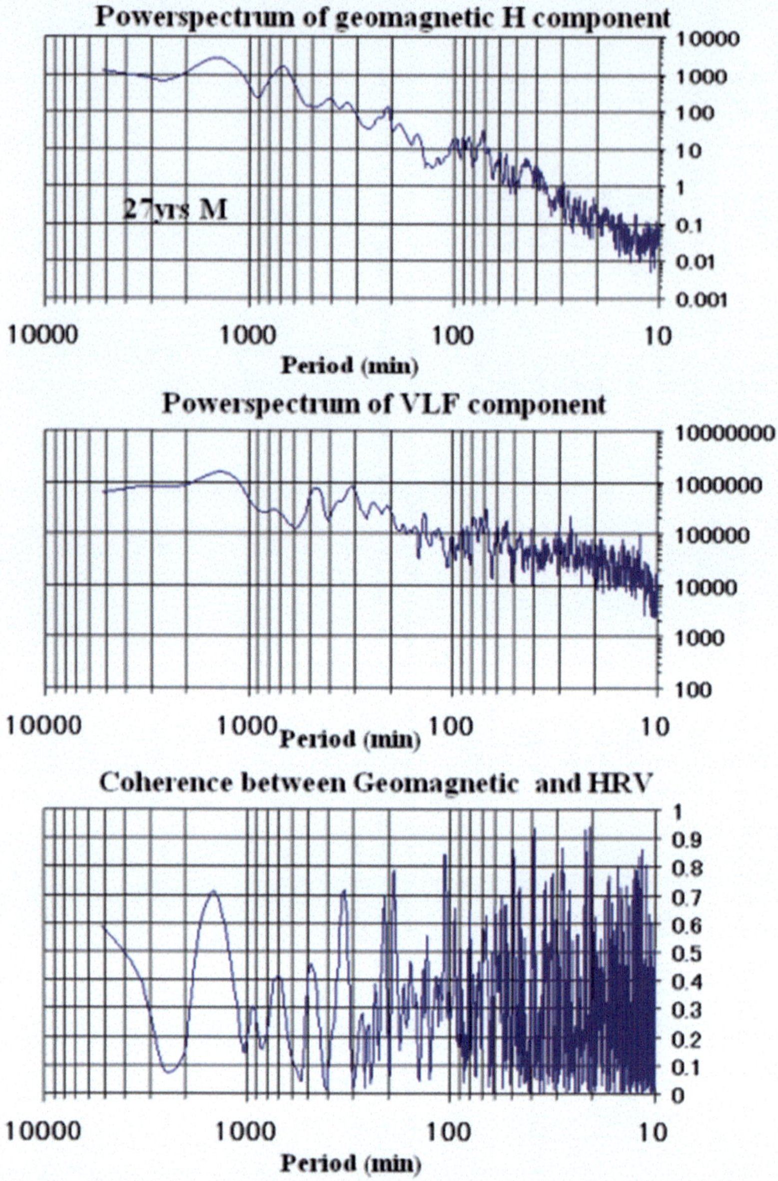

Fig. 14.7 An example of cross-spectral coherence (*bottom*) from H-component of the geomagnetics (*top*) to VLF-component of the HRV (*middle*)

significant group differences in average coherence were found at the circadian or circasemidian frequencies (in the period range from 10 to 34 hours).

Similarly, the group-averaged coherences between the geomagnetic D and the HRV "VLF" component were calculated. The average coherence of each group at

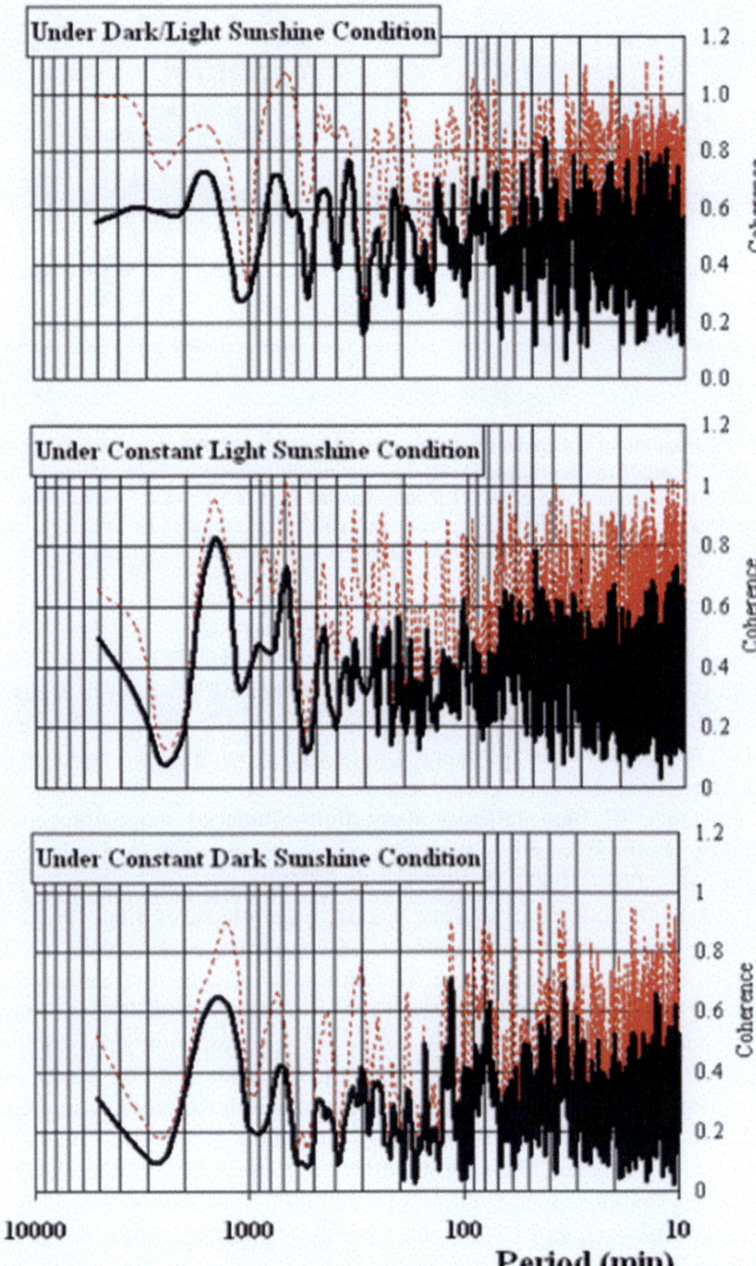

Fig. 14.8 Group-averaged coherence between geomagnetic H and VLF-component in HRV. *Top*, under D/L conditions (n = 3); *middle*, in L/L conditions (n = 3); and *bottom*, in D/D (n = 3)

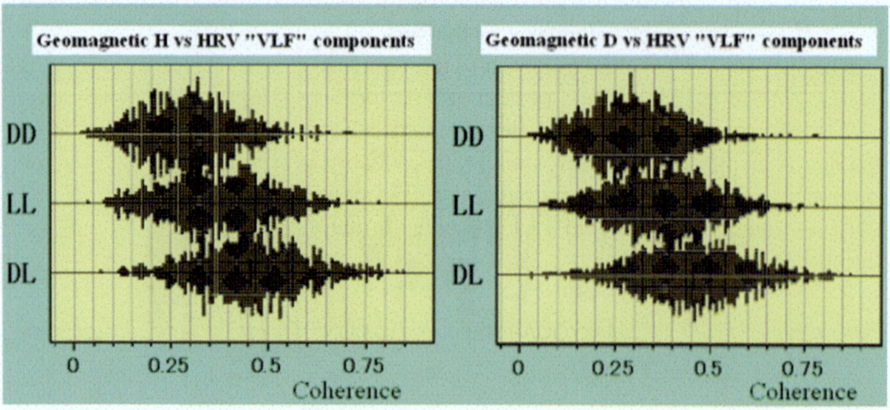

Fig. 14.9 Distribution of coherences between geomagnetics and HRV at frequencies coincident with the Pc 5′ band of the geomagnetic ultralow-frequency pulsations. *Left*: coherence between geomagnetic H-component and HRV VLF range under D/D, L/L, and D/L conditions. *Right*: coherence between geomagnetic D-component and HRV VLF range under D/D, L/L, and D/L conditions

frequencies coincident with Pc 5′ was 0.4551 ± 0.1451 (D/L), 0.3782 ± 0.1310 (L/L), and 0.2953 ± 0.1192 (D/D). The difference among groups was statistically significant ($F = 362.4$, $p < 0.000001$), the average during L/L being smaller than that during D/L ($p = 0.0007$) and the average during D/D being even smaller than that during L/L ($p = 0.0011477$) (Fig. 14.9, right).

The existence of light-darkness-alternation-influenced magnetoreception is supported by results from cross-spectral analyses. Group-averaged coherences at frequencies of 0.001–0.0001 Hz (0.25–5 hours) differed with statistical significance ($p < 0.000001$) among the three lighting conditions, on the average being smaller during L/L or D/D than during L/D. By contrast, no difference was found in the circadian-circasemidian range. The biological effects of so-called ultralow frequency (0.001–10 Hz) and extremely low frequency (10–300 Hz) magnetic fields have been investigated at several levels: human epidemiological studies, animal experiments, and cellular studies. Intermittent fields and components in the "ultralow frequency" band of the electromagnetic field have been considered to be more bioeffective than the predominantly sinusoidal power line fields in the "extremely low-frequency" band. The "ultralow frequencies" of the electromagnetic field are composed of 6 kinds of continuous pulsations, including Pc 1 (period range: 0.2–5 s), Pc 2 (5–10 sec), Pc 3 (10–45 sec), Pc 4 (45–150 sec), Pc 5 (150–600 sec), and Pc 5′ (10 minutes to 5 hours). The cross-spectral analysis results thus suggest that the magnetoreception mechanism may involve the Pc 5′ geomagnetic pulsations.

14.9 Altered Heart Rate Variability on Geomagnetically Disturbed Days and the Following Two Days in Subarctic

Halberg et al. [17] showed that mortality from myocardial infarction is increased on the day following a southward turn of the vertical component (Bz) of the interplanetary magnetic field, which has been associated with auroral geomagnetic storms. As a follow-up, we investigated for how long geomagnetic activity affected various measures of HRV. Associations between geomagnetic storms and HRV were observed not only on the day of a storm but also for 2 days immediately following a geomagnetically disturbed day. Alterations in HRV indices on each day were analyzed by Dunn test. Whereas changes in HR, NN interval, r-MSSD, and pNN50 were found to be statistically significant only during the day of the storm, changes in SDNNIDX (30-min) and in the power-law scaling of the power spectra ("β (beta)") remained statistically significant for one or 2 days following the day with a geomagnetic storm. One illustrative example is given in Fig. 14.10, which shows the concomitant 5-day record of the 1-min geomagnetic H-component and the 5-min β (beta) index (the slope of 1/f fluctuations of HRV). The lasting effect of geomagnetic storms on the power-law scaling of the power spectra was further checked for the case of "β $(\text{beta})_{3h}$" by ANOVA, using Bonferroni's correction. Results indicate that "β $(\text{beta})_{3h}$" became statistically significantly steeper not only on the day of the storm but also on the 2 days following a geomagnetic storm

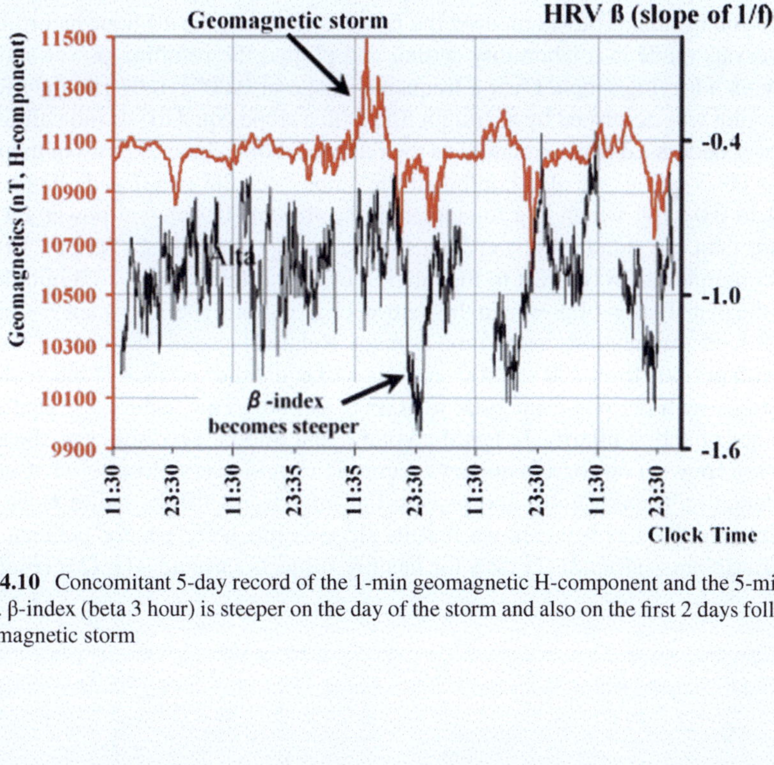

Fig. 14.10 Concomitant 5-day record of the 1-min geomagnetic H-component and the 5-min beta index. β-index (beta 3 hour) is steeper on the day of the storm and also on the first 2 days following a geomagnetic storm

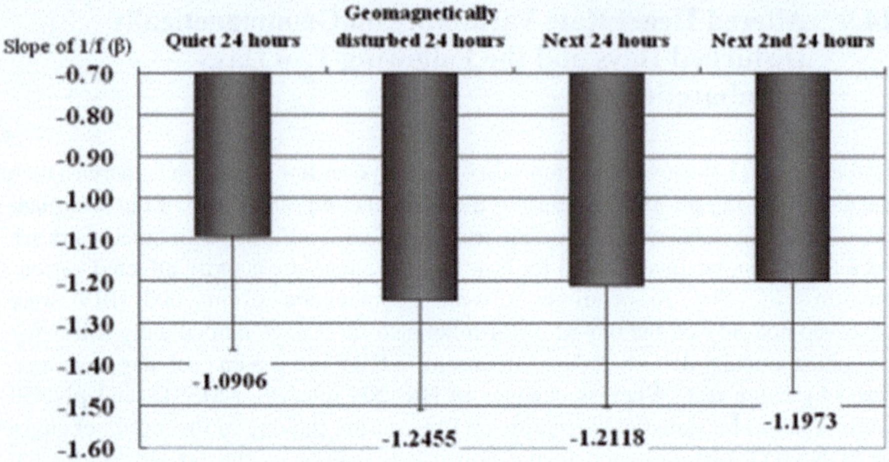

Fig. 14.11 Effect of geomagnetic disturbance on the fractal scaling of HRV (n=6)

($p<0.0001$) (Fig. 14.11), suggesting that geomagnetic disturbances have a lasting influence on the fractal scaling of HRV.

The slope (24 hours) of 1/f fluctuations of NN (power-law scaling of the power spectra, "β (beta)") on the day of a geomagnetic storm was also much steeper than on a geomagnetically quiet day. In 1982, Kobayashi and Musha [30] reported the frequency dependence of the power spectrum of RR-interval fluctuations in a healthy young man. They computed the power spectrum of a 10-hour recording of RR intervals made in a laboratory setting and plotted the resulting power and frequency on a log-log graph. Over a frequency range of 0.0001–0.02 Hz, they found that the plot was described by a straight line with a slope equal to −1, indicating that the power decreased approximately as the reciprocal of frequency, 1/f. On the basis of their observation, we also computed the slope over the two-decade band from 0.0001 to 0.01 Hz, which is a fundamentally different RR-interval power spectral measure than the standard power bands representing the "ULF," "VLF," "LF," or "HF" components. In contrast to the latter, the slope reflects not the magnitude but rather the distribution of power in this two-decade region. These frequency-domain features have implications in the time domain. Relatively rapid RR-interval oscillations with periods from 100 to 1000 seconds will equal the variance of much slower oscillations with periods from 1000 to 10,000 seconds. This feature is called scale invariance or self-similarity. In healthy young- and middle-aged subjects, the slope of the log (power) on log (frequency) regression line is very close to −1. Cardiac transplantation (denervation) and myocardial infarction shift the entire regression line down, and it has been shown that its slope is much steeper for patients with heart transplants (around −2) than for healthy subjects (around −1). Our results in

Alta show that "β $(\text{beta})_{24h}$" was −1.0738 (SD; 0.1155) during geomagnetically quiet conditions and that it was much steeper (average; −1.1838, SD; 0.1106) during a geomagnetic storm ($p < 0.005$ by paired *t*-test). Several authors suggested that this slope is a better predictor of all-cause mortality than the traditional power spectral bands. Thus, the concomitant longitudinal monitoring of geomagnetic activity and clinical data, especially the fractal scaling of HRV, may provide useful information for designing countermeasures for the prevention of adverse vascular events among patients at high cardiovascular disease risk.

14.10 Altered Heart Rate Variability Associated with Exposure to Artificially Generated "ULF" Waves Approximating Geomagnetic Pulsations During Geomagnetic Storms

To explore any influence of artificially generated "ULF" waves approximating geomagnetic pulsations during magnetic storms, two healthy volunteers were exposed for 8 hours to 50 nT ULF waves (0.0016 Hz) in a magnetically shielded room for 12 consecutive weekends. The exposure was carried out at one of six different circadian stages (i.e., 3:00–11:00 hour, 7:00–15:00 hour, 11:00–19:00 hour, 15:00–23:00 hour, 19:00–3:00 hour, and 23:00–7:00 hour), while subjects remained on a wooden bed inside the room for 24 hours, except for meals. The magnetically shielded room (interior dimensions: 3 m width × 3 m depth × 3 m height), which cuts off the "ELF" ("extremely low frequency"; 3–3000 Hz) and "VLF" ("very low frequency"; 3–30 KHz) electromagnetic field, was kept at a temperature of about 23 °C. All illuminations in the room were turned off during the night (i.e., 23:00–7:00 hour). The beds in the room were each located between a pair of custom-built Helmholtz coils 95 cm in radius with 12 turns (N) of enameled copper wire. The coils were connected to a function generator (Stanford Research Systems Model DS345 Function & Arbitrary Waveform Generator, Stanford Research Systems, Sunnyvale, CA) outside the room. Protection was provided to prevent the field from exceeding 50 nT with fuses and a breaker. Two investigators continuously monitored the current in the coils, using a Tektronix Model TDS210 Digital Oscilloscope (Tektronix, Beaverton, OR).

Systolic (S) and diastolic (D) BP were measured automatically every 30 minutes for 24 hours by ambulatory monitoring (A & D Co., Ltd, Tokyo). A Holter ECG record was obtained concomitantly. Alterations of HRV endpoints similar to those observed in a subarctic area in association with auroral geomagnetic storms were found in relation to the exposure to artificially generated "ULF" waves. Specifically, HRV indices ("ULF" and "VLF" components) were statistically significantly lower after exposure (4289.0 vs. 6550.6 msec^2, and 3466.0 vs. 3931.2 msec^2, $P < 0.001$) in comparison to sham exposure. The slope of 1/f fluctuations (power-law behavior,

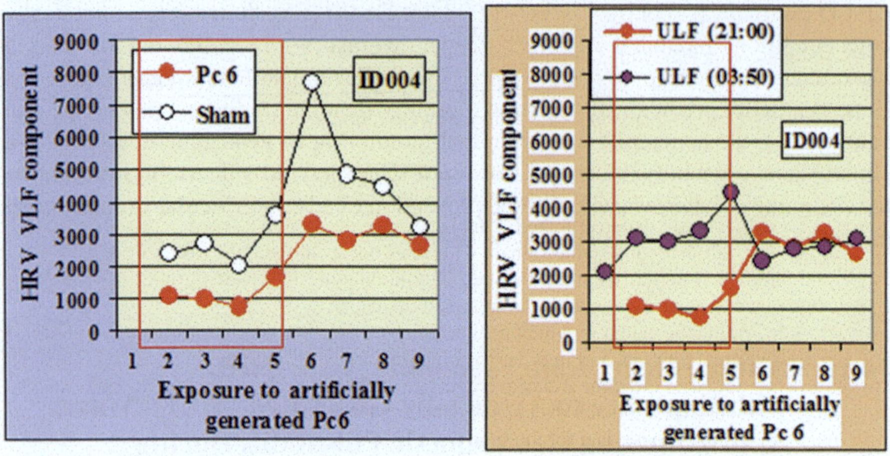

Fig. 14.12 *Left*: Decrease in heart rate variability (HRV "VLF" component) both during and after exposure to artificially generated "ULF" waves (Pc 6). *Right*: Difference in the response of HRV "VLF" to Pc 6 exposure at two different circadian stages (21:00 and 03:50). The effect of the electromagnetic field on HRV was smaller at night (03:50) than in the evening (21:00)

"β (beta)") of NN intervals was much steeper during and after actual than during and after sham exposure (–0.9462 vs. –0.7914, $p<0.005$, and –0.9416 vs. –0.8144, $p<0.0001$, respectively). SBP did not show any statistically significant alterations during real vs. sham exposure (115.0 vs. 115.3 mmHg), but DBP was statistically significantly higher during real than during sham exposure (74.0 vs. 71.7 mmHg, $p=0.044$). One illustrative example is shown in Fig. 14.12.

It is noteworthy that the alteration of the scaling properties of HRV continued after ULF exposure. Recently, not only did the analysis of time- and frequency-domain measures of HRV from 24-hour ambulatory ECG recordings provide prognostic information on patients after an acute myocardial infarction, but so did fractal analysis methods for the general population and for patients with impaired left ventricular function. An abnormal HRV predicts non-sudden fatal and nonfatal cardiac events as well as noncardiac causes of death. A number of new methods of nonlinear system theory ("chaos theory and fractals") have been developed to quantify the complex HR dynamics as a complement to the conventional measures of HRV. Increasing evidence shows that scaling properties of HRV, analyzed by methods of nonlinear system theory, may provide information about the risk of life-threatening arrhythmic events, complementing, as a feature of chronomics, traditional measures in different types of pathophysiological settings. Thus, results observed in the field in Alta, Norway, and in the laboratory by artificially generated "ULF" waves, suggest that prospective studies in the subarctic and at other latitudes should include fractal measures of HRV to further check on health hazards of natural, solar variability-driven temporal variations of the earth's magnetic field or of factors for which these variations constitute a proxy.

14.11 Conclusion

In clinical health, both HR and BP variabilities (HRV and BPV) are characterized by a broad time structure that includes the prominent circadians and also ultradian (notably about 8-hour and about 12-hour) and infradian (notably about-weekly, about-yearly, and about 10-yearly) changes, in addition to undergoing trends with growth, maturation, and aging. Alterations in these HRV and BPV chronomes are observed in the presence of disease and at times of increasing magnetic activity. We postulate that the physiological chronomes, such as those of HRV endpoints, have counterparts in our environment and that our genetic makeup in time may have evolved to adapt to and integrate with our cosmos. Future work should focus on how phenomena in the cosmos, including helio- and geomagnetics, can affect physiological chronomes, those of the HRV in particular.

Human beings have shielded themselves from the worst features of obvious extremes of weather and climate by building houses and have developed means, e.g., sulfonamides and antibiotics, to deal with at least some of the microbial contagions confronting us. These infections via the mutation of microorganisms causing them and by (steroidal?) changes in the host may be accelerated by solar activity. This is one more reason to explore the possibility of feedsidewards underlying phase response curves along genetically anchored built-in cycles from "then" (a billion years of evolution ago) as well as "now."

In conclusion, there is a need for a transdisciplinary combined archival, physiological, and demographic, always inferential statistical methodology in order to resolve multifactorial interactions underlying "contagions" of the body and of the mind, which latter may also be under the nonphotic influence of the sun or the galaxies, at least insofar as their cyclicity is concerned. They may lead to a heightened susceptibility of the masses to demagogues as well as to the demagogues themselves. We face the task of a scientific chronobioethic that should complement chronomics and chronobiology in everybody's service. It may not be coincidence that circadecadal changes in religious motivation [49] and homicide [20] are nearly in antiphase (200° apart).

References

1. Baevsky RM, Petrov VM, Chernikova AG. Regulation of autonomic nervous system in space and magnetic storms. Adv Space Res 1998;22:227–34.
2. Baevsky RM, Petrov VM, Cornélissen G, Halberg F, Orth-Gomer F, Akerstedt T, Otsuka K, Breus T, Siegelova J, Dusek J, Fiser B. Meta-analyzed heart rate variability, exposure to geomagnetic storms, and the risk of ischemic heart disease. Scr Med 1997;70:201–6.
3. Bigger JT Jr, Steinman RC, Rolnitzky LM, Fleiss JL, Albrecht P, Cohen RJ. Power law behavior of RR-interval variability in healthy middle-aged persons, patients with recent acute myocardial infarction, and patients with heart transplants. Circulation 1996;93:2142–51.
4. Bigger JT, Fleiss JL, Rolnitzky LM, Steinman RC. Frequency domain measures of heart period variability to assess risk late after myocardial infarction. J Am Coll Cardiol 1993;21:729–36.

5. Breus T, Cornélissen G, Halberg F, Levitin AE. Temporal associations of life with solar and geophysical activity. Ann Geophys 1995;13:1211–22.
6. Cornélissen G, Halberg F, Schwartzkopff O, Delmore P, Katinas G, Hunter D, Tarquini B, Tarquini R, Perfetto F, Watanabe Y, Otsuka K. Chronomes, time structures, for chronobioengineering for "a full life". Biomed Instrum Technol 1999;33:152–87.
7. Cornélissen G, Wendt HW, Guillaume F, Bingham C, Halberg F, Breus TK, Rapoport S, Komarov F. Disturbances of the interplanetary magnetic field and human pathology. Chronobiologia 1994;21:151–4.
8. Cornélissen G, Bakken E, Delmore P, Orth-Gomer K, Akerstedt T, Carandente O, Carandente F, Halberg F. From various kinds of heart rate variability to chronocardiology. Am J Cardiol 1990;66:863–8.
9. Cornélissen G, Breus TK, Bingham C, Zaslavskaya R, Varshitsky M, Mirsky B, Teibloom M, Tarquini B, Bakken E, Halberg F, International Womb-to-Tomb Chronome Initiative Group. Beyond circadian chronorisk: worldwide circaseptan-circasemiseptan patterns of myocardial infarctions, other vascular events, and emergencies. Chronobiologia 1993;20:87–115.
10. Cornélissen G, Halberg F, Breus T, Syutkina EV, Baevsky R, Weydahl A, Watanabe Y, Otsuka K, Siegelova J, Fiser B, Bakken EE. Non-photic solar associations of heart rate variability and myocardial infarction. J Atmos Solar-Terr Phys 2002;64:707–20.
11. Cornélissen G, Hillman D, Katinas GS, Rapoport S, Breus TK, Otsuka K, Bakken EE, Halberg F. Geomagnetics and society interact in weekly and broader multiseptans underlying health and environmental integrity. Biomed Pharmacother 2002;56:319s–26.
12. Cornélissen G, Halberg F. Introduction to chronobiology. Medtronic Chronobiology Seminar #7, April 1994, 52 pp. Library of Congress Catalog Card # 94-060580 z.umn.edu/halbergchronobiologyctr
13. Cornélissen G, Schwartzkopff O, Halberg F, Otsuka K, Watanabe Y. 7-day ambulatory monitoring for adults with hypertension and diabetes [letter]. Am J Kidney Dis 2001;37:878.
14. deBoer RW, Karemaker JM, Strackee J. Hemodynamic fluctuations and baroreflex sensitivity in humans: a beat-to-beat model. Am J Physiol 1987;253(3 Pt 2):H680–9.
15. Ghione S, Mezzasalma L, Del Seppia C, Papi F. Do geomagnetic disturbances of solar origin affect arterial blood pressure ? J Hum Hypertens 1998;12:749–54.
16. Gmitrov J, Ohkubo C. Geomagnetic field decreases cardiovascular variability. Electro-Magnetobiol 1999;18:291–303.
17. Halberg F, Breus TK, Cornélissen G, Bingham C, Hillman DC, Rigatuso J, Delmore P, Bakken E, International Womb-to-Tomb Chronome Initiative Group. Chronobiology in space. Keynote, 37th Annual Meeting of the Japan Society for Aerospace and Environmental Medicine, Nagoya, 8–9 Nov 1991. University of Minnesota/Medtronic Chronobiology Seminar Series, #1, Dec 1991, 21 pp. of text, 70 figures.
18. Halberg F, Cornélissen G, Halberg J, Fink H, Chen C-H, Otsuka K, Watanabe Y, Kumagai Y, Syutkina EV, Kawasaki T, Uezono K, Zhao ZY, Schwartzkopff O. Circadian Hyper-Amplitude-Tension, CHAT: a disease risk syndrome of anti-aging medicine. J Anti Aging Med 1998;1:239–59.
19. Halberg F, Cornélissen G, Halpin C, Burchell H, Watanabe Y, Kumagai Y, Otsuka K, Zaslavskaya R. Fleeting "monitor-", "conflict-", or "grief-associated" blood pressure disorders: MESOR-hypertension and circadian hyper-amplitude-tension (CHAT). Euro Rehab 1996;6:225–40.
20. Halberg F, Cornélissen G, Katinas G, Hillman D, Sampson M, Schwartzkopff O. Season's Appreciations 2000. Chronomics complement, among many other fields, genomics and proteomics. Neuroendocrinol Lett 2001;22:53–73.
21. Halberg F, Cornélissen G, Otsuka K, Schwartzkopff O, Halberg J, Bakken EE. Chronomics. Biomed Pharmacother 2001;55 Suppl 1:153–90.
22. Halberg F, Cornélissen G, Otsuka K, Schwartzkopff O. Chronodiagnosis-based chronotherapy: is blood pressure monitoring for weeks too costly to prevent long-term care for years? Gerontol Geriatr 1999;2:20–30.

23. Halberg F, Cornélissen G, Otsuka K, Watanabe Y, Katinas GS, Burioka N, Delyukov A, Gorgo Y, Zhao ZY, Weydahl A, Sothern RB, Siegelova J, Fiser B, Dusek J, Syutkina EV, Perfetto F, Tarquini R, Singh RB, Rhees B, Lofstrom D, Lofstrom P, Johnson PWC, Schwartzkopff O. International BIOCOS Study Group. Cross-spectrally coherent ~10.5- and 21-year biological and physical cycles, magnetic storms and myocardial infarctions. Neuroendocrinol Lett 2000;21:233–58.
24. Halberg F, Cornélissen G, Otsuka K, Watanabe Y, Woods MA, Lambert CR, Zaslavskaya R, Gubin D, Petukhova EY, Delmore P, Bakken E. Rewards in practice from chrono-meta-analyses "recycling" heart rate, ectopy, ischemia and blood pressure information. J Med Eng Technol 1997;21:174–84.
25. Halberg F, Cornélissen G, Wall D, Otsuka K, Halberg J, Watanabe Y, Siegelova J, Homolka P, Fiser B, Dusek J, Sánchez de la Peña S, Maggioni C, Delyukov A, Gorgo Y, Gubin D, Carandente F, Schaffer E, Borer K, Sonkowsky RP, Schwartzkopff O. Engineering and governmental challenge: 7-day/24-hour chronobiologic blood pressure and heart rate screening: Part I. Biomed Instrum Technol 2002;36:89–122.
26. Halberg F, Haus E, Cornélissen G. From biologic rhythms to chronomes relevant for nutrition. In: Marriott BM, editor. Not eating enough: overcoming underconsumption of military operational rations. Washington, DC: National Academy Press; 1995. p. 361–72.
27. Herold M, Cornélissen G, Loeckinger A, Koeberle D, Koenig P, Halberg F. About 8-hourly variation of circulating human endothelin-1 (ET-1) in clinical health. Peptides 1998;19:821–5.
28. Huikuri HV, Makikallio TH, Peng CK, Goldberger AL, Hintze U, Moller M. Fractal correlation properties of R-R interval dynamics and mortality in patients with depressed left ventricular function after an acute myocardial infarction. Circulation 2000;101:47–53.
29. Katinas GS, Halberg F, Cornélissen G, Hawkins D, Bueva MV, Korzhevsky DE, Sapozhnikova LR, Rhodus N, Schaffer E. About 8- and ~84-hour rhythms in endotheliocytes as in endothelin-1 and effect of trauma. Peptides 2001;22:647–59.
30. Kobayashi M, Musha T. 1/f fluctuation of heartbeat period. IEEE Trans Biomed Eng 1982;29:456–7.
31. La Rovere MT, Bigger Jr JT, Marcus FI, Mortara A, Schwartz PJ, ATRAMI Investigators. Baroreflex sensitivity and heart-rate variability in prediction of total cardiac mortality after myocardial infarction. Lancet 1998;351:478–84.
32. Li T-Y, Yorke JA. Period three implies chaos. Am Math Mon 1975;82:985–92.
33. Lipa BJ, Sturrock PA, Rogot E. Search for correlation between geomagnetic disturbances and mortality. Nature 1976;259:302–4.
34. Makikallio TH, Huikuri HV, Makikallio A, Sourander LB, Mitrani RD, Castellanos A, Myerburg RJ. Prediction of sudden cardiac death by fractal analysis of heart rate variability in elderly subjects. J Am Coll Cardiol 2001;37:1395–402.
35. Malin SRC, Srivastava BJ. Correlation between heart attacks and magnetic activity. Nature 1979;277:646–8.
36. Oinuma S, Kubo Y, Otsuka K, Yamanaka T, Murakami S, Matsuoka O, Ohkawa S, Cornélissen G, Weydahl A, Holmeslet B, Hall C, Halberg F, on behalf of the ICEHRV Group. Graded response of heart rate variability, associated with an alteration of geomagnetic activity in a subarctic area. Biomed Pharmacother 2002;56:284s–8.
37. Otsuka K, editor. Chronome & Janus-medicine: Heart Rate Variability (HRV) and BP Variability (BPV) from a viewpoint of chronobiology and ecology. Tokyo: Medical Review; 1998. 213 pp.
38. Otsuka K, Cornélissen G, Halberg F, Oehlerts G. Excessive circadian amplitude of blood pressure increases risk of ischaemic stroke and nephropathy. J Med Eng Technol 1997;21:23–30.
39. Otsuka K, Cornélissen G, Halberg F. Predictive value of blood pressure dipping and swinging with regard to vascular disease risk. Clin Drug Investig 1996;11:20–31.
40. Otsuka K, Cornélissen G, Halberg F. Circadian rhythmic fractal scaling of heart rate variability in health and coronary artery disease. Clin Cardiol 1997;20:631–8.

41. Otsuka K, Cornélissen G, Halberg F. Circadian rhythms and clinical chronobiology. Biomed Pharmacother 2001;55:7–18.
42. Otsuka K, Cornélissen G, Shinagawa M, Halberg F. Blood pressure variability assessed by semiautomatic and ambulatorily functional devices for home use. Clin Exp Hypertens 1999;21:729–40.
43. Otsuka K, Cornélissen G, Zhao ZY, Weydahl A, Delyukov A, Gorgo Y, Wang ZR, Perfetto F, Tarquini R, Kubo Y, Shinagawa M, Hotta N, Ishii T, Omori K, Watanabe Y, Nunoda S-I, Ohkawa S-i, Halberg F. Rhythm and trend elements in the time structure, chronome, of heart rate variability. Gerontol Geriatr 1999;2:31–48.
44. Otsuka K, Murakami S, Oinuma S, Kubo Y, Shinagawa M, Weydahl A, Cornélissen G, Halberg F. Altered heart rate variability on geomagnetically disturbed day and the following two days in subarctic. In: Mitro P, Pella D, Rybar R, Valocik G, editors. Proceedings, 2nd Congress on Cardiovascular Diseases, Kosice, 25–27 Apr 2002. Bologna: Monduzzi Editore; 2002. pp. 271–83.
45. Otsuka K, Oinuma S, Cornélissen G, Weydahl A, Ichimaru Y, Kobayashi M, Yano S, Holmeslet B, Hansen TL, Mitsutake G, Engebretson MJ, Schwartzkopff O, Halberg F. Alternating-light-darkness-influenced human electrocardiographic magnetoreception in association with geomagnetic pulsations. Biomed Pharmacother 2001;55 Suppl 1:63–75.
46. Phillips JB, Borland SC. Behavioural evidence for use of a light-dependent magnetoreception mechanism by a vertebrate. Nature 1992;359:142–144.
47. Rajaram M, Mitra S. Correlation between convulsive seizure and geomagnetic activity. Neurosci Lett 1981;24:187–91.
48. Simpson HW, Bartter FC, Halberg F, Gill JR, Delea C, Gullner H-G, Haus E. A circaoctohoran rhythm of urine epinephrine excretion in man. Prog Clin Biol Res 1990;341A:157–60.
49. Starbuck S, Cornélissen G, Halberg F. Is motivation influenced by geomagnetic activity? Biomed Pharmacother 2002;56 Suppl 2:289–97.
50. Stoupel E, Keret R, Assa S, Kaufman H, Shimshoni M, Laron Z. Secretion of growth hormone, prolactin and corticosteroids during different levels of geomagnetic activity. Neuroendocrinol Lett 1983;5:365–8.
51. Stoupel E, Wittenberg C, Zabludowski J, Boner G. Ambulatory blood pressure monitoring in patients with hypertension on days of high and low geomagnetic activity. J Hum Hypertens 1995;9:293–4.
52. Task Force of the European Society of Cardiology the North American Society of Pacing Electrophysiology. Heart rate variability. Standards of measurement, physiological interpretation, and clinical use. Circulation 1996;93:1043–65.
53. Tsuji H, Venditti FJ, Manders ES, Evans JC, Larson MG, Feldman CL, Levy D. Reduced heart rate variability and mortality risk in an elderly cohort. Framingham Heart Study Circ 1994;90:878–83.
54. Villoresi G, Breus TK, Iucci N, Dorman LI, Rapoport SI. The influence of geophysical and social effects on the incidences of clinically important pathologies (Moscow 1979–1981). Phys Med 1994;10:79–91.
55. Villoresi G, Kopytenko YA, Ptitsyna NG, Tyasto MI, Kopytenko EA, Iucci N, Voronov PM. The influence of geomagnetic storms and man-made magnetic field disturbances on the incidence of myocardial infarction in St Petersburg (Russia). Phys Med 1994;10:107–17.
56. Vladimirskii BM, Narmanskii VY, Temuriantz NA. Global rhythmics of the solar system in the terrestrial habitat. Biophysics 1995;40:731–6.
57. Watanabe Y, Cornélissen G, Halberg F, Otsuka K, Ohkawa S-I. Association by signatures and coherences between the human circulation and helio- and geomagnetic activity. Biomed Pharmacother 2001;55 Suppl 1:76–83.
58. Halberg F, Cornélissen G, Otsuka K, Schwartzkopff O. Can the nomenclature for HF and LF/HF ECG power be replaced in a chronome context? In: Otsuka K, editor. Chronome & Janus-medicine: heart rate variability (HRV) and BP variability (BPV) from a viewpoint of chronobiology and ecology. Tokyo: Kyowa; 1998. p. 126–34.
59. Panofsky H, Halberg II F. Thermo-variance spectra; simplified computational example and other methodology. Exp Med Surg 1961;19:323–38.

Chapter 15
A Transdisciplinary Near-Transyear in Sudden Cardiac Death Validates Its Counterpart in the Solar Wind and Geomagnetics

Abstract The results for SCD in Minnesota resemble those in Tokyo. They are also in keeping with an about-1.3-year component detected in a 10-year long record of atrial fibrillations, potentially fatal arrhythmia that may underlie SCD. These results imply the possible influence of nonphotic as well as of photic solar effects, at least in some geographic locations.

This possibility is in keeping with the finding of an about-10.5-year cycle characterizing the incidence of MI in Minnesota (1968–1996), accounting for a 5 % difference in mortality between years of maximal vs. minimal solar activity. The fact that nonphotic (e.g., magnetic) effects are latitude dependent may account in part for geographic differences observed in the patterns of incidence of SCD. Nonphotic influences may also stem from several sources, such as cosmic rays, solar flares, the solar wind, the interplanetary magnetic field, and/or geomagnetic disturbances as measured on earth, to name but a few.

For SCD, high geomagnetic activity is associated with a decreased incidence of SCD. For SCD, underlying factors may be explored in pertinent markers such as the incidence of potentially lethal arrhythmia recorded in some pacemaker-cardioverter-defibrillators, monitored longitudinally for the individual patient, before an untoward event occurs.

Keywords Sudden cardiac death • Circaseptan component • Transyears • Near-transyears • Geomagnetics

15.1 Introduction

This study aimed at reexamining, in different geographic/geomagnetic locations, the time patterns of SCD, in the strict sense, as a refined diagnosis according to the International Classification of Diseases, revision 10 (ICD10, code I46.1). Thereby, we wish to assess the relative prominence, gauged by a nonzero amplitude (A) of different spectral components (classified by the 95 % CI of their period not overlapping a period between 0.8- and 2.0-year length). Thus, we aim at exploring the relative importance of photic vs. nonphotic influences on SCD or, more specifically, of the seasons vs. unseen cosmic factors for which magnetic cycles may be a proxy, if not a cause, or both.

K. Otsuka et al., *Chronomics and Continuous Ambulatory Blood Pressure Monitoring*, DOI 10.1007/978-4-431-54631-3_15

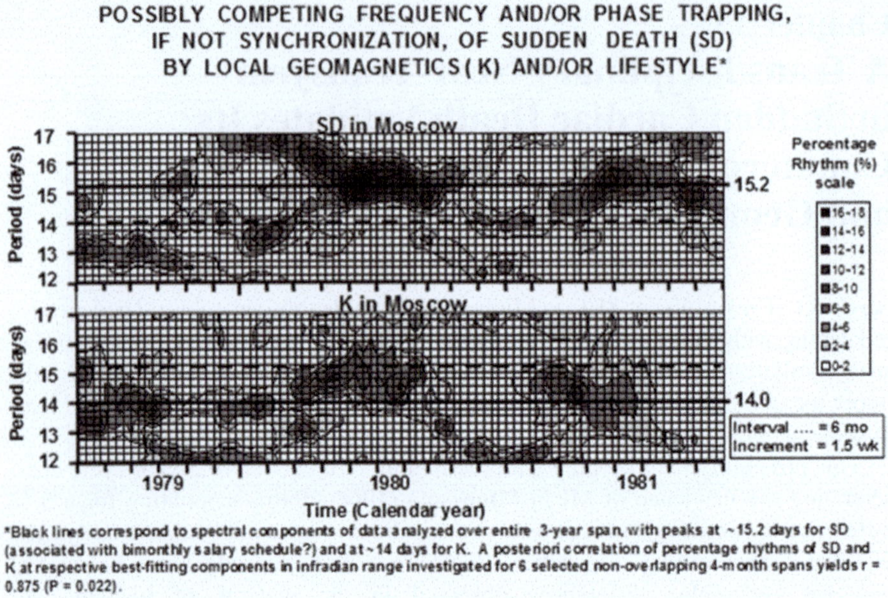

Fig. 15.1 Wobbliness of the about-2-week variation in "sudden cardiac deaths" (*top*) and the local index, K, of geomagnetic activity (*bottom*), revealed by gliding spectra wherein data in a 6-month interval are progressively displaced by a 1.5-week increment. Note that detection of the about-2-week component is not consistent throughout the 3-year span. A possible resonance with occasional frequency trapping between the multiseptans of the local geomagnetic activity index, K, and "sudden cardiac death" is suggested by the more prominently expressed (*darker*) about-half-monthly variation in "sudden cardiac deaths" observed when this component is also detected in the spectrum of K [3, 4] (© Halberg Chronobiology Center)

By 1991, it was apparent that the least-squares spectrum of SCD (as classified at the time, including MIs) already differed from that of MIs, strokes, cardiac arrhythmia, hypertensive episodes, and asthma [1, 2]. "SCD," in quotation marks (including presumably MIs), had a somewhat prominent circadiseptan component [2–4] (also shown in Fig. 15.1), in addition to a very prominent about-yearly (Fig. 15.1) and a lesser about-weekly (Fig. 15.2) component, which latter it shared with other conditions. A circadiseptan component of about 15.2 days may be interpreted as reflecting the influence at a given time and/or through evolution of a very wobbly magnetic cycle in the environment (see Fig. 15.1), e.g., in a global geomagnetic index, K_p, or a local such index, K [4]; cf. [5, 6], although the influence of factors of lifestyle, such as alcohol consumption on paydays, cannot be ruled out (Fig. 15.1) [3, 4].

Fig. 15.2 (continued) the rejection of the zero-amplitude assumption below the 2 % level of statistical significance (© Halberg Chronobiology Center). (**c**) Stacking of the data in Fig. 15.2a for a calendar year does not reveal any statistical significance in a zero-amplitude (no-rhythm) test. This would be expected if only a transyear (Fig. 15.2b), but no calendar-yearly component characterized the data (© Halberg Chronobiology Center)

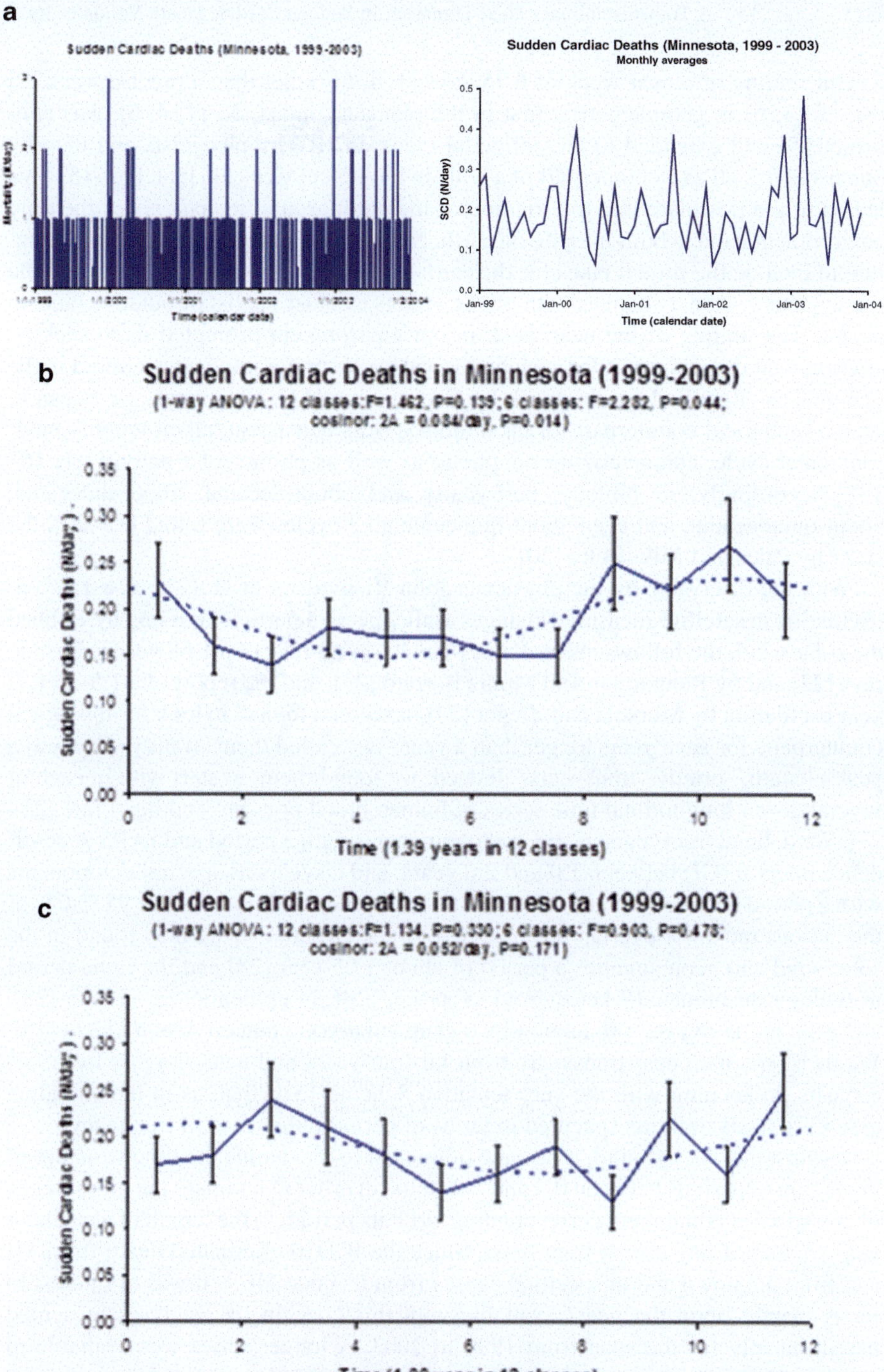

Fig. 15.2 (**a**) Incidence of sudden cardiac death, in the strict sense according to the International Classification of Diseases, 10th revision (ICD10), code I46.1, in Minnesota. The naked eye has difficulty in discerning a pattern, yet some peaks occur apparently at intervals longer than 1 year (© Halberg Chronobiology Center). (**b**) The fit of a 1.39-year cosine curve, a period resolved by linear-nonlinear rhythmometry, to the data from Fig. 1a stacked for this transyear allows

Our finding of a near week of 6.75 days [1, 4, 6] rather than a precise week [7] [8]; cf. [4, 6] in geomagnetics, first in the planetary index, K_p [1, 4, 6], was confirmed [9] and extended to the antipodal index aa [10] by physicists and then by ourselves [6], all as a counterpart of a built-in biological week [11]; cf. [12–15] This led us to the postulate that the rotation of the earth around its axis is not the only environmental cycle built into life on earth. Not even the extension of the time structure to include the time it takes for the earth to revolve once around the sun and the consequent seasonal changes with photic effects exhaust the list of putative built-in cycles. The finding of the near week in our environment prompted us to seek an environmental counterpart for each biological cycle, preferably one anchored in the genome, as suggested for the week by free-running [11]. Vice versa, we began to seek a biological counterpart for each more or less consistent, albeit wobbly environmental cycle, characterizing nonphotic as well as photic solar activity [1, 16–18]. Accordingly, in biology, half-yearly and about-decadal, about-didecadal, about-quindecadal, and even about-quincentennial cycles were found [18–20], the latter by Pales and Mikulecky [20].

With the discovery by the physicists John Richardson et al. [21] of a 1.3-year oscillation in satellite-measured changes of the speed of particles passing by, dubbed the solar wind, the follow-ups by Valdes-Galicia et al. [22] of a 1.68 year in cosmic rays [22] and by Rangarajan and Muniz Barreto [23], and the report of a 1.5- to 1.7-year oscillation by Mursula and Zieger [24], it seemed logical to look for biological counterparts for such years longer than a year – we called them "trans [= beyond] a year's length," briefly "transyears." Indeed, we found them, to start with in each of several dozen longitudinal time series of human blood pressure and heart rate [25–27]. We defined transyears as spectral components with a period and its 95 % confidence interval (CI) between 1.0 and 2.0 years, and cisyears as spectral components with a period between 0.5 and 1.0 year and added the condition that the 95 % CIs of the periods must not overlap the specified limits. Eventually, we also found in the solar wind and geomagnetics a period of about 1.05 year [28] and the same period in biology: in human 17-ketosteroid excretion [29], in epidemiological data [28], and even in the oxygen production by a giant eukaryotic unicell *Acetabularia* [19]. We then separated near-transyears from far-transyears and near-cisyears from far-cisyears, in keeping with the very tentative Scheme 15.1, that, as in the circadian case [27], leads to limits specified in terms of period rather than in frequency.

Against this background, it became of interest to reexamine the time structure of the region of periods between 0.3 and 2.0 years, in order to compare the prominence of any spectral component corresponding, with its period, to the length of a calendar year vs. that of any cis- or transyears. Since the ICD10 eliminated death from MI and from unknown and unspecified causes from its code I46.1, it was of interest to focus mainly upon the years available with this code in the analyses presented herein, mostly for the span from 1999 to 2003. A longer series was available to check on the assumptions, supported by other data [25], that 95 % CIs overlapping 2.0 years were probably due to the brevity of the series analyzed and that the 95 % CIs of the periods will become tighter with increasing series length, as had been the case for the transannual and for the decadal components in blood pressure and heart rate series.

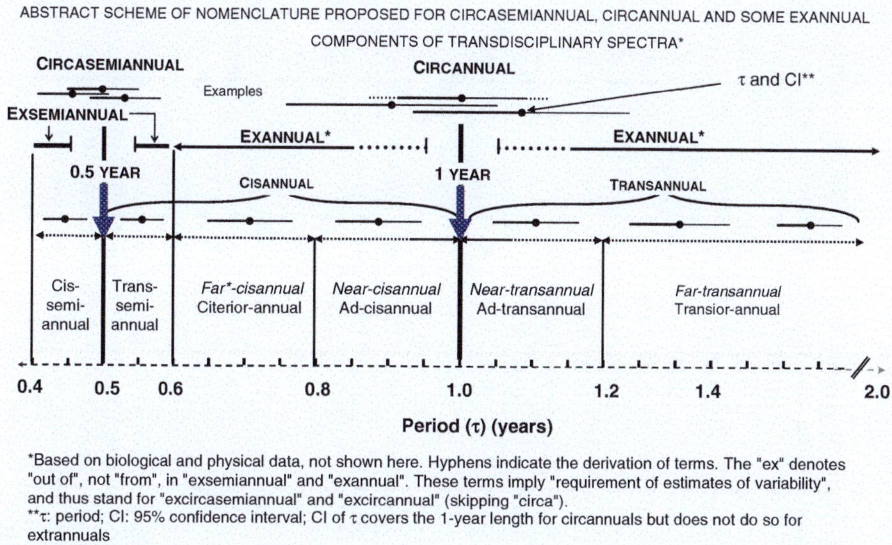

Scheme 15.1 Nomenclature of cycles with periods between 0.4 and 2.0 years

15.2 Epidemiology, Materials, and Methods

Data were first obtained from the Minnesota Department of Health through the courtesy of Ms. Judy Palermo. When a period corresponding to a far-transyear was readily apparent in these data, since the 95 % CI overlapped neither the 1.0- nor the 2.0-year period, as seen in Table 15.1, and no component with the length of a calendar year was noted, this was interpreted as a putative signature of a magnetic cycle corresponding in length to cycles described in the solar wind by Richardson et al. [21] and Mursula and Zieger [24]. This assumption was in keeping with the rule of "reciprocity of cycles in and around us" made for investigation of The Biosphere and the Cosmos, BIOCOS [15]. Added data were then solicited from Arkansas; North Carolina; the Czech Republic; Tbilisi, Republic of Georgia; and Hong Kong. The data covered all or part of the span from 1999 to 2003, with a longer SCD series also added from the Czech Republic, as were Czech and Minnesotan data on MI. Linear-nonlinear rhythmometry was used on each series [30–34]. We also looked for product-moment correlations among the epidemiological series themselves and between each of these series and the planetary geomagnetic disturbance index K_p. Superposed epochs in relation to events defined by a $K_p > 6$ were also applied. As yet, the correlation coefficients were small and the superposed epochs showed a minimum at $K_p > 6$, but minima were also found at timepoints distant from the $K_p > 6$ and hence are not shown herein.

Table 15.1 Geomagnetic/geographic differences among cycles with periods in the range of 0.8–2.0 years, characterizing the incidence of sudden cardiac death (SCD)[a,b] and myocardial infarction (MI)

Site	Span	T, Δt, N	SCD (N)	Period (y)	(95%CI)	Amplitude	(95%CI)	A(% MESOR)	P-value[c]
				------------- Transyear (TY) or Candidate Transyear (cTY) Detected -------------					
Minnesota	1999–2003	5y, 1d, 1826	343	<u>1.392</u> (TY)	(1.173, 1.611)	0.042	(0.00, 0.09)	22.0	0.014
Arkansas	1999–2003	5y, 1d, 1826	273	1.095	(0.939, 1.251)	0.032	(0.00, 0.07)	21.1	0.040
				<u>1.686</u> (cTY)	(1.293, 2.071)	0.031	(0.00, 0.07)	20.7	0.044
Czech Rep.	1999–2003	5y, 1d, 1826	1006	0.974	(0.856, 1.091)	0.078	(0.00, 0.16)	14.2	0.007
				<u>1.759</u> (cTY)	(1.408, 2.110)	0.077	(0.00, 0.15)	13.9	0.010
	1994–2003	10y, 1d, 3652	1792	<u>1.726</u> (TY)	(1.605, 1.848)	0.074	(0.02, 0.13)	15.1	<0.001
				1.000	(0.944, 1.056)	0.052	(0.00, 0.10)	10.6	0.010
				------------- Candidate Transyear Not Detected -------------					
North Carolina	1999–2003	5y, 1d, 1826	752	0.929	(0.834, 1.023)	0.069	(0.00, 0.14)	16.9	0.007
Tbillsi, Georgia	Nov99–2003	4.1y, 1d, 1505	130	0.988	(0.862, 1.114)	0.035	(0.00, 0.07)	40.7	0.007
Hong Kong	2001–2003	3y, 1m, 36	52	0.843	(0.651, 1.036)	0.022	(NS)	44.9	0.077
Myocardial Infarction (MI)									
Site	Span	T, Δt, N	MI (N)	Period (y)	(95%CI)	Amplitude	(95%CI)	A(% MESOR)	P-value[2]
				------------- Coexisting Year (Circannual) and Transyear (TY) -------------					
Czech Rep.	1999–2003	5y, 1d, 1826	52598	1.014	(0.989, 1.038)	2.85	(2.22, 3.48)	9.88	<0.001
				<u>1.354</u> (TY)	(1.252, 1.456)	1.35	(0.69, 2.02)	4.68	<0.001
	1994–2003	10y, 1d, 3652	115520	0.998	(0.988, 1.009)	3.03	(2.47, 3.60)	9.58	<0.001
				<u>1.453</u> (TY)	(1.417, 1.489)	1.91	(1.34, 2.49)	6.04	<0.001
				<u>1.15</u> (TY)	(1.116, 1.184)	1.23	(0.64, 1.82)	3.88	<0.001

[a]*With focus on transyears, with periods longer than 1.0 year*

[b]International Classification of Disease (ICD10) Code 146.1 excluding MI and sudden death of unknown or unspecified cause (except before 1999). T: Length of data series (y = years); T, Δt, N: record length; sampling interval (d=day, m=month); N: number of data (including 0s). Period and 95% confidence interval (CI) estimated by nonlinear least squares. In longer (10-y) series, a neartransyear (cycle with a period between 1.0 and 1.2y) is detected for MIs in addition to a fartransyear. Brevity of series and lack of ordering statistical significance qualify results from Hong Kong. Note that trans years are found in 3 of 6 locations ($P<0.05$ by linear least squares) with a relative amplitude >12 (% of MESOR)

[c]From linear least squares analysis, not corrected for multiple testing. Amplitude expressed in N/day

15.3 Transyear Rhythms in SCD and Mortality from MI

Table 15.1 shows that under sudden cardiac death in 3 of the 6 geographic/geomagnetic regions examined, there was only a spectral component of a calendar year in length. In these locations (North Carolina, Tbilisi, and Hong Kong), the point estimate of the period may differ slightly numerically from the exact year's length, but the 95 % CI in each case covers the precise calendar-year length. By contrast, in the data from Minnesota, there is only a transyearly component. Figure 15.2a shows the actual data: it is difficult to interpret these very low monthly incidences without analyses. Nonetheless, some peaks are easy to see and are more than a year distant from each other and are roughly in keeping with the period given in Table 15.1. Stacking of the data over this period in Fig. 15.2b results in a zero-amplitude test significant below the 2 % probability level. By contrast, as would be expected if a calendar-yearly component were absent, the fit of a 1-year cosine curve to data stacked along the calendar year in Fig. 15.2c does not allow the rejection of the zero circannual amplitude assumption (P=0.171).

In the data from Arkansas and the Czech Republic, there are components with the length of a calendar year, along with a probable candidate transyear. We speak of a candidate transyear when the period is between 1.0- and 2.0-year length and does not overlap 1 year, but overlaps the length of 2 years. In other words, in the 1999–2003 series, the condition of the 95 % CI of the period not overlapping 2 years is not (yet?) met by transyears in Arkansas and the Czech Republic in the data limited to 5 years, but it is met for the longer series from 1994 to 2003 from the Czech Republic. As noted, it seems likely from Table 15.1 and from individual series of blood pressure and heart rate measurements over decades (that are continuing) that the longer the series, the more it may be characterized by a transyearly component. Indeed the longer SCD series from the Czech Republic reveals a clear far-transyear, with no overlap of 2.0 years (an overlap is found with the shorter series from the same area). Mortality from MI, also included in Table 15.1, shows in turn during 1999–2003 a very prominent circannual component and a much less prominent transannual. In the longer series on MI, also summarized in Table 15.1, a near-transyear and a far-transyear (see Scheme 15.1) are both resolved.

15.4 Biomedical Background

The variability of a large data set on diagnoses made in the context of calls for an ambulance in Moscow, including 71,525 "SCDs" (that included some MIs) during a span of intensive solar activity from 1979 to 1981, had prompted us to look into any effects of a turn in the north–south vector, B_z, of the interplanetary magnetic field, as a measure of a magnetic storm [1]. There has been a long controversy about the merits or demerits of a "heliobiology." Its importance was very strongly supported in the Soviet Union and subsequently again fully accepted and highly regarded as a scientific area sui generis in present-day Russia [35–45]. The effect of

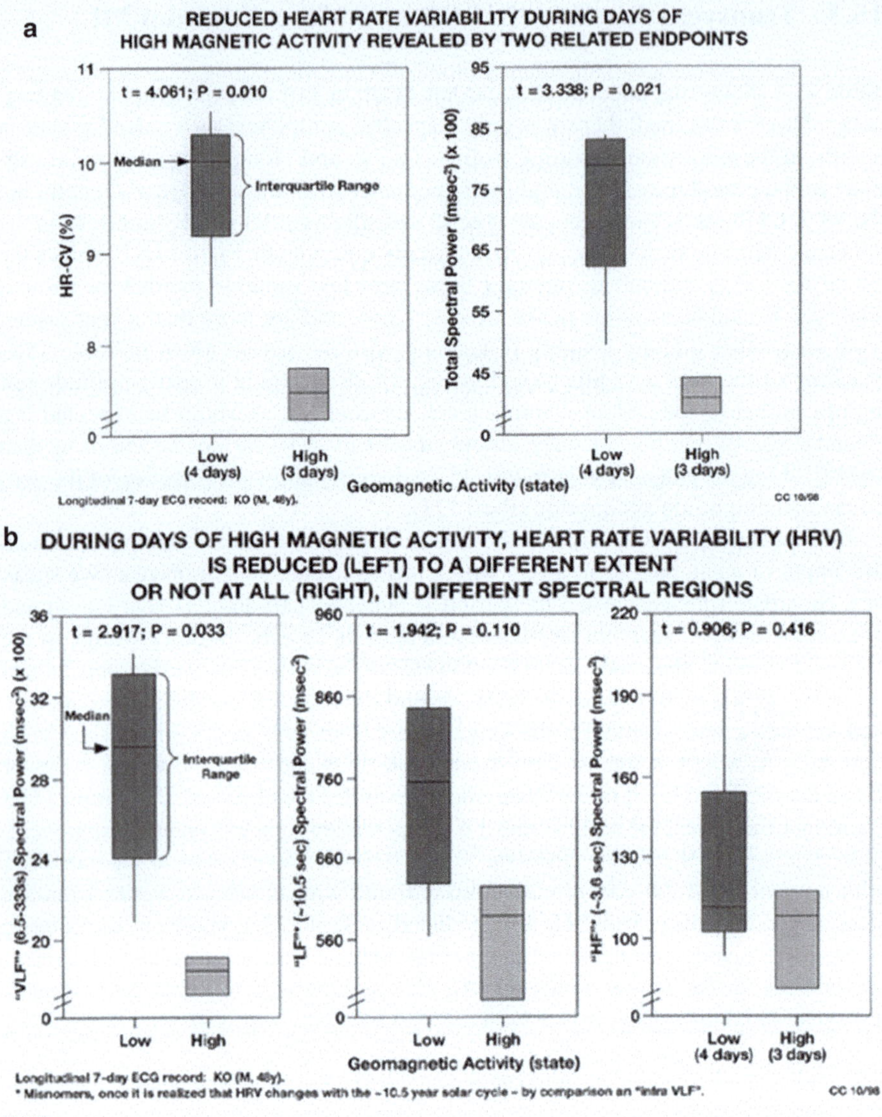

Fig. 15.3 (**a**) Reduced heart rate variability of a clinically healthy 48-year-old man (KO) on days of high magnetic activity vs. days of low magnetic activity, assessed longitudinally in beat-to-beat data recorded around the clock over a 7-day span and analyzed over consecutive 14.4-min intervals. Heart rate variability is gauged by two related endpoints: the coefficient of variation of RR intervals (time domain; *left*) and the total spectral power (frequency domain; *right*). The abnormal record prompted an inquiry by one of us (GC) into possible causes. The response by KO was first that nothing was unusual until the record of geomagnetics became available and a magnetic storm was found during the recording span (© Halberg Chronobiology Center). (**b**) During the span examined in Fig. 15.3a, during days of high magnetic activity, heart rate variability (HRV), in different spectral regions of a clinically healthy 48-year-old man (KO) on days of high magnetic activity vs. days of low magnetic activity, assessed longitudinally, is reduced (*left*) to different extents or

geomagnetic activity as such or as a proxy for solar effects upon human mortality from MI and other cardiovascular conditions is in keeping further with a summary of mortality from different causes in Europe in the 1930s [46–48] and in keeping with the continuing contributions of Stoupel [49, 50], Ertel [51, 52], Mikulecky [14, 53–55], Strestik and Prigancova [56], the Russian team leader Breus [57] with Villoresi et al. [58, 59], and ourselves [1, 15, 57, 60]. Superposed epochs revealed an effect of the magnetic storm [1]. A mechanism possibly underlying these associations was found by self-studies [60] in Fig. 15.3a, b and found in the laboratory as well [42, 61]. The strongest support came from an approach by subtraction and addition, in an individual or group [62, 63].

Geographic differences may underlie the failure of very thorough investigations in the USA by Feinleib et al. [64] and Lipa et al. [65] to confirm earlier findings, notably in Russian and German studies, and studies in Israel and Lithuania, led with proper documentation by Stoupel. Much more on the effect of magnetic storms or cycles upon human morphology, physiology and pathology have been the subject of nine conferences [19, 66, 67].

A special section of Table 15.1 reveals further that both yearly (or circannual) and transyearly components coexist in MI, but in that case, the transyear is substantially smaller in amplitude than the calendar-yearly component. It is the more remarkable that during the span investigated for the Minnesota SCDs, in the strict sense, there is only a transyearly and no calendar-yearly component detected, notably since for a much longer 29-yearly Minnesota series of MI, the seasonal component was much more prominent than an also-present signature of a non-photic, probably magnetic circadecadal component.

15.5 Caveats

Two thousand years ago, Hippocrates reportedly associated thoracic pain with sudden cardiac death [68]. In 1707 and again in 1718, Giovanni Maria Lancisi [69, 70]; cf. [71] published a scholarly, all-encompassing monograph on sudden death; and in 1917, Heinrich Ewald Hering published a book on "death within seconds, with special reference to fibrillation" [72]. The topic remains of interest as an entity in its own right, although the suddenness of the death is differently described by different authors, with the International Classification of Diseases, 10th revision, now referring to death within up to 1 hour from the start of symptoms to the terminal event

Fig. 15.3 (continued) not at all (*right*): the decrease in HRV is more pronounced for periods in the spectral region between 6.5 and 333 sec than in other spectral regions. This result has been confirmed on additional cases; it may point to some underlying physiological mechanism responsible for the physiological response to changes in magnetic activity other than the parasympathetic, usually identified with spectral power in the 3.5-sec region. *References to "high" and "low" frequency are misnomers once it is realized that HRV changes, with the ~10.5-year solar cycle, a frequency much lower than a very low frequency: an "infra VLF" [56, 57] (© Halberg Chronobiology Center)

[73–75]. The forensic pathologist, however, in the case of an unwitnessed death, may use the definition of sudden death for a person known to be alive and functioning 24 hours earlier, and this is acceptable to scholars in the field with proper qualification. "To the best of my knowledge" is accordingly included in the wording of death certificates. It seems reasonable in the future to focus on a subsample of cases (e.g., with cardioverter-defibrillators and/or other monitors) whose premortal records may be studied for antecedents, even when death was not witnessed.

With such qualifications, it must also be noted that the relative brevity of the series in Table 15.1 is a serious limitation, as is the restriction of analyses to only six sites. Almost certainly, as seen in many other instances, with an increase in length of a given series, a spectrum can be very greatly changed, and we must not be speculating herein. For physics, Mursula and Zieger [24] proposed that an about-1.3-year component changes to an about-1.5- to about-1.7-year oscillation with each subsequent solar cycle. In pursuing this possibility, a further analysis of a 131-year series of the geomagnetic aa index then revealed the near-transyear as well, and as in the case of Mursula and Zieger [24], these components were intermittent; our analyses in sections of fixed rather than solar cycle lengths do not suffice to rule in a relation to changes in the solar bipolarity [6, 25].

The data in Table 15.1, however, suffice to point to the need for further scrutiny of any regional differences, even within the same country (USA), where three different situations are found with North Carolina having only an about-yearly component; Arkansas having both components, the year and a candidate transyear, with nearly equal amplitude; and Minnesota having just a transyear in the limited data available.

Geographic/geomagnetic differences have also been documented for the circadidecadal phase of human neonatal body weight, with a near antiphase between 2,150,122 newborns in Minnesota and 1,166,206 newborns in Denmark [76]. Whether local geo- or atmosphero-magnetics and/or other factors account for this difference and others, including those in Table 15.1, remains an open question. It seems reasonable to suggest that one may withhold inferences in any aspects of heliobiology until multiple locations are scrutinized for as long a span as possible, for a given pattern; one must also define the length of the time series examined, underlying a given finding so that it can be further qualified when added longer time series become available. In the context of broader evidence, it seems reasonable to postulate that unseen nonphotic cycles that have their origin in the solar wind, for which geomagnetics are a proxy if not cause, contribute substantially but differently to the pattern of human sudden death and to body weight at birth and that this contribution is also dependent on local factors, possibly terrestrial and atmospheric electrical currents. The role of nonphotic cycles in transdisciplinary human affairs is a challenge, for which an atlas in space and time constitutes an indispensable requirement [18], as agreed upon in a series of international consensus meetings [19].

The task of providing the data is the opportunity of national governments and international agencies in a historical context. We have shown cycles of about-50-year length, including one in stroke incidence with similar periods in the Czech

Republic and in Minnesota, USA, and supporting data from Sweden, Slovakia, and Arkansas, USA [15, 67, 77]. Until governments and international agencies act, action by ministers of health like Bohumil Fiser [77] and/or responsible statisticians served us herein. Letters to US state health departments other than those of Minnesota, Arkansas, and South Carolina have remained unanswered for weeks and will remain so perhaps indefinitely. Government statisticians will have to realize that they are not only bookkeepers; their opportunities, if not mandates, should include the cost-free provision of dense (when indicated, preferably hourly) data as a minimum and the chronomic time series analysis by themselves of the accumulating records as an optimum. Health care can take stock of cycles which may differently affect MI and SCD in different geographic locations and to a different extent and in a different way in different variables [78]. Much added international research against this background could focus on many aspects of time structure, complementing focus on time of day [79–81] or rather circadian phase [82, 83]; cf. [84, 85], day of week or season [81], or rather circaseptan [2] and circannual phase [2, 82, 86, 87], cycles adding to the heart's rhythm [88] a broader historical context.

The importance of the circadians was shown in the laboratory by the 1950s, when their stage tipped the scale between death and survival in response to a cardiac drug (Fig. 15.4) [84, 85]. So do the stages of a much broader time structure to the human response to other stimuli as well (Figs. 15.5 and 15.6). By the 1970s, the complexity of both circadian and circannual patterns became time-microscopically apparent, and eventually, the complexity of the "season" emerges in Table 15.1 and Fig. 15.2.

Alchemy was historically important, but eventually it became chemistry. Oversimplifications have their historical *raison d'être,* such as the study of the effects of the time of day, time of week, or season. The students of both death statistics and pacemaker-cardioverter-defibrillators and similar corresponding as-one-goes recording devices have an opportunity by means of chronomic analyses to look at the complexities involved, using marker rhythms for body time, so that appropriate research into causation can lead to pertinent countermeasures.

The recording of geomagnetic changes was started by Gauss [89] and separately by Alexander von Humboldt [90]. It should be done, but more systematically in different locations aligned with biomedical data collection. Likewise, the period lengths of about-yearly changes here reported are based on too-sparse and too-short series from too-few locations. They suffice, however, to point to the need for the study of geographic/geomagnetic differences in environmental epidemiology. The time structures of epidemiologic data can become the topic of a budding chronoepidemiology if data are properly collected according to future chronobiologically specified ICD revisions. Until then, current ICD-criteria-based statistics of morbidity and mortality should be made available free of charge to those interested in analyzing them.

The impact of revising international classifications of disease has been commented upon [91–93] with the correct and important mandate "that trends over time can be analyzed." This latter article then looks at the impact of the changes in baselines *(sic)* for government targets to reduce mortality from circulatory diseases. In

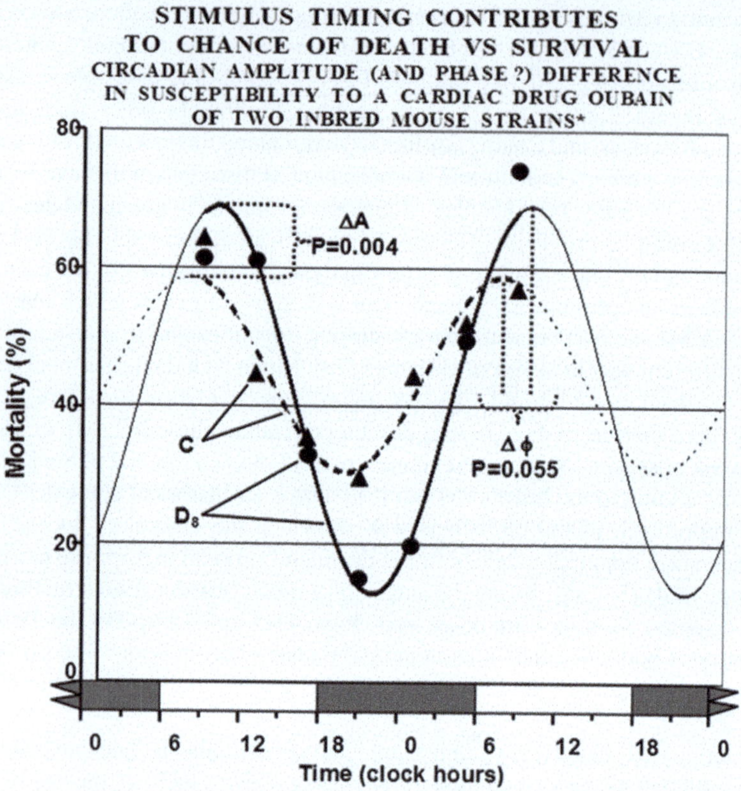

Strain	P	M	Amplitude (A)	Acrophase (φ**)
D_8	P < 0.001	40.0	28.1 (21.5; 34.8)	-135 (-119; -150)
C	P = 0.011	45.2	14.2 (7.3; 21.0)	-104 (- 73; -136)

C = Bagg albino, D_8 = subline 8 of Dilute Brown strain; oubain 0.15 or 0.5 mg/20 g of body weight, respectively. Thin lines in the graph are extrapolations of fitted cosine curves.
** In the table: 95% confidence limits in parentheses, φ in degrees (360° = 24 hours).
M = MESOR; P from zero-A (no-rhythm) test.
Data from F.Halberg et al., Minnesota Acad Sci Proc 1959; 27: 139-143.

Fig. 15.4 Susceptibility rhythms to ouabain of two inbred stocks of mice: a laboratory. Demonstration of the importance of circadian stage in tipping the scale between most animals dying vs. surviving from the same dose of the same cardiac drug, as a function solely of timing in relation to a synchronizing regimen of 12 hours of light alternating with 12 hours of darkness (© Halberg Chronobiology Center)

the body of the same article, we again find a seeming mandate, namely, the search for a "baseline" figure sought to ensure comparability. We can only cite for epidemiology what we wrote over half a century ago for physiology [94], namely, that we need not seek an imaginary baseline. We almost certainly confront a time structure consisting of (1) the trends that government statisticians have recognized, (2) probabilistic and other chaos, and (3) cyclicities. Moreover, the longer the series,

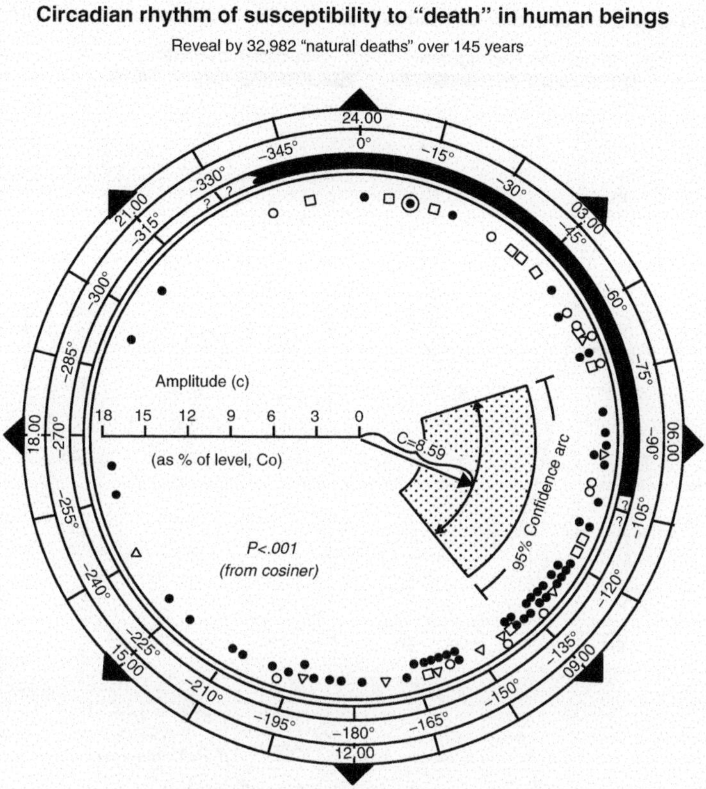

Mortality category	*No.of cases*	*No.of (C, φ)*	*Symbol for acrophase*
Pediatreic	21,673	12	○
Postsurgical	500	1	◉
Cardiovascular	7,644	9	△
Pulmonary	9,357	12	□
Cause unspecified	393,718	54	•
Cosinor of above	432,892	88	[cosinor symbol]

Fig. 15.5 Circadian rhythm of susceptibility to "death" in human beings. Polar plot summarizes the timing of individual acrophases for 88 time series representing nearly 433,000 deaths. Almost 65 % of the acrophases occurs between 04:00 and 11:00. Greatest overall sample mortality occurs at 07:12. *Inner shaded* portion indicates suspected nocturnal rest span of sample prior to death [84] (© Halberg Chronobiology Center)

the more likely it becomes that some of the trends are rhythms. When we deal with cyclicities and regard them as "baselines," the changes from baselines can be misinterpreted. Information on cyclicity is needed to test the effectiveness of interventions. Eventually, we must identify all of the cycles involved and any role played by the stages of all cycles before we can assess what is going on and why. Returning to

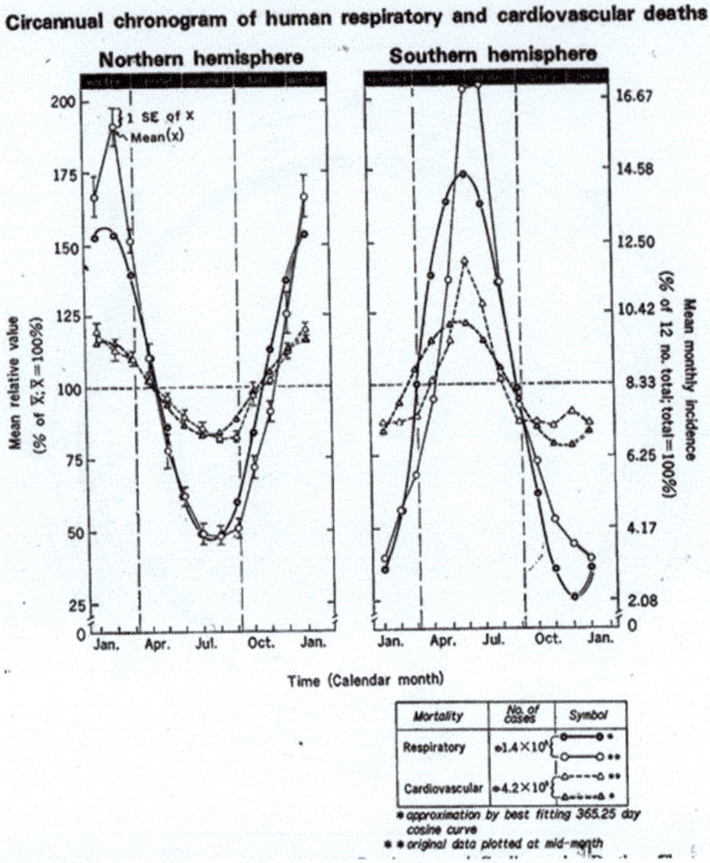

Fig. 15.6 Circannual chronogram of human respiratory and cardiovascular deaths. Chronograms of human mortality population rhythms reveal highest incidences during winter irrespective of hemisphere. Least-squares approximation of the monthly means by a 365.25-day cosine curve accurately reflects waveforms for the northern hemisphere. Slight discrepancies between original data and approximating curve for the southern hemisphere due to small number of samples available [84]. Table 15.1 suggests somewhat greater complexity (© Halberg Chronobiology Center)

our analogy, chemistry is more complex than alchemy, and chronopharmacodynamics is more complex than pharmacokinetics, but the FDA sees to it that we use usually the latter, not the former. Somebody will have to see to it that we use chronomes rather than the "baselines" for current health statistics.

15.6 Reciprocal Periods in Transdisciplinary Science

Reciprocal periods in spectra of different systems in and around us complement each other in several ways [15, 95]. They first support each other in inferential statistical tests of their reality. In a cosinor approach, based on prior evidence, they represent a proxy for prior evidence complementing that in a given variable itself, allowing the anticipation of the period and thus dispensing with the need for corrections required in the case of the testing of multiple periods. Equally important, they facilitate a subtraction-addition or remove-and-replace approach to the study of causal relations. While the built-in day and year are obvious photic periods in and around us, there are also nonphotic reciprocal periods in terrestrial vs. solar wind as well as sunspot-based magnetism and, of course, between cycles in physics vs. biology. Beyond the built-in biological week as a counterpart to a geomagnetic near week and a geomagnetic half year with a biological counterpart, there are biological counterparts for the solar about-decadal sunspot cycle, in natality, physiology, and pathology, including cardiovascular morbidity and mortality, notably of MI [1]. Figure 15.7 shows an excess of 220 MI cases/year at solar maximum, a change that compares unfavorably in extent with the patterns of the year and the week. For the case of MI, a far-transyear is a statistically significant peak, distinctly different from that corresponding to the calendar year but very small in amplitude.

The amplitude ratio of the transyear to the year for SCD can be the reverse of that for MI, as seen in Fig. 15.2. A peaklet at a period very slightly longer than a year has a counterpart in a slightly shorter symmetrical component, located at a similar distance from the precise year; thus, both probably represent a modulation by a component with a much lower frequency. By contrast, the component of a 1.15 near-transyear shown in Table 15.1 has no symmetrical counterpart and may be a component in its own right.

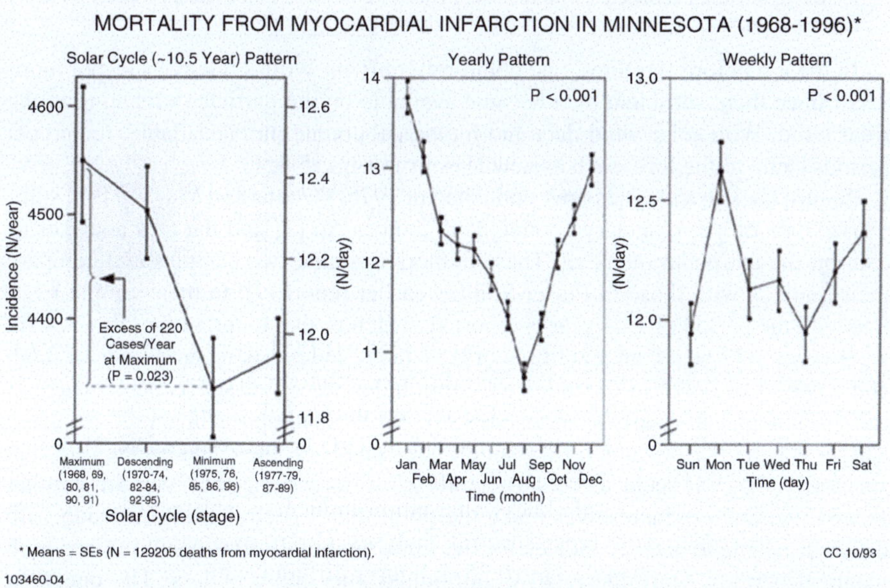

Fig. 15.7 Nonrandom patterns (means ± SEs) of incidences of mortality from myocardial infarction (MI) in Minnesota (1968–1996) along the scales of the week (*right*), year (*middle*), and an about-10.5-year sunspot activity cycle (*left*). By comparison with times of solar minimum, at times of solar maximum, there is an excess of 220 deaths per year due to MI. N = 129,205 deaths from MI (© Halberg Chronobiology Center)

Biologists need the caution of physicists who had found the quasi-persistence of oscillations with a period somewhat longer than a year in magnetism, including aurorae and cosmic rays [7, 96–99], and who, like Schwabe [100]; cf. [101], were unsure of the stability or persistence of the periods they resolved and so stated. In 1968, Ralph Shapiro marked his spectrum of the international magnetic character figure C_4 by peaks at 1.41 and 1.10 years in a time series covering the span from 1884 to 1964. Shapiro also deserves credit for testing the statistical significance of the peak and for reaching the 5 % level with one approach. Fraser-Smith confirmed this in 1972: he found that [7]:

> There was some indication of a 1.4-year line in the *(Ap)* spectrum, and it is a more obvious feature in the spectrum for the daily *Ap* data. An unexpected line occurs at a period of 1.09 years, with an amplitude not much smaller than that of the 1.43-year line. These two lines are all that are evident with periods greater than 6 months. Again, there is no trace of an annual line.

(Indeed, there was, and is, no such line in the spectrum of the geomagnetic index aa in the data reported by Bartels in 1932 after folding [97]. Folding is fine for visualizing the circasemiannual component, but must not be used to describe nonexistent yearly variation in the antipodal aa index in toto.

In any event, by 1972, both a near- and a far-transyear were clearly noted by two independent investigators [7, 96], but there was no component with a corresponding period in sunspots. In the discussion [7], Fraser-Smith observed:

> The *Ap* spectrum also has weak lines at periods of 1.43 and 1.09 years that may not be related to anything connected with the sun. (There are no lines in the sunspot spectrum at either period.)

In the era before satellites, one could rely only on Wolf's relative sunspot numbers. Since then, sufficient data became available on the particles streaming in the solar wind. With solar wind data and biological counterparts available, reciprocal periods came to the fore, such as new lessons in Fig. 15.8.

To turn back to a classic paper published in 1975, Delouis and Mayaud [98] summarize data on the planetary geomagnetic indices Ap [7] and the new analyses of data on the antipodal index aa. These authors examine two samples and compare their findings with those in Fraser-Smith's earlier report [7]. In their sample F1, a line with the period of 1.44 year is reported, which is also found in sample F2, with a period of 1.47 years, an amplitude of 0.51 in the index Ap, an amplitude of 0.62, and a period of 0.473 in the index aa. Amplitudes are expressed as percentages of the prominent half-yearly component [98]. While confidence intervals are given neither for the periods nor for the amplitudes involved, the peaks in the different spectra [7, 96, 98] seem to be sufficiently close to assume that they are similar. Indeed, the authors themselves make the point that "spectral analyses made with different magnetic activity indices for the same temporal series result in lines with identical periods. Only the relative amplitudes vary more or less. The question, however, is to know if all the lines … are truly significant." This question also applies to a period of 1.09 year in Ap, with an amplitude of 0.48, and to a period of 1.086 year, with an amplitude of 0.44. (Incidentally, Delouis and Mayaud deserve

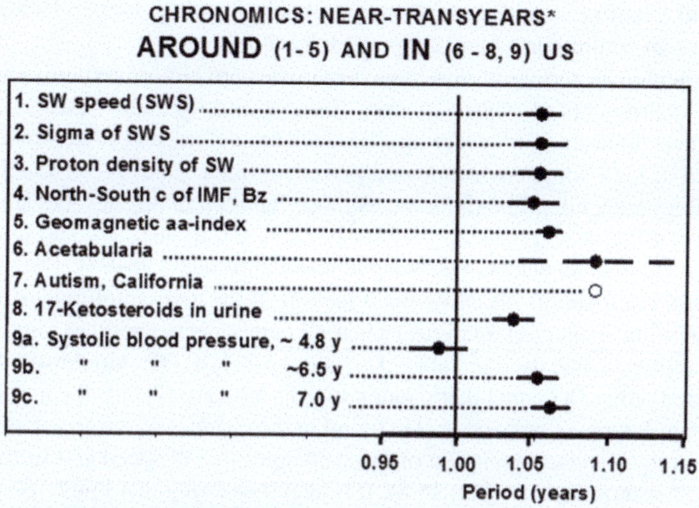

*as separate components and probably not as sidelobes of circannual variation, stemming from amplitude- and/or phase-modulation by components with a lower frequency.
* 1-4 data from OMNI 2 series (1963-2003): 1-3 = Solar wind (SW); 2 - Standard deviation of SW; 4 = Bz = North-South component of interplanetary magnetic field (IMF); 5 = 131 years (1868-1998); 6 = oxygen production (1980-1994); 7 = Quarterly incidence, 10.5 y (1994-2004); 8 = daily excretion for 15 y of urinary steroidal metabolites (17-ketosteroids, CH, October 25 1948 - October 22 1963); 9a = half-hourly record from March 31, 1998; from a man (GSK), 72 years old at start of measurements; 9b - same data series, extended to December 10, 2004; 9c - extended to April 1, 2005.
Circle = period length (point estimation); horizontal line crossing the circle = 95% confidence interval (CI) of spectral component; dashed line: CI of a spectral component not validated nonlinearly.

Fig. 15.8 Transdisciplinary near-transyears in solar and terrestrial biophysics and biology (© Halberg Chronobiology Center)

credit for demonstrating a yearly component in data from each of the antipodal stations when they were analyzed separately [98].)

To complete the story of the discovery of transyears in geophysics, Samuel M. Silverman and Ralph Shapiro in 1984 [99] examined Swedish auroral sightings for the span from 1721 to 1943: "an unexpected peak near 1.4 years is found … [it is] statistically significant at between the 5 % and 1 % level and found consistently in a sequence of spectra covering consecutive, overlapping, approximately 22-year data segments … The 1.4-year peak shows a strong modulation of about 65–68 years … The origin of this peak is unknown." Richardson cites some of these earlier papers when he reports on an oscillation of 1.3 years in the solar wind [21]; a moving spectrum suggests, in the hands of others [24] and our own [6, 19], that in all these cases, we are dealing with very wobbly spectral components. This wobbliness, documented in physics and in biology, can be checked for geomagnetics for a century and a half or so and in organisms probably for much longer. Here, the two disciplines can communicate and support each other.

Biomedicine can serve physics, since organisms are almost certainly older than the oldest instruments of physicists even if we accept the report that the Chinese used magnets thousands of years ago. Organisms before and after hominization have coded at least some environmental periods in their genes [15, 102]. The periods in biology may not all be the same as those reported in physics, but similarities exist between the near-transyear, the hourly excretion of steroids, and that in 131 years of aa and in other series of geomagnetics or aurorae [7, 96–98]. The biological and physical components all allow the rejection of the zero-amplitude assumption by linear-nonlinear spectral analyses (when all components are fitted jointly, a conservative approach also used in Tables 15.1 and 15.2) [28]. The similarity of periods is not limited only to a geomagnetic index such as aa or Ap [7, 98]. It gains in importance if a similar near-transyear is also found in the changes of the speed of the solar wind and in the standard deviation of these changes in SW speed and further in the solar wind's proton density (Fig. 15.8). It is also interesting that longer periods corresponding roughly to the 1.44 or 1.47 years [1, 96, 98, 99], as in geomagnetics, are also found in the temperature of the solar wind.

The observations gain further in interest when periods between near-transyears at about 1.05 year are found by us both in biology and physics, as well as far-transyears with periods and confidence intervals between 1.20 and 2.00 years. We need to take them seriously when near-transyears and far-transyears are also found to characterize the urinary excretion of the breakdown products of important hormones such as the steroids secreted by the two principal glands subservient to survival (the adrenal) and reproduction (the gonads). Their reality and importance may depend upon age and series length when they gain in prominence in the blood pressure of a physician-scientist (GSK) who monitored himself with few interruptions at half-hour intervals around the clock, starting at the age of 73 (Fig. 15.9). Analyses of the data are shown after 4.75 years (top) and again after nearly 7 years (bottom). What appeared to be a 1-year rhythm (because its 95 % confidence interval (CI) overlapped the calendar-year length) became a transyear in the 7-year record, wherein the CI of the period estimate near 1 year no longer overlapped the 1-year length.

We now turn from transyears to half-yearly components, long known in physics and rightly used by Delouis and Mayaud as references to estimate the prominence of other components [98]. In human SCD, there are spectral components, with a length on this side (= cis) of the 0.5-year length, with the CI of these components not overlapping the exact half year. Such half years differing from the precise half year, being shorter, were found for SCD in Hungary, the Czech Republic, and Minnesota and (only) during the corresponding spans also in geomagnetics and in the solar wind. Longer series of these physical environmental variables show no cis-half year, e.g., when the entire available series is being analyzed. Likewise, turning back to the bottom rows of Table 15.1, the geomagnetic index K_p and the solar wind speed show periods corresponding in length to one of those in Arkansas and the Czech Republic, and their CI overlaps even that of the putative Minnesotan transyear of SCD. But these similarities of periods are only hints: As yet the

Table 15.2 Chronomics of the geomagnetic antipodal index aa

The 3-hourly data series of the geomagnetic antipodal index aa, from 1868 to 1998 (131 years), was averaged over consecutive 2-day spans prior to analysis by linear-nonlinear rhythmometry. One set of analyses was carried out on 11 consecutive 11-year spans and a 10-year span. Least-squares spectra were computed separately for each span in the frequency range extending from 1 to 25 cycles per 11 years, using a 0.2 harmonic increment, and were summarized by population-mean cosinor. The root-mean-square amplitude spectrum was also computed. A second set of analyses was performed on the entire 131-year series, consisting of nonlinear least-squares analyses at each trial period, found to be statistically significant (P<0.05) by population-mean cosinor. The analyses yielded the following model, consisting of:
A linear increasing trend An about 10.93-year component (95 % confidence interval [CI]: 10.82, 11.04) with a double amplitude of 5.98 (95 % CI: 4.68, 7.28) An about 1.327-year component (95 % CI: 1.320, 1.334) with a double amplitude of 1.36 (95 % CI: 0.06, 2.64), i.e., the Richardson transyear An about 1.059-year component (95 % CI: 1.054, 1.063) with a double amplitude of 1.32 (95 % CI: 0.02, 2.62), i.e., the near-transyear A half-year component with a double amplitude of 4.70 (95 % CI: 3.40, 5.98)

In addition to the two main components with periods of about 11 years and about 0.5 year, both the near-(trans-) year and Richardson's ~1.3 transyear were found to contribute to the time structure of aa with statistical significance.

The estimated period length of the near-transyear is somewhat elusive, depending in part on the span considered for analysis. Most certainly, the width of the 95 % confidence intervals for the period estimate is greatly underestimated. This is because the analytical method assumes that we are dealing with stationary and stable signals. This does not seem to be the case, the signal changing its amplitude and period length with time, intermodulating in part with other nonstationary signals with lower frequencies such as the solar activity cycle.

In order to determine the extent to which results from our linear-nonlinear analyses can be relied upon for long-term prediction, and in order to compare our results with those published by physicists earlier, we analyzed the same 2-day average data of the antipodal geomagnetic index aa with the same linear-nonlinear approach over spans matching those considered in earlier publications

An extensive analysis of these data for the span from 1868 to 1998 (131 years) had shown the presence of a linear trend, a prominent half-year, and about 10.93-, 1.327-, and 1.059-year components. The 95 % CI of the near-transyear extended from 1.054 to 1.063 years. These results were compared to those of aa covering the span from 1868 to 1972 (105 years) considered by Fraser-Smith [7] and Delouis and Mayaud [98].

Using the same initial values in the same model, the linear trend and half-year remained more or less unchanged, the about-10.93-year cycle has a slightly longer period of 11.18 years, and both far- and near-transyears are no longer detected with statistical significance. After adjusting the initial period lengths of these components on the basis of the least-squares spectrum, the far-transyear is estimated to have a period of about 1.456 years (95 % CI: 1.446–1.466 years), almost 10 % longer than its estimate for a span covering 26 more years. Whereas this component is thus detected with statistical significance, the near-transyear is not, even when considering a period closer to the spectral peak, estimated at 1.072 years.

Also noteworthy is the finding of an about-1.162-year component (95 % CI: 1.149, 1.174 years) found for the 59-year span considered by Bartels (1872–1930). In this case, the period of the near-transyear is also about 10 % longer than its estimate for the longer 131-year span considered by us originally. It is validated nonlinearly in a model that adjust the initial value of the period to the spectral peak of the linear analysis when the far-transyear is excluded from the model. When the latter component is included in the model, the period estimate is about the same, but the zero-amplitude test is no longer rejected with statistical significance.

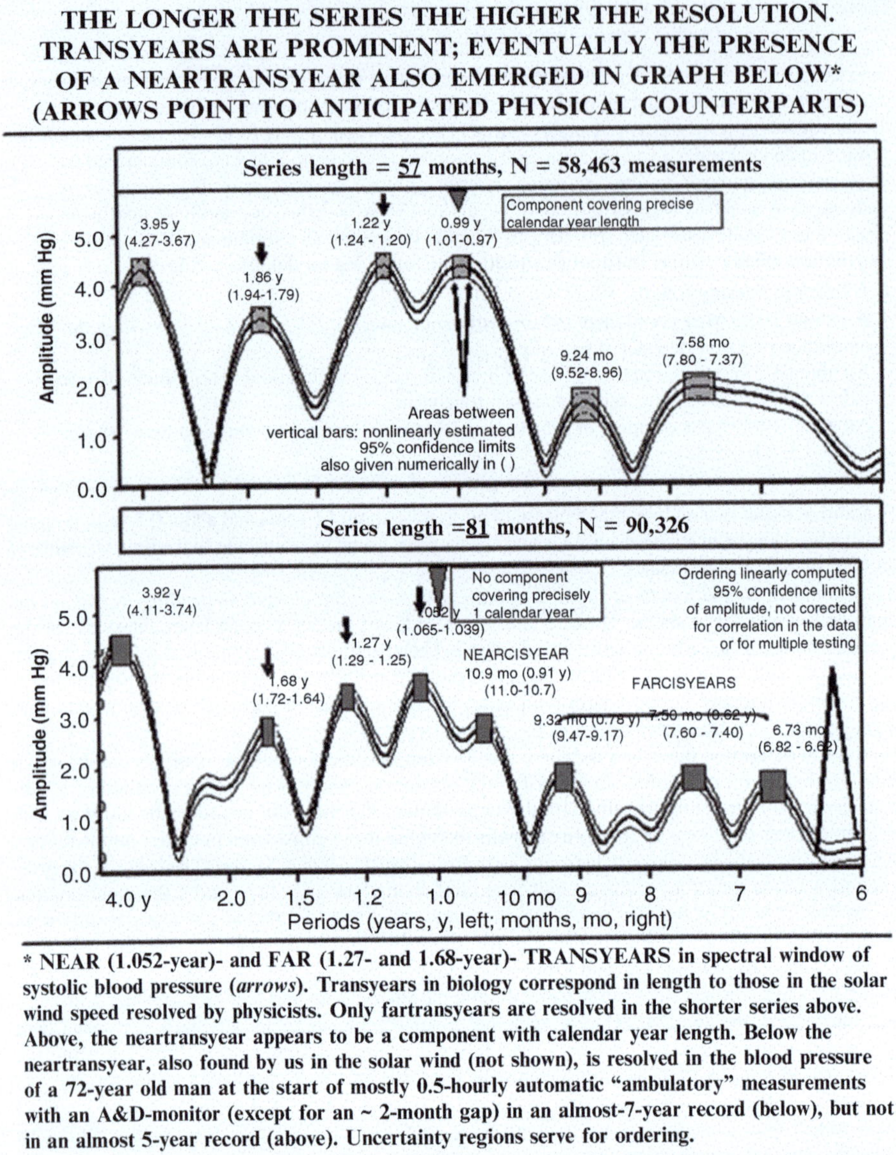

Fig. 15.9 Changing spectrum with age and increasing length of the time series. A circannual component (*top*) becomes transannual (*bottom*). Note at bottom that the 95 % confidence interval of the period does not overlap the 1.0-year length, as in the case on top (© Halberg Chronobiology Center)

geographic differences are based on much too small samples insofar as SCD is concerned, and larger samples from more geographic locations are indispensable, as are available for MI in Fig. 15.10.

In the pattern of SCD, when all available series are used, irrespective of geographic location, and the values of each series are expressed as a percentage of the

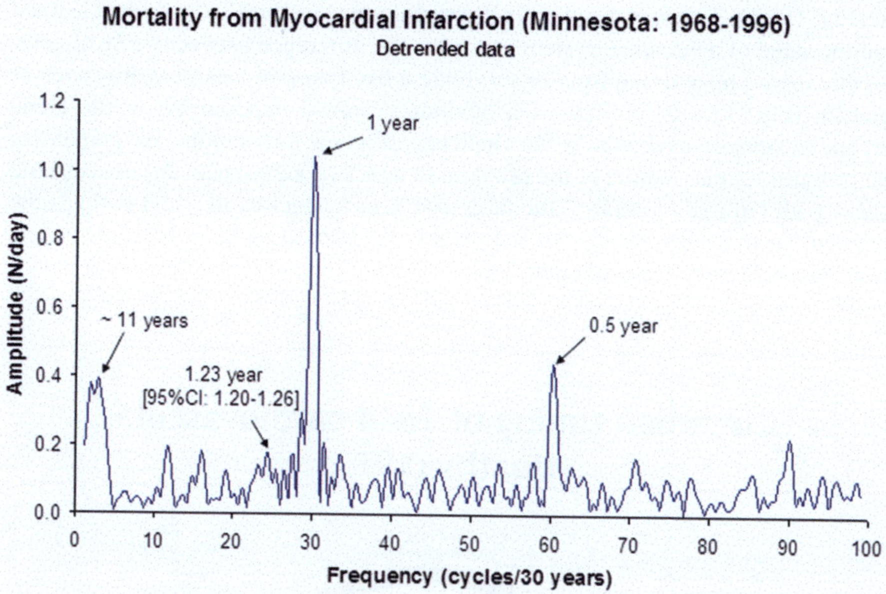

Fig. 15.10 Clear prominence, in a relatively long time series of myocardial infarctions, of yearly, probably climatic component over any magnetic ~11-year, 0.5-year, and transyearly components, the latter smaller yet with a nonzero amplitude (© Halberg Chronobiology Center)

series mean and then averaged, the calendar-year component also predominates and the transyear is not seen. This finding underlines the need for more extensive SCD data to find clues as to whether we are dealing only with random chance error or also, as we see for MI, with statistically significant components. Much controversy about heliobiology, apart from small numbers, an argument that has been laid to rest by large numbers for MI, may also stem from a changing relation to geo- and helio-magnetics, as shown by time-varying phase synchronizations and coherences [103].

Different patterns of human cardiovascular mortality may differ from each other; may also differ with time, as do solar-terrestrial phenomena; and probably differ with geographic location, a possibility that urgently requires the provision by different governments of data for analysis. These findings of interest to biomedicine may also serve physics since, as noted, the human time series can reveal dynamics that prevailed perhaps since hominization [102] and the *Acetabularia* time series allows us to take a look at perhaps 500,000,000 years ago. Mayaud [104] writes that an important solar noise is still superimposed upon the semiannual variation averaged over 100 years. It will be of interest with the aid of biological findings to see what other components can be lifted from the noise term now that biomedicine provides a 0.42-cis-semiannual component in SCD. These findings need to be aligned with the more precise half-yearly spectral component. The cis-semiannual, but not the average 0.5-year circasemiannual component, may contribute to SCD.

Over three decades ago, by objective curve fitting, the nonrandom distribution of human mortality along the 24-hour scale was confirmed, quantified, and extended to human mortality more generally in Fig. 15.5 and to cardiovascular death in particular (Fig. 15.11). This was then hardly surprising since by the 1950s, we had learned that the stage of a circadian system can make the difference between 75 % of comparable inbred mice dying from or surviving a fixed dose of a cardiac drug such as ouabain (Fig. 15.4). Carrying such laboratory animal experiments to the clinic beyond the circadian rhythm is the challenge of a new millennium. We emphasize the transdisciplinary nature of the problem to which, if geographic differences hold water, both Gilbert's *terrella* – the large and near magnet earth – and perhaps the huge but far magnet solar system and the still broader cosmos, far beyond a single fluid star, such as the sun, all contribute.

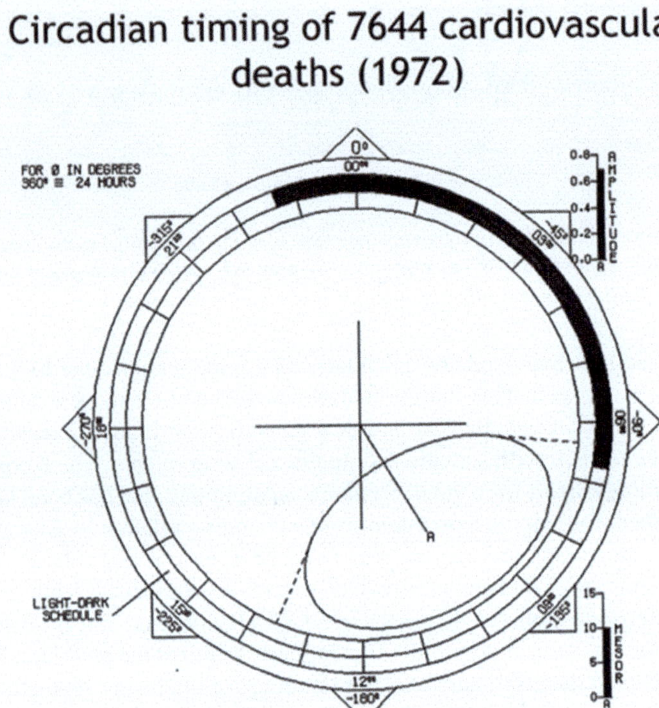

POPULATION MEAN COSINOR

KEY	P	N	PR	MESOR +-SE		AMPLITUDE *	ACROPHASE (Ø) *
A SMOLENSKY +	.006	9	50	10.00	.00	0.69 (.24 1.13)	-149° (-95 -204)

P: probability of hypothesis; amplitude=0
N: number of series used by mean cosinor technique to find ellipse
*Conservative 95% confidence limits ( ) derived from cosinor ellipse
Smolensky M, Halberg F, Sargent F II. Chronobiology of the life sequence. In: Itoh S, Ogata K, Yoshimura H, editors. Advances in Climatic Physiology. Tokyo: Igaku Shoin Ltd.; 1972. p. 281-318.

Fig. 15.11 Cosinor summary of the temporal distribution of deaths along the 24-hour scale involving the heart (© Halberg Chronobiology Center)

With a background in endocrinology, one realizes that the similarity of periods is nothing but a hint, notably of the importance of the phenomenon, whether or not the hint revolves around life vs. death from the same dose of the same agent in the case of the mouse or of a similar condition in the case of humans. Endocrinology also provides a remove-and-replace approach which can be and has been transferred to physics. We have such evidence [62] about the sun's influence upon a biological near week in human heart rate but not yet for a biological and physical near-half year and not for near years. As a recent report by Prof. Mikulecky and Dr. Florida notes, in the case of human birth in the Philippines, the transyear can also be larger in amplitude than the calendar year [105–107]. More data are needed and more are being recorded, but excluding some notable exceptions, this information is withheld by the inertia and bureaucracy of many governments.

Since the development of civilization, physicists have helped medicine, i.e., with imaging in space being part of everyday practice with X-ray, MRI, sonography, and innumerable other physical contributions. Since Gilbert and Schwabe, biomedicine has attempted to return the favor. The time for a transdisciplinary science has arrived, notably if we go beyond cardiovascular disease and start to focus on the human mind and spirit that intensely involves our circulation. An organism is a set of cycles; wherever you cut it, there is both a beginning and an end, with interactions in space-time in between. Chronomics images the cycles in and around us, some seen and very many largely unseen, except for aurorae.

Let us turn from the cycles, including different transyears in natality and death, at the extremes of extrauterine life, to those that form parts of time structures. These consist of many rhythms, of probabilistic and other chaos, as well as of trends that accompany us and can now be monitored in an increasingly cost-effective way. For a health care and for broader science, time-microscopically resolved structures can replace spotchecks or series examined by the naked eye without a transdisciplinary inferential statistical chronobiology and chronomics.

15.7 Conclusion

The as-one-goes epidemiological data analysis should become the mandate of government statisticians; their role should be extended beyond that of bookkeepers collecting data to establish imaginary baselines and selling data while so doing. They should realize that organisms are systems open to their environment and hence subject to influences by intrinsic as well as local, regional, and global climatic and other geographic factors. The latter include geomagnetic periodisms, some connected with the movement of the solid core of the earth, such as the free oscillation of about 1.18 years of Chandler [108, 109]. Mortality, morbidity, and natality statistics constitute an international treasure, not a commodity to be sold, and lessons learned from them may lead to valuable countermeasures. But before such countermeasures can be explored, the cycles and the broader time structures of data have to be analyzed with events explored as to any dependence upon temporal differences in extent or kind of response.

In the introduction to his *History of the Disorders of* [the classical] *Cardiac Rhythm*, Berndt Lüderitz [88] writes: "Arrhythmias are the main complication of ischemic heart disease, and they have been directly linked to the frequently arrhythmogenic sudden death syndrome, which is now presumed to be an avoidable 'electrical accident' of the heart." Conceivably, once the entire spectrum of rhythms is mapped, including yearly and perhaps transyearly and cis-half-yearly population rhythms, their mechanisms can be understood and can lead to countermeasures beyond current pacemaker-cardioverter-defibrillators. This remains a challenge for students of the cardiac rhythm spectrum who could clarify environmental effects near and far, so that indeed "electrical accidents" can be prevented in a rational way.

To serve health, departments by that name must mandate interpretation rather than just bookkeeping for establishing imaginary "baselines" and "trends" that may be in part cycles. Almost certainly, the circadian, circannual, transannual, and other rhythm stages, some demonstrated to tip the scale in response to a cardiac drug (Fig. 15.4) and in response to daily life (Fig. 15.6) [88], are hints of mechanisms contributing to, if not underlying, an "avoidable 'electrical accident' of the heart." Some mechanisms such as those of the heart are periodic at frequencies much lower than 1 Hz. Taking those lower frequencies into account can change the sensitivity of what we do by orders of magnitude [94, 110]. Implantable devices have already substantially lowered the incidence of SCD, and we may do even better.

The challenges of chronobiology and chronomics are much broader [111, 112] and warrant another series of publications complementing this volume and its predecessors [113–118]. As to specific applications, modern life in many settings is inconceivable without heating and cooling. If undue effects, including those of the magnetic environment, are clarified as contributing to the health of society as clues to chronobioethics, as well as to major vascular diseases, a further optimization of our environment may become possible.

References

1. Halberg F, Breus TK, Cornélissen G, et al. Chronobiology in space. Keynote, 37th Ann. Mtg. Japan Soc. for Aerospace and Environmental Medicine. Nagoya, November 8–9, 1991. University of Minnesota/Medtronic Chronobiology Seminar Series, #1, December 1991, 21 pp. of text, 70 figures.
2. Cornélissen G, Breus TK, Bingham C, et al. Beyond circadian chronorisk: worldwide circaseptan-circasemiseptan patterns of myocardial infarctions, other vascular events, and emergencies. Chronobiologia 1993;20:87–115.
3. Murakami S, Cornélissen G, Katinas G. et al. Circamultiseptan aspect of sudden death: competing socio-ecological synchronizers: alcohol and magnetics? Scripta medica 2005;78:67–74.
4. Cornélissen G, Hillman D, Katinas GS, et al. Geomagnetics and society interact in weekly and broader multiseptans underlying health and environmental integrity. Biomed Pharmacother 2002;56 Suppl 2:319s–26.
5. Cornélissen G, Halberg F, Obridko VN, et al. Quasi-11-year modulation of global and spectral features of geomagnetic disturbance. Biofizika 1998;43:677–80 [In Russian.].
6. Halberg F, Cornélissen G, Bingham C, et al. Season's appreciations 2001. Neuroendocrinol Lett 2002;23:170–87.

7. Fraser-Smith AC. Spectrum of the geomagnetic activity index Ap. J Geophys Res 1972;77:4209–20.
8. Karinen A, Mursula K, Ulich T, et al. Does the magnetosphere behave differently on weekends? Ann Geophys 2002;20:1137–42.
9. Roederer JG. Are magnetic storms hazardous to your health? Eos, Trans AGU 1995;76:441, 4–5.
10. Vladimirskii BM, Narmanskii VY, Temuriantz NA. Global rhythmics of the solar system in the terrestrial habitat. Biophysics 1995;40:731–6.
11. Halberg F, Engeli M, Hamburger C, et al. Spectral resolution of low-frequency, small-amplitude rhythms in excreted 17-ketosteroid; probable androgen induced circaseptan desynchronization. Acta Endocrinol (Kbh) 1965;50 Suppl 103:5–54.
12. Dérer L. Rhythm and proliferation with special reference to the six-day rhythm of blood leukocyte count. Neoplasma 1960;7:117–34.
13. Hildebrandt G, Bandt-Reges I. Chronobiologie in der Naturheilkunde: Grundlagen der Circaseptanperiodik Heidelberg: Haug; 1992. 102 pp.
14. Mikulecky M, Duris I. Chronocosmobiology and chronocosmomedicine in Slovakia: the legacy of Academician Ladislav Derer. In: Proceedings of the 3rd International Symposium of Chronobiology and Chronomedicine, Kunming, China, Oct. 7–12, 1998. p. 56–7.
15. Halberg F, Cornélissen G, Otsuka K, et al. Cross-spectrally coherent ~10.5- and 21-year biological and physical cycles, magnetic storms and myocardial infarctions. Neuroendocrinol Lett 2000;21:233–58.
16. Cornélissen G, Halberg F, Pöllmann L, et al. Circasemiannual chronomics: half-yearly biospheric changes in their own right and as a circannual waveform. Biomed Pharmacother 2003;57 Suppl 1:45s–54.
17. Halberg F, Cornélissen G, Otsuka K, et al. Chronomics. Biomed Pharmacother 2001;55 Suppl 1:153–90.
18. Halberg F, Cornélissen G, Schack B, et al. Blood pressure self-surveillance for health also reflects 1.3-year Richardson solar wind variation: spin-off from chronomics. Biomed Pharmacother 2003;57 Suppl 1:58s–76.
19. Halberg F, Cornélissen G, Regal P, et al. Chronoastrobiology: proposal, nine conferences, heliogeomagnetics, transyears, near-weeks, near-decades, phylogenetic and ontogenetic memories. Biomed Pharmacother 2004;58 Suppl 1:S150–87.
20. Pales E, Mikulecky M. Periodic emergence of great poets in the history of Arabia & Persia, China and Japan. Neuroendocrinol Lett 2004;25:169–72.
21. Richardson JD, Paularena KI, Belcher JW, et al. Solar wind oscillations with a 1.3-year period. Geophys Res Lett 1994;21:1559–60.
22. Valdés-Galicia JF, Pérez-Enríquez R, Otaola JA. The cosmic-ray 1.68-year variation: a clue to understand the nature of the solar cycle? Sol Phys 1996;167:409–17.
23. Rangarajan GK, Barreto LM. Long term variability in solar wind velocity and IMF intensity and the relationship between solar wind parameters & geomagnetic activity. Earth Planets Space 2000;52:121–32.
24. Mursula K, Zieger B. The 1.3-year variation in solar wind speed and geomagnetic activity. Adv Space Res 2000;25:1939–42.
25. Halberg F, Cornélissen G, Stoynev A, et al. Season's Appreciations 2002 and 2003. Imaging in time: the transyear (longer-than-the-calendar year) and the half-year. Neuroendocrinol Lett 2003;24:421–40.
26. Cornélissen G, Masalov A, Halberg F, et al. Multiple resonances among time structures, chronomes, around and in us. Is an about 1.3-year periodicity in solar wind built into the human cardiovascular chronome? Hum Physiol 2004;30(2):86–92.
27. Halberg F, Cornélissen G, Katinas G et al. In memoriam: Ion Baciu. Mutually supporting near-transyears in solar and terrestrial magnetics, microbial and cell biology, physiology and pathology. In: Cornélissen G, Kenner R, Fiser B, Siegelova J, editors. Proceedings of the Symposium: Chronobiology in Medicine. Dedicated to the 85th Anniversary of Professor Franz Halberg. Brno: Masaryk University; 2004. p. 78–86.

28. Halberg F, Cornélissen G, Panksepp J et al. Near-transyears in geophysics and autism. Abstract, Noninvasive methods in cardiology, 2.11.2004, Congress MEFA, Brno, Czech Republic, 2.-4.11.2004, p. 7.
29. Cornélissen G, Halberg F, Hamburger C et al. Chronomics: multiple distinct yearly cycles in urinary steroid metabolite excretion. Abstract, 2nd Int. Symp., Problems of Rhythms in Natural Sciences, Moscow, March 1–3, 2004. Moscow: Russian People's Friendship University; 2004. p. 17–8.
30. Halberg F, Cornélissen G, Katinas G et al. Transdisciplinary unifying implications of circadian findings in the 1950s. J Circadian Rhythms 2003; 1: 2. 61 pp. www.JCircadianRhythms.com/content/pdf/1740-3391/1/2.pdf
31. Halberg F. Chronobiology. Annu Rev Physiol 1969;31:675–725.
32. Halberg F. Chronobiology: methodological problems. Acta med rom 1980;18:399–440.
33. Cornélissen G, Halberg F. Chronomedicine. In: Armitage P, Colton T, editors. Encyclopedia of Biostatistics, v. 1. Chichester, UK: John Wiley & Sons Ltd;1998. 642–9.
34. Marquardt DW. An algorithm for least-squares estimation of nonlinear parameters. J Soc Ind Appl Math 1963;11:431–41.
35. Chizhevsky AL. Action de l'ionisation de l'atmosphère et de l'ionisation artificielle de l'air sur les organismes sains et les organismes malades. In: Piéry M, ed. Traité de Climatologie: Biologique et médicale. Tome premier. Paris: Masson et Cie; 1934. p. 662–73.
36. Chizhevsky AL (Fedynsky VV, ed). The Earth in the Universe. Translated from Russian [and edited by IRST staff]. NASA TT F-345. Jerusalem: Israel Program for Scientific Translations [available from US Dept of Commerce, Clearinghouse for Federal Scientific and Technical Information, Springfield, Virginia]; 1968. 280 pp.
37. Dubrov AP (Sinclair FL, trans; Brown FA Jr, translation ed). The geomagnetic field and life: geomagnetobiology. Plenum Press, New York, 1978, 318 pp.
38. Komarov FI, Breus TK. Rapoport SI et al [Biomedical effects of solar activity.]. Vestn Ross Akad Med Nauk 1994;11:37–50.
39. Zhadin MN. Review of Russian literature on biological action of DC and low-frequency AC magnetic fields. Bioelectromagnetics 2001;22:27–45.
40. Breus TK, Chibisov SM, Baevsky RM, et al. The chronostructure of heart biorhythms under effects of external environment factors. Moscow: Russian University of People's Friendship, Polygraph Service; 2002. 231 pp.
41. Breus TK, Rapoport SI. Magnitnie bure: Medico-biologicheskie i geofizicheskie aspekti. Moscow: Sovetskiy Sport Izdatel'stvo; 2003. 192 pp.
42. Chibisov S, Cornélissen G, Halberg F. Chronomics: circadian effects of magnetic storms in rabbits' circulation and transannual variations? Abstract, 2nd International Symposium, Problems of Rhythms in Natural Sciences, Moscow, March 1–3, 2004. Moscow: Russian People's Friendship University; 2004. p. 21–3.
43. Chibisov S, Halberg F, Cornélissen G et al. Amplitude changes of biorhythms as a sign of desynchronosis and magnetic storm. Abstract, 2nd International Symposium, Problems of Rhythms in Natural Sciences, Moscow, March 1–3, 2004. Moscow: Russian People's Friendship University; 2004. p. 493–6.
44. Alexandrov SI. Gamburtsev AG (Oleinic O, trans.) Atlas prirodnyh processov [Atlas of Natural Processes]. Vol. 1. Order and Chaos in Lithosphere and Other Spheres. Moscow: United Institute of the Physics of the Earth; 1994. 176 pp.
45. Alexandrov SI. Gamburtsev AG (Oleinic O, trans.) Atlas vremennykh variacij prirodnykh antropogennykh i social'nykh processov [Atlas of Temporal Variations in Natural, Anthropogenic and Social Processes]. Vol. 2: Cyclic dynamics in nature and society. Supplement. Moscow: Scientific World; 1999. 72 pp.
46. Düll T, Düll B. Über die Abhängigkeit des Gesundheitszustandes von plötzlichen Eruptionen auf der Sonne und die Existenz einer 27tägigen Periode in den Sterbefällen. Virchows Arch 1934;293:272–319.
47. Düll T, Düll B. Zusammenhänge zwischen Störungen des Erdmagnetismus und Häufungen von Todesfällen. Deutsch med Wschr 1935; 61: 95 (18. Januar).

48. Düll B, Düll T. Zur Frage solar-aktiver Einflusse auf die Psyche. Z gesamte Neurol und Psychiatr 1938; 162, N 3: 495–7
49. Stoupel E. Forecasting in Cardiology. New York: John Wiley & Sons; 1976. 141 pp.
50. Stoupel E, Domarkiene S, Radishauskas R et al. Links between monthly rates of four subtypes of acute myocardial infarction and their corresponding cosmophysical activity parameters. J Basic Clinical Physiology Pharmacology 2004;15(3–4):175–84.
51. Ertel S. Space weather and revolutions: Chizhevsky's heliobiological claim scrutinized. Stud Psychol 1996;39:3–22.
52. Ertel S. Bursts of creativity and aberrant sunspot cycles: hypothetical covariations. In: Nyborg H, editor. The Scientific Study of Human Nature: Tribute to Hans J. Eysenck at 80. New York: Elsevier Science; 1997. p. 491–508.
53. Mikulecky M, editor. The Moon and Living Matter. Kosice, Slovakia, September 23–25, 1993. Bratislava: Slovak Medical Society; 1993. 97 pp
54. Mikulecky M, editor. Sun, Moon and Living Matter. Bratislava, Slovakia, June 28-July 1, 1994. Bratislava: Slovak Medical Society; 1994. 159 pp
55. Mikulecky M, editor. Chronobiology & Its Roots in the Cosmos. High Tatras, Slovakia, September 2–6, 1997. Bratislava: Slovak Medical Society; 1997. 287 pp
56. Strestik J, Prigancova A. On the possible effect of environmental factors on the occurrence of traffic accidents. Acta Geodætica, Geophysica et Montanistica Hungarica 1986;21:155–66.
57. Breus T, Cornélissen G, Halberg F, et al. Temporal associations of life with solar and geophysical activity. Ann Geophys 1995;13:1211–22.
58. Villoresi G, Breus TK, Iucci N, et al. The influence of geophysical and social effects on the incidences of clinically important pathologies (Moscow 1979–1981). Phys Med 1994;10: 79–91.
59. Villoresi G, Kopytenko YA, Ptitsyna NG, et al. The influence of geomagnetic storms and man-made magnetic field disturbances on the incidence of myocardial infarction in St. Petersburg (Russia). Phys Med 1994;10:107–17.
60. Cornélissen G, Halberg F, Schwartzkopff O, et al. Chronomes, time structures, for chronobioengineering for "a full life". Biomed Instrum Technol 1999;33:152–87.
61. Chibisov SM, Cornélissen G, Halberg F. Magnetic storm effect on the circulation of rabbits. Biomed Pharmacother 2004;58 Suppl 1:S15–9.
62. Cornélissen G, Halberg F, Wendt HW, et al. Resonance of about-weekly human heart rate rhythm with solar activity change. Biologia (Bratisl) 1996;51:749–56.
63. Cornélissen G, Halberg F. Introduction to Chronobiology. Medtronic Chronobiology Seminar #7, April 1994, 52 pp. (Library of Congress Catalog Card #94-060580; https://sites.google.com/a/umn.edu/halbergchronobiologycenter/)
64. Feinleib M, Rogot E, Sturrock PA. Solar activity and mortality in the United States. Int J Epidemiol 1975;4:227–9.
65. Lipa BJ, Sturrock PA, Rogot E. Search for correlation between geomagnetic disturbances and mortality. Nature 1976;259:302–4.
66. Halberg F, Cornélissen G, Otsuka K, et al. Essays on chronomics spawned by transdisciplinary chronobiology: Witness in time: Earl Elmer Bakken. Neuroendocrinol Lett 2001;22:359–84.
67. Cornélissen G, Halberg F, Breus T, et al. Non-photic solar associations of heart rate variability and myocardial infarction. J Atmos Solar Terr Phys 2002;64:707–20.
68. Roskamm H, Neumann F-J, Kalusche D, Bestehorn HP, editors. Herzkrankheiten: Pathophysiologie / Diagnostik / Therapie. Berlin: Springer; 2004. p. 438–49, 446–8.
69. Lancisi GM. De subitaneis mortibus libri duo. Rome: Jo: Francisci Buagni; 1707. 243 pp.
70. Lancisi GM. Opera quae hactenus prodierunt omnia; dissertationibus nonnullis adhunc dum ineditis locupletata, & ab ipso auctore, recognita atque emendata. Collegit, ac in ordinem digessit Petrus Assaltus. (t. 1. De subitaneis mortibus libri duo. Dissertatio de nativis, deque adventitiis Romani coeli qualitatibus. Cui accedit Historia epidemiae rheumaticae. De noxiis paludum effluviis eorumque remediis libri duo. - t. 2. Dissertatio historica de bovilla peste. Disserationes variae.) Geneva: sumptibus J.A. Cramer & filii; 1718.

71. Snellen HA. History of Cardiology. Rotterdam: Donker; 1984. 191 pp.
72. Hering HE. Der Sekundenherztod mit besonderer Beruecksichtigung des Kammerflimmerns. Berlin: Springer; 1917.
73. Myerburg R, Castellanos A. Cardiac arrest and sudden death. In: Braunwald E, editor. Heart Disease. Philadelphia: Saunders; 1997. p. 742–79.
74. Zipes DP, Wellens HJJ. Sudden cardiac death. Circulation 1998;98:2334–51.
75. Waldecker B. Zur Epidemiologie des plötzlichen Herztodes. Notfall & Rettungsmedizin 2003;6:313–7.
76. Halberg F, Schwartzkopff O, Cornélissen G, et al. Rechnergestützte Wissenszweige treffen sich in München, November 29–30, 2002. Neuroendocrinol Lett 2003;24 Suppl 1:27–59.
77. Fiser B, Cornélissen G, Siegelova J, et al. Increase in stroke deaths after 1997 in the Czech Republic. Scr med (Brno) 2002;75:95–100.
78. Schwartzkopff O, Cornélissen G, Halberg F. Differences in the time structures (chronomes) of blood pressure (BP) and heart rate (HR). Abstract, 6° Convegno Nazionale de Cronobiologia, Chianciano, Italy, November 27–28, 1998. p. 31.
79. Muller J, Ludmer P, Willich S, et al. Circadian variation in the frequency of sudden cardiac death. Circulation 1987;75:131–8.
80. Peckova M, Fahrenbruch C, Cobb L, et al. Circadian variations in the occurrence of cardiac arrests: initial and repeat episodes. Circulation 1998;98:31–9.
81. Arntz H-R, Müller-Nordhorn J, Willich SN. Cold Monday mornings prove dangerous: epidemiology of sudden cardiac death. Curr Opin Crit Care 2001;7:139–44.
82. Smolensky M, Halberg F, Sargent II F. Chronobiology of the life sequence. In: Itoh S, Ogata K, Yoshimura H, editors. Advances in Climatic Physiology. Tokyo: Igaku Shoin Ltd.; 1972. p. 281–318.
83. Cornélissen G, Tamura K, Tarquini B, et al. Differences in some circadian patterns of cardiac arrhythmia, myocardial infarction and other adverse vascular events. Chronobiologia 1994;21:79–88.
84. Halberg F, Stephens AN. Susceptibility to ouabain and physiologic circadian periodicity. Proc Minn Acad Sci 1959;27:139–43.
85. Halberg F, Haus E, Stephens A. Susceptibility to ouabain and physiologic 24-hour periodicity. Fed Proc 1959;18:63.
86. Halberg F. Laboratory techniques and rhythmometry. In: Mills JN, editor. Biological Aspects of Circadian Rhythms. London/New York: Plenum Press; 1973. p. 1–26.
87. Reinberg A, Gervais P, Halberg F, et al. Mortalité des adultes: Rythmes circadiens et circannuels dans un hôpital parisien et en France. Nouv Presse Med 1973;2:289–94.
88. Lüderitz B. History of the Disorders of Cardiac Rhythm. Armonk, NY: Futura; 1995. 167 pp.
89. Gauss CF, Weber W, ed. Resultate aus den Beobachtungen des magnetischen Vereins im Jahre 1839. Mit 4 Steindrucktafeln. Leipzig, im Verlage der Weidmannschen Buchhandlung, 1840
90. Biermann K-R, editor. Briefwechsel zwischen Alexander von Humboldt und Carl Friedrich Gauß: Zum 200. Geburtstag von C.F. Gauß im Auftrage des Gauß-Komitees bei der Akademie der Wissenschaften der DDR. Berlin: Akademie-Verlag; 1977. 202 pp. + plates.
91. Griffiths C, Rooney C. The effect of the introduction of ICD-10 on trends in mortality from injury and poisoning in England and Wales. Health Stat Q 2003;19:10–21.
92. Rooney C, Griffiths C, Cook L. The implementation of ICD-10 for cause of death coding: some preliminary results from the bridge coding study. Health Stat Q 2002;13:31–41.
93. Griffiths C, Brock A, Rooney C. The impact of introducing ICD-10 on trends in mortality from circulatory diseases in England and Wales. Health Stat Q 2004;22:14–20.
94. Halberg F. Some physiological and clinical aspects of 24-hour periodicity. Journal Lancet (Minneapolis) 1953;73:20–32.
95. Halberg F, Cornélissen G, Wendt H et al. Chronomik von Komplementärsystemen der Naturwissenschaften und Medizin: Reziproke Zyklizitäten der Biosphäre und des Kosmos. In: Schröder W (editor). Arbeitskreis Geschichte der Geophysik und Kosmische Physik.

Meteorological and Geophysical Field Dynamics (A book to commemorate the centenary of the birth of Hans Ertel). Bremen: Wilfried Schröder/Science Edition/Arbk Geschichte Geophysik: 2004. p. 284–302.
96. Shapiro R. Interpretation of the subsidiary peaks at periods near 27 days in power spectra of geomagnetic disturbance indices. J Geophys Res 1967;72:4945–9.
97. Bartels J. L'activité du magnétisme terrestre et ses relations avec les phénomènes solaires. C R Cong Int Electr Paris 1932, 1932; 12: 77–88.
98. Delouis H, Mayaud PN. Spectral analysis of the geomagnetic activity index aa over a 103-year interval. J Geophys Res 1975;80:4681–8.
99. Silverman SM, Shapiro R. Power spectral analysis of auroral occurrence frequency. J Geophys Res 1983;88(A8):6310–6.
100. Schwabe H. Über die Flecken der Sonne. Astronomische Nachrichten 1838;15:244–8 (no. 350).
101. Schwabe H. Sonnen-Beobachtungen im Jahre 1843. Astronomische Nachrichten 1844;21:254–6 (no. 495).
102. Le Roy E. Les origines humaines et l'évolution de l'intelligence. Paris: Boivin & cie; 1928. vii, 375 p.
103. Halberg F, Cornélissen G, Bingham C, et al. Chronomics: Imaging in time by phase synchronization reveals wide spectral-biospheric resonances beyond short rhythms. ("Wenn man über kurze Rhythmen hinausgeht") In memoriam – lost future: Dr.-Ing. habil. Dr. rer. nat. Barbara Schack: 1952–2003. Neuroendocrinol Lett 2003;24:355–80.
104. Mayaud PN. Analyse d'une série centenaire d'indices d'activité magnétique: I. -- Recherche d'un effet lunaire et récurrence de 27 jours. Annales de Geophysique 1975; 31: 419–31.
105. Mikulecky M, Florida PL. Daily birth numbers in Davao, Philippines, 1993–2003: Halberg's transyear stronger than year. Abstract, 26th Seminar, Man in his terrestrial and cosmic environment, Upice, Czech Republic, May 17–19, 2005, unpaginated.
106. Cornélissen G, Halberg F, Mikulecky M, et al. Yearly and perhaps transyearly human natality patterns near the equator and at higher latitudes. Biomed Pharmacother 2005;59 Suppl 1:S117–22.
107. Cornelissen G, Halberg F, Mikulecky M, Florida P, Faraone P, Yamanaka T, Murakami S, Otsuka K, Bakken EE. Yearly and perhaps transyearly human natality patterns near the equator and at higher latitudes. Biomed Pharmacother 2005;59(Suppl 1):S117–22.
108. Chandler Jr SC. On the variation of latitude: VII. Astron J 1892;277:97–101.
109. Schuh H, Nagel S, Seitz T. Linear drift and periodic variations observed in long time series of polar motion. J Geod 2001;74:701–10.
110. Halberg F. Some correlations between chemical structure and maximal eosinopenia in adrenalectomized and hypophysectomized mice. J Pharmacol Exp Ther 1952;106:135–49.
111. Berry BL. Synchronous processes in the Earth's shells and their cosmic reasons. Vestnik Moskovskogo Universiteta, seriya Geografiya 1991; n. 1: 20–7. In Russian.
112. Berry BL. Basic systems of geosphere-biospheric cycles and the prediction of natural conditions. Biophysics 1993;37(3):328–41.
113. Otsuka K, editor. Proceedings of the 1st International Symposium Workshop on Circadian Rhythms and Clinical Chronobiology, 11 Nov 2000, Tokyo, Japan. Biomed Pharmacother 2001;55 Suppl 1:7s–190.
114. Otsuka K, editor. Proceedings of the 2nd International Symposium Workshop on Circadian Rhythms and Clinical Chronobiology, 17 Nov 2001, Tokyo, Japan. Biomed Pharmacother 2002; 56 Suppl 2: 231s-382.
115. Otsuka K, editor. Proceedings of the 3rd International Symposium Workshop on Circadian Rhythms and Clinical Chronobiology, 2002 Nov 9. Tokyo, Japan. Biomed Pharmacother 2003; 57 Suppl 1: 1s-198.
116. Otsuka K, editor. Proceedings of the 4th International Symposium Workshop on Circadian Rhythms and Clinical Chronobiology;2003 Nov 8. Tokyo, Japan. Biomed Pharmacother 2004; 58 Suppl 1: S1-188

117. Otsuka K, editor. Proceedings of the 5th International Symposium Workshop on Circadian Rhythms and Clinical Chronobiology; 2004 Nov 6, Tokyo, Japan. Biomed Pharmacother 2005; 59 Suppl 1: S1-262
118. Halberg F, Cornelissen G, Otsuka K, et al. Incidence of sudden cardiac death, myocardial infarction and far- and near-transyears. Biomed Pharmacother 2005;59 Suppl 1:S239–61.

Chapter 16
Diagnosing Vascular Variability Disorders and Vascular Variability Syndromes

Abstract Currently, 24-hour monitoring of blood pressure (BP) by ambulatory functioning devices is a gold standard, reserved for special cases of high BP, left uninterpreted in terms of its time structure. General reliance upon a single measurement (or a single 24-hour profile) of BP, however, has been dubbed "flying blind." Long-term BP monitoring, analyzed time-structurally (chronomically, from chronome = time structure), detects physiological-physical interactions. The chronobiological and chronomicinterpretation of ambulatory blood pressure monitoring (C-ABPM) records in the light of time-specified reference values derived from healthy peers matched by gender and age identifies vascular variability anomalies (VVAs) for an assessment of cardio-cerebro-reno-vascular disease risk. Even within the conventionally accepted normal range, VVAs have been associated with a statistically significant increase in risk. Long-term C-ABPM records help to "know ourselves," serving for relief of psychological and other strain once transient VVAs are linked to the source of a load, prompting adjustment of one's lifestyle for strain reduction.

Keywords Chronobiologically interpreted 24-hour/7-day BP and HR monitoring • Vascular variability anomalies • Vascular variability disorders • Vascular variability syndromes

16.1 Preamble

The current guidelines for diagnosing abnormal, notably high BP in a substantial segment of the population, have to be revised, according to a consensus meeting held at St. Anna Hospital, Masaryk University, Brno, Czech Republic, on October 6, 2008 (in the setting where instrumentation for beat-to-beat measurements of BP was developed) [1–3]. Specifications are needed for the minimal number of measurements, for how long and how often they should be taken, including their temporal placement. Methods are also needed for the assessment of dynamics, in keeping with a document originally prepared for this meeting by Dr. Germaine Cornélissen, Professor of Integrative Biology and Physiology at the University of Minnesota, revised in Brno by those undersigned. The terms "normotension" and "hypertension" can be replaced by the terms "MESOR-normotension" (MN) and "MESOR-hypertension" (MH),

K. Otsuka et al., *Chronomics and Continuous Ambulatory Blood Pressure Monitoring*, DOI 10.1007/978-4-431-54631-3_16

respectively, whenever the conditions for a chronobiologically interpreted 24-hour/7-day BP and HR monitoring (C-ABPM) are met. The term MH indicates only one of several vascular variability disorders (VVDs) that can combine to form sets of 2, 3, and n-component vascular variability syndromes (VVSs). For current health-care practice, the foregoing diagnoses do not include changes in day-night ratios (DNRs). DNRs and their alterations are routinely computed for research. Thus far for predicting outcomes, DNRs were all inferior to the parametric and nonparametric assessment of VVDs, including some that carry a risk higher than MH itself. In diagnosing MESOR-prehypertension, correctly identified with a chronobiologic approach, the DNR misled: the DNR was normal in patients with minimal change retinopathy and abnormal in the normal controls without any retinal involvement. Prediabetes was diagnosed chronobiologically, while the DNR failed to detect it. Any chronobiologically assessed consequences of MESOR-hypotension have yet to be assessed in terms of outcomes and remain beyond the scope of this consensus.

VVDs and VVSs derived from C-ABPM gauge an increased vascular disease risk, including conditions unnoticed in current practice, some of which may be treated. When the conditions for C-ABPM 24/7 are met, the term "diagnosing hypertension" can be replaced by the wider scope of the terms "diagnosis of VVDs and of VVSs."

16.2 Vascular Variability Disorders

Complementing other indices of risk studied, e.g., after a myocardial infarction [4, 5], the search for vascular variability disorders (VVDs) and vascular variability syndromes (VVSs), albeit aimed in particular at healthy subjects [3, 6–12] to detect earliest alterations, concerns a very large number of patients diagnosed conventionally as "hypertensive" [13–25] (about 72 million in the USA) [26]. The following diagnoses can be based upon a summary called a "sphygmochron," provided with accompanying materials [27–30] (Fig. 16.1).

(1) **M**ESOR-**n**ormotension (MN), when (a) all characteristics of a model fitted parametrically are within the limits of 90 % **p**rediction **i**ntervals (PI) [31] and (b) all endpoints obtained nonparametrically by stacking the data are acceptable (do not exceed thresholds) computed from data of peers matched by gender and age and, preferably, when possible, ethnicity and geography.

(2) **M**ESOR-**h**ypertension (MH), a chronobiologically validated elevated BP (Fig. 16.2-I). This term is used only if the diagnosis is based on the MESOR, obtained by the least-squares fit of cosine curves with anticipated (24- and 12-hour) periods, for comparison with ranges (90 % PIs) of acceptable MESORs characterizing data from clinically healthy peers matched by gender and age. The minimal time series length is a 24-hour/7-day record of data collected automatically at hourly or shorter intervals (or manually for 7 days at 4-hour intervals), analyzed both daily and for the week as a whole. It serves to rule out MH, if negative when analyzed parametrically for the weeklong record as a whole, but data collection is continued

when abnormality is found. MH is a condition where the BP-M is above the upper 95 % prediction limit of BP-Ms from clinically healthy peers matched by gender and age. BP elevation is noted when the hyperbaric index (extent of excess during 24 hours) exceeds the threshold of 50 mm Hg x hour during 24 hours. MH can be the sole VVD or it can coexist and/or alternate with other VVDs to constitute a VVS.

(3) CHAT (circadian hyper-amplitude-tension) or circadian BP overswing (excessive BP swing) is a condition characterized by a double amplitude of BP (BP-2A) derived from a 24/7 record exceeding the upper 95 % prediction limit of BP-2As from clinically healthy peers matched by gender and age (Fig. 16.2-II). CHAT can occur alone in MN and with a usual timing of the circadian BP rhythm, or it can coexist and/or alternate with other VVDs, such as complicating MH. C-ABPM is recommended to all patients with treated MH to ascertain that the elimination or reduction of MH is not a trade for the greater risk of CHAT (Fig. 16.3).

A measure of the extent of predictable change in this model, approximating the within-day BP change, the 24-hour BP-2A can be grossly underestimated by the DNR that does not account for alterations in the circadian acrophase and/or waveform. The DNR neither accounts for change with age in circadian BP characteristics, notably in terms of a postprandial BP dip, nor does it have reference values for differences between genders and is limited to a few aspects of within-day change, ignoring the many rhythms with frequencies other than circadian.

(4) BP ecphasia is a condition characterized by an odd timing of the circadian acrophase (ϕ) of BP but not of the ϕ of HR (BP ecphasia with HR euphasia) (Fig. 16.2-III). ϕ is a measure of the timing of overall high values recurring each day, measured as the lag from a given reference time (e.g., local midnight) of the peak in the 24-hour cosine curve of the model approximating the data. BP ecphasia is differentiated from the consequences of shift work that can be associated with an altered timing of both BP and HR (Fig. 16.1). It is defined by the BP-ϕ derived from a 24/7 record lying outside the 90 % PI of BP-ϕs from clinically healthy subjects matched by gender and age.

(5) Ecfrequentia is a statistically significant deviation of the period from 24 hours, assessed by the nonoverlap of 24 hours by the 95 % confidence interval of the period assessed by nonlinear least squares (Fig. 16.2-IV).

(6) Excessive pulse pressure (EPP) is defined by a difference between the systolic (S) and diastolic (D) BP-Ms in a 24/7 record above 60 mm Hg (until gender- and age-matched reference standards become available) (Fig. 16.2-V).

(7) Deficient HR variability (DHRV) is defined as a standard deviation (SD) of HR from a 24/7 record below 7.5 beats/minute (until gender- and age-matched reference standards become available) (Fig. 16.2-VI).

(8) Excessive pulse pressure product is a product of the MESORs of systolic blood pressure and heart rate larger (/100) larger than 100 ((mmHg.beats/min)/100) (until it can be further qualified in terms of gender and age) (Fig. 16.2-VII).

Several prospective and also much larger retrospective outcome studies have shown that VVDs such as CHAT carry a vascular disease risk that can be higher than the risk of MH, even among normotensive subjects [32] (Fig. 16.4). For

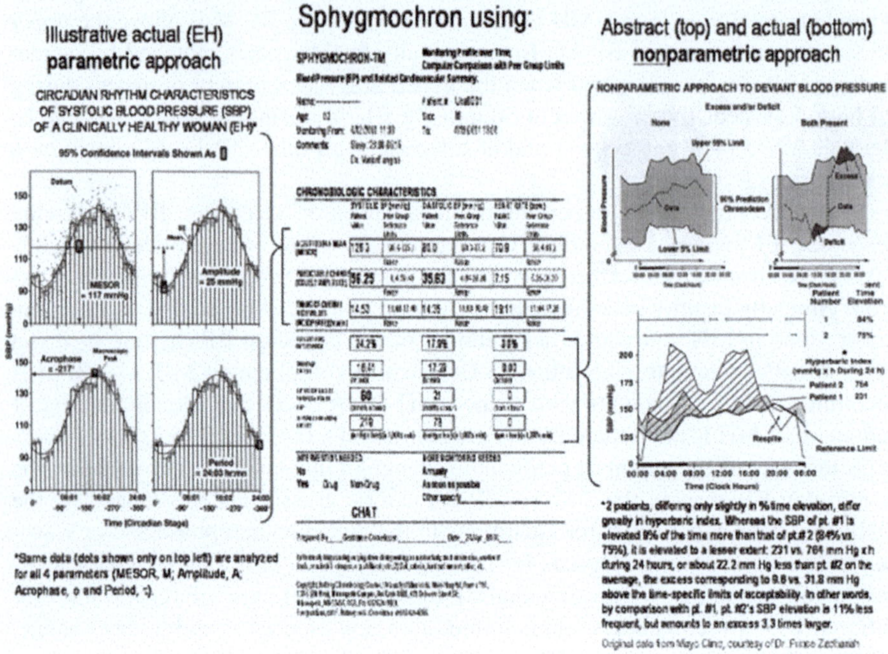

Fig. 16.1 Illustrative parametric (*left*) and nonparametric (*right*) approach brackets a sphygmochron (*middle*) from a MESOR-normotensive man with CHAT, a first tentative diagnosis requiring additional monitoring.

After data covering preferably at least 7 days of blood pressure (BP) and heart rate (HR) are downloaded from the e-mail into a computer for analysis, the following results are provided (since the 1990s and currently cost free from corne001@umn.edu) for the patient as well as for the care provider:

1. A list of actual measurements and the times at which they were obtained.
2. A plot of data as a function of time, shown together with the time-specified prediction intervals (PIs) of acceptability for systolic (S) and diastolic (D) BP and HR characteristics.
3. A data summary and a report of any BP and/or HR excess in consecutive 3-hour intervals. This part of the report may be accompanied by a "rhythmometric summary," which is just a more technical form from which the information is derived to prepare the

(4.) "sphygmochron." A sample "sphygmochron" (center) illustrates how results are being reported. First, above, the participant's name is kept confidential; a codename is used instead. Gender and age are listed, along with the date and time at which monitoring started, and for how long data were collected. The numerical report consists of two parts labeled "Characteristics" (parametric results) and "Indices of Deviation" (nonparametric results). In each case, results are shown for SBP (when the heart contracts) on the left, DBP (when the heart relaxes) in the middle, and HR on the right. Under "parametric results," a mathematical model of a smooth curve is fitted to the data to assess their circadian variation, which is primarily characterized by four numbers shown in the left-hand section of the graph, one of which, the period, covers with its uncertainty the precise 24 hours, so that the other 3 numbers are given from the fit of a 24-hour cosine curve. One characteristic, called the "MESOR," is the average value around which values fluctuate. It is very similar to the mean value, but yields more reliable results when the data are not collected at precisely regular intervals and has a smaller error when the data are equally spaced. Another characteristic, called the "double amplitude," is a measure of the predictable change occurring within a day, from the overall low values found usually during sleep to the high values during the daily active span. The third characteristic, called the "acrophase," is a measure of the time when overall example, the risk of vascular morbidity (cerebral ischemic event, coronary artery disease, nephropathy, and retinopathy) associated with CHAT alone (in MESOR-normotensive subjects) is twice as large as that associated with uncomplicated MH. The VVDs listed above are mostly independent and additive. When two or more coexist to constitute a VVS, the risk is usually larger than that of any one of the VVDs present alone. The risk of vascular morbidity in patients with three VVDs is much higher than the risk of patients with any one VVD alone (Fig. 16.5).

VVDs and/or VVSs detect earliest risk elevations, when these may be more readily reversed as prehypertension [6–10], prediabetes [11, 12] and, possibly with a focus on pulse pressure and obesity [13], a pre-metabolic syndrome. Hospitals with

Fig. 16.1 (continued) high values are likely to recur each day. For each of the three characteristics ("parameters"), the participant's value is compared to a range of acceptable values, derived from data provided by clinically healthy people of the same gender and age group as the participant. For instance, in the example shown here in sphygmochron, the average SBP, the DBP, and all other parameters are within the rectangles, indicating the range of acceptable values except for the double amplitudes of SBP and DBP. Under "nonparametric results," the participant's data are compared by computer with time-specified reference values, also derived from chronobiological archives on clinically healthy subjects matched by gender and age. For this analysis, all data are stacked over an idealized 24-hour day. Whenever a given person's profile exceeds the limits of acceptability of peers, the data are marked as being excessive or deficient. The "percentage time of elevation" reports the relative incidence of excessive values during a 24-hour day. It is common to have occasional high values, but in the example herein, there is reason for concern. The next item, the time of excess, becomes useful when drug treatment should be timed prior to the peak in excess.
Excessive values may either be barely above the limit or in turn can be very much higher than the limit. It is therefore important to express the extent of deviation by the "area under the curve," that is, the area between the values when they exceed the limit and this limit itself. Empirically, it has been shown that excess up to about 50 (mmHg x hour during 24 hours) may still be acceptable and accountable for by daily worries and/or physical activities. In the case summarized, the HBI is 60, in bold, and if confirmed in the next 7/24 profile, a reason for treatment.
On the *top right*, an abstract illustration of excess and deficit is accompanied below by 2 cases that are similar in terms of percent time elevation. They are very different in terms of hyperbaric index. In patient #2, although the percent time elevation is 9 % smaller than that in patient #1, the hyperbaric index is several times larger.
The "timing of excess" can be used as a guide to time the administration of non-drug or, if need be, of drug treatment once there is BP excess above 50 (mmHg x hour during 24 hours) and/or an elevation in MESOR, taking into account the chronopharmacokinetics of the drug prescribed. When, e.g., a tentative diagnosis of MESOR-normotension with CHAT is made, with insight into information provided on the questionnaire given to the participant with the monitor, as a first step, additional analyses may be carried out. Additional monitoring is recommended to check on any abnormality detected during the first monitoring, and, if confirmed, the need for intervention is reported to the person monitored so that it can be reported to the health-care provider. In one case summarized elsewhere, the follow-up 7-day monitoring showed that CHAT persisted for both SBP and DBP, while the MESORs were again acceptable. Thus, the diagnosis of CHAT with MESOR-normotension was confirmed. Consultation with a health-care provider was strongly and urgently recommended. In two cases of CHAT without an elevation of the BP MESOR, when such recommendations were ignored, catastrophic disease and high cost occurred, a myocardial infarction in a man [29, 30] or eclampsia in a pregnant woman with pressures of 115/67 mm Hg (SBP/DBP), leading to the delivery of a very premature boy hospitalized for 26 months at a cost of US$1 million [28]. © Halberg Chronobiology Center

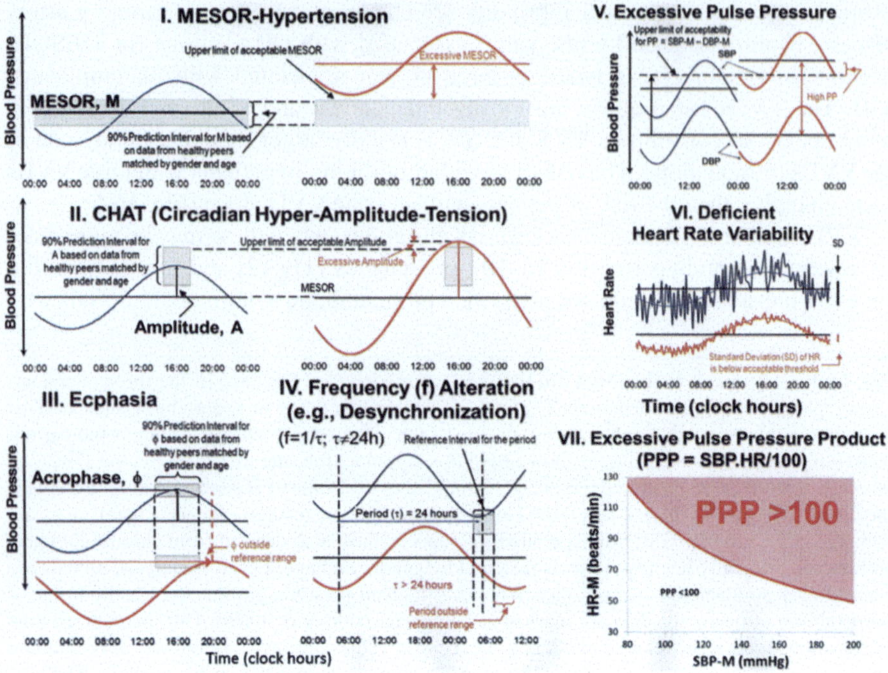

Fig. 16.2 Definitions of some vascular variability disorders (VVDs).
I. MESOR-hypertension (*top, left*, and *middle*) is an elevation of the blood pressure MESOR above the upper 95 % prediction limits of clinically healthy peers matched by gender and age.
II. CHAT (circadian hyper-amplitude-tension) (*middle, left*, and *middle*) is an elevation of the 24-hour amplitude of blood pressure above the upper 95 % prediction limit of clinically healthy peers matched by gender and age.
III. Ecphasia (*bottom left*) is a deviation of 24-hour acrophase from the 90 % prediction limits of clinically healthy peers matched by gender and age.
IV. Ecfrequentia (*bottom middle*) is a statistically significant deviation of the period from 24 hours, assessed by the nonoverlap of 24 hours by the 95 % confidence interval of the period assessed by nonlinear least squares.
V. Excessive pulse pressure (*top right*) is a difference of MESORs of systolic and diastolic blood pressure larger than 60 mmHg (until it can be further qualified in terms of gender and age).
VI. Deficient heart rate variability (*middle right*) is a (7-day) standard deviation of heart rate below 7.5 beats/min (until it can be further qualified in terms of gender and age).
VII. Excessive pulse pressure product (*bottom right*) is a product of the MESORs of systolic blood pressure and heart rate larger (/100) larger than 100 ((mmHg.beats/min)/100) (until it can be further qualified in terms of gender and age).
These VVAs can occur while most, if not all data lie within the conventional physiological range. Their presence has been associated with an increased risk of adverse cardiovascular events in several outcome studies. © Halberg Chronobiology Center

modern technology, equipped with hardware and software for automatic continuous self-surveillance, could make a change in health care by cost-effectively detecting the earliest changes when they precede severe vascular disease.

A manned website, for pre-habilitation, rather than only for rehabilitation, can provide self-helpers with analyses of their data collected with devices for ABPM

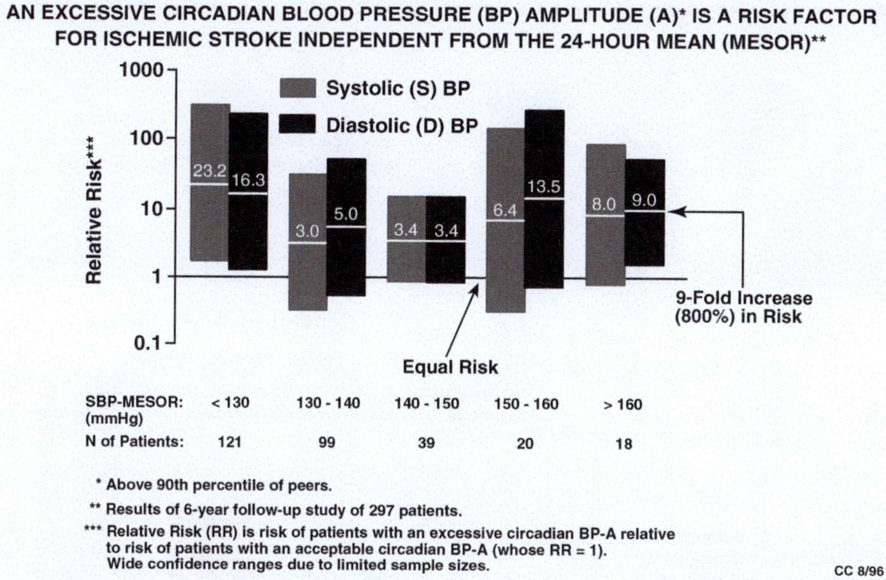

Fig. 16.3 Illustrative results supporting the need for continued surveillance and for a chronomic analysis of blood pressure series.
An excessive circadian amplitude of blood pressure (BP-A) is a risk factor for ischemic stroke independent from the 24-hour mean (MESOR). Finding that CHAT carries a very high risk even among MESOR-normotensives who do not need antihypertensive medication

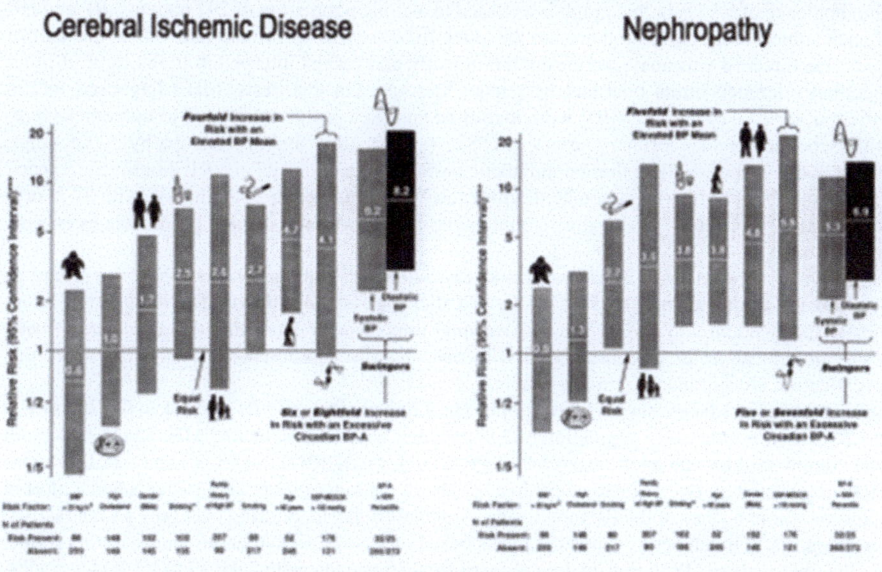

Fig. 16.4 Illustrative results supporting the need for continued surveillance and for a chronomic analysis of blood pressure series.
Left: Among risk factors, an excessive circadian BP amplitude (A) raises the risk of ischemic stroke most. *Right*: Among risk factors, an excessive circadian BP-A raises the risk of nephropathy most. Detection of abnormal circadian pattern of blood pressure (CHAT, "overswinging") associated with a risk of cerebral ischemia and nephropathy larger than other risks (including "hypertension") assessed concomitantly

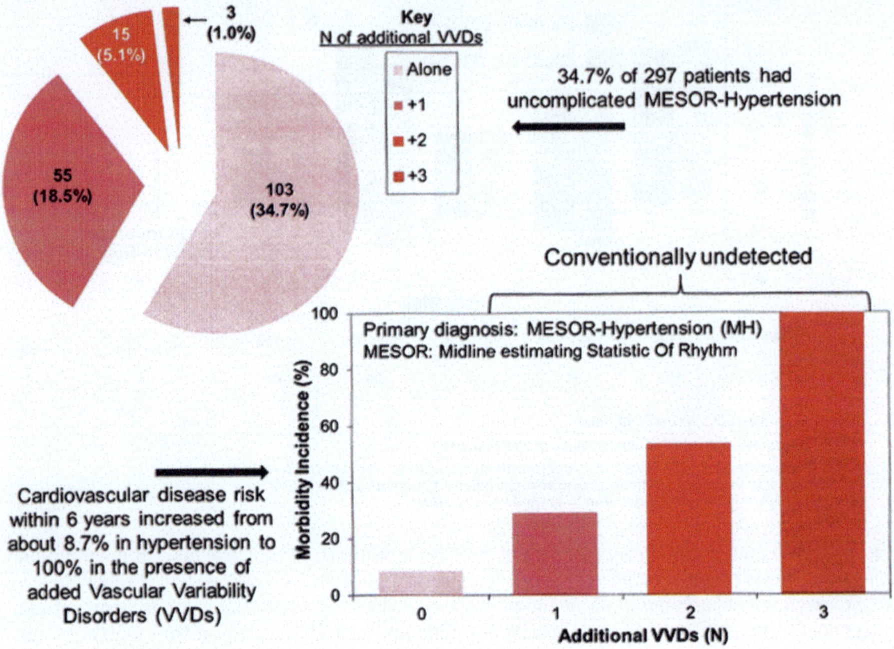

Fig. 16.5 Percentages of VVDs and VVSs missed in current practice.
The incidence of VVDs in this graph is assessed in a clinic population of 297 patients. BP and HR of each subject were monitored around the clock for 2 days at 15-minute intervals at the start of study. Each record was analyzed chronobiologically and results interpreted in the light of time-specified reference limits qualified by gender and age. On this basis, MH (diagnosed in 176 patients), EPP, CHAT, and DHRV were identified and their incidence related to outcomes (cerebral ischemic attack, coronary artery disease, nephropathy, and/or retinopathy). Outcomes, absent at the start of study in these nondiabetic patients, were checked every 6 months for 6 years, to estimate the relative risk associated with each VVD alone or in combination with 1, 2, or 3 additional VVDs, shown in columns complementing each circular display of incidences of variability disorders.
Earlier work showed that CHAT was associated with a risk of cerebral ischemic event and of nephropathy higher than MH (Fig. 16.4), and that the risks of CHAT, EPP, and DHRV were mostly independent and additive. It thus seemed important to determine the incidence of each VVD, present alone or in combination with one or more additional VVDs. Results from this investigation are summarized in this graph.
Results related to MH are shown in the graph. The 176 patients with MH are broken down into 103 (34.7 % of the whole study population of 297 patients) with uncomplicated MH, 55 (18.5 %) with MH complicated by one additional VVD (EPP, CHAT, or DHRV), 15 (5.1 %) and 3 (1.0 %) with MH complicated by two or three additional VVDs. In the latter group, all 3 patients had a morbid outcome within 6 years of the BP monitoring. Ambulatory BP monitoring over only 48 hours, used for diagnosis, is much better than a diagnosis based on casual clinic measurements, yet its results apply only to groups. With this qualification, of the 176 patients with MH, 73 (42.2 %) have additional VVDs that further increase their vascular disease risk, and that are not considered in the treatment plan of these patients since current practice does not assess these VVDs. This proportion may be smaller in a 7-day record (available for CHAT)

and data transfer, as is now done on a small scale by e-mail within the scope of the BIOCOS project (corne001@umn.edu).

16.3 Benefit of 7-Day/24-Hour ABP Monitoring

The self-helper in health care gets the needed information on his or her health status cost free, without requiring support from caregivers, unless the analyses suggest the need for a consultation. We here illustrate what can be gained by chronobiologically implemented ambulatory BP and HR self-monitoring, carried out automatically, as a start around the clock at hourly or shorter intervals for 7 days (C-ABPM 24/7 or at about 3- to 4-hour intervals manually, the ambulatory automatic instrument being much preferred).

Merits are compared with the status quo, limiting the use of ABPM to special cases, for one or a few days only, without any chronobiologic interpretation. Under such current practice, the proportion of cases that remain unrecognized and hence untreated corresponds to the darker red segments in the upper left corner of Fig. 16.5, corresponding to MH complicated by one, two, or three additional VVDs. It also corresponds to the lightest red part of Fig. 16.5, representing VVDs other than MH not complicated by any other VVD (including MH).

16.4 Background of 7-Day/24-Hour ABP Monitoring

The 7th Report of the Joint National Committee on Detection, Evaluation, and Treatment of High Blood Pressure (JNC-7) describes a SBP <120 mmHg and DBP<80 mmHg as normal. A SBP of 120–139 mmHg and DBP of 80–89 mmHg has been termed as prehypertension, while a SBP>140 mmHg and/or a DBP>90 mmHg is hypertension [33]. These guidelines are based on the (office) spot measurements of BP, which are suspect in the diagnosis of normotension, prehypertension, or hypertension. A "control" group can have many false-negative diagnoses; vice versa, a group of mild hypertensives may have false positives. There is evidence from 24/7 records of half-hourly data that "spotcheck hypertensives" (so misdiagnosed based only on a single 24-hour profile) can be MESOR-normotensive based on a 7-day/24-hour record and thus have some acceptable 7-day MESORs and vice versa. This observation may contribute to questionable results such as 48 % cures by a placebo when the false positives at entry into a clinical trial and the false negatives at its end are not assessed [27, 34].

Current practice no longer needs to rely on one or a few measurements of BP taken in the physician's office under standardized conditions with a mercury sphygmomanometer, interpreted against fixed limits applying to all adults 18 years or older [17, 33]. Thirty home measurements without indication of their temporal placement are required by the Austrian guidelines [18] to be interpreted in the light of a time-unspecified limit, as in the international guidelines [33]. A fixed limit for a rhythmically changing variable can make the diagnosis dependent on the time of day when the measurement is made, an abstract fact [20], documented in clinical practice as checked at the US National Institutes of Health [21]. Such limitations notwithstanding, treatment of an elevated BP has been related to a decline in the incidence of cardiovascular morbidity and mortality [22], yet there are a number of people who receive treatment they do not need while others, who need treatment, do not get it [35]. With rising health-care costs, any robust reduction in cardiovascular disease will be extremely helpful for those concerned about the cost of health care for preventing incapacitation and suffering.

Several improvements are directly within reach. It is widely accepted that BP is not constant but varies predictably, among others according to the individual's circadian rhythm usually of large amplitude [19]. Measuring BP around the clock is now readily feasible with ambulatory monitors without too much disturbance of sleep and the daily routine. Measurements from these monitors have also been shown to be superior to clinic measurements in terms of diagnosis and prognosis [24]. BP is also known to change as a function of age and to differ between men and women [25] and among individuals of the same gender, age, and ethnicity [16]. Not yet generally known is that in decades-long series, a number of newly discovered cyclicities have been mapped that all contribute to variability which has to be resolved. Their periods coexist with those of the environmental day and year or replace the calendar year, differing from a precise year. Some of the periods reflect different aspects of solar activity, including beat periods of nonradial solar rotations at different solar latitudes [36–48].

Accordingly, at variance with guidelines [33], an international project on the BIOsphere and the COSmos (BIOCOS) [29, 36, 37, 49] has derived time-specified reference values qualified by gender and age to interpret data from ambulatory monitors, collected around the clock, preferably for at least 7 days and in some cases for decades [44, 48]. A double-barreled approach has also been developed that consists of a parametric and nonparametric assessment of the data.

16.5 The Current Chronobiologic Approach

The current chronobiologic approach relies as a first step on the estimation of circadian rhythm characteristics (parameters), obtained by the least-squares fit of a two-component model consisting of cosine curves with periods of 24 and 12 hours (Fig. 16.1). The inclusion in the model of a second harmonic with a period of 12 hours accounts for the non-sinusoidal waveform of the circadian rhythm in BP (and HR). With 24/7 monitoring, half-weekly and weekly components can be included in the model, yet one has only 1 or 2 cycles available for assessment, which is the equivalent of taking the pulse for one or two cycles (or seconds). As longer series become available over years and decades, they serve for the detection of infradian alterations that may also signal prehypertension [48].

At this time, parametrically, about 3.5- and 7-day as well as circadian rhythm characteristics can be routinely estimated by the fitting of several cosines, their number depending on the length and density of the record. The parameters are an improved mean, the MESOR (M) and at each period fitted, the double amplitude (2A), and the acrophase (ϕ). For each of the infradian endpoints, improved reference standards (90 % PIs) will have to be computed from data of clinically healthy peers of each gender and several age groups, e.g., for about half-weekly and weekly as well as circadian (A, ϕ) pairs and for the PP and the HR-SD. In a complementary nonparametric approach, data from a 24/7 record are stacked over an idealized 24-hour day and are compared with the time-specified 90 % PIs of healthy peers matched by gender and age. The stacking can be done over periods different from 24 hours, if pertinent, but awaits the derivation of appropriate reference limits for non-24-hour components. Deviations from parametric and/or nonparametric limits guide the diagnosis of VVDs and VVSs, interpreted in the light of actual and/or proxy outcomes [32, 49–55].

Just as it is possible to treat MH, CHAT can also be treated in the absence as well as in the presence of MH non-pharmacologically or with antihypertensive drugs [27, 35, 49, 52, 53, 56–60]. Sometimes, in the case of iatrogenic CHAT, all it takes is to change the timing of administration of the same dose of the same treatment [35]. Treating CHAT may reduce cardiovascular morbidity and mortality by about 50 % [60].

Clinical studies have relied on groups or populations for diagnosis [61–63] or treatment, seeking the optimal timing of administration of a given drug. All subjects in a given group usually receive the same medication at the same circadian stage, irrespective of symptoms. In view of the different risks associated with the VVDs that all have to be optimized on an individualized basis, it becomes critical to monitor both for a chronodiagnosis and for finding the best timings of each treatment, varied for each patient systematically along the 24-hour scale, so as to validate that all VVDs are eliminated, or at least reduced as far as possible. Such linking of chronotherapy to the chronodiagnosis has been referred to as "chronotheranostics" [49]. The best time to administer a given antihypertensive agent may differ, e.g., among individuals [49] and/or among drugs used [56, 57], depending on whether the patient has CHAT or a small circadian double amplitude of BP. The decision to treat a patient with MH with felodipine or lercanidipine differs depending on whether the patient has an acceptable or a deficient HR variability since only one of the two drugs, lercanidipine, seems to be able to increase HRV [49]. By contrast, in a case with MH and CHAT, felodipine should be tried first since it reduces both the BP MESOR and the circadian double amplitude of BP. These chance findings require confirmation.

16.6 Comparison with Day-Night Ratios

Attempts have been made to simplify the assessment of a circadian rhythm by the computation of daytime and nighttime means and of day-night ratios (DNRs). Many studies have linked some cardiovascular pathologies to the DNR [e.g., 64–68]. Whenever the merits of the DNR have been compared with those of the circadian rhythm characteristics on the same data with outcomes, the chronobiologic characteristics were shown to be superior: in a 6-year prospective study of 297 patients in Japan [35, 49, 50]; in a smaller 7-year prospective study of dental patients in Minnesota [55]; in a study of 25 controls and 25 patients with minimal change retinopathy [7; cf. 6]; and in a study of 1,177 patients, using the left ventricular mass index as surrogate outcome measure [15, cf. 12].

The comparisons already completed were, without exception thus far, in favor of a chronobiologic approach. The latter allows the differentiation between undue changes in phase-only vs. amplitude-only which are assessed separately by the parametric approach while they are confounded in the DNR. Moreover, chronobiology uses added independent requirements for assessment, such as a much longer time series (of at least 7 days rather than 24 hours), and reference standards, some of which can already be qualified by gender and age, which as yet is not done for the DNR. Apart from a more thorough since less incomplete database for a circadian assessment, the chronobiologic approach requiring a weeklong record allows further the exploration of about half-weekly, near-weekly, and weekly cycles that cannot be assessed by a (highly variable) 24-hour monitoring-based DNR. The latter has served group assessment in studies based on populations, but does not allow inferences for the individual, for whom the goal of chronobiology is as dense as practical, and eventually a lifelong monitoring. This continued around-the-clock surveillance suggests already that infradian spectral alterations such as the replacement of a calendar year component by cycles longer than a year by about 4 months (far-transyears) may also be a harbinger to be considered for preventive treatment, decades before the onset of MESOR-hypertension [48]. The role of other infradian components in the spectra of HR or BP and of hormonal variables and in those of environmental variables near and far as yet is a topic for research.

A comparison of the relative merits of DNRs and circadian parameters derived from a chronobiologic approach should be continued on the large data sets with outcomes already collected via research grants from public sources by different investigations that included ABPM at the start. The refusal in the past of advocates of the DNR to carry out that comparison or allow BIOCOS to do so may be based on the misconception that the major source of changes in BP relates to the cycle of sleep, wakefulness, and activity which favors an analysis based on the waking-sleeping difference (dipping) and that the dipping classification system makes no assumption of an intrinsic circadian rhythm. A partly genetically anchored circadian variation in BP was documented in the clinic by a free-run of SBP from sleep-wakefulness and activity [69]. Beyond circadians, there is a broad spectrum of infradian rhythms that, their relatively small amplitude notwithstanding, may be of clinical [38–42] as well as basic interest, since they are signatures of beat periods of solar rotation in phenomena such as sudden cardiac death [44] as well as suicide [46] and hormones that influence BP.

A chronobiologic approach detects prehypertension and prediabetes. In the latter case, discrimination by means of the DNR fails. In a study of patients with minimal change retinopathy and healthy controls by Cugini et al. [6], differences were found in terms of circadian rhythm characteristics but not in terms of the DNR; the latter misled [6, 7].

Exploring circadian parameters as part of the dynamics of HR and BP can extend the range of rhythms from seconds and days [70] to infradians [71] and can explore the possibility of infradian signaling of prehypertension [48]. Whether infradian rhythm alterations have prognostic value will have to be investigated with automatic, ambulatorily usable instrumentation that is obtrusive and affordable. Data already collected on test pilots for decades document the possibility of the implementation of self-surveillance, with currently only slightly obtrusive instrumentation.

16.7 Affordability

Automatic ambulatorily usable monitors are available through a project on The BIOsphere and the COSmos, briefly BIOCOS (corne001@umn.edu), in exchange for the data, with a great cost reduction; cost is further reduced by the planned use of these tools on a friends-and-family basis. In systematic continuous use for at least 50 profiles/year (good for 5 years), the collection of a 7-day profile now costs less than US $3, allowing for the acquisition of an automatic monitor for each set of a family and friends, if not yet for an economical automatic monitor for each individual. Monitors become affordable as the demand for them increases. The implementation of the latter aim is planned by the Phoenix Study Group, composed of volunteering members of the Twin Cities Chapter of the Institute of Electrical and Electronics Engineers (http://www.phoenix.tc-ieee.org), along with a website (http://www.sphygmochron.org/) for free analyses in exchange for the data to be used for research.

16.8 Broader Implication

The example eventually set by benefit from VVD and VVS detection and treatment may be followed by a change from a health care based on spotcheck evidence (obtained in office visits or in one- or a few-day profiles without inferential statistical tests applied to the individual's data) to cost-effective inferential statistical personalized self-surveillance. Notably after the diagnosis of a VVD or VVS is made, treatment is prescribed by a care provider, and self-surveillance is again automatically implemented with sequential and parameter tests [72, 73] applied to time series of the given person. Beyond BP and HR, the dynamics of each variable have to be mapped, rather than assuming, as does homeostatic physiology, that there is the fata morgana of a "true" BP or a "true" HR or, more generally, that a de facto variable can be approximated as an imaginary constant by invoking control theory and homeostasis. The assessments of decades-long records of patients with MH have already demonstrated to medical practice that once the diagnosis of MH is made, there is no alternative to continuous monitoring of the treatment's efficacy or failure [74].

Self-surveillance was done by opinion leaders from the time of diagnosis of their VVDs for a lifetime by manual measurements [75, 76]. This is now feasible automatically as well as ambulatorily, during sleep and one's waking routine. Individualized test procedures are available to detect abnormality [72, 73] in order to do something about it. By monitoring, a relaxation or other treatment may be seen to lower the BP-M. But the desired lowering effect shown by control chart, assessing changes in BP-M that can be tracked to a given intervention, must not be accompanied by an increase in circadian double amplitude of BP that can lead to CHAT [49, 52, 53].

16.9 Summary

This consensus proposes the use in clinical practice of C-ABPM 24/7 to rule out a VVD (or VVS). If in turn a VVD is detected (and as long as it is not absent for a full week), the monitoring is continued so that the caregiver and receiver do not fly blind [77]. Currently, hundreds of millions of people are diagnosed (some rightly, others wrongly) to have "high BP." Only relatively few of them are recognized as false positives (white-coat effect) or false negatives (masked hypertension) under the misconception that there is a "true" BP. This current approach can be replaced by developing a system via a website for monitoring BP and HR on a large scale, as is already done on a smaller scale by BIOCOS worldwide, so that the diagnosis of a VVD or VVS can be made without undue cost by earliest public education in self-help. Finding out how best to treat is a major challenge. This focus on variability is not new: Ignaz Zadek recommended assessing variability as far back as 1880 in a doctoral thesis and then in 1881 [23]. In 1904, Theodore C. Janeway of Johns Hopkins University (Baltimore), the opinion leader of his time, wrote [78]: "…it is essential that a record of the pressure be made at frequent intervals at some time previous [presumably to an examination], to establish the normal level and the extent of the periodic variations. When this is done, it may be possible to demonstrate changes of small extent, which, lacking this standard for comparison, would be considered within the limits of normal variation." Janeway was right: even if he could not foresee that BP can undergo not only partly endogenous circadian rhythms, but also cycles longer than a year, driven by counterparts in the solar wind, yet genetically coded like circadians, since they persist when the solar driving at the given frequency is no longer detected.

And in 1974, Bartter [21] suggested writing about a patient whose BP was diagnosed differently by two physicians who saw him at different times of day: "By conventional standards, this patient is clearly normotensive every morning. Yet the blood pressure determined each day at 6 in the afternoon provides especially convincing evidence that this patient is a hypertensive. .... My plea today is that information contained in such curves [cosinor fits] becomes a routine minimal amount of information accepted for the description of a patient's blood pressure. The analysis of this information by cosinor should become a routine. It is essential that enough information be collected to allow objective characterization of a periodic phenomenon, to wit, an estimate of M [the time structure or chronome-adjusted mean, or MESOR], an estimate of [the amplitude] A itself, and finally an estimate of acrophase, φ [a measure of timing]. In this way, a patient can be compared with himself at another time, or under another treatment, and the patient can be compared with a normal or with another patient."

16.10 Conclusion

Today we have the wherewithal to implement Zadek's, Janeway's, Bartter's, and others' [10] extensively documented suggestions. We monitor garages around the clock to prevent crime. We monitor small rodents by telemetry to develop drugs. Let us use available technology to consider those who have to be reliably diagnosed, who need the drugs, and establish the approach to the diagnosis and treatment of each patient in an individualized inferential statistical way by now-available [72, 73, 79] and yet to be further extended software.

Based on chronobiologically interpreted C-ABPM 24/7, five VVDs can be diagnosed (MH, CHAT, BP ecphasia, EPP, and/or DHRV). Eventually, alterations of cycles in the ultradian and infradian domains may be found to be useful and should be added to a screen with repeated passes over the continuously accumulating data, first mapping and thereafter interpreting infradians with longer and longer periods.

In case of a VVD, to make the diagnosis reliable, 24-hour/7-day C-ABPM should be repeated for at least 7 added days; this approach may rule out transient VVDs. Alternatively, as long as a VVD persists, C-ABPM should be made continuous, depending on outcomes of C-ABPM 24/7. The effect of treatment should be controlled by C-ABPM with as-one-goes sequential tests [72], and, when the CUSUM in a control chart emerges from the decision interval, and parameter tests [73] indicate a change, treatment is accordingly introduced or modified.

16.11 Epilogue

Earlier consensus meetings were held in Brussels on March 17–18, 1995 [35], by the New SIRMCE Federation (the International Society for Research on Civilization Diseases and the Environment), followed by consensus meetings in Moscow on Oct. 10–12, 2005 (III International Conference, Civilization diseases in the spirit of V.I. Vernadsky, People's Friendship University of Russia) [80], in Baku, Azerbaijan, on September 24–27, 2007 (International Interdisciplinary Workshop, Natural Cataclysms and Global Problems of the Modern Civilization) [14], and in Sudak, Crimea, Ukraine, on October 1–6, 2007 (VII International Crimean Conference "Cosmos and biosphere") [81]. At all these consensus meetings, a 24-hour/7-day profile interpreted by a parametric and nonparametric approach was recommended to likely rule out abnormality once generally affordable automatic instrumentation for continuing observation becomes available on a large scale.

This consensus also strongly recommends that reference values be collected continuously from test pilots such as astronaut and cosmonaut candidates and their families. Thus, human BP and HR reference values from conception to old age (preferably to be monitored prospectively for successive generations) may become available. There is a need for institutes to accomplish the research on data collected preferably automatically by a website which provides the individual with analyses in exchange for using the information for both health care and research, including improved reference values and refined earliest indicators of disease on the one hand. On the other hand, the population morbidity and mortality, including crime, war, and terrorism, may serve to prevent these societal cataclysms.

Among telluric, atmospheric, and solar oscillations and their associations in physiology, psychology, sociology, epidemiology, criminology, and warfare, criteria of the cycles involved qualified by CIs (95 % confidence intervals) of their characteristics become measures amenable to a subtraction and addition approach occurring naturally, implemented by the sun and the broader cosmos. Just as we currently have heat and air conditioning and meet other extremes of the unduly seen, felt, tasted, or smelled environs, we need to understand the unseen and not consciously noticed, yet no less important nonphotic environmental risks [82, 83].

We live in the atmosphere of the sun. We learned of effects of magnetic storms and found geographic and temporal differences in association with the incidence patterns of various diseases. We find signatures of solar activity in everyday physiology. We need to learn more of the consequences of unseen nonphotic effects that can override the seasons and can replace them in the incidence patterns of sudden cardiac death, of suicide, and apparently also of terrorism, influencing public opinion as well. The tasks on hand are those of transdisciplinary science that ranges broadly from politics to philanthropy. This science will pay for itself in BP and HR monitoring that can fend off hard events like stroke for millions of people now treated for hypertension or in a prehypertensive state who actually have risks as VVSs greater than hypertension, which are currently unrecognized, yet could be treated if detected.

To meet the challenge of noninvasive cardiology, BIOCOS aims at the construction of a universal, multilingual website for both series in health care and research in science without disciplinary barriers. All interested are welcome to participate.

References

1. Penaz J. Indirect continuous recording of blood pressure in man. Physiol Bohemoslov 1970;19(4):341.
2. Penaz J, Honzikova N, Fiser B. Spectral analysis of resting variability of some circulatory parameters in man. Physiol Bohemoslov 1978;27(4):349–57.
3. Fiser B, Honzikova N, Penaz J. Power spectra of spontaneous variations of indirectly recorded blood pressure, heart rate and acral blood flow. Automedica 1978;2:143–7.

4. Honzikova N, Semrad B, Fiser B, Labrova R. Baroreflex sensitivity determined by spectral method and HRV and two-years mortality in patients after myocardial infarction. Physiol Res 2000;49:643–50.
5. Bailey JJ, Hodges M, Church TR. Decision to implant cardio- verter-defibrillator after myocardial infarction: the role of ejection fraction v. other risk factor markers. Med Decis Making 2007;27:151–60.
6. Cugini P, Cruciani F, Turri M, Regine F, Gherardi F, Petrangeli CM, Gabrieli CB. 'Minimal-change hypertensive retinopathy and arterial pre-hypertension', illustrated via ambulatory blood-pressure monitoring in putatively normotensive subjects. Int Ophthalmol 1998;22(3):145–9.
7. Cornélissen G, Cugini P, Siegelova J, Fiser B, Halberg F. Cugini's minimal change hypertensive retinopathy, resolved chronobiologically while dipping fails, supports the concept of pre-hypertension. In: Halberg F, Kenner T, Fiser B, Siegelova, J, editors. Proceedings of the Noninvasive Methods in Cardiology;2007; Brno: Department of Functional Diagnostics and Rehabilitation, Faculty of Medicine, Masaryk University (ISBN 978 80 7018 463 4); 2007. p. 55–61.
8. Kumagai Y, Shiga T, Sunaga K, Cornélissen G, Ebihara A, Halberg F. Usefulness of circadian amplitude of blood pressure in predicting hypertensive cardiac involvement. Chronobiologia 1992;19:43–58.
9. Halberg J, Halberg E, Hayes DK, Smith RD, Halberg F, Delea CS, Danielson RS, Bartter FC. Schedule shifts, life quality and quantity modeled by murine blood pressure elevation and arthropod lifespan. Int J Chronobiol 1980;7:17–64.
10. Carandente F, Halberg F. (Eds.) Chronobiology of blood pressure in 1985. Chronobiologia 1984;11(3):189–341.
11. Sánchez-de la Peña S, Gonzalez C, Cornélissen G, Halberg F. Blood pressure (BP), heart rate (HR) and non-insulin-dependent diabetes mellitus (NIDDM) chronobiology. Int J Cardiol 2004;97 Suppl 2:S14.
12. Gupta AK, Greenway FL, Cornélissen G, Pan W, Halberg F. Pre-diabetes is associated with abnormal circadian blood pressure variability. J Human Hypertension 2008; 22: 627–33. doi:10:1038/jhh.2008.32.
13. Cornélissen G, Siegelova J, Abramson J, Sundaram B, Mandel J, Holley D, Halberg F. Body mass index (BMI), pulse pressure (PP) and pre-metabolic syndrome. In: Halberg F, Kenner T, Fiser B, Siegelova J, editors. Proceedings of the Noninvasive Methods in Cardiology; Brno, Czech Republic 2008.p. 116–23.
14. Halberg F. Challenges from "60 years of translational Chronobiology. In: Kofler W editor. Proceedings of the International Interdisciplinary Workshop Natural Cataclysm and Global Problems of the Modern Civilization;2007 Sep 24–27; Baku, Azerbaijan; Transactions of the International Academy of Science H&E. Baku/Innsbruck: ICSD/IAS; 2007. p. 165–78.
15. Cornélissen G, Halberg F, Otsuka K, Singh RB, Chen CH. Chronobiology predicts actual and proxy outcomes when dipping fails. Hypertension 2007;49:237–9. doi:10.1161/01.HYP.0000250392.51418.64.
16. Halberg F, Cornélissen G, Otsuka K, Sánchezdela Peña S, Schwartzkopff O, Watanabe Y, Pati AK, Wall DG, Delmore P, Borer K, Beaty LA, Nolley ES, Adams C, Siegelova J, Homolka P, Dusek J, Fiser B, Prikryl P. Why and how to implement 7-day/247-day/24-hour blood pressure monitoring? Int J Geronto-Geriatrics 2005;8(8):1–31.
17. Hagen P, editor. Mayo Clinic guide to self-care: answers for every day health problems. Rochester, MN/Jacksonville, FL/Scottsdale, AZ: Mayo Clinic; 2003. p. 180–1.
18. Magometschnigg D. Definition und Klassifikation der Hypertonie. J Hyperton 2004;8(1):12–3.
19. Richardson DW, Honour AJ, Fenton GW, Stott FH, Pickering GW. Variation in arterial pressure throughout the day and night. Clin Sci 1964;26:445–60.

20. Cornélissen G, Halberg F. Impeachment of casual blood pressure measurements and the fixed limits for their interpretation and chronobiologic recommendations. Ann N Y Acad Sci 1996;783:24–46.
21. Bartter FC. Periodicity and medicine. In: Scheving LE, Halberg F, Pauly JE, editors. Chronobiology. Tokyo: Igaku Shoin Ltd.; 1974. p. 6–13.
22. Lenfant C. Reflections on hypertension control rates: a message from the director of the National Heart, Lung, and Blood Institute [Editorial]. Arch Intern Med 2002;162:131–2.
23. Zadek I. Die Messung des Blutdrucks am Menschen mittelst des Basch'chen Apparates. Z Klin Med 1881;2:509–51.
24. Verdecchia P, Clement D, Fagard R, Palatini P, Parati G. Blood pressure monitoring. Task Force III: Target-organ damage, morbidity and mortality. Blood Press Monit 1999;4:303–17.
25. Lakatta EG. Mechanisms of hypertension in the elderly. J Am Geriatr Soc 1989;37:780–90.
26. Rosenfeld I. Take control of your blood pressure. Parade, 2007 July 8. p. 16.
27. Halberg F, Cornélissen G, Halberg E, Halberg J, Delmore P, Shinoda M, Bakken E. Chronobiology of human blood pressure. Medtronic Continuing Medical Education seminars. 4th ed. Minneapolis: Medtronic Inc.; 1988. 242 pp.
28. Halberg F, Cornélissen G, Carandente F. Chronobiology leads toward preventive health care for all: cost reduction with quality improvement. A challenge to education and technology via chronobiology. Chronobiologia 1991;18:187–93.
29. Cornélissen G, Otsuka K, Halberg F. Blood pressure and heart rate chronome mapping: a complement to the human genome initiative. In: Otsuka K, Cornélissen G, Halberg F, editors. Chrono- cardiology and chronomedicine: humans in time and cosmos. Tokyo: Life Science Publishing; 1993. p. 16–48.
30. Cornélissen G, Halberg F, Gubin D, Carandente F, Halberg J, Zaslavskaya R, Syutkin V, Kumagai Y, Watanabe Y, Otsuka K. Carpe diem mensuratem: Fulminant CHAT, its recognition a chronobiologic path to preventing a myocardial infarction? Abstract, 4° Convegno Nazionale, Società Italiana di Cronobiollogia, Gubbio (Perugia), Italy, June 1–2, 1996. p. 35–36.
31. Nelson W, Cornélissen G, Hinkley D, Bingham C, Halberg F. Construction of rhythm-specified reference intervals and regions with emphasis on "hybrid" data, illustrated for plasma cortisol. Chronobiologia 1983;10:179–93.
32. Cornélissen G, Halberg F, Otsuka K, Singh RB. Separate cardiovascular disease risks: Circadian Hyper-Amplitude-Tension (CHAT) and an excessive pulse pressure. World Heart J 2008;1(3):223–32.
33. World Health Organization, International Society of Hypertension Writing Group. World Health Organization (WHO)/International Society of Hypertension (ISH) statement on management of hypertension. J Hypertension 2003; 21: 1983–92.
34. Wilcox RG, Mitchell JRA, Hampton JR. Treatment of high blood pressure: should clinical practice be based on results of clinical trials? Br Med J 1986;293:433–7.
35. Halberg F, Cornélissen G, International Womb-to-Tomb Chronome Initiative Group: Resolution from a meeting of the International Society for Research on Civilization Diseases and the Environment (New SIRMCE Confederation), Brussels, Belgium, March 17–18, 1995: Fairy tale or reality? Medtronic Chronobiology Seminar #8, April 1995, 12 pp. text, 18 figures. https://sites.google.com/a/umn.edu/halbergchronobiologycenter/
36. Halberg F, Cornélissen G, Otsuka K, Watanabe Y, Katinas GS, Burioka N, Delyukov A, Gorgo Y, Zhao ZY, Weydahl A, Sothern RB, Siegelova J, Fiser B, Dusek J, Syutkina EV, Perfetto F, Tarquini R, Singh RB, Rhees B, Lofstrom D, Lofstrom P, Johnson PWC, Schwartzkopff O. International BIOCOS Study Group. Cross-spectrally coherent ~10.5-and 21-year biological and physical cycles, magnetic storms and myocardial infarctions. Neuroendocrinol Lett 2000;21:233–58.
37. Halberg F, Cornélissen G, Regal P, Otsuka K, Wang ZR, Katinas GS, Siegelova J, Homolka P, Prikryl P, Chibisov SM, Holley DC, Wendt RW, Bingham C, Palm SL, Sonkowsky RP, Sothern RB, Pales E, Mikulecky M, Tarquini R, Perfetto F, Salti R, Maggioni Jozsa R, Konradov AA, Kharlitskaya EV, Revilla M, Wan CM, Herold M, Syutkina EV, Masalov AV, Faraone P, Singh

RB, Singh RK, Kumar A, Singh R, Sundaram S, Sarabandi T, Pantaleoni GC, Watanabe Y, Kumagai Y, Gubin D, Uezono K, Olah A, Borer K, Kanabrocki EA, Bathina S, Haus E, Hillman D, Schwartzkopff O, Bakken EE, Zeman M. Chronoastrobiology: proposal, nine conferences, heliogeomagnetics, transyears, nearweeks, near-decades, phylogenetic and ontogenetic memories. Biomed Pharmacother 2004;5 Suppl 1:S150–87.

38. Otsuka K, editor. Proceedings of the 1st International Symposium Work-shop on Circadian Rhythms and Clinical Chronotherapy 2000 Nov 11;, Tokyo, Japan; Biomed & Pharmacother; 2001; 55 (Suppl 1): 7s-190s.
39. Otsuka K, editor Proceedings of the 2nd International Symposium Workshop on Circadian Rhythms and Clinical Chronotherapy; 2001 Nov 17; Tokyo, Japan: Biomed & Pharmacother 2002; 56 (Suppl 2): 231s-382s.
40. Otsuka K, editor. Proceedings of the 3rd International Symposium Workshop on Circadian Rhythms and Clinical Chronotherapy; 2002 Nov 9; Tokyo, Japan: Biomed & Pharmacother 2003; 57 (Suppl 1): 1s-198s.
41. Otsuka K, editor. Proceedings of the 4th International Symposium Work-shop on Circadian Rhythms and Clinical Chronotherapy; 2003 Nov 8; Tokyo, Japan: Biomed & Pharmacother 2004; 58 (Suppl 1): S1-S188
42. Otsuka K, editor. Proceedings of the 5th International Symposium Work-shop on Circadian Rhythms and Clinical Chronotherapy; 2004 Nov 6; Tokyo, Japan: Biomed & Pharmacother 2005; 59 (Suppl 1): S1-S261
43. Cornélissen G, Masalov A, Halberg F, Richardson JD, Katinas GS, Sothern RB, Watanabe Y, Syutkina EV, Wendt HW, Bakken EE, Romanov Y. Multiple resonances among time structures, chronomes, around and in us. Is an about 1.3-year periodicity in solar wind built into the human cardiovascular chronome? Hum Physiol 2004;30(2):86–92.
44. Halberg F, Cornélissen G, Katinas G, Tvildiani L, Gigolashvili M, Janashia K, Toba T, Revilla M, Regal P, Sothern RB, Wendt HW, Wang ZR, Zeman M, Jozsa R, Singh RB, Mitsutake G, Chibisov SM, Lee J, Holley D, Holte JE, Sonkowsky RP, Schwartzkopff O, Delmore P, Otsuka K, Bakken EE, Czaplicki J, International BIOCOS Group. Chronobiology's progress: season's appreciations 2004–19 2005. Time-, frequency-, phase-, variable-, individual-, age- and site-specific chronomics. J Applied Biomedicine 2006; 4: 1–38.Available from http://www.zsf.jcu.cz/vyzkum/jab/4_1/halberg.pdf
45. Katinas GS, Halberg F, Cornélissen G, Sánchez de la Peña S, Czaplicki J, Siegelova J, BIOCOS project. C-ABPM reveals solar cis-halfyear and transyear signatures in human diastolic blood pressure. In: Halberg F, Kenner T, Fiser B, Siegelova J, editors. Proceedings of the Noninvasive Methods in Cardiology; Brno, Czech Republic, October 4–7, 2008. p. 144–6.
46. Halberg F, Cornélissen G, Berk M, Dodd S, Henry M, Wetterberg L, Nolley E, Beaty L, Siegelova J, Fiser B, Wolff C, BIOCOS project. Solar signatures in Australian suicide incidence: gender differences in prominence of photic vs. nonphotic spectral components. In: Halberg F, Kenner T, Fiser B, Siegelova J, editors. Proceedings of the Noninvasive Methods in Cardiology; Brno, Czech Republic, October 4–7, 2008, p.44-62
47. Cornélissen G, Tarquini R, Perfetto F, Otsuka K, Gigolashvili M, Halberg F. About 5-month cycle in human circulating melatonin: signature of weather in extraterrestrial space? Poster presentation Fourth UN/ESA/NASA/JAXA Workshop on the International Heliophysical Year 2007 and Basic Space Science: "First Results from the International Heliophysical Year 2007", Sozopol, Bulgaria, June 2–6, 2008
48. Watanabe Y, Cornélissen G, Halberg F, Hillman D, Czaplicki J, Sothern RB, Otsuka K, Siegelova J, BIOCOS project. Transyears, no calendar-year in blood pressure decades before MESOR-hypertension: normal or abnormal? In: Halberg F, Kenner T, Fiser B, Siegelova J, editors. Proceedings of the Noninvasive Methods in Cardiology; Brno, Czech Republic, October 4–7, 2008. p. 184–8.
49. Cornélissen G, Halberg F, Bakken EE, Singh RB, Otsuka K, Tomlinson B, Delcourt A, Toussaint G, Bathina S, Schwartzkopff O, Wang ZR, Tarquini R, Perfetto F, Pantaleoni GC,

Jozsa R, Delmore PA, Nolley E. 100 or 30 years after Janeway or Bartter, Healthwatch helps avoid "flying blind". Biomed Pharmacother 2004;58 Suppl 1:S69–86.
50. Otsuka K, Cornélissen G, Halberg F. Predictive value of blood pressure dipping and swinging with regard to vascular disease risk. Clin Drug Investig 1996;11:20–31.
51. Cornélissen G, Otsuka K, Chen C-H, Kumagai Y, Watanabe Y, Halberg F, Siegelova J, Dusek J. Nonlinear relation of the circadian blood pressure amplitude to cardiovascular disease risk. Scr med (Brno) 2000;73:85–94.
52. Halberg F, Cornélissen G, Wall D, Otsuka K, Halberg J, Katinas G, Watanabe Y, Halhuber M, Müller-Bohn T, Del more P, Siegelova J, Homolka P, Fiser B, Dusek J, Sanchez de la Peña S, Maggioni C, Delyukov A, Gorgo Y, Gubin D, Carandente F, Schaffer E, Rhodus N, Borer K, Sonkowsky RP, Schwartzkopff O. Engineering and governmental challenge: 7-day/24-hour chronobiologic blood pressure and heart rate screening: Part I. Biomedical Instrumentation & Technology 2002; 36: 89–122.
53. Halberg F, Cornélissen G, Wall D, Otsuka K, Halberg J, Katinas G, Watanabe Y, Halhuber M, Müller- Bohn T, Delmore P, Siegelova J, Homolka P, Fiser B, Dusek J, Sanchez de la Peña S, Maggioni C, Delyukov A, Gorgo Y, Gubin D, Carandente F, Schaffer E, Rhodus N, Borer K, Sonkowsky RP, Schwartzkopff O. Engineering and governmental challenge: 7-day/24-hour chronobiologic blood pressure and heart rate screening: Part II. Biomedical Instrumentation & Technology 2002; 36: 183–97.
54. Müller-Bohn T, Cornélissen G, Halhuber M, Schwartzkopff O, Halberg F. CHAT und Schlaganfall. Dtsch Apoth Ztg 2002;142:366–70.
55. Schaffer E, Cornélissen G, Rhodus N, Halhuber M, Watanabe Y, Halberg F. Blood pressure outcomes of dental patients screened chronobiologically: a seven-year follow-up. JADA 2001;132:891–9.
56. Watanabe Y, Cornélissen G, Watanabe M, Watanabe F, Otsuka K. Ohkawa S-i, Kikuchi T, Halberg F. Effects of autogenic training and antihypertensive agents on circadian and circaseptan variation of blood pressure. Clin Exp Hypertens 2003;25:405–12.
57. Tamura K, Kohno I, Saito Yuzo, Wakasugi K, Achiwa S, Imanishi Y, Cugini P, Halberg F. Antihypertensive individualized therapeutic strategy. Difesa Sociale 1991; 6: 109–24
58. Watanabe Y, Cornélissen G, Halberg F, Beaty L, Siegelova J, Otsuka K, Bakken EE. Harm vs. benefit from losartan with hydrochlorothiazide at different circadian times in MESOR hypertension or CHAT. In: Halberg F, Kenner T, Fiser B, Siegelova J, editors. Proceedings of the Noninvasive Methods in Cardiology; Brno, Czech Republic. October 4–7, 2008. p. 149–67.
59. Halberg F, Cornélissen G, Katinas G, Tvildiani L, Gigolashvili M, Janashia K, Toba T, Revilla M, Regal P, Sothern RB, Wendt HW, Wang ZR, Zeman M, Jozsa R, Singh RB, Mitsutake G, Chibisov SM, Lee J, Holley D, Holte JE, Sonkowsky RP, Schwartzkopff O, Delmore P, Otsuka K, Bakken EE, Czaplicki J, International BIOCOS Group. Chronobiology's progress: Part II, chronomics for an immediately applicable biomedicine. J Applied Biomedicine 2006; 4:73–86. Available from http://www.zsf.jcu.cz/vyzkum/jab/4_2/halberg2.pdf
60. Shinagawa M, Kubo Y, Otsuka K, Ohkawa S, Cornélissen G, Halberg F. Impact of circadian amplitude and chronotherapy: relevance to prevention and treatment of stroke. Biomed Pharmacother 2001;55 Suppl 1:125s–32s.
61. Lemmer B. The importance of circadian rhythms on drug response in hypertension and coronary heart disease --from mice and man. Pharmacol Therapeut 2006;111:629–51.
62. Lemmer B, Middeke M, Schaaf B, Eckes L. Prescribing practices and morning blood pressure control: results of a large-scale, primary-care study conducted in Germany. J Hum Hypertens 2008;22:295–7.
63. Middeke M, Lemmer B, Schaaf B, Eckes L. Prevalence of hypertension-attributed symptoms in routine clinical practice: a general practitioners-based study. J Hum Hypertens 2008;22:252–8.
64. O'Brien E. Ambulatory blood pressure measurement: a trove of hidden gems? Hypertension 2006;48:364–5.
65. Stolarz K, Staessen JA, O'Brien ET. Night-time blood pressure: dipping into the future? J Hypertens 2002;20:2131–3.

66. Jerrard-Dunne P, Mahmud A, Feely J. Circadian blood pressure variation: relationship between dipper status and measures of arterial stiffness. J Hypertens 2007;25:1233–9.
67. Giles TD. Circadian rhythm of blood pressure and the relation to cardiovascular events. J Hypertens Suppl 2006;24:S11–6.
68. Zweiker R, Eber B, Schumacher M, Toplak H, Klein W. "Non-dipping" related to cardiovascular events in essential hypertensive patients. Acta Med Austriaca 1994;21:86–9.
69. Halberg F, Good RA, Levine H. Some aspects of the cardiovascular and renal circadian system. Circulation 1966;34:715–7.
70. Hoyer D, Clairambault J. Rhythms from seconds to days: Physiological importance and therapeutic implications. IEEE Eng Med Biol Mag 2007;26(6):12–3.
71. Watanabe Y, Cornélissen G, Katinas G, Schwartzkopff O, Halberg F. Case report: a drug for damping circannual and transannual amplitudes of blood pressure and heart rate. Abstract, 2nd International Symposium, Problems of Rhythms in Natural Sciences, Moscow, March 1–3, 2004. Moscow: Russian People's Friendship University; 2004. p. 18–20.
72. Cornélissen G, Halberg F, Hawkins D, Otsuka K, Henke W. Individual assessment of antihypertensive response by self-starting cumulative sums. J Med Eng Technol 1997;21:111–20.
73. Bingham C, Arbogast B, Cornélissen Guillaume G, Lee JK, Halberg F. Inferential statistical methods for estimating and comparing cosinor parameters. Chronobiologia 1982;9:397–439.
74. Katinas GS, Cornélissen G, Otsuka K, Haus E, Bakken EE, Halberg F. Why continued surveillance? Intermittent blood pressure and heart rate abnormality under treatment. Biomed Pharmacother 2005;59 Suppl 1:S141–51.
75. Bartter FC, Delea CS, Baker W, Halberg F, Lee JK. Chronobiology in the diagnosis and treatment of mesor-hypertension. Chronobiologia 1976;3:199–213.
76. Levine H, Cornélissen G, Halberg F, Bingham C. Self-measurement, automatic rhythmometry, transmeridian flights and aging. In: Scheving LE, Halberg F, editors. Chronobiology: Principles and Applications to Shifts in Schedules. Alphen aan den Rijn, The Netherlands: Sijthoff and Noordhoff; 1980. p. 371–392.
77. Fossel M. Editor's Note (to Halberg F, Cornélissen G, Halberg J, Fink H, Chen C-H, Otsuka K, Watanabe Y, Kumagai Y, Syutkina EV, Kawasaki T, Uezono K, Zhao ZY, Schwartzkopff O. Circadian Hyper-Amplitude-Tension, CHAT: a disease risk syndrome of anti-aging medicine. J Anti-Aging Med 1998; 1: 239–59). J Anti-Aging Med 1998;1:239.
78. Janeway TC. The clinical study of blood pressure. New York: D. Appleton & Co.; 1904. 300 pp.
79. Refinetti R, Cornélissen G, Halberg F. Procedures for numerical analysis of circadian rhythms. Biological Rhythm Research 2007; 38 (4): 275–325. Available from http://dx.doi.org/10.1080/09291010600903692
80. Chibisov SM (summarized by). Resolution concerning chronobiology and chronomics. III International Conference, Civilization diseases in the spirit of V.I. Vernadsky, People's Friendship University of Russia, Moscow, Oct. 10–12, 2005, p.23–5.
81. Halberg F, Cornélissen G, Hillman D, Sothern RB, Nolley ES, Beaty LA, Schwartzkopff O, Otsuka K, Chibisov SM, Valenzi V, Pantaleoni G, Singh RB. Bakken's prehabilitation in the service of a budding chronoastrobiology. Invited lecture, VII International Crimean Conference "Cosmos and biosphere", Sudak, Crimea, Ukraine, October 1–6, 2007. p. 10–3.
82. Halberg F, Cornélissen G, Sothern RB, Katinas GS, Schwartzkopff O, Otsuka K. Cycles tipping the scale between death and survival. (="life"). Progress of Theoretical Physics 2008; Suppl 173: 153–81
83. Halberg F, Bakken EE, Katinas GS, Cornélissen G, Zaslavskaya RM, Blank MA, Syutkina EV, Breus TK, Watanabe Y, Masalov A, Chibisov SM. Chronoastrobiology: Vernadsky's future science? Benefits from spectra of circadians and promise of a new transdisciplinary spectrum of near matching cycles in and around us. Opening keynote, Proceedings of the III International Conference, Civilization diseases in the spirit of V.I. Vernadsky, People's Friendship University of Russia, Moscow; p. 4–22.

Chapter 17
Among-Day Blood Pressure Variability Assessed by 7-Day/24-Hour Ambulatory Blood Pressure Monitoring and Prognostic Value of Vascular Variability Anomalies

Abstract Currently, 24-hour monitoring of blood pressure (BP) by ambulatory functioning devices is a gold standard, reserved for special cases of high BP, left uninterpreted in terms of its time structure. General reliance upon a single measurement (or a single 24-hour profile) of BP, however, has been dubbed "flying blind" and is at variance with the documented need to meet requirements, stated repeatedly for over a century by opinion leaders, i.e., that we must evaluate periodic BP variations before a patient is examined. We review the relative merits of long-term BP monitoring, analyzed time-structurally (chronomically, from chronome = time structure). Among-day BP variability, including "masked non-dipping" (dipping on day 1, but non-dipping on at least 1 of the following 6 days), may underlie cardiovascular disease accompanying a key component of psychological depression.

Keywords 7-day/24-hour ambulatory BP monitoring • Among-day variability of circadian BP characteristics • Monday morning surge in BP • Disruption of circadian rhythm in systolic BP • Community-based comprehensive medical assessment

17.1 Within-Day and Among-Day Variability Assessed by 7-Day/24-Hour Ambulatory Blood Pressure Monitoring

Blood pressure (BP) being highly variable, ambulatory BP monitoring (ABPM) more thoroughly depicts the underlying BP behavior and predicts cardiovascular risk better than conventional measurements [1–4]. ABPM also assesses the extent of BP decrease ("dip") during sleep. "Non-dipping," defined as a day-night ratio (DNR) $<10\ \%$, is reportedly associated with target organ damage and cardiovascular events, independently of the overall BP mean [1, 2]. Day-to-day variability in circadian BP characteristics is often observed in 7-day ABPM records. Changes with age in the extent of day-to-day variability in 24-hour SBP mean and DNR are examined herein.

A greater morning BP surge on Mondays was found in Uraus, a rural northern Japanese town, as shown in Fig. 17.1 [5], but not in Tosa, a southern Japanese city. Geographic differences in changes with age in circadian features of BP are explored herein between these two cities and Tokyo.

K. Otsuka et al., *Chronomics and Continuous Ambulatory Blood Pressure Monitoring*, DOI 10.1007/978-4-431-54631-3_17

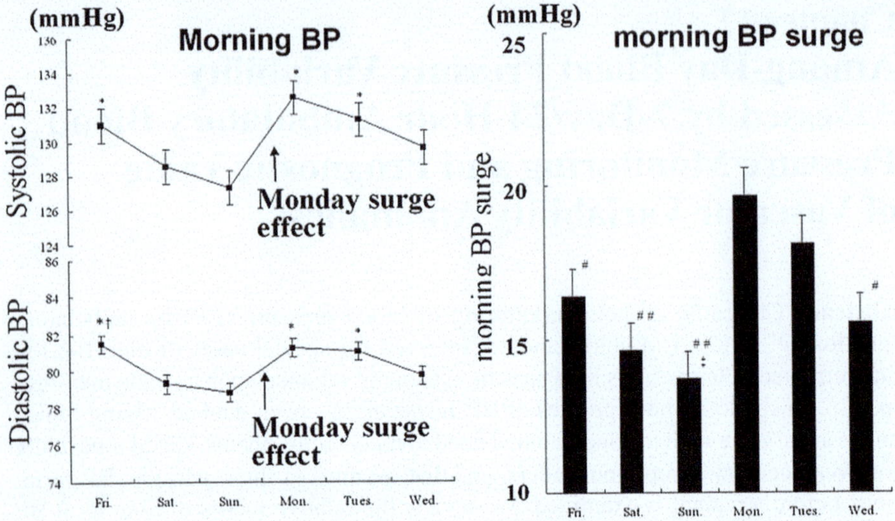

Fig. 17.1 Repeated ambulatory monitoring reveals a Monday morning surge in BP in a community-dwelling population (Am J Hypertens 2004;17:1179–83)

From the 644 community-dwelling subjects, living in Tokyo, Uraus (Hokkaido Prefecture), or Tosa (Kochi Prefecture) initially recruited, 609 subjects (368 women and 241men, 23–79 years of age) who provided ABPM records spanning at least 6 days were included in this study. As reported earlier [5–9], BP was measured oscillometrically at 30-min (07:00–22:00) or 60-min (22:00–07:00) intervals, using TM-2431 monitors (A&D Co., Japan), validated for accuracy [10]. Daily (24-hour) averages and DNR values were computed.

Hypertension (HT) was defined as a 24-hour BP mean above 130/80 mmHg. Non-dipping (ND) was defined as a DNR of systolic (S) BP <10 %. The 24-hour BP mean and DNR served to subdivide the 609 subjects into four groups. In one classification, groups I–IV consisted of consistent normotension (NT), NT on day 1 but HT on at least 1 of the following 6 days (masked "HT"), HT on day 1 but NT on at least 1 of the following 6 days ("masked NT"), and consistent HT, respectively. In the other classification, groups I–IV consisted of consistent dipping (DP), DP on day 1 but ND on at least 1 of the following 6 days ("masked ND"), ND on day 1 but DP on at least 1 of the following 6 days ("masked DP"), and consistent ND, respectively.

The incidence of consistent and inconsistent classification was compared by the Kruskal-Wallis test (significance at the 0.05 probability level) for subjects in three age groups (23–49 years, N=113; 50–64 years, N=257; 65–79 years, N=239), for all subjects, men and women separately, and subjects in each of the three cities.

Inconsistent classifications from day-to-day increased with age ($p<0.0001$), "masked NT" being more frequent than "masked HT." Results were significant for women ($p<0.0001$) but not for men ($p=0.2154$) (Table 17.1), for citizens of Uraus

Table 17.1 Gender difference of day-to-day variability in mean 24-hour blood pressure

			Day-to-day change of ABP HT within a week				
	Age (year)	Subject number	Persistent Normotension	Masked Hypertension	Masked Normotension	Persistent Hypertension	p-value
All subjects	23-49	113	56 (49.6 %)	18 (15.9%)	17 (15.0%)	22 (19.5%)	<0.0001
	50-64	257	76 (29.6 %)	40 (15.6 %)	69 (26.8%)	72 (28.0 %)	
	65-79	239	47 (19.7 %)	31 (13.0 %)	82 (34.3 %)	79 (33.1 %)	
	Total	609	179 (29.4 %)	89 (14.6 %)	168 (27.6 %)	173 (28.4 %)	
Women	23-49	64	42 (65.6 %)	9 (14.1 %)	3 (4.7 %)	10 (15.6 %)	<0.0001
	50-64	167	58 (34.7 %)	24 (14.4 %)	43 (25.7 %)	42 (25.1 %)	
	65-79	137	30 (21.9 %)	18 (13.1 %)	41 (29.9 %)	48 (35.0 %)	
	Total	368	130 (35.3 %)	51 (13.9 %)	87 (23.6 %)	100 (27.2 %)	
Men	23-49	49	14 (28.6 %)	9 (18.4 %)	14 (28.6 %)	12 (24.5 %9	0.21541
	50-64	90	18 (20.0 %)	16 (17.8 %)	26 (28.9 %)	30 (33.3 %)	
	65-79	102	17 (16.7 %)	13 (12.7 %)	41 (40.2 %)	31 (30.4 %)	
	Total	241	49 (20.3 %)	38 (15.8 %)	81 (33.6 %)	73 (30.3 %)	

(p=0.0007) and Tosa (p=0.0048) but not Tokyo (p=0.7201) (Table 17.2). Inconsistent DNR classifications also increased with age, overall, and for men and women separately (all p=0.0008) (Table 17.3), being significant in Uraus (p=0.0130) but not in Tosa (p=0.0907) or Tokyo (p=0.5364) (Table 17.4).

Depressive mood affects BP and BP variability [6, 7, 11], depressed citizens showing a more prominent circaseptan component in SBP, associated with subjective sleep disturbances. Aging is herein found to affect the extent of day-to-day BP variability, more so in women than in men, perhaps because day-to-day variability is already high in younger men. An age effect was not found in Tokyo, where the incidence of consistent HT was already high in the younger age group.

An age effect in DNR classification was significant for men and women but only in Uraus in northern Japan, where winters are cold. Results in Tosa are similar but not significant, perhaps because the clement weather of southern Japan favors a higher incidence of consistent dippers in all age groups. A busy and noisy lifestyle in Tokyo may account for the largest percentage of inconsistent day-to-day classification (80.2 %).

DNR classification is almost invariably more inconsistent than BP-mean classification. Habituation to the monitor may account for the numerically higher incidence of "masked-normotension" than "masked-hypertension," observed primarily in the elderly (p=0.0216).

In conclusion, aging affects day-to-day BP variability in the 24-hour BP mean and in the DNR. Several consensus meetings advocated 7-day/24-hour ABPM for a refined diagnosis to decide on a treatment plan, especially in the elderly. They proposed a chronobiological interpretation of the data in the light of time-specified reference values derived from healthy peers matched by gender and age to identify

Table 17.2 Difference of day-to-day variability in mean 24-hour blood pressure among three Japanese cities

			Day-to-day change of ABP HT within a week				
	Age (year)	Subject number	Persistent Normotension	Masked Hypertension	Masked Normotension	Persistent Hypertension	p-value
Tokyo	23-49	10	2 (20.0 %)	2 (20.0 %)	1 (10.0 %)	5 (50.0 %)	0.7201
	50-64	35	5 (14.3 %)	2 (5.7 %)	11 (31.4 %)	17 (48.6 %)	
	65-79	51	2 (3.9 %)	6 (11.8 %)	17 (33.3 %)	26 (51.0 %)	
	Total	96	9 (9.4 %)	10 (10.4 %)	29 (30.2 %)	48 (50.0 %)	
Tosa	23-49	53	25 (47.2 %)	7 (13.2 %)	9 (17.0 %)	12 (22.6 %)	0.0048
	50-64	107	33 (30.8 %)	24 (22.4 %)	24 (22.4 %)	26 (24.3 %)	
	65-79	126	27 (21.4 %)	18 (14.3 %)	40 (31.7 %)	41 (32.5 %)	
	Total	286	85 (29.7 %)	49 (17.1 %)	73 (25.5 %)	79 (27.6 %)	
Uraus	23-49	50	29 (58.0 %)	9 (18.0 %)	7 (14.0 %)	5 (10.0 %)	0.0007
	50-64	115	38 (33.0 %)	14 (12.2 %)	34 (29.6 %)	29 (25.2 %)	
	65-79	62	18 (29.0 %)	7 (11.3 %)	25 (40.3 %)	12 (19.4 %)	
	Total	227	85 (37.4 %)	30 (13.2 %)	66 (29.1 %)	46 (20.3 %)	

Table 17.3 Day-to-day variability in day-night ratio (nocturnal dipping) of blood pressure in women and men

			Day-to-day change of nocturnal BP dipping within a week				
	Age (year)	Subject number	Persistent Dipper	Masked Non-dipper	Masked Dipper	Persistent Non-dipper	p-value
All subjects	23-49	113	53 (46.9 %)	41 (36.3 %)	19 (16.8 %)	0 (0 %)	<0.0008
	50-64	257	92 (35.8 %)	116 (45.1 %)	45 (17.5 %)	4 (1.6 %)	
	65-79	239	68 (28.5 %)	103 (43.1 %)	61 (25.5 %)	7 (2.9 %)	
	Total	609	213 (35.0 %)	260 (42.7 %)	125 (20.5 %)	11 (1.8 %)	
Women	23-49	64	27 (42.2 %)	25 (39.1 %)	12 (18.8 %)	0 (0 %)	<0.0008
	50-64	167	60 (35.9 %)	75 (44.9 %)	29 (17.4 %)	3 (1.8 %)	
	65-79	137	39 (28.5 %)	56 (40.9%)	35 (25.5 %)	7 (5.1 %)	
	Total	368	126 (34.2 %)	156 (42.4 %)	76 (20.7 %)	10 (2.7 %)	
Men	23-49	49	26 (53.1 %)	16 (32.7 %)	7 (14.3 %)	0 (0 %)	<0.0008
	50-64	90	32 (35.6 %)	41 (45.6 %)	16 (17.8 %)	1 (1.1 %)	
	65-79	102	29 (28.4 %)	47 (46.1 %)	26 (25.5 %)	0 (0 %)	
	Total	241	87 (36.1 %)	104 (43.2 %)	49 (20.3 %)	1 (0.4 %)	

Table 17.4 Difference of day-to-day variability in day-night ratio (nocturnal dipping) of blood pressure among three Japanese cities

			Day-to-day change of nocturnal BP dipping within a week				
	Age (year)	Subject number	Persistent Dipper	Masked Non-dipper	Masked Dipper	Persistent Non-dipper	p-value
Tokyo	23-49	10	2 (20.0 %)	5 (50.0 %)	3 (30.0 %)	0 (0 %)	0.5364
	50-64	35	7 (20.0 %)	16 (45.7 %)	9 (25.7 %)	3 (8.6 %)	
	65-79	51	5 (9.8 %)	25 (49.0 %)	19 (37.3 %)	2 (3.9 %)	
	Total	96	14 (14.6 %)	46 (47.9 %)	31 (32.3 %)	5 (5.2 %)	
Tosa	23-49	53	29 (54.7 %)	18 (34.0 %)	6 (11.3 %)	0 (0 %)	0.0907
	50-64	107	41 (38.3 %)	50 (46.7 %)	15 (14.0 %)	1 (0.9 %)	
	65-79	126	50 (39.7 %)	48 (38.1 %)	26 (20.6 %)	2 (1.6 %)	
	Total	286	120 (42.0 %)	116 (40.6 %)	47 (16.4 %)	3 (1.0 %)	
Uraus	23-49	50	22 (44.0 %)	18 (36.0 %)	10 (20.0 %)	0 (0 %)	0.0130
	50-64	115	44 (38.3 %)	50 (43.5 %)	21 (18.3 %)	0 (0 %)	
	65-79	62	13 (21.0 %)	30 (48.4 %)	16 (25.8 %)	3 (4.8 %)	
	Total	227	79 (34.8 %)	98 (43.2 %)	47 (20.7 %)	3 (1.3 %)	

vascular variability anomalies (VVAs) for an assessment of cardio-, cerebro-, and renovascular disease risk [12]. Even within the conventionally accepted normal range, VVAs have been associated with a statistically significant increase in risk. Results herein corroborate the need for prolonged, preferably continuous ABPM.

17.2 Circadian Disruption of Blood Pressure Observed by 7-Day/24-Hour Ambulatory Blood Pressure Monitoring

Many biological variables are circadian periodic. Most of them are partly endogenous with a period of about 24 hours, coordinated by a biological circadian clock [13], also involved in the circadian coordination of blood pressure (BP) [14]. Disruption of circadian rhythms can seriously impact overall health and increase the risk of adverse cardiovascular outcomes [13]. In mammals, including humans, aging is associated with alterations in circadian rhythm characteristics, usually involving a phase advance and/or a reduced amplitude [15]. Age-related alterations of the circadian rhythm of BP measured oscillometrically for 7 days (TM-2431, A&D Co., Tokyo, Japan) are examined herein.

A comprehensive community-based medical assessment examined any association between adverse cardiovascular outcomes and disruption of the circadian BP rhythm. Subjects were fitted with an ambulatory monitor programmed to record BP and heart rate (HR) every 30 (07:00–22:00) or 60 (22:00–07:00) minutes for 7 days. From the 292 subjects living in a rural Japanese town (Tosa) initially recruited, 283 (170 women and 113 men), aged 39–74 years, who completed monitoring for 5 (N=3), 6 (N=4), or 7 (N=276) days, were included in the study.

The circadian period of BP was estimated by the Maximum Entropy Method, using the MemCalc software (GMS Co., Tokyo) [16]. A circadian rhythm with an estimated period in the range of 22.5–25.5 hours was found in 266 citizens. The average period (± standard deviation; SD) was 24.04 ± 0.46 hours (range from 22.52 to 25.17 hours). A model consisting of cosine curves with periods of 24 and 12 hours fitted by least squares to the data was used to estimate the MESOR (midline estimating statistic of rhythm) and the 24-hour amplitude and acrophase (time of predicted maximum) [12]. The acrophase (± SD) of systolic (S) and diastolic (D) BP and of heart rate (HR) averaged 13:57 ± 1:39, 13:55 ± 1:37, and 14:41 ± 1:51, respectively.

Subjects were subdivided into four age groups: I (35–44 years, N=34), II (45–54 years, N=44), III (55–64 years, N=79), and IV (65–74 years, N=126). The incidence of circadian SBP disruption increased with age ($p=0.0343$). The relationship may be J-shaped (Fig. 17.2), however, with 2 (5.9 %), 0, 2 (2.5 %), and 13 (10.3 %) cases of disrupted SBP rhythm in groups I–IV, respectively (increased incidence statistically significant only for group IV vs. groups I–III, $p=0.035$).

SBP was also higher in group IV than in younger groups (7-day MESOR, daytime and nighttime means, office and morning home measurements), despite the fact that elderly subjects were more likely to be treated with antihypertensive medication (Table 17.5). Whereas there was no difference in the circadian amplitude of SBP or DBP among the four groups, all acrophases were statistically significantly advanced with increasing age ($p<0.0001$) (Fig. 17.3).

Although there were no intergroup differences in depressive mood, subjective quality of life (health and happiness), or times of falling asleep and getting up, sleep duration was longer in the elderly (Fig. 17.4).

As compared to subjects with an about 24-hour SBP variation, subjects with a disrupted circadian rhythm had an earlier acrophase (12:12 ± 3:36 vs. 13:57 ± 1:39; $p<0.0001$), smaller double amplitude (12.5 ± 7.7 vs. 30.5 ± 10.6; $p<0.00001$), as shown in Table 17.6 and Fig. 17.5, and a smaller nightly BP dip (7.0 ± 11.6 % vs. 19.5 ± 10.0 %; $p<0.0001$). They were older (65.0 vs. 59.9 years, $p=0.0440$) and included a larger proportion of subjects on antihypertensive medication ($p=0.0179$).

In conclusion, aging was found to be associated with changes in circadian rhythm characteristics of BP and with an increased incidence of rhythm disruption [17]. We already reported that day-to-day variation in nightly BP dip was more frequent in the elderly [18], especially in depressive citizens [6, 7, 11]. Herein, we showed that the effects of aging may become more pronounced when combined with the effect of antihypertensive medication. Elsewhere, we showed that treatment at an unsuitable time can disturb the circadian BP rhythm [19]. While several hurdles still need to be overcome, we propose that treatment timing should be optimized on an individualized basis in view of the great increase of disrupted circadian BP rhythm among elderly people treated with antihypertensive drugs. This task is

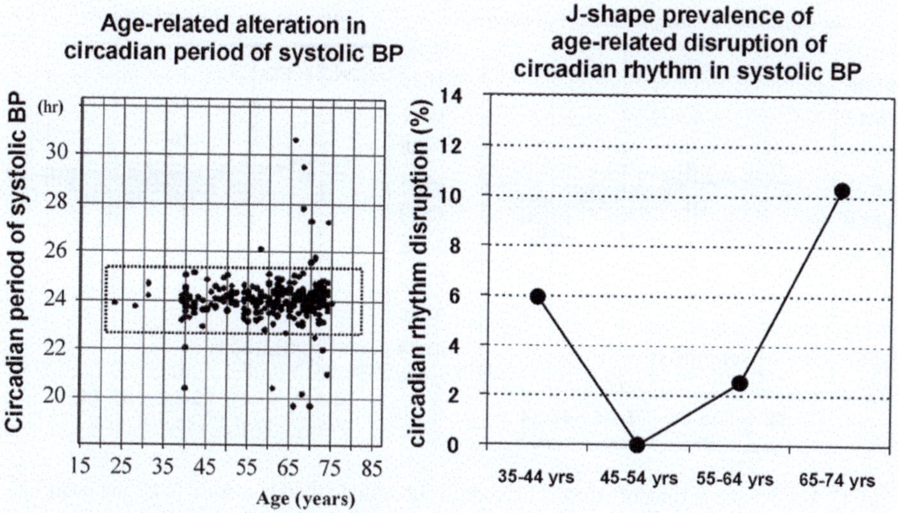

Fig. 17.2 The incidence of circadian SBP disruption increased with age

Table 17.5 The J-shaped prevalence of age-related disruption of circadian rhythm in systolic BP

	Total (N=283)	Group I (35-44 yrs, N=34)	Group II (45-54 yrs, N=44)	Group III (55-64 yrs, N=79)	Group IV (65-74 yrs, N=126)	p value (a)
Subjects with disrupted circadian SBP rhythm (N, %)	17 (6.0 %)	2 (5.9 %)	0 (0 %)	2 (2.5 %)	13 (10.3 %)	0.0343##
Age (year)	60.3 (10.3)	40.7 (1.4) ***	50.1 (2.9) ***	59.9 (2.8) ***	69.4 (2.7)	<0.0001
Women (%)	40.0	44.1	54.6	70.9	59.5	0.0453##
Anti-hypertension medication (%)	25.4	0	15.9	25.3	34.9	0.0002##
Geriatric Depression Scale (range 0-15) (±SD)	3.8 (3.1)	4.7 (3.5)	3.3 (3.0)	3.5 (3.4)	3.9 (3.0)	0.2930
modified Patient Health Questionnaire-2 (range, 0-2) (±SD)	0.58 (0.73)	0.58 (0.72)	0.60 (0.77)	0.43 (0.66)	0.66 (0.75)	0.3256
Subjective QOL of health (%)	68.5 (20.4)	75.1 (18.3)	67.0 (22.1)	69.8 (20.9)	66.7 (19.9)	0.2692
Subjective QOL of happiness (%) (±SD)	70.8 (19.9)	75.5 (19.7)	74.5 (17.9)	71.0 (22.0)	68.5 (19.2)	0.2494
Time for falling asleep (min.) (±SD)	24.2 (24.2)	20.4 (22.9)	19.9 (19.1)	25.8 (29.4)	25.9 (23.0)	0.4648
Time of Getting up (hh:mm) (±SD)	6:31 (0:41)	6:34 (0:41)	6:28:0 (0:32)	6:30 (0:36)	6:31 (0:46)	0.9495
Sleep duration (hh:mm) (±SD)	7:51 (1:08)	7:07 (0:39) ***	7:07 (0:45) ***	7:45 (0:57) ***	8:21 (1:11)	<0.0001
SBP-MESOR (mmHg) (±SD)	127.4 (13.5)	120.8 (14.0) **	128.6 (16.7)	126.1 (10.7)	129.5 (13.3)	0.0063
DBP-MESOR (mmHg) (±SD)	78.2 (7.8)	77.2 (9.0)	81.9 (10.0) ***	78.6 (6.6)#	76.8 (7.0)	0.0022
HR-MESOR (bpm) (±SD)	71.3 (7.6)	74.2 (8.0) *	72.6 (6.8)	71.2 (7.0)	70.1 (8.0)	0.0225
SBP-Amplitude (mmHg) (±SD)	29.4 (11.4)	26.6 (8.4)	29.5 (8.3)	31.3 (11.7)	28.9 (12.6)	0.2086
DBP-Amplitude (mmHg) (±SD)	18.4 (7.0)	19.9 (4.7)	19.9 (6.6)	18.7 (6.5)	17.4 (7.9)	0.1044
HR-Amplitude (bpm) (±SD)	16.7 (6.5)	15.9 (6.1)	14.7 (6.2) *	16.4 (6.3)	17.8 (6.6)	0.0395
SBP-Acrophase (hh:mm) (±SD)	13:50 (1:52)	14:34 (1:14) ***	14:44 (1:07) ***	13:50 (2:00)	13:21 (1:59)	<0.0001
DBP-Acrophase (hh:mm) (±SD)	13:31 (1:43)	14:31 (1:05) ***	14:32 (1:34) ***	14:02 (1:41) *	13:21 (1:48)	<0.0001
HR-Acrophase (hh:mm) (±SD)	14:39 (1:48)	15:35 (1:55) ***	15:17 (2:10) **	14:36 (1:16)	14:14 (1:48)	<0.0001
SBP 7-day Average (mmHg) (±SD)	130.1 (13.6)	123.5 (14.6) **	131.5 (16.9)	128.9 (10.7)	132.1 (13.2)	0.0078
SBP 7-daytime Average (mmHg) (±SD)	136.4 (14.4)	128.9 (15.8) ***	137.8 (17.3)	135.7 (11.6)	138.5 (13.9)	0.0053
SBP 7-nighttime Average (mmHg) (±SD)	111.6 (14.5)	104.9 (12.6) ***	110.9 (16.7)	109.7 (12.4) #	114.8 (14.6)	0.0018
DBP 7-day Average (mmHg) (±SD)	80.0 (7.9)	79.5 (9.2)	83.8 (10.2) ***	80.3 (6.6)#	78.4 (7.1)	0.0013
Office systolic BP (mmHg) (±SD)	140.2 (21.5)	130.8 (27.6) *	141.5 (20.4)	138.3 (19.0)	143.3 (21.2)	0.0206
Office diastolic BP (mmHg) (±SD)	85.8 (12.7)	85.3 (17.1)	90.9 (14.7) **	87.8 (11.2) **	82.6 (10.8)	0.0006
Office Pulse (bpm) (±SD)	71.3 (10.6)	75.3 (10.8)	71.2 (8.6)	71.9 (10.7)	69.8 (11.0)	0.0523
Morning home systolic BP (mmHg) (±SD)	131.6 (16.2)	121.8 (16.8)***	127.5 (16.0) **	130.0 (13.2) **	136.9 (16.0)	<0.0001
Morning home diastolic BP (mmHg) (±SD)	81.7 (9.4)	79.2 (12.6)	84.1 (9.8)	82.8 (8.4)	80.9 (8.7)	0.0629
Morning home pulse (bpm) (±SD)	66.9 (8.7)	71.2 (10.1) **	68.6 (7.7) #	66.5 (7.3)	65.2 (9.0)	0.0020
Evening home systolic BP (mmHg) (±SD)	124.3 (14.6)	119.7 (18.6)	124.5 (16.5)	124.8 (12.7)	125.3 (13.8)	0.2606
Evening home diastolic BP (mmHg) (±SD)	75.3 (9.3)	75.3 (13.4)	79.1 (10.6) ***	76.9 (6.9) ***	73.1 (8.4)	0.0007
Evening Home pulse (bpm) (±SD)	69.6 (9.0)	73.5 (9.8) **	71.5 (8.5) #	68.6 (8.1)	68.3 (9.3)	0.0083
SD of morning systolic BP for 30 days (mmHg) (±SD)	10.2 (2.9)	9.0 (3.3) *	9.8 (3.1)	10.2 (2.4)	10.7 (3.0)	0.0289
SD of evening systolic BP for 30 days (mmHg) (±SD)	11.1 (3.6)	10.0 (5.0)	10.4 (3.8)	11.7 (3.3)	11.3 (3.1)	0.0782

(a) chi-square test for percentages, and 1-way ANOVA for numerical values.

# p<0.05 from Student t-test (comparisin vs. group IV), * p<0.01 from Student t-test (comparison vs. IV), ** p<0.005 from Student t-test (comparison vs. group IV), *** p<0.001 from Student t-test (comparison vs. group IV), ## p-value from chi-square test.

SBP = systolic blood pressure, DBP = diastolic blood pressure, HR = heart rate, MESOR = midline-estimating statistics of rhythm, QOL = quality of life, SD: Standard Deviation

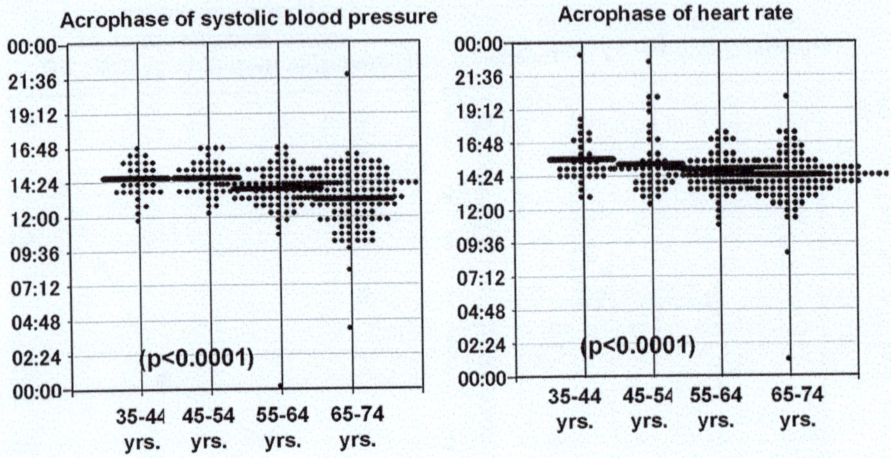

Fig. 17.3 Advance in time of circadian acrophase of systolic blood pressure (*left*) and heart rate (*right*)

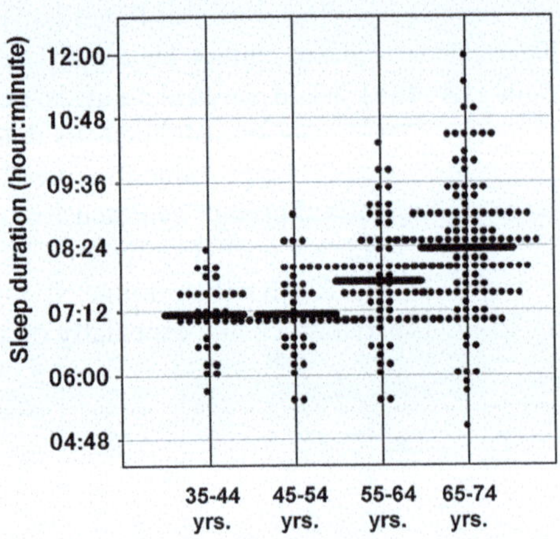

Fig. 17.4 Duration in bed for sleep increases with age

greatly facilitated by advances in technology in the last few decades and by the availability of statistical methods applicable to the individual patient. Parameter tests [20] and cumulative control charts [21] have been used to determine the best treatment time for the given patient, the personalized chronotherapy being tailored to the particular chronodiagnosis. This approach includes the screening for vascular variability anomalies to detect an elevated BP and other abnormal patterns of BP (and HR) variability.

Table 17.6 Characteristic factors associated with a disrupted circadian rhythm group of systolic BP

	Disrupted circadian rhythm group (n=18)		Normal circafdian rhythm group (n=265)		Student t-test
	Mean	SD	Mean	SD	P-value
Male (1)/Femal (0)	0.278	0.461	0.408	0.492	N.S.
Age (years)	65.0	10.1	59.9	10.3	0.04401
Body mass Index (BMI)	23.9	2.7	24.0	3.0	N.S.
Classification of HT(0; normotensin, 1; HT without medication, 3; HT with medication	1.278	0.752	0.819	0.791	0.01792
GDS-15 (0~15)	4.7	3.1	3.7	3.1	N.S.
Sense of sadness (0,1)	0.39	0.51	0.22	0.41	0.00480
Decline in interest (0,1)	0.46	0.52	0.35	0.48	N.S.
PHQ-2 (0~2)	0.85	0.90	0.56	0.72	N.S.
Subjective QOL of health (%)	66.4	17.0	68.6	20.6	N.S.
Subjective QOL of mood (%)	70.5	15.7	71.1	18.2	N.S.
Subjective QOL of family relationship (%)	75.2	16.4	82.6	18.7	N.S.
Subjective QOL of relationship with friends (%)	77.8	17.5	80.4	18.9	N.S.
Subjective QOL of finantial satisfaction (%)	50.9	21.6	55.3	23.3	N.S.
Subjective QOL of daily life (%)	59.3	19.9	66.3	22.2	N.S.
Subjective QOL of happiness (%)	58.1	11.3	71.7	20.1	0.01049
Time of getting up (hour:minute)	6:14	0:42	6:31	0:42	N.S.
Sleep duration (hour:minute)	7:51	0:57	7:49	1:16	N.S.
Time for falling asleep(minute)	17.7	8.0	24.7	24.9	N.S.
Office Systolic BP (mmHg)	141.9	24.6	140.1	21.3	N.S.
Office Diastolic BP (mmHg)	80.2	10.3	86.1	12.8	N.S.
Office Pulse (bpm)	72.0	12.9	71.2	10.5	N.S.
Systolic BP 7-day Average (mmHg)	136.0	14.8	129.7	13.5	N.S.
Systolic BP 7-daytime Average (mmHg)	137.8	15.1	136.3	14.4	N.S.
Systolic BP 7-night Average (mmHg)	127.0	13.4	110.5	14.0	<0.00001
Diastolic BP 7-day average (mmHg)	79.6	7.4	79.8	8.8	N.S.
Systolic BP-MESOR (mmHg)	134.9	14.7	126.9	13.3	0.01497
Diastolic BP-MESOR (mmHg)	78.7	7.1	78.2	7.9	N.S.
Heart Rate-MESOR (bpm)	72.1	8.3	71.3	7.6	N.S.
Systplic BP-Amp (mmHg)	12.5	7.7	30.5	10.6	<0.00001
Diastolic BP-Amp (mmHg)	9.4	5.0	19.1	6.7	<0.00001
Heart Rate-Amp (bpm)	18.5	5.9	16.5	6.5	N.S.
Systolic BP-Acro	12:12	3:36	13:57	1:39	0.00010
Diastolic BP-Acro	13:01	2:52	13:54	1:36	0.03285
Heart Rate-Acro	14:10	0:57	14:41	1:51	N.S.
Systolic BP MESOR-HT (0,1)	0.33	0.49	0.16	0.37	N.S.
Diastolic BP MESOR-HT (0,1)	0.11	0.32	0.11	0.31	N.S.
EPP >60mmHg (0,1)	0.33	0.49	0.08	0.27	0.00020
CHAT Syatolic BP (0,1)	0	0	0.29	0.45	0.01720
CHAT Diastolic BP (0,1)	0	0	0.09	0.29	N.S.
Ecphasia (0,1)	0.06	0.24	0	0.06	N.S.
DHRV (0,1)	0	0	0.04	0.20	N.S.
Number of VVDs	0.83	0.99	0.72	0.99	N.S.
anyVVD (0,1)	0.50	0.51	0.44	0.50	N.S.
HBP Systolic BP (morning) (mmHg)	141.8	19.4	130.9	15.8	0.00706
HBP Diasolic BP (morning) (mmHg)	82.3	7.1	81.7	9.6	N.S.
HBP Pulse (morning) (bpm)	70.1	10.3	66.7	8.6	N.S.
HBP Systolic BP (evening) (mmHg)	130.3	15.2	123.9	14.5	N.S.
HBP Diasolic BP (evening) (mmHg)	74.5	7.5	75.4	9.5	N.S.
HBP Pulse (evening) (bpm)	69.1	10.6	69.6	8.9	N.S.

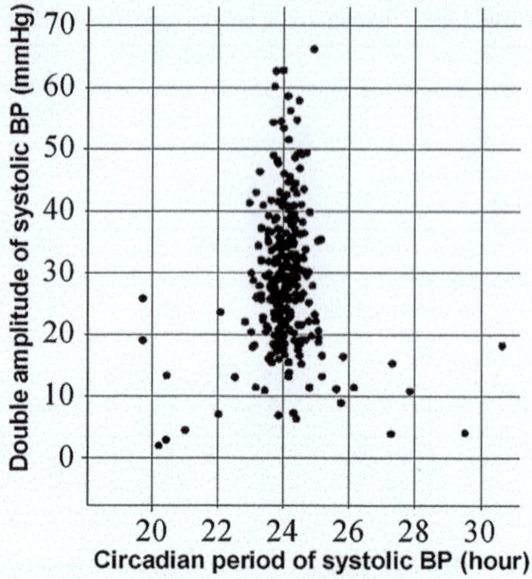

Fig. 17.5 Decrease in circadian amplitude along with circadian disruption of blood pressure

17.3 The 7-Day/24-Hour Is Superior to the 1-Day/24-Hour Ambulatory Blood Pressure Monitoring to Predict Cardiovascular Outcome

Long-term prognosis is much greater for ambulatory BP (ABP) monitoring than office measurements [22–24], because it includes short-term BP variability like white-coat hypertension, masked hypertension, and various responses during exercise or laboratory stress. A relationship between within-day variability of BP and prognosis has been established, but among-day variability based on home BP or visit-to-visit office BP measurements has recently attracted much attention [11, 22, 24, 25]. Thus, we examined whether 7-day/24-hour ABP offers any advantage in prognostic implications of cardiovascular (CV) diseases as compared to 1-day/24-hour ABP monitoring.

A community-based comprehensive medical assessment (CMA), including 7-day/24-hour ABP monitoring, was performed in Uraus town, Hokkaido, located in northern Japan. Participants were 124 women and 82 men, aged 24–79 years (average 57.1 years). Each provided an ABP record spanning at least 6 days. As reported earlier [5–9, 11, 12, 17–19, 25–27], BP was measured oscillometrically at 30-min (07:00–22:00) or 60-min (22:00–07:00) intervals, using TM-2431 monitors (A&D Co., Japan), validated for accuracy [10].

A CV outcome was defined as a new onset of CV disease, including infarction, effort angina, and myocardial infarction. Subjects were followed-up prospectively for about 5 years (average: 1633 days) from April 2001 to October 2005. Event-free survival from CV disease was analyzed by the Kaplan-Meier method, and its statistical significance was assessed by the log-rank (Cox-Mantel) and Mantel-Haenszel tests. The Student *t*-test was used to compare variable characteristics between subjects diagnosed as normotensive (NT) or hypertensive (HT), based on 7-day ABP monitoring. A P-value below 0.05 was considered statistically significant. This study was approved by the Medical Ethics Committee of Tokyo Women's Medical University (Clinical Study #2912 "Health assessment of community-dwelling elderly in Japan"). Written informed consent was obtained from all participants regarding data collection, analysis, and publication.

Hypertension (average SBP/DBP >130/80 mmHg) vs. normotension was defined in two ways, depending on whether the diagnosis relied on the first 24-hour (1-day) or on the entire 7-day ABP monitoring. There were 101 citizens with 1-day HT and 105 with 1-day NT, and there were 77 citizens with 7-day HT and 129 with 7-day NT. We compared the prognostic value of 7-day vs. 1-day categorization.

CV diseases were observed in nine subjects. Results from the Kaplan-Meier analysis are shown in Fig. 17.6. The event-free ratio was significantly lower when comparing 7-day HT vs. 7-day NT (right). No statistically significant difference was found, however, when comparing 1-day HT vs. 1-day NT (left).

Cox proportional hazards regression analysis was not able to show a statistical significant difference between 1-day HT and 1-day NT, but, as compared to 7-day NT, 7-day HT statistically significantly predicted CV outcome. The corresponding relative risk (RR) and 95 % confidence interval (CI) was 5.9 (1.20–28.1, $p=0.0274$). CV outcome was also affected by age (RR for 5 years: 1.6, CI: 1.1–2.5, $p=0.0186$). The age-adjusted RR of 7-day HT was 4.9 (CI: 1.01–23.6, $p=0.0480$).

Accordingly, we assessed characteristics of subjects diagnosed with 7-day HT ($N=77$) and compared them with those of 7-day NT subjects ($N=129$). Results are summarized in Table 17.7. Although there were no statistically significant differences between the two groups in terms of depressive mood, subjective quality of life of health and happiness, and sleep quality, 7-day HT subjects were older (59.8 vs. 55.5 years, $p=0.0087$), were more likely to be treated with antihypertensive drugs

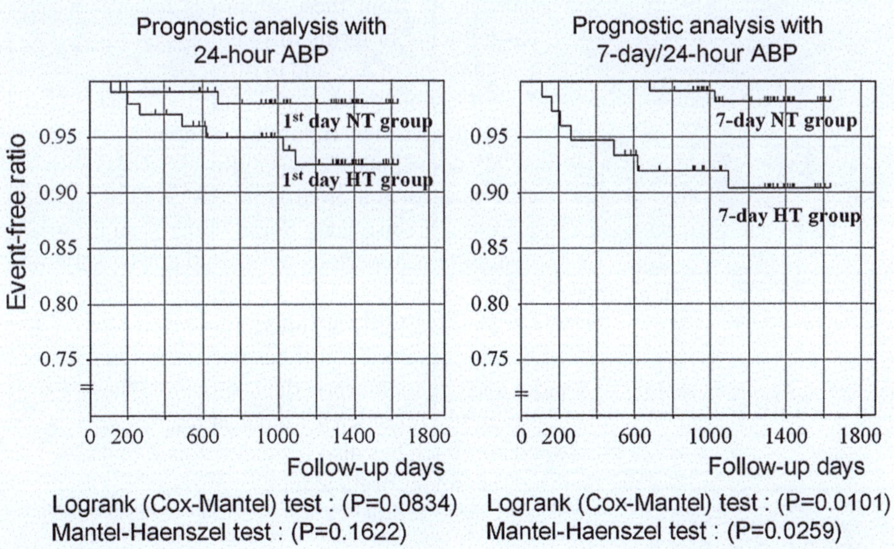

Fig. 17.6 Improved predictability of cardiovascular outcome from 7-day/24-hour vs. 1-day/24-hour ABP monitoring

(p<0.0001), and had a larger arterial stiffness measured by baPWV (1673.9 vs. 1455.3 cm/s, p<0.0001). Their day-night ratio of SBP was smaller, whether it was determined conventionally or chronobiologically (p<0.0001). Their circadian amplitude of SBP (p=0.0443) and DBP (p=0.0090) was smaller. Their incidence of excessive pulse pressure was higher (24.68 % vs. 0.78 %, p<0.0001), as was their incidence of any vascular variability disorder (72.73 % vs. 20.94 %, p<0.0001), as compared to the 7-day NT subjects (Table 17.7).

Table 17.7 Comparison between citizens diagnosed as normotensive (NT) or hypertensive (HT) based on 7-day/24-hour ambulatory BP monitoring. Results from a community-based medical assessment

	Total (n=206)		7-day NT (n=129)		7-day HT (n=77)		Student t-test
	Mean	SD	Mean	SD	Mean	SD	P-value
Age (years)	57.1	11.2	55.5	11.8	59.8	9.6	0.0087
Sex (M/F)	0.40	82/124	0.34	44/85	0.49	38/39	0.0439
Body Mass Index (BMI)	24.5	2.9	23.9	2.9	25.6	2.6	0.0001
N of subjects: NT/untreated HT/treated HT	78/50/78		78/19/32		0/31/46		<0.0001
GDS-15	4.1	3.1	4.0	3.1	4.3	3.2	NS
Subjective QOL of health (%)	64.9	17.7	66.5	17.7	62.3	17.6	NS
Subjective QOL of happiness (%)	74.0	19.5	75.7	19.9	71.2	18.7	NS
Time of getting up	6:07	0:48	6:07	0:49	6:07	0:46	NS
Sleep duration (hour:minute)	7:52	1:01	7:46	1:02	8:03	0:59	NS
Follow-up days	1283.3	328.0	1292.3	287.9	1268.2	387.6	NS
Total death (ratio, number)	0.019	4/202	0.016	2/127	0.026	2/75	NS
Cardiovascular outcome: Yes/No (ration, number)	0.044	9/197	0.016	2/127	0.091	7/70	0.0271
Incidence of cancer onset: Yes/No (ratio, number)	0.024	5/201	0.031	4/125	0.013	1/76	NS
Office SBP (mmHg)	140.0	23.9	131.7	20.5	154.0	22.8	<0.0001
Office DBP (mmHg)	85.5	13.1	81.8	11.6	91.6	13.2	<0.0001
Office Heart Rate (bpm)	71.8	10.2	71.9	10.3	71.8	10.0	NS
SBP 7-day Average (mmHg)	128.0	14.6	119.7	8.3	141.8	12.1	<0.0001
SBP 24-hour (day 1) Average (mmHg)	129.9	16.4	121.1	10.2	144.5	14.1	<0.0001
Conventional SBP dip (day-night ratio) (%)	17.6	9.9	19.8	9.4	13.9	9.8	<0.0001
Chronobiological SBP dip (day-night ratio) (%)	19.1	11.2	21.6	10.6	14.8	10.9	<0.0001
Conventional DBP dip (day-night ratio) (%)	17.3	7.0	19.1	6.8	14.1	6.2	<0.0001
Conventional HR dip (day-night ratio) (%)	17.0	7.8	16.8	8.6	17.3	6.3	NS
SBP MESOR from 7-day/24-hour ABP (mmHg)	126.5	14.7	117.5	8.1	141.5	10.4	<0.0001
DBP MESOR 7-day/24-hour ABP (mmHg)	77.3	8.5	73.0	5.9	84.5	7.1	<0.0001
HR MESOR 7-day/24-hour ABP (bpm)	69.6	6.9	69.7	7.2	69.2	6.3	NS
SBPAmp 7-day/24-hour ABP (mmHg)	25.0	10.6	26.1	10.5	23.1	10.4	0.0443
DBPAmp 7-day/24-hour ABP (mmHg)	16.5	6.8	17.5	6.6	14.9	6.8	0.0090
HR Amp 7-day/24-hour ABP (bpm)	15.9	6.2	15.8	6.7	16.1	5.3	NS
SBPAcro 7-day/24-hour ABP	13:47	2:25	14:01	2:00	13:25	2:59	NS
DBPAcro 7-day/24-hour ABP	13:47	1:40	13:55	1:37	13:35	1:43	NS
HRAcro 7-day/24-hour ABP	14:06	2:12	13:58	2:27	14:19	1:40	NS
SBP PTE 7-day/24-hour ABP (mmHg)	12.2	23.2	1.4	4.3	30.3	29.9	<0.0001
DBP PTE 7-day/24-hour ABP (mmHg)	6.8	14.9	1.2	3.0	16.1	21.0	<0.0001
HR PTE 7-day/24-hour ABP (bpm)	0.5	2.2	0.7	2.7	0.1	0.8	NS
SBP MESOR HT (%, Yes/No)	19.90	41/165	0.78	1/128	51.95	40/37	<0.0001
DBP MESOR HT (%, Yes/No)	13.11	27/179	0.78	1/128	33.77	26/51	<0.0001
EPP >60mmHg (%, Yes/No)	9.71	20/186	0.78	1/128	24.68	19/58	<0.0001
SBP CHAT (%, Yes/No)	15.53	32/174	17.83	23/106	11.69	9/68	NS
DBP CHAT (%, Yes/No)	4.23	11/195	7.75	10/119	1.30	1/76	0.0462
Ecphasia (%, Yes/No)	1.46	3/203	0	0/129	3.90	3/74	0.0239
DHRV (%, Yes/No)	0.05	1/205	0.08	1/128	0	0/77	NS
anyVVD (%, Yes/No)	67.48	83/123	20.94	27/102	72.73	56/21	<0.0001
baPWV max (cm/sec)	1541.1	342.3	1455.3	323.1	1673.9	330.5	<0.0001
ABIminimum	1.08	0.09	1.08	0.09	1.09	0.10	NS

GDS; geriatric depression score, QOL; quality of life, SBP; systolic blood pressure, DBP; diastolic blood pressure, HR; heart rate, Amp; circadian amplitude, Acro; circadian acrophase, PTE; percent time of elevation, EPP; excessive pulse pressure, CHAT; circadian hyper-amplitude tension, DHRV; decreased heart rate variability, VVD; vascular variability disorder, baPWV; brachial ankle pulse wave velocity, ABI; ankle brachial index

Conventional dip was calculated as a dipping ration of nighttime BP/daytime BP, Chronobiological dip was done as a dipping ratio of nighttime BP/ 24-hour average BP

In conclusion, whereas clear advantages of 24-hour ABP over casual office BP measurements have been established to assess CV disease risk, the present results indicate that 24-hour ABP is not enough. Whereas 24-hour ABP may be less affected from any novelty pressor effect [8], it does not take into consideration among-day BP variability in the day-night ratio (masked non-dipper) [11, 12, 18, 19, 25, 27], in the BP MESOR (masked ABP hypertension) [6, 9, 12, 17, 19, 28, 29], or in episodic BP increases in the morning (morning BP surge) [26, 29] or on Monday mornings (masked Monday morning surge) [5, 29], as we have reported. Results herein corroborate the need for prolonged, preferably continuous ABPM [12, 27, 29].

17.4 Prognostic Value of Combination Scoring of Vascular Variability Anomalies and Among-Day Variability Assessed by 7-Day/24-Hour Ambulatory Blood Pressure Monitoring

We have already reported that 7-day/24-hour ambulatory blood pressure (ABP) monitoring makes it possible to more precisely diagnose target organ damage and assess cardiovascular (CV) disease risk [3, 4, 12, 27–29]. Herein, we confirm the prognostic advantage of 7-day/24-hour ABP monitoring when it is used to assess among-day variability and a combination of vascular variability anomalies (VVAs), even in a small community study. VVAs are abnormalities of the circadian pattern of systolic (S) BP, diastolic (D) BP, and/or pulse (HR).

A community-based comprehensive medical assessment (CMA) was performed in Uraus town, in northern Japan, including 7-day/24-hour ABP monitoring. Each of 206 citizens (124 women and 83 men) 24–79 years of age (average: 57.1 years) provided an ABPM record spanning at least 6 days. BP was measured oscillometrically every 30 min (07:00–22:00) or 60 min (22:00–07:00), using TM-2431 monitors (A&D Co., Japan), validated for accuracy.

A CV outcome was defined as a new onset of CV disease, including infarction, effort angina, and myocardial infarction. Subjects were followed-up prospectively for about 5 years (average: 1633 days) from April 2001 to October 2005. Event-free survival from CV disease was analyzed by the Kaplan-Meier method, and its statistical significance was assessed by the Log-rank (Cox-Mantel) test. One-way ANOVA and Student t-tests were used to compare variable characteristics among subjects classified according to their 7-day/24-hour ABP as being at mild, moderate, or high CV risk. A P-value below 0.05 was considered statistically significant. This study was approved by the Medical Ethics Committee of Tokyo Women's Medical University (Clinical Study #2912: "Health assessment of community-dwelling elderly in Japan"). Written informed consent was obtained from all participants regarding data collection, analysis, and publication.

Each ABP record was screened for VVAs (1, SBP MESOR-HT; 2, DBP MESOR-HT; 3, SBP CHAT; 4, DBP CHAT; 5, BP ecphasia; 6, excessive pulse pressure (EPP); and 7, deficient heart rate variability (DHRV)), where HT and NT stand for hypertension and normotension, respectively, CHAT stands for circadian hyper-amplitude-tension (too large a 24-hour amplitude of BP), ecphasia is an odd 24-hour phase of BP but not of HR, EPP is a pulse pressure above 60 mmHg, and DHRV is a standard deviation of HR less than 7.5 bpm. Each VVA was counted as 1 point on the risk factor score. Masked non-dipping and masked ABP HT were also counted as 1 point. Persistent normotension (NT), masked HT, masked NT, and persistent HT were counted as 0, 1, 2, and 3 points, respectively. Persistent dipping, masked non-dipping, masked dipping, and persistent non-dipping were also counted as 0, 1, 2, and 3 points, respectively. The risk factor score of each citizen was determined by summing points.

Among the 206 citizens, 28, 39, 31, 38, 25, 18, 15, 9, and 3 had scores of 0, 1, 2, 3, 4, 5, 6, 7, and 8, respectively. They were subdivided into three groups at mild (scores 0 or 1; N=67), moderate (scores 2 or 3; N=69), or high (scores 4–8; N=70) CV risk.

CV diseases were observed in nine subjects. Results from the Kaplan-Meier analysis are shown in Fig. 17.1. The event-free ratio was significantly lower in the high-risk group compared to both the mild- and moderate-risk groups (Fig. 17.7). From the viewpoint of a community-based CMA, variable characteristics were assessed in each group and compared among them (Table 17.8). When one-way ANOVA results were statistically significant, the high-risk group was compared to the mild and moderate-risk groups by the Student *t*-test. Results in Table 17.8 stem from the comparison between the high- and moderate-risk groups.

Subjects in the high-risk group (1) were older (p=0.0005), (2) had a higher BMI (p=0.0001), and (3) were more likely to be hypertensive (p<0.0001), to be treated with antihypertensive drugs (p<0.0001), and to take more than 3 drugs (p=0.0007). (4) Their office and ABP measurements were higher (p<0.0001) and

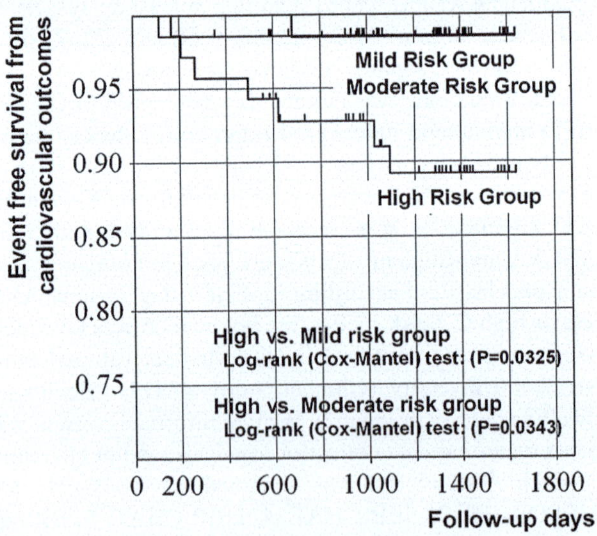

Fig. 17.7 Predictability of the risk factor score assessed by vascular variability anomalies based on 7-day/24-hour ABP monitoring

Table 17.8 Characteristics of three risk groups of citizens predicting of cardiovascular outcomes

		Mild Risk Group (n=67)		Moderate Risk Group (n=69)		High Risk Group (n=70)		1-way ANOVA or chi-square test (p-value)
		Mean	SD	Mean	SD	Mean	SD	
Outcome	Follow-up days	1324.7	280.0	1280.5	315.9	1246.4	378.5	0.3767
	Total death (1,0) (%)	1/66 (0.015)		2/67 (0.029)		1/69 (0.014)		0.7990
	Cardiovascular events (1,0) (%)	1/66 (0.015)		1/78 (0.014)		7/63 (0.10)		0.0179
	Cancer occurrence (1,0) (%)	2/65 (0.030)		2/67 (0.029)		1/69 (0.014)		0.7995
Background characteristics and office BP profiles	Sex (M/F)	19/48 (0.284)		33/36 (0.478)		30/40 (0.429)		0.0553
	Age (years)	52.9	11.5	58.7	10.3	59.7	10.7	0.0005
	Body Mass Index	23.3	2.4	24.9	3.2	25.4	2.6	0.0001
	Anti-hypertension medication (1,0)	10/57 (0.149)		29/40 (0.420)		39/31 (0.5579		<0.0001
	Classes of drug therapy (0/1/2/3/4)	57/5/2/2/1		40/15/4/6/2		30/19/4/13/1		0.0097
	Groups of treatment (0/1/2)	57/0/10		20/20/29		1/30/39		<0.0001
	Office BP (systolic BP) (mmHg)	124.5	18.4	142.1	21.5	152.8	22.7	<0.0001
	Office BP (diastolic BP) (mmHg)	77.9	10.4	87.4	11.8	90.8	13.4	<0.0001
	Office BP (pulse) (bpm)	72.0	9.3	72.6	10.1	70.9	11.1	0.6010
Grading of Risk for cardiovascular outcome from a viewpoint of 7-day/24-hour ABP monitoring	Systolic BP MESOR HT (1,0) (%)	0/67 (0)		0/69 (0)		41/29 (58.6)		<0.0001
	Diastolic BP MESOR HT (1,0) (%)	0/67 (0)		3/66 (4.3)		24/46 (34.3)		<0.0000
	EPP >60mmHg (1,0) (%)	0/67 (0)		1/68 (1.4)		19/51 (27.1)		<0.0001
	CHAT Systolic BP (1,0) (%)	6/61 (9.0)		16/53 (23.2)		10/60 (14.3)		0.0681
	CHAT Diastolic BP (1,0) (%)	1/66 (1.5)		5/64 (7.2)		5/65 (7.1)		0.2336
	Ecphasia (1,0) (%)	0/67 (0)		0/69 (0)		3/67 (4.3)		0.0520
	DHRV (1,0) (%)	0/67 (0)		0/69 (0)		1/69 (1.4)		0.3768
	anyVVA (1,0) (%)	6/61 (9.0)		20/49 (29.0)		57/13 (81.4)		<0.0001
	Number of VVAs (0/1/2/3/4)	61/6/0/0/0		49/15/5/0/0		13/21/28/6/2		<0.0001
	Marked non-dipper (0/1/2/3)	36/31/0/0		22/34/12/1		15/25/28/2		<0.0001
	Marked 24h ABP HT (0/1/2/3)	65/2/0/0		12/25/31/1		1/1/28/40		<0.0001
Indices of 7-day/24-hr ABP monitoring	Systolic BP MESOR (mmHg)	112.8	6.8	124.7	8.2	141.3	11.7	<0.0001
	Diastolic BP MESOR (mmHg)	70.7	5.1	77.6	6.5	83.5	8.1	<0.0001
	Heart Rate MESOR (bpm)	70.1	6.5	70.0	7.4	68.6	6.7	0.3420
	Systolic BP Amplitude (mmHg)	26.1	8.0	25.3	11.4	23.6	11.8	0.3568
	Diastolic BP Amplitude (mmHg)	17.8	5.1	16.4	6.9	15.4	7.8	0.1163
	Heart Rate Amplitude (bpm)	16.7	5.5	15.9	6.7	15.1	6.3	0.3299
	Systolic BP Acrophase (hour:minute)	14:16	1:06	13:33	2:23	13:34	3:14	0.1538
	Diastolic BP Acrophase (hour:minute)	14:06	1:09	13:33	1:53	13:44	1:49	0.1586
	Heart Rate Acrophase (hour:minute)	14:05	2:16	14:01	1:44	14:12	2:32	0.8898
	Systolic BP PTE (%)	0.3	1.0	3.1	5.5	32.6	30.4	<0.0001
	Diastolic BP PTE (%)	0.4	1.3	3.1	7.1	16.4	21.4	<0.0001
	Heart Rate PTE (%)	0.7	2.5	0.7	2.8	0.2	0.9	0.2632
Characteristics of the 1st day 24-hour ABP	24-hour average of Systolic BP	116.2	7.6	128.1	12.5	144.7	13.5	<0.0001
	24-hour average of Pulse Pressure	43.5	5.3	48.4	7.8	59.4	10.3	<0.0001
	SD of pulse over 24 hours	12.5	4.9	11.3	4.9	9.7	3.4	0.0015
	1st day effect BP≧8 (1,0)	5/62 (0.075)		13/56 (0.188)		19/51 (0.271)		0.0108
	1st daytime effect BP≧10 (1,0)	17/50 (0.254)		20/49 (0.290)		15/55 (0.214)		0.5908
	1st night effect BP≧5 (1,0)	14/53 (0.209)		22/47 (0.319)		27/43 (0.386)		0.0773
	Dipping ratio of Systolic BP	21.9	7.1	17.9	9.5	13.1	10.8	<0.0001
	Dipping ratio of Diastolic BP	23.0	8.9	16.8	9.6	16.3	11.7	0.0002
	Dipping ratio of Heart rate	16.4	13.5	13.8	12.4	14.8	9.3	0.4503
Indices of arterial stiffness	baPWV (cm/sec)	1369.2	311.9	1548.1	271.1	1682.1	368.0	<0.0001
	ABI	1.08	0.07	1.08	0.10	1.09	0.10	0.8379
Depressive mood and quality of life	GDS-15 (0-15)	3.5	2.7	4.9	2.8	4.6	3.6	0.0272
	Subjective QOL of health (%)	69.3	16.1	64.2	19.6	61.4	16.4	0.0322
	Subjective QOL of mood (%)	72.6	17.6	68.5	19.8	66.0	17.1	0.1101
	Subjective QOL of family relationships (%)	82.8	18.3	79.7	19.5	77.6	19.8	0.2853
	Subjective QOL of relationships with friends (%)	81.1	16.6	77.0	19.7	76.5	18.2	0.2850
	Subjective QOL of financial satisfaction (%)	64.3	24.7	62.7	27.6	62.6	22.1	0.9081
	Subjective QOL of life (%)	71.7	23.0	71.1	22.6	68.2	19.9	0.6089
	Subjective QOL of happiness (%)	76.4	19.6	75.5	19.8	70.1	19.0	0.1307
Sleeping habits	Time of getting-up (hour:minute)	6:09	0:48	6:05	0:49	6:08	0:47	0.8948
	Time of falling asleep (hour:minute)	22:47	1:05	22:19	1:02	22:26	0:49	0.0233
	sleep duration (hour:minute)	7:30	0:52	8:07	1:08	7:59	0:56	0.0065
	Feeling of good sleep (0,1)	1.2	0.4	1.2	0.4	1.3	0.5	0.3430
	Time for falling asleep (minute)	17.5	11.5	19.9	16.9	24.0	22.7	0.1492
	Duration of snoring (month)	4.2	2.5	4.7	2.6	4.4	2.6	0.7885
	Waking not rested (0,1)	1.655	0.739	1.672	0.778	1.627	0.720	0.9365
Dietary habits	Salt intake a day (g)	13.0	1.8	14.2	2.4	13.4	1.8	0.0099
	Eating cooked rice a day (cup)	2.4	1.0	2.5	0.8	2.5	0.8	0.4411
	Drinking miso soup a day (cup)	1.3	0.7	1.5	0.7	1.6	0.7	0.0580
	Vegetables	10.3	7.3	9.9	4.6	8.5	3.2	0.1970
	Fruits	4.1	3.1	4.8	3.3	5.2	4.2	0.3195

(5) their PTE (percentage time elevation above time-specified reference values qualified by gender and age) was higher ($p<0.0001$). They were more likely to have (6) SBP and (7) DBP MESOR-HT ($p<0.0001$), (8) EPP ($p<0.0001$), or (9) any VVA ($p<0.0001$) (Table 17.8). (10) Their score on the subjective QOL of health was also lower ($p=0.0322$).

In conclusion, results herein show that in addition to a higher 24-hour BP average, an abnormal circadian BP profile and among-day BP variability constitute risk factors impacting CV outcome corroborating the need for prolonged, preferably continuous ABPM.

References

1. Shimada K, Kawamoto A, Matsubayashi K, et al. Silent cerebrovascular disease in the elderly. Correlation with ambulatory pressure. Hypertension 1990;16:692–9.
2. Verdecchia P, Porcellati C, Schillaci G, et al. Ambulatory blood pressure. An independent predictor of prognosis in essential hypertension. Hypertension 1994;24:793–801.
3. Otsuka K, Cornelissen G, Halberg F. Predictive value of blood pressure dipping and swinging with regard to vascular disease risk. Clin Drug Investig 1996;11:20–31.
4. Otsuka K, Cornélissen G, Halberg F, et al. Excessive circadian amplitude of blood pressure increases risk of ischemic stroke and nephropathy. J Med Eng Technol 1997;21:23–30.
5. Murakami S, Otsuka K, Kubo Y, et al. Repeated ambulatory monitoring reveals a Monday morning surge in blood pressure in a community-dwelling population. Am J Hypertens 2004;17:1179–83.
6. Shinagawa M, Otsuka K, Murakami S, et al. Seven-day (24-h) ambulatory blood pressure monitoring, self-reported depression and quality of life scores. Blood Press Monit 2002;7:69–76.
7. Otsuka K, Yamanaka G, Shinagawa M, et al. Chronomic community screening reveals ~31% depression, elevated blood pressure and infradian vascular rhythm alteration. Biomed Pharmacother 2004;58:48s–55.
8. Otsuka K, Okajima K, Yamanaka T, et al. Ageing and the novelty pressor effect in men on the first day of 7-day/24-hour ambulatory blood pressure monitoring. J Am Geriatr Soc 2014;62:1602–5.
9. Otsuka K, Cornelissen G. Alcohol consumption and vascular variability anomalies (VVAs) and disorders (VVDs) detected by C-ABPM. Open Nutraceut J 2012;5(Suppl 1-M1):81–9.
10. Palatini P, Frigo G, Bertolo O, et al. Validation of the A&D TM-2430 device for ambulatory blood pressure monitoring and evaluation of performance according to subjects' characteristics. Blood Press Monit 1998;3:255–60.
11. Okajima K, Yamanaka G, Oinuma S, et al. Even mild depression is associated with among-day blood pressure variability, including masked non-dipping assessed by 7-d/24-h ambulatory blood pressure monitoring. Clin Exp Hypertens 2015;37(5):426–32. doi:10.3109/10641963.2015.1013114.
12. Halberg F, Powell D, Otsuka K, et al. Diagnosing vascular variability anomalies, not only MESOR-hypertension. Am J Physiol Heart Circ Physiol 2013;305:H279–94.
13. Richards J, Gumz ML. Mechanism of the circadian clock in physiology. Am J Physiol Regul Integr Comp Physiol 2013;304:R1053–64.
14. Curtis AM, Cheng Y, Kapoor S, et al. Circadian variation of blood pressure and the vascular response to asynchronous stress. Proc Natl Acad Sci USA 2007;104:3450–5.
15. Halberg F. Chapter 13: Biologic rhythms, hormones, and aging. In: Vernadakis A, Timiras P, editors. Hormones in development and aging. New York: Spectrum; 1982. p. 451–76.
16. Saito K, Koyama A, Yoneyama K, et al., editors. A recent advance in time series analysis by maximum entropy method: applications to medical and biological sciences. Sapporo: Hokkaido University Press; 1994. p. 406.
17. Otsuka K, Okajima K, Oinuma S, et al. Aging and circadian disruption of blood pressure observed using 7-day/24-hour ambulatory blood pressure monitoring. J Am Geriatr Soc 2014;62:2213–5.
18. Otsuka K. Seven-day (24-hour) ambulatory blood pressure monitoring and frequently observed day-to-day difference in the elderly. Jpn J Geriatr 2009;46:488–92.
19. Watanabe Y, Halberg F, Otsuka K, Cornelissen G. Toward a personalized chronotherapy of high blood pressure and a circadian overswing. Clin Exp Hypertens 2013;35:257–66.
20. Bingham C, Arbogast B, Cornelissen G, et al. Inferential statistical methods for estimating and comparing cosinor parameters. Chronobiologia 1982;9:397–439.
21. Cornelissen G, Halberg F, Hawkins D, et al. Individual assessment of antihypertensive response by self-starting cumulative sums. J Med Eng Technol 1997;21:111–20.

22. Task Force for the Management of Arterial Hypertension of the European Society of Hypertension and of the European Society of Cardiology. 2013 ESH/ESC Guidelines for the Management of Arterial Hypertension. J Hypertens 2013;31:1281–357.
23. Hermida RC, Smolensky MH, Ayala DE, et al. 2013 ambulatory blood pressure monitoring recommendations for the diagnosis of adult hypertension, assessment of cardiovascular and other hypertension-associated risk, and attainment of therapeutic goals. Chronobiol Int 2013;30:355–410.
24. Shimamoto K, Ando K, Fujita T, et al. The Japanese Society of Hypertension guidelines for the management of hypertension (JSH 2014). Hypertens Res 2014;37:362–87.
25. Okajima K, Otsuka K, Oinuma S, et al. Aging and within- and between-day variability assessed using 7-day/24-hour ambulatory blood pressure monitoring. J Am Geriatr Soc 2014;62:2440–2.
26. Murakami S, Otsuka K, Kono T, et al. Impact of outdoor temperature on prewaking morning surge and nocturnal decline in blood pressure in a Japanese population. Hypertens Res 2011;34:70–3.
27. Cornelissen G, Halberg F, Otsuka K, Singh RB, Chen CH. Chronobiology predicts actual and proxy outcomes when dipping fails. Hypertension 2007;49:237–9.
28. Halberg F, Cornelissen G, Otsuka K, et al. Extended consensus on need and means to detect vascular variability disorders (VVDs) and vascular variability syndromes (VVSs). Int J Gerontol Geriatr 2008;11:119–46.
29. Otsuka K. Clinical chronomedicine; glocalization in time and space. Tokyo: Nakayama Shoten Co. Ltd.; 2013. p. 325 (in Japanese).

Chapter 18
Case Presentations I. Within-Day BP Variations Observed Using 7-Day/24-Hour Ambulatory BP Monitoring

Abstract In this chapter, we introduce case presentations of the circadian and circaseptan (i.e., about 7 day) profiles of blood pressure (BP), including "ambulatory BP (ABP) normotension," "white-coat hypertension," "BP morning surge," and "Monday BP surge," from a viewpoint of 7-day/24-hour ABP monitoring.

Keywords Ambulatory blood pressure normotension • Chronodesm • Circadian period of systolic BP • Maximum Entropy Method (MEM) • Least-squares analysis • White-coat hypertension syndrome

18.1 A Case Report of "Ambulatory BP Normotension"

An example of "ambulatory blood pressure (ABP) normotension" based on 7-day/24-hour ambulatory blood pressure (ABP) monitoring is this 39-year-old non-obese (BMI = 20.4 kg/m^2) woman (ID number: Tosa0195). She usually goes to bed around 00:00 and gets up around 07:35. She feels like she sleeps well on most days and wakes up well rested in the morning. She does not have a depressive mood and shows a null score on the two-item Patient Health Questionnaire (PHQ-2).

Her office BP was 126/86 mmHg, and her home BP measurements averaged over 30 days were 117.0/77.0 mmHg (morning) and 122.6/80.5 mmHg (evening). Time series of systolic BP (SBP), diastolic BP (DBP), and pulse (HR) are shown in Fig. 18.1. A model consisting of cosine curves with periods of 24 and 12 hours was fitted by least squares to her data [1]. The MESOR (midline estimating statistic of rhythm) estimates of SBP, DBP, and HR from the 7-day/24-hour ABP record were 109.7 mmHg, 70.0 mmHg, and 76.0 bpm, respectively.

Average circadian profiles of SBP and HR are shown in Fig. 18.2. Each time-specified point, shown as a closed square, represents the average of measurements at that time of day during the 7 days of monitoring. The depicted line graphs in Fig. 18.2 thus illustrate averaged circadian profiles of SBP (left) and HR (right).

Time-specified reference limits, chronodesms, have been computed for each variable in each group as 90 % prediction limits, derived from ABP monitoring of a non-hypertensive reference population. Thus, Fig. 18.2 depicts not only circadian

K. Otsuka et al., *Chronomics and Continuous Ambulatory Blood Pressure Monitoring*, DOI 10.1007/978-4-431-54631-3_18

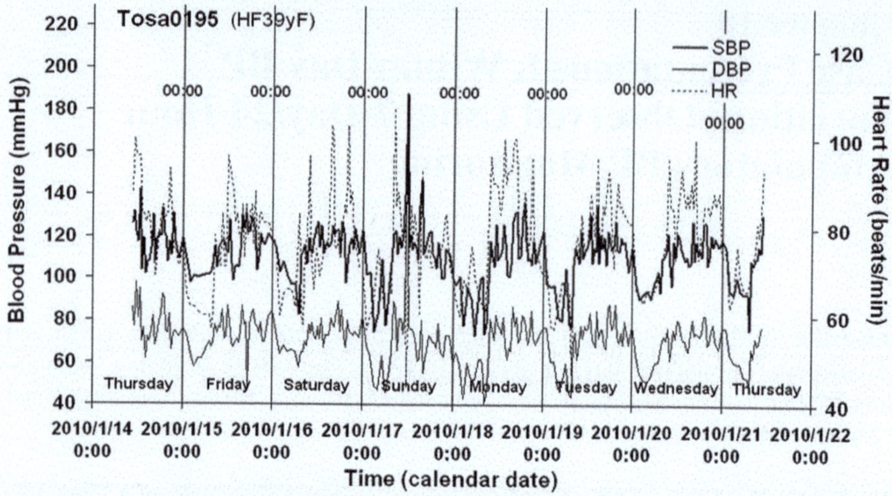

Fig. 18.1 An example of normotension. Time plots of the 7-day/24-hour record of a 39-year normotensive woman: SBP (*thick continuous line*), DBP (*thin continuous line*), and HR (*dashed line*). These variables fluctuate somewhat predictably, following a circadian rhythm, suggesting a regular rest-activity schedule

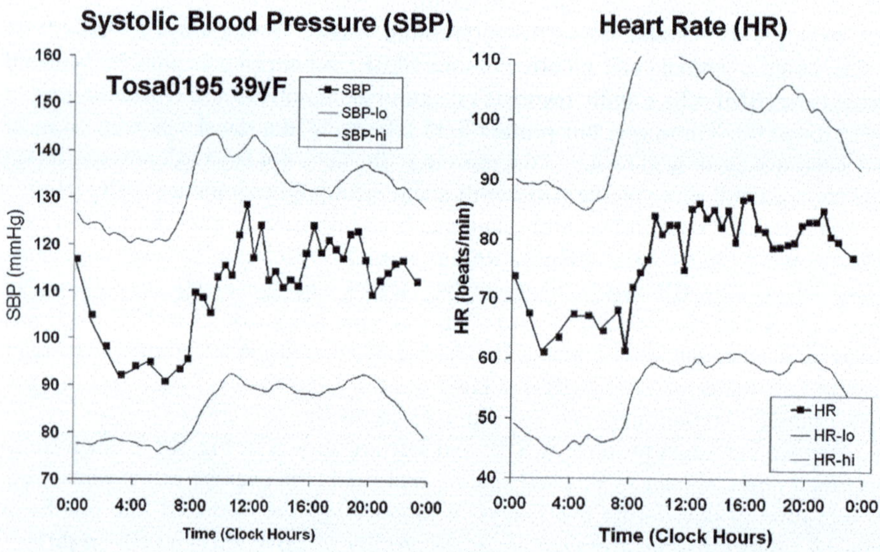

Fig. 18.2 Circadian profiles of systolic BP (*left*) and heart rate (HR) (*right*). 24-hour patterns of systolic BP (*left*) and heart rate (HR) (*right*) of a normotensive woman, averaged over the 7 days of monitoring (*polygonal line* graph with closed *squares*), are shown along with corresponding time-specified 90 % prediction limits (*continuous thin lines*). Both variables vary in the *middle* portion of the respective chronodesms

profiles of the 39-year-old normotensive woman but also the 90 % prediction limits (i.e., upper and lower thresholds) derived from data provided by clinically healthy women of her age group, shown as the two continuous curves. It can be seen that the circadian profiles of SBP and HR of this subject are completely located within the reference limits, drawn in the central portion of the reference range for most of the 24-hour cycle.

Parameters of circadian profiles were calculated by the chronobiological method are. Her double 24-hour amplitudes of SBP and DBP were 24.39 and 17.51 mmHg, respectively, both within acceptable limits. Circadian acrophases of SBP and DBP were 15:23 and 15:28, respectively, both also within the 90 % prediction limits (between ~12:00 and ~16:00).

The circadian period of SBP was estimated by the Maximum Entropy Method (MEM), using the MemCalc software (GMS Co., Tokyo) [2], which requires the data to be equidistant. Outliers were first deleted from the original SBP time series (middle section of Fig. 18.3). Data were then made equidistant by averaging and linear interpolation (bottom section of Fig. 18.3).

The MEM spectrum is shown on the middle left of Fig. 18.4, which shows a prominent circadian component with a period of 1.0033 day, together with two

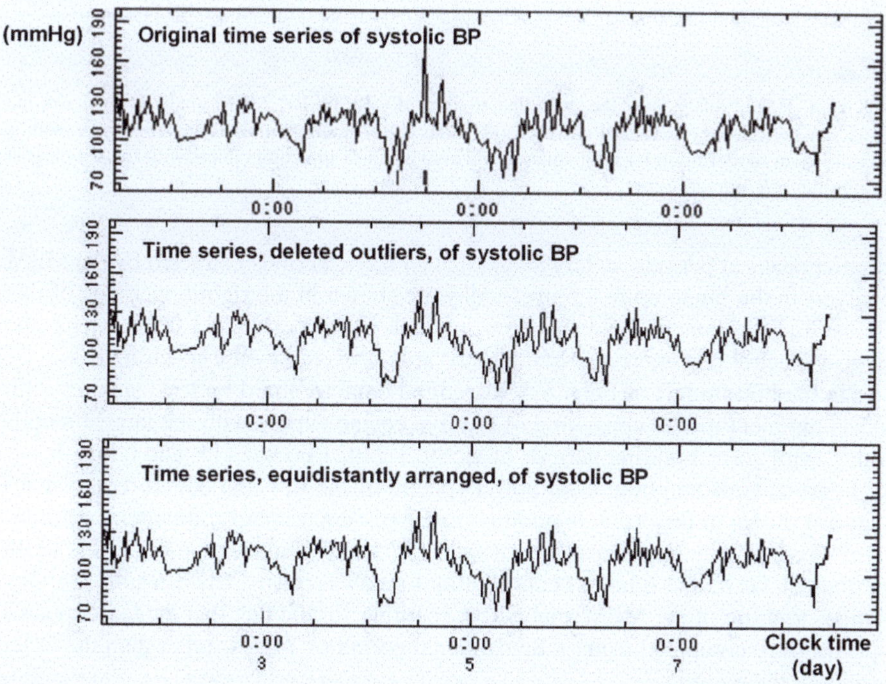

Fig. 18.3 Three versions of the time series of systolic BP. *Upper*: original data of systolic BP. *Middle*: edited data after mathematically excluding outliers. *Lower*: equidistantly arranged (averaged and linearly interpolated) data for analysis by the Maximum Entropy Method, using the MemCalc software (GMS Co., Tokyo)

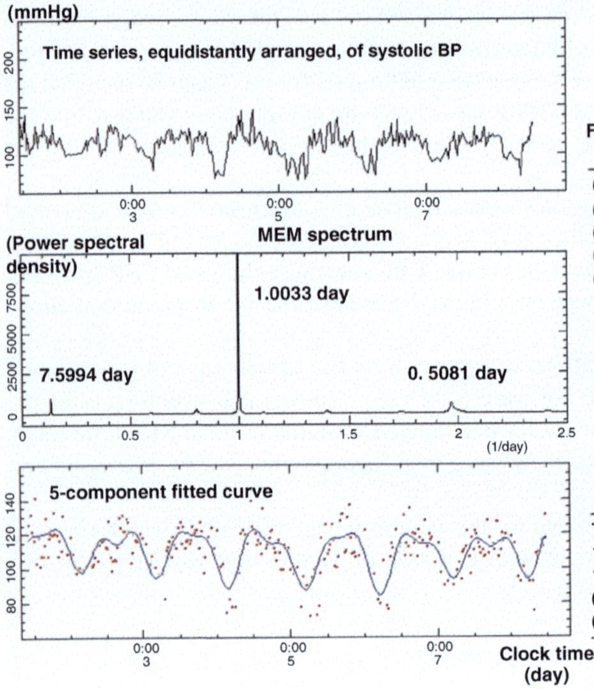

Tosa0195 39yF

Frequency (/ day)	Period (days)	Area
0.132	7.5994	4.7711
0.334	2.9984	0.7412
0.594	1.6841	1.5554
0.800	1.2492	4.4788
0.997	1.0033	74.5997
1.111	0.8997	2.1568
1.399	0.7150	3.8575
1.735	0.5764	2.7067
1.968	0.5081	18.8695
2.406	0.4156	2.9151

Period (days)	Amplitude	Acro-phase
7.599	2.81	2-01:32
1.249	2.92	2-00:57
1.003	13.11	1-16:09
0.715	2.66	1-13:07
0.508	5.75	1-21:32

Fig. 18.4 Period analysis of the circadian rhythm of systolic BP. MEM analysis estimates the circadian period of systolic BP to be 1.0033 day (*left middle part*). Data are shown with a 5-component model fitted by least squares and consisting of the components with the five largest amplitudes (*right lower part*)

smaller peaks at periods of 7.5994 and 0.5081 days. Periods extracted by the MEM analysis in the range up to 2.5 cycles/day are shown in the right upper side of this figure. Ten components are listed here, and that with a period of 1.0033 day has the largest spectral power, as assessed by the "area" covered by the spectral peak.

On the bottom left of Fig. 18.4, the edited data are fitted by least squares with a 5-component model consisting of the five cosine curves with the largest amplitudes, with corresponding periods of 7.599, 1.249, 1.003, 0.715, and 0.508 days.

Least-squares amplitude and power spectra of the raw and edited (equidistant) data are shown in Fig. 18.5. In addition to a weak circaseptan component, a prominent circadian rhythm is readily apparent, three additional harmonic terms (with periods of 12, 8, and 6 hours) contributing to its waveform. These results are very similar to those of the MEM analysis, accounting for the fact that the least-squares spectra were computed using a fundamental period of 7 days rather than the exact length of the record.

Figure 18.6 shows that the results do not differ greatly between the analysis of the original (raw) data and the edited record where outliers were deleted and data linearly interpolated to render the time series equidistant. Only small differences in the relative prominence of the 12-hour and 6-hour components are observed in the least-squares spectra of the raw and edited data. As seen from Fig. 18.7, models

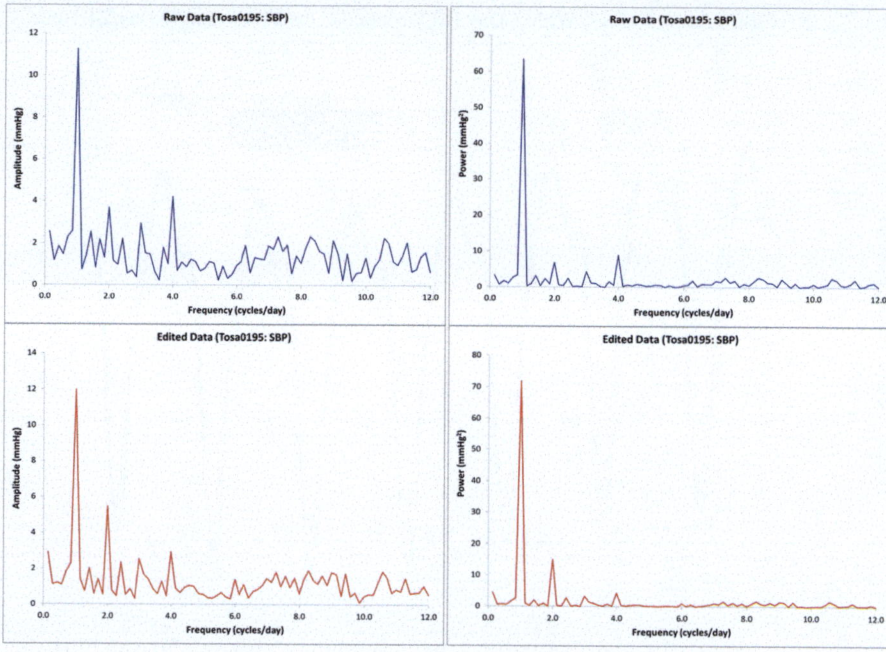

Fig. 18.5 Least-squares amplitude (*left*) and power (*right*) spectra of the original (*top*) and edited (*bottom*) SBP data. Note similarity of results with those of the MEM analysis

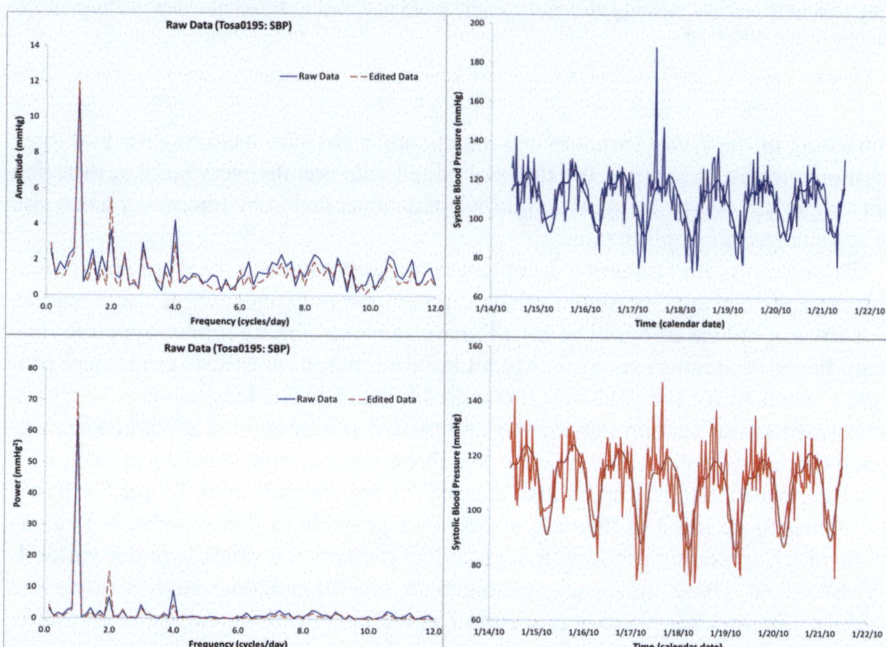

Fig. 18.6 Comparison of least-squares amplitude (*top, left*) and power (*bottom, left*) spectra from raw and edited SBP data shows only small differences. Signals are reconstructed on the basis of the five largest components (with periods of 7 days, 24, 12, 8, and 6 hours) for the raw (*top, right*) and edited (*bottom, right*) data

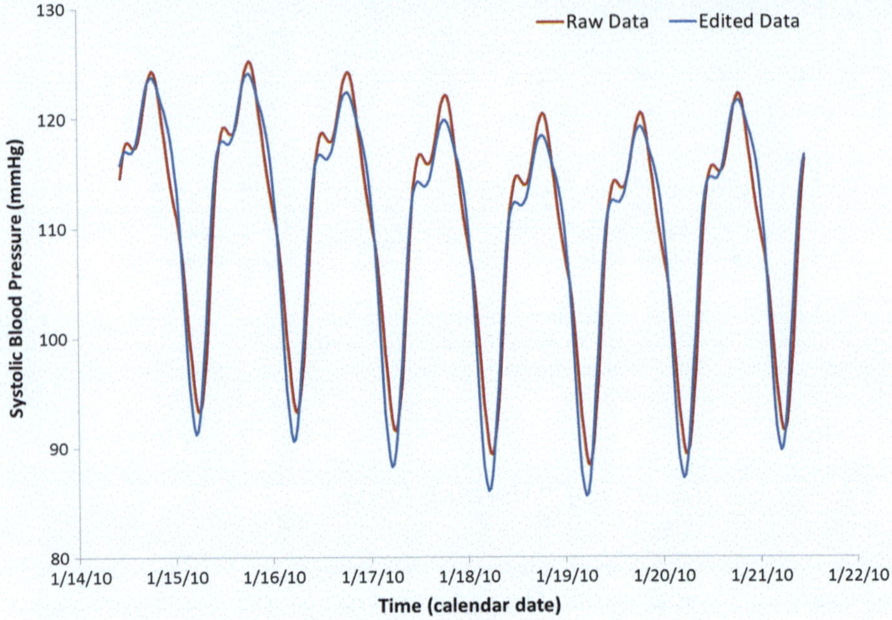

Fig. 18.7 Comparison of 5-component models obtained for the original and edited data shows only small differences, notably during the night when data had to be interpolated because of the change in sampling rate

consisting of the 7-day variation and the circadian rhythm, including the first three harmonic terms, describing the raw and edited data are also very similar, differing only slightly, notably during the night when data needed to be interpolated because of the change in sampling rate.

These results illustrate two complementary approaches to analyze blood pressure records for an assessment of anticipated components such as the circadian and circaseptan variations. The MEM analysis yields a more refined spectral view than the least-squares spectrum. Model building based on MEM results may provide a better fit for the data over the available record. The least-squares spectrum performed while accounting for the anticipated presence of a 24-hour synchronized circadian rhythm and a 7-day synchronized biological week, in turn, provides estimates that may apply also outside the observation span. By further fixing the model to be fitted to the data, it becomes possible to derive reference values from which screening for abnormal circadian patterns identifies vascular variability disorders. These diagnoses, validated in several outcome studies, represent useful guidelines for primary prevention as well as for treatment optimization by timing (chronotherapy).

18.2 Case Report 1 of "White-Coat Hypertension"

In this chapter, we introduce four cases of "white-coat hypertension" from the viewpoint of 7-day/24-hour ambulatory blood pressure (ABP) monitoring.

18.2.1 *Case Report 1 of "White-Coat Hypertension"*

The first example of "white-coat hypertension" is a 40-year-old non-obese (BMI = 22.6 kg/m^2) man (ID number: Tosa0282). His weight was 64.5 kg and height 169 cm. He goes to bed usually around 23:30 and gets up around 06:40. He falls asleep within 10 min and believes he sleeps well most days. Whereas he does not wake up at night and has no disturbances of waking too early, he does not feel rested upon awakening. He does not smoke and usually does not drink alcohol, but his intake of meat is frequent and he seldom eats fish. His intake of vegetables is somewhat below average. He eats eggs regularly every day, but does not drink milk. Once or twice a week, he eats something within 2 hours of going to bed.

He feels healthy now, but does not take care of his health. He does not have a depressive mood and has something to live for. Scores on the Geriatric Depression Score 15 (GDS-15) and PHQ-2 were 5 and 0, respectively. Subjective quality of life (QOL) was 85 % (health) and 65 % (happiness).

This subject does not take any antihypertensive medications. His office BP was 155/94 mmHg and his home BP measurements averaged over 30 days were 113.4/70.8 mmHg (morning) and 105.7/58.5 mmHg (evening). Time series of systolic BP (SBP), diastolic BP (DBP), and heart rate measured by pulse (HR) are shown in Fig. 18.8. A model consisting of cosine curves with periods of 24 and 12 hours was fitted by least squares to his data [1]. The MESOR (midline estimating statistic of rhythm) estimates of SBP, DBP, and HR from the 7-day/24-hour ABP record were 114.6 mmHg, 71.4 mmHg, and 74.1 bpm, respectively.

Average circadian profiles of SBP and HR of this subject are shown in Fig. 18.9. They are well within the reference limits despite the episodic BP elevation (white-coat effect) on the first day.

The circadian profiles are characterized by their double 24-hour amplitudes of 30.55 mmHg (SBP) and 25.16 mmHg (DBP), both within acceptable limits, and by their 24-hour acrophases of 14:26 (SBP) and 14:32 (DBP), also within acceptable limits (between ~12:00 and ~16:00).

The circadian period of SBP was estimated by the Maximum Entropy Method (MEM), using the MemCalc software (GMS Co., Tokyo), which requires the data to be equidistant. Outliers were deleted from the original SBP time series (middle section of Fig. 18.10). Data were then made equidistant (bottom section of Fig. 18.10).

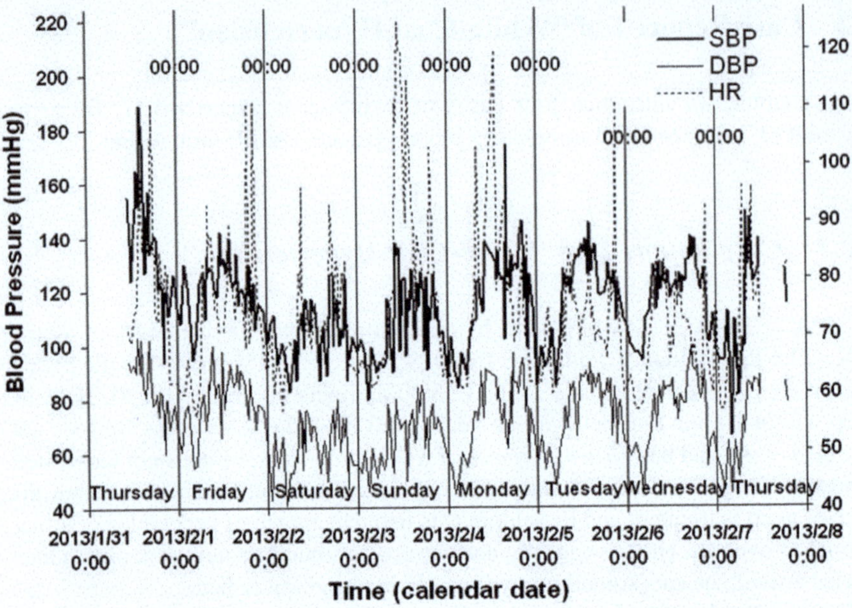

Fig. 18.8 An example of white-coat hypertension. Time plots of the 7-day/24-hour records of SBP (*thick continuous line*), DBP (*thin continuous line*), and HR (*dashed line*) of a 40-year-old man with white-coat hypertension. When the 7-day/24-hour ABP monitoring session was started at 11:00 in front of a doctor on January 31, 2013, his blood pressure and pulse were 155/94 mmHg and 71 bpm, suggesting mild hypertension, but measurements dropped thereafter within the normal range, where they fluctuated periodically, following a circadian rhythm, suggesting a regular rest-activity schedule. High systolic BP on the first day of monitoring can readily be seen

The MEM spectrum is shown on the middle left of Fig. 18.11, which shows a prominent circadian component with a period of 1.0213 day, together with three smaller peaks at periods of 5.9442, 3.8640, and 0.4761 days. Periods extracted by the MEM analysis in the range up to 2.5 cycles/day are shown in the right upper side of this figure. Ten components are listed here, and that with a period of 1.0213 day has the largest spectral power, as assessed by the "area" covered by the spectral peak.

On the bottom left of Fig. 18.11, the edited data are fitted by least squares with a 5-component model consisting of the five cosine curves with the five largest amplitudes, with corresponding periods of 5.944, 3.864, 1.358, 1.021, and 0.476 days (right lower part of Fig. 18.11).

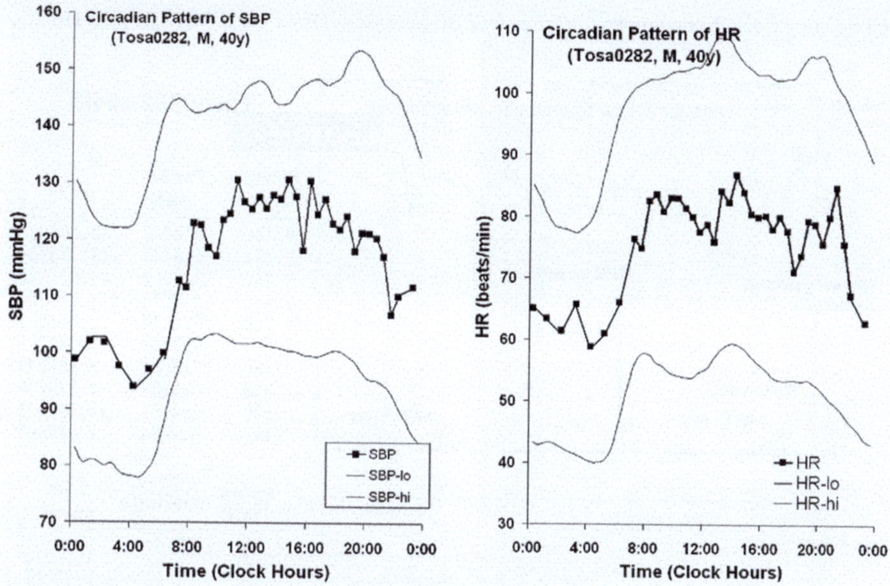

Fig. 18.9 Circadian profiles of systolic BP (*left*) and heart rate (HR) (*right*). 24-hour patterns of systolic BP (*left*) and heart rate (HR) (*right*) of this subject with white-coat hypertension, averaged over the 7 days of monitoring (*polygonal line* graph with closed *squares*), are shown along with time-specified 90 % prediction limits (*continuous thin lines*). Both variables vary in the *middle* portion of their respective chronodesms

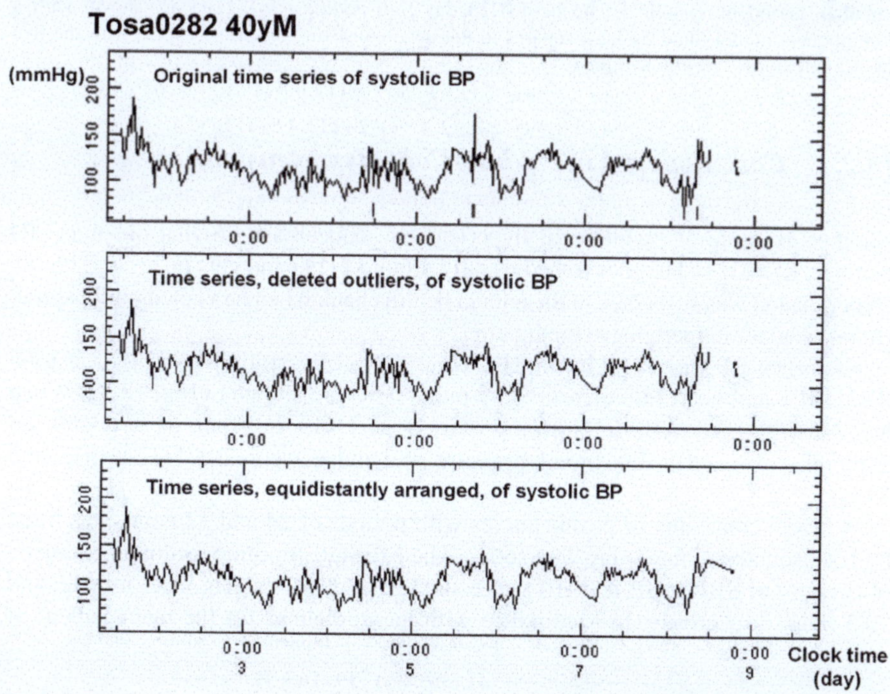

Fig. 18.10 Three versions of the time series of systolic BP. *Upper*: original data of systolic BP. *Middle*: edited data after mathematically excluding outliers. *Lower*: equidistantly arranged time series for analysis by the Maximum Entropy Method, using the MemCalc software (GMS Co., Tokyo)

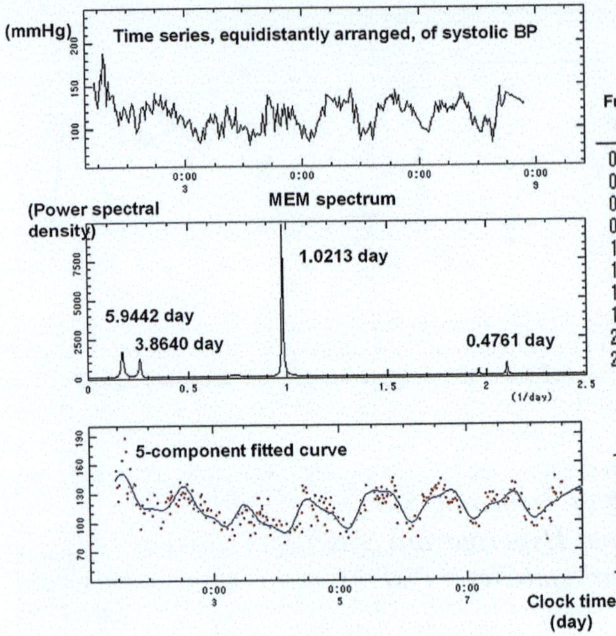

Tosa0282 40yM

Frequency (/ days)	Period (days)	Area
0.168	5.9442	39.304030
0.259	3.8640	23.353234
0.736	1.3582	8.811140
0.979	1.0213	116.643054
1.232	0.8119	3.051327
1.441	0.6939	2.242762
1.600	0.6251	0.597240
1.959	0.5104	4.573290
2.100	0.4761	11.738071
2.485	0.4024	1.576665

Period (days)	Amplitude	Acro-phase
5.944	8.44	1-12:02
3.864	6.71	1-15:29
1.358	3.43	1-14:01
1.021	14.54	1-12:30
0.476	4.55	1-14:06

Fig. 18.11 Period analysis of the circadian rhythm of systolic BP. MEM analysis estimates the circadian period of systolic BP to be 1.0213 day (*left middle part*). Data are shown with a 5-component model fitted by least squares to the data, consisting of cosine curves with the five largest amplitudes (*right lower part*)

18.2.2 Case Report 2 of "White-Coat Hypertension"

Another example of "white-coat hypertension" concerns a 76-year-old man (ID number: KOts078). He goes to bed usually around 21:30 and gets up around 05:00. Hypertension was suspected at the annual health check when he visited our hospital. He takes no antihypertensive medication.

His office BP was 185/110 mmHg. Time series of systolic BP (SBP), diastolic BP (DBP), and heart rate measured by pulse (HR) are shown in Fig. 18.12. When his 7-day/24-hour ABP monitoring started at 15:13 in front of a doctor on September 19, 2001, his blood pressure and pulse were 203/111 mmHg and 76 bpm.

A model consisting of cosine curves with periods of 24 and 12 hours was fitted by least squares to the 7-day data series. The MESOR (midline estimating statistic of rhythm) of SBP, DBP, and HR was estimated as 135.1 mmHg, 90.1 mmHg, and 70.3 bpm, respectively. In accordance with the guidelines for the management of

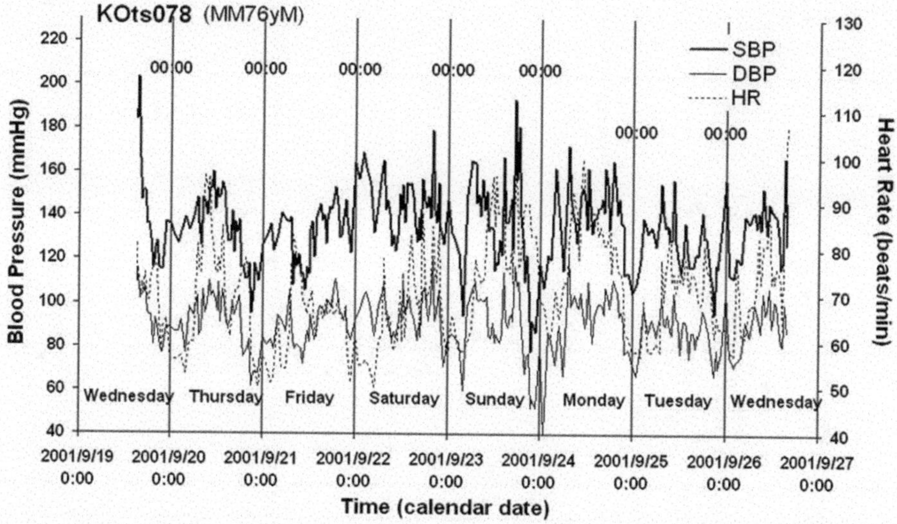

Fig. 18.12 An example of white-coat hypertension syndrome. Time plots of the 7-day/24-hour records of SBP (*thick continuous line*), DBP (*thin continuous line*), and HR (*dashed line*) of a 76-year-old man with white-coat hypertension. When the 7-day/24-hour ABP monitoring session was started at 15:13 in front of a doctor, his blood pressure and pulse were 203/111 mmHg and 76 bpm, coinciding with grade 3 hypertension, but these values decreased thereafter, getting close to the normal range where they fluctuate rhythmically as a function of his rest-activity cycle. Nocturnal rise is observed on Friday night, and a non-dipping pattern is seen on Tuesday night (SBP and DBP day-night ratios: −20.6 % and 1.5 %, respectively). A low day-night ratio also prevails for the entire profile, suggesting the presence of "masked non-dipping"

high blood pressure in Japan, published in 2014 [3], hypertension is defined for 24-hour ABP as ≧130 (SBP) and/or ≧80 mmHg (DBP). Thus, hypertension is diagnosed in this case. However, the circadian profiles of SBP and HR of this subject are located within the 90 % chronodesmic limits most of the time, as depicted in Fig. 18.13 along the 24-hour scale. Thus, this case should be diagnosed as white-coat hypertension syndrome [4].

The double amplitudes of SBP and DBP estimated from the fitted model are 14.34 and 14.88 mmHg, respectively, both within acceptable limits. The corresponding circadian acrophases of SBP and DBP were 11:12 and 12:19, respectively, suggesting a slight degree of phase advance in SBP along with an aging effect [5].

The circadian period of SBP was estimated by the Maximum Entropy Method (MEM), using the MemCalc software (GMS Co., Tokyo), which requires the data to

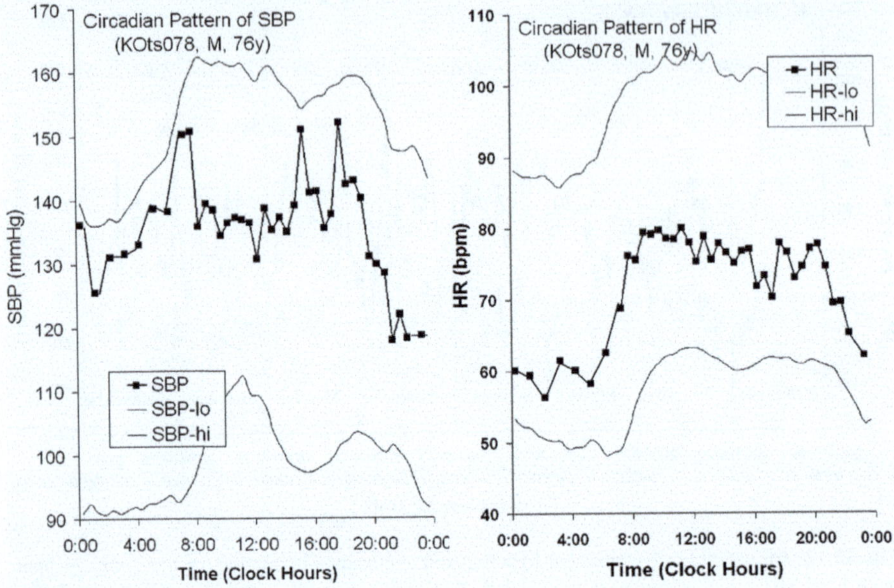

Fig. 18.13 Circadian profiles of systolic BP (*left*) and heart rate (HR) (*right*). 24-hour patterns of systolic BP (*left*) and heart rate (HR) (*right*) of this subject with white-coat hypertension, averaged over the 7 days of monitoring (*polygonal line* graph with *closed squares*), are shown along with time-specified 90 % prediction limits (*continuous thin lines*). Although both patterns are located well within the acceptable limits, the SBP profile is biased toward the *upper part* of the chronodesm, and a mild morning surge is also observed. This subject's BP is particularly high at night, close to the *upper* limit of the chronodesm, in keeping with a non-dipping pattern

be equidistant. Outliers were deleted from the original SBP time series (middle section of Fig. 18.14). Data were then made equidistant (bottom section of Fig. 18.14).

The MEM spectrum is shown on the middle left of Fig. 18.15, which shows a prominent circasemidian component with a period of 0.5133 day, together with three smaller peaks at periods of 2.4326, 0.9963, and 0.8223 days. Periods extracted by the MEM analysis in the range up to 2.5 cycles/day are shown in the right upper side of this figure. Twelve components are listed here, and that with a period of 0.5133 day has the largest spectral power, as assessed by the "area" covered by the spectral peak.

On the bottom left of Fig. 18.4, the edited data are fitted by least squares with a 5-component model consisting of the five cosine curves with the five largest

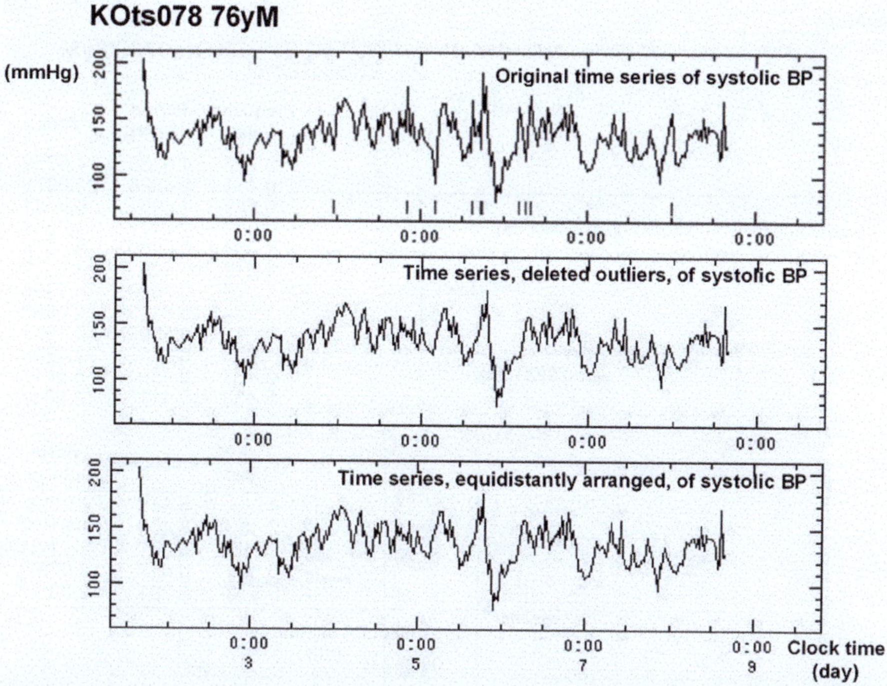

Fig. 18.14 Three versions of the time series of systolic BP. *Upper*: original data of systolic BP. *Middle*: edited data after mathematically excluding the outliers. *Lower*: equidistantly arranged time series for analysis by the Maximum Entropy Method, using the MemCalc software (GMS Co., Tokyo)

amplitudes, with corresponding periods of 4.604, 2.433, 0.996, 0.822, and 0.513 days (right lower part of Fig. 18.15). The most prominent component with a period of 0.513 day cannot be readily seen in the fitted mode, contrasting with results from the MEM spectral analysis.

In conclusion, the clinical diagnosis derived from the 7-day/24-hour ABP monitoring consists of (1) a prominent white-coat effect, (2) a mild ABP hypertension, (3) a "masked non-dipping" hypertension, and (4) a circadian disruption of systolic BP. Accordingly, we started to treat by antihypertensive medicine with a calcium channel blocker (CCB) (benidipine, 4 mg, once daily), together with guidance regarding lifestyle, focusing especially on an erratic sleeping habit.

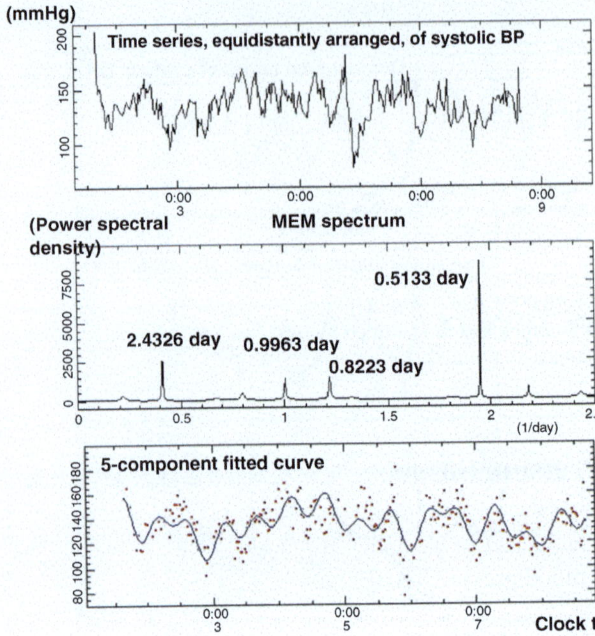

KOts078 76yM

Frequency (/ day)	Period (days)	Area
0.217	4.6039	14.9789
0.411	2.4326	29.0975
0.669	1.4951	6.4441
0.801	1.2477	14.4688
1.004	0.9963	18.4316
1.216	0.8223	24.0327
1.326	0.7539	5.1496
1.620	0.6173	0.6855
1.812	0.5520	2.1971
1.948	0.5133	39.7723
2.181	0.4584	8.6617
2.433	0.4111	12.5037

Period (days)	Amplitude	Acro-phase
4.604	6.444	4-18:35
2.433	8.406	1-21:32
0.996	6.081	2-11:51
0.822	6.017	2-09:03
0.513	8.701	2-03:17

Fig. 18.15 Period analysis of the circadian rhythm of systolic BP. MEM analysis shows a weak circadian component for systolic BP with a period of 0.9963 day (*left middle part*). Data are shown with a 5-component model fitted by least squares to the data, consisting of cosine curves with the five largest amplitudes (*right lower part*)

18.2.3 *Case Report 3 of "White-Coat Hypertension"*

The third example of "white-coat hypertension" is a 77-year-old non-obese (BMI = 23.0 kg/m^2) man (ID number: KOts063). He goes to bed usually around 22:00 and gets up around 07:00. He takes benidipine (4 mg daily) as antihypertensive medication at about 08:00 after breakfast. He does not have a depressive mood, his scores on the GDS-15 and PHQ-2 scales being 5 and 0 points, respectively. His subjective quality of life (QOL) of health was fairly low, however, at 35 %, and his subjective QOL of happiness was 65 %.

His office BP was 148/94 mmHg. Time series of systolic BP (SBP), diastolic BP (DBP), and heart rate measured by pulse (HR) are shown in Fig. 18.16. When starting his 7-day/24-hour ABP monitoring at 13:00 in front of a doctor on March 30, 2001, his blood pressure and pulse were 183/93 mmHg and 53 bpm. At the time of his 7-day/24-hour ABP monitoring, he attended the Shogi game at his town on Sunday and won his game 4 days, from Sunday to Wednesday, but lost the championship on Thursday after a close competition. Thus, this time series includes not only a white-coat phenomenon on the first day but also mental stress related to the close competition lasting for 5 days.

The circadian profiles of SBP and HR of this subject are shown in Fig. 18.17. SBP differs greatly between day and night, exceeding the normal range for the

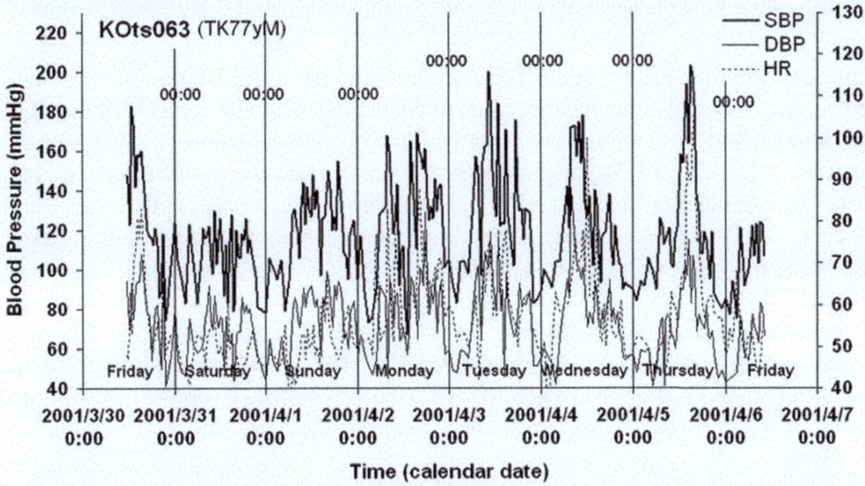

Fig. 18.16 Another example of white-coat hypertension syndrome complicated by circadian overswinging hypertension. Time plots of the 7-day/24-hour records of SBP (*thick continuous line*), DBP (*thin continuous line*), and HR (*dashed line*) in a 77-year-old man. When the 7-day/24-hour ABP monitoring was started at 13:00 in front of a doctor on March 30, 2001, his blood pressure and pulse were 183/93 mmHg and 53 bpm, suggesting grade 3 hypertension. Values were lower the next day, within the acceptable range. Differences of BPs between day and night are large from the 4th to 7th day. Data fluctuate rhythmically, following mostly his rest-activity cycle

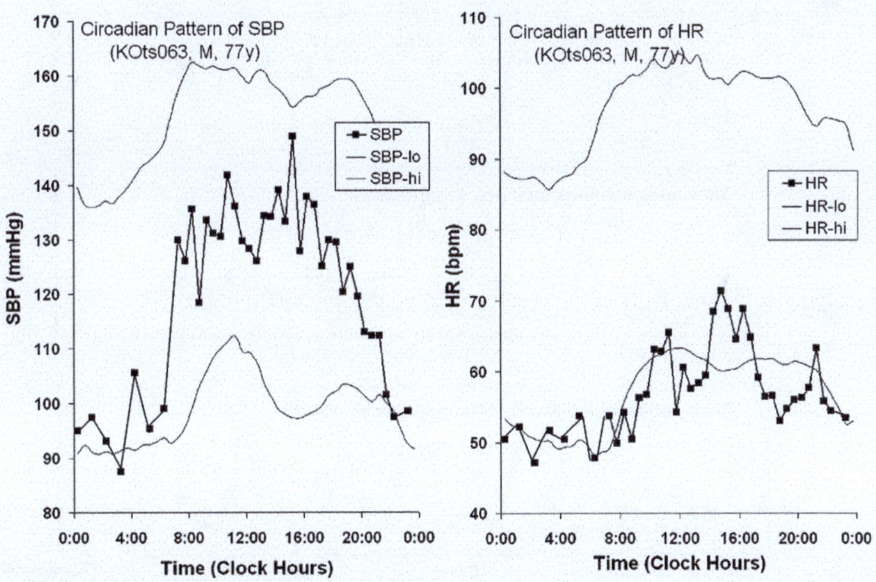

Fig. 18.17 Circadian profiles of systolic BP (*left*) and heart rate (HR) (*right*). 24-hour patterns of systolic BP (*left*) and heart rate (HR) (*right*) of this subject with white-coat hypertension, averaged over the 7 days of monitoring (*polygonal line* graph with *closed squares*), are shown along with time-specified 90 % prediction limits (*continuous thin lines*). Within the chronodesms, SBP swings greatly and HR varies only a little

double 24-hour amplitude. The MESOR, estimated from the 2-component model fitted to the 7-day/24-hour ABP record, of SBP, DBP and HR was 116.8 mmHg, 71.5 mmHg, and 56.2 bpm, respectively. The circadian acrophase of SBP and DBP was 13:13 and 13:34, respectively. All parameters were within acceptable limits. The double amplitude of SBP and DBP, however, was excessive, reaching 44.92 and 30.19 mmHg, respectively. Such an over-swinging circadian pattern, characterized by an excessive amplitude, is called circadian hyper-amplitude-tension (CHAT).

The circadian period of SBP was estimated by the Maximum Entropy Method (MEM), using the MemCalc software (GMS Co., Tokyo), which requires the data to be equidistant. Outliers were deleted from the original SBP time series (middle section of Fig. 18.18). Data were then made equidistant (bottom section of Fig. 18.18).

The MEM spectrum is shown on the middle left of Fig. 18.19, which displays a prominent circadian component with a period of 1.0056 day, together with three smaller peaks at periods of 6.3192, 0.5397, and 0.4756 days. Periods extracted by the MEM analysis in the range up to 2.5 cycles/day are shown in the right upper side

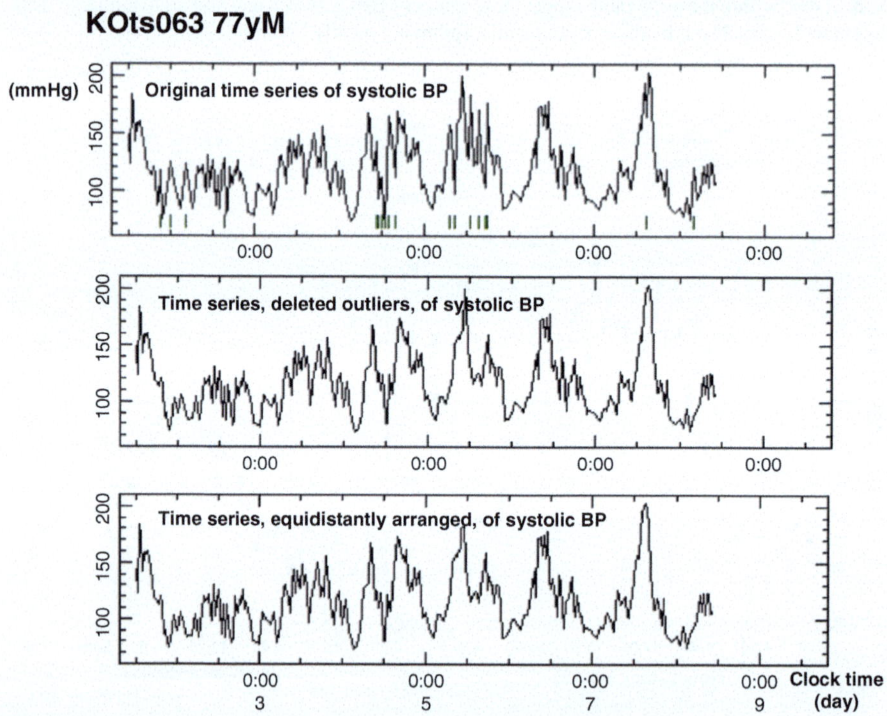

Fig. 18.18 Three versions of the time series of systolic BP. *Upper*: original data of systolic BP. *Middle*: edited data after mathematically excluding outliers. *Lower*: equidistantly arranged time series for analysis by the Maximum Entropy Method, using the MemCalc software (GMS Co., Tokyo)

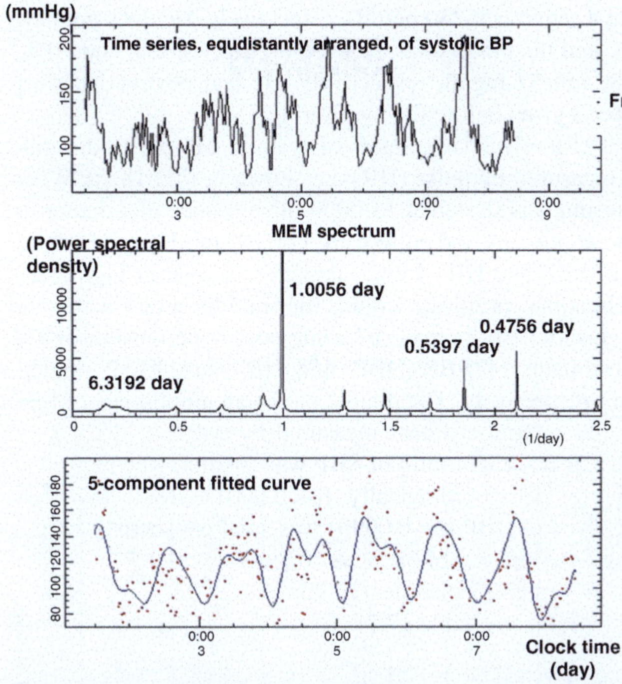

KOts063 77yM

Frequency (/ day)	Period (days)	Area
0.158	6.3192	33.40
0.219	4.5716	22.88
0.489	2.0467	12.86
0.705	1.4175	18.29
1.906	1.1034	52.52
1.994	1.0056	234.90
1.287	0.7772	18.54
1.477	0.6770	18.34
1.692	0.5909	16.92
1.853	0.5397	32.21
2.103	0.4756	33.74
2.263	0.4419	6.55
2.478	0.4035	12.40

Period (days)	Amplitude	Acro-phase
6.319	9.01	5-12:08
1.103	9.00	1-21:18
1.006	26.64	1-13:13
0.540	8.94	1-14:26
0.476	9.11	1-12:13

Fig. 18.19 Period analysis of the circadian rhythm of systolic BP. MEM analysis estimates the circadian period of systolic BP to be 1.0056 day (*left middle part*). Data are shown with a 5-component model fitted by least squares to the data, consisting of cosine curves with the five largest amplitudes (*right lower part*)

of this figure. Thirteen components are listed here, and that with a period of 1.0056 day has the largest spectral power, as assessed by the "area" covered by the spectral peak.

On the bottom left of Fig. 18.19, the edited data are fitted by least squares with a 5-component model consisting of the five cosine curves with the five largest amplitudes, with corresponding periods of 6.319, 1.103, 1.0068, 0.540, and 0.476 days (right lower part of Fig. 18.19).

18.2.4 Case Report 4 of "White-Coat Hypertension"

The fourth case of "white-coat hypertension" is a 61-year-old man (ID number: KOts001). He goes to bed usually at about 23:00 and gets up at about 07:40. Hypertension is treated with a long-acting nifedipine tablet (20 mg daily) as antihypertensive medication taken at about 08:00 after breakfast. Plasma renin activity and aldosterone concentration were 6.1 ng/ml/hour (normal range: 0.34–3.5 ng/ml/

hour) and 110 pg/ml (normal range: 35–240 pg/ml), respectively. He does not have diabetes, but dyslipidemia, and his blood total cholesterol, triglyceride, and HDL-cholesterol were 241, 308, and 57 mg/dl, respectively. He had obstructive sleep apnea syndrome (OSAS) for 3 years before the monitoring.

His office BP was 154/102 mmHg. Time series of systolic BP (SBP), diastolic BP (DBP), and heart rate measured by pulse (HR) are shown in Fig. 18.20. When 7-day/24-hour ABP monitoring was started at 15:00 in the presence of a doctor on November 28, 2000, his blood pressure and pulse were 171/103 mmHg and 60 bpm.

The circadian profiles of SBP and HR of this subject are shown in Fig. 18.21, which depicts normal circadian profiles, within the middle portion of the chronodesms, despite an episodic BP increase of white-coat hypertension on the first day. The MESOR estimates of SBP, DBP, and HR were 130.2 mmHg, 69.2 mmHg, and 65.9 bpm, respectively. The double circadian amplitudes of SBP

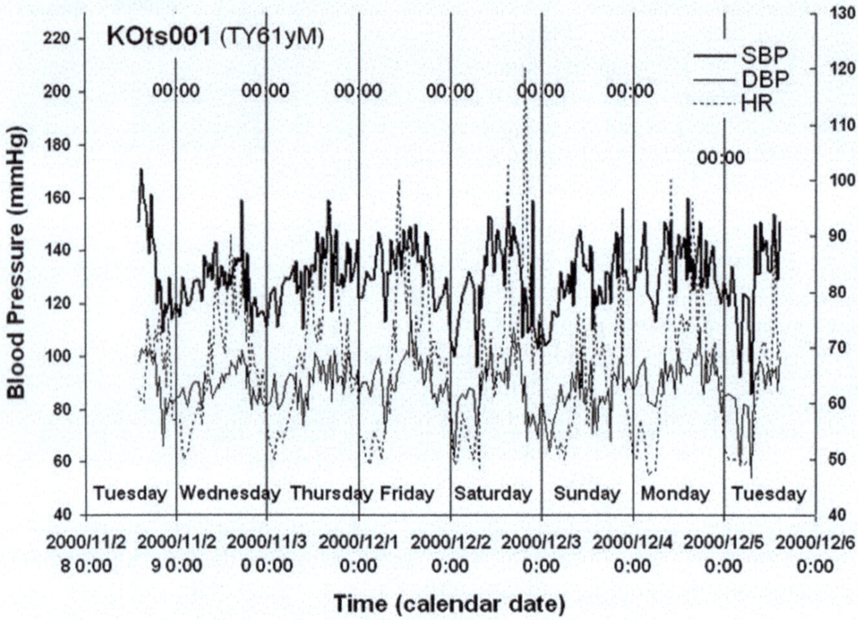

Fig. 18.20 An example of white-coat hypertension syndrome with obstructive sleep apnea. Time plots of the 7-day/24-hour records of SBP (*thick continuous line*), DBP (*thin continuous line*), and HR (*dashed line*) of a 76-year-old man with white-coat hypertension. When the 7-day/24-hour ABP monitoring session was started at 15:00 in front of a doctor, his blood pressure and pulse were 171/103 mmHg and 60 bpm, coinciding with grade 2 hypertension. But measurements decreased thereafter, becoming close to the normal range and fluctuating rhythmically as a function of his rest-activity cycle. A "non-dipping" pattern is observed on Thursday and Sunday nights (day-night ratio: 2.9 % and 0.9 %, respectively), corresponding to "masked non-dipper" hypertension. He is suffering from obstructive sleep apnea and receives CPAP treatment, but he did not use it on Thursday and Sunday

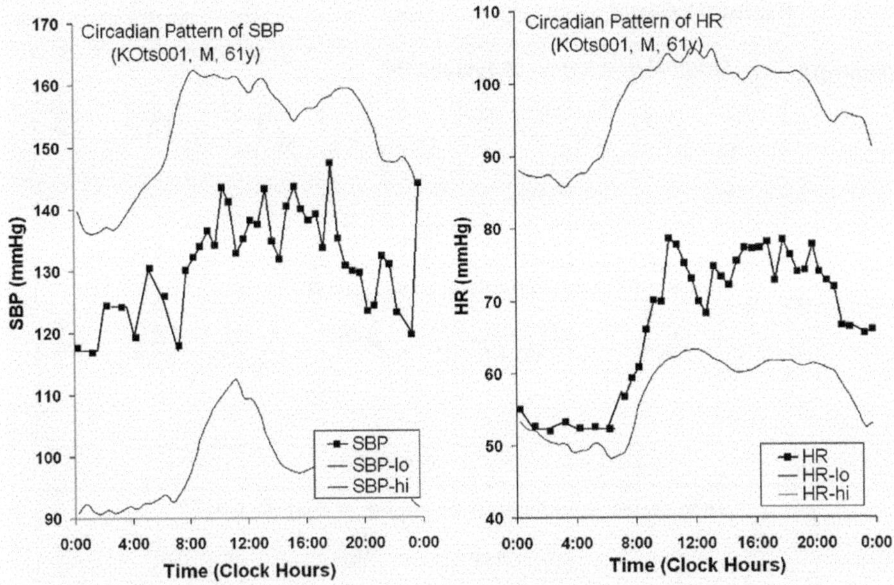

Fig. 18.21 Circadian profiles of systolic BP (*left*) and heart rate (HR) (*right*). 24-hour patterns of systolic BP (*left*) and heart rate (HR) (*right*) of this subject with white-coat hypertension, averaged over the 7 days of monitoring (*polygonal line* graph with *closed squares*), are shown along with time-specified 90 % prediction limits (*continuous thin lines*). Both variables vary in the *middle* portion of their respective chronodesms

and DBP were 19.15 and 12.28 mmHg, respectively, within acceptable limits. The circadian acrophases of SBP and DBP were 13:38 and 14:26, respectively, within acceptable limits (between ~12:00 and ~16:00).

The circadian period of SBP was estimated by the Maximum Entropy Method (MEM), using the MemCalc software (GMS Co., Tokyo), which requires the data to be equidistant. Outliers were deleted from the original SBP time series (middle section of Fig. 18.22). Data were then made equidistant (bottom section of Fig. 18.22).

The MEM spectrum is shown in the middle part of the left side of Fig. 18.23, which shows a prominent circadian period of 0.9861 day, together with three smaller peaks at periods of 3.0736, 0.7772, and 0.5233 days. Periods extracted by the MEM analysis in the range up to 2.5 cycles/day are shown in the right upper side of this figure. Eleven components are listed here, and that with a period of 0.9861 day

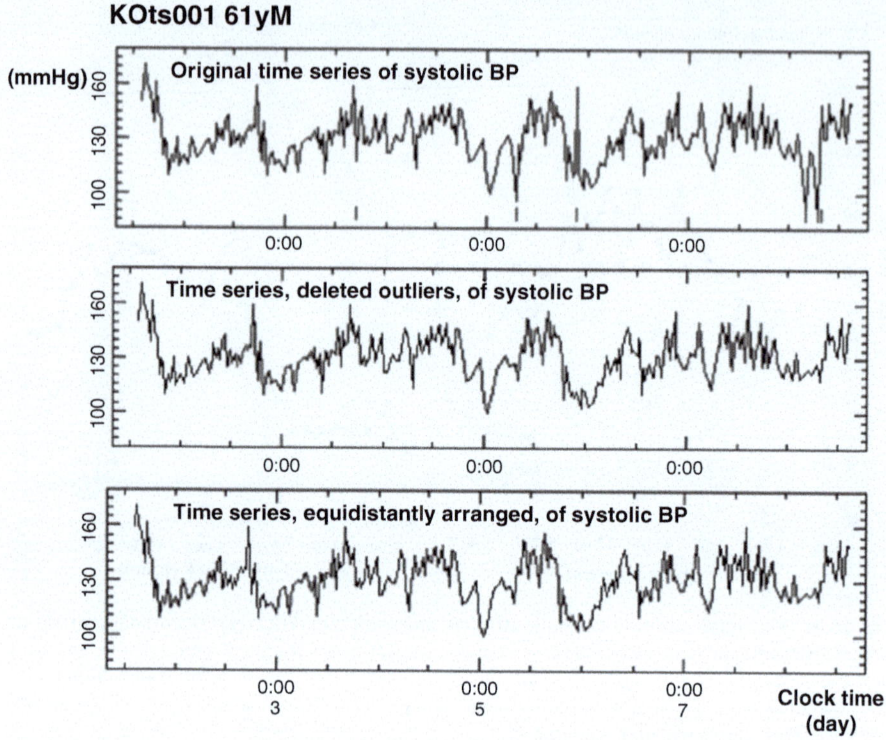

Fig. 18.22 Three versions of the time series of systolic BP. *Upper*: original data of systolic BP. *Middle*: edited data after mathematically excluding outliers. *Lower*: equidistantly arranged time series for analysis by the Maximum Entropy Method, using the MemCalc software (GMS Co., Tokyo)

has the largest spectral power, as assessed by the "area" covered by the spectral peak.

On the bottom left of Fig. 18.23, the edited data are fitted by least squares with a 5-component model consisting of the five cosine curves with the five largest amplitudes, with corresponding periods of 3.074, 0.986, 0.777, 0.523, and 0.240 days (right lower part of Fig. 18.23).

In conclusion, the clinical diagnosis reached after the 7-day/24-hour ABP monitoring consists of (1) a prominent white-coat effect, (2) a mild ABP hypertension treated by antihypertensive medicine, and (3) a "masked non-dipping" hypertension present in the absence of CPAP treatment. Thus, the recommendation is to continue a combination therapy with antihypertensive medicine for hypertension and with CPAP treatment for OSAS.

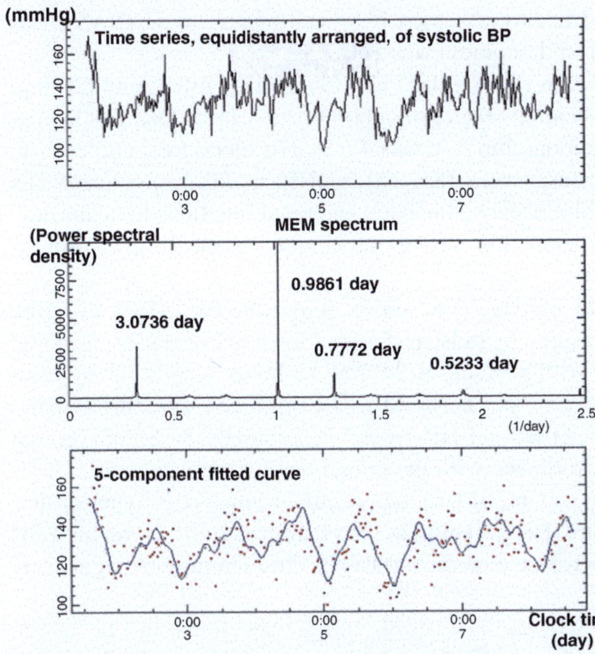

KOts001 61yM

Frequency (/ day)	Period (days)	Area
0.122	8.2130	0.4098
0.325	3.0736	11.0350
0.579	1.7274	2.7201
0.758	1.3199	3.3572
1.014	0.9861	47.5000
1.287	0.7772	9.9217
1.462	0.6839	3.1729
1.701	0.5880	2.9342
1.911	0.5233	5.6901
2.108	0.4745	0.8353
2.472	0.4046	1.2769

Period (days)	Amplitude	Acro-phase
3.074	5.057	4-07:52
0.986	9.687	2-13:57
0.777	4.565	2-09:03
0.523	3.265	1-14:56
0.240	2.399	1-19:53

Fig. 18.23 Period analysis of the circadian rhythm of systolic BP. MEM analysis estimates the circadian period of systolic BP to be 0.9861 day (*left middle part*). Data are shown with a 5-component model fitted by least squares to the data, consisting of cosine curves with the five largest amplitudes (*right lower part*)

18.3 Blood Pressure Morning Surge

In this chapter, we introduce four cases of "morning surge" determined on the basis of 7-day/24-hour ambulatory blood pressure (ABP) monitoring.

18.3.1 Case Report 1 of "Morning Surge"

The first example of "morning surge" is a 64-year-old non-obese (BMI=22.8 kg/m^2) man (ID number: KOts012). His weight was 62.0 kg and height 165 cm. He usually went to bed at about 20:30 and woke up at about 05:30. He felt he slept well most days. He did not smoke or drink alcohol, and he frequently ate vegetables and fruits. He usually did not eat snacks between meals and tried not to eat too many fatty foods. He used to drink two cups of milk every morning. He went for a walk every morning, taking out his dog, about as large as himself.

He felt healthy and took good care of his health. He had something to live for and did not have a depressive mood. His scores on the GDS-15 and PHQ-2 scales were zero. His subjective QOL of health, assessed by visual analog scale (VAS), was 90 % and his subjective QOL of happiness was 100 %.

He was treated with valsartan (80 mg/day) and long-acting nifedipine (20 mg/day) as antihypertensive medication taken around 08:00 after breakfast. His fasting glucose was 97 mg/dl, and hemoglobin A1c was 4.6 %. His blood total cholesterol, triglyceride, and HDL-cholesterol were 185, 121, and 52 mg/dl, respectively. His biochemical examination of blood showed hepatic and renal functions to be normal. He had a family history of hypertension, stroke (father and brothers), and cardiovascular sudden death (brothers).

His office BP was 177/111 mmHg. Time series of systolic BP (SBP), diastolic BP (DBP), and heart rate measured by pulse (HR) are shown in Fig. 18.24. A model consisting of cosine curves with periods of 24 and 12 hours was fitted by least squares to the data collected over 7 days. The MESOR (midline estimating statistic of rhythm) estimates of SBP, DBP, and HR were 149.2 mmHg, 88.5 mmHg, and 71.2 bpm, respectively. In accordance with the guidelines for management of high blood pressure in Japan published in 2014, in the case of 24-hour ABP, hypertension is defined as SBP ≧130 and/or DBP ≧80 mmHg. Thus, this case still shows high BP despite being on two antihypertensive medications. His chronobiologic diagnosis is MESOR-hypertension.

The circadian profiles of SBP and HR of this subject are shown in Fig. 18.25. The double circadian amplitudes of SBP and DBP were 10.24 and 7.57 mmHg,

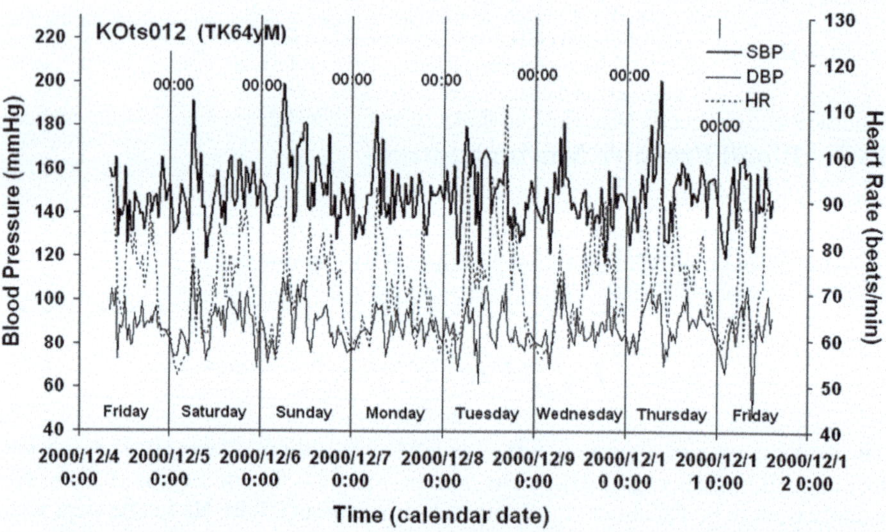

Fig. 18.24 An example of morning surge, complicated by MESOR-hypertension and non-dipping. Time plots of the 7-day/24-hour records of SBP (*thick continuous line*), DBP (*thin continuous line*), and HR (*dashed line*) of a 64-year-old man. Prominent BP peaking is noted in the morning. It is repeated every day, but its extent differs from day to day, apparently following a weekly pattern

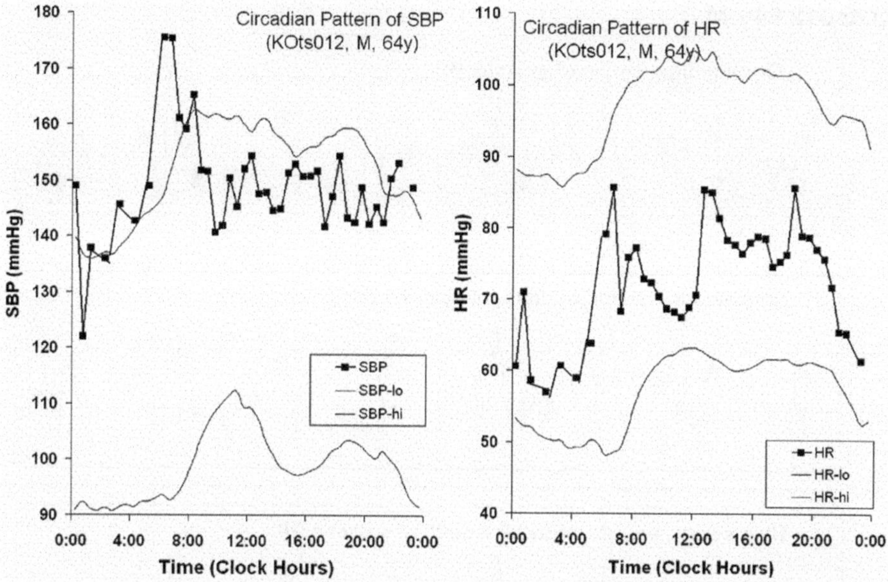

Fig. 18.25 Circadian profiles of systolic BP (*left*) and heart rate (HR) (*right*). 24-hour patterns of systolic BP (*left*) and heart rate (HR) (*right*) of a hypertensive man with BP morning surge, averaged over the 7 days of monitoring (*polygonal line* graph with *closed squares*), are shown along with time-specified 90 % prediction limits (*continuous thin lines*). Whereas the 24-hour pattern of HR is located in the *middle* of the chronodesm, well within acceptable limits, the 24-hour pattern of SBP is close to the *upper* limit of the chronodesm in the daytime and exceeds it at night, indicating a hypertensive non-dipping pattern. The marked elevation of BP in the morning (morning surge) is a distinguishing feature of this profile

respectively, both fairly lower than those of clinically healthy peers, albeit still within acceptable limits, suggesting a hypertensive non-dipping pattern. In addition, a marked BP elevation in the morning is a distinct characteristic of this subject. Whereas the circadian acrophase of HR at 14:10 was within acceptable limits (~12:00 to ~16:00), the acrophases of SBP and DBP were 08:20 and 11:51, respectively, somewhat earlier than those of matched healthy peers, probably owing to a prominent morning surge of systolic BP.

Unfortunately, this subject died 2 years after this monitoring session, on a bitingly cold winter morning when he went for a walk with his big dog.

The circadian period of SBP was estimated by the Maximum Entropy Method (MEM), using the MemCalc software (GMS Co., Tokyo), which requires the data to be equidistant. Outliers were deleted from the original SBP time series (middle section of Fig. 18.26). Data were then made equidistant (bottom section of Fig. 18.26).

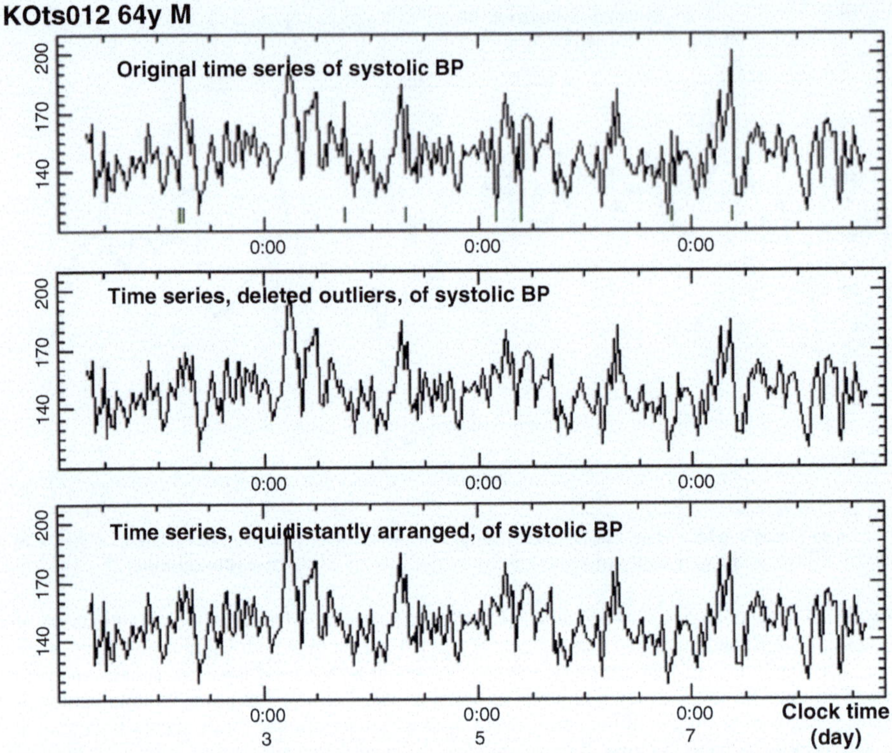

Fig. 18.26 Three versions of the time series of systolic BP. *Upper*: original data of systolic BP. *Middle*: edited data after mathematically excluding outliers. *Lower*: equidistantly arranged time series for analysis by the Maximum Entropy Method, using the MemCalc software (GMS Co., Tokyo)

The MEM spectrum is shown on the middle left of Fig. 18.27, which shows a prominent circadian period of 1.0378 day, together with two other peaks with periods of 2.2522 and 0.4979 days. Periods extracted by the MEM analysis in the range up to 2.5 cycles/day are shown in the right upper side of this figure. Eleven components are listed here, and that with a period of 1.0378 day has the largest spectral power, as assessed by the "area" covered by the spectral peak. Although it is a very small peak, a component with a period of 7.7228 days may relate to a weekly pattern in the extent of the morning surge.

On the bottom left of Fig. 18.27, the edited data are fitted by least squares with a 5-component model consisting of the five cosine curves with the five largest amplitudes, with corresponding periods of 2.252, 1.038, 0.498, 0.333, and 0.203 days (right lower part of Fig. 18.27).

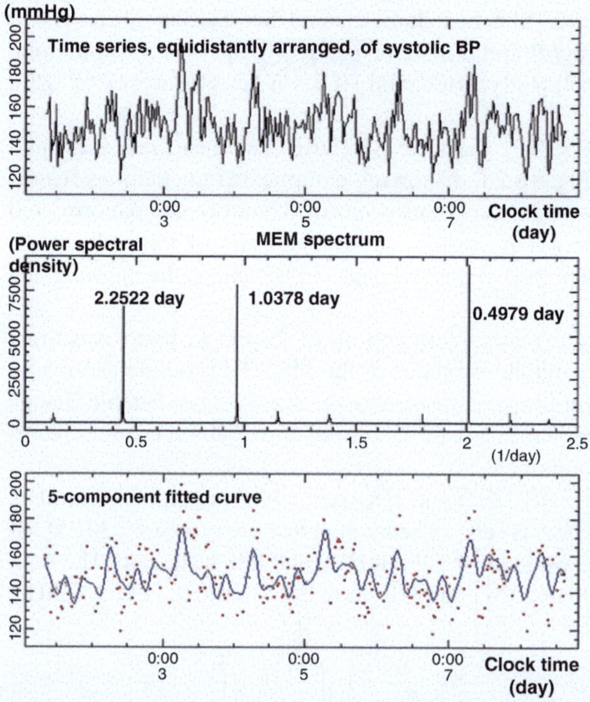

KOts012 64y M

Frequency (/ day)	Period (days)	Area
0.129	7.7228	4.7213
0.444	2.2522	14.9219
0.657	1.5215	1.0404
0.964	1.0378	18.1875
1.150	0.8695	3.9410
1.382	0.7235	4.2403
1.802	0.5549	1.2017
2.008	0.4979	7.7072
2.195	0.4555	2.6030
2.370	0.4220	1.7730
2.739	0.3651	0.5970

Period (days)	Amplitude	Acro-phase
2.252	5.615	3-03:16
1.038	6.033	2-05:32
0.498	4.228	1-19:04
0.333	6.623	1-14:32
0.203	3.205	1-10:28

Fig. 18.27 Period analysis of the circadian rhythm of systolic BP. MEM analysis estimates the circadian period of systolic BP to be 1.0378 day (*left middle part*). Data are shown with a 5-component model fitted by least squares to the data, consisting of cosine curves with the five largest amplitudes (*right lower part*)

18.3.2 Case Report 2 of "Morning Surge"

Another example of "morning surge" was found in a 60-year-old woman (ID number: KOts030) who is slightly overweight (BMI = 26.4 kg/m^2). She goes to bed regularly at 23:00 and gets up at 08:00. She feels she does not sleep very well. She usually takes care of her health and has something to live for. She has a depressive mood, however, and her scores on the GDS-15 and PHQ-2 scales were 11 and 2, respectively. Her subjective QOL of health assessed by the visual analog scale (VAS) was 50 %, and her subjective QOL of happiness was 80 %.

She does not smoke and usually does not drink, and her intake of vegetables and fruits is frequent. She usually tries not to eat too many fatty foods. She has a cup of miso soup every morning, but prefers light seasoning. She goes for a walk every morning and evening.

She is treated for hypertension with a three-drug combination, but her office BP still was 172/82 mmHg. She takes a calcium channel blocker (CCB) and an angiotensin II receptor blocker (ARB) twice daily at 08:30 and 18:30 and an alpha-blocking agent at 22:30. She has diabetes and her fasting glucose and hemoglobin A1c were 172 mg/dl and 8.5 %, respectively. She has dyslipidemia, and her blood total cholesterol, triglyceride, and HDL-cholesterol were 245, 226, and 64 mg/dl, respectively.

Time series of systolic BP (SBP), diastolic BP (DBP), and heart rate measured by pulse (HR) are shown in Fig. 18.28, displaying complicated fluctuations featuring morning and evening BP surges, hypertensive masked non-dipping patterns, and excessive pulse pressure. Circadian profiles of SBP and HR of this subject are shown in Fig. 18.29, which depicts two marked peaks in BP, one in the morning and one in the evening.

A model consisting of cosine curves with periods of 24 and 12 hours was fitted by least squares to the data, yielding estimates of the MESOR (midline estimating statistic of rhythm), double amplitude (a measure of the extent of predictable change within a cycle), and acrophase (a measure of the timing of overall high values recurring in each cycle) of each component.

The MESOR estimates of SBP, DBP, and HR were 139.9 mmHg, 73.0 mmHg, and 52.3 bpm, respectively. These results indicate the presence of systolic MESOR-hypertension, excessive pulse pressure (66.9 mmHg), and low heart rate. The double circadian amplitudes of SBP, DBP, and HR were 24.44 mmHg, 13.90 mmHg,

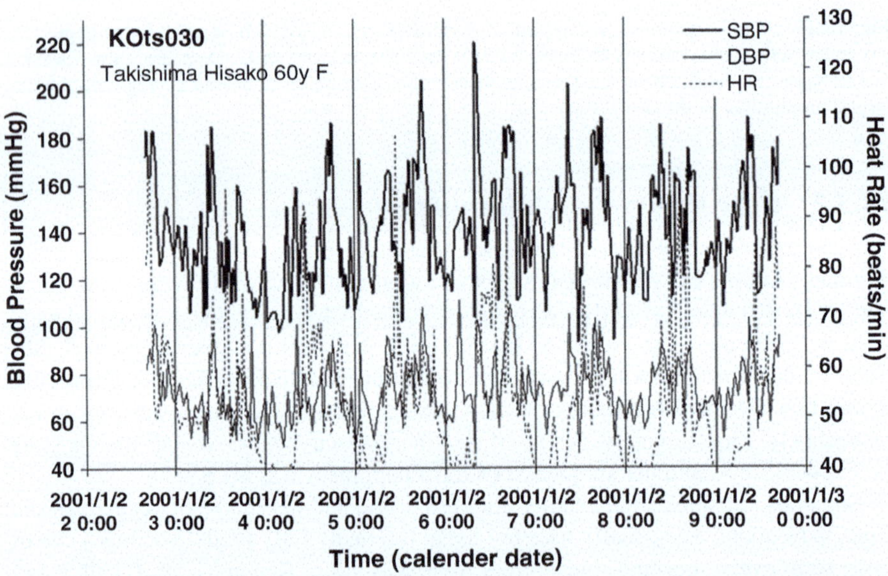

Fig. 18.28 An example of systolic BP MESOR-hypertension complicated by morning and evening BP surges, hypertensive masked non-dipping, and excessive pulse pressure. Time plots of the 7-day/24-hour records of SBP (*thick continuous line*), DBP (*thin continuous line*), and HR (*dashed line*) in a 60-year-old hypertensive woman. BP fluctuations are ambiguous even though she follows a regular rest-activity cycle, i.e., going to bed at 23:00 and getting up at 08:00

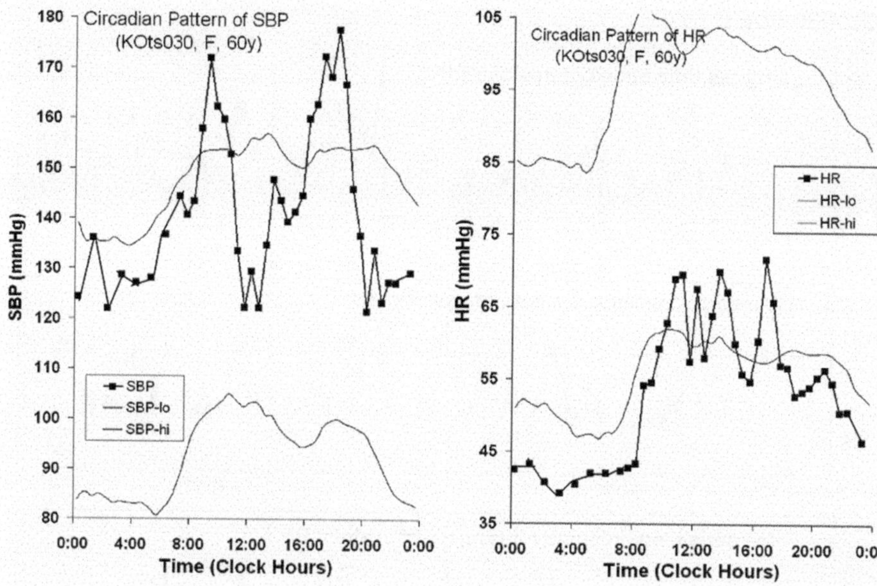

Fig. 18.29 Circadian profiles of systolic BP (*left*) and heart rate (HR) (*right*). 24-hour patterns of systolic BP (*left*) and heart rate (HR) (*right*) of this subject, averaged over the 7 days of monitoring (*polygonal line* graph with *closed squares*), are shown along with time-specified 90 % prediction limits (*continuous thin lines*). Apart from a clear morning surge in SBP, a prominent evening SBP surge is also apparent. Notably during the night, HR is below the *lower* limit of the chronodesm

and 24.98 bpm, respectively, within their respective acceptable ranges. The circadian acrophase of SBP, DBP, and HR was 13:21, 14:09, and 14:38, respectively, within acceptable limits (between ~12:00 and ~16:00).

The circadian period of SBP was estimated by the Maximum Entropy Method (MEM), using the MemCalc software (GMS Co., Tokyo), which requires the data to be equidistant. Outliers were deleted from the original SBP time series (middle section of Fig. 18.30). Data were then made equidistant (bottom section of Fig. 18.30).

The MEM spectrum is shown on the middle left of Fig. 18.31, which shows a weak circadian component with a period of 0.9710 day. It has a wide bandwidth, and its power spectral density is very small. Two other spectral peaks corresponding to periods of 3.6645 and 0.5030 days are also small. Periods extracted by the MEM analysis in the range up to 2.5 cycles/day are shown in the right upper side of this figure. Ten components are listed here, and that with a period of 0.9710 day has the largest spectral power, as assessed by the "area" covered by the spectral peak, suggesting that the circadian periodicity persisted even though it was ambiguous.

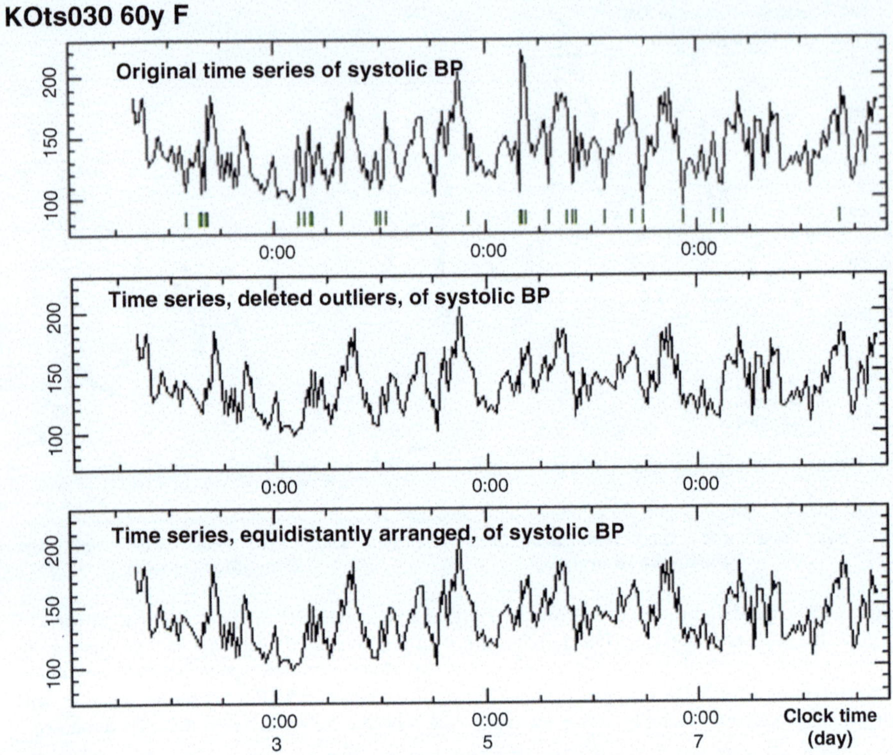

Fig. 18.30 Three versions of the time series of systolic BP. *Upper*: original data of systolic BP. *Middle*: edited data after mathematically excluding outliers. *Lower*: equidistantly arranged time series for analysis by the Maximum Entropy Method, using the MemCalc software (GMS Co., Tokyo)

On the bottom left of Fig. 18.31, the edited data are fitted by least squares with a 5-component model consisting of the five cosine curves with the five largest amplitudes, with corresponding periods of 6.346, 3.664, 0.971, 0.503, and 0.330 days (right lower part of Fig. 18.31).

In conclusion, the clinical diagnosis of this subject based on a chronobiologically interpreted 7-day/24-hour ABP monitoring record consists of (1) BP MESOR-hypertension, (2) marked BP morning surge, (3) distinct BP evening surge, (4) hypertensive "masked non-dipping" pattern, (5) excessive pulse pressure (66.9 mmHg), and (6) low heart rate. Her diabetes had been poorly controlled on variable insulin regimens. Thus, we had a consultation with her diabetologist to make lifestyle recommendations, in terms of diet in relation to her diabetes and of mood improvement. She followed the recommended treatment well, and fortunately she is in good health at the time of this writing.

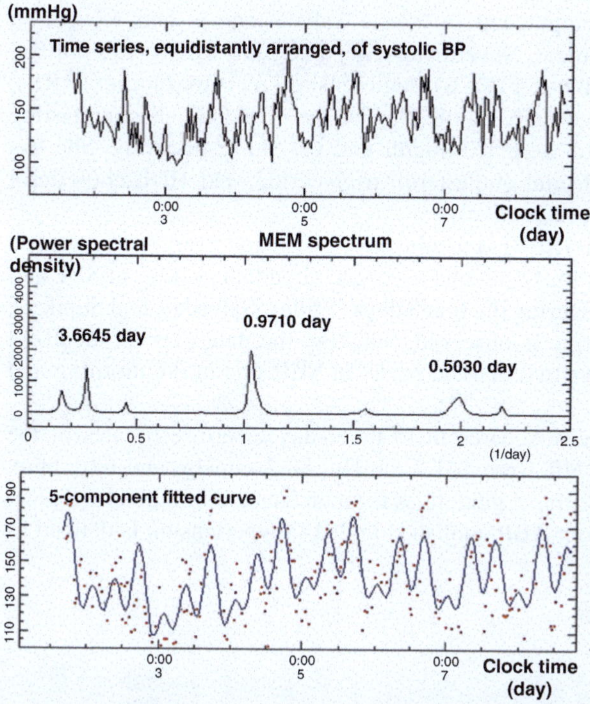

KOts030 60y F

Frequency (/ day)	Period (days)	Area
0.158	6.3458	24.0928
0.273	3.6645	38.5213
0.454	2.2005	8.2570
0.798	1.2531	2.4917
1.030	0.9710	84.1764
1.420	0.7043	3.1035
1.553	0.6437	10.3844
1.988	0.5030	43.9097
2.189	0.4568	10.1196
2.564	0.3900	7.0555

Period (days)	Amplitude	Acro-phase
6.346	8.59	6-09:07
3.664	6.87	4-23:35
0.971	12.33	2-15:41
0.503	9.00	1-17:44
0.330	11.93	1-17:53

Fig. 18.31 Period analysis of the circadian rhythm of systolic BP. MEM analysis estimates the circadian period of systolic BP to be 0.9710 day (*left middle part*). Data are shown with a 5-component model fitted by least squares to the data, consisting of cosine curves with the five largest amplitudes (*right lower part*)

18.3.3 Case Report 3 of "Morning Surge"

This case of "morning surge" concerns a 63-year-old obese (BMI = 28.4 kg/m^2) woman (ID number: KOts090). She goes to bed regularly at 21:30 and gets up at 08:00. She feels that she does not sleep well and wakes up not rested. She usually takes care of her health, but does not have many interests in her life. She has a depressive mood, and her scores on the GDS-15 and PHQ-2 scales were 10 and 2, respectively. Her subjective QOL of health, assessed by a visual analog scale (VAS), was 45 %, and her subjective QOL of happiness was 63 %. She experiences panic episodes accompanied by an increase in BP and tachycardia, bringing her frequently to the emergency room.

She does not smoke and usually does not drink alcohol. She often eats vegetables and fruits. She usually tries not to eat too many fatty foods. She has a cup of miso soup every morning, but prefers light seasoning. She goes for a walk every morning and evening.

She has hypertension complicated by aortic valve stenosis and receives a three-drug combination treatment. Nevertheless, her office BP was 168/99 mmHg. She takes a calcium channel blocker (CCB), angiotensin II receptor blocker (ARB), and beta-blocking agent once daily at 08:30. She has diabetes, and her fasting glucose and hemoglobin A1c were 152 mg/dl and 6.7 %, respectively. She has dyslipidemia, and her blood total cholesterol, triglyceride, and HDL-cholesterol were 265, 255, and 44 mg/dl, respectively.

Time series of systolic BP (SBP), diastolic BP (DBP), and heart rate measured by pulse (HR) are shown in Fig. 18.32, and averaged circadian profiles of SBP and HR are shown in Fig. 18.33. During the first 3 days (Friday, Saturday, and Sunday), a prominent morning BP surge is observed. Stacking the data over an idealized 24-hour days clearly displays two marked peaks in SBP, one in the morning and one in the evening.

From the 2-component least-squares fit to the 7-day record, estimates of the MESOR of SBP, DBP, and HR were 141.7 mmHg, 80.2 mmHg, and 67.5 bpm, respectively. In accordance with the guidelines for management of high blood pressure in Japan based on 24-hour ABP published in 2014, hypertension is defined as

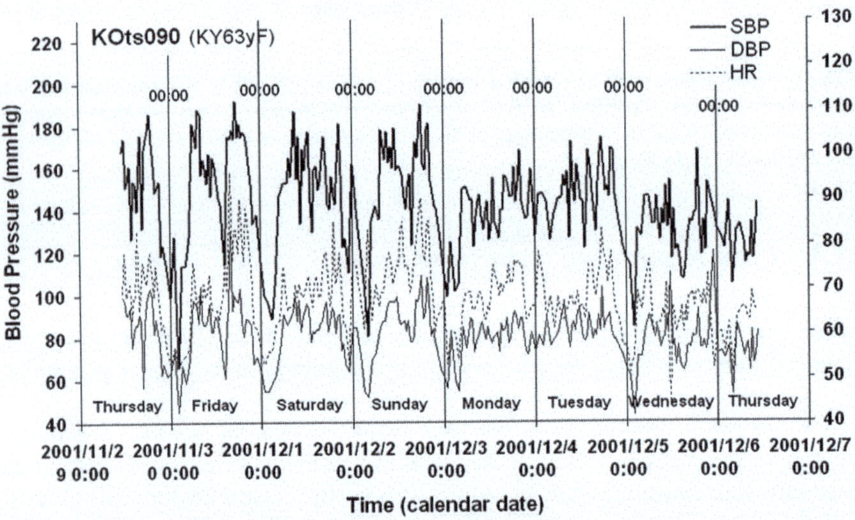

Fig. 18.32 An example of morning surge complicated by SBP MESOR-hypertension, masked ABP normotension (Wednesday), and hypertensive masked non-dipping (Monday and Wednesday night). Time plots of the 7-day/24-hour records of SBP (*thick continuous line*), DBP (*thin continuous line*), and HR (*dashed line*) of a 63-year-old woman. During the first 3 days (Friday, Saturday, and Sunday), a prominent morning BP surge is observed, with a daytime dip seen on Thursday, Friday, and Sunday, coincident with afternoon naps. On Monday and Wednesday night, non-dipping hypertension occurs, coincident with insomnia owing to a strong depressive mood. Whether afternoon naps should be avoided in subjects with a depressive mood to prevent non-dipping hypertension deserves further investigation

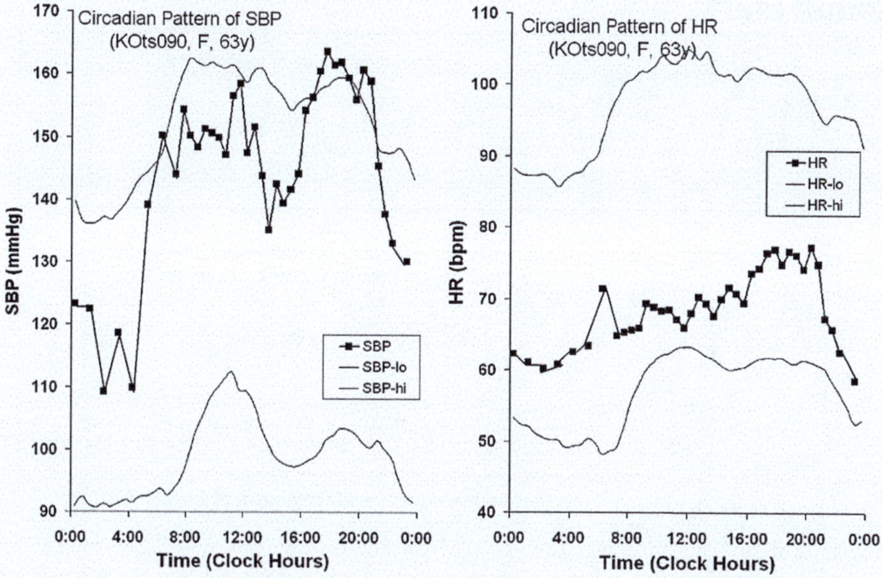

Fig. 18.33 Circadian profiles of systolic BP (*left*) and heart rate (HR) (*right*). 24-hour systolic BP (*left*) and heart rate (HR) (*right*) patterns of a subject with BP morning surge, averaged over the 7 days of monitoring (*polygonal line* graph with *closed squares*), are shown along with time-specified 90 % prediction limits (*continuous thin lines*). Not only can a clear morning BP surge be seen, an even more prominent evening BP surge is observed. HR remains close to the *lower* limit of the chronodesm

SBP ≧130 mmHg and/or DBP ≧80 mmHg. Thus, this subject is still diagnosed as hypertensive despite treatment with three antihypertensive medications.

The circadian acrophase of SBP, DBP, and HR was 14:27, 13:57, and 15:37, respectively. The double amplitudes of SBP and DBP were 31.38 and 18.54 mmHg, respectively, all within acceptable limits.

The circadian period of SBP was estimated by the Maximum Entropy Method (MEM), using the MemCalc software (GMS Co., Tokyo), which requires the data to be equidistant. Outliers were deleted from the original SBP time series (middle section of Fig. 18.34). Data were then made equidistant (bottom section of Fig. 18.34).

The MEM spectrum is shown on the middle left of Fig. 18.35, which shows a prominent circadian period of 1.0517 day, together with five peaks with periods of 10.4850, 0.9696, 0.6377, 0.5109, and 0.4908 days. Periods extracted by the MEM analysis in the range up to 2.5 cycles/day are shown in the right upper side of this figure. Thirteen components are listed here, and the one with a period of 1.0517 day has the largest spectral power, as assessed by the "area" covered by the spectral peak. MEM spectral analysis suggests a splitting of the circadian periodicity likely related to the nonstationarity of the data.

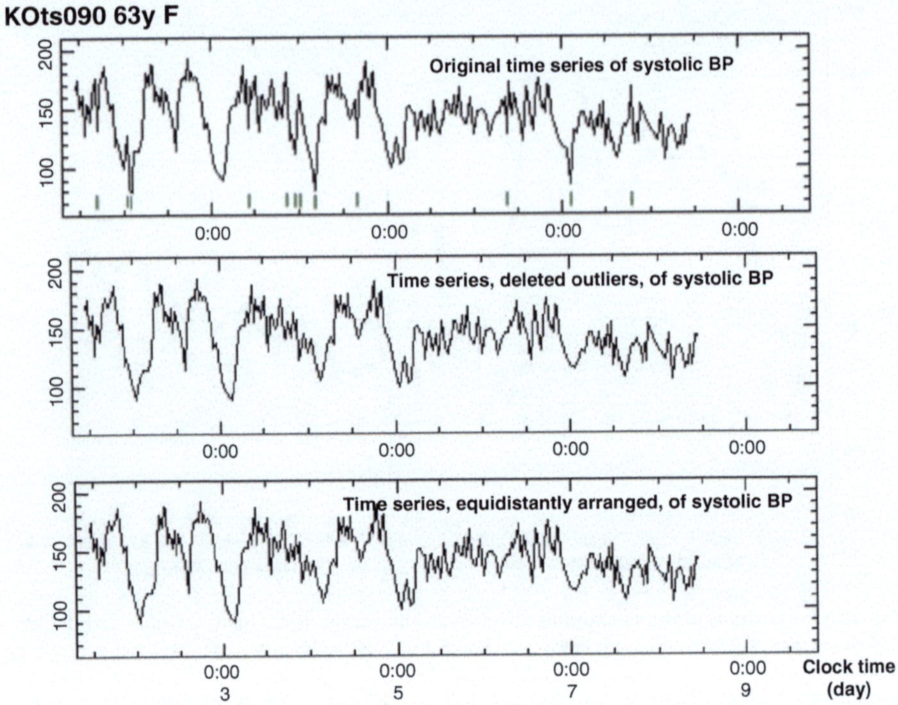

Fig. 18.34 Three versions of the time series of systolic BP. *Upper*: original data of systolic BP. *Middle*: edited data after mathematically excluding outliers. *Lower*: equidistantly arranged time series for analysis by the Maximum Entropy Method, using the MemCalc software (GMS Co., Tokyo)

On the bottom left of Fig. 18.35, the edited data are fitted by least squares with a 5-component model consisting of the five cosine curves with the five largest amplitudes, with corresponding periods of 10.485, 1.052, 0.970, 0.511, and 0.491 days (right lower part of Fig. 18.35). Splitting of the circadian periodicity is further suggested by the similar amplitudes of the 1.052-day and 0.970-day components with close periods (14.53 and 12.49 mmHg, respectively).

In conclusion, the clinical diagnosis based on the chronobiologically interpreted 7-day/24-hour ABP monitoring consists of (1) SBP MESOR-hypertension, (2) marked BP morning surge, (3) even more pronounced BP evening surge, (4) intermittent hypertensive "masked non-dipping," and (5) intermittent masked ABP normotension. There is indeed a drastic difference in the time course of SBP during the first and last 3 days of the monitoring session, likely related to occasional anxiety and sleep disturbances. Accordingly, her panic episodes and depressive mood were treated with antidepressant medicines, and dietary recommendations were made to treat her obesity. She is still presenting with resistant hypertension, episodic symptoms of anxiety, and sleep disturbances with depressive mood, however.

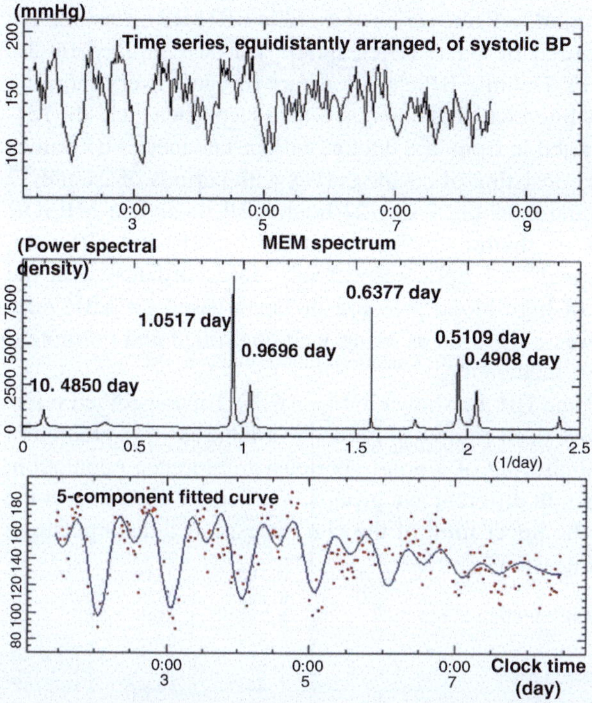

KOts090 63y F

Frequency (/ day)	Period (days)	Area
0.0954	10.4850	23.570
0.3720	2.6882	14.387
0.5284	1.8923	9.572
0.6451	1.5500	4.401
0.9508	1.0517	99.367
1.0313	0.9696	57.620
1.1731	0.8524	0.791
1.5680	0.6377	15.870
1.7666	0.5661	9.222
1.9573	0.5109	67.499
2.0377	0.4908	44.292
2.4075	0.4154	8.719
2.5928	0.3857	6.978

Period (days)	Amplitude	Acro-phase
10.485	8.80	3-16:18
1.052	14.53	1-13:24
0.970	12.49	1-13:03
0.511	11.70	1-19:19
0.491	9.36	1-19:25

Fig. 18.35 Period analysis of the circadian rhythm of systolic BP. MEM analysis estimates the circadian period of systolic BP to be 1.0517 day (*left middle part*). Data are shown with a 5-component model fitted by least squares to the data, consisting of cosine curves with the five largest amplitudes (*right lower part*)

18.3.4 Case Report 4 of "Morning Surge": Morning Surge with White-Coat Hypertension Syndrome Observed by 7-Day/24-Hour Ambulatory BP Monitoring

The fourth example of "morning surge," complicated by marked white-coat hypertension, is a 68-year-old non-obese (BMI = 24.4 kg/m^2) man (ID number: KOts048). He goes to bed regularly at 21:00 and gets up at 07:00. He feels like he does not sleep so well and wakes up not rested. He usually takes care of his health, has something to live for, and does not have a depressive mood. His scores on the GDS-15 and PHQ-2 scales were 2 and 1, respectively. His subjective QOL of health, assessed by the visual analog scale (VAS), was 67 %, and his subjective QOL of happiness was 82 %.

He is treated with a CCB (benidipine, 4 mg/day) as antihypertensive medication, taken once daily at about 08:00 after breakfast. He does not have diabetes, and his blood total cholesterol, triglyceride, and HDL-cholesterol were 151, 289, and 31 mg/dl, respectively. His biochemical examination of blood showed that his hepatic and renal functions were normal.

His office BP was 157/98 mmHg. Time series of systolic BP (SBP), diastolic BP (DBP), and heart rate, measured by pulse (HR), monitored on February 16–24, 2001, are shown in Fig. 18.36. Definite BP elevations are repeated every morning when he gets up, and a clear white-coat phenomenon is also seen when the 7-day/24-hour ABP monitoring was started in front of a doctor, with instantaneous BP values of 231/102 mmHg. A model consisting of cosine curves with periods of 24 and 12 hours was fitted by least squares to the 7-day/24-hour ABP data. The MESOR (midline estimating statistic of rhythm) estimates of SBP, DBP, and HR were 129.8 mmHg, 77.5 mmHg, and 75.1 bpm, respectively. In accordance with the guidelines for management of high blood pressure in Japan based on ABP, published in 2014, this subject was diagnosed as being well controlled and normotensive when on antihypertensive medication.

Circadian profiles of SBP and HR are shown in Fig. 18.37. Time-specified reference limits (chronodesms) have been computed for each variable as 90 % prediction limits, derived from ABP monitoring of a non-hypertensive reference population, separately for men and women in different age groups. Marked BP elevation in the morning, when BP exceeds the upper limit of the chronodesm, is a distinguishing feature of this subject (left, Fig. 18.37).

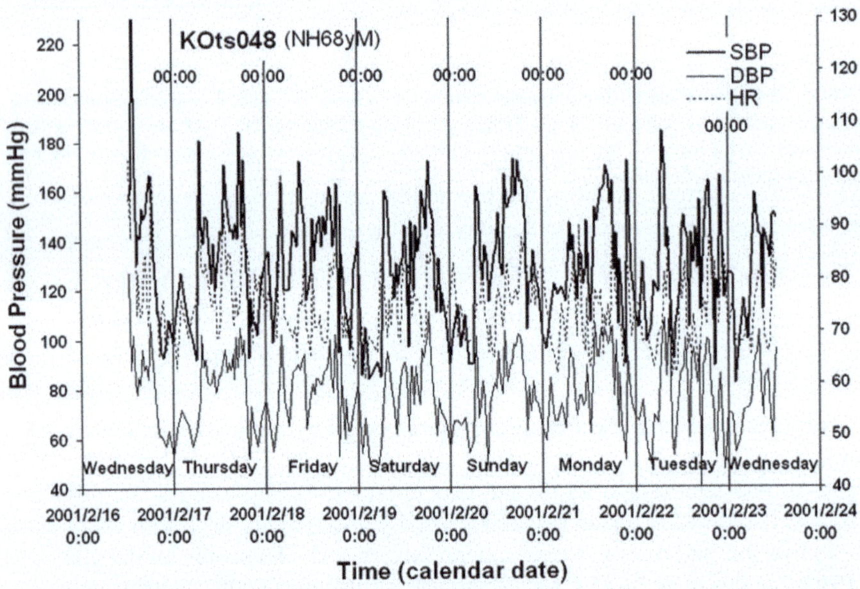

Fig. 18.36 An example of morning surge complicated by white-coat hypertension syndrome. Time plots of the 7-day/24-hour records of SBP (*thick continuous line*), DBP (*thin continuous line*), and HR (*dashed line*) of a 77-year-old man. When the 7-day/24-hour ABP monitoring session was started at 13:30 in front of a doctor, his blood pressure and pulse were 231/102 mmHg and 125 bpm, suggesting white-coat hypertension. Measurements dropped within the acceptable range the next day, where they fluctuated periodically according to a circadian rhythm (suggesting a regular rest-activity cycle), with a prominent BP morning surge

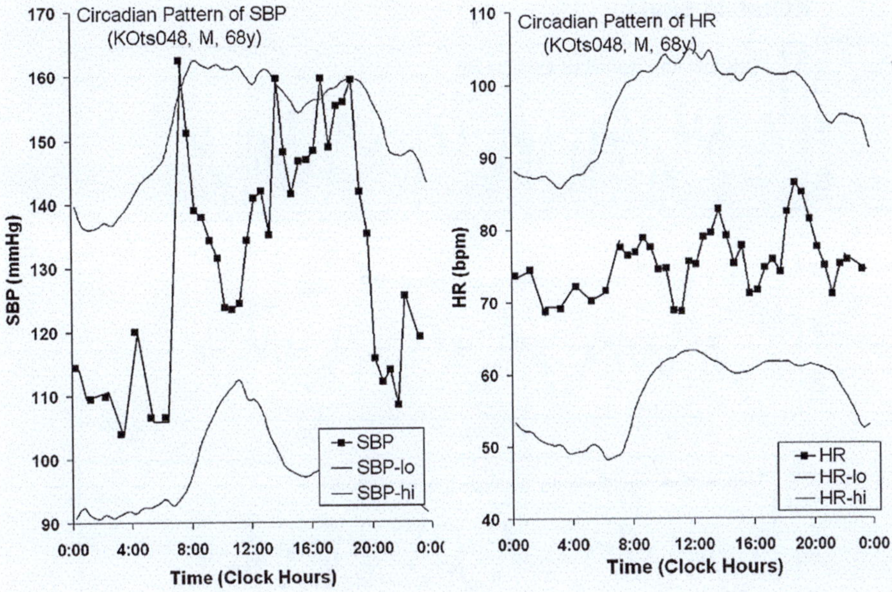

Fig. 18.37 Circadian profiles of systolic BP (*left*) and heart rate (HR) (*right*). 24-hour patterns of systolic BP (*left*) and heart rate (HR) (*right*) of a man with BP morning surge, averaged over the 7 days of monitoring (*polygonal line* graph with *closed squares*), are shown along with time-specified 90 % prediction limits (*continuous thin lines*). Although SBP averaged 129.8 mmHg and was mostly within acceptable limits, this subject shows a prominent BP morning surge, as seen from the small and brief excursion of SBP outside the chronodesm just before 08:00. On the other hand, HR is located well within the chronodesmic limits

Based on the fitted models, the circadian acrophases of SBP, DBP, and HR were 13:51, 13:41, and 15:55, respectively, all within acceptable limits (between ~12:00 and ~16:00). The double amplitude of SBP was 37.87 mmHg, above the upper limit of acceptability. This condition is known as circadian hyper-amplitude-tension (CHAT). The double amplitude of DBP was 24.69 mmHg, within acceptable limits.

The circadian period of SBP was estimated by the Maximum Entropy Method (MEM), using the MemCalc software (GMS Co., Tokyo), which requires the data to be equidistant. Outliers were deleted from the original SBP time series (middle section of Fig. 18.38). Data were then made equidistant (bottom section of Fig. 18.38).

The MEM spectrum is shown on the middle left of Fig. 18.39, which shows a prominent circadian period of 0.9995 day, together with three clear, albeit smaller peaks with periods of 2.7772, 0.5202, and 0.4497 days. Periods extracted by the MEM analysis in the range up to 2.5 cycles/day are shown in the right upper side of this figure. Eleven components are listed here, and that with a period of 0.9995 day has the largest spectral power, as assessed by the "area" covered by the spectral peak.

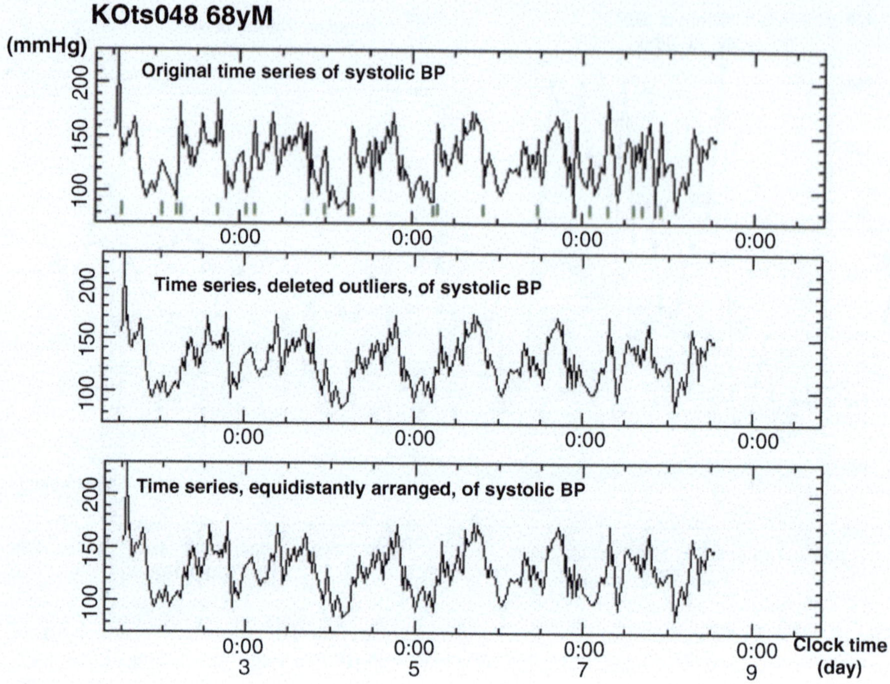

Fig. 18.38 Three versions of the time series of systolic BP. *Upper*: original data of systolic BP. *Middle*: edited data after mathematically excluding outliers. *Lower*: equidistantly arranged time series for analysis by the Maximum Entropy Method, using the MemCalc software (GMS Co., Tokyo)

On the bottom left of Fig. 18.39, the edited data are fitted by least squares with a 5-component model consisting of the five cosine curves with the five largest amplitudes, with corresponding periods of 0.999, 0.520, 0.329, 0.250, and 0.199 days (right lower part of Fig. 18.39).

In conclusion, the clinical diagnosis based on the chronobiologically interpreted 7-day/24-hour ABP monitoring consists of (1) hypertension treated by CCB, (2) marked BP morning surge, (3) SBP-CHAT, and (4) distinct white-coat hypertensive effect. After this monitoring session, an aneurysm of the abdominal aorta was discovered. We had a consultation with a specialist of cardiovascular surgery, but unfortunately, the aneurysm ruptured before the operation. An emergency operation was successful and the medical treatment for hypertension was continued for several years. This subject, however, passed away in the morning during a hard 2012 winter, suffering from a slowly progressively decreased renal function.

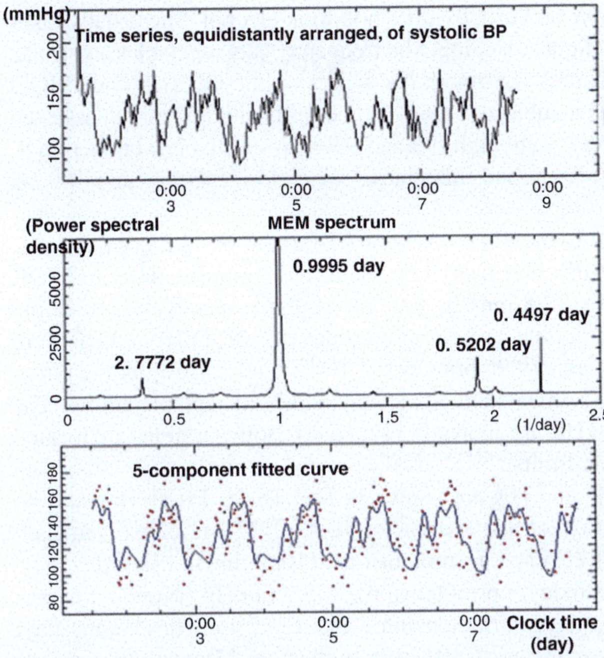

KOts048 68yM

Frequency (/ day)	Period (days)	Area
0.165	6.0643	1.96
0.361	2.7722	11.17
0.552	1.8115	6.61
0.723	1.3830	7.07
1.001	0.9995	234.67
1.237	0.8081	7.98
1.436	0.6962	2.84
1.744	0.5734	2.69
1.922	0.5202	22.41
2.008	0.4979	8.60
2.224	0.4497	6.43

Period (days)	Amplitude	Acro-phase
0.999	20.95	1-14:25
0.520	6.87	1-14:16
0.329	6.42	1-17:04
0.250	5.10	1-12:54
0.199	5.14	1-13:28

Fig. 18.39 Period analysis of the circadian rhythm of systolic BP. MEM analysis estimates the circadian period of systolic BP to be 0.9995 day (*left middle part*), which is the most prominent component. Data are shown with a 5-component model fitted by least squares to the data, consisting of cosine curves with the five largest amplitudes (*right lower part*)

18.4 Monday Blood Pressure Surge

In this chapter, we introduce two cases of "Monday BP surge" from the viewpoint of 7-day/24-hour ambulatory blood pressure (ABP) monitoring.

18.4.1 Case Report 1 of "Monday Blood Pressure Surge"

The first example of "Monday BP surge" is a 62-year-old slightly obese (BMI = 29.2 kg/m^2) woman (ID number: Tosa0257). She is not taking any antihypertensive medication. Her office BP and pulse were 135/86 mmHg and 73 bpm. Her home BP and pulse averaged over 30 days were 116.9/71.1 mmHg and 62.3 bpm in the morning and 129.2/73.4 mmHg and 72.2 bpm in the evening.

She has a regular sleep-wake cycle and goes to bed regularly at 23:00 and gets up at 07:00. She falls asleep within 10 min and feels that she sleeps well most days, waking rested. She does not smoke and usually does not drink alcohol. She usually takes care of her health and prefers light seasoning. She frequently eats vegetables and fruits. She has, however, a slight depressive mood, her scores on the GDS-15 and PHQ-2 being 7 and 1, respectively. Her subjective QOL indices of health and mood, assessed by a visual analog scale (VAS), were high (93 and 95 %, respectively), but her QOL indices for financial satisfaction and happiness were low (53 % and 48 %, respectively).

Based on the model fitted to the 7-day/24-hour ABP monitoring record, systolic BP (SBP), diastolic BP (DBP), and heart rate measured by pulse (HR) MESOR estimates were 123.3 mmHg, 72.8 mmHg, and 71.3 bpm, respectively, indicating MESOR-normotension. The double 24-hour amplitudes of SBP, DBP, and HR were 23.32 mmHg, 17.35 mmHg, and 19.95 bpm, respectively, and corresponding circadian acrophases were 12:41, 13:38, and 15:49, all within acceptable limits. The circadian profiles of SBP and HR are shown in Fig. 18.40. Both variables are located well within the chronodesmic limits.

Time series of SBP, DBP, and HR are shown in Fig. 18.41. From Thursday to Wednesday, the 24-hour average of SBP assumes values of 122.3 mmHg (Thursday; normotension), 128.6 mmHg (Friday; normotension), 110.5 mmHg (Saturday; normotension), 121.8 mmHg (Sunday; normotension), 135.9 mmHg (Monday; hypertension), 126.0 mmHg (Tuesday; normotension), and 133.2 mmHg (Wednesday; hypertension). As the 24-hour mean of SBP was highest on Monday, this case is illustrative of a Monday BP surge.

The day-night ratio of SBP indicates a "masked dipping pattern," alternating between non-dipping, dipping, and extreme dipping patterns. It changed from 1 day to another, assuming values of 4.0 % (Thursday night; non-dipping), 9.2 % (Friday night; non-dipping), 3.8 % (Saturday night; non-dipping), 21.3 % (Sunday night; extreme dipping), 19.9 % (Monday night; dipping), 0.1 % (Tuesday night; non-dipping), and 23.5 % (Wednesday night; extreme dipping).

The circadian period of SBP was estimated by the Maximum Entropy Method (MEM), which requires the data to be equidistant. Outliers were deleted from the original SBP time series (middle section of Fig. 18.42). Data were then made equidistant (bottom section of Fig. 18.42).

The MEM spectrum is shown on the middle left of Fig. 18.43, which shows a prominent peak corresponding to the circadian component with a period of 1.0391 day, together with a small circasemidian peak and three other peaks at periods of 7.2466, 2.4354, and 1.6132 days. Periods extracted by the MEM analysis in the range up to 2.5 cycles/day are shown in the right upper side of this figure. Eleven components are listed here, and that with a "period" of 1.0391 day has a large spectral power, as assessed by the "area" covered by the spectral peak.

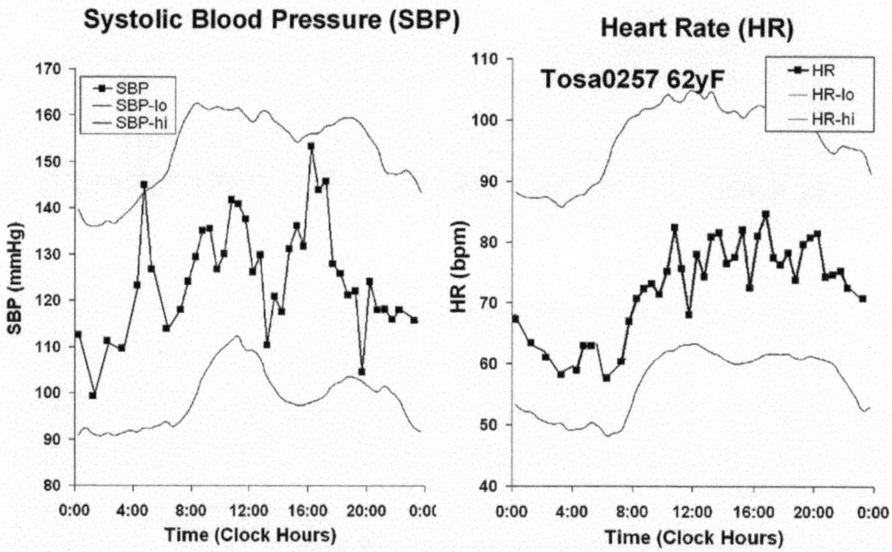

Fig. 18.40 Circadian profiles of systolic BP (*left*) and heart rate (HR) (*right*). 24-hour systolic BP and HR patterns of a 62-year-old woman, averaged over the 7 days of monitoring (*polygonal line* graph with closed *squares*), are shown along with time-specified prediction limits (*continuous thin lines*). Both SBP and HR are located inside their chronodesms

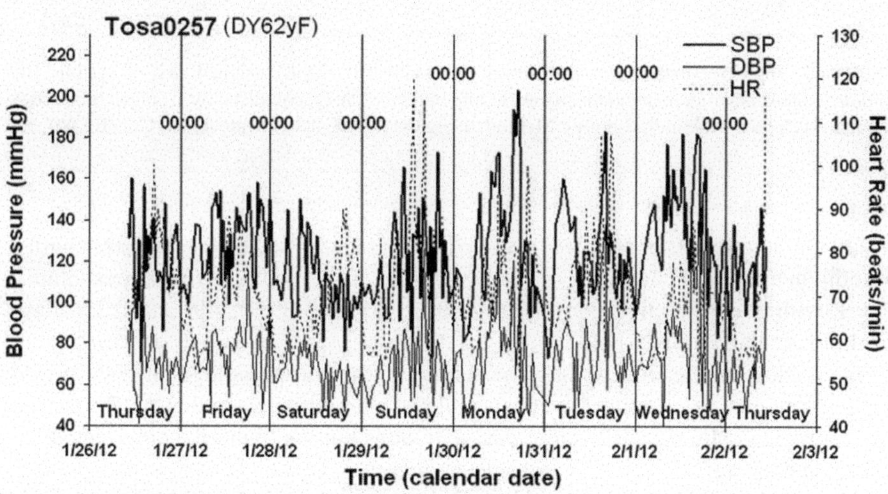

Fig. 18.41 An example of "Monday BP surge." Time plots of the 7-day/24-hour records of SBP (*thick continuous line*), DBP (*thin continuous line*), and HR (*dashed line*) of a 62-year-old woman. An increase in BP is clearly seen on the 5th day (Monday), in keeping with the diagnosis of Monday BP surge

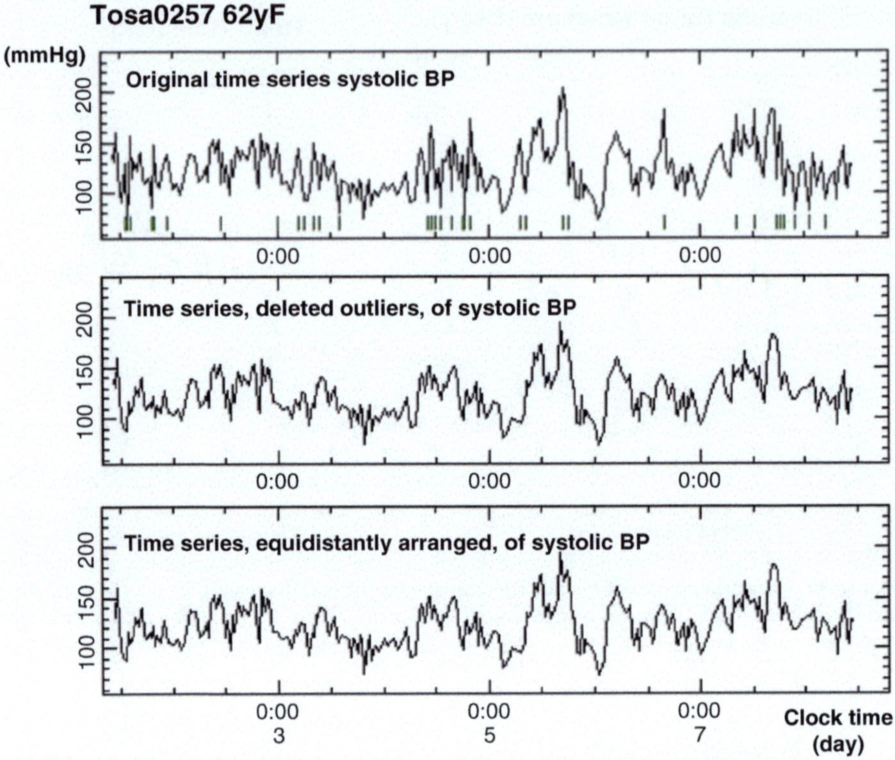

Fig. 18.42 Three versions of the time series of systolic BP. *Upper*: original data of systolic BP. *Middle*: edited data after mathematically excluding outliers. *Lower*: equidistantly arranged time series for analysis by the Maximum Entropy Method, using the MemCalc software (GMS Co., Tokyo)

On the bottom left of Fig. 18.43, the edited data are fitted by least squares with a 5-component model consisting of the five cosine curves with the five largest amplitudes, with corresponding periods of 2.435, 1.613, 1.039, 0.549, and 0.250 days (right lower part of Fig. 18.43).

In conclusion, the clinical diagnosis based on the chronobiologically interpreted 7-day/24-hour ABP monitoring consists of (1) marked between-day ABP variability, including Monday BP surge and (2) non-dipping pattern complicated by "masked dipping."

Accordingly, we recommended that she strictly improve her lifestyle regarding salt intake. We also started antihypertensive treatment. We selected a low-dose CCB focusing on improving the between-day ABP variability. Under this treatment plan, she now seems to be in good health.

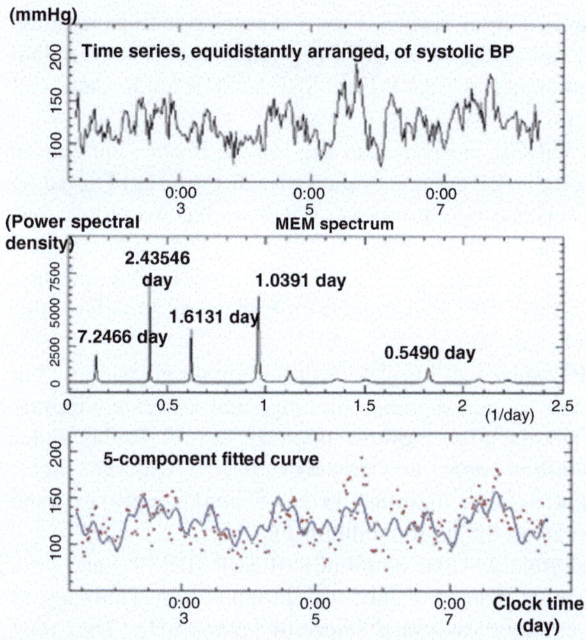

Tosa0257 62yF

Frequency (/ day)	Period (days)	Area
0.000	-	1.071
0.138	7.2466	22.293
0.411	2.4354	31.594
0.620	1.6131	32.111
0.962	1.0391	76.359
1.122	0.8912	23.218
1.361	0.7350	11.831
1.506	0.6641	11.869
1.822	0.5490	23.627
2.076	0.4817	10.471
2.225	0.4494	10.780
2.440	0.4099	5.039

Period (days)	Amplitude	Acro-phase
2.435	7.07	2-18:04
1.613	7.01	2-18:25
1.039	11.00	2-10:21
0.549	6.86	1-18:20
0.250	8.29	1-16:15

Fig. 18.43 Period analysis of the circadian rhythm of systolic BP. MEM analysis estimates the circadian period of systolic BP to be 1.0391 day (*left middle part*). Data are shown with a 5-component model fitted by least squares to the data, consisting of cosine curves with the five largest amplitudes, the 1.039-day component being very prominent (*right lower part*)

18.4.2 Case Report 2 of "Monday Blood Pressure Surge"

The second example of "Monday BP surge" is a 57-year-old non-obese (BMI = 21.0 kg/m^2) woman (ID number: Tosa0213). She does not take any antihypertensive medication. Her office BP and pulse were 141/86 mmHg and 73 bpm, and her home BP and pulse averaged over 30 days were 125.2/74.6 mmHg and 61.8 bpm in the morning and 120.9/70.7 mmHg and 63.7 bpm in the evening.

She goes to bed at different times and gets up at extremely irregular times. She needs more than 120 min to fall asleep. She does not sleep well most days and wakes up not rested. She does not smoke and usually does not drink alcohol. She usually takes care of her health and her seasoning is average. She has a depressive mood, and her scores on the GDS-15 and PHQ-2 scales were was 5 and 1, respectively. Her subjective QOL indices of health, mood, family relationship, relationship with friends, financial satisfaction, life satisfaction, and happiness were 75 %, 52 %, 68 %, 70 %, 50 %, 46 %, and 68 %, respectively.

Based on the model fitted to the 7-day/24-hour ABP monitoring record, systolic BP (SBP), diastolic BP (DBP), and heart rate measured by pulse (HR) MESOR estimates were 121.1 mmHg, 66.4 mmHg, and 66.2 bpm, respectively, indicating MESOR-normotension. The double 24-hour amplitudes of SBP, DBP, and HR were

34.99 mmHg, 15.27 mmHg, and 7.6 bpm, respectively, and corresponding circadian acrophases were 15:09, 15:13, and 17:02, all within acceptable limits. The circadian profiles of SBP and HR are shown in Fig. 18.44. Both SBP and HR are located well within the chronodesmic limits.

Time series of SBP, DBP, and HR are shown in Fig. 18.45. From Thursday to Wednesday, the 24-hour average of SBP assumes values of 116.1 mmHg (Thursday; normotension), 134.3 mmHg (Friday; hypertension), 123.3 mmHg (Saturday; normotension), 123.5 mmHg (Sunday; normotension), 130.5 mmHg (Monday; hypertension), 127.9 mmHg (Tuesday; normotension), and 121.2 mmHg (Wednesday; normotension). The 24-hour mean of SBP was highest on Friday and Monday, indicating Monday BP surge with Friday surge.

The day-night ratio of SBP shows a "masked extreme-dipping" pattern, but varies between reverse dipping to non-dipping, dipping, and extreme dipping, assuming values of −3.0 % (Thursday night; reverse dipping), 35.6 % (Friday night; extreme dipping), 8.5 % (Saturday night; non-dipping), 18.5 % (Sunday night; dipping), 30.6 % (Monday night; extreme dipping), 38.9 % (Tuesday night; extreme dipping), and 30.1 % (Wednesday night; extreme dipping).

Acceptable limits for the double 24-hour amplitude of SBP (BP-2A) are from 3.27 to 37.15 mmHg for healthy women in her age group. From Thursday to Wednesday, her daily BP-2A estimates assumed values of 6.93 mmHg (Thursday;

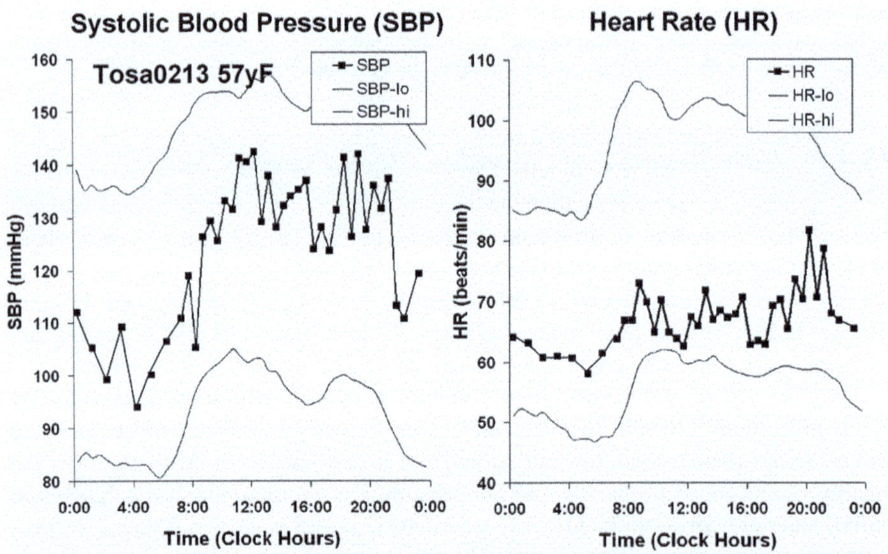

Fig. 18.44 Circadian profiles of systolic BP (*left*) and heart rate (HR) (*right*). 24-hour systolic BP and HR patterns of a 57-year-old woman, averaged over the 7 days of monitoring (*polygonal line* graph with *closed squares*), are shown along with time-specified prediction limits (*continuous thin lines*). Both SBP and HR are located inside their chronodesms. Both SBP and HR are located in the *middle* of their respective chronodesms

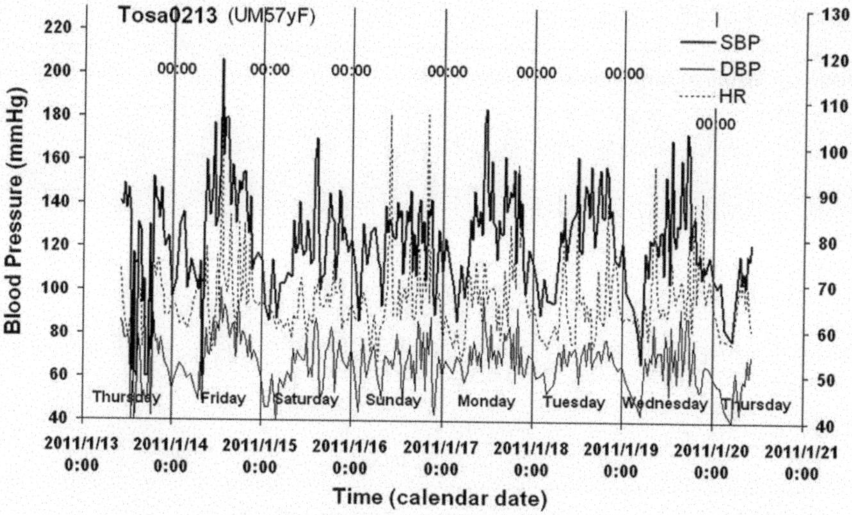

Fig. 18.45 An example of "Monday BP surge." Time plots of the 7-day/24-hour records of SBP (*thick continuous line*), DBP (*thin continuous line*), and HR (*dashed line*) of a 57-year-old woman. A clear increase in BP is observed on the 2nd day (Friday) and on the 5th day (Monday), indicating Monday BP surge with Friday surge

acceptable), 67.21 mmHg (Friday; CHAT), 14.43 mmHg (Saturday; acceptable), 27.03 mmHg (Sunday; acceptable), 47.55 mmHg (Monday; CHAT), 52.84 mmHg (Tuesday; CHAT), and 48.33 mmHg (Wednesday; CHAT). It is excessive on Friday, Monday, Tuesday, and Wednesday, when SBP-CHAT is diagnosed.

The circadian period of SBP was estimated by the Maximum Entropy Method (MEM), which requires the data to be equidistant. Outliers were deleted from the original SBP time series (middle section of Fig. 18.46). Data were then made equidistant (bottom section of Fig. 18.46).

The MEM spectrum is shown on the middle left of Fig. 18.47, which shows a prominent peak corresponding to the circadian component with a period of 1.0011 day, together with a smaller circasemidian peak (0.5108 day). Periods extracted by the MEM analysis in the range up to 2.5 cycles/day are shown in the right upper side of this figure. Nine components are listed here, and that with a "period" of 1.0011 day has the largest spectral power, as assessed by the "area" covered by the spectral peak.

On the bottom left of Fig. 18.47, the edited data are fitted by least squares with a 5-component model consisting of the five cosine curves with the five largest amplitudes, with corresponding periods of 3.750, 1.565, 1.173, 1.001, and 0.325 days (right lower part of Fig. 18.47).

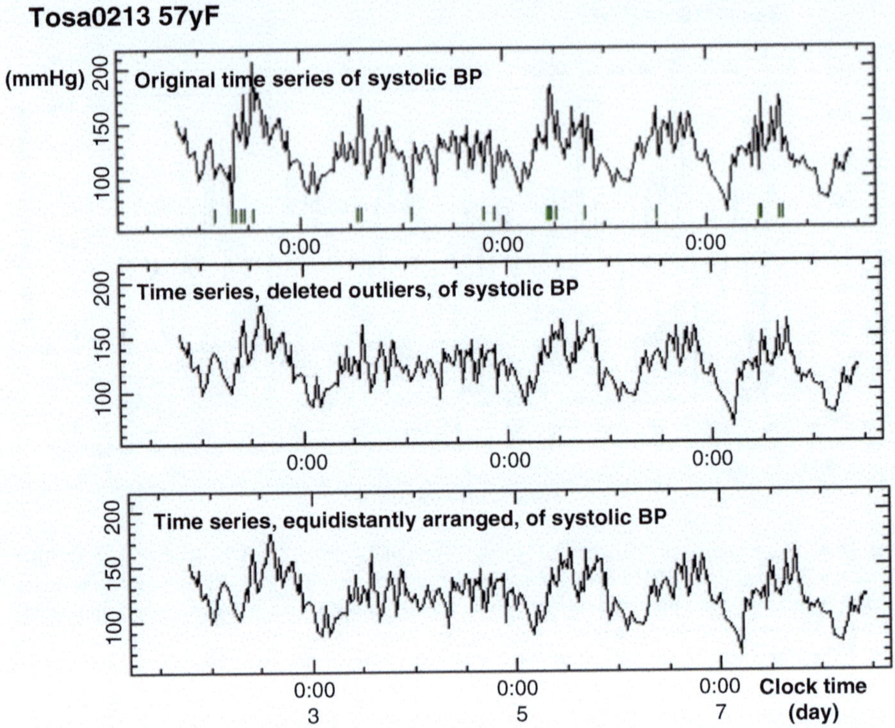

Fig. 18.46 Three versions of the time series of systolic BP. *Upper*: original data of systolic BP. *Middle*: edited data after mathematically excluding outliers. *Lower*: equidistantly arranged time series for analysis by the Maximum Entropy Method, using the MemCalc software (GMS Co., Tokyo)

In conclusion, the clinical diagnosis based on the chronobiologically interpreted 7-day/24-hour ABP monitoring consists of (1) basically MESOR-normotension, (2) Monday BP surge with Friday surge, (3) "masked extreme dipping" complicated by large between-day variation in the day-night ratio (from extreme to reverse dipping), and (4) SBP-CHAT on Friday, Monday, Tuesday, and Wednesday. Accordingly, we recommended her to strictly improve her lifestyle, focusing on salt intake and on observing a regular sleep-wake cycle. We also started antihypertensive treatment with a low-dose of CCB and antidepressive medications to improve her mood and minimize her sleep disturbances. Under this plan, she now seems to be in good health.

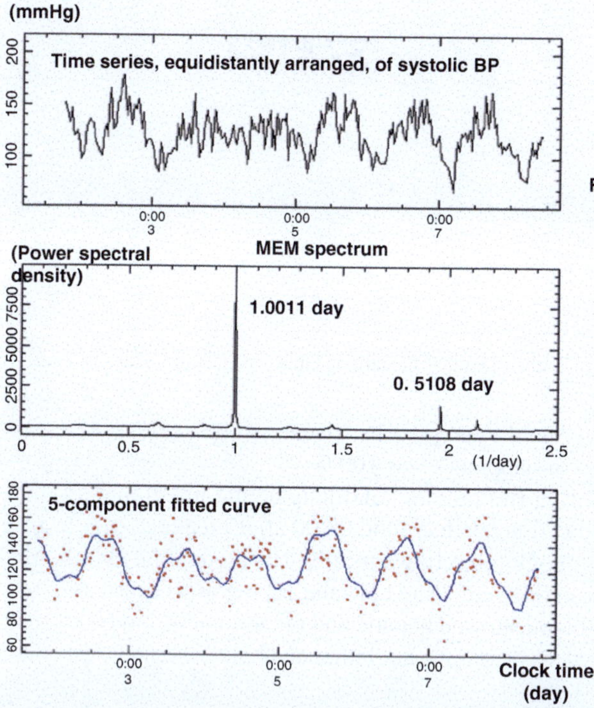

Tosa0213 57yF

Frequency (/ day)	Period (days)	Area
0.267	3.7503	17.07
0.639	1.5651	14.40
0.853	1.1727	10.73
0.999	1.0011	181.91
1.258	0.7952	4.82
1.452	0.6886	6.33
1.958	0.5108	8.70
2.129	0.4697	6.79
2.425	0.4124	3.03

Period (days)	Amplitude	Acro-phase
3.750	4.93	1-23:16
1.565	5.88	2-10:51
1.173	5.10	1-19:44
1.001	19.24	2-15:32
0.325	4.68	1-20:28

Fig. 18.47 Period analysis of the circadian rhythm of systolic BP. MEM analysis estimates the circadian period of SBP to be 1.0011 day (*left middle part*). Data are shown with a 5-component model fitted by least squares to the data, consisting of cosine curves with the five largest amplitudes, the 1.001-day component being very prominent (*right lower part*)

References

1. Halberg F, Powell D, Otsuka K, et al. Diagnosing vascular variability anomalies, not only MESOR-hypertension. Am J Physiol Heart Circ Physiol 2013;305:H279–94.
2. Saito K, Koyama A, Yoneyama K, et al., editors. A recent advance in time series analysis by maximum entropy method: applications to medical and biological sciences. Sapporo: Hokkaido University Press; 1994. p. 406.
3. The Japanese Society of Hypertension. The Japanese Society of Hypertension guidelines for the management of hypertension (JSH 2014). Hypertens Res 2014;37:253–390.
4. Kario K, Pickering TG. White-coat hypertension or white-cost hypertension syndrome: which is accompanied by target organ damage? Arch Intern Med 2000;160:3497–8.
5. Otsuka K, Okajima K, Oinuma S, et al. Aging and circadian disruption of blood pressure observed using 7-day/24-hour ambulatory blood pressure monitoring. J Am Geriatr Soc 2014;62:2213–5.

Chapter 19
Case Presentations II. Vascular Variability Syndrome

Abstract In this chapter, we introduce case presentations of vascular variability anomalies, including the Midline-Estimating Statistics Of Rhythm (i.e., MESOR)-hypertension and circadian hyper-amplitude-tension (i.e., CHAT), BP ecphasia, excessive pulse pressure, and "vascular variability syndrome," from a viewpoint of 7-day/24-hour ABP monitoring.

Keywords MESOR (Midline-Estimating Statistics Of Rhythm)-hypertension • Circadian hyper-amplitude-tension • Ecphasia • Excessive pulse pressure • Vascular variability syndrome

19.1 MESOR-Hypertension

In this chapter, we introduce three cases of "Midline-Estimating Statistics Of Rhythm (MESOR) hypertension" based on 7-day/24-hour ambulatory blood pressure (ABP) monitoring. The MESOR is a rhythm-adjusted mean that takes into consideration the rhythmic variation in the data, approximated by cosine curve(s) (left, Fig. 19.1). MESOR-hypertension is a chronobiologically validated elevation in BP (right, Fig. 19.1).

19.1.1 Case Report 1 of "MESOR-Hypertension"

The first example of "MESOR-hypertension" is a 56-year-old woman (ID number: KOts098). Her office BP and HR were 160/93 mmHg and 75 bpm, but she does not take any antihypertensive medication. Time series of systolic BP (SBP), diastolic BP (DBP), and heart rate measured by pulse (HR) are shown in Fig. 19.2. A model consisting of cosine curves with periods of 24 and 12 hours fitted by least squares to the data was used to estimate the MESOR and the amplitude and acrophase (time of predicted maximum) of each component. The MESORs of SBP, DBP, and HR, estimated from the 7-day/24-hour ABP measurements, were 152.9 mmHg, 82.2 mmHg, and 70.0 bpm, respectively. Thus, she is diagnosed as having SBP

K. Otsuka et al., *Chronomics and Continuous Ambulatory Blood Pressure Monitoring*, DOI 10.1007/978-4-431-54631-3_19

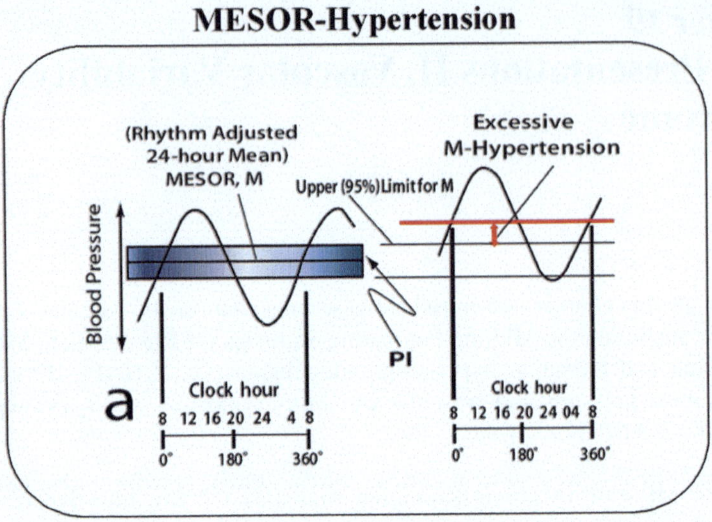

Fig. 19.1 MESOR-hypertension. MESOR-hypertension (MH) is a chronobiologically validated elevation in BP (*right*, Fig. 19.1), which can be systolic (S-MH), diastolic (D-MH), or both (SD-MH), or mean arterial (MA-MH), demonstrated parametrically. Although this graph is demonstrated with a single component model, the data are usually fitted with a two-component model consisting of cosine curves with periods of 24 and 12 hours. Acceptable limits for the MESOR have been derived as 90 % prediction limits (PI) from ABP obtained from clinically healthy peers matched by gender and age. "MESOR-hypertension" (MH) is used only if the diagnosis is based on the MESOR obtained by the least squares fit of cosine curves with anticipated (24- and 12-hour) periods, for comparison with ranges (90 % P!s) of acceptable MESORs characterizing data from clinically healthy peers matched by gender and age. The minimal time series length is a 24-hour/7-day record of data collected automatically at hourly or shorter intervals, analyzed both daily and for the week as a whole. It serves to rule out MH if negative when analyzed parametrically for the weeklong record as a whole, but data collection is continued when abnormality is found. MH is a condition where the BP-M is above the upper 95 % prediction limit of BP-Ms from clinically healthy peers matched by gender and age. BP elevation is noted when the hyperbaric index (extent of excess during 24 hours) exceeds the threshold of 50 mmHg x hour during 24 hours. MH can be the sole vascular variability disorder (VVD) or it can coexist and/or alternate with other VVDs to constitute a vascular variability syndrome (VVS)

MESOR-hypertension. The difference between the MESORs of SBP and DBP was 70.7 mmHg. As it is above 60 mmHg, she is also diagnosed as having an excessive pulse pressure (EPP), another vascular variability disorder (VVD).

The 24-hour double amplitudes of SBP, DBP, and HR were 28.42 mmHg, 13.00 mmHg, and 19.82 bpm, respectively, all within acceptable limits. The circadian acrophases of SBP, DBP, and HR were 13:12, 12:18, and 15:58, respectively, all within acceptable limits (between ~12:00 and ~16:00).

The 24-hour SBP pattern, as shown in the left side of Fig. 19.3, exceeds the upper limit of the chronodesm during most of the time along the 24-hour scale. On the average over 7 days, SBP remains above 130 mmHg, indicative of persistent 24-hour ABP hypertension. By contrast, HR is located well within the chronodesmic limits.

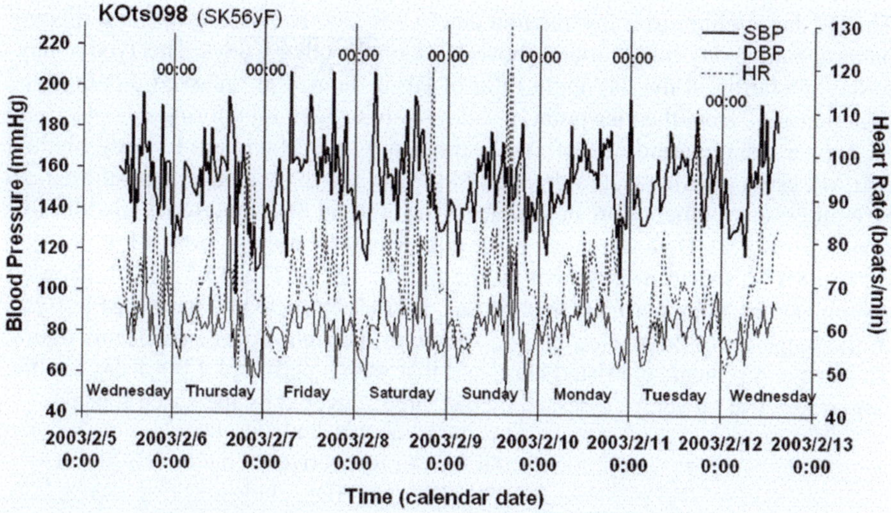

Fig. 19.2 An example of SBP MESOR-hypertension. Time plots of the 7-day/24-hour records of SBP (*thick continuous line*), DBP (*thin continuous line*), and HR (*dashed line*) of a 56-year-old woman. SBP MESOR-hypertension is complicated by excessive pulse pressure (EPP) and masked dipping-type hypertension (nocturnal BP displays a non-dipping pattern on the 1st day, but a dipping pattern prevails during the subsequent 6 days)

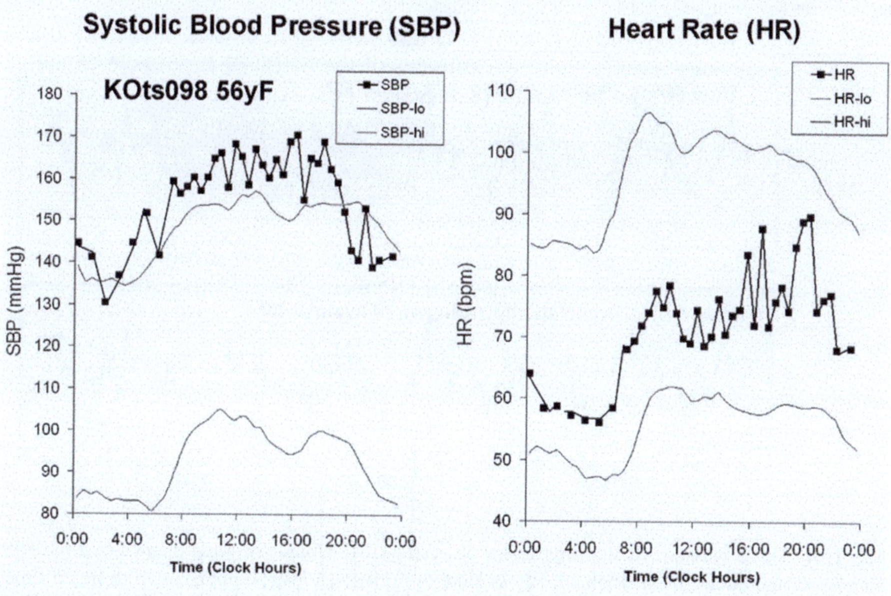

Fig. 19.3 Circadian profiles of systolic BP (*left*) and heart rate (HR) (*right*) plotted in the light of circadian chronodesms (*thin lines*). The average 24-hour SBP pattern is elevated during most of the time, exceeding the upper chronodesmic limit

The BP day-night ratio is low the first day (5.8 %), corresponding to a non-dipping pattern, but the day-night ratio is above 10 % on the other 6 days. This type of day-to-day alteration in the day-night ratio of BP is defined as "masked dipping-type hypertension" from the viewpoint of 7-day/24-hour ABP monitoring.

The circadian period of SBP was estimated by the Maximum Entropy Method (MEM), using the MemCalc software (GMS Co., Tokyo), which requires the data to be equidistant. Outliers were deleted from the original SBP time series (middle section of Fig. 19.4). Data were then made equidistant (bottom section of Fig. 19.4).

The MEM spectrum is shown on the middle left of Fig. 19.5, which shows a prominent circadian period of 1.0061 day. Periods extracted by the MEM analysis in the range up to 2.5 cycles/day are shown in the right upper side of this figure. Thirteen components are listed here, and that with a period of 1.0061 day has the largest spectral power, as assessed by the "area" covered by the spectral peak.

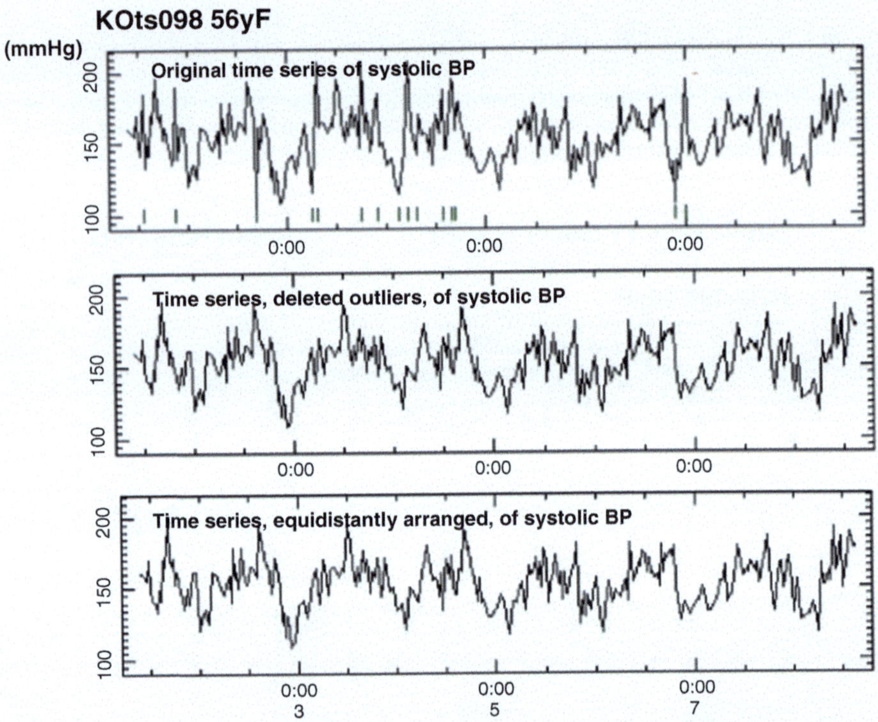

Fig. 19.4 Three versions of the time series of systolic BP. *Upper*: original data of systolic BP. *Middle*: edited data after mathematically excluding outliers. *Lower*: equidistantly arranged time series for analysis by the Maximum Entropy Method, using the MemCalc software (GMS Co., Tokyo)

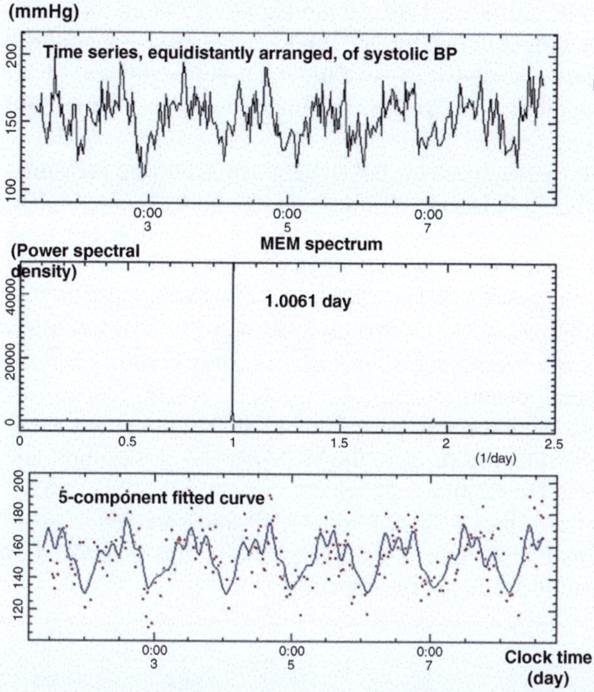

KOts098 56yF

Frequency (/ day)	Period (days)	Area
0.000	–	0.17
0.221	4.5217	1.83
0.453	2.2081	2.08
0.618	1.6177	1.21
0.755	1.3244	2.20
0.994	1.0061	103.29
1.062	0.9420	4.38
1.318	0.7587	4.25
1.487	0.6723	4.27
1.683	0.5942	0.75
1.935	0.5169	9.61
2.042	0.4897	2.41
2.269	0.4407	1.62
2.439	0.4101	7.93

Period (days)	Amplitude	Acro-phase
1.003	14.29	1-12:44
0.517	4.52	1-16:25
0.410	3.95	1-09:49
0.231	3.65	1-10:50
0.188	3.38	1-11:31

Fig. 19.5 Period analysis of the circadian rhythm of systolic BP. MEM analysis estimates the circadian period of systolic BP to be 1.0061 day (*left middle part*). Data are shown with a five-component model fitted by least squares to the data, consisting of cosine curves with the five largest amplitudes (*right lower part*)

On the bottom left of Fig. 19.5, the edited data are fitted by least squares with a five-component model consisting of the five cosine curves with the five largest amplitudes, with corresponding periods of 1.006, 0.517, 0.410, 0.231, and 0.188 days (right lower part of Fig. 19.5).

In conclusion, the clinical diagnosis based on the chronobiologically interpreted 7-day/24-hour ABP monitoring consists of (1) SBP MESOR-hypertension (152.9 mmHg), (2) excessive pulse pressure (70.7 mmHg), (3) persistent 24-hour ABP hypertension, and (4) "masked dipping"-type hypertension. This patient thus has moderate to severe hypertension. Accordingly, we immediately started treatment with a long-acting CCB (benidipine, 8 mg twice daily).

19.1.2 Case Report 2 of "MESOR-Hypertension"

The second example of "MESOR-hypertension" is a 48-year-old non-obese (BMI = 21.2 kg/m^2) woman (ID number: KOts051). She is treated with a long-acting calcium channel blocker (CCB) and an angiotensin II receptor blocker (ARB) once daily, but her office BP and HR still were 176/110 mmHg and 75 bpm. She takes care of her health and has something to live for, but she has a depressive mood; her PHQ-2 score was 2 points. Her subjective quality of life (QOL) of health was 75 %, and her subjective QOL of happiness was 70 %. On average, she sleeps 8 hours and 50 min, perhaps a little too long.

Time series of systolic BP (SBP), diastolic BP (DBP), and heart rate measured by pulse (HR) are shown in Fig. 19.6. During all 7 days, the 24-hour average of SBP is invariably above 130 mmHg, indicative of persistent ABP hypertension. Her day-night ratio of BP varies from day to day, however. On the first day, it is 9.8 % (non-dipping), and on days 2–7, it changes to 18.2 % (dipping), 14.7 % (dipping), 20.9 % (extreme dipping), 18.9 % (dipping), 20.4 % (extreme dipping), and 18.9 % (dipping). Her "masked dipping-type hypertension" includes not only non-dipping and dipping but also extreme-dipping patterns.

A model consisting of cosine curves with periods of 24 and 12 hours was fitted by least squares to the data, yielding estimates of the MESOR, and of the amplitude and acrophase (time of predicted maximum) of each component. MESORs of SBP, DBP, and HR, estimated from the 7-day/24-hour ABP measurements, were 153.9 mmHg, 98.5 mmHg, and 71.4 bpm, respectively. She is thus diagnosed as having both systolic and diastolic MESOR-hypertension.

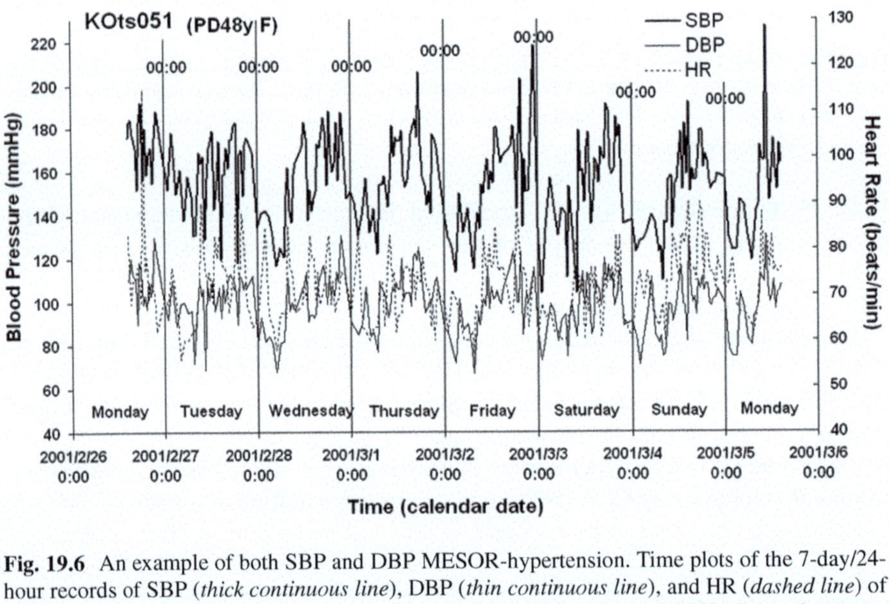

Fig. 19.6 An example of both SBP and DBP MESOR-hypertension. Time plots of the 7-day/24-hour records of SBP (*thick continuous line*), DBP (*thin continuous line*), and HR (*dashed line*) of a 48-year-old woman. MESOR-hypertension is complicated by "masked dipping"-type hypertension (non-dipping on the 1st day, but dipping or extreme dipping on subsequent days)

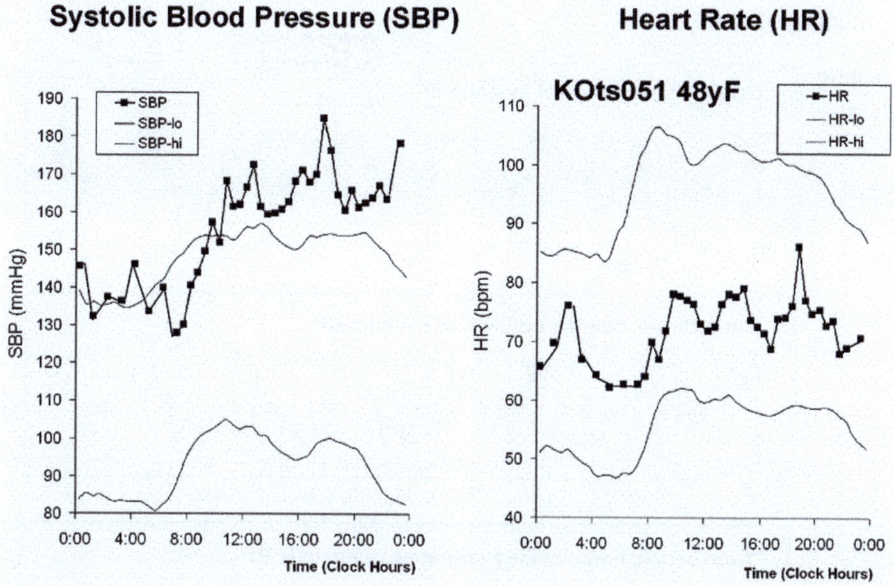

Fig. 19.7 Circadian profiles of systolic BP (*left*) and heart rate (HR) (*right*) plotted in the light of circadian chronodesms. The average 24-hour SBP pattern is elevated above the upper chronodesmic limit almost during the entire 24-hour period

The 24-hour SBP pattern, as shown on the left side of Fig. 19.7, lies mostly above the upper limit of the chronodesm. By contrast, HR varies well within the acceptable limits. The circadian acrophases of SBP, DBP, and HR were 16:40, 16:32, and 16:06, respectively, slightly phase-delayed with respect to acrophases of clinically healthy peers (between ~12:00 and ~16:00). The 24-hour double amplitudes of SBP, DBP, and HR were 36.39 mmHg, 21.92 mmHg, and 10.97 bpm, respectively, all within acceptable limits.

The circadian period of SBP was estimated by the Maximum Entropy Method (MEM), using the MemCalc software (GMS Co., Tokyo), which requires the data to be equidistant. Outliers were deleted from the original SBP time series (middle section of Fig. 19.8). Data were then made equidistant (bottom section of Fig. 19.8).

The MEM spectrum is shown on the middle left of Fig. 19.9, which shows a prominent circadian period of 0.9925 day, together with two other clear peaks corresponding to periods of 9.7226 and 0.5648 days. Periods extracted by the MEM analysis in the range up to 2.5 cycles/day are shown in the right upper side of this figure. Ten components are listed here, and the one with a period of 0.9925 day has the largest spectral power, as assessed by the "area" covered by the spectral peak.

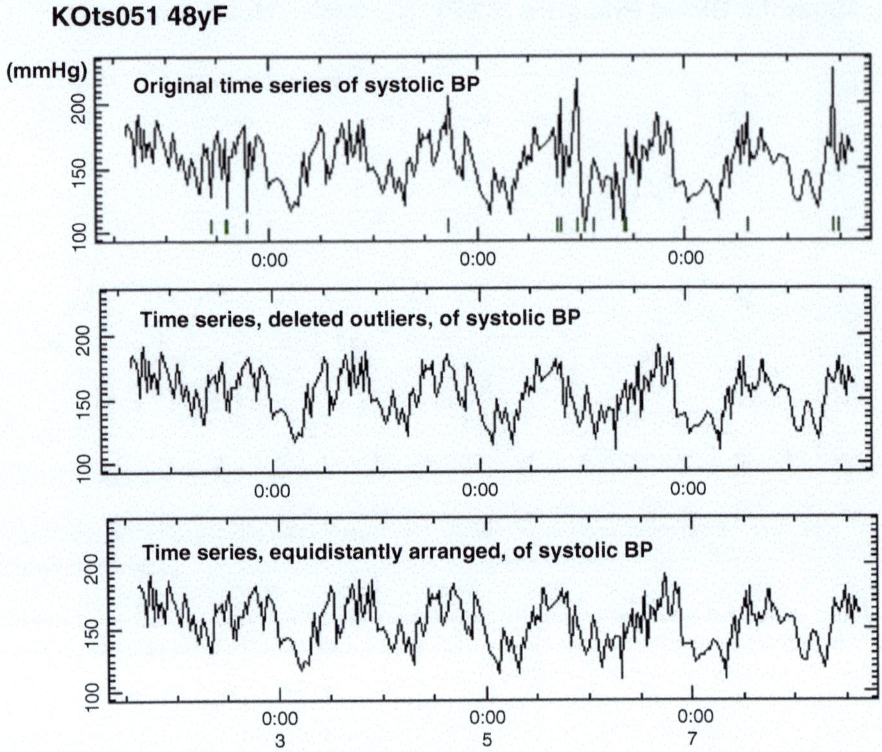

Fig. 19.8 Three versions of the time series of systolic BP. *Upper*: original data of systolic BP. *Middle*: edited data after mathematically excluding outliers. *Lower*: equidistantly arranged time series for analysis by the Maximum Entropy Method, using the MemCalc software (GMS Co., Tokyo)

On the bottom left of Fig. 19.9, the edited data are fitted by least squares with a five-component model consisting of the five cosine curves with the five largest amplitudes, with corresponding periods of 9.723, 2.136, 0.992, 0.565, and 0.250 days (right lower part of Fig. 19.9).

In conclusion, the clinical diagnosis based on the chronobiologically interpreted 7-day/24-hour ABP monitoring consists of (1) both SBP and DBP MESOR-hypertension (153.9/98.5 mmHg), (2) delayed phase of the circadian SBP rhythm, (3) persistent ABP hypertension, and (4) "masked dipping"-type hypertension, including non-dipping, dipping, and extreme-dipping patterns on different days. Results suggest that this subject has a moderate hypertensive condition. Accordingly, we reiterated lifestyle recommendations focusing on an improvement of her depressive mood. This treatment seems to work well and she appears to be in good health now.

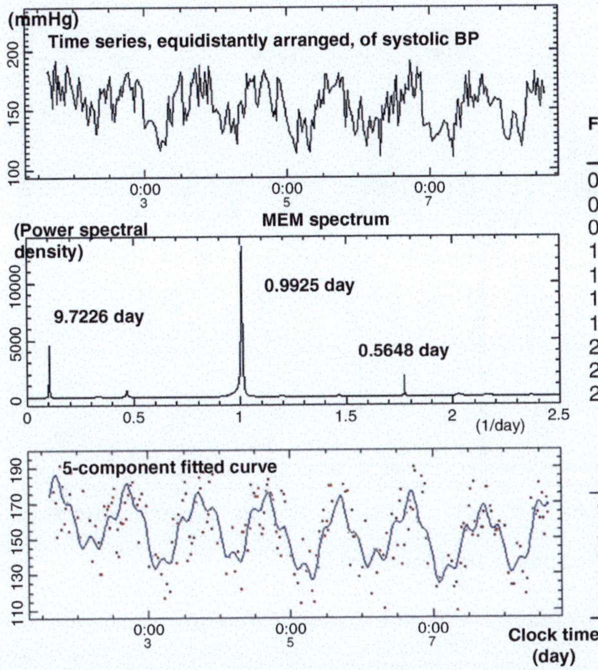

KOts051 48yF

Frequency (/ day)	Period (days)	Area
0.1029	9.7226	15.47
0.3341	2.9928	2.04
0.4682	2.1360	7.21
1.0076	0.9925	168.92
1.1957	0.8363	1.60
1.4648	0.6827	1.73
1.7705	0.5648	5.86
2.0259	0.4936	3.76
2.1621	0.4625	3.27
2.3648	0.4229	1.30

Period (days)	Amplitude	Acro-phase
9.723	4.81	1-21:00
2.136	3.81	2-02:04
0.992	17.90	1-17:10
0.565	3.05	1-16:29
0.250	5.20	1-16:34

Fig. 19.9 Period analysis of the circadian rhythm of systolic BP. MEM analysis estimates the circadian period of systolic BP to be 0.9925 day (*left middle part*). Data are shown with a five-component model fitted by least squares to the data, consisting of cosine curves with the five largest amplitudes (*right lower part*)

19.1.3 Case Report 3 of "MESOR-Hypertension"

The third example of "MESOR-hypertension" is a 36-year-old mildly obese (BMI = 26.7 kg/m^2) woman (ID number: KOts058). She has a regular rest-activity cycle and goes to bed usually at about 23:00 and gets up around 06:00. She is treated with a three-drug combination, a long-acting calcium channel blocker (CCB), an angiotensin II receptor blocker (ARB), and a β-blocker, taken once daily at about 08:00 after breakfast. Nevertheless, her office BP and HR were 145/96 mmHg and 96 bpm. She takes care of her health and has something to live for. She does not have a mild depressive mood, and her PHQ-2 score was 1 point. Her subjective quality of life (QOL) was low: her QOL of health was 65 %, and her QOL of happiness was 50 %. She does not have diabetes; her fasting glucose and hemoglobin A1c were 90 mg/dl and 4.6 %, respectively. She has a mild dyslipidemia, her blood total cholesterol, triglyceride, and HDL-cholesterol being 183, 150, and 68 mg/dl, respectively.

Time series of systolic BP (SBP), diastolic BP (DBP), and heart rate measured by pulse (HR) are shown in Fig. 19.10. During the 7 days of monitoring, the 24-hour average of SBP is invariably above 130 mmHg, indicating persistent ABP hyperten-

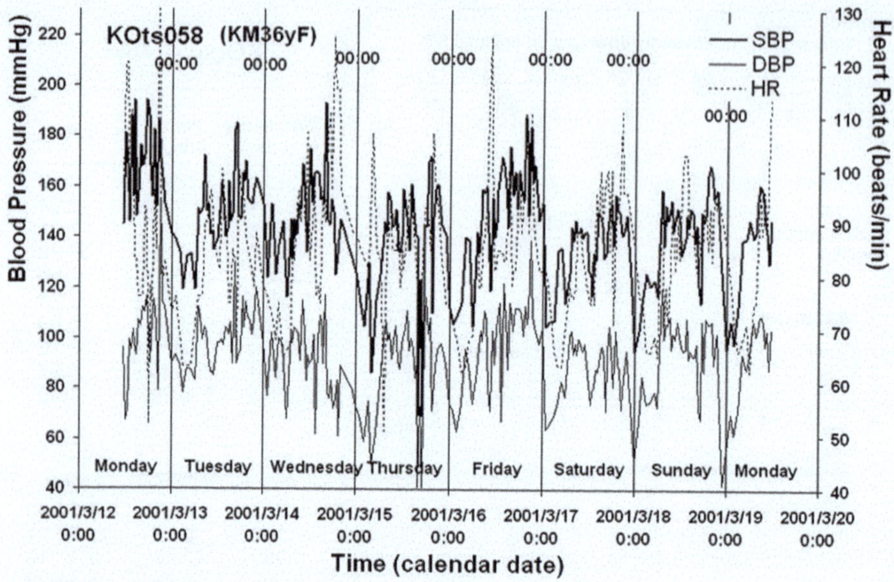

Fig. 19.10 An example of both SBP and DBP MESOR-hypertension. Time plots of the 7-day/24-hour records of SBP (*thick continuous line*), DBP (*thin continuous line*), and HR (*dashed line*) of a 36-year-old woman. MESOR-hypertension is complicated by "masked non-dipping"-type hypertension (extreme-dipping on the 1st day, but non-dipping pattern on the 2nd day)

sion. Her BP day-night ratio varies from day to day, however. On the first day, it is 21.3 % (extreme dipping), but on days 2–7, it changes to 8.6 % (non-dipping), 26.6 % (extreme dipping), 13.3 % (dipping), 21.9 % (extreme dipping), 19.3 % (dipping), and 21.0 % (extreme dipping). She shows a "masked non-dipping-type hypertension," including not only non-dipping but also dipping and extreme-dipping patterns on different days.

A model consisting of cosine curves with periods of 24 and 12 hours was fitted by least squares to the data to estimate the MESOR and the amplitude and acrophase (time of predicted maximum) of each component. MESOR estimates of SBP, DBP, and HR, calculated based on the 7-day/24-hour ABP measurements, were 142.0 mmHg, 91.3 mmHg, and 84.3 bpm, respectively. She is thus diagnosed as having both systolic and diastolic MESOR-hypertension.

The 24-hour SBP pattern, as shown in the left side of Fig. 19.11, exceeds the upper limit of the chronodesm during most of the time along the 24-hour scale. By contrast, HR is located well within the chronodesmic limits. The circadian acrophase of SBP, DBP, and HR was 15:28, 14:13, and 16:13, respectively. Acrophases of systolic and diastolic BP were within acceptable limits (between ~12:00 and ~16:00). The 24-hour double amplitudes of SBP and DBP were 30.52 and 17.01 mmHg, respectively, both within acceptable limits.

A model consisting of cosine curves with periods of 24 and 12 hours was fitted by least squares to the data to estimate the MESOR and the amplitude and acro-

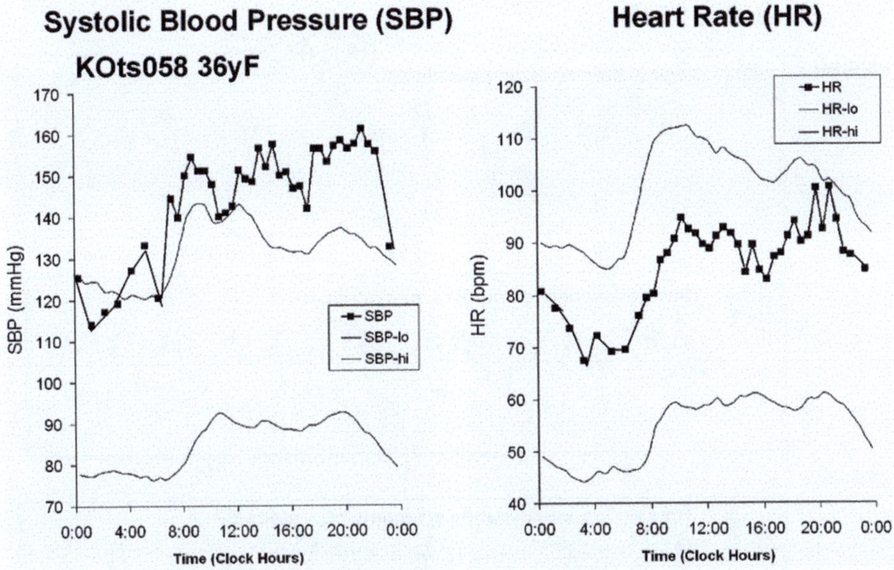

Fig. 19.11 Circadian profiles of systolic BP (*left*) and heart rate (HR) (*right*) plotted in the light of circadian chronodesms (*thin lines*). The average 24-hour SBP pattern is elevated during most of the time, exceeding the upper chronodesmic limit. The circadian profile of HR, however, is located within the chronodesmic limits

phase (time of predicted maximum) of each component. The MESORs of SBP, DBP, and HR, based on the 7-day/24-hour ABP measurements, were 153.9 mmHg, 98.5 mmHg, and 71.4 bpm, respectively. This subject is thus diagnosed as having both systolic and diastolic MESOR-hypertension.

The 24-hour SBP pattern, as shown in the left side of Fig. 19.11, exceeds the upper limit of the chronodesm during most of the time along the 24-hour scale. By contrast, HR is located well within the chronodesmic limits. The circadian acrophase of SBP, DBP, and HR was 15:28, 14:13, and 16:13, respectively, mostly within acceptable limits (between ~12:00 and ~16:00). The 24-hour double amplitude of SBP, DBP, and HR was 36.39 mmHg, 21.92 mmHg, and 10.97 bpm, respectively, all within acceptable limits.

The circadian period of SBP was estimated by the Maximum Entropy Method (MEM), using the MemCalc software (GMS Co., Tokyo), which requires the data to be equidistant. Outliers were deleted from the original SBP time series (middle section of Fig. 19.12). Data were then made equidistant (bottom section of Fig. 19.12).

The MEM spectrum is shown on the middle left of Fig. 19.13, which shows a prominent circadian period of 0.9869 day, together with three other clear peaks corresponding to periods of 10.4674, 3.3828, and 0.4970 days.. Periods extracted by the MEM analysis in the range up to 2.5 cycles/day are shown in the right upper side of this figure. Ten components are listed here, and that with a period of 0.9925 day

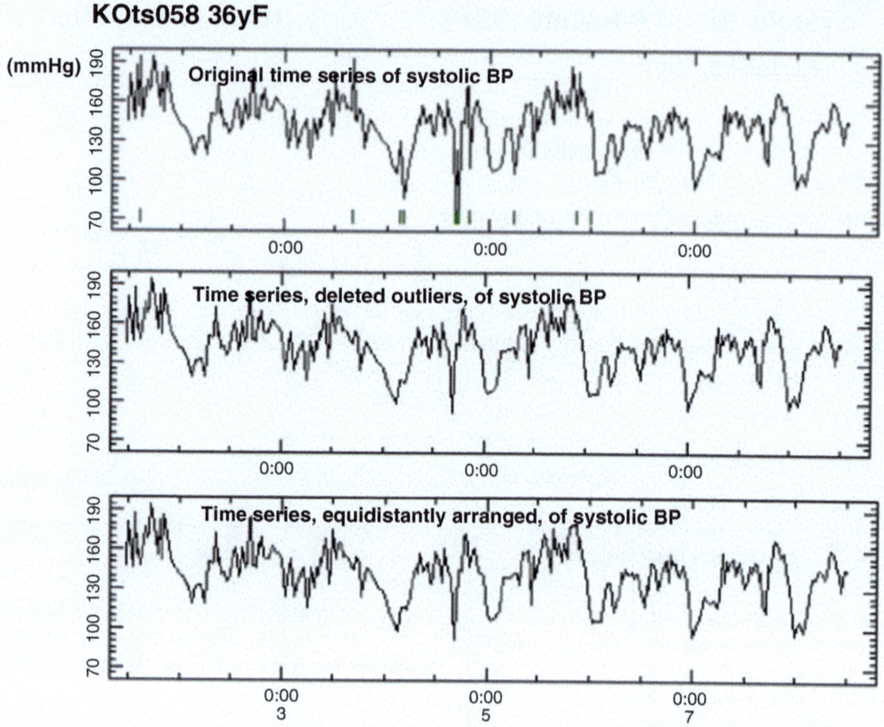

Fig. 19.12 Three versions of the time series of systolic BP. *Upper*: original data of systolic BP. *Middle*: edited data after mathematically excluding outliers. *Lower*: equidistantly arranged time series for analysis by the Maximum Entropy Method, using the MemCalc software (GMS Co., Tokyo)

has the largest spectral power, as assessed by the "area" covered by the spectral peak.

On the bottom left of Fig. 19.13, the edited data are fitted by least squares with a five-component model consisting of the five cosine curves with the five largest amplitudes, with corresponding periods of 10.467, 3.383, 2.079, 0.987, and 0.497 days (right lower part of Fig. 19.13).

In conclusion, the clinical diagnosis based on the chronobiologically interpreted 7-day/24-hour ABP monitoring consists of (1) both SBP and DBP MESOR-hypertension (142.0/91.3 mmHg), (2) persistent ABP hypertension, and (3) "masked non-dipping"-type hypertension, including extreme-dipping, dipping, and non-dipping patterns on different days. This subject is still moderately hypertensive despite being on three antihypertensive medications. Accordingly, we reiterated recommendations for improving her lifestyle, focusing on salt intake and on improving her mood. The treatment seems to be working well and she is now healthy and normotensive.

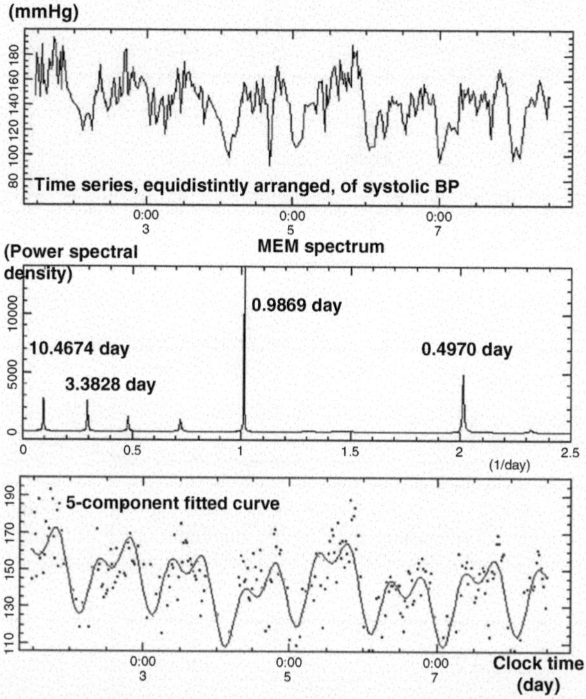

KOts058 36yF

Frequency (/ day)	Period (days)	Area
0.0955	10.4674	19.94
0.2956	3.3828	19.50
0.4811	2.0787	12.22
0.7207	1.3875	12.05
1.0132	0.9869	119.33
1.3071	0.7651	3.28
1.4791	0.6761	3.85
1.8087	0.5529	0.54
2.0119	0.4970	63.28
2.3193	0.4312	6.00

Period (days)	Amplitude	Acro-phase
10.467	6.06	11-13:35
3.383	6.28	2-07:12
2.079	5.89	3-09:56
0.987	14.93	1-16:20
0.497	10.65	1-20:40

Fig. 19.13 Period analysis of the circadian rhythm of systolic BP. MEM analysis estimates the circadian period of systolic BP to be 0.9869 day (*left middle part*). Data are shown with a five-component model fitted by least squares to the data, consisting of cosine curves with the five largest amplitudes (*right lower part*)

19.2 Blood Pressure Circadian Hyper-Amplitude-Tension (CHAT)

In this chapter, we introduce three cases of "circadian hyper-amplitude-tension (CHAT)" from the viewpoint of 7-day/24-hour ambulatory blood pressure (ABP) monitoring. CHAT is a circadian overswing (excessive BP swing), a condition characterized by a double 24-hour amplitude of BP (BP-2A), derived from a 24-hour/7-day record, exceeding the upper 95 % prediction limit of BP-2A estimates from clinically healthy peers matched by gender and age (Fig. 19.14).

CHAT can occur alone in MESOR-normotension and with a usual timing of the circadian BP rhythm, or it can coexist and/or alternate with other vascular variability disorders (VVDs), such as complicating MESOR-hypertension.

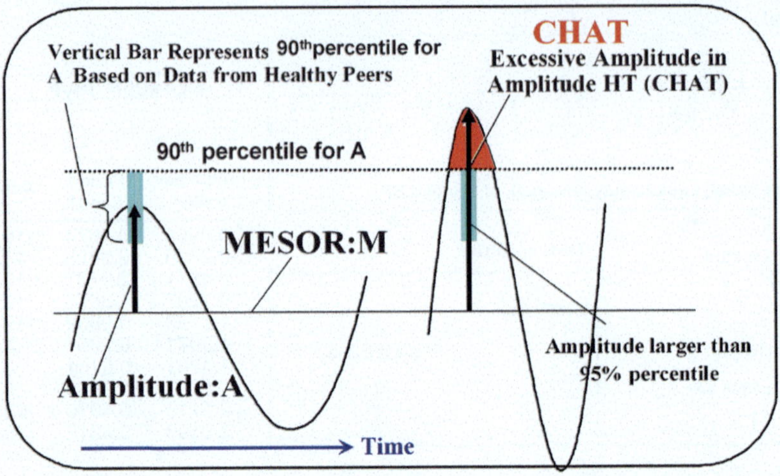

Fig. 19.14 CHAT (circadian hyper-amplitude-tension). CHAT means circadian over-swinging (excessive BP swing), which is a condition characterized by a double amplitude of BP exceeding the upper 95 % prediction limit of double amplitudes of BP from clinically healthy peers matched by gender and age

19.2.1 Case Report 1 of "CHAT"

The first example of "CHAT" is a 61-year-old non-obese (BMI = 21.6 kg/m^2) woman (ID number: Tosa0275). Her weight was 52.5 kg and height 156 cm. She goes to bed usually around 23:30 and gets up at 07:20. She falls asleep within 10 min and feels that she sleeps well most days. She does not smoke and usually does not drink alcohol. She eats meat only rarely, but frequently eats fish with an average intake of vegetables. She prefers light seasoning.

She feels healthy and takes care of her health. She has something to live for and has no depressive mood. Her scores on the GDS-15 and PHQ-2 scales were 6 and 0 points, respectively. Her subjective quality of life (QOL) of health, assessed by a visual analog scale (VAS), was 85 %, but her subjective QOL of happiness was a little low (65 %).

She does not take any antihypertensive medication. Her office BP was 127/83 mmHg, and her home BP measurements averaged over 30 days were 130.9/81.6 mmHg in the morning and 120.4/76.9 mmHg in the evening. Time series of systolic BP (SBP), diastolic BP (DBP), and heart rate measured by pulse (HR) are shown in Fig. 19.15. A model consisting of cosine curves with periods of 24 and 12 hours was fitted by least squares to the data to estimate the MESOR (Midline-Estimating Statistic Of Rhythm) and the amplitude and acrophase (time of predicted maximum) of each component. MESOR estimates of SBP, DBP, and HR, estimated from the 7-day/24-hour ABP measurements, were 125.1 mmHg, 77.6 mmHg, and 73.1 bpm, respectively. In accordance with the guidelines for management of high

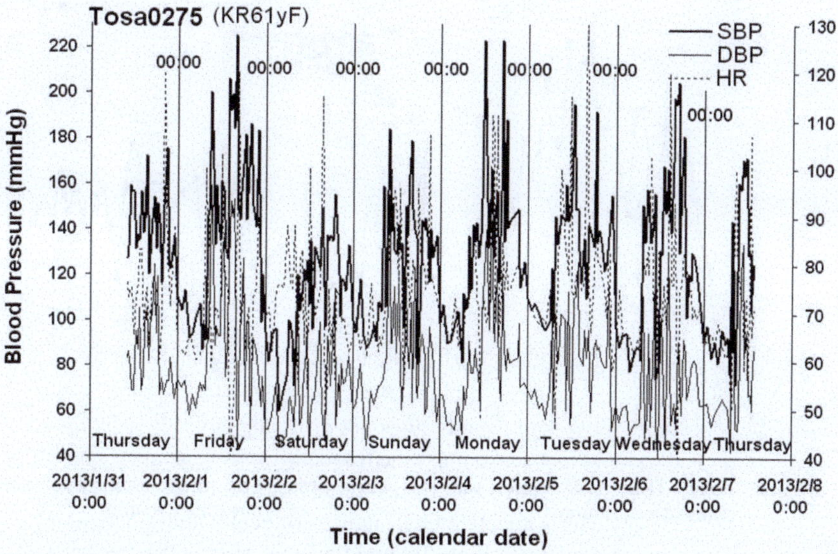

Fig. 19.15 Both SBP CHAT and DBP CHAT occur in masked ABP normotension with a usual timing of the circadian BP rhythm. Time plots of the 7-day/24-hour records of SBP (*thick continuous line*), DBP (*thin continuous line*) and HR (*dashed line*) of a 61-year-old woman. The 24-hour average of her SBP (in mmHg) varies from day to day, as follows: 132.7 (Thursday; hypertensive), 135.3 (Friday, hypertensive), 119.7 (Saturday; normotensive), 126.3 (Sunday; normotensive), 137.6 (Monday; hypertensive), 130.4 (Tuesday; hypertensive), and 127.4 (Wednesday; normotensive)

blood pressure based on ABP in Japan published in 2014, she is normotensive; her 24-hour SBP and 24-hour DBP averages are lower than 130 mmHg and 80 mmHg, respectively.

The circadian profile of SBP shows an excessive BP swing, that is, extreme dipping at night and overshooting above a time-varying threshold in the daytime (Fig. 19.16, left). The circadian profile of HR is mostly located in the center of the chronodesmic limits (Fig. 19.16, right).

Based on the fitted model, the circadian acrophases of SBP, DBP, and HR were 15:28, 14:42, and 14:59, respectively, within acceptable limits (between ~12:00 and ~16:00). The double circadian amplitudes of SBP and DBP were 60.23 and 32.40 mmHg, both exceeding the upper 95 % prediction limit in health. Daily estimates of her SBP double amplitude (BP-2A) changed from Thursday to Wednesday, assuming values of 49.48, more than 100, 37.15, 48.29, 60.61, 48.52, and 75.73 mmHg, respectively, all exceeding the upper 95 % prediction limit of BP-2A (34.00 mmHg in this case, estimated from clinically healthy women of her age group).

The circadian period of SBP was estimated by the Maximum Entropy Method (MEM), using the MemCalc software (GMS Co., Tokyo), which requires the data to be equidistant. Outliers were deleted from the original SBP time series (middle section of Fig. 19.17). Data were then made equidistant (bottom section of Fig. 19.17).

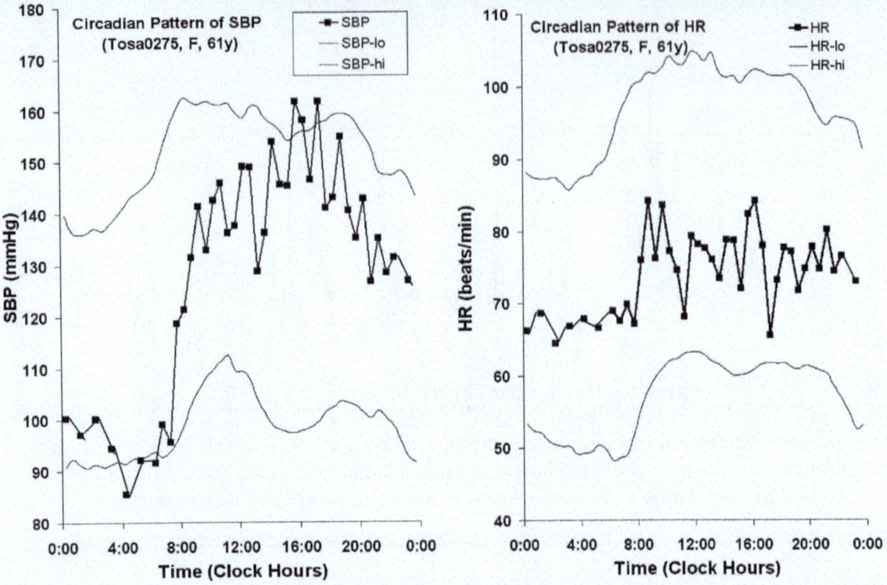

Fig. 19.16 Circadian profiles of systolic BP (*left*) and heart rate (HR) (*right*). 24-hour systolic BP (*left*) and heart rate (HR) (*right*) patterns of a woman with CHAT, averaged over the 7 days of monitoring (*polygonal line* graph with closed *squares*), are shown along with time-specified prediction limits (*continuous thin lines*)

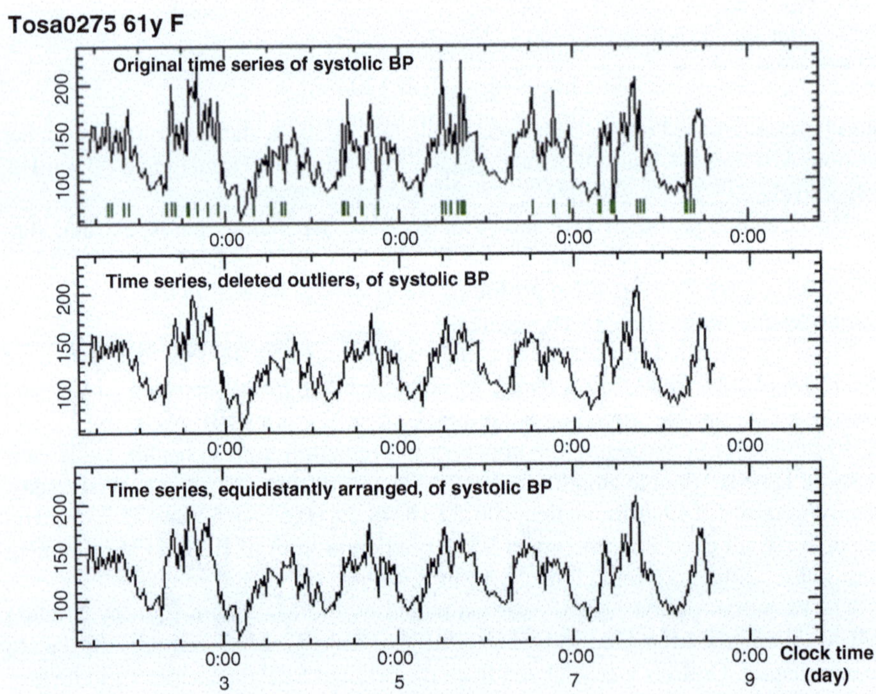

Fig. 19.17 Three versions of the time series of systolic BP. *Upper*: original data of systolic BP. *Middle*: edited data after mathematically excluding outliers. *Lower*: equidistantly arranged time series for analysis by the Maximum Entropy Method, using the MemCalc software (GMS Co., Tokyo)

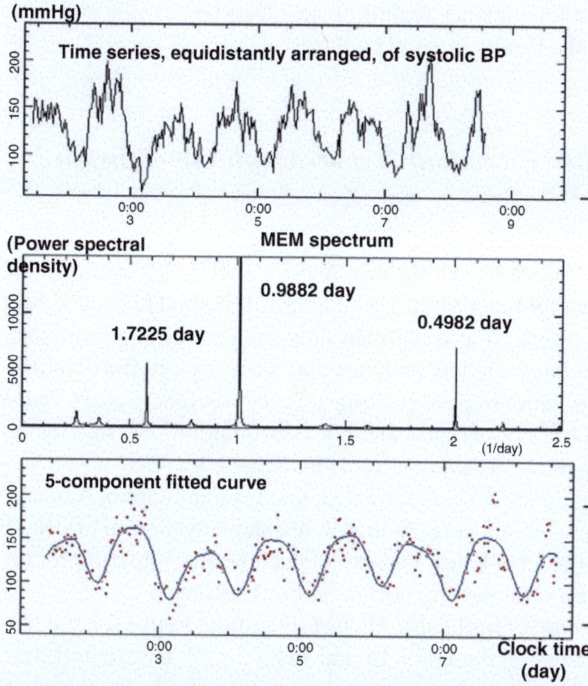

Tosa0275 61y F

Frequency (/ day)	Period (days)	Area
0:000	–	0.76
0:256	3.9102	19.24
0:361	2.7664	16.50
0:581	1.7225	31.12
0:790	1.2654	11.45
1.012	0.9882	436.12
1.409	0.7098	11.41
1.604	0.6234	4.43
2.007	0.4982	25.11
2.227	0.4491	5.29
2.490	0.4017	5.13

Period (days)	Amplitude	Acro-phase
3.910	6.41	2-03:02
2.766	4.77	2-05:30
1.723	7.71	2-13:55
0.988	29.77	1-16:16
0.498	7.02	1-21:36

Fig. 19.18 Period analysis of the circadian rhythm of systolic BP. MEM analysis estimates the circadian period of systolic BP to be 0.9882 day (*left middle part*). Data are shown with a five-component model fitted by least squares to the data, consisting of cosine curves with the five largest amplitudes, the 0.988 day being the most prominent (*lower right part*)

The MEM spectrum is shown on the middle left of Fig. 19.18, which shows a prominent circadian period of 0.9882 day, together with two clear peaks with periods of 1.7225 and 0.4982 days. Periods extracted by the MEM analysis in the range up to 2.5 cycles/day are shown in the right upper side of this figure. Ten components are listed here, and that with a period of 0.9882 day has the largest spectral power, as assessed by the "area" covered by the spectral peak.

On the bottom left of Fig. 19.18, the edited data are fitted by least squares with a five-component model consisting of the five cosine curves with the five largest amplitudes, with corresponding periods of 3.910, 2.766, 1.723, 0.988, and 0.498 days (right lower part of Fig. 19.18).

In conclusion, the clinical diagnosis based on the chronobiologically interpreted 7-day/24-hour ABP monitoring consists of (1) both SBP and DBP CHAT, (2) persistent SBP CHAT, and (3) masked ABP normotension, with a usual timing of the circadian BP rhythm. We have reported that CHAT is associated with a very high risk of ischemic stroke and other catastrophic disease. This subject also had dyslipidemia; her blood LDL-cholesterol and HDL-cholesterol were 159 and 51 mg/dl, respectively (LDL/HDL ratio = 3.12), although she did not have diabetes. Her ECG and CAVI were normal. After giving her lifestyle recommendations, focusing notably on salt intake, and providing her with guidance regarding her sleep-wake cycle and how to improve her dyslipidemia, she is now in good health.

19.2.2 Case Report 2 of "Circadian Hyper-Amplitude-Tension (CHAT)"

Another example of "CHAT" is a 63-year-old non-obese (BMI = 22.2 kg/m^2) man (ID number: Tosa0292). His weight was 59.0 kg and height 163 cm. He goes to bed usually at about midnight and gets up at 07:30. He falls asleep within 15 min and feels that he sleeps well most days. He has diabetes and a history of effort angina, which was treated by coronary aorta bypass grafting surgery when he was 58 years old. He stopped smoking and his occasional alcohol consumption. He frequently eats fish and his intake of vegetables is average, as is his seasoning.

His diabetes and dyslipidemia are now well treated, and his hemoglobin A1c and LDL-/HDL-cholesterol were 6.7 % and 115/79 mg/dl, respectively. Serum uric acid was 4.5 mg/dl. He has no atrial fibrillation and no left ventricular hypertrophy on ECG, and his CAVI was 8.0 m/s, suggesting normal arterial stiffness.

He feels healthy and takes care of his health. He has something to live for and has no depressive mood. Both scores on the GDS-15 and PHQ-2 scales were null. His subjective quality of life (QOL) of happiness, however, assessed by a visual analog scale (VAS), was low (47 %), while his subjective QOL of health was 85 %.

Coronary artery disease and hypertension are treated by angiotensin II receptor blocker (ARB) once daily. Nevertheless, his office BP remains high at 153/95 mmHg, and his home BP measurements averaged over 30 days were 137.8/85.0 mmHg in the morning and 116.9/74.1 mmHg in the evening.

Time series of systolic BP (SBP), diastolic BP (DBP), and heart rate measured by pulse (HR) are shown in Fig. 19.19. MESOR estimates of SBP, DBP, and HR, estimated from the 6-day/24-hour ABP measurements, were 126.8 mmHg, 79.8 mmHg, and 64.8 bpm, respectively. In accordance with the guidelines for management of high blood pressure based on ABP in Japan, published in 2014, he is normotensive (24-hour SBP and DBP averages are lower than 130 mmHg and 80 mmHg, respectively). However, day-to-day analyses indicated "masked ABP normotension," as the 24-hour average of SBP changed from day to day: 141.5 (Saturday; hypertensive), 135.0 (Sunday; hypertensive), 125.8 (Monday; normotensive), 144.9 (Tuesday; hypertensive), 125.1 (Wednesday; normotensive), and 124.5 mmHg (Thursday; normotensive).

The circadian profile of SBP shows an overswing, with extreme dipping at night and overshooting above a time-varying threshold in the daytime (Figs. 19.19 and 19.20, left). The circadian profile of HR is normal, however, as shown on the right side of Fig. 19.20.

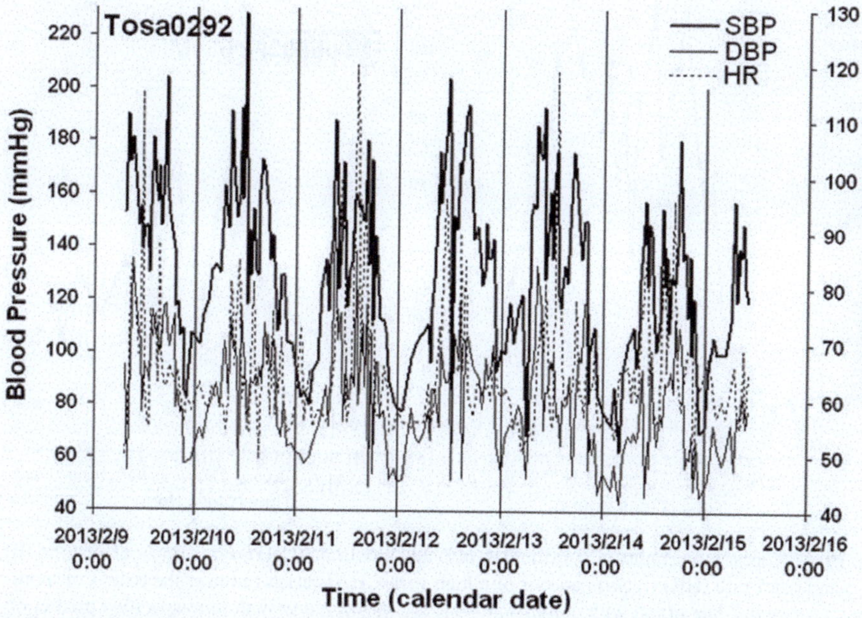

Fig. 19.19 Both SBP CHAT and DBP CHAT complicated by "masked ABP normotension." Time plots of the 6-day/24-hour records of SBP (*thick continuous line*), DBP (*thin continuous line*), and HR (*dashed line*) of a 63-year-old man. His 24-hour average of SBP changed from day to day between 124.5 and 144.9 mmHg. Day-to-day analyses also showed that his double 24-hour amplitude (BP-2A) of SBP varied greatly: 67.98 (Saturday), 78.08 (Sunday), 72.25 (Monday), 66.45 (Tuesday), 82.14 (Wednesday), and 44.25 mmHg (Thursday), all exceeding the upper 95 % prediction limit of BP-2A in healthy peers matched by gender and age. His SBP day-night ratio indicates extreme dipping every day (from Saturday to Thursday: 24.9 %,41.1 %,37.5 %,33.5 %,46.4 %, and 31.9 %)

The circadian acrophases of SBP, DBP, and HR were 13:08, 13:28, and 13:42, respectively, all within acceptable limits, but the double amplitudes of SBP and DBP were 66.23 and 35.43 mmHg, both exceeding the upper limit of acceptability. Day-to-day analyses yielded estimates of the double amplitude of SBP (SBP-2A) which varied from Saturday to Thursday: 67.98, 78.08, 72.25, 66.45, 82.14, and 44.25 mmHg, respectively, all exceeding the upper 95 % prediction limit of BP-2A estimated from clinically healthy peers matched by gender and age (34.00 mmHg in this case).

The circadian period of SBP was estimated by the Maximum Entropy Method (MEM), using the MemCalc software (GMS Co., Tokyo), which requires the data to be equidistant. Outliers were deleted from the original SBP time series (middle section of Fig. 19.21). Data were then made equidistant (bottom section of Fig. 19.21).

The MEM spectrum is shown on the middle left of Fig. 19.22, which shows a prominent circadian period of 1.0375 day, together with two clear peaks with periods of 3.2436 and 0.5016 days. Periods extracted by the MEM analysis in the range

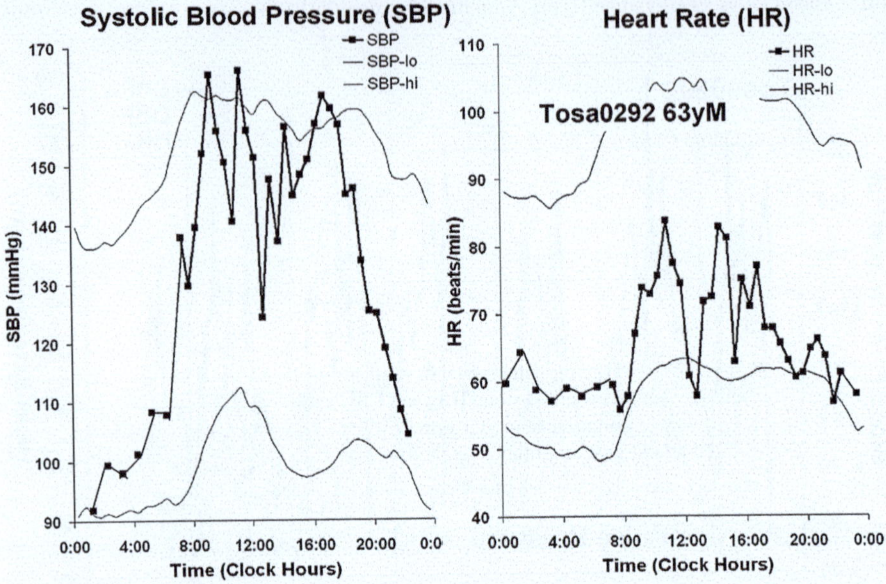

Fig. 19.20 Circadian profiles of systolic BP (*left*) and heart rate (HR) (*right*). 24-hour systolic BP (*left*) and heart rate (HR) (*right*) patterns of a man with CHAT, averaged over the 6 days of monitoring (*polygonal line* graph with *closed squares*), are shown along with time-specified prediction limits (*continuous thin lines*)

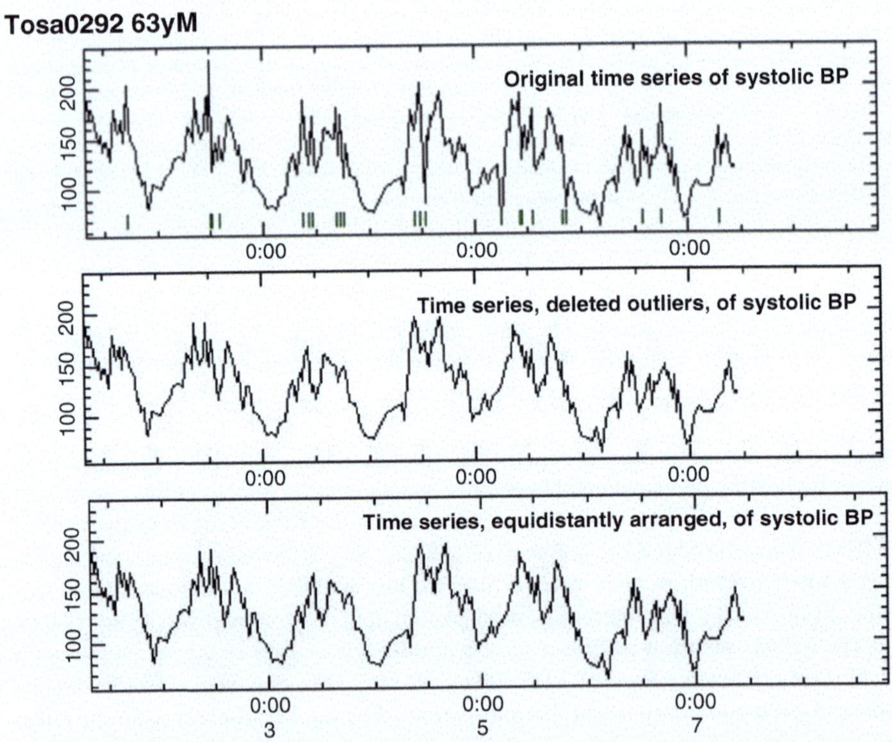

Fig. 19.21 Three versions of the time series of systolic BP. *Upper*: original data of systolic BP. *Middle*: edited data after mathematically excluding outliers. *Lower*: equidistantly arranged time series for analysis by the Maximum Entropy Method, using the MemCalc software (GMS Co., Tokyo)

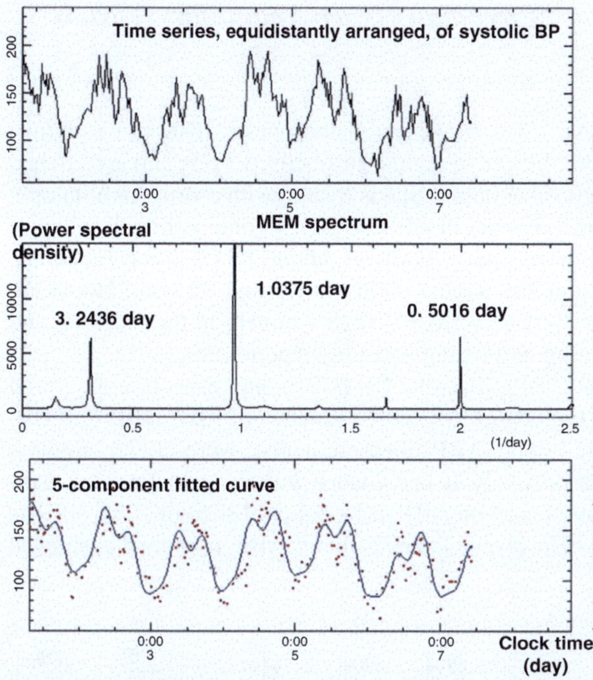

Tosa0292 63yM

Frequency (/ day)	Period (days)	Area
0.151	6.6196	35.32
0.308	3.2436	99.59
0.495	2.0199	3.99
0.964	1.0375	479.95
1.101	0.9080	18.68
1.352	0.7399	7.39
1.659	0.6029	8.24
1.994	0.5016	43.27
2.235	0.4475	11.75
2.498	0.4003	13.27

Period (days)	Amplitude	Acro-phase
6.620	4.46	2-22:18
3.244	14.17	1-17:45
1.038	30.44	1-10:37
0.502	9.86	1-18:43
0.338	7.71	1-08:30

Fig. 19.22 Period analysis of the circadian rhythm of systolic BP. MEM analysis estimates the circadian period of systolic BP to be 1.0375 day (*left middle part*). Data are shown with a five-component model fitted by least squares to the data, consisting of cosine curves with the five largest amplitudes, the 1.038 day being the most prominent one (*right lower part*)

up to 2.5 cycles/day are shown in the right upper side of this figure. Ten components are listed here and that with a period of 1.0375 day has the largest spectral power, as assessed by the "area" covered by the spectral peak.

On the bottom left of Fig. 19.22, the edited data are fitted by least squares with a five-component model consisting of the five cosine curves with the five largest amplitudes, with corresponding periods of 6.620, 3.244, 1.038, 0.502, and 0.388 days (right lower part of Fig. 19.22).

In conclusion, the clinical diagnosis based on the chronobiologically interpreted 6-day/24-hour ABP monitoring consists of (1) both SBP and DBP CHAT, (2) persistent SBP CHAT, and (3) masked ABP normotension, with a usual timing of the circadian BP rhythm. We have reported that CHAT is associated with a very high risk of ischemic stroke and other catastrophic disease. In addition, this subject had office hypertension and morning hypertension, assessed by home BP monitoring. Accordingly, we reviewed his lifestyle and made recommendations, focusing on salt intake, sleep-wake cycle, and how to improve the management of his diabetes. In addition, after starting a combination therapy for his hypertension and CAD with a CCB (benidipine, 4 mg/day), he now seems to be in good health.

19.2.3 Case Report 3 of "Circadian Hyper-Amplitude-Tension (CHAT)"

The third example of "CHAT" is a 61-year-old non-obese (BMI = 21.8 kg/m^2) woman (ID number: Tosa0206). She has a regular rest-activity cycle and goes to bed usually at about 23:30 and gets up at 06:40. She is treated with 3-drug combination regimen, a long-acting calcium channel blocker (CCB), an angiotensin II receptor blocker (ARB), and a β-blocker, once daily at about 08:00 after breakfast. Nevertheless, her office BP and HR were 138/96 mmHg and 95 bpm. Home BP measurements averaged over 30 days were 144.0/90.4 mmHg in the morning and 127.4/79.7 mmHg in the evening, indicating morning hypertension.

Time series of systolic BP (SBP), diastolic BP (DBP), and heart rate measured by pulse (HR) are shown in Fig. 19.23. MESOR estimates of SBP, DBP, and HR, estimated from the 7-day/24-hour ABP measurements, were 127.1 mmHg, 81.6 mmHg, and 71.0 bpm, respectively. In accordance with the guidelines for management of high blood pressure based on ABP in Japan, published in 2014, she has mild diastolic hypertension. However, 7-day/24-hour ABP monitoring showed

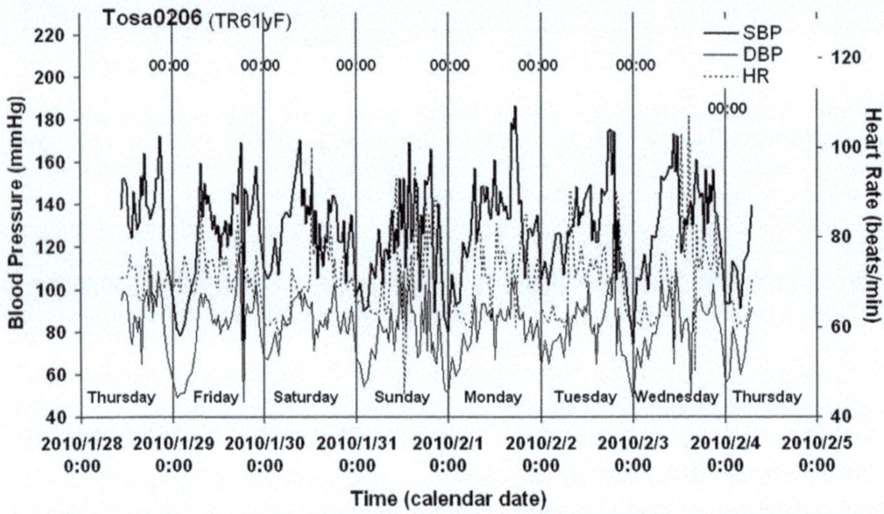

Fig. 19.23 SBP CHAT complicated by "masked ABP normotension." Time plots of the 7-day/24-hour records of SBP (*thick continuous line*), DBP (*thin continuous line*), and HR (*dashed line*) of a 61-year-old woman. The 24-hour SBP average varied from day to day between 123.6 and 134.7 mmHg. Day-to-day analyses yielded estimates of the double 24-hour amplitude of SBP (SBP-2A) of 52.30 (Thursday; CHAT), 14.76 (Friday), 32.07 (Saturday), 38.20 (Sunday; CHAT), 38.12 (Monday; CHAT), 46.40 (Tuesday; CHAT), and 41.56 mmHg (Wednesday; CHAT), exceeding the upper 95 % prediction limit of SBP-2A on 5 days (all days, except Friday and Saturday). The day-night ratio of SBP indicates the presence of extreme dipping on all days, except Friday and Monday: 38.2 % (Thursday, extreme dipping), 13.0 % (Friday; dipping), 23.5 % (Saturday; extreme dipping), 20.6 % (Sunday; extreme dipping), 18.7 % (Monday; dipping), 21.3 % (Tuesday; extreme dipping), and 27.5 % (Wednesday; extreme dipping)

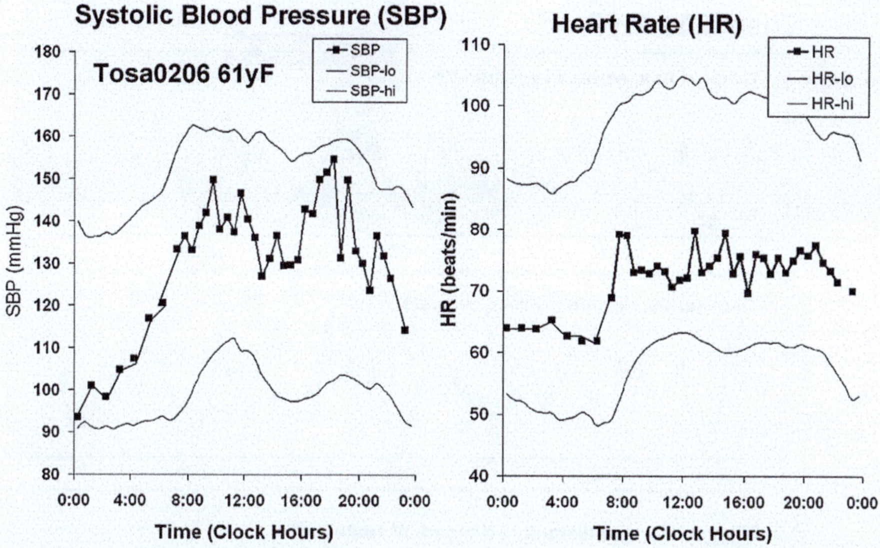

Fig. 19.24 Circadian profiles of systolic BP (*left*) and heart rate (HR) (*right*) plotted with respect to their respective circadian chronodesms. 24-hour systolic BP (*left*) and heart rate (HR) (*right*) patterns of a woman with intermittent systolic CHAT, averaged over the 7 days of monitoring (*polygonal line* graph with *closed squares*), are shown along with time-specified prediction limits (*continuous thin lines*). Whereas overall, the double circadian amplitude of SBP is excessive, the circadian profile of HR is well within acceptable limits

"masked SBP normotension," as her 24-hour average of SBP varied from day to day, assuming values of 132.3 (Thursday; hypertension), 132.2 (Friday; hypertension), 123.6 (Saturday; normotension), 127.9 (Sunday; normotension), 134.7 (Monday; hypertension), 132.0 (Tuesday; hypertension), and 134.2 mmHg (Wednesday; hypertension). Her circadian profile of SBP shows an excessive BP swing with respect to the time-varying thresholds (Fig. 19.24, left), while her circadian HR profile is within acceptable limits (Fig. 19.24, right).

The circadian acrophases of SBP, DBP, and HR were 13:54, 13:52, and 15:06, respectively, all within acceptable limits, but the double 24-hour amplitudes (BP-2A) of SBP and DBP were 37.24 and 22.72 mmHg. SBP-2A exceeding the upper limit of acceptability in clinically healthy peers matched by gender and age (34.00 mmHg in this case). Day-to-day analyses yielded estimates of the double 24-hour amplitude of SBP (BP-2A) that varied from Thursday to Wednesday, assuming values of 52.30, 14.76, 32.07, 38.20, 38.12, 46.40, and 41.56 mmHg, respectively, indicating "intermittent CHAT" on Thursday, Sunday, Monday, Tuesday, and Wednesday.

The circadian period of SBP was estimated by the Maximum Entropy Method (MEM), using the MemCalc software (GMS Co., Tokyo), which requires the data to be equidistant. Outliers were deleted from the original SBP time series (middle section of Fig. 19.25). Data were then made equidistant (bottom section of Fig. 19.25).

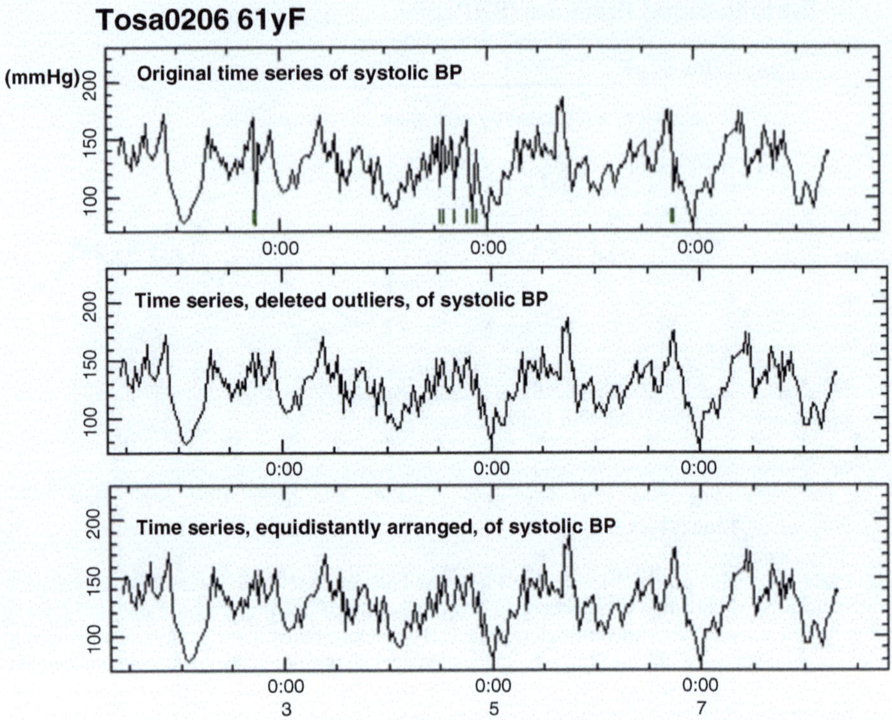

Fig. 19.25 Three versions of the time series of systolic BP. *Upper*: original data of systolic BP. *Middle*: edited data after mathematically excluding outliers. *Lower*: equidistantly arranged time series for analysis by the Maximum Entropy Method, using the MemCalc software (GMS Co., Tokyo)

The MEM spectrum is shown on the middle left of Fig. 19.26, which shows a prominent circadian period of 0.9921 day, together with four clear peaks with periods of 5.0233, 2.2459, 0.5474, and 0.4972 days. Periods extracted by the MEM analysis in the range up to 2.5 cycles/day are shown in the right upper side of this figure. Twelve are listed here and that with a period of 0.99215 day has the largest spectral power, as assessed by the "area" covered by the spectral peak.

On the bottom left of Fig. 19.26, the edited data are fitted by least squares with a five-component model consisting of the five cosine curves with the five largest amplitudes, with corresponding periods of 2.246, 0.992, 0.547, 0.497, and 0.214 days (right lower part of Fig. 19.26).

In conclusion, the clinical diagnosis based on the chronobiologically interpreted 7-day/24-hour ABP monitoring consists of (1) intermittent SBP CHAT and (2) masked ABP normotension, with a usual timing of the circadian BP rhythm. This subject also has morning hypertension. Accordingly, we provided lifestyle recommendations, focusing on salt intake, exercise, and sleep habits.

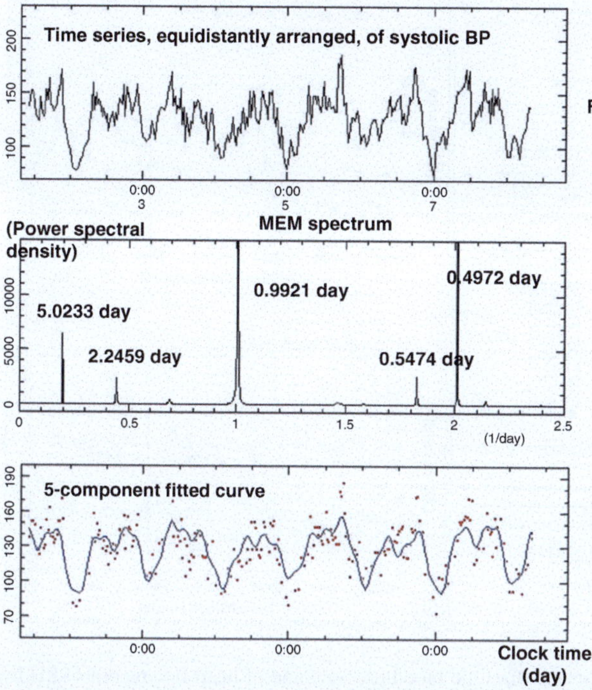

Tosa0206 61yF

Frequency (/ day)	Period (days)	Area
0.199	5.0233	5.06
0.445	2.2459	15.92
0.548	1.8247	3.68
0.691	1.4464	8.27
1.008	0.9921	180.27
1.272	0.7859	1.50
1.466	0.6820	8.31
1.584	0.6311	4.39
1.827	0.5474	14.86
2.011	0.4972	67.04
2.142	0.4669	4.96
2.394	0.4177	0.91

Period (days)	Amplitude	Acro-phase
2.246	6.15	3-06:35
0.992	18.72	1-14:31
0.547	5.07	1-19:44
0.497	10.99	1-20:01
0.214	4.30	1-15:37

Fig. 19.26 Period analysis of the circadian rhythm of systolic BP. MEM analysis estimates the circadian period of systolic BP to be 0.9921 day (*left middle part*). Data are shown with a five-component model fitted by least squares to the data, consisting of cosine curves with the five largest amplitudes, the 0.992 day being the most prominent one (*right lower part*)

19.2.4 Case Report 4 of "Circadian Hyper-Amplitude-Tension (CHAT)"

In 2009, FH was a 90-year-old man treated for MESOR-hypertension (MH) by daily dosage of two tablets of Hyzaar (losartan 50 mg, hydrochlorothiazide 12.5 mg) and two tablets of metoprolol (50 mg) on awakening in the morning. FH had a history of MH complicated by intermittent CHAT, mostly on single days but with occasional weekly CHAT [1, 2]. FH used his ABPM from A&D (Tokyo, Japan) and another from Colin Electronics (Komaki, Japan) for decades.

FH had been monitoring his BP for the previous 22 years [3, 4]. In 2009, he used the TM-2421 monitor manufactured by A&D (Tokyo, Japan). On a weekly basis, FH's data were analyzed by sphygmochron (Fig. 19.27), a summary of results from curve-fitting, with the combined fit of 24-and 12-hour cosine curves to the 7-day series as a whole (top) and a summary of the data carried out after the stacking of the values from different days along a single 24-hour scale (bottom). Figure 19.28 shows the analysis of 24-hour sections of the weekly record (only the day-to-day 24-hour results are shown in this figure and are summarized routinely for a week in a sphygmochron in Fig. 19.27).

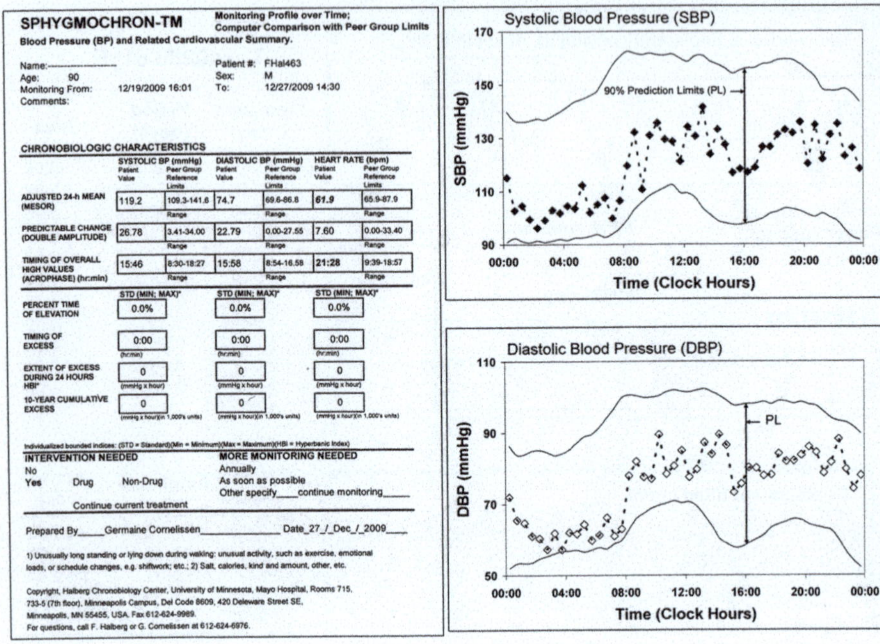

SPHYGMOCHRON-TM Monitoring Profile over Time; Computer Comparison with Peer Group Limits
Blood Pressure (BP) and Related Cardiovascular Summary.

Name:————
Age: 90
Monitoring From: 12/19/2009 16:01
Comments:
Patient #: FHal463
Sex: M
To: 12/27/2009 14:30

CHRONOBIOLOGIC CHARACTERISTICS

	SYSTOLIC BP (mmHg) Patient Value	Peer Group Reference Limits	DIASTOLIC BP (mmHg) Patient Value	Peer Group Reference Limits	HEART RATE (bpm) Patient Value	Peer Group Reference Limits
ADJUSTED 24-h MEAN (MESOR)	119.2	109.3-141.6 Range	74.7	69.6-86.8 Range	***61.9***	65.9-87.9 Range
PREDICTABLE CHANGE (DOUBLE AMPLITUDE)	26.78	3.41-34.00 Range	22.79	0.00-27.55 Range	7.60	0.00-33.40 Range
TIMING OF OVERALL HIGH VALUES (ACROPHASE) (hr:min)	15:46	8:30-18:27 Range	15:58	8:54-16.58 Range	**21:28**	9:39-18:57 Range

	SYSTOLIC STD (MIN; MAX)*	DIASTOLIC STD (MIN; MAX)*	HEART RATE STD (MIN; MAX)*
PERCENT TIME OF ELEVATION	0.0%	0.0%	0.0%
TIMING OF EXCESS	0:00 (hr:min)	0:00 (hr:min)	0:00 (hr:min)
EXTENT OF EXCESS DURING 24 HOURS HBI*	0 (mmHg x hour)	0 (mmHg x hour)	0 (mmHg x hour)
10-YEAR CUMULATIVE EXCESS	0 (mmHg x hour)(in 1,000's units)	0 (mmHg x hour)(in 1,000's units)	0 (mmHg x hour)(in 1,000's units)

Individualized bounded indices: (STD = Standard)(Min = Minimum)(Max = Maximum)(HBI = Hyperbanic Index)

INTERVENTION NEEDED
No
Yes Drug Non-Drug
Continue current treatment

MORE MONITORING NEEDED
Annually
As soon as possible
Other specify____continue monitoring_____

Prepared By_____Germaine Cornelissen______________ Date_27_/_Dec_/_2009____

1) Unusually long standing or lying down during waking: unusual activity, such as exercise, emotional loads, or schedule changes, e.g. shiftwork; etc.; 2) Salt, calories, kind and amount, other, etc.

Copyright, Halberg Chronobiology Center, University of Minnesota, Mayo Hospital, Rooms 715, 733-5 (7th floor), Minneapolis Campus, Del Code 8609, 420 Deleware Street SE, Minneapolis, MN 55455, USA. Fax 612-624-9989.
For questions, call F. Halberg or G. Cornelissen at 612-624-6976.

Fig. 19.27 Summaries of systolic and diastolic blood pressure and of heart rate record of FH by curve fitting (*left*) and stacking (*left* and *right*). These summaries of the 7-day record as a whole are acceptable, but see Fig. 19.2. Heart rate abnormalities (*bold*) have been present for a long time (© Halberg Chronobiology Center)

An elevation of the BP-2A on a single day is seen in Fig. 19.28 in the absence of a change in BP-M, occurring when FH was a host on Dec. 24 but was absent when he was a guest at a party in the evening of the following day. Figure 19.27 summarizes and Fig. 19.28 shows in detail that the subject at the time was MESOR-normotensive, due to treatment, as subsequently suggested early in 2010 by a prompt, within-48-hour rise in the BP-M when the treatment with Hyzaar and metoprolol was discontinued for 2 days (Table 19.1). This response was much shorter than the gradual month-long change after discontinuing atenolol in another case [5]. Certainly, the documentation of a rebound's temporal characteristics depends, i.e., on the drug(s) and on the subject; one must not generalize from the results in Table 19.1 that are suggestive but too short and not replicated. FH's overall BP-M, the 24-hour 2A of BP, and the circadian timing are all within peer-group limits during the span from December 19 to 27 summarized as a whole in Fig. 19.27. The MESOR of heart rate (HR) and its circadian timing are outside limits, an alteration of long standing, first detected after FH's second coronary artery bypass surgery in 1997 (the first was in 1981). A 24-hour CHAT is smoothed out in Fig. 19.27, yet indeed it was present as Fig. 19.28 shows when FH was the host of a dinner party looking after guests' desires on December 24, but not on December 25, when he was a guest at a similar party and could retire into a corner and have fun in writing in the

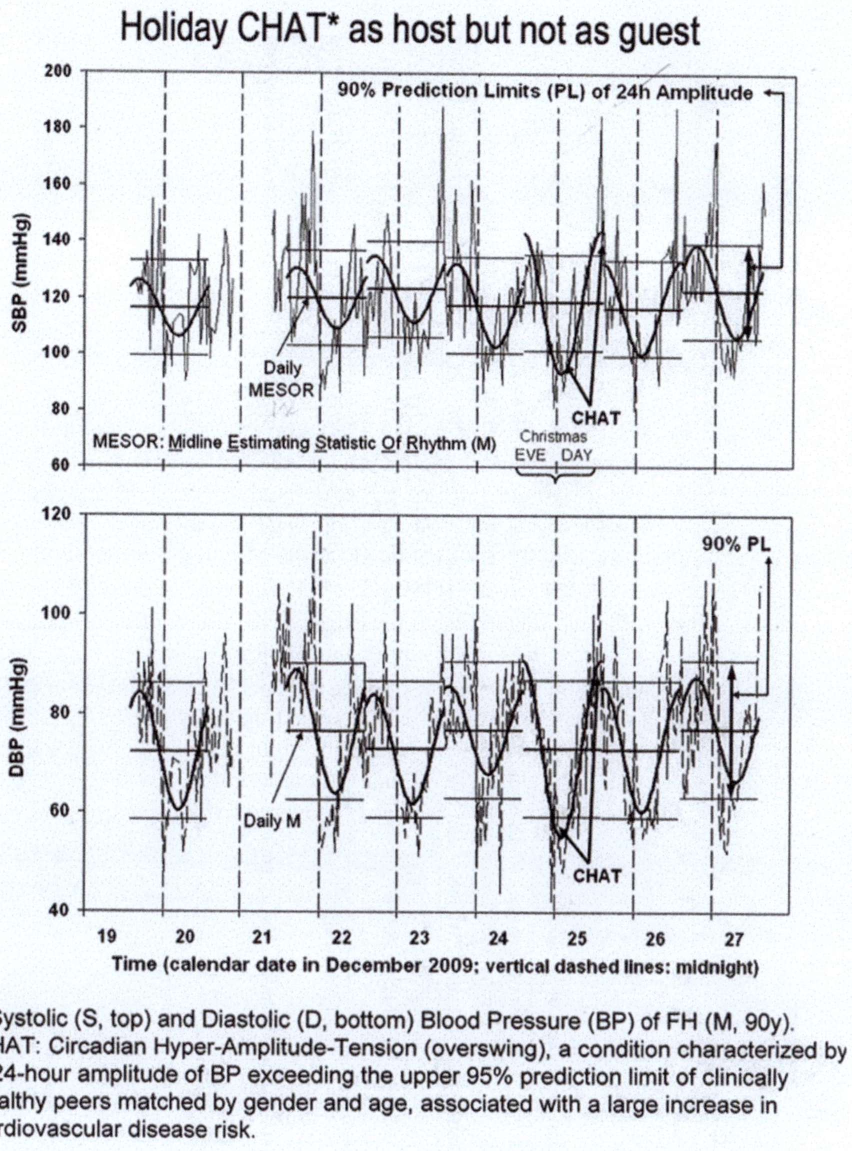

Fig. 19.28 Among-day variation of CHAT observed by 7-day/24-hour ambulatory blood pressure monitoring. FH attended parties on both Christmas Eve (host) and Christmas Day (guest). As host, he could not follow his favorite task of writing; in the capacity of a guest, he took the liberty of unobtrusively doing some professional homework. His blood pressure during the latter, but not the former role, was acceptable. Situational 24-hour transient CHAT, as here recorded, was limited to a single 24-hour span; it is smoothed out by the fit of a fit of a 24-hour cosine to 7 days of observations as a whole (© Halberg Chronobiology Center)

Table 19.1 Comparison of rhythm parameters of an elderly man on 2 days with (January 8–9) or without (January 9–10) hypotensive medication[a]

	Systolic BP (mmHg)		Test of equality of parameters[b]	Diastolic BP (mmHg)		Test of equality of parameters	Heart Rate (beats/min)	Test of equality of parameters	
Variable (units)	Jan 8–9	Jan 9–10	F-value (P-value)	Jan 8–9	Jan 9–10	F-value (P-value)	Jan 8–9	Jan 9–10	F-value (P-value)
MESOR	122	131.8	5.246 (0.025)	80.1	89	10.923 (0.002)	68.8	81.2	6.774 (0.011)
24-hour Amplitude	14.6	20.9	1.142 (0.289)	10.4	13.8	0.822 (0.368)	11.8	5.2	1.063 (0.306)
24-hour Acrophase	−254	−240	0.479 (0.491)	−216	−238	1.236 (0.270)	−35	−45	0.027 (0.872)

[a]Post hoc ergo propter hoc reasoning is not excluded. The longer record during the preceding months consists mostly of MESORs comparable to that before treatment or lower ones

[b]Bingham et al. [28]

context of "homework" that he had brought along. Since CHAT constitutes a risk of stroke greater than that of hypertension [6; cf. 7, 8, 9], it seems desirable to eliminate it as much as possible. FH attempted this weekly by adjusting the kind and timing of drug therapy.

Figure 19.29a shows the original data underlying an association of CHAT (fit not shown) with visitors who were friendly, one a prospective collaborator, the other an earlier provider of important data for analysis. FH was unaware of the fact that the arrival on March 7 of one and on March 10 of the other was preceded by spikes in BP by hours or a day, respectively. Figure 19.29b shows that all results were within the acceptable range, yet the swing within that range was excessive, as seen in Fig. 19.29c. CHAT was present while the visitors were around.

As part of a study in Tosa City, 26 inhabitants of Tosa Town in Kochi provided 7-day (24 of them) or 6-day (2 of them) half-hourly around-the-clock records. All ABPM recordings were obtained with the TM-2430 monitor from A&D (Tokyo, Japan).

Table 19.2 shows that, in each of 26 records, 24 of 7-day and 2 of 6-day length, a chronobiologic analysis happened to detect transient abnormality when these series were examined for each day separately. A vascular variability abnormality for a day or so is a physiological response in most people [8, 10]. Until unobtrusive, affordable instrumentation becomes generally available for long-term strain monitoring, for the limited purpose of cardiovascular disease risk monitoring, a 1-day VVD can be neglected. A transient VVD of up to 3 days length can likely be washed out by the fit to the 7-day series as a whole, as seen in Table 19.3. The results in Tables 19.2 and 19.3 underline the need for a minimal 7-day profile analyzed as a whole. When this fit of the cosine model to the 7-day record as a whole is abnormal, a VVD is tentatively considered but certainly may not be diagnosed based on data for a single week (even if it persisted in the series analyzed as a whole). This 7-day VVD may also represent a physiological response. The need for replicating the 7-day profile is seen in Table 19.4. Seven of the 26 profiles from Tosa City (~27 %) show a VVD in analyses of the records as a whole.

Figure 19.30 summarizes mortality during December and January in Minnesota in data stacked in these 2 months from 29 consecutive years. It compares mortality between December 1 and 23 vs. that between December 24 and January 1 vs. that between January 2 and 31, irrespective of the year. A one-way analysis of variance shows, for original and relative data, statistical significance ($P<0.05$ but $P>0.01$). Surprisingly, Christmas and New Year's Day are local minima. In dealing with thousands, we confirm the lesson from the millions that there are more deaths during the holiday season, but on the holidays themselves, we find local minima. These meta-analyses should constitute a warning to draw inferences from populations without also considering geographic and many other differences, just as decisions to treat a VVD should be considered for the individual patient rather than based on population results alone.

Phillips et al. [11], based on a data set of 53 million from the USA, followed up on Kloner et al.'s sample of 220,000 cases from Los Angeles [12], reporting that cardiac mortality was higher during Christmas and New Year's Day, listing the holi-

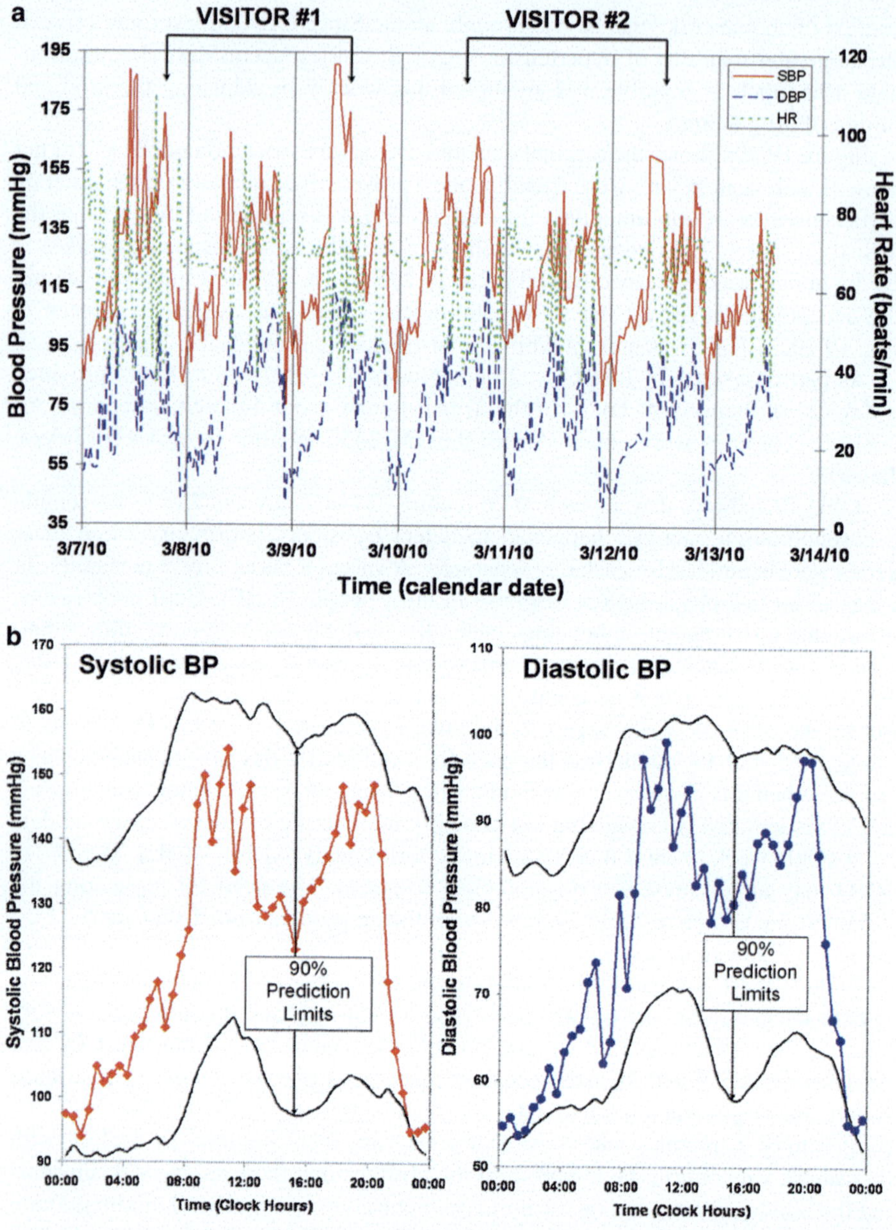

Fig. 19.29 (**a**) Time series data from March 7, 2010, to March 14, 2010, shows "visitor CHAT." Visitor CHAT: original data revealing both 7-day CHAT and CHAT on a number of days during which the subject took care of friendly visitors (© Halberg Chronobiology Center). (**b**) Circadian patterns of systolic (*left*) and diastolic (*right*) blood pressure in FH (M, 90y) (© Halberg Chronobiology Center). (**c**) CHAT was present during two friendly visits within 1 week. All results of systolic and diastolic blood pressure in FH (M, 90y) are within the acceptable range, but the swing within the range was excessive (© Halberg Chronobiology Center)

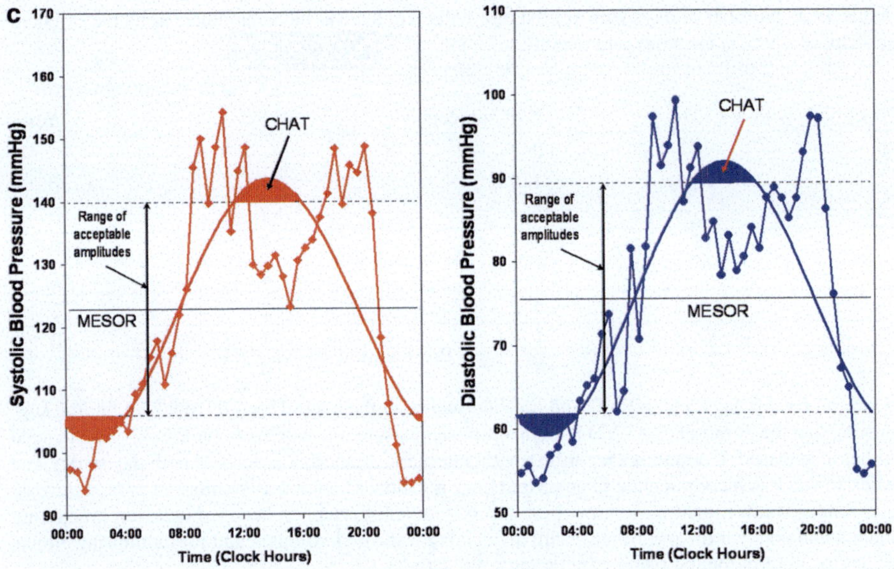

Fig. 19.29 (continued)

Table 19.2 In weeklong records from 26 subjects (4M, 22F) 39–74 years of age, there is none without a vascular variability disorder (VVD) on at least 1 day[a]

		Daily analyses							Daily analyses (%)							
	Ndays	S-MH	D-MH	EPP	S-CHAT	D-CHAT	DHRV	anyVVD	S-MH	D-MH	EPP	S-CHAT	D-CHAT	DHRV	anyVVD	
S01-0185	7	1		1	2	3	1	6	14.29		14.29	28.57	42.86	14.29	85.71	
S02-0186	7	-	-	-	-	2	-	2	-	-	-	-	28.57	-	28.57	
S03-0187	7	-	-	-	2	2	-	2	-	-	-	28.57	28.57	-	28.57	
S04-0188	7	1	1	-	2	3	1	5	14.29	14.29	-	28.57	42.86	14.29	71.43	
S05-0189	6	-	1	-	2	3	-	3	-	16.67	33.33	-	50.00	-	50.00	
S06-0190	7	6	3	5	4	3	-	7	85.71	42.86	71.43	57.14	42.86	-	100.00	
S07-0191	7	-	-	-	4	-	2	4	-	-	-	57.14	-	28.57	57.14	
S08-0192	7	1	2	-	2	2	-	4	14.29	28.57	-	28.57	28.57	-	57.14	
S09-0193	7	4	-	5	4	2	6	7	57.14	-	71.43	57.14	28.57	85.71	100.00	
S10-0194	7	-	3	-	3	2	-	4	-	42.86	-	42.86	28.57	-	57.14	
S11-0195	7	-	-	-	1	-	-	1	-	-	-	14.29	-	-	14.29	
S12-0196	7	-	-	-	-	-	2	2	-	-	-	-	-	28.57	28.57	
S13-0197	7	-	-	-	1	-	-	1	-	-	-	14.29	-	-	14.29	
S14-0198	7	-	-	-	3	3	-	3	-	-	-	42.86	42.86	-	42.86	
S15-0199	6	1	3	-	1	1	-	4	16.67	50.00	-	16.67	16.67	-	66.67	
S16-0200	7	2	-	-	1	-	1	2	28.57	-	14.29	-	14.29	28.57		
S17-0201	7	-	-	2	1	-	1	2	-	-	28.57	14.29	-	14.29	28.57	
S18-0202	7	-	-	-	-	-	1	1	-	-	-	-	-	14.29	14.29	
S19-0203	7	-	-	-	5	4	-	5	-	-	-	71.43	57.14	-	71.43	
S20-0204	7	-	-	-	1	2	-	2	-	-	-	14.29	28.57	-	28.57	
S21-0205	7	-	-	-	4	3	-	5	-	-	-	57.14	42.86	-	71.43	
S22-0206	7	-	-	-	5	2	4	7	-	-	-	71.43	28.57	57.14	100.00	
S23-0207	7	-	-	-	1	1	-	2	-	-	-	14.29	14.29	-	28.57	
S24-0208	7	-	-	-	4	3	-	4	-	-	-	57.14	42.86	-	57.14	
S25-0209	7	1	5	-	-	1	-	5	14.29	71.43	-	-	14.29	-	71.43	
S26-0210	7	-	-	-	2	-	2	4	-	-	-	28.57		28.57	57.14	
Nsubjects:		8	7	4	22	18	10	26							52.29	mean
(%)		30.77	26.92	15.38	84.62	69.23	38.46	100.00							26.91	SD
															5.28	SE

[a]Abnormality on at least 1 day is found in each weeklong record, sometimes as more than a single VVD. The three subjects with abnormality on all 7 days also have an abnormal 7-day record. The number of abnormal days in a weeklong record as such does not tell whether the 7-day record is also abnormal. Longer, preferably lifelong monitoring is recommended along with a qualification of the extent of severity of each VVD, beyond the percentage time of their relative occurrence here tabulated from daily analyses

Table 19.3 Number of days with at least one VVD will have to be complemented by the extent of severity of VVD(s) each day and overall[a]

N of days with a VVD	N of subjects	VVD in weeklong record? No	Yes
0	0		
1	3	3	0
2	7	7	0
3	2	2	0
4	6	4	2
5	4	2	2
6	1	1	0
7	3	0	3

[a]One can have 6 days with a VVD and no abnormality in the overall record (row 7), a finding suggesting that the duration of a VVD is an incomplete measure, the extent of abnormality also in need of being assessed. Continuous recording with automatic analyses carried out both day to day and week to week is recommended in any event once a validated, unobtrusive monitor is miniaturized and rendered affordable in a chronobiologic system subservient to health monitoring, including mental and other strain assessment, with dividends for medical research and for monitoring effects of natural environmental factors, including solar activity

days as a risk factor for death. Saposnik et al. [13] ask whether birthdays predispose to vascular events. A situational hypertension was described by Uexküll and Wick [14] in 1962, and Ettinger et al. [15] had referred to the "holiday heart" as alcohol-associated cardiac rhythm disorder. A holiday effect on traffic accidents is documented [16].

Based on a study completed within 3 days, Karl Hecht et al. reported a "Post-New Year's Eve Hypertension [17]," reviewing the literature on what they describe as "holiday celebration stress." By spotchecks between 11:00 and 13:00, these authors applied a BP relaxation test [18–20].

The subject sits comfortably with eyes closed for 10 min with conscious mental concentration on relaxed breathing. During the 10 min, an automatic, noiseless instrument measures BP at 1-min intervals. The difference between the initial value (I) and the relaxation value (R), i.e., the lowest value among the last 5 BP measurements, determines the extent of relaxation. Hecht et al. [17] examined 11 "normotensive" subjects on 12/31/06 and for ten of them (one behaved differently and was excluded) found a mean of systolic (S) BP of 131 mmHg, reduced to 117 mmHg by relaxation, which pressures increased to 152 and 136 mmHg on 1/1/07, being 140 mmHg and 120 mmHg on 1/2/07. These studies focused on most but not all individuals and drew inferences only on the basis of a presumably "true resting relaxed BP" irrespective of time, presumably a BP that can be explored at any convenient time chosen for testing. The one reported exception suffices to indicate that the test is not applicable to all individuals and groups and, perhaps, also not to the same individual at all times, since the same stimulus [21] or the same drug dose ([22]; see table in [23]) can have very different effects at various circadian stages. The question may be raised whether the relaxation test outcomes may have differed at times other than those used by Hecht et al. [17], a probability deserving further

Table 19.4 Vascular variability disorders (VVDs) usually combine into syndromes (VVSs)[a]

	Entire (6- to 7-day) record								
	N days	S-MH	D-MH	EPP	S-CHAT	D-CHAT	DHRV	anyVVD	N of VVDs
S01-0185	7								
S02-0186	7								
S03-0187	7								
S04-0188	7	1		1	1			1	3
S05-0189	6								
S06-0190	7								
S07-0191	7								
S08-0192	7								
S09-0193	7	1		1			1	1	3
S10-0194	7								
S11-0195	7								
S12-0196	7								
S13-0197	7								
S14-0198	7								
S15-0199	6								
S16-0200	7								
S17-0201	7								
S18-0202	7								
S19-0203	7				1	1		1	2
S20-0204	7								
S21-0205	7								
S22-0206	7				1			1	1
S23-0207	7								
S24-0208	7				1			1	1
S25-0209	7							1	1
S26-0210	7		1						
Nsubjects:		2	1	2	5	1	1	7	
(%)		7.69	3.85	7.69	19.23	3.85	3.85	26.92	

[a]Weeklong records show abnormality in over one fourth of subjects investigated, a reason for recommending a follow-up 7-day profile once abnormality is detected in first 7-day record
S systolic blood pressure, *D* diastolic blood pressure, *MH* MESOR-hypertension, *EPP* excessive pulse pressure (>60 mmHg), *CHAT* circadian hyper-amplitude-tension, *DHRV* deficient heart rate variability (HR-SD<7.5 bpm)

study. Certainly the use of time-specified reference values, or chronodesms [24–26], can be recommended when BP elevations or drops are considered, and these limits will have to be specified beyond gender and age for ethnicity and social class and with reference to special test conditions.

Instead of generalizing from a relaxation test, we here introduce a dividend from long-term BP monitoring: the concept of a situational CHAT, to match and, as the case may be, possibly complement chronobiologically diagnosed situational MESOR-hypertension, by the elevation of the double circadian amplitude (2A) of

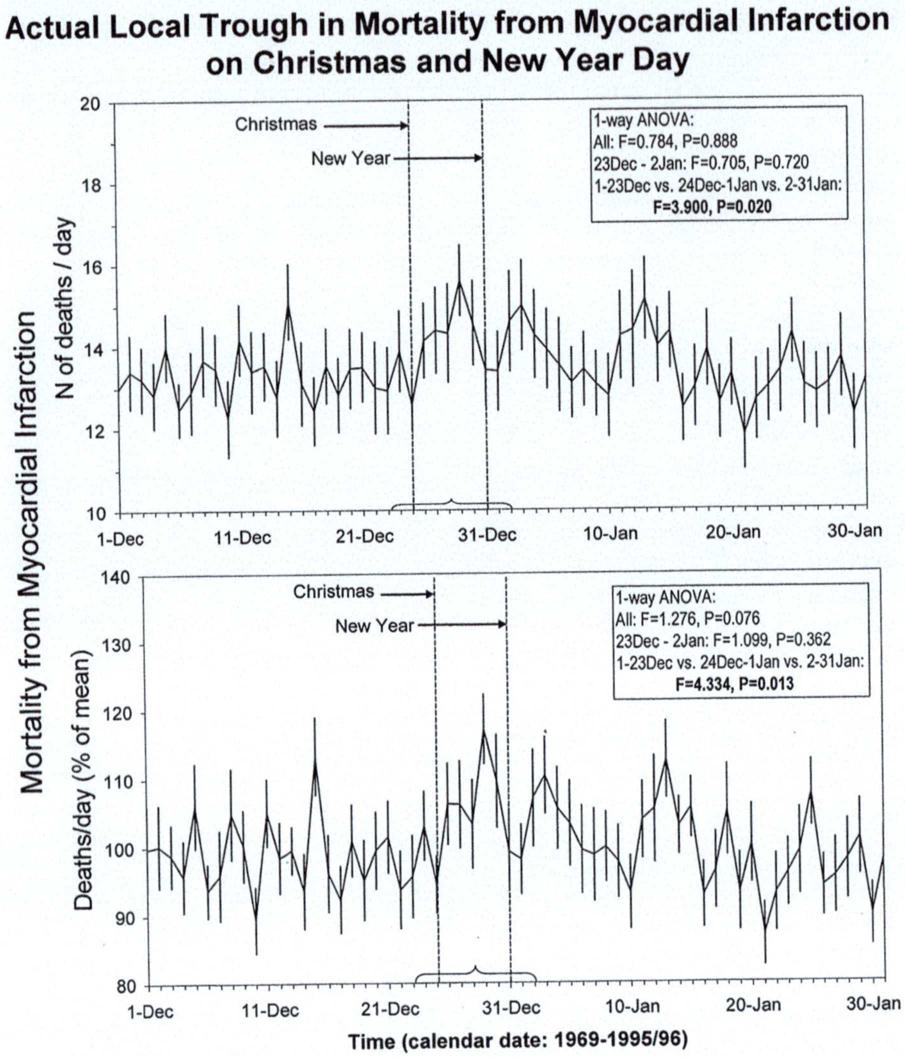

Fig. 19.30 The inference from a much larger database of previous studies is confirmed for Minnesota, insofar as a higher mortality during the Christmas-to-New Year's Day season is concerned, yet the data are in keeping with the assumption that specific holidays as such are not necessarily detrimental, even though the highest mortality in a data set of 53 million deaths was found on Christmas and New Year's Day. On those days, mortality from myocardial infarction in a 29-year record in Minnesota was at a local minimum, whether the original number of deaths per day (*top*) or relative data (*bottom*) are considered. We here propose chronobiologically interpreted blood pressure and heart rate monitoring to explore whether, for a given subject, a certain event such as a holiday, birthday, or symposium is desirable (benetensive) or undesirable (maletensive). (We remember anecdotally a prominent scientist who left the podium at a lecture and was diagnosed as having a myocardial infarction. The same scientist died at the end of his invited presentation at another conference.) (© Halberg Chronobiology Center)

the BP above the upper 95 % prediction limit, derived from gender- and age-matched clinically healthy peers [26].

In conclusion, BP and HR can be surveyed in everybody from childhood and preferably from birth at, e.g., half-hour intervals around the clock 24/7/365/lifetime, once affordable, unobtrusive, miniaturized (watch-size) devices become available within a chronobiologic system of analyses (provided automatically from the Internet, without the need for a care provider) as long as no indication for treatment exists.

19.3 Blood Pressure Ecphasia

In this chapter, we introduce two cases of "BP ecphasia" from the viewpoint of 7-day/24-hour ambulatory blood pressure (ABP) monitoring. Ecphasia is a condition characterized by an odd timing of the circadian acrophase (ϕ) of BP but not of the ϕ of HR (BP ecphasia with HR euphasia). The circadian acrophase, ϕ, is a measure of the timing of overall high values recurring each day, measured as the lag from a given reference time (e.g., local midnight) of the peak in the 24-hour cosine curve of the model approximating the data.

BP ecphasia is differentiated from the consequences of shift work that can be associated with an altered timing of both BP and HR. It is defined by the BP-ϕ derived from a 7-day/24-hour record lying outside the 90 % prediction interval (PI) of BP-ϕs from clinically healthy subjects matched by gender and age (Fig. 19.31).

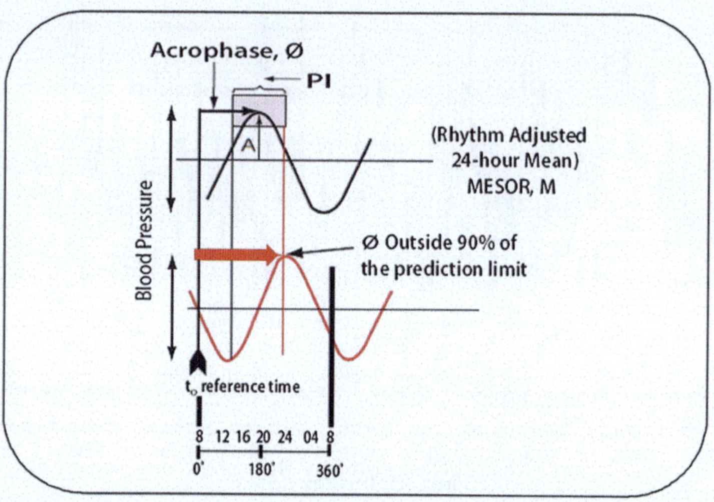

Fig. 19.31 SBP, DBP, or both ecphasia. Ecphasia means an odd timing of the circadian acrophase (ϕ) of BP but not of the ϕ of HR (BP ecphasia with HR euphasia), ϕ lying outside the 90 % prediction interval (PI) of BP-ϕs from clinically healthy peers matched by gender and age

19.3.1 *Case Report 1 of "BP Ecphasia (Odd Timing of the Circadian Rhythm of BP but Not of That in HR)"*

The first example of "ecphasia" is a 63-year-old slightly obese (BMI = 28.0 kg/m^2) man (ID number: Tosa0220). His weight was 81.0 kg and height 170 cm. He uses a wheelchair because of a spinal cord injury when he was young. He has a regular sleep-wake cycle and goes to bed at about midnight and gets up at 08:00. He falls asleep within 10 min and claims to sleep well most days. He does not smoke and usually does not drink alcohol. He rarely eats meat but eats fish frequently with an average intake of vegetables. He prefers moderate seasoning.

He has a depressive mood. His geriatric depression score 15 (GDS-15) was 9; his subjective quality of life (QOL) of health, assessed by a visual analog scale (VAS), was 5 %; and his subjective QOL of happiness was 17 %. While his subjective QOL of family relations was 92 %, his QOL of relations with friends was only 66 %.

He does not take any antihypertensive medication. His office BP was 155/96 mmHg, and his home BP measurements averaged over 30 days were 128.3/78.5 mmHg in the morning and 127.5/79.1 mmHg in the evening. Time series of systolic BP (SBP), diastolic BP (DBP), and heart rate measured by pulse (HR) are shown in Fig. 19.32. A model consisting of cosine curves with periods of 24 and 12 hours was fitted by least squares to the data to estimate the MESOR (Midline-Estimating Statistic Of Rhythm) and the amplitude and acrophase (time of predicted

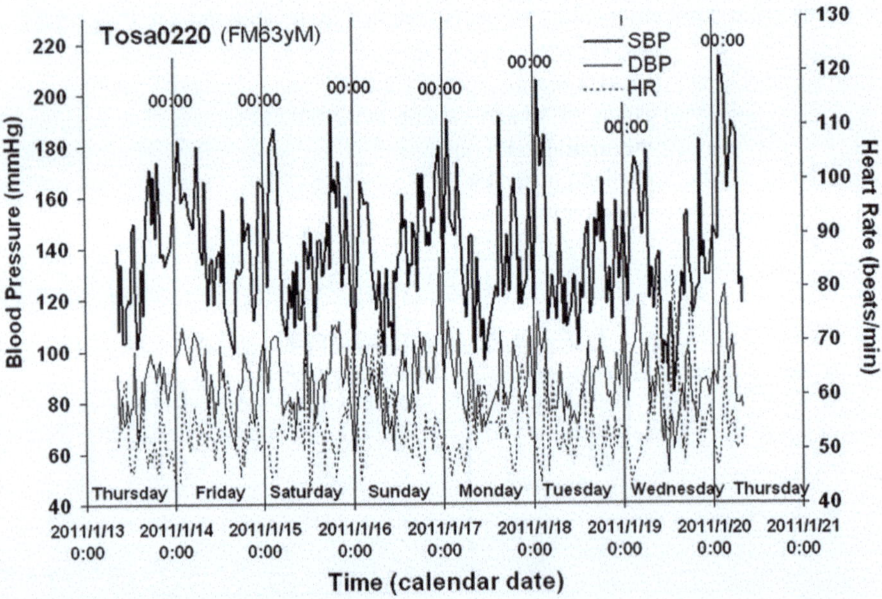

Fig. 19.32 SBP and DBP ecphasia (odd timing of the circadian rhythm of BP). Time plots of the 7-day/24-hour records of SBP (*thick continuous line*), DBP (*thin continuous line*), and HR (*dashed line*) of a 61-year-old man

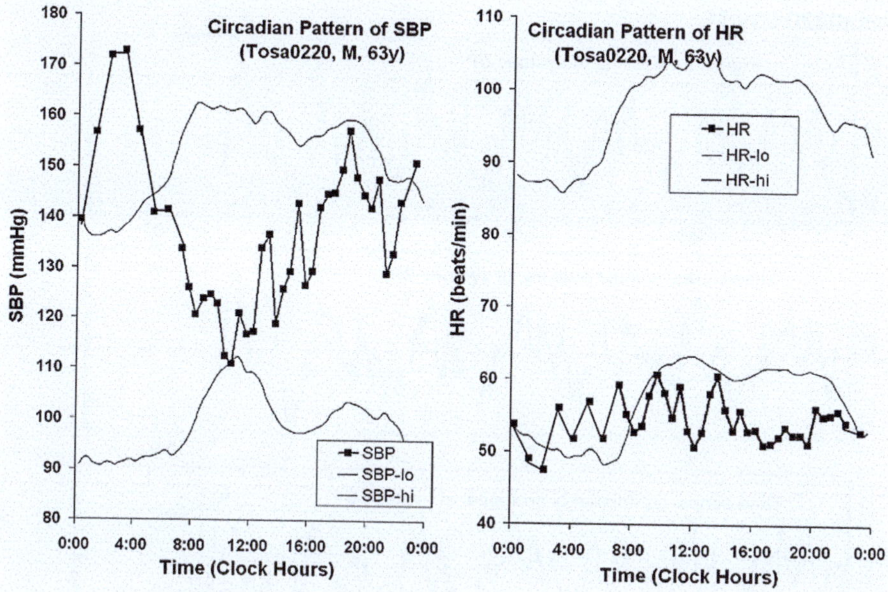

Fig. 19.33 Circadian profiles of systolic BP (*left*) and heart rate (HR) (*right*). 24-hour systolic BP (*left*) and heart rate (HR) (*right*) patterns of a man with BP ecphasia, averaged over the 7 days of monitoring (*polygonal line* graph with *closed squares*), are shown along with time-specified prediction limits (*continuous thin lines*)

maximum) of each component. MESOR estimates of SBP, DBP, and HR, estimated from the 7-day/24-hour ABP measurements, were 139.5 mmHg, 88.4 mmHg, and 54.0 bpm, respectively. In accordance with the guidelines for management of high blood pressure based on ABP in Japan, published in 2014, he is hypertensive, as his 24-hour SBP and 24-hour DBP averages are above 130 mmHg and 80 mmHg, respectively.

The circadian profile of SBP shows "ecphasia," that is, an odd timing of the circadian rhythm of BP with respect to the chronodesm (Fig. 19.33, left). The circadian profile of HR lies at the lower part of the chronodesm (Fig. 19.33, right).

Based on the fitted model, the double 24-hour amplitudes of SBP and DBP were 32.71 and 16.35 mmHg, respectively, but that of HR was 3.24 bpm, which means a deficient HR variability (DHRV). The circadian acrophases of SBP, DBP, and HR were 00:03, 23:49, and 10:40, respectively, SBP and DBP peaking at night rather than in the afternoon. HR also showed a small deviation from the usual circadian timing, which may relate to the DHRV.

The circadian period of SBP was estimated by the Maximum Entropy Method (MEM), using the MemCalc software (GMS Co., Tokyo), which requires the data to be equidistant. Outliers were deleted from the original SBP time series (middle section of Fig. 19.34). Data were then made equidistant (bottom section of Fig. 19.34).

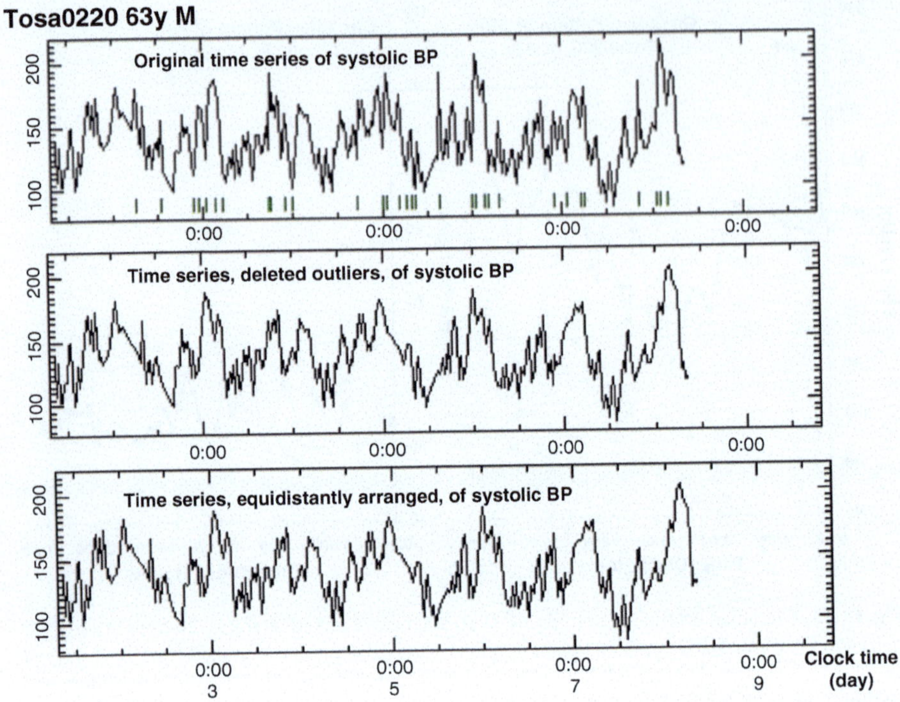

Fig. 19.34 Three versions of the time series of systolic BP. *Upper*: original data of systolic BP. *Middle*: edited data after mathematically excluding outliers. *Lower*: equidistantly arranged time series for analysis by the Maximum Entropy Method, using the MemCalc software (GMS Co., Tokyo)

The MEM spectrum is shown on the middle left of Fig. 19.35, which shows a prominent circadian period of 1.0239 day, together with several peaks, notably one at a period of 0.4979 day. Periods extracted by the MEM analysis in the range up to 2.5 cycles/day are shown in the right upper side of this figure. Twelve components are listed here, and that with a period of 1.0239 day has the largest spectral power, as assessed by the "area" covered by the spectral peak.

On the bottom left of Fig. 19.35, the edited data are fitted by least squares with a five-component model consisting of the five cosine curves with the five largest amplitudes, with corresponding periods of 1.183, 1.024, 0.498, 0.360, and 0.333 days (right lower part of Fig. 19.35).

In conclusion, the clinical diagnosis based on the chronobiologically interpreted 7-day/24-hour ABP monitoring consists of (1) both SBP and DBP ecphasia, (2) mild SBP and DBP MESOR-hypertension, and (3) DHRV. Office BP also showed mild hypertension. Accordingly, we made lifestyle recommendations, focusing on salt intake and the sleep-wake cycle. In addition, we started antihypertensive treatment with a calcium channel blocker (CCB) (benidipine, 4 mg/day). The subject now seems to be in good health.

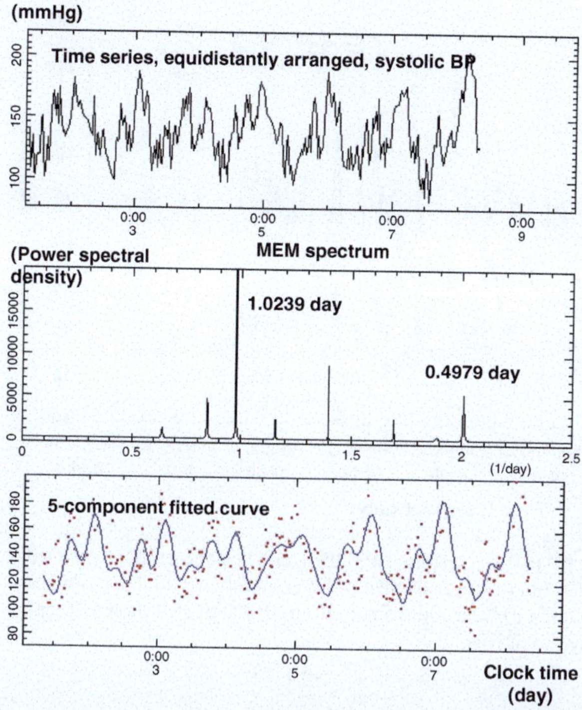

Tosa0220 63y M

Frequency (/ day)	Period (days)	Area
0.1069	9.3510	3.68
0.3064	3.2637	2.48
0.6399	1.5627	12.36
0.8452	1.1831	33.77
0.9767	1.0239	181.63
1.1570	0.8643	10.97
1.3974	0.7156	11.38
1.6935	0.5905	9.02
1.8882	0.5296	9.72
2.0085	0.4979	46.24
2.1677	0.4613	4.34
2.5393	0.3938	8.25

Period (days)	Amplitude	Acro-phase
1.183	8.72	2-06:43
1.024	19.12	1-22:35
0.498	8.81	1-15:55
0.360	5.79	1-16:29
0.333	6.03	1-10:29

Fig. 19.35 Period analysis of the circadian rhythm of systolic BP. MEM analysis estimates the circadian period of systolic BP to be 1.0239 day (*left middle part*). Data are shown with a five-component model fitted by least squares to the data, consisting of cosine curves with the five largest amplitudes, the 1.0239 day being the most prominent one (*right lower part*)

19.3.2 Case Report 2 of "BP Ecphasia (Odd Timing of the Circadian Rhythm of BP but Not of That in HR)"

The second example of "ecphasia" is a 68-year-old slightly obese (BMI = 25.7 kg/m^2) woman (ID number: KOts092). She is suffering from major depression and her PHQ-2 score was 2. She takes antidepressant medicines. Her sleep-wake cycle is a little irregular. She has 7–8 hours sleep, but does not feel she sleeps well and is not rested when she wakes. She needs more than 120 min to fall asleep and wakes up 2–3 times at night, and her time to get up in the morning is irregular. She does not smoke and usually does not drink alcohol. She rarely eats meat, but frequently eats fish, with average intake of vegetables. She prefers moderate seasoning.

She is treated with benidipine (4 mg/day), which is a long-acting calcium channel blocker. Her office BP was 109/73 mmHg. Time series of systolic BP (SBP), diastolic BP (DBP), and heart rate measured by pulse (HR) are shown in Fig. 19.36.

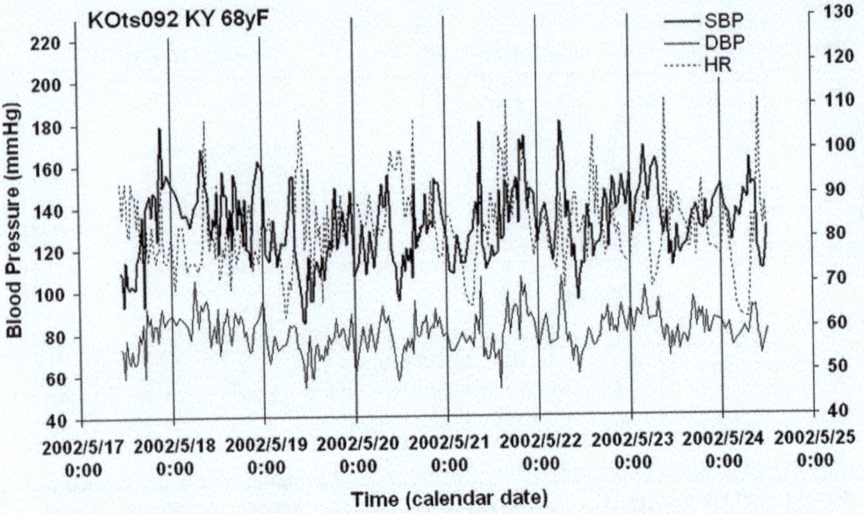

Fig. 19.36 Time series of systolic BP (SBP), diastolic BP (DBP), and heart rate (HR) from Friday to Friday. Time plots of the 7-day/24-hour records of SBP (*thick continuous line*), DBP (*thin continuous line*), and HR (*dashed line*) of a 61-year-old woman. Both SBP and DBP show BP ecphasia (odd timing of the circadian rhythm of BP)

A model consisting of cosine curves with periods of 24 and 12 hours was fitted by least squares to the data to estimate the MESOR (Midline-Estimating Statistics Of Rhythm) and the amplitude and acrophase (time of predicted maximum) of each component. MESOR estimates of SBP, DBP, and HR, estimated from the 7-day/24-hour ABP measurements, were 131.9 mmHg, 80.2 mmHg, and 82.3 bpm, respectively. In accordance with the guidelines for management of high blood pressure based on ABP in Japan, published in 2014, she still has mild hypertension, as her 24-hour SBP and DBP means are above 130 mmHg and 80 mmHg, respectively.

However, her day-to-day analyses showed "masked ABP normotension," as her 24-hour average of SBP varied, assuming values of 132.5 (Friday; hypertension), 134.7 (Saturday; hypertension), 120.0 (Sunday; normotension), 125.9 (Monday; normotension), 136.7 (Tuesday; hypertension), 134.9 (Wednesday; hypertension), and 133.2 mmHg (Thursday; hypertension).

The circadian profile of SBP shows "ecphasia," that is, an odd timing of the circadian rhythm of BP with respect to the chronodesm (Fig. 19.37, left). The circadian profile of HR lies in the upper part of the chronodesm (Fig. 19.37, right).

Based on the fitted model, the circadian acrophases of SBP, DBP, and HR were 00:27, 23:49, and 14:56, respectively, SBP and DBP peaking at an odd time along the circadian scale. HR, however, peaks in the afternoon, as anticipated (between ~12:00 and ~16:00). The double 24-hour amplitudes of SBP, DBP, and HR were

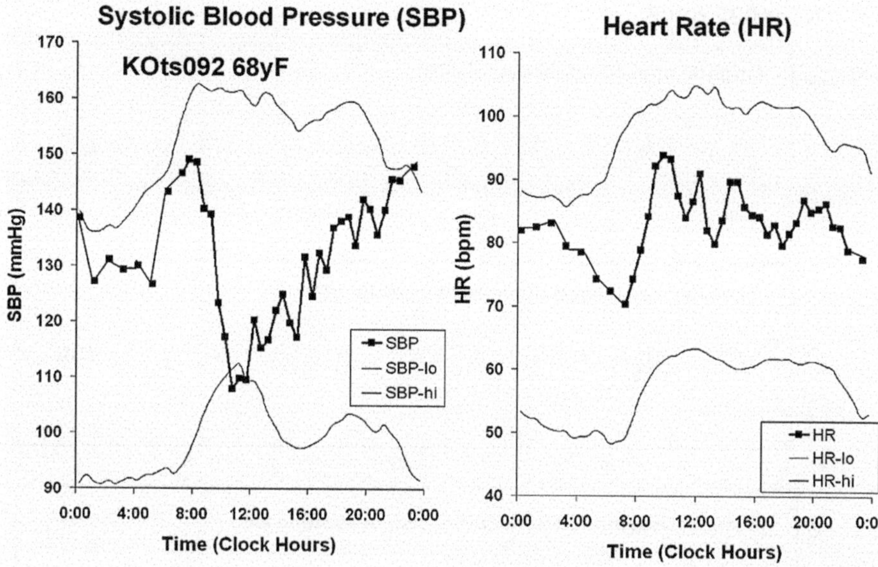

Fig. 19.37 Circadian profiles of systolic BP (*left*) and heart rate (*HR*) (*right*). 24-hour systolic BP (*left*) and heart rate (HR) (*right*) patterns of a woman, averaged over the 7 days of monitoring (*polygonal line* graph with *closed squares*), are shown along with time-specified prediction limits (*continuous thin lines*). Ecphasia is observed for SBP

17.14 mmHg, 7.03 mmHg, and 8.33 bpm, respectively. From a viewpoint of chronomedicine, she is diagnosed as having SBP and DBP ecphasia.

The circadian period of SBP was estimated by the Maximum Entropy Method (MEM), using the MemCalc software (GMS Co., Tokyo), which requires the data to be equidistant. Outliers were deleted from the original SBP time series (middle section of Fig. 19.38). Data were then made equidistant (bottom section of Fig. 19.38).

The MEM spectrum is shown on the middle left of Fig. 19.39, which shows a distinct circadian period of 1.0129 day, together with a prominent peak with a period of 0.4993 day. Periods extracted by the MEM analysis in the range up to 2.5 cycles/day are shown in the right upper side of this figure. Eleven components are listed here, and those with periods of 1.0129 and 0.4993 days have the largest spectral power, as assessed by the "area" covered by the spectral peak.

On the bottom left of Fig. 19.39, the edited data are fitted by least squares with a five-component model consisting of the five cosine curves with the five largest amplitudes, with corresponding periods of 5.956, 1.696, 1.013, 0.499, and 0.331 days (lower right part of Fig. 19.39).

In conclusion, the clinical diagnosis based on the chronobiologically interpreted 7-day/24-hour ABP monitoring consists of (1) both SBP and DBP ecphasia, with

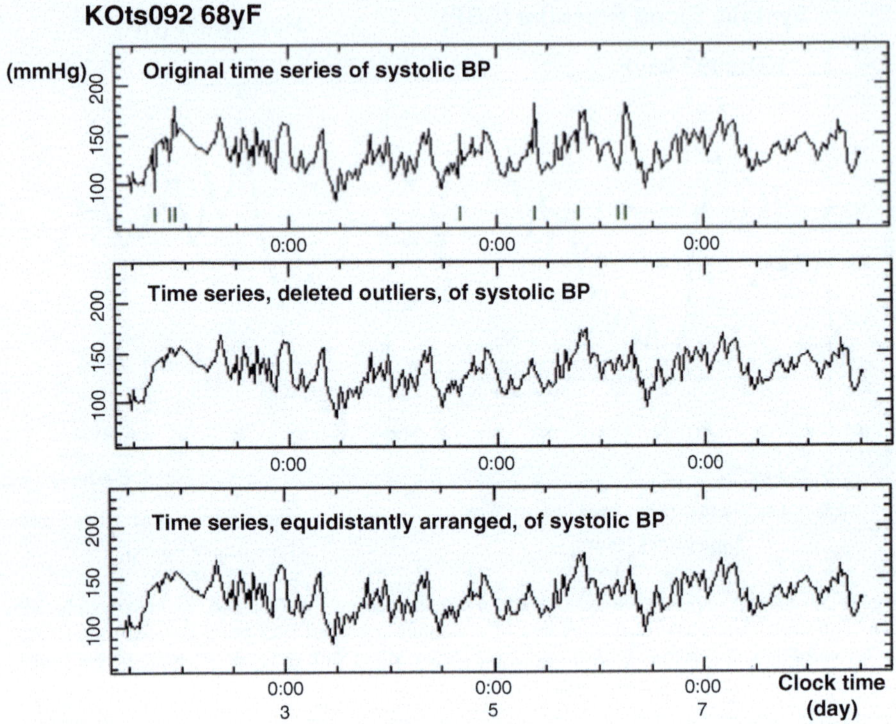

Fig. 19.38 Three versions of the time series of systolic BP. *Upper*: original data of systolic BP. *Middle*: edited data after mathematically excluding outliers. *Lower*: equidistantly arranged time series for analysis by the Maximum Entropy Method, using the MemCalc software (GMS Co., Tokyo)

normal timing of the circadian HR rhythm, (2) hypertensive condition despite antihypertensive medication, and (3) masked ABP normotension.

This subject needed more than 120 min to fall asleep, and she woke up 2–3 times a night. Her sleep disturbance was associated with nocturnal hypertension. She had excessive sleepiness during the daytime, and her circadian BP profiles showed a daytime dip together with a BP lowering effect of the antihypertensive medication in the morning. Circadian profiles of SBP and DBP showed BP ecphasia, probably exacerbated by treatment at the wrong time. Accordingly, we changed the time of treatment from the morning to the evening. In addition, we increased the dose of antidepressive medicine, which improved her depressive mood. She now has a good quality of sleep and is in good health. This example shows the importance of a personalized chronotherapy of high BP and a circadian ecphasia of BP as shown in our previous report [27].

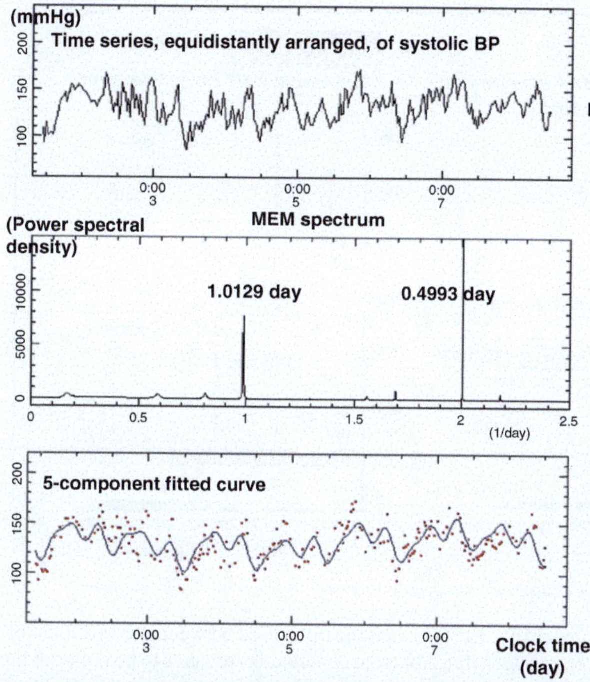

KOts092 68yF

Frequency (/ day)	Period (days)	Area
0.168	5.9557	34.586
0.399	2.5061	12.089
0.590	1.6959	16.611
0.805	1.2418	12.841
0.987	1.0129	43.496
1.294	0.7729	3.369
1.556	0.6425	3.967
1.690	0.5918	4.837
2.003	0.4993	36.902
2.178	0.4592	3.180
2.369	0.4222	0.395

Period (days)	Amplitude	Acro-phase
5.956	7.758	7-05:18
1.696	5.435	2-09:51
1.013	9.409	1-22:55
0.499	8.738	1-19:31
0.331	5.785	1-15:46

Fig. 19.39 Period analysis of the circadian rhythm of systolic BP. MEM analysis estimates the circadian and circasemidian periods of systolic BP to be 1.0129 and 0.4993 days (*left middle part*). Data are shown with a five-component model fitted by least squares to the data, consisting of cosine curves with the five largest amplitudes (*right lower part*)

19.4 Excessive Pulse Pressure

In this chapter, we introduce a case of "excessive pulse pressure (EPP)" from the viewpoint of 7-day/24-hour ambulatory blood pressure (ABP) monitoring. EPP is defined by a difference between the systolic (S) and diastolic (D) BP MESORs in a 7-day/24-hour record above 60 mmHg (until gender- and age-matched reference standards become available), as shown in Fig. 19.40.

19.4.1 Case Report 1 of "Excessive Pulse Pressure (EPP)"

An example of "EPP" is a 66-year-old non-obese (BMI = 23.7 kg/m^2) woman (ID number: KOts084). She has a regular sleep-wake cycle and goes to bed at about 23:00 and gets up at 07:30. She falls asleep within 20 min and sleeps well most

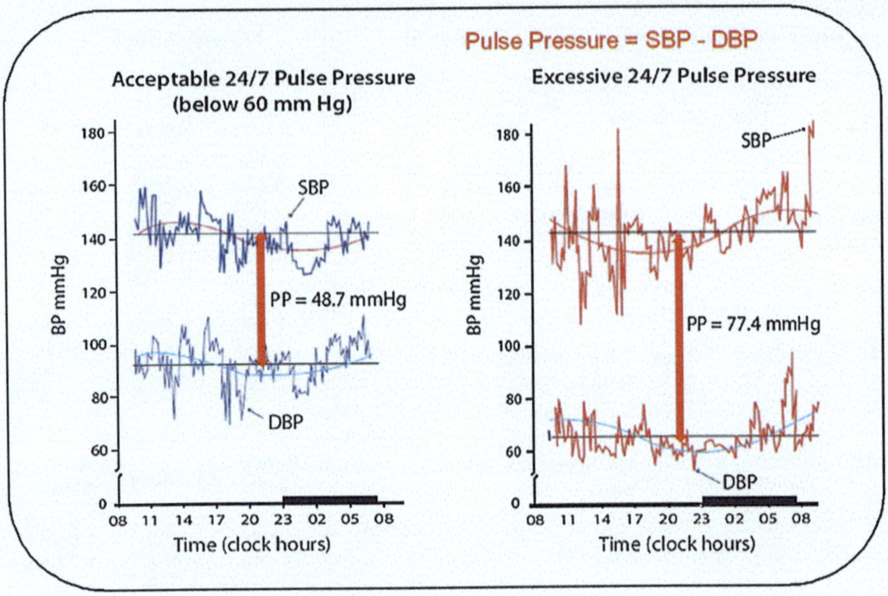

Fig. 19.40 Excessive pulse pressure (EPP). EPP is when the difference between the MESORs of SBP and DBP for adults exceeds 60 mmHg, a threshold that (eventually) remains to be replaced by reference values from clinically healthy peers with disease-free long-life outcomes, specified further by gender, age, ethnicity, and geography

days. She does not take any antihypertensive medication. Her office BP was 167/87 mmHg and pulse was 61 bpm. Time series of systolic BP (SBP), diastolic BP (DBP), and heart rate measured by pulse (HR) are shown in Fig. 19.41.

MESOR estimates from the 7-day/24-hour ABP measurements of SBP, DBP, and HR were 156.8 mmHg, 79.8 mmHg, and 49.4 bpm, respectively. The difference between the systolic and diastolic BP MESORs was 77.8 mmHg, which is above 60 mmHg. Hence, she is diagnosed as having an excessive pulse pressure (EPP) with SBP MESOR-hypertension. According to the guidelines for management of high BP based on ABP in Japan, published in 2014, she is also diagnosed as SBP hypertensive, as her 24-hour SBP is above 130 mmHg, but her 24-hour DBP is below 80 mmHg.

Day-to-day analyses showed "persistent ABP hypertension," as each day, her 24-hour average of SBP was above 130 mmHg, assuming values of 154.1 (Friday), 154.7 (Saturday), 157.4 (Sunday), 158.5 (Monday), 164.4 (Tuesday), 155.3 (Wednesday), and 157.8 mmHg (Thursday). By contrast, her 24-hour average of DBP showed intermittent hypertension, as it varied from Friday to Thursday, assum-

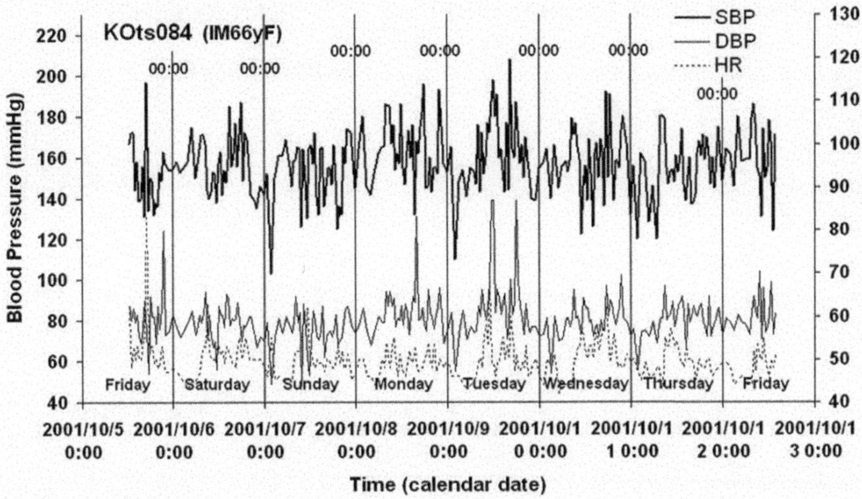

Fig. 19.41 An example of "excessive pulse pressure (EPP)" from the viewpoint of 7-day/24-hour ABP monitoring. Time plots of the 7-day/24-hour records of SBP (*thick continuous line*), DBP (*thin continuous line*), and HR (*dashed line*) of a 61-year-old woman

ing values of 80.4 (hypertension), 76.7 (normotension), 77.5 (normotension), 82.4 (hypertension), 86.5 (hypertension), 79.9 (normotension), and 81.9 mmHg (hypertension). Thus, daily pulse pressure from Friday to Thursday was 73.7, 78.0, 79.9, 76.1, 77.9, 75.4, and 75.9 mmHg, respectively, invariably above 60 mmHg.

Weekly variation of the day-night ratio of SBP showed "persistent non-dipping," but the dipping pattern varied from 1 day to another, assuming values of (from Friday to Thursday) −4.7 % (Friday night; riser), 7.5 % (Saturday night; non-dipping), −3.3 % (Sunday night; riser), 9.6 % (Monday night; non-dipping), 9.3 % (Tuesday night; non-dipping), 7.9 % (Wednesday night; non-dipping), and −2.7 % (Thursday night; riser).

The circadian profile of SBP shows "MESOR-hypertension with a non-dipping pattern" (Fig. 19.42, left), and the circadian profile of HR shows a low HR lying below the lower 5 % chronodesmic limit (Fig. 19.42, right).

Based on the fitted model, the double 24-hour amplitudes of SBP, DBP, and HR were 7.79 mmHg, 8.84 mmHg, and 6.04 bpm, respectively, all showing a dampened circadian variation. The circadian acrophases of SBP, DBP, and HR were 12:18, 14:12, and 14:47, respectively, all within acceptable limits.

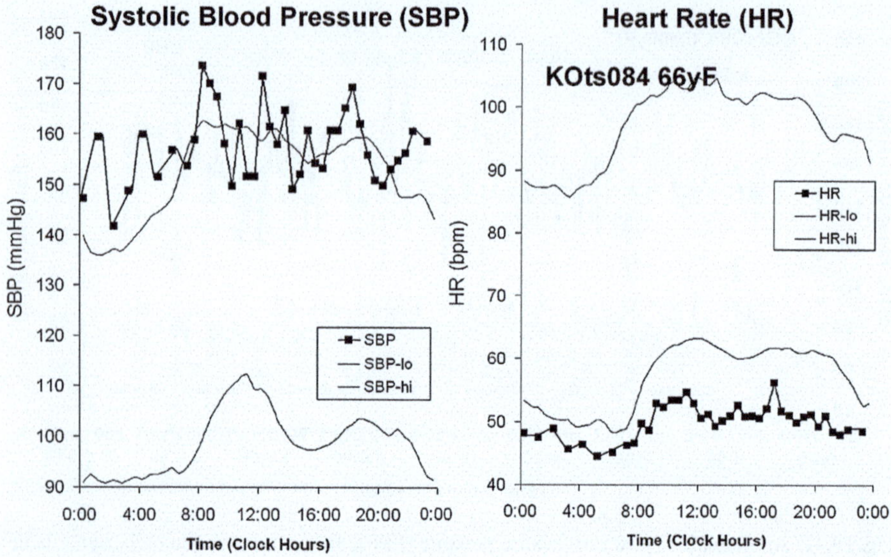

Fig. 19.42 Circadian profiles of systolic BP (*left*) and heart rate (HR) (*right*). 24-hour systolic BP (*left*) and heart rate (HR) (*right*) patterns of a woman with excessive pulse pressure, averaged over the 7 days of monitoring (*polygonal line* graph with *closed squares*), are shown along with time-specified prediction limits (*continuous thin lines*). The 24-hour SBP pattern is indicative of non-dipping hypertension, and the HR pattern remains below the lower 5 % time-varying threshold along the entire 24-hour scale

The circadian period of SBP was estimated by the Maximum Entropy Method (MEM), using the MemCalc software (GMS Co., Tokyo), which requires the data to be equidistant. Outliers were deleted from the original SBP time series (middle section of Fig. 19.43). Data were then made equidistant (bottom section of Fig. 19.43).

The MEM spectrum is shown on the middle left of Fig. 19.44, which shows a disrupted circadian rhythm: instead of the usual circadian variation, there are several peaks, in the ultradian range of the spectrum. The major peaks correspond to periods of 4.0430, 0.6809, 0.5827, or 0.4641 days. Periods extracted by the MEM analysis in the range up to 2.5 cycles/day are shown in the right upper side of this figure. Twelve components are listed here, but there was no major prominent component.

On the bottom left of Fig. 19.44, the edited data are fitted by least squares with a five-component model consisting of the five cosine curves with the five largest amplitudes, with corresponding periods of 1.152, 1.017, 0.583, 0.393, and

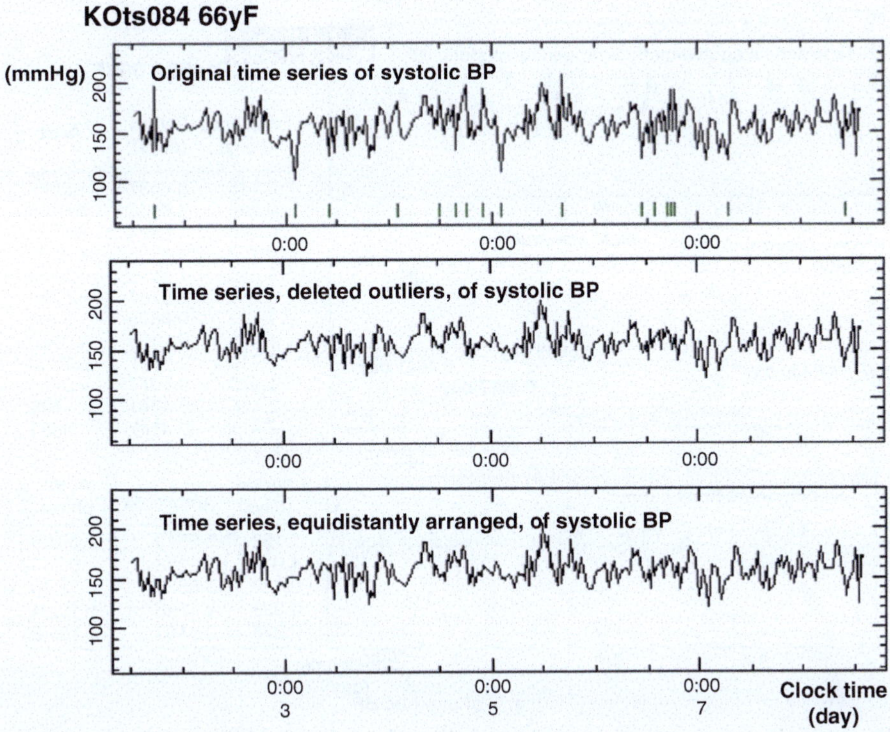

Fig. 19.43 Three versions of the time series of systolic BP. *Upper*: original data of systolic BP. *Middle*: edited data after mathematically excluding outliers. *Lower*: equidistantly arranged time series for analysis by the Maximum Entropy Method, using the MemCalc software (GMS Co., Tokyo)

0.228 days (lower right of Fig. 19.44). None of the detected components had an amplitude greater than 10 mmHg.

In conclusion, the clinical diagnosis based on the chronobiologically interpreted 7-day/24-hour ABP monitoring consists of (1) SBP MESOR-hypertension; (2) EPP; (3) persistent ABP hypertension; (4) persistent "non-dipping" pattern, changing from non-dipping to riser from 1 day to another; and (5) disrupted circadian rhythm of SBP. Accordingly, we recommended this subject to strictly improve her lifestyle, focusing on salt intake and daily exercise. We also started antihypertensive treatment with a calcium channel blocker (CCB) (benidipine, 4 mg/day). Treatment has been going well and she is now in good health.

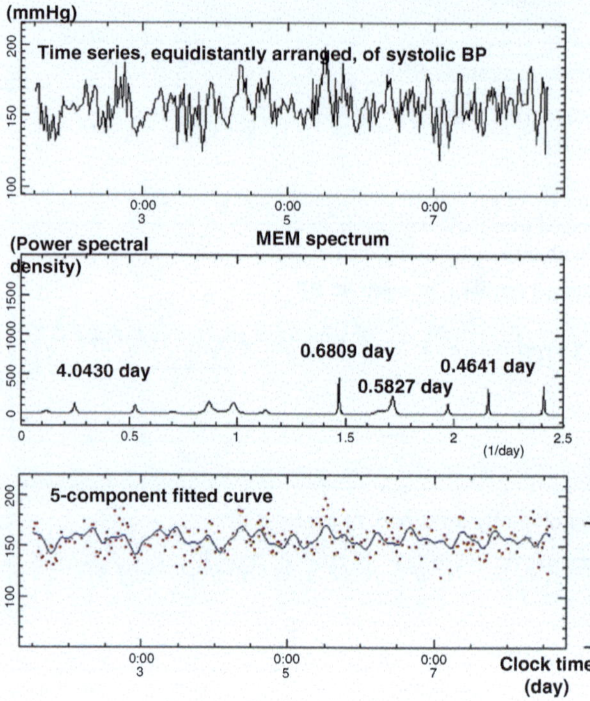

KOts084 66yF

Frequency (/ day)	Period (days)	Area
0.115	8.6697	1.3647
0.247	4.0430	2.6195
0.528	1.8924	2.2717
0.702	1.4236	1.2207
0.868	1.1515	7.0885
0.984	1.0167	6.5755
1.132	0.8836	1.8401
1.469	0.6809	4.0780
1.716	0.5827	9.2396
1.970	0.5077	2.0460
2.155	0.4641	2.7625
2.411	0.4147	3.8581

Period (days)	Amplitude	Acro-phase
1.152	3.5513	2-05:05
1.017	3.2278	2-09:45
0.583	3.8325	1-12:46
0.393	4.4398	1-13:04
0.228	2.9163	1-15:42

Fig. 19.44 Period analysis of the circadian rhythm of systolic BP. MEM analysis shows a disrupted circadian rhythm of systolic BP (*left middle part*). Data are shown with a five-component model fitted by least squares to the data, consisting of cosine curves with the five largest amplitudes. No clear circadian periodicity can be demonstrated (*right lower part*)

19.5 Among-Day Variability of Vascular Variability Anomaly

19.5.1 An Example of Vascular Variability Syndrome (VVS) Consisting of Four Kinds of Vascular Variability Anomaly (VVA)

An example of vascular variability syndrome (VVS) is a 63-year-old non-obese (BMI = 23.5 kg/m^2) man (ID number: KOts019). He takes antihypertensive medication (CCB, benidipine 4 mg daily), but his office BP and pulse were still 157/107 mmHg and 76 bpm, respectively.

He usually takes care of his health and his seasoning is average. He does not have a depressive mood, and his scores on the GDS-15 and PHQ-2 scales were 2 and 0, respectively. His subjective QOL indices of health, mood, family relationship, relationship with friends, financial satisfaction, life satisfaction, and happiness were 90 %, 90 %, 100 %, 90 %, 80 %, 90 %, and 90 %, respectively.

Based on the model fitted to the 7-day/24-hour ABP monitoring record, systolic BP (SBP), diastolic BP (DBP), and heart rate measured by pulse (HR) MESOR estimates were 150.1 mmHg, 89.2 mmHg, and 66.6 bpm, respectively, indicating both SBP and DBP MESOR-hypertension. As the difference between the SBP and DBP MESORs is 60.9 mmHg, he has an "excessive pulse pressure (EPP)," another vascular variability anomaly (VVA). The double 24-hour amplitudes of SBP, DBP, and HR were 40.80 mmHg, 22.84 mmHg, and 15.036 bpm, respectively. Because the 95 % prediction limit for the double amplitude of SBP extends from 3.41 to 34.00 mmHg in healthy men of his age group, he is also diagnosed with SBP CHAT. Corresponding circadian acrophases were 14:04, 13:47, and 14:59, all within acceptable limits. These results are summarized in the sphygmochron of Fig. 19.45.

The circadian profiles of SBP and HR are shown in Fig. 19.46. SBP varies close to the upper chronodesmic limit, exceeding this limit during the daytime, an indica-

Name:--------------------
Age: 63
Monitoring From: 1/12/2001 13:13
Comments: Sleep: 23:30-06:00
Rx:Amlodipine 2.5mg at 06:30

Patient #: KOts019
Sex: M
To: 1/19/2001 13:30

MESOR-Hypertension

Systolic CHAT

CHRONOBIOLOGIC CHARACTERISTICS

	SYSTOLIC BP (mmHg) Patient Value	Peer Group Reference Limits	DIASTOLIC BP (mmHg) Patient Value	Peer Group Reference Limits	HEART RATE (bpm) Patient Value	Peer Group Reference Limits
ADJUSTED 24-hour MEAN (MESOR)	150.1	109.3-141.6 Range	89.2	69.6-86.8 Range	66.6	65.9-87.9 Range
PREDICTABLE CHANGE (DOUBLE AMPLITUDE)	40.80	3.41-34.00 Range	22.84	0.00-27.55 Range	15.03	0.00-33.40 Range
TIMING OF OVERALL HIGH VALUES (ACROPHASE) (hr:min)	14:04	8:30-18:27 Range	13:47	8:54-16:58 Range	14:59	9:39-18:57 Range

	STD (MIN;MAX)*	STD (MIN;MAX)*	STD (MIN;MAX)*
PERCENT TIME OF ELEVATION	41.3%	9.4%	0.0%
TIMING OF EXCESS	14:58 (hr:min)	17:42 (hr:min)	0:00 (hr:min)
EXTENT OF EXCESS DURING 24 HOURS HBI*	66 (mmHg x hour)	4 (mmHg x hour)	0 (mmHg x hour)
10-YEAR CUMULATIVE EXCESS	240 (mmHg x hour)(in 1.000's units)	13 (mmHg x hour)(in 1.000's units)	0 (mmHg x hour)(in 1.000's units)

Individualized bounded indices: (STD = Standard)(Min = Minimum)(Max = Maximum)(HBI = Hyperbanic Index)

Fig. 19.45 Sphygmochron summarizing parametric and nonparametric results of the 7-day record interpreted in the light of reference data in clinically healthy peers matched by gender and age

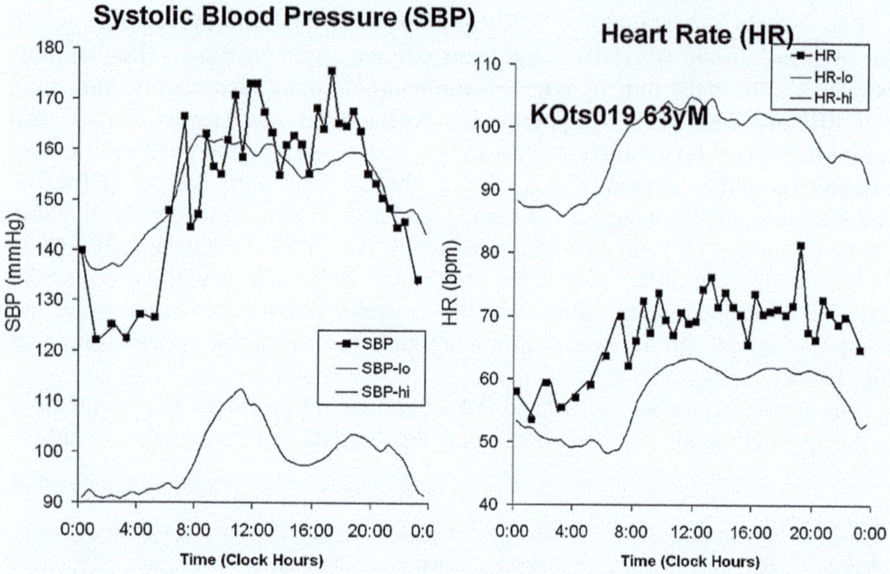

Fig. 19.46 Circadian profiles of systolic BP (*left*) and heart rate (HR) (*right*). 24-hour systolic BP and HR patterns of a 63-year-old man, averaged over the 7 days of monitoring (*polygonal line* graph with *closed squares*), are shown along with time-specified prediction limits (*continuous thin lines*). SBP varies mostly above the upper limit of the chronodesm during the daytime. HR varies well within the chronodesmic limits, closer to the lower than the upper limit

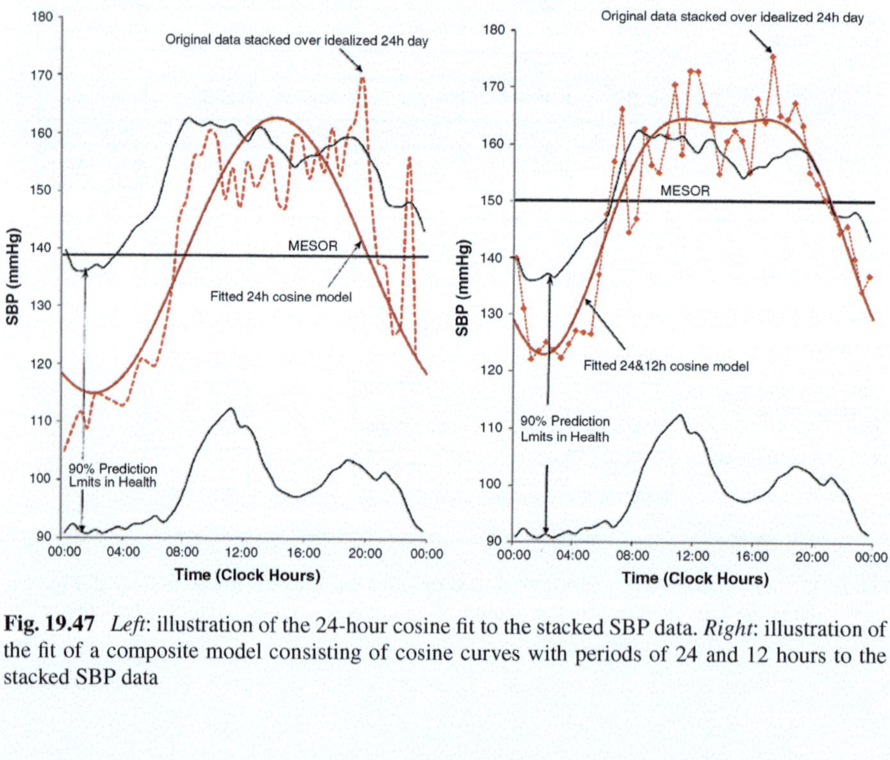

Fig. 19.47 *Left*: illustration of the 24-hour cosine fit to the stacked SBP data. *Right*: illustration of the fit of a composite model consisting of cosine curves with periods of 24 and 12 hours to the stacked SBP data

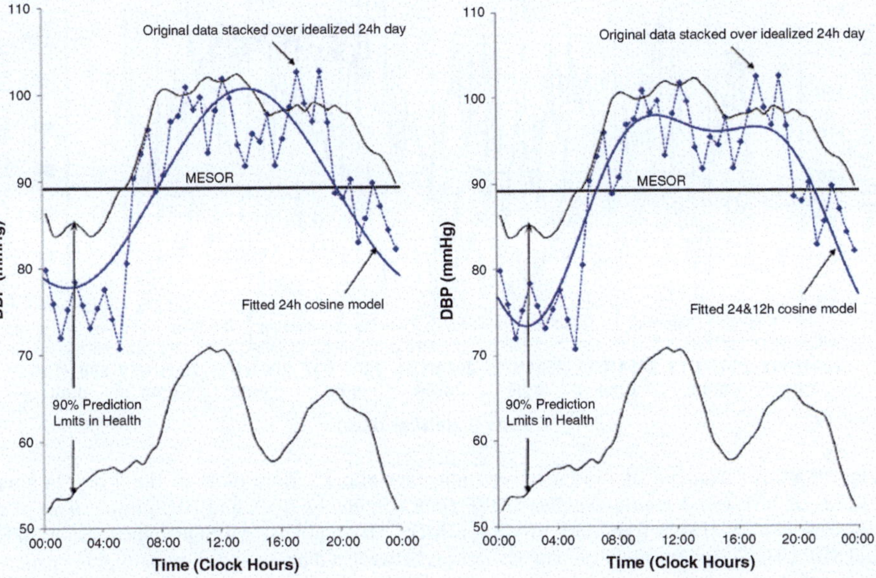

Fig. 19.48 *Left*: illustration of the 24-hour cosine fit to the stacked DBP data. *Right*: illustration of the fit of a composite model consisting of cosine curves with periods of 24 and 12 hours to the stacked DBP data

tion of SBP MESOR-hypertension. Figure 19.47 illustrates the fit of a 24-hour cosine curve (left) and of a composite model consisting of cosine curves with periods of 24 and 12 hours (right) to the stacked SBP data. Corresponding results for DBP are shown in Fig. 19.48.

Time series of SBP, DBP, and HR are shown in Fig. 19.49. From Friday to Thursday, the 24-hour average of SBP varies from day to day, assuming values of 149.2 mmHg (Friday; hypertension), 159.0 mmHg (Saturday; hypertension), 145.7 mmHg (Sunday; hypertension), 153.3 mmHg (Monday; hypertension), 162.6 mmHg (Tuesday; hypertension), 151.6 mmHg (Wednesday; hypertension), and 155.6 mmHg (Thursday; hypertension). The 24-hour mean of SBP was above 130 mmHg all 7 days, indicating persistent hypertension.

The day-night ratio of SBP is consistently above 10 %, indicating a "persistent dipping" pattern, but it varied from day to day between dipping and extreme dipping, assuming values of 29.8 % (Friday night; extreme dipping), 29.6 % (Saturday night; extreme dipping), 20.5 % (Sunday night; extreme dipping), 18.8 % (Monday night; dipping), 15.1 % (Tuesday night; dipping), 19.1 % (Wednesday night; dipping), and 29.6 % (Thursday night; extreme dipping).

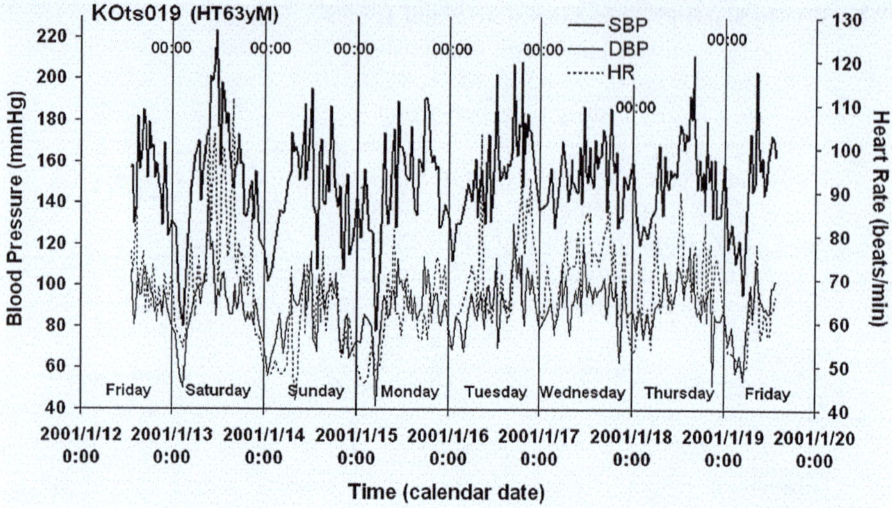

Fig. 19.49 An example of "vascular variability syndrome." Time plots of the 7-day/24-hour records of SBP (*thick continuous line*), DBP (*thin continuous line*), and HR (*dashed line*) of a 63-year-old man. He has four kinds of VVA, namely, SBP and DBP MESOR-hypertension, EPP, and SBP CHAT. His day-night ratio of SBP varies between a dipping and extreme-dipping pattern from 1 day to another

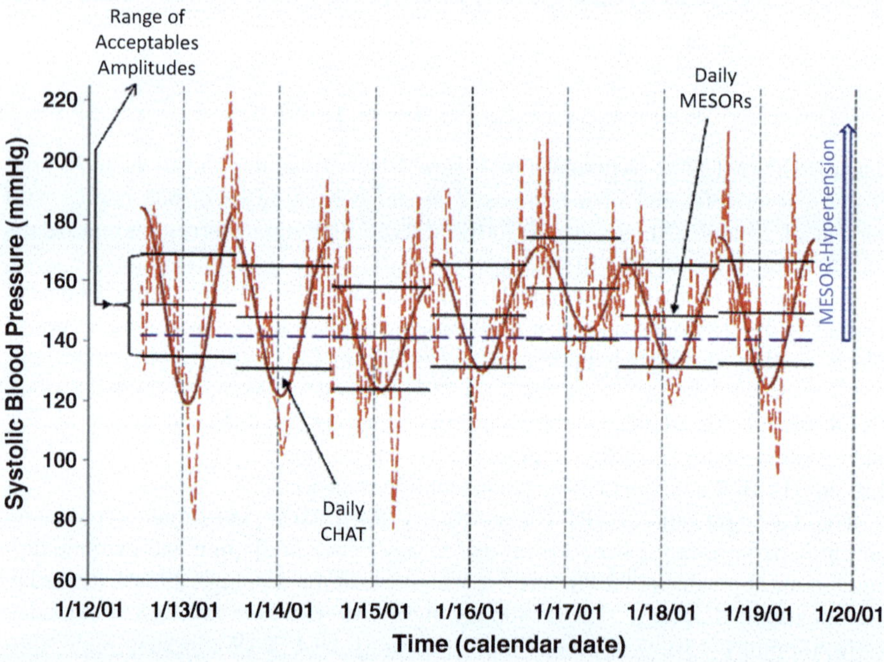

Fig. 19.50 Illustration of the day-to-day changes in the MESOR of systolic blood pressure. MESOR-hypertension is found on 6 of the 7 days of monitoring. It is close to the upper limit of acceptability on the only day when it is not excessive. The circadian amplitude of systolic blood pressure is excessive (CHAT) on 5 of the 7 days of monitoring

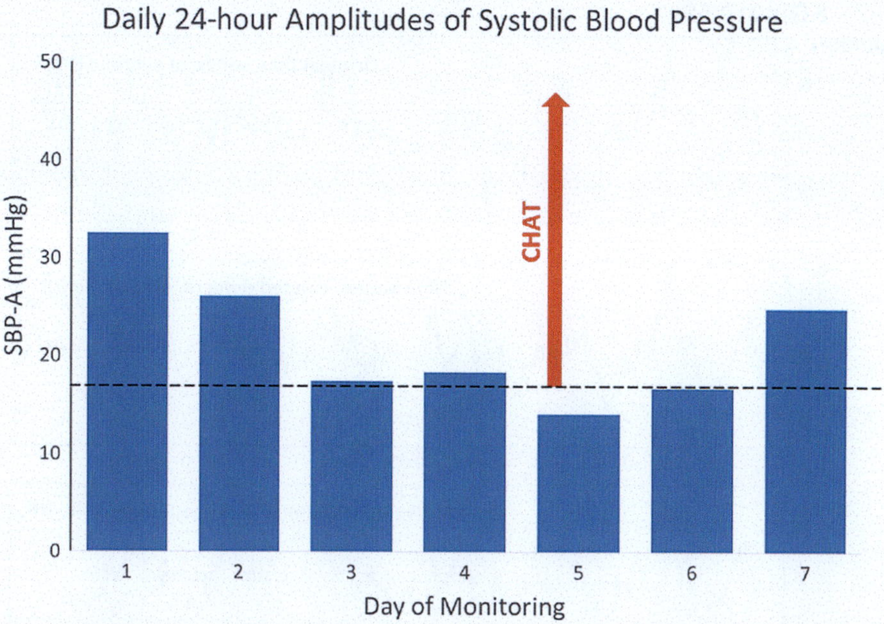

Fig. 19.51 Systolic CHAT occurs on 5 of the 7 days of monitoring, as indicated by the 24-hour amplitude of SBP exceeding the threshold in health (*horizontal dashed line*)

Day-to-day changes in the MESOR and 24-hour amplitude of SBP are illustrated in Fig. 19.50. MESOR-hypertension is present on 6 of the 7 days and is close to the acceptable limit on the day when it is not excessive. Systolic CHAT is observed on 5 of the 7 days of monitoring, as also shown in Fig. 19.51.

The circadian period of SBP was estimated by the Maximum Entropy Method (MEM), which requires the data to be equidistant. Outliers were deleted from the original SBP time series (middle section of Fig. 19.52). Data were then made equidistant (bottom section of Fig. 19.52).

The MEM spectrum is shown on the middle left of Fig. 19.53, which shows a prominent peak corresponding to the circadian component with a period of 1.0005 day, together with smaller peaks corresponding to periods of 1.7505, 1.2063, and 0.8579 days. Periods extracted by the MEM analysis in the range up to 2.5 cycles/day are shown in the right upper side of this figure. Twelve components are listed here, and that with a "period" of 1.0005 day has the largest spectral power, as assessed by the "area" covered by the spectral peak.

On the bottom left of Fig. 19.53, the edited data are fitted by least squares with a five-component model consisting of the five cosine curves with the five largest

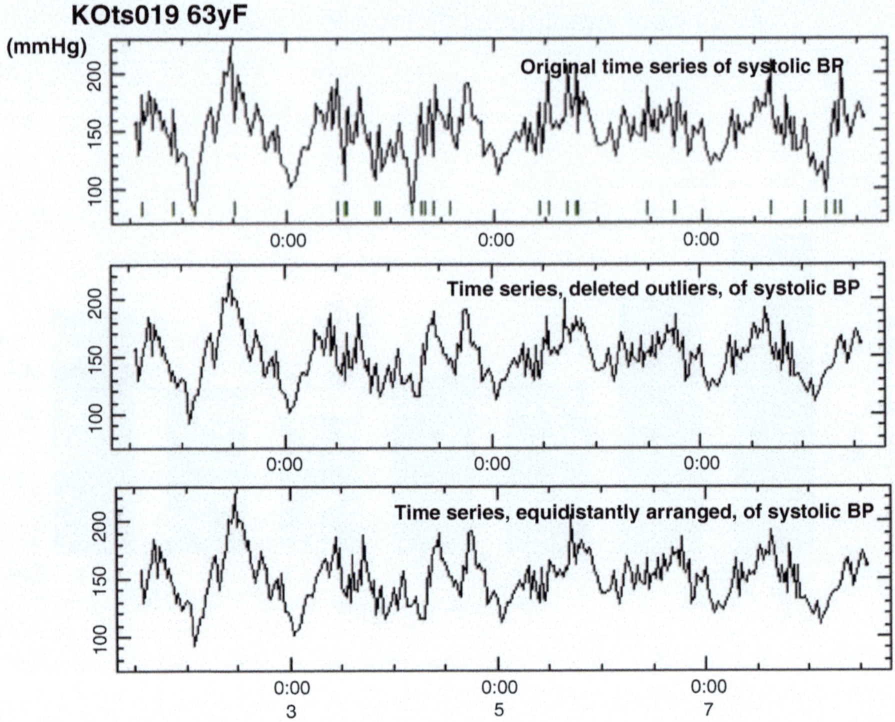

Fig. 19.52 Three versions of the time series of systolic BP. *Upper*: original data of systolic BP. *Middle*: edited data after mathematically excluding outliers. *Lower*: equidistantly arranged time series for analysis by the Maximum Entropy Method, using the MemCalc software (GMS Co., Tokyo)

amplitudes, with corresponding periods of 1.206, 1.001, 0.858, 0.502, and 0.385 days (right lower part of Fig. 19.53).

In conclusion, the clinical diagnosis based on the chronobiologically interpreted 7-day/24-hour ABP monitoring consists of (1) both SBP and DBP MESOR-hypertension, (2) persistent ABP hypertension, (3) excessive pulse pressure, (4) SBP CHAT, and (5) between-day ABP variability of the day-night ratio of SBP (between dipping and extreme-dipping pattern). Despite being on antihypertensive treatment (benidipine 4 mg once daily), he was still diagnosed with four VVAs or a VVS.

Accordingly, we made strict dietary recommendations, especially on salt intake. We also started a combination therapy of CCB and ARB. Under this plan, he now seems to be in good health.

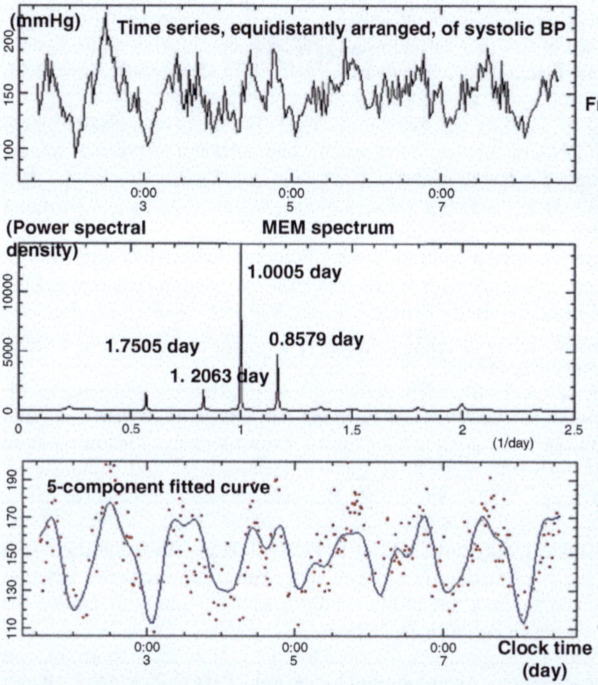

KOts019 63yM

Frequency (/ day)	Period (days)	
0.227	4.4132	9.46
0.309	3.2329	5.22
0.571	1.7507	9.48
0.829	1.2063	15.20
0.999	1.0005	168.61
1.166	0.8579	40.23
1.359	0.7359	9.77
1.507	0.6634	4.13
1.799	0.5557	6.29
1.994	0.5016	18.51
2.092	0.4780	3.58
2.331	0.4290	2.82

Period (days)	Amplitude	Acro-phase
1.206	5.43	2-08:40
1.001	17.73	1-14:06
0.858	7.90	1-15:31
0.502	5.59	1-19:38
0.385	5.11	1-17:44

Fig. 19.53 Period analysis of the circadian rhythm of systolic BP. MEM analysis estimates the circadian period of SBP to be 1.0005 day (*left middle part*). Data are shown with a five-component model fitted by least squares to the data, consisting of cosine curves with the five largest amplitudes, the 1.001-day component being very prominent (*right lower part*)

References

1. Halberg F, Cornélissen G, Otsuka K, Sanchez de la Peña S, Schwartzkopff O, Watanabe Y, Pati AK, Wall DG, Delmore P, Borer K, Beaty LA, Nolley ES, Adams C, Siegelova J, Homolka P, Dusek J, Fiser B, Prikryl P. Why and how to implement 7-day/24-hour blood pressure monitoring? Gerontol Geriatr Int J Gerontol ChronomeGeriatr 2005;8(1):1–31.
2. Halberg F, Katinas GS, Cornélissen G, Sanchez de la Peña S. Continued self-surveillance for treating a frequent vascular variability syndrome. Gerontol Geriatr Int J Gerontol ChronomeGeriatr 2008;11:147–54.
3. Halberg F, Cornélissen G, Katinas G, Tvildiani L, Gigolashvili M, Janashia K, Toba T, Revilla M, Regal P, Sothern RB, Wendt HW, Wang ZR, Zeman M, Jozsa R, Singh RB, Mitsutake G, Chibisov SM, Lee J, Holley D, Holte JE, Sonkowsky RP, Schwartzkopff O, Delmore P, Otsuka K, Bakken EE, Czaplicki J. International BIOCOS Group. Chronobiology's progress: season's appreciations 2004–2005. Time-, frequency-, phase-, variable-, individual-, age- and site-specific chronomics. J Appl Biomed 2006;4:1–38.

4. Halberg F, Cornélissen G, Siegelova J, Homolka P, Dusek J, Fiser B. Differences in time structural environmental congruence of vascular variables in an elderly man. In: Halberg F, Kenner T, Fiser B, Siegelova J, editors. Proceedings, Noninvasive Methods in Cardiology, Brno, 7–10 July 2009. pp. 271–80. http://web.fnusa.cz/files/kfdr2009/sbornik_2009.pdf
5. Scheving LE, Tsai TH, Pauly JE, Halberg F. Circadian ACTH 1–17 effects on murine tritiated thymidine incorporation into DNA of thymus, bone marrow and spleen; also on total splenic RNA and DNA. Chronobiologia 1984;11:109–30.
6. Otsuka K, Cornélissen G, Halberg F. Predictive value of blood pressure dipping and swinging with regard to vascular disease risk. Clin Drug Investig 1996;11:20–31.
7. Halberg F, Cornélissen G. International Womb-to-Tomb Chronome Initiative Group: resolution from a meeting of the International Society for Research on Civilization Diseases and the Environment (New SIRMCE Confederation), Brussels, 17–18 Mar 1995: Fairy tale or reality? Medtronic Chronobiology Seminar #8, Apr 1995, 12 pp. text, 18 figures. https://sites.google.com/a/umn.edu/halbergchronobiologycenter/
8. Halberg F, Cornélissen G, Otsuka K, Katinas GS, Schwartzkopff O, Halpin C, Mikulecky M, Revilla M, Siegelova J, Homolka P, Dusek J, Fiser B, Singh RB. Chronomics detects altered vascular variabilities constituting risks greater than hypertension: with an illustrative case report. In: Mitro P, Pella D, Rybar R, Valocik G, editors. Proceedings, 2nd Congress on Cardiovascular Diseases, Kosice, 25–27 Apr 2002. Bologna: Monduzzi Editore; 2002. pp. 223–58.
9. Halberg F, Cornélissen G, Otsuka K, Siegelova J, Fiser B, Dusek J, Homolka P, Sanchez de la Peña S, Singh RB, BIOCOS project. Extended consensus on need and means to detect vascular variability disorders (VVDs) and vascular variability syndromes (VVSs). Leibniz-Online Nr. 5; 2009. (http://www.leibniz-sozietaet.de/journal). 35 pp.
10. Cornélissen G, Halberg F, Wall D, Siegelova J, Zaslavskaya RM. How long to screen: ice hockey game and transient circadian hyper-amplitude-tension, *CHAT*. Scr Med (Brno) 1997;70:189–98.
11. Phillips DP, Jarvinen JR, Abramson IS, Phillips RR. Cardiac mortality is higher around Christmas and New Year's than at any other time: the holidays as a risk factor for death. Circulation 2004;110:3781–8.
12. Kloner RA, Poole WK, Perritt RL. When throughout the year is coronary death most likely to occur? A 12-year population-based analysis of more than 220,000 cases. Circulation 1999;100:1630–4.
13. Saposnik G, Baibergenova A, Dang J, Hachinski V. Does a birthday predispose to vascular events? Neurology 2006;67:300–4.
14. Uexküll TV, Wick E. Die Situationshypertonie. Arch Kreislaufforschung 1962;39:236–71.
15. Ettinger PO, Wu CF, De La Cruz Jr C, Weisse AB, Ahmed SS, Regan TJ. Arrhythmias and the "holiday heart": alcohol-associated cardiac rhythm disorders. Am Heart J 1978;95:555–62.
16. Arnold R, Cerrelli EC. Holiday effect on traffic fatalities. Washington, DC: US Department of Transportation, National Highway Traffic Safety Administration; 1987. Publication DOT-HS-807-115.
17. Hecht K, Scherf H-P, Yilmaz Y, Meffert P. Post-Silvesterparty-Hypertonus. J Hyperton 2007;11:6–10.
18. Hecht K, Vogt W-E, Fietze I. Beziehung zwischen Insomnien und arterieller Hypotonie. Pneumologie 1991;24:1996–9.
19. Hecht K, Andler S, Breinl S, Lander HJ, Stück M. Objektive Kontrolle der Selbstentspannungsfähigkeit anhand von Zeitreihenmessungen des Blutdrucks und der elektrodermalen Aktivität (EDA). In: Hecht K, Scherf H-P, König O, editors. Stress durch Überforderung und Unterforderung. Berlin/Milow: Schibri Verlag; 2001. p. 253–72.
20. Scherf H-P, Hecht K, Yilmaz Y, Meffert P. Fehlgemessener oder realer Hypertoniker? J Hyperton 2006;10:6–11.
21. Lee MS, Lee JS, Lee JY, Cornélissen G, Otsuka K, Halberg F. About 7-day (circaseptan) and circadian changes in cold pressor test (CPT). Biomed Pharmacother 2003;57 Suppl 1:39s–44.

22. Watanabe Y, Cornélissen G, Halberg F, Beaty L, Siegelova J, Otsuka K, Bakken EE. Harm vs. benefit from losartan with hydrochlorothiazide at different circadian times in MESOR-hypertension or CHAT. In: Halberg F, Kenner T, Fiser B, Siegelova J, editors. Proceedings, Noninvasive Methods in Cardiology, Brno, 4–7 Oct 2008. pp. 149–67.
23. Cornélissen G. Profiles in science: Franz Halberg. J Exp Ther Oncol 2008;7(4):257–62.
24. Halberg F, Lee JK, Nelson WL. Time-qualified reference intervals – chronodesms. Experientia (Basel) 1978;34:713–6.
25. Halberg F, Nelson W, Lee JK, Levine H, Halberg J, Halberg E, Bartter FC, Delea C, Haus E, Halberg F, Kawasaki T, Omae T, Scheving LE. Individualized cosinor-based chronodesms of systolic blood pressure. Chronobiologia 1978;5:96.
26. Nelson W, Cornélissen G, Hinkley D, Bingham C, Halberg F. Construction of rhythm-specified reference intervals and regions, with emphasis on "hybrid" data, illustrated for plasma cortisol. Chronobiologia 1983;10:179–93.
27. Watanabe Y, Halberg F, Otsuka K, Cornelissen G. Toward a personalized chronotherapy of high blood pressure and a circadian overswing. Clin Exp Hypertens 2013;35:257–66.
28. Bingham C, Arbogast B, Guillaume GC, Lee JK, Halberg F. Inferential statistical methods for estimating and comparing cosinor parameters. Chronobiologia 1982;9:397–439.

Chapter 20
Case Presentations III: Among-Day Variability of Circadian Profiles of 24-Hour ABP

Abstract In this chapter we introduce case presentations of the circaseptan (i.e., about 7-day) profiles of ambulatory blood pressure (ABP), including "masked ABP hypertension," "masked ABP normotension," "ABP masked non-dipper," "ABP masked dipper," "persistent non-dipper," and "riser hypertension," from a viewpoint of 7-day/24-hour ABP monitoring.

Keywords Among-day variability of 24-hour ambulatory BP • Masked ambulatory BP hypertension • Masked ambulatory BP normotension • Ambulatory BP masked non-dipper • Ambulatory BP masked dipper

20.1 Masked Ambulatory Blood Pressure Hypertension

We have investigated not only the circadian blood pressure (BP) profile [1–10] but also several kinds of circaseptan, i.e., about 7-day profile of ambulatory BP (ABP), including between-day and among-day variability [11–30]. In this chapter we introduce a case of "masked ambulatory blood pressure (ABP) hypertension" from the viewpoint of 7-day/24-hour ABP monitoring. "Masked ABP hypertension" is defined as normotension (NT) on day 1 but hypertension on at least one of the following 6 days. Hypertension is defined as a 24-hour BP mean above 130/80 mmHg.

20.1.1 Case Report 1 of "Masked Ambulatory Blood Pressure Hypertension"

An example of "masked ABP hypertension" is this 41-year-old nonobese (BMI = 23.8 kg/m^2) man (ID number: KOts100). He visited us for a health check. He goes to bed regularly at 23:30 and gets up at 07:30, but he does not sleep well. His office BP was 129/99 mmHg and pulse was 71 betas per minute (bpm). He has a history of cerebellum bleeding when he worked in China 3 years ago and takes three antihypertensive medications, including CCB and ARB.

K. Otsuka et al., *Chronomics and Continuous Ambulatory Blood Pressure Monitoring*, DOI 10.1007/978-4-431-54631-3_20

Time series of systolic BP (SBP), diastolic BP (DBP), and heart rate measured by pulse (HR) are shown in Fig. 20.1. From Saturday to Friday, the 24-hour average of SBP assumes values of 111.6 mmHg (Saturday; NT), 115.9 mmHg (Sunday; NT), 121.3 mmHg (Monday; NT), 116.8 mmHg (Tuesday; NT), 130.2 mmHg (Wednesday; hypertension), 131.9 mmHg (Thursday; hypertension), and 115.5 mmHg (Friday; NT). The 24-hour mean of SBP was above 130 mmHg on Wednesday and Thursday, indicating the presence of masked ABP hypertension. The corresponding 24-hour averages of DBP were 82.6 mmHg (Saturday), 81.3 mmHg (Sunday), 85.7 mmHg (Monday), 84.4 mmHg (Tuesday), 90.4 mmHg (Wednesday), 92.3 mmHg (Thursday), and 83.2 mmHg (Friday), indicating persistent DBP hypertension.

Based on the model fitted to the 7-day/24-hour ABP monitoring record, MESOR estimates of SBP, DBP, and HR were 118.0 mmHg, 83.7 mmHg, and 79.9 bpm, respectively, indicating DBP MESOR-hypertension. The double 24-hour amplitudes of SBP, DBP, and HR were 13.23 mmHg, 9.68 mmHg, and 13.39 bpm, respec-

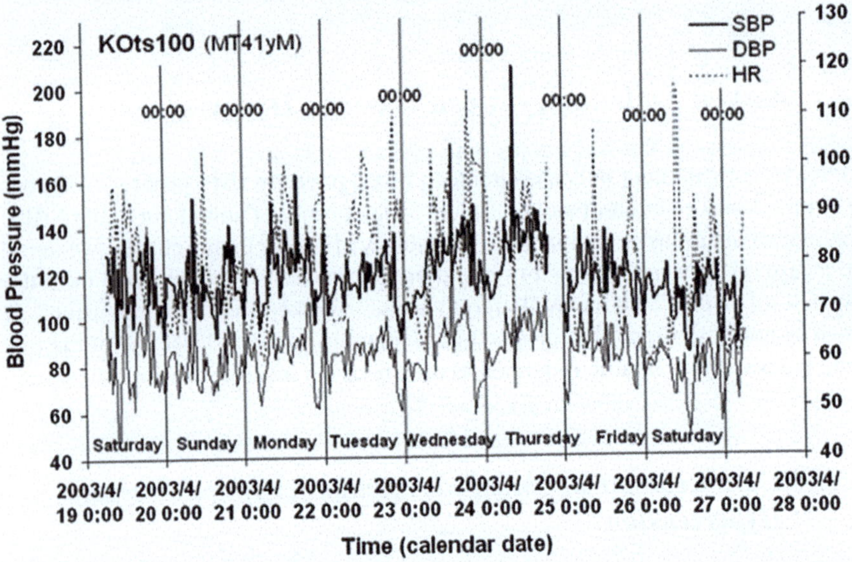

Fig. 20.1 An example of "masked ambulatory blood pressure hypertension." Time plots of the 7-day/24-hour records of SBP (*thick continuous line*), DBP (*thin continuous line*), and HR (*dashed line*) of a 41-year-old man. The day-night ratio of SBP varies from day to day, assuming values of 1.9 % (Saturday; non-dipping), −0.5 % (Sunday; reverse dipping), 10.8 % (Monday; dipping), 11.2 % (Tuesday; dipping), 13.2 % (Wednesday; dipping), 15.5 % (Thursday; dipping), and 5.6 % (Friday; non-dipping), indicative of a masked dipping pattern of SBP

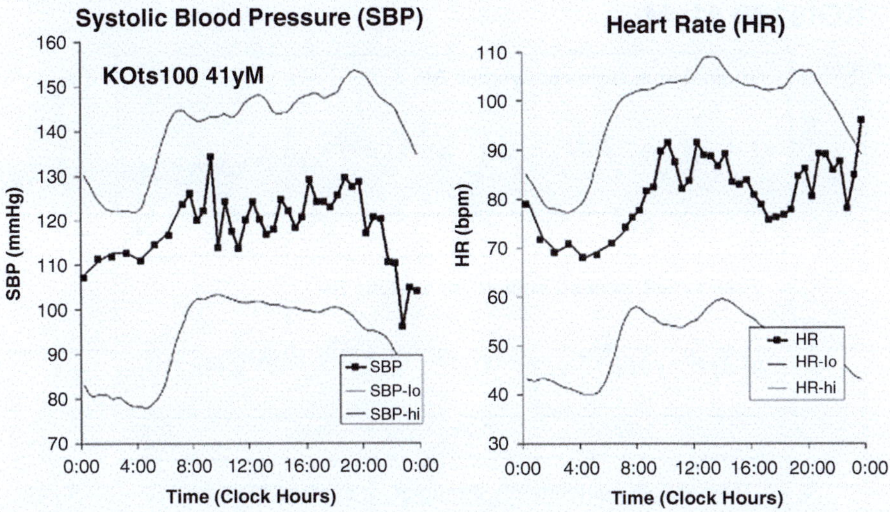

Fig. 20.2 Circadian profiles of systolic BP (*left*) and heart rate (HR) (*right*). 24-hour profiles of systolic BP and heart rate (HR) of a 41-year-old man, averaged over the 7 days of monitoring (*polygonal line* graph with *closed squares*), are shown along with time-specified prediction limits (*continuous thin lines*). Both SBP and HR patterns vary in the *middle* of the chronodesmic range

tively, and corresponding circadian acrophases were 13:36, 13:00, and 15:27, all within acceptable limits.

The circadian profiles of SBP and HR are shown in Fig. 20.2. Both SBP and HR vary well within the chronodesmic limits, but their circadian pattern is somewhat unusual with evening values of SBP much lower and HR values much higher than those found at other circadian stages.

The circadian period of SBP was estimated by the Maximum Entropy Method (MEM), which requires the data to be equidistant. Outliers were deleted from the original SBP time series (middle section of Fig. 20.3). Data were then made equidistant (bottom section of Fig. 20.3).

The MEM spectrum is shown on the middle left of Fig. 20.4, which shows a clear but weak circadian component with a period of 1.0358 day, together with two other peaks at periods of 0.8853 and 0.5017 days. Periods extracted by the MEM analysis in the range up to 2.5 cycles/day are shown in the right upper side of this figure. Eleven components are listed here, and that with a "period" of 1.0358 day has the second largest spectral power, as assessed by the "area" covered by the spectral peak, the most prominent component having a period of 0.5017 day.

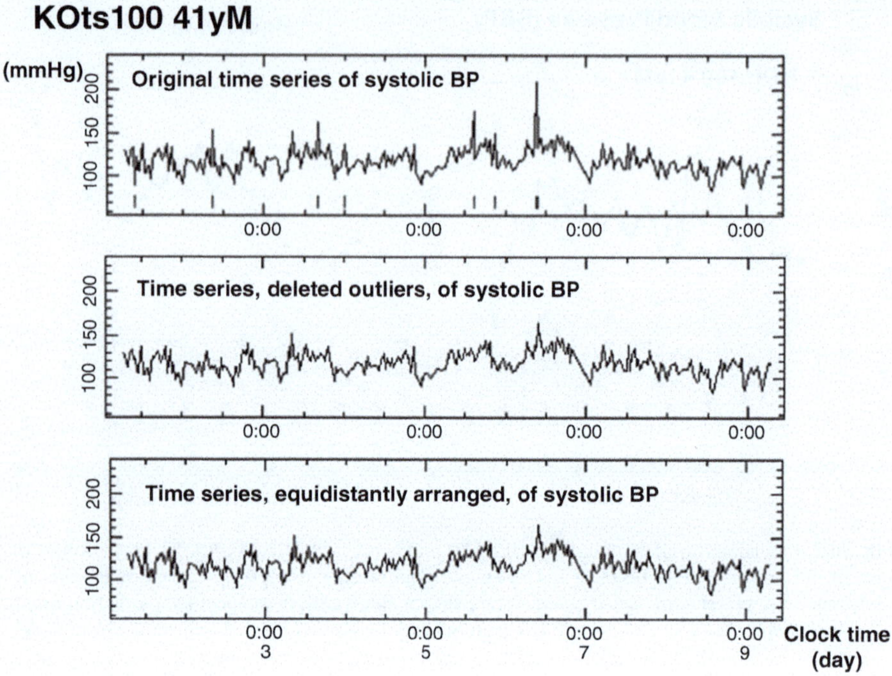

Fig. 20.3 Three versions of the time series of systolic BP. *Upper*: Original data of systolic BP. *Middle*: Edited data after mathematically excluding outliers. *Lower*: Equidistantly arranged time series for analysis by the Maximum Entropy Method, using the MemCalc software (GMS Co., Tokyo)

On the bottom left of Fig. 20.4, the edited data are fitted by least squares with a five-component model consisting of the five cosine curves with the five largest amplitudes, with corresponding periods of 5.891, 2.486, 1.036, 0.885, and 0.502 days (right lower part of Fig. 20.4). None of these components has an amplitude larger than 10 mmHg.

In conclusion, the clinical diagnosis based on the chronobiologically interpreted 7-day/24-hour ABP monitoring consists of (1) masked ABP hypertension; (2) DBP MESOR-hypertension; (3) persistent DBP hypertension; (4) masked dipping pattern, changing from non-dipping or reverse dipping to dipping from one day to another; and (5) weak circadian rhythm of SBP. Accordingly, we recommended that him to strictly improve his lifestyle, focusing on salt intake. Under this treatment plan, he now seems to be in good health.

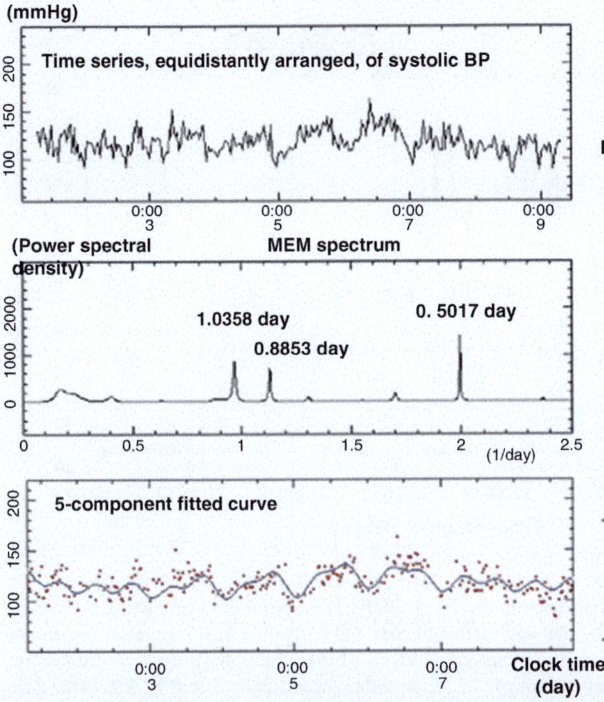

KOts100 41yM

Frequency (/ day)	Period (days)	Area
0.170	5.8911	28.7493
0.402	2.4862	5.3825
0.630	1.5871	1.0486
0.882	1.1334	3.0060
0.965	1.0358	18.0778
1.130	0.8853	9.4644
1.304	0.7667	1.9574
1.552	0.6442	0.8527
1.700	0.5883	3.5901
1.993	0.5017	11.5950
2.367	0.4225	0.6938

Period (days)	Amplitude	Acro-phase
5.891	6.667	6-02:04
2.486	2.125	3-14:08
1.036	6.540	1-10:39
0.885	4.982	1-23:55
0.502	4.824	1-18:14

Fig. 20.4 Weak circadian rhythm of systolic BP detected by the MEM spectral analysis and the least-squares fit of cosine curves to the data. MEM analysis shows a weak circadian rhythm of systolic BP (*left middle part*). Data are shown with a five-component model fitted by least squares to the data, consisting of cosine curves with the five largest amplitudes, showing no prominent circadian component (*right lower part*)

20.2 Masked Ambulatory Blood Pressure Normotension

In this chapter we introduce a case of "masked ambulatory blood pressure (ABP) normotension" from the viewpoint of 7-day/24-hour ABP monitoring. "Masked ABP normotension" is defined as hypertension (HT) on day 1 but normotension on at least one of the following 6 days. Hypertension is defined as a 24-hour BP mean above 130/80 mmHg.

20.2.1 Case Report 1 of "Masked Ambulatory Blood Pressure Normotension"

The first example of "masked ABP normotension" is a 75-year-old nonobese (BMI = 22.9 kg/m^2) woman (ID number: Ura0256). Her office BP was 144/97 mmHg while on two antihypertensive medications, CCB and ARB.

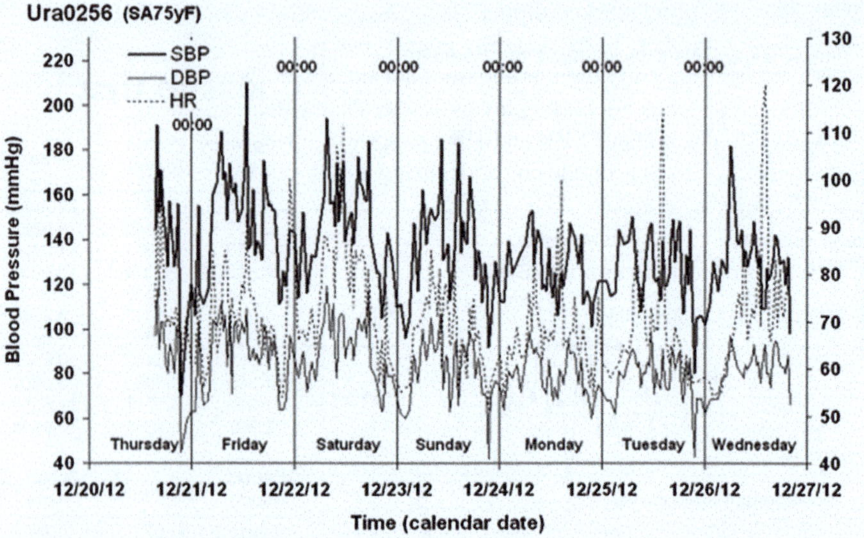

Fig. 20.5 An example of "masked ambulatory blood pressure hypertension." Time plots of the 6-day/24-hour records of SBP (*thick continuous line*), DBP (*thin continuous line*), and HR (*dashed line*) of a 75-year-old woman. The day-night ratio of SBP varies from 1 day to another, assuming values of 23.6 % (Thursday; extreme dipping), 14.6 % (Friday; dipping), 26.3 % (Saturday; extreme dipping), 14.9 % (Sunday; dipping), 4.2 % (Monday; non-dipping), and 12.3 % (Tuesday; dipping), indicating the presence of a masked non-dipping pattern of SBP

Time series of systolic BP (SBP), diastolic BP (DBP), and heart rate measured by pulse (HR), covering 6 days, are shown in Fig. 20.5. From Thursday to Tuesday, 24-hour averages of SBP assumed values of 140.8 mmHg (Thursday; HT), 148.3 mmHg (Friday; HT), 143.7 mmHg (Saturday; HT), 137.7 mmHg (Sunday; HT), 126.6 mmHg (Monday; normotension), and 125.7 mmHg (Tuesday; normotension). The 24-hour mean of SBP was below 130 mmHg on Monday and Tuesday, indicating masked ABP normotension. The corresponding 24-hour means of DBP also showed masked normotension, assuming values of 87.6 mmHg (Thursday; HT), 89.2 mmHg (Friday; HT), 88.3 mmHg (Saturday; HT), 83.4 mmHg (Sunday; HT), 78.3 mmHg (Monday; normotension), and 77.8 mmHg (Tuesday; normotension), the 24-hour mean of DBP dropping below 80 mmHg on Monday and Tuesday.

Based on the model fitted to the 6-day/24-hour ABP monitoring record, MESOR estimates of SBP, DBP, and HR were 134.3 mmHg, 82.5 mmHg, and 70.0 bpm, respectively. As shown in Fig. 20.5, the 24-hour means of SBP/DBP vary from 1 day to another, assuming values of 144.0/88.0 mmHg (Thursday; both SBP and DBP MESOR-HT), 146.3/90.0 mmHg (Friday; both SBP and DBP MESOR-HT), 135.9/82.3 mmHg (Saturday; normotension), 127.4/79.0 mmHg (Sunday; normotension), 126.6/77.3 mmHg (Monday; normotension), and 126.1/78.2 mmHg (Tuesday; normotension), indicating SBP and DBP MESOR-HT with masked normotension.

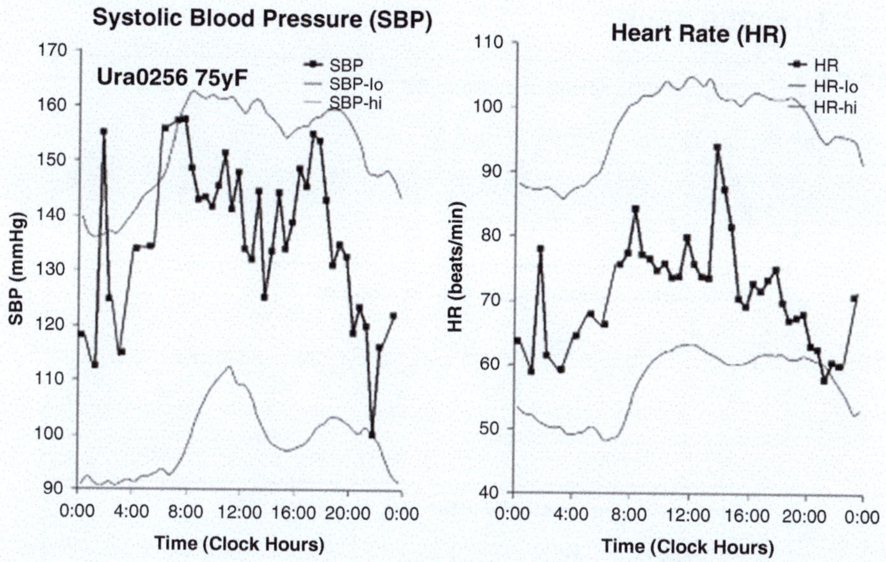

Fig. 20.6 Circadian profiles of systolic BP (*left*) and heart rate (HR) (*right*). 24-hour systolic BP and HR patterns of a 75-year-old woman, averaged over the 6 days of monitoring (*polygonal line* graph with *closed squares*), are shown along with time-specified prediction limits (*continuous thin lines*)

The double 24-hour amplitudes of SBP, DBP, and HR were 29.47 mmHg, 20.56 mmHg, and 18.21 bpm, respectively, all within acceptable limits. Gender- and age-matched limits for the circadian amplitude of SBP in health are 3.41–34.0 mmHg. The circadian amplitude of SBP is acceptable on some days but excessive (CHAT) on others, assuming values from 1 day to another of 51.43 mmHg (Thursday; CHAT), 35.74 mmHg (Friday; CHAT), 36.80 mmHg (Saturday; CHAT), 14.65 mmHg (Sunday; acceptable), 13.94 mmHg (Monday; acceptable), and 29.69 mmHg (Tuesday; acceptable), indicating intermittent CHAT.

She goes to bed regularly at 21:30 and gets up at 05:00, but her circadian acrophases of SBP, DBP, and HR were 10:51, 11:10, and 11:47, respectively, slightly earlier than usually found in clinical healthy women of her age group. The circadian profiles of SBP and HR are shown in Fig. 20.6. SBP varies mostly within the chronodesmic limits, peaking slightly earlier than in other clinically healthy women her age. HR varies well within the chronodesmic limits.

The circadian period of SBP was estimated by the Maximum Entropy Method (MEM), using the MemCalc software (GMS Co., Tokyo), which requires the data to be equidistant. Outliers were deleted from the original SBP time series (middle section of Fig. 20.7). Data were then made equidistant (bottom section of Fig. 20.7).

The MEM spectrum is shown on the middle left of Fig. 20.8, which shows a circadian component with a period of 0.9631 day, together with a prominent peak at a period of 10.5764 days and a weaker peak at a period of 0.4998 day. Periods extracted by the MEM analysis in the range up to 2.5 cycles/day are shown in the right upper side of this figure. Nine components are listed here, and that with a

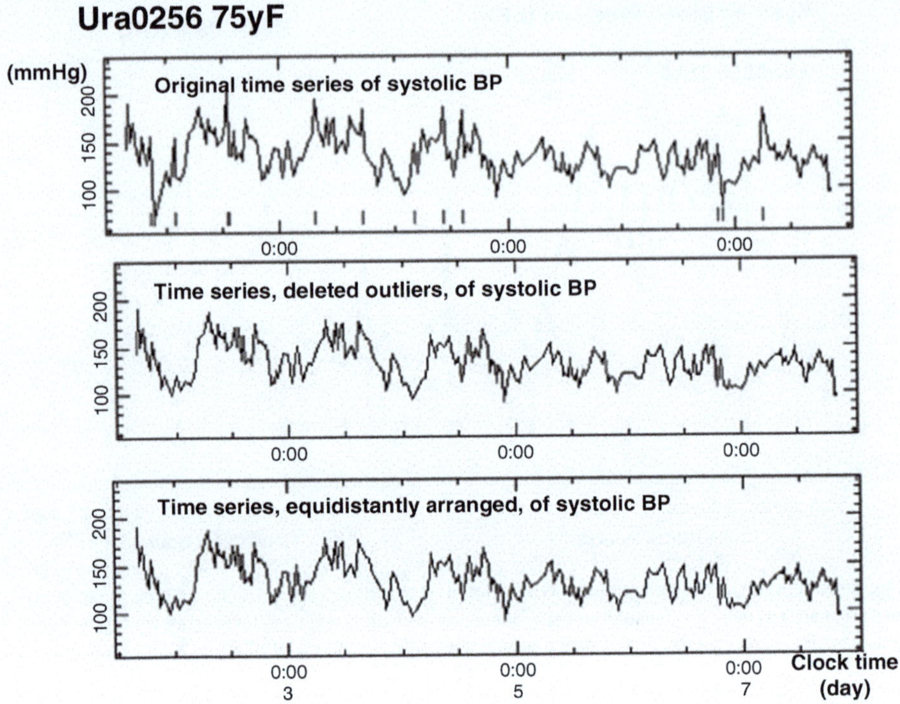

Fig. 20.7 Three versions of the time series of systolic BP. *Upper*: Original data of systolic BP. *Middle*: Edited data after mathematically excluding outliers. *Lower*: Equidistantly arranged time series for analysis by the Maximum Entropy Method, using the MemCalc software (GMS Co., Tokyo)

"period" of 10.58 days has the largest spectral power, as assessed by the "area" covered by the spectral peak.

On the bottom left of Fig. 20.8, the edited data are fitted by least squares with a five-component model consisting of the five cosine curves with the five largest amplitudes, with corresponding periods of 10.576, 0.963, 0.714, 0.500, and 0.338 days (lower right of Fig. 20.8).

In conclusion, the clinical diagnosis based on the chronobiologically interpreted 6-day/24-hour ABP monitoring consists of (1) masked ABP normotension, (2) intermittent CHAT, (3) masked non-dipping pattern, and (4) a small circadian phase advance of SBP. Accordingly, we increased the dose of the calcium channel blocker (CCB) and made lifestyle recommendations to her, focusing on salt intake. Under this treatment plan, she now seems to be in good health.

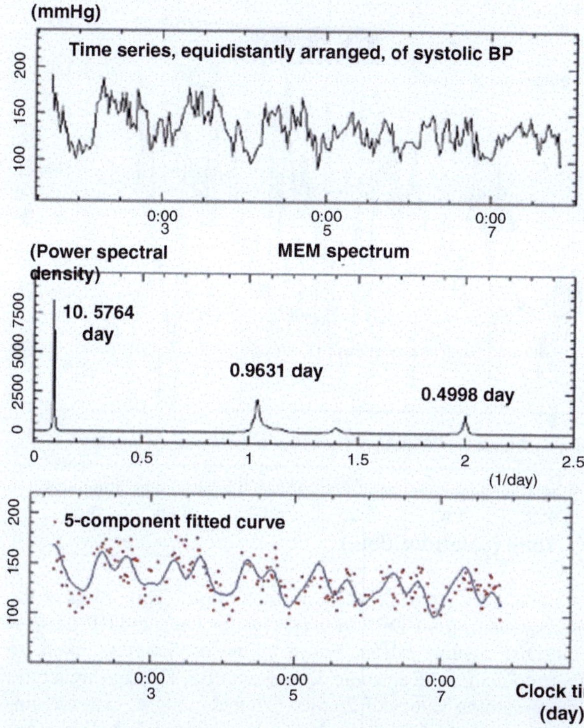

Ura0256 75yF

Frequency (/ day)	Period (days)	Area
0.0946	10.5764	53.99
0.3104	3.2219	1.36
0.6058	1.6507	2.92
0.8385	1.1927	4.21
1.0384	0.9631	107.90
1.4004	0.7141	17.87
1.5960	0.6266	3.01
2.0008	0.4998	36.01
2.2592	0.4426	3.28

Period (days)	Amplitude	Acro-phase
10.576	9.65	1-19:01
0.963	13.37	2-12:45
0.714	4.37	2-08:16
0.500	8.08	1-17:59
0.338	5.91	1-23:33

Fig. 20.8 Period analysis of the circadian rhythm of systolic BP. MEM analysis estimates the circadian period of systolic BP to be 0.9631 day (*left middle part*). Although the circadian rhythm is clearly detected, an even more prominent infradian component also characterizes SBP. Data are shown with a five-component model fitted by least squares to the data, consisting of cosine curves with the five largest amplitudes. Superimposed on the circadian variation is a prominent 10.5764-day component (*right lower part*)

20.2.2 Case Report 2 of "Masked Ambulatory Blood Pressure Normotension"

The second example of "masked ABP normotension" is a 66-year-old nonobese (BMI = 24.5 kg/m^2) man (ID number: Tosa0131). His weight was 70.5 kg and height 170 cm. His office BP was 171/111 mmHg and home BP measurements averaged over 30 days were 149.4/96.9 mmHg in the morning and 145.3/94.0 mmHg in the evening. He was not yet on antihypertensive treatment.

He has a regular rest-activity cycle, usually going to bed around 21:30 and getting up at 06:30. He falls asleep within 10 min and claims to sleep well most days. He does not snore and wakes up rested. He says he prefers light seasoning but has two cups of miso soup every morning and eats pickled salted vegetables at least

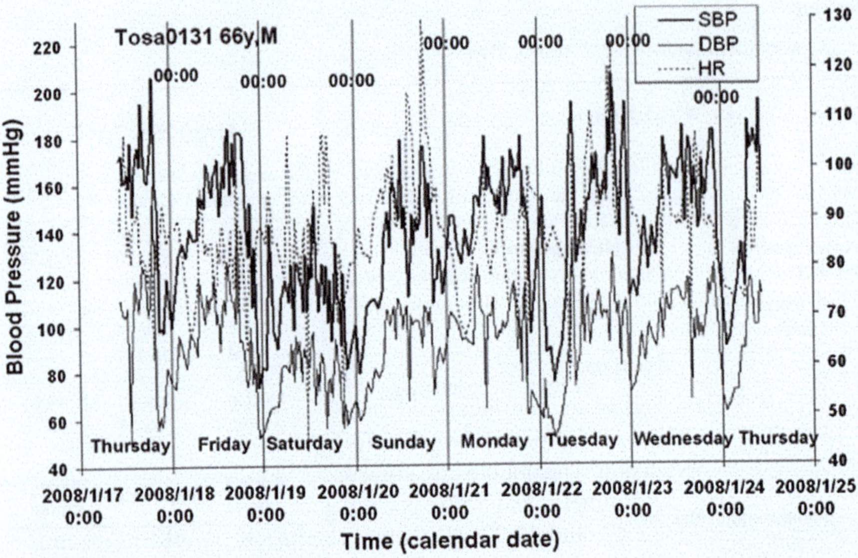

Fig. 20.9 An example of "masked ambulatory blood pressure normotension." Time plots of the 7-day/24-hour records of SBP (*thick continuous line*), DBP (*thin continuous line*), and HR (*dashed line*) of a 66-year-old man. SBP and DBP assume extremely low values on Saturday, likely an effect of alcohol. Increases in heart rate coincide with episodic decreases in BP. The day-night ratio of SBP varies from 1 day to another, assuming values of 26.6 % (Thursday night; extreme dipping), 39.5 % (Friday night; extreme dipping), 15.5 % (Saturday night; dipping), 8.4 % (Sunday night; non-dipping), 37.8 % (Monday night; extreme dipping), 22.6 % (Tuesday night; extreme dipping), and 32.7 % (Wednesday night; extreme dipping), indicating the presence of masked non-dipping on Sunday night

twice a day. He does not smoke but usually drinks more than 2 go (360 ml) of "Japanese sake" every evening. Not rarely he ends up drinking too much.

He does not have a depressive mood, and his scores on the GDS-15 and PHQ-2 scales were 1 and 0, respectively. He maintains that he usually takes care of his health and has something to live for. His subjective QOL indices of health, mood, family relationships, relationships with friends, financial satisfaction, life satisfaction, and happiness, assessed by visual analog scale (VAS), were 96 %, 100 %, 99 %, 99 %, 84 %, 95 %, and 96 %, respectively.

Time series of systolic BP (SBP), diastolic BP (DBP), and heart rate measured by pulse (HR) covering 7 days are shown in Fig. 20.9. From Thursday to Wednesday, the 24-hour average of SBP assumed values of 151.7 mmHg (Thursday; hypertension), 141.0 mmHg (Friday; hypertension), 115.3 mmHg (Saturday; normotension), 142.7 mmHg (Sunday; hypertension), 146.3 mmHg (Monday; hypertension), 152.6 mmHg (Tuesday; hypertension), and 154.9 mmHg (Wednesday; hypertension). The 24-hour mean of SBP was below 130 mmHg on Saturday, indicating masked ABP normotension.

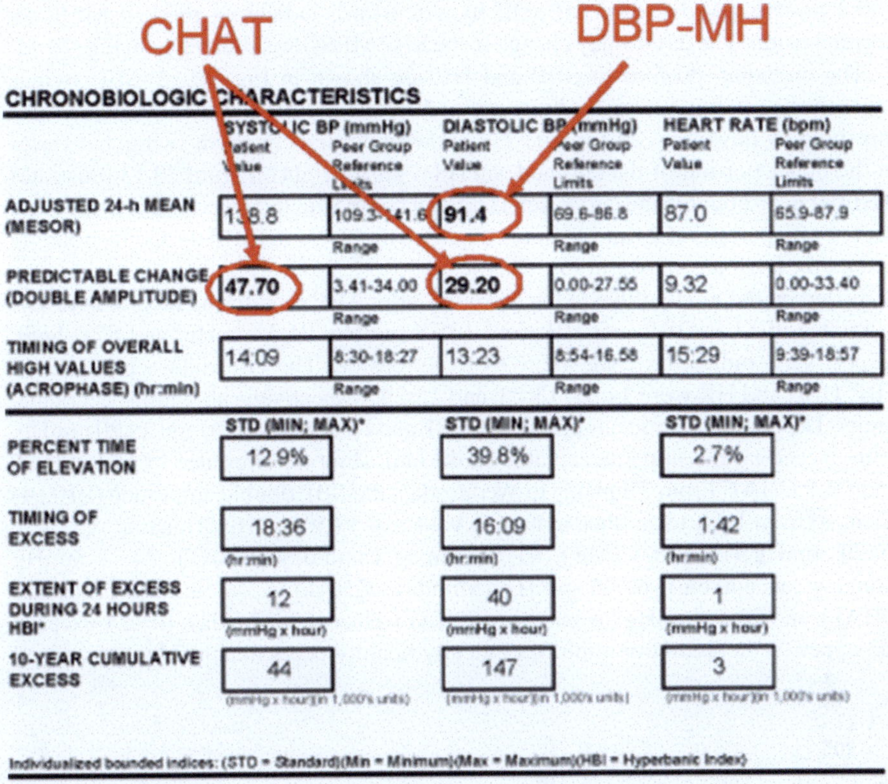

CHRONOBIOLOGIC CHARACTERISTICS

	SYSTOLIC BP (mmHg) Patient Value	Peer Group Reference Limits	DIASTOLIC BP (mmHg) Patient Value	Peer Group Reference Limits	HEART RATE (bpm) Patient Value	Peer Group Reference Limits
ADJUSTED 24-h MEAN (MESOR)	138.8	109.3-141.6 Range	**91.4**	69.6-86.8 Range	87.0	65.9-87.9 Range
PREDICTABLE CHANGE (DOUBLE AMPLITUDE)	**47.70**	3.41-34.00 Range	**29.20**	0.00-27.56 Range	9.32	0.00-33.40 Range
TIMING OF OVERALL HIGH VALUES (ACROPHASE) (hr:min)	14:09	8:30-18:27 Range	13:23	8:54-16.58 Range	15:29	9:39-18:57 Range

	STD (MIN; MAX)*	STD (MIN; MAX)*	STD (MIN; MAX)*
PERCENT TIME OF ELEVATION	12.9%	39.8%	2.7%
TIMING OF EXCESS	18:36 (hr:min)	16:09 (hr:min)	1:42 (hr:min)
EXTENT OF EXCESS DURING 24 HOURS HBI*	12 (mmHg x hour)	40 (mmHg x hour)	1 (mmHg x hour)
10-YEAR CUMULATIVE EXCESS	44 (mmHg x hour)(in 1,000's units)	147 (mmHg x hour)(in 1,000's units)	3 (mmHg x hour)(in 1,000's units)

Individualized bounded indices: (STD = Standard)(Min = Minimum)(Max = Maximum)(HBI = Hyperbaric Index)

Fig. 20.10 Sphygmochron summarizing parametric and nonparametric results of the 7-day record interpreted in the light of reference data in clinically healthy peers matched by gender and age

The sphygmochron is shown in Fig. 20.10. This two-barreled approach includes the least-squares fit of a model consisting of cosine curves with periods of 24 and 12 hours (results of the 24-hour component are listed with the MESOR for each of the three variables), and the computer comparison of the subject's data stacked over an idealized 24-hour day with time-specified reference values qualified by gender and age. The MESOR and the circadian amplitude and acrophase are further evaluated in the light of 90 % prediction limits derived from clinically healthy subjects of the same gender and age group as the subject. As seen in Fig. 20.10, the MESOR of DBP is above the upper 95 % prediction limit in health. Accordingly, diastolic MESOR-hypertension is diagnosed. Similarly, CHAT is diagnosed as the double amplitude of both SBP and DBP is above the upper limit of acceptability.

The MESOR (rhythm-adjusted mean), interpreted in the light of gender- and age-matched reference limits, varies between 113.5 and 141.6 mmHg, assuming values of 146.7 mmHg (Thursday; MESOR-hypertension), 133.0 mmHg (Friday; MESOR-normotension), 113.5 mmHg (Saturday; MESOR-normotension), 141.4 mmHg (Sunday; MESOR-normotension), 137.9 mmHg (Monday; MESOR-normotension), 149.8 mmHg (Tuesday; MESOR-hypertension), and

150.3 mmHg (Wednesday; MESOR-hypertension), indicating masked MESOR-normotension. The day-to-day change in the SBP MESOR is illustrated in Fig. 20.11.

The circadian profiles of SBP and HR are shown in Fig. 20.12. SBP swings excessively (extreme dipping by night and overshooting above the time-varying threshold by day) (Fig. 20.12, left). HR shows a normal circadian swing, but varies in the upper portion of the chronodesm (Fig. 20.12, right). Figure 20.13 illustrates the fit of a 24-hour cosine curve (left) and of a composite model consisting of cosine curves with periods of 24 and 12 hours (right) to the stacked SBP data. Corresponding results for DBP are shown in Fig. 20.14.

Based on the model fitted to the 7-day/24-hour ABP monitoring record, MESOR estimates of SBP, DBP, and HR were 138.8 mmHg, 91.4 mmHg, and 87.0 bpm, respectively, indicating DBP MESOR-hypertension. The circadian acrophases of SBP, DBP, and HR were 14:09, 13:23, and 15:29, respectively, all within acceptable limits. The double 24-hour amplitudes of SBP and DBP were 47.70 and 29.20 mmHg, both exceeding the upper acceptable limit, indicating the presence of SBP CHAT and DBP CHAT. From Thursday to Wednesday, the SBP double amplitude (BP-2A) changed from 1 day to another, assuming values of 59.50 mmHg (Thursday; CHAT), 79.79 mmHg (Friday; CHAT), 41.50 mmHg (Saturday; CHAT), 24.33 mmHg (Sunday; acceptable), 67.46 mmHg (Monday; CHAT), 37.77 mmHg (Tuesday; CHAT), and 52.89 mmHg (Wednesday; CHAT). Except for Sunday, BP-2A exceeds the upper 95 % prediction limit of clinically healthy peers matched by gender and

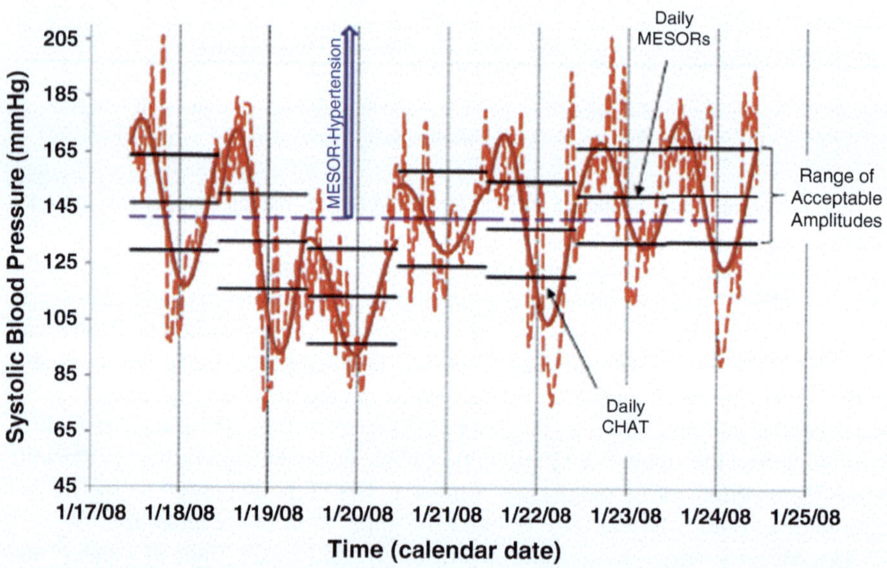

Fig. 20.11 Analysis of day-to-day variation of the 24-hour BP profile showing "masked ambulatory BP normotension." As a result of the day-to-day change in the MESOR of systolic blood pressure, MESOR-hypertension is found on some days, whereas MESOR-normotension is found on other days. Similarly, changes from day to day in the circadian amplitude of systolic blood pressure result in the diagnosis of CHAT on some but not on all days

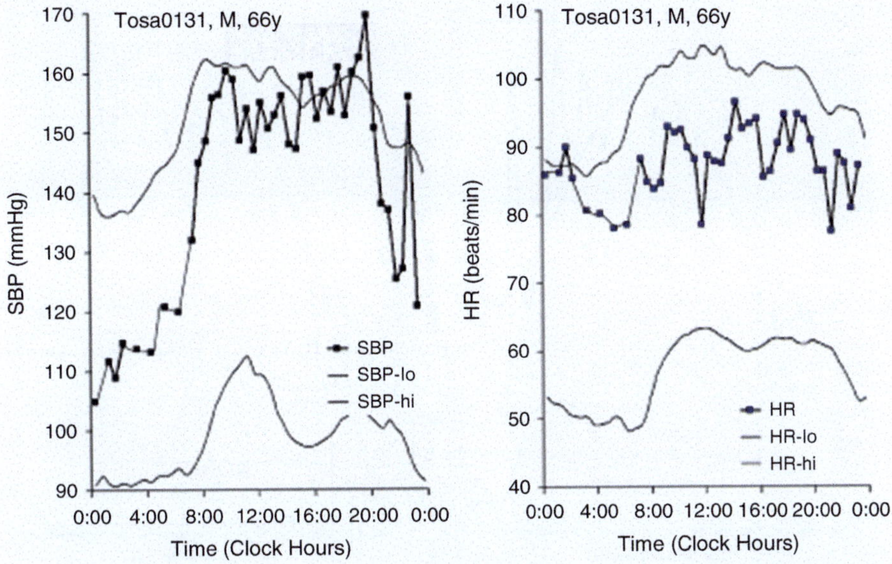

Fig. 20.12 Averaged circadian profiles of systolic BP (*left*) and heart rate (HR) (*right*). 24-hour systolic BP and HR patterns, averaged over the 7 days of monitoring (*polygonal line* graph with *closed squares*), are shown along with time-specified prediction limits (*continuous thin lines*)

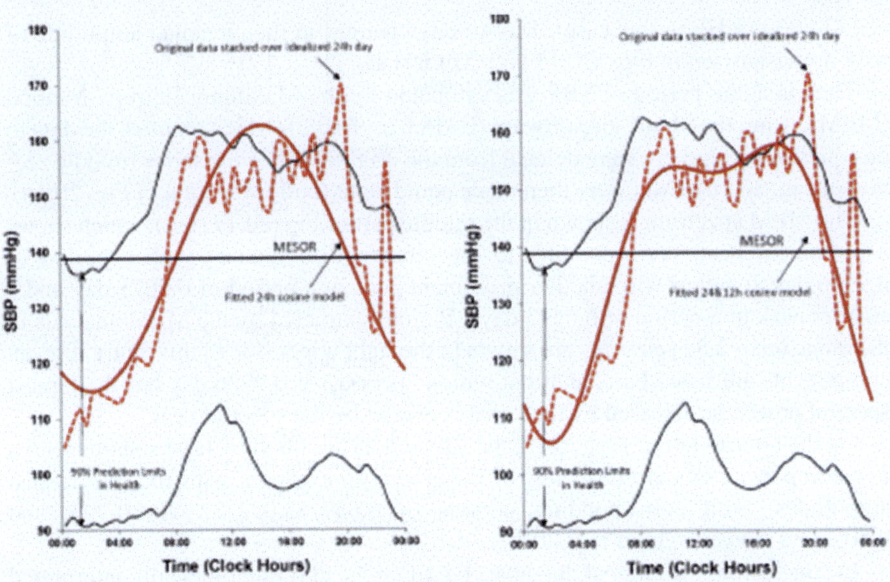

Fig. 20.13 *Left*: Illustration of the 24-hour cosine fit to the stacked SBP data. *Right*: Illustration of the fit of a composite model consisting of cosine curves with periods of 24 and 12 hours to the stacked SBP data

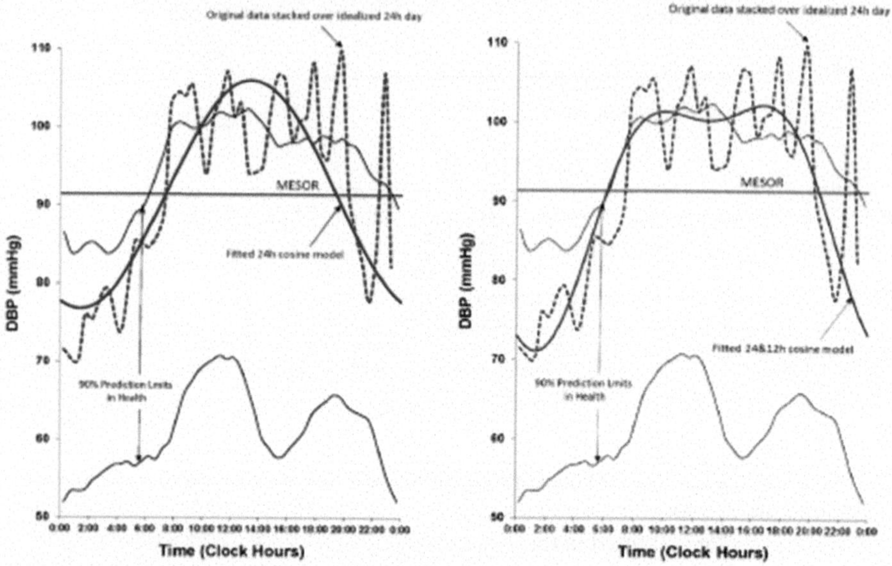

Fig. 20.14 *Left*: Illustration of the 24-hour cosine fit to the stacked DBP data. *Right*: Illustration of the fit of a composite model consisting of cosine curves with periods of 24 and 12 hours to the stacked DBP data

age (34.00 mmHg in this case). Day-to-day changes in the circadian amplitude of SBP are visualized in Figs. 20.11 and 20.15.

The circadian period of SBP was estimated by the Maximum Entropy Method (MEM), using the MemCalc software (GMS Co., Tokyo), which requires the data to be equidistant. Outliers were deleted from the original SBP time series (middle section of Fig. 20.16). Data were then made equidistant (bottom section of Fig. 20.16).

The MEM spectrum is shown in the middle part of Fig. 20.17 (left), which shows a prominent peak corresponding to the circadian component with a period of 0.9926 day, together with another prominent peak at a period of 0.5071 day and a smaller peak at a period of 6.6592 days. Periods extracted by the MEM analysis in the range up to 2.5 cycles/day are shown in the right upper side of this figure. Eleven components are listed here, and that with a "period" of 0.9926 day has the largest spectral power, as assessed by the "area" covered by the spectral peak.

On the bottom left of Fig. 20.17, the edited data are fitted by least squares with a five-component model consisting of the five cosine curves with the five largest amplitudes, with corresponding periods of 6.659, 3.024, 2.520, 0.993, and 0.507 days (lower right of Fig. 20.17).

In conclusion, the clinical diagnosis based on the chronobiologically interpreted 7-day/24-hour ABP monitoring consists of (1) masked ABP normotension (or masked MESOR-normotension), showing a much lower 24-hour average of ABPs on Saturday, with a usual timing of the circadian BP rhythm; (2) masked non-dipping pattern on Sunday night; (3) both SBP and DBP CHAT; (4) persistent SBP

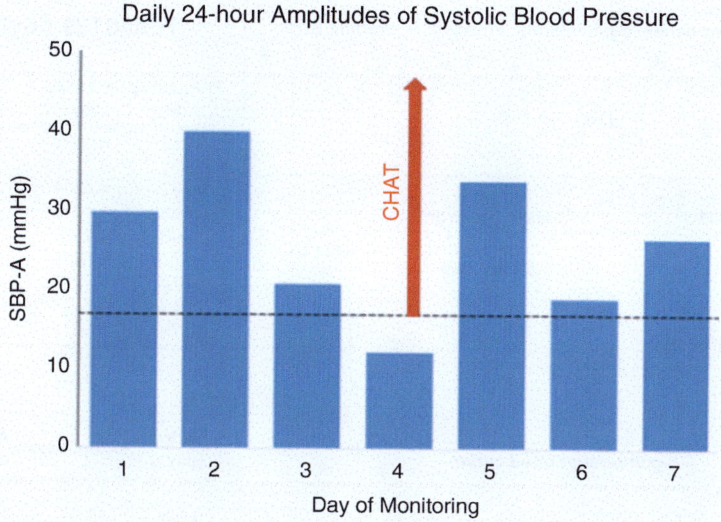

Fig. 20.15 Systolic CHAT occurs on 6 of the 7 days of monitoring, the 24-hour amplitude of SBP largely exceeding the threshold in health (*horizontal dashed line*) on some days

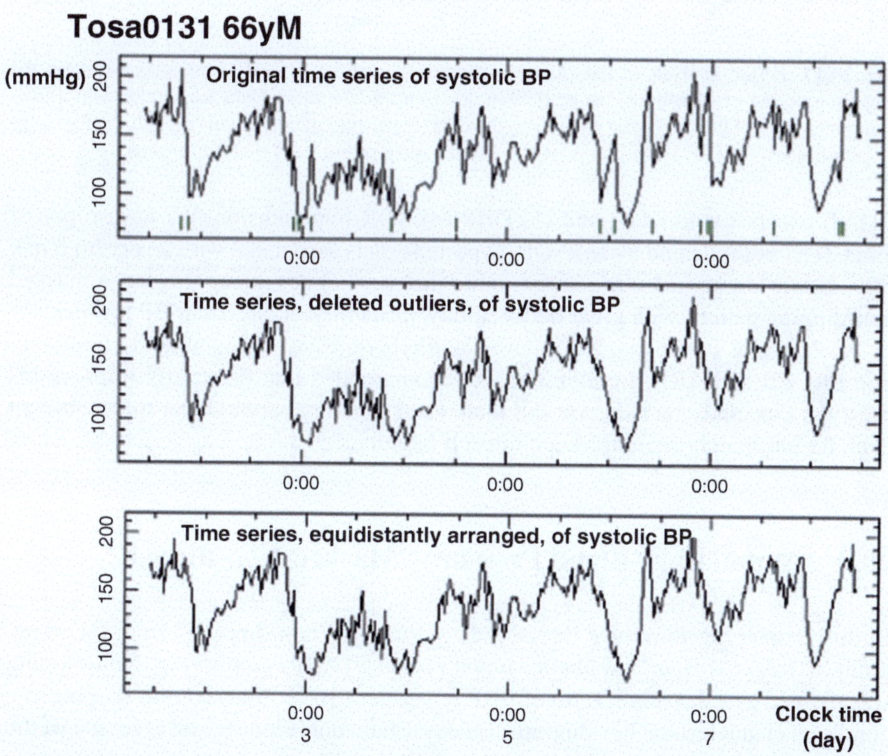

Fig. 20.16 Three versions of the time series of systolic BP. *Upper*: Original data of systolic BP. *Middle*: Edited data after mathematically excluding outliers. *Lower*: Equidistantly arranged time series for analysis by the Maximum Entropy Method, using the MemCalc software (GMS Co., Tokyo)

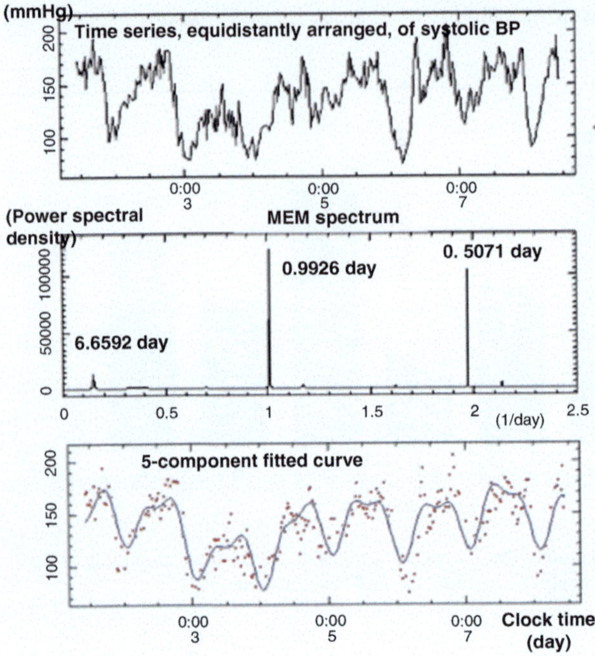

Tosa0131 66yM

Frequency (/ day)	Period (days)	Area
0.150	6.6592	127.00
0.331	3.0236	44.91
0.397	2.5204	50.90
0.698	1.4333	24.56
1.008	0.9926	287.59
1.173	0.8524	40.84
1.375	0.7273	6.05
1.619	0.6178	9.60
1.972	0.5071	64.34
2.136	0.4683	14.86
2.408	0.4153	1.55

Period (days)	Amplitude	Acro-phase
6.659	14.83	7-01:51
3.024	8.94	2-08:04
2.520	8.60	2-06:25
0.993	24.79	1-14:39
0.507	10.91	1-18:27

Fig. 20.17 Period analysis of the circadian rhythm of systolic BP. MEM analysis estimates the circadian period of systolic BP to be 0.9926 day (*left middle part*). Data are shown with a five-component model fitted by least squares to the data, consisting of cosine curves with the five largest amplitudes, the 0.993-day component being the most prominent (*right lower part*)

CHAT, except for Sunday; and (5) DBP MESOR-hypertension. We have reported that CHAT complicated by MESOR-hypertension is associated with a very high risk of ischemic stroke and other catastrophic diseases. This subject also has a masked non-dipping pattern with great between-day variability of circadian BP profiles.

Accordingly, we immediately started antihypertensive treatment with a combination of CCB and ARB. In addition, we recommended that he strictly improve his lifestyle, especially focusing on salt intake and drinking habit. Under this treatment plan, he fortunately seems to keep himself healthy.

20.3 Ambulatory Blood Pressure "Masked Non-dipper"

In this chapter we introduce three cases of "masked non-dipping" from the viewpoint of 7-day/24-hour ambulatory blood pressure (ABP) monitoring. Non-dipping was defined as a day-night ratio of SBP lower than 10 %. Masked non-dipping is a condition characterized by "dipping" on day 1 but "non-dipping" on at least 1 of the following 6 days.

20.3.1 Case Report 1 of "Masked Non-dipper"

The first example of "masked non-dipping" is a 61-year-old nonobese (BMI = 24.9 kg/m^2) woman (ID number: Tosa0251). Her weight is 56.0 kg and height 150 cm. Although she has cervical spondylosis and knee joint pain, she feels healthy and has an easy time every day. She does housework for 4 hours and walks to do her shopping for 2 hours daily. She does not smoke and usually does not drink alcohol. She rarely eats meat but frequently eats fish and vegetables. She prefers her food lightly seasoned.

She does not have a depressive mood and has something to live for. Her scores on the GDS-15 and PHQ-2 scales were 2 and 1, respectively. Her subjective QOL of health, mood, family relationships, relationships with friends, financial satisfaction, and happiness were 95 %, 85 %, 85 %, 85 %, 70 %, 90 %, and 95 %, respectively. She has a family history of hypertension (parents) and stroke (parents and elderly brother).

She does not take any antihypertensive medication. Her office BP was 135/80 mmHg and her home BP measurements averaged over 30 days were 120.4/80.6 mmHg in the morning and 126.8/83.4 mmHg in the evening. Time series of systolic BP (SBP), diastolic BP (DBP), and heart rate measured by pulse (HR) are shown in Fig. 20.18. A model consisting of cosine curves with periods of 24 and

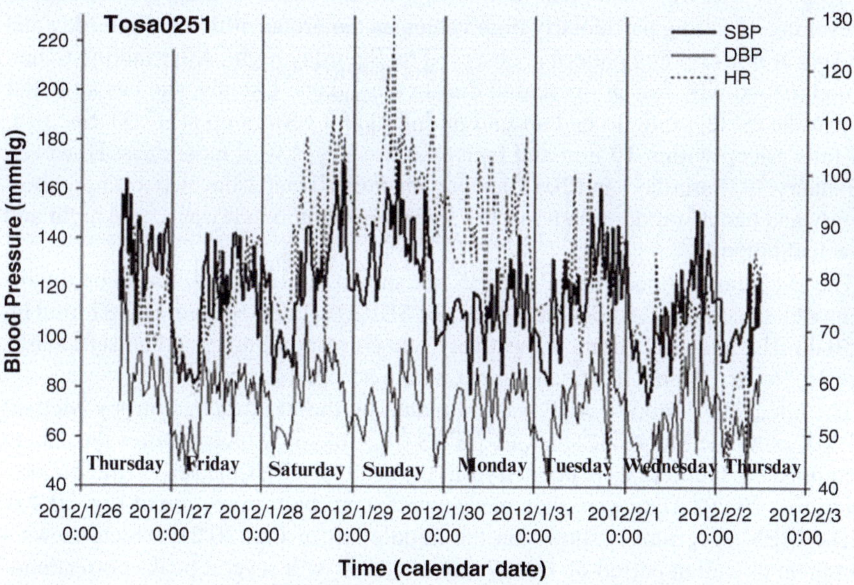

Fig. 20.18 An example of "masked non-dipping." Time series of SBP, DBP, and heart rate (HR) are shown from Friday to the next Friday. Time plots of the 7-day/24-hour records of SBP (*thick continuous line*), DBP (*thin continuous line*), and HR (*dashed line*) of a 61-year-old woman. An increase in BP in the evening is invariably observed, which reaches particularly high values on the 3rd and 4th day (Saturday and Sunday, January 29 and 30)

12 hours was fitted by least squares to the data to estimate the MESOR (Midline-Estimating Statistics Of Rhythm) and the amplitude and acrophase (time of predicted maximum) of each component.

MESOR estimates of SBP, DBP, and HR, estimated from the 7-day/24-hour ABP measurements, were 117.8 mmHg, 73.5 mmHg, and 80.9 bpm, respectively. In accordance with the guidelines for management of high blood pressure, in the case of 24-hour ABP, in Japan, published in 2014, she is diagnosed as being normotensive, as her 24-hour SBP and DBP averages were below 130 mmHg and 80 mmHg, respectively.

However, day-to-day analyses showed "masked ABP hypertension," as the 24-hour average of SBP changed from one day to another, assuming values of 124.7 (Thursday; normotension), 118.1 (Friday; normotension), 136.2 (Saturday; hypertension), 125.3 (Sunday; normotension), 111.3 (Monday; normotension), 117.5 (Tuesday; normotension), and 113.7 mmHg (Wednesday; normotension).

Variation in the day-night ratio of SBP during the weeklong record indicated the presence of "masked non-dipping," the dipping pattern varying from 1 day to another (from extreme dipping to dipping and non-dipping), and the day-night ratio assuming values of 35.4 % (Thursday night; extreme dipping), 14.4 % (Friday night; dipping), 7.3 % (Saturday night; non-dipping), 25.9 % (Sunday night; extreme dipping), 12.4 % (Monday night; dipping), 25.6 % (Tuesday night; extreme dipping), and 12.3 % (Wednesday night; dipping).

From this 7-day/24-hour ABP monitoring, we can see an evident BP increase in the evening, reaching particularly high values on the 3rd and 4th day (Saturday and Sunday). A non-dipping pattern is observed on Saturday night. After talking to her, we understood why her BP increased during the 2 days. Usually, she has a regular sleep-wake cycle, going to bed around midnight and getting up at 07:00. She usually falls asleep within 10 min and feels like she sleeps well most days. However, on January 29 (Saturday) and 30 (Sunday), she took a brief 2-day trip to a neighboring city and had an outdoor bath in a hot spring, even though it was a cold night and snow had fallen.

The circadian profiles of SBP and HR are shown in Fig. 20.19. Based on the fitted model, the double 24-hour amplitudes of SBP, DBP, and HR were 32.37 mmHg, 23.50 mmHg, and 12.27 bpm, respectively, and corresponding circadian acrophases were 16:56, 16:19, and 16:46, all almost within acceptable limits.

The circadian period of SBP was estimated by the Maximum Entropy Method (MEM), using the MemCalc software (GMS Co., Tokyo), which requires the data to be equidistant. Outliers were deleted from the original SBP time series (middle section of Fig. 20.20). Data were then made equidistant (bottom section of Fig. 20.20).

The MEM spectrum is shown on the middle left of Fig. 20.21, which shows a prominent circadian period of 1.0137 day, together with several peaks corresponding to periods of 6.1118, 2.6474, and 0.4954 days, among others. Periods extracted by the MEM analysis in the range up to 2.5 cycles/day are shown in the right upper side of this figure. Twelve components are listed here and that with a period of 1.0137 day has the largest spectral power, as assessed by the "area" covered by the spectral peak.

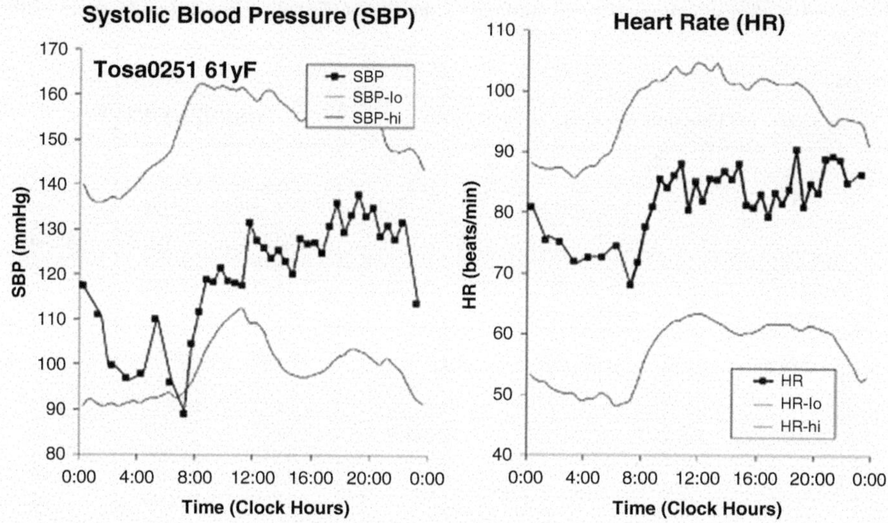

Fig. 20.19 Circadian profiles of systolic BP (*left*) and heart rate (HR) (*right*). 24-hour patterns of systolic BP and HR, averaged over the 7 days of monitoring (*polygonal line* graph with closed *squares*), are shown along with time-specified 90 % prediction limits (*continuous thin lines*). Both patterns are well within the chronodesmic limits, except for SBP in the morning (around 07:30)

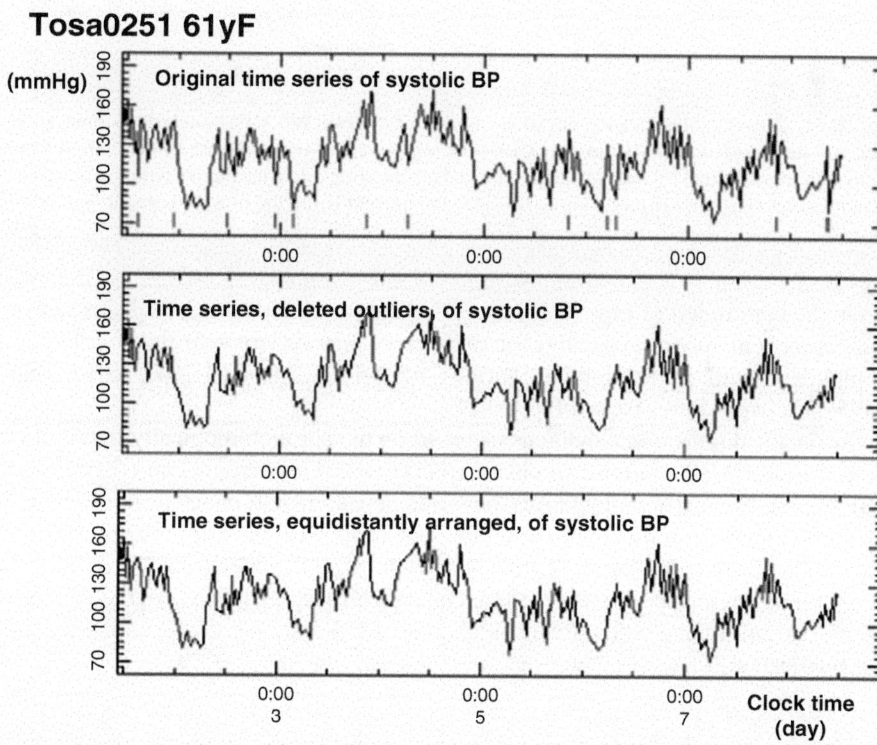

Fig. 20.20 Three versions of the time series of systolic BP. *Upper*: Original data of systolic BP. *Middle*: Edited data after mathematically excluding outliers. *Lower*: Equidistantly arranged time series for analysis by the Maximum Entropy Method, using the MemCalc software (GMS Co., Tokyo)

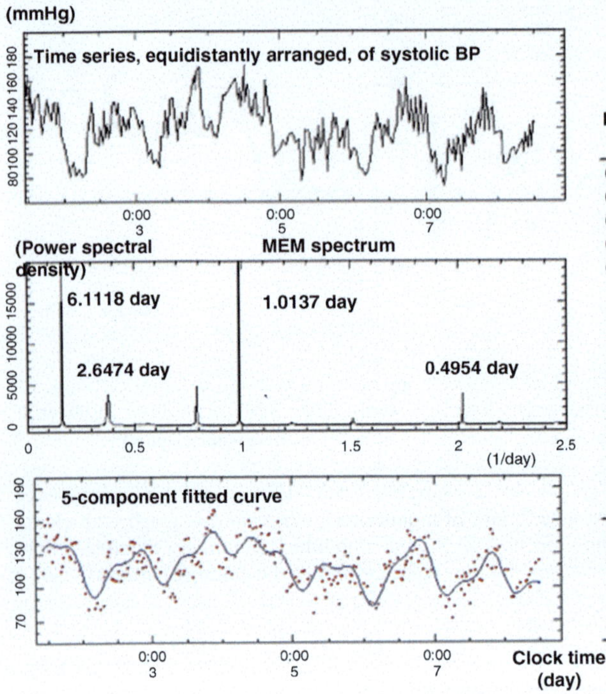

Tosa0251 61yF

Frequency (/ day)	Period (days)	Area
0.164	6.1118	53.40
0.378	2.6474	57.18
0.563	1.7774	12.44
0.792	1.2633	29.74
0.987	1.0137	108.07
1.229	0.8134	5.16
1.379	0.7253	2.54
1.511	0.6617	8.36
1.832	0.5459	2.29
2.018	0.4954	17.78
2.187	0.4573	5.08
2.458	0.4069	7.51

Period (days)	Amplitude	Acro-phase
6.112	11.40	3-19:38
2.647	10.41	1-13:11
1.263	8.62	1-15:22
1.014	13.50	1-15:27
0.495	6.79	1-21:45

Fig. 20.21 Period analysis of the circadian rhythm of systolic BP. MEM analysis estimates the circadian period of systolic BP to be 1.0137 day (*left middle part*). Data are shown with a five-component model fitted by least squares to the data, consisting of cosine curves with the five largest amplitudes (*right lower part*), the 1.0137-day component being the most prominent

On the bottom left of Fig. 20.21, the edited data are fitted by least squares with a five-component model consisting of the five cosine curves with the five largest amplitudes, with corresponding periods of 6.112, 2.647, 1.263, 1.014, and 0.495 days (right lower part of Fig. 20.21).

In conclusion, the clinical diagnosis based on the chronobiologically interpreted 7-day/24-hour ABP monitoring consists of (1) masked non-dipping and (2) masked ABP hypertension. These diagnoses, however, depend in part on environmental factors, as evidenced by effects observed in association with the brief winter trip outdoor including the hot-spring bath on a wintery night when snow was falling. Accordingly, we made lifestyle recommendations, focusing on environmental conditions and salt intake. She remains in good health despite not taking antihypertensive medication.

20.4 Case Report 2 of "Masked Non-dipper"

The next example of "masked non-dipping" is a 68-year-old nonobese (BMI = 24.3 kg/m^2) woman (ID number: KOts080). She feels healthy and has an easy time every day. She has a regular sleep-wake cycle, going to bed around midnight and getting up at 07:00. She believes she sleeps well most days.

She does not take any antihypertensive medication. Her office BP was 158/115 mmHg. Time series of systolic BP (SBP), diastolic BP (DBP), and heart rate measured by pulse (HR) are shown in Fig. 20.22. Based on the fitted model, MESOR estimates of SBP, DBP, and HR, estimated from the 7-day/24-hour ABP measurements, were 132.8 mmHg, 88.4 mmHg, and 72.1 bpm, respectively, indicating the presence of DBP MESOR-hypertension. In accordance with the guidelines for management of high BP, in the case of 24-hour ABP, in Japan published in 2014, she was also diagnosed with diastolic hypertension, as her 24-hour DBP average is above 80 mmHg.

Day-to-day analyses indicate the presence of "persistent ABP hypertension," the 24-hour average of SBP assuming values of 133.8 (Wednesday), 143.4 (Thursday), 135.2 (Friday), 136.3 (Saturday), 133.6 (Sunday), 132.9 (Monday), and 135.7 mmHg (Tuesday). Day-to-day changes in the day-night ratio of SBP indicate the presence

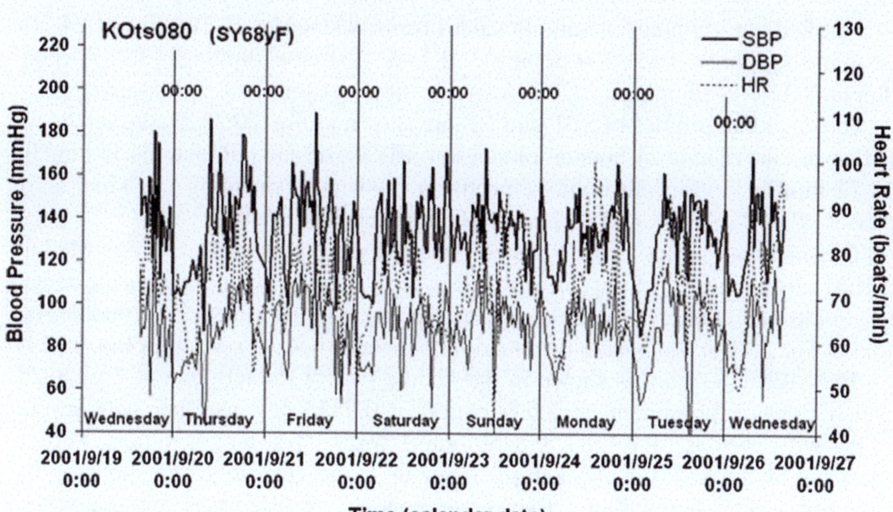

Fig. 20.22 An example of "masked non-dipping." Time series of SBP, DBP, and heart rate (HR) are shown. Time plots of the 7-day/24-hour records of SBP (*thick continuous line*), DBP (*thin continuous line*), and HR (*dashed line*) of a 68-year-old woman. A non-dipping pattern occurs on Saturday night (4th day)

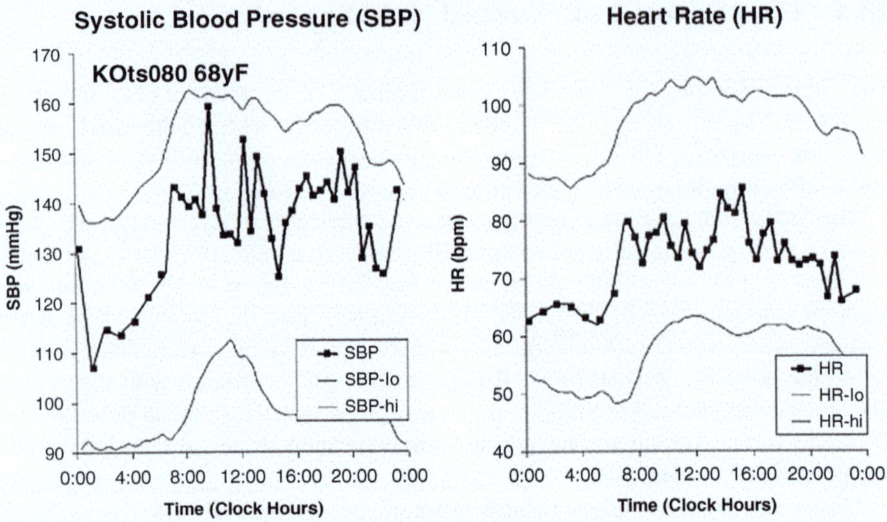

Fig. 20.23 Circadian profiles of systolic BP (*left*) and heart rate (HR) (*right*). 24-hour patterns of systolic BP and HR, averaged over the 7 days of monitoring (*polygonal line* graph with *closed squares*), are shown along with time-specified 90 % prediction limits (*continuous thin lines*). Both patterns are well within the chronodesmic limits

of "masked non-dipping," assuming values from Wednesday to Tuesday of 24.7 % (extreme dipping), 16.2 % (dipping), 21.1 % (extreme dipping), 3.2 % (non-dipping), 16.7 % (dipping), 23.2 % (extreme dipping), and 19.2 % (dipping).

The circadian profiles of SBP and HR are shown in Fig. 20.23. Based on the fitted model, the double 24-hour amplitudes of SBP, DBP, and HR were 24.13 mmHg, 18.77 mmHg, and 15.30 bpm, respectively, with corresponding circadian acrophases of 14:08, 13:34, and 13:25, all within acceptable limits.

The circadian period of SBP was estimated by the Maximum Entropy Method (MEM), using the MemCalc software (GMS Co., Tokyo), which requires the data to be equidistant. Outliers were deleted from the original SBP time series (middle section of Fig. 20.24). Data were then made equidistant (bottom section of Fig. 20.24).

The MEM spectrum is shown in the middle part of the left side of Fig. 20.25, which shows a prominent circadian period of 0.9855 day, together with several peaks with periods of 9.1383, 0.8450, and 0.5029 days, among others. Periods extracted by the MEM analysis in the range up to 2.5 cycles/day are shown in the right upper side of this figure. Twelve components are listed here and that with a period of 0.9855 day has the largest spectral power, as assessed by the "area" covered by the spectral peak.

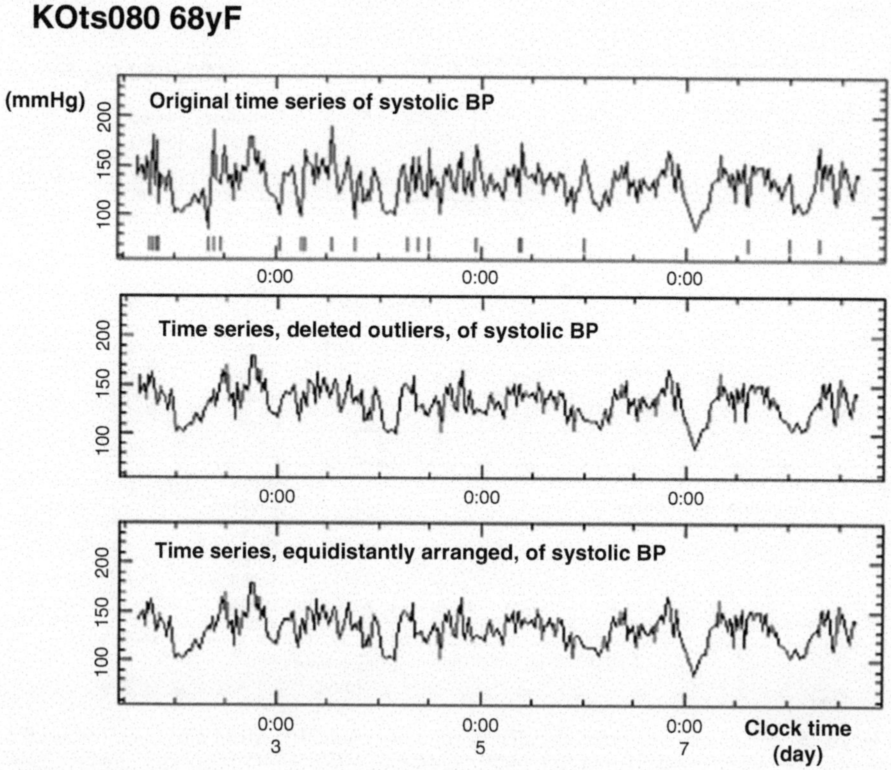

Fig. 20.24 Three versions of the time series of systolic BP. *Upper*: Original data of systolic BP. *Middle*: Edited data after mathematically excluding outliers. *Lower*: Equidistantly arranged time series for analysis by the Maximum Entropy Method, using the MemCalc software (GMS Co., Tokyo)

On the bottom left of Fig. 20.25, the edited data are fitted by least squares with a five-component model consisting of the five cosine curves with the five largest amplitudes, with corresponding periods of 9.138, 2.104, 0.985, 0.845, and 0.503 days (lower right of Fig. 20.25).

In conclusion, the clinical diagnosis based on the chronobiologically interpreted 7-day/24-hour ABP monitoring consists of (1) masked non-dipping, (2) persistent ABP hypertension (24-hour average of SBP is consistently above 130 mmHg), and (3) DBP MESOR-hypertension. Accordingly, we recommended her to strictly improve her lifestyle, focusing on salt intake, and we started antihypertensive treatment with a CCB (benidipine, 4 mg once daily). Under treatment, this subject seems to be keeping herself healthy.

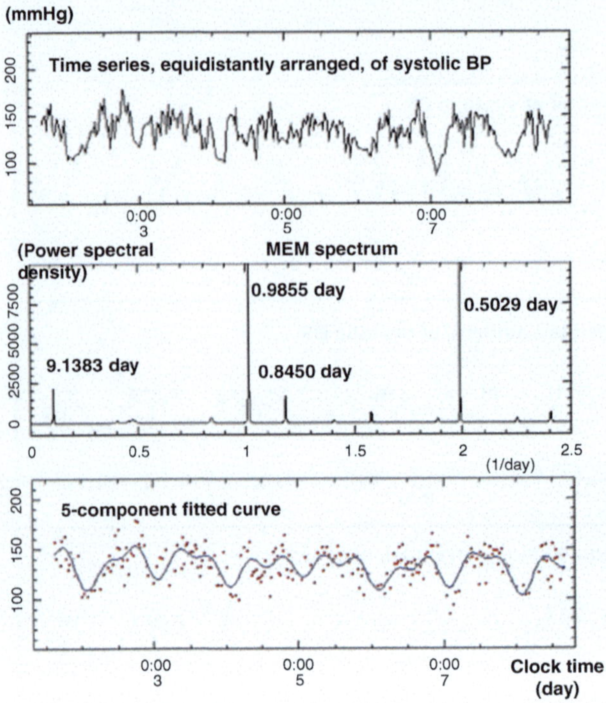

KOts080 68yF

Frequency (/ day)	Period (days)	Area
0.109	9.1383	7.861
0.399	2.5063	4.911
0.475	2.1038	11.514
0.839	1.1920	6.424
1.015	0.9855	70.765
1.183	0.8450	10.211
1.407	0.7106	2.395
1.576	0.6346	3.475
1.885	0.5306	3.618
1.988	0.5029	32.175
2.251	0.4443	3.279
2.406	0.4157	5.847

Period (days)	Amplitude	Acro-phase
9.138	3.78	2-15:12
2.104	5.29	3-03:13
0.985	11.29	2-14:57
0.845	4.23	1-17:12
0.503	7.79	1-19:12

Fig. 20.25 Period analysis of the circadian rhythm of systolic BP. MEM analysis estimates the circadian period of systolic BP to be 0.9855 day (*left middle part*). Data are shown with a five-component model fitted by least squares to the data, consisting of cosine curves with the five largest amplitudes (*right lower part*), the 0.985-day component being the most prominent (*right lower part*)

20.5 Case Report 3 of "Masked Non-dipper"

The third example of "masked non-dipping" is a 74-year-old nonobese (BMI = 21.3 kg/m^2) man (ID number: Tosa0255). His weight was 58.0 kg and height 164.5 cm. Girth of abdomen was 78.5 cm. He has hypertension since he was 70 years of age and is treated with antihypertensive medication. He does not have diabetes mellitus. He has had pollakisuria since he was 60 years of age and takes medication for prostate hypertrophy. Within the 3 months prior to the monitoring, he had several episodes of getting tongue-tied or feeling tightness in the chest.

He has a family history of hypertension (father and elderly brother) and his father died from stroke at age 69. He goes to bed regularly at 22:00 and gets up at 06:30.

He sleeps well without any sleeping tablets. He falls asleep within 10 min, but sometimes he has an episode of sleep paralysis when he is going to sleep. He has been snoring for about 6 years, but seldom dreams. He wakes up for urination more than three times a night (around 01:00, 03:00, and 05:00), but he can fall asleep immediately thereafter. He has neither disturbances of waking too early nor waking not rested. He has no symptoms of nocturnal angina, leg cramps, or restless legs. He usually does not take a nap during the daytime.

He tries to get physical activity and usually jogs more than 40 min five times a week and walks for 60 min to do his shopping every day. He has been smoking 20 cigarettes a day for 50 years since he was 25 years of age.

He regularly takes three meals a day, consuming an average amount of vegetables, fruits, and meat. He usually tries not to eat too many fatty foods. He claims to prefer light seasoning, but has three cups of miso soup every morning and eats pickled salted vegetables more than twice a day. He drinks more than two glasses of Japanese sake every evening. Although he is conscious of nutrition, he often ends up drinking too much.

He usually takes care of his health and has something to live for. He does not have a depressive mood and his scores on the GDS-15 and PHQ-2 scales were 1 and 0, respectively. However, his subjective QOL measures of mood, health, and happiness assessed by a visual analog scale (VAS) were 45 %, 45 %, and 53 %, respectively.

Inquiring about his view of ideal aging, he answered (1) to remain healthy one's entire life, (2) to age well without having any regrets throughout one's life, (3) to be able to fulfill daily living necessities, and (4) to be able to look after oneself for one's entire life.

SpO_2 was 97 % and ECG showed normal sinus rhythm (65 beats per minute), with no ST-T changes indicative of myocardial ischemia or left ventricular hypertrophy. CTR of chest X-ray was 44.9 %, showing neither calcification of aorta nor particular abnormal findings. He had dyslipidemia and his blood LDL-cholesterol was 141 mg/dl, but his HDL-cholesterol and triglyceride were normal, assuming values of 55 and 91 mg/dl, respectively. Urinalysis was fine.

His office BP and HR were 140/88 mmHg and 84 bpm while on antihypertensive medication. His BP and HR on supine and standing positions were 146/86 mmHg (73 bpm) and 132/84 mmHg (87 bpm). Ankle brachial index was normal (left 1.16; right 1.22), but CAVI was 11.2 m/s, suggesting moderately increased arterial stiffness.

Home BP measurements, averaged over 30 days, were 123.7/85.6 mmHg in the morning and 112.4/73.6 mmHg in the evening. Time series of systolic BP (SBP), diastolic BP (DBP), and heart rate measured by pulse (HR) are shown in Fig. 20.26. A model consisting of cosine curves with periods of 24 and 12 hours was fitted by least squares to the data to obtain estimates of the MESOR (Midline-Estimating Statistics Of Rhythm) and of the amplitude and acrophase (time of predicted maximum) of each component. MESOR estimates of SBP, DBP, and HR, estimated from the 7-day/24-hour record, were 135.4 mmHg, 86.6 mmHg, and 74.9 bpm, respectively. In accordance with the guidelines for management of high blood pressure in Japan published in 2014, he is normotensive based on home BP (<135/85 mmHg), but hypertensive based on ABP (>130/80 mmHg).

Day-to-day analyses show "masked ABP normotension," as the 24-hour average of SBP/DBP varied from day to day, assuming values of 139.0/89.5 (Thursday),

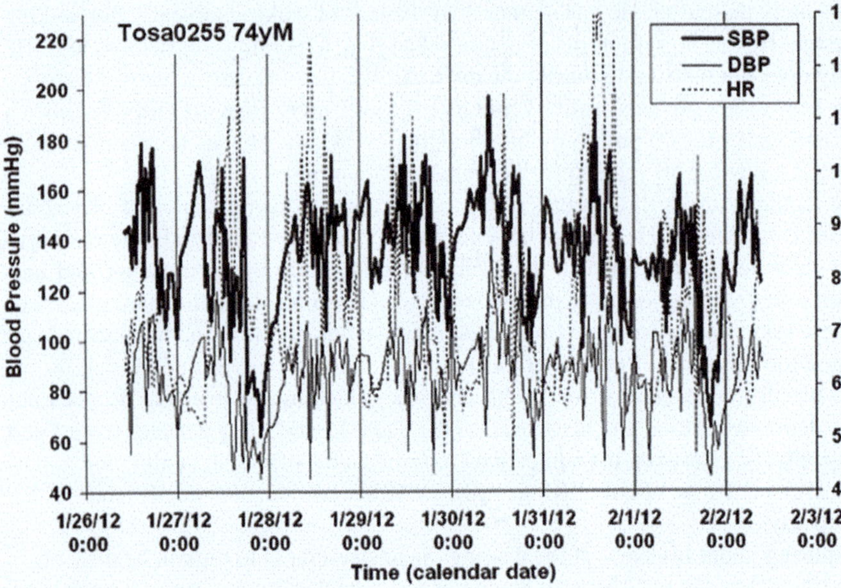

Fig. 20.26 An example of "masked non-dipping." Time plots of the 7-day/24-hour records of SBP (*thick continuous line*), DBP (*thin continuous line*), and HR (*dashed line*) of a 74-year-old man. A non-dipping pattern is observed on the 3rd (Saturday night) and 6th (Tuesday night) day

121.5/76.8 (Friday), 143.3/89.8 (Saturday), 146.2/88.7 (Sunday), 141.8/92.9 (Monday), 136.3/88.3 (Tuesday), and 125.7/85.0 mmHg (Wednesday). Day-to-day changes in the day-night ratio of SBP showed "masked non-dipping" on the 3rd (Saturday night) and 6th (Tuesday night) day, as shown in Fig. 20.26.

He usually drinks alcohol and had too much to drink on two occasions during the monitoring span, on the 2nd (Friday evening) and 7th (Wednesday evening) day, which is reflected in Fig. 20.1 by a distinct increase in HR, followed by an excessive BP dipping.

The circadian profiles of SBP and HR are shown in Fig. 20.27. Based on the fitted model, the double circadian amplitudes of SBP, DBP, and HR were 27.68 mmHg, 17.807 mmHg, and 27.48 bpm, respectively; all were within acceptable limits. Corresponding circadian acrophases were 10:23, 10:28, and 14:23, showing a slight phase advance of the circadian rhythms of SBP and DBP.

The circadian period of SBP was estimated by the Maximum Entropy Method (MEM), using the MemCalc software (GMS Co., Tokyo), which requires the data to be equidistant. Outliers were deleted from the original SBP time series (middle section of Fig. 20.28). Data were then made equidistant (bottom section of Fig. 20.28).

The MEM spectrum is shown on the middle left of Fig. 20.29, which shows a prominent circadian period of 0.9837 day, together with several peaks with periods of 4.8810, 1.6116, and 0.4958 days, among others. Periods extracted by the MEM analysis in the range up to 2.5 cycles/day are shown in the right upper side of this

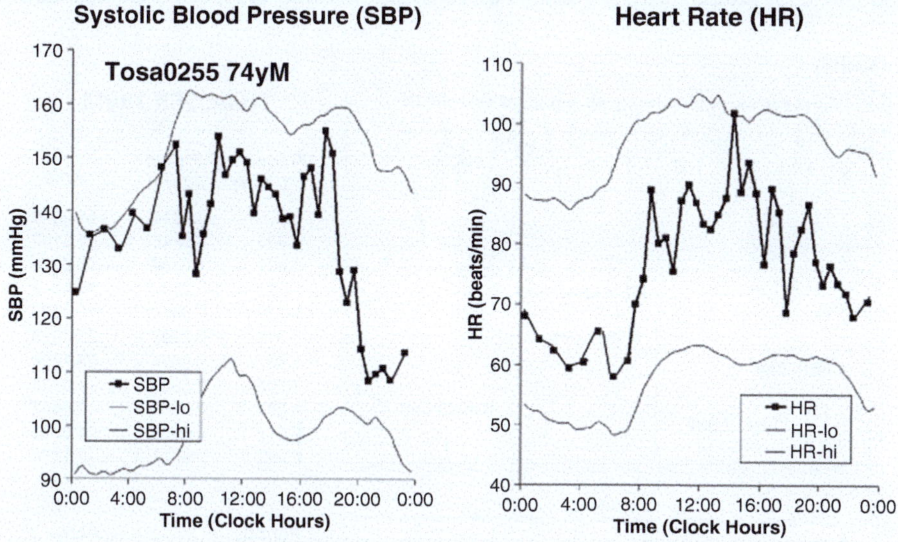

Fig. 20.27 Circadian profiles of systolic BP (*left*) and heart rate (HR) (*right*). 24-hour systolic BP (*left*) and heart rate (HR) (*right*) patterns, averaged over the 7 days of monitoring (*polygonal line* graph with *closed squares*), are shown along with time-specified prediction limits (*continuous thin lines*). While BP is high-normal during the night, HR is well within the chronodesmic limits

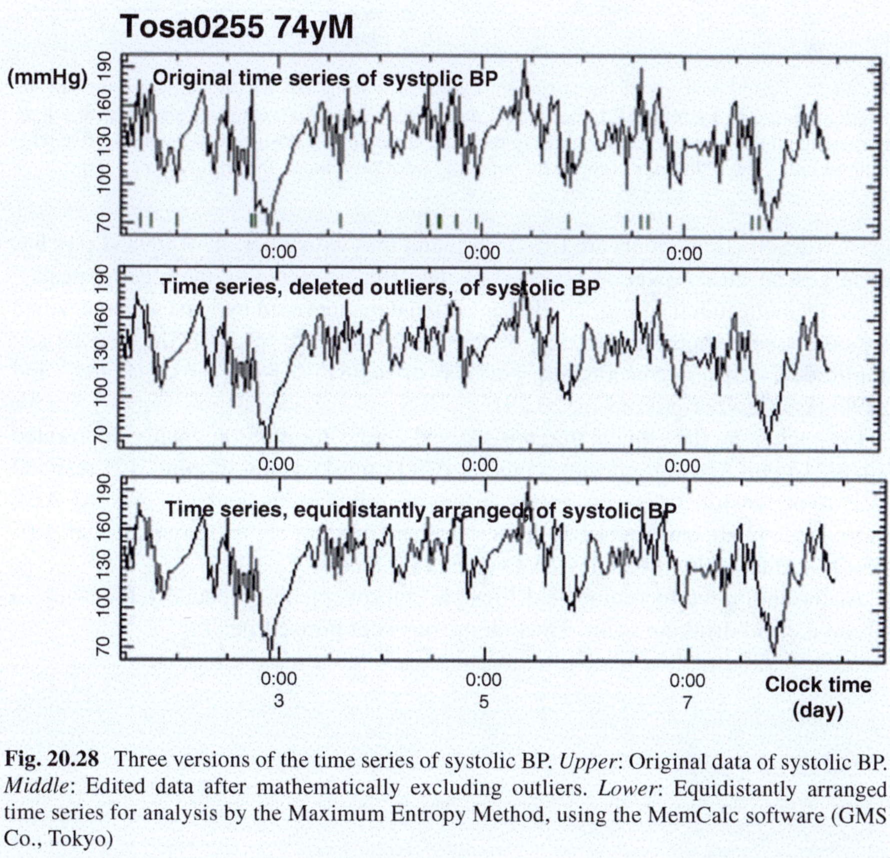

Fig. 20.28 Three versions of the time series of systolic BP. *Upper*: Original data of systolic BP. *Middle*: Edited data after mathematically excluding outliers. *Lower*: Equidistantly arranged time series for analysis by the Maximum Entropy Method, using the MemCalc software (GMS Co., Tokyo)

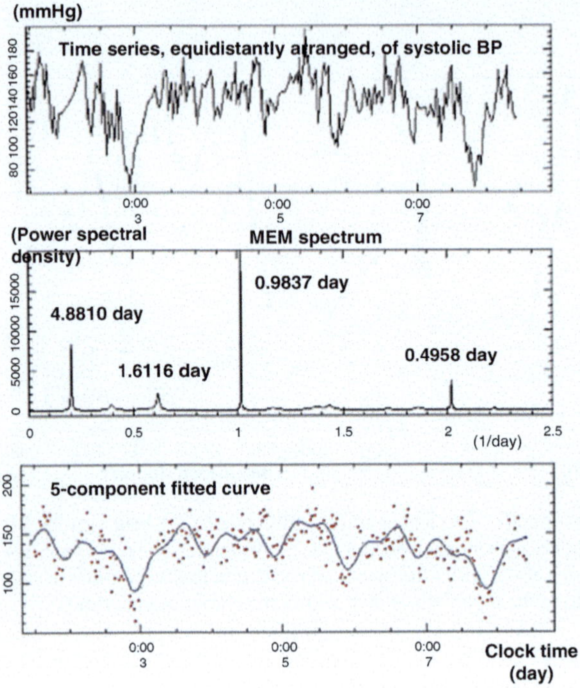

Tosa0255 74yM

Frequency (/ day)	Period (days)	Area
0.205	4.8810	65.2345
0.393	2.5446	24.5192
0.621	1.6116	53.6975
1.017	0.9837	74.9760
1.170	0.8550	12.0528
1.371	0.7293	23.2325
1.437	0.6958	23.7758
1.708	0.5855	4.3624
1.858	0.5383	9.3547
2.017	0.4958	24.3992
2.222	0.4501	2.7574

Period (days)	Amplitude	Acro-phase
4.881	10.96	5-02:25
2.545	5.36	1-13:45
1.612	10.40	2-01:16
0.984	13.08	1-11:44
0.496	7.99	1-16:06

Fig. 20.29 Period analysis of the circadian rhythm of systolic BP. MEM analysis estimates the circadian period of systolic BP to be 0.9837 day (*left middle part*). Data are shown with a five-component model fitted by least squares to the data, consisting of cosine curves with the five largest amplitudes, the 0.984-day component being the most prominent one (*right lower part*)

figure. Eleven components are listed here, and that with a period of 0.9837 day has the largest spectral power, as assessed by the "area" covered by the spectral peak.

On the bottom left of Fig. 20.29, the edited data are fitted by least squares with a five-component model consisting of the five cosine curves with the five largest amplitudes, with corresponding periods of 4.881, 2.545, 1.612, 0.984, and 0.496 days (lower right of Fig. 20.29).

In conclusion, the clinical diagnosis based on the chronobiologically interpreted 7-day/24-hour ABP monitoring consists of (1) masked non-dipping, (2) residual ABP hypertension while on antihypertensive medication, and (3) masked ABP normotension. He has high LDL-cholesterol and moderately increased arterial stiffness, as indicated by the high CAVI value (11.2 m/s).

Accordingly, we recommended lifestyle improvements, especially focusing on salt intake and drinking habit. Thus far, he has kept himself healthy.

20.6 Ambulatory Blood Pressure "Masked Dipper"

In this chapter we introduce two cases of "masked dipping" from the viewpoint of 7-day/24-hour ambulatory blood pressure (ABP) monitoring. Non-dipping was defined as a day-night ratio of SBP less than 10 %. Masked dipping is a condition characterized by "non-dipping" on day 1 but "dipping" on at least 1 of the following 6 days.

20.6.1 Case Report 1 of "Masked Dipper"

The first example of a "masked dipper" is a 67-year-old nonobese (BMI = 23.3 kg/m^2) woman (ID number: KOts082). She goes to bed regularly at 22:30 and gets up at 06:30, but she frequently dreams and does not sleep well. While her office BP was 151/82 mmHg, she was not taking any antihypertensive medication.

The circadian profiles of SBP and HR are shown in Fig. 20.30. SBP is close to the upper part of the chronodesm, and during the daytime, readings sometimes

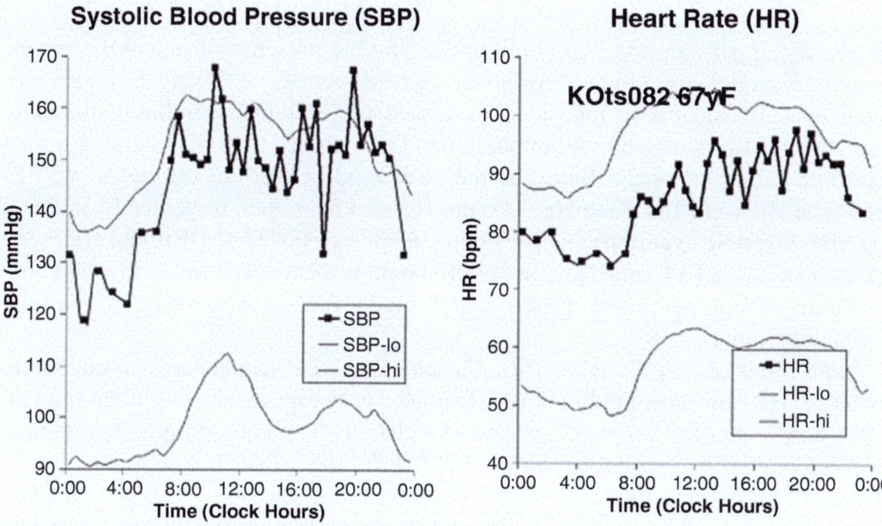

Fig. 20.30 Circadian profiles of systolic BP (*left*) and heart rate (HR) (*right*). 24-hour systolic BP and HR patterns, averaged over the 7 days of monitoring (*polygonal line* graph with *closed squares*), are shown along with time-specified prediction limits (*continuous thin lines*). SBP remains close to the upper limit of the chronodesm, exceeding it several times during the daytime. HR varies well within the chronodesmic limits, albeit in the high-normal range in late afternoon

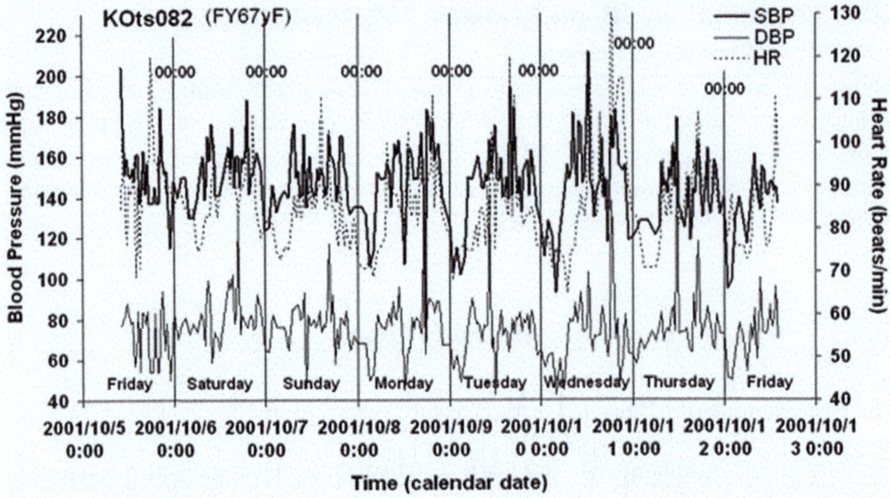

Fig. 20.31 An example of "masked dipping." Time plots of the 7-day/24-hour records of SBP (*thick continuous line*), DBP (*thin continuous line*), and HR (*dashed line*) of a 67-year-old woman are shown from Friday to the next Friday. Nocturnal BP dipping varies greatly from day to day, alternating between non-dipping, dipping, and extreme-dipping patterns

exceed the upper chronodesmic limit, indicating the presence of mild SBP hypertension. A model consisting of cosine curves with periods of 24 and 12 hours was fitted by least squares to the data to estimate the MESOR (Midline-Estimating Statistics Of Rhythm) and the amplitude and acrophase (time of predicted maximum) of each component. Based on the fitted model, MESOR estimates of SBP, DBP, and HR were 144.5 mmHg, 74.6 mmHg, and 85.4 bpm, respectively, indicating SBP MESOR-hypertension. The double amplitudes of SBP, DBP, and HR were 25.44 mmHg, 13.44 mmHg, and 17.38 bpm, respectively, and corresponding circadian acrophases were 14:22, 14:03, and 16:25, respectively, all within acceptable limits.

Time series of systolic BP (SBP), diastolic BP (DBP), and heart rate measured by pulse (HR) are shown in Fig. 20.31. Day-to-day changes in the day-night ratio of SBP show a "masked dipping" pattern, as it changed from non-dipping to dipping, or extreme dipping, assuming values of 5.4 % (Friday night; non-dipping), 9.4 % (Saturday night; non-dipping), 14.9 % (Sunday night; dipping), 23.6 % (Monday night; extreme dipping), 24.% (Tuesday night; extreme dipping), 20.1 % (Wednesday night; extreme dipping), and 14.3 % (Thursday night; dipping).

The circadian period of SBP was estimated by the Maximum Entropy Method (MEM), using the MemCalc software (GMS Co., Tokyo), which requires the data to be equidistant. Outliers were deleted from the original SBP time series (middle section of Fig. 20.32). Data were then made equidistant (bottom section of Fig. 20.32).

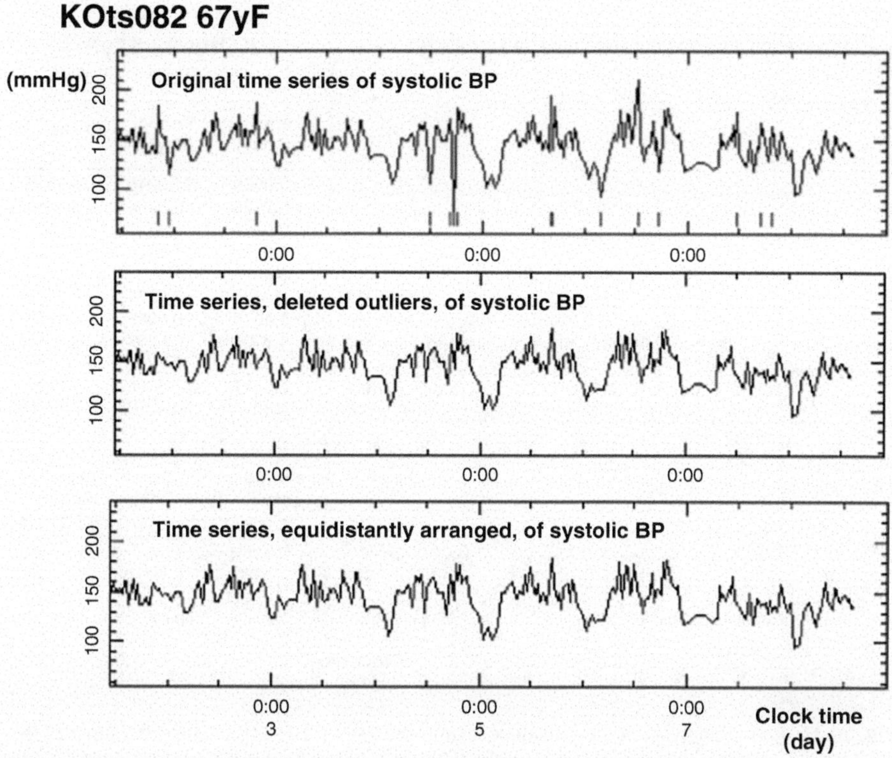

Fig. 20.32 Three versions of the time series of systolic BP. *Upper*: Original data of systolic BP. *Middle*: Edited data after mathematically excluding outliers. *Lower*: Equidistantly arranged time series for analysis by the Maximum Entropy Method, using the MemCalc software (GMS Co., Tokyo)

The MEM spectrum is shown on the middle left of Fig. 20.33, which shows a prominent peak corresponding to a circasemidian component with a period of 0.4955 day, together with a distinct circadian component with a period of 1.0308 day and several other peaks, notably corresponding to periods of 11.0040 and 0.5305 days. Periods extracted by the MEM analysis in the range up to 2.5 cycles/day are shown in the right upper side of this figure. Eleven components are listed here, and those with periods of 0.4955 and 1.0308 day have the largest spectral power, as assessed by the "area" covered by the spectral peaks.

On the bottom left of Fig. 20.33, the edited data are fitted by least squares with a five-component model consisting of the five cosine curves with the five largest amplitudes, with corresponding periods of 11.004, 1.031, 0.927, 0.590, and 0.496 days (lower right of Fig. 20.33).

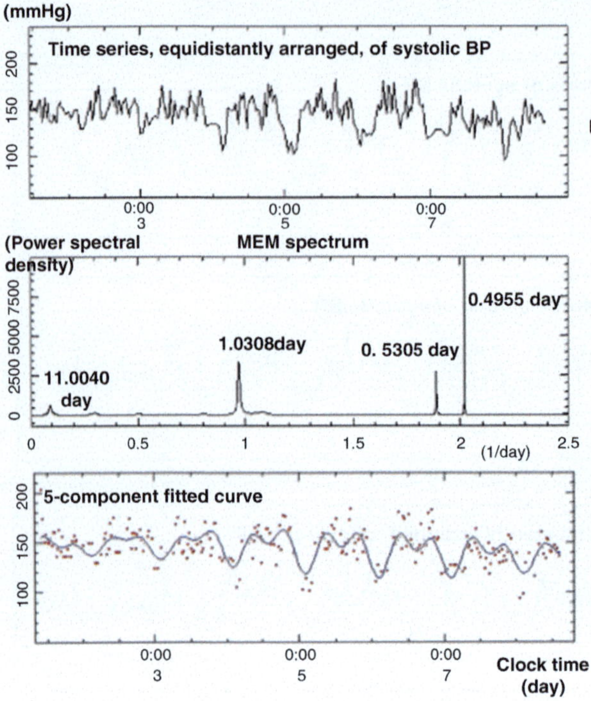

KOts082 67yF

Frequency (/ day)	Period (days)	Area
0.0909	11.0040	14.9912
0.2986	3.3484	5.5589
0.5046	1.9816	2.6824
0.8050	1.2422	3.0803
0.9701	1.0308	60.4574
1.0783	0.9274	12.4499
1.4378	0.6955	0.4672
1.7339	0.5767	0.7186
1.8851	0.5305	13.3493
2.0181	0.4955	40.6937
2.4874	0.4020	2.8176

Period (days)	Amplitude	Acro-phase
11.004	5.546	3-04:31
1.031	9.669	1-11:19
0.927	5.169	1-21:14
0.530	5.064	1-14:13
0.496	8.820	1-21:14

Fig. 20.33 Period analysis of the circadian rhythm of systolic BP. MEM analysis estimates the circadian and circasemidian periods of systolic BP to be 1.0308 and 0.4955 day (*left middle part*). Data are shown with a five-component model fitted by least squares to the data, consisting of cosine curves with the five largest amplitudes, the 0.4955-day and 1.031-day components having the two largest amplitudes (*right lower part*)

In conclusion, the clinical diagnosis of this subject, based on the chronobiologically interpreted 7-day/24-hour ABP monitoring, is SBP MESOR-hypertension complicated by a "masked dipping" pattern. Accordingly, we recommended improvements in her lifestyle, notably her sleeping habit and salt intake. We also started antihypertensive treatment with a CCB, once daily. She has been keeping herself healthy since then.

20.6.2 Case Report 2 of "Masked Dipper"

Another example of "masked dipping" is a 64-year-old obese (BMI=26.3 kg/m^2) man (ID number: KOts091) with diabetes mellitus. His office BP was 135/97 mmHg while on four antihypertensive medications, including CCB and ARB. From the analysis of 7-day/24-hour ABP records, MESOR estimates of systolic BP (SBP),

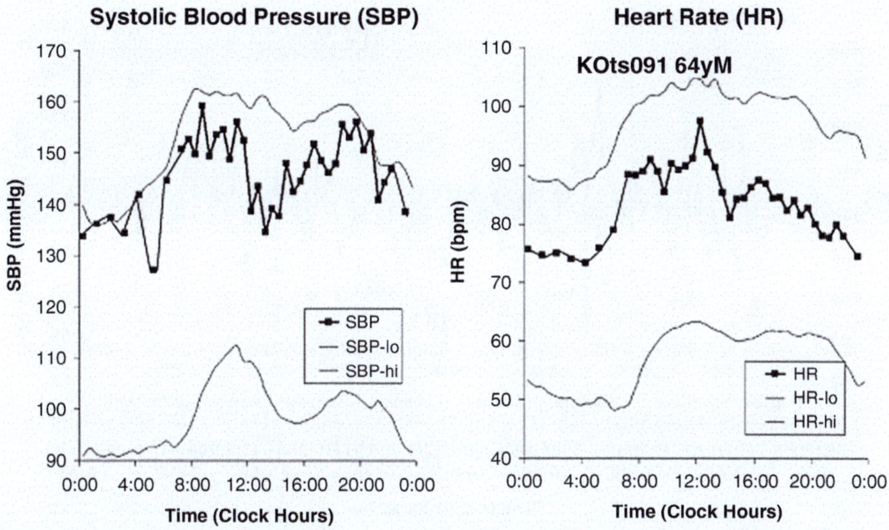

Fig. 20.34 Circadian profiles of systolic BP (*left*) and heart rate (HR) (*right*). 24-hour systolic BP and HR patterns, averaged over the 7 days of monitoring (*polygonal line* graph with closed *squares*), are shown along with time-specified prediction limits (*continuous thin lines*). SBP varies close to the upper limit of the chronodesm, in keeping with SBP MESOR-hypertension. Whereas, on the average, SBP remains mostly within the chronodesmic limits, the MESOR exceeds the upper 95 % prediction limit in health. HR varies approximately in the midportion of the chronodesm

diastolic BP (DBP), and heart rate measured by pulse (HR) were 144.5 mmHg, 101.1 mmHg, and 82.2 bpm, respectively, indicating the presence of both SBP and DBP MESOR-hypertension. The double circadian amplitudes of SBP, DBP, and HR were 10.57 mmHg, 5.35 mmHg, and 15.69 bpm, respectively, with corresponding acrophases of 13:43, 12:42, and 12:22, all within acceptable limits. The circadian profiles of SBP and HR are shown in Fig. 20.34. SBP varies close to and on both sides of the upper limit of the chronodesm, in keeping with mild SBP hypertension.

Time series of SBP, DBP, and HR are shown in Fig. 20.35. Day-to-day changes in the day-night ratio of SBP showed "masked dipping," differing from 1 day to another between non-dipping and dipping, assuming values of 4.9 % (Thursday night; non-dipping), 2.4 % (Friday night; non-dipping), 9.0 % (Saturday night; non-dipping), 14.3 % (Sunday night; dipping), 11.3 % (Monday night; dipping), 4.0 % (Tuesday night; non-dipping), and 14.3 % (Wednesday night; dipping). Whereas a non-dipping pattern occurs on days 1–3 and 6, on days 4, 5, and 7, a dipping pattern is observed (Fig. 20.35). Accordingly, this subject is diagnosed as having SBP MESOR-hypertension complicated by a "masked dipping" pattern.

The circadian period of SBP was estimated by the Maximum Entropy Method (MEM), using the MemCalc software (GMS Co., Tokyo), which requires the data to

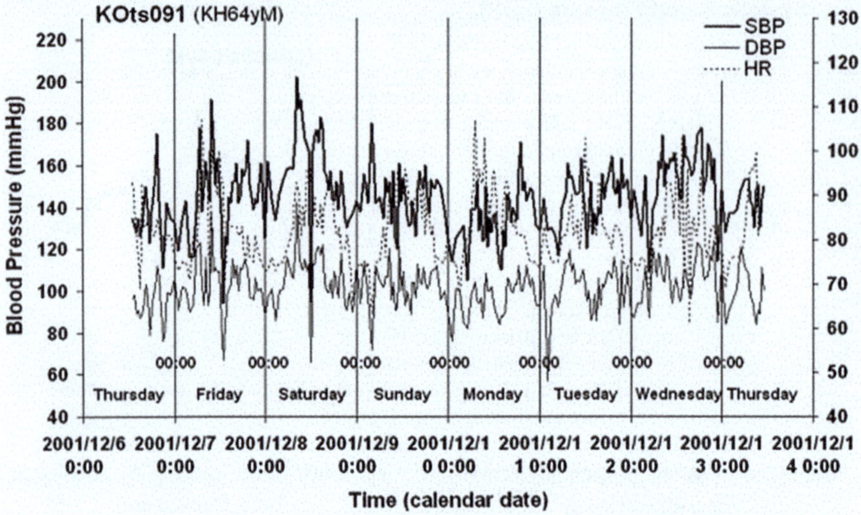

Fig. 20.35 Another example of "masked dipping." Time plots of the 7-day/24-hour records of SBP (*thick continuous line*), DBP (*thin continuous line*), and HR (*dashed line*) of a 64-year-old man are shown from Thursday to Wednesday. The 24-hour average values of SBP change from 1 day to another, assuming values of 137.0 (Thursday), 149.9 (Friday), 156.1 (Saturday), 140.0 (Sunday), 136.8 (Monday), 145.0 (Tuesday), and 154.4 mmHg (Wednesday), indicating "persistent ABP hypertension." Episodic increases in SBP are observed both on the 3rd (Saturday) and 7th (Wednesday) day. A non-dipping pattern is found on days 1–3 and 6, but a dipping pattern occurs on days 4, 5, and 7, indicating the presence of "masked dipping"

be equidistant. Outliers were deleted from the original SBP time series (middle section of Fig. 20.36). Data were then made equidistant (bottom section of Fig. 20.36).

The MEM spectrum is shown in the middle part of the left side of Fig. 20.37, which shows a prominent peak corresponding to circasemiseptan component with a period of 3.9741 days, together with a distinct peak at the circadian period of 1.0242 day and several other peaks corresponding to periods of 1.4252, 0.8381, 0.5718, and 0.4936 days, among others. Periods extracted by the MEM analysis in the range up to 2.5 cycles/day are shown in the right upper side of this figure. Ten components are listed here, and that with a period of 3.9741 days has the largest spectral power, as assessed by the "area" covered by the spectral peak.

On the bottom left of Fig. 20.37, the edited data are fitted by least squares with a five-component model consisting of the five cosine curves with the five largest amplitudes, with corresponding periods of 3.974, 1.425, 1.024, 0.572, and 0.494 days (lower right of Fig. 20.37).

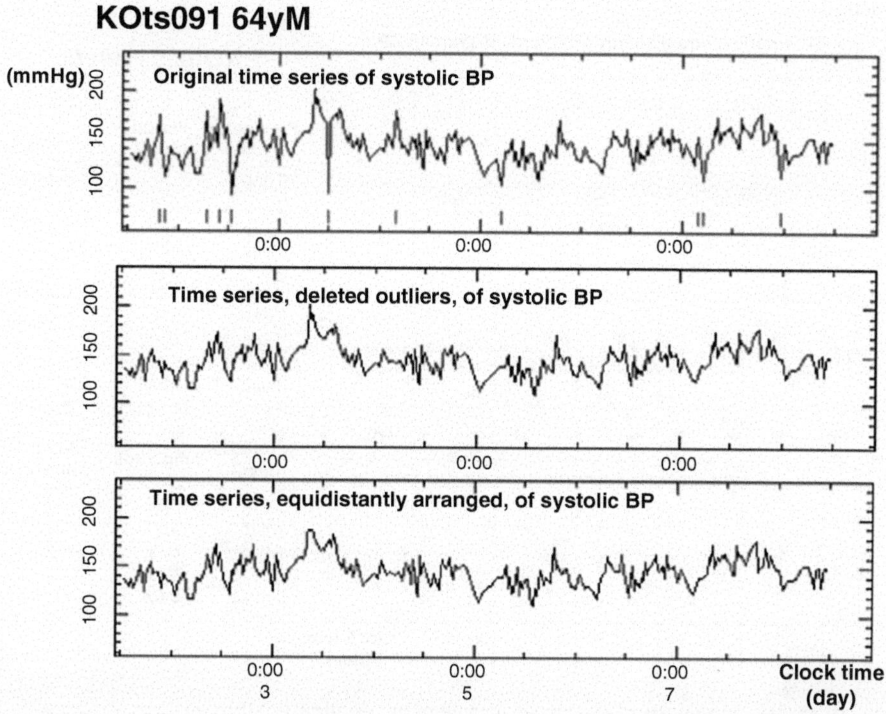

Fig. 20.36 Three versions of the time series of systolic BP. *Upper*: Original data of systolic BP. *Middle*: Edited data after mathematically excluding outliers. *Lower*: Equidistantly arranged time series for analysis by the Maximum Entropy Method, using the MemCalc software (GMS Co., Tokyo)

In conclusion, the clinical diagnosis of this subject, based on the chronobiologically interpreted 7-day/24-hour ABP monitoring, consists of (1) both SBP and DBP MESOR-hypertension; (2) masked dipping, changing from a non-dipping to a dipping pattern from day to day; (3) persistent ABP hypertension, with episodic increases in the 24-hour ABP; and (4) the presence of a prominent half-weekly component exceeding the circadian SBP rhythm in prominence. This subject also has diabetes mellitus. His fasting glucose and hemoglobin A1c were 222 mg/dl and 7.5 %, respectively. Accordingly, we recommended that he strictly improve his lifestyle, focusing on his diet (including salt intake) and merits of daily exercise. As he did not follow the treatment recommendations very well, as this is written he has developed an aneurysmal change of the abdominal aorta.

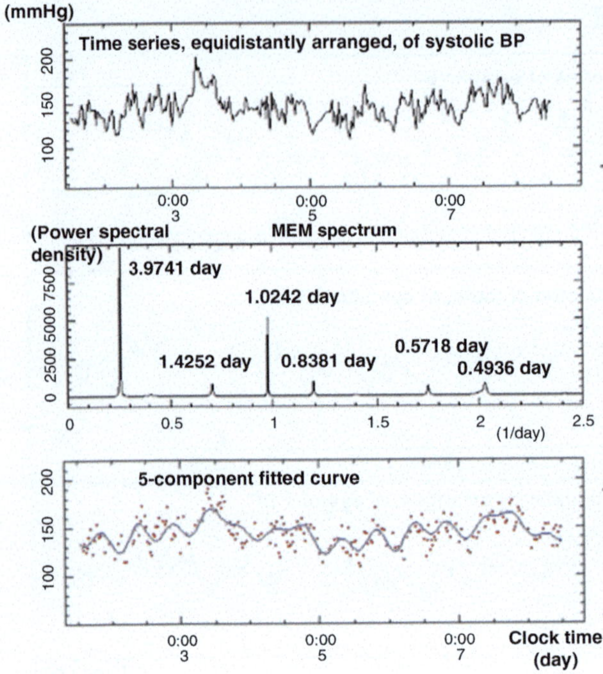

KOts091 64yM

Frequency (/ day)	Period (days)	Area
0.000	-	0.6138
0.252	3.9741	52.8332
0.406	2.4636	6.5258
0.702	1.4252	10.5039
0.976	1.0242	18.7178
1.193	0.8381	9.4245
1.398	0.7154	2.5806
1.565	0.6392	1.4663
1.749	0.5718	10.6294
2.026	0.4936	22.1005
2.412	0.4145	2.2961

Period (days)	Amplitude	Acro-phase
3.974	10.56	3-12:07
1.425	4.21	1-22:09
1.024	6.62	2-13:02
0.572	4.15	1-19:34
0.494	6.63	1-21:17

Fig. 20.37 Period analysis of the circadian rhythm of systolic BP. MEM analysis estimates the circadian and circasemiseptan periods of systolic BP to be 1.0242 and 3.9741 days (*left middle part*). Data are shown with a five-component model fitted by least squares to the data, consisting of cosine curves with the five largest amplitudes, the 1.0242-day and 3.974-day components having the two largest amplitudes (*right lower part*)

20.7 Ambulatory Blood Pressure "Persistent Non-dipper"

In this chapter we introduce three cases of "persistent non-dipping" from the viewpoint of 7-day/24-hour ambulatory blood pressure (ABP) monitoring. Non-dipping is defined as a day-night ratio of SBP lower than 10 %. Persistent non-dipping is a condition such that the nocturnal non-dipping pattern remains consistent each day throughout the 7-day record.

20.7.1 Case Report 1 of "Persistent Non-dipper"

The first example of "persistent non-dipping" is a 50-year-old nonobese (BMI = 22.4 kg/m^2) woman (ID number: KOts086). She visited us for a health check. She goes to bed regularly at 00:30 and gets up at 06:30, but she frequently has nightmares and does not sleep well. Her office BP was 132/81 mmHg, but she

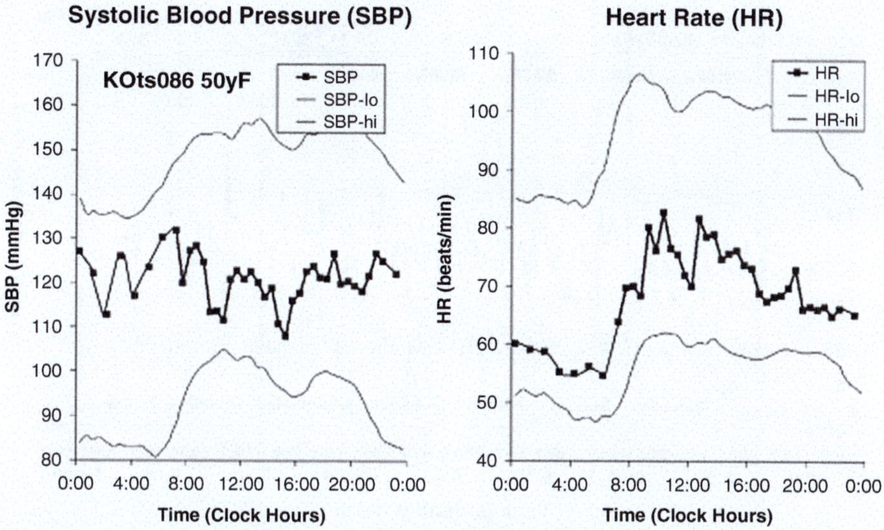

Fig. 20.38 Circadian profiles of systolic BP (*left*) and heart rate (HR) (*right*). 24-hour systolic BP and HR patterns, averaged over the 7 days of monitoring (*polygonal line* graph with closed *squares*), are shown along with time-specified prediction limits (*continuous thin lines*). While SBP varies well within the chronodesmic limits, the day-night difference is small. A model consisting of cosine curves with periods of 24 and 12 hours fitted by least squares to the data showed an odd timing of the circadian acrophase, which occurs at 02:52, indicating ecphasia. HR also varies in the middle portion of the chronodesm but shows a clear circadian rhythm with an acrophase at 13:45 in the afternoon as anticipated

has a fast pulse of about 100 beats per minute (bpm). She does not take any antihypertensive medication.

Based on the model fitted to the 7-day/24-hour ABP monitoring record, MESOR estimates of systolic BP (SBP), diastolic BP (DBP), and heart rate measured by pulse (HR) were 121.2 mmHg, 77.6 mmHg, and 67.1 bpm, respectively. The double 24-hour amplitudes of SBP, DBP, and HR were 5.28 mmHg, 1.40 mmHg, and 19.27 bpm, respectively, suggesting a small day-night difference of BPs. Corresponding circadian acrophases were 02:52, 02:20, and 13:45, indicating the presence of both SBP and DBP ecphasia (i.e., an odd timing of the circadian acrophase of BP but not of the acrophase of HR). The circadian profiles of SBP and HR are shown in Fig. 20.38. While SBP varies in the middle portion of the chronodesm, it is hard to discern a clear circadian pattern by the naked eye.

Time series of SBP, DBP, and HR are shown in Fig. 20.39. During all 7 days, the 24-hour average of SBP remained below 130 mmHg, indicating persistent ABP normotension. The day-night ratio of BP also remains consistently below 10 %, indicating persistent non-dipping, albeit alternating with reverse dipping. Day-to-day day-night ratios assumed values of 5.8 % (non-dipping), −17.3 % (reverse dipping),

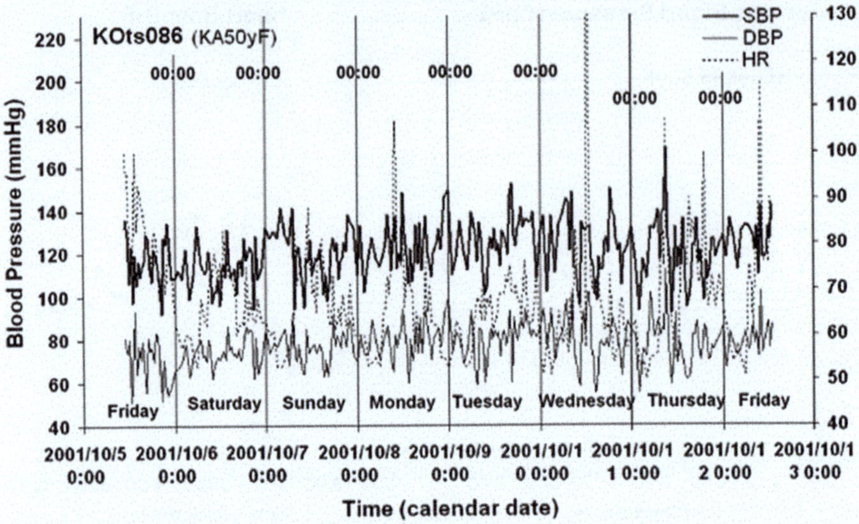

Fig. 20.39 An example of "persistent non-dipping" from the viewpoint of 7-day/24-hour ABP monitoring. Time plots of the 7-day/24-hour records of SBP (*thick continuous line*), DBP (*thin continuous line*), and HR (*dashed line*) of a 50-year-old woman

–0.3 % (reverse dipping), 0.6 % (non-dipping), 5.5 % (non-dipping), 6.8 % (non-dipping), and –9.3 % (reverse dipping) from days 1 to 7.

In accordance with international guidelines, including the Japanese guidelines for 24-hour ABP monitoring, this subject is diagnosed as being hypertensive by night as her nighttime BP averages more than 120/70 mmHg. As shown by day-to-day analyses, nighttime averages of SBP vary, alternating between nighttime hypertension and normotension. From Friday to Thursday, nighttime averages of SBP assume values of 109.8 mmHg (Friday night; normotension), 131.8 mmHg (Saturday night; hypertension), 116.0 mmHg (Sunday night; normotension), 121.5 mmHg (Monday night; hypertension), 120.8 mmHg (Tuesday night; hypertension), 112.5 mmHg (Wednesday night; normotension), and 126.3 mmHg (Thursday night; hypertension). Accordingly, she was diagnosed as having "masked nocturnal hypertension."

The circadian period of SBP was estimated by the Maximum Entropy Method (MEM), using the MemCalc software (GMS Co., Tokyo), which requires the data to be equidistant. Outliers were deleted from the original SBP time series (middle section of Fig. 20.40). Data were then made equidistant (bottom section of Fig. 20.40).

The MEM spectrum is shown on the middle left of Fig. 20.41, which shows a disrupted circadian rhythm. Instead of a prominent circadian peak, the 1.0295-day component is only a secondary peak with a small amplitude. There are several peaks in the ultradian and infradian ranges, the most prominent one corresponding to a period of 0.8560 day. Periods extracted by the MEM analysis in the range up to 2.5 cycles/day are shown in the right upper side of this figure. Twelve components are listed here, with no prominent circadian "period."

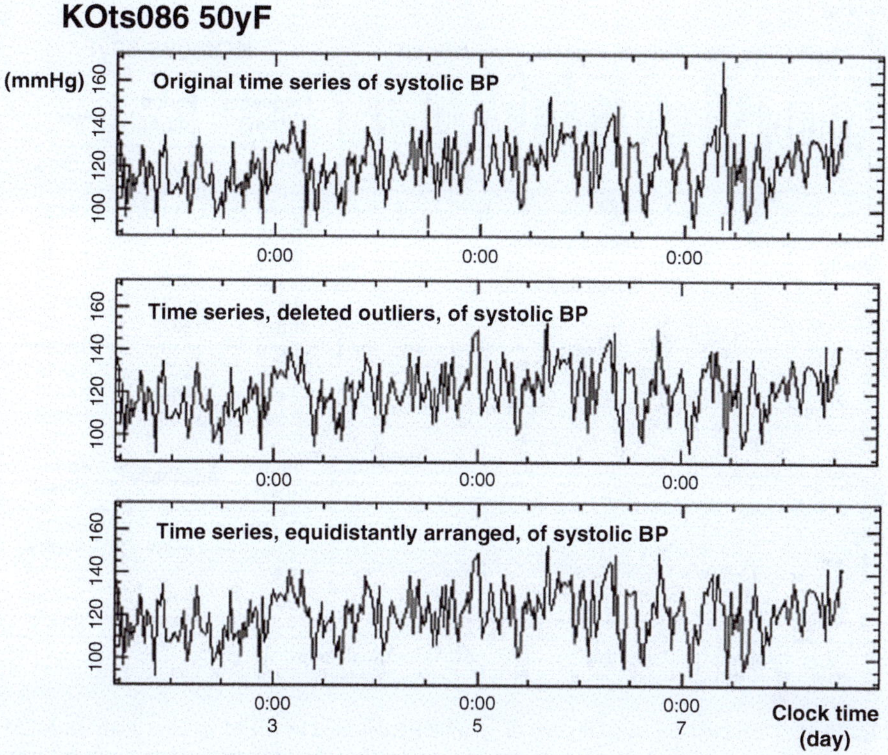

Fig. 20.40 Three versions of the time series of systolic BP. *Upper*: Original data of systolic BP. *Middle*: Edited data after mathematically excluding outliers. *Lower*: Equidistantly arranged time series for analysis by the Maximum Entropy Method, using the MemCalc software (GMS Co., Tokyo)

On the bottom left of Fig. 20.41, the edited data are fitted by least squares with a five-component model consisting of the five cosine curves with the five largest amplitudes, with corresponding periods of 7.600, 2.661, 0.597, 0.258, and 0.232 days (lower right of Fig. 20.41). None of the detected components has an amplitude larger than 10 mmHg.

In conclusion, the clinical diagnosis of this subject, based on the chronobiologically interpreted 7-day/24-hour ABP monitoring, consists of (1) masked nocturnal hypertension; (2) persistent "non-dipping" pattern, alternating between a non-dipping and a reverse-dipping pattern from 1 day to another; (3) SBP and DBP ecphasia; and (4) disrupted circadian rhythm of SBP. Accordingly, we recommended that she strictly improve her lifestyle, focusing on her sleeping habit and salt intake, and to adopt an exercise routine. We also started antihypertensive treatment with a calcium channel blocker (CCB) (benidipine, 4 mg once daily). Under this treatment plan, she is now in good health.

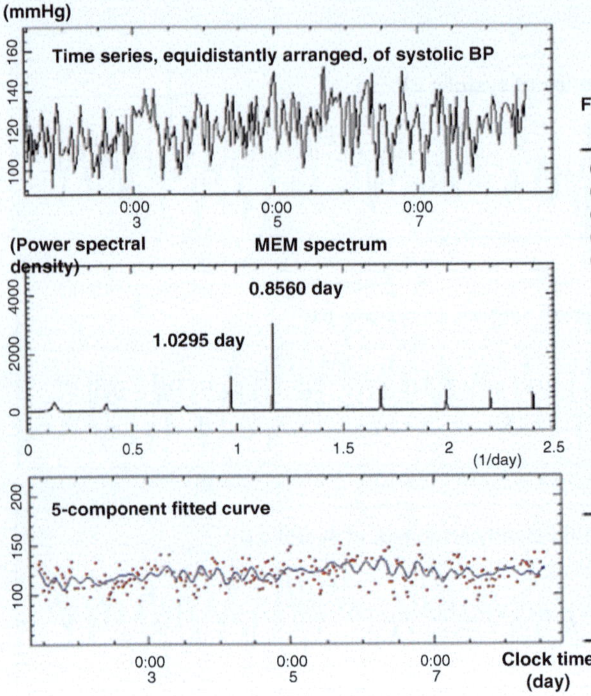

KOts086 50yF

Frequency (/ day)	Period (days)	Area
0.132	7.6003	12.0903
0.376	2.6608	5.4887
0.582	1.7195	1.3722
0.745	1.3427	2.6049
0.971	1.0295	4.1146
1.168	0.8560	4.1423
1.499	0.6672	0.9623
1.675	0.5970	5.5798
1.818	0.5501	0.9390
1.987	0.5033	4.0412
2.196	0.4553	2.4350
2.399	0.4168	1.5966

Period (days)	Amplitude	Acro-phase
7.600	4.4005	6-00:47
2.661	3.4187	3-05:15
0.597	3.1362	2-01:08
0.253	3.3898	1-16:50
0.232	3.4496	1-15:55

Fig. 20.41 Disrupted circadian rhythm of systolic BP detected by MEM spectral analysis. MEM analysis shows a disrupted circadian rhythm of systolic BP (*left middle part*). Data are shown with a five-component model fitted by least squares to the data, consisting of cosine curves with the five largest amplitudes. A circadian component cannot be discerned by the naked eye (*right lower part*)

20.7.2 Case Report 2 of "Persistent Non-dipper"

The second example of a "persistent non-dipper" is a 67-year-old nonobese (BMI = 23.7 kg/m^2) man (ID number: KOts007). His weight was 63.0 kg and height 163 cm. He goes to bed usually at 22:00 and gets up at 07:00. He needs 30–60 min to fall asleep and does not sleep well most days. Because of several nocturnal awakenings and disturbances and instances of waking too early, he wakes up not feeling well rested. He does not smoke and usually does not drink alcohol. He frequently eats fish and rarely meat, his intake of vegetables being somewhat below average. He prefers light seasoning. He regularly has eggs every day, but does not drink milk.

He feels healthy now and takes care of his health. He has a depressive mood and does not have many interests in life. His scores on the GDS-15 and PHQ-2 scales were 13 and 2 points, respectively. His subjective QOL indices of health, mood, family relationships, relations with friends, financial satisfaction, life satisfaction,

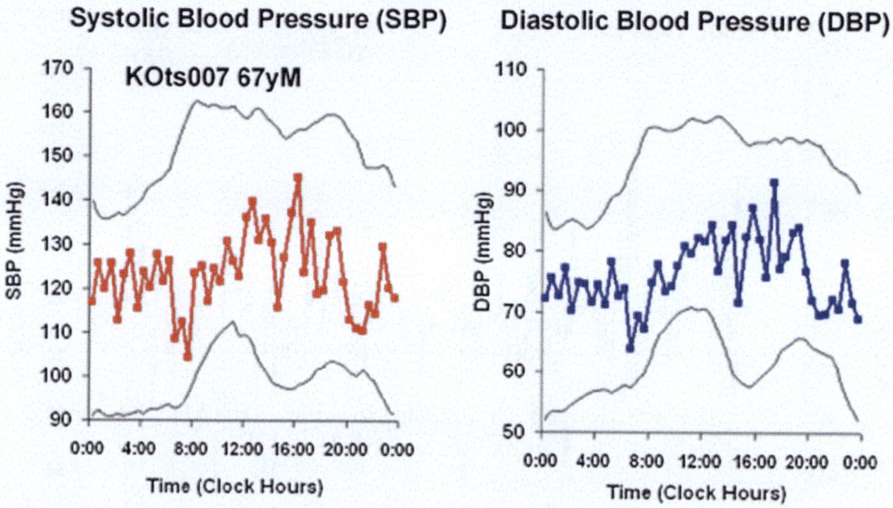

Fig. 20.42 Circadian profiles of systolic (*left*) and diastolic (*right*) BP. 24-hour systolic and diastolic BP patterns, averaged over the 6 days of monitoring (*polygonal line* graph with *closed squares*), are shown along with time-specified prediction limits (*continuous thin lines*). Both are within acceptable limits during the entire 24-hour period

and happiness, assessed by VAS, were 47 %, 50 %, 75 %, 49 %, 82 %, 63 %, and 48 %, respectively.

His office BP was 156/95 mmHg, despite being on four antihypertensive drugs, including CCG and ARB. Based on model fitting to the 6-day/24-hour ABP data, MESOR estimates of systolic BP (SBP), diastolic BP (DBP), and heart rate measured by pulse (HR) were 123.4 mmHg, 76.1 mmHg, and 57.5 bpm, respectively. The double 24-hour amplitudes of SBP, DBP, and HR were 11.9 mmHg, 10.5 mmHg, and 20.0 bpm, respectively, suggesting a small day-night difference of BPs. Corresponding circadian acrophases were 14:09, 14:49, and 15:48, all within the acceptable range. Circadian profiles of SBP and DBP are shown in Fig. 20.42.

Time series of SBP, DBP, and HR are shown in Fig. 20.43. During each of the 6 days of monitoring, the 24-hour SBP average is invariably below 130 mmHg, indicating persistent ABP normotension. The day-night ratio of SBP is also invariably below 10 %, indicating a persistent non-dipping pattern. From day 1 to 6, the day-night ratio of SBP assumes values of 6.2 % (non-dipping), 8.4 % (non-dipping), 6.5 % (non-dipping), −2.2 % (reverse dipping), 9.9 % (non-dipping), and 5.3 % (non-dipping).

On a daily basis, the nighttime SBP average varied from Friday to Thursday, assuming values of 126.5 mmHg (Friday night; hypertension), 127.6 mmHg (Saturday night; hypertension), 117.1 mmHg (Sunday night; normotension), 125.3 mmHg (Monday night; hypertension), 106.5 mmHg (Tuesday night;

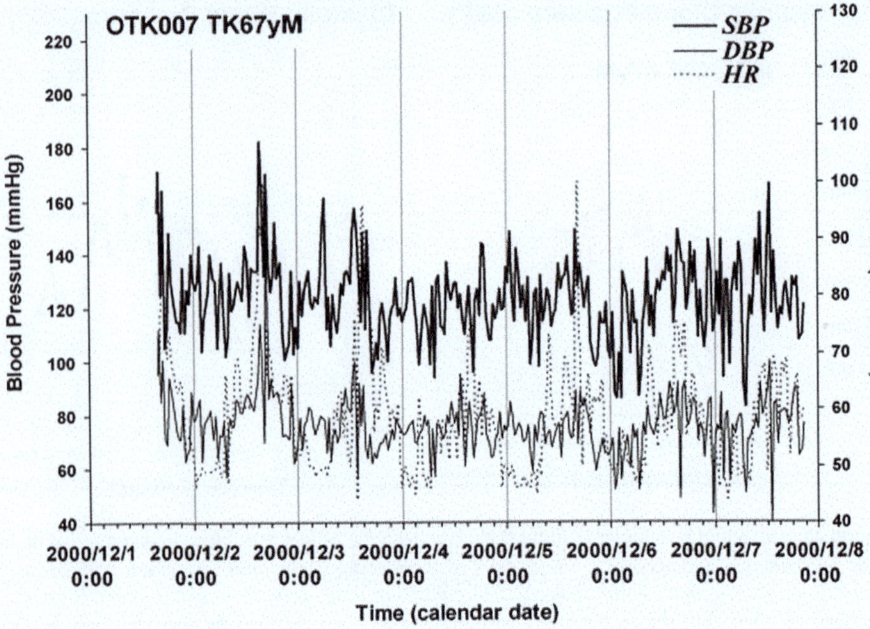

Fig. 20.43 An example of "persistent non-dipping" present during all 6 days of ABP monitoring. Time plots of the 6-day/24-hour records of SBP (*thick continuous line*), DBP (*thin continuous line*), and HR (*dashed line*) of a 67-year-old man. When 6-day/24-hour ABP monitoring was started at 15:47 in front of a doctor on December 1, 2000, BP and pulse were 171/95 mmHg and 96 bpm, respectively, suggesting white-coat hypertension, as BP dropped thereafter within the acceptable range, except for occasional BP spikes on the 2nd, 3rd, and 6th day

normotension), and 123.0 mmHg (Wednesday night; hypertension). Thus, he was diagnosed with "masked nocturnal normotension."

The circadian period of SBP was estimated by the Maximum Entropy Method (MEM), using the MemCalc software (GMS Co., Tokyo), which requires the data to be equidistant. Outliers were deleted from the original SBP time series (middle section of Fig. 20.44). Data were then made equidistant (bottom section of Fig. 20.44).

The MEM spectrum is shown on the middle left of Fig. 20.45, which shows a disrupted circadian rhythm. Instead of a prominent circadian peak, the 0.9779-day component has an amplitude smaller than several other peaks in the infradian and ultradian frequency ranges, notably corresponding to periods of 6.3222 and 0.4898 days. Periods extracted by the MEM analysis in the range up to 2.5 cycles/day are shown in the right upper side of this figure. Nine components are listed here, with none of the components accounting for a major proportion of the variance.

On the bottom left of Fig. 20.45, the edited data are fitted by least squares with a five-component model consisting of the five cosine curves with the five largest

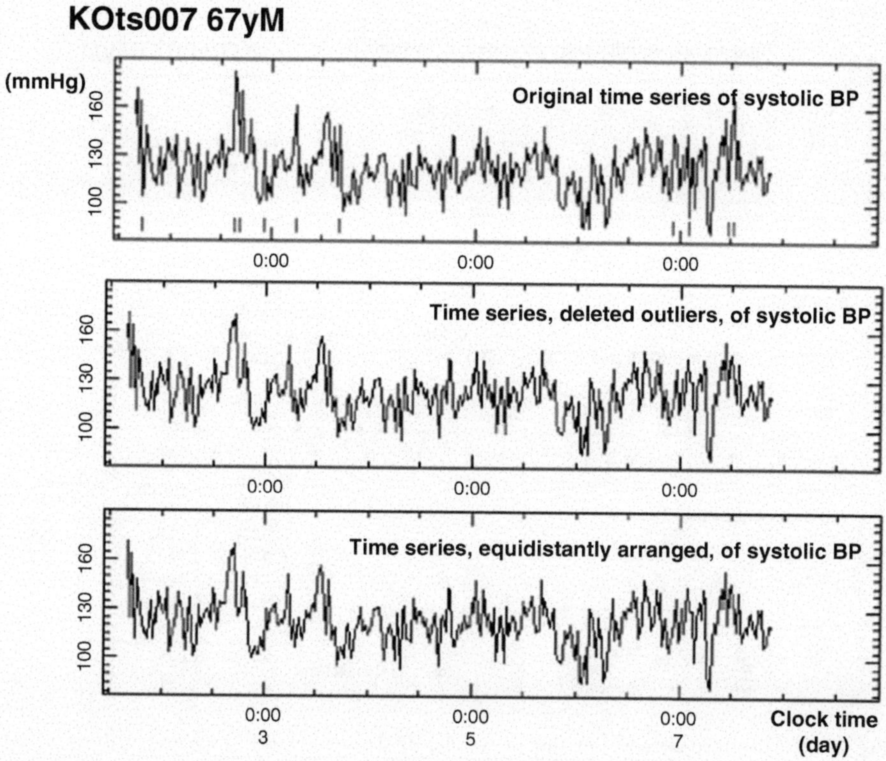

Fig. 20.44 Three versions of the time series of systolic BP. *Upper*: Original data of systolic BP. *Middle*: Edited data after mathematically excluding outliers. *Lower*: Equidistantly arranged time series for analysis by the Maximum Entropy Method, using the MemCalc software (GMS Co., Tokyo)

amplitudes, with corresponding periods of 6.322, 0.978, 0.792, 0.490, and 0.391 days (lower right of Fig. 20.45). None of the components had an amplitude larger than 10 mmHg.

In conclusion, the clinical diagnosis of this subject, based on the chronobiologically interpreted 6-day/24-hour ABP monitoring, consists of (1) nocturnal hypertension with masked normotension, (2) persistent "non-dipping" pattern changing from non-dipping to reverse dipping, (3) white-coat hypertension with episodic BP spikes, and (4) disrupted circadian rhythm of SBP. He was suffering from a strongly depressed mood and a moderately disturbed quality of sleep. We recommended him to improve his lifestyle, focusing on his sleeping habit and salt intake. We also prescribed soporific and antidepressive medications. Under this treatment plan, he now seems to be in good health.

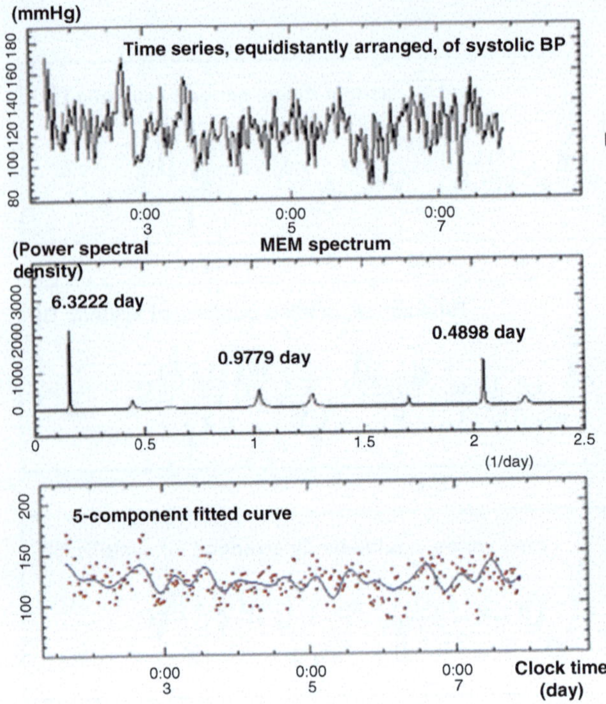

KOts007 67yM

Frequency (/ day)	Period (days)	Area
0.158	6.3222	8.7944
0.444	2.2530	6.3020
0.615	1.6255	3.3233
0.832	1.2013	1.2533
1.023	0.9779	14.5910
1.263	0.7920	13.5169
1.704	0.5867	3.5034
2.042	0.4898	14.8351
2.232	0.4480	8.1840

Period (days)	Amplitude	Acro-phase
6.322	4.2382	2-00:32
0.978	4.7113	2-15:20
0.792	4.4996	1-19:06
0.490	5.6093	2-03:27
0.391	4.0189	1-23:41

Fig. 20.45 Disrupted circadian rhythm of systolic BP detected by MEM spectral analysis. MEM analysis shows a disrupted circadian rhythm of systolic BP (*left middle part*). Data are shown with a five-component model fitted by least squares to the data, consisting of cosine curves with the five largest amplitudes. A circadian rhythm cannot be discerned by the naked eye (*right lower part*)

20.7.3 Case Report 3 of "Persistent Non-dipper"

The third example of "persistent non-dipping" is a 64-year-old nonobese (BMI = 22.1 kg/m^2) woman (ID number: KOts071). Her office BP was 173/112 mmHg despite the fact that she was on four antihypertensive medications, including CCB (long-acting nifedipine, 20 mg, twice daily) and ARB (candesartan, 8 mg, once daily at 09:30).

She goes to bed usually at 22:00 and gets up regularly at 05:00. She falls asleep within 10 min and sleeps well most days. She does not smoke and usually does not drink alcohol. She prefers light seasoning and takes care not to add too much salt.

She feels fairly healthy and takes care of her health. She does not have a depressive mood and has something to live for. Her scores on the GDS-15 and PHQ-2 scales were 5 and 0 points, respectively. Her subjective QOL indices of health, mood, family relationships, relationships with friends, financial satisfaction, life

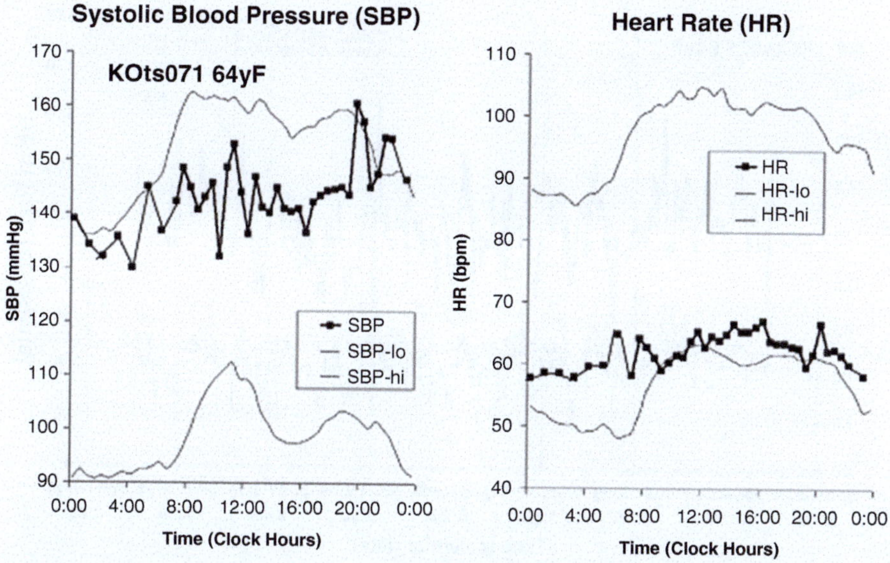

Fig. 20.46 Circadian profiles of systolic BP (*left*) and heart rate (HR) (*right*). 24-hour systolic BP and HR patterns, averaged over the 7 days of monitoring (*polygonal line* graph with closed *squares*), are shown along with time-specified prediction limits (*continuous thin lines*). SBP varies in the upper part of the chronodesm with a small day-night difference. HR varies in the lower part of the chronodesm, also with a small circadian variation. These patterns may be influenced by the twice-daily administration of an alpha-beta-blocking agent (arotinolol, 10 mg at 09:30 and 21:00)

satisfaction, and happiness, assessed by VAS, were 65 %, 65 %, 95 %, 75 %, 65 %, 75 %, and 70 %, respectively.

Based on the least-squares fit of a two-component model consisting of cosine curves with periods of 24 and 12 hours on the 7-day/24-hour ABP data, estimates of the MESOR of systolic BP (SBP), diastolic BP (DBP), and heart rate measured by pulse (HR) were 142.6 mmHg, 83.3 mmHg, and 61.8 bpm, respectively, indicating SBP MESOR-hypertension. The double 24-hour amplitudes of SBP, DBP, and HR were 8.24 mmHg, 8.10 mmHg, and 6.07 bpm, respectively, suggesting small day-night differences of BPs and HR. Corresponding circadian acrophases were 17:25, 16:18, and 14:23, respectively. Circadian profiles of SBP and HR are shown in Fig. 20.46. SBP varies inside the chronodesm, but a clear circadian pattern cannot be discerned by the naked eye.

Time series of SBP, DBP, and HR are shown in Fig. 20.47. During each of the 7 days of monitoring, the 24-hour average of SBP was consistently above 130 mmHg, indicating persistent ABP hypertension. The BP day-night ratio was also consistently below 10 %, indicative of a persistent non-dipping pattern. From the first (Friday) to the last (Thursday) day, the day-night ratio of SBP assumed values of

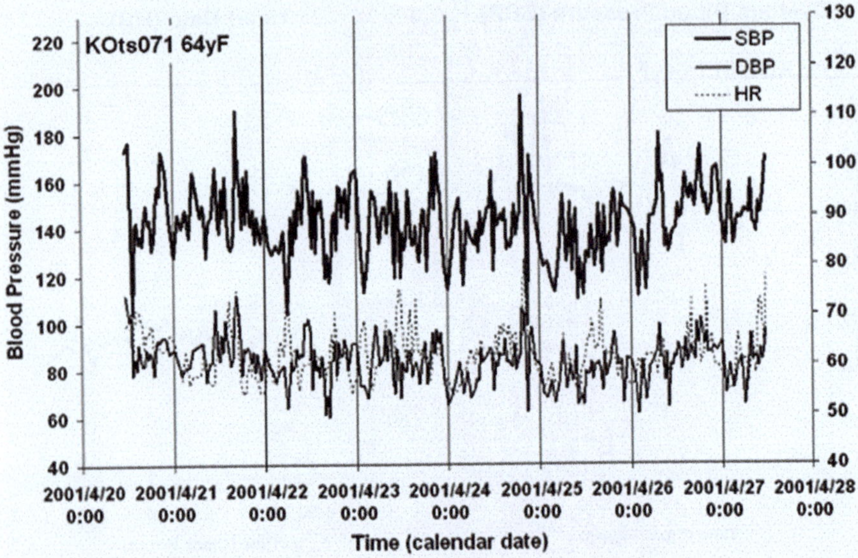

Fig. 20.47 An example of "persistent non-dipping." Time plots of the 7-day/24-hour records of SBP (*thick continuous line*), DBP (*thin continuous line*), and HR (*dashed line*) of a 64-year-old woman

1.0 %, 9.8 %, 0.3 %, 6.2 %, 9.9 %, 0.1 %, and 2.4 %. Daily nighttime averages of SBP showed persistent nocturnal hypertension, being invariably above 130 mmHg, albeit they varied from 1 day to another, assuming values of 140.3 mmHg (Friday night), 134.2 mmHg (Saturday night), 143.2 mmHg (Sunday night), 130.5 mmHg (Monday night), 126.5 mmHg (Tuesday night), 132.5 mmHg (Wednesday night), and 146.5 mmHg (Thursday night).

The circadian period of SBP was estimated by the Maximum Entropy Method (MEM), using the MemCalc software (GMS Co., Tokyo), which requires the data to be equidistant. Outliers were deleted from the original SBP time series (middle section of Fig. 20.48). Data were then made equidistant (bottom section of Fig. 20.48).

The MEM spectrum is shown on the middle left of Fig. 20.49, which shows a disrupted circadian rhythm. Instead of a prominent circadian peak, the 1.0273-day component has an amplitude smaller than several other peaks in the infradian and ultradian frequency ranges. Periods extracted by the MEM analysis in the range up to 2.5 cycles/day are shown in the right upper side of this figure. Twelve components are listed here, with none of the components accounting for a major proportion of the variance.

On the bottom left of Fig. 20.49, the edited data are fitted by least squares with a five-component model consisting of the five cosine curves with the five largest

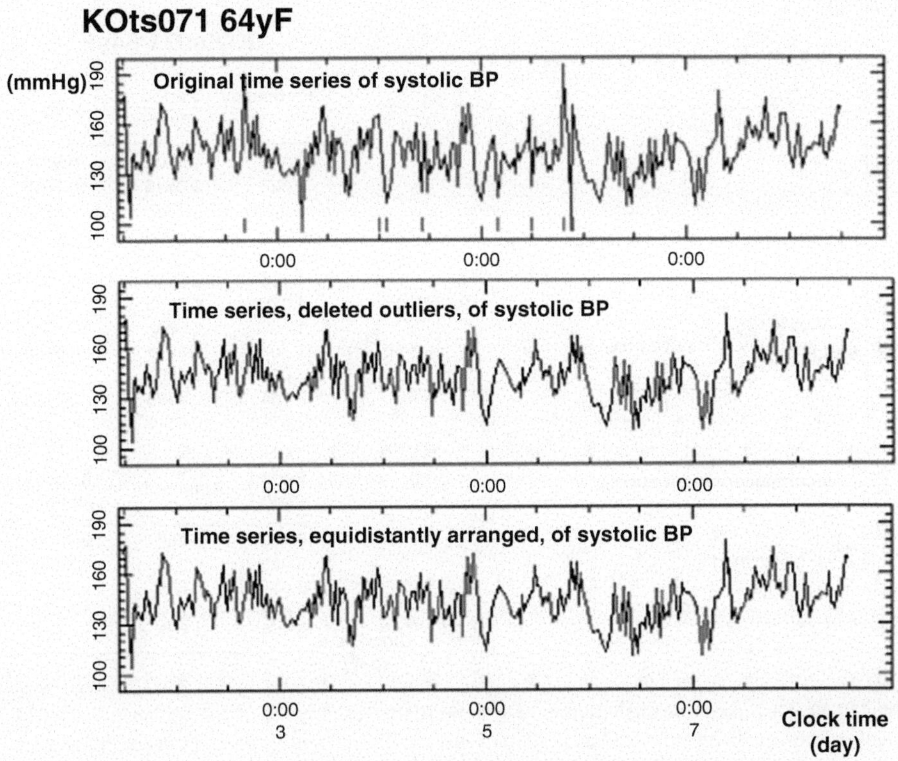

Fig. 20.48 Three versions of the time series of systolic BP. *Upper*: Original data of systolic BP. *Middle*: Edited data after mathematically excluding outliers. *Lower*: Equidistantly arranged time series for analysis by the Maximum Entropy Method, using the MemCalc software (GMS Co., Tokyo)

amplitudes, with corresponding periods of 6.160, 3.055, 0.495, 0.415, and 0.339 days (lower right of Fig. 20.49). None of the components had an amplitude larger than 10 mmHg.

In conclusion, the clinical diagnosis of this subject, based on the chronobiologically interpreted 7-day/24-hour ABP monitoring, consists of (1) hypertension despite treatment with antihypertensive medications; (2) SBP MESOR-hypertension; (3) "persistent non-dipping" pattern, along with persistent nocturnal hypertension; (4) disrupted circadian rhythm of SBP; and (5) insufficient HR variability. Accordingly, we recommended to strictly improve her lifestyle, focusing on salt intake, and we changed the timing of administration of medications from twice daily to once daily in the morning. Under this treatment plan, she now seems to be in good health.

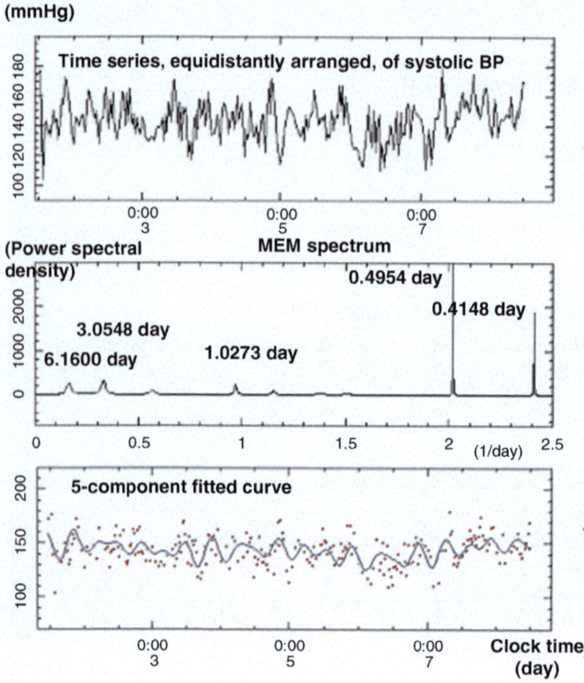

KOts071 64yF

Frequency (/ day)	Period (days)	Area
0.162	6.1600	9.5620
0.327	3.0548	10.7698
0.567	1.7635	4.1856
0.745	1.3423	0.3658
0.973	1.0273	4.8308
1.156	0.8650	2.8347
1.376	0.7265	1.7202
1.508	0.6632	1.8098
1.702	0.5874	0.5148
2.019	0.4954	15.9527
2.134	0.4686	0.4137
2.411	0.4148	6.7115

Period (days)	Amplitude	Acro-phase
6.160	3.789	2-16:54
3.055	4.313	1-19:53
0.495	5.819	1-21:32
0.415	3.847	1-20:09
0.339	4.841	1-11:36

Fig. 20.49 Disrupted circadian rhythm of systolic BP detected by MEM spectral analysis. MEM analysis shows a disrupted circadian rhythm of systolic BP (*left middle part*). Data are shown with a five-component model fitted by least squares to the data, consisting of cosine curves with the five largest amplitudes. A circadian rhythm cannot be discerned by the naked eye (*right lower part*)

20.8 Riser (Reverse-Dipping) Hypertension

In this chapter we introduce two cases of "riser (reverse-dipping) hypertension" from the viewpoint of 7-day/24-hour ABP monitoring. A "riser" (or "reverse dipper") is a subject whose day-night ratio of SBP is negative (below 0 %).

20.8.1 Case Report 1 of "Riser Hypertension"

The first example of "riser hypertension" is a 75-year-old nonobese (BMI = 22.8 kg/m^2) woman (ID number: KOts085). She came for a health check, especially in relation to her BP. She goes to bed at 23:00 and gets up regularly at 06:30, but she does not sleep well. Her office BP was 170/113 mmHg in the absence of antihypertensive medication.

Based on the model fitted to the 7-day/24-hour ABP monitoring record, systolic BP (SBP), diastolic BP (DBP), and heart rate measured by pulse (HR) MESOR estimates were 159.1 mmHg, 91.1 mmHg, and 66.7 bpm, respectively. The double 24-hour amplitudes of SBP, DBP, and HR were 25.26 mmHg, 6.40 mmHg, and 13.21 bpm, respectively. Corresponding circadian acrophases were 3:13, 3:59, and 14:27, indicative of SBP and DBP ecphasia (odd timing of the circadian rhythm of BP but not of HR).

The circadian profiles of SBP and HR are shown in Fig. 20.50. SBP varies close to the upper chronodesmic limit during the day and markedly exceeds this limit at night. SBP is higher during the night than during the day. This ("reverse-dipping") pattern is that of a "riser." HR varies in the lower portion of the chronodesm, showing a usual circadian variation. As the difference between the MESORs of SBP (159.1 mmHg) and DBP (91.1 mmHg) is larger than 60 mmHg (159.1–91.1 = 68.0 mmHg), she is also diagnosed as having an "excessive pulse pressure (EPP)." Thus, she has four vascular variability anomalies (VVAs), namely, SBP MESOR-hypertension, DBP MESOR-hypertension, EPP, and BP ecphasia.

Time series of SBP, DBP, and HR are shown in Fig. 20.51. From Friday to Thursday, daily differences between daytime and nighttime averages of SBP assume values of −4.6 (153.6–158.2) mmHg (Friday), −55.5 (146.8–202.3) mmHg (Saturday), −3.4 (158.9–162.3) mmHg (Sunday), −28.8 (157.0–185.8) mmHg (Monday), −21.0 (144.3–165.3) mmHg (Tuesday), 15.7 (160.7–145.0) mmHg (Wednesday), and −48.6 (132.4–181.0) mmHg (Thursday). A reverse-dipping pattern is present during 6 of the 7 days of monitoring.

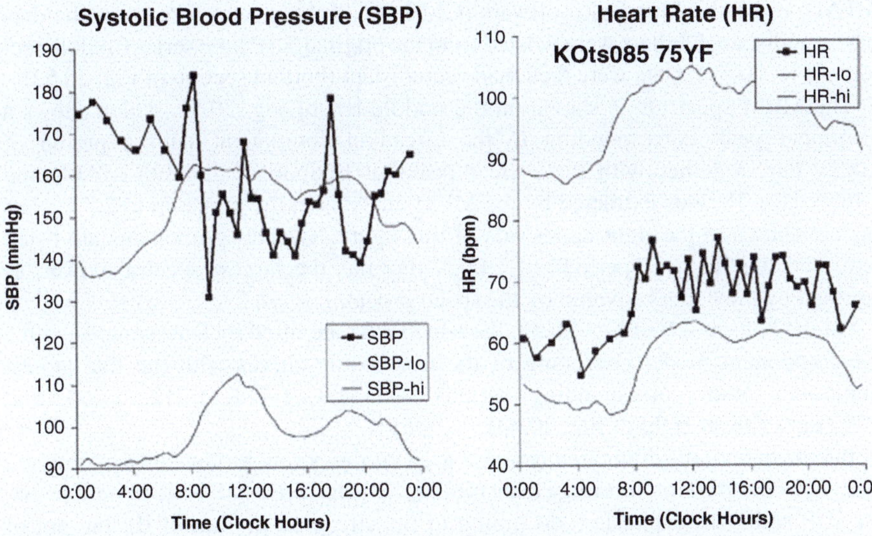

Fig. 20.50 Circadian profiles of systolic BP (*left*) and heart rate (HR) (*right*), averaged over the 7 days of monitoring (*polygonal line* graph with closed *squares*), are shown along with time-specified prediction limits (*continuous thin lines*). SBP is elevated most of the time, but particularly at night when it is higher than during the day, in contrast to HR, indicative of reverse-dipping ("riser hypertension")

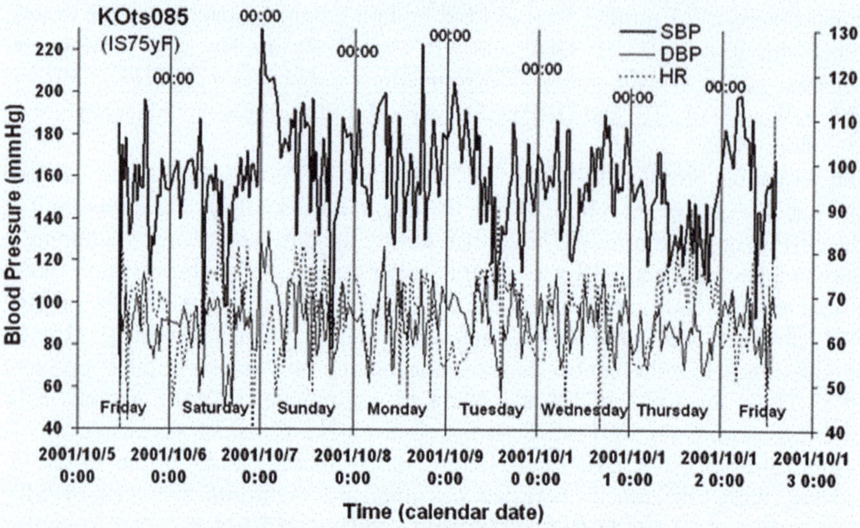

Fig. 20.51 An example of "riser hypertension" from the viewpoint of 7-day/24-hour ABP monitoring. Time plots of the 7-day/24-hour records of SBP (*thick continuous line*), DBP (*thin continuous line*), and HR (*dashed line*) of a 75-year-old woman

The circadian period of SBP was estimated by the Maximum Entropy Method (MEM), using the MemCalc software (GMS Co., Tokyo), which requires the data to be equidistant. Outliers were deleted from the original SBP time series (middle section of Fig. 20.52). Data were then made equidistant (bottom section of Fig. 20.52).

The MEM spectrum is shown on the middle left of Fig. 20.53, which shows a prominent peak corresponding to the circadian component with a period of 1.0029 day, together with three clear peaks at periods of 5.9689, 1.7437, and 0.4686 days. Periods extracted by the MEM analysis in the range up to 2.5 cycles/day are shown in the right upper side of this figure. Twelve components are listed here, and that with a "period" of 1.0029 day has the largest spectral power, as assessed by the "area" covered by the spectral peak.

On the bottom left of Fig. 20.53, the edited data are fitted by least squares with a five-component model consisting of the five cosine curves with the five largest amplitudes, with corresponding periods of 5.969, 1.744, 1.212, 1.003, and 0.725 days (lower right of Fig. 20.53).

In conclusion, the clinical diagnosis based on the chronobiologically interpreted 7-day/24-hour ABP monitoring consists of (1) SBP and DBP MESOR-hypertension, (2) SBP and DBP ecphasia (odd timing of the circadian rhythm of BP but not of HR), (3) persistent ABP hypertension, (4) "riser" hypertension (reverse-dipping pattern), and (5) "excessive pulse pressure (EPP)." In view of her severe "riser" hypertension coexisting with other VVAs, she is diagnosed as having a "Vascular Variability Syndrome" (VVS), i.e., multiple coexisting VVAs. Accordingly, we immediately started a combination therapy (CCB and ARB) and made lifestyle recommendations, focusing on salt intake and sleeping habits. Under this treatment

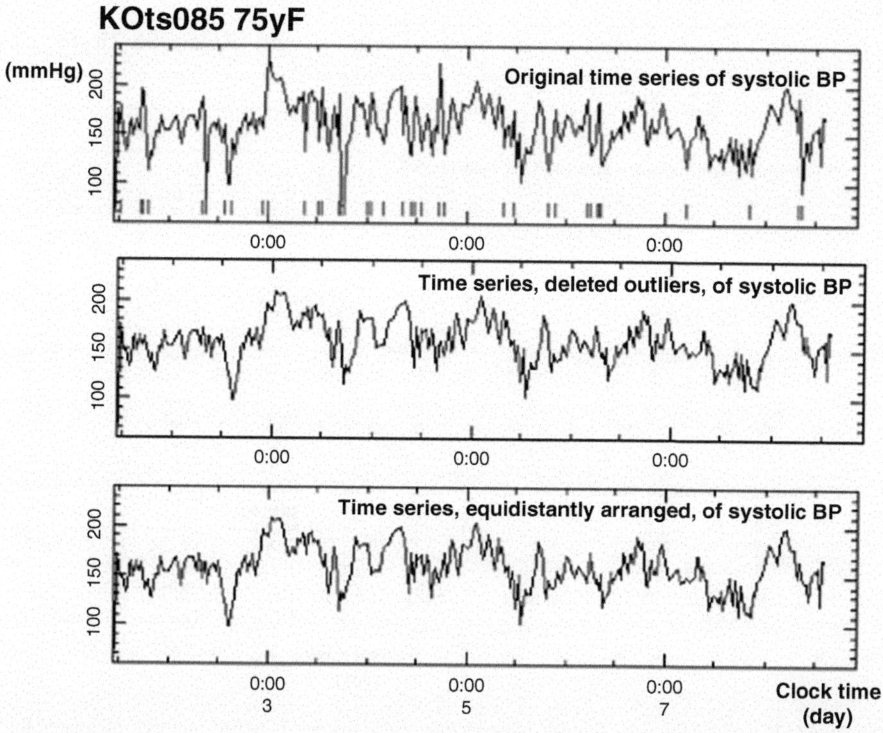

Fig. 20.52 Three versions of the time series of systolic BP. *Upper*: Original data of systolic BP. *Middle*: Edited data after mathematically excluding outliers. *Lower*: Equidistantly arranged time series for analysis by the Maximum Entropy Method, using the MemCalc software (GMS Co., Tokyo)

plan, this patient did well for 4 years, but suffered from cerebral bleeding and passed away 5 years after this monitoring.

20.8.2 Case Report 2 of "Riser Hypertension"

The second example of "riser hypertension" is a 65-year-old nonobese (BMI=22.1 kg/m^2) woman (ID number: KOts099). Her office BP was 172/99 mmHg while she was on four antihypertensive medications, including CCB and ARB. She goes to bed at 22:00, sleeps well and gets up regularly at 05:00. She does not smoke and usually does not drink alcohol. She prefers light seasoning and takes care not to take too much salt.

She feels fairly healthy and takes care of her health. She has something to live for and does not have a depressive mood. Her scores on the GDS-15 and PHQ-2 scales were 5 and 0 points, respectively. Her subjective QOL indices of health, mood, family relationships, relationships with friends, financial satisfaction, life satisfaction, and happiness, assessed by VAS, were 67 %, 64 %, 94 %, 73 %, 65 %, 77 %, and 74 %, respectively.

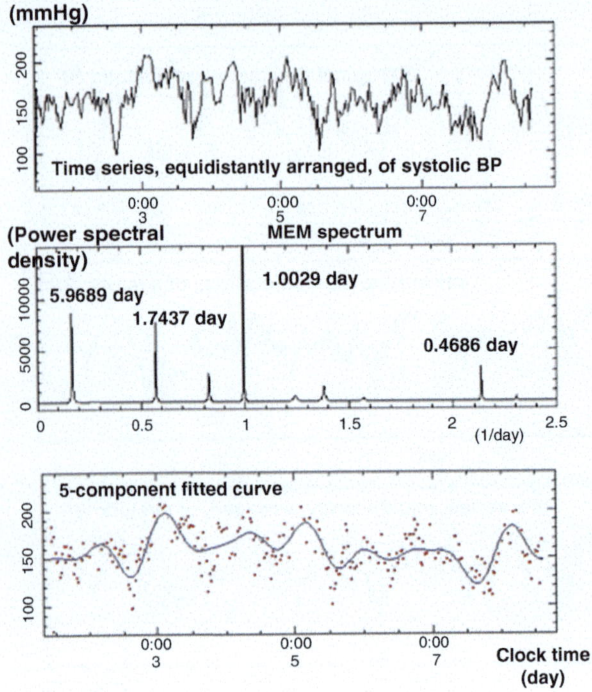

KOts085 75yF

Frequency (/ day)	Period (days)	Area
0.168	5.9689	62.5301
0.277	3.6038	7.8107
0.574	1.7437	42.6148
0.825	1.2121	25.5115
0.997	1.0029	81.2342
1.240	0.8067	13.9454
1.379	0.7254	27.2414
1.566	0.6384	9.1101
1.777	0.5627	4.3817
1.907	0.5245	2.3408
2.134	0.4686	17.7455
2.304	0.4341	4.7602

Period (days)	Amplitude	Acro-phase
5.969	11.64	4-01:40
1.744	9.34	1-11:25
1.212	5.39	1-23:09
1.003	12.84	2-03:05
0.725	6.92	1-13:50

Fig. 20.53 Period analysis of the circadian rhythm of systolic BP. MEM analysis estimates the circadian period of systolic BP to be 1.0029 day (*left middle part*). Data are shown with a five-component model fitted by least squares to the data, consisting of cosine curves with the five largest amplitudes, the 1.003-day component being the most prominent (*right lower part*)

Based on the model fitted to the 7-day/24-hour ABP monitoring record, MESOR estimates of systolic BP (SBP), diastolic BP (DBP), and heart rate measured by pulse (HR) were 153.5 mmHg, 85.8 mmHg, and 61.2 bpm, respectively. As the difference between the MESORs of SBP (153.5 mmHg) and DBP (85.8 mmHg) is larger than 60 mmHg (153.5–85.8 = 67.7 mmHg), she is also diagnosed as having an "excessive pulse pressure (EPP)." The double 24-hour amplitudes of SBP, DBP, and HR were 12.47 mmHg, 6.01 mmHg, and 10.55 bpm, respectively, suggesting a small day-night difference for BPs and HR. In addition, corresponding circadian acrophases were 07:15, 11:04, and 15:03, indicative of both SBP and DBP ecphasia (odd timing of the circadian rhythm of BP but not of HR).

The circadian profiles of SBP and HR are shown in Fig. 20.54. SBP varies close to the upper limit of the chronodesm and markedly exceeds this limit during the night when it is higher than during the day. This is the pattern of a "riser." HR varies close to the lower limit of the chronodesm and undergoes a clear circadian variation, albeit of small amplitude.

Time series of SBP, DBP, and HR are shown in Fig. 20.55. From Wednesday to Tuesday, daily differences between the daytime and nighttime averages of SBP assumed values of −8.5 (157.0–165.4) mmHg (Wednesday night; riser), −4.2 (156.6–

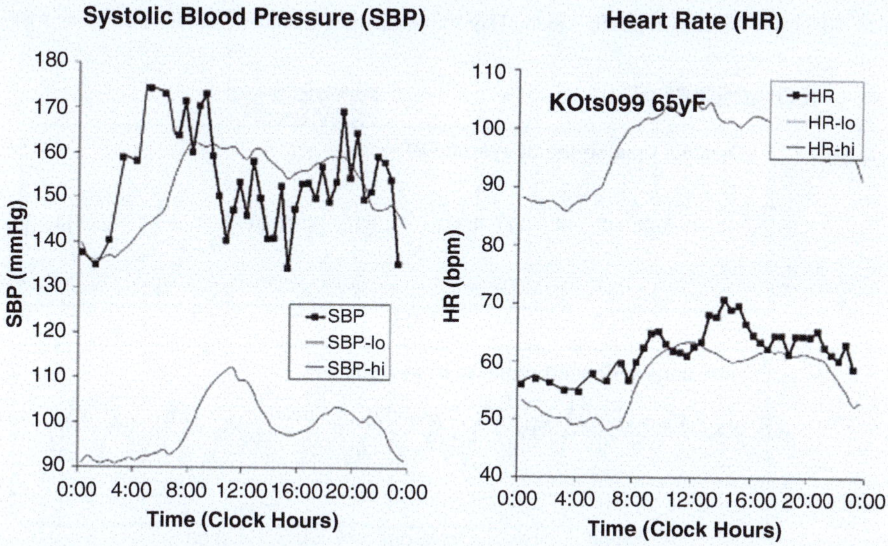

Fig. 20.54 Circadian profiles of systolic BP (*left*) and heart rate (HR) (*right*), averaged over the 7 days of monitoring (*polygonal line* graph with *closed squares*), are shown along with time-specified prediction limits (*continuous thin lines*). SBP is elevated most of the time, particularly during the night, when it markedly exceeds the upper limit of the chronodesm, indicative of "riser hypertension" ("reverse-dipping pattern")

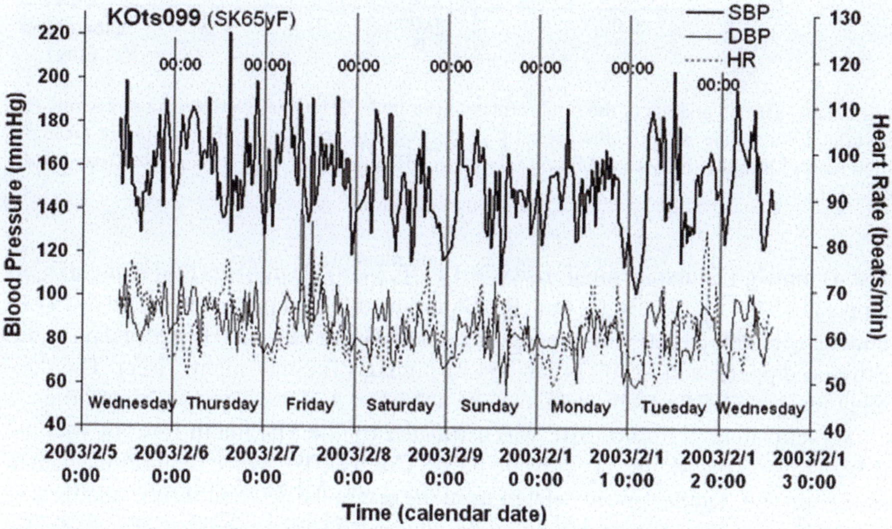

Fig. 20.55 Another example of "riser hypertension" from the viewpoint of 7-day/24-hour ABP monitoring. Time plots of the 7-day/24-hour records of SBP (*thick continuous line*), DBP (*thin continuous line*), and HR (*dashed line*) of a 65-year-old woman. The day-night ratio of SBP changes from day to day, corresponding to patterns of a riser (Wednesday), riser (Thursday), non-dipper (Friday), dipper (Saturday) non-dipper (Sunday), extreme dipper (Monday), and riser (Tuesday). She is also diagnosed with "MESOR-hypertension" and "excessive pulse pressure (EPP)." The circadian acrophase of SBP was 07:15, suggesting SBP ecphasia (odd timing of the circadian rhythm of SBP but not of HR)

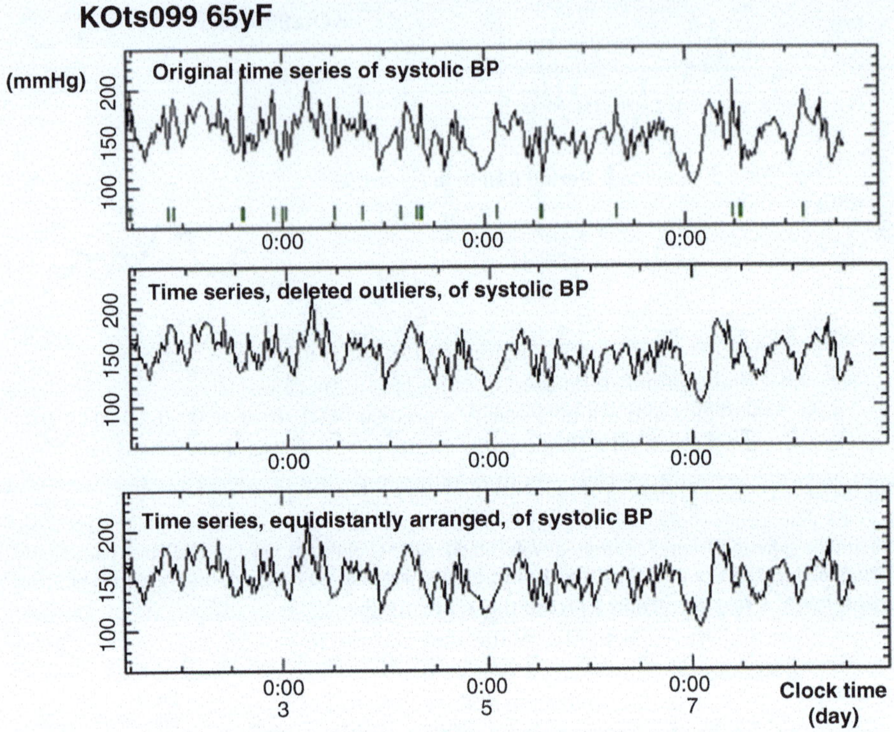

Fig. 20.56 Three versions of the time series of systolic BP. *Upper*: Original data of systolic BP. *Middle*: Edited data after mathematically excluding outliers. *Lower*: Equidistantly arranged time series for analysis by the Maximum Entropy Method, using the MemCalc software (GMS Co., Tokyo)

152.4) mmHg (Thursday night; riser), 8.1 (156.3–148.2) mmHg (Friday night; non-dipper), 15.7 (152.3–136.6) mmHg (Saturday night; dipper), 8.5 (149.5–141.0) mmHg (Sunday night; non-dipper), 32.4 (150.0–117.6) mmHg (Monday night; extreme dipper), and −5.0 (149.8–154.8) mmHg (Tuesday night; riser). The day-night ratio of SBP finds this subject to be a riser on 4 of the 7 days of monitoring.

The circadian period of SBP was estimated by the Maximum Entropy Method (MEM), using the MemCalc software (GMS Co., Tokyo), which requires the data to be equidistant. Outliers were deleted from the original SBP time series (middle section of Fig. 20.56). Data were then made equidistant (bottom section of Fig. 20.56).

The MEM spectrum is shown on the middle left of Fig. 20.57. Although there is a clear peak corresponding to the circadian component, with a period of 1.0506 day, three other peaks with periods of 8.4521, 0.4961, and 0.4616 days are more prominent. Periods extracted by the MEM analysis in the range up to 2.5 cycles/day are shown in the right upper side of this figure. Ten components are listed here, and that with a "period" of 8.4521 days has the largest spectral power, as assessed by the "area" covered by the spectral peak.

On the bottom left of Fig. 20.57, the edited data are fitted by least squares with a five-component model consisting of the five cosine curves with the five largest amplitudes, with corresponding periods of 8.452, 1.051, 0.681, 0.496, and 0.462 days (lower right of Fig. 20.57).

In conclusion, the clinical diagnosis based on the chronobiologically interpreted 7-day/24-hour ABP monitoring consists of (1) SBP MESOR-hypertension, (2) SBP ecphasia (odd timing of the circadian rhythm of BP but not of HR), (3) persistent ABP hypertension, (4) "riser" hypertension (reverse-dipping pattern), and (5) "excessive pulse pressure (EPP)." She is diagnosed with a "Vascular Variability Syndrome" since more than a single abnormality is detected. Accordingly, we immediately increased the dose of CCB and ARB medications, and we strictly recommended lifestyle modifications, focusing on salt intake and sleeping habits. Under this treatment plan, she now seems to be in good health.

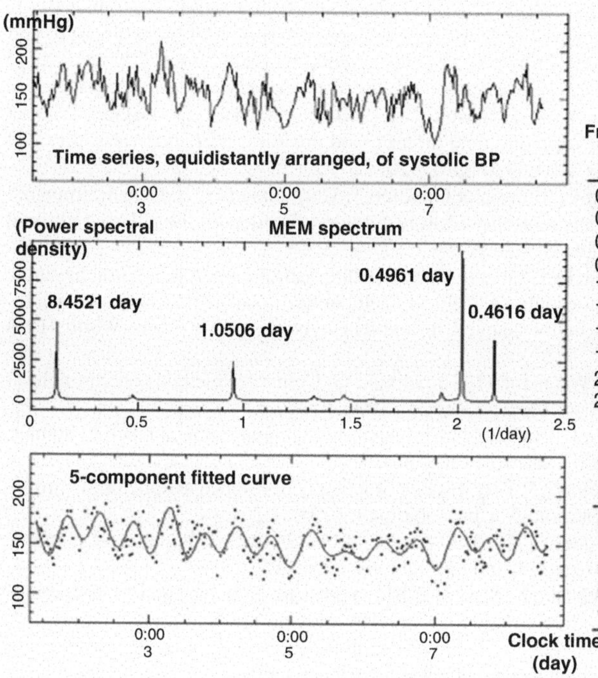

KOts099 65yF

Frequency (/ day)	Period (days)	Area
0.118	8.4521	42.2179
0.476	2.0987	6.1789
0.638	1.5685	0.4561
0.952	1.0506	26.8899
1.328	0.7532	7.8863
1.468	0.6811	11.9085
1.597	0.6263	3.3728
1.922	0.5204	11.8611
2.016	0.4961	68.1876
2.116	0.4616	20.8439

Period (days)	Amplitude	Acro-phase
8.452	9.97	1-19:23
1.051	7.13	2-04:58
0.681	5.09	1-22:37
0.496	12.53	1-19:13
0.462	4.57	1-20:47

Fig. 20.57 Period analysis of the circadian rhythm of systolic BP. MEM analysis estimates the circadian period of systolic BP to be 1.0506 day (*left middle*), but the circasemidian component with a period of 0.4961 day is the most prominent. Data are shown with a five-component model fitted by least squares to the data, consisting of cosine curves with the five largest amplitudes, the 0.496-day component being the most prominent (*right lower*)

References

1. Otsuka K, Watanabe H. Circadian amplitude of diastolic blood pressure gauges family history of high blood pressure in nurses. Chronobiologia 1988;15:266–7.
2. Otsuka K, Kitazumi T, Matsubayashi K, Kawamoto A, Sadakane N, Chikamori T, Kuzume O, Shimada K, Ogura H, Ozawa T. Age-related alterations in the circadian pattern of blood pressure. Am J Noninvas Cardiol 1989;3:159–65.
3. Otsuka K, Cornelissen G, Aslanian N, Helou B, Watanabe H, Holte JE, Halberg F. Circadian period of human blood pressure and heart rate in clinical health under ordinary conditions. In: Proceedings of the 2nd Annual IEEE Symposium on Computer Based Medical Systems. Washington, DC: IEEE Computer Society Press; 1989. pp. 206–19.
4. Otsuka K, Watanabe H, Cornelissen G, Shinoda M, Uezono K, Kawasaki T, Halberg F. Gender, age and circadian blood pressure variation of apparently healthy rural vs. metropolitan Japanese. Chronobiologia 1990;17:253–65.
5. Nakajima S, Otsuka K, Yamanaka T, Omori K, Kubo Y, Toyoshima T, Watanabe Y, Watanabe H. Ambulatory blood pressure and postprandial hypotension. Am Heart J 1992; 124(6):1669–71.
6. Otsuka K, Halberg F. Circadian profiles of blood pressure and heart rate of apparently healthy metropolitan Japanese. Front Med Biol Eng 1994;6(2):149–55.
7. Otsuka K, Cornélissen G, Matsuoka O, Kubo Y, Murakami S, Shinagawa M, Yamanaka T, Nishimura Y, Hotta N, Nunoda S-i, Ohkawa S-i, Halberg F. Mapping of blood pressure and heart rate variability: a model for pediatricians. Neuroendocrinol Lett 2003;24 Suppl 1:157–64.
8. Mitsutake G, Otsuka K, Oinuma S, Ferguson I, Cornélissen G, Wanliss J, Halberg F. Does exposure to an artificial ULF magnetic field affect blood pressure, heart rate variability and mood? Biomed Pharmacother 2004;58 Suppl 1:S20–7.
9. Otsuka K, Hotta N, Takasugi E, Yamanaka G, Kubo Y, Cornélissen G, Halberg F. Impaired cardiovascular function predicts mortality from respiratory disorders in the elderly [editorial comment]. Hypertens Res 2008;31:589–91.
10. Nakaoka T, Norboo T, Kawasaki T, Okumiya K, Matsubayashi K, Singh RB, Cornelissen G, Halberg F, Otsuka K. Chronoecological "glocal" (global and local) comprehensive health-watch at high altitude: men better adapted to ambient hypoxia than women. World Heart J 2012;4(2/3):165–75.
11. Otsuka K, editor. Halberg F, Watanabe H, Cornelissen G. Computer methods on chronobiology and chronomedicine. Tokyo: Medical Review; 1992. p. 297.
12. Otsuka K, editor. Cornelissen G, Halberg F. Chronocardiology and chronomedicine: humans in time and cosmos. Tokyo: Life Science; 1993. p. 138.
13. Otsuka K, Cornélissen G, Halberg F. Broad scope of a newly developed actometer in chronobiology, particularly chronocardiology. Chronobiologia 1994;21:251–64.
14. Otsuka K, Cornélissen G, Halberg F. Predictive value of blood pressure dipping and swinging with regard to vascular disease risk. Clin Drug Investig 1996;11:20–31.
15. Suzuki M, Guilleminault C, Otsuka K, Shiomi T. Blood pressure "dipping" and "non-dipping" in obstructive sleep apnea syndrome patients. Sleep 1996;19(5):382–7.
16. Otsuka K, Cornelissen G, Halberg F. Circadian amplitude-hypertension as a risk factor of ischemic stroke and nephropathy. Vrach 1997;4:9–11 [In Russian].
17. Otsuka K, Cornélissen G, Halberg F, Oehlert G. Excessive circadian amplitude of blood pressure increases risk of ischemic stroke and nephropathy. J Med Eng Technol 1997;21:23–30.
18. Shinagawa M, Otsuka K, Murakami S, Kubo Y, Cornélissen G, Matsubayashi K, Yano S, Mitsutake G, Yasaka K, Halberg F. Seven-day (24-h) ambulatory blood pressure monitoring, self-reported depression and quality of life scores. Blood Press Monit 2002;7(1):69–76.
19. Otsuka K, Cornélissen G, Schwartzkopff O, Bakken EE, Halberg F, Burioka N, Katinas GS, Kane R, Regal PJ, Schaffer E, Sonkowsky R, Patterson R, Engebretson M, Brockway B, Wang ZR, Delmore P, Halpin C, Sarkozy S, Wall D, Halberg J. Clinical chronobiology and chronome-

geriatrics: at variance with recommendations of subsequent guidelines, yet focusing indeed on pre-hypertension in the physiological range. Biomed Pharmacother 2003;57 Suppl 1:164s–98.

20. Otsuka K, Yamanaka G, Shinagawa M, Murakami S, Yamanaka T, Shibata K, Yano S, Ishizuka S, Singh RB, Cornélissen G, Halberg F. Chronomic community screening reveals about 31 % depression, elevated blood pressure and infradian vascular rhythm alteration. Biomed Pharmacother 2004;58 Suppl 1:48s–55.
21. Otsuka K, Halberg F, Cornélissen G, Katinas G, Singh RB. Preface. Chronomics based on long-term vascular monitoring leads to scientific and clinical findings. Clin Exp Hypertens 2004;27(2–3):1–5.
22. Murakami S, Otsuka K, Kubo Y, Shinagawa M, Yamanaka T, Ohkawa S, Kitaura Y. Repeated ambulatory monitoring reveals a Monday morning surge in blood pressure in a community-dwelling population. Am J Hypertens 2004;17:1179–83.
23. Murakami S, Otsuka K, Kubo Y, Shinagawa M, Matsuoka O, Yamanaka T, Nunoda S-I, Ohkawa S-I, Kitaura Y. Weekly variation of home and ambulatory blood pressure and relation between arterial stiffness and blood pressure measurements in community-dwelling hypertensives. Clin Exp Hypertens 2005;27(2–3):231–9.
24. Otsuka K. Seven-day (24-hour) ambulatory blood pressure monitoring and frequently observed day-to-day differences in the elderly. Jpn J Geriatr 2009;46(6):488–92.
25. Murakami S, Otsuka K, Kono T, Soyama A, Umeda T, Yamamoto N, Morita H, Yamanaka G, Kitaura Y. Impact of outdoor temperature on prewaking morning surge and nocturnal decline in blood pressure in a Japanese population. Hypertens Res 2011;34:70–3.
26. Otsuka K, Cornelissen G. Alcohol consumption and vascular variability anomalies (VVAs) and disorders (VVDs) detected bu C-ABPM. Open Nutraceuticals J 2012;5(Suppl 1-M1):81–9.
27. Otsuka K, Okajima K, Yamanaka T, Oinuma S, Taniwaki Y, Otsuka K, Cornelissen G. Aging and the novelty pressor effect in men on the first day of 7-day/24-hour ambulatory blood pressure monitoring. J Am Geriatr Soc 2014;62:1602–5.
28. Otsuka K, Okajima K, Oinuma S, Sasaki J, Yamanaka T, Takahashi Y, Kokaji A, Kitamura K, Cornelissen G. Aging and circadian disruption of blood pressure observed using 7-day/24-hour ambulatory blood pressure monitoring. J Am Geriatr Soc 2014;62:2213–5.
29. Okajima K, Otsuka K, Oinuma S, Sasaki J, Yamanaka T, Takahashi Y, Cornelissen G. Aging and within- and between-day variability assessed using 7-day/24-hour ambulatory blood pressure monitoring. J Am Geriatr Soc 2014;62:2440–2.
30. Okajima K, Yamanaka G, Oinuma S, Kikuchi T, Yamanaka T, Otsuka K, Cornelissen G. Even mild depression is associated with among-day blood pressure variability, including masked non-dipping assessed by 7-day/24-hour ambulatory blood pressure monitoring. Clin Exp Hypertens 2015;35:257–66.

Chapter 21
"What Is Life?" Cycles Tipping the Scale Between Death and Survival

Abstract Around and in living matter, a system of transdisciplinary cycles with common average periods is being quantified with point and interval estimates of parameters. The cycles' characteristics constitute useful information once their uncertainties are assessed. The cycles in space climate are critical in discussing global warming. The cycles' periods are described as congruent when their CIs (95 % confidence intervals) overlie or overlap and the amplitudes' CIs' lower limits are positive. Some congruent cycles in organisms, counterparts of the environmental day and the seasons, relate to electromagnetic radiation in the visible domain; these are the usually environmentally synchronized socio-photo-thermoperiodisms (photics). The biosphere also resonates with or is pulled or driven by nonstationary, environmental nonphotic cycles (nonphotics) – particle emissions from the sun and the wider cosmos, cosmoheliogeomagnetics, ultraviolet flux, gravitation, acoustics, and others.

Nonphotics, a set of in part transdisciplinarily novel spectral components, can be intermittent; when present, they coexist and compete with signatures of photic cycles, monitored, e.g., in blood pressure (BP) and heart rate (HR) for vascular disease detection and prevention. Nonphotics of, e.g., about (~) 1 week (circaseptans), ~17 months (transyears), characterize mood and performance, modulate, and sometimes override society's (photic) schedules, even in dying suddenly unintentionally or by one's own will or at the hand of others. Nonphotics persist but are damped in physiology or in terrorism when their environmental counterpart is not detected.

Biospheric monitoring complements the records from purely physical tools for surveilling the variable sun, validating in the biosphere the reality of intermittent, aeolian environmental spectral components that can be more consistent than their physical counterparts once they are coded in genes, indicating the relevant associations of space weather with human health and ecology. We invite readers, as a dividend from self-surveillance for preventive health care, to monitor, on themselves, solar variability, a transdisciplinary gauge of individuals' and societies' health and disease, including violence, notably terrorism.

Keywords Aeolian • Chronomics • Congruence • Transdisciplinary

K. Otsuka et al., *Chronomics and Continuous Ambulatory Blood Pressure Monitoring*, DOI 10.1007/978-4-431-54631-3_21

21.1 Introduction

These complementary materials extend the scope of a presentation on "What is life?" [1], prompted, in the wake of Erwin Schrödinger's book of the same title [2] by Schrödinger's colleague, the physicist Arthur March [3], to whose memory this chapter is dedicated. In 1946, we answered "What is life?" with "periodic waves," and we do so again in 2007 [4] with observations from the senior author's (FH) lifetime and that of many others. Further background to this summary, presented at the Hideki Yukawa 100-year Memorial Seminar "What is life?," at the Yukawa Institute of Theoretical Physics in Kyoto, Japan, October 15–20, 2007, is found in cartography, clinical and basic, in an invited presentation, under the title "60 years of translational chronobiology," to open a symposium on "circadian rhythms from animals to humans" at Experimental Biology 2007 in Washington DC on April 29, 2007 [4], earlier at the European Geosciences Union General Assembly in Vienna, April 15–20, 2007 [5] and later at the International Academy of Science symposium on "Natural cataclysms and global problems of modern civilization," Baku, Azerbaijan, September 24–27, 2007 [4], and at the VII International Crimean Conference "Cosmos and biosphere," Sudak, Ukraine, October 1–6, 2007 [6].

21.2 Preamble

Suitbert Ertel [7] summarized a schism between East and West that remains pertinent today:

> At an Annual Meeting of the American Meteorological Society, 1926, an American participant delivered a paper written by Alexander L. Chizhevsky, a Russian scholar. Its bombastic title ("The influence of cosmic factors upon the behavior of organized human masses as well as upon the universal historical process") appeared laughable. The author claimed that occurrences of social unrest, rebellions, upheavals, revolutions are significantly correlated with solar activity, i.e., with the ups and downs of magnetic turbulence of the sun….
>
> Chizhevsky was imprisoned by the KGB from 1942–1958. His claim was oppressed; in Western countries, it was ridiculed or simply ignored. There are exceptions. At least one historian, William S. Langer, President of the American Historical Association and outstanding figure in his field, suggested that "We may … be on the threshold of a new era when the historian will have to think … even in cosmic terms" [7].

As of 2007, there are indeed many exceptions to Ertel's [7] claim that Western science dismisses Chizhevsky's theories (Astbury, personal communication; [8–42]). In post-Soviet Russia, Chizhevsky is revered, with a most recent meeting in Sudak, Crimea, Ukraine, dedicated to the 110th anniversary of his birth [43].

In conclusion, his review with meta-analyses in support of Chizhevsky's theses concerning not only "infections" of the mind but also epidemics ([44–47]; cf. [7], [14]) notes that "researchers in chronobiology/chronomedicine and in biogeomagnetics … are not far from setting the stage."

Negative findings, however, have been made in the West with impeccable methods on large databases (Table 21.1) albeit over hardly more than a single sunspot cycle [48, 49]. The chronobiologic stage was perhaps set:

- When temporal and geographic differences could be documented in several mortality endpoints investigated (Table 21.2, Fig. 21.1a, b) [50, 51]
- When the presence vs. absence of an about-7-day (circaseptan) component in solar activity was accompanied by an amplification vs. damping of a persistent

Table 21.1 The problem of heliobiology in the "West": negative results on millions of people in 2 North American papers[a]

1. From Feinleib et al. [48]:		
1964–1966 (3 years) daily; 1964–1971 (8 years) monthly		
Daily deaths		Estimated totals 1964–1971 (8 years)
All		14,493,120
Coronary heart disease		4,383,000
Stroke		1,577,880
2. From Lipa et al. [49]:		
1962–1966 (5 years) = 1826 days		
	Coronary heart disease	Stroke
Total (1962–1966)	2,750,880	1,002,246

[a]Totals estimated based on infradians in the original papers on highest and lowest numbers/day that led to an estimated value/day multiplied by the number of days [1] or based on weekly means [2]

Table 21.2 Differences among three consecutive solar cycles found by fit of a 10.5-year cosine curve to daily mortality from myocardial infarction in Minnesota[a]

Cycle N	P (A = 0)	Average ± SE (cases/day)	Amplitude (A) (95% CI) (cases/day)	Relative amplitude	Phase of maximum (95% CI) (degrees)[b]
1	0.006	12.0 ± 0.1	0.36 (0.18; 0.53)	0.030	−135 (−105; −164)
2	<0.001	12.4 ± 0.1	0.76 (0.55; 0.96)	0.061	−145 (−132; −158)
3	0.830	12.1 ± 0.1	0.07 ( )	0.006	−144 ( )
Test of equality of parameters					
Parameter(s)			ndf	F	P
Average			(2, 20)	11.296	0.0005
Amplitude (A)			(2, 20)	12.820	0.0003
Phase of maximum (ϕ)			(2, 20)	0.231	0.7954
(A,ϕ)			(4, 20)	6.762	0.0013
(M,A,ϕ)			(6, 20)	9.586	<0.0001

[a]In the data analyzed after removal of a linearly decreasing trend. Average computed as a rhythm-adjusted mean (MESOR, *m*idline-*e*stimating *s*tatistic *o*f *r*hythms); amplitude: measure of extent of predictable change within one cycle; phase of maximum expressed in (negative) degrees, with 360° equated to 10.5 yr; phase reference: 00:00 on July 1, 1967

[b]95% CI: 95% confidence interval

a **GEOGRAPHIC LOCATION and DIFFERENCES in OUTCOMES in PARTLY CONTEMPORANEOUS DATA* LIMITED TO FRACTIONS OF A SOLAR CYCLE; CONTROVERSY RESOLVED BY MINNESOTAN DATA** COVERING SEVERAL SOLAR CYCLES**

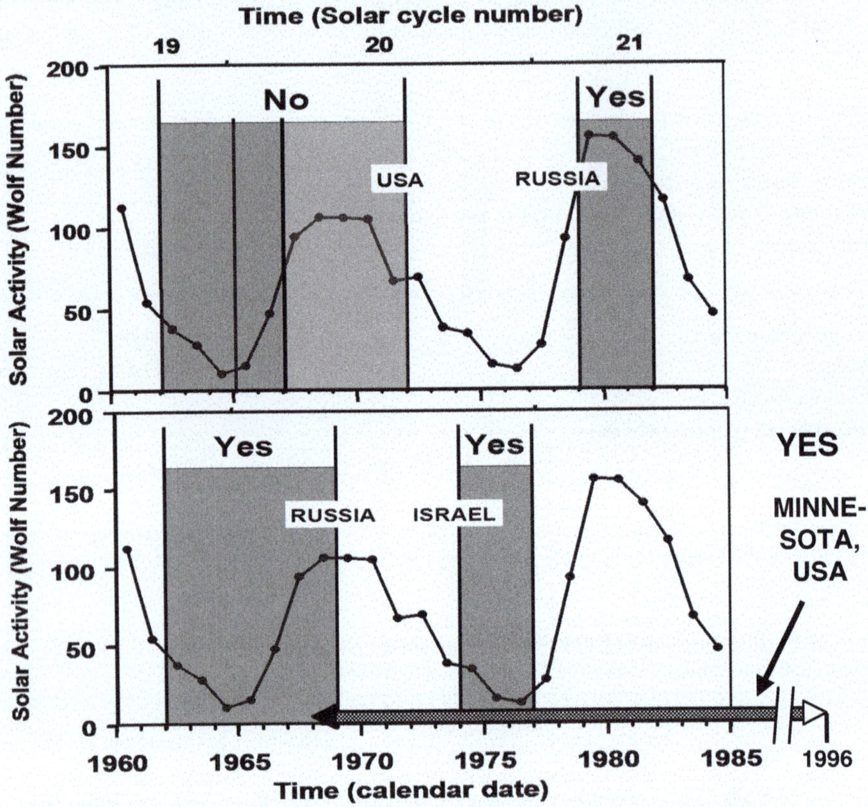

*Association of magnetic storms and mortality from myocardial infarction (MI) not detected in the USA (1962-1966 daily data and only monthly to 1971) but detected in Russia (including morbidity) in 1979-1981 (top) and earlier. *Bottom*: association also detected in 1962-1968 in Russia and in 1974 –1976 in Israel, at minima of Schwabe (~10.5-y) and Hale (~21-y) cycles, unless data in the late sixties in Russia contribute the outcome at the preceding minimum. **Minnesotan data over several solar cycles (1968 –1996) document excess of 220 MI/year during solar maxima vs. solar minima (*black horizontal arrow* along abscissa.

Fig. 21.1 (**a**) Resolution of a controversy by the recognition of time- and site-specific changes that in the light of accumulating evidence can be anticipated insofar as such variability in results is repeatedly documented. The analogy of blood pressure measurement over a single circadian cycle comes to mind: such a spotcheck may lead to a lifetime of unneeded treatment when, e.g., in several repeated visits to a care provider's office, a hypertensive patient may happen to be caught on several measurements (notably in the afternoon) with occasional normotensive values and hence will not be treated until a stroke occurs. By the same token, a mostly normotensive patient may be excited, the spotchecks may show "hypertension," and the patient may be treated unnecessarily for a lifetime. Furthermore, an individual with blood pressure overswinging limited to the normal range may not be recognized as being at a very high risk of stroke by limitation to a single cycle

b

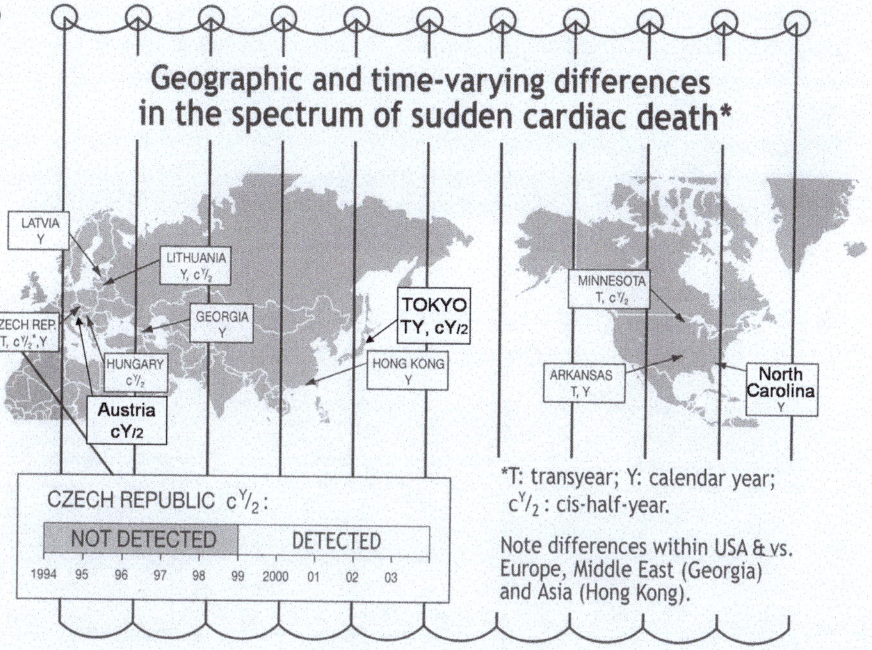

Fig. 21.1 (continued) or again he or she may be misdiagnosed and treated unnecessarily. Thus, studies that cover a single solar cycle may not be conclusive and may miss a relation that can be documented as being off and on aeolian elsewhere or over 2 of 3 solar cycles (in Minnesota, USA). Taking evidence over a single solar cycle is equivalent to taking the pulse for one second. The graph also shows geographic differences in the same solar cycle, as does Fig. 21.1b (© Halberg Chronobiology Center) (**b**) This figure shows initial limited data now further extended to additional geographic and temporal differences for the case of sudden cardiac death in the form of rhythmic patterns with different frequencies characterizing this condition at different times and in different geographic locations. A curtain of uncertainty, because of limited available data, hides any time- and geographic (geomagnetic or dip-magnetic) site specificity of various spectral aspects of sudden cardiac death. Thus, we find a transyear in Minnesota with a cis-half year (cY/2) and both a calendar year and a transyear in Arkansas and the Czech Republic: at the latter site a cis-half year, corresponding in length to an also-transient component of hard solar flares (Rieger et al., 1984; Ballester et al., 2002) is detected after but not before 1999 (Halberg et al., 2005). In 1998, Peter V. Minorsky reevaluated latitudinal differences in coconut palms' foliar spiral direction (FSD). He hypothesized that latitude-dependent biases in FSD, a non-Mendelian trait, will be associated with a temporally varying component of the earth's magnetic field. Earth currents, which are measurable in trees, bias the difference of auxin (or auxin transport proteins) in young embryos, such that left-handed palm trees (LHP) are produced preferentially in the Northern Hemisphere and right-handed ones (RHP) in the Southern Hemisphere. He built his observations on a previously collected database by Davis and Davis, revealing that the ratio of LHP and RHP/total was better correlated with magnetic (dip) latitude than with geographic or geomagnetic (centered dipole) latitude. Minorsky and Bronstein (2005) referred to this classical case of morphological antisymmetry in which dextral and sinistral forms are not inherited and are equally common within a species, proposing that Pc1-induced earth currents may bias the diffusion of morphogens in coconut palm embryos, thereby giving rise in asymmetries of FSD. Whether other geographic differences in sudden cardiac death may also relate to any magnetic latitude deserves scrutiny. It is noteworthy in any event that cardiac arrhythmias can also transiently reveal a transyear or a cis-half year, each in a different solar Schwabe cycle stage (Halberg et al. 2006a) (© Halberg Chronobiology Center)

circaseptan rhythm of HR in RBS, a clinically healthy subject documented by about five daily self-measurements made for several decades, in keeping with other decade-long series from four other individuals investigated (Fig. 21.2a, b) [52, 53]

- When the solar wind speed lost an about 17-month broad band transyear cycle and when, with a transyear lag, the same transyear was partly lost and the remainder damped in the systolic blood pressure of an elderly man (Fig. 21.3) [54].

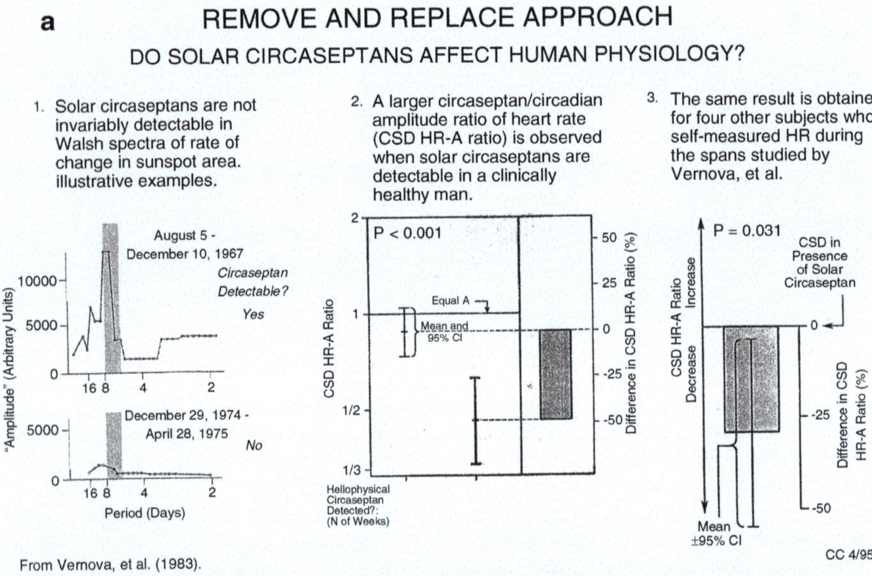

Fig. 21.2 (**a**) Average decrease in circaseptan-to-circadian amplitude ratio of heart rate for five subjects (*right*) is shown for one of the subjects (*middle*) during a span (V) when no solar circaseptan is detected in Walsh spectra vs. four other spans (I–IV) during which a component near 1 cycle in 6–8 days is detected in Walsh spectra (Vernova et al., 1983). The actual Walsh spectra of the rate of change in sunspot area during spans I (*top*) and V (*bottom*) are shown on the left. In data of subject RS (*middle*), amplitude ratios log-transformed to render the distribution closer to a normal one were computed during consecutive nonoverlapping adjacent 7-day intervals for comparison by Student's t-test. All data collected during and for one week after a trip associated with a transmeridian flight involving at least four time zones were excluded, as were data collected during illness (usually no more than a cold or flu). Even though nonoverlapping adjacent weekly intervals served to compute the circaseptan-to-circadian amplitude ratios, it may be argued that some lingering serial correlation may remain. The smaller amplitude ratio persists ($P<0.001$) when the Student's t-test is repeated using weekly intervals separated by 2-day gaps to reduce any serial correlation (© Halberg Chronobiology Center). (**b**) Assigning the average circaseptan amplitude of heart rate in each span for each subject (expressed as a percentage of the mean computed across all spans for that subject) to the midpoint of each span, the fit of an 11.6-year cosine curve to the total of 15 values pooled over the five subjects is statistically significant ($P=0.018$ for the circaseptan amplitude and $P=0.016$ for the circaseptan-to-circadian amplitude ratio), with maxima occurring around May 1971 and May 1970, respectively. A statistically significant 11.6-year component is also found for the circadian (rhythm-adjusted) mean ($P=0.031$) and for the circadian amplitude

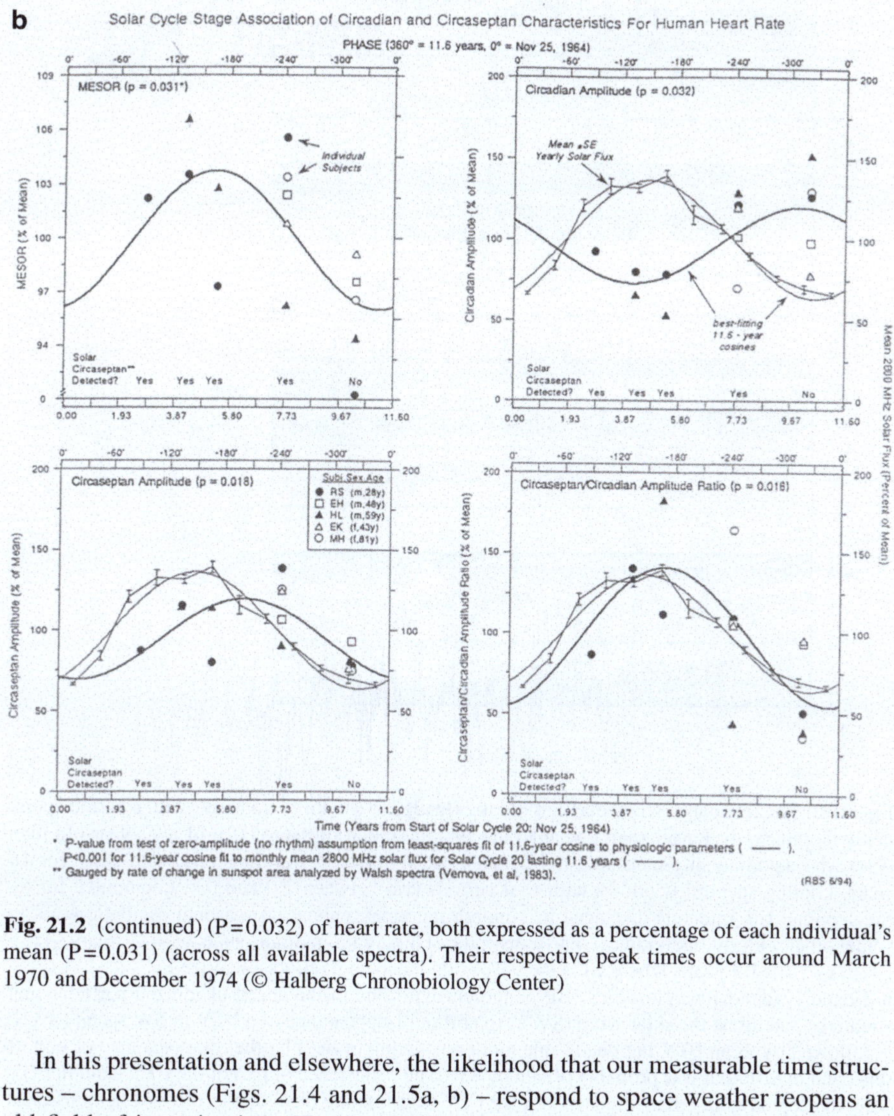

Fig. 21.2 (continued) (P=0.032) of heart rate, both expressed as a percentage of each individual's mean (P=0.031) (across all available spectra). Their respective peak times occur around March 1970 and December 1974 (© Halberg Chronobiology Center)

In this presentation and elsewhere, the likelihood that our measurable time structures – chronomes (Figs. 21.4 and 21.5a, b) – respond to space weather reopens an old field of investigations [7, 14, 43–45, 47–49], perhaps pertinent to setting the course of history, as Chizhevsky [44] put it insofar as the hard-to-define "good" and "bad," such as religious motivation and terrorism, are involved. With inferential statistics implemented by computers on time series that were not available in the precomputer, pre-satellite, pre-molecular biology era, methods now allow us to follow a tree of life [55] with a figurative microscopy and telescopy in time. We can say that nothing in transdisciplinary science can be fully understood without a record of the broad spectrum of cycles in the cosmos that constitutes a quantifiable (variable but not random) spectrum (Fig. 21.5b) and thus is a hardly dispensable background to any experiment or investigation in time, replacing fictitious

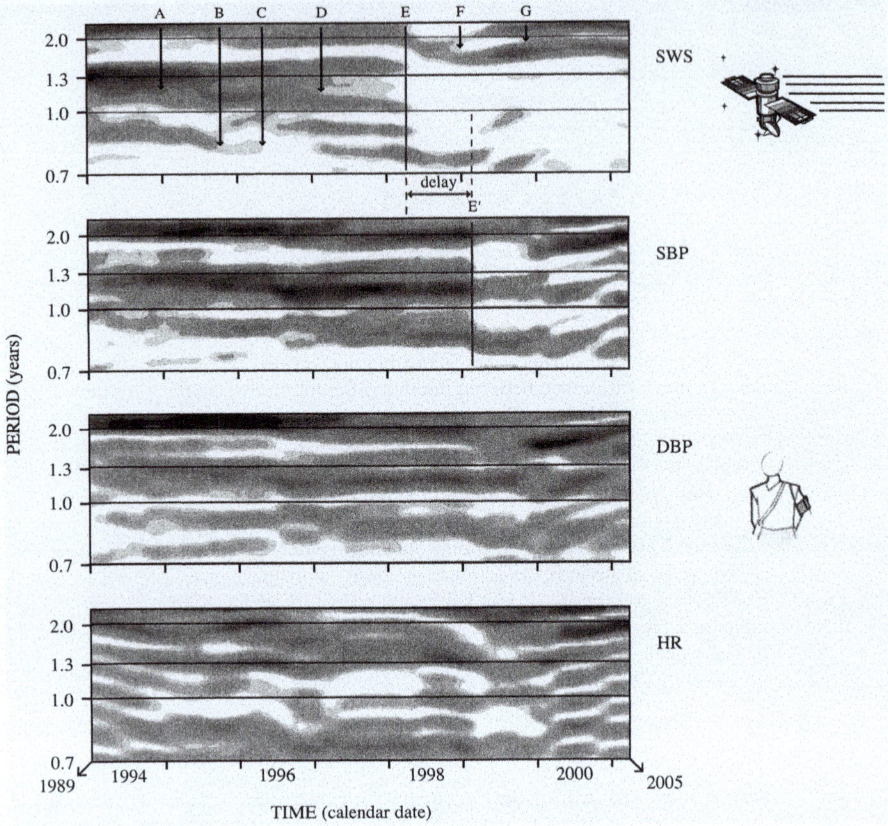

Fig. 21.3 The resonance with solar activity is extended to a wobbly transyear with a period statistically significantly longer than a year in the systolic blood pressure (BP) of a 72-year-old man (FH) who automatically measured his BP and HR at about 30-min intervals, with gaps, for 16 years. Gliding spectral windows with trial periods (τs) between 2.0 and 0.4 y computed for 8y overlapping data intervals displaced in 1-month increments show no seasonal oscillations, as in some other elderly individuals. Far-transyears (FTY, with periods such as 1.2 years $\leq [\tau - \text{CI}] < [\tau + \text{CI}] < 1.9$ years, where CI is the 95% confidence interval) characterize FH's data. A spectral band is found for systolic (S) BP congruent with one in the speed of ionized particles that stream by satellites: the solar wind (SW). An abrupt disappearance of a FTY in SW speed (SWS) is followed (with an FTY lag due to one more resonating cycle?) by the disappearance of part of this band in SBP that was probably driven by SW. The persistence of part of the FTY band in FH's SBP, in keeping with other evidence of a BP FTY free-running from that of the SW, demonstrates an evolutionary acquisition of a solar FTY oscillation that may reflect some of the sun's dynamics for billions of years past. The circumstance that the remaining part of the transyear band is damped is in keeping with the assumption that it was also driven by SW. Only minor coincident changes are seen in diastolic BP (DBP) or heart rate (HR) (bottom 2 sections). Aeolian rhythms in gliding spectra of SWS and SBP change in frequency (smoothly [A] or abruptly [B,C,D], bifurcating [D,F], and rejoining [G], they also change in amplitude (B)) (up to disappearing [C,E] and reappearing). During a nearly 16-year span, there are no consistent components with a period averaging precisely 1 year in the three physiologic variables, probably an effect of advancing age. The finding of FTY patterns, along with a cycle of about 5 months, i.e., a cis-half year in sudden cardiac death, in some but not all geographic locations (Fig. 21.1b) shows the importance of the earth in modulating solar effects. Chance and *post hoc ergo propter hoc* reasoning cannot be ruled out, but the circadian story is extended to infradians, beyond the fact, underlining their importance, that both components can account for the difference between life and death (Halberg et al. 54) (© Halberg Chronobiology Center)

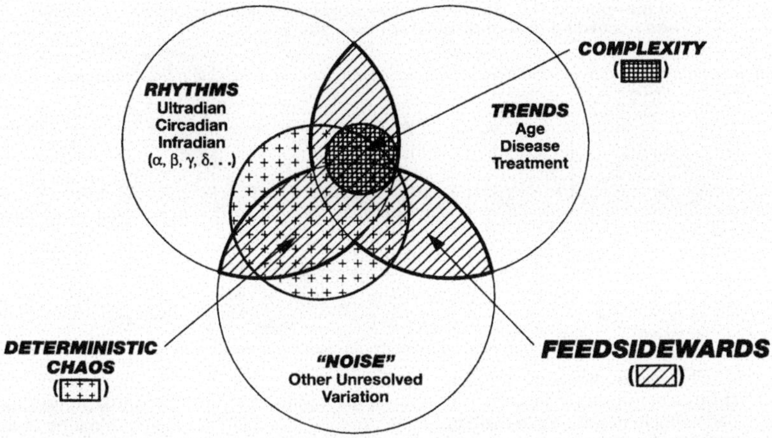

Fig. 21.4 Chronome: time structure consisting of multifrequency rhythms covering frequencies over 18 orders of magnitude, elements of deterministic and other chaos, trends in chaotic and rhythmic endpoints, and any other, as-yet unresolved variability (© Halberg Chronobiology Center)

spotcheck-based "baselines." Accordingly, we reconsider a statement attributed to Max Planck that the experiment carried out critically by a specialized researcher must yield the same result at each time and in each place. With respect to "experiments of nature," we follow, in BIOCOS, the reasons underlying the need for studying "the influence of the sun's emanations" on earth by an international effort given by Sir Edward Appleton, the discoverer of the ionosphere, in introducing a book by Harlan True Stetson [56]:

- First, the earth is round, not flat; hence, solar radiations impinge unequally in different regions.
- Second, the earth rotates and interesting solar features may occur at any time; hence, observers are needed at different terrestrial longitudes so that no effects are missed.
- Third, the earth is a large magnet in its own right, so that electrical particles from the sun are constrained to travel in curved tracks that may even have effects on the side of the earth's atmosphere further from the sun (at night). Magnetic qualities are very different in different locations on earth, and hence they modify solar effects with unequal intensity at different terrestrial latitudes.

With the use of these three points, Sir Edward was a master in inviting international cooperation to find information critical to communication science as well as for the fields of bioheliogeomagnetics or what has also been dubbed chronoastrobiology [15]. In part for the same fundamental reasons, there are geographic and temporal differences among the signatures of photic vs. nonphotic phenomena, e.g., in incidence patterns of natality and mortality as well as of events in between.

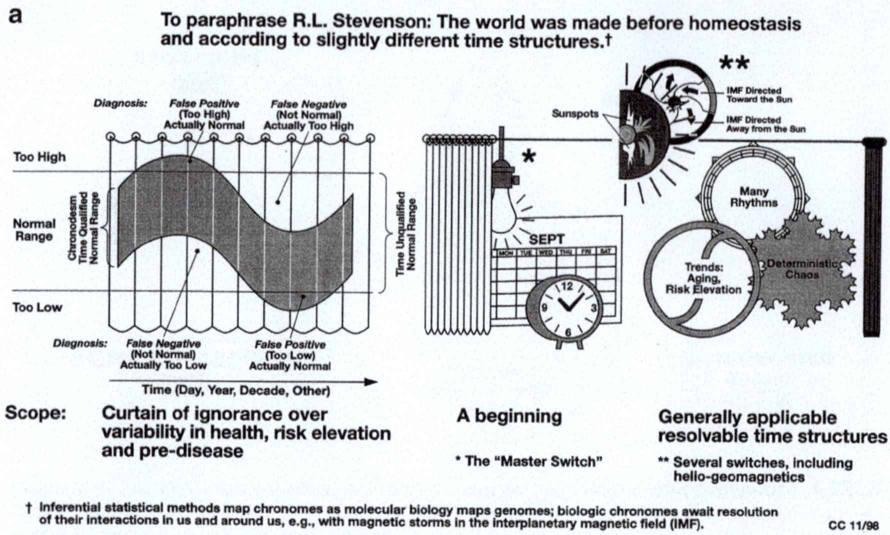

Fig. 21.5 (**a**) From the mistaken assumption of a master clock serving regulation for constancy, we turn to recognition of an integrative internal-external collateral hierarchy for physiological coordination. Homeostasis postulates that physiological processes remain largely within certain ranges in health and seeks departures from such "normal values" as baselines to diagnose overt disease. Thereby, variability within the normal range, however, is often dealt with as if it were narrow, random, or trivial, the body striving for at least a relative "constancy." The alternative, rather than complement, to this status quo, on the left of this figure, is learning about the rules of rhythmic and chaotic variations that take place within the "usual value" ranges that led to the postulation of a "biological clock" that would enable the body to keep track of time. By removal and replacement experiments, a "clock" was located first in the adrenal, shown to be responsible for some but not for other circadians that persisted after brain ablation. The fact that single cells and even bacteria are genetically coded for a spectrum of rhythmic variation indicates that the concept of "clock" needs extension. The concept of figurative calendars followed, aligned with clocks and further a biological week, a 2-week cycle, one of five and another of 6 months, and many more cycles, including some of ~1, ~2, and ~5 decades' lengths, and other rules were found in variability, such as deterministic chaos and the long-known trends, some of which may turn out to be cycles (see Fig. 21.4). When the giant alga Acetabularia, a prominent model of a circadian "clock," is placed into continuous light, its spectrum of electrical activity recorded by Hellmuth Broda (Cornélissen et al., 1986) reveals the largest amplitude for a component of about 1 week rather than of 1 day. When over a decade of studies on this alga are pooled, an about 10-year cycle emerges in the data set as a whole. The alignment of spectral components and chaos and trends in and around us has also begun. Long-term longitudinal, but not yet entire, lifetime monitoring of critical variables complements current linked cross-sectional (hybrid) reference values required for preventive health and environmental care. Changes occurring within the usual value range are resolvable by chronomics as longer and still longer cycles, with a (predictable multifrequency) rhythmic element that allows us to measure the dynamics of everyday life, in order to obtain, e.g., warnings before the fait accompli of disease, so that prophylactic measures can be instituted in a timely way, to detect heretofore largely undetected or unquantified environmental effects or internal changes, which are silent not only to the person involved at high risk but also to conventional health care and science. These vascular variability disorders are detected by cost-free analyses in the BIOCOS project (corne001@umn.edu). On the top right, an abstract sketch is an often-used snapshot reminiscent of the more complex sector structure of the interplanetary magnetic field, shown by three arrows in opposite directions. The fourth arrow is covered by another circle showing solar flares.

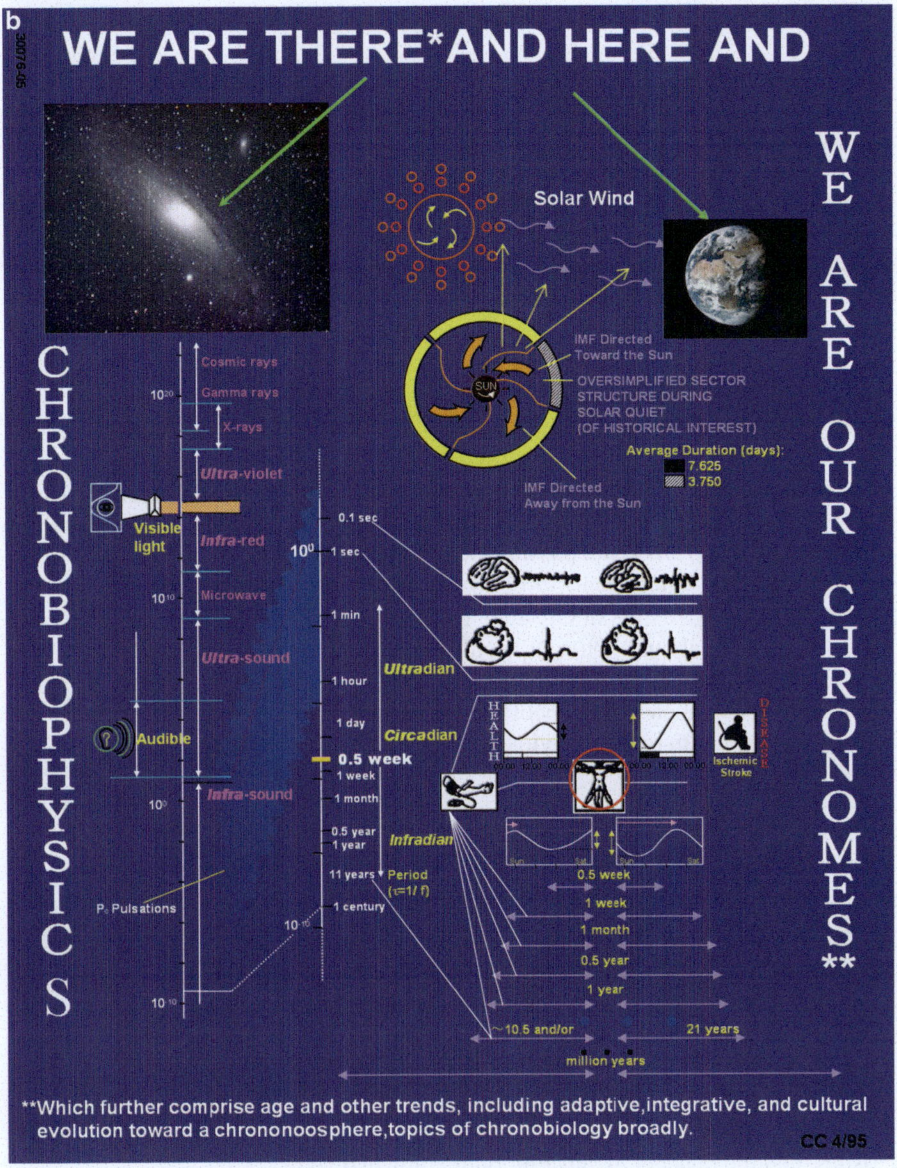

Fig. 21.5 (continued) The parameters of the solar wind and of solar flares are much more variable than visualized when first described. The three *circles* below represent a chronome (cf. Fig. 21.4); the right-hand section as a whole symbolizes associations of helio- and geomagnetic variability, not only with myocardial infarctions and strokes but also with suicide and terrorism that are accumulating and are just the tip of the iceberg, with a profound effect of magnetism (recognized by Gilbert in 1600), apparent in the human ECG, notably in the auroral region (Otsuka et al. 2001). In external-internal interactions, a broad spectrum of rhythms (in both the environment and in living matter) organizes deterministic and other chaos and trends. Trends pursued long enough may become low-frequency cycles, information that may serve, e.g., for the detection of any risk elevation and for the taking of timely action (© Halberg Chronobiology Center). (**b**) The chronomes of variables in us, which came about as a function of chronomes of variables around us and were eventually coded genetically, await further exploration in health care and broader science, including focus on mechanisms underlying adaptive, integrative, and cultural evolution toward a chrononoosphere, topics of chronobiology broadly (© Halberg Chronobiology Center)

21.3 General Background

Part of a vast, diverse, relatively recent literature on cycles in us and in the cosmos attempts to approach the open system called an organism, as a dynamo and a magnet, living on the earth, a larger magnet [57], in the atmosphere of the sun [58], within galaxies and universes, with magnetic storms causing blackouts in cities [59] and relating to as yet irreversible ones in human hearts. Organisms are open to their environment magnetically, electrically, mechanically, and in many other ways yet to be elucidated. We do not know the extent to which they resonate with or depend upon sets of processes near and far that are not consciously seen, heard, tasted, touched, or otherwise knowingly registered. We are, willy-nilly, an integral, continuously informed part of the environment, by which we also happen to be energized. What we can do to learn more about what Frank A. Brown Jr. [60] called subtle influences is to remove and preferably also replace parts of our body in the laboratory or, when this is not possible, let the cosmos do the subtracting and adding (i.e., of a spectral component) to observe what happens in ourselves. Herein, we succinctly summarize uses of circadian mapping (Fig. 21.6a) as a precedent for the merits of concern about photics in health care; we then turn to maps about photics (Fig. 21.6b) and nonphotics in and around us (Fig. 21.6c–g), to what we learned about methods for their study, about remaining problems, and about their relevance thus far, notably with respect to violence.

21.4 Remove and Replace = Subtraction and Addition in the Study of Circadian Rhythms

Removing a gland was a time-honored approach to finding out what the removed gland does, by looking for deficiencies or other alterations and seeing further what replacement does. This was how endocrinology was reportedly founded by Arnold Adolph Berthold [61, 62].[1] Two cockerels with their testicles removed remained capons (subtraction); two other castrated cockerels that received testicular transplants (remove, subtraction; replace, addition) developed secondary sexual characteristics, and the concept of substitution therapy evolved.

---

[1] The subtraction-addition (removal and replacement) approach applied, in the spirit of important predecessors, to the male gonad by Arnold Adolph Berthold (1803–1861), the founder of endocrinology. Berthold was inspired by Scots surgeon John Hunter (1728–1793), who transplanted cock testes into the abdominal cavity of a hen. Berthold castrated four cockerels; two remained capons (removal = subtraction) but two had both testes transplanted into the abdominal cavity of the other (removal and replacement = subtraction and addition). The latter regained their secondary sexual characteristics of growing and aggressiveness. The removal and replacement also applied to other endocrine glands – the adrenal, thyroid, and pancreas – and led to substitution treatment pioneered by Charles Édouard Brown-Sequard (1817–1861).

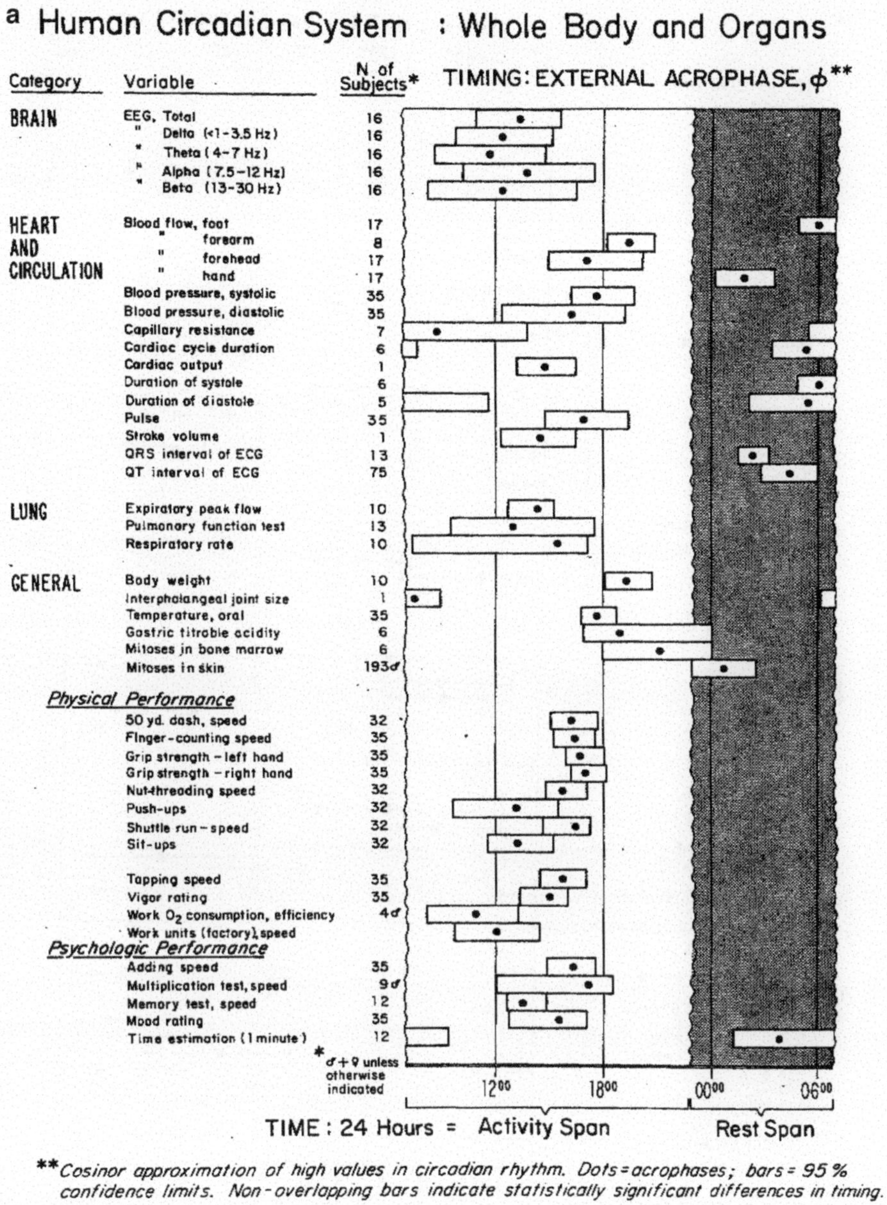

Fig. 21.6 (**a**) Illustrative circadian phase map suggesting a synchronization of several aspects of human physiological and psychological performance with differences in timing. Summary of authors' and others' published data (Halberg, 1969) (© Halberg Chronobiology Center). (**b**) Human circannual rhythms in urine and blood, yielding endpoints for the assessment of the risk of developing several diseases of our civilization (Halberg, 1983). Note, however, many indications of timing without horizontal lines that indicate uncertainties computed as 95% confidence intervals. When circannual phases are followed for decades in the same individual, they can reveal very great variability, possibly resulting from beats between photic and nonphotic, calendar year vs.

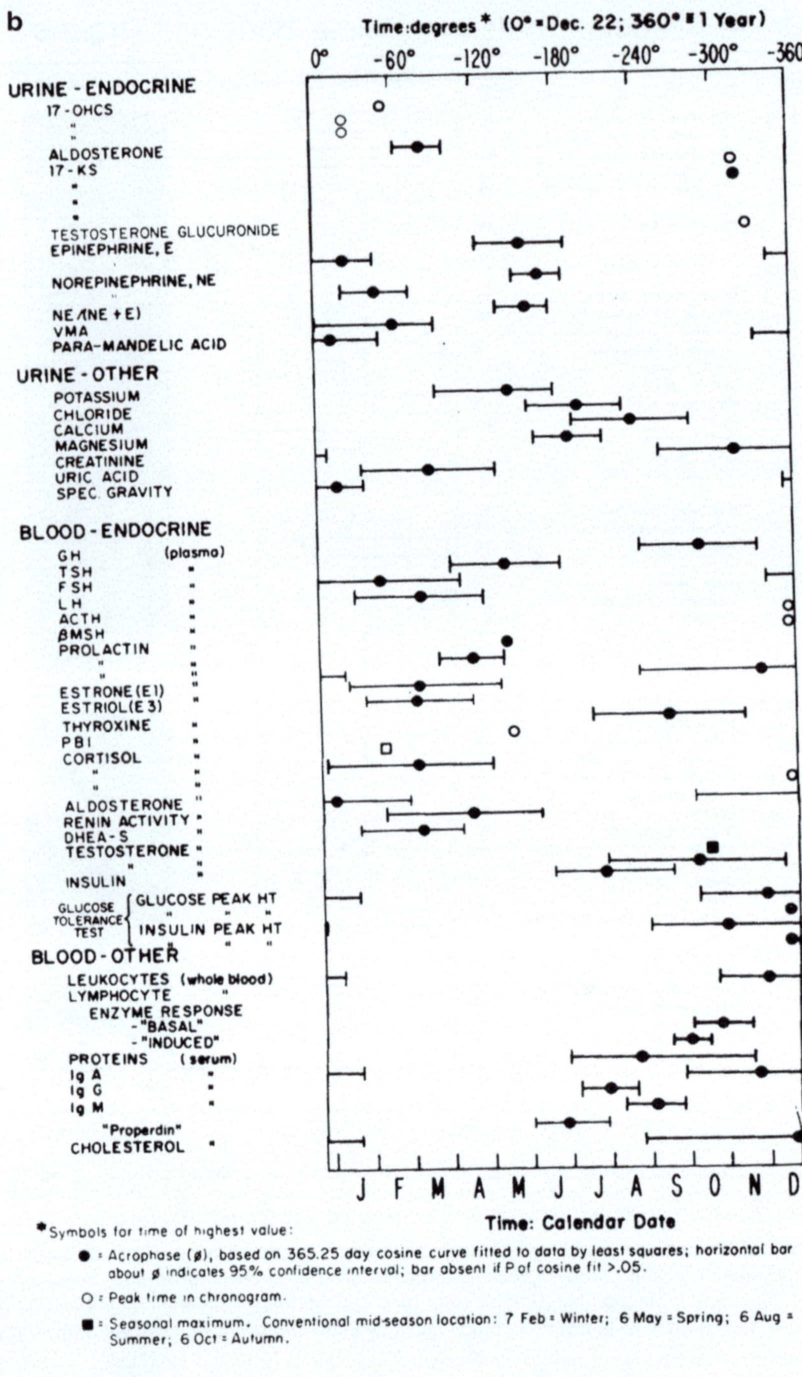

Fig. 21.6 (continued) cis- and transyear mechanisms (© Halberg Chronobiology Center). (**c**) Start of a partial circasemiannual acrophase map (© Halberg Chronobiology Center). (**d**) Near-transyears around [1–3] and in [4–9] organism(s). It is indispensable to specify the calendar span

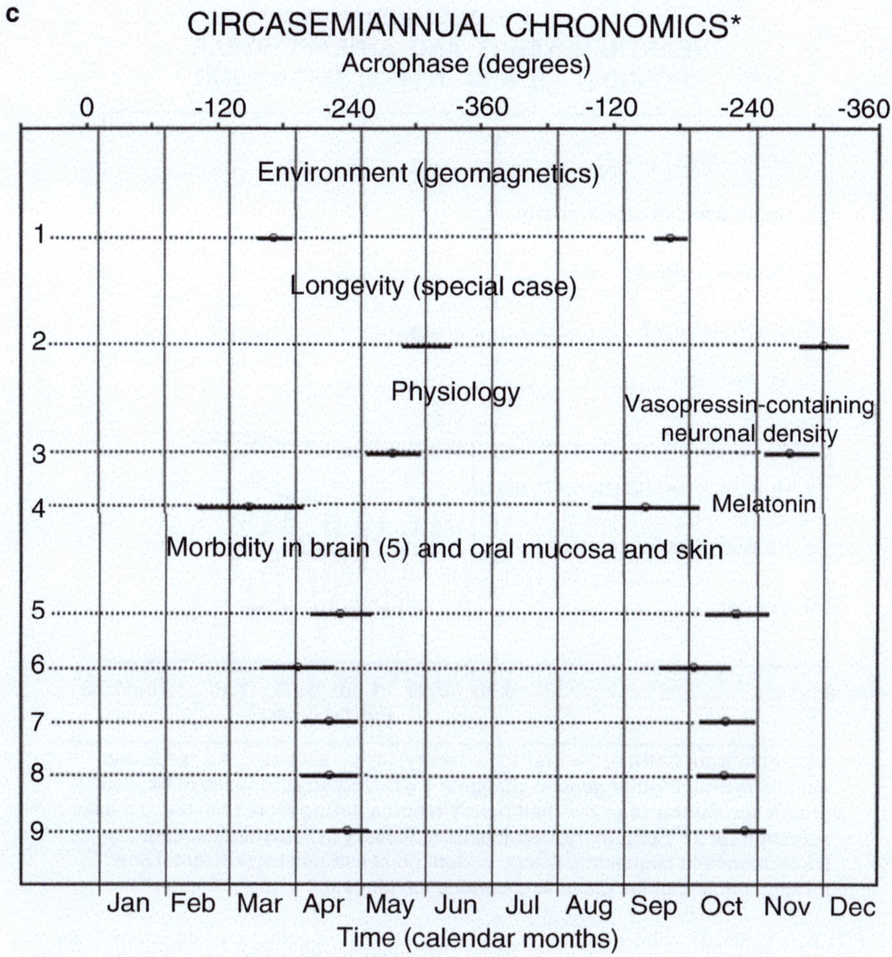

* Doubly plotted acrophases of a half-yearly spectral component from fit of model consisting of cosine curves with periods of 1.0 and 0.5 year fitted jointly by cosinor. Key to series numbers given in separate table.

** 360° = period length (6 months); 0° = December 22.

Fig. 21.6 (continued) and thereby also the length of a time series upon which the resolution of a (set of) spectral component(s) (c) with period(s) (τ) rests. As to mechanisms, it seems pertinent that the only endocrine record covering 15 years of observation of an adult's 17-ketosteroid excretion reveals globally no yearly component, only a near- and two far-transyears, the latter not shown. The finding of physical and biological near-transyears, defined by 1.00 year (y) $<\tau<$ 1.20 y, is mutually reinforcing. These near-transyears in us, acquired by evolution, reinforce the reality of those around us from which they likely derive. We all stacked data in the past over a year and other periods such as a day; yet whenever possible, we should get an estimate of the period involved before stacking! This was a sine qua non well over 50 years ago for circadians and has become one for circannuals and in particular for aeolians, such as the near-transyear, here depicted as calendar-time-location- and series-length-dependent near-transyears (Halberg et al. 2004a and b). It remains to be scrutinized whether near-transyears are separate components and probably not as sidelobes of circannual variation (not stemming from amplitude and/or phase modulation by components

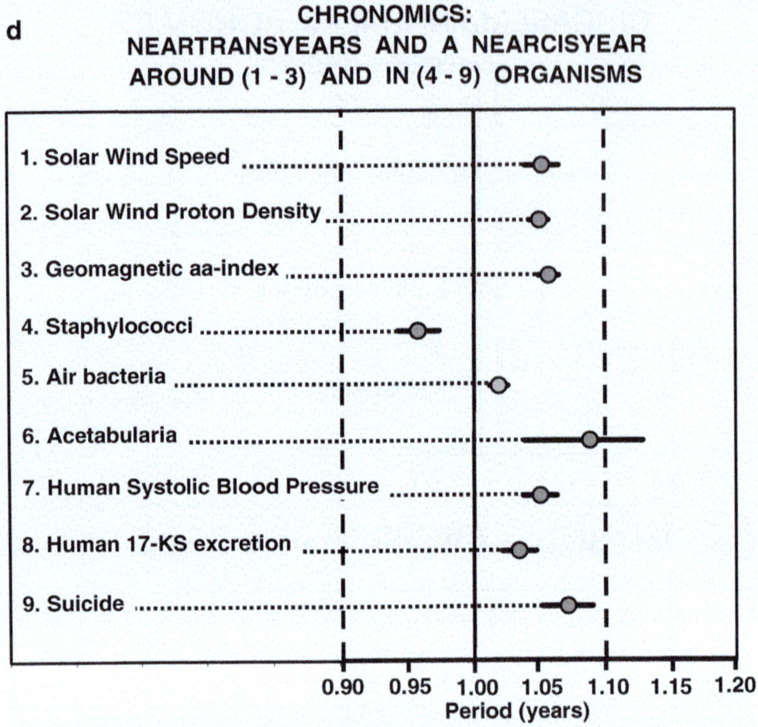

1-2 = data from OMNI 2; 3 = first 131 years (y); 4, 5 = sectoring, i.e., probable mutations and/or other genetic changes; 6 = Circadian acrophase of oxygen production during 14 y; 7 = ~half-hourly records during more than 6 y; 8 = daily excretion for 15 y of urinary steroidal metabolites (17-ketosteroids, CH); 9 = incidence in Minnesota. Circle: period, point estimation; horizontal line through circle: conservative 95% confidence interval.

Fig. 21.6 (continued) with a lower frequency) (© Halberg Chronobiology Center). (**e**) Phase map of ~20-year cycles. Note relative in-phaseness shown by overlapping 95% confidence intervals between criminality and religious motivation, among other variables: a transdisciplinary challenge (© Halberg Chronobiology Center). (**f**) A phase map of about 10-year cycles is illustrated here. Since studies are often carried out over relatively short spans, it is more important to map long-term cycles, so that the time horizon thus provided may serve to guide the design of shorter-term studies. This is one of the immediately useful chronomics that maps the broad time structures in biota and aligns them with environmental chronomes (© Halberg Chronobiology Center). (**g**) Population studies also point to the presence of ~50-year cycles in hardly trivial "diseases" of society, such as homicides and wars, almost certainly involving the human brain, as does the ~20-year cycle in religious motivation (© Halberg Chronobiology Center)

In 1950, Franz Halberg found, as others did before him, very large variations along the 24-hour scale in counts of cells staining with a pink dye (eosin): the eosinophil cells [63]. The eosinophil counts were depressed by certain adrenocortical hormones. In the euphoria of the finding that the adrenal hormone made lame people walk again, adrenal glands were removed in human research looking for under-

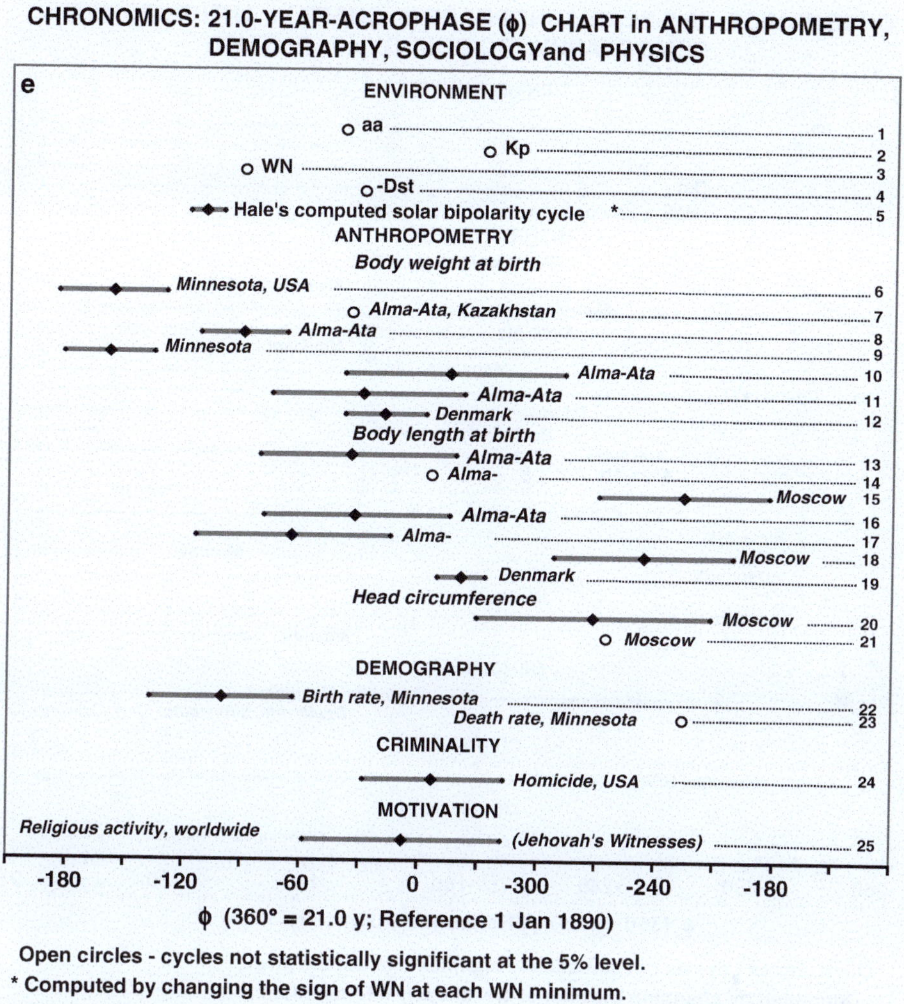

Fig. 21.6 (continued)

lying mechanisms. The opportunity revealed that a cycle in eosinophil count was lost in patients without adrenals, as it was lost in adrenalectomized mice [64]. The finding held for the circadian rhythm of eosinophil cell counts; in the circadian rhythm of serum iron, there were only changes of the amplitude, A, and the acrophase, ϕ, after adrenalectomy, but no abolition (Fig. 21.7a) (bottom and top, respectively). Some decades later, the ablation of the suprachiasmatic nuclei abolished the circadian rhythm in water drinking, but not that in alcohol drinking or in DNA labeling in the digestive tract (Fig. 21.7b) or core temperature (Fig. 21.7c).

The team of Jerome W. Conn (of Conn syndrome fame) confirmed the finding in adrenocortical insufficiency and showed in addition that replacement with a single large dose at the right time but not with divided smaller doses of oral cortisone rein-

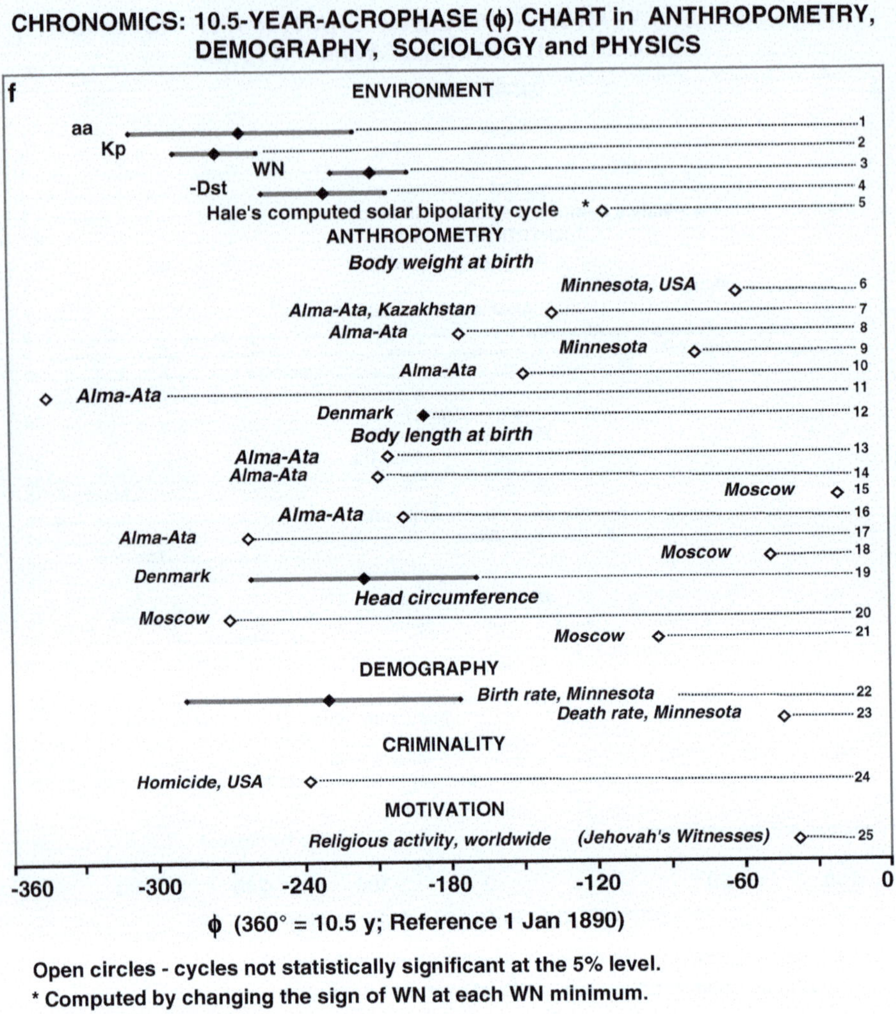

Fig. 21.6 (continued)

duced the rhythm [65]. The approach by "remove" (subtraction) was thus again complemented by "replace" (addition). The approach by subtraction was extended in the earliest 1950s beyond gland removal to study other mechanisms of the eosinophil rhythm. When the eyes, receptors of a synchronizing lighting cycle, were "subtracted" by surgery or genetics (in ZRD mice born blind), a previously 24-hour synchronized cycle, while it did not disappear, was free-running from the 24-hour routine, with a built-in period near 1-day length, but not of precisely 24 hours (Fig. 21.7d and e). A circadian desynchronization was already documented by periodograms (that took a week to compute on a desk calculator and another week to

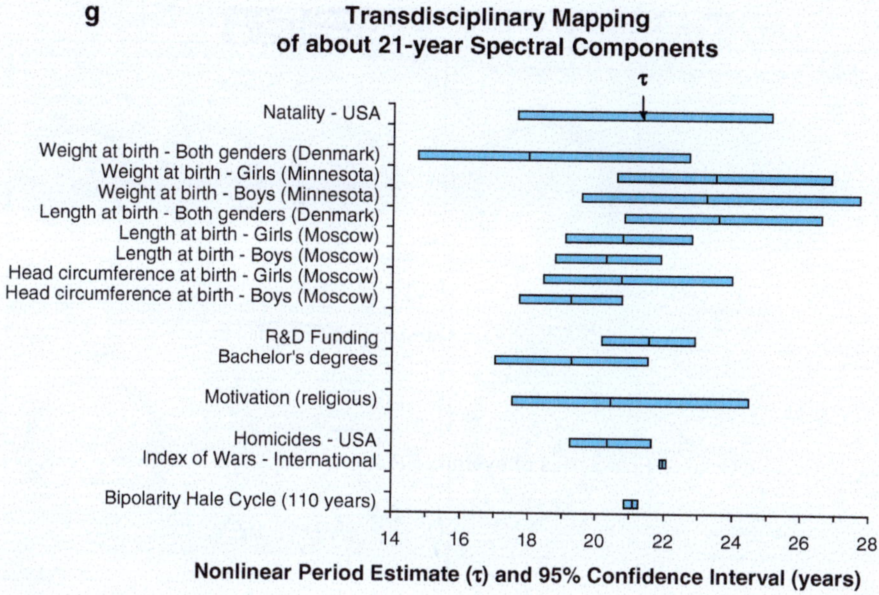

Fig. 21.6 (continued)

check) and were the topic of a discussion with Julius Bartels in 1953, to indicate that his harmonic dial could also be used flexibly for fitting anticipated desynchronized environmental cycles and not necessarily a precise 24-hour day [66].

"Circadian" was coined – in the early 1950s [67] and endorsed by international nomenclature committees [68] – both to indicate partial endogenicity and (another reason for "circa" that is too often forgotten) to refer to the also-encountered variability inherent in the cycle's characteristics. "Circadian" was intended to replace "diurnal," a term used in biomedicine to distinguish activity or morbidity patterns during the daily photofraction from those during the daily scotofraction, as diurnality vs. nocturnality, respectively. While they are obvious signatures of the photic day, circadians are also influenced nonphotically by natural and artificial electric fields [69] and by subtle environmental factors in caves [70]. Circadian timing accounts for the difference between ~75 % survival and ~75 % death from the same dose of a cardiac drug, among others, under controlled laboratory conditions as a function of timing (Fig. 21.7f). Alterations of circadian characteristics result, e.g., in vascular variability disorders (VVDs) that represent a risk of stroke which is greater than that of hypertension and can be treated (Fig. 21.7g).

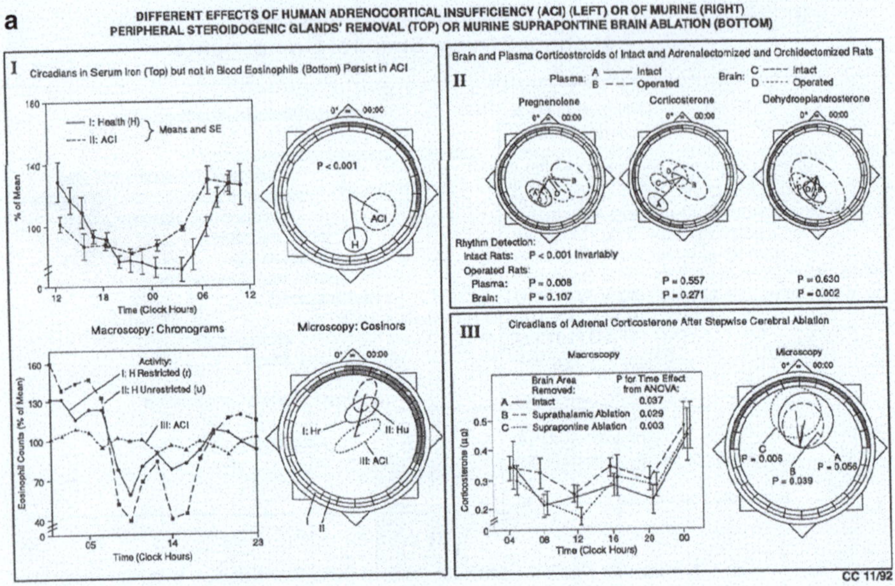

Fig. 21.7 (**a**) Chronobiologic self-experimentation extended to groups of medical students and staff served to show that the adrenal glands are an important source of circadian rhythmicity. For instance, a circadian component of variation in the circulating eosinophil count could not be demonstrated for patients with adrenocortical insufficiency (I *bottom*), whereas in healthy people with either restricted or unrestricted activity (I *top*), this rhythmicity was not only demonstrable but also amplified by enhanced motor activity. Whereas for patients with Addison's disease a circadian rhythm in eosinophil counts could not be statistically validated (polar cosinor displays both), circadians persisted, however, for serum iron (polar plots on *top*, not covering the pole, i.e., *center*) (P<0.001). Self-experimentation then leads to studies in the laboratory. Whereas in intact rats a circadian rhythm could invariably be detected for several steroidal hormones (pregnenolone, corticosterone, and dehydroepiandrosterone), in adrenalectomized and orchidectomized rats, a circadian rhythm remained demonstrable for plasma pregnenolone and for brain dehydroepiandrosterone (II). Circadians of adrenal corticosterone also persisted on the day after stepwise cerebral ablation (III) (© Halberg Chronobiology Center). (**b**) Circadian rhythm alteration rather than obliteration after lesioning of suprachiasmatic nuclei (SCN). By eyeballing alone of Section IIA, the circadian rhythm in telemetered core temperature, each measurement shown by a dot, is clearly seen in data from a sham-operated control on the left and seems to be lost in the rat with a SCN lesion on the right of this section, whether one examines squeezed values in the top row or stretched values in the second row (of dots). A circadian rhythm in temperature for individual animals is also displayed in Section I *top*, with a smaller within-day change of lesioned animals (IB) as compared to controls (IA). This finding is also seen after averaging in Section IIB (*bottom*). Time microscopy, in section IIC, apart from quantifying the rhythm by cosinor, reveals, by a *shorter arrow*, a great amplitude lowering after a bilateral SCN ablation and a phase advancement seen as an earlier vector. Section IIC also validates the persisting rhythm by the nonoverlap of the center (or pole) of the graph, by the error ellipse representing a 95% confidence region: the removal of the SCN is compatible with the persistence of a statistically highly significant circadian rhythm in core temperature quantified with its parameters and their uncertainties, after histologically validated bilateral SCN ablation. When the ablation unintentionally, as discovered at postmortem, was unilateral (U), the circadian amplitude was enhanced (Section I, *bottom*), a finding suggesting a subtractive coupling between the two SCN. Section III is in keeping with the speculation of an effect by lunar factors upon the "free-running" (or rather lunar?) period of about 24.8 hours found in controls

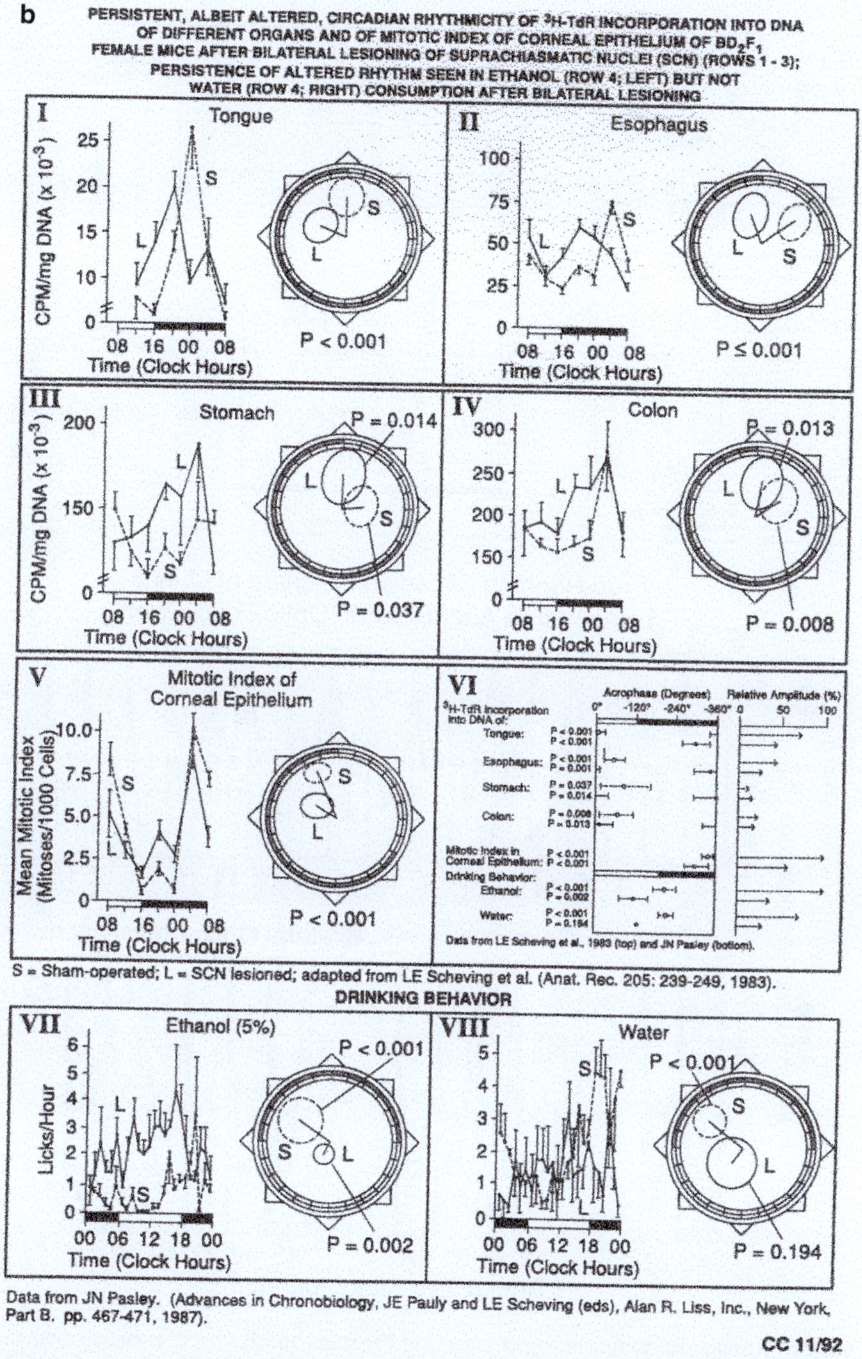

Fig. 21.7 (continued) or unilaterally ablated animals at that light intensity. If this should be in part a lunar effect, it is lost in animals subjected to bilateral (B) suprachiasmatic lesions and apparently tightened by unilateral (U) lesions, a possibility requiring further experimental scrutiny (© Halberg

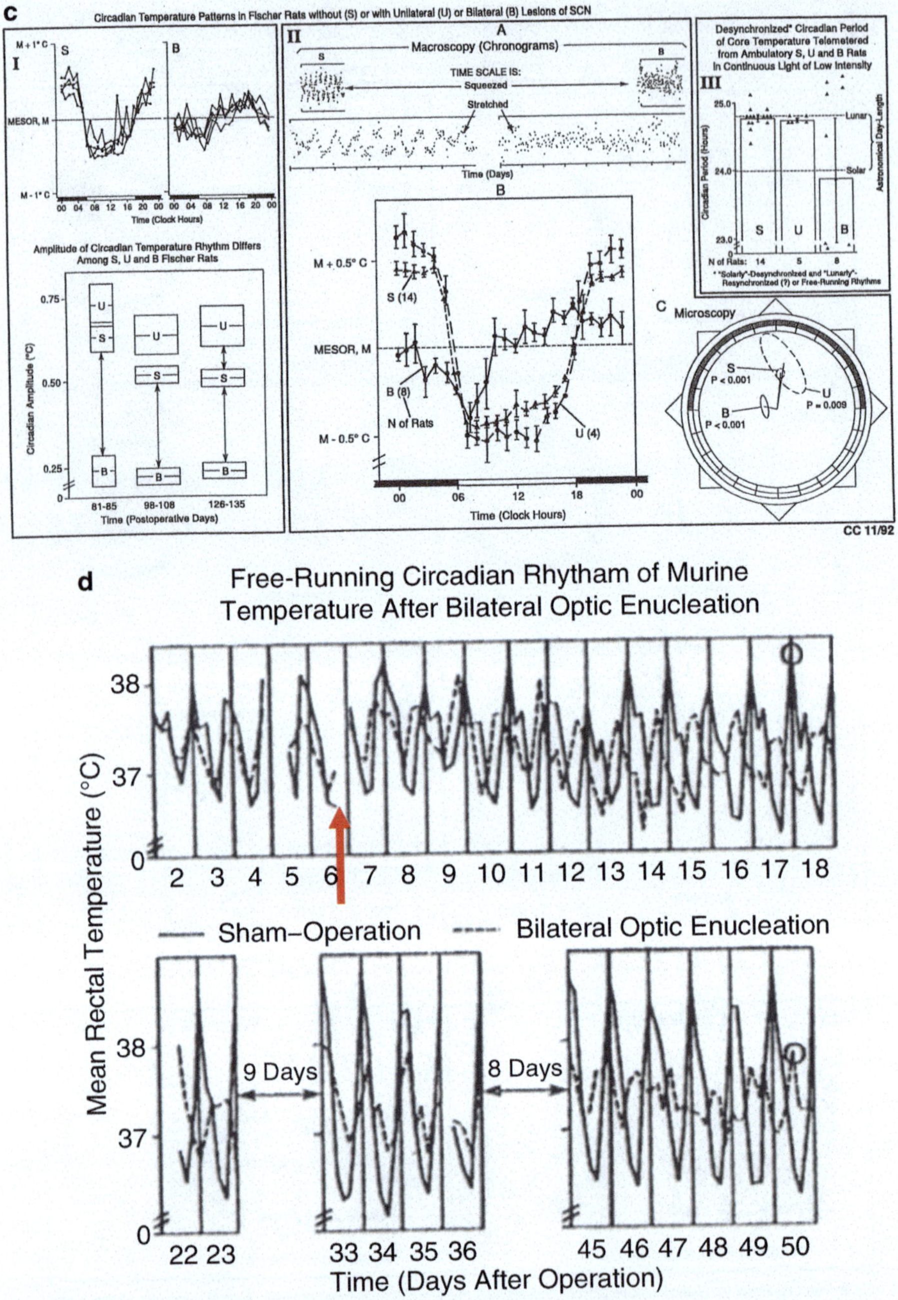

Fig. 21.7 (continued) Chronobiology Center). (**c**) Persistent, albeit altered, circadian rhythmicity of ^{3}H-TdR incorporation into DNA of different organs and of mitotic index of corneal epithelium of BD2F$_1$ female mice after bilateral lesioning of suprachiasmatic nuclei (SCN) (rows 1–3); persistence of altered rhythm seen in ethanol (row 4; *left*) but not water (row 4; *right*) consumption

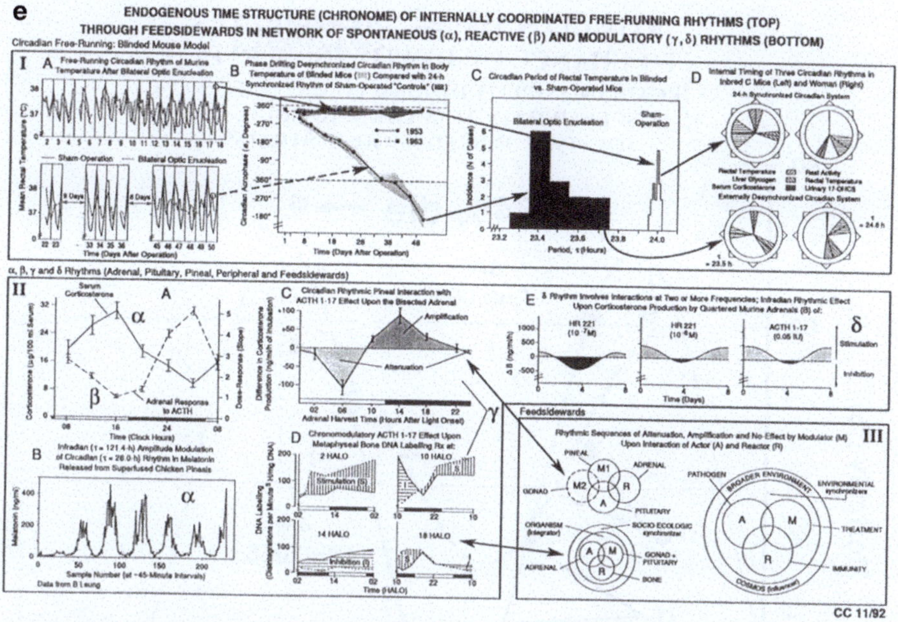

Fig. 21.7 (continued) after bilateral lesioning. Data from J.N. Pasley (Pasley et al., 1987). The SCN coordinates a collateral hierarchy that can be quantified in terms of amplitude and phase: the major effect of bilateral SCN ablation is thus far invariably, comparably to the behavior of core temperature, an advance in phase for the SCN-lesioned (L) animals in 8 cases out of 8, with a reduction in amplitude, except for DNA labeling in the stomach and colon, which may respond to food directly rather than by the SCN, Section III. Section VI shows a microscopic phase and amplitude chart summarizing the finding in the other sections (I–V, VII, and VIII). This chart extends the scope of the SCN lesson learned from studies of core temperature (Fig. 21.7b) to a number of other variables studied as marker rhythms: rather than being a master clock leading to the loss of all rhythms when ablated bilaterally, the SCN is compatible with the rhythm's persistence in several of the variables investigated, except for a few important ones like locomotor activity, water drinking, and blood pressure. A subjective time-macroscopic interpretation-based impression that led to the master clock illusion is replaced by the objective quantification of a mechanism for period, phase, and amplitude in a network with a collateral rather than up-down hierarchy (© Halberg Chronobiology Center). (**d**) Macroscopic circadian desynchronization in mice after bilateral optic enucleation (dashed line connecting dots) visualizing the need for objective inferential statistical time-microscopic analyses added in sections B and C of Fig. 21.7e. While sham-operated controls (solid line connecting open circles) show a daily peak at or close to the vertical 20:30 lines, peak temperatures of blind animals already on day 6 seem to diverge, rising while those of the other group are falling. A graph of the original finding of this separation of the two groups and the decision based thereon to continue measurement every 4 hours around the clock was interpreted as "paranoia." (At the time, in the precomputer era, the provision of a periodogram on desk calculators took a week, and its checking another week. Indeed, today the approach with a moving fit of a 24-hour cosine curve, rapidly implemented on a computer, is preferred to inspection that has to eliminate "misbehaving rogues" and remains subjective.) On the average, on top, peaks in temperature of the blinded group occur earlier and earlier, but there are uncertainties in such eyeballing. A transient antiphase at 22 to 23 days after blinding is readily seen. If, around that stage after blinding, 2-timepoint checks are carried out on the 2 groups, opposite results can be obtained on mice with and without eyes and later when they are again in phase, 2-timepoint checks show no difference, a puzzle readily resolved by an objective quantification of the rhythm characteristics. The need for this microscopy in time becomes obvious, notably if an inference is desired as soon

f

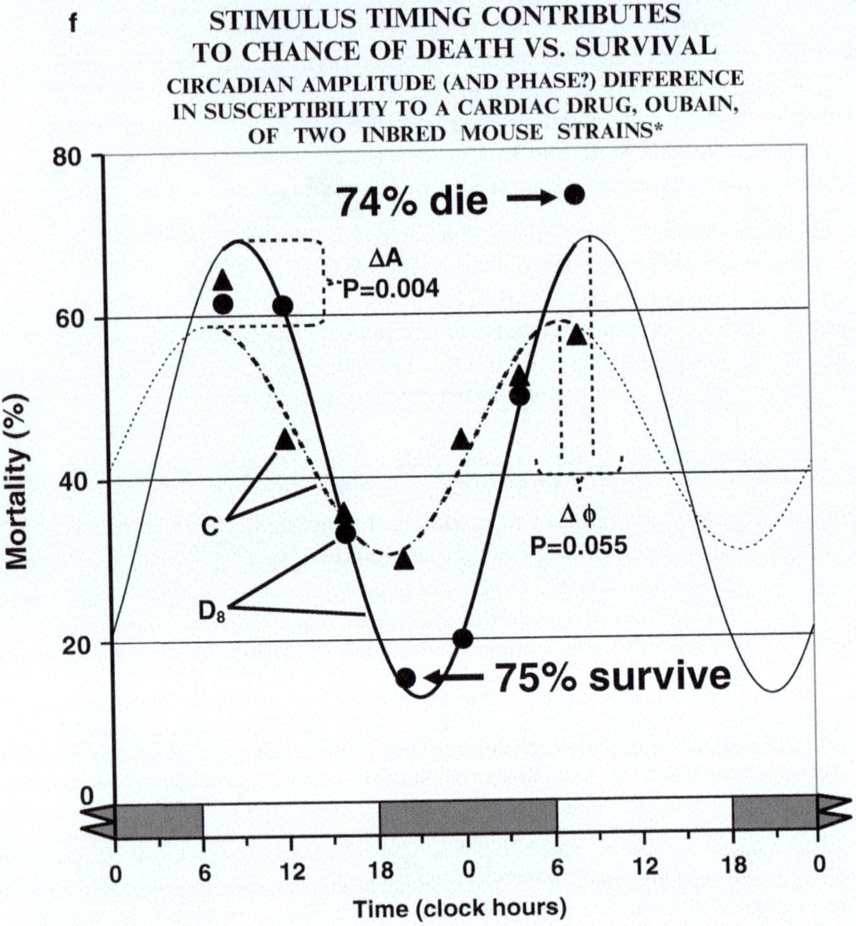

Strain	P	M	Amplitude (A)	Acrophase (ϕ**)
D_8	P < 0.001	40.0	28.1 (21.5; 34.8)	-135 (-119; -150)
C	P = 0.011	45.2	14.2 (7.3; 21.0)	-104 (-73; -136)

*** C = Bagg albino, D_8= subline 8 of Dilute Brown strain; oubain 0.15 or 0.5 mg/20 g of body weight, respectively. Thin lines in the graph are extrapolations of fitted cosine curves. ** In the table: 95% confidence limits in parentheses, in degrees (360° = 24 hours). M = MESOR; P from zero-A (no-rhythm) test.**
Data from F.Halberg et al., Minnesota Acad Sci Proc 1959; 27:139-43.

Fig. 21.7 (continued) as possible with an estimate of uncertainty. (Circadian desynchronization also characterizes congenitally blind ZRD mice) (© Halberg Chronobiology Center). (**e**) *Top*, section I: Desynchronization of circadian rhythm in core temperature of mice after blinding, seen time-macroscopically in IA (much better in Fig. 21.7d), here leads, in IB, to time microscopy with a chronobiologic serial section showing a different time course of the core temperature acrophases, ϕ, with early separation of the two groups by nonoverlapping 95% confidence intervals ϕ in IC, to a summary of individual periodograms that form two separate distributions, and in ID to time

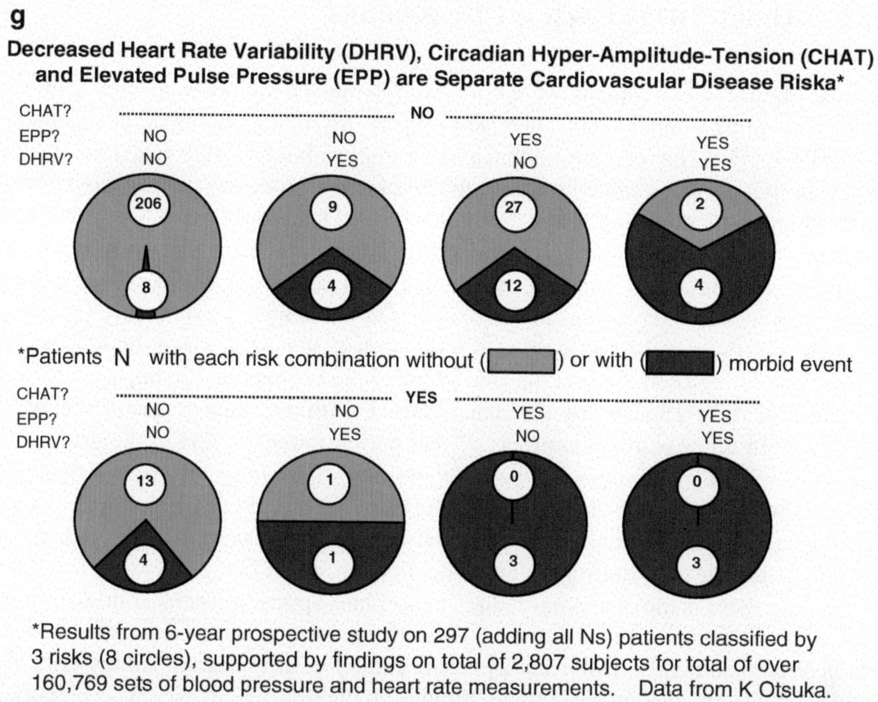

Fig. 21.7 (continued) relations among three variables in a 24-hour synchronized (*top*) or free-running (*bottom*) system (of mice, left, and of a human, *right*). Section II shows a spontaneous (α) rhythm in circulating corticosterone of mice in antiphase with the slope of an in vitro response rhythm to ACTH, a reactive (β) rhythm. The components of the chronome (time structure) are internally coordinated through feedsidewards in a network of rhythms that are more or less spontaneous (α), others primarily reactive (β) or modulatory at a single mapped frequency, such as a circadian (γ), IIC and IID, or at multiply mapped (δ) frequencies, IIE (© Halberg Chronobiology Center). (**f**) Susceptibility rhythm to ouabain in mice of two stocks. Twenty ($\mp$1) male mice (about 5 months of age)/timepoint/stock. Note daily change in mortality from 15% to 75% in D8 mice by comparison to change from 30% to 65% in C mice. Note also greater susceptibility of C mice (given smaller dose). Objective quantification is provided in the cosinor summary at the *bottom* (© Halberg Chronobiology Center). (**g**) As compared to patients with an acceptable variability of blood pressure and heart rate (*top left*) who have 4% cardiovascular morbidity after 6 years, the presence of a single vascular variability disorder (VVD) such as DHRV or EPP alone (*top middle*) greatly increases the risk, even more so when both conditions coexist (*top right*) to become a vascular variability syndrome, VVS; the added diagnosis of a circadian blood pressure overswing, CHAT (*bottom*), further increases risk. VVDs or VVSs can constitute prehypertension (Watanabe et al., 1997), in the absence of an elevation of the average of around-the-clock measures of blood pressure, while they can also complicate a high blood pressure, i.e., MESOR-hypertension; the latter, because of (always to be checked) variability, is also a VVD in its own right (© Halberg Chronobiology Center)

21.5 About-7-Day Cycles: Circaseptans

Almost-ubiquitous circadians were soon complemented by other spectral components including about (not precisely)-7-day rhythms, dubbed circaseptans (Fig. 21.8a) [71]. The recognition that a free-running about-weekly schedule resides in part in living matter came first from the breakdown products (metabolites) of the very adrenocortical and gonadal hormones (17-ketosteroids, 17-KS), which are important for survival and reproduction (Fig. 21.8b), the circadian story repeated itself. An apparently free-running rhythm in 17-KS had a period close to yet different from precisely 7 days (Fig. 21.8a, b) [71]. On weekends, the subject (CH) drank an extra cup of tea and a bottle of beer, a schedule that synchronized the volume he excreted and measured, but not the steroid metabolites contained in his urine which he determined for 15 years by the same method, with the same reagents bought at the outset. In other words, the precise 7-day period found for CH in the urine volume in which the metabolites had been determined, coexisted with a period of about (but different from precisely) 7 days in the excreted steroidal metabolite and was also different (desynchronized) from another near-7-day period in the planetary geomagnetic index Kp during the urine's collection span [71].

Second, there is the observation that circaseptans appear after the administration of a single stimulus that carries no 7-day information, as in the case of a partial hepatectomy studied by Hübner [72] in rats and by us after human kidney [73, 74] or heart [75] transplantation. Single stimulus induction of circaseptans, or rather their manifestation after a single stimulus, traced back to antiquity, was also the topic of a classic book on circaseptans by Gunther Hildebrandt and Ingrid Bandt-Reges [76].

Third, as to ontogeny, a dominant weekly pattern, much more prominent than the also-present circadian component in the same data, was seen in a survey of more than 181 profiles of blood pressure, each profile on a separate baby, each measured every 30 minutes each for 48 hours on different 2 consecutive days of the first week of life, and in several hundred additional longitudinal monitorings for several weeks during the first months of life; by the fourth week of life, the circadian component usually dominated the human spectrum (Fig. 21.8c). In a crayfish, the circaseptan dominated the activity pattern at the age of 6 months. Aeolian circaseptans persisted into adolescence, albeit they came and went and were only quasi-persistent.

Fourth, again as to ontogeny, the circaseptans regain relative prominence with elderly age, concurrently with a circadian decrease in amplitude [77].

Fifth, as to phylogeny, the giant unicell *Acetabularia acetabulum*, standardized in light and darkness alternating every 12 hours, thereafter released into continuous light, after signal averaging of electrical activity, shows a circaseptan rhythm with a free-running near-one-week period, with a much larger amplitude than that of the also free-running circadian rhythm (Fig. 21.8d). Assuming that this alga has not changed in the interim, this eukaryote may tell us about the environmental cycles it adapted to when it developed, perhaps 500 million years ago [78] at a time that physical instruments cannot yet readily trace.

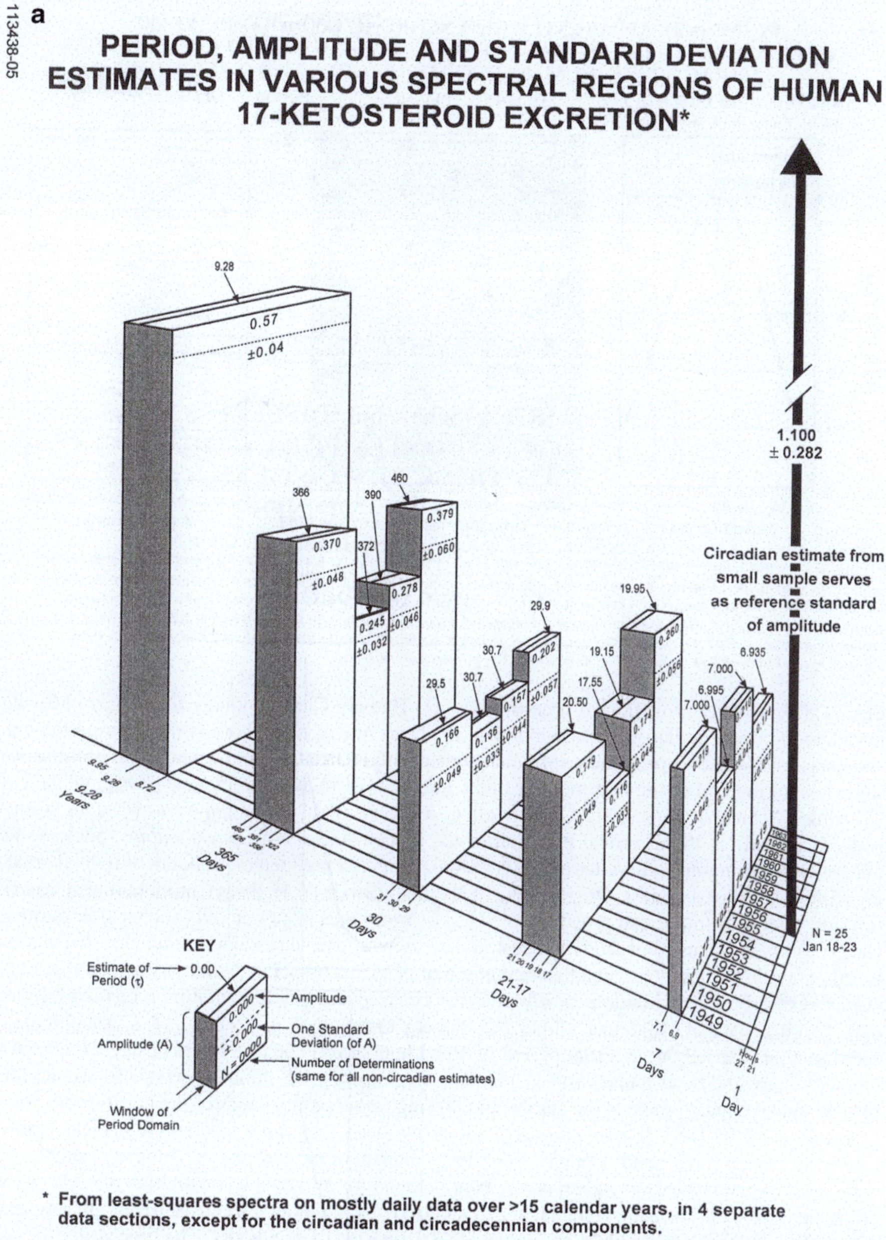

Fig. 21.8 (**a**) Spectrum of urinary 17-ketosteroid excretion by a clinically healthy man measured over 15 calendar years. Estimates of periods, amplitudes, and standard deviations in six of the many more spectral regions are shown. Analyses in 4 spectral regions by least-squares spectra are on 4 separate sections. A number of cycles with long periods characterize this one-and-the-same biological variable (and other body functions). Reproducibility is seen when four sections are analyzed separately. For added components, see Fig. 21.8b (© Halberg Chronobiology Center). (**b**) Spectral windows with transyears are shown globally (*left* and *right*) and are resolved as a

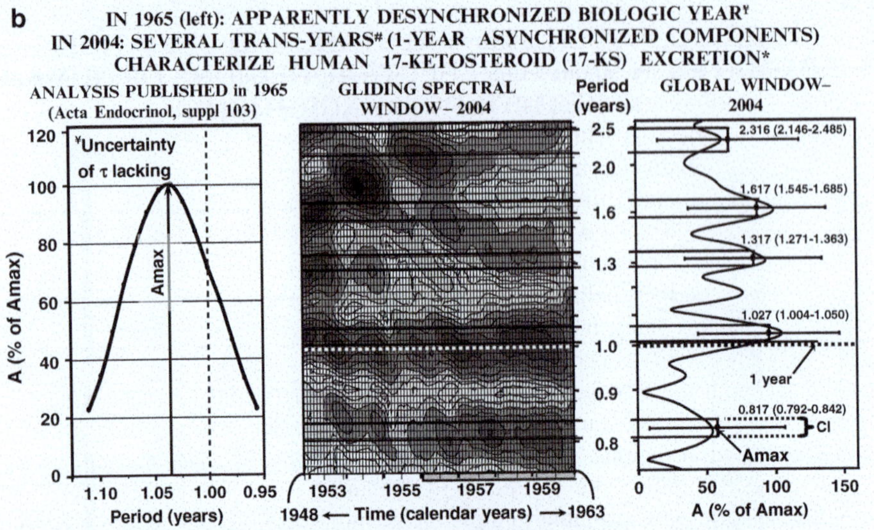

#Transyears are oscillations with 95% confidence intervals (CIs) of their period (τ) between 1 and 2 years, overlapping neither of these τs nor one another. Cisyears are oscillations with periods between 1 and 0.5 year with CIs overlapping neither the 1-year nor the 0.5-year τ, as seen for a τ of ~ (but not) = 0.8 y. *~ 15-year record, daily data (with gaps) from a clinically healthy man (CH, 43 – 58 y), N = 3719. Gliding interval = 8 years (y), increment = 1 month, longest trial τ = 2.265 y, shortest τ = 0.75 y, harmonic increment = 0.05. Darker shading corresponds to larger amplitude (A). Horizontal shaded bar starting in 1956 indicates span of intermittent androgen Rx; data during Rx omitted from analysis.

Fig. 21.8 (continued) function of time (middle) (© Halberg Chronobiology Center). (**c**) Moving infradian-to-circasemidian spectral window of the heart rate of a boy born at term during the first 40 weeks of life shows prominent infradians postnatally. Circadian-over-circaseptan prominence starts toward the end of the first month of life: side view of relative prominence. Original data of Yoshihiko Watanabe (© Halberg Chronobiology Center). (**d**) Circaseptan-over-circadian prominence observed for the electrical potential of the unicell *Acetabularia acetabulum* following its release into continuous light, after prior synchronization to a regimen of 12 hours of light alternating with 12 hours of darkness. Original data of Hellmuth Broda (© Halberg Chronobiology Center). (**e**) Time course of tumor cell growth pooled across four kinds of tumor cells shown after normalization for weeks 2 and 3 of study to illustrate circaseptan-over-circadian prominence. Irradiation shortly after the circaseptan-circadian maximum of β-ATP is associated with larger tumor cell kill (lesser survival) than irradiation shortly after the circaseptan-circadian minimum of β-ATP (middle). Original data of Waldemar Ulmer (© Halberg Chronobiology Center). (**f**) Chronobiologic serial sections carried out at a trial period of 164.4 hours (*left*) and 168 hours (7 days) (*right*) for blood pressure and heart rate measured ambulatorily, mostly at 15-minute intervals for 16 months by a woman 67 years of age at the start of monitoring (*top 3 panels*) and for the north–south component (Bz) of the IMF (*bottom panel*). Shading (by *bracketing dots*) indicates statistical significance and its size corresponds to a 95% confidence interval of the acrophases. A mostly statistically significant (when shaded) circadian component is found for all variables, with both fits. The acrophase curve of Bz is horizontal (stable) at a trial period of 164.4 hours, whereas the acrophase curve of blood pressure is more or less horizontal at a trial period of 168 hours. The time course of the heart rate phases drifts with the 168-hour fit. The physiological variables are differently synchronized, blood pressure mainly by the societal week and heart rate pulled by the 164.4-hour period of Bz. In the case of heart rate, however, a small phase shift occurs more or less at the middle of the monitoring span. This shift in circadian acrophase of heart rate coincides with a similar shift in circaseptan acrophase of Bz from the interplanetary magnetic field. The left half of this figure shows that blood pressure clearly drifts away from 164.4 hours, further validating the 7-day-synchronized circaseptan component in blood pressure. In the case of heart rate, there is also a

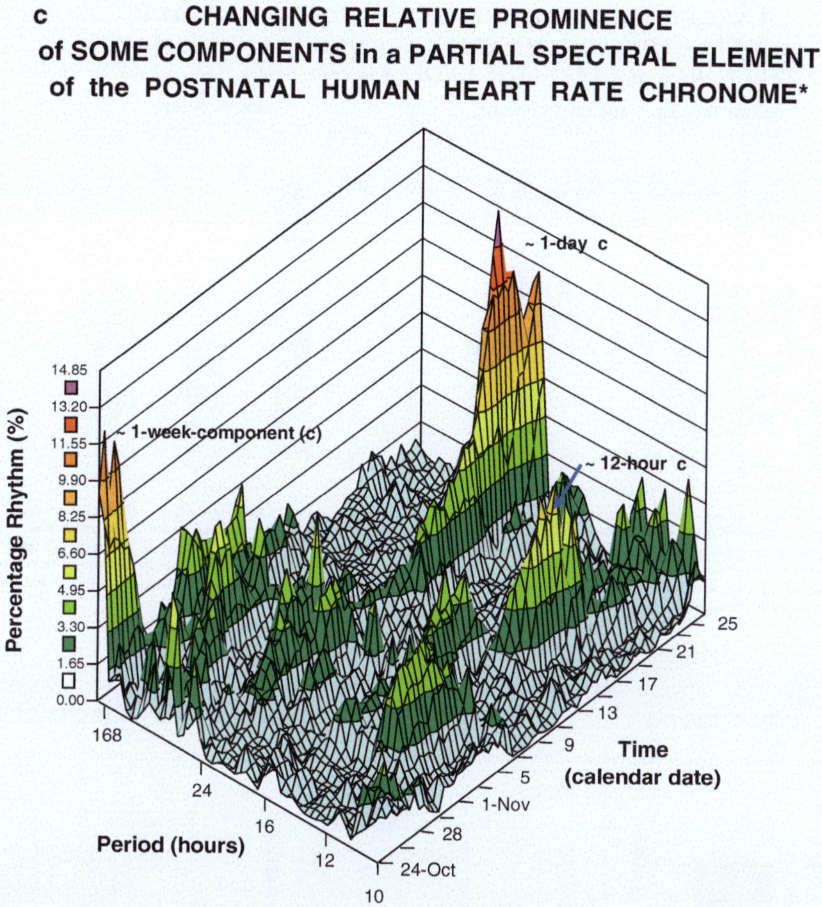

*** in a healthy boy, born 19.10.1992, whose heart rate was measured at mostly 30-minute intervals from 20.10 for the ensuing 40 days, and analyzed as a moving spectrum in separate weekly intervals, displaced in 12-hourly increments through the data set. An initially greater relative prominence of infradians (see ~ 1 week c, *left*), shown by height and color, corresponding to a larger percentage rhythm, contrasts with the relative prominence of circadians and circasemidians in later weeks of life, while any ultradians with still higher frequencies and any trends and chaos, two other chronome elements, are here unassessed.**

Fig. 21.8 (continued) transient drift of the acrophase of the 164.4-hour component for several months, but it is not as pronounced or consistent as that observed for blood pressure. Moreover, there are spans when the 164.4-hour acrophase seems to lock and remain stationary and thus consistent with this period at the beginning of the series and again in the second half of the monitoring span. These are spans when a stable 164.4-hour acrophase is also observed for Bz. From these observations, it appears that Bz may pull heart rate and that there is occasional phase locking as well as frequency-locking between these two variables. It should be noted further that the phase difference between heart rate and Bz at the beginning of the series is about 90°, whereas in the second half of the series, the phase difference is much smaller; this means that when there is frequency-locking between the two variables, this may occur with a different phase difference. Whether or not these phase differences and phase locking follow a predictable pattern such as being dependent upon the stage of the solar activity cycle remains to be further investigated. These results are only observational at this stage, and a mechanism or model accounting for these observations has not yet

d **LARGER *ABOUT-WEEKLY* THAN *ABOUT-DAILY* CYCLE in ELECTRICAL POTENTIAL of a UNICELLULAR ALGA *ACETABULARIA ACETABULUM* EVOLVED 500 MILLION YEARS AGO***

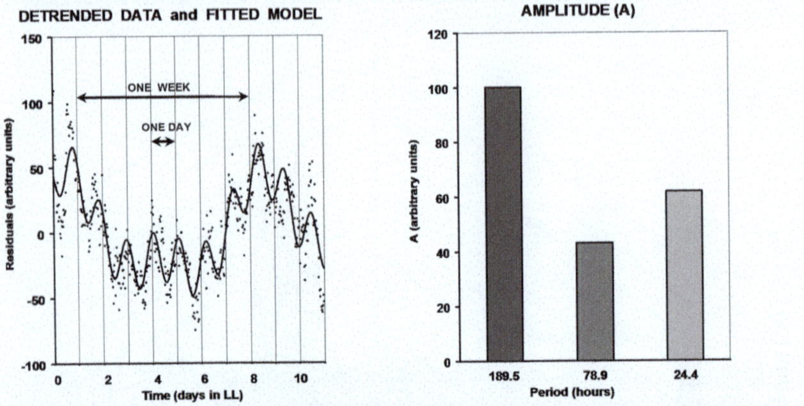

*Nonlinear spectral analysis on signal-averaged data from 20 single cells, released (zero time) into continuous light (LL), after prior standardization in light and darkness alternating at 12-hour intervals (LD12:12) for up to one week. Total number of observations: 38,578; experimental span: 376 days. Note a more prominent amplitude (A) for a component with a period near a week, than the As of the about daily and about half-weekly components (all free-running). The circaseptan A is eqated to 100 and the other As are expressed as percentage of the circaseptan. Metaanalyzed data of Dr. Sigrid Berger, Dr. Lübbo von Lindern and the late Dr. Hans-GeorgSchweiger.

e **LARGER *ABOUT-WEEKLY THAN ABOUT_DAILY* CYCLE IN GROWTH OF TUMOR CELLS***

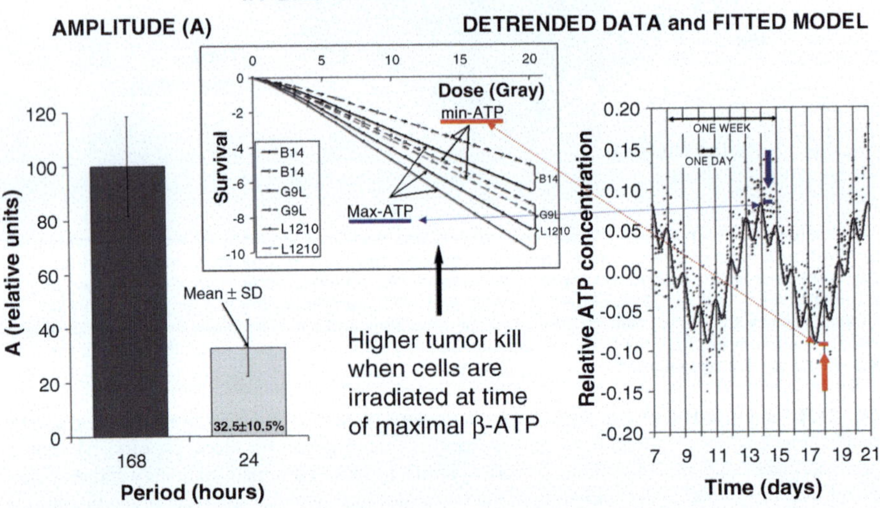

*Nonlinear spectral analysis of data pooled from four kinds of tumor cells, each cultured at a pH of 6.9 (experimental #1) and 7.3 (experimental #2). Data shown for weeks 2-3 (total 6 weeks, exhibitng persistent cycles of similar relative prominence).

Fig. 21.8 (continued) been developed. One avenue that seems promising may be to consider the possibility that there is more than a single circaseptan component. This was the case for geomagnetic pulsations in Antarctica as well as in geomagnetic indices during some but not necessarily all spans investigated, since recording of aa or Kp started: one component was close to 7 days and the other

f

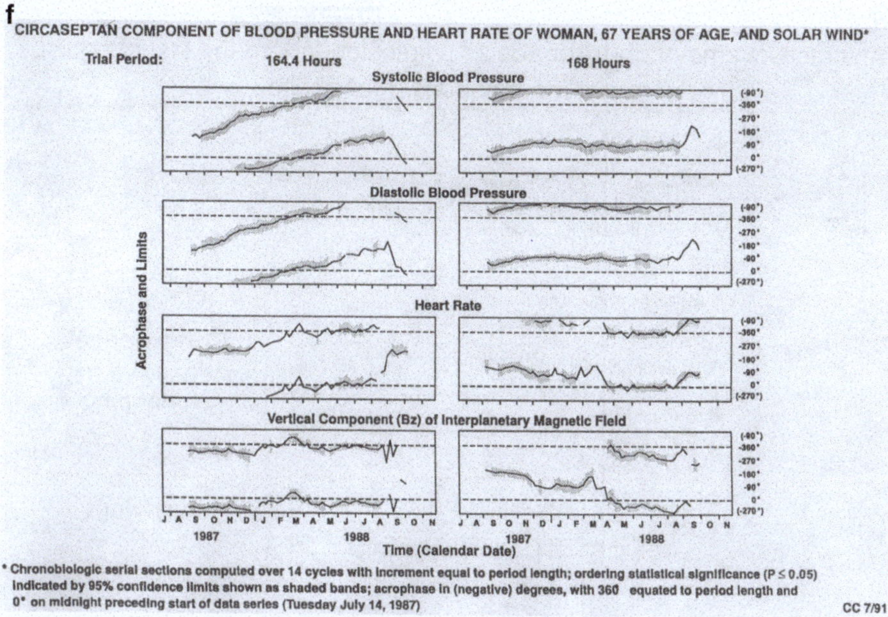

Fig. 21.8 (continued) was slightly shorter than 7 days. When two components with close periods are present jointly, they may pull during some spans, but not during others, depending upon the relative instantaneous phase difference between the two different components, e.g., whether they are in phase or out of phase (© Halberg Chronobiology Center). (**g**) Congruence can characterize characteristics such as the period or phase of a spectral component in 2 or more time series of the same variables or of different variables within tissues, organs, organ systems, and disciplines and, most interestingly, when they are transdisciplinary, as are transyears (© Halberg Chronobiology Center). (**h**) Transdisciplinary congruence in the circaseptan spectral region (© Halberg Chronobiology Center). (**i**) Combination of gliding spectral window on top with special focus on the behavior of two selected periods (*middle* and *bottom*), with the time course of the phase validating the 6.77 period by a more or less horizontal trajectory of phases in only part of the record, but invalidating a precise 7-day periodicity since no dots bracket any horizontal time course of phases, and the small initial section with dots shows a gradual advance (© Halberg Chronobiology Center)

Sixth, like *Acetabularia*, tumor cell cultures, as spheroids or as monolayer, also show a major circaseptan peak in β-ATP concentration, with a superposed circadian in amplitude that is barely one-third of the circaseptan (Fig. 21.8e). While *Acetabularia* shows free-running circadians and circaseptans, the tumor cells are 1- and 7-day synchronized. While *Acetabularia* shows a dispersion of periods after a first cycle or so, the tumor cells remain 1- and 7-day synchronized for 6 consecutive cycles. Tumor cell kill by radiation was much better at the circadian and circaseptan peak of the β-ATP marker, as compared to radiation administered at the circaseptan and circadian minimum [79, 80].

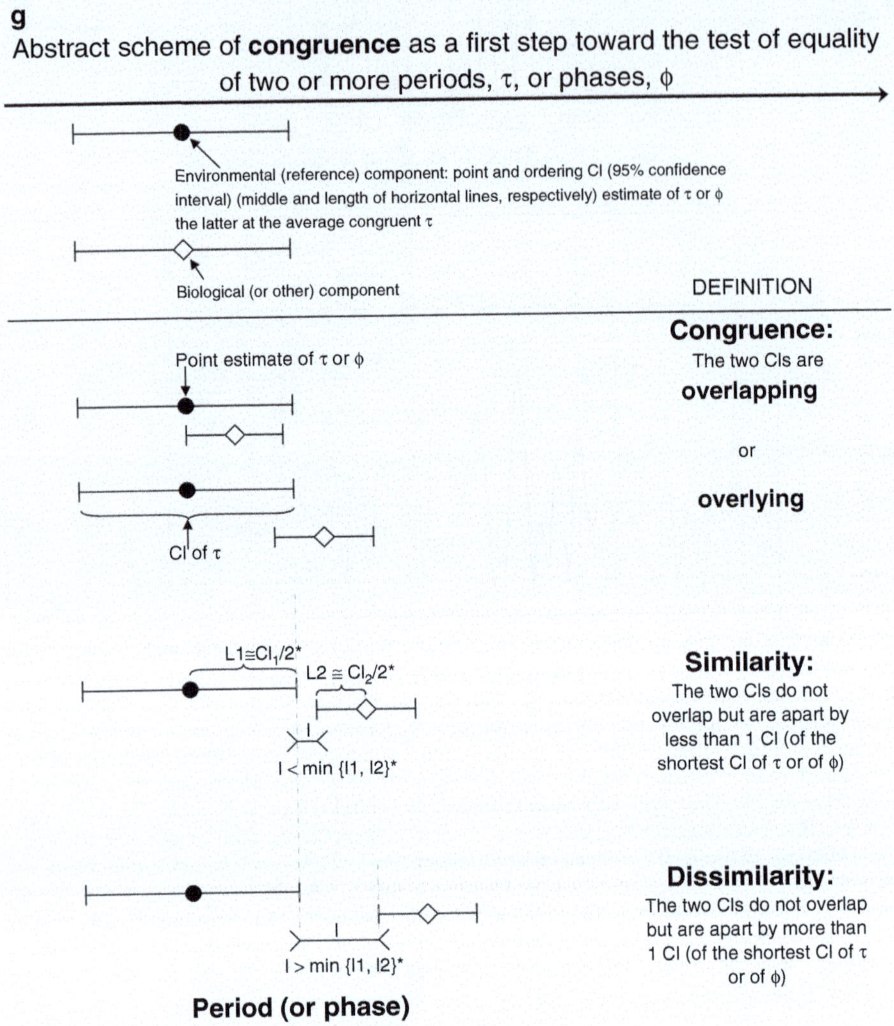

Wobbly nature of some natural physical environmental and biological components qualify the assumptions underlying the inferential statistical approach, according to Marquardt (J SIAM 1963; 11: 431-41; cf. Halberg F. Acta med rom 1980; 18: 399-440). Only the congruence of anticipated components can be meaningfully assessed to approximate the test of H_0: τ_1 (e.g., environmental τ) = τ_2 (e.g., biological τ) [= ... = τ_k (e.g., sociological τ)] , or the test of H_0: $\phi_1 = \phi_2$ [= ... = ϕ_k].

* l : distance between non-overlapping CIs of τs or ϕs; l1 and l2: distances between point estimates of τ or ϕ and end of its CI proximal to another CI of τ or ϕ being considered.

Fig. 21.8 (continued)

21.6 Remove and Replace Revisited, with Circaseptans

About-7-day rhythms opened a new chapter in chronomics, the aligned recording of variables in and around us. Again a remove-and-replace approach proved to be helpful, except that this time the sun, not the surgeon (in blinding) or genetics (in the

h CONGRUENCE IN CERTAIN ENVIRONMENTAL CIRCASEPTAN SPECTRAL COMPONENTS AND IN SOME CARDIOVASCULAR COUNTERPARTS DURING 1998 - 2005*

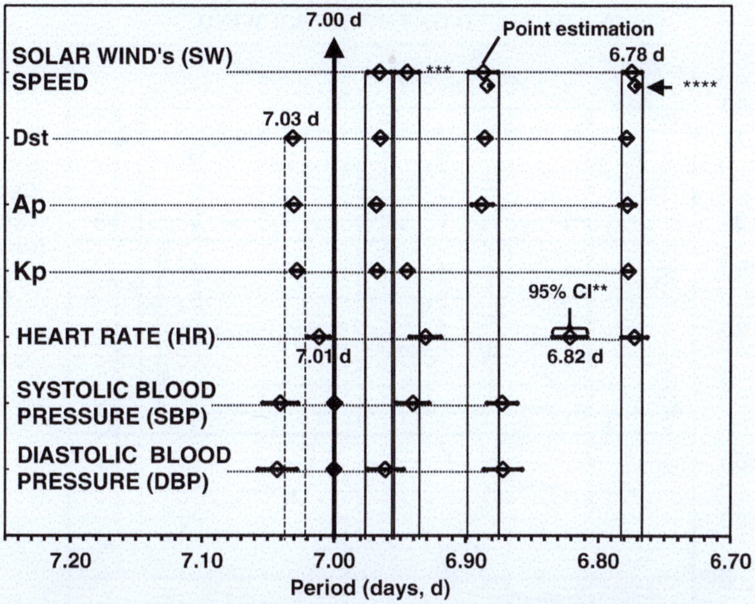

*All peaks are statistically significant (P<0.001) by linear-nonlinear least squares cosinor spectra (not corrected for multiple testing). HR, SBP and DBP (N=124,263 each) - half-hourly records of GSK, a 72-year old man at start of around-the-clock monitoring, Data: SW (N=69,845) hourly values from http://omniweb.gsfc.nasa.gov/, Dst, Ap and Kp 3-hourly data ((N=23,376 each) from http;//spidr.ngdc.noaa.gov/.
**CI = confidence interval. ***Two separate spectral peaks without CI-overlapping.
****All available daily SW data during 1963 - 2005.

Fig. 21.8 (continued)

case of congenitally blind ZRD mice), did the removal and the cosmos the replacement: when solar activity had 7-day components in its (Walsh) spectra (computed in St. Petersburg) [81], human HR (in Minnesota and Connecticut) had these components amplified in its spectrum and vice versa when these components were lacking in the sun's record, the HR's corresponding components were damped, but not lost (Fig. 21.2a, 2b); they were also partly built in [53, 71]. This coding of the circaseptans in our genes is compatible with resonance; hence, it is hardly surprising to find an occasional pull of a biological circaseptan acrophase by a geomagnetic cycle with a proximal frequency, e.g., for HR but not for BP in Fig. 21.8f.

Before us, Fraser-Smith [82] found a precise 7-day component in the geomagnetic planetary index Ap. We confirmed him, of course [83], for the precise data accumulated by the time when he did the analyses, but in the longer series of the planetary geomagnetic index Kp, which we subsequently analyzed many years later [84], the major period of a component in the circaseptan spectral region was of

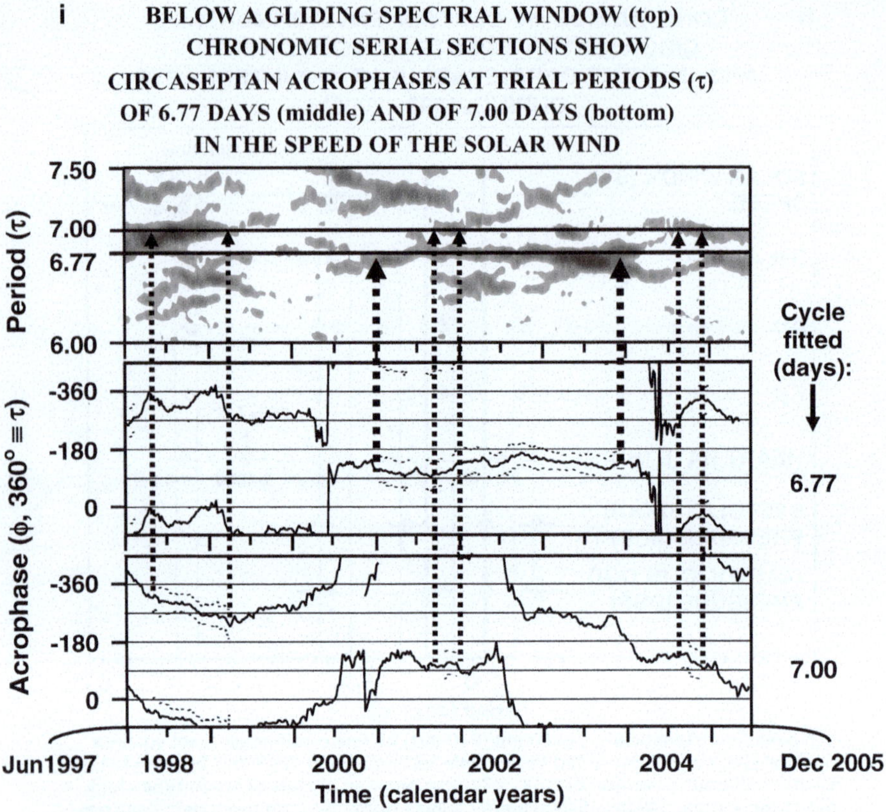

Data from http://omniweb.gsfc.nasa.gov/html/ow_data.html.
In gliding spectral window, interval = 1 year, increment = 1 week; shaded areas show percentage of rhythm (from 1.2 to 6.8 %). Statistical significance seen as 95% confidence intervals of ϕs shown as dots bracketing curves (*middle* and *bottom*). Reference time: 00:00 on Dec 21, 1997. *Dotted arrows* indicate correspond ence between circaseptan components in the phase (*middle* and *bottom*) and period (*top*) domains. 23-rd solar cycle began in May 1996, its maximum was in Apr 2000 and the cycle still continued descending after the end of record.

Fig. 21.8 (continued)

~6.75 days [83, 85, 86]. Others subsequently confirmed this component in Kp, noting its changes with solar cycle stage [87], and extended it to the antipodal geomagnetic index aa [88], as we did subsequently, in the context of extensive studies on biomedical circaseptans, notably by the school of Gunther Hildebrandt [76, 86, 89], among others [74, 90, 91].

At this point, it seemed worthwhile to look for congruence, defined for aeolians in view of overlying or overlapping CIs (95% confidence intervals) of two or more periods or phases, as visualized in the abstract Fig. 21.8g. By congruence, we test for the correspondence of periods and only in this restricted (rather than literal) sense for the reciprocity of two or more periods. Congruences were found among time series of the same variable, organ, organ system, or organism, also within a biological discipline such as physiology or within areas of physics, and furthermore transdisciplinarily among variables of many different disciplines such as cosmo-helio- or geophysics vs. physiology, pathology, and epidemiology, e.g., Fig. 21.8h.

21.7 Differential Bioenvironmental Circaseptan Congruences

The transdisciplinary Fig. 21.8h summarizes data from an elderly man, GSK (with MESOR-hypertension complicated by an intermittent circadian BP overswing), who performed rigorous half-hourly monitoring with few gaps for 10 years, aligned with changes during the same span in a few variables of his terrestrial and more remote environment. These results stem from global (again in the sense only of "based on the entire time series") spectral windows computed for a narrow about-7-day region of the spectrum; multiple components are seen in physics and physiology, each with a CI (95% confidence interval) shown as the length of a horizontal line through the point estimate of a period, the latter shown as a diamond. The top row shows periods found in the speed of the solar wind (SWS). Four SWS components are congruent with one or another of the periods detected (in the next 3 rows) in three related geomagnetic indices, the equatorial Dst and the planetary Ap and Kp. As noted with the abstract Fig. 21.8g, congruence refers to no more than overlapping or overlying CIs of a characteristic such as the period and not necessarily to causal relations. The global summary in Fig. 21.8h cannot tell whether any particular component was present only during different sections of the series rather than during the entire span. In view of aeolian behavior, any intermittent presence of one or the other, if not of most components, is hardly surprising.

Chronobiologic serial sections, aligned with a gliding spectral window of the SWS in Fig. 21.8i, support caution by revealing that we are dealing with a component drifting in frequency that is 6.77-days long, present consistently (while phases are bracketed by dots representing CIs) only part-time (during the descending phase of solar cycle 23). In this approach, the fit of the wrong period of 7.00 days, also shown, reveals no consistent statistical significance during a relatively long section of the span investigated and hence serves to rule out a precise 7-day component. (When phases are bracketed by dots, they are advancing, indicating a period shorter than that fitted.) By contrast, the fit of two correct periods in the same span examined suggests the concomitant presence of two separate wobbly circaseptan components in human HR, one slightly longer, the other somewhat shorter than 7 days (not shown).

Congruence in Fig. 21.8h of a 7.03-day component among the three geomagnetic indices is not, however, shared by SWS. This is hardly surprising since no more than a plot (not shown) of Kp vs. Wolf numbers reveals differences as well as similarities between these variables that, if then available, may have prevented General Sir Edward Sabine from saying that the behavior of sunspots and geomagnetic activity is identical, an impression that may well be gained from viewing the much shorter time series of the two variables that were available during his time.[2] Furthermore, some environmental signatures in HR and BP differ. There is congruence for the 6.78-day component found in SWS and all three geomagnetic variables with HR, but not with BP. There is no congruence of HR or SW for an about-7.03-day component found in the geomagnetic indices and in both systolic and diastolic BP. But spurious differences that can come about from many sources, of course, cannot be ruled out, just as congruence in itself does not rule in a relationship.

With this qualification, differential congruences between some but not all variables in and around us may occur at different periods, at different times. They are found at frequencies that are very close to each other. It seems likely that the many close spectral peaklets may be all manifestations of the same physical phenomena. They may be figurative splashes of the "boiling" hydrodynamo sun, as Phil Regal put it in looking at gliding spectra, and as Thomson et al. [92] described the "more than ten million normal modes" of the sun. But more distant and thus well-separated differential congruences are also seen in the para-annual and multi-annual range for GSK and others.

21.8 Cis-Half Years

Rieger et al. [93] documented an about-5-month (~0.42-year) cycle in hard solar flares found to be a wobbly phenomenon by Ballester et al. (2002) for which we found congruence in sudden cardiac death (Fig. 21.1b) and in several kinds of cardiac arrhythmias, again as an intermittent phenomenon, dubbing it a "cis-half year" since "cis" denotes "on this side" of the half year or year.

21.9 Circasemiannuals

In their treatise on geomagnetism, Chapman and Bartels [94] write:

> The 6-monthly wave is, of course, not restricted to *C* [international daily magnetic character-figures], but is also the significant physical characteristic of the annual variation of the u_1-measure of [magnetic] activity ... on the other hand, no reason to attribute special significance to the 12-monthly wave has so far been found. [This is an important statement in 2007 as well, since, in the long available time series of the geomagnetic index, aa, we find

[2] The similarities prompted Sabine, on April 6, 1852, to announce that the sun's 11-year sunspot cycle was "absolutely identical" to the earth's 11-year magnetic cycle. This statement must be qualified by differences found among certain indices of terrestrial magnetism on the one hand and a coronal index and Wolf numbers on the other hand, revealed among others by time-varying phase synchronization to a degree differing among geomagnetic indices.

a near-transyear rather than a circannual component, as seen in Fig. 21.9a.] The 6-monthly sine wave in u_1 has a fairly constant phase, but its amplitude in disturbed years is more than twice as great as in quiet years.

[In the previous paragraph these authors state that:] In the corresponding summation-dial … the decisive preponderance of the directions indicating maxima near the equinoxes excludes all similarity with a "random-walk", and illustrates the reality of the 6-monthly wave, which has just been quantitatively proved by the κ-test.

a PROMINENT CIRCASEMIANNUAL COMPONENT MOSTLY IN THE ABSENCE OF A CALENDAR YEAR, SHOWING BETTER CONSISTENCY THAN THE TRANSYEARS APPARENT MAINLY AFTER 1920 IN THE ANTIPODAL GEOMAGNETIC INDEX aa

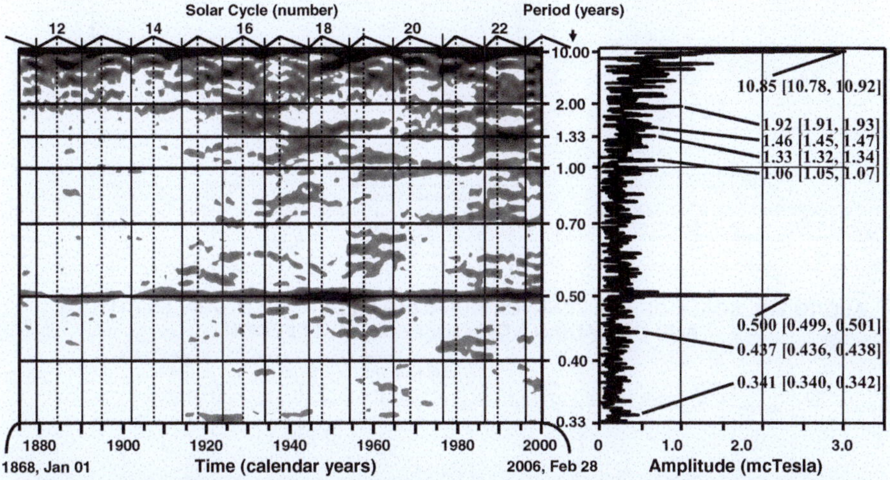

~137 years (y) of daily data from http://www.geomag.bgs.ac.uk.aaindex.html. (N= 50,463); gliding spectrum computed with interval = 14 y (= longest period here fitted), increment = 1 y, harmonic increment = 0.25; *darker shading* corresponds to larger amplitude; vertical continuous or dotted lines = Wolf No minima or maxima, respectively. *Right*: 95% confidence intervals (CIs) of spectral peaks in global window are lacking due to very small size but are given in [ ]. {During 1963-2003, sunspot cycle averages 10.90 [10.72, 11.07] years, differing from 8.39 [8.07, 8.72]-year cycle in solar wind speed.}

Fig. 21.9 (**a**) Transyears in the antipodal geomagnetic index aa are intermittent, wobbly when present as compared to a prominent half-yearly component, the latter commented upon by Chapman and Bartels, and as the same authors implied, for most of the time examined, no yearly component is seen (© Halberg Chronobiology Center). (**b**) At 43.47°N latitude, circulating melatonin may undergo an about-yearly rhythm during the daylight hours and an about half-yearly rhythm during the hours of darkness (© Halberg Chronobiology Center). (**c**) At 65.00°N in Oulu, Finland, human circulating melatonin is characterized by a circasemiannual pattern at noon (© Halberg Chronobiology Center). (**d**) Observations on the equatorial geomagnetic index, Dst, indicating a storm, followed by quiet (*top*), are aligned with results on the hypothalamus, Ht (*middle*), that may be activated via the storm directly and/or via the pineal, Pin, which latter gland may be damped if not exhausted (*bottom*). The storm also involved the adrenal cortex, a neuroendocrine network. These observations on rats can be aligned with findings of a decreased human urinary excretion of melatonin during magnetic storms, in keeping with a damping of pineal function, including its dynamics, gauged by the lowered amplitude and MESOR of the circadian rhythm of the in situ melatonin, presumably as a consequence of the storm. This "experiment of nature" offers a putative mechanism that cannot be dissected in vivo in humans. Nonetheless, post hoc ergo propter hoc reasoning and confounding by any unassessed effects upon circaseptan rhythms cannot be excluded and will require rebuttal or confirmation. A solution may be most readily available if

b POSSIBLE SPECTRAL VARIANCE TRANSPOSITION FROM A 1.0-YEAR (SUNSHINE-RELATED?) TO A 0.5-YEAR (MAGNETICS-RELATED?) MODULATION OF CIRCULATING MELATONIN (MEL) BY NIGHT*

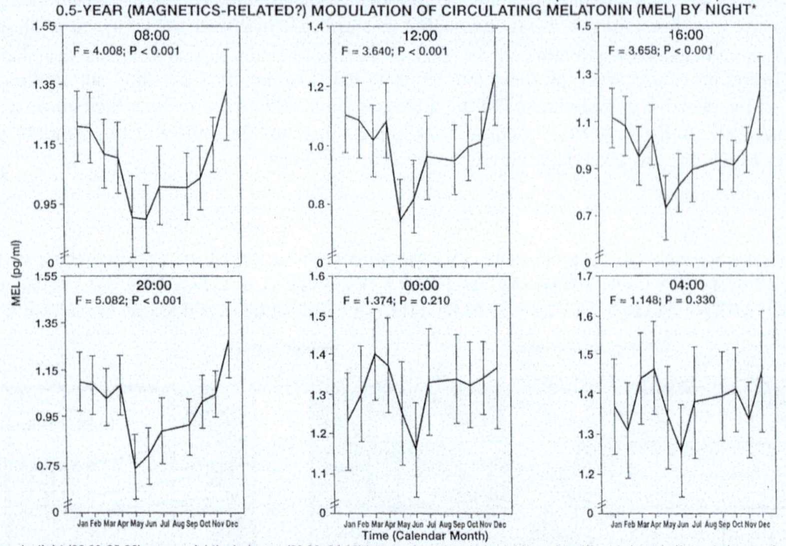

* During daylight (08:00-20:00) versus nightly darkness (00:00; 04:00) hours, about-yearly (sunshine-related?) and about-half-yearly (magnetics-related?) effects are seen; one-way analyses of variance on $\log_{10}$-transformed data: means and 95% confidence limits; data collected at 4-hour intervals for 24 hours on 172 subjects studied between 13 October 1992 and 15 December 1995, in Florence, Italy, 43.47 °N. CC 5/97

c ALONG THE SCALE OF THE YEAR, CIRCULATING MELATONIN REVEALS LATITUDE AND CIRCADIAN STAGE-DEPENDENT EFFECTS*

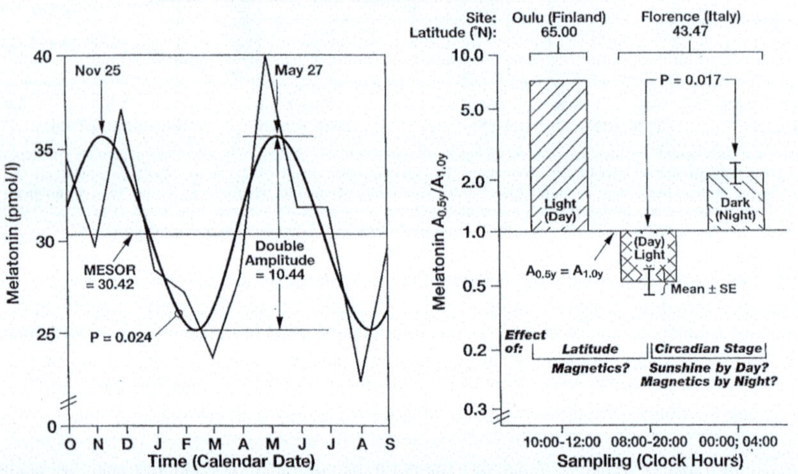

* Half-yearly (possibly magnetics-related) response pattern is seen at 65.00 °N in Oulu, Finland (left half of graph) when measured around noon; accordingly, the log-ratio of the half-yearly to yearly amplitudes exceeds unity (first bar in right half of graph) whereas this ratio (original data not shown) is much smaller than unity (second bar in right half of graph) during waking hours (including noon) in Florence, Italy, at 43.47 °N, where the amplitude ratio exceeds unity only during nightly darkness (third bar in right half of graph); original data from 172 patients of B. Tarquini, each sampled every 4 hours for 24 hours; data in Oulu are from 11 healthy men sampled monthly for one year (Martikainen et al., Acta Endocr. 109:446-450, 1985). CC 5/97

Fig. 21.9 (continued) investigators of pineal function check the presence or absence of a magnetic storm during their studies; if they find days of storms or of extreme magnetic quiet, they can compare data from magnetically extreme days with those on quiet days by a cosinor computation of parameters. In any event, original data or circadian parameters with calendar dates can be sent to corne001@umn.edu for further analyses and/or integration with the results summarized herein and elsewhere (Jozsa et al. [16]) (© Halberg Chronobiology Center). (**e**) When, according to Scheme 1,

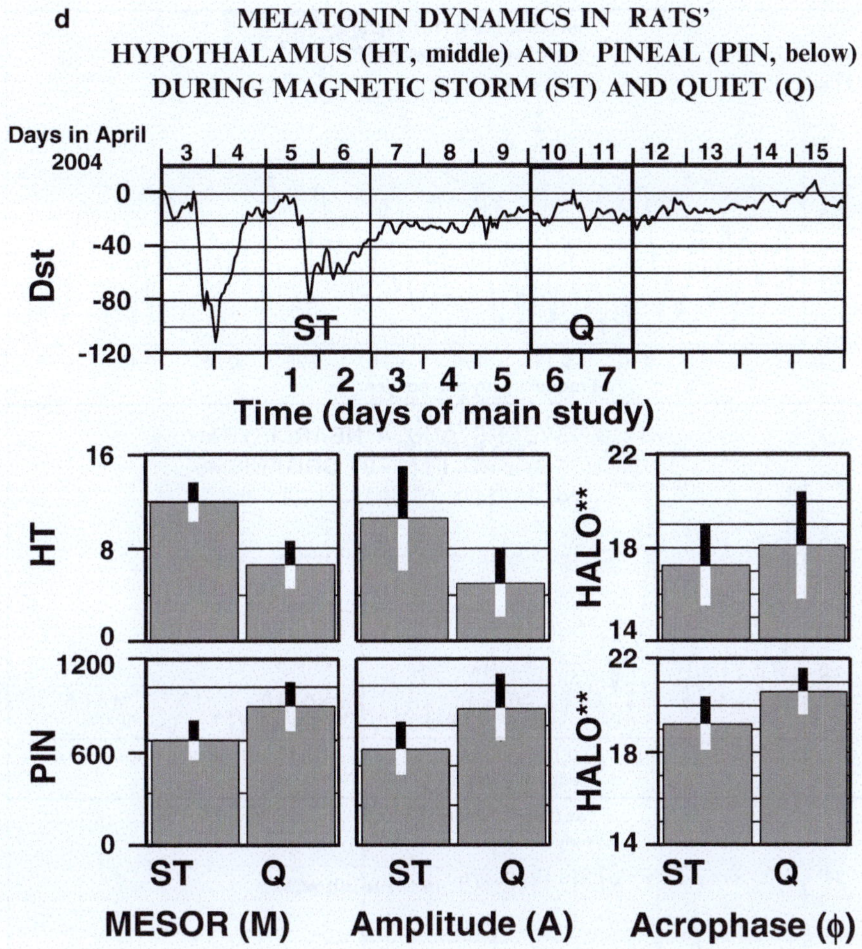

ST =2 days of the second part of a moderate double magnetic storm, Q = 2 days of magnetic quiet, gauged by the geomagnetic equatorial disturbance index Dst in nT (plotted, *top*) of -115 and -80 and Kp (a planetary magnetic disturbance index of slightly above 6 in arbitrary units in each storm (Kp not shown).
*** HALO = hours after light on. Vertical lines straddling bars are 95% confidence intervals. P < 0.05 in each comparison, except for ϕ in hypothalamus.**

Fig. 21.9 (continued) the percentage rhythm (PR) computed by analyses of consecutive 1-year data sections is analyzed, the longest nonlinearly validated period differs from that in a time series consisting of MESORs (*middle*) or circannual amplitudes (*bottom*). The longest major period could be a harmonic of the Hale cycle in PR on *top*; the major period in the *middle*, perhaps a harmonic of the Schwabe cycle; and that at the *bottom* could correspond to Markov and Sivaraman's (1989) global solar cycle (© Halberg Chronobiology Center). (**f**) Transdisciplinarily applicable methods illustrated for Wolf's relative sunspot numbers (IA-D), terrestrial telescope-assessed solar magnetism (IIA-C), and solar wind speed (IIIA-C)*. Gliding spectral windows of amplitude, IA, and of percentage rhythm, IB. A time series of sunspot numbers from 1500 to 1995 was kindly provided

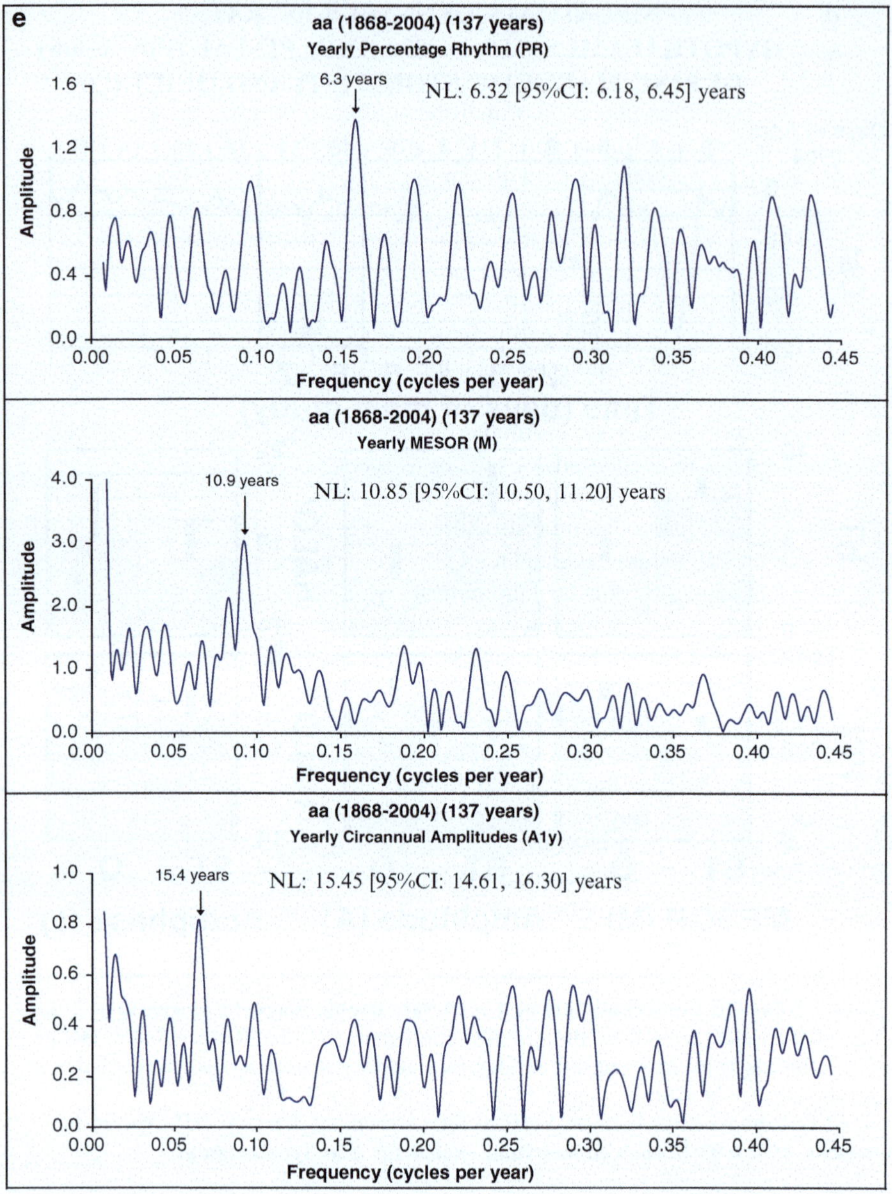

Fig. 21.9 (continued) by Dr. Jaroslav Strestik (Strestik and Mikulecky 1994). The percentage rhythm, but not the amplitude, visualized the persistence of the circadecadal Schwabe cycle during the Maunder minimum. Other illustrative gliding spectral (amplitude) windows, ID, II, IIIB, or a gliding period with its 95% confidence interval, IC, of solar activity, including those of terrestrial telescope recorded solar magnetism, II, are of interest as such and methodologically by showing the trade-off in choosing the interval for analysis between the resolution of dynamic behavior in time vs. frequency (II). The resolution in frequency increases with the length of the interval analyzed (II *bottom* to *top*). Note variability in original data illustrated for the solar wind speed, IIIA measured by satellites, revealing clear peaks by global spectral analysis, IIIC, albeit in time we are dealing with a very wobbly (aeolian) oscillation, IIIB, that is difficult to discern with the unaided

f

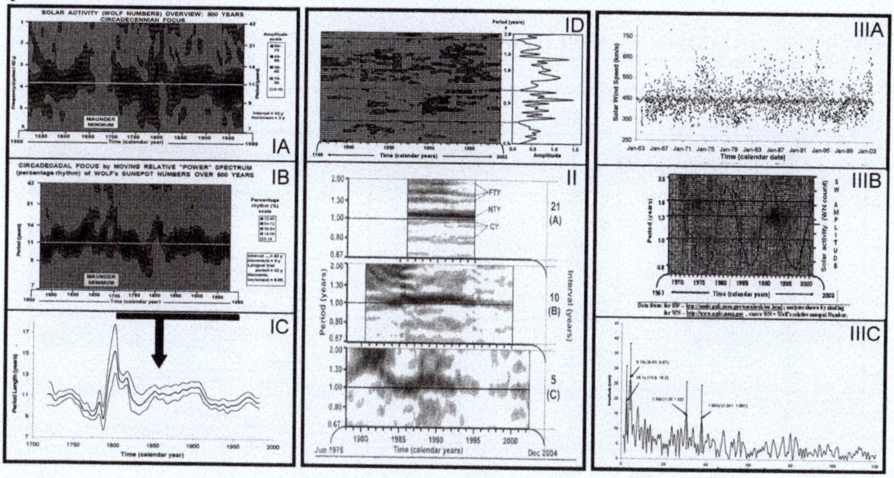

g

TIME-VARYING ABOUT -YEARLY SPECTRAL REGION OF ANTIPODAL MAGNETIC DISTURBANCE INDEX aa* : WAXING AND WANING IN AMPLITUDE WITH AEOLIAN CHANGES IN FREQUENCY

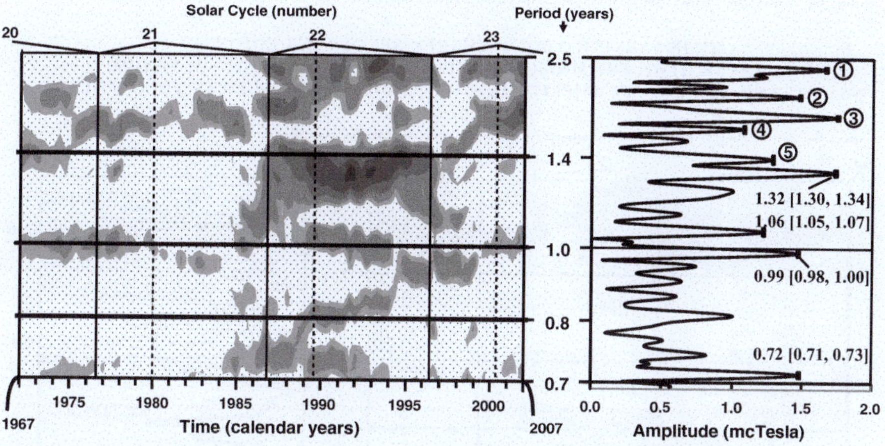

* Weekly averages of data from http://www.geomag.bgs.ac.uk.aaindex.html, during 1967 - 2006 (N=2088). Gliding spectrum computed with interval = 10 y, increment = 4 months, harmonic increment = 0.05; darker shading corresponds to larger amplitude. Shading begins at ordering P=0.001. In global spectrum 95% confidence intervals (CIs) are shown as bars at peaks with statistically significant ordering P < 0.01. CIs for peaks identified by circled numbers: 1 – 2.31 [2.27 – 2.35], 2 – 1.92 [1.89 – 1.95], 3 – 1.71 [1.69 – 1.74], 4 – 1.61 [1.58 – 1.64], 5 – 1.39 [1.37 – 1.42].

Fig. 21.9 (continued) eye when the results of an anticipated 1.30-year cosine curve are fitted to the data, for comparison with an also-fitted precise 1-year trial period, IIIA. A gliding spectral window, IIIB, shows the variability on hand in keeping with variability reported by moving spectra for the solar wind earlier by Mursula and Zieger (2000) and by ourselves. During Maunder's minimum, the about 11-year solar activity cycle persists, as shown by the percentage rhythm (IB), albeit with a much reduced amplitude (IA), the period also varies widely (IC). Also shown by gliding window are wobbly transyears resolved in complementary global window (right, ID). Deviation from precise calendar year needs resolution better achieved with longer interval progressively displaced throughout time series (IIA-C). Relatively minuscule transyears in solar wind speed, revealed in chronogram (IIIA) and least-squares window (IIIC), change as a function of time but not necessarily as a function of solar cycle stage (IIIB) (© Halberg Chronobiology Center).

h TIME-VARYING ABOUT-YEARLY SPECTRAL REGION
IN SOLAR WIND's SPEED*
WAXING AND WANING IN AMPLITUDE WITH AEOLIAN CHANGES IN FREQUENCY

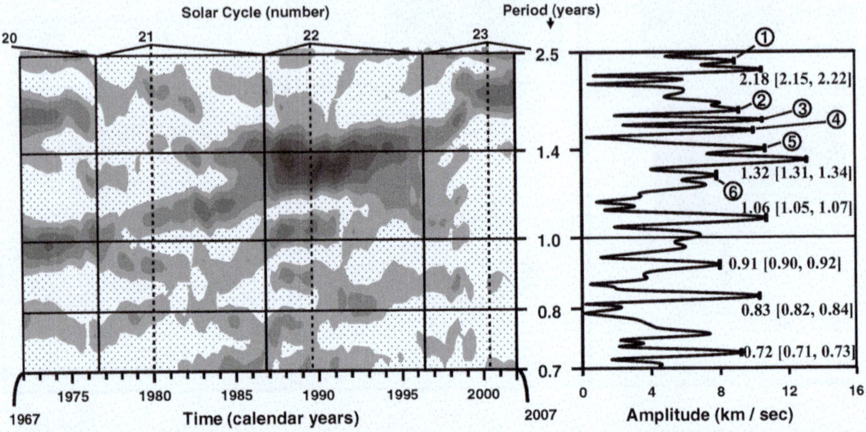

* Weekly averages of data from http://omniweb.gsfc.nasa.gov during 1967 - 2006 (N=1958). Gliding spectrum: interval = 10 y, increment = 4 months, harmonic increment = 0.05; darker shading corresponds to larger amplitude. Shading begins at ordering P=0.001. In global spectrum 95% confidence intervals (CIs) are shown as bars at peaks with statistically significant ordering P < 0.01. CIs for peaks identified by circled numbers: 1 – 2.33 [2.28 – 2.37], 2 – 1.69 [1.67 – 1.72], 3 – 1.60 [1.58 – 1.62], 4 – 1.52 [1.50 – 1.54], 5 – 1.39 [1.37 – 1.41], 6 – 1.24 [1.23 – 1.26].

i TIME-VARYING ABOUT -YEARLY SPECTRAL REGION
OF 1-MINUTE TIME ESTIMATION IN HEALTHY MAN* DURING ~ 35 YEARS:
WAXING AND WANING IN AMPLITUDE WITH AEOLIAN CHANGES IN FREQUENCY

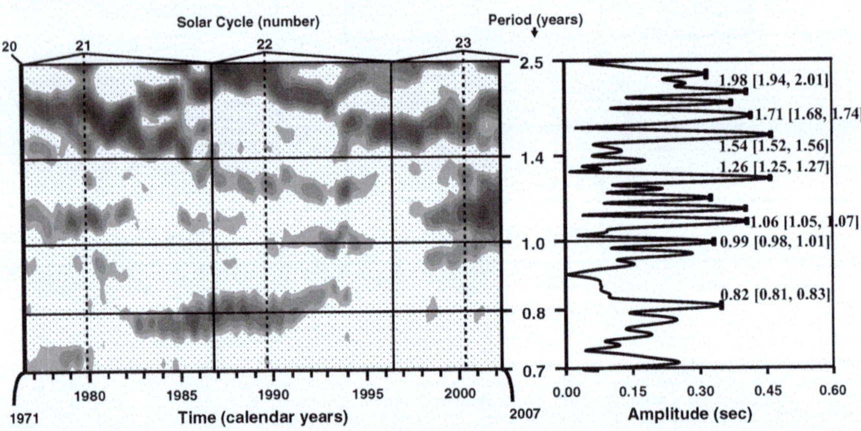

*Weekly averages of data from RBS, clinically healthy man, 21-y old at start of self-measuring 5 – 7 times per day during 1971 - 2006 (N=1758). Gliding spectrum computed with interval = 10 y, increment = 4 months, harmonic increment = 0.05; darker shading corresponds to larger amplitude. Shading begins at ordering P=0.001. In global spectrum 95% confidence intervals (CIs) are shown as bars at peaks with statistically significant ordering P < 0.01.

Fig. 21.9 (continued) (**g**) Time course (*left*) and global summary (*right*) during 4 decades of wobbly, transyearly, and cisyearly behavior of the geomagnetic index aa during span of study by a clinically healthy man, RBS. Note prominent amplification during the years from about the mid-1980s through the mid-1990s (© Halberg Chronobiology Center). (**h**) Time course (*left*) and global summary (*right*) during 4 decades of wobbly, transyearly, and cisyearly behavior of solar wind speed during span of study by a clinically healthy man, RBS. Note prominent amplification during the years from ~1985 to ~1995 (© Halberg Chronobiology Center). (**i**) Time course (*left*) and global summary (*right*) during 3.5 decades of wobbly, transyearly, and cisyearly behavior

j TIME-VARYING ABOUT -YEARLY SPECTRAL REGION
OF SYSTOLIC BLOOD PRESSURE IN HEALTHY MAN* DURING ~40 YEARS:
WAXING AND WANING IN AMPLITUDE WITH AEOLIAN CHANGES IN FREQUENCY

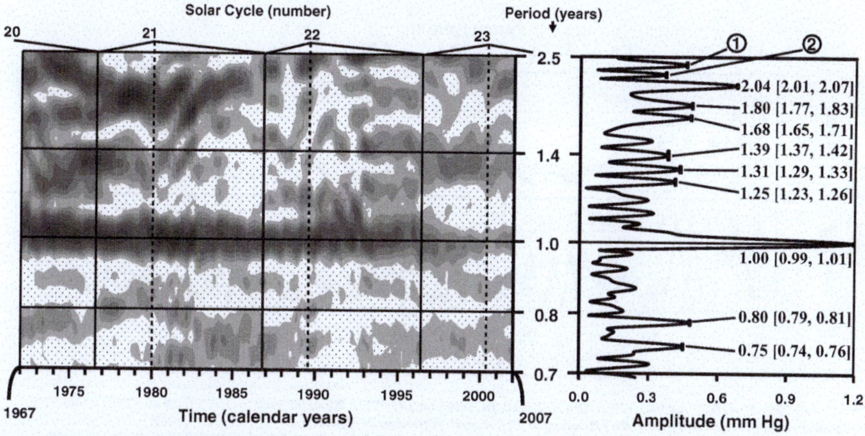

* Weekly averages of data from RBS, clinically healthy man, 21-y old at start of self-measuring 5 – 7 times per day during 1967 - 2006 (N=1962). Gliding spectrum computed with interval = 10 y, increment = 4 months, harmonic increment = 0.05; darker shading corresponds to larger amplitude. Ordering shading begins at P=0.001. In global spectrum 95% confidence intervals (CIs) are shown as bars at peaks with statistically significant ordering P < 0.01. CIs for peaks identified by circled numbers: 1 – 2.35 [2.30 – 2.40], 2 – 2.19 [2.15 – 2.24].

k TIME-VARYING ABOUT -YEARLY SPECTRAL REGION
OF HEART RATE IN HEALTHY MAN* DURING ~40 YEARS:
WAXING AND WANING IN AMPLITUDE WITH AEOLIAN CHANGES IN FREQUENCY

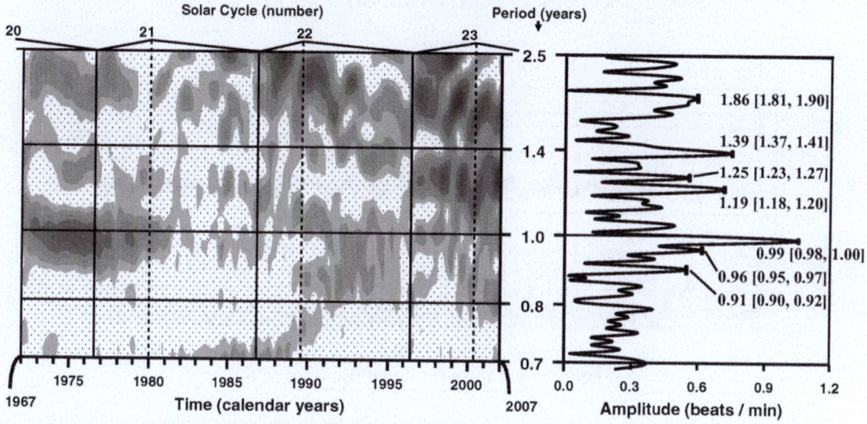

* Weekly averages of data from RBS, clinically healthy man, 21-y old at start of self-measuring 5 – 7 times per day during 1967 - 2006 (N=1978). Gliding spectrum computed with interval = 10 y, increment = 4 months, harmonic increment = 0.05; darker shading corresponds to larger amplitude. Ordering shading begins at P=0.001. In global spectrum 95% confidence intervals (CIs) are shown as bars at peaks with statistically significant ordering P < 0.01.

Fig. 21.9 (continued) of the estimation of 1 minute by a clinically healthy man, RBS. Note a fainter band around a trial period of 1.3 years between 1985 and 1995, possibly mimicking the more prominent bands in Fig. 9g and h (© Halberg Chronobiology Center). (**j**) Time course (*left*) and global summary (*right*) during 4 decades of wobbly, transyearly, and cisyearly behavior of systolic blood pressure of a clinically healthy man, RBS. Note, by contrast, consistently prominent circannual periodism dominating, without interruption, in the gliding spectral window on the left and towering over other components in the global summary on the right (© Halberg Chronobiology Center). (**k**) Time course (*left*) and global summary (*right*) during 4 decades of a consistent

l DECADAL, MULTI- & EXTRA-ANNUAL CONGRUENCE OR SIMILARITY GAUGED BY OVERLAPPING OR OVERLYING VS CONTIGUOUS CIs¶ OF PERIOD, τ, IN SPECTRA OF SOLAR WIND SPEED (SWS) AND SYSTOLIC BLOOD PRESSURE (SBP) MESOR (M)*

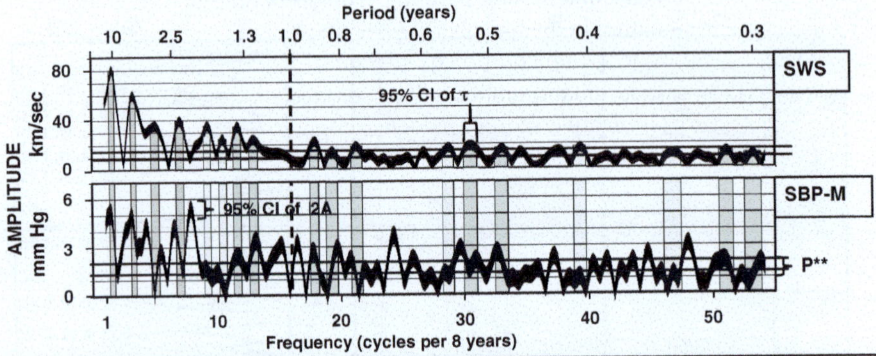

*SWS: daily data from ftp://nssdcftp.gsfc.nasa.gov/spacecraft_data/omni, N= 5272. BP data: daily averages from Dec 1989 to Jul 2006 monitored ~ every 30 min (with gaps) by FH, a man of 70 y at start of records. N = 2684.
**Horizontal lines indicate ordering significance at the 0.001 and 0.05 levels only as a first approximation; until more robust methods become available, that are not dependent upon the assumptions of independence, normality and homogeneity of variance, a transdisciplinary congruence of periods and a "remove – replace" approach remain the criteria of importance. Congruence between components with amplitudes differing from zero with an ordering P betweein 0.001 and 0.05 are questionable.
¶ CI = 95% confidence interval of τ (of each component fitted separately).

m DECADAL, MULTI- & EXTRA-ANNUAL CONGRUENCE OR SIMILARITY GAUGED BY OVERLAPPING OR OVERLYING VS CONTIGUOUS CIs¶ OF PERIOD, τ, IN SPECTRA OF SOLAR WIND SPEED (SWS) AND HEART RATE (HR) MESOR (M)*

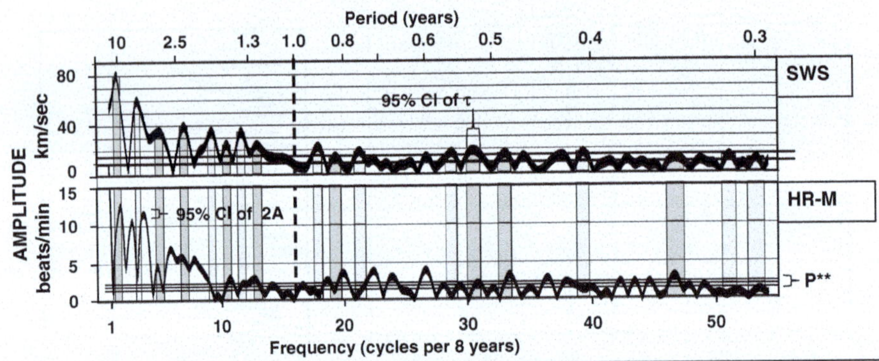

*SWS: daily data from ftp://nssdcftp.gsfc.nasa.gov/spacecraft_data/omni, N= 5272. HR data: daily averages from Dec 1989 to Jul 2006 monitored ~ every 30 min (with gaps) by FH, a man of 70 y at start of records. N = 2447.
**Horizontal lines indicate ordering significance at the 0.001 and 0.05 levels only as a first approximation; until more robust methods become available, that are not dependent upon the assumptions of independence, normality and homogeneity of variance, a transdisciplinary congruence of periods and a "remove – replace" approach remain the criteria of importance. Congruence between components with amplitudes differing from zero with an ordering P betweein 0.001 and 0.05 are questionable.
¶ CI = 95% confidence interval of τ (of each component fitted separately).

Fig. 21.9 (continued) circannual rhythm of heart rate of a clinically healthy man, RBS. Note also intermittent transyearly and cisyearly bands, but the circannual component is globally most prominent, not quite as in Fig. 21.9j (by contrast to Fig. 21.9g, h and i) (© Halberg Chronobiology Center). (**l**) Alignment of spectral windows of the speed of the solar wind (*top*) with those of the MESOR of systolic blood pressure of an elderly man, FH (© Halberg Chronobiology Center). (**m**) Alignment of spectral windows of the speed of the solar wind (*top*) with those of the MESOR of heart rate of an elderly man, FH (© Halberg Chronobiology Center). (**n**) Alignment of spectral

n MULTI- & EXTRA-ANNUAL CONGRUENCE GAUGED BY OVERLAPPING OR OVERLYING CIs¶ OF PERIOD, τ, IN SPECTRA OF SOLAR WIND SPEED (SWS), SYSTOLIC BLOOD PRESSURE (SBP) MESOR (M) AND CIRCADIAN DOUBLE AMPLITUDE (2A)*

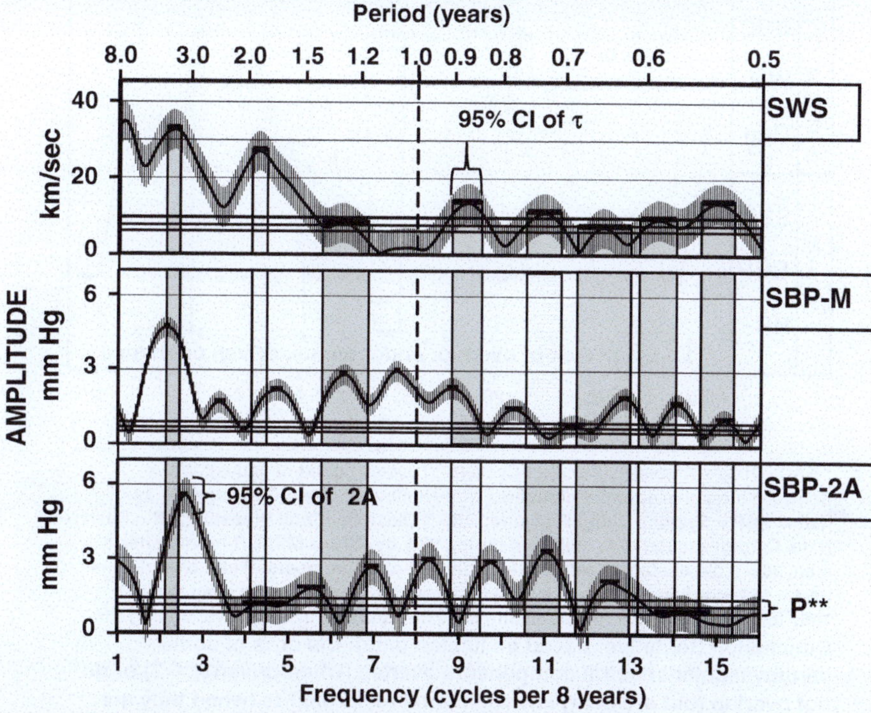

*** SWS: daily data from ftp://nssdcftp.gsfc.nasa.gov/spacecraft_data/omni, N=3013. BP data: daily averages from Mar 1998 to Jul 2006 monitored ~ every 30 min (with few gaps) by GSK, a man of 72 y at start of records. Original N = 133,561, daily averages = 2774.**
**** Horizontal lines indicate ordering significance at the 0.05, 0.01 and 0.001 levels only as a first approximation; until more robust methods become available, that are not dependent upon the assumptions of independence, normality and homogeneity of variance, a transdisciplinary congruence of periods and a "remove – replace" approach remain the criteria of importance. Congruence above ordering P=0.001 is questionable.**
¶ CI = 95% confidence interval (each component fitted separately).

Fig. 21.9 (continued) windows of the speed of the solar wind (*top*) with those of the systolic blood pressure MESOR (*middle*) and double amplitude of an elderly man, GSK (© Halberg Chronobiology Center). (**o**) Differential circadecadal and multidecadal congruence. In the multidecadal range of an adult man's, RBS's, circulation during 4 decades, his longest period in heart rate is congruent with the Schwabe cycle (WN) and the next longest period with the Hale cycle (BSC), whereas the longest period in blood pressure (differs from that in heart rate and) is congruent with a period in the Hale cycle; the next peak, the blood pressure, is congruent with yet another period in the Schwabe cycle. Further resolution in time of multidecadal prominence is not possible because of the limited (40-year) length of the available time series, but the results are in keeping with lock-ins of different physiological variables into different frequencies of the same environmental variables (© Halberg Chronobiology Center)

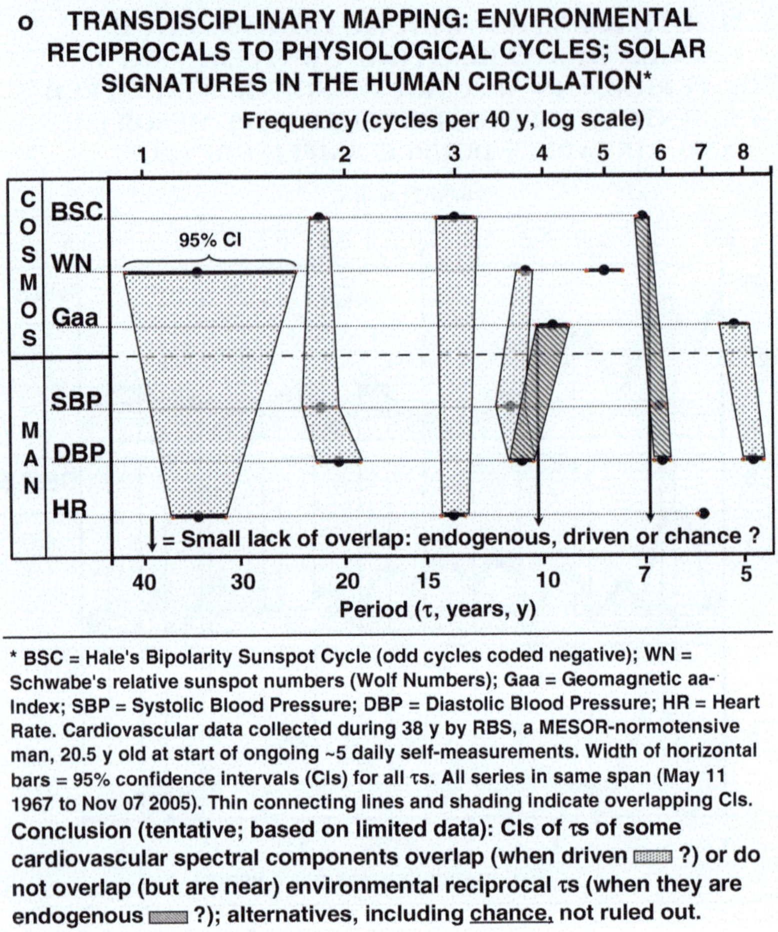

Fig. 21.9 (continued)

Against this background, Armin Grafe [95] not only recognized the solar origin of the physical half year but had looked into its biological associations (albeit not yet with congruences) to learn about a half-year pattern in human epilepsy (personal communication). Subsequently, Russell and McPherron [96] modeled this cycle, which indeed has physiological and pathological counterparts (Fig. 21.6c). A half-yearly cycle in circulating melatonin, restricted at middle geographic latitudes to nightly values [97] (Fig. 21.9b) found at noon at high latitudes [98] (Fig. 21.9c), renders this variable of particular interest as a mechanism underlying effects of terrestrial magnetism on mammals [16, 30, 99, 100]. Both the midline-estimating statistic of rhythms (MESOR) and the circadian amplitude of the rhythm in melatonin content in situ are increased in the hypothalamus and dampened in the pineal by magnetic storms in the rat (Fig. 21.9d) [16]. Human nocturnal melatonin excretion at all latitudes investigated is also decreased during magnetic storms [99, 100].

21.10 Circannual Variation

Julius Bartels, the late mathematician turned opinion-leading geophysicist, advocated the use of inferential statistics. He did not, however, refrain from stacking data to describe a yearly variation. A half-yearly pattern, but no annual variation, is seen (in Fig. 21.4 of Bartels [101]). Zieger and Mursula [102] describe an annual variation in near-earth solar wind speed and in the geomagnetic activity, Kp, along with an IMF sector component with a scholarly review of the literature. Yearly data of the aa index, which greatly antedates Kp, were analyzed over the span from 1868 to 2004 (137 years), fitting a 1-year cosine curve by least squares to each separate calendar year's data, to obtain estimates of the MESOR and circannual amplitude and acrophase as well as a measure of the coefficient of determination or percentage rhythm. Three new, broader time series (N = 137 each) were then formed of the MESOR, circannual amplitude, and percentage rhythm, according to the flowchart in Scheme 21.1. Results are shown in Fig. 21.9e, revealing differences in the most prominent spectral peak with a period of about 6.3 years for the percentage rhythm, 10.9 years for the MESOR, and 15.4 years for the circadian amplitude. The following components are linearly statistically significant:

1. The percentage rhythm (PR) is characterized by components with periods of about 10.4 years ($P = 0.037$), 6.3 years ($P = 0.002$), 5.2 years ($P = 0.030$), 3.4 years ($P = 0.024$), and 3.1 years ($P = 0.022$).
2. The MESOR is characterized by components with periods of 12.1 years ($P = 0.014$) and 10.9 years ($P < 0.001$). The latter component corresponds to the solar activity cycle, expected to also characterize the average aa values (or the yearly MESORs).
3. The circannual amplitude is characterized by components with periods of 72.1 years ($P = 0.020$) and 15.4 years ($P = 0.003$). The latter component is the most prominent and is seen to modulate the circannual amplitude of aa. In a cosinor periodogram of daily aa data, it would be seen as a double peak at frequencies of 1.065 and 0.935 cycles per year, the latter corresponding to a near-transyear component often seen in spectra of helio- and geomagnetic indices, albeit not invariably as a symmetrical peak bracketing the year, as would be anticipated if it corresponded only to a modulation in amplitude, phase, or frequency (Fig. 21.9a–e).
4. Analyses of the circannual acrophases are more difficult to carry out as it is a circular variable. After expanding the phase in such a way as to always choose a given year's phase so that it differs from the previous year's phase by no more than 180 degrees, the phase sequence is characterized by a long-term trend that can be approximated by a third-order polynomial, which in turn may perhaps correspond to an about-350-year cycle described by the late Ronald Bracewell [103].[3]

[3]When Bracewell (1953) suggested that, in keeping with "Hale's well-known discovery, based on observations of the magnetic polarity of sunspots, the sunspot cycle lasts twenty-two, not eleven, years ...," he added that "there is no known difference in the terrestrial effects between successive

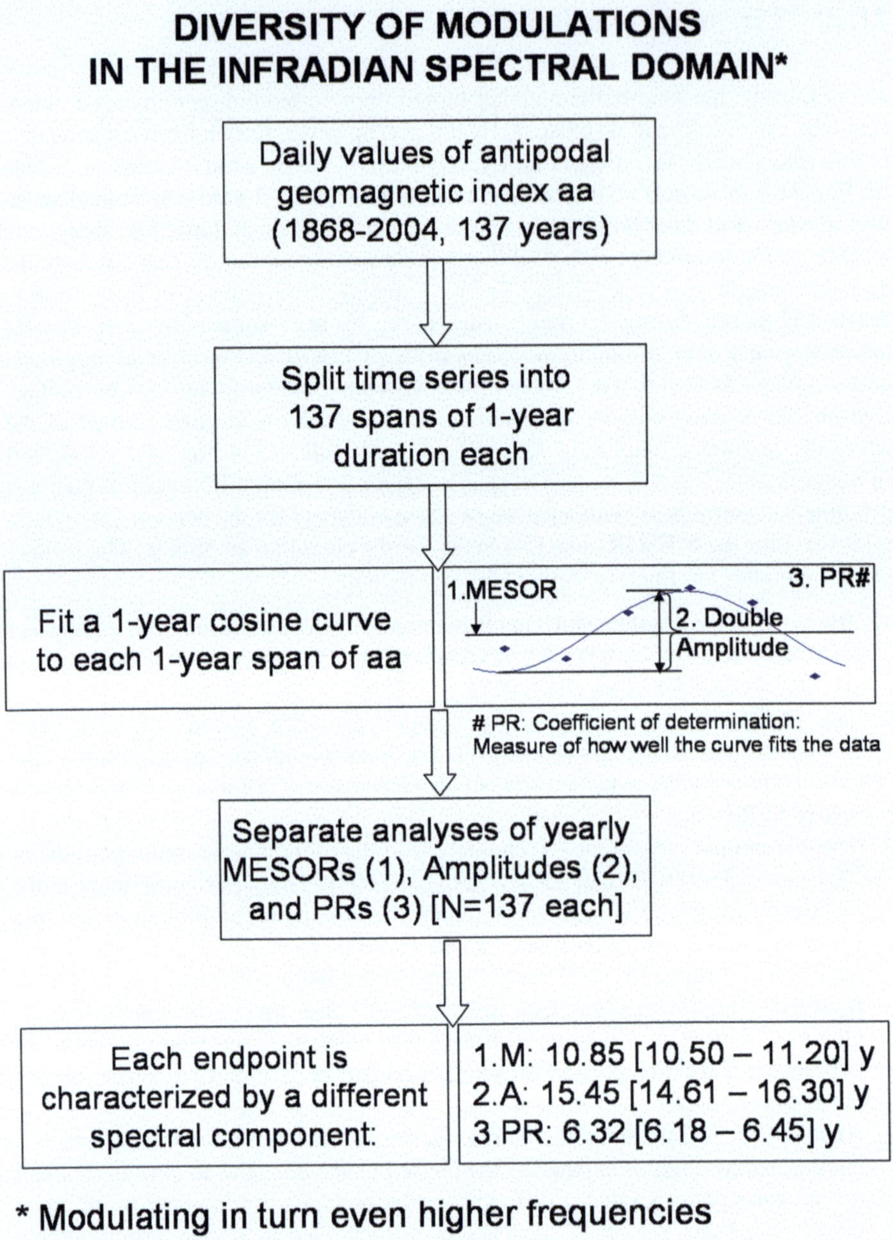

Scheme 21.1 Flowchart describing different analytical steps

Residuals from this model are characterized by components with periods of about 72.1 years ($P<0.001$), most prominently 38.1 years ($P<0.001$), 26.9 years ($P<0.001$), 21.4 years ($P=0.004$), and probably also 11.1 years ($P=0.071$), perhaps corresponding to the Schwabe cycle. Biological circannuals and their modulations of circadian rhythms have been described by Halberg et al. [40, 104].

21.11 Cis- and Transyears

Cis-years and cycles longer than (trans = beyond) a year, the transyears, are para-annual phenomena; the danger of artifacts looms large in data with gaps, as decimations of time series show and as further visualized elsewhere [105].

When Gonzalez and Gonzalez [106] discovered an ~1.3-year oscillation in data from the IMF and Richardson et al. [107] found it in the speed of the solar wind, we had already postulated the desirability to look for reciprocal periods [108] (in the sense of congruence) in biomedicine and physics and, even more broadly, based on the congruence of circaseptans. Mclntosh et al. [109] had reported a 600-day (1.66-year) periodicity in solar coronal holes, and Pap et al. [110] had reviewed periodicities of solar irradiance and solar activity indices shorter than a year, including periods of 8–11 months, about 150–157 days (~0.42 year, as reported by Rieger), and of 51 and 13.5 days. Szabo et al. [111] thereafter found the 1.3-year oscillation again in the interplanetary magnetic field (97; cf. 98). Gazis et al. [112] extended the observation span retrospectively beyond that by Richardson et al. [107] and report a lack of the 1.3-year periodicity during the two preceding solar cycles, while Zieger and Mursula [102] reported on an annual variation in Kp, which we extended to aa for the span examined by them (not shown), but not for the longer series of aa data as a whole (Fig. 21.9a). Mursula and Zieger [113] furthermore report an alternation between an ~1.3-year and an ~1.5- to 1.7-year spectral component by moving spectra computed from data on Kp, available since 1932, as well as from data on solar wind speed dating back only from 1963, while, of course, Fig. 21.9a documents a much longer span.

Prabhakaran Nayar [114] thoroughly reviewed the pertinent literature of physical far-transyears and with his own wavelet analyses uncovered a 33-year variation with a biological published counterpart [105], another example of transdisciplinarily congruent environmental and biological periods, where at least the physical component

---

eleven-year cycles." In 2002, however, Cornélissen et al. reported differences among 3 consecutive solar cycles found by the fit of a 10.5-year cosine curve to daily mortality for myocardial infarction in Minnesota (Table 21.2) to the point that the zero-amplitude assumption during 2 consecutive cycles could be demonstrated with P-values of 0.006 and less than 0.001; the next cycle the P was 0.830. In each solar cycle, the average incidence of myocardial infarctions was the same of 12.0, 12.4, and 12.1 cases/day; only the amplitude dropped by an order of magnitude. Even more persuasive for considering differences between Hale and Schwabe cycles are differences in the frequencies that may resonate with Hale and Schwabe cycles in the human BP and HR, discussed as bioenvironmental congruences of periods.

is independently validated and confirmed, biologically (or vice versa, as the case may be). We had searched and found, in dozens of long biologic time series covering years, most of them decades, what we called physiological far-transyears in the spectral region of ~1.3 years [115, 116], as well as other far-transyears with periods (τs) in the range of 1.2 years ≤ [τ – CI] < [τ + CI] < 1.9 years, e.g., Fig. 21.9a, and further near-transyears (1.00 year < [τ – CI] < [τ + CI] < 1.2 years) (Fig. 21.6d) [15]. Far-transyears in population statistics followed [15], and their scope was soon extended beyond cardiac and psychiatric pathology to natality in the Philippines [117] and in Brazil [118], epilepsy [21] and stroke in Slovakia [20], and, in the absence of a calendar-year component, to the worldwide pattern of terrorist activity [4, 119].

All of these findings rest on the shoulders of geophysicists who had earlier found spectral components longer than a year: Shapiro [120], Fraser-Smith [82], Delouis and Mayaud [121], and Silverman and Shapiro [122]. Corresponding results are also available from investigations of cosmic rays [123, 124]. Hence, it was no surprise to find transyears in a variety of transdisciplinary time series, including one in helioseismics at the solar equator, discovered by Howe et al. [125], and such a component (not commented upon) among others apparent in a spectrum of geoseismics published by Duma and Vilardo [126].

With the antipodal index aa, as did others, we went back to 1868 (when that index was initiated) and found that the far-transyears could not be consistently detected (with statistical significance) in the late nineteenth and early twentieth centuries (Fig. 21.9a) in keeping with the more general demonstration by Clilverd et al. [127] that from solar cycle 14 on, solar activity was increased. Wolf's relative sunspot numbers allow us to go further back in viewing solar variability, with its uncertainty as CI (Fig. 21.9f, Sections IA-ID) [54]. Methodology (Fig. 21.9f) also visualizes for telescope-assessed solar magnetism the role of the interval used for analysis in the transyear region. Both long and short intervals have to be used to resolve a frequency with lower frequency and to scrutinize its behavior in time, respectively, with shorter intervals, in Fig. 21.9f, section II. Section IIIA in Fig. 21.9f shows how small the ~1.3 transyear is in the overall variability, Section IIIB resolves the transyear's time course, and Section IIIC shows its global spectrum. For the span of four decades, Fig. 21.9g and h shows combined gliding (left) and global (right) environmental spectral windows for alignment with psychophysiological ones in Fig. 21.9i–k.

For biomedicine in turn, we found far-transyear oscillations not only in each of dozens of decade-long time series of systolic and diastolic BP and HR but also in variables such as a mental performance (Fig. 21.9i) as well as in core temperature, mood, and vigor and (only in certain geographic locations) in sudden cardiac death, as documented in detail.

The different course as a function of time of nonphotic spectral components in the environment (Fig. 21.9g and h) can be compared with differences during the same span for variables of three psychophysiological variables in Fig. 21.9i–k for the same person, RBS: photics and nonphotics coexisting in RBS, with photic dominance in the circulation (Fig. 21.9j, k), while a mental function's fainter about-1.3-year transyear band (Fig. 21.9i) mimics the behaviors of the solar wind speed and of geomagnetics during the same span (again in the absence of any dominant 1-year synchronized component) rather than showing the prominent circannual band found in the circulation of the same subject. Also note the absence of a 1-year synchronized component in the spectra of vascular variables of two elderly men in Fig. 21.9l–n. There is an age-dependent variance transposition from the extracircadian into the circadian range early in human and other life and a circadian-to-extracircadian variance transposition documented late in human ontogeny [77].

21.12 Multidecadal Congruence

Congruence is also found in the multidecadal range in RBS (Fig. 21.9o and Table 21.3) [128, 129], who had self-measured a dozen of his psychophysiological functions on the average about 5 times a day (2–7 times/day) for, by now, 40 years. The longest periods in systolic and diastolic BP and HR have overlapping or overlying CIs of their periods with different periods of the various environmental variables, each with distinct congruences. Competing effects of the sun and the earth, among others, influence different physiological variables. If one biological variable does not lock into the solar wind's pattern at one frequency, it may still lock into one in geomagnetics if it happens to be a nearby frequency or vice versa [66]. The large but remote sun and the smaller but closer earth, or, otherwise put, helio- or geomagnetics, respectively, may influence and transiently synchronize different human variables at several differing nonphotic as well as photic frequencies. Multifrequency congruence with the physical environment in the spectrum of several individuals remains to be examined on series from additional subjects in several, including additional, spectral regions, among which the ultradians are also important. In the experimental laboratory, the EEG and ECG can be continuously recorded, and human ECGs are now available to ambulatory monitoring to be coordinated with concurrent continuous geomagnetic monitoring at the same site.

21.13 Biospheric Consequences of the Subtraction of a Solar Far-Transyear

The global mapping in the longest available series and in series of corresponding length and density, when feasible, of each new transdisciplinary spectral component does not only contribute useful information. It should prompt us to apply a remove-and-replace approach, as soon as we find an opportunity offered by the environment, in particular the variable sun. In this context, we welcome the aeolian nature, among others, of the solar wind speed (Fig. 21.9f, h) and of geomagnetics (Fig. 21.9g). The sun is likely to offer the future generations occasions to observe what happens when (with the resolution used (Fig. 21.9f, h)), for example, a

Table 21.3 Differential congruence (corresponding periods) of some (orderingly) statistically significant spectral peaks in helio- and/or geomagnetics, signatures of the magnetoperiodic environment (E)[a]

Environment (E)		Clinically healthy man (RBS)		
Solar wind speed (S)	Geomagnetism, aa index (G)	Mental function (M): 1-minute estimation	Systolic blood pressure (B)	Heart rate (H)
	1.92 (1.89, 1.95)	1.98 (1.94, 2.01)		
		1.85 (1.82, 1.88)	1.80 (1.77, 1.83)	1.86 (1.81, 1.90)
1.69 (1.67, 1.72)	1.71 (1.69, 1.74)	1.71 (1.68, 1.74)	1.68 (1.65, 1.71)	
1.60 (1.58, 1.62)	1.61 (1.58, 1.64)			
1.52 (1.50, 1.54)		1.54 (1.52, 1.56)		
1.39 (1.37, 1.41)	1.39 (1.37, 1.42)		1.39 (1.37, 1.42)	1.39 (1.37, 1.41)
1.32 (1.31, 1.34)	1.32 (1.30, 1.34)		1.31 (1.29, 1.33)	
1.24 (1.23, 1.26)		1.26 (1.25, 1.27)	1.25 (1.23, 1.26)	1.25 (1.23, 1.27)
				1.19 (1.18, 1.20)
		1.12 (1.11, 1.13)		
1.06 (1.05, 1.07)	1.06 (1.05, 1.07)	1.06 (1.05, 1.07)		
	0.99 (0.98, 1.00)	0.99 (0.98, 1.01)	1.00 (0.99, 1.01)	0.99 (0.98, 1.00)
				0.96 (0.95, 0.97)
0.91 (0.90, 0.92)				0.91 (0.90, 0.92)
0.83 (0.82, 0.84)		0.82 (0.81, 0.83)	0.80 (0.79, 0.81)	
			0.75 (0.74, 0.76)	
0.72 (0.71, 0.73)	0.72 (0.71, 0.73)			

[a]In the region of trial periods between 2.0 and 0.7 years for the estimation (by counting) of 1 minute (about five times a day over 35 years) by a clinically healthy man, RBS. Congruence is defined as periods, with overlying or overlapping ordering 95% confidence intervals (detected with the nonlinearly extended cosinor [Refinetti et al., 2007] by Marquardt's [1963] algorithm)

transyear disappears and thereafter eventually reappears as a statistically significant entity. A transyear was present as a broad band around 1.3 years in SWS during about the first half of a 16-year record (a span during which an elderly person, FH, happened to monitor his systolic BP). This SWS-band first bifurcated and then disappeared from the record of the solar wind (Fig. 21.3 (top)) [54]. In the second row of this figure, with a lag of about 1 transyear, part of a similar broad band component of ~1.3 years in SBP was also lost. Does the lag of a transyear in the partial loss of the 1.3-year band in BP indicate a physiological resonance of this variable for an added cycle? Does the partial loss indicate driving? Does the partial, albeit damped persistence mean partial endogenicity? More cases are certainly needed to reapply the "subtraction" (and to seek "addition") at more environmental counterparts of biological cycles than a transyear and several circaseptans.

21.14 Do Unseen, Very Weak Magnetic Mechanisms Contribute to Terrorism?

Wendt [43] has reported "entangled" (as he puts it, in keeping with Schrödinger) empirical connections between emotional or attentional states and quantum indeterminate processes: he examined interplanetary magnetic field (IMF) polarity, collective emotions, and entropy changes of random event generators, using data from Princeton's Global Consciousness Project (http://noosphere.princeton.edu), a large-scale network-type system by Nelson et al. with Jahn et al. [130] and Radin [131]. These experimenters reported small changes in the output of up to 70 widely distributed random event generators (REG) whenever groups of individuals experienced life-threatening situations or participated in other collective activities. As of now, the findings are based on nearly 250 events and span solar cycle 23. The results are challenging whether they are viewed in the context of physics, as to entropy, or in that of (mass) psychology or in a chronophysiological perspective. Wendt analyzed the 1998–2007 Princeton data in the framework of the interplanetary magnetic field's (IMF) polarity that had been accessed since 1996 by the ACE satellite at the L1 libration point about 1.5 106 km from earth. Associations have now been reported between (a) entropy changes of the generator output when it coincides with (and sometimes anticipates) the groups' emotional arousal and (b) a composite measure of IMF polarity during the prior ~27-day solar rotation. Wendt [42] found statistically highly significant correlations in the IMF for human violence such as bombings, assassinations, and generally warlike acts. Such situations also reflect the essential lack of control on the part of victims and witnesses.

By contrast, correlations between REG deviations and near-earth polarity status were not statistically significant for fatal accidents such as from transportation and natural catastrophes. Large-scale celebrations, prayer, and meditation were not accompanied in Wendt's [42] study by statistically significant correlations. Cycles can play havoc with statistically highly significant correlations [132]. Hence, to check Wendt's findings, the spectral structure of international terrorist acts was examined with limited focus, namely, to determine the extent to which the incidence of terrorism may be influenced not only (if at all) by socio-photo-thermoperiodisms with a 1.0-year period but also by nonphotic magnetoperiodisms usually characterized by wobbly transyears with an about 1.3 year or similar period, which we had detected earlier in suicide [133]. Data on violence from the Global Consciousness Project database were analyzed by linear-nonlinear least-squares spectra, as the entire available series shown in Fig. 21.10a on top and again after excluding data on the left side of a gap seen in this figure, to check on the extent to which the large break may have contributed artifacts [106].

Table 21.4 summarizes the results obtained on the original data by the fit of a 1.3-year cosine curve with a one-parameter approach [134], since the more rigorous nonlinear approach did not resolve a transyear. Tabulated analyses on the data as a whole as well as on the shorter series, after omitting the limited information from the time before the gap, were nearly the same, suggesting that the break did not play a critical role. The results are also similar whether or not a linear or quadratic trend is added to the model. While we usually use the conservative approach, although the one-parameter approach is often recommended (e.g., Bingham, personal communication, and Marquardt, 1963 [135]), the 1-parameter period of 1.35 years agrees very well with the outcome of subsequent analyses made on the *M*emorial *I*nstitute for the *P*revention of *T*errorism's (MIPT) larger Terrorism Knowledge Base, to which Grigoryev and Vladimirskii kindly called our attention. These authors allowed us insight into their original scholarship.

Figure 21.10a (bottom) shows no peak at the trial period of a calendar year in the incidence pattern of violence, only a transyear [135], in keeping with results on the incidence patterns of sudden cardiac death (ICD10) in Minnesota and Tokyo, but not overall worldwide, and in keeping with the occurrence of both a transyear and a calendar-yearly component in sudden cardiac death in Arkansas and the Czech Republic. A near-transyear and a far-transyear are found with a calendar-yearly component and a prominent cis-year in the incidence of suicides in Minnesota with the amplitude of the transyear more prominent than that of the calendar year [132].

The absence of a calendar-yearly component also applies to our subsequent analyses of the MIPT data. It was also used to derive the monthly incidence of international terrorist acts between February 1968 and March 2007 (~39 years). The numbers of events per month were converted into numbers per day by dividing the monthly counts by the number of days in each month to account for the different numbers of days in different months and thereby avoid any spurious yearly cycle. The time series thus obtained was analyzed by the linear-nonlinear extended cosinor [136, 137]. Specifically, a least-squares periodogram was computed in the frequency range of one cycle in 40 years to about 3 cycles per year. Periods of identified anticipated spectral peaks were used as initial value(s) for the nonlinear analysis, yielding point and 95% confidence interval (CI) estimates of the respective period length(s). Further insight into the spectral region from 4.0 years to 0.67 year was obtained by a gliding spectral window with a 0.05 harmonic increment, using an interval of 8 years progressively displaced by 1 month throughout the time series [119].

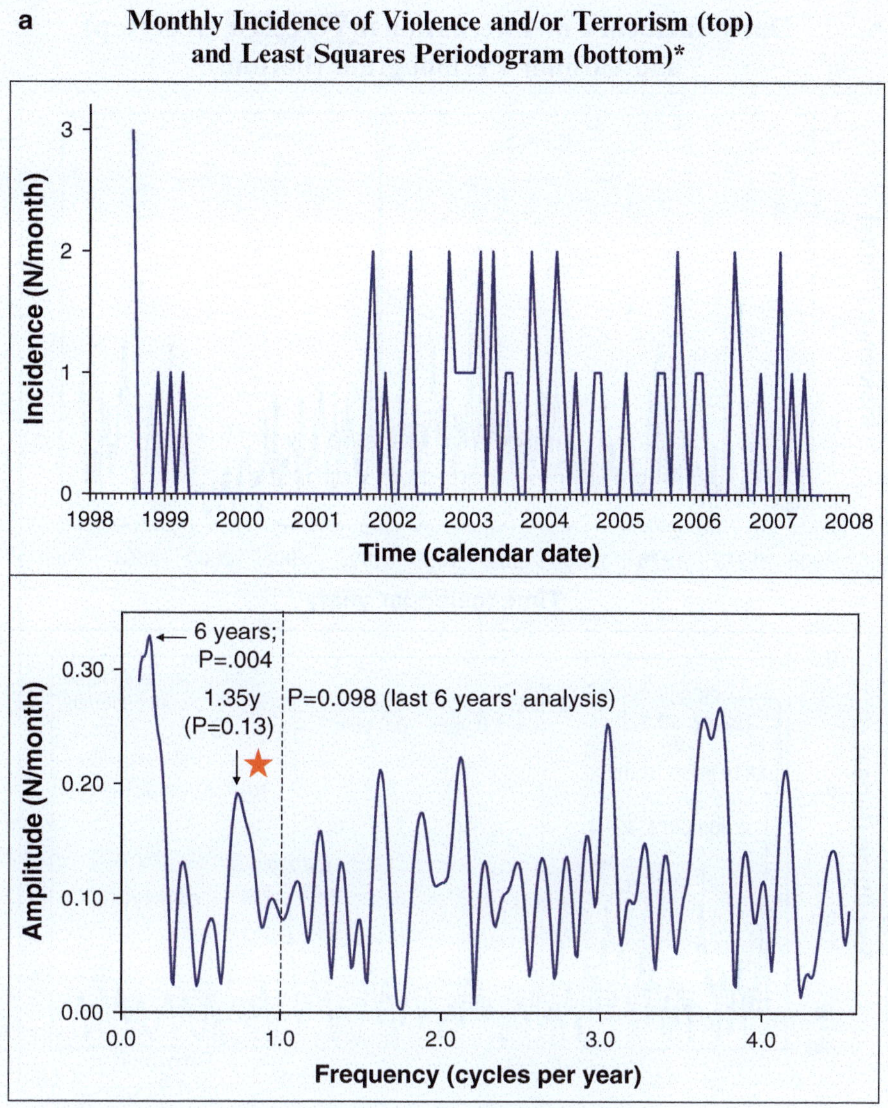

* 51 events from Global Consciousness database (http://noosphere.princeton.edu/). Period validated by check on a larger (N=10,240) and longer (39 years) database (MIPT).

Fig. 21.10 (**a**) In the absence of a calendar year component, a transyear of ~1.3 years is detected in data from the Global Consciousness Project database (http://noosphere.princeton.edu) (© Halberg Chronobiology Center). (**b**) In the absence of a calendar year component, a transyear of ~1.3 years is detected in data from the Memorial Institute for the Prevention of Terrorism's (MIPT) Terrorism Knowledge Base (© Halberg Chronobiology Center)

b **Daily Incidence of International Terrorist Acts (top) and Cosinor Periodogram (bottom)***

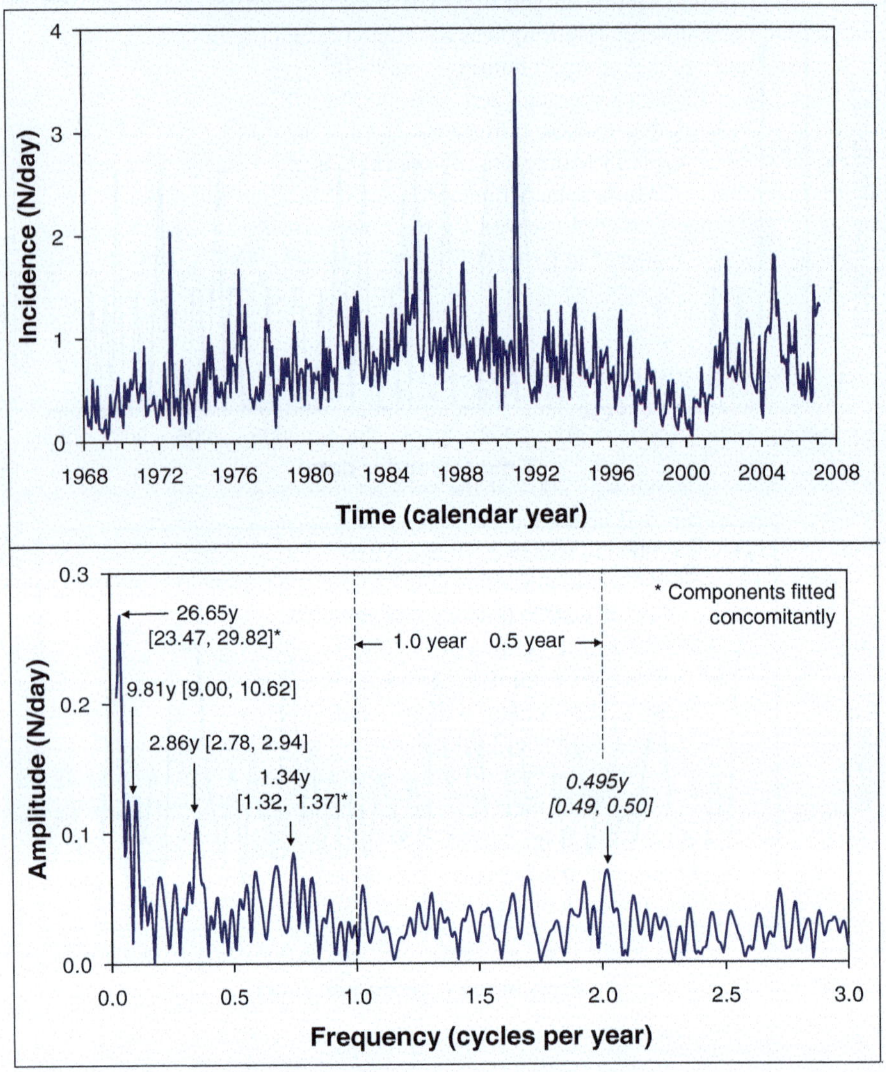

* From February 1968 to March 2007, daily incidence computed from monthly totals (adjusted for differences in the number of days per month). Data from MIPT Terrorism Knowledge Base (© Halberg Chronobiology Center)

Fig. 21.10 (continued)

Table 21.4 Periods found in a time series consisting of acts of terror with several investigated models[a]

Span analyzed	Model used	Period (years) (95% CI)	Amplitude (N/month) (1-parameter limits)
All data	1.3-year cosine curve	1.354 (1.248, 1.459)	0.19 (0.00, 0.38)
	+ linear trend	1.350 (1.177, 1.523)	0.20 (0.01, 0.39)
	+ quadratic trend	1.364 (1.183, 1.545)	0.21 (0.02, 0.40)
Last 6 years	1.3-year cosine curve	1.324 (1.086, 1.562)	0.26 (0.02, 0.49)
	+ linear trend	1.353 (1.083, 1.623)	0.26 (0.02, 0.50)

[a]Global Consciousness Project (http://noosphere.princeton.edu). The authors are greatly indebted for the contribution of Dr. Roger Nelson (Nelson et al. [174])

The daily incidence of international terrorist acts, assessed monthly from February 1968 to March 2007, is illustrated in Fig. 21.10b (top), with a least-squares periodogram shown at the bottom. The presence of an about-1.3-year transyear component found in the Global Consciousness database but not by the analyses tabulated with text in English by Grigoryev and Vladimirskii is seen yet again, with an amplitude much larger than that of any (absent) annual or semiannual cycle. Linearly, the zero-amplitude (no-rhythm) assumption cannot be rejected for the calendar year ($P>0.50$), whereas it is rejected for the transyear (ordering $P=0.004$), in agreement with Grigoryev and Vladimirskii. Three other peaks are identified linearly, with periods of about 25 years, about 9.5 years, and about 2.85 years. Nonlinearly, the transyear is validated when the major about-25-year component is included in the model. Results show a relatively narrow (yet conservative) CI extending from 1.32 to 1.37 years and a CI for the amplitude not overlapping zero. Whether the estimated 26.65-year cycle (CI: 23.47, 29.82 years) with an amplitude much larger (0.27, CI: 0.18, 0.36) or the estimated 9.81-year cycle (CI: 9.00, 10.62 years) with an amplitude (0.13, CI: 0.05, 0.21) larger than the transyear may be "pulled" or otherwise influenced by the Hale and Schwabe cycles of solar activity, respectively, remains to be determined on the basis of longer records. Albeit wobbly in frequency and amplitude, the transyear with a broadband period centering around ~1.3 years, albeit it remains mostly statistically significant throughout the whole record, is most prominent during the years that also show a prominent band in solar wind speed (Fig. 21.9h) in the geomagnetic index aa (Fig. 21.9g) as seen in the gliding spectral window.

Separate lines of evidence are thus almost concomitantly obtained (within a few weeks) with the results herein all associating human violence with the cosmos' nonphotic periodisms rather than with photoperiodisms, thereby offering perhaps an approach to the prehabilitation of social ills, so that eventually countermeasures are developed to prevent their potentially cataclysmic consequences. Toward this goal, rather than only for prediction of when an alert is indicated, it seems pertinent that among a number of variables self-measured by a clinically healthy man (RBS), namely, the estimation of 1 minute (Fig. 21.9i) was characterized by transyears, predominating over any calendar-year, with an intensified amplitude during the years when not only terrorist activity but solar wind speed and the geomagnetic index aa were all amplified. This as-yet time-macroscopic result awaits further scrutiny, not only in gliding spectra of more individuals but also with time-varying cross-spectral coherences of phase synchronizations [138]. All of these results should prompt experimentation concerning aggression, including focus on looking for time-varying coherence and phase synchronization in the ultradian and infradian spectral regions, where Roederer [139] has aligned, as yet only theoretically, the corresponding spectra in geomagnetic pulsations with those in physiology.

According to Freund [140–142], it seems pertinent that some of the nonphotic transdisciplinary, probably magnetic associations with human affairs, may be mediated by a coupling between the ionosphere and solid earth, with vortices in the former resulting in telluric currents in the latter, generating a horizontal torque to which human physiology may be responsive. We are probably dealing with selected frequency windows for which there may be special resonating mechanisms as postulated and investigated for frequencies much higher than those considered herein earlier by others [143].

21.15 Qualifications

Robert K. Adair [144] writes:

> An examination of the physical interaction of [extremely low frequency (ELF) electromagnetic] fields with the body shows that such interactions are too weak to have a significant effect on human biology at the cell level. Because of the high electrical conductivity of tissues, the coupling of external electric fields in air to tissue in the body is such that the effects of the internal fields on cells is smaller than thermal noise. Static magnetic fields smaller than the earth's field of 50 μT and varying fields weaker than the 4-μT 60-Hz fields that are equivalent in effect to that from walking in the earth's field, cannot be expected to generate significant biological effects. Moreover, the interactions of such weak fields at the cell level are also small compared to thermal noise.

Such reservations concerning biological effects of weak extremely low-frequency electromagnetic fields are shared by other physicists [145] and were briefly reviewed by Roederer [87] in the context of putative health hazards from magnetic storms. The order of magnitude is indeed a critical consideration as to underlying mechanisms. Adair felt that for biological effects, only that amount of energy matters, which is higher than thermal noise. The possibility that some very low-intensity signals at certain more or less specific frequencies may be selectively enforced by cellular mechanisms could herein be taken into account. Such mechanisms include chemical reactions with feedsideward coupling ([146]; cf. [147]), chemical waves, and chemical clocks, which underlie complex reactions obeying hypercycles, the latter defined as circular sequences of cyclic events.

There can also be magnetic interactions: Every magnetic field is not only characterized by the field's strength but also by its local orientation (gauged by components of a vector). Thus, the geomagnetic field may also vary in its strength as a function of geographic location on earth and of the magnetic terrestrial properties at that site. The local field strength may overlap with that of other neighboring fields. Hence, at a certain place, only that component which is independent of all other influences can show a specific relevance for some effects, such as periodic chemical reactions of charged reacting molecules. The resulting motion might be – besides the Brownian motion – a superimposition of the magnetic force leading to oscillations. As an analogy, in a beehive, bees can move randomly according to a Brownian motion, yet as a whole they can undergo rather slow oscillations like a pendulum.

Ion cyclotron resonances can also be considered. The influence of the geomagnetic field on Ca^{++} or Mg^{++} ions is rather similar to the foregoing considerations relating to magnetic interactions. In a vacuum, it is certainly necessary that a large diameter of about 80 m be available to observe a rotation under the Lorentz force (with the Larmor frequency). But in a cellular medium, some constraints prevent this large-diameter rotation. This implies that the ions get continuous impulses with more or less the same frequencies that must be in congruence with the frequencies of other processes so that amplification may result. If purely stochastic motions are not necessarily, and not likely, the only dominating processes in cells, the tracks of ions and biomolecules (e.g., proteins, hormones) may show – in addition to the indeterminate directions of Brownian motions – some preferred directions resulting in frequencies that can be anticipated to produce order, which is a (usually somewhat wobbly) spectral structure rather than maximal entropy.

The precise nature of any environmental synchronizers and influences, directly responsible for all of the associations described, remains unknown. This fact does not detract from the observation that nonphotic cycles (the week, the cis-half year, the half year, cis- and transyears, and many others) can occasionally have a much larger amplitude than a photic one (the day or the year). We here add more evidence on the relative prominence of nonphotic cycles in health and disease and even in terrorism.

It seems justified, however, to extrapolate to psychophysiological mechanisms as a basis for rational countermeasures to violence, to suicide and sudden cardiac death, among others. The pineal and the hypothalamus as well as the circulation may all be involved (Fig. 21.11) [128]. As to the next steps, the availability of satellites to probe the interplanetary magnetic field has allowed the use of a subtractive approach in the longitudinal decade-long around-the-clock study of over a dozen individuals' physiology and of one's psychology [129] as well. Whenever nature happens to permit it, the approach by remove and replace is best applied to other time series on social ills as well and to their association with the solar wind, the existence of which was already reportedly suspected by Kepler, who noted that comets' tails point away from the sun. They may point to a sooner or later manipulable undesirable aspect of ecosystems such as ours that happen to be in the atmosphere of the sun [58].

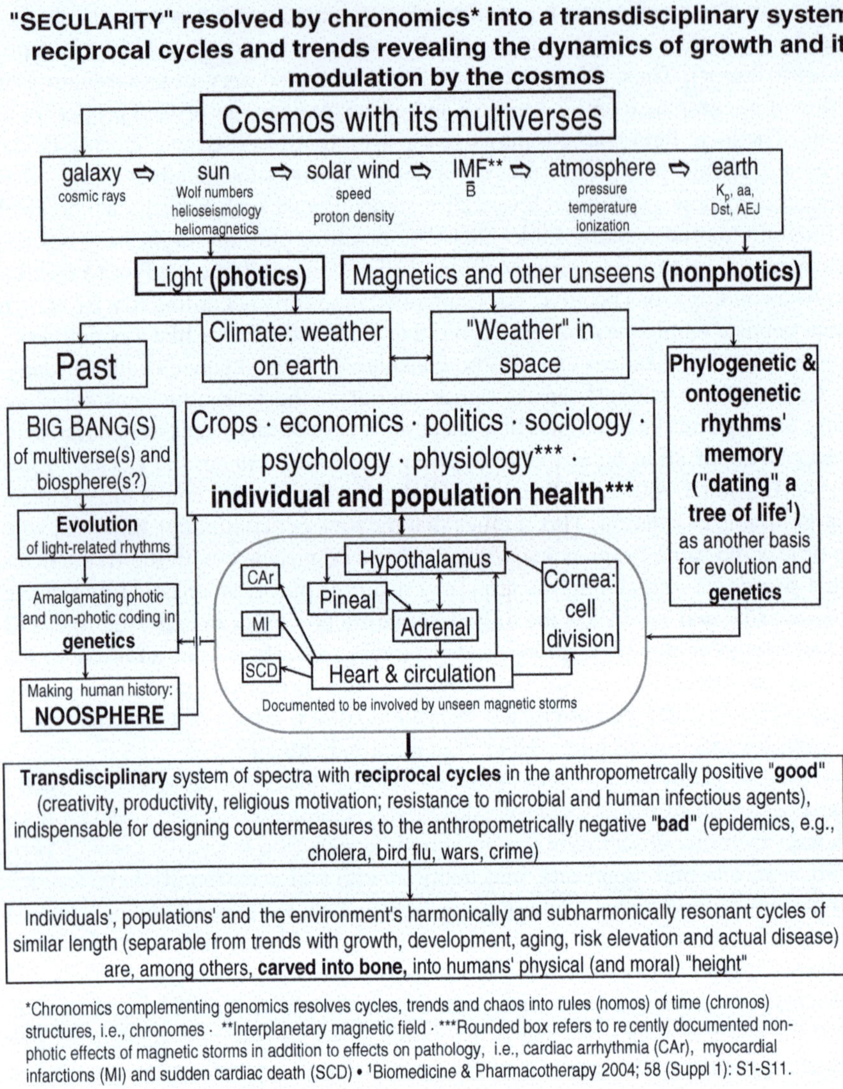

Fig. 21.11 The time has come to replace "secularity" whenever possible with chronomes (Fig. 21.4) (in the abstract) and by the many maps here presented with their as-yet tentative uncertainties. A review of the underlying mechanisms, a topic of intense study in Russia and Ukraine, is beyond our scope; some pertinent illustrative Russian papers are cited. As to studies by BIOCOS, there is the hint of an effect at the cellular level from a lower amplitude of the circadian rhythm of corneal mitosis of rats on the day of a magnetic storm (Halberg et al., 2006c), and the studies by Jozsa et al. [16] hint at the effects of a magnetic storm on the endocrines. Effects upon the circulation in health and disease lead to this consideration of photic and nonphotic environmental effects coexisting and competing with each other awaiting further testing (© Halberg Chronobiology Center)

The sun and the cosmos have many "modes," but biospheric evolution may tell us about those that have been around consistently and were coded in our genes since they have survival value. Some of these associations differ with geographic location, and such differences may be reported in the Russian text of Grigoryev and Vladimirskii on terrorism.

Just as we shield – heat and cool – against the sun's photic and thermic effects, we face the task of resolving "entangled" consequences of unseen, probably largely magnetics that call for an international effort. The IMF may influence not only myocardial infarction, sudden cardiac death, and suicide, whether or not there is also interdigitated information with both collective emotions and changes in the output entropy of random event generators [42]. Certainly the absence of a calendar-yearly component in two databases examined concomitantly, one of them by different investigators and the presence of transyears in both, and our finding that the transyears in terrorism mimic nonphotics in helio- and geomagnetics and the finding of a similar transyearly signature in an important mental function all point to an association of aggression with the physics of the cosmos and render unlikely artifacts often found under circumstances dealing with several periodic functions when these differ in frequency or phase [148].

Much other mapping is alluded to elsewhere. Cycles of 500 years, among many others, are mapped for climate change in the coloring of stalagmites. Again the spectrum of cycles must be considered in this context of lively debates, too often requiring the realization that, albeit they were aeolians, in global summaries of the longest available series, some congruent behaviors deserve further study and may help recognize what consciousness, if not life, really is.

21.16 Conclusion

Separate lines of evidence obtained almost concomitantly with the results herein all associate human violence with magnetoperiodisms rather than with the calendar year's rather prominent socio-photo-thermoperiodisms (notably in Minnesota, "the icebox of the USA"). These results may be a foot in the door for an approach to the reduction, if not prevention, "prehabilitation," of social ills, if countermeasures can be developed against potentially cataclysmic consequences [135].

For the prevention of the individual's cardiovascular ills [105, 149, 150], the Phoenix Project of the Institute of Electrical and Electronics Engineers in Minnesota (http://www.phoenix.tc-ieee.org) advocates a website (http://www.sphygmochron.org/) that can monitor solar variability, as BIOCOS (corne001@umn.edu) now does, by providing free time-series analyses for all participants in self-health care who donate their data (e.g., time series of BP and HR) for, e.g., monitoring solar variability (Fig. 21.9l–n).

Dokutchaev deserves credit for emphasis on the need for transdisciplinarity, suggesting that finding laws in biology only, as Darwin did, or in chemistry only, as Lavoisier did, was not enough ([151]; cf. [51]). In his, Galileo's and Humboldt's transdisciplinary sense, we can try to go beyond what is regular [152]; even when the endpoints are highly variable, we need to render them measurable in all their aspects now dealt with in artificially separated disciplines, usually without any uncertainty estimate, e.g., in the solar wind of the physicist as in the patient's blood pressure of a physician. We need to realize not only that we measure in time and only thus meaningfully but also that whatever we single out for our assessing endeavor, there will always remain, in the background and very often in the foreground, an immense residual, dynamic complementary system that is not only intra- or interdisciplinary but transcends more than a few disciplinary boundaries [87, 153, 154].

We conclude theoretically that nothing in transdisciplinary science, including disciplines dealing with human affairs, can be understood except in the light of the cosmos (read: space weather) and the biosphere and a noosphere that have evolved in it. In that information base, biospheric evolution stands tall, but not alone, and remains incomplete until a broad spectrum of transdisciplinary cycles, the predictable element in time structures, is mapped and thus can be taken into account. Those very cycles may help physicists backtrack to see with what time structure the solar wind blew billions of years ago, when other tools [148] may not suffice.

21.17 Epilogue

21.17.1 History: The Soul, Chronomics, and the Pineal

The involvement of this gland in a magnetic storm (Fig. 21.9d) [16] is a reminder, in a historical context, that interest in the pineal gland as the "house" of the soul *(animus/anima)* started, according to the late Brunetto Tarquini [155], presumably with the legendary Yellow Emperor (said to have reigned from 2697 to 2597 BC) [156]. Other scholars of this time have failed to list the gland in their index (e.g., Huang Ti Nei Ching Su Wen, 1949). In the Vedas, the religious teachings of India, the pineal gland was one of the chakras, the centers of vital energy. This chakra was the "door to perfect peace and harmony." In the Vedic pantheon, one of the most potent deities was Varona, the god of the nocturnal sky, which directs the course of the sun.

Herophilus of Alexandria (325–280 BC), who taught anatomy and physiology during the reign of the first two Ptolemys, mentioned the pineal gland as an entity that is functionally involved in the coordination of the flow of thought (*psychikon*): as a sphincter between the ventricles of the brain and the nerves, so did Galen by indicating that the soul is in the center of the brain, in the pineal gland, rather than in the heart as Aristotle claimed. Herophilus's lasting fame stems first and foremost more generally from studies on the brain, the eyes, and the circulation. He described the nerves, the *calamus scriptorius* (or *calamus herophilii*), the occipital bone, the retina, and the duodenum. Herophilus may have been first to count the beats of the pulse, with an ingenious water clock, and thereby to diagnose even amorous passion, if not the healing, subtle energies of love, interpreted by a psychiatrist [157] in the light of subtle electromagnetic energy fields.

In the Confessions of St. Augustine (354–430 AD), a broad concept of the soul, *anima*, was equated with the Greek *psyche* and Hebrew *nephes*: the latter means "life," including everything that contributes to life. Tarquini speculated whether the teachings of the Vedas became known in Europe through the Templars, the monk-soldiers of St. Bernard, eventually mostly annihilated by Philip "the Beautiful" of France, except for those retained by the initiators of Freemasonry [158–161]. From the Templars may stem the kiss of the behind, which should excite or awaken the serpent Kundalini, a cosmic force which resides in the roots of the dorsal spine and in the sex glands and which once awakened reaches the pineal gland, an explicit reference relating this gland to the cosmos. The Templars are credited with the recognition of the "third eye" in the pineal gland, providing a direct vision of time and space.

In the seventeenth century, René Descartes (1596–1650) wrote: "There is a small gland in the brain, the pineal, in which the soul exercises its function in a more peculiar way than in all other parts" (Descartes 1664/2003). He thus maintained, interpretations to the contrary notwithstanding, that spirituality cannot be localized in any one part of the body; but he considered the pineal gland as the organ where the mind or anima *(res cogitans)* meets the material, the fleeting mortal body *(rex extensa)*. Elsewhere, we have reviewed the current evidence supporting the proposition that the pineal modulates the effect of space weather [155].

Several interdigitating and partly interwoven fields are centered around the pineal gland, where interactions between life and the cosmos may partly lie. One of these areas, chronobiology, concerns the rules of intermodulations within organisms far beyond the circadian system. Another field, chronomics, focuses upon individuals' adjustments and adaptations of species as features of the integration of organisms, as open systems with their ever-changing, predictable only insofar as cyclic, environment. Life has coded time structure in genes, but organisms still respond to remote drummers. The pineal gland may play a role as one of the windows not only for photic and nonphotic effects of the sun but also for other galactic effects. Barriers to these fields exist only in the mind as deeply rooted conventions, such as homeostasis. A third area, chronobioethics, pertains to concerns about spirituality, if not the soul, in purely physical terms. Once we realize that it is possible to walk through intra- and interdisciplinary walls, which are only there by our own deeply rooted biases, we arrive at the verge of contributing to progress in science and ethics and thus to their concomitant interdigitated pursuit.

Sooner or later, we may have to tear down the barrier between science and spirituality and approach both objectively. Next, we may tear down the barriers within the sciences, such as those among disciplines as different as physics and biology. Barriers between integrationists and reductionists then fade not only when the pineal gland is at issue. The physicist Giovanni Gregori (of the Istituto di Acustica O.M. Corbino – CNR, Rome, Italy) noted (in a refereeing communication) that "according to the current evidence, it is reasonable guessing that a large fraction of the Earth's crust ought to contain a large amount of recycled biomass. That is, the biosphere played a fundamental role in the history of the Earth: it is not just a matter of geophysics and geochemistry! Therefore, geophysicists and geochemists ought to get acquainted with the observational perspectives related to the biosphere." To this last, we add "and vice versa." This recognition may serve the dwindling numbers of pinealists, whose historic background is the search for the soul. To vitalists who believe that the soul leaves the body with its last breath, the soul is the anima.

Cicero's *animus mundi*, however, may be used as a preferred alternative in focusing on relations among cycles and events viewed as time structures in and around us. To resolve these chronomes (portmanteau'd from *chronos*, time, and *nomos*, rule), we must abandon the current general position with respect to physiological variation in the normal range, which corresponds to that in preatomic physics. It was then believed that the "a-tom" was the smallest particle that could not be further split. Breaking the atom opened the door to a new universe of particles governed by new forces and physical laws. The field of nuclear physics evolved and brought with it new knowledge, a new energy source, and a wealth of practical applications [52]. Indeed, the analogy applies to splitting of the normal range into the time structures of everyday physiology. From picking different times of day and seasons for study, a transdisciplinary science, chronobiology, emerged to search objectively for the physiology of cognition, if not of the *animus mundi*, by focus on photic cycles, followed by resolving transdisciplinarily new, nonphotic ones.

21.17.2 Render Measurable What as Yet Is Not: Emotion, Mind, and Spirit

In this context, several inextricably interwoven fields are challenging to scholars studying the pineal gland, whether they regard this gland as the window to light, the latter in turn viewed as a "master" switch [162, 163], and/or also accept the pineal gland as a keyhole to geomagnetics, another switch at noon at high latitudes and only by night at lower latitudes [164, 165], and/or as a putative site of consciousness, emotions, and the mind, if not of the soul, already considered from the viewpoints of (eventually we hope chrono-) molecular biology [166] or from that of an eminent crystallographer and physicist [167]. Whatever the focus, or for all of the foregoing purposes, chronobiology is an indispensable tool, needed to resolve the rules of intermodulations of rhythms and other elements of time structure within organisms [52, 154] and the subtle factors [60, 154] or subtle (magnetic field) energies [69, 167] that may be involved with life and the environment [7, 44].

In the course of evolution, organisms may have coded many of the cycles and events of their proximal and distant habitat in their genes, as for the circadian change of human heart rate [168, 169]. From this viewpoint, life on earth is a living fossil, telling us perhaps about the cosmos of the past [8, 84, 149, 154]. On the other hand, the biosphere is still responsive today to the very cycles and events that also characterize its current environment [149].

Chronobioethics [170–172] is most pertinent to those concerned about the mind and spirit, if not soul, whether they come from philosophy, molecular biology, or integrative physics [130, 173, 174]. Tiller [167, 175] and Tiller, Dibble, and Kohane [176] provide a dual four-space frame, complementing the physical four dimensions of space-time with a mathematically underpinned reciprocal four-space which they embedded into emotions (a nine-space) that in turn is embedded in the frame of mind (a ten-space), while the embedding frame for all this is the frame of spirit (Scheme 1 in Halberg et al. [156]).

21.17.3 *Let Us Learn That We Do Not Know Even What We Should Know* (Discamus Nos Nescire Etiam Quid Nobis Sciendam Sit)

According to Tiller [167] reportedly [157] "… the entire human species seems to be part of a vast organism." We single out instead, as a system to be analyzed, the even vaster body of our planet earth, by using its planetary magnetic disturbance indices Kp and Dst, in line with Gilbert's [57] *magnum magnes est orbis terrestris* (our earthly orb is a great magnet). We look further into the problem of obtaining from spaceborne vehicle measures of the interplanetary magnetic field (to paraphrase Gilbert by writing magnum magnes est orbis solaris) only to find the putative effects of the cosmos that prompt us to view our physiology as reflecting a *magnum magnis est orbis cosmi* or *orbis mundi* if not *orbis multiversi*, once we realize that helioseismological probing of the interior of the sun can provide cycles of 1.3-year length [125].

R.D. Nelson et al. [174] "describe data from a global network of physical random number generators that shows unexpected structure apparently associated with major world events." Deviations from randomness after September 11, 1991 are particularly noteworthy, as is the endeavor of the Princeton group as a whole [130, 173, 174]. Another entangled perspective of correlations of continuous random data with major world events is provided by Hans W. Wendt, who introduced the interplanetary magnetic field (IMF) into the interaction of consciousness and physical systems. He found a very high correlation between the deviations of Nelson's global consciousness, human violence, and the polarity of the IMF [42] that actually prompted our looking for and finding transyears in the same database, soon extended by others (Grigoryev and Vladimirskii) and by us [119, 174] to another longer database. As a further check of Wendt's [42] and Nelson et al.'s [174] findings, we analyzed four more variables of the IMF. The waxing and waning and drifting of these transyears mimics that in terrorism, in keeping with the assumption that a perhaps quantum mechanism is involved in the *animus mundi* and can be influenced by human violence. That other motivations may also be influenced environmentally is apparent from studies on religious motivation [177].

More specifically, to eventually build a scientific bridge to religion by examining whether nonphotic solar effects may influence motivation, invaluable yearly worldwide data over half a century on activities from 1950 to 1999 by Jehovah's Witnesses on behalf of their church from over 113 geographic locations were analyzed chronobiologically. The nonlinear least-squares fit of a model consisting of a linear trend and a cosine curve with a trial period of 21.0 years, numerically approximating that of the Hale cycle during the span examined, validated the about-21.0-year component in about 70% of the data series, with the nonoverlap of zero by the 95% confidence interval of the amplitude estimate. Estimates of MESOR, amplitude, and period were further regressed with geomagnetic latitude. The about-21.0-year amplitude tends to be larger at low and middle than at higher latitudes, and the resolution of the about-21.0-year cycle, gauged by the width of 95% confidence intervals for the period and amplitude, is higher (the 95% confidence intervals are statistically significantly smaller) at higher than at lower latitudes. Near-matches of periods in solar activity and human motivation hint that the former may influence the latter, while the dependence on latitude constitutes evidence that geomagnetic activity may affect certain brain areas involved in motivation, just as it was earlier found that it is associated with effects on the electrocardiogram and in anthropometry [30, 178].

We are open to our environment mechanically, electrically, magnetically, electromagnetically, and in many other ways yet to be elucidated. We do not know the extent to which life also depends upon sets of cyclic processes near and far that are not consciously seen, heard, tasted, touched, or otherwise knowingly registered. We are, willy-nilly, an integral, continuously informed part of the environment, by which we also happen to be energized. We share rhythms with the environment from which life (read: rhythms) evolved. Thus, the origin and essence of life, as a result of evolution, by natural selection interacting with the cosmos, is the origin and essence of rhythms, including their use as the alternation of wakefulness (read emerging self-awareness and ever broader consciousness) and sleep. Consciousness then is no longer the "last refuge of obscurantists in science," but emerges as an aspect of self-awareness [179], as does pain [180] and eventually fear. To cite Gell-Mann [179]:

> Now do an experiment with a great ape. Make a mark on one of his hands. He will wipe it off, or try to. Then make a mark on his forehead and let him look in a mirror. If he is a gorilla, he will not attempt to wipe it off, presumably not knowing that he is seeing himself in the mirror. However, a chimpanzee (or, presumably, a bonobo or pygmy chimp) will try to wipe the mark off his forehead, and will succeed if the mark is not indelible.
>
> A related property of the mind, somewhat more advanced, is knowing that others of the same species have the same sorts of needs, reactions, thoughts, and so on as one's self.

In summary, in the past 61 years, cycles in and around life were documented with detailed examples that eventually crossed disciplinary barriers. First, rhythms with different frequencies stood out in different biological variables and systems. Next, cycles with various frequencies were quantified in the same variable, such as urinary 17-ketosteroid excretion. Eventually sets of spectra with congruent environmental and biospheric components, not only in populations but also in individuals, were found. In some cases, subtraction or complete removal of a cycle in the environment was associated with damping but not a complete loss of the biological cyclic counterpart. A host of other cycles, with higher and lower frequencies in consciousness and behavior of individuals and of populations in a hard-to-define spirituality, await similar scrutiny with remove-and-replace approaches.

Wendt's correlations [42] prompted further analyses of four variables from the IMF – plasma β, flow pressure, proton temperature, and the σ of proton temperature. In gliding spectra, these aspects of the IMF, like the also-analyzed geomagnetic indices Kp and Dst (and aa studied earlier), all more or less mirrored the pattern found in terrorism. Many roads lead to purely physical aspects of consciousness; the one followed here, and recommended for further approaches by all those who responded to the brilliant preparation and organization of Prof. Masatoshi Murase's meeting is: "Measure, in any **partial system** of interest at a given time, everything pertinent that is measurable and render measurable what is pertinent but not yet measurable, as simply as possible, but not simpler, in time and hence meaningfully, taking into account chronomic (time-structural) maps, a partly quantified **complementary system** of common transdisciplinary, partly non-stationary and often invisible nonphotic cycles, interacting with the visible, less unstationary since usually synchronized signature in the biosphere of the photic, now also societal day and season" [Amplified from "Omnia propria ex **systemate partiali** metire quæcumque licet et propria immensa, quam simplicissime sed non simplicius, ad mensuram tempestive et ergo significative redige, reddens rationem tabulae chronomicae ad **systema complementare** quantificatum" [181].]

References

1. Halberg F. Chapter on "Medizin" in: Jahrbuch der Internationalen Hochschulwochen des Österreichen College. Salzburg: Igonta Verlag; 1946. p. 336–51.
2. Schrödinger E. What is life? Cambridge: Cambridge University Press; 1945. p. 91.
3. Moore W. Schrödinger: life and thought. Cambridge: Cambridge University Press; 1992. p. 525.
4. Halberg F. Challenges from "60 years of translational chronobiology". Uchenyerye zeaniski Tavricheskogo natsional'iogo Universiteta im V.I. Vernadskogo, Seriya "Biologia, khaimiya" 2007;20(19):107–22; Kofler W, editor. Proceedings, International Interdisciplinary Workshop, Natural Cataclysms and Global Problems of the Modern Civilization. Baku, Azerbaijan, Sep 24–27, 2007. Transactions of the international academy of science H&E. Baku/Innsbruck: ICSD/IAS; 2007. p. 165–78.
5. Halberg F, Cornélissen G, Sothern R, Mikulecky M, Kovac M, Florida P, Watanabe Y, Otsuka K, Mazaudier C, Schroeder W. Solar wind's ~15-month cycle's signature in the human blood circulation: partly built-in, partly driven? Geophys Res Abstr 2007a;9:10986. SRef-ID: 1607-7962/gra/EGU2007-A-10986. p. 2.
6. Halberg F, Cornélissen G, Hillman D, Sothern RB, Nolley ES, Beaty LA, Schwartzkopff O, Otsuka K, Chibisov SM, Valenzi V, Pantaleoni G, Singh RB. Bakken's prehabilitation in the service of a budding chronoastrobiology. Invited lecture, VII International Crimean Conference "Cosmos and Biosphere", Sudak, Crimea, Ukraine, Oct 1–6, 2007b. pp. 10–3.
7. Ertel S. Space weather and revolutions: Chizhevsky's heliobiological claim scrutinized. Stud Psychol 1996;39:3–22.
8. Breus T, Cornélissen G, Halberg F, Levitin AE. Temporal associations of life with solar and geophysical activity. Ann Geophys 1995;13:1211–22.

9. Breus TK, Chibisov SM, Baevsky RM, Schebzuhov KV. The chronostructure of heart biorhythms under effects of external environment factors. Moscow: Russian University of People's Friendship, Polygraph Service; 2002. p. 231.
10. Breus TK, Pimenov KYU, Cornélissen G, Halberg F, Syutkina EV, Baevsky RM, Petrov VM, Orth-Gomer K, Akerstedt T, Otsuka K, Watanabe Y, Chibisov SM. The biological effects of solar activity. Biomed Pharmacother 2002;56 Suppl 2:273s–83s.
11. Breus TK, Rapoport SI. Magnitnie bure: Medico-biologicheskie i geofizicheskie aspekti. Moscow: Sovetskiy Sport Izdatel'stvo; 2003. p. 192.
12. Chibisov SM, Cornélissen G, Halberg F. Magnetic storm effect on the circulation of rabbits. Biomed Pharmacother 2004;58 Suppl 1:S15–9.
13. Dorman LI, Iucci N, Villoresi G. The use of cosmic rays for continuous monitoring and prediction of some dangerous phenomena for the earth's civilization. Astrophys Sp Sci 1993;208:55–68.
14. Ertel S. Influenza pandemics and sunspots – easing the controversy. Naturwissenschaften 1994;82:308–10.
15. Halberg F, Cornélissen G, Regal P, Otsuka K, Wang ZR, Katinas GS, Siegelova J, Homolka P, Prikryl P, Chibisov SM, Holley DC, Wendt RW, Bingham C, Palm SL, Sonkowsky RP, Sothern RB, Pales E, Mikulecky M, Tarquini R, Perfetto F, Salti R, Maggioni C, Jozsa R, Konradov AA, Kharlitskaya EV, Revilla M, Wan CM, Herold M, Syutkina EV, Masalov AV, Faraone P, Singh RB, Singh RK, Kumar A, Singh R, Sundaram S, Sarabandi T, Pantaleoni GC, Watanabe Y, Kumagai Y, Gubin D, Uezono K, Olah A, Borer K, Kanabrocki EA, Bathina S, Haus E, Hillman D, Schwartzkopff O, Bakken EE, Zeman M. Chronoastrobiology: proposal, nine conferences, heliogeomagnetics, transyears, near-weeks, near-decades, phylogenetic and ontogenetic memories. Biomed Pharmacother 2004;58 Suppl 1:S150–87.
16. Jozsa R, Halberg F, Cornélissen G, Zeman M, Kazsaki J, Csemus V, Katinas GS, Wendt HW, Schwartzkopff O, Stebelova K, Dulkova K, Chibisov SM, Engebretson M, Pan W, Bubenik GA, Nagy G, Herold M, Hardeland R, Hiither G, Poggeler B, Tarquini R, Perfetto F, Salti R, Olah A, Csokas N, Delmore P, Otsuka K, Bakken EE, Allen J, Amory-Mazaudier C. Chronomics, neuroendocrine feedsidewards and the recording and consulting of nowcasts – forecasts of geomagnetics. Biomed Pharmacother 2005;59 Suppl 1:S24–30.
17. Kanabrocki E, Sothern RB, Cornélissen G, Halberg F. Chronomics of aging: schwabe cycles in clinical biochemistry and physiology. Abstract (Read by Title) 14, Proceedings, 5th International Workshop on Chronoastrobiology and Chronotherapy, Matsubayasi K editor, Division of Human-Nature Dynamics, Center for Southeast Asian Studies, Nov 6, 2004. pp. 70–3.
18. Kashulin PA, Kalacheva NV. Short-term cycles in terrestrial ecosystems and possible biological significance of quasi-biennial oscillations. In: "Physics of Auroral Phenomena", Proceedings XXIX Annual Seminar, Apatity, 2006. pp. 295–98.
19. Kofler W editor. Natural cataclysms and global problems of the modern civilization (Proceedings, International Interdisciplinary Workshop, Baku, Azerbaijan, Sep 24–27, 2007). Special edition, Transaction of the International Academy of Science H&E. Baku/Innsbruck: ICSD/IAS; 2007. p. 759.
20. Kovac M, Mikulecky M. Secular rhythms and Halberg's paraseasonality in the time occurrence of cerebral stroke. Bratisl Lek Listy 2005;106(2):423–7.
21. Kovac M, Mikulecky M. Time sequence of epileptic attacks from the point of view of possible lunisolar connections. International Conference on the Frontiers of Biomedical Science: Chronobiology. Chengdu, China, Sep 24–26, 2006, pp. 175–79.
22. Lushnov MS. Simultaneity of long-term fluctuations of cosmogeophysical parameters and dynamics of biochemical parameters of people. International Congress, Coherence 2004, Rome, 12–13 Feb 2004. p. 3.
23. Martynyuk VS, Kalinovsky PS, Tseisler YV. Influence of 8-Hz magnetic field on the binding of chloroform to proteins. Biophysics 2004;49 Suppl 1:S17–21.
24. Martynyuk VS, Panov DA. Surfactant properties of natural phospholipids in media treated with extremely low frequency magnetic field. Biophysics 2004;49 Suppl 1:S22–5.

25. Mikulecky M, editor. Sun, moon and living matter. Bratislava: Slovak Medical Society; 1994. p. 159.
26. Mikulecky M, editor. The moon and living matter. Kosice/Bratislava: Slovak Medical Society; 1993. p. 97.
27. Mikulecky M, editor. Chronobiology & its roots in the cosmos. High Tatras/Bratislava: Slovak Medical Society; 1997. p. 287.
28. Minorsky PV. Latitudinal differences in coconut foliar spiral direction: a re-evaluation and hypothesis. Ann Bot 1998;82:133–40.
29. Minorsky PV. Solar-terrestrial effects on bean seed imbibition. P30006, Plant Biology and Botany 2007 Joint Congress. Chicago, Illinois: American Society of Plant Biologists; Jul 7–11,2007.http://2007.botanyconference.org/engine/search/index.php?func=detail&aid=1348.
30. Otsuka K (editor). Proceedings, 1st International Symposium Workshop on Circadian Rhythms and Clinical Chronotherapy. 11 Nov 2000, Tokyo, Japan. Biomed Pharmacother 2001; 55(Suppl 1):7s–190s.
31. Otsuka K (editor). Proceedings, 2nd International Symposium Workshop on Circadian Rhythms and Clinical Chronotherapy. 17 Nov 2001, Tokyo, Japan. Biomed Pharmacother 2002;56(Suppl 2):231s–382s.
32. Otsuka K (editor). Proceedings, 3rd International Symposium Workshop on Circadian Rhythms and Clinical Chronotherapy. 9 Nov 2002, Tokyo, Japan. Biomed Pharmacother 2003;57 (Suppl 1):1s–198s.
33. Otsuka K (editor). Proceedings, 4th International Symposium Workshop on Circadian Rhythms and Clinical Chronotherapy. 8 Nov 2003, Tokyo, Japan. Biomed Pharmacother 2004;58 (Suppl 1):S1–S188.
34. Otsuka K (editor). Proceedings, 5th International Symposium Workshop on Circadian Rhythms and Clinical Chronotherapy. 6 Nov 2004, Tokyo, Japan. Biomed Pharmacother 2005;59 (Suppl 1):S1–S261.
35. Ouattara F, Amory-Mazaudier C, Legrand JP, Simon P. On the geomagnetic activity change from 1900 to 2000. Geophys Res Abstr 2007;9/04849. SRef-ID: 1607-7962/gra/EGU2007-A-04849.
36. Stoupel E. Forecasting in cardiology. New York: John Wiley & Sons; 1976. p. 141.
37. Temuryants NA, Martynyuk VS, Chuyan EN, Minko VA, Brusil IA. Changes in the infradian rhythmicity of blood lymphocyte dehydrogenases in rats exposed to an extremely low frequency variable magnetic field. Biophysics 2004;49 Suppl 1:S26–31.
38. Vladimirskii BM, Chizhevsky AL. Heliobiology on the eve of the XXI century: results, problems, and prospects. Biophysics 1992;43:532–6. Works of A.L. Chizhevsky on solar-terrestrial links.
39. Watanabe Y, Cornélissen G, Halberg F. Thousands of blood pressure and heart rate measurements at fixed clock hours may mislead. Neuroendocrinol Lett 2003;24:339–40.
40. Watanabe Y, Cornélissen G, Halberg F, Bingham C, Siegelova J, Otsuka K, Kikuchi T. Incidence pattern and treatment of a clinical entity, overswinging or circadian hyperamplitudetension (CHAT). Scr Medica (Brno) 1997;70:245–61.
41. Watanabe Y, Nintcheu-Fata S, Katinas G, Cornélissen G, Otsuka K, Hellbrügge T, Schwartzkopff O, Bakken E, Halberg F. Methodology: partial moving spectra of postnatal heart rate chronome. Neuroendocrinol Lett 2003;24 Suppl 1:139–44.
42. Wendt HW. Interplanetary magnetic field (IMF) polarity, collective emotions and entropy changes of random event generators. Bulletin of the People's Friendship University of Russia, proceedings, 8th International Congress "Health and Education Millennium", Moscow, Russia, Nov 14–17, 2007, pp. 81–4.
43. Vladimirskii B, Shnol S (chairpersons, international program committee). Proceedings, VII International Crimean Conference "Cosmos and Biosphere", Sudak, Crimea, Ukraine, Oct 1–6, 2007. Simferopol, Ukraine: science-center.net. p. 200.
44. Chizhevsky AL. Astronomy, psychology and history. Moscow: M.A. Institute; 1921. p. 78.
45. Chizhevsky AL. Les épidémies et les perturbations électromagnétiques du milieu extérieur. Paris: Éditions Hippocrate; 1938. p. 239.

46. Chizhevsky AL (de Smitt VP, trans and condensed). Physical factors of the historical process. Cycles 1971;22:11–27. http://cyclesresearchinstitute.org/cycles-history/chizhevsky1.pdf.
47. Sigel F (Dreier W, Lerche D, Übers, Göring H, Wissenschaftl. Red. der deutschsprachigen). Schuld ist die Sonne. Thun/Frankfurt am Main: Harri Deutsch; 1979. p. 215.
48. Feinleib M, Rogot E, Sturrock PA. Solar activity and mortality in the United States. Int J Epidemiol 1975;4:227–9.
49. Lipa BJ, Sturrock PA, Rogot E. Search for correlation between geomagnetic disturbances and mortality. Nature 1976;259:302–4.
50. Cornélissen G, Halberg F, Breus T, Syutkina EV, Baevsky R, Weydahl A, Watanabe Y, Otsuka K, Siegelova J, Fiser B, Bakken EE. Non-photic solar associations of heart rate variability and myocardial infarction. J Atmos Solar Terr Phys 2002;64:707–20.
51. Halberg F, Bakken EE, Katinas GS, Cornélissen G, Zaslavskaya RM, Blank MA, Syutkina EV, Breus TK, Watanabe Y, Masalov A, Chibisov SM. Chronoastrobiology: Vernadsky's future science? Benefits from spectra of circadians and promise of a new transdisciplinary spectrum of near-matching cycles in and around us. Opening keynote, Proceedings, III International Conference, Civilization Diseases in the Spirit of V.I. Vernadsky, People's Friendship University of Russia, Moscow, Oct. 10–12, 2005, pp. 4–22.
52. Cornélissen G, Halberg F, Introduction to chronobiology. Medtronic Chronobiology Seminar #7; Apr 1994. pp. 52. (Library of Congress Catalog Card #94-060580; http://halbergchronobiologycenter.umn.edu).
53. Cornélissen G, Halberg F, Wendt HW, Bingham C, Sothern RB, Haus E, Kleitman E, Kleitman N, Revilla MA, Revilla Jr M, Breus TK, Pimenov K, Grigoriev AE, Mitish MD, Yatsyk GV, Syutkina EV. Resonance of about-weekly human heart rate rhythm with solar activity change. Biologia (Bratisl) 1996;51:749–56.
54. Halberg F, Cornélissen G, Katinas G, Tvildiani L, Gigolashvili M, Janashia K, Toba T, Revilla M, Regal P, Sothern RB, Wendt HW, Wang ZR, Zeman M, Jozsa R, Singh RB, Mitsutake G, Chibisov SM, Lee J, Holley D, Holte JE, Sonkowsky RP, Schwartzkopff O, Delmore P, Otsuka K, Bakken EE, Czaplicki J, International BIOCOS Group. Chronobiology's progress: season's appreciations 2004–2005. Time-, frequency-, phase-, variable-, individual-, age- and site-specific chronomics. J App Biomed 2006;4:1–38. http://www.zsf.jcu.cz/vyzkum/jab/4_1/halberg.pdf.
55. Halberg F, Otsuka K, Katinas G, Sonkowsky R, Regal P, Schwartzkopff O, Jozsa R, Olah A, Zeman M, Bakken EE, Cornélissen G. A chronomic tree of life: ontogenetic and phylogenetic 'memories' of primordial cycles - keys to ethics. Biomed Pharmacother 2004;58 Suppl 1:S1–S11.
56. Stetson HT. Sunspots in action. New York: Ronald Press Co.; 1947. p. 252.
57. Gilbert W. De magnete, magneticisque corporibus, et de magno magnete tellure; physiologia noua, plurimis et argumentis, et experimentis demonstrata [On magnetism, magnetic bodies, and the great magnet Earth]. London: Short; 1600. p 204.
58. Kamide Y. We reside in the sun's atmosphere. Biomed Pharmacother 2005;59 Suppl 1:S1–4.
59. Lanzerotti LJ. Space weather effects on communications. In: Daglis IA, editor. Space storms and space weather hazards. Kluwer, Dordrecht: NATO Science Series; 2001. p. 313–34.
60. Brown Jr FA. Response to pervasive geophysical factors and the biological clock problem. Cold Spr Harb Symp Quant Biol 1960;25:57–71.
61. Berthold AA. Transplantation der Hoden. Arch. Anat. Physiol. Wissenschr. Med 1849;42–6; (Quiring DP, Trans, 1944). The Transplantation of Testes. Bulletin of the History of Medicine 1944;16:339–401.
62. Berthold AA. Fragment einer Selbstbiographie von Arnold Adolph Berthold. Dtsch Arch Gesch Med Med Geogr 1880;3:74–100.
63. Halberg F, Visscher MB. Regular diurnal physiological variation in eosinophil levels in five stocks of mice. Proc Soc Exp Biol Med (NY) 1950;75:846–7.
64. Halberg F, Visscher MB, Flink EB, Berge K, Bock F. Diurnal rhythmic changes in blood eosinophil levels in health and in certain diseases. Journal Lancet (Minneapolis) 1951;71:312–9.

65. Kaine HD, Seltzer HS, Conn JW. Mechanism of diurnal eosinophil rhythm in man. J Lab Clin Med 1955;45(2):247–52.
66. Halberg F. Historical encounters between geophysics and biomedicine leading to the Cornélissen-series and chronoastrobiology. In: Schröder W, editor. Long- and short-term variability in sun's history and global change. Bremen: Science Edition; 2000. p. 271–301.
67. Halberg F, Cornélissen G, Katinas G, Syutkina EV, Sothern RB, Zaslavskaya R, Halberg F, Watanabe Y, Schwartzkopff O, Otsuka K, Tarquini R, Perfetto P, Siegelova J. Transdisciplinary unifying implications of circadian findings in the 1950s. J Circ Rhy 2003;1:2–61. www.JCircadianRhythms.com/content/pdf/1740-3391-2-3.pdf.
68. Halberg F. Physiologic 24-hour periodicity; general and procedural considerations with reference to the adrenal cycle. Z. Vitam Horm-u Fermentforsch 1959;10:225–96.
69. Wever RA. The circadian system of man: results of experiments under temporal isolation. New York: Springer-Verlag; 1979. p. 276.
70. Halberg F, Cornélissen G, Sonkowsky RP, Lanzoni C, Galvagno A, Montalbini M, Schwartzkopff O. Chrononursing (chronutrics), psychiatry and language. New Trends Exp Clin Psychiatr 1998;14:15–26.
71. Halberg F, Engeli M, Hamburger C, Hillman D. Spectral resolution of low-frequency, small-amplitude rhythms in excreted 17-ketosteroid; probable androgen induced circaseptan desynchronization. Acta Endocrinol (Kbh) 1965;50(Supp 103):5–54.
72. Hübner K. Kompensatorische Hypertrophie, Wachstum und Regeneration der Rattenniere. Ergeb Allg Pathol Pathol Anat 1967;100:1–80.
73. Ratte J, Halberg F, Kühl JFW, Najarian JS. Circadian variation in the rejection of rat kidney allografts. Surgery 1973;73:102–8.
74. DeVecchi A, Halberg F, Sothern RB, Cantaluppi A, Ponticelli C. Circaseptan rhythmic aspects of rejection in treated patients with kidney transplant. In: Walker CA, Winget CM, Soliman KFA, editors. Chronopharmacology and chronotherapeutics. Tallahassee: Florida A & M University Foundation; 1981. p. 339–53.
75. Liu T, Cavallini M, Halberg F, Cornélissen G, Field J, Sutherland DER. More on the need for circadian, circaseptan and circannual optimization of cyclosporine therapy. Experientia 1986;42:20–2.
76. Hildebrandt G, Bandt-Reges I. Chronobiologie in der Naturheilkunde: Grundlagen der Circaseptanperiodik. Heidelberg: Haug; 1992. p. 102.
77. Gubin D, Cornélissen G, Halberg F, Gubin G, Uezono K, Kawasaki T. The human blood pressure chronome: a biological gauge of aging. In Vivo 1997;11:485–94.
78. Berger S, Kaever MJ. Dasycladales: an illustrated monograph of a fascinating algal order. Stuttgart: Thieme-Verlag; 1992. p. 247.
79. Cornélissen G, Ulmer W, Malkova I, Syutkina EV, Masalov A, Hardeland R, Linthe HJ, Grafe A, Otsuka K, Siegelova J, Fiser B, Halberg F. Geographically proximal versus global geomagnetic reference values for biological studies. In: Halberg F, Kenner T, Fiser B, Siegelova J, editors. Proceedings, Noninvasive Methods in Cardiology 2007, Brno, Czech Republic, Nov 11–14, 2007. Brno: Department of Functional Diagnostics and Rehabilitation, Faculty of Medicine, Masaryk University; 2007a. (ISBN 978 80 7018 463 4). pp. 46–54.
80. Ulmer W, Cornélissen G, Revilla M, Siegelova J, Dusek J, Halberg F. Circadian and circaseptan dependence of the beta-ATP peak of four different cancer cell cultures: implications for chronoradiotherapy. Scr Medica (Brno) 2001;74:87–92.
81. Vernova YS, Pochtarev VI, Ptitsyna NG, Tyasto MI. Short-period variations in the rate of change of solar activity as a geosensitive parameter. Geomagn Aeron 1983;23:425–7.
82. Fraser-Smith AC. Spectrum of the geomagnetic activity index Ap. J Geophys Res 1972;77:4209–20.
83. Halberg F, Cornélissen G, Bingham C, Hillman D, Katinas G, Sampson M, Revilla M, Prikryl Sr P, Prikryl Jr P, Sanchez de la Pena S, Gonzalez C, Amory-Mazaudier C, Bouvet J, Barnwell F, Maggioni C, Sothern RB, Wang ZR, Schwartzkopff O, Bakken E. Season's appreciations 2001. Neuroendocrinol Lett 2002;23:170–87.

84. Halberg F, Breus TK, Cornélissen G, Bingham C, Hillman DC, Rigatuso J, Delmore P, Bakken E, International womb-to-tomb chronome initiative group, 1991. Chronobiology in space. Keynote, 37th Annual Meeting Japan Society for Aerospace and Environmental Medicine, Nagoya 8–9 Nov, 1991. University of Minnesota/Medtronic Chronobiology Seminar Series, #1; Dec 1991, p. 21 of text, 70 figures.
85. Hillman DC. Physiologic 7- and 3.5-day patterns in health and disease revealed by free-run and single-stimulus induction. PhD thesis, University of Minnesota, May 1993,p. 279.
86. Cornélissen G, Hillman D, Katinas GS, Rapoport S, Breus TK, Otsuka K, Bakken EE, Halberg F. Geomagnetics and society interact in weekly and broader multiseptans underlying health and environmental integrity. Biomed Pharmacother 2002;56(Supp 2):319s–26s.
87. Roederer JG. Are magnetic storms hazardous to your health? Eos, Transactions. Am Geophys Un 1995;76:441. 444–5.
88. Vladimirskii BM, Narmanskii VYA, Temuriantz NA. Global rhythmics of the solar system in the terrestrial habitat. Biophysics 1995;40:731–6.
89. Cornélissen G, Halberg F. The biological week and broader time structures (chronomes): in memory of Gunther Hildebrandt. Percept Mot Skills 2000;90:579–86.
90. Cornélissen G, Halberg F, Tarquini B, Mainardi G, Panero C, Cariddi A, Sorice V, Cagnoni M. Blood pressure rhythmometry during the first week of human life. In: Tarquini B, editor. Social diseases and chronobiology: Proceedings III International Symposium Social Diseases and Chronobiology, Florence, Nov. 29, 1986. Bologna: Società Editrice Esculapio; 1987. pp. 113–22.
91. Campbell J. Winston Churchill's afternoon nap. New York: Simon and Schuster; 1986. p. 432.
92. Thomson DJ, Lanzerotti LJ, Vernon III FL, Lessard MR, Smith LTP. Solar modal structure of the engineering environment. Proc IEEE 2007;95:1085–132. doi:10.1109/JPROC.2007.894712.
93. Rieger A, Share GH, Forrest DJ, Kanbach G, Reppin C, Chupp EL. A 154-day periodicity in the occurrence of hard solar flares? Nature 1984;312:623–5.
94. Chapman S, Bartels J. Geomagnetism. 3rd ed. Oxford: Clarendon Press; 1962. p. 1049.
95. Grafe A. Einige charakterische Besonderheiten des geomagnetischen Sonneneruptionseffektes. Geofis Pura e Appl 1958;40:172–9.
96. Russell CT, McPherron RL. Semiannual variation of geomagnetic activity. J Geophys Res 1973;78:92–108.
97. Tarquini B, Cornélissen G, Perfetto F, Tarquini R, Halberg F. Chronome assessment of circulating melatonin in humans. In Vivo 1997;11:473–84.
98. Martikainen H, Tapanainen J, Vakkuri O, Leppaluoto J, Huhtaniemi I. Circannual concentrations of melatonin, gonadotrophins, prolactin and gonadal steroids in males in a geographical area with a large annual variation in daylight. Acta endocrinol (Cph) 1985;109:446–50.
99. Burch JB, Reif JB, Yost MG, Keefe TJ, Pitrat CA. Reduced excretion of a melatonin metabolite in workers exposed to 60 Hz magnetic fields. Am J Epidemiol 1999;150:27–36.
100. Weydahl A, Sothem RB, Cornélissen G, Wetterberg L. Geomagnetic activity influences the melatonin secretion at latitude 70°N. Biomed Pharmacother 2001;55:57–62.
101. Bartels J. L'activité du magnétisme terrestre et ses relations avec les phénomènes solaires. In: de Valbreuze R, editor. Comptes rendus des travaux de la onzième section et de la treizième section, Congrès international d'électricité, Paris 1932. Paris: Gauthier-Villars; 1932. p. 77–88.
102. Zieger B, Mursula K. Annual variation in near-earth solar wind speed: evidence for persistent north–south asymmetry related to solar magnetic polarity. Geophys Res Lett 1998;25: 841–4.
103. Bracewell R. The sunspot number series. Nature 1953;179:649–50.
104. Halberg F, Cornélissen G, Haus E, Northrup G, Portela A, Wendt H, Otsuka K, Kumagai Y, Watanabe Y, Zaslavskaya R. Clinical relevance of about-yearly changes in blood pressure and the environment. Int J Biometeorol 1996;39:161–75.

105. Halberg F, Cornélissen G, Katinas G, Tvildiani L, Gigolashvili M, Janashia K, Toba T, Revilla M, Regal P, Sothern RB, Wendt HW, Wang ZR, Zeman M, Jozsa R, Singh RB, Mitsutake G, Chibisov SM, Lee J, Holley D, Holte JE, Sonkowsky RP, Schwartzkopff O, Delmore P, Otsuka K, Bakken EE, Czaplicki J. International BIOCOS Group. Chronobiology's progress: season's appreciations 2004–2005. Time-, frequency-, phase-, variable-, individual-, age- and site-specific chronomics. J App Biomed 2006;4:1–38. http://www.zsf.jcu.cz/vyzkum/jab/4_1/halberg.pdf.
106. Gonzalez ALC, Gonzalez WD. Periodicities in the interplanetary magnetic field polarity. J Geophys Res 1987;92(A5):4357–75.
107. Richardson JD, Paularena KI, Belcher JW, Lazarus AJ. Solar wind oscillations with a 1.3-year period. Geophys Res Lett 1994;21:1559–60.
108. Halberg F, Cornélissen G, Otsuka K, Watanabe Y, Katinas GS, Burioka N, Delyukov A, Gorgo Y, Zhao ZY, Weydahl A, Sothern RB, Siegelova J, Fiser B, Dusek J, Syutkina EV, Perfetto F, Tarquini R, Singh RB, Rhees B, Lofstrom D, Lofstrom P, Johnson PWC, Schwartzkopff O, International BIOCOS Study Group. Cross-spectrally coherent ~10.5- and 21-year biological and physical cycles, magnetic storms and myocardial infarctions. Neuroendocrinol Lett 2000;21:233–58.
109. McIntosh PS, Thompson RJ, Willock EC. A 600-day periodicity in solar coronal holes. Nature 1992;360:322–4.
110. Pap J, Tobiska WK, Bouwer SD. Periodicities of solar irradiance and solar activity indices, I. Sol Phys 1990;129:165–9.
111. Szabo A, Lepping RP, King JH. Magnetic field observations of the 1.3-year solar wind oscillation. Geophys Res Lett 1995;22:1845–8.
112. Gazis PR, Richardson JD, Paularena KI. Long-term periodicity in solar wind velocity during the last three solar cycles. Geophys Res Lett 1995;22:1165–8.
113. Mursula K, Zieger B. The 1.3-year variation in solar wind speed and geomagnetic activity. Adv Space Res 2000;25:1939–42.
114. Prabhakaran Nayar SR. Periodicities in solar activity and their signature in the terrestrial environment. Goa: ILWS Workshop; 2006. p. 9.
115. Cornélissen G, Masalov A, Halberg F, Richardson JD, Katinas AS, Sothem RB, Watanabe Y, Syutkina EV, Wendt HW, Bakken EE, Romanov Y. Multiple resonances among time structures, chronomes, around and in us. Is an about 1.3-year periodicity in solar wind built into the human cardiovascular chronome? Hum Physiol 2004;30(2):86–92.
116. Halberg F, Cornélissen G, Wendt HW, Schwartzkopff O, Berger S, Schweiger HG. Erweiterung transdisziplinärer photisch-nichtphotischer Spektren: Sich gegenseitig stützende Nah- und/oder Ferntransjahre in Sonnenmagnetik, Sonnenwind, Interplanetarmagnetik, Erdmagnetik und Biologie. Drei Magneotzyklizitäten im Autismus (Zusammenfassung). In: Kautzleben H, Hamel J, editors. "Albert Einstein in Berlin": Kolloquium am 17. März 2005. Sitzungsberichte der Leibniz-Sozietät; 2005; 78/79:209–14.
117. Mikulecky M, Florida PL. Daily birth numbers in Davao, Philippines, 1993–2003: Halberg's transyear stronger than year. Abstract, 26th Seminar, Man in his Terrestrial and Cosmic Environment, Upice, Czech Republic; May 17–19, 2005.
118. Mikulecky M. Reanalyza natality v jizni brazilii – opet dominuje Halbergova parasezonalita. International Conference on the Frontiers of Biomedical Science: Chronobiology. Chengdu, China; Sep 24–26, 2006, pp. 188–93.
119. Cornélissen G, Halberg F, Wendt HW, Sothern RB, Chibisov SM, Kulikov SI, Agarwal RK. Weak magnetoperiodism rather than socio-photo-thermoperiodism characterizes human terrorism detection of about 1.3-year aeolian transyear but not precise 1.0-year cycle. Proceedings, 8th International Congress "Health and Education Millennium". Moscow, Russia; Nov 14–17, 2007c, pp. 77–80.
120. Shapiro R. Interpretation of the subsidiary peaks at periods near 27 days in power spectra of geomagnetic disturbance indices. J Geophys Res 1967;72:4945–9.
121. Delouis H, Mayaud PN. Spectral analysis of the geomagnetic activity index aa over a 103-year interval. J Geophys Res 1975;80:4681–8.

122. Silverman SM, Shapiro R. Power spectral analysis of auroral occurrence frequency. J Geophys Res 1983;88:6310–6.
123. Cornélissen G, Halberg F, Kudela K, Chibisov, SM (presenter). Transdisciplinary transyears and some of their uncertainties in cosmic rays. In: Proceedings, International Symposium, Problems of Ecological and Physiological Adaptation, People's Friendship University of Russia, Moscow, 30–31 Jan 2007. Moscow: People's Friendship University of Russia; 2007b. pp. 519–24.
124. Valdes-Galicia JF, Pérez-Enriquez R, Otaola JA. The cosmic-ray 1.68-year variation: a clue to understand the nature of the solar cycle? Sol Phys 1996;167:409–17.
125. Howe R, Christensen-Dalsgaard J, Hill F, Komm RW, Larsen RM, Schou J, Thompson MJ, Toomre J. Dynamic variations at the base of the solar convection zone. Science 2000;287(5462):2456–60.
126. Duma G, Vilardo G. Seismicity cycles in the Mt. Vesuvius area and their relation to solar flux and the variations of the earth's magnetic field. Phys Chem Earth 1998;23(9–10):927–31.
127. Clilverd MA, Clark TDG, Clarke E, Rishbeth H. Increased magnetic storm activity from 1868 to 1995. J Atmos Solar Terr Phys 1998;60:1047–56.
128. Halberg F, Cornélissen G, Katinas G, Chibisov S, Holley D, Czaplicki J, Otsuka K, Wang ZR, Bakken EE. Cycles in the biosphere in the service of solar-terrestrial physics and vice versa. Proceedings, International Conference on the Frontiers of Biomedical Science: Chronobiology. Chengdu, China; Sep 24–26, 2006b. pp. 36–39.
129. Sothern RB, Katinas GS, Cornélissen G, Czaplicki J, Halberg F. Differential congruence of periods in helio- and/or geomagnetics and in human psychophysiology. PS–074, Proceedings, 2nd World Congress of Chronobiology. Tokyo, Japan; Nov 4–6, 2007. p. 98.
130. Jahn RG, Dunne BJ, Nelson RD, Dobyns YH, Bradish GJ. Correlations of random binary sequences with pre-stated operator intention: a review of a 12-year program. J Sci Explor 1997;11:345–67.
131. Radin D. Entangled minds: extrasensory experiences in a quantum reality. New York: Paraview Pocket Books; 2006. p. 368.
132. Cornélissen G, Halberg F, Toward a "chron-sensus" on neuroimmunomodulation, with "modulation" operationally and inferentially defined. In: Fabris N, Jankovic BD, Markovic BM, Spector NH, editors. Ontogenetic and phylogenetic mechanisms of neuroimmunomodulation: from molecular biology to psychosocial sciences. Ann NY Acad Med Sci 1992;650: 60–7.
133. Cornélissen G, Halberg F. Chronomics of suicides and the solar wind. Br J Psychiatry 2006;189:567–8.
134. Marquardt DW. An algorithm for least-squares estimation of nonlinear parameters. J Soc Ind Appl Math 1963;11:431–41.
135. Halberg F, Cornélissen G, Sothern RB, Chibisov SM, Wendt HW. Do unseen, very weak magnetic mechanisms contribute to terrorism in wobbly spectral windows? Bulletin of the People's Friendship University of Russia, Proceedings, 8th international congress "health and education millennium", Moscow, Russia; Nov 14–17, 2007c. pp. 63–6.
136. Cornélissen G, Halberg F. Chronomedicine. In: Armitage P, Colton T, editors. Encyclopedia of biostatistics. 2nd ed. Chichester: John Wiley & Sons Ltd; 2005. p. 796–812.
137. Halberg F. Chronobiology: methodological problems. Acta Medica Romana 1980;18:399–440.
138. Halberg F, Cornélissen G, Bingham C, Witte H, Ribary U, Hesse W, Petsche H, Engebretson M, Geissler H-G, Weiss S, Klimesch W, Rappelsberger P, Katinas G, Schwartzkopff O. Chronomics: Imaging in time by phase synchronization reveals wide spectral-biospheric resonances beyond short rhythms. ("Wenn man über kurze Rhythmen hinausgeht") In memoriam – lost future: Dr.-Ing. habil. Dr. rer. nat. Barbara Schack: 1952–2003b. Neuroendocrinol Lett 2003;24:355–80.
139. Roederer JG. Effects of natural magnetic field disturbances on biota. Space Med Med Eng (Chn) 1996;9:7–16.
140. Freund FT. When the earth speaks: understanding earthquake signals. #426, First European Conference on Earthquake Engineering and Seismology. Geneva, Switzerland; 3–8 Sep 2006.

141. Freund FT. Pre-earthquake signals. Part I: deviatoric stresses turn rocks into a source of electric currents. Nat Hazard Earth Syst Sci 2007;7:1–7.
142. Freund FT. Pre-earthquake signals. Part II: flow of battery currents in the crust. Nat Hazard Earth Syst Sci 2007;7:1–6.
143. Adey WR. Potential therapeutic applications of nonthermal electromagnetic fields: ensemble organization of cells in tissue as a factor in biological field sensing. In: Rosch PJ, Markov M, editors. Bioelectromagnetic medicine. New York: Marcel Dekker; 2004. p. 1–15.
144. Adair RK. Constraints on biological effects of weak extremely low-frequency electromagnetic fields. Phys Rev A 1991;43:1039–48.
145. Bennett WR. Cancer and power lines. Phys Today 1994;24(4):23–9.
146. Ulmer W, Cornélissen G, Halberg F. Physical chemistry and the biologic week in the perspective of chrono-oncology. In Vivo 1995;9:363–74.
147. Halberg F, Cornélissen G, Katinas GS, Watanabe Y, Otsuka K, Maggioni C, Perfetto F, Tarquini R, Schwartzkopff O, Bakken EE. Feedsidewards: intermodulation (strictly) among time structures, chronomes, in and around us, and cosmo-vasculo-neuroimmunity. About ten-yearly changes: what Galileo missed and Schwabe found. In: Conti A, Maestroni GJM, McCann SM, Sternberg EM, Lipton JM, Smith CC, (editors). Neuroimmunomodulation (Proceedings 4th International Congress, International Society for Neuroimmunomodulation, Lugano, Switzerland, Sep 29-October 2, 1999). Ann NY Acad Sci 2000b;917:348–376.
148. Usoskin IG, Mursula K, Solanki S, Schüssler M, Alanko K. Reconstruction of solar activity for the last millennium using 10Be data. Astron Astrophys 2004;413:745–51.
149. Cornélissen G, Halberg F, Schwartzkopff O, Delmore P, Katinas G, Hunter D, Tarquini B, Tarquini R, Perfetto F, Watanabe Y, Otsuka K. Chronomes, time structures, for chronobioengineering for "a full life". Biomed Instrum Technol 1999;33:152–87.
150. Vladimirskii BM, Martynyuk V, Cornélissen G, HalbergF. Proposed resolution: consensus at Sudak (Crimea); 6 Oct 2007. http://www.science-center.net/files/sudak-consensus2.pdf.
151. Bailes KE. Science and Russian culture in an age of revolutions: V.I. Vernadsky and his scientific school, 1863–1945. Bloomington/Indianapolis: Indiana University Press; 1990. p. 238. Dokuchaev, p. 182–3.
152. Halberg F, Cornélissen G, Schwartzkopff O, Bakken EE. Cycles in the biosphere in the service of solar-terrestrial physics? In: Schroeder W, editor. Case studies in physics and geophysics. Bremen: Wilfried Schroeder/Science Edition, 2006c. pp. 39–87. (Beiträge zur Geophysik und Kosmischen Physik/Joumal for the History of Geophysics and Cosmical Physics, Special issue, 2006/2. ISSN 1615–2824)
153. Halberg F. Some physiological and clinical aspects of 24-hour periodicity. Journal Lancet (Minneapolis) 1953;73:20–32.
154. Halberg F. Chronobiology. Annu Rev Physiol 1969;31:675–725.
155. Halberg F, Cornélissen G, Conti A, Maestroni G, Maggioni C, Perfetto F, Salti R, Tarquini R, Katinas GS, Schwartzkopff O. The pineal gland and chronobiologic history: mind and spirit as feedsidewards in time structures for prehabilitation. In: Bartsch C, Bartsch H, Blask DE, Cardinali DP, Hrushesky WJM, Mecke W, editors. The pineal gland and cancer: neuroimmunoendocrine mechanisms in malignancy. Heidelberg: Springer; 2001. p. 66–116.
156. Huang Ti Nei Ching Su Wen (The Yellow Emperor's Classic of Internal Medicine) (Veith I, trans). Berkeley: University of California Press; 1949. p 260.
157. Pecci EF. Foreword. In: Tiller WA, editor. Science and human transformation: subtle energies, intentionality and consciousness. Walnut Creek: Pavior; 1997. p. xiii–xix.
158. Barber M. The new knighthood: a history of the order of the temple. Cambridge: Cambridge University Press; 1994. p. 465.
159. Demurger A. Vie et mort de l'ordre du Temple. Paris: Seuil; 1989. p. 437.
160. Partner P. The murdered magicians: the templars and their myth. Oxford: Oxford University Press; 1982. p. 209.
161. Seward D. The monks of war: the military religious orders. London: Penguin; 1995. p. 416.
162. Pennisi E. Multiple clocks keep time in fruit fly tissues. Science 1997;278:1560–1.

163. Plautz JD, Kaneko M, Hall JC, Kay SA. Independent photoreceptive circadian clocks throughout drosophila. Science 1997;278:1632–5.
164. Cornélissen G, Tarquini B, Syutkina EV, Siegelova J, Halberg F. *Cherchez le* feedsideward: contradictory results versus predictable outcomes in infradian, notably circaseptan, meta-chronanalytic perspective. Cancer Biother Radiopharm 1997;12:419–20.
165. Halberg F, Cornélissen G, Tarquini B, Grafe A, Syutkina EV, Otsuka K, Watanabe Y, Siegelova J, Sanchez de la Pena S, Carandente F, Schwartzkopff O. Pineals, cancers and feedsidewards in the Biosphere and the Cosmoi (BIOCOS). Cancer Biother Radiopharm 1997;12:421–2.
166. Crick F. The astonishing hypothesis: the scientific search for the soul. New York: Scribners; 1994. p. 317.
167. Tiller WA. Science and human transformation: subtle energies, intentionality and consciousness. Pavior: Walnut Creek; 1997. p. 316.
168. Halberg F. *Quo vadis* basic and clinical chronobiology: promise for health maintenance. Am J Anat 1983;168:543–94.
169. Hanson BR, Halberg F, Tuna N, Bouchard Jr TJ, Lykken DT, Cornélissen G, Heston LL. Rhythmometry reveals heritability of circadian characteristics of heart rate of human twins reared apart. Cardiologia 1984;29:267–82.
170. Cornélissen G, Bingham C, Siegelova J, Fiser B, Dusek J, Prikryl P, Sonkowsky RP, Halberg F. Cardiovascular disease risk monitoring in the light of chronobioethics. Chronobiologia 1994;21:321–5.
171. Cornélissen G, Zaslavskaya RM, Kumagai Y, Romanov Y, Halberg F. Chronopharmacologic issues in space. J Clin Pharmacol 1994;34:543–51.
172. Halberg F, Bingham C, Siegelova J, Fiser B, Dusek J, Prikryl P, Sonkowsky RP, Cornélissen G. "Cancer marker chronomes" assessed in the light of chronobioethics. Chronobiologia 1994;21:327–30.
173. Jahn RG. The persistent paradox of psychic phenomena: an engineering perspective. Proc IEEE 1982;70:136–79.
174. Nelson RD, Radin DI, Shoup R, Bancel PA. Correlations of continuous random data with major world events. Found Phys Lett 2002;15(6):1–13.
175. Tiller WA. Towards a predictive model of subtle domain connections to the physical domain aspect of reality: the origins of wave-particle duality, electric-magnetic monopoles and the mirror principle. J Sci Explor 1999;13:41–67.
176. Tiller WA, Dibble Jr WE, Kohane MJ. Towards objectifying intention via electronic devices. Subtle Energ Energ Med 1999;8:103–23.
177. Starbuck S, Cornélissen G, Halberg F. Is motivation influenced by geomagnetic activity? Biomed Pharmacother 2002;56 Suppl 2:289s–97s.
178. Halberg F, Cornélissen G, Otsuka K, Syutkina EV, Masalov A, Breus T, Viduetsky A, Grafe A, Schwartzkopff O. Chronoastrobiology: neonatal numerical counterparts to Schwabe's 10.5 and Hale's 21-year sunspot cycles. In memoriam Boris A. Nikityuk. Int J Prenat Perinat Psychol Med 2001;13:257–80.
179. Gell-Mann M. Consciousness, reduction and emergence: some remarks. Ann NY Acad Sci 2001; 929 (Cajal and consciousness: Scientific Approaches to Consciousness on the Centennial of Ramón y Cajal's Textura):41–9.
180. Pitts GC. An evolutionary approach to pain. Perspect Biol Med 1994;37:275–84.
181. Halberg F, Cornélissen G, Stoynev A, Ikonomov O, Katinas G, Sampson M, Wang ZR, Wan CM, Singh RB, Otsuka K, Sothern RB, Sothern SB, Sothern MI, Syutkina EV, Masalov A, Perfetto F, Tarquini R, Maggioni C, Kumagai Y, Siegelova J, Fiser B, Homolka P, Dusek J, Uezono K, Watanabe Y, Wu JY, Sonkowsky R, Schwartzkopff O, Hellbrügge T, Spector NH, Baciu I, Hriscu M, Bakken E. Season's appreciations 2002 and 2003. Imaging in time: the transyear (longer-than-the-calendar year) and the half-year. Neuroendocrinol Lett 2003;24:421–40.

Chapter 22
Astro-glocal Spatially and Temporally (*Glo*bal and Lo*cal*) Comprehensive Health-Watch Especially at High Altitude

Abstract On August 1, 2010, an entire hemisphere of the sun erupted. Filaments of magnetism snapped and exploded, shock waves raced across the stellar surface, and billion-ton clouds of hot gas billowed into space. A coronal mass ejection (CME) headed directly for earth. The huge solar storm triggered unusual northern and southern aurorae appearing on the night of August 3. Geomagnetic turbulence in Japan was observed on August 4. The CME could have driven disastrous flooding occurring in Ladakh on August 5 and 6, 2010, in association with an annual acrophase of environmental temperature recorded by us there.

Chronoecological health-watch in Ladakh, using "glocal comprehensive assessment" to study the human circulation, autonomic nervous system activity, and health quality as a physiological system at high altitude, complements a chronoecological study in Japanese towns. Ladakh is a very arid region of east Kashmir, adjacent to Tibet, at an altitude of 2500–4600 m between the Karakoram and the Himalayan ranges. Ladakh residents were checked out annually from 2004 to 2010.

Space weather affected the citizens' health, as demonstrated in this investigation, at least around an 11-year cycle. Thus, an astro-glocal assessment, in space and time, is recommended as background for diagnosis and treatment, especially at high altitudes. We can ask whether one of the longer solar, or galactic geological, cycles can override the conditions of the citizens' health as well as the earth weather especially in Ladakh recorded during the minimum in a Horrebow-Schwabe about 11-year sunspot cycle.

Keywords Chronoecological health-watch • High-altitude environments • Huge solar storm • Disastrous floods in Ladakh • Glocal assessment in space and time

22.1 Introduction

Ladakh is one of the most remote regions of India, with the Karakoram to the northwest, the Himalayas in the southwest, and the Trans-Himalayas at its core. A sparsely populated area of Indian Himalaya is here, at 2500–4600 m altitude, consisting mainly of arid desert. High-altitude environments are generally harsh and fragile. They have little oxygen, low pressure, cold temperature, and strong ultraviolet radiation, and the weather in the mountains is very changeable. Certainly

K. Otsuka et al., *Chronomics and Continuous Ambulatory Blood Pressure Monitoring*, DOI 10.1007/978-4-431-54631-3_22

individuals living there at high altitude showed changeable SpO2 values as well as severe hypoxemia, to which women seem to be more sensitive than men. Meteorological and broader climatological environments should affect the health of citizens in Ladakh. There have been no precise reports there, however. Thus, we started the weather and climate monitoring and tried to make an interdisciplinary approach for the relation between the mountains' weather and the citizens' health.

22.2 Chronoecological Health-Watch in Ladakh

We have continued free health screening of a chronoecological health-watch annually since 2001–2010. The first visit to Ladakh was 2001, coincident with a span of high solar activity, that is in the solar maximum, and the sixth visit of 2008 was close to a solar minimum. Space weather could have a strong effect on the high-altitude environment, and in this investigation we assessed a relation between solar activity and the health of citizens in Ladakh.

A chronoecological health-watch in Ladakh, using "glocal comprehensive assessment (GCA)" to study the human circulation, autonomic nervous system activity, and health quality as a physiological system at high altitude, complements a chronoecological study in Japanese towns. Ladakh is a very arid region of east Kashmir, adjacent to Tibet, at an altitude of 2500–4600 m between the Karakoram and the Himalayan ranges.

In association with the 23rd solar cycle, 3418 Ladakh residents (13–92 years, average 51.4 years of age, 1428 men and 1990 women), i.e., 549 citizens in 2004 (13–82 years, average 51.1 years of age), 461 citizens in 2005 (19–89 years, average 54.9 years of age), 447 citizens in 2007 (20–86 years, average 47.8 years of age), 164 citizens in 2008 (18–92 years, average 55.7 years of age), 788 citizens in 2009 (19–92 years, average 53.1 years of age), 420 citizens in 2010 (in June and July) (19–85 years, average 55.9 years of age), and 589 citizens in 2010 (in August and September) (22–91 years, average 55.7 years of age), were examined annually from 2004 to 2010.

High-altitude environments have generally less oxygen, lower pressure, cold temperature, and strong UV radiation. From June 2009 to September 2011, we monitored the climate at Ladakh, North India (34°27′, 76°49′, 3808 m altitude), every 10 min. Air temperature (°C), relative humidity (%), air pressure (hPa), wind direction (degree), wind speed (m/s), and rainfall (mm) were measured by using a complex sensor (Vaisala CVS-WXT520). Upward/downward long/short wave radiation (W/m^2) was assessed by an instrument of Hukseflux CHF-NR01 and UV radiation (305–385 nm, W/m^2) by KIPP&ZONEN CUV4 [1]. Rhythmicity and 1/f fluctuations of the time series of measurements were analyzed by the Maximum Entropy Method (MemCalc software Ver2.0, GMS, Tokyo), also investigated in chronomic serial sections with a 24-hour and a separate 7-day cosine fit, by fitting polynomials and by the nonlinearly extended cosinor.

Examinations of health-watch were as follows: pulse oximetry (SpO_2), blood pressure (BP), heart rate (HR), respiration rate (RR), and body mass index (BMI). BP and HR were measured twice in each of the sitting, supine, and standing positions using a semiautomated BP device (UA-767PC, A&D Co, Ltd., Tokyo, Japan). Aortic stiffness of *c*ardio-*a*nkle *v*ascular *i*ndex (CAVI) of right and left ankles [2, 3] was measured twice using a VaSera instrument (Fukuda Denshi, Tokyo). For the glocal comprehensive assessment, we recorded sleep habit, depressive mood questionnaire and cognitive function of the Kohs block design test and the time estimation test. The Up & Go test measured, in seconds, the time it took the subject to stand up from a chair, walk a distance of 3 m, turn, walk back to the chair, and sit down again. Functional Reach, used to evaluate balance, represents the maximal distance a subject can reach forward beyond arm's length while maintaining a fixed base of support in the standing position. Manual dexterity was assessed using a panel with combinations of 10 hooks (hook-on), 10 big buttons (button-on-and-off), and 5 small buttons (button-on-and-off). The total manual dexterity time in sec., defined as the button score (Button), was calculated by adding the average times for one hook-on and one big or small button-on-and-off.

All data from Ladakh were analyzed with the Statistical Software for Windows (StatFlex Ver.5.0, Artec, Osaka, http://www.statflex.net) and some of these and other data from controls in Minnesota, by methods in the Halberg Chronobiology Center (University of Minnesota, Minneapolis, MN). Student's t-tests and one-way analyses of variance (ANOVA) served for the comparison of two or more groups. Significance was considered at $p<0.05$.

22.3 Association of Blood Pressure in Ladakh and Solar-Terrestrial Cycle

The first visit to Ladakh was 2001 at the sunspot maximum. From then on, solar activity decreased for 6–7 years and the 23rd solar cycle became the sunspot minimum in 2008. The eighth visit was in 2010 and we have examined citizens in Ladakh from the solar maximum to the minimum (Fig. 22.1).

Chronoecological health-watch showed a variation of the systolic BP readings as follows: in 2004 (135.1 ± 22.1 mmHg, $n=546$), in 2005 (130.6 ± 25.2 mmHg, $n=433$), in 2007 (129.4 ± 20.1 mmHg, $n=444$), in 2008 (128.8 ± 21.2 mmHg, $n=164$), in 2009 (128.0 ± 22.3 mmHg, $n=786$), and in 2010 (130.0 ± 23.6 mmHg, $n=418$). It was compared with the 23rd solar cycle and their fitted models were surprisingly similar (Fig. 22.2a), yet differing between men and women, with the men apparently resonating with their systolic blood pressure and the women with their diastolic blood pressure (Fig. 22.2b). Systolic but much less so, if at all, diastolic blood pressure was damped when a 1.3-year far-transyear component was no longer detected in the spectrum of the solar wind (4).

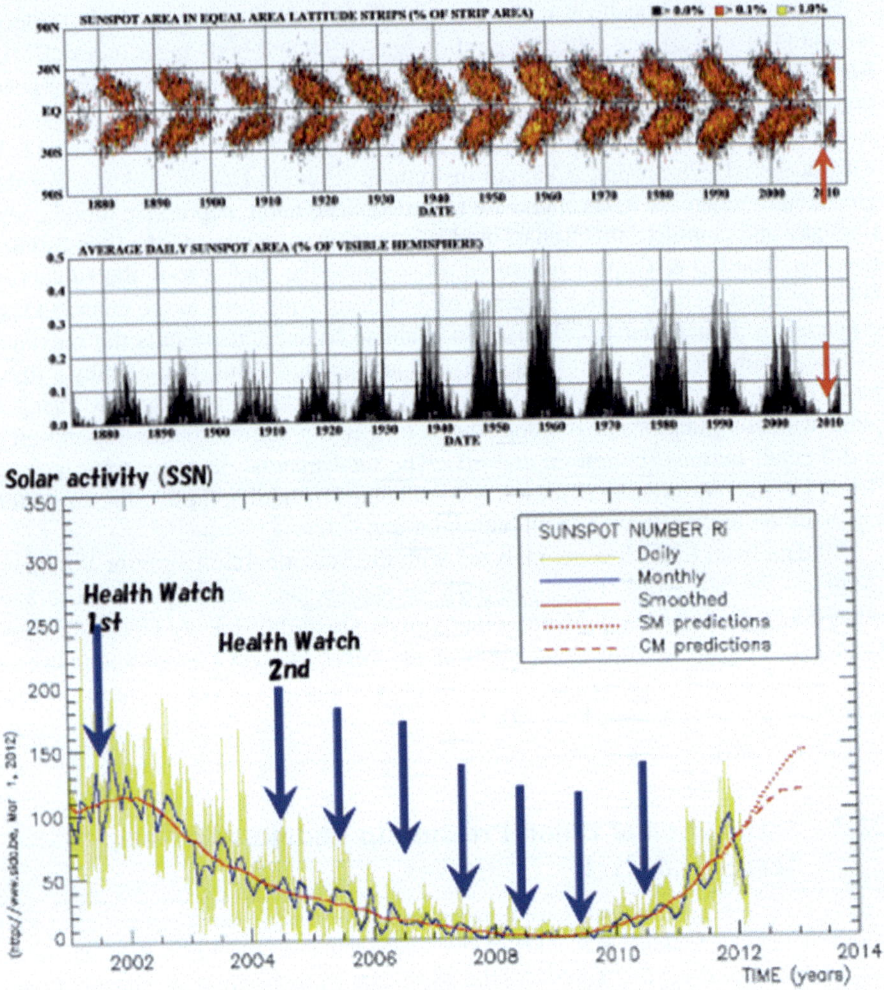

Fig. 22.1 Solar cycle and chronoecological health-watch in Ladakh. *Top*: Time series of the solar cycle. *Bottom*: Timing of eight times of visit to Ladakh for chronoecological health-watch since 2004–2010 (*blue* ↓), which coincides with the 23rd solar cycle (to December 2007) and with the 24th one (from January 2008). *Red* ↓ shows the timing of disastrous flooding in Ladakh

Figures 22.2c, d are from a man and a woman monitored longitudinally in Minnesota during the same or much of the same span. Both individuals (Fig. 22.2c, d) and a population such as that in Ladakh can monitor the sun, but such studies must be extended before drawing comparative inferences. It seems clear, however, that measurements in populations, preferably much longer than that in Ladakh, can mirror solar activity, and hence routine clinic and home measurements may well be tested worldwide for this possible use.

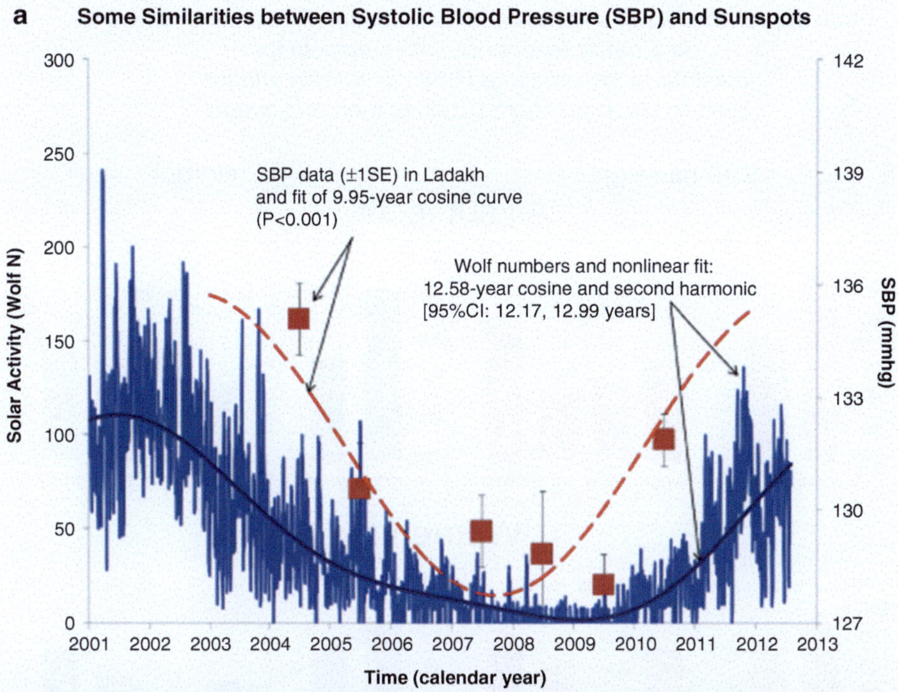

Fig. 22.2 (**a**) Some similarities between systolic blood pressure (SBP) and sunspots. Daily values of Wolf numbers from January 1, 2001, to July 31, 2012, were analyzed by the linear-nonlinear extended cosinor. Using a trial period of 10 years, nonlinearly, the period estimate was about 17 years but converged to 12.58 (95 % CI: 12.17, 12.99) years when adding a second harmonic in the model, as shown here (*blue curve*). Using a similar model (10-year trial period, with addition of a second harmonic) for the SBP data from Ladakh, the period estimate is about 9.95 years, but the 95 % CIs of the amplitude of both the fundamental and second harmonic cover zero, likely because the time series covers a shorter span of 7 years, shorter than a single solar activity cycle. The dashed red curve was obtained by using the parameters from the linear cosinor corresponding to a trial period of 9.95 years, with the qualification that linearly, the best fitting period is longer than 15 years, as was the case for Wolf numbers in the absence of a second harmonic term. (**b**) Yearly mean values of systolic and diastolic blood pressure of all Ladakh's residents as well as of women and men considered separately. Results are shown as bar graphs assigned to the year of data collection, taking into consideration the fact that no data were collected in 2006. This added consideration slightly changes the R^2 values and the equations of the fitted second-order polynomial curves (© Halberg Chronobiology Center.) (**c**) A plot of an elderly man's (FH's) SBP data from 2001 to 2011. The weekly mean values were fitted nonlinearly with a cosine curve using 10 years as a trial period. The model converges to a period of 9.36 years, similar to that found for the residents of Ladakh. There is, however, a large phase difference between the SBP data in Ladakh and those of FH (© Halberg Chronobiology Center.) (**d**) A plot of a woman's (GC's) SBP data from 2004 to 2011 (no data collected between 2001 and 2004). The weekly mean values were fitted nonlinearly with a cosine curve using 10 years as a trial period. The model converges to a period of 11.91 years. The 95 % CI of the period is quite wide in view of the very short span of data available to reach statistical significance (© Halberg Chronobiology Center)

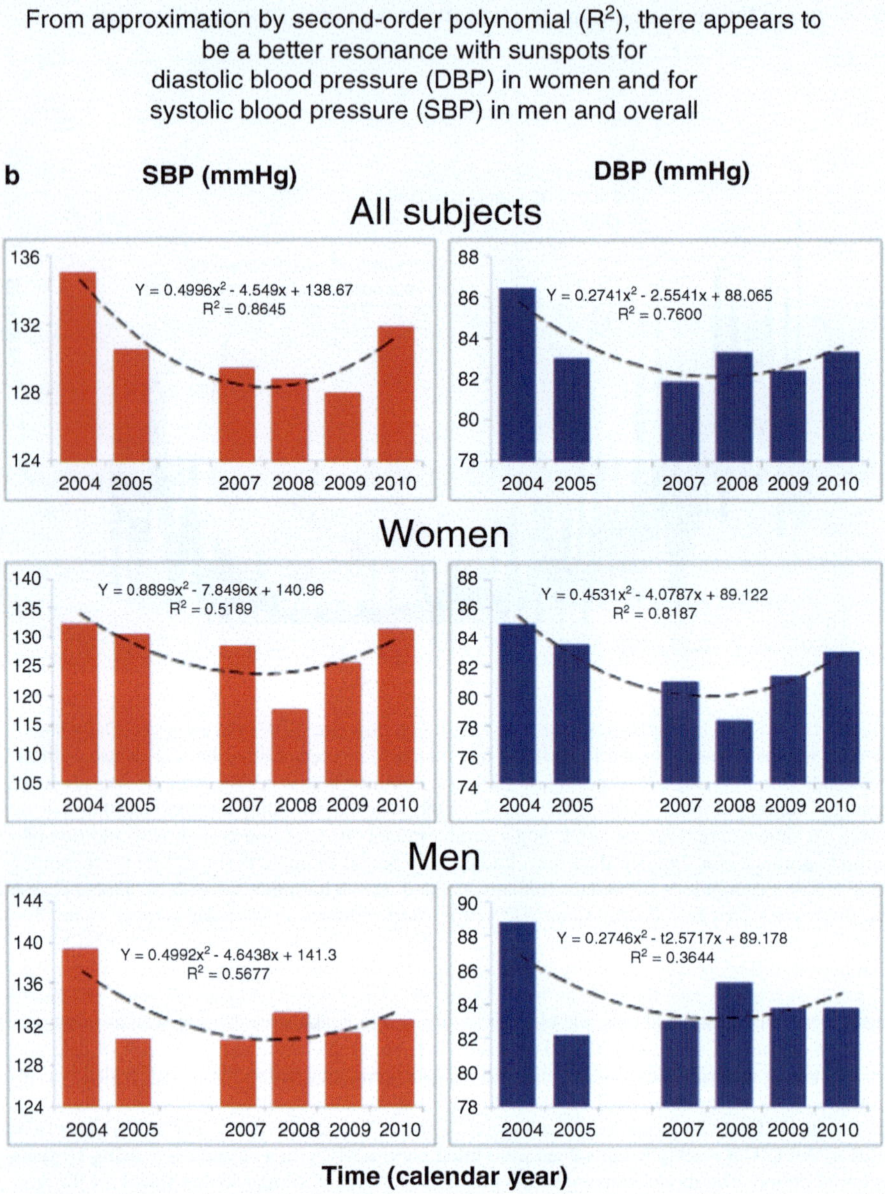

Fig. 22.2 (continued)

22.4 Coronal Mass Ejection (CME) on August 1, 2010

On August 1, 2010, an entire hemisphere of the sun erupted. Filaments of magnetism snapped and exploded, shock waves raced across the stellar surface, and billion-ton clouds of hot gas billowed into space [4–8]. A coronal mass ejection (CME) headed directly for earth (Fig. 22.3a, b). The huge solar storm triggered

c

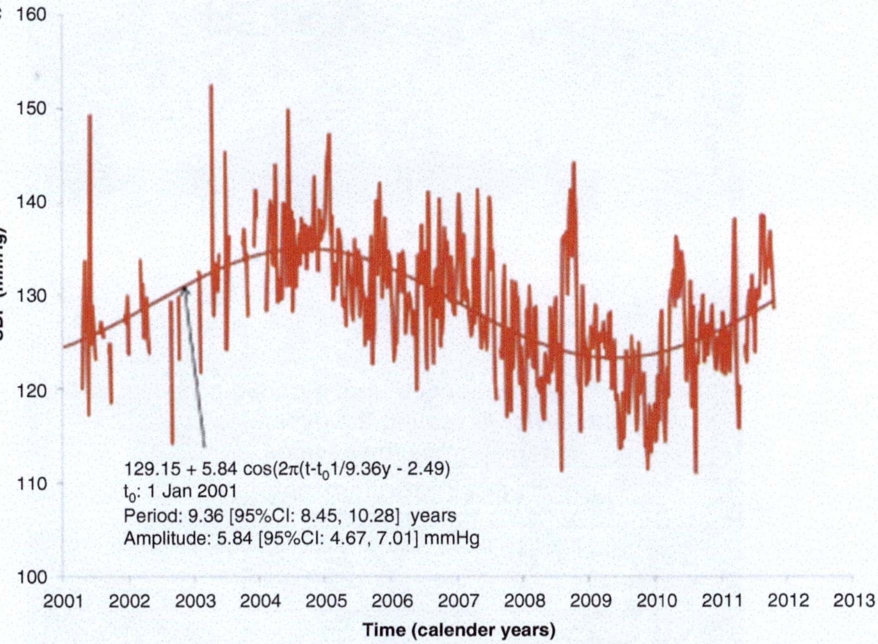

d

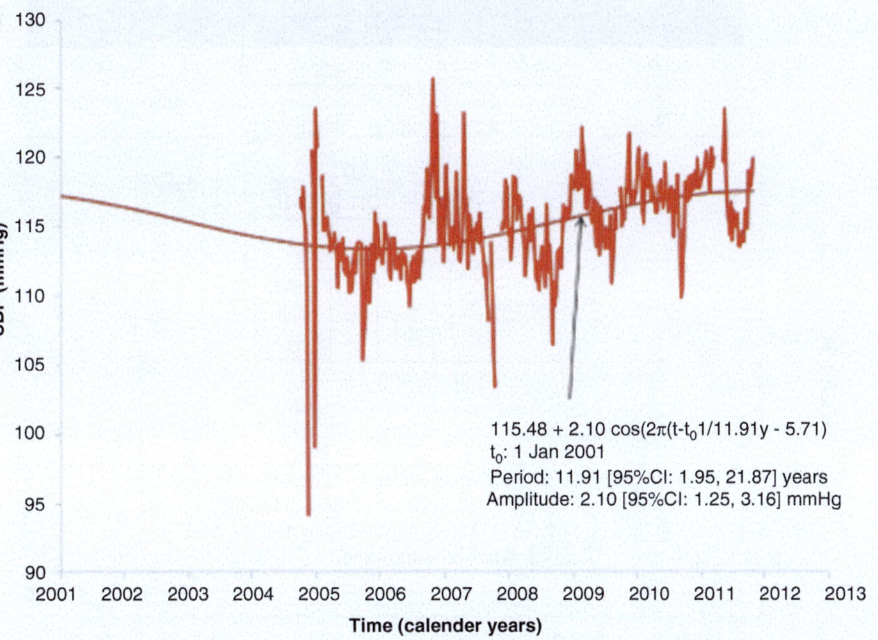

Fig. 22.2 (continued)

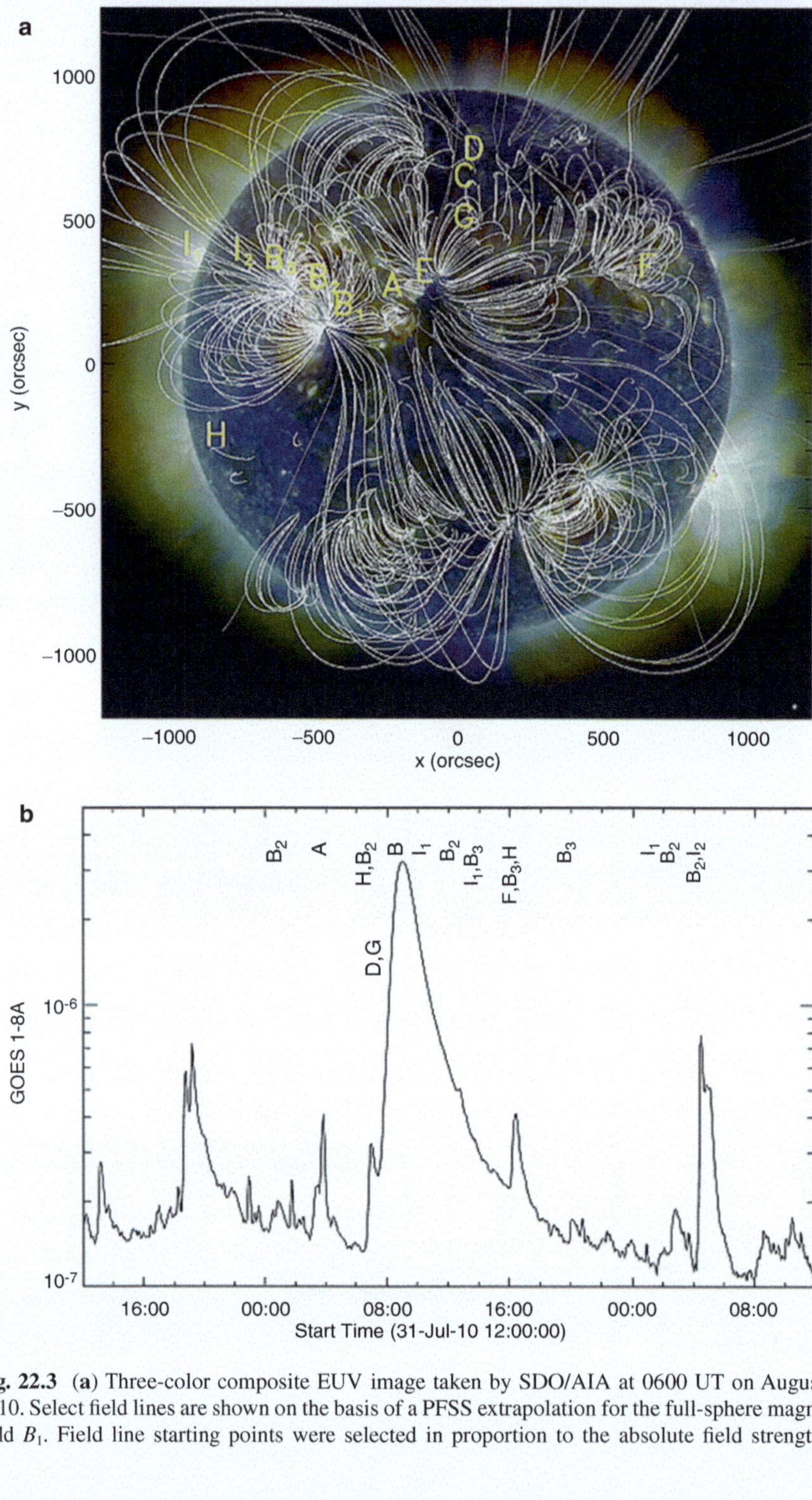

Fig. 22.3 (**a**) Three-color composite EUV image taken by SDO/AIA at 0600 UT on August 1, 2010. Select field lines are shown on the basis of a PFSS extrapolation for the full-sphere magnetic field B_1. Field line starting points were selected in proportion to the absolute field strength in

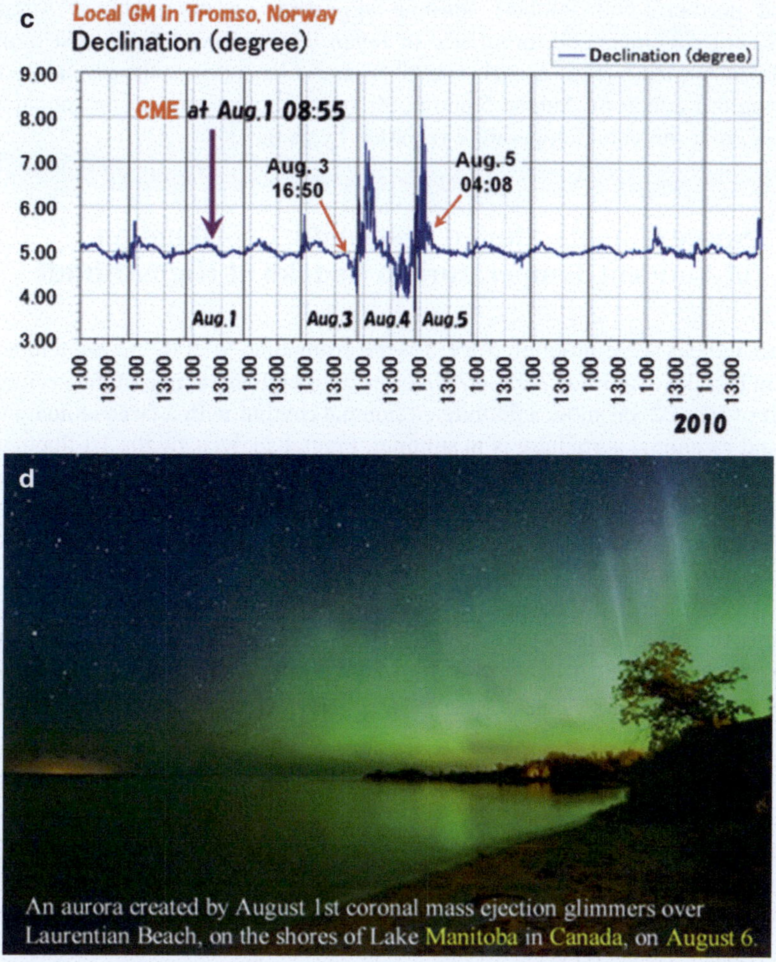

Fig. 22.3 (continued) the full-sphere map at the resolution of the PFSS model; the same starting positions were used in all images and movies in this study. The earth is shown to scale in the lower right corner (Schrijver and Title [4]). (**b**) A time line of the sun eruption on August 1, 2010, when an entire hemisphere of the sun erupted. Long-range magnetic couplings between solar flares and coronal mass ejections observed by Solar Dynamics Observatory (SDO) and the two STERO spacecraft are summarized in this figure. GOES 1–8 Å light curve for a 2-day interval centered on August 1, 2010; the plot's vertical scale ranges from the flux levels equivalent to B1 up to C5 (Schrijver and Title [4]). (**c**) Time series of geomagnetic activity observed in Tromsø, Norway. Please note the long-lasting geomagnetic turbulence, which continued from 16:50 on August 3 to 04:08 on August 5. (**d**) The huge solar storm triggered unusual northern and southern aurorae appearing on the night of August 3 and also in Canada on August 6. Aurorae happen when energized particles from the sun wash over earth and flow down the planet's magnetic field lines toward the poles. Along the way, the charged particles bang into nitrogen, oxygen, and other atoms in our atmosphere. The charged solar particles give earth's atmospheric atoms an energy boost, which then gets released as light, producing the shimmering curtains of greens, reds, blues, and other colors

unusual northern and southern aurorae appearing on the night of August 3 (Fig. 22.3c). Geomagnetic turbulence in Japan was observed on August 4 and in Canada on August 6 (Fig. 22.3d). The CME could have driven disastrous flooding occurring in Ladakh on August 5 and 6, 2010, in association with an annual acrophase of environmental temperature recorded by us there.

22.5 Space Climate Altered Systematic Coordination of Ambient Temperature in Ladakh at High Altitude

Weather indices were monitored for 2 years from June 9, 2009, to September 11, 2011, in Ladakh at altitude 3808 m. Time series of ambient temperature, seen on the top left of Fig. 22.4a, show a definite circannual rhythm with a large annual amplitude, and its annual acrophase is in summer. Fractal analysis by the 1/f fluctuations shows a definite disruption in August which precedes disastrous flooding occurring in Ladakh on August 5 and 6, 2010, as shown on the right of Fig. 22.4a.

Ambient temperature investigated in chronomic serial sections with a 7-day cosine fit also shows a change of circaseptan rhythmicity as seen in the third upper panel of Fig. 22.4b. From June to winter in 2009, ambient temperature showed a stable circaseptan rhythm in Ladakh (at an altitude of 3808 m). From the winter to summer, the circaseptan starts to drift (advance) in phase. However, around July to August, the circaseptan rhythmicity changes coincident with a change of space weather.

22.6 Disastrous Flooding Caused by the Huge CME on August 1, 2010

Disastrous floods on August 5 followed the CME on August 1, 2010, seen in serial sections of rainfall (Fig. 22.5a, b). Meteorological observation disclosed a disruption of 1/f fractal scaling of the environmental temperature before the floods in Ladakh, whereas serial sections showed the circannual acrophase. An annual health-watch shows, as compared to Japanese subjects, that residents in Ladakh slept less (7.8 vs. 6.9 hours, $p<0.001$) and had a lower SpO2 (84.4 % vs. 96.0 %, $p<0.00001$), decreased parasympathetic activity (lower LF and HF; $p<0.05$), higher sympathetic activity (increased LF/HF ratio, $p<0.05$), and higher diastolic blood pressure (BP) and heart rate ($p<0.00001$) [3].

Floods on August 5 affected the health of Ladakh citizens as shown in Table 22.1. Systolic and diastolic BPs increased from 130.1/81.7 to 135.2/85.7 mmHg ($p<0.005$), and arterial stiffness CAVI increased from 7.72 to 8.22 m/s ($p<0.0001$). The percentage of citizens suffering from depressive mood increased from 5.3 % to 8.8 % ($p<0.0001$).

We advocate aligned biospheric and space-terrestrial weather monitoring as a feature of both health care and a unified science. To learn whether more than one solar cycle stage may be considered in relation to human cataclysms, we use an illustrative example asking (only rhetorically!) whether one of the many longer than ~11-year solar or galactic cycles can lead to cataclysms [9, 10].

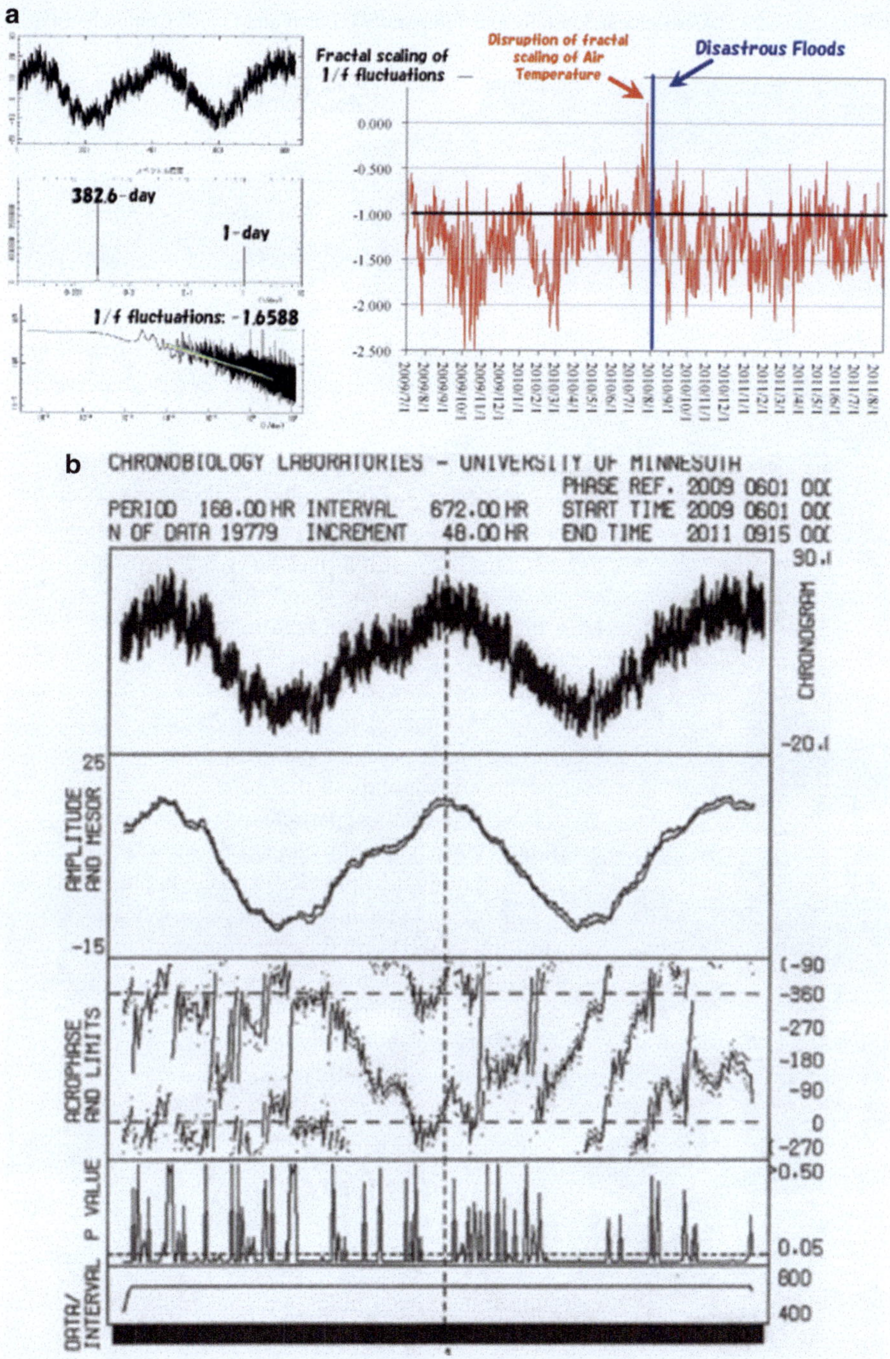

Fig. 22.4 (**a**) Disturbed fractal scaling of the ambient temperature precedes flood disasters in Ladakh. *Left Top*: Time series data of ambient temperature from June 9, 2009, to September 11, 2011, in Ladakh at high altitude (3898 m). *Left Middle*: MEM spectrum shows two rhythms of 382.6 day and 1.00 day periodicities. *Left Bottom*: Fractal analysis of total data of the ambient temperature. *Right*: Time series of its 1/f fractal scaling of the 90-day records at an increment of 7 days. (**b**) Circannual rhythm of ambient temperature and its fluctuating circaseptan rhythmicity

Fig. 22.5 (**a**) Mountain and country scenery in Ladakh before disastrous floods. (**b**) Flood disaster in Ladakh on August 5, 2010

Table 22.1 Chronoecological health-watch before and after the disastrous flooding in Ladakh

	2005 (control 1)		2010 before flood		2010 after flood		Statistics: after vs. before		Statistics: after vs. control 1	
	Mean	S.D.	Mean	S.D.	Mean	S.D.	t-value	P-value	t-value	P-value
Number of subjects	461		420		590					
Altitude (m)	3775 (3390–4590)		3089 (2500–3800)		3673 (3200–3800)		–16.08	<0.0001	3.634	<0.0005
Age (year)	54.9	15.3	55.9	15.2	55.7	14.1	0.254	N.S.	–0.821	N.S.
Range (year)	19–89		19–85		22–91					
Gender	0.356		0.438		0.381		1.812	N.S.	–0.853	N.S.
BMI	21.9	3.1	22.3	3.4	23.4	3.9	–4.737	<0.0001	–6.831	<0.0001
Syst BP (mm Hg)	130.6	25.2	130.1	23.6	135.2	24.3	–3.335	<0.001	–2.969	<0.005
Diast BP (mm Hg)	83.0	14.3	81.7	14.7	85.7	13.7	–4.352	<0.0001	–2.953	<0.005
Pulse rate (bpm)	79.9	16.5	77.8	14.7	77.9	13.8	–0.131	N.S.	2.111	<0.05
SpO_2 (%)	86.8	5.8	86.3	6.3	91.9	3.6	–17.645	<0.0001	–17.201	<0.0001
CAVI (m/sec)	8.38	2.04	7.72	1.68	8.22	1.51	–0.4801	<0.0001	1.318	N.S.
ABI	1.130	0.121	1.120	0.107	1.153	0.098	–4.568	<0.0001	–2.845	<0.005
Functional reach (cm)	*	*	25.0	1.0	28.5	8.4	–4.612	<0.0001	*	*
Depression Q1	0.020	0.139	0.014	0.339	0.113	0.317	χ^2; 32.0	<0.000001	χ^2; 32.9	<0.000001
Depression Q2	0.015	0.123	0.053	0.224	0.088	0.287	χ^2; 4.38	<0.05	χ^2; 25.3	<0.000001

The CME could have driven disastrous flooding occurring in Ladakh on August 5 and 6, 2010, in association with an annual acrophase of environmental temperature recorded by us there and also reported in Pakistan and in Gansu, China. A heat wave and smog were presumably associated with the CME in Moscow on August 2–9.

We can ask (again only rhetorically) whether one of the longer solar or galactic geological cycles can override the conditions of the minimum in a Horrebow-Schwabe about 11-year sunspot cycle [11, 12]. In 2010 [13] we were presumably near the end of one of the ~30 million-year cycle, possibly a Hans Stille geological cycle as the sun passes through the galactic equator area (personal communication from Elchin Khalilov. But we cannot sufficiently resolve the stage of myriadennian biospheric possible coperiodisms [14] of ~37.8 million years that may have peaked or reached a critical stage during a minimum of the undecennian sunspot cycle stage [11, 12]. Only prospective monitoring of ourselves, preferably longitudinally (Fig. 22.2c, d) and around us, in the steps of others, can provide an answer to this speculation.

An astro-glocal (in space and time) assessment is recommended as background for diagnosis and treatment, especially at high altitude. Both a relatively short (about 11-year) and long (up to and beyond about 30- or 37-million-year) cycle [9] can be monitored in the future in us, as soon as we transfer the technology now available to monitor pressure in our tires to humans at all ages from womb to tomb (whether or not we drive).

Cycles in the Occurrence of Major Earthquakes

Major Earthquakes

Major earthquakes may not occur randomly, being characterized by cycles along different time scales that may help understand underlying mechanisms and allow the development of precautionary measures to minimize their impact on populations and their infrastructures, whether or not the cycles also serve the prediction and thus the evasion of cataclysms.

The case often made against the predictability of earthquakes (Edgescience 2011; 7: 12–4) is built upon the fact that fractures and failure in bedrock deep beneath the Earth are notoriously intractable. Any cycles found to characterize the occurrence of earthquakes are, of course, only statistical by nature and may not be accurate enough for actual prediction.

Between January 1900 and June 2010, the dates of 331 earthquakes (primary magnitude $\geq$1.0; MMI 1.0; deaths $\geq$10) were listed in the NOAA (NESDIS/NGDC) website for Natural Hazards (accessed on 26 July, 2011). A tsunami followed in 102 cases. The monthly incidence of earthquakes and tsunamis was determined and analyzed by the extended cosinor (Acta med rom 1980; 18: 399–440).

(continued)

(continued)

Least squares spectra in the range of one cycle per 100 years and 5 cycles per year identified components with periods of about 50, 11.5, 1.44, and 0.41 year(s), all validated by nonlinear least squares when considered separately. Whereas the about 50-year component is less prominent in tsunamis, it is also detected with statistical significance, along with an about 12-year cycle. An about 2-year component characterizes tsunamis more prominently, along with a large-amplitude about 0.38-year cycle that has a period somewhat shorter than that characterizing all earthquakes. Fitted separately, it is validated nonlinearly, as is an about 2-year component, not present in the spectrum of earthquakes. Fitted concomitantly by nonlinear least squares, all four components have 1-parameter limits for the amplitude non-overlapping zero.

A possible connection between sunspots and earthquakes was postulated in 1920 by Alter (Science 1920; 51: 486–7). A short period of about 14.8 months is also reported, corresponding to a transyear, along with a longer period of about 78 years. Zhang (Acta Seismologica Sinica 1998; 11: 495–500) also reports that earthquakes occur frequently around minima of solar activity. Variations in electric tension underground have been postulated to underlie the triggering of earthquakes (Nat Hazards Earth Syst Sci 2007; 7: 1–6). It should be noted, however, that results herein point to a period of about 12.2 years, with a 95% confidence interval for the period ranging from 11.1 to 13.3 years, corresponding to a cycle length somewhat longer than that characterizing sunspots during the same 110-year span, estimated at 10.52 [95% CI: 10.48, 10.56] years. The largest earthquakes may not be evenly distributed throughout the 110-year-long seismic record. A big-quake cluster was identified that ran from 1950 to 1965, including 7 of the 9 greatest quakes of the twentieth century. It was followed by a 36-year span without a quake of magnitude 8.4 or greater (Science 2011; 332: 411), preceding a second cluster that may have started with the magnitude-8.4 quake off Peru in 2001. These observations are well in keeping with the about 50-year cycle detected in the data analyzed herein and found also in physiology, pathology and societal upheavals. The influence of solar activity on seismic and volcanic activity has also been extensively studied by Elchin Khalilov and his school (Trans Int Acad Sci H & E 2008; 3: 217–40). Khalilov deserves much credit for his endeavors aimed at forecasting strong earthquakes by means of a specially-designed detector of super-long gravitational waves. According to his estimation, gravitational signals can precede strong earthquakes by 5–15 days, and can be detected in about 90% of occurrences, even when epicenters are as far away as 1,000 km from the monitoring station (Natural Cataclysms and Global Problems of the Modern Civilization. Baku-Innsbruck, ICSD/IAS 2007, pp. 51–69).

(continued)

(continued)

Physiological Changes in Relation to the 2011 East Japan Earthquake

It is investigated whether systolic (S) and/or diastolic (D) blood pressure (BP) and/or heart rate (HR) changed in relation to the magnitude 9.0 Tohoku earthquake and tsunami of 2011.

Bracketing the 12 May 2008 Wenchuan magnitude 8.0 earthquake in China, locomotor activity of mice monitored automatically around-the-clock was found to be dramatically decreased in 6 of the 8 animals on day 3 before the earthquake, the behavioral change lasting 6 days (Bioelectromagnetics 2009; 30: 613–20). Abnormal behavior in fish, snakes, dogs, frogs and mice before an earthquake has been primarily anecdotal, however. In humans, the literature has focused on consequences from cataclysms rather than on any precursors (J Am Coll Cardiol 1997; 29: 926–33). This investigation draws on the ongoing self-surveillance of BP and HR for health monitoring by several individuals to seek any changes that may have either followed or perhaps preceded the onset of an earthquake.

Weeklong around-the-clock profiles of SBP, DBP and HR from 13 Japanese in Tokyo at the time of the earthquake, from one of them during 11 prior years, and from 11 similar individuals monitored during prior years in March were analyzed by the fit of a two-component model consisting of cosine curves with periods of 24 and 12 hours, separately for each calendar day and also for each day relative to the time of occurrence of the earthquake. The circadian parameters were normalized for each profile and the relative values analyzed by one-way analysis of variance (ANOVA) across all subjects in each of the 3 groups, testing for any day-to-day differences. As the 2011 earthquake occurred on a Friday, Friday was chosen as the reference "event" day to account for any weekly pattern known to characterize BP and HR as well as the incidence patterns of adverse cardiovascular events. Controlling for the entire pertinent spectrum of rhythms would have required additional decade-long records from subjects who happened to be exposed to an earthquake.

Based on analyses of 7 calendar days, day 4 being the day when the earthquake occurred, the 12-hour amplitude of SBP was decreased 2 days prior to EQ ($F=3.903$, $P=0.002$), a feature not observed in the controls. The standard deviation of HR was larger before the earthquake and started decreasing on day 2 after the earthquake ($F=2.486$, $P=0.040$). Particularly noteworthy is the steady increase in the MESOR of SBP, reaching a peak on the first day after the earthquake ($F=3.157$, $P=0.013$). Patterns in the controls were not statistically significant and showed primarily no noticeable change.

Comparing patterns of the different endpoints among the three datasets, characteristic features associated with the 2011 earthquake, not seen in the controls, include a steady increase in the MESOR of both SBP and HR preceding the earthquake and peaking one day after the disaster and a drop in the

(continued)

(continued)

SD of HR starting on the second day after the earthquake. Whereas the latter effect may be a response to the disaster and may contribute to an increase in cardiovascular disease risk, in keeping with earlier reports, the increase in both SBP and HR preceding the earthquake is a novel finding in humans, akin to earlier observations of precursor signs made in animals. Continuous physiological surveillance could serve not only for primary prevention in healthcare but also as a way to monitor the antecedents of cataclysms in the environment. Monitoring on a broader scale in different geographic locations can yield signals that may eventually serve as a warning. If found to be amenable to generalization in different populations at different times, the data may be particularly welcome to those seismologists who hold the view that earthquakes are unpredictable. The biosphere needs to be tested for use as a very sensitive frequency-dependent magnetometer.

References

1. Yatagai A. Meteorological observation launched at Domkar, Ladakh. Himal Study Monogr 2010;11:116–26 (in Japanese with English abstract).
2. Shirai K, Utino J, Otsuka K, et al. A novel blood pressure-independent arterial wall stiffness parameter: cardio-ankle-vascular index (CAVI). J Atheroscler Thromb 2006;13:101–7.
3. Nakaoka T, Kawasaki T, Okumiya K, Norboo T, Otsuka K, Matsubayashi K, Singh RB, Cornélissen G, Halberg F. Chronoecological "glocal" (global and local) health-watch: men better adapted to ambient hypoxia than women. World Heart J 2012;4(2/3):165–75.
4. Schrijver CJ, Title AM. Long-range magnetic couplings between solar flares and coronal mass ejections by SDO and STEREO. J Geophys Res 2011;116, A04108.
5. Li T, Zhang J, Zhang Y, Yang S. Three-dimensional reconstruction of an erupting filament with solar dynamics observatory and STEREO observations. Astrophys J 2011;739:43. doi:10.1088/0004-637X/739/1/43.
6. Temmer M, Vrsnak B, Rollett T et al. Characteristics of kinematics of a coronal mass ejection during the 2010 August 1 CME-CME interaction event. arXiv 2012; 1202.0629vl [astro-ph SR].
7. Cowen R. Linked solar eruptions explained. Nature 2011;509. doi:10.1038/news.
8. Torok T, Panasenco O, Titov VS et al. A model for magnetically coupled sympathetic eruptions. arXiv 2011; 1108.2069vl [astro-ph.SR].
9. Halberg F, Cornélissen G, Katinas G, Tvildiani L, Gigolashvili M, Janashia K, et al. Chronobiology's progress: season's appreciations 2004–2005. Time-, frequency-, phase-, variable-, individual-, age- and site-specific chronomics. J Appl Biomed 2006;4:1–38. http://www.zsf.jcu.cz/vyzkum/jab/4_1/halberg.pdf
10. Halberg F, Cornélissen G, Sothern RB, Czaplicki J, Schwartzkopff O. Thirty-five-year climatic cycle in heliogeophysics, psychophysiology, military politics, and economics. Izv Atmos Oceanic Phys 2010;46(7):844–64. (Backtranslation from Geophysical Processes and Biosphere 2009;8(2): 13–42.
11. Thiele TN. De Macularum Solis antiquioribus quibusdam observationibus Hafniae institutis. Astronomische Nachrichten 1859;50:259–61.

12. Schwabe H. Sonnen-Beobachtungen im Jahre 1843. Astronomische Nachrichten 1844;21: 254–6 (no. 495).
13. Khain VE, Khalilov EN. Cycles in geodynamic processes: their possible nature. Moscow: Scientific World; 2009. 520 pp.
14. Cornélissen G, Bakken EE, Sonkowsky RP, Halberg F. A 38-million-year cycle among myriadennians in the diversity of oceanic genera. Abstract, III international conference, civilization diseases in the spirit of V.I. Vernadsky, People's Friendship University of Russia, Moscow, 10–12 Oct 2005. pp. 47–9.

Chapter 23
Chronobioethics and Alerting Chronobiologic Ambulatory Blood Pressure Monitoring (C-ABPM; Stress/Strain Test) for Preventing Cardiovascular and for Understanding and Avoiding/Evading Societal and Natural Cataclysms by Personalized and Generalized Chronouspheres

Unawares, we swim in a magnetic sea of weather in space
Which with rain and shine on earth is in a continuous race.
Components of both competitors in the spectrum abound
With amplitudes that differ in one or the other round;
Sometimes only some photics or only some nonphotics are seen;
In sudden cardiac death in Minnesota near and far transyears are keen.
The calendar year, lost for sudden death in the icebox of the nation,
Is found further south at a North Carolina, USA station.
Phenomena interact at more than one geographic site globally
Wrangling here and there, now and then, in space and time, locally.
Hence analyzing everything in segments and in its entirety
Requires a conceptual and methodological temporo-spatial glocality.
Everywhere in the biosphere there are aeolian cycles galore;
Structures in time as in space unfold more and more.
Chronomics and genetics are inextricably interdigitated;
From monitoring oneself multi-decadal cycles originated.

This paper is a posthumous work of Professor Franz Halberg, prepared in cooperation with Germaine Cornelissen, Lyazzat Gumarova, Francine Halberg, Waldemar Ulmer, Weihong Pan, Rita Jozsa, Ram B Singh, Jarmila Siegelova, Yoshihiko Watanabe, Shiyu Hong, Jinyi Wu, Othild Schwartzkopff, Hans Wendt, and Kuniaki Otsuka.

This report with a backbone based on self-surveillance whenever possible by chronobiologic ambulatory blood pressure monitoring (C-ABPM) summarizes a lifetime's search for controls (cycles) along different time scales (chronobiology and chronomics). My endeavors in health care started with successfully fighting an epidemic of typhus (supporting material). I acted in the steps of Alexander Leonidovich Chizhevsky without knowing this in this context and in many other ways in which I was in the position of Molière's M. Jourdain like most of my contemporaries: "Par ma foi! il y a plus de quarante ans que je dis de la prose *[chronobiology]* sans que j'en susse rien, et je vous suis le plus obligé du monde de m'avoir appris cela" (My goodness, for over 40 years I've spoken prose [chronobiology/chronomics]?! I'm most grateful to you in all the world for telling me so!) It is good for all of us, at any age, to learn that we are in the sun's atmosphere and affected by it in many ways, some detailed herein.

K. Otsuka et al., *Chronomics and Continuous Ambulatory Blood Pressure Monitoring*, DOI 10.1007/978-4-431-54631-3_23

Famous poets, physicians and historians emerge with
Mikulecky's semimillennians
Complementing Vernadsky's noosphere and Chizhevsky's
undecennians,
Develop them into a time-structured chronousphere;
"Know thyself" and learn how you fare in the sun's
atmosphere.
Accompanying our best wishes for the now-past holidays
With delays we summarize working/fun goals of our ways:
Personalized and generalized chronouspheres are facts, not
fiction
In physics as in the biosphere, yet beyond any single
discipline's jurisdiction.
Self- and others' surveillance has the chronousphere as the
yield
Against ignorance of what the cosmos does to us a shield.
The chronousphere, the time structured realm of the mind,
Could instill love, remove hate, and prevent us from flying to
our cosmos blind.
This is how complementary art and science and concern about
well-being began
And ever since a unified science and culture remains the goal of
woman and man.

I backtrack to my days in Boston (1948–1949), when I used 2.8 mg of cortisone in a test that, with chronobiology in Minnesota, could be done with one microgram. In the interim, I have had help by very many and should have more coauthors here who made this summary and a more extensive bio-bibliography in Folia anthropologica (2012; 12: 5–134) possible. Earl E. Bakken, the developer of the first implantable cardiac pacemaker, who was there from the beginning in Minnesota, celebrates his 89th birthday on January 10, and I trust he will welcome seeing Larry A. Beaty continue what he started as an engineer, serving medicine with a view of prehabilitation as well as rehabilitation. Many others are coauthors in the appended recent bibliography and/or in earlier publications available from halbergchronobiologycenter.umn.edu.

This summary of personalized chronouspheres based on C-ABPM is dedicated to pioneers in self-assessment, with best wishes for continued healthy productivity, to Miroslav Mikulecky, emeritus head of internal medicine at Comenius University, Bratislava, Slovakia, who as professor of internal medicine wrote a textbook of inferential statistics for his medical students, had organized a series of meetings at which he documented lunisolar associations in clinical medicine, and has demonstrated a seleno-solar (double tidal) rhythm in his own cardiac function, complementing his fellow countryman Jan Evangelista Purkinje's contributions, also on himself, by self-study.

It is also dedicated to Ram Bahadur Singh, professor of cardiology at Halberg Hospital and Research Institute, Moradabad, India, and president emeritus of the International Colleges of Cardiology and of Nutrition, who excelled in doing on himself an optimization of timing exercise, showing that chronobiology and chronomics can be of particular use in developing areas.

It is further dedicated to Kuniaki Otsuka, President, Daini Hospital, Tokyo Women's Medical University, who led me to glocality in space that required a complement in time, who realized in two Japanese cities the monitoring by C-ABPM of the elderly and who plans on the first clinical department of chronomics complementing the many chronobiology and sleep disorder clinics worldwide, including cost-effective focus on the cosmos in health care.

Abstract Humanity needs transdisciplinary science for spirituality assessment as a critical step to optimizing ethics. Genetics has shed light on the biosphere, technology upon the noosphere, in which, the achievements of the information age notwithstanding, no obvious spinoff for morality is yet apparent.

A scientific optimization may become possible in ethics to the extent to which any reproducible since cyclic features of spirituality and of criminality become measurable. Should either the "good" or the "bad" or both be found to be at least passively influenced by cyclic physical environmental factors, as is putatively the case, these aspects of behavior may eventually become actively manipulable, perhaps utilizable for human survival.

Toward this goal, we have focused elsewhere on outcomes as different as religious motivation, if not spirituality, criminality, and atrocity, and on the underlying dynamics as states at least partly within the brain and in the brain's relations with the circulation of blood, the heart, and the broad neuroendocrines.

We introduce in this chapter that chronomics has already mapped time structures in religious behavior that can lead to a study of underlying geographic/geomagnetic latitude-associated mechanisms.

Keywords Chronobiologic ambulatory blood pressure monitoring • Time structures in space-terrestrial weather • Chrono-cloud system • Natural and social cataclysms • Environmental-biospheric coperiodisms • Chronouspheres

---

It is dedicated as well to Robert B. Sothern (the coauthor of a fine textbook on biological rhythms to which he added independent scholarship and objective quantification) and to his parents, Samuel B. and the late Margaret I. Sothern. Their family autorhythmometry allowed the concomitant demonstration of changes with age in the para-annual region of the spectrum with interindividual differences in timing of the (concomitantly recorded) transitions from dominance of the signature of the calendar year, that is, from weather on earth to the prominence in physiology of periods found in solar magnetism as near-transyears and in interplanetary magnetism as far-transyears, ending with the absence of a detectable calendar-year component with greatly advancing age. It seems noteworthy that an association of human mentality with helio- and geomagnetism was found in the language of frequencies in the infradian range of a novel transdisciplinary spectrum of cycles in us, many but not all matching cycles around us, some free-running, insofar as this terminology, for which Earl E. Bakken is co-responsible, applies to a partly endogenous component that coexists, in several cases examined, with other spectral components.

I also fondly remember many late friends and coauthors, in particular Frederic C. Bartter of Bartter syndrome, Director of the Clinical Center at NIH, who self-measured blood pressure until a lethal stroke, and Howard Levine, professor of medicine at the University of Connecticut and chief of medicine at the New Britain [Conn.] General Hospital whom I met while I fought a typhus epidemic in Wels, Upper Austria, while he was on the staff of a US Army field hospital, and who practiced a battery of self-measurements until the very day of his death from amyotrophic lateral sclerosis, setting in this way and many others, an example for all of us.

23.1 Preamble

In keeping with Lord Kelvin's "to measure is to know" [1], in retrospect, I have been measuring what I could and trying to render measurable what as yet was not measurable: *Omnia metire quæcumque licet et immensa ad mensuram redige.* With Professor Robert Sonkowsky's (of Classics at the University of Minnesota) advice on nomenclature, I soon inserted *tempestive* before *redige* to indicate that time is critical in any measurement. Next, further quantification of the indispensability of multiple time scales in what we measure and what happens more broadly became apparent: "Measure, in a partial system, everything pertinent that is measurable and render measurable what is pertinent but as yet is not measurable, as simply as possible, but not simpler, in time and space, globally and locally, i.e., glocally, and hence meaningfully, taking into account chronomic maps, a described and quantified *complementary system*" *[Omnia propria ex systemate partiali metire quæcumque licet et propria immensa, quam simplicissime sed non simplicius, ad mensuram tempestive et ergo significative redige, reddens rationem tabulae chronomicae ad systema complementare descriptum et quantificatum]* [2]. Spatiotemporal glocality applies as a concept and as a method. In space, we best consider, beyond terrestrial, an inadvertent cosmic "globalization." (To "think globally, act locally," we add "in keeping with what occurs invisibly yet with major effects on us, cosmically." What affects us unknowingly in space-time is being deciphered in a new language of periods, τs, or frequencies, 1/τs.) For this purpose, we need analyses not only of entire data series (globally, i.e., of the longest available) but also of sections systematically varied in length (locally in time). In so doing, we find cycles in us that lead us to environmental coperiodisms and vice versa and can start analyzing the consequences of the environmental change and possibly effects unknowingly exerted by us on our habitat. Thus, we arrive, i.e., at a personalized sphere of the mind, which "sleeps" each day but is not cyclically inactive during each (sleep) rest span, while also following other photic and thermic cycles like those of the day and the year, with what I try to introduce and document here, an ongoing invisible competition among photics and nonphotics.

What is new is the measurability of nonphotic cycles, characterized by nonstationarities in space and time, such as, among others, the natural near week(s) and its (their) harmonics and subharmonics, the sun's near-27-day rotation around its axis, competing lunar periods, the solar flares' ~5-month pattern, the geomagnetic half year, the societal as well as solar calendar week, solar magnetism's slightly (by days or weeks) or greatly (by months, in the case of solar wind speed) longer than yearly spectral components (near- and far-transyears), as well as circabiennial (i.e., once every 2 years), parasemidecadal, decadal, multidecadal, semimillennian, and even myriadennian components (Fig. 23.1a). One or more of these nonphotics wrangle with and can replace the photic year, albeit more rarely the societally and environmentally usually synchronized biological day. Climate cycles of 50 or 60 years [3] have biospheric counterparts, i.e., in human anthropometry, stroke incidence, and economics, confirming, extending, and qualifying Kondratiev's contributions by their uncertainties. Cycles of about 500 years found by Miroslav Mikulecky and Emil Pales in the emergence of prominent historians, poets, and physicians on different

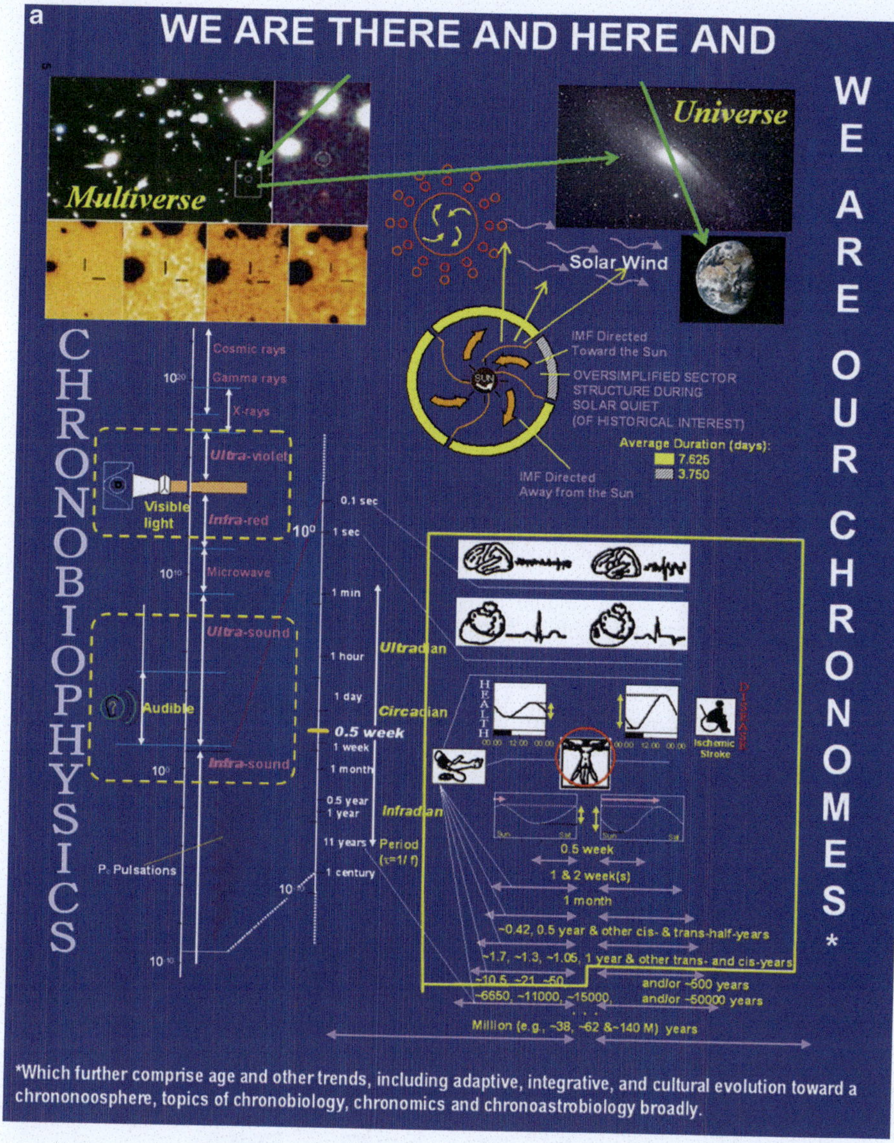

Fig. 23.1 (**a**) We are there (part of galaxies) and here (on earth), with multifrequency cycles in us and around us near and far. Biological cycles tend to have environmental counterparts. Some cycles detected in biology prompted their discovery in the environment. A near-transyear in solar magnetism recorded by terrestrial telescopes was found by us after its discovery by C-ABPM in an elderly man and in the excretion of steroidal metabolites of another man and even in the oxygen production of a giant unicell. We found a near-week in terrestrial magnetism earlier as a coperiodism of a societally desynchronized free-running near-week in steroidal metabolite excretion. Vice versa, a far-transyear of about 1.3 years in solar wind speed and an about-154-day periodicity in solar flares led to biospheric counterparts in the human blood circulation in health and in incidence patterns of disease. A transdisciplinary spectrum of coperiodisms thus evolved as the backbone in time of human science, art, and even broader culture. As humanity starts to intentionally alter the crust of the earth, the cycles provide a perspective in time to, among others, Vernadsky's sphere of the mind, rendering it into a generalized chronousphere, accumulating by transfer from books, journals, and museums to the Internet. Can we systematize the cycles and render them into an alerting system to prevent personal catastrophes and to understand and eventually prevent or evade

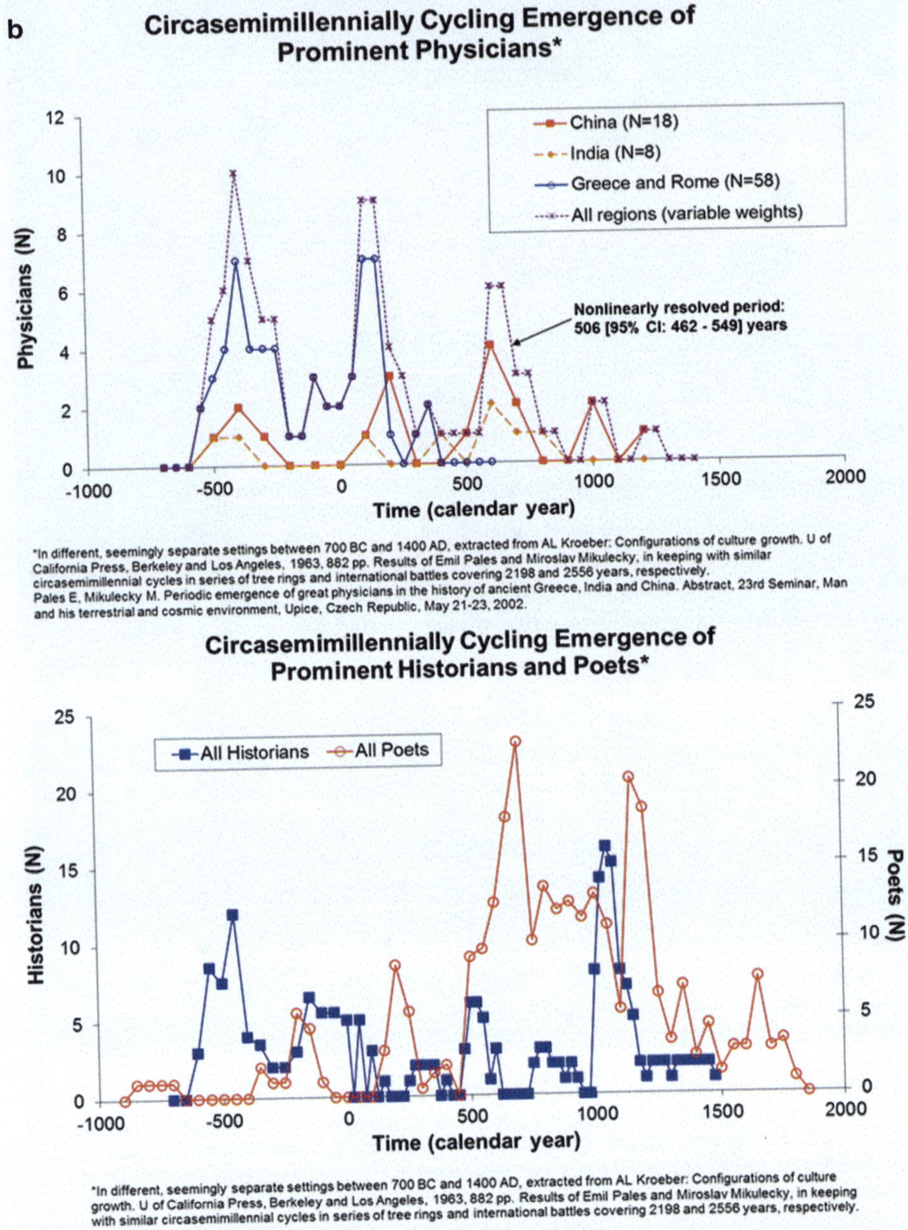

Fig. 23.1 (continued) societal and natural cataclysms, respectively? © Halberg Chronobiology Center. (**b**) About-500-year cycles in the emergence of outstanding minds: physicians (*top*) and historians and poets (*bottom*). © Halberg Chronobiology Center. (**c**) Circasemimillennian periods aligned with other roughly congruent cyclic features, among others, as parts of a generalized chronousphere: a transdisciplinary spectrum of cycles in international battles, tree growth, and climate. © Halberg Chronobiology Center. (**d**) Important findings can be made by longitudinal monitoring, as shown by Santorio Santorio who discovered insensible perspiration by carrying out all his activities on a scale hung from the ceiling of a room in his house

c

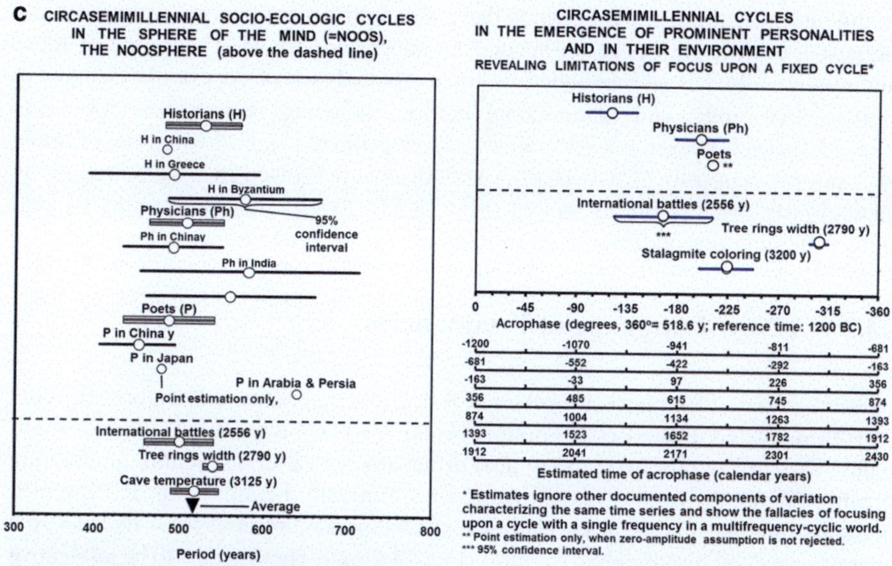

d

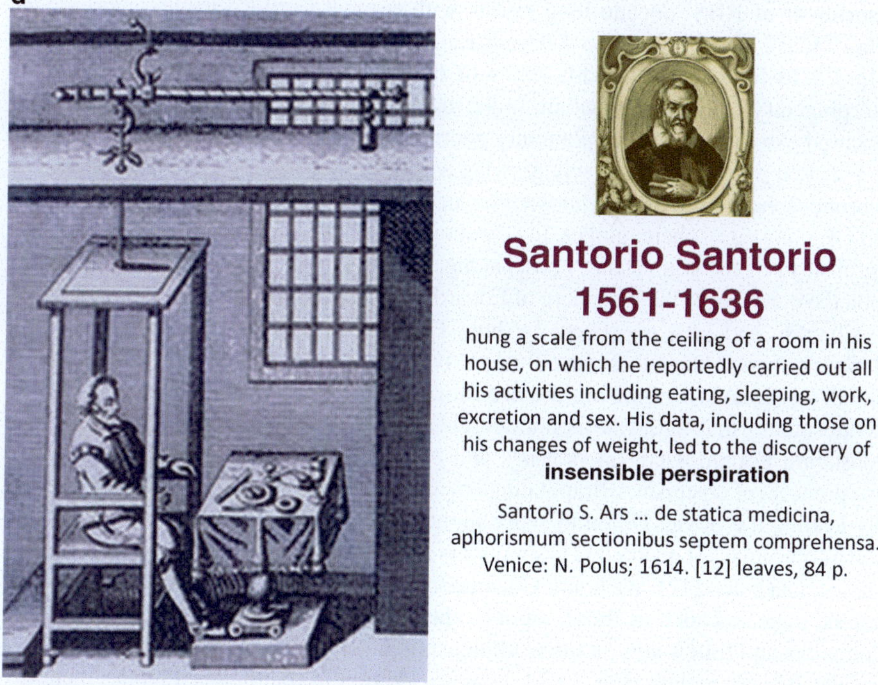

Fig. 23.1 (continued)

continents, prevailing at a time when there was no intercontinental communication, suggest a cosmic influence on special human minds [4–6]. The scope of these findings is validated and extended in our meta-analyses of coperiodic changes in climate, tree rings, and international battles, including periods of ~500 years (Fig. 23.1b, c), a glance at a generalized chronousphere [7]. Following the example of Santorio Santorio (1561–1636), who discovered insensible perspiration by meticulously monitoring his weight (Fig. 23.1d), longitudinal monitoring of vital signs is helping our understanding of influences from space-terrestrial weather.

23.2 A New Language of Frequencies

Interindividual differences in personalized chronouspheres differ for MESORs (rhythm-adjusted means) vs. 24-hour (FH and YW) or 7-day (WRB) double amplitudes (2As in Fig. 23.2a), as they also differ among variables, within and among organ systems. As to the time structure of a clinically healthy man, RBS, multidecadals modulate, in a set of 11 different variables monitored for 4 decades, the characteristics of his circadian rhythms (Fig. 23.2b, c). There seems to be a selective assortment of RBS' decade-long cycles with respect to those in the environment (Fig. 23.2d). The 3-month to 1.9-year range of a spectrum of shared frequencies (Fig. 23.2e–h) reveal that odds ratios of the association of mentality with either interplanetary or earth magnetism more than match those of the (in physics) well-accepted association of interplanetary geomagnetism (Fig. 23.2h) [8].

We are trying to build a slowly growing edifice of transdisciplinary coperiodisms as a chronousphere (from Gk *chronos*=time, in our case cycles, the Attic Gk *nous*=mind, and Gk *sphairos*=sphere, globe). In "chronousphere," we use "nous" for portmanteauing, instead of the equivalent "noos," along with "chronos" and "sphairos." We build upon the concepts of the biosphere of Eduard Suess, of hominization of Pierre Teilhard de Chardin, and of a noosphere by him, Edouard le Roy, and Vladimir Ivanovich Vernadsky, who, like others, used the derivation of "noosphere" from "noos" (also "mind" in Greek), recognizing the need for a sphere of the human creative mind or broader culture that, as did the atomic age, changes the physical environment. Vernadsky, Konstantin Eduardovich Tsiolkovsky, and Alexander Leonidovich Chizhevsky were also pioneers of cosmism, extrapolating beyond the sphere of available data yet eventually leading the Soviet Union to space flight. Critics [9] include questionable [10] as well as objective yet guarded [11] supporters and overwhelmingly strong confirmers on a novel data batch [12]. We honor Chizhevsky's genius by meta-analyses of invaluable data he collected such as those, among others, in Fig. 23.3a–c, prepared by Lyazzat Gumarova and follow-ups on his work in using glocal methods for analysis, including wavelets for orientation (Fig. 23.3d, e) as an addition to (but not as a replacement of) combined global and gliding spectra, aligned with serial sections [13].

In referring to a personal chronousphere, we aim to complement the generalized sum total of humanity's contributions to culture, including science, thus far, to which each individual can contribute measurements if not concepts or rules, the latter the privilege of a few outstanding and, as such, outlying personalities

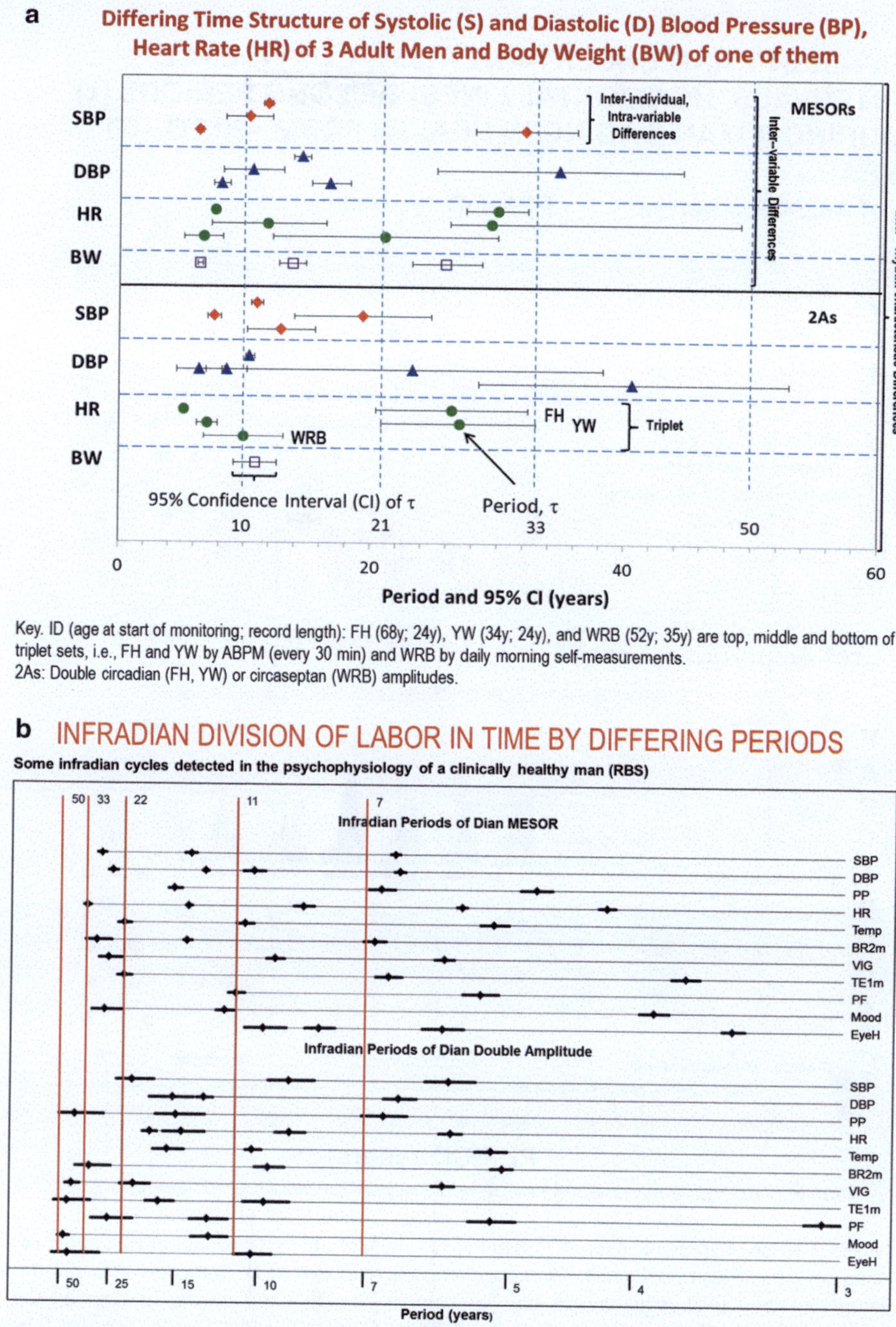

Fig. 23.2 (**a**) Inter- and intraindividual differences among periods and their uncertainties, CIs (95 % confidence intervals) found in chronobiologically interpreted ambulatory blood pressure and pulse monitoring by two men (FH and YW) and once-daily measurements of the same variables and body weight by a third man (WRB), separately for MESORs and for double 24-hour (FH and YW) or 7-day (WRB) amplitudes (2As): parts of personalized chronouspheres, revealing inter- and intraindividual and inter- and intra-variable and even inter-rhythm characteristics (MESOR, Midline-Estimating Statistic Of Rhythm, vs. amplitude) differences – selective

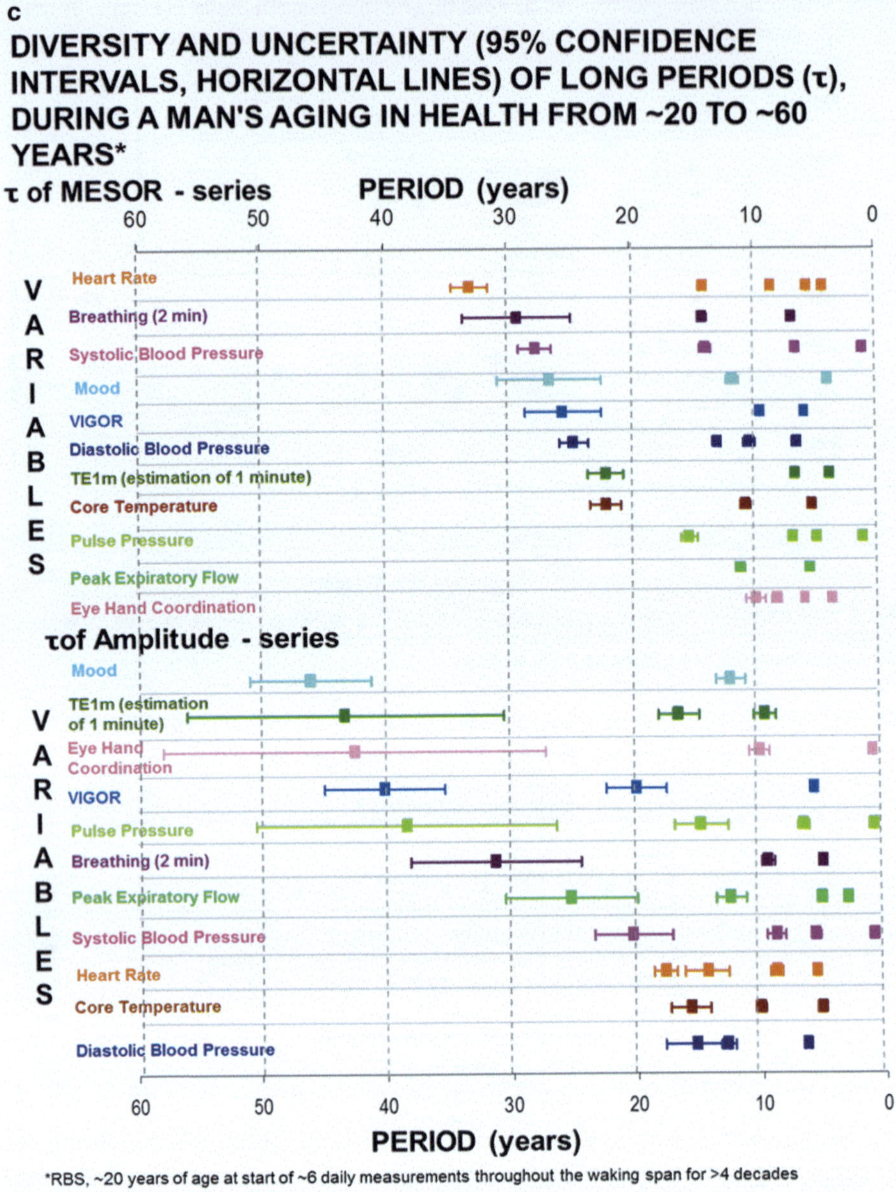

Fig. 23.2 (continued) assortments (cf. also Fig. 23.2b) © Halberg Chronobiology Center. (**b**) Personalized self-measured and self-rated chronousphere of Robert B. Sothern (RBS): illustration of intraindividual acongruences and congruences, the latter revealed by overlapping CIs (95 % confidence intervals) of periods in the multidecadal spectral region, separately for time series consisting of MESORs (*upper 11 rows*) and of 24-hour amplitudes (*lower 11 rows*). SBP, systolic blood pressure; DBP, diastolic blood pressure; PP, pulse pressure; HR, heart rate; Temp, oral temperature; BR2m, breathing rate (2 minutes); VIG, self-rated vigor; TE1m, 1-minute time estimation; PF, peak (expiratory) flow; Mood, self-rated mood; EyeH, eye-hand coordination; horizontal bars have 95 % confidence intervals of nonlinearly estimated periods © Halberg Chronobiology Center. (**c**) Another view of RBS' acongruences and congruences, the latter revealed by overlaps of CIs (95 % confidence intervals) of periods, shown as horizontal bars, some within the width of the symbol giving the point estimate of the period, replotted from Fig. 23.2b.

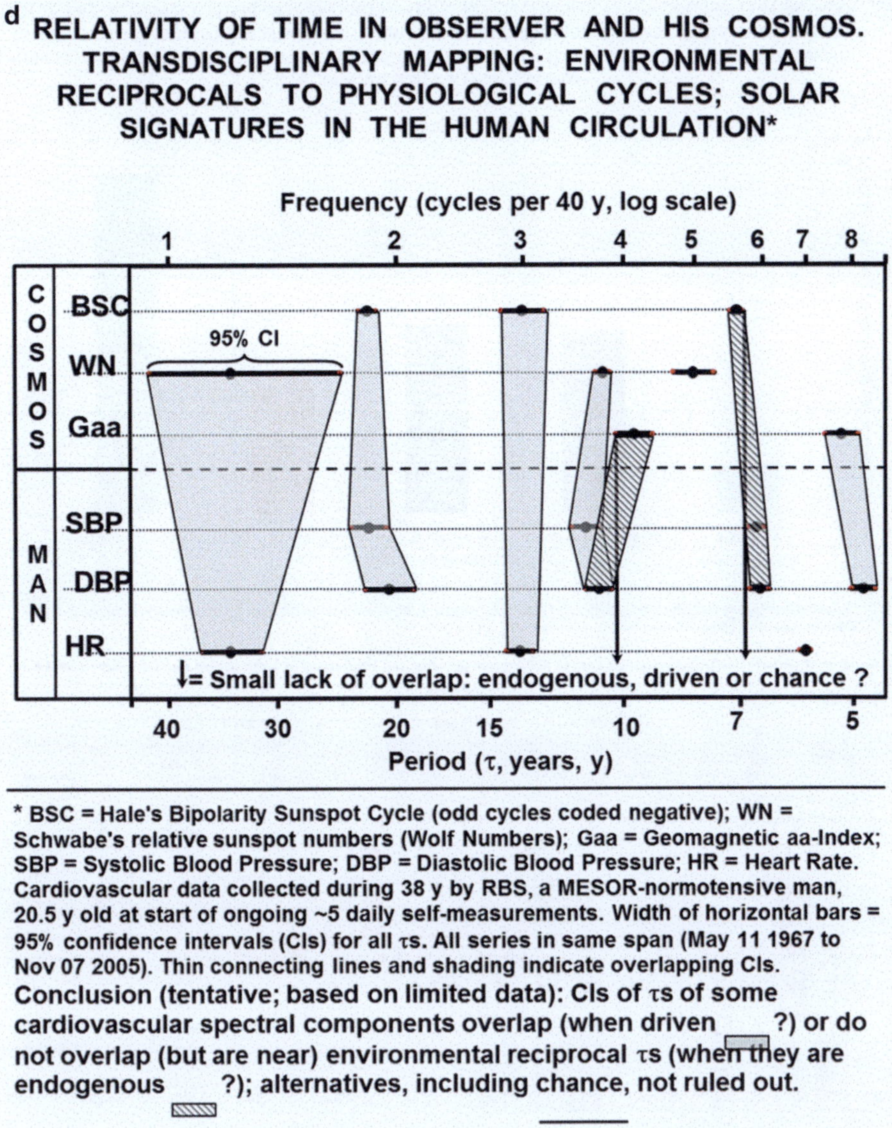

Fig. 23.2 (continued) © Halberg Chronobiology Center. (**d**) Selective assortment in RBS of heart rate with Wolf numbers' paratridecadal component and of blood pressure with sunspot (didecadal) bipolarity. © Halberg Chronobiology Center. (**e**) Para-annual congruences found in RBS, with environmental magnetism, solar wind speed (SWS), an approximation of interplanetary magnetism, and the antipodal geomagnetic index (aa). © Halberg Chronobiology Center. (**f**) An anticipated influence of the nonphotic environment (gauged by SWS and aa) on human mentality, gauged by 1-minute time estimation, was assessed by means of the congruence of periods of their spectral components (defined by overlap of the 95 % confidence intervals of the periods, in the frequency range of one cycle in 2.5 years to three cycles per year). The biological data stem from 40 years of self-ratings by RBS. © Halberg Chronobiology Center. (**g**) An anticipated influence of the antipodal index of geomagnetic disturbance (aa) and of solar wind speed, i.e., of the nonphotic environment on human psychophysiology, was assessed by means of the congruence of periods of their spectral components (defined by overlap of the 95 % confidence intervals of the periods, in the frequency range of one cycle in 2.5 years to three cycles per year) (see also Fig. 23.2d–f).

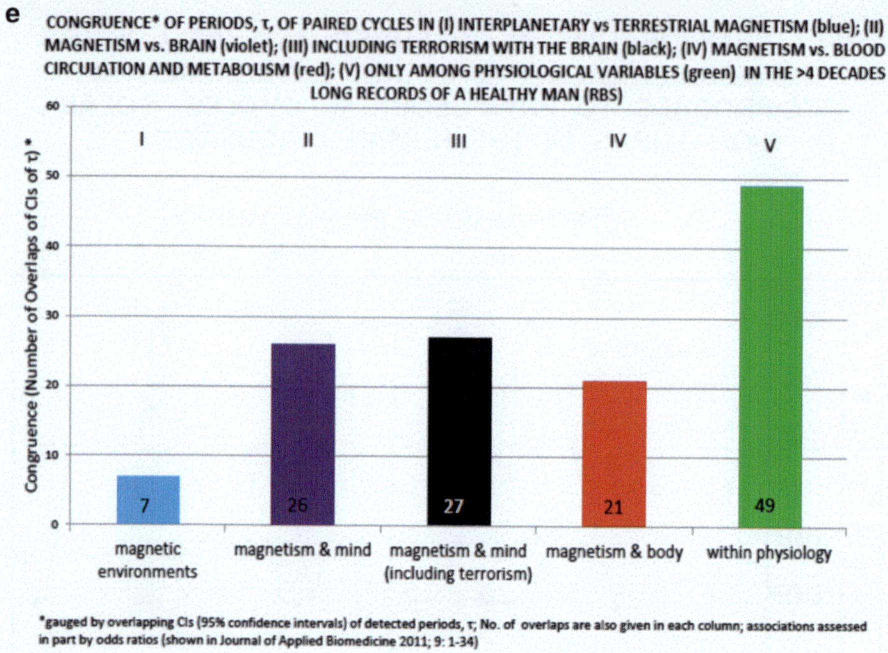

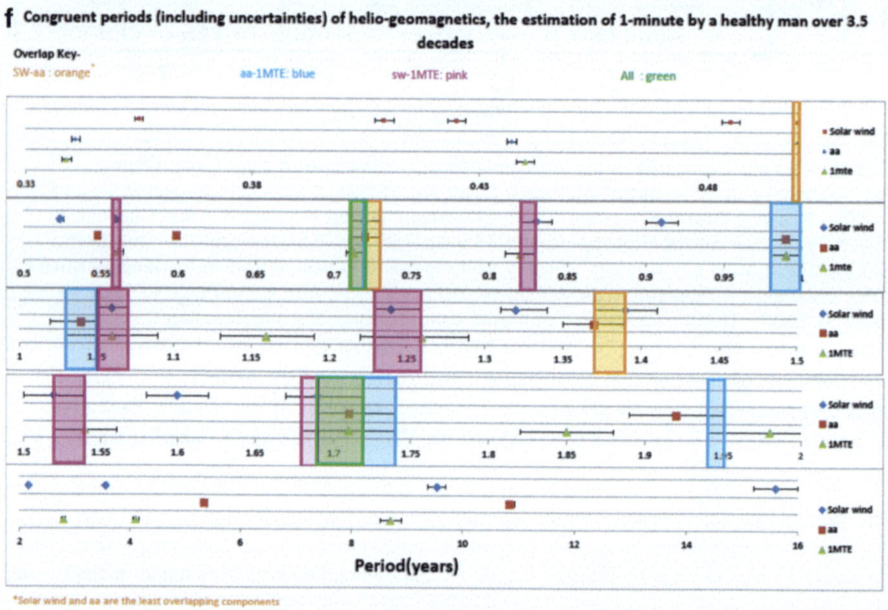

Fig. 23.2 (continued) The biological data stem from 40 years of self-measurements of oral temperature (Temp), systolic blood pressure (SBP), diastolic blood pressure (DBP), and heart rate (HR) and of ratings of mood and vigor and the time estimation of 1 minute by counting (1MTE), performed about 6 times a day by a clinically healthy man, RBS (J Appl Biomed 2011; 9: 63–102, J Appl Biomed 2011; 9: 1–34). © Halberg Chronobiology Center. (**h**) Congruences (assessed by means of odds ratios based on the noncentral hypergeometric distribution) found for 1MTE and for several other variables self-measured by RBS more than equal that of the known association of helio- and geomagnetism (*bottom*, last column on right of dashed vertical line in *blue*). Mental functions (*full red*) show numerically higher congruence than somatic functions (*green*).

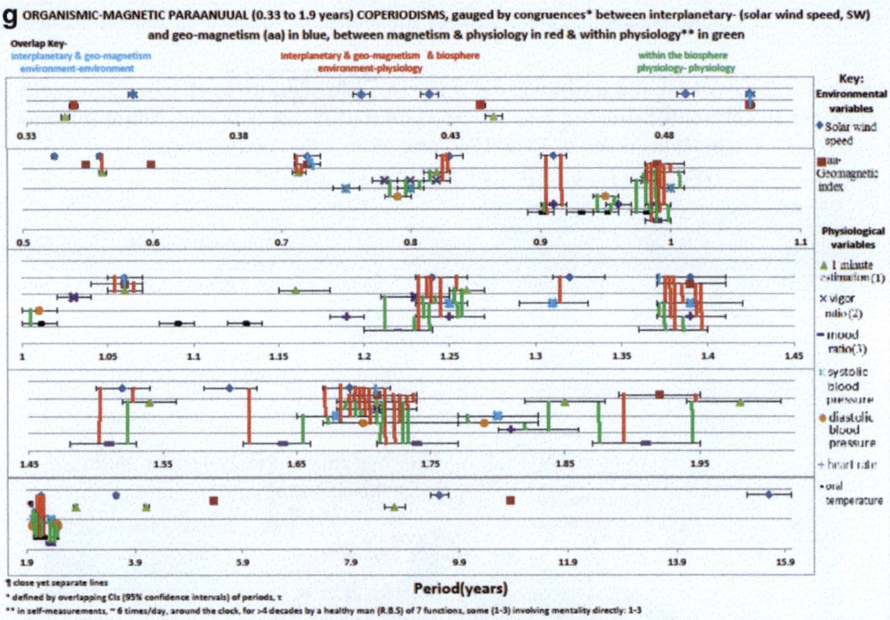

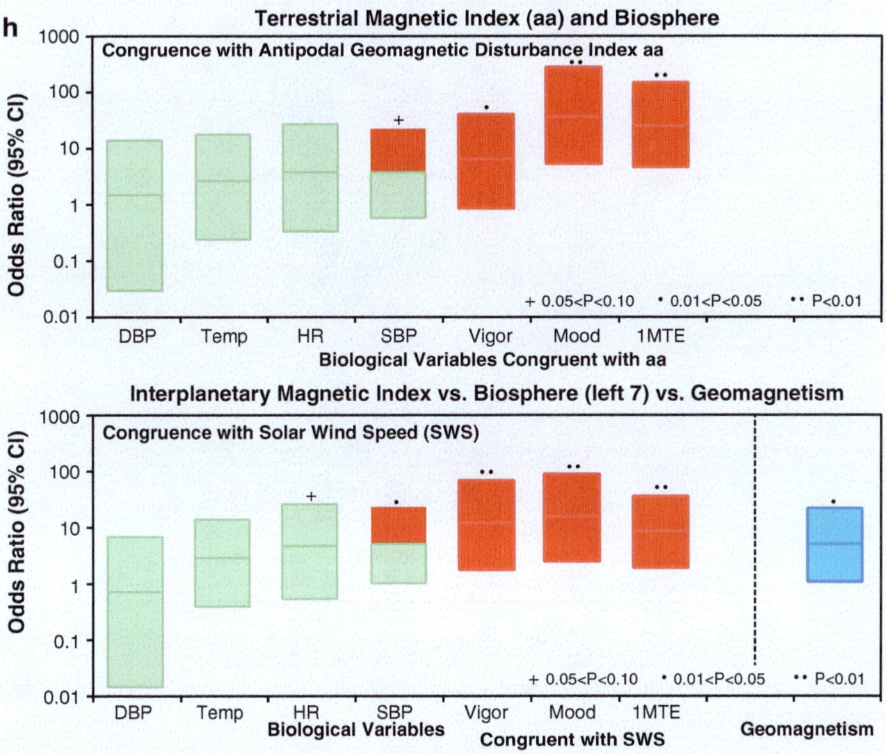

Fig. 23.2 (continued) Among the latter, systolic blood pressure (SBP) is most responsive, constituting a proxy for the mental functions. P-values are based on the noncentral Fisher hypergeometric distribution, with 95 % confidence intervals computed by Fisher's exact test, used since the null hypothesis was rejected in some yet not all cases. © Halberg Chronobiology Center

a Geographic decadal and multidecadal differences in signatures of space and earth weather in communicable (top) and para-annual differences in non-communicable (bottom) disease, 1823-1926 and 1999-2003, respectively

DECADAL and MULTI-DECADAL ASSORTMENT

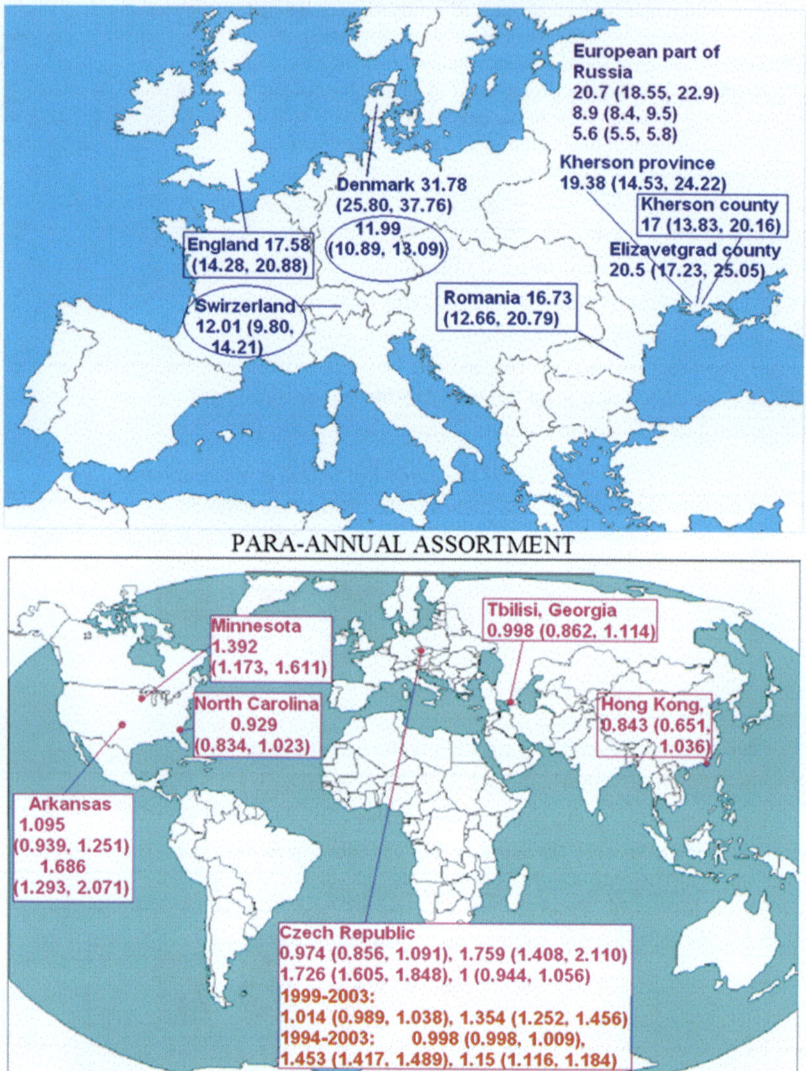

——— Diphteria, ——— Cholera,
——— Sudden Cardiac Death, ——— Myocardial Infarction*

* longer observation span reveals a near-transyear, in addition to a calendar-year and a far-transyear component.

Fig. 23.3 (**a**) Different cycles, given in years with their CIs (95 % confidence intervals) in parentheses, can characterize the same noncommunicable (*bottom*) or communicable (*top*) diseases in various, sometimes neighboring geographic locations, as is the case for diphtheria and cholera (*top*) and for myocardial infarction and sudden cardiac death (*bottom*). Cycles (in years) detected in croup, diphtheria, and relapsing fever in five European nations between 1860 and 1910,

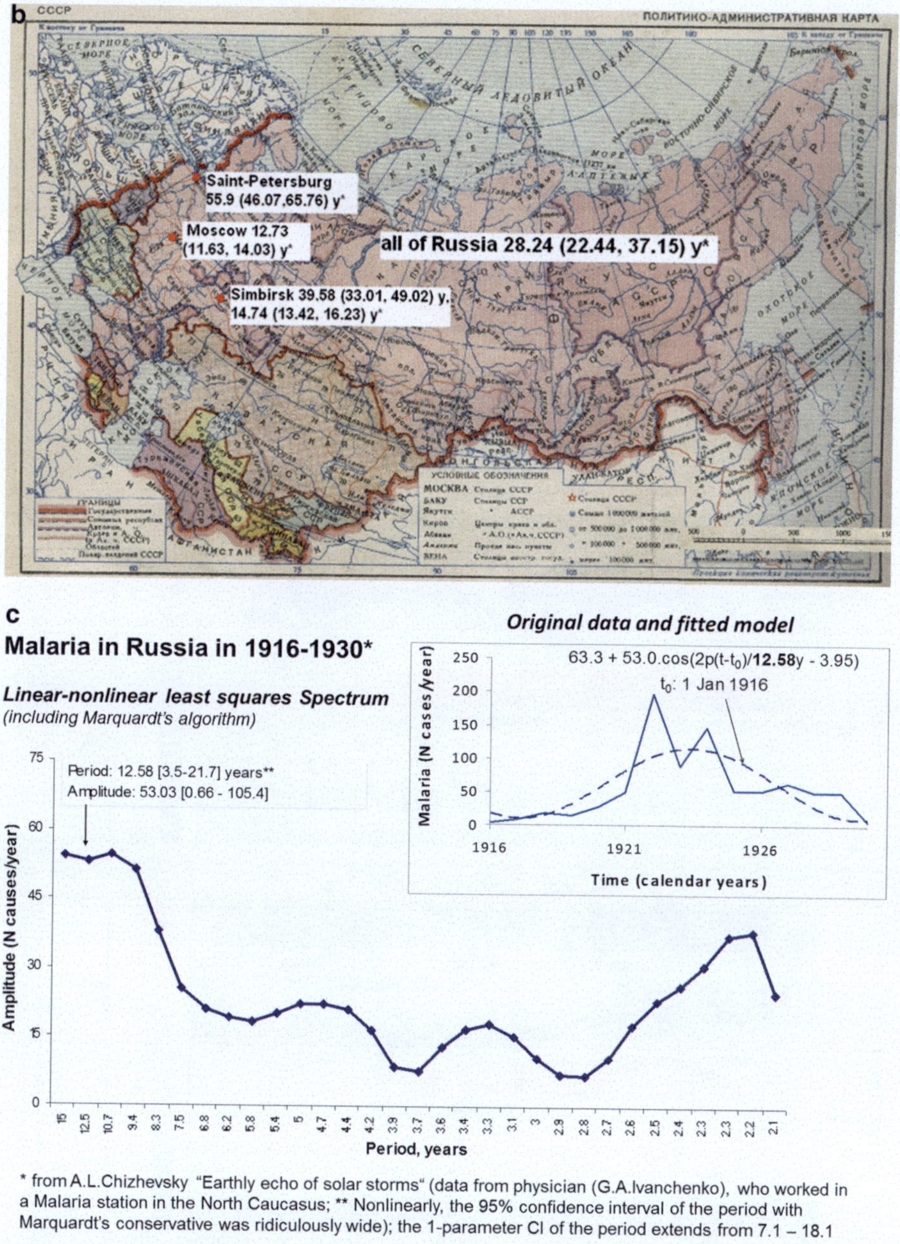

Fig. 23.3 (continued) in 104 years of incidence of cholera in Russia (1823–1926), and mortality from cholera in India (1901–1961) show a spectrum of cycles that are probably a feature of atavistic built-in decadal and multidecadal resonance, the importance of which Chizhevsky recognized, perhaps too succinctly as an echo of the sun. Analyses of communicable diseases by Lyazzat Gumarova. © Halberg Chronobiology Center. (**b**) Numbers in this graph indicate periods (in years) with their CIs (95 % confidence intervals) found by Lyazzat Gumarova's meta-analyses in Chizhevsky's compilation of statistics. Note that an about-11-year cycle found in Moscow coexists

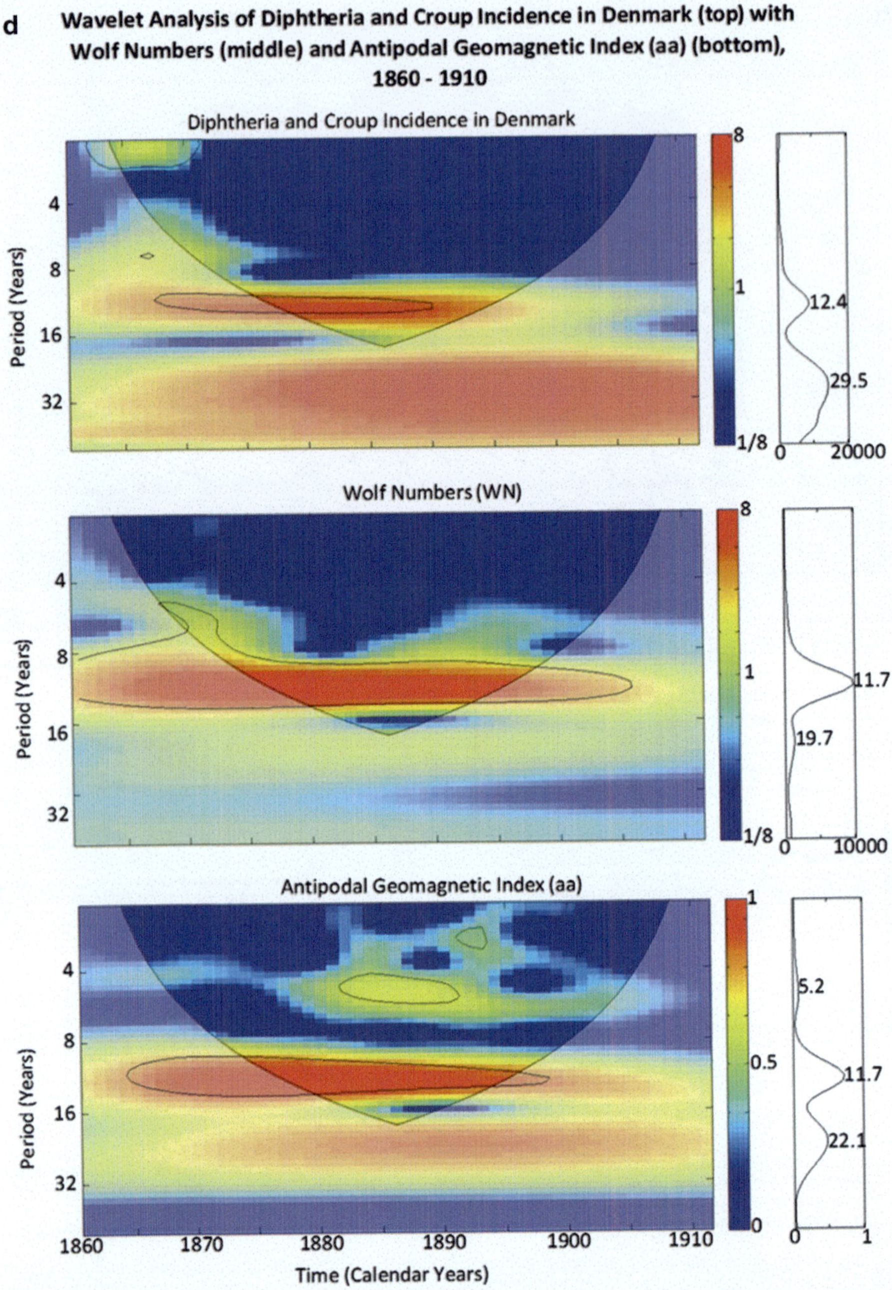

Fig. 23.3 (continued) with other cycle lengths in other geographic locations in Russia, a fact requiring a glocal concept and, critically, a spatially (as well as temporally) glocal procedure. © Halberg Chronobiology Center. (**c**) Meta-analysis by Lyazzat Gumarova of a cycle in malaria in data published by Chizhevsky. A period seen by his naked eye has a very wide uncertainty as apparent from the 95 % confidence interval in the linearly-nonlinearly extended cosinor analysis. The data published by Chizhevsky (*upper right*) encouraged him to write about this periodicity.

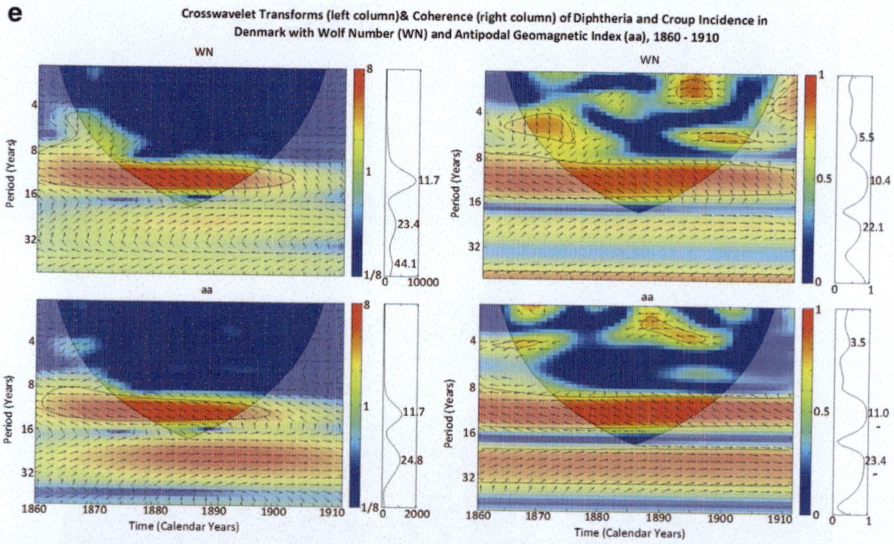

Fig. 23.3 (continued) His genius, based on inspection, was right: a periodicity could be validated by rejecting the zero-amplitude assumption by Marquardt's conservative approach, albeit the period's CI is very wide, prompting the question: Should one always provide an uncertainty? Even Carl Friedrich Gauss, the king of mathematicians, relied in some cases on inspection alone [14]. In most cases, however, the meta-analysis used and advocated herein and more generally is critical to check the impression gained by the naked eye, of a possible periodicity, and allows more rigorous comparisons with other data, notably when a multidecadal length of the cycle prevents insistence on its replication. © Halberg Chronobiology Center. (**d**) Putative time- and geographic area-varying reflection of past and/or present solar variability, dependent perhaps on terrestrial interactions, seen here only for Denmark by wavelet analysis of incidence patterns of two infectious diseases (pooled data) centuries ago when they were rampant (*top row*), of Wolf sunspot numbers (*middle*), and of the antipodal geomagnetic index aa (*bottom*), with peaks indicated by color (key), and corresponding periods listed in the spectrum (*right*, next to the color key) © Halberg Chronobiology Center. (**e**) Cross-wavelet transforms show a strong association of the incidence of two infectious diseases (pooled) with WN and aa at a period of about 11.7 years (within the cone of influence). Coherence (*right*) is also found within the cone of influence at periods of about 10.4 and 11.0 years. These periods are near those validated for the incidence of diphtheria and croup by the extended nonlinear cosinor (not shown). There are also statistically significant differences, some within the cone of influence between the associations (among others) of rampant pandemics with solar vs. earth magnetism in other geographic areas (not shown) © Halberg Chronobiology Center

(e.g., Fig. 23.1b, c) [7]. We emphasize that in a sphere of the mind, above all, one need to know oneself in keeping with the Greek *gnothi seauton* (perhaps Pythagoras') precept above the portico of the temple at Delphi and the Latin *cognosce te ipsum.* We can try to explore our mental and emotional state of well-being to start with by monitoring blood pressure and heart rate that can be recorded automatically as one goes and soon analyzed automatically, thanks to Larry A. Beaty (Fig. 23.4a), as markers (documented as such for mental functions such as mood, vigor, and the estimation of one minute; Fig. 23.2e, h).

Fig. 23.4 (**a**) Larry A. Beaty, de facto honorary professor of the Minnesota branch of the project on The BIOsphere and the COSmos (BIOCOS). As such, and as a member of the Phoenix Study Group, composed of volunteering members of the Twin Cities chapter of the Institute of Electrical and Electronics Engineers (http://www.phoenix.tc-ieee.org), he demonstrated by dint of engineering that affordable blood pressure monitors can be constructed at a cost below $100. He represented us this year at an international meeting in Moscow (Space Weather Effects on Humans in Space and on Earth, 4–8 June 2012), at one in Al Ahsa, Saudi Arabia (4th International Conference on Advanced Cardiac Sciences: King of Organs, 18–21 November 2012), and at a national IEEE meeting in Houston, Texas. © Halberg Chronobiology Center. (**b**) Illustrative parametric (*left*) and nonparametric (*right*) approach bracket a sphygmochron (*middle*) from a MESOR-normotensive man with circadian hyper-amplitude-tension (CHAT), a condition characterized by the 24-hour amplitude of blood pressure (BP) exceeding the upper 95 % prediction limit of clinically healthy peers matched by gender and age. A sample "sphygmochron" (*center*) illustrates how results are being reported. The numerical report consists of two parts labeled "characteristics" (parametric results) and "indices of deviation" (nonparametric results). Under "parametric results," a mathematical model of a smooth curve is fitted to the data to assess their circadian variation. The "MESOR" is the average value around which values fluctuate. It is very similar to the mean value, but yields more reliable results when the data are not collected at precisely regular intervals, and has a smaller error when the data are equally spaced. The "double amplitude" is a measure of the predictable change occurring within a day. The "acrophase" is a measure of the time when overall high values are likely to recur each day. For each of the three characteristics ("parameters"), the participant's value is compared to a range of acceptable values, derived from data provided by clinically healthy people of the same gender and age group as the participant. In this case, the double amplitudes of SBP and DBP exceed the reference limits, leading to the diagnosis of CHAT. Under "nonparametric results," the participant's data are compared by computer with time-specified reference values, also derived from chronobiological archives on clinically healthy subjects matched by gender and age. For this analysis, all data are stacked over an idealized 24-hour day. Whenever a given person's profile exceeds the limits of acceptability of peers, the data are marked as being excessive or deficient. The "percentage time of elevation" reports the relative incidence of excessive values during a 24-hour day. It is common to have occasional high values, but in the example herein, there is reason for concern. The next item, the

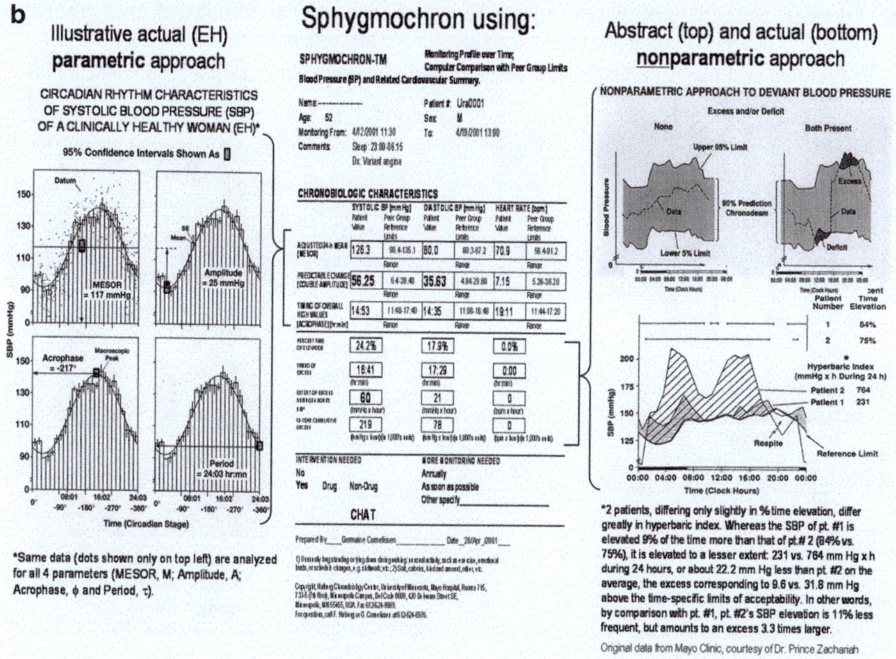

Fig. 23.4 (continued) time of excess, becomes useful when drug treatment should be timed prior to the peak in excess. Since excessive values may either be barely above the limit or be very much higher than the limit, the "area under the curve" delineated by the values when they exceed the limit and this limit itself conveys the severity of excess. Empirically, it has been shown that excess up to about 50 (mmHg×hour during 24 hours) may still be acceptable and accountable for by daily worries and/or physical activities. In the case summarized, the HBI is 60, in *bold*, and, if confirmed in the next 7/24 profile, a reason for treatment © Halberg Chronobiology Center. (**c**) The Phoenix Group of volunteering electrical and electronic engineers from the Twin Cities chapter of the Institute of Electrical and Electronics Engineers (http://www.phoenix.tc-ieee.org) is planning on developing an inexpensive, cuffless automatic monitor of blood pressure and on implementing the BIOCOS concept of an international website and data center (www.sphygmochron.org) for the analysis of time series collected with these instruments aligned with information, among others, from space weather stations, epidemiological data from health departments, statistics on crime rates from police departments, and trends in economics for research in a unified science and art. Concept validation by such varied data collection and analysis is found in some of the appended recent bibliography © Halberg Chronobiology Center. (**d**) The many dividends visualized by Dean and Professor Yinjing Guo, with Dr. Jinyi Wu and Bing Wu, from a chronomic cloud system with a data center for service in self-surveillance to individuals by an as-one-goes lifetime's strain test and as background for caregivers and for research on understanding, preventing, and/or avoiding or evading natural and human-made cataclysms on an even broader scale by mining epidemiologic, sociologic, and environmental variables, utilizing current technology for smart homes and smart transportation, the latter examining, e.g., a driver's alertness by physiological monitoring as well as trying to take over some of the driver's tasks © Halberg Chronobiology Center. (**e**) Operations plan for 21-day/8-rat Biosatellite: sequence of lighting regimens to be instituted after lift-off at times determined by results from "as-you-go" analyses of the data. Design prompted by financial restraints: the government's contractor (General Electric) had to provide a system for eight rats that were to be in extraterrestrial space for 21 days. Apparently, the extension of the hardware to last for another day and/or to include another rat had an added cost of $1 million (before inflation) in a budget of $100 million (so we were told). The hybrid design [15] was the result. The study never flew, but it served to develop physiological telemetry of core temperature with motor activity at first and a design with

c Preventive and curative health care can yield the dividend of biomedical monitoring of space weather by time-structural analyses of ambulatory blood pressure and heart rate series[1]

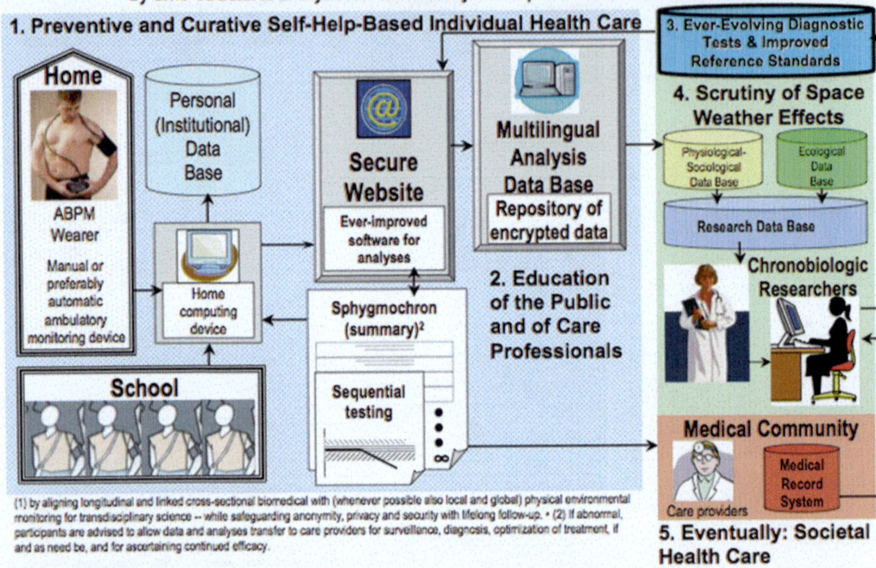

Modified from Figure 1 (Phoenix Architecture) in Adams C Privacy requirements for low-cost chronomedical systems. Int Conf on the Frontiers of Biomedical Science: Chronobiology, Chengdu, China, September 24-26, 2006, p. 64-69, originally with Larry A. Beaty (www.sphygmochron.org) of the Phoenix Project (www.phoenix.tc-ieee.org/).

d Broadening of scope of alerting system

ALERTING for understanding and eventual prevention or evasion of personal, natural and societal cataclysms, the latter including terrorism*

***Via an international chronomics platform (website or cloud system) for worldwide service to individuals and nations**

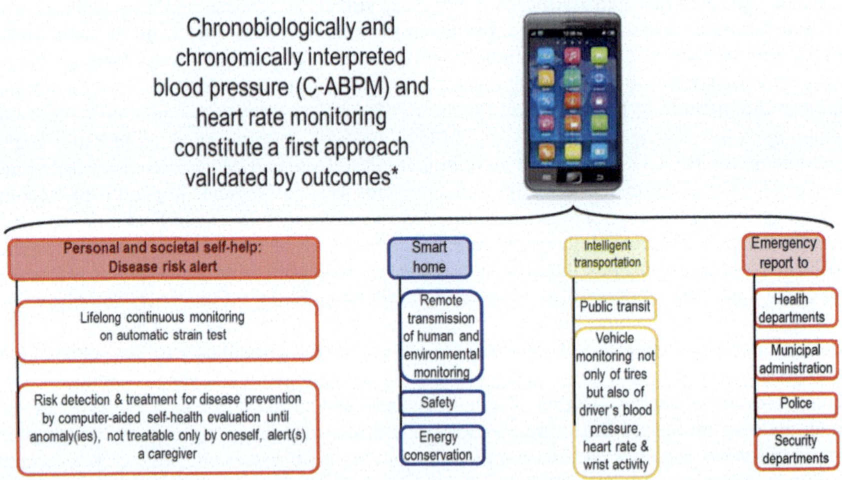

*Hypertension 2007; 49: 237-239; Geronto-Geriatrics: Int J Gerontology-ChronomeGeriatrics 2008; 11 (14): 119-146.
¶With local and national inputs.

Fig. 23.4 (continued) many applications on earth as well. The same as-one-goes analyses serve for assessing rhythm characteristics in experimental work on earth and in the clinic in order to detect seemingly spontaneous rhythm alterations and those after changes in routine or drug treatment. Further demonstrations of a hybrid design combining meta-analyses done longitudinally on individuals and transversely (with but one observations/individual) on one group are shown elsewhere © Halberg Chronobiology Center

e

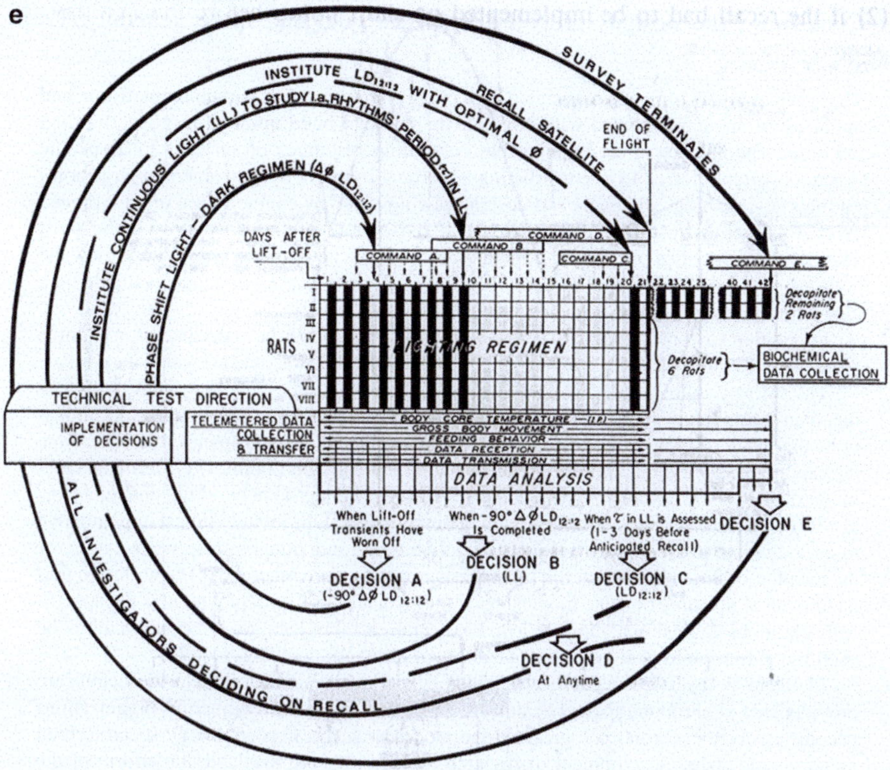

Fig. 23.4 (continued)

More specifically, we can seek to measure automatically a personalized sphere of the human mind by the extent of rhythm alteration in C-ABPM analyzed as one goes, not only for personal health when data are analyzed by sphygmochron (Fig. 23.4b) but also as a way to learn about space-terrestrial weather influences on our physiology. It seems desirable to update one's recent state of mind on a daily, weekly, and monthly basis, whether or not one can extend one's horizon, to become an outstanding creative mind – we wish in more individuals and more frequently than a few every 500 years, now that intercontinental communication is possible. One's experience and acquired knowledge about oneself and others can be placed into the context of information on past and present events more generally, i.e., into a pool of information in a generalized chronousphere on the Internet. This double (personalized and generalized) chronousphere could become a basis for trying to improve one's health and that part of the environment in which one lives and beyond, as far as oneself and/or others can reach by glocal pooling of contributions by ever-improved human minds. Contemporary technology (Fig. 23.4c, d) permits the processing of a monitored personalized chronousphere, notably when combining longitudinal records from different individuals in hybrid designs such as those used in a Biosatellite program (Fig. 23.4e), as planned by Dr. Jinyi Wu and Bing Wu with Yinjing Guo, dean and professor, Shandong University of Science and Technology, Shandong, China.

Concerning the designs proposed by Dean Guo and more generally, Larry Beaty wrote:

> The services layers typically associated with cloud computing (infrastructure, platform, and application) are implemented with components that can build upon each other. Building upon and specializing a set of available computing resources or services to a specific domain is often referred to as constructing a 'vertical solution' by developers of large computing systems; block diagrams and design drawings used by engineers are often arranged vertically, with infrastructure components at the bottom of the drawing and the most abstract functionalities of multiple domain-specific applications at the top. We can view the collection of *services specific to chronobiologically-based healthcare* made available by a cloud computing system as one such vertical solution.
>
> In addition to construction of vertical solutions, services can be specialized for multiple types of users within a specific domain, or for subsets of the functions needed within the domain.
>
> The infrastructure-as-a-service capability of a chronobiology cloud system (networking, data storage, raw compute capability) could be made available to technology organizations that want to build systems with those building blocks. Data schemas that encourage the capture, transmission, and storage of chronobiologically/chronomically-useful data that might otherwise be ignored by system developers could underlie networking and data storage infrastructure services. Useful schema components might include timestamps relative to an individual's 'local body time' (i.e., relative to the individual's sleep/wake cycle), or the location of the individual at the times measurements were recorded. (The ability to find an individual in a time of acute medical need for medical services deployment is not specifically chronobiological in nature, but a *history* [time series] of location information might be useful in chronobiological healthcare analyses or chronobiological research.)
>
> Platform-as-a-service capability might not have much applicability to the healthcare provider side of chronobiological healthcare, but would be useful if numerous organizations become involved in chronobiological/chronomic research. The chronobiology-specific cloud system could speed development of research tools by making available a platform of the basic tools along with computer programming languages or computer program development systems for building specialized research tools of interest.
>
> A potentially more powerful application of cloud computing concepts to chronobiology/chronomics occurs at the software-as-a-service level. If software that performs chronobiological healthcare data collection and analyses is made available as a service, the "user interface" is separated from the core analysis and data storage mechanisms in a way that allows specialization of public services without the loss of efficiency usually associated with duplicate implementation of capabilities at the infrastructure and application software layers. Different implementations of websites for physicians with different specialties would be built upon the same software, tapping into capabilities that are useful for that specialty, but ignoring other capabilities, through customization and/or configuration. Customizable or configurable software services can be designed to ensure consistency and data integrity across a collection of applications that appear different to their users while using common algorithms and massive data storage, an important consideration in any system design with centralization (or at least standardization) as feature of major importance. Despite the different user interfaces, the system functions would still be available to all implementation at no extra development or operational cost.
>
> For illustration, a simple example of a software service that might be made available at this level of abstraction would be a recording of and time-sensitive analysis of a stream of data. When used by a cardiologist, the time-sensitive analysis capability might note that a person's heart rate has dropped below or has remained above acceptable preset values and automatically notify emergency medical personnel and/or care providers. The same time-sensitive analysis capability used by another physician to measure physical activity to encourage a patient to exercise more would most likely have the automatic notification capability configured off to prevent emergency personnel from being called just because the patient kept exercising. Yet both systems would be using the same soft-

ware to record the primary data elements (heart rate or physical activity) along with secondary data elements typically useful for chronobiological analyses (e.g., timestamp, location), with the reasonable expectation that later data mining activities will be more useful without incurring extra cost.

Cloud systems have a risk that is somewhat unique: sharing of computing resources for efficiency (a major characteristic of cloud technology) can result in insufficient resources at a time of need. A *critical* healthcare system based on cloud technology that is expected to automatically deploy and guide emergency response personnel to patients in need during an epidemic could fail if, for example, it were sharing computing resources with computerized entertainment (e.g., movie delivery, shopping) systems that were also processing larger-than-average loads due to large segments of the population staying home from work and school. Non-cloud systems around the world already see such failures (e.g., telephone networks clogged with personal phone calls between worried family members during a natural disaster while emergency personnel try to use the same networks for coordination of resources and response); the cloud system architecture shares computing resources in a way that non-cloud system do not, introducing another potential failure point. Careful 'disaster' planning and testing can prevent this type of problem, and must explicitly be included in critical cloud system design and ongoing operation. One common mechanism to prevent resource problems is to negotiate a contractually-enforced priority system with the other users/owners of the shared computing resources, such that critical functions of a healthcare cloud system would be guaranteed computing resources in time of need.

The services provided in a cloud-based chronobiological healthcare system for China go beyond simply 'Sensoring China.' Collecting data from sensors is not particularly useful without also analyzing it. But it is not likely that all of the useful analysis services can be predicted in advance, presuming there are continued advancements in chronobiologic/chronomic research. Therefore, the chronobiological healthcare cloud system must be designed to evolve over time; it would be just too ironic for a *chrono*biology system to not have to adapt to the surrounding environment as it changes. The current state of computer hardware and software development is that such changes are painstakingly (and perhaps expensively) developed by engineers. The mechanics of such changes are usually straightforward (from an engineer's point of view), but ensuring that a long-lived project has leaders with predictive vision that evolves and might not be so straightforward at the outset, and should be explicitly planned for in a national or international scale healthcare system.

23.3 Recent Findings and Events

In Kaluga, Russia, a statue was recently erected in honor of Chizhevsky (http://www.youtube.com/watch?v=TfRyD-RmHrU), recognizing him as the pioneer of heliobiology in the pre-space era. He did not have the means (computers, inferential statistics, and satellites) for detecting and documenting the pervading transdisciplinary spectrum of solar, interplanetary, geo-, and biospheric cycles; yet he visionarily and eloquently described the very many consequences in the human mind and body of aeolian disappearing and/or reappearing nonphotic rhythms. We can honor his legacy if we extend the scope of one of our major summaries of this year's synthesis (Fig. 23.5a) into many more personalized chronouspheres. We could all try to know ourselves, by recognizing an unseen strain, consisting in part of contributions by the cosmos, that like single, most of the time, not in themselves lethal, stimuli along the circadian scale also, can make the difference along infradian scales between death and survival, in the cases of human sudden cardiac death, suicide or crime, and terrorism. It has been shown in a circadian context that noises, such as the ringing of bells, and doses of drugs that are perfectly tolerated at one phase of a circadian cycle can kill a few hours earlier or later. The same likely applies to the effects of solar flares and magnetic storms in certain stages of the infradian susceptibility rhythms, as shown in Fig. 23.5a–c. The effects of the cosmos on mortality are currently forgotten by the Western scientific community at large, yet were prominent in data reported with the proper interpretation to the French Academy of Science by 1922. These data are meta-analyzed in Fig. 23.5d, documenting that sunspots are associated with an increase in mild or severe symptoms [16], including mortality [17]. The astronomer and philanthropist Joseph Vallot and the physicians Maurice Faure and Gaston Sardou [16, 17] made their discoveries concomitantly with Chizhevsky's studies of the cosmos and human affairs [18, 19], including natality and mortality [20–25], which were confirmed in the 1930s by Traute and Bernhard Düll [26], among many others, notably in Russia [27], also in Israel by Eliahu Stoupel [28, 29], and much later by ourselves [30], with just two exceptions [31, 32] that can be anticipated in aeolian nonstationary phenomena [33].

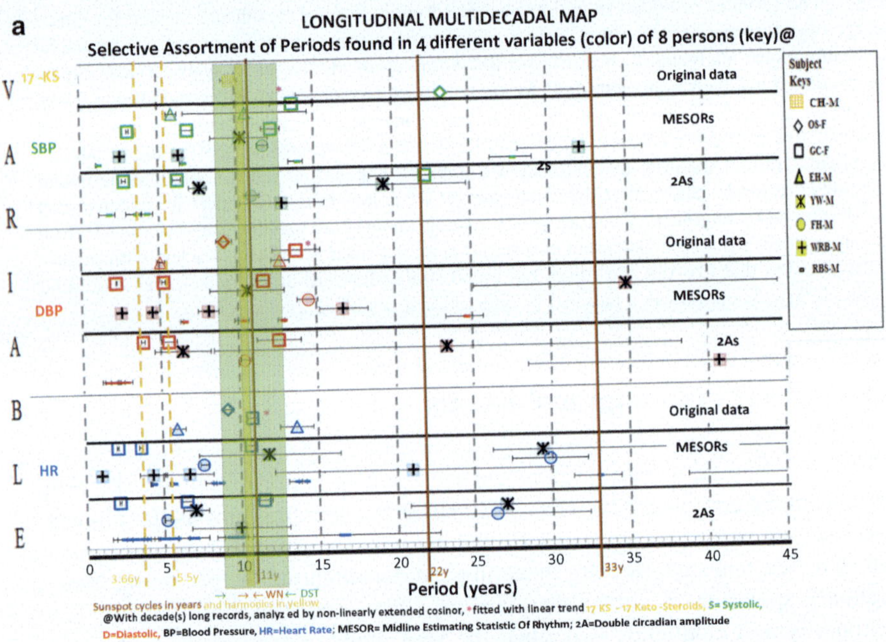

Fig. 23.5 (**a**) Periods of cycles in eight persons with their uncertainties, CIs (95 % confidence intervals) shown as horizontal lines, detected by the nonlinearly extended cosinor, rejecting the assumption of "no rhythm" with the period fitted or one near it, extending the scope of a critical assortment of periods shown in Fig. 23.2a–c. Series from 15 up to 40 years in length assess the time structure of systolic (S) and diastolic (D) blood pressure (BP) and heart rate (HR) and of metabolism (17-KS) of eight biomedical scientists, including six physicians, four of them under hypotensive treatment. Note cluster around 10 years, with a vertical shaded yellow band indicating Wolf numbers during the prior 100 years with their uncertainty and a vertical shaded green band

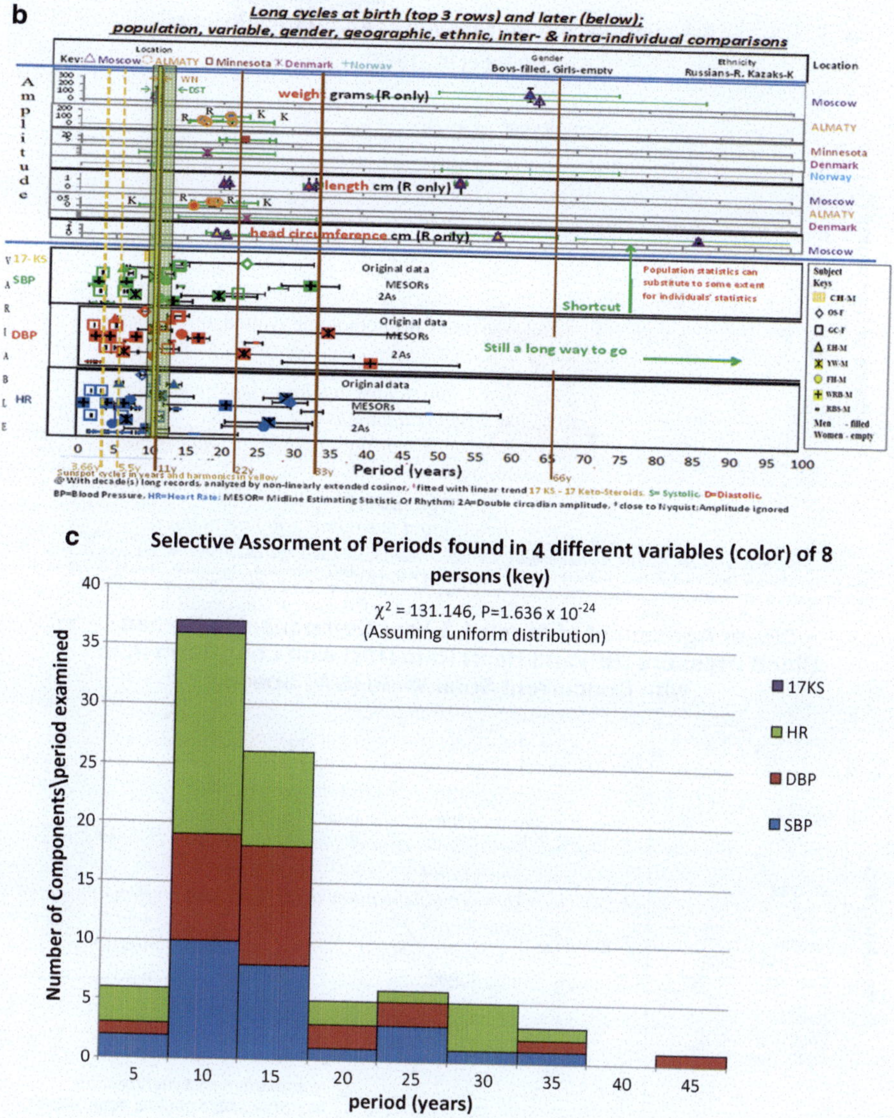

Fig. 23.5 (continued) giving the CIs of the period of the equatorial geomagnetic index Dst during the past 27 years, a rough illustration of environmental periods that we happened to have computed. The environmental counterparts of the span corresponding to each subject's record are not shown here. Periods in the adult human circulation are aligned with those found in neonatal anthropometry in Fig. 23.5b © Halberg Chronobiology Center. (**b**) The graph in Fig. 23.5a is aligned with time series, including one of 110 years in length, on body weight, body length, and head circumference at birth, revealing, i.e., trans-semicentennian cycles (trans = beyond, i.e., longer than half a century, i.e., 50 years). © Halberg Chronobiology Center. (**c**) Histogram of nonrandom distribution of periods found in Fig. 23.5a, b showing a peak around 10 years. © Halberg Chronobiology Center. (**d**) Solid evidence that solar magnetism (sunspots) can represent a stress resulting in enhanced symptoms quantifiable on a population basis was presented by 1922 [16, 17]. Severe

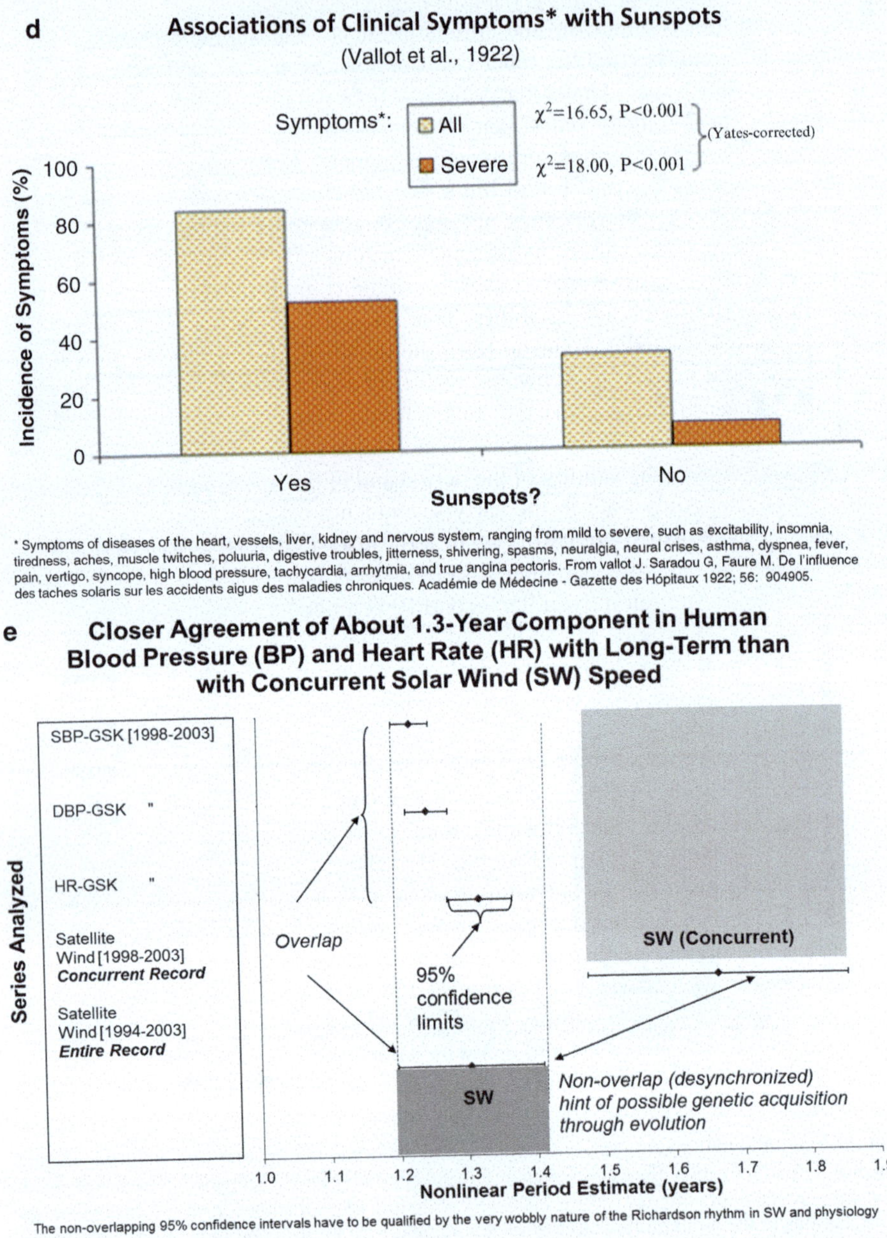

Fig. 23.5 (continued) symptoms included death. Self-surveillance, notably of systolic blood pressure (Fig. 23.2h), may detect this strain as a vascular variability anomaly (VVA) on an individualized basis. © Halberg Chronobiology Center. (**e**) The about-1.3-year component of blood pressure and heart rate of GSK (M, 72–77 years of age) recorded during 1998–2003 has a period closer to the solar wind speed period characterizing the entire available record from the wind satellite than that for the concurrent 5-year span. Physiological variables may resonate with nonphotic environmental cycles that may have entered the genetic code during evolution. From Cornelissen et al. [34] © Halberg Chronobiology Center

At several meetings in 2012 in Moscow, Russia; Al Ahsa, Saudi Arabia; and Houston, Texas, USA, Germaine Cornelissen and/or Larry Beaty demonstrated, in a pool of many transdisciplinary variables' meta-analyses, cycles amenable to a cartography to be carried out globally and locally (we reemphasize glocally) in time, space, concept, and method. Culture, art, and science can benefit from a systematic mapping of the cycles in a chronousphere in and around us, representing surprisingly prominent putative signatures of the past as well as present behavior of the sun coded in our genes (Fig. 23.5e) [2, 13]. The pooling of the results of mapping that started with circadian cartography complements the findings in other meta-analyses of records from many long professional lifetimes of others' productivity, e.g., that of Raymond Holder Wheeler [35], revealing decadal, multidecadal, and still longer cycles in various affairs relating to the health of human individuals and societies. R. B. Singh, professor of Cardiology in Moradabad, India, and emeritus president of the International Colleges of Cardiology and of Nutrition, demonstrated that a 7-day C-ABPM is feasible in outlying areas; his self-study of activity joined a distinguished list of scientist physicians who were both experimenters and probands in their research [36]. Diversity in space – genetics – developed by Johann Gregor Mendel in a pea patch in Brno, Czech Republic, now has a counterpart at St. Anna Hospital, also in Brno, under the guidance of Prof. MUDr. Jarmila Siegelova, whose hundreds of 7-day C-ABPM focus, as Prof. Singh did, on exercise chronomics, documenting, among others, that exercise at the wrong time can harm [37]. Jarmila's yearly proceedings summarize our decade-long cooperation [38–47]. In 2012, our contributions from Minnesota were given by Skype. Germaine Cornelissen and Larry Beaty had a splendid resonance, as the following newspaper report conveys broader interest (http://www.saudigazette.com.sa/index.cfm?method=home.regcon&contented=20121120143419).

23.4 Long-term Diverse Data Collection Opens Medical Frontiers

We were fortunate to have Lyazzat Gumarova of Almaty, Kazakhstan, join us for 2012. Her analyses of Chizhevsky's data (e.g., Fig. 23.3a–c), her fluency in Russian, and her scholarship allowed us a new excursion into calendars in other areas of the world that recognized soli-lunar events and to the data or specifications in the contributions of Alexander Leonidovich Chizhevsky and Vladimir Ivanovich Vernadsky. She applied our glocal analyses which separately provide uncertainties for (global) spectra of a time series as a whole and for (local) serial sections to the treasure of data compiled in Russian. Lyazzat also contributed analyses of blood pressure and heart rate records collected in Kazakhstan by her colleague Dr. Almagul Mansharipova of the Kazakhstan-Russian Medical University in Almaty (Fig. 23.6a, b). Among the vascular variability anomalies (VVAs) detected in this population, circadian hyper-amplitude-tension (CHAT) deserves special scrutiny since it has been associated with a large increase in cardiovascular disease risk in several outcome

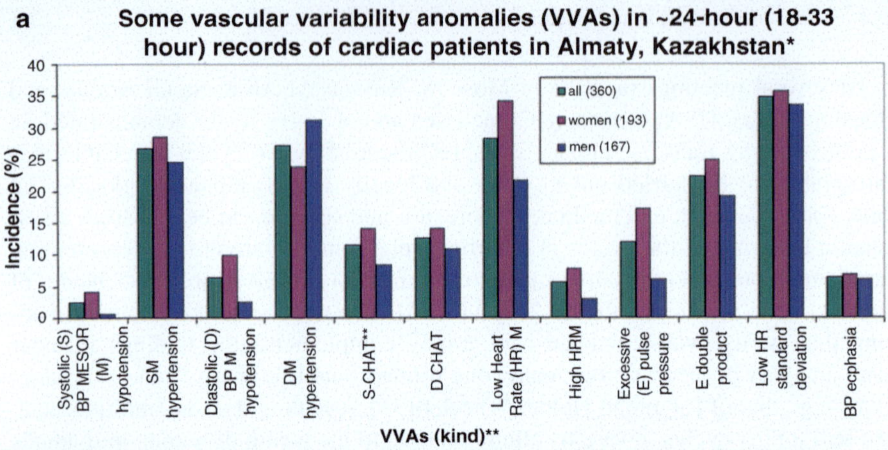

*** Above 90% of patients were hospitalized during the monitoring; ** CHAT, short for Circadian Hyper-Amplitude-Tension;**

See also Halberg F, Cornelissen G, Otsuka K, Siegelova J, Fiser B, Dusek J, Homolka P, Sanchez de la Pena S, Singh RB, BIOCOS project. Extended consensus on need and means to detect vascular variability disorders (VVDs) and vascular variability syndromes (VVSs). Geronto-Geriatrics: Int J Gerontology-ChronomeGeriatrics 2008; 11 (14): 119-146 AND Leibniz-Online Nr. 5, 2009 (http://www2.hu-berlin.de/leibniz-sozietaet/journal/archiv_5_09.html). 35pp, AND World Heart J 2010; 2 (4): 279-305.

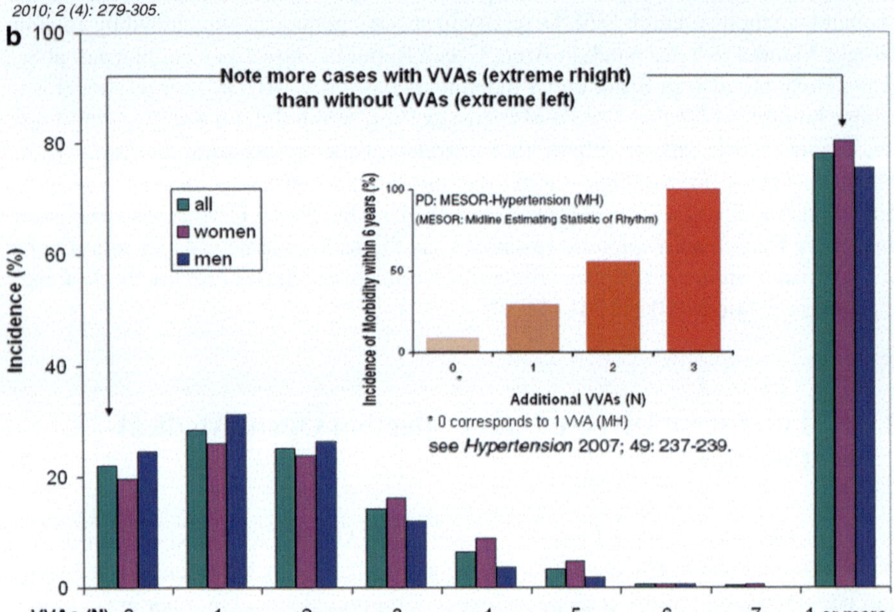

Fig. 23.6 (**a**) Incidence of some vascular variability anomalies (VVAs) in about-24-hour ABPM records obtained in routine hospital practice during 2010–2012. Analysis by Lyazzat Gumarova © Halberg Chronobiology Center. (**b**) In a hospital population (Almaty), the proportion of patients found to be free of any vascular variability anomaly (VVA) is less than that of patients with a single VVA or multiple VVAs in an about-24-hour ABPM record. The complication of MESOR-hypertension by one or several VVAs was associated with an increased risk of an adverse event in an outcome study of 297 Japanese patients (insert) © Halberg Chronobiology Center. (**c**) Circadian hyper-amplitude-tension (CHAT), an excessive circadian amplitude of blood pressure, was observed to precede MESOR-hypertension in the experimental laboratory (*left*, experiments by Erna and Julia Halberg), in clinical research (*middle*, data from Dr Yuji Kumagai), and in clinical practice (*right*, data from Dr Yoshihiko Watanabe) © Halberg Chronobiology Center

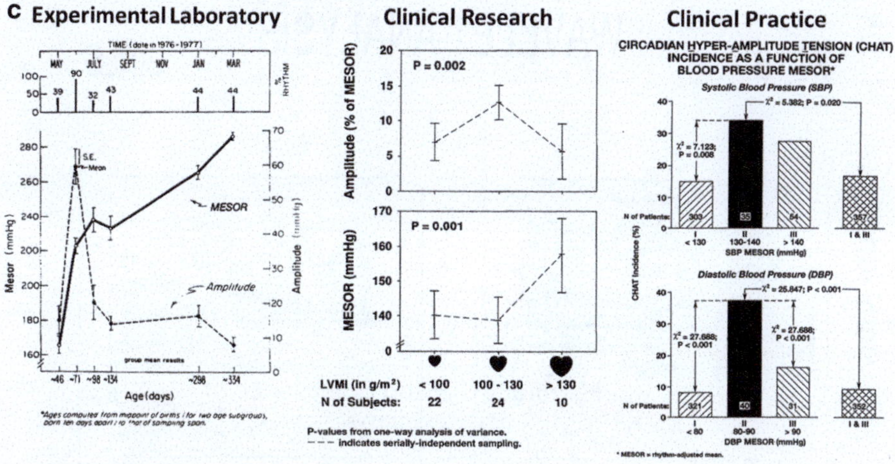

Fig. 23.6 (continued)

studies. Its presence may also be a precursor of MESOR-hypertension, as shown in the experimental laboratory, clinical research, and practice (Fig. 23.6c).

We thank Stefano Sello of Enel Research, Pisa, Italy, for including wavelets of anthropometric data for orientation and comparison of time structures in space-terrestrial weather (Fig. 23.7a–c). We also thank Shiyu Hong for analyzing longitudinal blood pressure records (Fig. 23.7d) by adding glocal wavelets (Fig. 23.7e–h) and by comparing cross-wavelet coherences of the incidence of myocardial infarction and international terrorist acts with Wolf numbers as a gauge of solar activity (Fig. 23.7i). Figure 23.7a–i illustrates how wavelet analysis can be a useful addition to chronobiometry; albeit, it should not replace the combined local (gliding spectra) and global (linear-nonlinear least-squares spectra) rhythmometric analyses (Fig. 23.7j) and/or spectrograms (Fig. 23.7k, l), notably when they are also aligned with chronobiologic serial sections, as often done whenever the record is sufficiently long, as apparent from the appended update of our recent bibliography. Figure 23.7a–l shows the relative merits of wavelets and other procedures, emphasizing the need to use them in combination, notably when numerical estimates of uncertainties are required. Wavelets have been used, for instance, as building blocks to be included for quick orientation in a strain test, where part of the strain may be the unfavorable space weather. Paul J. Rosch, clinical professor of medicine and psychiatry at the New York Medical College and president of the American Institute of Stress and provocative author of regular interesting and candidly critical newsletters that report on how stress and strain relate to current health care, put this endeavor into a historical perspective [48] (as a former associate of Hans Selye) and may place it into the context of magnetism, to which he contributed his authoritative handbook covering the development of medico-biological aspects of that topic as well [49].

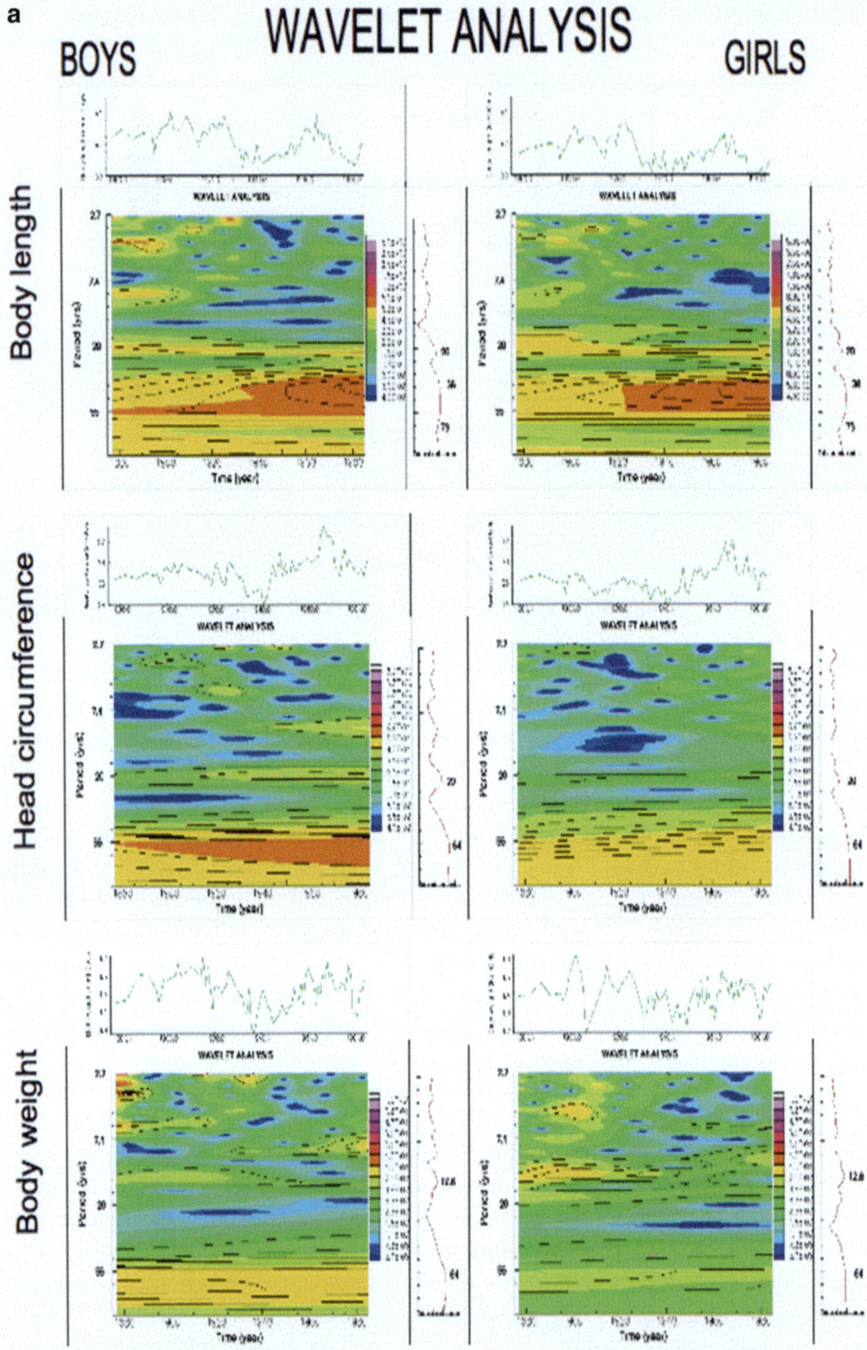

Fig. 23.7 (**a**) Morlet wavelets of the late Boris Nikityuk's anthropometric data reveal time course, among others, of the signatures of long periods, including after 1930 a paratridecadal Brückner-Egeson-Lockyer (BEL) signature as well, that was missed by earlier analyses but could be ascertained when tested by the nonlinearly extended cosinor © Halberg Chronobiology Center.

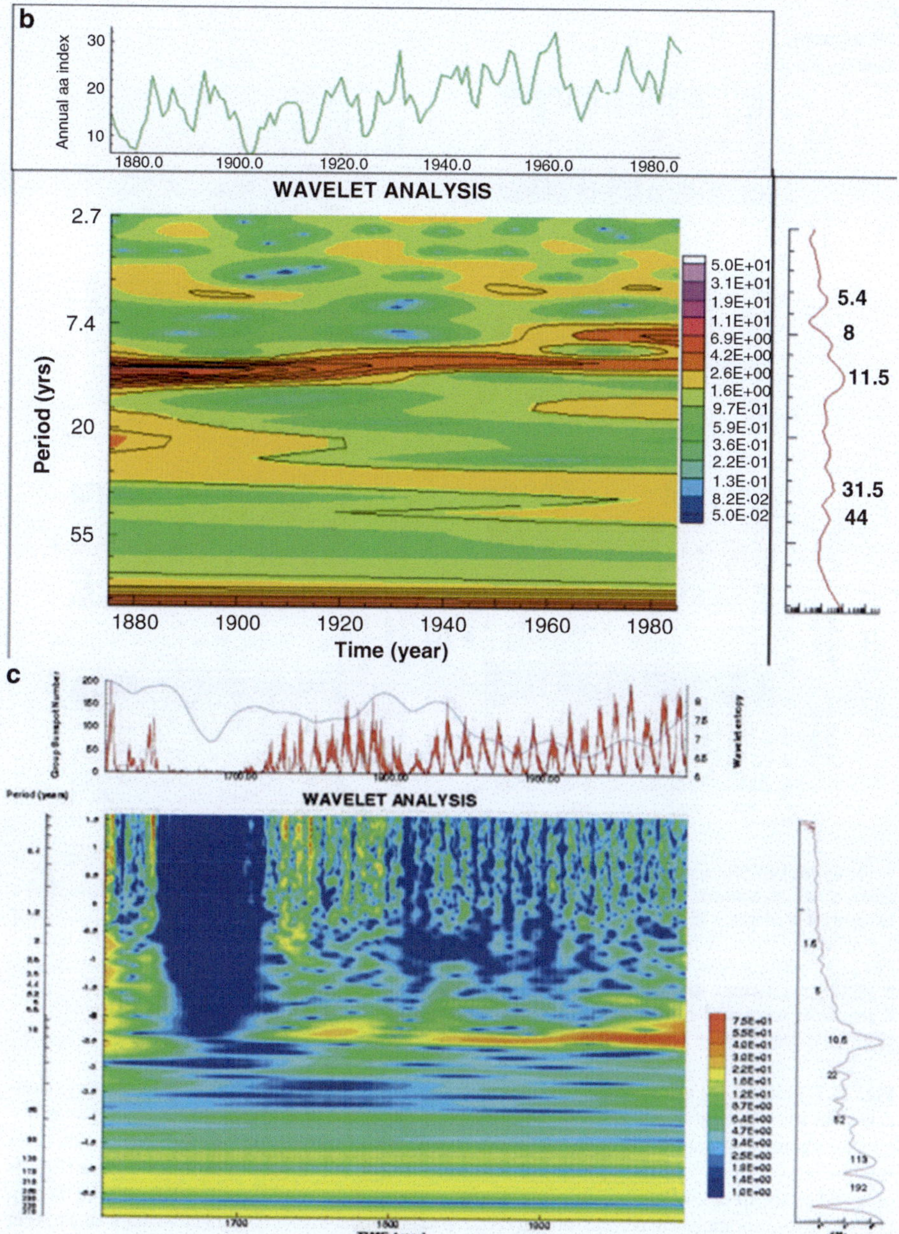

Fig. 23.7 (continued) (**b**) Morlet wavelets reveal, in the antipodal geomagnetic index aa, the time course of periods centered on 5.4 (8–11.5) years and 31.5–44.0 years. Note the broad yellow band starting near the about-20-year Hale cycle and the near 30-year Brückner-Egeson-Lockyer (BEL) cycle, periodicities putatively associated in part with changes with long-term solar activity. © Halberg Chronobiology Center. (**c**) With the particular time-frequency parameters used, an about-10- to

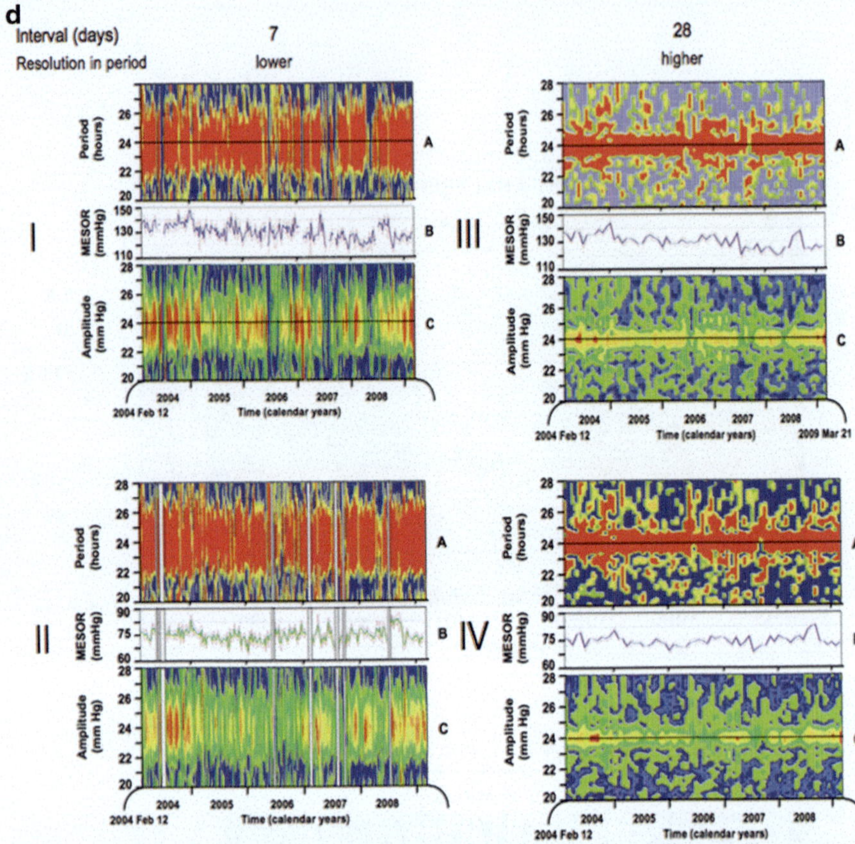

*In FH, treated MESOR-hypertensive man 84 years of age at start of observation, monitored every ~30 minutes with few gaps (original data analyzed, N=67,598). Gliding interval and increment = 7 (left) or 28 (right) days, longest trial period (τ) = 28 hours (h), shortest τ = 20.

A: Probability of rhythm, without account for any pink spectrum; colors correspond to: blue >0.05, green 0.05-0.01, yellow 0.01-0.001, red 0.001-0.0001, violet <0.0001.

B: MESOR: red sidelines: 99.9% confidence corridor; dotted line: upper 95% prediction limit.

C: amplitudes: colors correspond to: dark green 1.9-3.8, light green 3.8-5.7, yellow 5.7-11.4, orange 11.4-17.1, red 17.1-22.8 mm Hg (red = CHAT = **C**ircadian **H**yper-**A**mplitude-**T**ension. © Halberg.

Fig. 23.7 (continued) about-11-year band is seen in group sunspot numbers [50, 51], but any paratridecadal trace is largely missed, a demonstration of the need for analyses with different time-frequency parameters and/or different methods. © Halberg Chronobiology Center. (**d**) A circadian rhythm of systolic (I and III) and diastolic (II and IV) blood pressure of an elderly man (FH) is assessed by the fit of 24-hour cosine curves to data sections of 7 (I and II) and 28 (III and IV) days in order to assess anomalies in records smoothed so as to eliminate transients limited to a few days. Note the large red spot at the very end of row C, showing the recurrence of CHAT (brief for circadian hyper-amplitude-tension). Need for continued blood pressure monitoring: CHAT recurs after its long absence, even in smoothed 28-day records analyzed. Continuous monitoring is required once a vascular variability disorder (VVD), such as a high blood pressure, and/or a circadian overswing (CHAT) is diagnosed. It should be noted, however, that CHAT is only a risk factor, and hence there should be no dilemma in reconciling its presence in some subjects with their long survival, as is the

e

Nine Consecutive Weekly Circadian-Circasemiseptan Profiles in 2008
FH, 89-year old man

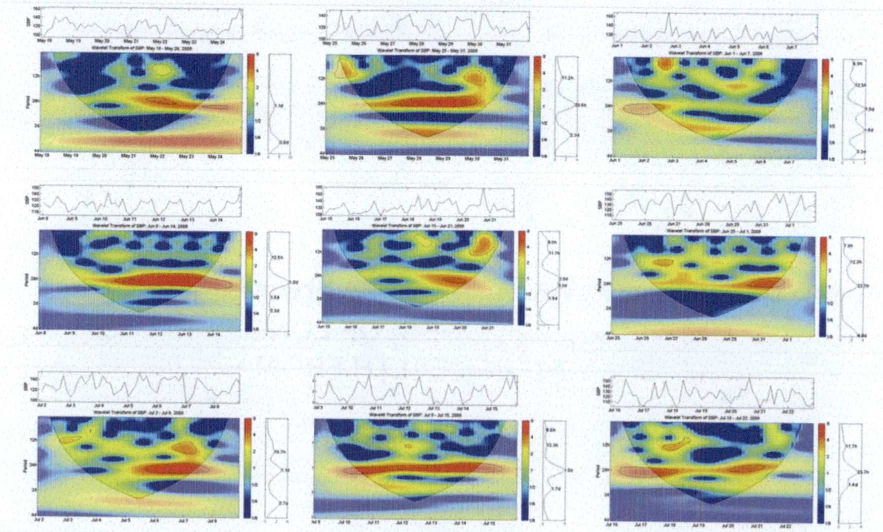

f

Eight 'Consecutive' Weekly Circadian-Circasemiseptan Profiles in 2011
FH, 92-year old man

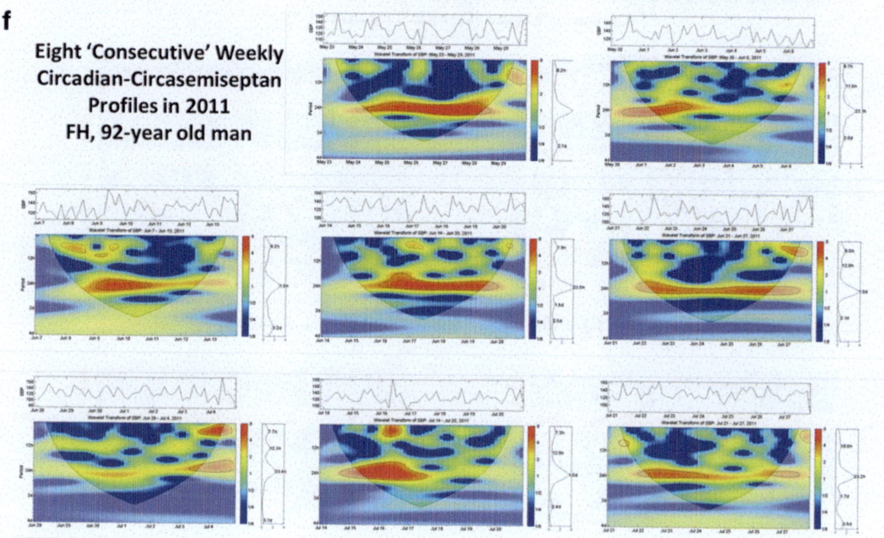

Fig. 23.7 (continued) case of FH. In some prospective clinical trials based on 48-hour ABPM, CHAT represented a risk of morbid cardiovascular event within 6 years numerically larger than hypertension. New trials are needed to determine the merit of reducing or eliminating CHAT in curbing adverse outcomes. © Halberg Chronobiology Center. (**e**) Wavelets of systolic blood pressure of a 92-year-old man (FH) measured around the clock for 9 consecutive weeks in 2008, focusing on the circadian-circasemiseptan time structure. Data bracket the week of June 15–21 when the circadian amplitude was not prominent. Focus on extracircadian frequency anomalies is overdue but requires more longitudinal reference data in health. © Halberg Chronobiology Center. (**f**) Wavelets of systolic blood pressure of a 92-year-old man (FH) measured around the clock for 8 consecutive weeks in 2011, focusing on the circadian-circasemiseptan time structure. Data bracket

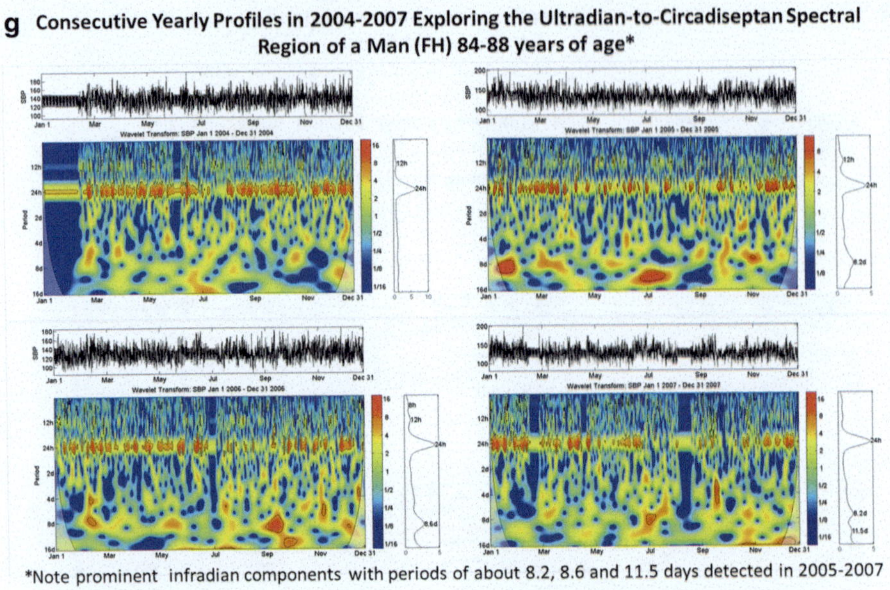

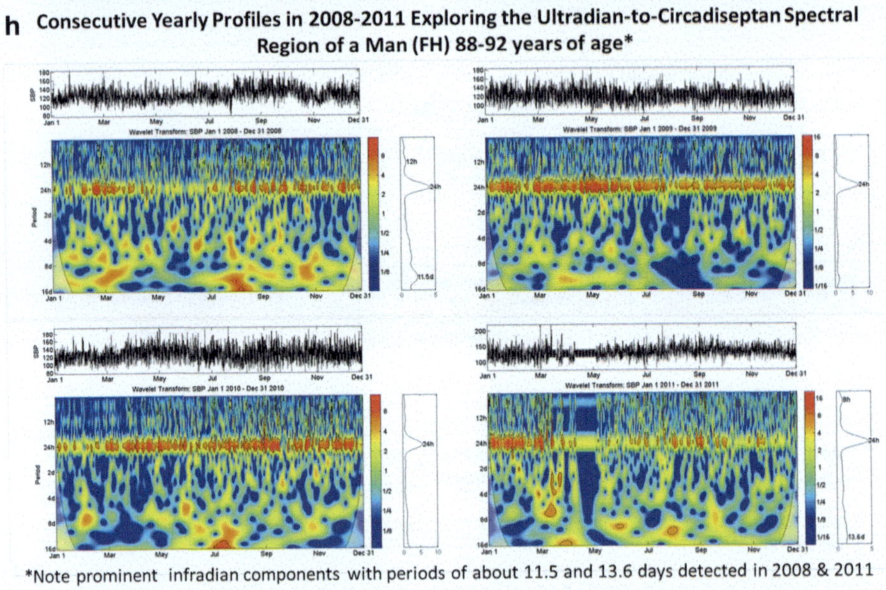

Fig. 23.7 (continued) the week of June 21–27 when systolic CHAT was present. © Halberg Chronobiology Center. (**g**) Wavelets of systolic blood pressure of an elderly man (FH) measured around the clock during 2004–2007 (with short interruptions), focusing on the circannual-circadiseptan time structure, could be automatically checked at any intervals prompted by as one-goes analyses. © Halberg Chronobiology Center. (**h**) Wavelets of systolic blood pressure of an elderly man (FH) measured around the clock during 2008–2011 (with short interruptions), focusing on the circannual-circadiseptan time structure, could be automatically checked at any intervals prompted by as one-goes analyses. © Halberg Chronobiology Center. (**i**) Cross-wavelet coherence with Wolf numbers (WN, gauging solar activity, *top row*) differs between the incidence of myocardial infarc

i

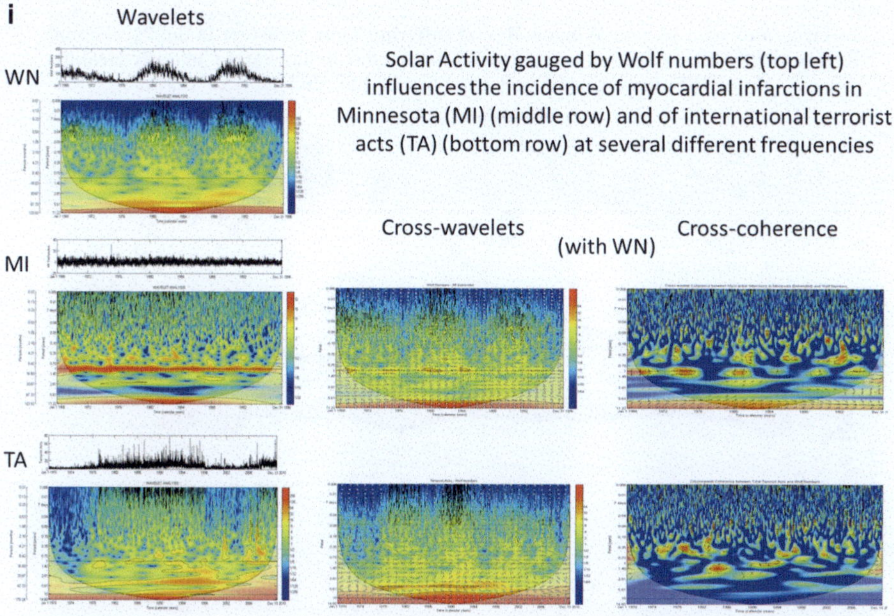

j PROMINENT CIRCASEMIANNUAL COMPONENT MOSTLY IN THE ABSENCE OF A CALENDAR YEAR, SHOWING BETTER CONSISTENCY THAN THE TRANSYEARS APPARENT MAINLY AFTER 1920 IN THE ANTIPODAL GEOMAGNETIC INDEX aa

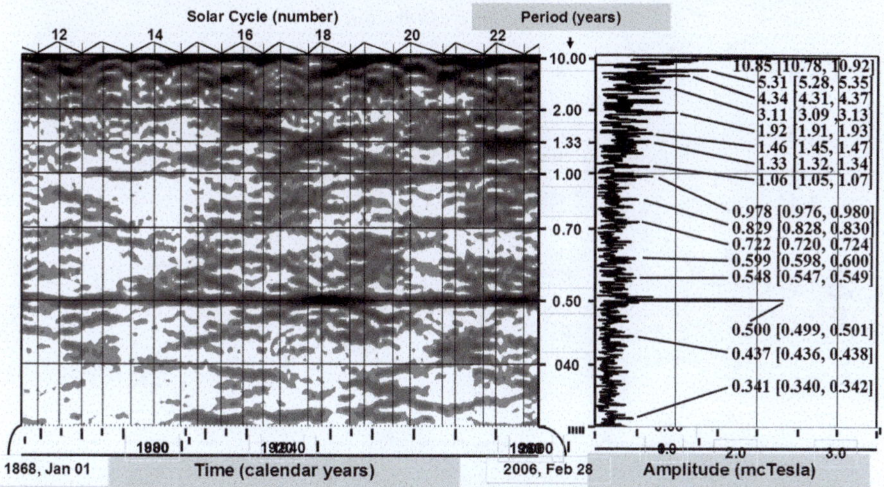

~137 years (y) of daily data from http://www.geomag.bgs.ac.uk.aaindex.html. (N= 50,463); gliding spectrum computed with interval = 14 y (= longest period here fitted), increment = 1 y, harmonic increment = 0.25; darker shading corresponds to larger amplitude; vertical continuous or dotted lines = Wolf No minima or maxima, respectively. Right: 95% confidence intervals (CIs) of spectral peaks in global window are lacking due to very small size but are given in [ ]). {During 1963-2003, sunspot cycle averages 10.90 [10.72, 11.07] years, differing from 8.39 [8.07, 8.72]-year cycle in solar wind speed.}

Fig. 23.7 (continued) tions in Minnesota (MI, *middle row*) and the incidence of international terrorist acts (TA, from the total Global Terrorism Database, *bottom row*). Whereas a low-frequency component is visible in the wavelet spectra of all three variables (*left*) and in the cross-wavelets of MI and TA with WN (*middle*), only MI (but not TA) shows coherence in this region. © Halberg Chronobiology Center. (**j**). Global and gliding (local) spectrum of antipodal geomagnetic index,

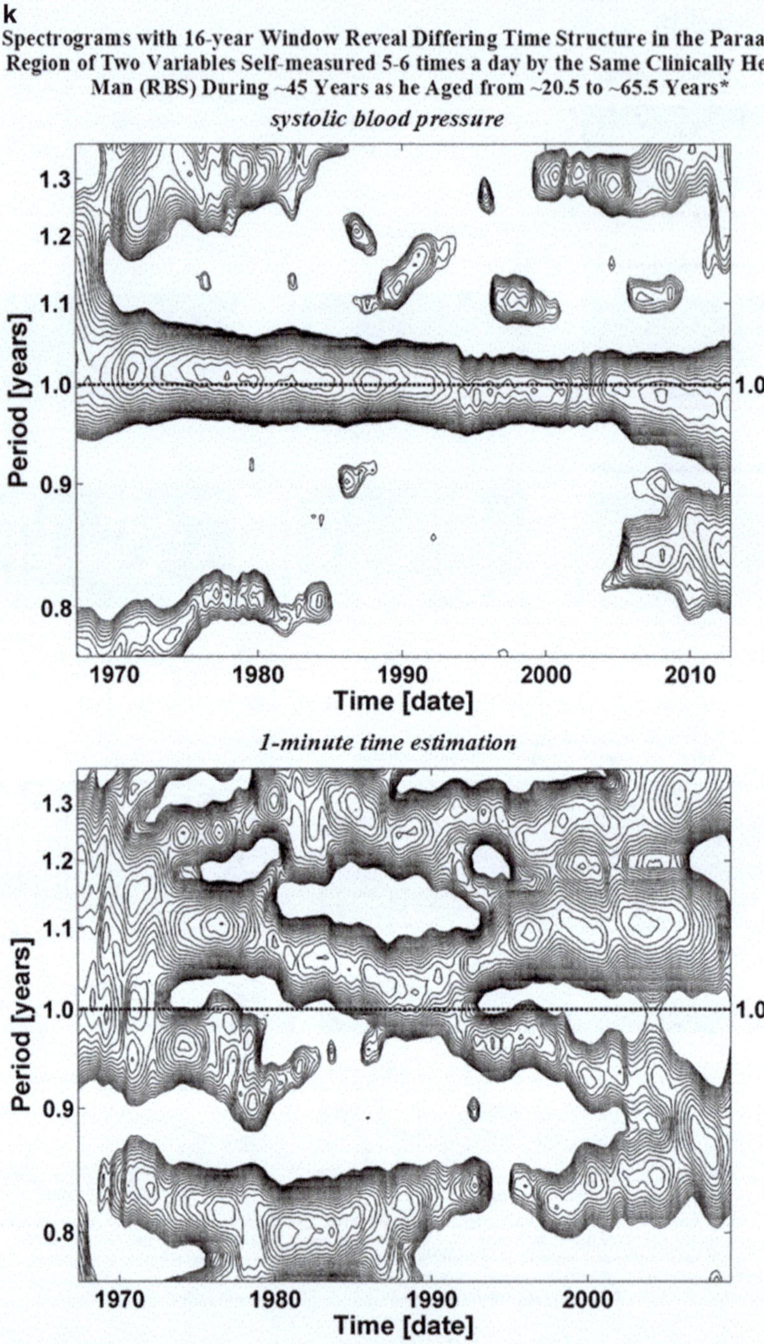

Fig. 23.7 (continued) aa. © Halberg Chronobiology Center. (**k**) Using a 16-year window, spectrograms of systolic blood pressure (*top*) and 1-minute time estimation (*bottom*) were obtained from 7-day averages of self-measurements 5 to 6 times a day by a clinically healthy man (RBS) collected over about 45 years. Whereas systolic blood pressure is primarily characterized by a prominent

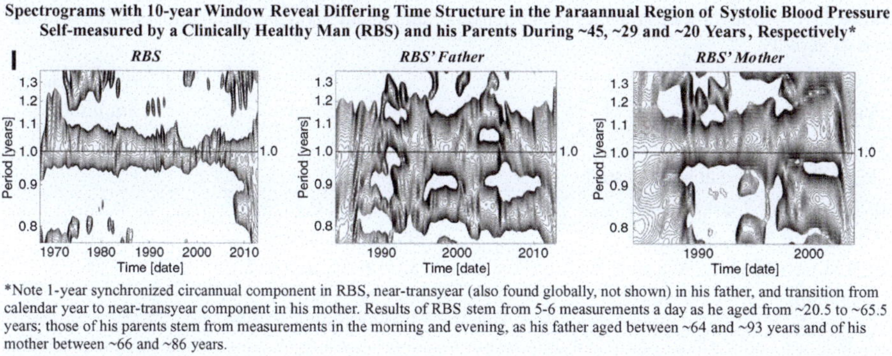

Fig. 23.7 (continued) circannual variation, the time structure of 1-minute time estimation is less stable (more "aeolian"). A transyear with a period of about 1.2 to 1.3 years predominates over the circannual variation for 1-minute time estimation © Halberg Chronobiology Center. (**I**) Spectrograms of systolic blood pressure are compared between RBS and his parents. Data analyzed are 7-day averages of self-measurements taken about 5 to 6 times a day by RBS and twice a day (morning and evening) by his parents, collected over about 45 years (RBS), 29 years (RBS's father), and 20 years (RBS's mother). Using the same 10-year window bracketing the calendar year, differences are readily seen among the three records: RBS has a sturdy and stable circannual variation. The calendar year is accompanied by a near-transyear in his father's record. His mother's record shows a slightly desynchronized but stable year at the beginning of the record, but with a period starting to lengthen later on to become a near-transyear © Halberg Chronobiology Center

Automatic passes over the accumulating data from sections of decade-long C-ABPM have already served to track the coming, going, and reappearing VVAs such as CHAT (Fig. 23.7d) to summarize endeavors toward eliminating them, which were only transiently successful, as can be seen within graphs by the appearance and recurrence of red denoting the undesirable condition. With wavelets, Fig. 23.7e–h explores the extracircadian domain in an elderly man. For anomalies to be detected therein, a first focus is possible in the ultradian and up to about a circasemiseptan region of the spectrum (Fig. 23.7e, f), but reference values need to be collected before any, e.g., cis-half-yearly anomalies can be defined and diagnosed in the extracircadian range. Figure 23.7g, h models future automatic yearly screenings, or screenings at other intervals, with a length depending on the number of cycles required in health to reliably assess a given (e.g., infradian) component of interest, such as a quinmensal, which, along with a transyear, is reflected in sudden cardiac death as well as in cardiovascular variables and also, as we found out on ourselves, in wrist activity (Table 23.1).

We continued to learn with and from the Japanese team of Dr. Kuniaki Otsuka, Professor, Tokyo Women's Medical University, and President, Daini Hospital, Tokyo, with the data covering invariably 7 days/person continuing to come from the cities of Uraus and Tosa where he implemented what former Roseville Mayor Dan Wall visualized for Minnesota (but his successor did not continue) as a prehabilitation of the elderly by detecting VVAs by C-ABPM and treating them properly (Fig. 23.8a, b) in applying such goals of chronobiology and chronomics in elders' everyday life [52–56].

Table 23.1 About-5-month cycles (quinmensals)[a] and a candidate *far-transyear* in human motor (wrist) activity

Period, τ (years, y)	Amplitude (count[c])	% of 24-hour amplitude
Days, d, CI[b]		
OS, F, 89 y of age at start of study		
0.7728 y (0.6675, 0.8781)	0.459 (0.145, 0.773)	9.50
0.4160 y (0.3664, 0.4655)	**0.305 (0.024, 0.585)**	**6.31**
0.2324 y (0.2125, 0.2522)	0.252 (−0.01, 0.515)	5.22
24.00 hour	4.83 (4.58, 5.09)	–
GC, F, 59 y of age at start of study		
0.4100 y (0.3920, 0.4279)	**0.555 (0.389, 0.721)**	**9.07**
24.00 hour	6.12 (5.95, 6.29)	–
FH, M, 92 y of age at start of study		
1.860 y (1.527, 2.193)	0.990 (0.806, 1.170)	18.00
0.4536 y (0.4179, 0.4892)	**0.481 (0.275, 0.687)**	**8.75**
0.3367 y (0.3237, 0.3496)	0.699 (0.493, 0.905)	12.71
24.00 hour	5.50 (5.30, 5.70)	–

[a]Originally also referred to as cis-half years (cis = on the near side, i.e., shorter than), with counterparts (coperiodisms) in solar flare incidence
[b]CI: 95 % confidence interval
[c]Count from digital integration with a unit from Respironics
Note relatively small infradian amplitudes as compared to the circadian amplitude

It is a pleasure to plan with our new (Yinjing Guo, Dean and Professor, Shandong University of Science and Technology) and long-time (Jinyi Wu and his son Bing Wu) colleagues in the People's Republic of China to bring chronobiologic/chronomic health care to their country. They aim to establish a chrono-cloud system and to deliver chronobiologically interpreted services into Chinese community health care, as originally scheduled (Fig. 23.4c), yet possibly in a broader perspective (Fig. 23.4d).

BIOCOS welcomes Dr. Abdullah al-Abdulgader, director of the Prince Sultan Cardiac Center in Al Ahsa, Saudi Arabia, and the organizer of King of Organs cardiology conferences. On Paul Rosch's advice, he gave chronobiology a prominent place and a full day of lectures on his program and had a special symposium on chronomics on another day. From his hospital, 7-day records of C-ABPM are already being contributed to BIOCOS on a scale that we trust he can enlarge. Abdullah monitors nearby magnetic activity and heart rate variability in his setting, as a station of an international coherence project guided by Dr. Rollin McCraty (Director of Research, Institute of HeartMath, Boulder Creek, California), who initiated such combined monitoring, and plans on doing this continuous surveillance of "coherence" among people on a large scale. In Rollin's data, we have already found nonphotic signatures [57], among other results [58–61]. Gravitational recording is the main task of Prof. Elchin Khalilov, head of the project on geocataclysms [62], who came to the Al Ahsa meeting. His worldwide activity aimed at predicting

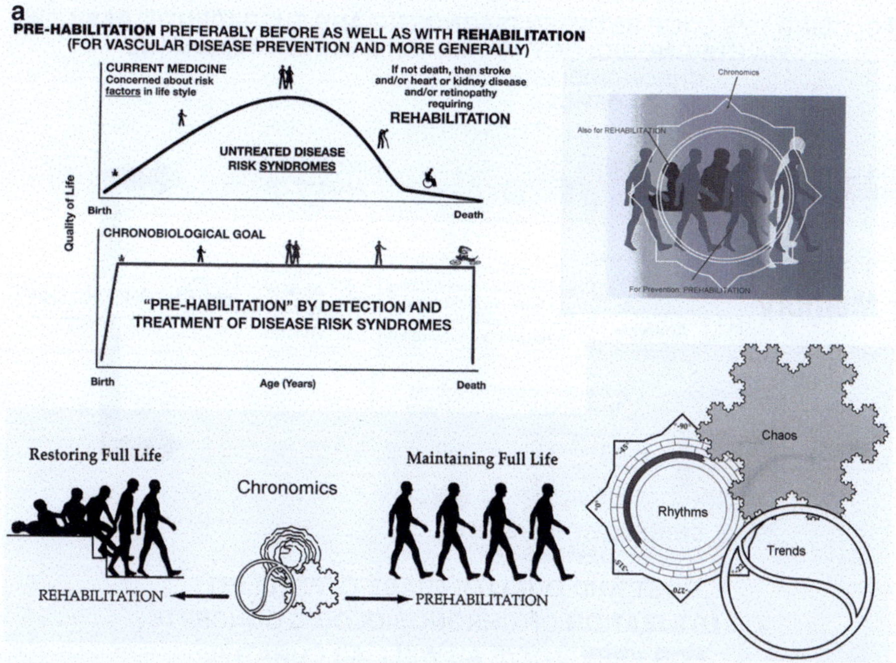

Fig. 23.8 (**a**) *Top left*: Prehabilitation, preferably before as well as after or with rehabilitation. By the early detection of disease risk syndromes in the individual patient, countermeasures for primary prevention can be instituted, whether by non-pharmacologic or pharmacologic means. *Top right*: For prehabilitation, the development of affordable, unobtrusive instrumentation for monitoring blood pressure and heart rate is needed, as are databases of records in health combined with outcomes from prospective and/or retrospective trials for a chronobiologic interpretation of the results and, when needed, for treatment optimized by timing (chronotherapy), preferably adjusted to the individual patient, accounting for the chronodiagnosis (chronotheranostics). Such a chronomic approach provides the tools serving both prehabilitation and rehabilitation. *Bottom right*: Chronomics helps quantify elements of chronomes (multifrequency rhythms, chaos, and trends) in us and around us. *Bottom left*: Applications of chronomes extend focus to include along with rehabilitation (*left part*), prehabilitation, i.e., the timely institution of prophylactic measures (*right part*). © Halberg Chronobiology Center. (**b**) *Top*: Closing the loop between diagnostic and therapeutic devices via chronobiology may lead to chronomedicine. The addition of a memory for data storage and for chronobiologic data analysis as one goes, according to the principles of (repeated passes for) windowing (i.e., analyses in spectral regions of progressively lower frequencies), compacting (in the form of the characteristics of the rhythm with the just-analyzed frequency), and recycling for the analyses on longer and longer series and the (preferably automatic) interpretation of the results, in the light of time-specified reference standards, could provide a bridge between monitoring endeavors aimed at screening, diagnosis and prognosis, and instrumentation for the delivery of treatment. In thus closing the diagnostic and therapeutic loops, a step could be made toward the online optimization of treatment by timely and timed treatment according to rhythms. *Bottom*: Investment into physiological monitoring and education in chronobiology, to detect warning signs indicative of an elevated risk, rather than only of the fait accompli of disease, can prompt preventive intervention with the goal of avoiding the crippling of catastrophic diseases, also a major drain of financial resources. By placing added emphasis on prevention, health-care costs could decrease, while the quality of care is improved. © Halberg Chronobiology Center

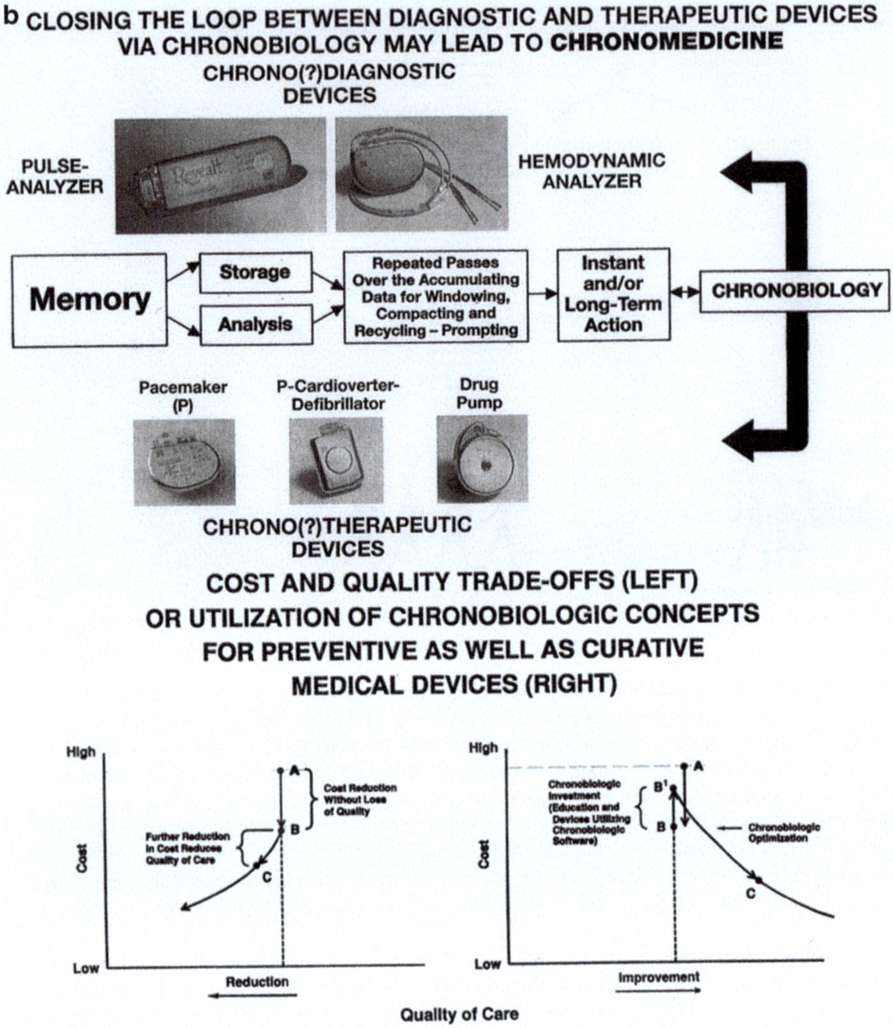

Fig. 23.8 (continued)

earthquakes could take the lead in an international chronomic alerting system, from alignment with BIOCOS, considered in Saudi Arabia if not in Azerbaijan, and preferably in both locations.

Like a group of scholars at Princeton University, we ask, also like Rollin, about any effect on the environment exerted by the feelings and emotions of large groups of people and vice versa, a topic considered further by Hans Wendt (Prof.psychol. em., Prof. sc. nat. [hon.], A.v. Humboldt Behavioral Geomedicine, St. Paul, Minnesota), who in turn found a putative effect of the reaction of masses of people to stirring events entangled with the cosmos (Fig. 23.9a–c) [63]. The effect of human emotions has to be disentangled from that of the cosmos [64].

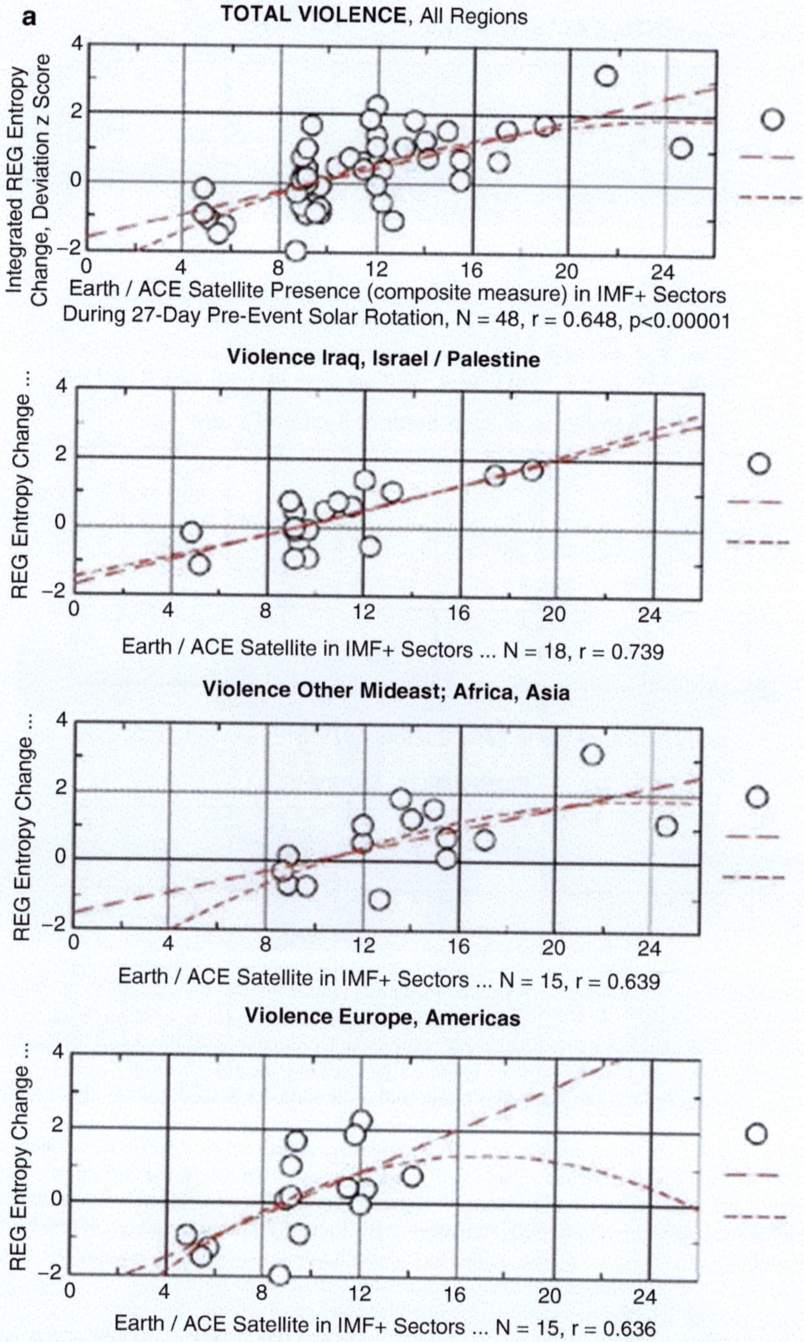

Fig. 23.9 (**a**) Correlations with a gauge of the polarity of the interplanetary magnetic field (IMF) and deviations from randomness "entangled" with "violence from below," qualified by lack of consideration for periodicity in the phenomena assessed. Approach and analyses by Hans W. Wendt. (**b**) Scatterplots of deviation from randomness vs. IMF polarity "entangled" with

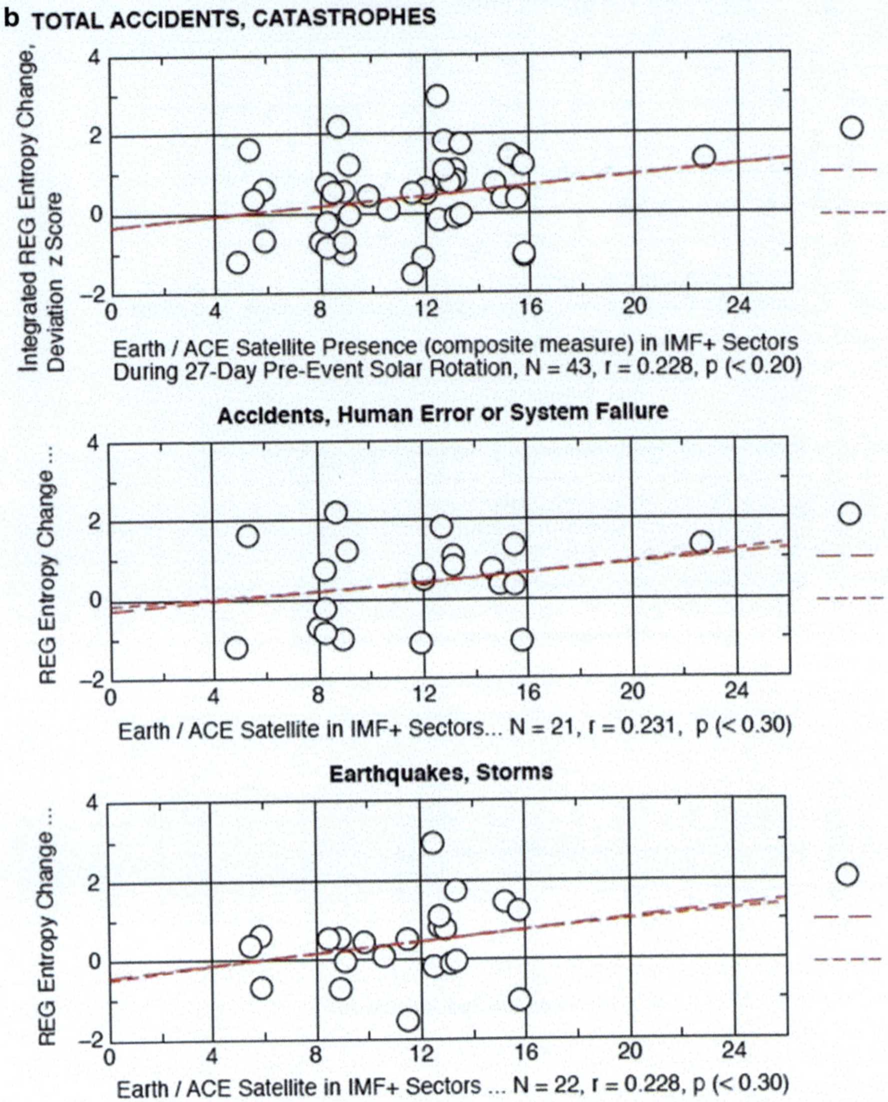

Fig. 23.9 (continued) the incidence of accidents and other catastrophes. Approach and analyses by Hans W. Wendt. (**c**) Inward-directed acts such as meditation and prayer in the context of a putative correlation between the output of random event generators and the interplanetary magnetic field of borderline and otherwise qualified statistical significance. Approach and analyses by Hans W. Wendt

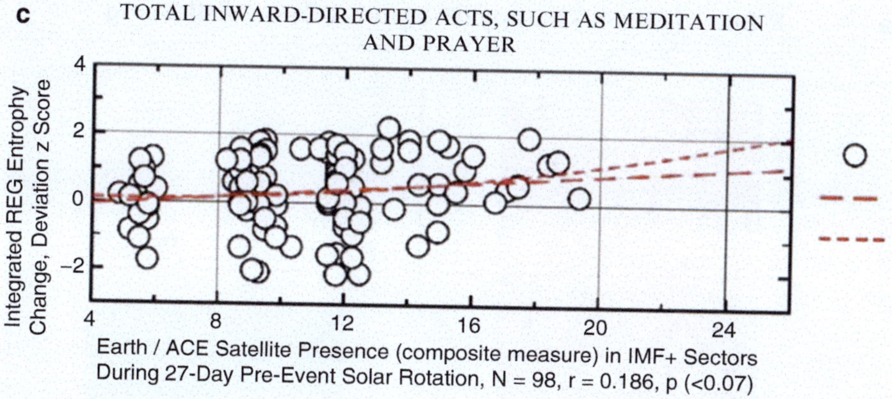

Fig. 23.9 (continued)

Cycles characterize both gravity and magnetism and can be monitored as such to seek information on whether and, if so, what seems likely, how both these unseen features of our environment interact, if indeed their effects can be separated. It was therefore a pleasure for Abdullah to welcome Acad. Elchin Khalilov (Chairman, InterGeoTethys, Baku, Azerbaijan) with the possibility of adding gravitational to biological and magnetic surveillance. Abdullah, Elchin, and Rollin willing, these endeavors could lead, in combination with BIOCOS, to an international alerting system to better understand and avoid natural and social cataclysms, with the very data accumulating as harbingers prompting action to prevent personal adverse cardiovascular events. In this context, endeavors from the Phoenix Study Group, composed of volunteering members of the Twin Cities chapter of the Institute of Electrical and Electronics Engineers (http://www.phoenix.tc-ieee.org), founded and maintained by Ellis Nolley with Larry Beaty's outstanding activities [65–67] and with Chris Adams' broad scenarios [68], are promising. A chronomic data center (corne001@umn.edu) at the University of Minnesota has been analyzing presently accumulating as well as published data (Figs. 23.3a–e, 23.6a–b, and 23.7a–i), and the wherewithal for the broader implementation of this activity is part of the endeavor visualized by El, Larry, and Chris and worked on in the laboratory with engineering students by Larry.

The year 2012 continued to show the palliative success of Vera Brandes' SANOSON music therapy (Fig. 23.10a) in a case of twice-yearly depression where drugs had failed for two decades [69]. Patient/investigator Mrs. Judy Finley documented the wrangling of several coexisting periods in her cardiovascular system, mental state, and 11,702 endocrine determinations (on the same person) of six hormones sampled at 4-hour intervals around the clock along with other variables (Fig. 23.10b and Table 23.2). For Judy, it was possible to demonstrate a double tidal period in variables as different as self-ratings of vigor (twice), urine volume, and heart rate (Fig. 23.10c) in the computer output (not input). In the wrangling of her coexisting components, the double tidal period

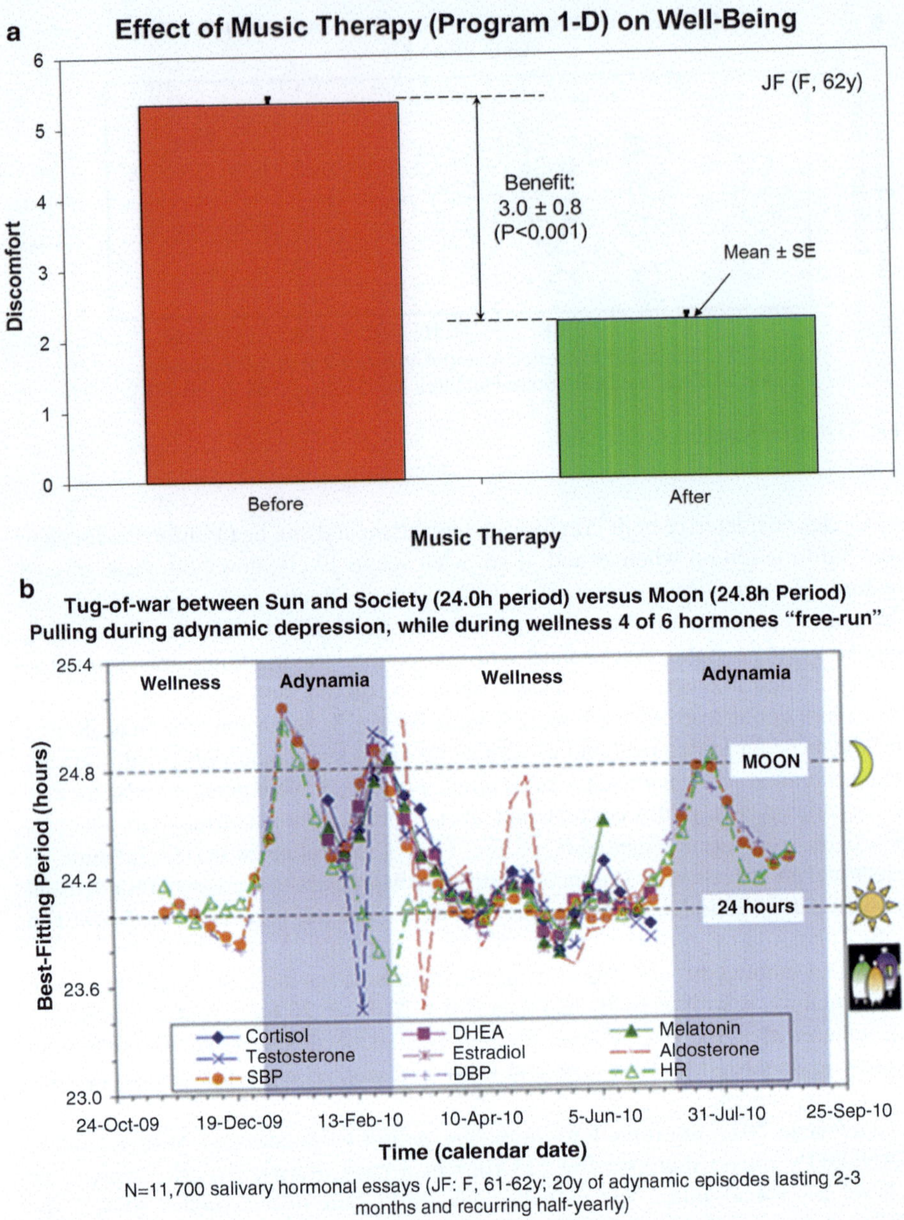

Fig. 23.10 (**a**) Listening to specially composed music (SANOSON) alleviates, for at least 1 hour, adynamic depressive symptoms of a selenosensitive woman (JF) with depression episodes lasting 2–3 months at a time, recurring twice yearly for the past 21 years. © Halberg Chronobiology Center. (**b**) Wrangling between the double tide on the one hand and society/solar light and temperature on the other hand in JF's endocrines as in her circulation, with the pull of the double tides lengthening the dominant circadian period during depression. © Halberg Chronobiology Center. (**c**) The heart rate of JF exhibits a double tidal period in the computer output (not input) of analysis. © Halberg Chronobiology Center. (**d**) Several circadian and circasemidian components

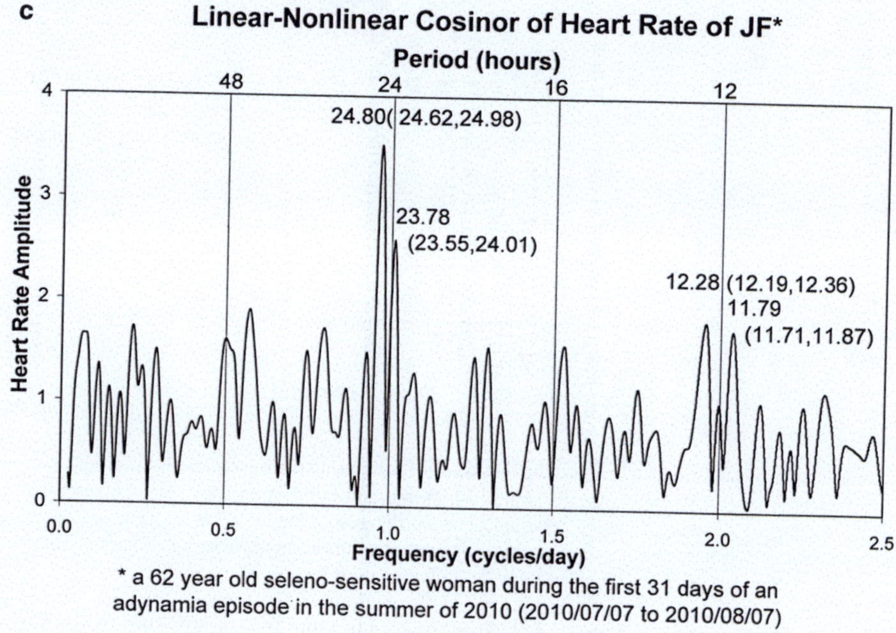

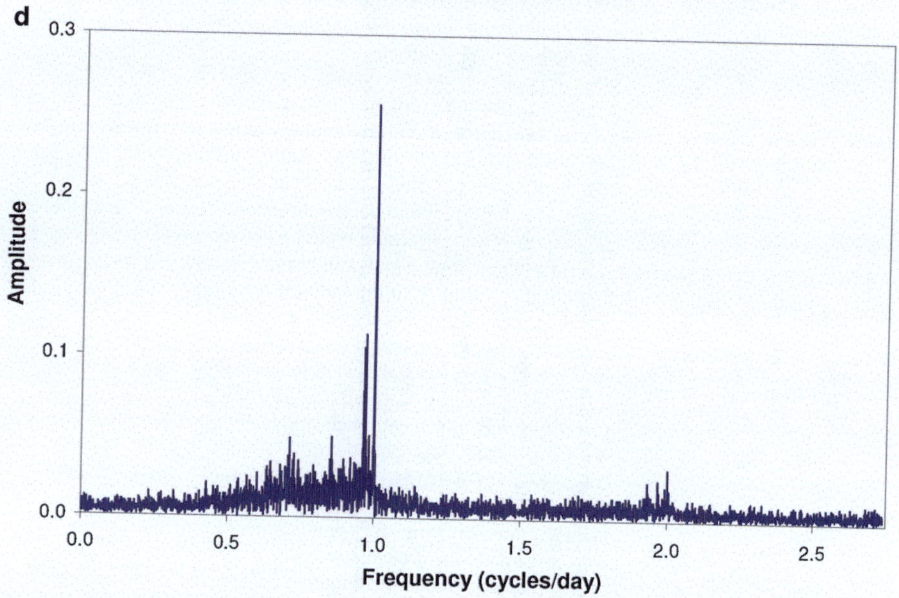

Fig. 23.10 (continued) characterize the rest-activity schedule of a healthy man (JC) on a self-selected routine. In addition to a 24-hour synchronized component, an about-24.8-hour (double tidal) component is present together with an about-24.4-hour cycle (compromise between earth and moon?) and an about-24.2-hour (free-running?) component, all resolved by linear-nonlinear cosinor. © Halberg Chronobiology Center

Table 23.2 Recorded recurrent adynamia. Start-end dates and length (weeks) of spans of adynamia when recorded in winter (W) and summer (S). f: date of full moon; n: date of new moon

	2002	2003	2004	2005	2006	2007	2008	2009	2010	2011	2012
W	-	01.18 f 02.17 f (4)	01.12 02.20 n (5)	01.05 02.15 (5)	12.25 n 02.14 n (7)	01.03 f 03.03 f (8)	01.08 n 02.21 f (6)	12.26 n 02.23 n (8)	01.02 f 02.28 f (8)	01.04 n 03.05 n (8)	01.04 05.15 (18)
S	06.09 08.07 (8)	06.15 f 08.12 f (8)	06.17 n 08.18 n (8)	06.20 f 08.09 (7)	06.04 08.28 (11)	05.31 f 08.12 n (10)	? 08.18 f (?)	06.09 f 08.08 f (8)	07.08 09.16 (10)	07.07 09.13[a] (10)	07.03 09.22 (11)

[a]"I have concluded that Sept 1 was a false end to the downtime. I had five functional days, but the last eight [i.e., 09.06–09.13] have been horrible again!" (JF, personal communication, 14 September 2011)

has the larger amplitude (wins) during episodes of depression in alternation with a larger 24-hour amplitude between episodes of adynamia. Judy's many contributions thus far constitute a major investment by her into science as well as health and so are the many chemical and numerical analyses. We hope she resumes vascular, urine, wrist activity, and possibly salivary surveillance with self-ratings of mood and vigor.

Dr. John Costella's data during the 3 years on a self-selected activity/rest span are long enough and sufficiently consistent on a relatively societally independent living routine to allow the concomitant demonstration of four coexisting periods in the same healthy person [70] (Table 23.3 and Fig. 23.10d), including components related to lunar tides. From his own analysis of his sleep record, he concluded, as a physical scientist, that he was selenosensitive. The 24.00-hour predominates over the also prominent 24.84-hour period and that the latter's amplitude is less than one half of that of the societal-solar period. The statistically significant tidal period of 12.41 hours also has a smaller amplitude than the 12.0-hour period in John. He deserves credit for demonstrating, if not multiple oscillators necessarily, than coexisting multiple circadians in health in sleep-wakefulness. Multiple coexisting circadian oscillators have been documented in plants [71] and considered broadly [72] and should be extended not only to the extracircadian domain but also to seek and examine environmental coperiodisms. Against this background of several coexisting circadian periods in depression [73] and in selenosensitive health on a self-selected routine of living, non-24-hour sleep disorder can be viewed [74, 75]. The question could be asked whether, given long time series of months or years, allowing their resolution, an effect of several circadian periods, including double tidal periods and compromise periods, should be sought. Isolation studies in caves have yielded compromise periods about halfway between 24.00 and 24.84 hours [76–79] or, outright, the double tidal period [80, 81]. Noteworthy in this context is the finding of a very tiny peak at 24.8 hours in health on a societal routine during 45 years in a clinically healthy man's (RBS) vigor, who had been studied in Rütger Wever's bunker [82], where his oral temperature in our meta-analyses showed a double tidal period. In RBS' 44.5-year record of self-rated vigor, the 24.00-hour component has an amplitude, A, of 1.0, while that at 24.8 hours has an A of 0.018; yet, the point estimate of neither A has a CI (95 % confidence interval) overlapping zero. If it is not an artifact of very large numbers (this will be checked further) and if this component is found in a few other variables of RBS and/or in other subjects, it would lead to tests of the hypothesis that the effect of the double tides is ever present at least in one variable in some other individuals. This is in keeping with a near-tidal component in another subject (FH) with decade-long data (Table 23.4). A tidal or double tidal component, even if minuscule, could lead to "butterfly effects" that can struggle with and can overcome societal effects but coexists as an added component.

We congratulate Dieter Broers of the International Council for Scientific Development (Erlangen, Germany) on his award from Amazon and on the successes of his film (Solar Revolution) based on interviews about what the cosmos does. Another complementary film (Resonance: Beings of Frequency by James

Table 23.3 Periods, amplitudes, and acrophases[a] found in sleep-wakefulness on a largely self-selected schedule by JC[b]

Period (CI[a])			
Name	Length (hours)	Amplitude (CI) Acrophase (CI)	
Solar day	24.001 (24.000, 24.001)	0.25443 (0.2337, 0.2751)	−293° (−288, −287)
Lunar day	24.836 (24.833, 24.838)	0.10606 (0.0854, 0.1268)	−233° (−222, −244)
Compromise?	24.432 (24.427, 24.436)	0.05643 (0.0357, 0.0772)	−250° (−229, −271)
Free-running?	24.260 (24.252, 24.268)	0.03253 (0.0118, 0.0533)	−10° (−334, −47)
Half-day	11.999 (11.997, 12.001)	0.03204 (0.0113, 0.0527)	−356° (−319, −33)
Tide	12.414 (12.410, 12.418)	0.01728 (0.0034, 0.0380)	−199° (−131, −268)

[a]With uncertainties (CI, 95 % confidence interval)
[b]A healthy man, 23 years of age at start of recording from February 19, 1990, to March 11, 1993 with linear-nonlinear rhythmometry by the extended cosinor. Local analyses by separate serial sections with 24.0- and 24.8-hour periods (in 1-week intervals) reveal the dominance, in alternation, of these 2 periods in the data

Table 23.4 A minuscule signature with a tiny amplitude, A, of the double tidal period, τ, of ~24.84 hours in the vigor[a] of a healthy man (RBS) on a 24-hour routine of living

Component	τ (CI) (hours)	A (CI) (arbitrary)	Acrophase (CI, rounded)
Societal solar	24.000(24.000, 24.000)	1.00324(0.9851, 1.0214)	−220^0 (−219^0 , −221^0)
Double tidal	24.829(24.828, 24.830)	0.01804(0.0027, 0.0334)	−340^0 (−291^0, −28^0)

Reference value: 00:00 on January 1, 1967
[a]Self-ratings, ~ 5 to 6 times a day in a 44.5-year record by a 20- to 66-year-old man (RBS). Note that the 24.83-hour component's amplitude is <2 hundredths of the amplitude of the 24-hour synchronized component, with limits in further decimals, when the two components are fitted together by the nonlinearly extended cosinor. Apart from side lobes, computed to lie at 24.20, 24.13, 24.07, 23.93, 23.87, and 23.80 hours in a 1-year section of the series. Not here tabulated are a period of 24.20 hours, corresponding perhaps to an endogenous circadian component, and periods of 24.40 and 24.54 hours, which could correspond to a wrangling of multiple components, found in the original spectrum.

Qualifications, among others, stem from (1) the violation of the independence assumption when so many data are analyzed (74413 ratings covering 44.5 years from 196707261218 to 201212311722) and the facts that (2) a small change in the trial period inserted into the nonlinear program used can make a difference in the program's solution; (3) the program does not print out enough decimals to give a CI for each amplitude and period tested; (4) the nonoverlapping of zero by the amplitude strictly was applied only to vigor (chance ?) out of a total of 14 variables tested, although the overlap was small in two other variables, temperature, and mood; and (5) no adjustment was made for multiple testing. It is most pertinent that the double tidal period was prominent in RBS during isolation from society in the late Ruetger Wever's bunker in Andechs, Bavaria. The most important consideration is that each of the five peaklets that stand out above the spectrum could be anticipated. Furthermore, a 12.45-hour period, related perhaps to the average period of the tide of 12.4 hours, was found in the heart rate of another subject (FH) during a span from 2004 to 2011. The wrangling or coexistence seen in selenosensitive persons between the double tidal 24.8-hour and 24.0-hour periods (JF and JC) may also occur at around 12.4 vs. 12.0 hours in people on a conventional societal-solar 24.0-hour average routine.

Russell; www.youtube.com/watch?v=5vb9R0x_0NQ) and a future one focused on data such as those of Prof. Rita Jozsa (Fig. 23.11 a–o) summarized elsewhere [13] could well introduce a quantification of the response of the human brain to its cosmos in the broader context of a transdisciplinary spectrum of cycles and their use as a multipurpose alerting system.

Figures 23.2a–h were prepared to amplify on text sent for Dieter's presentation in Al Ahsa, should he have chosen to refer to the association (well accepted in physics) between helio- and geomagnetism being more than matched by that between either helio- or geomagnetism on the one hand and several gauges of human mentality, such as mood, vigor, and 1-minute estimation, on the other hand. Franz Daxecker, emeritus professor of ophthalmology at the University of Innsbruck, Austria, who had written books about Galileo's opponent Christoph Scheiner [83, 84; cf. 85–87] compiled sunspot counts from the latter's publications, making it possible for Dr. Dewayne Hillman, beyond his other contributions, to assess sunspot cycles of 12 years in Scheiner's data. One solar cycle was thus found to have started coincident with the start of the Thirty Years' War (1618–1648) (Fig. 23.12a, b and Table 23.5) [88]. Franz Daxecker further scrutinized Scheiner's Latin texts for references to periodicity in sunspots and found hints that Scheiner, a thorough physicist who made observations for two decades, not only considered periodicity as such but may have been aware of the variability of a cycle in sunspots. It may have been only the negative attitude of astronomers toward periodicity that prevented Scheiner from publishing explicitly about variability. This disregard of cycles is itemized in the presentation by Manuel J. Johnson [89], the president of the Royal Astronomical Society, of a gold medal to Schwabe, well over two centuries later. It is likely to have prevailed earlier, as it also exists today. It may have been justified by the failure (then and now of a majority of investigators reporting periodicities) to provide hypothesis testing of the validity of the period and of the uncertainties of the parameters – understandably only in the early seventeenth century, before Gauss' introduction of least squares, but hardly so in 2012, 292 years after Gauss [90]. In our computer era, 95 % confidence intervals could accompany each amplitude to check the "no-cycle" assumption (by a 95 % confidence interval's overlap of zero) and, if this is rejected, to provide the 95 % confidence interval of the periods (as well as amplitudes). Such an overdue approach served for an inferential statistical presentation of periods in Table 23.5, with wide uncertainties because the data in Fig. 23.12a or b are sparse.

Germaine Cornelissen [91] had already demonstrated by meta-analyses the existence of a statistically significant about-10- to about-11-year cycle in sunspot numbers in (1) Samuel Heinrich Schwabe's data up to 1838 [92] and up to 1893 [93] (Fig. 23.12c) and in (2) Christian Pedersen Horrebow's data (Fig. 23.12d; [94]). She also found (3) a historically captivating transyear in data on the strength of the signal in radio communications as well as in Wolf sunspot numbers (Fig. 23.12e–g) in data taken off graphs published in 1930 by Harlan True Stetson of MIT [95]. Germaine found the periods listed in Fig. 23.12f for sunspots, one corresponding to a transyear reported half a century later from the same institution for solar wind speed by Richardson et al. [96] and one with a period of about 0.42 year (154 days),

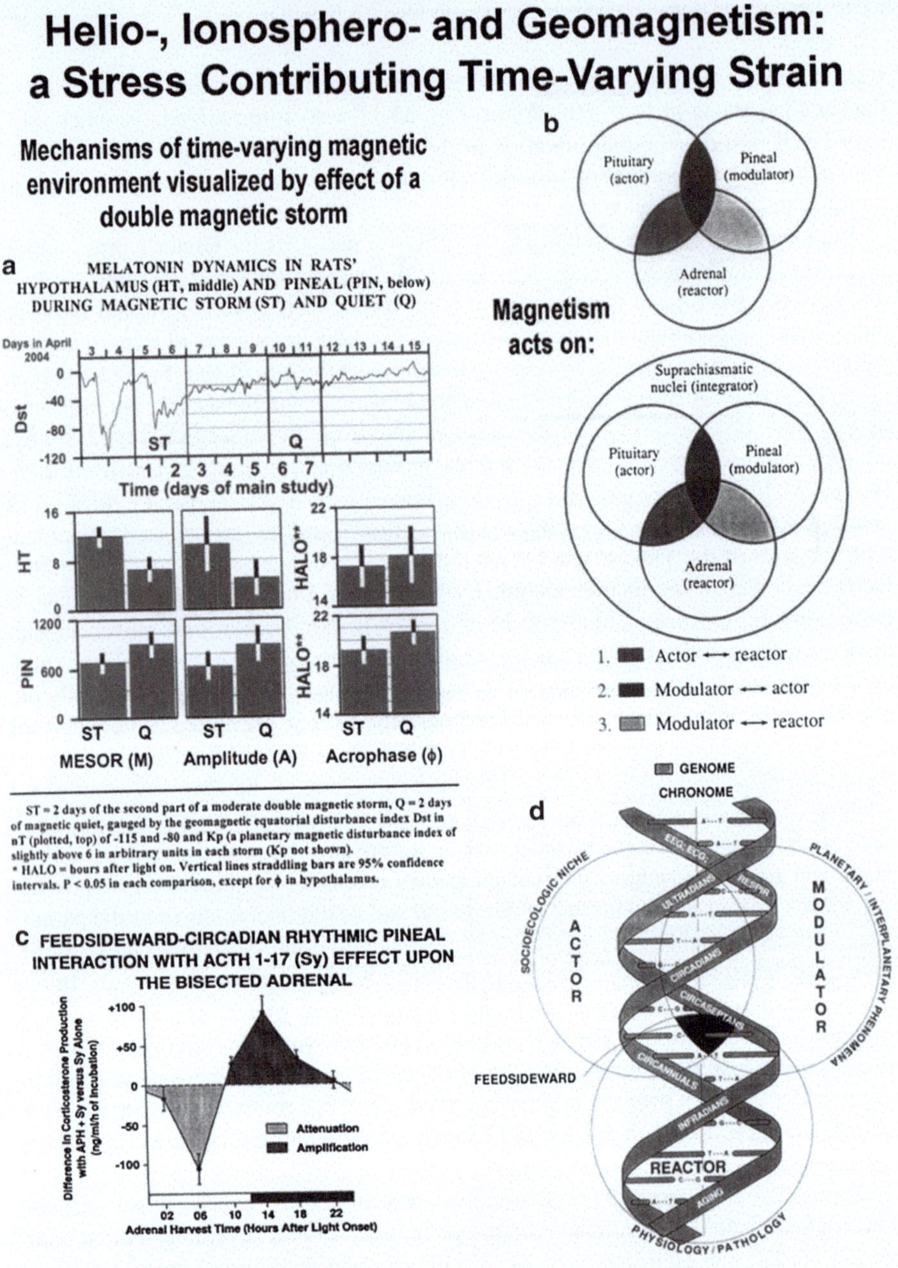

Fig. 23.11 (**a**) Magnetic storm, gauged by equatorial geomagnetic index Dst (*top*), is associated with an increase in MESOR and circadian amplitude of murine hypothalamic (HT) melatonin, a decrease in MESOR and circadian amplitude of murine pineal (Pin) melatonin, and an advance in circadian acrophase of both pineal and hypothalamic melatonin. (**b**) The pineal, pituitary, and adrenal are involved in feedsideward interactions, consisting of the response of a reactor to an actor being modified (modulated) by a third entity, the modulator, all entities being influenced by the environment. (**c**) Example of feedsideward interaction, where the addition of pineal homogenate to harvested adrenals changes their corticosterone production when stimulated by ACTH (produced

a putative signature of solar flares, again over half a century before it was reported by Rieger et al. in 1984 [97], as predicted by Charles Wolff [98] and confirmed by many other physicists (Table 23.6). Accordingly, we started to call attention to Stetson, an American scientific pioneer, as well as a popularizer (as Wikipedia has him), publishing, among other books, hard data on the topic of "Man and the Stars" [95], as well as to Scheiner and further to Charles Egeson [99, 100] and paid a first tribute to Bernhard and Gertraud Düll for providing the concept (Fig. 23.12h) [101; cf. 102], as well as data that, by their own analyses, confirmed by our meta-analyses, allowed them to differentiate the effect of space and terrestrial weather on suicide or other deaths from sensory or mental illness, on the one hand, vs. cardiovascular, respiratory, or all other diseases, on the other hand. Signatures of the cosmos with validated statistical significance of anticipated periods indeed have been found to characterize a host of variables related to physiology, pathology, and socio-ecology (Fig. 23.12i).

In retrospect, many pioneers acted in keeping with advice from Sir Norman Lockyer, the codiscoverer of helium and founder of the journal *Nature*: Surely in Meteorology, as in Astronomy, the thing to hunt down is a cycle, and if that is not to be found in the temperate zone, then go to the frigid zones or to the torrid zone to look for it, and if found, then above all things, and in whatever manner, lay hold of, study it, record it and see what it means [103]. Following his advice, a new spectrum of cycles emerged in the biosphere, providing the as-yet missing cyclic temporal frame to Vernadsky's explicitly linear (in time) noosphere [104] ("noosphere" = of the human mind), previously structured only in space by genetics. This chrono-noosphere was dubbed a chronousphere and documented by C-ABPM and meta-analyses of time series on individuals' and societies' health and disease [13, 105].

In Fig. 23.13a, we summarize all available periods in a range of long cycles found by meta-analyses of data collected during several professional lifetimes of others as well as my own, yielding overwhelming evidence for the transdisciplinary ubiquity of environmental-biospheric coperiodisms.[1] Cycles in human affairs, including bodily, economic, and military-political health and the latter's impairment, i.e., crime, terrorism (Table 23.7), and war, match periods in interplanetary and earth magnetism, as the peaks at about 20 and 10 years suggest when series covering hundreds and a very few thousands of years are included in a histogram (Fig. 23.13a–c).

Fig. 23.11 (continued) by the pituitary); there is a lack of effect, attenuation, or amplification depending on the circadian stage of the adrenals when pineal homogenate is added to ACTH. When a magnetic storm reaches an organism, its effect depends on the network's stage at exposure time. (**d**) Chronomodulation, a modulator (space-terrestrial weather) alters the (socio-)ecological conditions in the habitat (such as the synchronizing effect of the lighting regimen, the actor in the animal room) defining the state of the organismic hypothalamic-pineal, adrenal, and vascular network (the reactor). © Halberg Chronobiology Center

[1] Detected by rejecting with the nonlinearly extended cosinor, the "no-cycle" assumption (for the period fitted or a period near that fitted), with other methods, is used but in one exception.

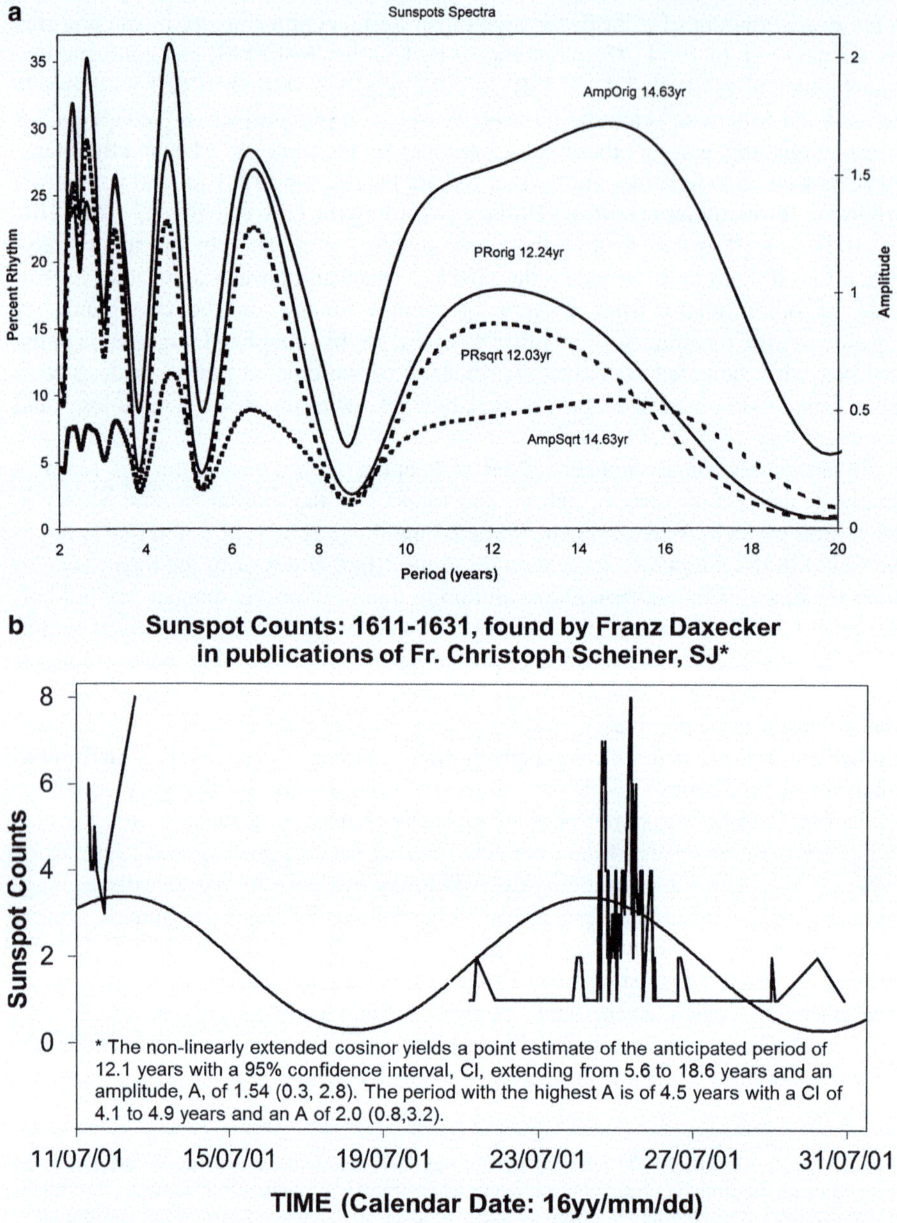

Fig. 23.12 (**a**) Least-squares periodogram of original and square root-transformed data on sunspot counts from Scheiner. © Halberg Chronobiology Center. (**b**) Fit to the original sunspot counts by Steiner of a 12.1-year cosine curve resolved by nonlinear least squares and any other components shown in Table 23.5. © Halberg Chronobiology Center. (**c**) With yearly counts of sunspot groups from 1826 to 1837, Schwabe did not explicitly refer to periodicity. Only after a second series spanning the years 1838 to 1843 did he (macroscopically) recognize a "periodicity" in print in 1844. Gauss' least squares were available in 1801. If the least-squares-based cosinor is applied separately to the 1826–1837 data, a statistically significant cycle can be demonstrated, as for his entire time series. © Halberg Chronobiology Center. (**d**) *Left*: Yearly sunspot data (with interruptions)

c Waiting for nearly 2 complete cycles before writing explicitly about a periodicity*

Periodicity ? NOT YET IN 1838 (1)

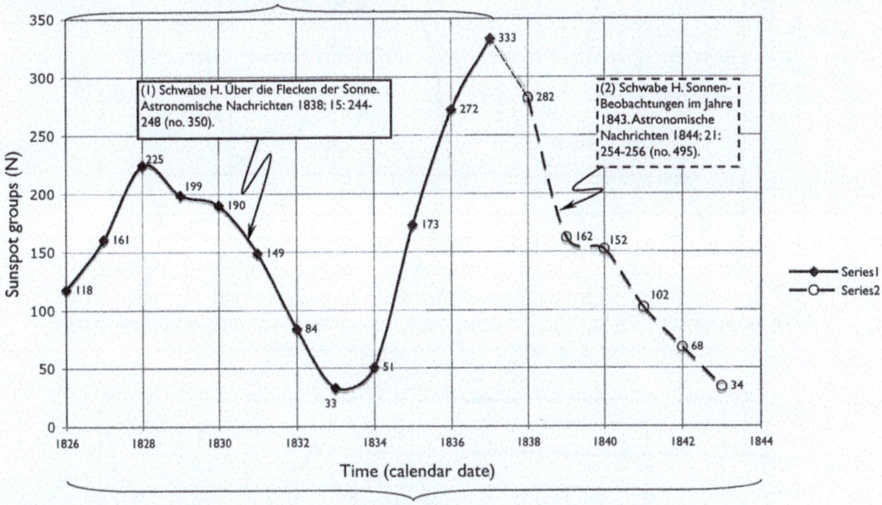

Periodicity ? YES in 1844 (2) AFTER SERIES 2 WAS ADDED

*Yes, for time-macroscopic eyeballing. "Vergleicht man nun die Zahl der Gruppen und der fleckenfreien Tage mit einander, so findet man, dass die Sonnenflecken eine **Periode** von ungefähr 10 Jahren hatten ..." (Schwabe 1844)

d

Horrebow's Data from 18th Century (top) Allow Detection (middle) and Model Fitting (bottom) of Circadecadal Sunspot Cycle

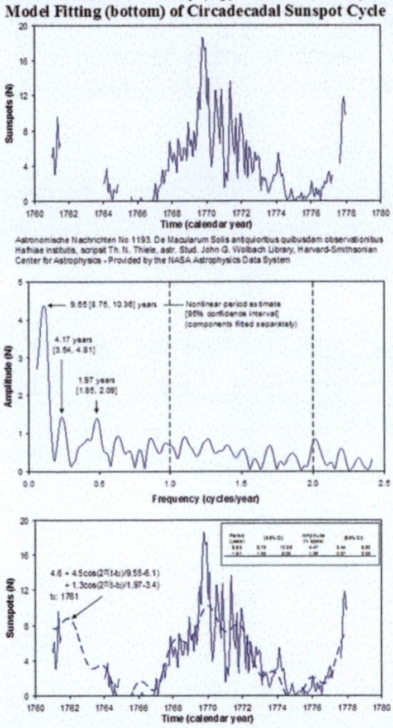

Using Gauss' Least Squares, Schwabe Could Have Validated a Periodicity in Sunspots with His First 9 Years of Data *

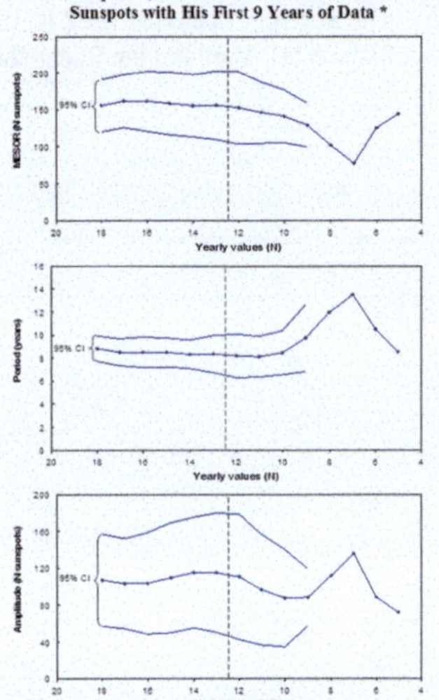

* Time series of yearly sunspot numbers (1826-1837 and 1838-1843) were combined for the nonlinear fit of a single cosine curve with a trial period of 11 years (that could not then but can now be anticipated) to all 18 values and after the progressive removal of the latest value. Nonlinear estimates of the MESOR (top), period (middle) and amplitude (bottom) with their respective 95% confidence intervals (95%CI) (when the zero-amplitude assumption could be rejected with P<0.05) shown as a function of the decreasing number of data indicate that relatively stable estimates are obtained with 9 or more years of data.

Fig. 23.12 (continued) from Horrebow (published by Thiele) (*top*) are characterized by three spectral peaks corresponding to periods of about 9.55, 4.17, and 1.97 years validated nonlinearly when considered separately (*middle*). The composite 2-component model is fitted to the data (*bottom*). *Right*: Results from nonlinear analyses of Schwabe's data analyzed over the entire 18-year

e

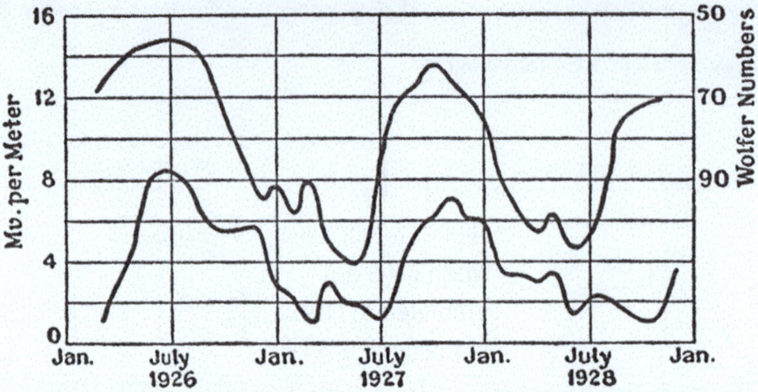

Radio reception for 1926 to 1928. The upper curve is the inverted curve of sunspot numbers. The lower curve is the strength of radio signals measured in microvolts.

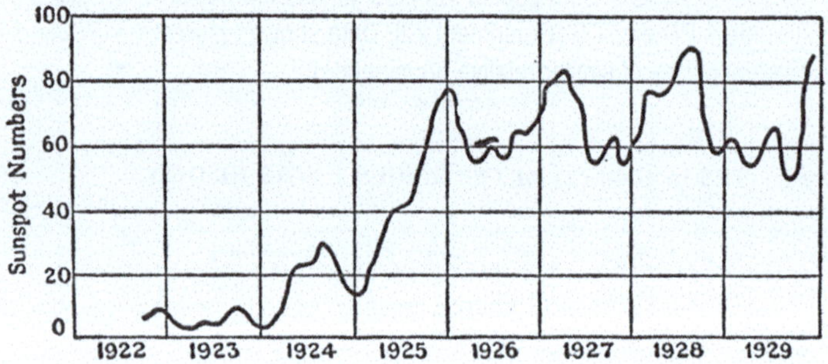

The sunspot cycle for 1922 to 1929 shows a 15-month secondary fluctuation.
From Stetson HT. Man and the Stars. New York: Whittlesey House, McGraw-Hill; 1930. p. 158.

f

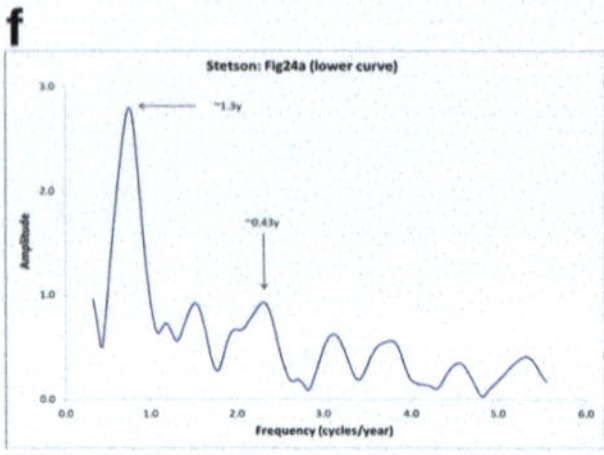

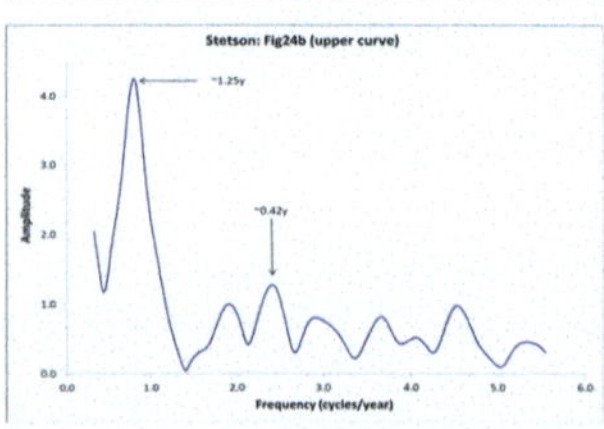

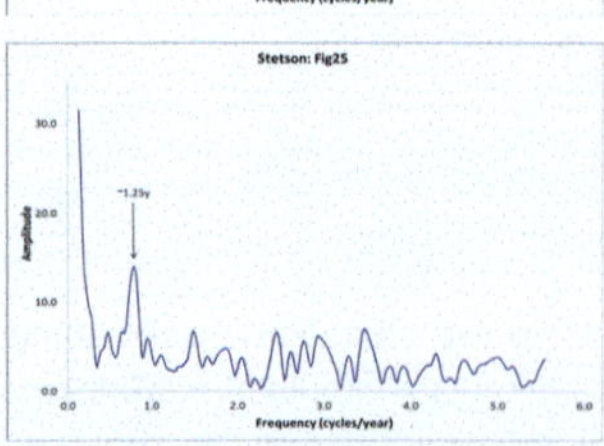

		Period	[95% CI]		Amplitude	[95% CI]		
Fig24a	Orig	1.33	1.25	1.41	2.77	2.00	3.54	
		0.44	0.41	0.47	0.87	0.13	1.60	
Fig24b	Orig	1.25	1.17	1.32	4.26	3.03	5.49	
Fig25	Orig	1.26	1.16	1.36	14.01	2.74	25.29	
	+LinTrend	1.43	1.31	1.55	9.52	2.58	16.46	
	+QuaTrend	1.38	1.31	1.45	10.75	5.63	15.88	
	+TerTrend	1.34	1.27	1.41	11.29	5.81	16.77	
		0.42	0.40	0.44	3.79	-1.70	9.28	
						1.31	6.27	1p

Fig. 23.12 (continued) span and on progressively shorter series obtained by removing each time the last yearly value, in an attempt to determine what is the shortest series suitable for an

g

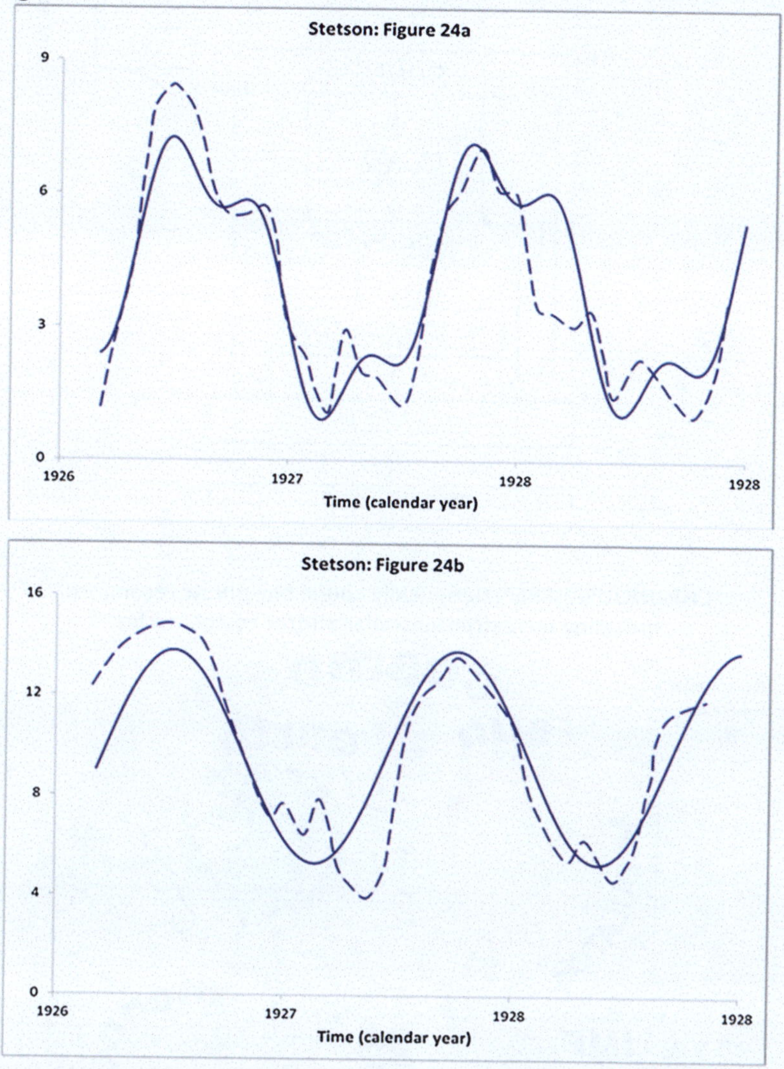

Fig. 23.12 (continued) assessment of the solar activity cycle at the time. Even with only the first 9 yearly data, it is possible to derive point and 95 % confidence intervals for all parameters. Results shown for the MESOR (*top*), period length (*middle*), and amplitude (*bottom*). © Halberg Chronobiology Center. (**e**) Original data from Harlan True Stetson: his Figs. 23.23 (a: lower curve; b: upper curve) and 23.25. Like Maurice Faure, Gaston Sardou and Joseph Vallot, he notes by the unaided eye a similar time course of the strength of the signal in radio communication and the number of sunspots and writes about implications for human affairs. (**f**) Least-squares spectra of data taken off (Figs. 23.23 and 23.24) of HT Stetson (see original published graphs in Fig. 23.12e). Validation by nonlinear least squares of transyear and cis-half-year components fitted separately or with a trend approximated by first- (LinTren), second- (QuaTrend), or third-order (TerTrend) polynomials. © Halberg Chronobiology Center. (**g**) Models resolved by linear-nonlinear rhythmometry fitted to data taken off published graph by HT Stetson. A transyear is detected with statistical significance for both curves in his published Fig. 23.23, accompanied by a cis-half year in the case of his lower curve (strength of radio signals). © Halberg Chronobiology Center. (**h**) Dülls' view of influence of space-terrestrial weather on human affairs. (**i**) Signatures of the cosmos with validated statistical significance of anticipated periods characterize: a. dozens of decade-long time series of human blood pressure and heart rate; b. other physiology and psychology, including human mental

h

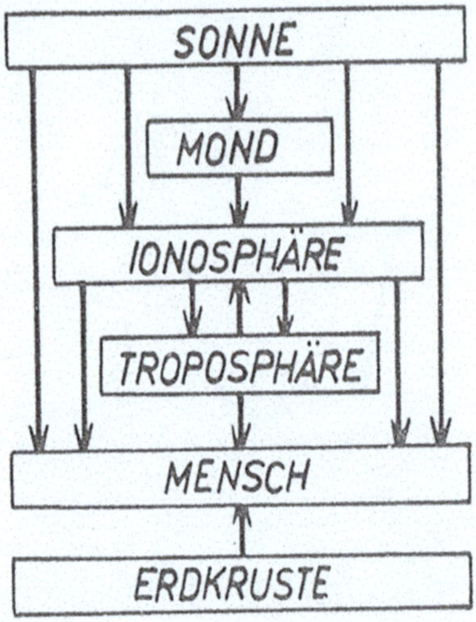

i

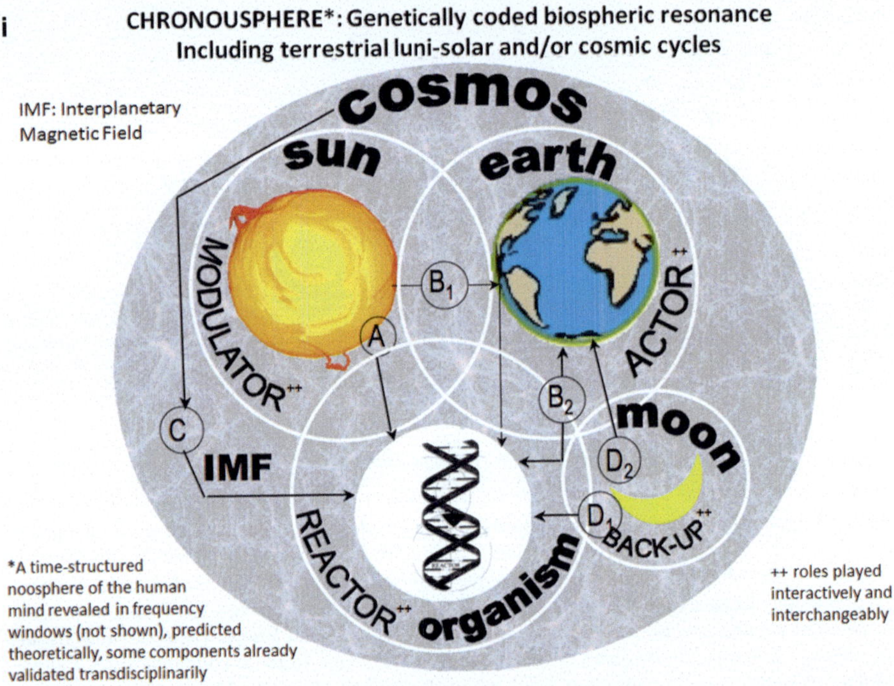

Fig. 23.12 (continued) functions; c. religious proselytism; d. suicide; e. sudden cardiac death; f. terrorist activity for the past 39 or 41 years; g. 2,556 years of international battles compiled by Raymond Holder Wheeler; h. military expenditures for training in nonmedical science; i. degrees earned; j. Gallup Polls; and k. political and military actions in nearly 200 years, meta-analyzed from the much broader treasure of data compiled by Alexander Leonidovich Chizhevsky. © Halberg Chronobiology Center

Table 23.5 Periods, τ, and corresponding amplitudes, A, each with its CI (95 % confidence interval) of sparse sunspot counts compiled from publications by Christoph Scheiner[a]

τ (years)	A (counts)
2.35 (2.24, 2.46)	1.37 (0.63, 2.10)
2.67 (2.53, 2.80)	1.39 (0.70, 2.08)
3.30 (3.10, 3.50)	1.47 (0.60, 2.33)
4.51 (4.11, 4.91)	2.04 (0.81, 3.27)
6.52 (5.43, 7.60)	1.60 (0.62, 2.58)
12.10 (5.60, 18.59)	1.54 (0.32, 2.76)

[a]Found by the nonlinearly extended cosinor as separately fitted components, one at a time, in sparse sunspot counts, compiled by Franz Daxecker from publications by Christoph Scheiner, SJ. Period of 12 years was in the range anticipated from the analyses of original counts (above) and after their square root transformation (latter results are similar; not shown). A period longer than 11 years was also suggested by our analysis of an extrapolation of available sunspot data to the span from 1611 to 1631 (the extrapolation was by Strestik J, Mikulecky M. An attempt to predict the Wolf sunspot numbers for the next decades using those observed so far [106]). Background in Daxecker F. Sonnenfleckenzyklen bei Christoph Scheiner. Eine retrospektive chronobiologische Untersuchung, Sammelblatt des Historischen Vereins Ingolstadt, in Druck, 2013

Table 23.6 Point estimates of a cis-half-year (~154-day) periodicity in solar flares, sometimes given with estimates of uncertainty and hypothesis testing

Period (in days except where noted)	Reference
154.3	Wolff CL. The rotational spectrum of g-modes in the sun. Astrophys J 1983; 264: 667–76.
154	Rieger A, Share GH, Forrest DJ, Kanbach G, Reppin C, Chupp EL. A 154-day periodicity in the occurrence of hard solar flares? Nature 1984; 312: 623–25.
158	Kiplinger AL, Dennis BR, Orwig LE. Detection of a 158-day periodicity in the solar hard X-ray flare rate. Bull Am Astronom Soc 1984; 16: 891.
152	Bogart RS, Bai T. Confirmation of a 152-day periodicity in the occurrence of solar flares inferred from microwave data. Astrophys J 1985; 299: L51-L55.
152–158	Dennis BR. Solar hard X-ray bursts. Solar Physics 1985; 100: 465–90.
152	Bai T, Sturrock PA. The 152-day periodicity of the solar flare occurrence rate. Nature 1987; 327: 601–04.
Near 155	Lean JL, Brueckner GE. Intermediate-term solar periodicities -- 100–500 days. Astrophys J 1989; 337: 568–78.
152	Özgüç A, Ataç T. Periodic behavior of solar flare index in solar cycles 20 and 21. Solar Physics 1989; 123: 357–65.
154 (±0.6)	Bai T, Cliver EW. A 154 day periodicity in the occurrence rate of proton flares. Astrophys J 1990; 363: 299–309.
Near 155	Carbonell M, Ballester JL. A short-term periodicity near 155 day in sunspot areas. Astron Astrophys 1990; 238: 377–81.

(continued)

Table 23.6 (continued)

Period (in days except where noted)	Reference
153	Dröge W, Gibbs K, Grunsfeld JM, Meyer P, Newport BJ, Evenson P, Moses D. A 153 day periodicity in the occurrence of solar flares producing energetic interplanetary electrons. Astrophys J Suppl Ser 1990; 73: 279–83. Applying Rayleigh test for periodicity.
155	Lean J. Evolution of the 155-day periodicity in sunspot areas during solar cycles 12 to 21. Astrophys J 1990; 363: 718–27.
155	Silverman SM. The 155-day solar period in the sixteenth century and later. Nature 1990; 347: 365–367. "[A]lthough values between 150 and 160 days have been reported, I refer to it here as the 155-day period, for convenience"
154	Bai T, Sturrock PA. The 154-day and related periodicities of solar activity as subharmonics of a fundamental period. Nature 1991; 350: 141–3.
154	Kile JN, Cliver EW. A search for the 154 day periodicity in the occurrence rate of solar flares using Ottawa 2.8 GHz burst data, 1955–1990. Astrophys J 1991; 370: 442–8.
152–158	Verma VK, Joshi JC, Uddin W, Paliwal DC. Search for a 152–158 days periodicity in the occurrence rate of solar flares inferred from spectral data of radio bursts. Astron Astrophys Suppl Ser 1991; 90: 83–7.
154	Bai T. The 77 day periodicity in the flare rate of cycle 22. Astrophys J 1992; 388: L69-L72.
Near 155	Carbonell M, Ballester JL. The periodic behaviour of solar activity: the near 155-day periodicity in sunspot areas. Astron Astrophys 1992; 255: 350–62.
152	Verma VK, Joshi JC, Paliwal DC. Study of periodicities of solar nuclear gamma ray flares and sunspots. Solar Physics 1992; 138: 205–08.
154	Bai T, Sturrock PA. Evidence for a fundamental period of the sun and its relation to the 154 day complex of periodicities. Astrophys J 1993; 409: 476–86.
~153	Cane HV, Richardson IG, von Rosenvinge TT. Interplanetary magnetic field periodicity of ~153 days. Geophys Res Lett 1998; 25: 4437–40.
158	Oliver R, Ballester JL, Baudin F. Emergence of magnetic flux on the Sun as the cause of a 158-day periodicity in sunspot areas. Nature 1998; 394: 552–3 | doi:10.1038/29012
Near 158	Ballester JL, Oliver R, Baudin F. Discovery of the near 158 day periodicity in group sunspot numbers during the eighteenth century. Astrophys J 1999; 522: L153-L156.
151–155	Lou YQ. Rossby-type wave-induced periodicities in flare activities and sunspot areas or groups during solar maxima. Astrophys J 2000; 540: 1102–8.
153.9	Hady AA. Analytical studies of solar cycle 23 and its periodicities. Planetary and Space Science 2002; 50: 89–92.
156	Krivova NA, Solanki SK. The 1.3-year and 156-day periodicities in sunspot data: wavelet analysis suggests a common origin. Astron Astrophys 2002; 394: 701–6.
Near 160	Ballester JL, Oliver R, Carbonell M. The near 160 day periodicity in the photospheric magnetic flux. Astrophys J 2002; 566: 505–11.

(continued)

Table 23.6 (continued)

Period (in days except where noted)	Reference
Near 5 months	Han Yanben, Han Yonggang. Time variation of the near 5-month period of sunspot numbers. Chinese Sci Bull 2002; 47 (23): 1967–73. "Many scholars … found similar periods of solar activity from other observations of the Sun. However, these periods are different, such as about 152-, 154-, 156-day, etc. Here we name it a near 5-month period (N5MP) since it is not definite."
154	Prabhakaran Nayar SR, Radhika VN, Revathy K, Ramadas V. Wavelet analysis of solar, solar wind and geomagnetic parameters. Solar Phys 2002; 208: 359–373. "The global wavelet spectra of these parameters …exhibit the presence of a variety of prominent quasi periods around 16 years, 10.6 years, 9.6 years, 5.5 years, 1.3 years, 180 days, 154 days …"
153	Bai T. Periodicities in solar flare occurrence analysis of cycles 19–23. Astrophys J 2003; 591: 406–15.
Near 160	Ballester JL, Oliver R, Carbonell M. Return of the near 160 day periodicity in the photospheric magnetic flux during solar cycle 23. Astrophys J 2004; 615: L173-L176.
151–158	Knaack R, Stenflo JO, Berdyugina SV. Evolution and rotation of large-scale photospheric magnetic fields of the Sun during cycles 21–23: Periodicities, north–south asymmetries and r-mode signatures. Astron Astrophys 2005; 438: 1067–82.
~152	Chowdhury P, Ray PC. Periodicities of solar electron flare occurrence: analysis of cycles 21–23. Mon Not Roy Astronom Soc 2006; 373: 1577–89.
155	Chowdhury P, Ray PC, Ray S. Periodicity of 155 days in solar electron fluence. Ind J Phys 2008; 82: 95–104.
150–160	Dimitropoulou M, Moussas X, Strintzi D. Enhanced Rieger-type periodicities' detection in X-ray solar flares and statistical validation of Rossby waves' existence. Mon Not R Astron Soc 2008; 386: 2278–84.
158	Javaraiah J, Ulrich RK, Bertello L, Boyden JE. Search for short-term periodicities in the sun's surface rotation: a revisit. Solar Phys 2009; 257: 61–9.
155	Vaquero JM, Trigo RM, Vazquez M, Gallego MC. 155-day periodicity in solar cycles 3 and 4. New Astronomy 2009. doi: 10.1016/j.newast.2009.11.004
155–160	Zaqarashvili TV, Carbonell M, Oliver R, Ballester JL. Magnetic Rossby waves in the solar tachocline and Rieger-type periodicities. arXiv:0911.4591v1 [astro-ph.SR] 24 Nov 2009
154	Fischbach E, Jenkins JH, Buncher JB, Gruenwald JT, Sturrock PA, Javorsek D II. Evidence for solar influences on nuclear decay rates. arXiv:1007.3318v1 [hep.ph] 20 Jul 2010

This incomplete list (added contributions are cited in the references provided) suffices to indicate the variable nature of the period being discussed. Specifications of the decadal solar (Horrebow-Schwabe) cycle in which they are found point indirectly to the intermittency of the components being discussed

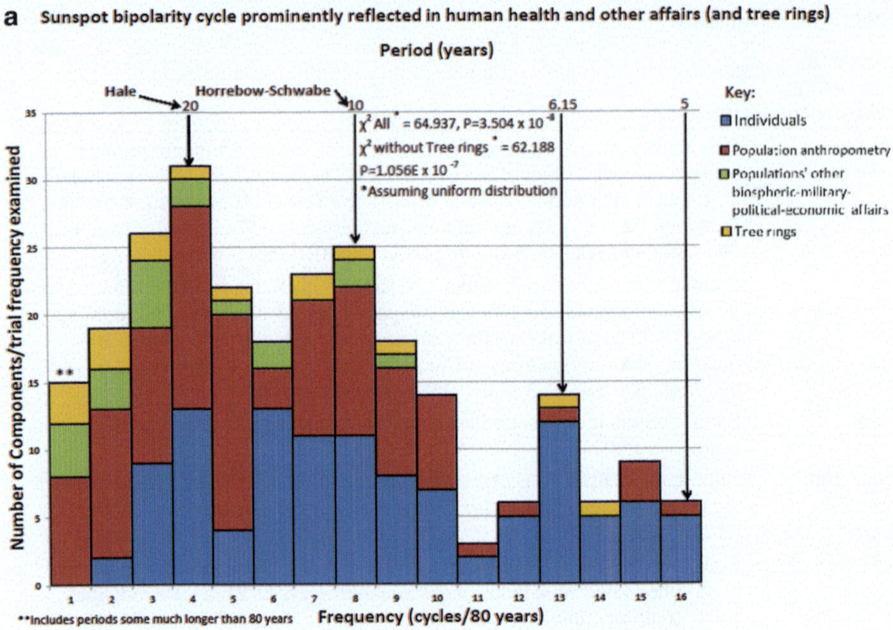

Fig. 23.13 (**a**) As meta-analyses of longer time series are added to the pool summarized in Fig. 23.5c, the distribution of best-fitting periods of long cycles in individuals and populations confirms a peak of cycles with periods of about 10 years, the Horrebow-Schwabe cycle of sunspot counts, and reveals an even greater peak at 20 years, near the Hale sunspot bipolarity cycle (the sunspots change polarity, with the start of each new cycle). While the didecadal and decadal peaks are prominent, there are also cycles with other periods, in keeping with the possibility that some spectral components may free-run without being driven by a concurrent environmental counterpart, a possibility that exists also for (congruent by definition) coperiodisms, i.e., in case of biospheric and environmental cycles with overlapping CIs of their periods. © Halberg Chronobiology Center. (**b**) Both near- and far-transyears are detected with statistical significance in a 28.8-year record of self-measurements by a man (SBS) 64 years of age at start of monitoring, in the absence (of detection) of a calendar year despite the large annual variations in temperature and day length in Minnesota, where SBS lives. Note similarity and differences in terms of period lengths and 95 % confidence intervals among different variables measured concomitantly. A clear separation of coexisting near- and far-transyears is noteworthy in SBS and in his wife's (MIS) data in Fig. 23.13c. SBS began using antihypertensive medication on Mar 1, 1988, about 4 years after the start of self-measurements on Feb 27, 1984. © Halberg Chronobiology Center. (**c**) Both near- and far-transyears are detected with statistical significance in a 19.7-year record of self-measurements by a woman (MIS) 66 years of age at start of monitoring, in the absence of a calendar year detectable by the method used, despite the large annual variations in temperature and day length in Minnesota. Note similarity and differences in terms of period lengths and 95 % confidence intervals among different variables measured concomitantly in the same person's blood circulation. MIS began using antihypertensive medication at the start of self-measurements on Feb 27, 1984. She was institutionalized with a diagnosis of progressive dementia from Dec 31, 2000, to Nov 11, 2003, during which family members continued to measure her blood pressure twice a day. © Halberg Chronobiology Center

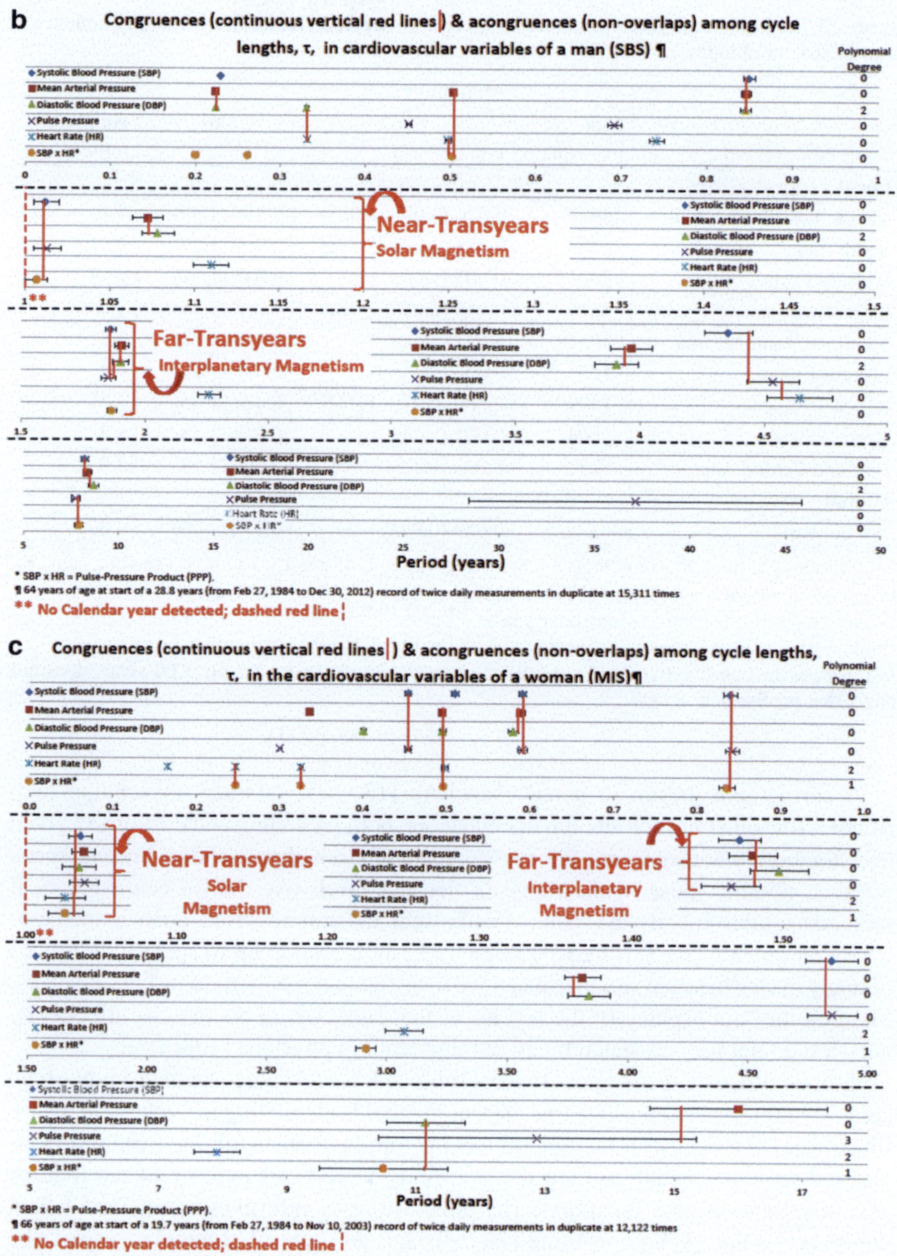

Fig. 23.13 (continued)

Table 23.7 Mostly transyear prevalence in spectra of terrorism incidence gauged by annual vs. para-annual amplitudes

Geographic region	N of incidents	Amplitude[a]		Amplitude ratio[b]
		$\tau_1 \equiv 1$ year	$\tau_2 > 1.061$, < 1.75	
Local				
Middle East/Persian Gulf	16977	0.026 (0.996–1.023)	0.216 (1.39)	8.85
South Asia	6464	0.073 (1.006–1.122)	0.069 (1.47)	0.91
Western Europe	5801	0.007 (valley[c])	0.045 (1.71)	6.43
Latin America	3816	0.016 (0.985–1.022)	0.047 (1.24)	2.94
Southeast Asia/Oceania	3051	0.021 (valley)	0.059 (1.72)	2.81
Eastern Europe	1652	0.005 (valley)	0.032 (1.18)	6.4
Africa	1499	0.009 (0.986–1.024)	0.021 (1.05)	2.33
North America	604	0.005 (valley)	0.011 (1.58)	2.2
East and Central Asia	259	0.002 (valley)	0.006 (1.68)	3.0
Global				
Whole world	40122	0.089 (0.973–1.007)	0.251 (1.70)	2.82

[a]Amplitudes (A, measures of extent of length) at periods, τ, of exactly 1 year (τ_1) and at a peak, τ_2, in a spectral region longer than 1.0 year and shorter than 1.75 year (transyears)
[b]Ratios of As of transyears vs. calendar years
[c]Valley: no confidence regions according to Marquardt for any As overlapping 1 year
In this case the conservative 95 % confidence interval of period near exactly 1.00 years, does not cover that precise year length

A cartography begun in detail elsewhere [13, 105] provides a planning map (if not a sine qua non) for all human endeavors, both preceding and during any given investigation, whether or not the cycles are signatures of past and/or present space weather, notably those in the range of decades. Well over half a century ago, it seemed useful to specify the spans of daily light and darkness for studies in the laboratory on rodents [107]. It may be immensely more important to specify the stages of many cycles in space weather when we investigate not only rodents but ourselves, living in the “atmosphere of the [cycles of the] sun,” where we may be affected by what we do not see , as stated by an opinion leading physicist [108, 109].

Among others, I also seemed to have followed, for 64 years, the path Sir Norman suggested for his meteorology and astronomy, fields abounding in cycles that, as we illustrated [13], deal with factors influencing various human affairs, from individuals’ and societies’ health to their disease. Sixty-three years ago, I realized that for any investigation of a biospheric rhythmic variable, information on what I then dubbed circadian cycles constituted an indispensable control, if blunders were to be avoided (Fig. 23.14a–d) [107]. I could postulate, but not fully anticipate, on the positive side, that the rewards of the gains of more energy release from atomic fission and fusion may be matched by gains in human health and happiness from splitting the normal range and integrating information gathered at different levels and different frequencies (Fig. 23.1a, 23.5c, and 23.13a).[2] I was, however, sufficiently awed by the lessons from Fig. 23.14a–d to devote my whole professional lifetime to the topic and to give priority thereto at the risk of one's job, already at Harvard and repeatedly in Minnesota, where I began research, thanks to a major mentor, Maurice B. Visscher, and was fortunate to be supported by him, honest disagreements notwithstanding, and by my late wife Erna.

The realization of the nonsense (opposite) inferences one can draw from the same data in replications some hours or days apart in the case of failing to consider circadian rhythms came most clearly and concomitantly from solving two puzzles, preventing me from publishing blunders – one can expect if differences in phase (mice feeding at different times) (Fig. 23.14a) (and much less so in Fig. 23.14b), or frequency (Fig. 23.14c, d) (after blinding of mice), are not considered in what we are doing. In the earliest 1950s, I compared a group of blinded mice (presumably no longer locked into the synchronizer of a 12-hourly alternation of light and darkness after loss of the transducers of the light, the eyes) with sham-operated controls, for which the lighting regimen prevailed as a synchronizer (clock time giver). The results up to day 6 in Fig. 23.14d[3] were shown to the well-meaning mentor, to whom I pointed out that at the last measurement on day 6, blinded mice had a core temperature increase, while the rectal temperature of controls still decreased, in keeping with a probable intergroup difference in period. He dismissed the observation of a shortening of period as a chance phenomenon and, presumably with the good intention of preventing further loss of my time, equated my intent to pursue "circadian" free-running to "paranoia." (The mentor failed to realize the implications of the fact of comparing two groups with a slightly different period; Fig. 23.14c and Sect. 1 in Fig. 23.14d.) By that time, I had already lost my mouse room at the university and yet had found an opportunity to do that work at the Cambridge State Hospital. Thus, at odds with the mentor's wishes, we kept measuring (Fig. 23.14d, Sect. 1). It took until day 22 to see by the unaided eye that the two groups reached (a transient) antiphase, and of course now we are convinced to continue measurements around the clock for years, manually with much help from motivated patients at the State Hospital. Neither the mentor's valued judgment nor the community of medical or scientific investigators today might fully assess the shortcomings of a time-

---

[2] This term, under the mantle of homeostasis, currently implies equating "physiological" with "negligible if not trivial," in summarizing every laboratory or clinical determination by looking only at whether values are inside or outside the range of a homeostatic "physiological" variability.

[3] With respect to the study that led me to propose the "free-running" (on the basis of Earl E. Bakken's comparison with endogenous rhythms with a free-running oscillator), I was forbidden by my department head, a helpful mentor who gave me a chance to come to Minnesota, to work on the effect of human bilateral optic enucleation and lost my university animal laboratory. After handing in my resignation (supported by my wife), I went ahead and did the study anyway in an old paint room at a state hospital ~60 miles from the Twin Cities. I resigned again when I was forbidden to study adrenalectomized patients, feeling it was a unique opportunity in the history of medicine (in the absence of institutional review boards). Seeing my determination, the department head relented and gratefully accepted coauthorship on reports of results, as seen in my bibliography from the 1950s.

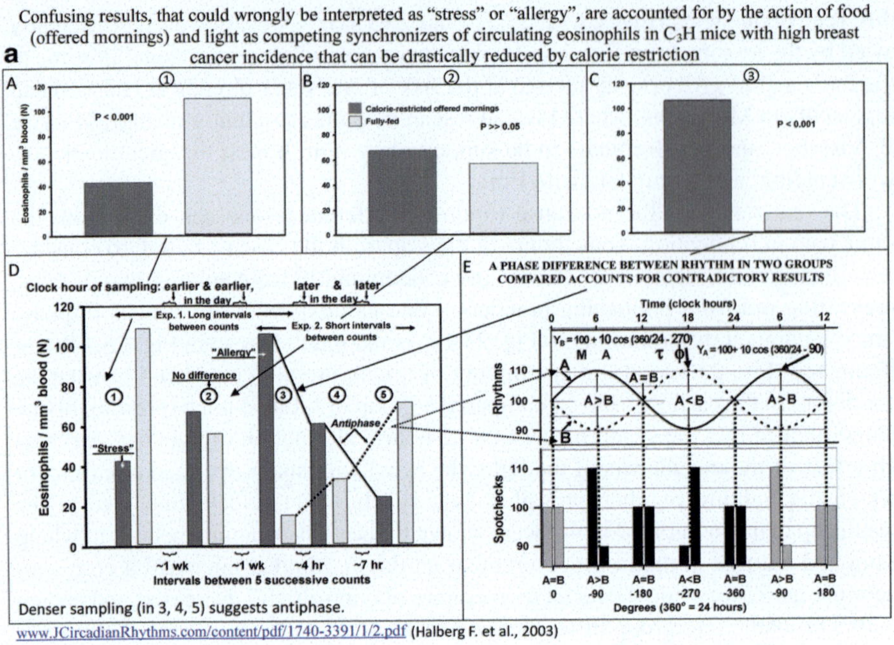

Fig. 23.14 (**a**) Importance of rhythms in assessing intervention effects, illustrated in relation to stress or allergy. A. Eosinophil counts seem to be lowered by fasting (and/or stress), when a 50 % reduction in dietary carbohydrates and fats (with proteins, vitamins, and minerals similar to control group) was fed in the morning to C_3H mice (*dark column*). (In this model, the naturally high incidence of breast cancer is lowered by a diet reduced in calories and by ovariectomy, not shown.) The result could have been interpreted as an adrenocortical activation and then assessed by eosinophil depression, with applications for treating breast cancer and for prolonging life. Steroids that depress eosinophil cell counts and perhaps mitoses could be a mechanism through which caloric restriction and ovariectomy act in greatly reducing cancer incidence. This tempting inference was never published. (**b**). In view of the importance of this finding for the etiology of cancer, results were replicated on a larger group of animals; one week later, a follow-up study with more animals starting at an earlier clock hour yielded confusing results, showing no statistically significant difference between the two groups of mice. (**c**). After another week, another study starting at an even earlier clock hour yielded results opposite to those in the first experiment when considered alone. These findings in C in themselves could have been interpreted as an allergic response, certainly contrary to the "stress" response in A. D. Sampling at intervals of a few hours in the third study, the stages called 4 and 5 hinted at the reason for the confusion: by sampling at different clock hours, two groups of mice were found to be characterized by a circadian rhythm with different phases. Opposite effects thus became predictable. E. Abstract illustration of two circadian rhythms in antiphase. Differences in opposite direction or no effect are then anticipated from sampling at different clock hours. © Halberg Chronobiology Center. (**b**) Effect of food restriction on circulating eosinophil counts in mice. Follow-up study on Fig. 23.10a, with a phase difference greatly reduced by offering the restricted diet in the evening. A. Even after $\log_{10}$ transformation of the data expressed as percentage of mean, great interindividual variability is apparent in the raw data. B. Plots of timepoint mean for mice in each group reveal different circadian patterns. C–E. Parameter tests quantify differences, indicating that calorie restriction is associated with a lower MESOR (C–D), a larger circadian amplitude (D–E), and only a slight difference in acrophase (**d**). The difference in acrophase in this study, where calorie-restricted mice were fed in the evening, is much smaller

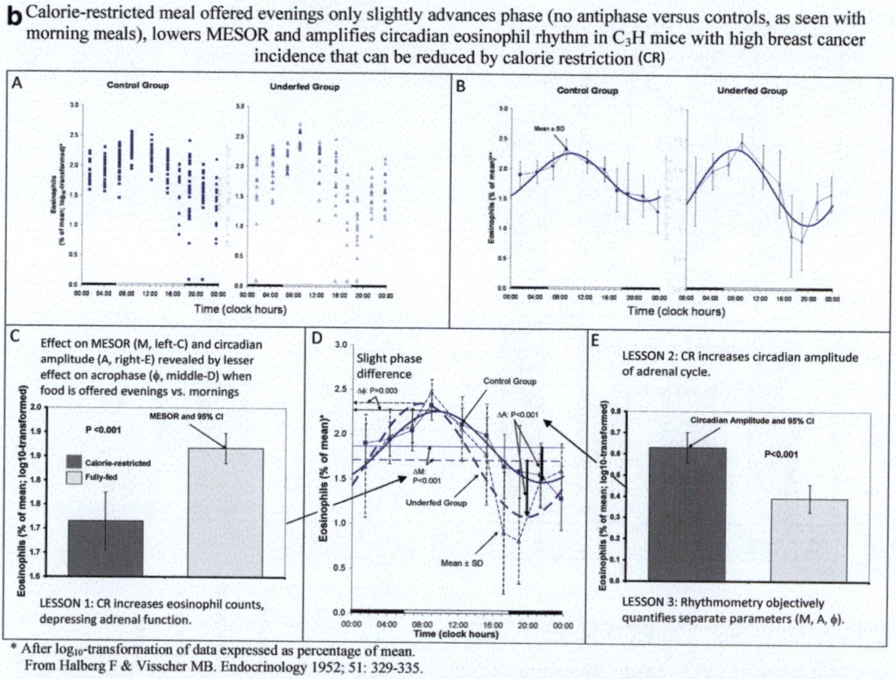

Fig. 23.14 (continued) than the almost antiphase observed in prior studies (Fig. 23.10a), where calorie-restricted mice were fed in the evening. © Halberg Chronobiology Center. C. Importance of assessing the rhythm's period. Whereas circadian rhythms are usually synchronized by environmental schedules, they may free-run, for instance, in mice without eyes, assuming a period that deviates slightly but statistically significantly from precisely 24 hours. When free-running occurs, results from two-timepoint studies may yield spurious results. *Top left*: Illustration for circulating eosinophil counts of mice sampled twice a day, during daytime and nighttime, after sham operation or bilateral optic enucleation. Results from 4 different (non-consecutive) days post-operation are shown. Whereas sham-operated mice display a consistent pattern of higher daytime and lower nighttime counts, blinded mice exhibit erratic patterns, the nighttime counts being sometimes higher than, sometimes lower than, and sometimes almost the same as daytime counts. *Bottom left*: Abstract illustration of two components with slightly different periods accounts for the changing relation between the two curves depending on whether they are in or out of phase with one another. *Right*: Time course of an intergroup difference between synchronized control and desynchronized experimental animals – when comparisons are made 24 hours apart, at one or other clock hour. A given physiologic function is assumed to be circadian periodic in both groups compared. One function could be 24-hour synchronized, while the other could differ in period by 160 minutes (case 3). On the plot, both functions start out in phase at 06. The difference between the two groups undergoes large changes with time, simply as a function of the particular clock hour chosen for observation. © Halberg Chronobiology Center. D. *Top*, section I: Desynchronization of circadian rhythm in core temperature of mice after blinding, seen time macroscopically in IA, here leads, in IB, to time microscopy with a chronobiologic serial section showing a different time course of the core temperature acrophases, ϕ, with early separation of the two groups by nonoverlapping 95 % confidence intervals of ϕ in IC, to a summary of individual periodograms that form two separate distributions, and in ID to time relations among three variables in a 24-hour synchronized (*top*) or free-running (*bottom*) system (of mice, *left*, and of a human, *right*). Section II shows a spontaneous (alpha) rhythm in circulating corticosterone of mice in antiphase with the slope of

c

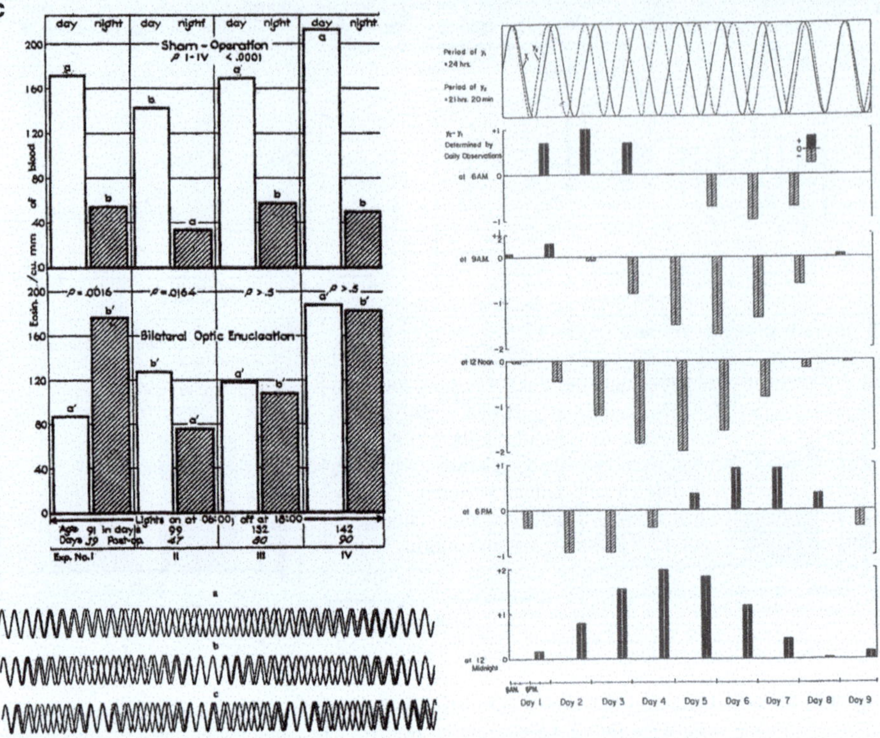

Fig. 23.14 (continued) an in vitro response rhythm to ACTH, a reactive (beta) rhythm. The components of the chronome (time structure) are internally coordinated through feedsidewards in a network of rhythms that are more or less spontaneous (alpha), others primarily reactive (beta) or modulatory at a single mapped frequency, such as a circadian (gamma), IIC, and IID, or at multiply mapped (delta) frequencies, IIE. The effect of one entity (the actor) upon a second (the reactor), such as the pituitary acting upon the adrenal cortical corticosterone production, may be influenced, predictably insofar as rhythmically, by a third entity such as melatonin (the modulator) at the level of the pituitary, and the same melatonin may act directly upon the adrenal. Reproducible sequences of attenuation, no effect, and amplification, the time-qualified feedsidewards, replace time-unqualified feedbacks and feedforwards (IIC to E). In sections II and III, feedsidewards include the interaction of a modulator (such as ACTH) upon an actor (such as adrenocortical corticosterone production) acting upon DNA labeling in bone (the reactor). The roles played by endocrines can and do change in various feedsidewards that replace time-unqualified feedbacks and feedforwards. Chronomolecular mapping of circadian acrophases has also begun. © Halberg Chronobiology Center. (**e**). Correlation analysis should be avoided when comparing variables sharing the same cycle: depending on the phase difference between the two variables being compared (none, anti-phase, or quadrature), spurious results are obtained as a statistically significant positive (*left*) or negative (*middle*) association, or the lack thereof (*right*). © Halberg Chronobiology Center

d

ENDOGENOUS TIME STRUCTURE (CHRONOME) OF INTERNALLY COORDINATED FREE-RUNNING RHYTHMS (TOP) THROUGH FEEDSIDEWARDS IN NETWORK OF SPONTANEOUS (α), REACTIVE (β) AND MODULATORY (γ, δ) RHYTHMS (BOTTOM)

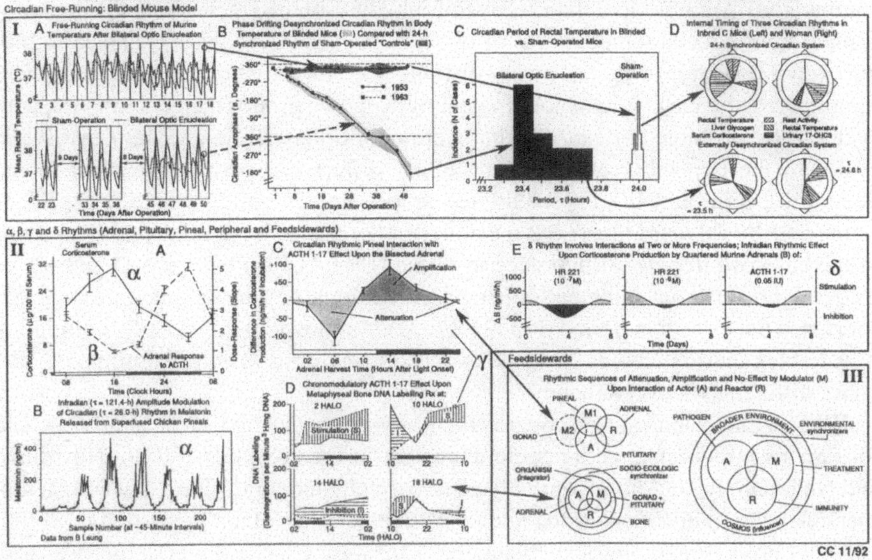

F09

e

ALL THESE NONSENSE CORRELATIONS OCCUR WHEN THE TWO VARIABLES SHARE A CYCLE OF THE SAME PERIOD BUT WITH DIFFERENT PHASE RELATIONS

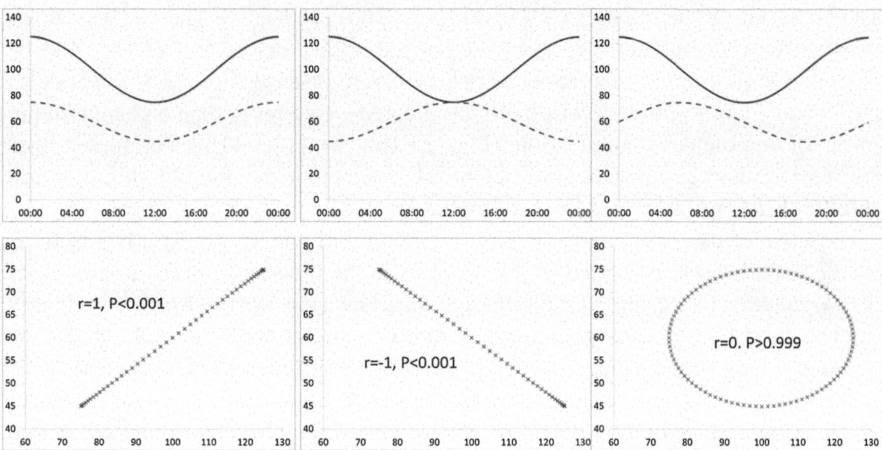

A difference in phase of the circadian rhythm in eosinophil counts brought about by competing synchronizers (light and food) contributed to the birth of chronobiology. Test statistics other than the Pierson product-moment correlation are advocated so that nonsense correlations and inferences drawn from them are avoided.

Fig. 23.14 (continued)

unqualified arbitrary "baseline," for instance, in order to gain a reference base for comparisons. With no less than 6 days of mostly 4-hourly measurements made for comparing the rectal temperatures of two groups of mice, without consideration of the overall behavior of cycles, a very distinguished investigator misinterpreted free-running as a chance phenomenon. In dealing with two groups being compared, with cycles with the same two periods that were out of phase (Fig. 23.14a) or with two cycles in the two other groups that slightly differed in period (Fig. 23.14c and Section I in Fig. 23.14d), one could necessarily find "no intergroup difference" or statistically highly significant but meaningless "opposite" results, a few hours apart, in the case of circadian differences in phase (Fig. 23.14a) (sections D and E), or weeks apart, after an intervention, such as blinding, that in the mice investigated shortened the circadian period (Fig. 23.14c, d). As shown in Fig. 23.14e, correlation of variables sharing a common cycle can lead to spurious results determined by the phase difference between them.

When Otto Hahn, the discoverer of fission, saw the first nuclear process and foresaw its importance, all he could utter was "If this is true …." If not Hiroshima and Nagasaki, the heretofore unrealized release of energy may have dawned on him (personal communication by the late Erika Cremer, then Hahn's assistant in Berlin, and later professor of physical chemistry in Innsbruck). The senior author was also in awe when he visualized the vast amount of misinformation made possible without the study of cycles, realizing that for finding mechanisms, just as the atom must be split to release energy (Fig. 23.15a) (top), so must the physiological range be split (Fig. 23.15b), to start within the circadian domain for new diagnoses, e.g., of vascular variability anomalies (VVAs) (Fig. 23.16a) [110–114]. These VVAs, notably when they persist and become vascular variability disorders (VVDs), can be associated with dramatic increases in risk of severe disease (Fig. 23.15a) (middle) that, in some cases, can be cost-effectively treated by no more than a change in the timing of hypotensive medication (Fig. 23.16b, c) [110–115]. Parameter tests (Fig. 23.16b) and cumulative sum (CUSUM) control charts (Fig. 23.16b, d, e) are useful tools to optimize the timing of treatment for the individual patient [115].

The broader use of the circadian chronotherapy of cancer can be advocated; as timed chemotherapy, it has led to 300 % gains in the laboratory in the treatment of L1210 leukemia [116], and as radiotherapy timed by a marker rhythm, it has already doubled the 2-year disease-free survival rate of patients with perioral cancers by circadian focus alone (Fig. 23.16f) [117, 118]. Further research on endocrine manipulation based on feedsideward mechanisms is warranted in view of results in the experimental laboratory (Fig. 23.16g). Chronoradiotherapy remains a major challenge, as does a further circaseptan optimization (Fig. 23.16h) that led Waldemar Ulmer to the atomic [119] and eventually to the quantum level to account for his prominent circaseptans, the phase of which affected the outcome of radiotherapy [120]. A cycle with an about-7-day (circaseptan) period in the heart rate (HR) of five humans resonates with solar activity; it is a coperiodism insofar as the HR's amplitude, A, at the about-7-day period is amplified in the presence of corresponding periods in rate of change of sunspot area and damped but not lost in the latter's absence. It is likely built into living matter by evolution (Fig. 23.16i) [121].

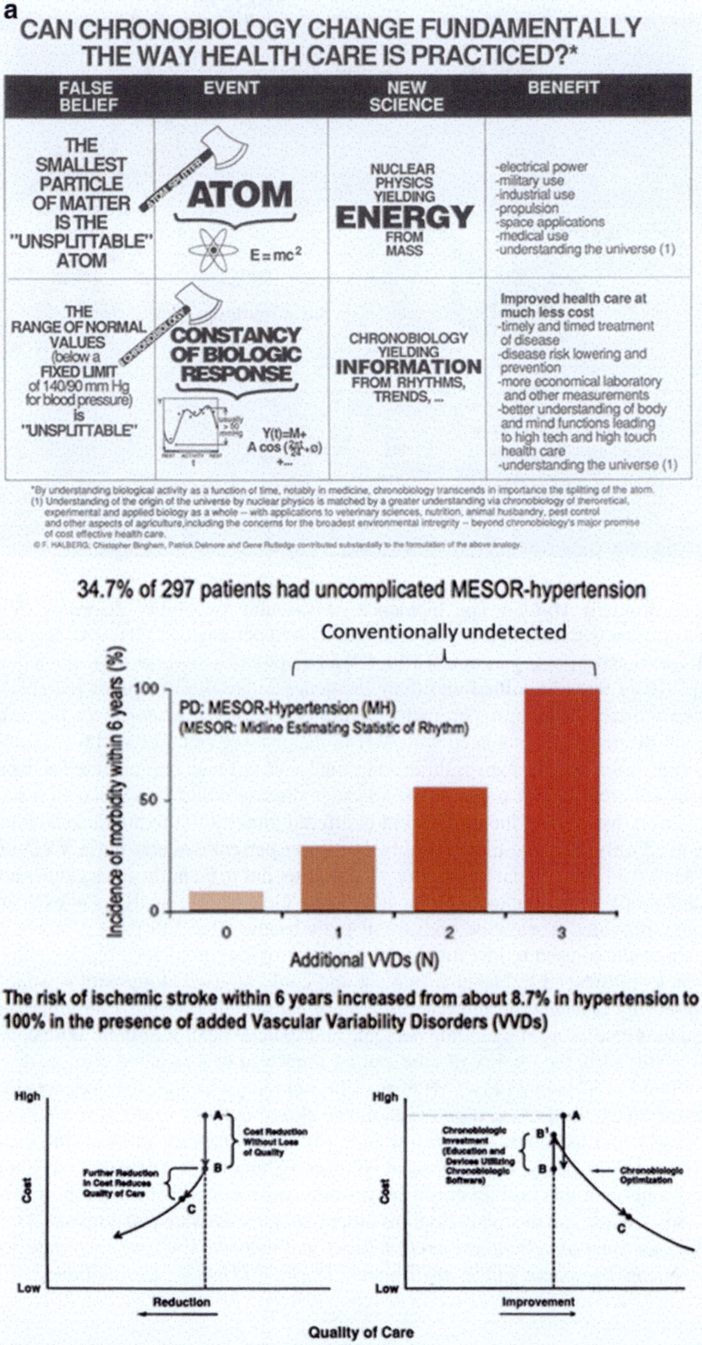

Fig. 23.15 (**a**) *Top*: Resolving chronomes (broad time structures) within the physiological range, viewed as a realm of relative monitoring (by the misconception of homeostasis), can be compared to splitting the atom, once thought to be the smallest indivisible particle of matter.

b

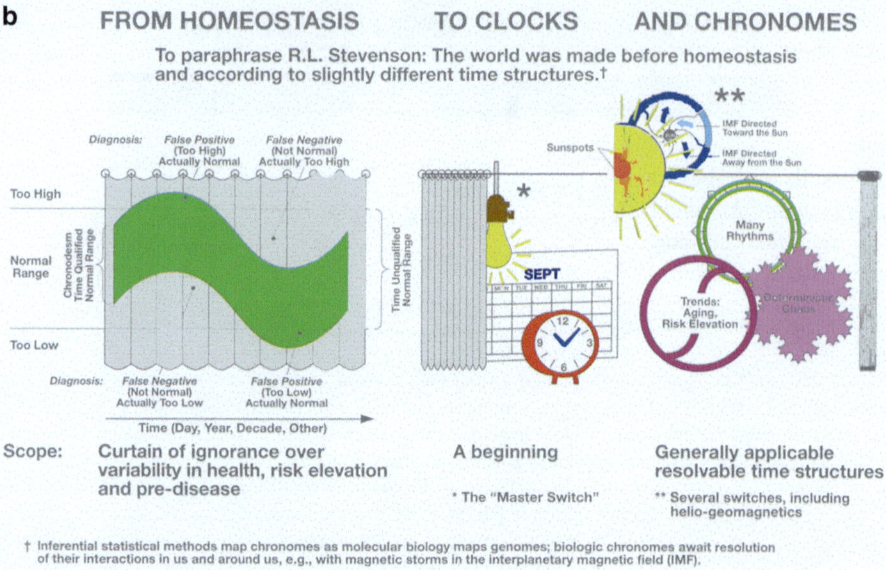

Fig. 23.15 (continued) *Middle*: The incidence of vascular variability disorders (VVDs) was assessed in a clinic population of 297 patients. MESOR-hypertension (MH) was diagnosed in 176 patients. MH, excessive pulse pressure (EPP), CHAT (a circadian overswing), and a deficient HR variability (DHRV) were identified and their incidence related to outcomes (cerebral ischemic attack, coronary artery disease, nephropathy, and/or retinopathy), assessed for the next 6 years. Cardiovascular disease risk associated with MH in the absence of other VVDs is much less than when it is complicated by one, two, or three additional VVDs. Once implemented across the board rather than in selected patient populations, vascular disease could be curbed to a much larger extent at relatively low cost if the monitoring is offered directly to the public and care providers become involved only after the more reliable, but never perfect detection of a VVD in a second survey of 7 days. An international multilingual data center has to be built, among other approaches, to educate and motivate the public at large, and to provide cost-free analyses in exchange for the data, as is now provided worldwide on a small scale by the BIOCOS project (corne001@umn.edu). The data could be used to investigate the greater problem of societal disease, such as crime and terrorism, as influenced by the environment and could be used along with seismic and other physical monitoring to detect the antecedents, e.g., of natural cataclysms. A multipurpose alerting system could thus result for safeguarding personal and societal health. *Bottom*: With education and self-help in health care, the quality of care can be improved at a reduced cost, as demonstrated with the "million US $" baby [122]. © Halberg Chronobiology Center. (**b**) Concepts like homeostasis constitute an excuse for a curtain of ignorance closed over the range of physiological variation (*left*). Clocks and calendars (*middle*), if and only if they replace as measured dynamic new endpoints (rather than being misinterpreted as a complement or alternative to an imaginary homeostasis), can open that curtain only partly, since coexisting structures such as deterministic and other chaos, trends, and the many rhythms other than the circadian and circannual components in and around us must also be considered. Clocks and calendars are no substitute for mapped coperiodisms in the biosphere and its environment (*right*). © Halberg Chronobiology Center

How this and other coperiodisms with a period longer than 28 hours, infradians, come about along the scale of fast ultradian cycles was considered, up to circaseptans and their harmonics and subharmonics by Ulmer, who found stunning circaseptans in cancer cell lines in vitro. From a physical viewpoint, several kinds of molecular processes could be involved: (1) diffusion of charged molecules in magnetic fields with the oscillating period for Ca^{2+} and K^{+} being of about 1 week [119] and (2) quantum mechanical perturbation and its resonance dominators, characterizing specific interactions among constituents, that permit the treatment of molecular processes by circuits with characteristic resonances and beat frequencies, which result from the primarily fast physical processes. As examples, Waldemar considered the tunneling processes between DNA base pairs (H bonds) and the decomposition of ATP [120]. At all levels, there is a need to estimate uncertainties, with the qualification that glocal methods of analysis are particularly promising for circadians, now documented by well over 67,000 articles in the PubMed database alone as of December 2012, and for which we here demonstrate two separate multifrequency oscillatory systems, one for the MESOR and another for the circadian amplitude of each human variable tested (Fig. 23.2a, b).

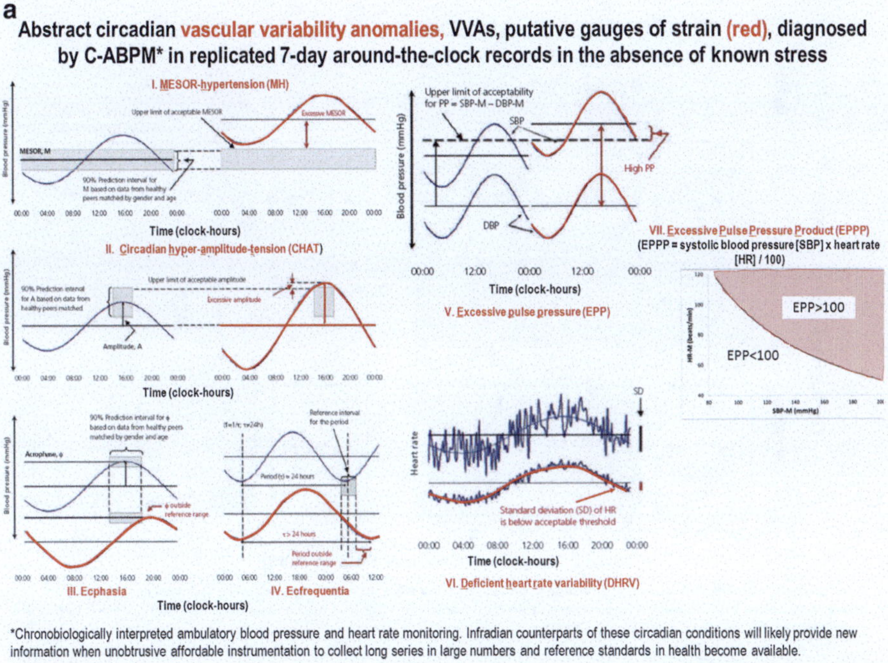

Fig. 23.16 (**a**) Abstract limits for blood pressure MESOR (*top left*), amplitude (*left*, row 2), acrophase and period (*left, bottom*), pulse pressure (*middle top*), heart rate variability (*middle bottom*), and pulse pressure product (PPP) (*right*). Deviations from these norms, derived from data collected on clinically healthy subjects (all eventually to be qualified by gender, age, and end-of-life outcomes) (*MH* MESOR-hypertension, *CHAT* circadian hyper-amplitude-tension, *EPP* ecphasia and ecfrequentia; excessive pulse pressure, *DHRV* deficient heart rate variability, *EPPP* excessive PPP),

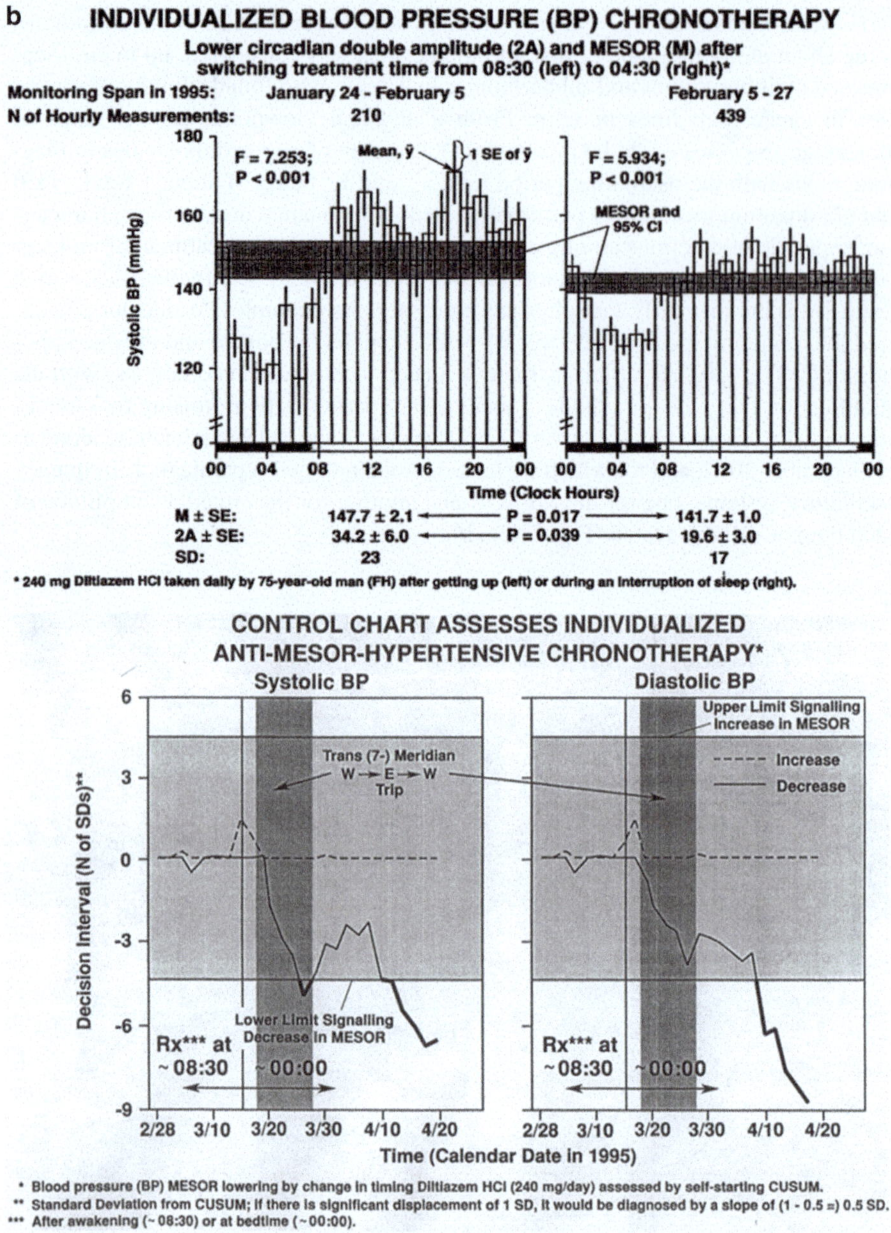

Fig. 23.16 (continued) have been associated with increased cardiovascular disease risk in several outcome studies (Fig. 23.15a) (*middle*). © Halberg Chronobiology Center. (**b**) *Upper left*: The usefulness of a chronobiologic approach is supported by the demonstration that treatment with antihypertensive drugs can be optimized by timing. The same dose of the same drug can have different effects on the MESOR and circadian amplitude of the same patient's blood pressure when it is administered daily at a different circadian stage, as seen by the naked eye and documented by

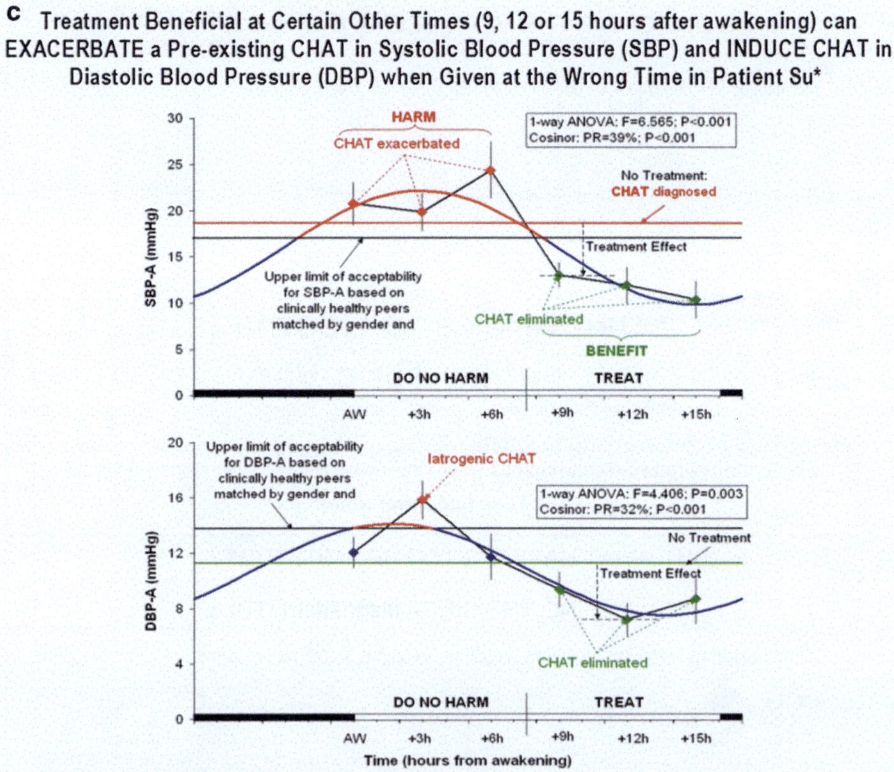

v*Su, M, 66y, treated with Losartan (50 mg) and hydrochlorothiazide (12.5 mg). Each point represents 1 week of half-hourly around-the-clock monitoring after ~1 month on a given treatment time. After this graph was completed, treatment at AW+15h yielded a SBP-A of 10.36 ± 2.09 mmHg and a DBP-A of 8.69 ± 1.85 mm Hg. By cosinor, a circadian stage-dependent response to treatment could be shown to account, with statistical significance for both endpoints, for 39% and 32% of the overall variance, respectively.

Fig. 23.16 (continued) parameter tests. *Lower right*: The chronobiologic approach includes inferential statistical methods for the rigorous assessment of intervention effects applicable to the individual patient, e.g., detecting whether the lowering in the MESOR of systolic (*left*) and diastolic (*right*) blood pressure can be associated with the time when the change in treatment timing was instituted. © Halberg Chronobiology Center. (**c**) A popular drug, if prescribed without personalized surveillance, can induce a vascular variability anomaly (VVA) such as circadian hyper-amplitude-tension (CHAT). A change in the time when the drug is taken can make the same dose of the same drug in the same person beneficial or vice versa. At one administration time (before noon), Hyzaar induces CHAT in diastolic blood pressure and exacerbates a preexisting CHAT in systolic blood pressure (*red*). At another time of administration, Hyzaar eliminates a preexisting VVA (*green*). These opposite effects were found in tests at six medication times, each administered for about a month, with half-hourly surveillance of blood pressure during the last week of each span. These differences occur as a function of the timing of the drug's use along the scale of 24 hours. © Halberg Chronobiology Center. (**d**) Changing timing of medication (ΔRx) during consecutive spans shows efficacy of treatment. An empirical approach to chronotherapy: immediately after diagnosis, one should ascertain that the treatment is effective. Optimization of treatment effects by timing can be achieved for the individual patient by systematically changing, e.g., advancing the time of treatment. Successful treatment of MESOR-hypertension assessed by a self-starting cumulative sum

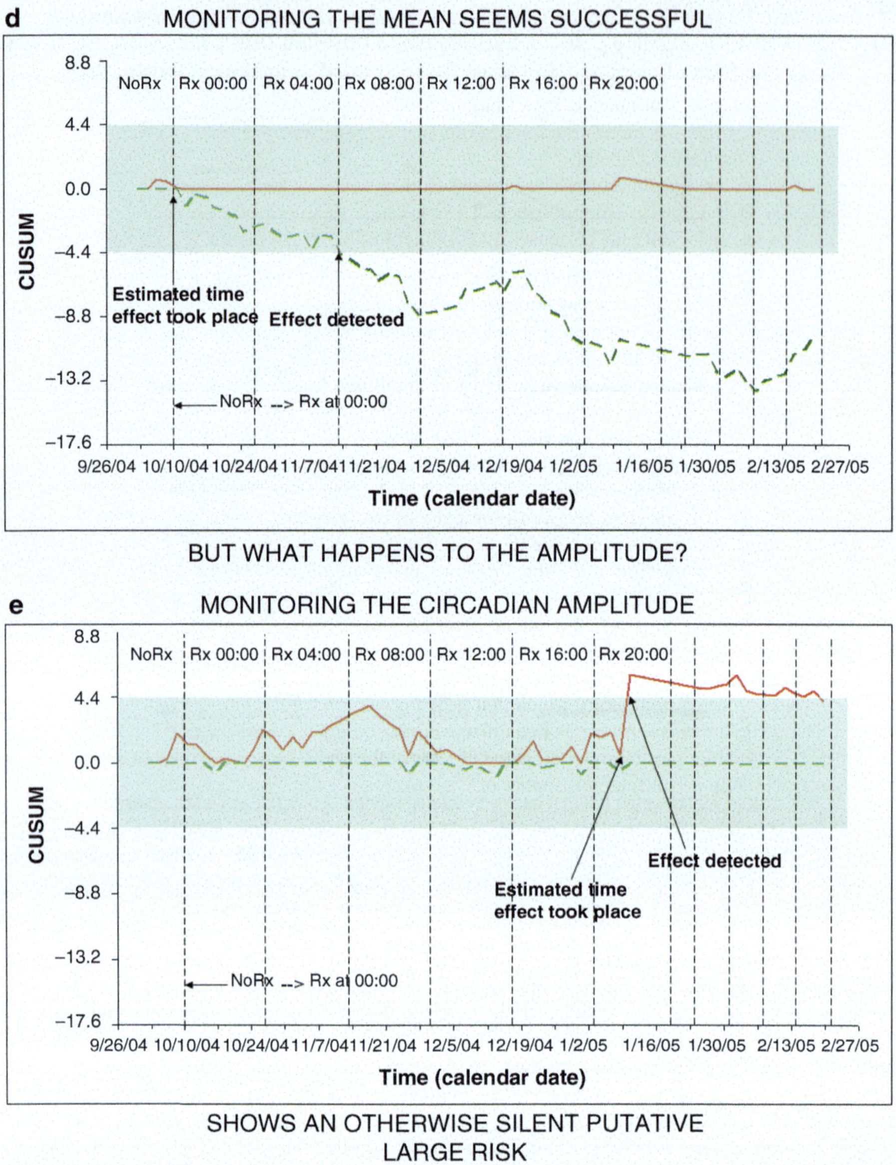

Fig. 23.16 (continued) (CUSUM) control chart. To optimize his hypotensive treatment (Rx), a just-diagnosed 24-year-old individual (TT) switched his Rx first every 17 days by 4 hours and then mostly at shorter intervals. Note statistically significant decrease in MESOR, evidenced by the breakout outside the decision interval of the negative CUSUM line. With continued Rx, the blood pressure MESOR leaves the decision interval, indicating a statistically significant decrease in overall blood pressure. © Halberg Chronobiology Center. (**e**) Changing timing of medication (ΔRx) during consecutive spans shows risk of iatrogenic circadian hyper-amplitude-tension (CHAT). An empirical approach to chronotherapy: immediately after diagnosis, one should ascertain that one does not induce CHAT by inappropriate timing of antihypertensive medication. In this

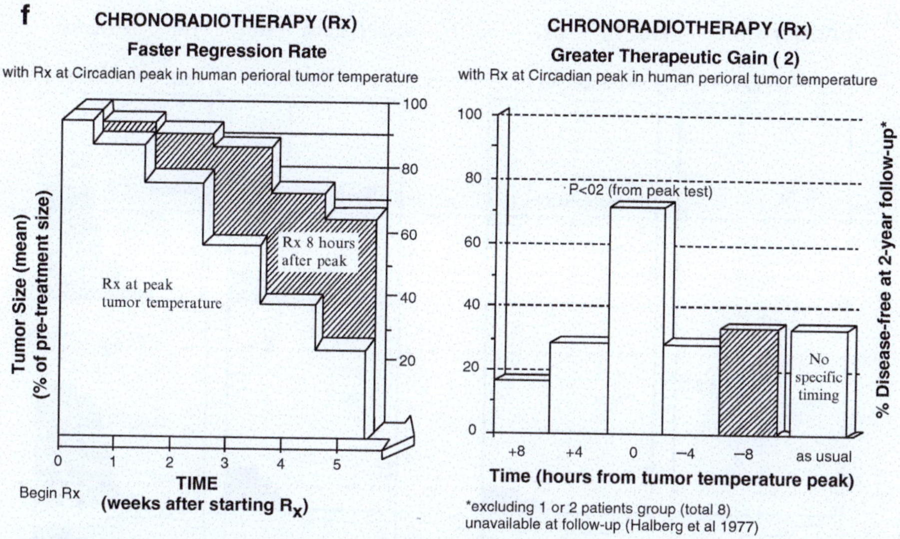

Fig. 23.16 (continued) 24-year-old man (TT) who advanced the time of treatment by 4 hours every 17 days initially and at shorter intervals thereafter, treatment in the evening was associated with iatrogenic CHAT, raising the question whether the risk of MESOR-hypertension may not have been traded for the even higher risk of stroke that CHAT represents. Iatrogenic CHAT, induced by treatment at 20:00 daily, was silent to office visits. TT may have traded benefit (lowering of the MESOR of blood pressure, Fig. 23.16d) for something worse (CHAT). This danger applies to some hypertensives (who tend to have a large circadian amplitude of blood pressure) to whom treatment time is not specified (or is specified for bedtime) by the care provider, as was the case for TT. A few others who took hypertensive medication at bedtime were also found to have CHAT. The figure shows the assessability of otherwise undetected harm by as-one-goes sequential analysis. © Halberg Chronobiology Center. (**f**) Treatment timing at peak tumor temperature leads to faster tumor regression (*left*) and to more than doubling of the 24-month disease-free survival rate (0 on abscissa, *right*), as compared to reference groups treated 8 or 4 hours before or after the tumor temperature peak (+8, +4, −4, −8 on abscissa, *right*) or "as usual" (last column, *right*). Studies implemented by B.D. Gupta and Akhil Deka, designed during a PL-480 program sponsored by the USA. © Halberg Chronobiology Center. (**g**) Breast cancer incidence (*black areas*; staircase starting at bottom) and death from other causes (*top*) in each of several experimental subgroups from two inbred strains. Breast cancers induced or promoted by pituitary isografts were reduced in incidence or prevented (in mice without the Bittner agent) by the addition of a hypothalamus to the ectopic pituitary. Laboratory results then await follow-up concerning the prevention of certain endocrine cancers of the mouse breast. © Halberg Chronobiology Center. (**h**) Circaseptan optimization may be equally as important as circadian chronotherapy, as seen in a 9L glioma (rat tumor model), and may be more readily scheduled in the clinic. © Halberg Chronobiology Center. (**i**) An about-7-day spectral component in the heart rate of five men is less prominent when the solar wind loses its counterpart of corresponding length. Implied, but not shown, is the persistence in the biosphere of an about-7-day component that can be amplified (driven) by a reciprocal component in solar activity. The approach by remove and replace in the study of circadian rhythms is here extended to the circaseptan [123]. © Halberg Chronobiology Center

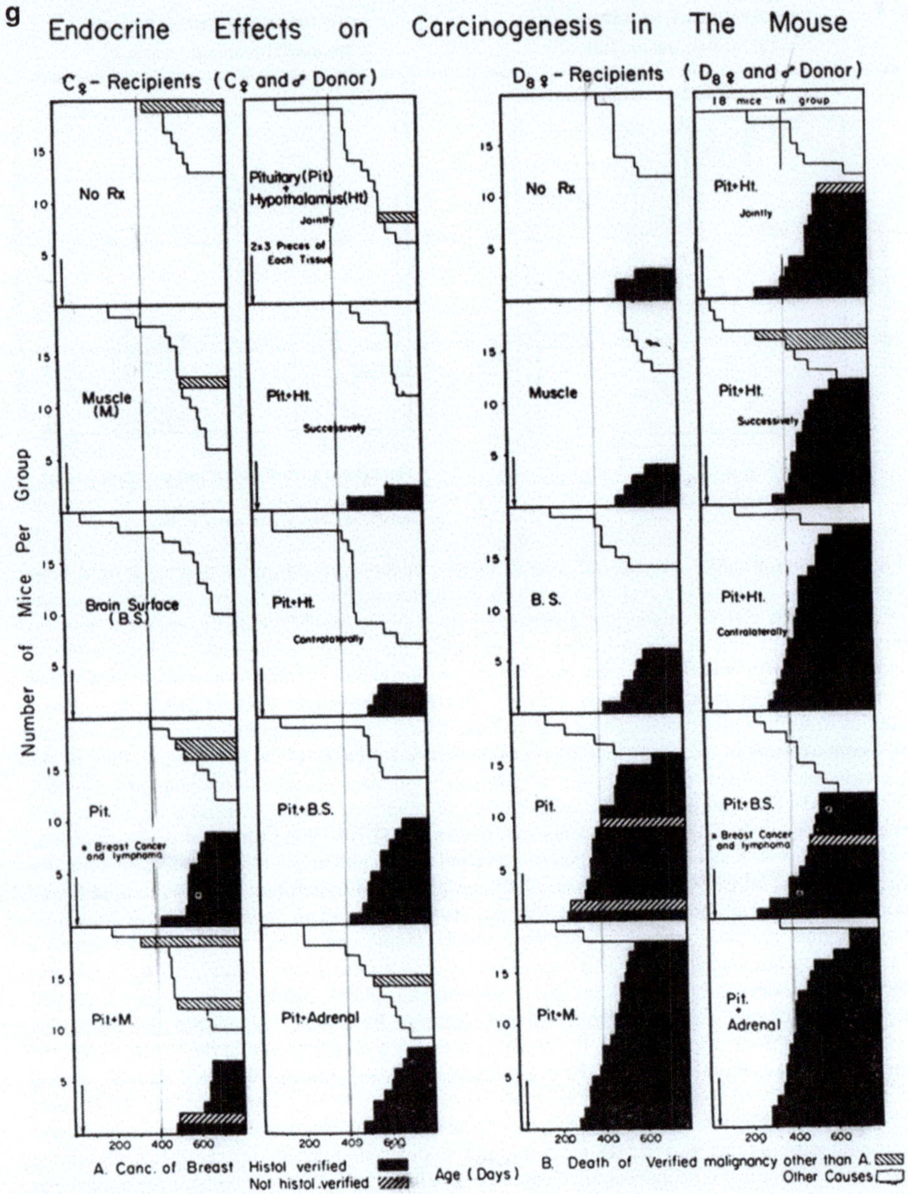

Fig. 23.16 (continued)

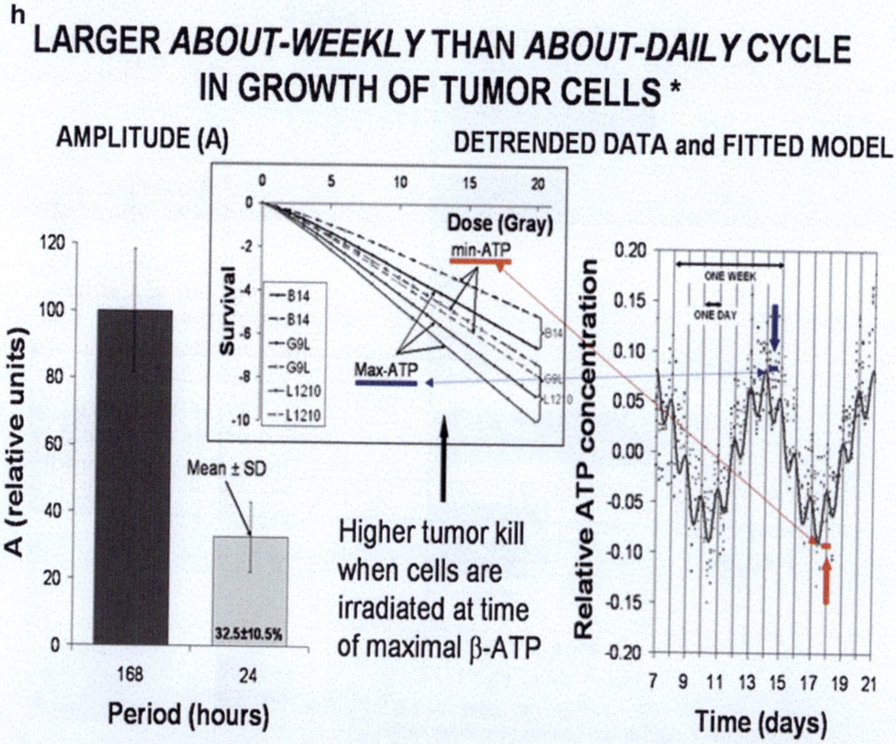

* Nonlinear spectral analysis of data pooled from four kinds of tumor cells, each cultured at a pH of 6.9 (experiment #1) and 7.3 (experiment #2). Data shown for weeks 2-3 (total 6 weeks, exhibiting persistent cycles of similar relative prominence).

Fig. 23.16 (continued)

In 1953, I wrote [124]: "… it is not surprising to encounter reference to 'the great variation in circulating eosinophil counts, even in normal conditions, which makes it impossible to establish a base-line for the study of the physiopathology of these cells. It is indeed impossible to establish a base-line, at least a straight base-line, for eosinophil counts in normal conditions, because there is none in normal conditions" (nor is there a base*line* for the many variables investigated in the interim) [13]. On the positive side of studying changes within the normal range, not yet exploited, however, are circadian VVAs and VVDs (Fig. 23.16a) that become apparent once we remove the curtain of

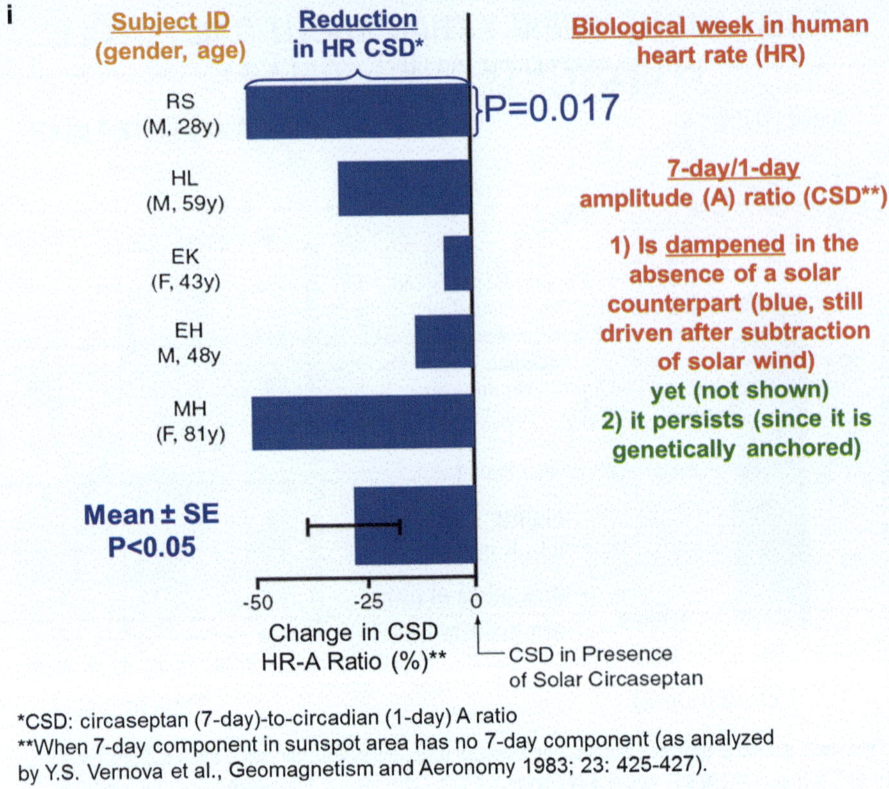

Fig. 23.16 (continued)

ignorance drawn by concepts of relative constancy or homeostasis over the range of everyday physiology (Fig. 23.15b) [8]. A highly elevated risk of severe cardiovascular disease (Fig. 23.15a, middle) goes unrecognized today, in millions of people; yet, if it is iatrogenic in origin, it can sometimes be treated simply and cost-effectively by no more than a change in the timing of hypotensive medication (Fig. 23.16c). Apart from imaginary baselines, we also learned in 1953 that work in the normal range allows much more sensitive responses than reliance on tests with endpoints outside that range. What had taken milligrams in work with an endpoint outside the normal range (eliminating a circadian variation in circulating eosinophil count by cortisone at Harvard) could now be assayed with one or a few micrograms (Fig. 23.17a [125 vs. 124; cf. 126]).

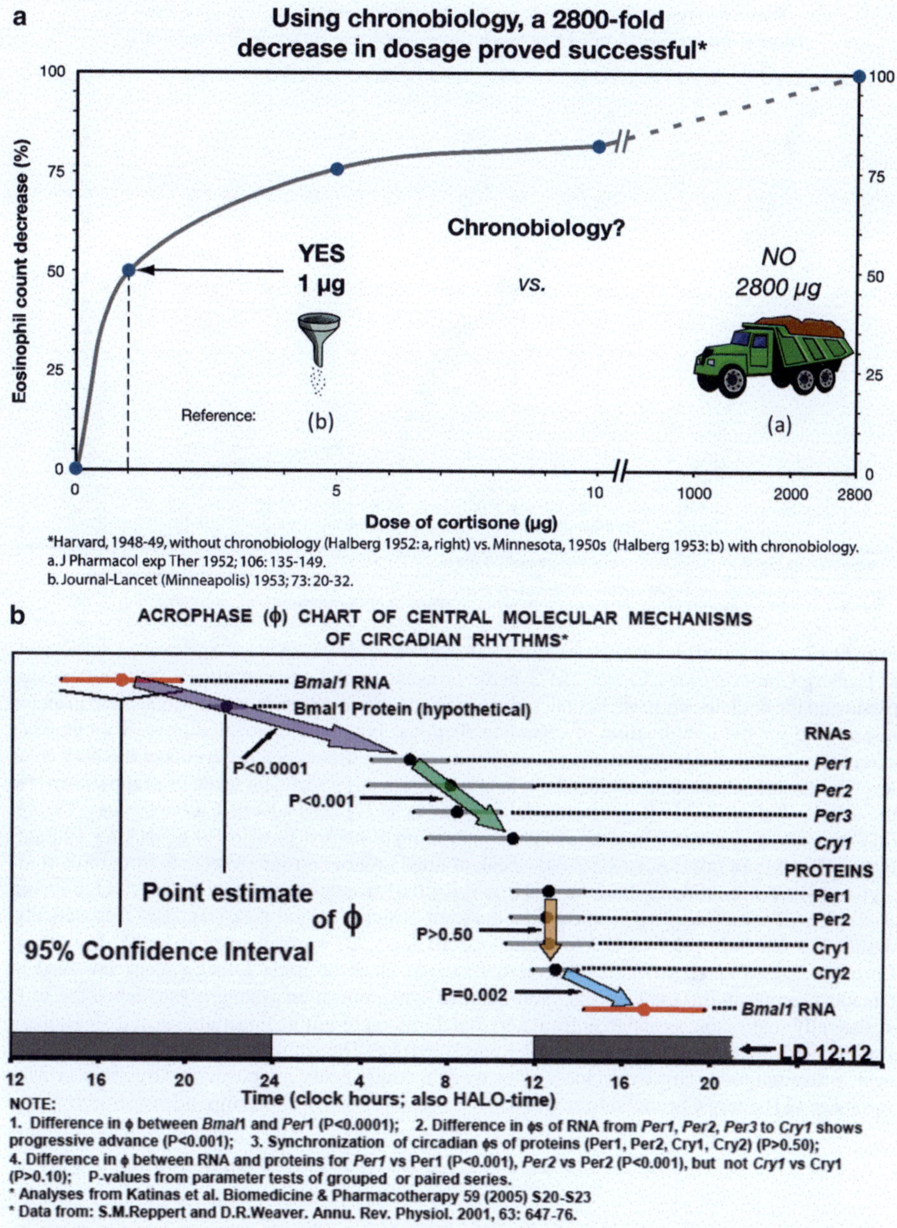

Fig. 23.17 (**a**) Gain in sensitivity from taking rhythms into account. Circulating eosinophil cell depressing activity was detected with 1 μg of cortisone when tested in the ascending phase of the circadian rhythm in the count of these blood cells. The increase in sensitivity of the assay is depicted by comparing the minuscule dose on the left, taking into account the rhythm, with the dose depicted as a truckload on the right, when the rhythm was "eliminated" by depressing the cell count to zero for 24 hours by 2.8 mg of cortisone used as the standard in my original bioassay in Boston.

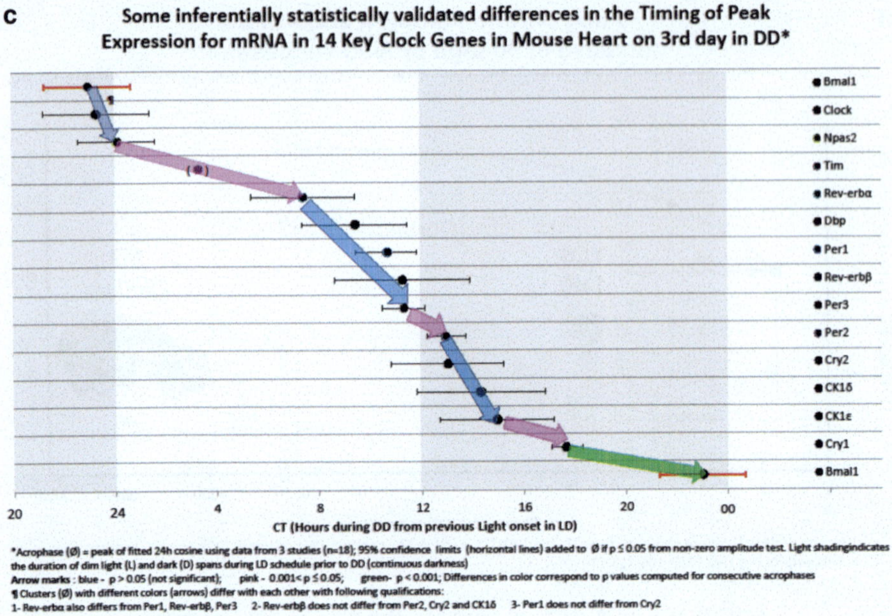

Fig. 23.17 (continued) Using chronobiology, a 2800-fold decrease in dosage proved successful. © Halberg Chronobiology Center. (**b**) Scheme of events at the molecular level in both the cytoplasm and the nucleus, presented as the sequence in activity of different genes, RNAs, and proteins responsible for the coordination of circadian rhythms. Inferential statistical tests of a circadian acrophase chart in the hypothalamus intended to visualize the scheme of events at the molecular level in the suprachiasmatic nuclei of the mouse (meta-analyses of data taken of graph drawn) by the scholarship of S. M. Reppert and D. R. Weaver [127]. © Halberg Chronobiology Center. (**c**) Circadian cardiac acrophase chart indicating timing of peak expression of mRNA for 14 putative clock genes in mice. Acrophase (ϕ) = peak of fitted 24-hour cosine using data from three studies ($n = 18$); 95 % confidence limits added to ϕ if $P \leq 0.05$ from nonzero amplitude test. *Lighter* and *darker gray bars* along x-axis indicate the duration of light (L) and dark (D) spans, respectively, during LD schedule prior to DD (continuous darkness); CT = circadian time [128]. © Halberg Chronobiology Center. (**d**) Circadian acrophase (ϕ) chart of some (clock) genes involved in circadian timing in different organs. Note relative synchronization among organs when the fit is statistically significant, as apparent from horizontal bars representing uncertainties of ϕ. © Halberg Chronobiology Center (Data from Toru Takumi [129]). (**e**) The circadian rhythm is usually prominent, with amplitudes larger than most other spectral components, as shown here for some urinary variables. © Halberg Chronobiology Center

d

Peak expression of mRNA for 14 clock genes in each of 7 peripheral organs of mice

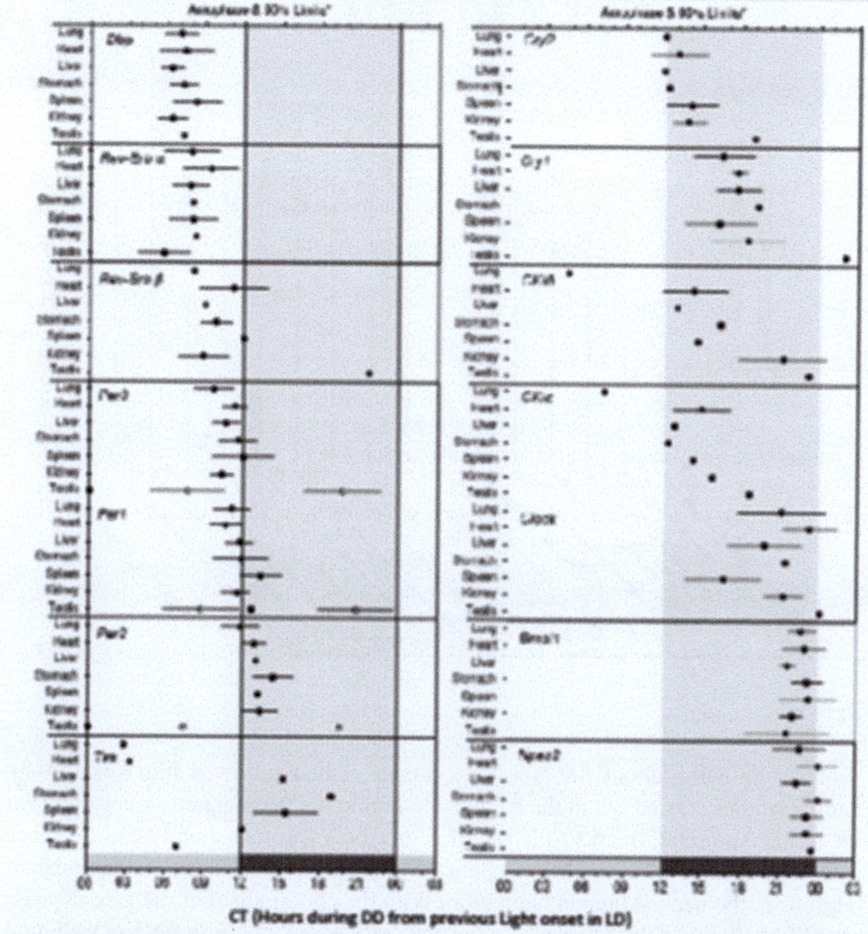

Acrophase (Ø) = peak of fitted 24h cosine using data from 3 studies (n = 18); 95% confidence limits added to Ø if p≤0.05 from non-zero amplitude test. Lighter and darker gray bars along x-axis indicate the duration of light (L) and dark (D) spans, respectively, during LD schedule prior to DD (continuous darkness in, the condition to which the mice had been exposed for two days before & on a third day of sampling); CT= circadian time. Analyses from Sothern et al. Data from Yamamoto T et al. BMC Mol Biol 2004; 5: 18(9pp)

Fig. 23.17 (continued)

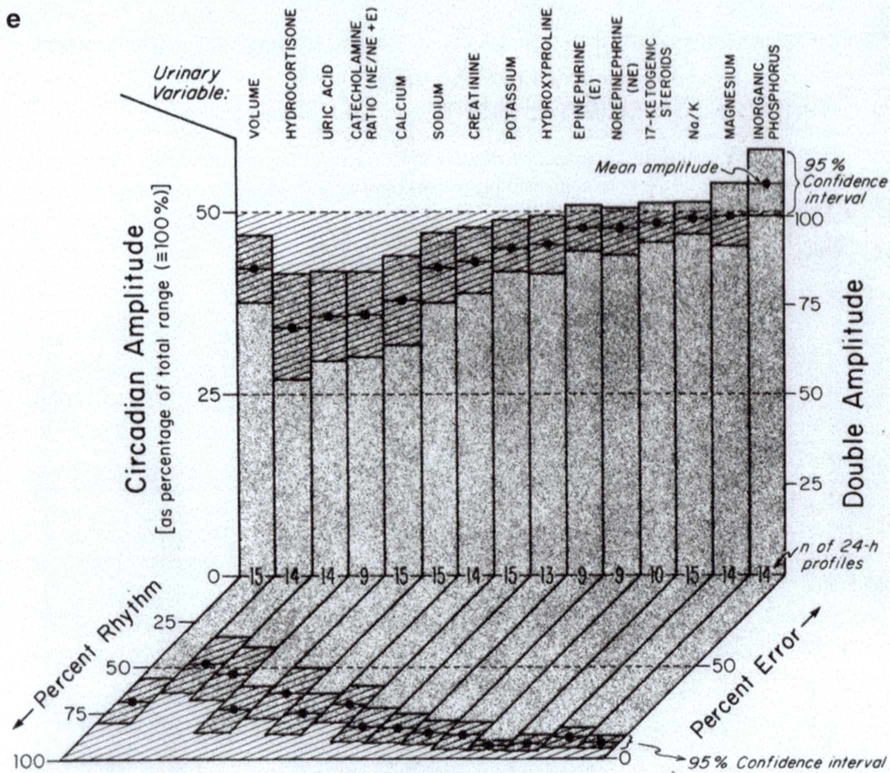

Fig. 23.17 (continued)

Not yet knowing about Sir Norman's advice, I had followed him since 1949, including studies carried out in the frigid zone to find the geomagnetic week and near week in the Antarctic, thanks to Mark Engebretson's (professor and head of Space Science, Augsburg College, Minneapolis, Minnesota) stand-alone magnetometer, 600 km from the nearest human habitation. With the coperiodism of the geomagnetic near week available in the Antarctic, in the planetary geomagnetic index Kp, we found the putative origin for the atavistic biospheric near week. It is hence not surprising that the inheritance of the week could be demonstrated with Dana Johnson (professor of Neonatology, University of Minnesota), by a comparison of diastolic blood pressure, heart rate, and body weight of mono- vs. dizygotic human twins [130], with the former vs. the latter revealing statistically significantly tighter estimates of cycle parameters. Emergenic inheritance of the circadian characteristics of heart rate had already been demonstrated in a Minnesotan study of twins reared apart [131]. Molecular biology has uncovered some of the genes underlying the circadian system, found not only in the suprachiasmatic nuclei but also in peripheric organs such as the heart (Fig. 23.17b–d). The circadian amplitude is often most prominent, as illustrated in Fig. 23.17e for some urinary variables. In the torrid zone, Miroslav Mikulecky found novel transyears with a period longer than a year in natality in Davao, Philippines [132, 133], extending by far the biospheric detection of about-16-month cycles [134]

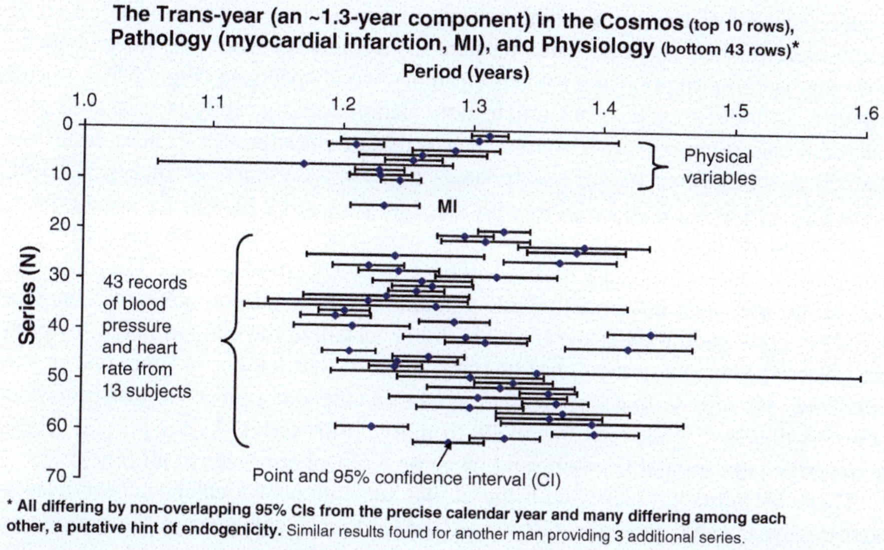

Fig. 23.18 Far-transyears around us (top 10) and incidence of myocardial infarctions in Minnesota (MI) and in 43 blood pressure and heart rate records from 13 subjects who provided longitudinal records covering several years. © Halberg Chronobiology Center

beyond sudden cardiac death and cardiac arrhythmia [135, 136], as well as blood pressure and heart rate, where it was originally found by us (Fig. 23.18) [2].

Major new lessons were learned by monitoring ourselves. The data underlying Fig. 23.5a were collected by eight adults, four of them at 30-minute intervals, with interruptions, around the clock for several (up to 4.5) decades [13]. The histogram of periods in Fig. 23.5c reveals a peak at 10 years and that in Fig. 23.13a on a larger set of periods detailed elsewhere adds a more prominent peak at two decades [13]. It seems that (as with circadians and circaseptans; [123, 137]) about-20- and/or about-10-year and/or paratridecadal cycles of about 35 years, notably in heart rate among other multidecadal solar signatures, are nearly ubiquitous in the physiologic time series approximating that length and were broadly but not always precisely of the length of an environmental coperiodism (Fig. 23.5e). There is a selective assortment of periods, whether or not they are putative genetically coded signatures of the variable cycles in sunspot bipolarity or count.

In my bibliography [3,725 titles as of August 19, 2015] are meta-analyses by the nonlinearly extended cosinor [13] of human population phenomena, among many others, from anthropometry on newborns (Fig. 23.7a) to morbidity and death of adults, 400 years of economic and 173 years of military-political affairs, including 2556 years of international battles and 2189 years of tree rings [138]. In successive summaries of cycles in the human blood circulation and metabolism (the last one in Fig. 23.5a), a nonrandom distribution with a peak around a period of 10 years became more prominent as data accumulated (Fig. 23.5c). The sample in hand is small, however, and hence qualifies generalization, notably since some of the time series cover only a few cycles. But when the cardiovascular data were aligned with our meta-analyses of more and some much longer series, a major peak emerged around periods of 20 years, yet that at 10 years remained (Fig. 23.13a) [13]. As data accumulate further, we might also prepare histograms for near- and far-transyears and other specific regions of a transdisciplinary broad spectrum (Fig. 23.1a), to call attention to their presence, but much more detailed cartography is indicated glocally. All classifications are tentative, as was that of circadians [13]. Sometimes we find clear separations of coexisting near- and far-transyears, as in Fig. 23.13b. Figure 23.13b and c is in keeping with separate coexisting biospheric counterparts to cycles in solar magnetism (the near-transyears) or in solar wind speed (the far-transyears). These results are global and have to be specified as such. They are useful, as are averages but no substitute for temporally local complements, such as serial sections and/or spectrograms. The latter visualize the aeolian nature of both the calendar-year component and of transyears, so that locally in time, one or the other may prevail, so that a specification as to a calendar-year vs. a near-transyear component is best made retrospectively from long time series. As one goes, it is best to describe para-annual behavior and insist on a glocal approach at all times.

There are other problems with the global separation of a calendar year from a near-transyear (1.00 year $<[\tau-CI]<[\tau+CI]<1.20$ years, where τ is the period) as well as from a far-transyear (1.2 years $\leq[\tau-CI]<[\tau+CI]<1.9$ years, where τ is the period). When the confidence interval of a far-transyear overlaps the lower 1.2-year limit, we may refer to a candidate far-transyear or imply this by omitting the prefix "far" or "near." This can also be done in the case of a near-transyear, when its 95 % confidence interval (CI) exceeds the in this case upper 1.2-year limit.

Among many others, about 10 (decadals), about 20 (didecadals), and about 30–40 (paratridecadals) years [13] putatively reflect signatures of past Horrebow-Schwabe sunspot counts, Hale sunspot bipolarity, and Brückner-Egeson-Lockyer paratridecadal cycles, respectively. Further environmental signatures, among others, include biospheric coperiodism to about-50-year cycles in the incidence of strokes in medicine, of earthquakes in seismology, and of Kondratiev's semicentennian in economics. Trans-semicentennians, corresponding to about-60-year and longer sunspot and terrestrial climate cycles, are prominent in the human newborn (Fig. 23.7a). All decadals and multidecadals may eventually serve as reference cycles to be consulted in further investigations, a sine qua non equally for those concerned about human-made climate or human aging, to cite but two areas where these maps are indispensable.

Characteristics of long cycles, even at shared frequencies, can differ in phase in different geographic locations and/or among transdisciplinary variables. Studies on twins support the heritability of about-24-hour and about-7-day cycles. A remove-and-replace approach, where the removal and replacement of a spectral component is done by weather in extraterrestrial and/or terrestrial space, lends further evidence for an influence of the environment, near and far, on human physiology and pathology (Fig. 23.19a, b), as documented in different geographic locations [30, 33, 139–143].

Prospective approaches on twins that have yielded positive results concerning the heritability of (emergenic, in the case of about 24-hour) circadian and of about-7-day cycles are much more lengthy (and hence more difficult) to implement in investigations of about-10-, about-20-, and about-35-year cycles. Even if restricted to a neonatal population study for a single cycle, they take a decade or more as a minimum and they are complicated by steep age trends, particularly from infancy to adolescence. It seems unlikely that the sunspots or other cosmic variables may soon loose (remove) a decadal cycle, but one can study what happens in the biosphere coincident with changes in sunspot cycle characteristics, as did Sir Norman's son William J.S. Lockyer [144]. The sunspot cycle's period and amplitude can differ drastically from one cycle to the next with a paratridecadal pattern found by Egeson, as well as shortly after him by Edward Brückner [145] and W.J.S. Lockyer.

The time course through childhood and adolescence of human body mass and skin fold as a function of age may also be modulated by a paratridecadal BEL, as Germaine's most recent analyses of data kindly sent by the neurologist Dr. Botond Buda of Szombathely, Hungary, suggest, not only by an inspection of plots but also by inferential statistical analyses, another story for another day to be told by those who collected the data. Whoever sets up new studies will have to check by denser sampling over a much longer span in adulthood with mono- and dizygotic twins included on the putative heritability of paratridecadal (about 35 year) body mass and skin fold or other variables of interest to those concerned with obesity, in a chronocosmoauxology [146–149].

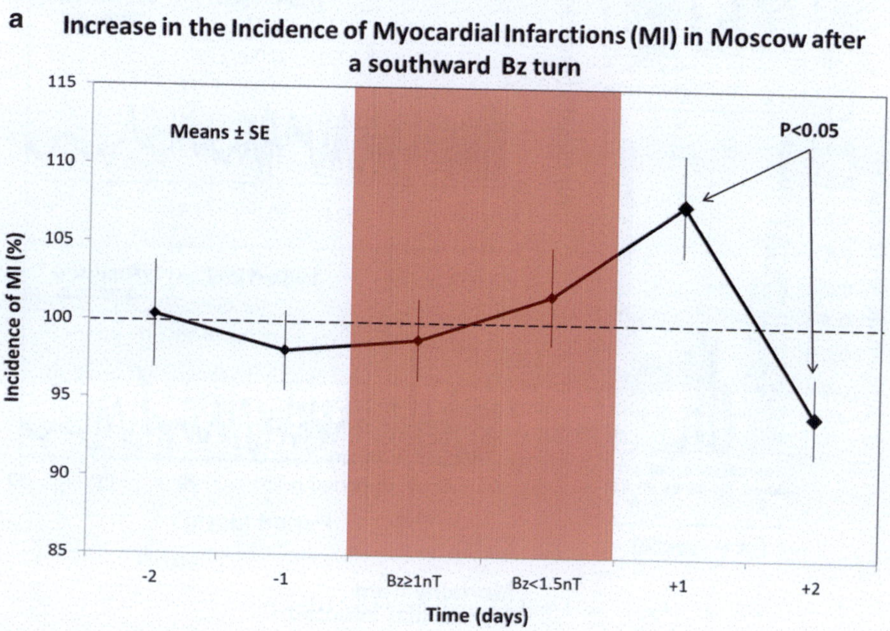

Fig. 23.19 (**a**) In a database of 6.3 million calls for an ambulance in Moscow, Russia, during 1979–1981, superposed epoch analysis reveals an about 7 % increase in the incidence of myocardial infarctions on the day following a magnetic storm, gauged by the southward turn of the vertical component of the interplanetary magnetic field, Bz. Data of each 6-day epoch are expressed as a percentage of the 6-day mean in order to reduce the variance due to the large-amplitude circannual rhythm. © Halberg Chronobiology Center. (**b**) An about-7-day component in mortality from myocardial infarction in Tbilisi, Republic of Georgia, is detected with statistical significance during 1980–1982 and again during 1989–1992 when solar activity is high but not during 1984–1987 when solar activity is low. © Halberg Chronobiology Center

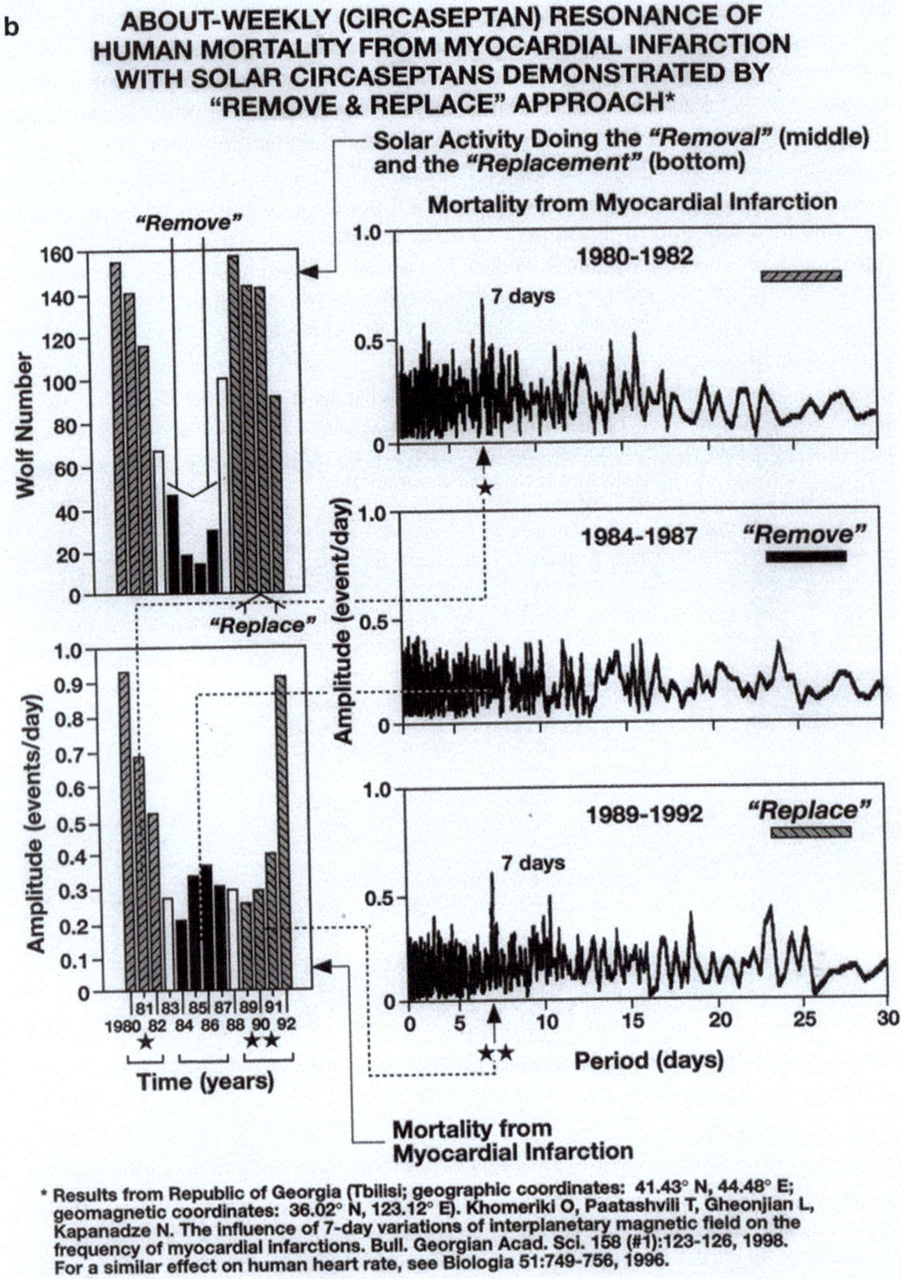

Fig. 23.19 (continued)

Whatever the underlying mechanisms, long cycles, because of their demonstrated near ubiquity, with their diverse uncertainties, require mapping at all ages. Thus, in the one case examined, Yoshihiko Watanabe's son Fumihiko, a transyear had differing time courses in blood pressure vs. heart rate during adolescence. While the transyear's period shortened in heart rate, it lengthened with only an occasional coexistence of a calendar year in systolic and diastolic blood pressure from 11 up to 20 years of age (Fig. 23.20a, b). Trends with age tend to be more pronounced in the very young and the very old. We map indispensable control information, pertinent equally to all investigators, to their referees, and to editors and publishers in all disciplinary and transdisciplinary endeavors in time. Examples of personalized chronouspheres are recommended in particular for the elderly, in whom we found a quinmensal signature of the solar cycle in wrist activity as well, which might turn out to be the least obtrusive way of infradian surveillance (Table 23.1).

a

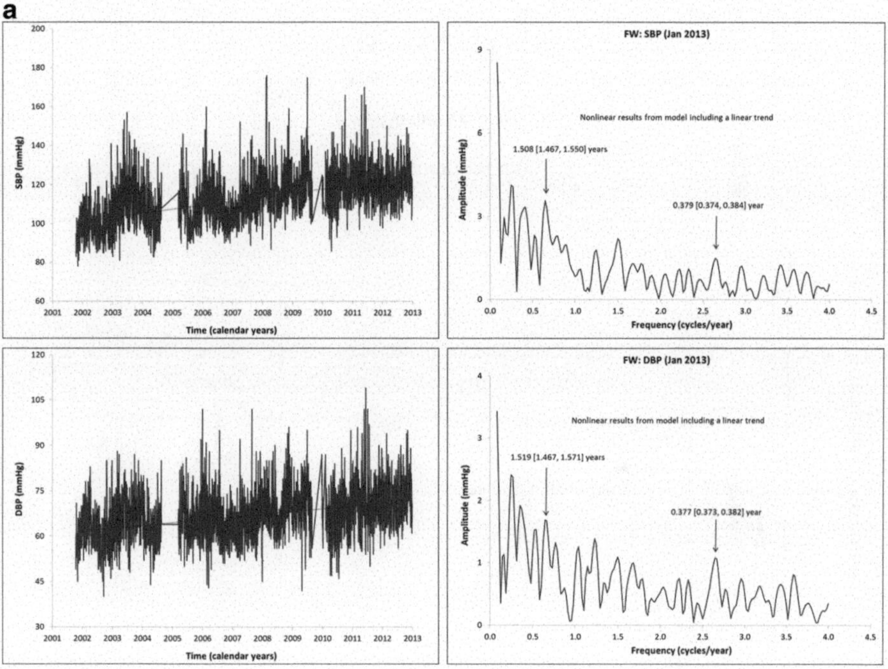

Fig. 23.20 (**a**) In the absence of a (detected) calendar-year component, the systolic (S) and diastolic (D) blood pressure (BP) of a clinically healthy boy (FW) now 20 years of age, measured upon awakening for over 11 years, is characterized by a far-transyear and a cis-half year, validated by nonlinear least squares. © Halberg Chronobiology Center. (**b**) Measurements of heart rate (HR) upon awakening by a clinically healthy boy (FW) for over 11 years (October 2001–December 2012) are characterized by a near-transyear (not a far-transyear, statistically significant when the series was four years shorter) and a 1-year synchronized component, by contrast to the time structure of blood pressure measured concomitantly (see Fig. 23.20a). © Halberg Chronobiology Center

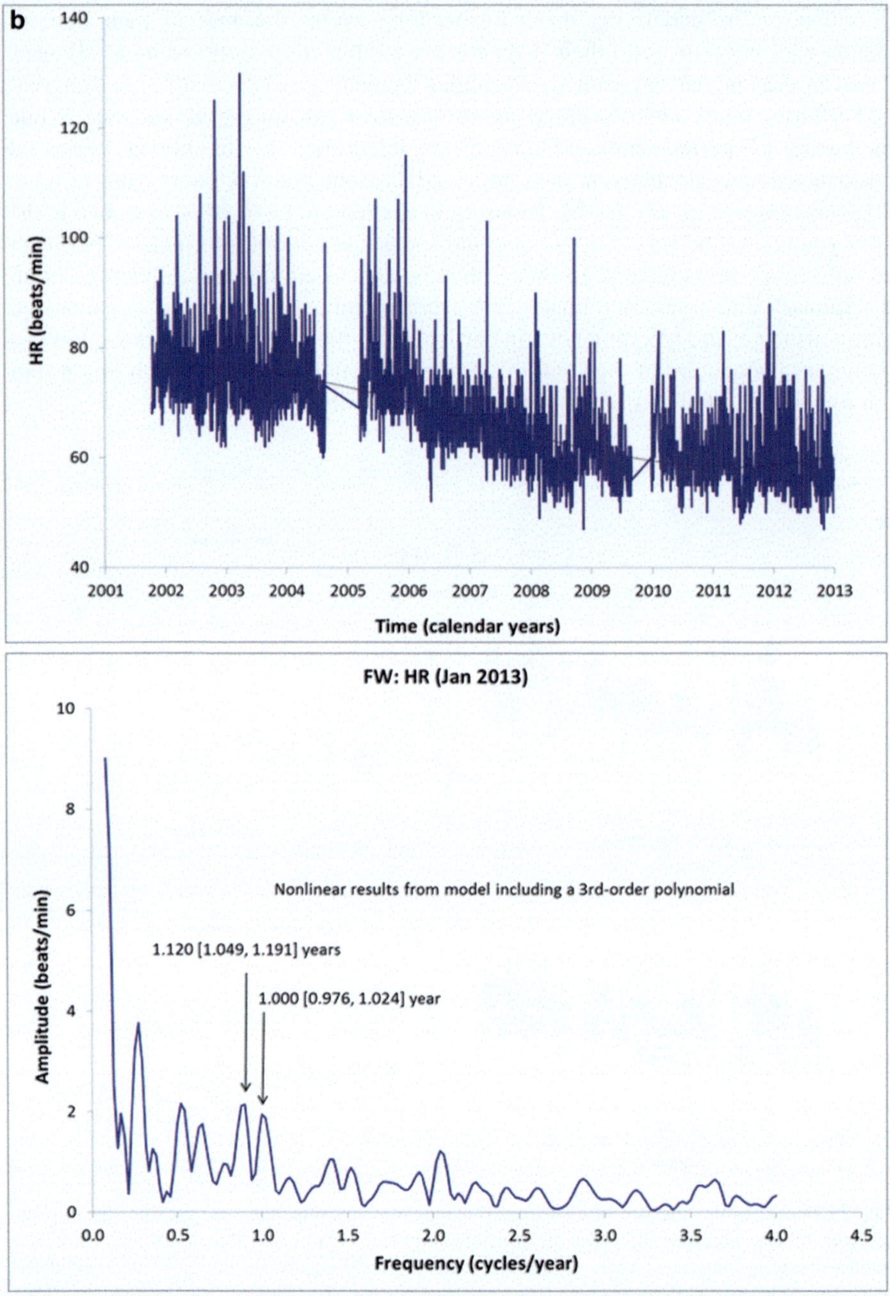

Fig. 23.20 (continued)

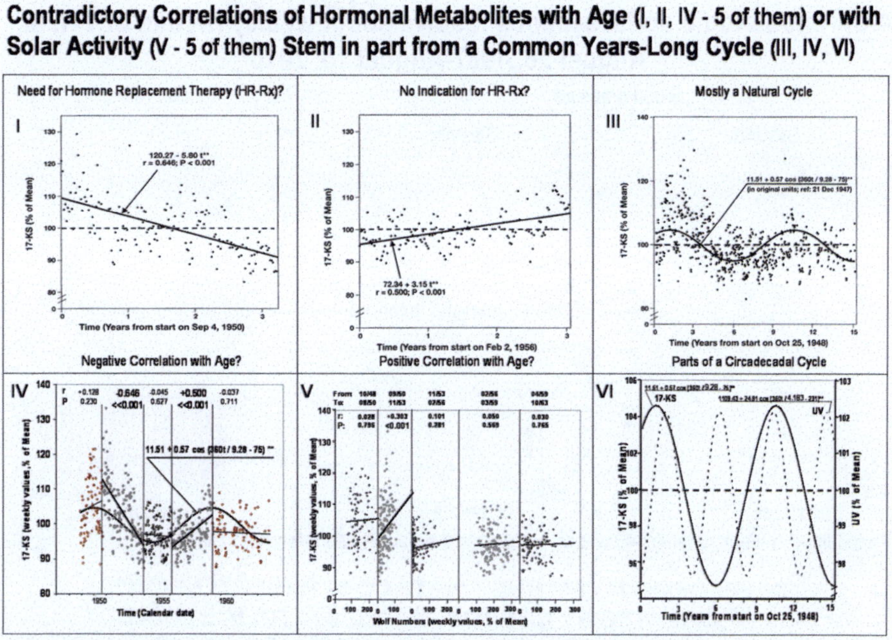

Fig. 23.21 A clinically healthy male medical professional (CH), seeing just the steady decrease in his urinary 17-ketosteroid excretion (17-KS), without knowing that it was part of an about-10-year cycle, indeed considered the need for androgen substitution therapy (*upper left graph*). He may not have realized that a decreasing trend was a transient, followed by an increase (*upper middle graph*), the increase and decrease being but part of a readily seen cycle (*upper right graph*) that when related to age (*lower left*) or to sunspot numbers (*lower middle*) can lead, during different subspans, to contradictory, even opposite correlations (when examined only as a trend, in the absence of prior information). The circadecennian period of 17-KS and the period of urine volume differ (*extreme bottom right*). There are circadecennian, circaseptan, and further transyearly differences between the periods of 17-KS and the volume of urine in which it was determined and even differences in phase at the same period (not shown). © Halberg Chronobiology Center

As documented for circadians (in Fig. 23.14a–e), the likelihood of blunders also applies to those who ignore infradians, notably decadals and multidecadals, as does most of the community of investigators of aging and practicing geriatrics or specialists in climate, who focus on at least 10-year cycles. Even ignoring a decadal cycle as such can also lead to wrong inferences in the clinic, e.g., to unwarranted treatment. Section 1 of Fig. 23.21 shows the data underlying a misdiagnosed androgen insufficiency that was treated, an unjustified action since a decrease in androgen excretion was just part of a decadal cycle and was followed by a spontaneous increase over several years [150]. Some nonsense results also stem from correlations of variables sharing a common cycle, as shown in Fig. 23.14e.

The tasks still lie ahead of us to develop reference values needed, for instance, to distinguish a decadal cycle from the onset of hypertension (Fig. 23.22a–d). If we gauge strain by C-ABPM, we act at least in keeping with the Voice of America and other audio materials (http://www.voanews.com/content/blood-pressure-patterns-could-offer-critical-health-clues/1563748.html OR http://www.youtube.com/watch?v=m-eILBZeTYg). There is a repeated professional consensus to do so.

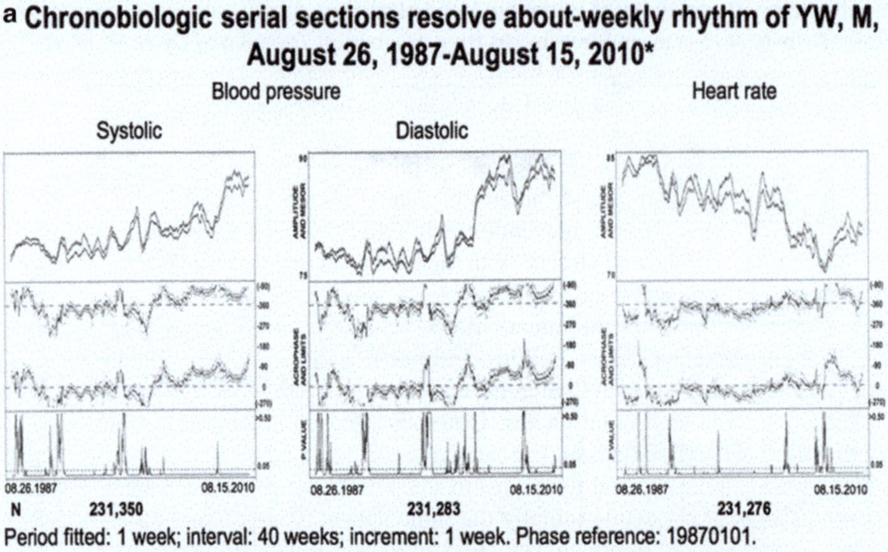

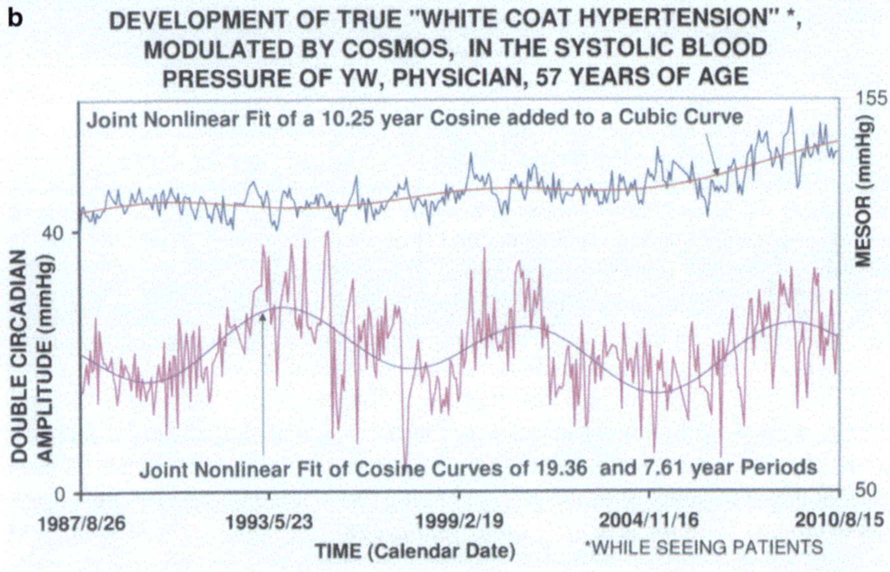

Fig. 23.22 (**a**) Stable weekly variation in blood pressure and heart rate underlies the consistently lower blood pressure on weekends than on work days in a cardiologist (YW) monitored around the clock during 23 years, as he develops MESOR-hypertension. © Halberg Chronobiology Center. (**b**) The presence of cycles with very long periods complicates the diagnosis of MESOR-hypertension, as the increase in systolic blood pressure may be no more than a (transient) stage of the long cycle(s). © Halberg Chronobiology Center. (**c**) The presence of cycles with very long periods complicates the diagnosis of MESOR-hypertension, as the increase in diastolic blood pressure may be no more than a (transient) stage of the long cycle(s). © Halberg Chronobiology Center. (**d**) Systolic blood pressure (SBP) self-measurements by EH (M, 38.5 years of age at start), averaged over consecutive 2-day intervals, were analyzed by the extended cosinor. The least-

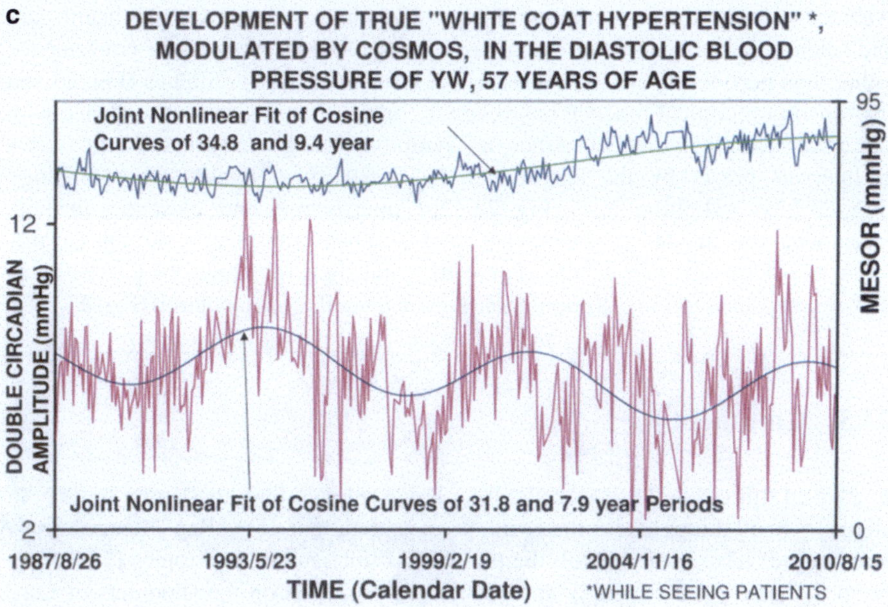

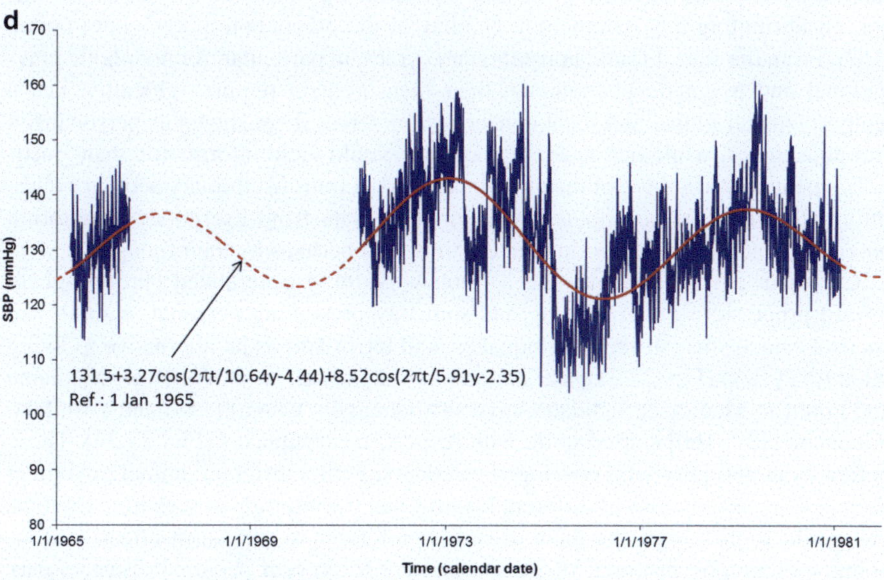

Fig. 23.22 (continued) squares spectrum revealed the presence of components with periods of about 12 and 6 years, resolved by nonlinear least squares. Nonlinear results were used to estimate the fitted model shown with the 2-day means. As for YW (see Fig. 23.22a), in the early 1970s, an increase in SBP could have been interpreted as the development of MESOR-hypertension; continued monitoring suggests that it was also contributed by the presence of a cycle with a very long period. © Halberg Chronobiology Center

We need to assess the variability within the physiological range. Current health care and science can change from being based on spotchecks at times of convenience rather than pertinence to a cyber-based cost-effective self-surveillance (Fig. 23.15a (bottom)) in a clinical practice based largely on self-help by self-surveillance with summaries locally in time when there are symptoms and globally on the entire series at intervals preset by the length and the noisiness of cycles. Chronobiology (Fig. 23.22a) and chronomics (Fig. 23.22b), microscopes and telescopes in time, reveal what the unaided eye cannot see, such as a near-transyear in variables as different as solar magnetism recorded on earth, vital signs monitored by C-ABPM, or the oxygen consumption of a population of giant unicells (Acetabularia) (Fig. 23.22c).

23.5 Comment

In a world full of cycles, it is difficult to derive rules or laws that apply linearly in all places to all things at all times, without considerations of cycling. Numerous different minds have so dealt with the problem of time and related topics [151–156]. From the start, I unknowingly approached the problems on the shoulders of Lord Kelvin, by not only measuring but also by rendering measurable what as yet was not, misattributing this statement to Galileo, as did others before me whom I cited [157]. From the start, I found nonstationary cycles, in particular, for nonphotic gravitational and magnetic phenomena; their measurement requires glocality, i.e., a temporo-spatial global and local concept of ourselves, here shown as personalized chronouspheres, while our accumulating data would yield information that constitutes a generalized sphere of the mind in a world of transdisciplinary aeolian cycles. Information may be contributed with input by data from everybody monitoring pressure in blood vessels, as done in car tires, and perhaps by their thoughts.

Cycles in any event constitute a framework for a generalized chronousphere toward progress in understanding and someday preventing personal, natural, and societal cataclysms (strokes, earthquakes, and terrorism) as an alternative to flying blind (to cycles) (Fig. 23.23), as Zhengrong Wang (dean, School of Basic Medicine and Forensic Medicine, Sichuan University, Chengdu, China) [158] and Yoshihiko Watanabe [159, 160] started to do with respect to earthquakes (Fig. 23.24). Those before us in antiquity who developed calendars, with a different animal governing each year in periodicities of different lengths, had the intuition to seek long rhythms but not the tools to measure them. It seems preferable to add a cost-effective chronomic cartography in time. An alerting system to prevent personal disasters may also serve to help understand and eventually to avoid or evade societal and natural cataclysms. There is no choice for health care and science but to monitor all we can as we also add to spheres of the mind [13].

Different arts and sciences, broadly physics and biology, in particular, are inseparable. Their tools and findings, cycles in particular, complement each other, in sharp disagreement with Jean Jacques d'Ortous de Mairan, who did not pursue his wonderful discovery of a persistent near-24-hour rhythm in the leaf movement of the sensitive plant because it was outside his professional field, by contrast, for example, to

a

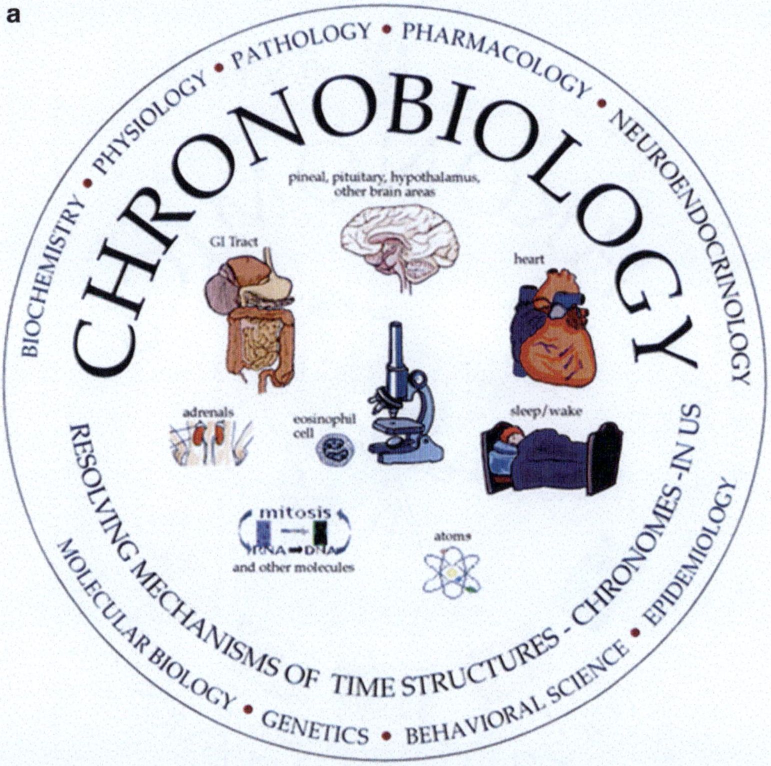

Fig. 23.23 (**a**) A computer-implemented virtual "microscope" (chronobiology) resolves cycles in ourselves, revealing, in different stages, sometimes opposite effects of the same stimulus, such as a hypotensive drug that can harm or yield benefit at different cycle stages, prompting marker rhythm-guided chronotherapy that also applies to X-ray irradiation, doubling the 2-year disease-free survival rate of patients with perioral cancer. Broadly, chronobiology replaces fictitious "baselines" in the life sciences, as the CONTROL information in BIOSCIENCE. © Halberg Chronobiology Center. (**b**) A computer-implemented virtual "telescope" resolves associations among reciprocal cycles in and around us that represent control information replacing fictitious "baselines" in any scientific or other human endeavor. © Halberg Chronobiology Center. (**c**) Linear-nonlinear rhythmometry on data consisting of circadian acrophases of Acetabularia's oxygen evolution recorded around the clock for up to 1 week in LD12:12 in 279 different experiments carried out over a 14-year span. Since the 24-hour acrophases from all LD12:12 cultures concentrated on one side of the circle, the circle was opened opposite the circular mean, and the 24-hour acrophases were taken effectively as deviations (in fractions of a cycle, i.e., in degrees/360°) from this mean. In the graphed results, the transyear with an estimated period of 1.31 years is more prominent (i.e., has a numerically larger amplitude) than the yearly component, with an estimated period of 0.98 year and a 95 % confidence interval (CI) overlapping 1.0 year, likely reflecting seasonal influences. The 1.31-year transyear is also more prominent than another transannual component with an estimated period of 1.09 years (and a CI not overlapping 1.0 year) corresponding perhaps to geomagnetic disturbances that usually are characterized by a component with a period differing slightly from precisely one year (for instance, the geomagnetic disturbance index Kp recorded from 1932 to 2003 is characterized by such a component, with a period of 1.056 years and a CI spanning from 1.05 to 1.06 years). Note the even greater prominence of a 9.24-year component, corresponding numerically in period length to another nonphotic, wobbly Schwabe cycle in relative sunspot numbers. Width of horizontal bars at spectral peaks indicates CIs of the periods determined by nonlinear least squares. Amplitudes shown on the ordinate represent predictable changes in circadian acrophase at corresponding frequencies. Not shown is the consistently greater amplitude of the transyear vs. calendar year for the other circadian parameters tested (24-hour amplitude and MESOR, the chronome-adjusted average): in each case, the transyear to calendar-year amplitude ratio is larger than unity. © Halberg Chronobiology Center

b

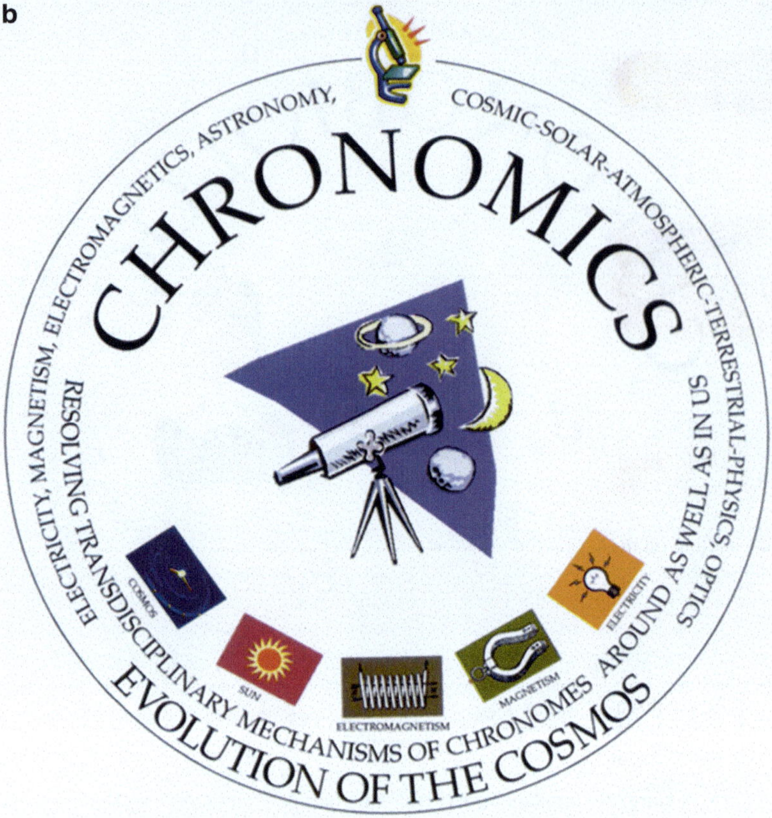

c **An About-Yearly Component is Neither the Sole nor the Largest Infrannual Variation Characterizing *Acetabularia*'s Oxygen Evolution***

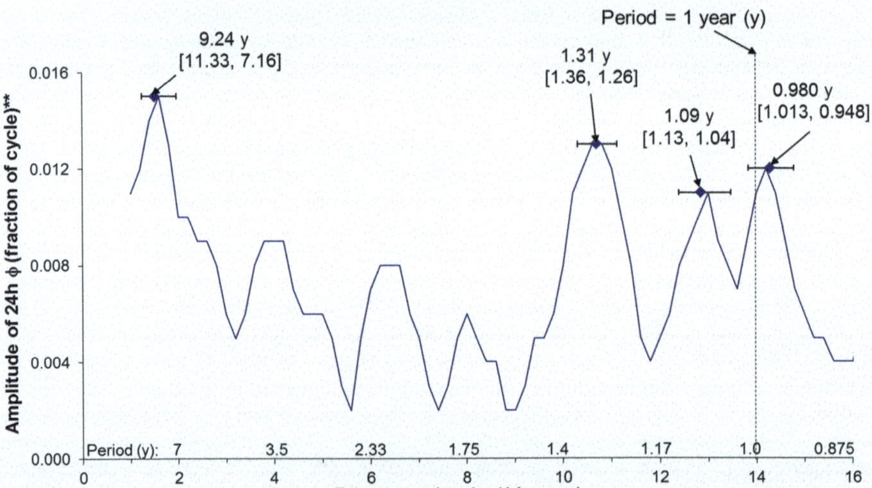

* In 14 years of experiments on these giant unicells kept in LD12:12, each cell contributing one of 279 estimates of the circadian acrophase based on up to a week's recording analyzed by cosinor. From Max Planck Institute for Cell Biology archives (HG Schweiger, director; courtesy of Sigrid Berger). Period lengths and 95% confidence intervals (CIs) assessed by nonlinear least squares. **Indicating prominence of change in circadian acrophase (ϕ) at frequency shown on abscissa.

Fig. 23.23 (continued)

Fig. 23.24 Acting on the basis of a conventionally interpreted (chronobiologically uninterpreted) 24-hour blood pressure record, circadian hyper-amplitude-tension (CHAT) and other vascular variability anomalies (VVAs) remain unrecognized by both those administering treatment and those receiving it. The illustration of Pieter Brueghel's The Parable of the "Blind Leading the Blind" is reproduced by kind permission of the Fototeca della Soprintendenza of the BAS PSAE and of the Polo Museale of the City of Naples. More broadly, C-ABPM can be an ongoing strain test and a harbinger of severe disease and may even serve for understanding and eventually avoiding social and natural cataclysms. Alternatively, everybody in current health care forces the consequences of flying blind to VVAs and to other harbingers of severe cardiovascular disease. © Halberg Chronobiology Center

Giovanni Gregori [161] and Juan Roederer [162], who represent a transdisciplinary perspective with the tearing down of disciplinary barriers in his own work as a physicist concerned with music. Before them, Samuel Heinrich Schwabe crossed from pharmacy over botany to astronomy to report a cycle in sunspots. In so doing, he did for physics what another amateur botanist, Androsthenes, a ship's captain, did for biology. Androsthenes found that the leaves of the tamarind tree closed by night and opened by day, what we now call nyctinasty. While, as noted, Schwabe, in counting sunspots up to 1838, did not dare to mention a cycle perhaps because of an unfavorable intellectual climate regarding cycles in the nineteenth century, he explicitly reported a periodicity in 1844 [93]. But he did not extrapolate at either time to the importance of the solar activity cycle on earth. While he deserves credit for caution (indeed final conclusions must not be drawn from the finding of variations with data covering not quite two cycles), it is nonetheless most interesting that in the very year of 1838, when Schwabe published the first 12 years of his observations (without even daring to suggest a cycle in the face of an obvious up-and-down trend), Clarke described an about-10- to about-11-year trade cycle [146]; cf. [147] in his data on economics. But the intellectual climate was not favorable, as apparent from Manuel

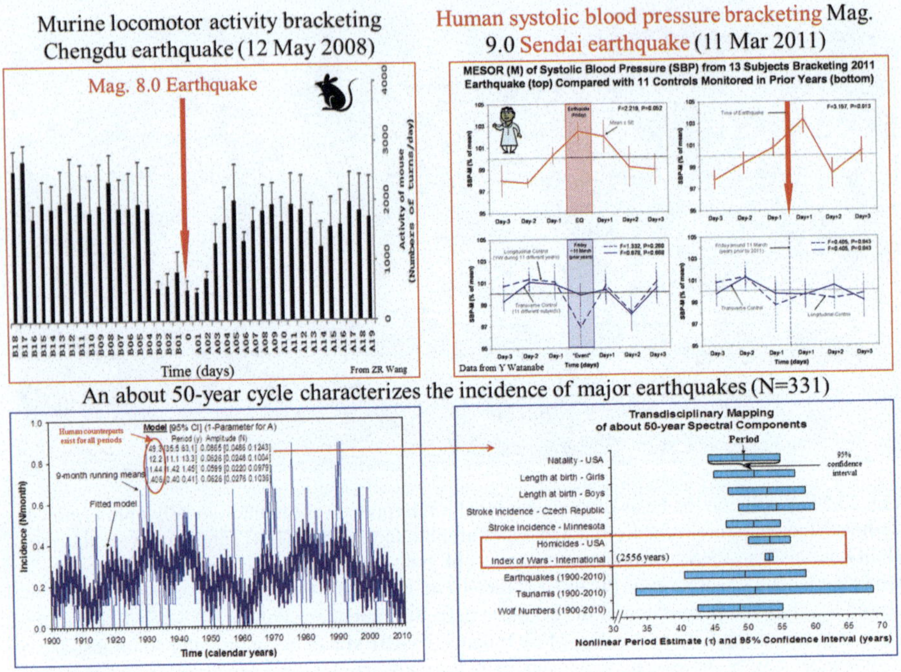

Fig. 23.25 Proposed biospheric contribution to the understanding, if not prediction of earthquakes. *Upper left*: Locomotor activity of mice telemetered around the clock was statistically significantly decreased starting 3 days prior to the magnitude 8.0 earthquake in Chengdu, China, on May 12, 2008. *Upper right*: Human systolic blood pressure started increasing 2 days prior to the magnitude 9.0 earthquake in Sendai, Japan, on March 11, 2011, documented on the basis of week-long records of around-the-clock ambulatory data from 13 Japanese. Similar records from longitudinal and transverse controls differ in their time course, suggesting that the trend observed before the earthquake was related to it rather than being a feature of an anticipated weekly pattern. *Lower left*: The monthly incidence of major earthquakes since 1900 is characterized by the presence of cycles with periods of about 49.3, 12.2, 1.44, and 0.41 year(s), given with their uncertainties in parentheses. *Lower right*: The most prominent about-50-year cycle is also documented in physiology, pathology, societal upheavals, and space weather. Nonlinearly estimated periods are displayed with their 95 % confidence intervals shown as the length of corresponding horizontal bars. © Halberg Chronobiology Center

J. Johnson's (the president of the Royal Astronomical Society) address on presenting the Society's medal to Schwabe in 1857 [89]:

> M. Schwabe is a gentleman resident in … Dessau, the capital of the Duchy of Anhalt Dessau, who, having no professional duties, has devoted much time and attention to scientific studies, and has attained considerable reputation in other departments of knowledge than that in which he is known to this Society. Among other accomplishments, I am informed that M. Schwabe is an excellent botanist, and has composed, though, I believe, not published, a flora of the neighbourhood of Dessau.
>
> It was in 1826 that he entered upon those researches which are now to engage our attention. I am not aware of the motive that induced him — whether any particular views had suggested themselves to his own mind—or whether it was a general desire of investigating, more thoroughly than his predecessors had done, the laws of a remarkable phenomenon,

which it had long been the fashion to neglect. [Author's note: He did a pointed description of the view that periodic phenomena were not profitable to study, perhaps conveying the thought of all those who seek solutions, mathematical or biomedical, without accounting for time.] He could hardly have anticipated the kind of result at which he has arrived; at the same time we cannot imagine a course of proceeding better calculated for its detection, even if his mind had been prepared for it, than that which he has pursued from the very commencement of his career. Assuredly if he entertained such an idea, it was not borrowed from the authorities of the last century [author's note: including Galileo and Newton], to whom the solar spots were objects of more attention than they have been of late years.

Nulla constanti temporum lege apparent aut evanescunt, says Keill in 1739. — Introduct. ad. Physic. Astronom. p. 253.

Il est manifeste par ce que nous venons de rapporter qu'il n'y a point de règle certaine de leur formation, ni de leur nombre et de leur figure, says Cassini II. in 1740. — Elém. d'Astron. vol. i, p. 82.

Il semble qu'elles ne suivent aucune loi dans leurs apparitions, says Le Monnier in 1746. — Instit. Astron. p. 83.

Solar spots observe no regularity in their shape, magnitude, number, or in the time of their appearance or continuance, says Long in 1764. — Astron. vol. ii., p. 472.

Les apparitions des tâches du soleil n'ont rien de régulier," says Lalande in 1771. — Astron., vol. iii, § 3131, 2d edit.

And Delambre's opinion may be inferred from a well-known passage in the third volume of his *Astronomy* (p. 20), published in 1814, where treating of the solar spots he says, "Il est vrai qu'elles sont plus curieuses que vraiment utiles."

I cite these passages to show, whatever might have been Schwabe's own view of the matter, that a periodicity of the spots was not a current idea at the time he entered upon the subject, and therefore that he has not merely developed a law for which men's minds were prepared, but that he has been, to all intents and purposes, the discoverer of it. The first suggestion I have met with (and that a very vague one) of anything of the kind occurs in the article "Sonnenflecken", in Gehler's *Physikalisches Wörterbuch,* written by the elder Littrow, and published in 1836, long after Schwabe had been at work.

It was in 1826, as I have said, that Schwabe began his observations, but I am not aware of any published results before those in No. 350 of the *Astronomische Nachrichten,* which appeared in April 1838. Here he gives a summary of twelve years' labour, and though at this time he must have begun to foresee the issue, for the indications of periodicity are plain, he makes no remark, nor does the memoir appear to have attracted the attention of astronomers.

From this time commence his annual contributions to the *Astronomische Nachrichten;* but it was not until the end of 1843, when he had passed through two periods of maximum and minimum, that he very modestly remarks [in 1844] that his observations heretofore had given indications of periodicity which that year's result tended to confirm. Still the subject attracted little attention. As far as I can discover, the only person who had taken it up was Julius Schmidt, the present indefatigable Director of the Observatory at Olmütz, then residing near Hamburg. But the philosopher of Dessau was not disconcerted; he went on accumulating fresh proofs of his great discovery, which, when announced in 1851, by Alexander Von Humboldt in the third volume of his immortal Cosmos, came upon the world with all the freshness of novelty, though in reality the secret had been revealed eight years before. [Our addition: Or even earlier by Horrebow, and perhaps with his 19 years of observation Scheiner encountered a cycle even earlier] …

The exact period Schwabe does not pretend to have determined. That it is liable to perturbation is evident. During twenty-seven years of the series the results were extremely regular; during the last three years they have shown symptoms of disturbance. The epoch of minimum, which, consistently with earlier indications, should have happened in 1853, did not occur until 1856.

After making a number of other comments on Schwabe's observations of sunspots, Johnson continued [89]:

> There is, however, one more topic too closely connected with our subject, and too remarkable to pass unnoticed.
>
> You are aware, gentlemen, it has been long known that the intensity of magnetic declination is subject to a daily variation; and this variation was also known to be in some way connected with the sun, attaining its maximum western limit when that body is at its upper and lower culmination, and its maximum eastern limit about 6 o'clock in the morning and evening. These are not the exact times, but they are sufficiently near for our purpose. About the year 1850, Professor Lamont announced that this variation was again liable to another variation, which observed a period, from maximum to minimum, and from minimum to maximum, of about 10 years. He was led to this result by a discussion of his own observations at Munich. It was shortly after fully confirmed by our colleague, General Sabine, by the observations made at the Magnetic Observatories of Toronto and Hobarton. But in the course of the latter discussions General Sabine detected another important circumstance which had escaped Lamont, viz., that the periods of maximum variation of the declination-needle corresponded exactly with those of the maximum frequency of the solar spots, and vice versa, the minimum of the one with the minimum of the other. Hardly had General Sabine's paper been read before the Royal Society on March 18, 1852, when intelligence was received that two Swiss physicists, Professor Gautier of Geneva, and Professor Wolf of Berne, had arrived at the same conclusion, independently of each other, from a perusal of Lamont's results.
>
> All the observations I have mentioned were made during nearly the same time — between the years 1840 and 1851 — comprehending only a single period of change. Obviously, therefore, in this state of the inquiry it was of the highest importance to obtain a trustworthy series made at another time and under different circumstances. Though hardly to be expected, it happened, most fortunately, that such a series was at hand. The diurnal variation was one of those subjects which, many years ago, had particularly interested the vigorous mind of the lamented Arago; and among his papers were found the records of a most laborious course of experiments, conducted with all the care which no one knew better than he how to bestow on a delicate investigation. These experiments, extending from 1820 to 1831, have been rigorously discussed by M. Thoman, and exhibit the prevalence of the same law that obtains in the later series. Nor does the evidence of connexion rest here. In a recent paper sent to the Royal Society, General Sabine has shown that at Toronto (the only series hitherto fully discussed) all the magnetic elements subjected to observation undergo similar variations. In the face of these coincidences, to doubt the relationship between the two phenomena seems to me to be almost as unreasonable as to doubt the influence of the moon on the tides of the ocean. [Our addition: This statement will also have to be qualified by differences among solar, interplanetary and terrestrial magnetism's associations in the biosphere.]
>
> M. Schwabe's service has been the discovery of an important astronomical fact, falling strictly within the cognizance of this Society. His merits are in no way affected, though, no doubt, the value of his discovery is enhanced, by this remarkable episode. No longer is its scope confined to the disclosure of a physical peculiarity in the constitution of the sun. It promises to be the means of revealing the prevalence of a principle, throughout the solar system, co-extensive with gravitation, and of establishing another link in the chain of analogies binding Earth with other worlds, the ultimate effect of which on the progress of physical research no one can foresee.

We trust Johnson would be pleased to see transdisciplinary coperiodisms being mapped as a new dimension of the human mind. In keeping with the advice from Sir Norman Lockyer, this 93.5-year-old has, with apologies for redundancies, taken the opportunity to summarize his addition, a personalized and generalized chronousphere, the personalized version starting with C-ABPM, as a strain test. Also, we may all contribute our data to BIOCOS (and eventually to an international UN-WHO [and also state]-operated data center[s]), so that, after a life-span's self-surveillance as a whole, with instructions in our will, for its executor, she/he will specify causes of death and also antecedents – to be entered into a transdisciplinary chronousphere, or whatever an entity in its own right may be called that deals with cycles. For sure this entity will include maps of the dynamics of "normal ranges," so that we eventually obtain age-specified reference values from the healthy long-lived among us. In their light, we shall not only to see what our cosmos does to us. We will have the opportunity to develop countermeasures against the undesirable – hate, aggression, and terrorism – as a start of a societal as well as individualized health care, in keeping with Manuel Johnson's and Yohsuke Kamide's vision [108, 109], all related to the fact that we live in the atmosphere of the sun and are affected by it.

23.6 Precedent, Courtesy of Larry A. Beaty for Phoenix

"In the United States, the Firestone recall in the late 1990s (which was linked to more than 100 deaths from rollovers following tire tread-separation), pushed the Clinton administration to legislate the TREAD Act. The Act mandated the use of a suitable TPMS technology in all light motor vehicles (under 10,000 pounds), to help alert drivers of severe under-inflation events. This act affects all light motor vehicles sold after September 1, 2007.

The issue isn't so much a question of whether we could use the same pressure monitor in people, but rather why was it important and practical to put continuous pressure monitoring into cars before we put it into presumably more valuable people? What can we learn from the history?

It would seem that the public didn't demand tire pressure monitors from the automobile manufacturers until the "tragedy" was publicized by the mass media. This suggests that one way to get people interested in chronobiologic/chronomic interpretation of blood pressure is to get somebody to start counting and publishing "death-and-disablement-by-VVA" statistics. That is, how many people died last year with a detectable, treatable VVA? This is different than measuring the cost of treating people in the mainstream health-care system and holding that number up in front of government agencies or insurance companies, or even the general public, claiming there's a "savings" to be had. Deaths mean more than dollars to many people. Figure 23.15a (middle) is the case in point. According to Dr. Thomas R. Frieden, director of the Centers for Disease Control, "We're talking about $131 billion in health care costs annually and 1,000 deaths/day, today and every day this year" that can be attributed to uncontrolled hypertension [163].

23.7 Conclusion

Congruences among the periods of two time series, defined by the overlap of their 95 % confidence intervals, are a first step for the study of cosmic-biospheric associations. Investigations of the biospheric consequences of the subtraction (removal) vs. addition (replacement) of an environmental frequency are the next step. Supporting

"SECULARITY" resolved by chronomics* into a transdisciplinary spectrum of coperiodisms (reciprocal cycles) revealing the dynamics of growth and its modulation by competing SPACE and TERRESTRIAL WEATHER

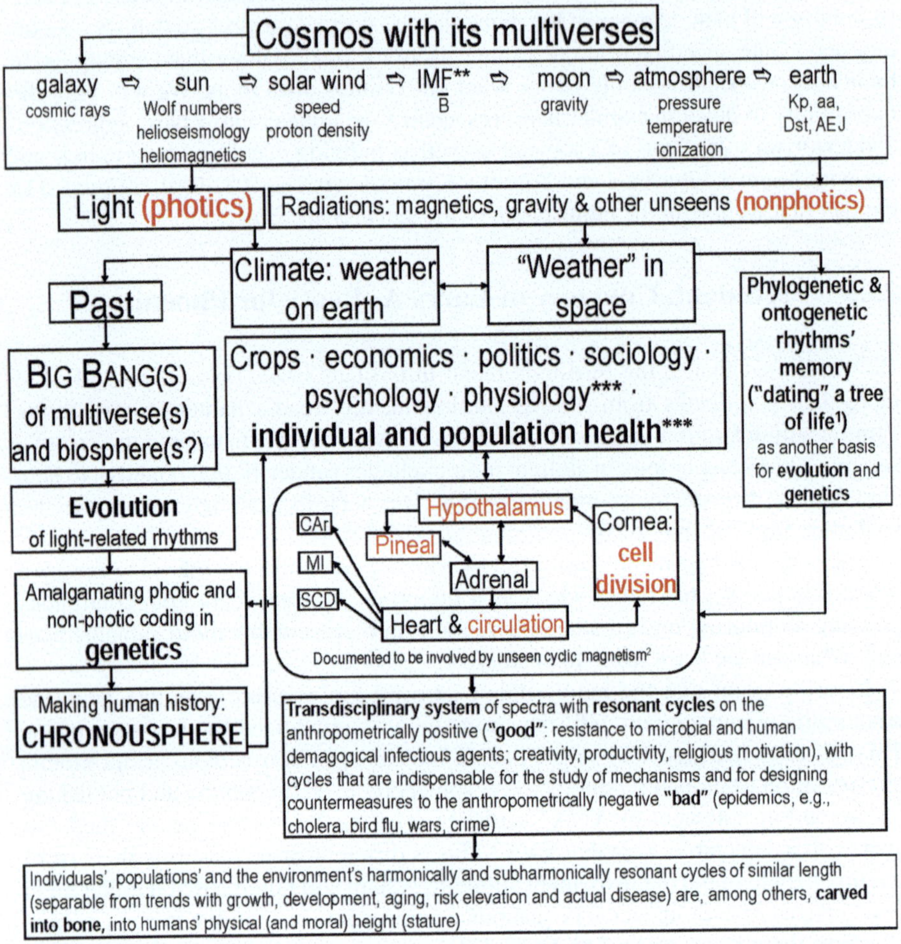

*Chronomics complementing genomics resolves cycles, trends and chaos into rules (nomos) of time (chronos) structures, i.e., chronomes · **Interplanetary magnetic field · ***Rounded box refers to recently documented non-photic effects of magnetic storms in addition to effects on pathology, i.e., cardiac arrhythmia (CAr), myocardial infarctions (MI) and sudden cardiac death (SCD) · [1]Biomedicine & Pharmacotherapy 2004; 58 (Suppl 1): S1-S11. [2]J Applied Biomedicine 2006; 4: 1-38 and 73-86.

Scheme 23.1 The time has come to replace "secularity" with chronome maps

evidence can be no more than a statistically significant increase in the amplitude of the environmental cycle that is followed by a statistically significant increase in the biospheric cycle's amplitude and vice versa. The (damped) persistence of a trans-year in an individual's systolic blood pressure or in the extent of terrorism encountered in a population speaks for a genetically coded infradian auto-resonance. The mechanisms involved are alluded to in Scheme 23.1 [13].

We advocate replicating congruence studies over decades of physiological data on populations of self-monitoring subjects: a replication on the same individual is lengthy and involves interactions with age. Congruence and addition/subtraction studies are desirable on all infradian periods, preferably by womb-to-tomb monitoring, as done in car tires.

Rather than only an echo, a genetically coded auto-resonance in mammals, including humans, triggered by magnetic storms (Fig. 23.11) (Section A), acting on the neuroendocrine (Sections A–C) and, at the level of the cell, nucleic acids, in particular, is likely a mechanism via feedsidewards [164]. A surveillable by C-ABPM, transmitted by cellphone to a website or cloud system, could underlie an as-one-goes stress-strain test telling us what we liked and disliked by the appearance and disappearance of VVAs. VVAs can also serve as an alerting system (Fig. 23.24) to predict and develop countermeasures not only against too much or too little light or heat, but also against too much or too little nonphotics, and thus against personal, societal and even natural cataclysms.

References

1. Thomson W (Lord Kelvin). Popular Lectures and Addresses. London: Macmillan and Co.; 1889. "In physical science the first essential step in the direction of learning any subject is to find principles of numerical reckoning and practicable methods for measuring some quality connected with it. I often say that when you can measure what you are speaking about, and express it in numbers, you know something about it; but when you cannot measure it, when you cannot express it in numbers, your knowledge is of a meagre and unsatisfactory kind; it may be the beginning of knowledge, but you have scarcely in your thoughts advanced to the state of *Science*, whatever the matter may be." p. 73–4.
2. Halberg F, Cornelissen G, Stoynev A, Ikonomov O, Katinas G, Sampson M, Wang ZR, Wan CM, Singh RB, Otsuka K, Sothern RB, Sothern SB, Sothern MI, Syutkina EV, Masalov A, Perfetto F, Tarquini R, Maggioni C, Kumagai Y, Siegelova J, Fiser B, Homolka P, Dusek J, Uezono K, Watanabe Y, Wu JY, Sonkowsky R, Schwartzkopff O, Hellbrügge T, Spector NH, Baciu I, Hriscu M, Bakken E. Season's appreciations 2002 and 2003. Imaging in time: the transyear (longer-than-the-calendar year) and the half-year. Neuroendocrinol Lett 2003; 24:421–40.
3. Mazzarella A, Scafetta N. Evidences for a quasi-60-year North Atlantic Oscillation since 1700 and its meaning for global climate change. Theor Appl Climatol 2012;107:599–609.
4. Pales E, Mikulecky M. Periodic emergence of great physicians in the history of ancient Greece, India and China. Abstract, 23rd Seminar. Man in his Terrestrial and Cosmic Environment, Upice, 21–23 May 2002.
5. Mikulecky M. Solar activity, revolutions and cultural prime in the history of mankind. Neuroendocrinol Lett 2007;28:749–56.
6. Kroeber AL. Configurations of culture growth. Berkeley: University of California Press; 1944. 882 pp.
7. Halberg F, Cornelissen G, Regal P, Otsuka K, Wang ZR, Katinas GS, Siegelova J, Homolka P, Prikryl P, Chibisov SM, Holley DC, Wendt RW, Bingham C, Palm SL, Sonkowsky RP, Sothern RB, Pales E, Mikulecky M, Tarquini R, Perfetto F, Salti R, Maggioni C, Jozsa R, Konradov AA, Kharlitskaya EV, Revilla M, Wan CM, Herold M, Syutkina EV, Masalov AV, Faraone P, Singh RB, Singh RK, Kumar A, Singh R, Sundaram S, Sarabandi T, Pantaleoni GC, Watanabe Y, Kumagai Y, Gubin D, Uezono K, Olah A, Borer K, Kanabrocki EA, Bathina

S, Haus E, Hillman D, Schwartzkopff O, Bakken EE, Zeman M. Chronoastrobiology: proposal, nine conferences, heliogeomagnetics, transyears, near-weeks, near-decades, phylogenetic and ontogenetic memories. Biomed Pharmacother 2004;58 Suppl 1:S150–87.

8. Lord of Time. London: Science without Borders/International Publishing House SWB; 2011. 45 p.
9. Hagemeister M. Russian cosmism in the 1920s and today. In: Rosenthal BG, editor. The occult in Russian and Soviet culture. Ithaca: Cornell UP; 1997. p. 185–202.
10. Fiks AP, Alexander L. Tchijevsky: "exquisite poet-philosopher", "creator of new sciences" or "charlatan". Caduceus 1994;10:179–84.
11. Young GM. The Russian cosmists: the esoteric futurism of Nikolai Fedorov and his followers. Oxford/New York: Oxford University Press; 2012. 280 pp.
12. Ertel S. Space weather and revolutions: Chizhevsky's heliobiological claim scrutinized. Studia Psychologica 1996;39:3–22.
13. Halberg F, Cornelissen G, Katinas GS, Hillman D, Otsuka K, Watanabe Y, Wu J, Halberg F, Halberg J, Sampson M, Schwartzkopff O, Halberg E. Many rhythms are control information for whatever we do: an autobiography. Folia Anthropologica 2012;12:5–134. http://ttk.nyme.hu/blgi/Knyvek%20kiadvnyok/FOLIA%20ANTHROPOLOGICA/folia12.pdf.
14. Gauss CF, Weber W. Resultate aus den Beobachtungen des magnetischen Vereins im Jahre 1839. Herausgegeben von Gauss CF, Weber W. Mit 4 Steindrucktafeln. Leipzig, im Verlag der Weidmannschen Buchhandlung, 1840.
15. Halberg F, Nelson W, Runge WJ, Schmitt OH, Pitts GC, Tremor J, Reynolds OE. Plans for orbital study of rat biorhythms. Results of interest beyond the Biosatellite program. Space Life Sci 1971;2:437–71.
16. Vallot J, Sardou G, Faure M. De l'influence des taches solaires: sur les accidents aigus des maladies chroniques. Gazette des Hôpitaux 1922;56:904–5.
17. Sardou G, Faure M. Les taches solaires et la pathologie humaine. La Presse médicale 1927;18(2 Mars):284–5.
18. Chizhevsky [Tchijevsky] AL (de Smitt VP, trans and condensed). Physical factors of the historical process. Cycles 1971;22:1–27. http://www.cyclesresearchinstitute.org/cycles-history/Chizhevsky1.pdf
19. Chizhevsky AL. (Gazenko O, foreword; Korzhuyev PA et al., Eds.) Earth echo of solar storms. Moscow: Thought; 1976. 367 p. with illustrations.
20. Chizhevsky AL. La natalité et la tension de l'activite solaire. La Côte d'Azur médicale (Toulon) 1930;11(9):230.
21. Chizhevsky AL. Les périodes solaires et la mortalité. La Côte d'Azur médicale (Toulon) 1930;11(9):232.
22. Chizhevsky AL. Influences des variations de l'activité solaire sur la natalité et sur la mortalité. Bulletin de l'Academie du Var (Toulon) 1931;98:6
23. Chizhevsky AL. L'activité solaire et la mortalité. Bulletin de l'Academie du Var (Toulon) 1931;98.
24. Chizhevsky AL. Action de l'activité périodique solaire sur la mortalité générale. In: Piéry M, editor. Traité de Climatologie: Biologique et médicale, vol. II. Paris: Masson et Cie; 1934. p. 1042–5.
25. Chizhevsky AL. L'activité corpusculaire, électromagnétique et périodique du Soleil et l'électricité atmosphérique, comme régulateurs de la distribution dans la suite des temps des maladies épidémiques et de la mortalité générale. Acta Medica Scandinavica 1937;91(6):491–522.
26. Düll T, Düll B. Über die Abhängigkeit des Gesundheitszustandes von plötzlichen Eruptionen auf der Sonne und die Existenz einer 27tägigen Periode in den Sterbefällen. Virchows Archiv 1934;293:272–319.
27. Gnevyshev MN, Novikova KF. The influence of solar activity on the earth's biosphere: part I. J Interdiscipl Cycle Res 1972;3:99–104.
28. Stoupel E. Forecasting in cardiology. New York: Wiley; 1976. 141 pp.
29. Stoupel E, Abramson E, Sulkes J. The effect of environmental physical influences on suicide: how long is the delay? Arch Suicide Res 1999;5:241–4.

30. Halberg F, Breus TK, Cornelissen G, Bingham C, Hillman DC, Rigatuso J, Delmore P, Bakken E. International Womb-to-Tomb Chronome Initiative Group: Chronobiology in space. Keynote, 37th Annual Meeting of the Japan Society for Aerospace and Environmental Medicine, Nagoya, 8–9 Nov 1991. University of Minnesota/Medtronic Chronobiology Seminar Series, #1 Dec 1991, 21 pp. of text, 70 figures.
31. Feinleib M, Rogot E, Sturrock PA. Solar activity and mortality in the United States. Int J Epidemiol 1975;4:227–9.
32. Lipa BJ, Sturrock PA, Rogot E. Search for correlation between geomagnetic disturbances and mortality. Nature 1976;259:302–4.
33. Cornelissen G, Halberg F, Breus T, Syutkina EV, Baevsky R, Weydahl A, Watanabe Y, Otsuka K, Siegelova J, Fiser B, Bakken EE. Non-photic solar associations of heart rate variability and myocardial infarction. J Atmos Solar-Terr Phys 2002;64:707–20.
34. Cornélissen G, Masalov A, Halberg F, Richardson JD, Katinas GS, Sothern RB, Watanabe Y, Syutkina EV, Wendt HW, Bakken EE, Romanov Y. Multiple resonances among time structures, chronomes, around and in us. Is an about 1.3-year periodicity in solar wind built into the human cardiovascular chronome? Hum Physiol 2004;30(2):86–92.
35. Wheeler RH. War 599 B.C.–1950 A.D. Indexes of international and civil war battles of the world. New York: Foundation for the Study of Cycles; 1951. 15 pp.
36. Singh RB, Halberg F, Siegelova J, Cornelissen G. What is the best time to exercise? In: Halberg F, Kenner T, Fiser B, Siegelova J, editors. Noninvasive Methods in Cardiology. Brno: Faculty of Medicine, Masaryk University; 2012. p. 163–4.
37. Homolka P, Cornelissen G, Homolka A, Siegelova J, Halberg F. Exercise-associated transient circadian hypertension (CHAT)? Abstract, III International Conference, Civilization Diseases in the Spirit of V.I. Vernadsky, People's Friendship University of Russia, Moscow, 10–12 Oct 2005, p. 419–21.
38. Halberg F, Kenner T, Fiser B, editors. Proceedings, Symposium: the importance of chronobiology in diagnosing and therapy of internal diseases. Brno: Department of Functional Diagnostics and Rehabilitation, Faculty of Medicine, Masaryk University; 2002. p. 206.
39. Halberg F, Kenner T, Siegelova J, editors. Proceedings, Symposium: chronobiological analysis in pathophysiology of cardiovascular system. Brno: Department of Functional Diagnostics and Rehabilitation, Faculty of Medicine, Masaryk University; 2003. p. 186.
40. Cornelissen G, Kenner R, Fiser B, Siegelova J, editors. Proceedings, Symposium: Chronobiology in Medicine. Brno: Department of Functional Diagnostics and Rehabilitation, Faculty of Medicine, Masaryk University; 2004. 146 pp.
41. Noninvasive Methods in Cardiology, Brno, Czech Republic, September 14, 2005. Brno: Department of Functional Diagnostics and Rehabilitation, Faculty of Medicine, Masaryk University; 2005. 25 pp.
42. Halberg F, Kenner T, Fiser B, Siegelova J, editors. Proceedings, Noninvasive Methods in Cardiology. Brno: Department of Functional Diagnostics and Rehabilitation, Faculty of Medicine, Masaryk University; 2007. ISBN 978 80 7018 463 4.
43. Halberg F, Kenner T, Fiser B, Siegelova J, editors. Proceedings, Noninvasive Methods in Cardiology. Brno: Faculty of Medicine, Masaryk University; 2008. 304 pp.
44. Halberg F, Kenner T, Fiser B, Siegelova J. editors. Proceedings, Noninvasive Methods in Cardiology, Brno: Faculty of Medicine, Masaryk University; 2009. 402 pp. http://web.fnusa.cz/files/kfdr2009/sbornik_2009.pdf
45. Halberg F, Kenner T, Fiser B, Siegelova J, editors. Noninvasive Methods in Cardiology. Brno: Faculty of Medicine, Masaryk University; 2010. 164 pp.
46. Halberg F, Kenner T, Fiser B, Siegelova J, editors. Noninvasive Methods in Cardiology. Brno: Faculty of Medicine, Masaryk University; 2011. 200 pp.
47. Halberg F, Kenner T, Fiser B, Siegelova J, editors. Noninvasive Methods in Cardiology. Brno: Faculty of Medicine, Masaryk University; 2012. 180 pp.
48. Halberg F, Cornelissen G, Schwartzkopff O, Otsuka K, Rosch PJ. Strain test for chronotherapy of pre- and MESOR-hypertension, involving Vascular Variability Anomalies/Disorders (VVAs/VVDs). In: Proceedings, 4th International Conference on Advanced Cardiac Sciences:

King of Organs 2012, Prince Sultan Cardiac Center Eastern Province, Al Ahsa, 18–21 Nov 2012. p. 117–41.
49. Rosch PJ, Markov MS, editors. Bioelectromagnetic medicine. New York: Marcel Dekker; 2004. 850 pp.
50. Hoyt DV, Schatten KH. Group sunspot numbers: a new solar activity reconstruction. Sol Phys 1998;179(1):189–219.
51. Hoyt DV, Schatten KH. Group sunspot numbers: a new solar activity reconstruction. Sol Phys 1998;181(2):491–513.
52. Otsuka K. Proceedings, 1st International Symposium Workshop on Circadian Rhythms and Clinical Chronotherapy. Biomed Pharmacother 2001;55 Suppl 1:7s–190s.
53. Otsuka K. Proceedings, 2nd International Symposium Workshop on Circadian Rhythms and Clinical Chronotherapy. Biomed Pharmacother 2002;56 Suppl 2:231s–382s.
54. Otsuka K. Proceedings, 3rd International Symposium Workshop on Circadian Rhythms and Clinical Chronotherapy. Biomed Pharmacother 2003;57 Suppl 1:1s–198s.
55. Otsuka K. Proceedings, 4th International Symposium Workshop on Circadian Rhythms and Clinical Chronotherapy. Biomed Pharmacother 2004;58 Suppl 1:S1–S188.
56. Otsuka K. Proceedings, 5th International Symposium Workshop on Circadian Rhythms and Clinical Chronotherapy. Biomed Pharmacother 2005;59 Suppl 1:S1–S261.
57. Cornelissen G, McCraty R, Atkinson M, Halberg F. Gender differences in circadian and extra-circadian aspects of heart rate variability (HRV). Time series (not spotcheck) evidence-based holistic approach to health, 1st International Workshop of the TsimTsoum Institute, Krakow, 3 Jun 2010, p. 27–8.
58. Hillman D, Finley J, Halberg F, Cornelissen G, Thomas F, Ilyia E, McCraty R, Singh RB, Kino T, Chrousos G. Data during sleep are indispensable for chronobiologically-interpreted ambulatory blood pressure monitoring (C-ABPM). Time series (not spotcheck) evidence-based Holistic approach to health, 1st International Workshop of the TsimTsoum Institute, Krakow, 3 Jun 2010, p. 29–31.
59. McCraty R, Cornelissen G, Atkinson M, Halberg F. Gender differences in various endpoints of the conventional range of heart rate variability (HRV). Time series (not spotcheck) evidence-based holistic approach to health, 1st International Workshop of the TsimTsoum Institute, Krakow, 3 Jun 2010, p. 24–5.
60. Finley J, Halberg F, Cornelissen G, Hillman D, Thomas F, Ilyia E, McCraty R, Kino T, Chrousos G. Chronobiologically-interpreted ambulatory blood pressure monitoring (C-ABPM) in every home. Time series (not spotcheck) evidence-based holistic approach to health, 1st International Workshop of the TsimTsoum Institute, Krakow, 3 Jun 2010, p. 12–3.
61. Cornelissen G, Halberg F, Hillman D, McCraty R, Finley J, Thomas F, Kino T. Abstract modeling of events in a recurrent half-yearly adynamia with internal and/or external, e.g., circadian desynchronization. Time series (not spotcheck) evidence-based holistic approach to health, 1st International Workshop of the TsimTsoum Institute, Krakow, 3 Jun 2010, p. 10–1.
62. Proceedings, Natural Cataclysms and Global Problems of the Modern Civilization. London: SWB International Publishing House; 2012. 621 pp.
63. Wendt HW. Interplanetary magnetic field (IMF) polarity, collective emotions and entropy changes of random event generators. In: Proceedings, 8th International Congress "Health and Education Millennium." Moscow: Bulletin of the People's Friendship University of Russia; 2007, 14–17 Nov 2007. p. 81–4.
64. Radin D. Entangled minds: extrasensory experiences in a quantum reality. New York: Paraview Pocket Books; 2006. 368 pp.
65. Beaty LA. The status quo of chronobiologic 7-day/24-hour monitoring of blood pressure and heart rate. Int J Gerontol ChronomeGeriatrics 2008;11(14):155.
66. Beaty LA. Web-enabled software to complement affordable automatic hardware: Phoenix website. Int J Gerontol ChronomeGeriatrics 2008;11(14):156–8.
67. Beaty LA, Cornelissen G, Adams C, Nolley E, Halberg F. Wide-scale cost-effective chronobiologically-interpreted blood pressure for home and clinic. Poster presentation, 2012 IEEE Healthcare Innovation Conference; 2012. Houston, 7–9 Nov 2012.

68. Adams C. Privacy requirements for low-cost chronomedical systems. International Conference on the Frontiers of Biomedical Science: Chronobiology, Chengdu, 24–26 Sept 2006. p. 64–9.
69. Halberg F, Cornelissen G, Hillman D, Ilyia E, Cegielski N, el-Khoury M, Finley J, Thomas F, Brandes V, Kino T, Papadoupoulou A, Chrousos GP, Costella JF, Mikulecky M. Multiple circadian periods in a lady with recurring episodes of adynamic depression: case report. In: Halberg F, Kenner T, Fiser B, Siegelova J, editors. Noninvasive Methods in Cardiology. Brno: Faculty of Medicine, Masaryk University; 2011. p. 45–67.
70. Costella JF, Halberg F, Hillman D, Mikulecky M, Cornelissen G. Four circadian and two circasemidian periods in sleep-wakefulness of a man on a self-selected routine. Folia Anthropologica 2012;11:51–3.
71. Hennessey TL, Field CB. Evidence of multiple circadian oscillators in bean plants. J Biol Rhythms 1992;7:105–13.
72. Rosenwasser AM, Adler NT. Structure and function in circadian timing systems: evidence for multiple coupled circadian oscillators. Neurosci Biobehav Rev 1986;10:431–48.
73. Halberg F, Cornelissen G, Hillman D, Ilyia E, Cegielski N, el-Khoury M, McCraty R, Strestik J, Finley J, Thomas F, Kino T, Sanchez de la Pena S, Chrousos GP, de Meester F, Singh RB, Mikulecky M. Half-yearly recurrent adynamic loss of 24-hour synchronization. Int J Geronto-Geriatrics 2010;13:35–62.
74. Uchiyama M, Shibui K, Hayakawa T, Kamei Y, Ebisawa T, Tagaya H, Okawa M, Takahashi K. Larger phase angle between sleep propensity and melatonin rhythms in sighted humans with non-24-hour sleep-wake syndrome. Sleep 2002;25(1):83–8.
75. Uchiyama M, Okawa M, Shibui K, Kim K, Tagaya H, Kudo Y, Kamei Y, Hayakawa T, Urata J, Takahashi K. Altered phase relation between sleep timing and core body temperature rhythm in delayed sleep phase syndrome and non-24-hour sleep-wake syndrome in humans. Neurosci Lett 2000;294(2):101–4.
76. Halberg F, Siffre M, Engeli M, Hillman D, Reinberg A. Etude en libre-cours des rythmes circadiens du pouls, de l'alternance veille-sommeil et de l'estimation du temps pendant les deux mois de séjour souterrain d'un homme adulte jeune. CR Acad Sci (Paris) 1965;260:1259–62.
77. Reinberg A, Halberg F, Ghata J, Siffre M. Spectre thermique (rythmes de la température rectale) d'une femme adulte avant, pendant et après son isolement souterrain de trois mois. CR Acad Sci (Paris) 1966;262:782–5.
78. Siffre M, Reinberg A, Halberg F, Ghata J, Perdriel G, Slind R. L'isolement souterrain prolongé. Etude de deux sujets adultes sains avant, pendant et après cet isolement. Presse méd 1966;74:915–9.
79. Halberg F, Reinberg A. Rythmes circadiens et rythmes de basses fréquences en physiologie humaine. J Physiol (Paris) 1967;59:117–200.
80. Halberg F, Cornelissen G, Sonkowsky RP, Lanzoni C, Galvagno A, Montalbini M, Schwartzkopff O. Chrononursing (chronutrics), psychiatry and language. New Trends Exp Clin Psychiatry 1998;14:15–26.
81. Halberg F, Cornelissen G, Sothern RB, Barnwell F, Cegielski N, Ilyia E, Siegelova J. The moon's and the genes' tides and double tides pulling the biosphere. In: Halberg F, Kenner T, Fiser B, Siegelova J, editors. Noninvasive Methods in Cardiology. Brno: Faculty of Medicine, Masaryk University; 2010. p. 23–45.
82. Wever RA. The circadian system of man: results of experiments under temporal isolation. New York: Springer; 1979. 276 pp.
83. Daxecker F. Briefe des Naturwissenschaftlers Christoph Scheiner SJ an Erzherzog Leopold V. von Österreich-Tirol 1620–1632. Innsbruck: Veröffentlichung der Universität Innsbruck; 1995. Hsg.: Universität Innsbruck, Bd. 207. ISBN 3-901249-21-4.
84. Daxecker F. Das Hauptwerk des Astronomen P. Christoph Scheiner SJ "Rosa Ursina sive Sol": eine Zusammenfassung. Ber Nat-Med Ver Innsbruck 1996; Suppl 13:82 Seiten. ISBN 3-7030-0300-6.

85. Daxecker F, Subaric L (Hg). Christoph Scheiners "Sol ellipticus". Veröffentlichung der Universität Innsbruck 1998; Bd. 226. ISBN 3-901249-39-7.
86. Daxecker F. The physicist and astronomer Christoph Scheiner: biography, letters, works. Veröffentlichung der Universität Innsbruck 2004;246. ISBN 3-901249-69-9.
87. Daxecker F. Der Physiker und Astronom Christoph Scheiner. Innsbruck: Universitätsverlag Wagner 2006: 160 S. ISBN 3-7030-0424-X.
88. Daxecker F. Sonnenfleckenzyklen bei Christoph Scheiner. Eine retrospektive chronobiologische Untersuchung, Sammelblattdes Historischen Vereins Ingolstadt. 2013;121:105–9.
89. Johnson MJ. [Address delivered by the President, M.J. Johnson, Esq., on presenting the Medal of the Society to M. Schwabe.] Monthly Notices, Royal Astronomical Society 1857;17:126–32.
90. Gauss CF. Theorematis fundamentalis in doctrine de residuis quadraticis demonstrationes et ampliationes novae. Commentationes societatis regiae scientiarum gottingensis recentiores 1820;4: cl. math:3–20.
91. Cornelissen G, Halberg F, Sonkowsky R, Siegelova J, Homolka P, Dusek J, Fiser B. Meta-analysis of Horrebow's and Schwabe's scholarship with a view of sampling requirements. In: Proceedings, Noninvasive Methods in Cardiology, Brno, 7–10 Jul 2009. p. 141–58. http://web.fnusa.cz/files/kfdr2009/sbornik_2009.pdf
92. Schwabe H. Über die Flecken der Sonne. Astron Nachr 1838;15:244–8 (no. 350).
93. Schwabe H. Sonnen-Beobachtungen im Jahre 1843. Astron Nachr. 1844;21:254–6 (no. 495).
94. Thiele TN. De Macularum Solis antiquioribus quibusdam observationibus Hafniae institutis. Astron Nachr 1859;50:259–61.
95. Stetson HT. Man and the stars. New York: Whittlesey House, McGraw-Hill; 1930. 221 pp.
96. Richardson JD, Paularena KI, Belcher JW, Lazarus AJ. Solar wind oscillations with a 1.3-year period. Geophys Res Lett 1994;21:1559–60.
97. Rieger E, Share GH, Forrest DJ, Kanbach G, Reppin C, Chupp EL. A 154-day periodicity in the occurrence of hard solar flares? Nature 1984;312:623–5.
98. Wolff CL. The rotational spectrum of g-modes in the sun. Astrophys J 1983;264:667–76.
99. Egeson C. Egeson's weather system of sun-spot causality: being original researches in solar and terrestrial meteorology. Sydney: Turner & Henderson; 1889. 63 pp.
100. Halberg F, Cornelissen G, Bernhardt K-H, Sampson M, Schwartzkopff O, Sonntag D. Egeson's (George's) transtridecadal weather cycling and sunspots. Hist Geo Space Sci 2010;1:49–61.
101. Düll B, Düll T. Erd- und Sonnenphysikalische Vorgänge in ihrer Bedeutung für Krankheits- und Todauslosung. Nosokomeion 1938;9:103–20.
102. Schwartzkopff O, Düll-Pfaff N, Halberg F, Freytag EM, Freytag J, Gumarova L, Halberg F, Cornelissen G. Cardiovascular/overall mortality vs. suicide differ by the Dülls' 27-day coperiodisms' phases. In: Proceedings, 4th International Conference on Advanced Cardiac Sciences: King of Organs 2012, Prince Sultan Cardiac Center Eastern Province, Al Ahsa, Saudi Arabia, 18–21 Nov 2012. p. 268–9.
103. Lockyer N. Contributions to solar physics, London: Macmillan; 1874. 676 pp. The thing to hunt down is a cycle, p. 424.
104. Vernadsky VI. Philosophical thought of naturalist. Moscow: Science; 1988. 520 p.
105. Halberg F, Cornelissen G, Gumarova L, Halberg Francine, Ulmer W, Hillman D, Siegelova J, Watanabe Y, Hong S, Otsuka K, Wu J, Lee JY, Schwartzkopff O, Wendt H. Integrated and as-one-goes analyzed physical, biospheric and noetic monitoring: preventing personal disasters by self-surveillance may help understand natural cataclysms: a chronousphere (chrononoösphere). London: SWB International Publishing House; 2012. 106 pp.
106. Mikulecky ME. Sun, Moon and living matter. Bratislava, June 28–July 1, 1994. Bratislava: Slovak Medical Society; 1994. p. 163–9.
107. Halberg F. Chronobiology. Annu Rev Physiol 1969;31:675–725.
108. Kamide Y. What human beings cannot see can exist: a message from recent studies of solar-terrestrial relationships. Biomed Pharmacother 2003;17:19–23.

109. Kamide Y. We reside in the sun's atmosphere. Biomed Pharmacother 2005;59 Suppl 1:S1–4.
110. Halberg F, Cornelissen G, International Womb-to-Tomb Chronome Initiative Group. Resolution from a meeting of the international society for research on civilization diseases and the environment (New SIRMCE Confederation), Brussels, 17–18 Mar 1995: Fairy tale or reality? Medtronic Chronobiology Seminar #8, April 1995, 12 pp. text, 18 figures. http://halbergchronobiologycenter.umn.edu
111. Halberg F, Cornelissen G, Otsuka K, Siegelova J, Fiser B, Dusek J, Homolka P, Sanchez De La Peña S, Singh RB, BIOCOS project. Extended consensus on need and means to detect vascular variability disorders (VVDs) and vascular variability syndromes (VVSs). Int J Gerontol ChronomeGeriatrics. 2008;11(14):119–46; Leibniz-Online Nr. 5, 2009 (http://www2.hu-berlin.de/leibniz-sozietaet/journal/archiv_5_09.html). 35 pp; and World Heart J 2010;2(4):279–305.
112. Halberg F, Cornelissen G, Hillman D, Beaty L, Hong S, Schwartzkopff O, Watanabe Y, Otsuka K, Siegelova J. Chronobiologically interpreted ambulatory blood pressure monitoring in health and disease. Global Adv Health Med 2012;1(2):64–88.
113. Watanabe Y, Cornelissen G, Halberg F, Beaty L, Siegelova J, Otsuka K, Bakken EE. Harm vs. benefit from losartan with hydrochlorothiazide at different circadian times in MESOR-hypertension or CHAT. In: Halberg F, Kenner T, Fiser B, Siegelova J, editors. Proceedings, Noninvasive Methods in Cardiology, Brno, 4–7 Oct 2008. p. 149–67. http://web.fnusa.cz/files/kfdr2008/sbornik_2008.pdf
114. Watanabe Y, Halberg F, Otsuka K, Cornelissen G. Toward a personalized chronotherapy of high blood pressure and a circadian overswing. Clin Exp Hypertens 2013;35(4):257–66.
115. Halberg F, Cornelissen G, Katinas G, Tvildiani L, Gigolashvili M, Janashia K, Toba T, Revilla M, Regal P, Sothern RB, Wendt HW, Wang ZR, Zeman M, Jozsa R, Singh RB, Mitsutake G, Chibisov SM, Lee J, Holley D, Holte JE, Sonkowsky RP, Schwartzkopff O, Delmore P, Otsuka K, Bakken EE, Czaplicki J, International BIOCOS Group. Chronobiology's progress: part II, chronomics for an immediately applicable biomedicine. J Appl Biomed 2006;4:73–86. http://jab.zsf.jcu.cz//4.1/halberg.pdf.
116. Halberg F, Nelson W, Cornelissen G, Haus E, Scheving LE, Good RA. On methods for testing and achieving cancer chronotherapy. Cancer Treat Rep 1979;63:1428–30.
117. Halberg F, Cornelissen G, Wang ZR, Wan C, Ulmer W, Katinas G, Singh R, Singh RK, Singh R, Gupta BD, Singh RB, Kumar A, Kanabrocki E, Sothern RB, Rao G, Bhatt MLBD, Srivastava M, Rai G, Singh S, Pati AK, Nath P, Halberg F, Halberg J, Schwartzkopff O, Bakken E, Shastri VK. Chronomics: circadian and circaseptan timing of radiotherapy, drugs, calories, perhaps nutriceuticals and beyond. J Exp Ther Oncol 2003;3:223–60.
118. Halberg Francine, Halberg J, Halberg E, Halberg Franz. Chronobiology, radiobiology and steps toward the timing of cancer radiotherapy. In: Goldson AL, editor, Kaiser H, series editor. Cancer growth and progression, vol. 9, ch. 19. Dordrecht: Kluwer; 1989. p. 227–53.
119. Ulmer W, Cornelissen G, Halberg F. Physical chemistry and the biologic week in the perspective of chrono-oncology. In vivo 1995;9:363–74.
120. Ulmer W, Cornelissen G, Halberg F. Interaction among (quantum mechanical) resonance-coupled electromagnetic circuits relevant to a natural week. World Heart J 2012;4(1):35–70.
121. Halberg F, Cornelissen G, Wilson D, Singh RB, De Meester F, Watanabe Y, Otsuka K, Khalilov E. Chronobiology and chronomics: detecting and applying the cycles of nature. Biologist 2009;56(4):209–14.
122. Halberg F, Cornelissen G, Carandente F. Chronobiology leads toward preventive health care for all: cost reduction with quality improvement. A challenge to education and technology via chronobiology. Chronobiologia 1991;18:187–93.
123. Cornélissen G, Halberg F, Wendt HW, Bingham C, Sothern RB, Haus E, Kleitman E, Kleitman N, Revilla MA, Revilla Jr M, Breus TK, Pimenov K, Grigoriev AE, Mitish MD, Yatsyk GV, Syutkina EV. Resonance of about-weekly human heart rate rhythm with solar activity change. Biologia (Bratisl) 1996;51:749–56.
124. Halberg F. Some physiological and clinical aspects of 24-hour periodicity. Journal-Lancet (Minneapolis) 1953;73:20–32.

125. Halberg F. Some correlations between chemical structure and maximal eosinopenia in adrenalectomized and hypophysectomized mice. J Pharmacol Exp Ther 1952;106:135–40.
126. Godlowski ZZ. The fate of eosinophils in hormonally induced eosinopenia and its significance. J Endocrinol 1952;8:102–25.
127. Reppert SM, Weaver DR. Molecular analysis of mammalian circadian rhythms. Annu Rev Physiol 2001;63:647–76. doi:10.1146/annurev.physiol.63.1.647.
128. Sothern RB, Cornélissen G, Yamamoto T, Takumi T, Halberg F. Time microscopy of circadian expression of cardiac clock gene mRNA transcription: chronodiagnostic and chrono-therapeutic implications. La Clinica Terapeutica 2009;160(2):e25–34.
129. Sothern RB, Yamamoto T, Cornélissen G, Takumi T, Halberg F. Central and peripheral circadian clock genes, their statistical analysis for rhythms, and relationship to health and disease. Scripta medica (Brno) 2009;82:133–63.
130. Cornelissen G, Engebretson M, Johnson D, Otsuka K, Burioka N, Posch J, Halberg F. The week, inherited in neonatal human twins, found also in geomagnetic pulsations in isolated Antarctica. Biomed Pharmacother 2001;55 Suppl 1:32s–50s.
131. Hanson BR, Halberg F, Tuna N, Bouchard Jr TJ, Lykken DT, Cornelissen G, Heston LL. Rhythmometry reveals heritability of circadian characteristics of heart rate of human twins reared apart. Cardiologia 1984;29:267–82.
132. Mikulecky M, Florida PL. Daily birth numbers in Davao, Philippines, 1993–2003: Halberg's transyear stronger than year. Abstract, 26th Seminar, Man in his Terrestrial and Cosmic Environment, Upice, 17–19 May 2005 unpaginated.
133. Cornelissen G, Halberg F, Mikulecky M, Florida P, Faraone P, Yamanaka T, Murakami S, Otsuka K, Bakken EE. Yearly and perhaps transyearly human natality patterns near the equator and at higher latitudes. Biomed Pharmacother 2005;59 Suppl 1:S117–22.
134. Kovac M, Mikulecky M. Secular rhythms and Halberg's paraseasonality in the time occurrence of cerebral stroke. Bratisl Lek Listy 2005;106(2):423–7.
135. Halberg F, Cornelissen G, Katinas G, Tvildiani L, Gigolashvili M, Janashia K, Toba T, Revilla M, Regal P, Sothern RB, Wendt HW, Wang ZR, Zeman M, Jozsa R, Singh RB, Mitsutake G, Chibisov SM, Lee J, Holley D, Holte JE, Sonkowsky RP, Schwartzkopff O, Delmore P, Otsuka K, Bakken EE, Czaplicki J, International BIOCOS Group. Chronobiology's progress: season's appreciations 2004–2005. Time-, frequency-, phase-, variable-, individual-, age- and site-specific chronomics. J Appl Biomed 2006;4:1–38. http://jab.zsf.jcu.cz//4.1/halberg.pdf.
136. Hamamatsu A, Cornelissen G, Otsuka Ku, Halberg F, Chibisov S (presenter). Linear-nonlinear rhythmometry documents a transyear and a cishalfyear in sudden cardiac death (ICD 10, code I46.1) in Tokyo. In: Proceedings, International Symposium, Problems of Ecological and Physiological Adaptation, People's Friendship University of Russia, Moscow, 30–31 Jan 2007. Moscow: People's Friendship University of Russia; 2007. p. 542–5.
137. Cornelissen G, Otsuka K, Halberg F. Remove and replace for a scrutiny of space weather and human affairs. In: Grigoriev AI, Zeleny LM, editors. Proceedings, international conference, space weather effects in humans: in space and on earth. Moscow: Space Research Institute; 2013. p. 508–38.
138. Halberg F, Cornelissen G, Sothern RB, Czaplicki J, Schwartzkopff O. Thirty-five-year climatic cycle in heliogeophysics, psychophysiology, military politics, and economics. Izv Atmos Oceanic Phys. 2010;46(7):844–64 (Backtranslation from Geophysical Processes and Biosphere 2009;8(2):13–42.).
139. Khomeriki O, Paatashvili T, Gheonjian L, Kapanadze N, Invia N. The influence of 7-day variations of interplanetary magnetic field on the frequency of myocardial infarctions. Bull Georgian Acad Sci 1998;158(#1):123–6.
140. Amiranashvili AG, Cornelissen G, Amiranashvili V, Gheonjian L, Chikhladze VA, Gogua RA, Matiashvili TG, Paatashvili T, Kopytenko YuA, Siegelova J, Dusek J, Halberg F. Circannual and circadecennian changes in mortality from cardiovascular causes in Tbilisi, Republic of Georgia (1980–1992). MEFA 8th International Fair of Medical Technology and Pharmacy, Brno, 7–10 Nov 2000, abstract 16.

141. Cornelissen G, Halberg F, Gheonjian L, Paatashvili T, Faraone P, Watanabe Y, Otsuka K, Sothern RB, Breus T, Baevsky R, Engebretson M, Schröder W. Schwabe's ~10.5- and Hale's ~21-year cycles in human pathology and physiology. In: Schröder W, editor. Long- and short-term variability in sun's history and global change. Bremen: Science Edition; 2000. p. 79–88.
142. Amiranashvili AG, Cornelissen G, Amiranashvili V, Gheonjian L, Chikhladze VA, Gogua RA, Matiashvili TG, Paatashvili T, Kopytenko YA, Siegelova J, Dusek J, Halberg F. Circannual and circadecennian stages in mortality from cardiovascular causes in Tbilisi, Republic of Georgia (1980–1992). Scripta Medica 2002;75:255–60.
143. Halberg F, Cornelissen G, Gheonjian L, Schwartzkopff O. Myocardial infarctions, sudden cardiac deaths and wrangling influences of space and terrestrial weather. In: Proceedings, 4th International Conference on Advanced Cardiac Sciences: King of Organs 2012, Prince Sultan Cardiac Center Eastern Province, Al Ahsa, 18–21 Nov 2012. p. 108–16.
144. Lockyer WJS. The solar activity 1833–1900. Proc Roy Soc Lond 1901;68:285–300.
145. Brückner E. Klimaschwankungen seit 1700 nebst Beobachtungen über die Klimaschwankungen der Diluvialzeit. Wien und Olmütz: E. Hölzel; 1890. 324 pp. (Penck A, Hrsg. Geographische Abhandlungen, Band IV.)
146. Clarke H. On the political economy and capital of joint stock banks. Railway Mag. 1838;27:288–93. Cf. also Clarke H. Physical economy: a preliminary inquiry into the physical laws governing the periods of famines and panics. Railw Regist 1847; 155–69.
147. Herschel WXIII. Observations tending to investigate the nature of the sun, in order to find the causes or symptoms of its variable emission of light and heat; with remarks on the use that may possibly be drawn from solar observations. Royal Soc Lond Philos Trans 1801;91:265–318.
148. Halberg F, Cornelissen G, Salti R, Perfetto R, Tarquini R, Stagi S, Hillman DC, Katinas GS, Hoogerwerf WA, Carandente F, Otsuka K, Czaplicki J, Chibisov SM, Scheving LA, Syutkina EV, Masalov A, Mitsutake G, Wang ZR, Wan CM, Schwartzkopff O, Bakken EE. Chronoauxology. Chronomics: trends and cycles in growth and cosmos rather than secularity. In: Proceedings, 10th Auxology Congress: Human Growth in Sickness and in Health, Florence, 4–7 Jul 2004. Florence: Edizioni Centro Studi Auxologici; 2010. 92 pp.
149. Halberg F, Cornelissen G, Otsuka K, Syutkina EV, Masalov A, Breus T, Viduetsky A, Grafe A, Schwartzkopff O. Chronoastrobiology: neonatal numerical counterparts to Schwabe's 10.5 and Hale's 21-year sunspot cycles. In memoriam Boris A. Nikityuk. Int J Prenat Perinat Psychol Med 2001;13:257–80.
150. Halberg F, Cornelissen G, Watanabe Y, Otsuka K, Fiser B, Siegelova J, Mazankova V, Maggioni C, Sothern RB, Katinas GS, Syutkina EV, Burioka N, Schwartzkopff O. Near 10-year and longer periods modulate circadians: intersecting anti-aging and chronoastrobiological research. J Gerontol A Biol Sci Med Sci 2001;56:M304–24.
151. Boskovic RJ. Criticism of Newton's alleged proof of absolute motion. From explanatory notes to Philosophiae recentioribus versibus tradita a Benedicto Stay Libri Decem. Romae 1755; III. p. 464 in Capek M. The concepts of space and time. Dordrecht-Boston: D. Reidel Publ.; 1976. p. 289–90.
152. Poincaré H. La mesure du temps. Revue de métaphysique et de morale 1898;6:1–13.
153. Fraser JT. Of time, passion and knowledge. New York: G. Braziller; 1975. 529 pp.
154. Macey SL, editor. Encyclopedia of time. New York: Garland Publishing; 1994. 699 pp.
155. Einstein A. Über den Äther. Verhandlungen der Schweizerischen naturforschenden Gesellschaft 1924;105(2):85–93.
156. Grafarend EW. Von A. Einstein über H. Weyl und E. Cartan zur Quanten-Gravitation: Vortrag in der Klasse für Naturwissenschaften am 13. Oktober 2011. Sitzungsberichte der Leibniz-Sozietät der Wissenschaften zu Berlin 2012;113:15–21.
157. Halberg F. Chapter on "Medizin" in: Jahrbuch der Internationalen Hochschulwochen des Österreichen College. Salzburg: Igonta Verlag; 1946. p. 336–51.
158. Li Y, Liu Y, Jiang Z, Guan J, Yi G, Cheng S, Yang B, Fu T, Wang Z. Behavioral change related to Wenchuan devastating earthquake in mice. Bioelectromagnetics 2009;30:613–20.
159. Watanabe Y, Halberg F, Otsuka K, Cornelissen G. Physiology and earthquakes, focusing on that in 2011 in East Japan. In: Proceedings, Natural Cataclysms and Global Problems of the

Modern Civilization, Istanbul, 19–21 September 2011. London: SWB International Publishing House; 2012. p. 260–4.

160. Halberg F, Cornelissen G, Schwartzkopff O, Khalilov E, Khalilov T, Damirov F, Wang Z, Watanabe Y, Otsuka K, Siegelova J, Al-Abdulgader AA. Preventive cardiology concerns avoidance of personal and societal health-related and natural cataclysms. In: Halberg F, Kenner T, Fiser B, Siegelova J, editors. Noninvasive Methods in Cardiology. Brno: Faculty of Medicine, Masaryk University; 2011. p. 33–44.
161. Gregori GP. Galaxy – sun – earth relations: the origin of the magnetic field and of the endogenous energy of the earth, with implications for volcanism, geodynamics and climate control, and related items of concern for stars, planets, satellites, and other planetary objects. A discussion in a prologue and two parts. Beiträge zur Geschichte der Geophysik und Kosmischen Physik, Band 3, Heft 3. 2002. 471 pp. ISNN: 1615–2824 - NE: Gregori, Giovanni P. © Science Edition, Arbeitskreis Geschichte Geophysik/W. Schröder, Germany 2002.
162. Roederer JG. Tearing down disciplinary barriers. Eos Trans Am Geophys Union 1985;66:681. 684–5.
163. Rosch PJ. Why treating hypertension has become a dismal fiasco. Health and Stress (The Monthly Newsletter of the American Institute of Stress) Oct 2003.
164. Halberg F, Cornelissen G, Katinas GS, Watanabe Y, Otsuka K, Maggioni C, Perfetto F, Tarquini R, Schwartzkopff O, Bakken EE. Feedsidewards: intermodulation (strictly) among time structures, chronomes, in and around us, and cosmo-vasculo-neuroimmunity. About ten-yearly changes: what Galileo missed and Schwabe found. In: Conti A, Maestroni GJM, McCann SM, Sternberg EM, Lipton JM, Smith CC. (Eds.) Ann NY Acad Sci 2000;917:348–376.

Closing Remarks

Evidence-based medicine is now strongly recommended in precedence to clinical medicine. It is often difficult, however, to assess the abrupt change from the stable state of disease to deterioration for the worse or almost impossible to predict sudden cardiac death. One may ask what is lacking to achieve these goals. First, current medicine does not have an interest in the science of time, and, second, it is rare to assess an individual comprehensively, taking into consideration the local characteristics.

In quest for a better medicine, we applied "chronomics" to detect disease at an early stage by longitudinal follow-ups of clinical and lifestyle data to assess changes in health status. Since 2000, we examined community-dwelling subjects from the viewpoint of a "glocal" (combined global and local) comprehensive assessment in several villages.

Herein, we conclude this book by Franz Halberg's words, which had been ready for this book 2 years ago.

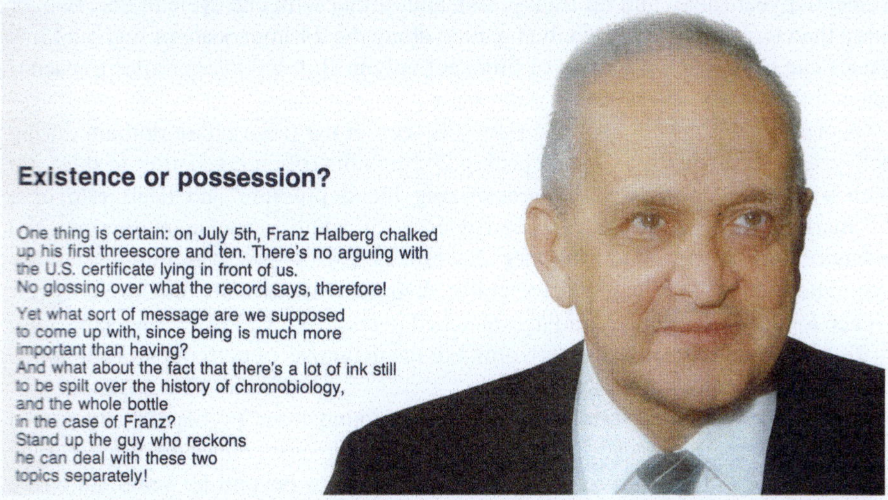

Existence or possession?

One thing is certain: on July 5th, Franz Halberg chalked up his first threescore and ten. There's no arguing with the U.S. certificate lying in front of us. No glossing over what the record says, therefore!

Yet what sort of message are we supposed to come up with, since being is much more important than having? And what about the fact that there's a lot of ink still to be spilt over the history of chronobiology, and the whole bottle in the case of Franz? Stand up the guy who reckons he can deal with these two topics separately!

(from CHRONOBIOLOGIA, Volume XVI, October-December 1989)

K. Otsuka et al., *Chronomics and Continuous Ambulatory Blood Pressure Monitoring*, DOI 10.1007/978-4-431-54631-3

Toward a Chronomically Interpreted Home ABPM

The desire prevails in current everyday medical practice for the fiction of a stable, "true" blood pressure measurement, that is, of "the average pressure over long periods of time" [1–6]. It is mistakenly assumed that, in a long record, blood pressure is time invariant, yet variations above and below fixed threshold (such as 140/90 or 135/85 mm Hg) are commonly seen from one day to another, from one week to another, from one month or even longer spans to the next span of similar duration, even for the 24-hour rhythm-adjusted mean value (MESOR). For instance, in the light of chronobiologic reference values, half-yearly summaries of half-hourly around-the-clock measurements, in a case studied for decades, changed from MESOR-hypertensive in one span to normotensive in the next such span. This fact and very many others [7–9] question a diagnosis based on office measurements with one or a few 24-hour ambulatory blood pressure monitorings (ABPM), the current false platinum standard [1–6]. The officially recommended goal of current WHO or national guidelines remains today to estimate a "true fixed" (yet imaginary!) pressure and to compare it with a fixed (gender-, age-, and time-invariant) threshold as a reference standard. The consequence is that false-negative and false-positive diagnoses abound and calling them white-coat hypertension or masked hypertension does not alter the number of misdiagnoses and hence of mistreatments or of failures to treat when this is indicated.

The foregoing status quo ignores the predictable variability due to the many cycles in and around us, apparent to the unaided eye in any around-the-clock and along-the-years record. Measurements are usually lower by night than by day; they vary further, not only with the seasons but often more so with frequencies found in the solar wind, shorter than a half year (cis-half years or quinmensals) or longer than a calendar year (near- and far-transyears), and further with one cycle in one, two, or more than three decades. Even the location along the 24-hour scale of high and low values can change as a function of time, not only in shift workers or after transcontinental flights but also in severe depression.

A clearly lengthened circadian period is seen in the (acro)phase domain during half-yearly recurring depressions lasting 2–3 months (often coexisting in this case with a 24-hour component), characterizing blood pressure and heart rate of a 61-year-old woman (JF) who experiences a circadian vascular, endocrine, and behavioral ecfrequentia during her downtimes [7]. VVAs (vascular variability anomalies) and VVDs (vascular variability disorders) are actually diagnosed by reference to data from gender- and age-matched peers ([8]; cf. [7]). For the same case of JF, if the values during sleep are omitted, the diagnosis of ecfrequentia cannot be made [9].

One's view changes further when one has found trans (= beyond)-tridecadal cycles (of about 35 years) in sets of around-the-clock series covering decades. One set of about-6-daily measurements extends from 20 to beyond 60 years, an entire clinically healthy adulthood; another begins with an acceptable blood pressure, to arrive, after two decades, at a high average blood pressure (MESOR-hypertension).

The MESOR, estimated by curve fitting, is usually more accurate and more precise than the arithmetic mean. In several other long series covering decades, one finds a mostly acceptable MESOR of blood pressure under treatment yet with occasional VVAs if not VVDs, all findings that must be detected to be corrected, if a high risk of severe cardiovascular disease associated with VVDs is to be avoided.

Thus, previously mysterious nonphotic, unseen features of human life become readily apparent under the virtual telescope of chronomics, the study of interactions among time structures (chronomes) in and around us, approached with glocality in time as in space. Glocality in space [10] means that globally valid approaches can be applied to find differences locally geographically, as Prof. Otsuka does for health surveillance in cities in Hokkaido and Kochi, as well as within the Arctic Circle and at high altitudes in the Indian subcontinent [11–15].

For a methodologic temporal glocality, we can prospectively as well as retrospectively look globally at ever longer cycles with repeated passes over the same data until we dispose of an entire 4-decade-long time series for assessing any transtridecadal cycle, while the same data for blood pressure and heart rate are also analyzed in weekly and longer sections of the time series as-one-goes, locally in time for their circadian and infradian characteristics. The sections may serve the individual for severe disease prevention, while the same blood pressures and heart rates are further pooled and aligned with time series of terrestrial and extraterrestrial weather to find, e.g., an about-35-year cycle, off and on, in rainfall and environmental temperature, important in human affairs, bearing on problems of global warming, economics, and military politics. Further nonphotic cycles include those of about (~) a week, ~ a month, 5 months, 6 months, or 16 months, as well as decadal, didecadal, or transtridecadal cycles. Quinmensals, semiannuals, and transannual cycles, along with circadians and circannuals, can all be monitored, e.g., by systolic blood pressure as are mental functions, to study mechanisms, e.g., of terrorism on a population basis [16], while initially the same data served, as noted, for the detection and personalized treatment of VVAs in 7-day around-the-clock records as a start. These 7-day profiles are to be repeated once an abnormality is found, and, if a VVD is confirmed, continuous, e.g., half-hourly, monitoring is advocated, which may serve everybody anyhow, once affordable, cuff-free unobtrusive instrumentation becomes available for automatic transfer into a multilingual also automatic website that does on a large scale what is being done by a project BIOCOS for all comers free of charge.

Against the background of these thoughts, it is a pleasure to introduce a book on blood pressure by Dr. Kuniaki Otsuka and Prof. Germaine Cornelissen. We do this with deep personal affection and continue to learn from him, first from 297 cases that served to define and compare five of the major VVAs, including MESOR-hypertension and a circadian overswing or circadian hyper-amplitude-tension (CHAT). In Asians, if several of these VVDs persist and coexist, they raise cardiovascular disease risk within 6 years to nearly 100% [17]. Second, we owe Prof. Otsuka the original introduction of a glocal approach that is global and local in space [10–15] and can be glocal in time.

For the practitioner of medicine and the educated lay person, the late Thomas G. Pickering, recipient of the prestigious Franz Volhard Award, asked in 2008 rightly ([1]; cf. [2–6]): "Should doctors still measure blood pressure?" [1]. Pickering also advocated, as do we, an international data base of ABPM studies. His father, Sir George White Pickering, Regius Professor of Medicine at Oxford University [18], had delivered the first Volhard lecture in 1972. Concern about blood pressure was in the Pickering family time structure (chronome; 1–6, 18), if not in their genome. (the Volhards – Jakob, the father [a chemist] who synthesized creatinine, and Franz, the son [a clinician] who preceded Goldblatt in focusing scholars of blood pressure on the kidney – had the kidney in their chronomes.) As clinicians, both Sir George and Tom Pickering eventually (Tom was a fast learner [1 vs. 2–6]) helped focus attention on the indispensability of ABPM and Tom also on modern data bases. Hence, this book is dedicated to their memories.

Pressure is being monitored continuously in the tires of modern cars to avoid accidents and in small animals to develop drugs. If there is a popular demand stemming from those who read Prof. Otsuka's book, affordable, unobtrusive technology will become available to continuously monitor the individual's pressure, to allow one to know oneself, to learn from VVA analysis what one likes and dislikes, a problem as yet unsolved [19], and further to treat VVAs, preferably before they become VVDs and VVSs, as well as when these latter conditions already prevail. A classification by dipping, its positive aspects on large groups notwithstanding, is no substitute; it may mislead [20–22].

We need not fly blind [23]: the conventional blood pressure cuff must be removed from the physician's office; instead, home ABPM can be interpreted chronomically (C-ABPM) today by BIOCOS (corne001@umn.edu) and eventually by an automatic multilingual website; it is immediately feasible for a personalized health care, e.g., for stroke prevention and further for providing data for monitoring the cosmos and eventually for the development of rational methods to prevent social diseases such as crime and terrorism. What is good for cars' tires and drug development is recommended for those who drive the cars and who may wish to avoid the need for drugs by early prevention as well as for those who wish to use these drugs with monitoring of the treatment of VVDs, as needed.

Franz Halberg, M.D., Dr. h.c. (mult.):
Montpellier
Ferrara
Tyumen
Brno
L'Aquila
Moscow (People's Friendship University of Russia)
Professor of Laboratory Medicine and Pathology,
Physiology, Biology, Bioengineering and Oral Medicine
Co-director, Halberg Chronobiology Center
University of Minnesota
US MAIL: Campus Mail Code 8609

OFFICE: 715 Mayo Memorial Building
420 Delaware St. S.E.
Minneapolis, MN 55455, USA
PHONE +1 (612) 624.6976, 624.3144 OR 626.5404
FAX +1 (612) 624.9989
URL http://halbergchronobiologycenter.umn.edu

References

1. Pickering TG, Gerin W, Schwartz JE, Spruill TM, Davidson KW. Franz Volhard lecture: should doctors still measure blood pressure? The missing patients with masked hypertension. J Hypertens 2008;26:2259–67.
2. Pickering T, Schwartz J, Verdecchia P, Imai Y, Kario K. An international database of prospective ambulatory blood pressure monitoring studies. Blood Press Monit 2003;8:147–9.
3. Pickering TG. White coat hypertension. Proceedings, III Meeting on "Conflicting Aspects in the Clinical Approach to Hypertension", Montecassino, Oct. 21–22, 1988. pp. 1–2.
4. Pickering TG. Masked hypertension and white-coat hypertension. In: Proceedings, 59th annual meeting, Japan Society of Neurovegetative Research. Tokyo; Nov 1–3, 2006. p. 32.
5. Cornélissen G, Chen CH, Halberg F. Predictive value of blood pressure variability: merits of circadian parameters versus dipping patterns. N Engl J Med 2006 [Aug 14]; 355; 8: 850; [Letter in response to Pickering TG, Shimbo D, Haas D. Ambulatory blood pressure monitoring. N Engl J Med 2006; 354: 2368–74. Reply by Pickering TG, Shimbo D, Haas D. N Engl J Med 2006 [Aug 14]; 355;8: 850–1.]
6. Kario K, Thomas G. Pickering – a great mentor. Blood Press Monit 2010;15:82–4.
7. Halberg F, Cornélissen G, Cegielski N, Hillman D, Halberg F, Schwartzkopff O, McCraty R, Finley J, Thomas F, Kino T, Chrousos G, Sonkowsky RP, El-Khoury M, Ilyia E. Circadian dysfrequentia of cortisol, melatonin, DHEA, testosterone and estradiol. In: Noninvasive Methods in Cardiology. Brno: Faculty of Medicine, Masaryk University, 2010, Brno, Czech Republic.
8. Halberg F, Cornelissen G, Otsuka K, Siegelova J, Fiser B, Dusek J, Homolka P, Sanchezdela Pena S, Singh RB, BIOCOS project. Extended consensus on need and means to detect vascular variability disorders (VVDs) and vascular variability syndromes (VVSs). Geronto-Geriatr: Int J Gerontol-Chronom Geriatr 2008;11(14):119–46.
9. Hillman D, Finley J, Halberg F, Cornélissen G, Thomas F, Ilyia E, McCraty R, Singh RB, Kino T, Chrousos G. Data during sleep are indispensable for chronobiologically interpreted ambulatory blood pressure monitoring (C-ABPM). Time Series (Not Spotcheck) Evidence-Based Holistic Approach to Health, 1st International Workshop of the TsimTsoum Institute, Krakow, Poland; June 3, 2010, p. 28–30.
10. Otsuka K, Cornélissen G, Norboo T, Takasugi E, Halberg F. Chronomics and "glocal" (combined global and local) assessment of human life. Progress of Theoretical Physics 2008; Suppl. 173: 134–52
11. Otsuka K, Oinuma S, Cornélissen G, Weydahl A, Ichimaru Y, Kobayashi M, Yano S, Holmeslet B, Hansen TL, Mitsutake G, Engebretson MJ, Schwartzkopff O, Halberg F. Alternating-light-darkness-influenced human electrocardiographic magnetoreception in association with geomagnetic pulsations. Biomed Pharmacother 2001;55 Suppl 1:63s–75.
12. Otsuka K, Cornélissen G, Schwartzkopff O, Bakken EE, Halberg F, Burioka N, Katinas GS, Kane R, Regal PJ, Schaffer E, Sonkowsky R, Patterson R, Engebretson M, Brockway B, Wang ZR, Delmore P, Halpin C, Sarkozy S, Wall D, Halberg J. Clinical chronobiology and chronome-geriatrics: at variance with recommendations of subsequent guidelines, yet focusing

indeed on pre-hypertension in the physiological range. Biomed Pharmacother 2003;57 Suppl 1:164s–98.
13. Otsuka K, Yamanaka G, Shinagawa M, Murakami S, Yamanaka T, Shibata K, Yano S, Ishizuka S, Singh RB, Cornélissen G, Halberg F. Chronomic community screening reveals about 31% depression, elevated blood pressure and infradian vascular rhythm alteration. Biomed Pharmacother 2004;58 Suppl 1:S48–55.
14. Otsuka K, Cornélissen G, Weydahl A, Holmeslet B, Hansen TL, Shinagawa M, Kubo Y, Nishimura Y, Omori K, Yano S, Halberg F. Geomagnetic disturbance associated with decrease in heart rate variability in a subarctic area. Biomed Pharmacother 2001;55:51–6.
15. Otsuka K, Norboo T, Otsuka Y, Higuchi H, Hayajiri M, Narushima C, Sato Y, Tsugoshi T, Murakami S, Wada T, Ishine M, Okumiya K, Matsubayashi K, Yano S, Chogyal T, Angchuk D, Ichihara K, Cornélissen G, Halberg F. Effect of aging on blood pressure in Leh, Ladakh, a high-altitude (3524 m) community, by comparison with a Japanese town. Biomed Pharmacother 2005;59 Suppl 1:S54–7.
16. Halberg F, Cornélissen G, Sothern RB, Katinas GS, Schwartzkopff O, Otsuka K. Cycles tipping the scale between death and survival (= "life"). Progress of Theoretical Physics 2008; Suppl. 173:153–81.
17. Otsuka K, Cornélissen G, Halberg F. Predictive value of blood pressure dipping and swinging with regard to vascular disease risk. Clin Drug Investig 1996;11:20–31.
18. Richardson DW, Honour AJ, Fenton GW, Stott FH, Pickering GW. Variation in arterial pressure throughout the day and night. Clin Sci 1964;26:445–60.
19. Halberg F, Cornélissen F, Halberg Francine, Kessler T, Otsuka K. Measuring mental strain by duration of blood pressure overswing (CHAT): case report. World Heart J 2010; 2(2): 141–67.
20. Gupta AK, Greenway FL, Cornélissen G, Pan W, Halberg F. Prediabetes is associated with abnormal circadian blood pressure variability. J Hum Hypertens 2008;22:627–33. doi:10:1038/jhh.2008.32.
21. Cugini P, Cruciani F, Turri M, Regine F, Gherardi F, Petrangeli CM, Gabrieli CB. 'Minimal-change hypertensive retinopathy' and 'arterial pre-hypertension', illustrated via ambulatory blood-pressure monitoring in putatively normotensive subjects. Int Ophthalmol 1999;22(3):145–9.
22. Cornélissen G, Halberg F, Beaty L, Kumagai Y, Halberg E, Halberg J, Lee J, Schwartzkopff O, Otsuka K. Cugini's syndrome in statu nascendi: oratio contra morem prevalentem et pro chronobiologica ratione ad pressione sanguinis curandam. Clin Ter 2009;160(2):e13–24.
23. Fossel M. Circadian hyper-amplitude-tension, CHAT: a disease risk syndrome of anti-aging medicine. J Anti-Aging Med 1998;1:239.

The manufacturer's authorised representative in the EU is Springer Nature Customer Service Centre GmbH, Europaplatz 3, 69115 Heidelberg, Germany. If you have any concerns regarding our products, please contact ProductSafety@springernature.com

Printed and bound by CPI Group (UK) Ltd, Croydon, CR0 4YY
25/09/2026
02222862-0003